Michael D. Johnson

Human
Biology

CONCEPTS AND CURRENT ISSUES

Eighth Edition

Acquisitions Editor: Star MacKenzie Burruto
Project Managers: Mae Lum and Brett Coker
Program Manager: Anna Amato
Developmental Editor: Susan Teahan
Editorial Assistant: Maja Sidzinska
Executive Editorial Manager: Ginnie Simione Jutson
Editor-in-Chief: Beth Wilbur
Program Management Team Lead: Michael Early
Project Management Team Lead: David Zielonka
Production Management: Lumina Datamatics, Inc.

Copyeditor: Lumina Datamatics, Inc.
Compositor: Lumina Datamatics, Inc.
Design Manager: Derek Bacchus
Interior & Cover Designer: tani hasegawa (TTEye)
Illustrators: Imagineering
Rights & Permissions Project Manager: Donna Kalal
Rights & Permissions Management: Candice Velez (QBS Learning)
Photo Researchers: Pat Holl (QBS Learning) and tani hasegawa (TTEye)
Manufacturing Buyer: Stacey Weinberger
Executive Marketing Manager: Lauren Harp

Cover Photo Credit: Alfred Pasieka/Science Photo Library/Getty Images

Library of Congress Cataloging-in-Publication Data

Johnson, Michael D.
  Human biology : concepts and current issues / Michael D. Johnson. — Eighth edition.
    pages cm
  Includes index.
  ISBN 978-0-13-404243-5 — ISBN 0-13-404243-3
  1. Human biology—Textbooks.   I. Title.
  QP34.5.J645 2017
  612—dc23
                                                            2015026650

4 17

www.pearsonhighered.com

ISBN 10: 0-134-04243-3;
ISBN 13: 978-0-134-04243-5 (Student Edition)

ISBN 10: 0-134-15400-2;
ISBN 13: 978-0-134-15400-8 (Books a la Carte Edition)

ABOUT THE AUTHOR

Dr. Michael D. Johnson spent most of his youth in the fields and forests of rural Washington, observing nature. He earned his B.S. degree in zoology from Washington State University and then moved east to earn a Ph.D. in physiology from the University of Michigan. After completing a Postdoctoral Research Fellowship at Harvard Medical School, he joined the faculty of West Virginia University, where he remained for most of his career.

In 2001, Dr. Johnson moved to the Middle East, where he served first as the founding dean of Oman Medical College in the Sultanate of Oman and then as associate dean for premedical education at Weill Cornell Medical College in Qatar. In both positions, he directed the premedical education of students from more than 25 countries. He returned to the United States in 2011 to focus on his writing.

Dr. Johnson received several teaching awards during his career, including the West Virginia University Foundation Outstanding Teacher award and the Distinguished Teacher Award of the School of Medicine. He is a member of the American Physiological Society, the Human Anatomy and Physiology Society, the National Association of Biology Teachers, and the American Association for the Advancement of Science.

Whether teaching undergraduates or medical students, Dr. Johnson has always had a keen interest in instilling in students an appreciation of science. He seeks to show students how the advancement of scientific knowledge sometimes raises unforeseen ethical, political, economic, and social issues for all of us to discuss and solve. Through this book, he encourages students to become scientifically literate so that they will feel comfortable making responsible choices as consumers of science.

You can contact Dr. Johnson at mdj5412@gmail.com with comments or questions about this book.

CONTENTS

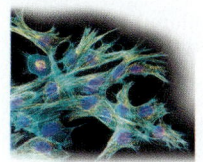
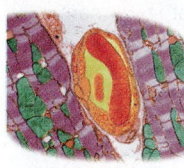

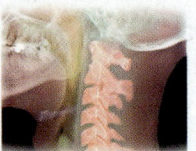

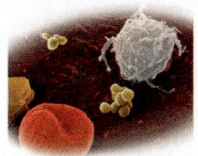

9 The Immune System and Mechanisms of Defense   187

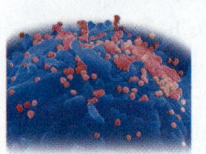

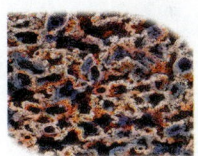

10 The Respiratory System: Exchange of Gases 219

11 The Nervous System: Integration and Control 243

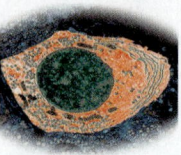

15 The Urinary System 351

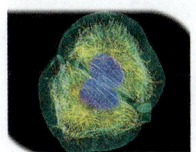

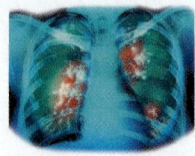
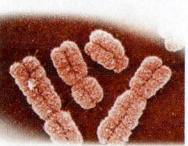

22 Evolution and the Origins of Life 503

23 Ecosystems and Populations 521

24 Human Impacts, Biodiversity, and Environmental Issues 541

Current Issue Global Warming and Global Climate Change   542

Should childhood vaccinations be mandatory for school attendance? Are genetically modified organisms (GMOs) a good or a bad thing? How will our future be affected by global warming and global climate change, and what, if anything, should we be doing about these phenomena? Are organic foods better for you than conventional foods?

Questions such as these seem to come up almost daily. Those of us who find these questions and the news stories about them fascinating—and yes, even exciting!—have an obligation to help others understand science and the impact it has on their lives. Science is too much fun and far too important to be left to scientists.

New to This Edition

Changes to this edition are designed to encourage students who do not have a strong background in science to become actively engaged in the course. Improved pedagogy helps students focus their learning, directs their attention to key concepts and current issues in biology, and encourages thoughtful analysis and critical thinking.

- **New organization to the chapter opening material.** To help the student develop an organized approach to a chapter's content, each chapter opener now includes an outline of the main headings and a list of the key concepts to be covered.
- **Addition of a "connections" passage.** The initial section of text in each chapter includes a "connections" passage, delineated by a chain-link icon, that provides the student with a sense of how a chapter's specific topic interrelates to the overall subject of human biology, biology in general, and the larger world.
- **New ways to access MJ's BlogInFocus entries.** To rouse students' interest in the science they encounter in their everyday lives, once again incorporated into each chapter are references to the author's blog. With this edition, the MJ's BlogInFocus is more accessible, as students can now view the blog entries via three different ways: directly with their smartphones by scanning a QR code, online by typing a URL into a search engine, or by visiting the MasteringBiology Web site. Each chapter includes two to four MJ's BlogInFocus references. It is hoped that these references to the author's Web site will encourage students to further explore science related topics that are of particular interest to them.
- **Refreshed Visual Content.** To revitalize the visual content, 120 new photos replace images from the previous edition, and 16 figures are new. More than 30 other figures have been improved from the previous edition.
- **The use of numbered steps.** Where complex processes are described, numbered step icons, ❶ , ❷ , ❸ ,

and so forth, are included both in the text and in the accompanying figure. These correlating step icons will help students follow the logical sequence of events as those events unfold within a complex process.

- **Updated Features, Graphs, Tables, and Text.** Key features of this text are currency and accuracy. Time-dependent data has been updated with the latest information available. The updated text includes eleven new or extensively updated *Current Issue* features, three new Health & Wellness features, and more than 60 new MJ's BlogInFocus entries.

The Focus Is On the Student

This book is written for students who do not yet have a strong background in science so that they, too, might share in the joy and wonder of science. Every effort is made to make the book accurate and up to date while keeping it inviting, accessible, and easy to read. The look and feel of the text is intentionally like that of a news magazine, peppered with short features likely to be of interest to the student and designed with a strong visual appeal.

Each chapter begins with an outline of the main topic headings and a list of key concepts to be covered. Next, a *Current Issue* feature highlights a recent controversy or ethical/social/political issue related to topics to be covered in the chapter. In the introductory section of each chapter, a new "connections" passage helps the student understand just how the topic of the chapter fits into the bigger picture of human biology and the larger world.

Students are naturally curious about how their own bodies work and human diseases and disorders. We capitalize on this curiosity with Health & Wellness features that highlight timely health topics. In addition, organ system chapters generally conclude with a section covering the more common human diseases and disorders.

Once again, a key feature of the book is MJ's BlogInFocus, brief references to a blog Web site written by Dr. Johnson in support of this text. The URL is www.humanbiologyblog.blogspot.com. Two to four MJ's BlogInFocus entries per chapter highlight recent discoveries or news items relevant to the subject of each chapter. Most of the blog entries have an additional embedded URL that takes the student directly to a news source or research paper. We hope that MJ's BlogInFocus entries and the author's blog will encourage curious students to dig a little deeper into topics that interest them. New to this edition are the means by which students can access the blog entries. Students can now get to the blog in any one of three ways: They can scan a QR code, type a URL into a search engine, or visit Pearson's MasteringBiology Web site.

To help students assess whether or not they understand the material, check questions throughout the text allow the students to test their understanding as they go along. Finally, at the end of each chapter is a range of question types, from concept review to recall to application, each designed to test the student's knowledge of facts as well as stimulate their critical thinking skills.

Unifying Themes Tie the Subjects Together

Several unifying themes in biology hold the chapters together. Homeostasis, the state of dynamic equilibrium in which the internal environment of an organism is maintained fairly constant, is one of those recurrent themes. The concept of homeostasis ties in with another recurrent theme: Structure and function are related. Structure/function relationships are the very core of the study of anatomy and physiology, and both of these fields in turn rely on the most unifying concept in all of biology: evolution. Only in the context of evolution can anatomy and physiology be fully understood; without the concept of evolution, very little in biology makes sense.

A predominant theme of this book is that each of us has choices to make—choices that will affect ourselves, other humans, and the entire planet. Should all children be vaccinated against childhood diseases? Should we spend time and money preparing for a pandemic that may never occur? Will we be willing and able to slow the rate of global warming? Is it important that we save other species from extinction, and if so, how should we go about it? Students are encouraged to formulate their own views on these and other topics so that they will feel comfortable with related choices they make.

The Organization Fits the Course

This book was designed to accommodate the fairly standard format for college courses in human biology. There are chapters that introduce science and chemistry, chapters that cover basic human biology from cells through the human organ systems, and finally, chapters on evolution, ecosystems and populations, and human impacts on the environment.

With such broad coverage, however, there is never enough time to teach all that is interesting, exciting, and relevant about human biology in one semester. Fortunately, because each chapter was written to stand on its own, this book allows for a certain degree of flexibility. Instructors wishing to emphasize the basics of human anatomy and physiology or focus on the medical aspects of human biology could omit or de-emphasize the last two chapters. Instructors should also feel free to present the organ system chapters in a different order if they feel more comfortable doing so. Within chapters, sections on diseases and disorders could be omitted or considered optional. Those interested in a more molecular or cellular approach might want to give greater emphasis to Chapters 2–4 and 17–21 and move more quickly through the organ systems chapters. Those more interested in the broader picture of where humans came from and how humans fit into the world order may want to allow sufficient time for the last three chapters, even if it means that they must move quickly or selectively through the organ system chapters. All of these approaches are equally valid.

However much you cover, dig in and enjoy your course!

Michael D. Johnson

KEEP **CURRENT** IN BIOLOGY

Through his teaching, his textbook, and in his online blog, award-winning teacher Michael D. Johnson sparks your interest by connecting basic biology to real-world issues relevant to your life.

"I hope the blog will stimulate students to go beyond the required reading, leading them to discover and explore subjects of personal interest. When this happens, students will ultimately be learning because they want to, not because they have to, and they'll be more comfortable with science and with biology."

—**Michael Johnson**, Author of *Human Biology: Concepts and Current Issues*

BlogInFocus in-text references appear at applicable points within the chapter and direct you to the blog that provides up-to-date insights on important issues in the news. The blog is updated 3–4 times per month.

NEW! Three options for accessing Michael Johnson's BlogInFocus entries: You may scan a QR code using a smartphone, type the URL (www.humanbiologyblog. blogspot.com) into a search engine, or log into your MasteringBiology subscription.

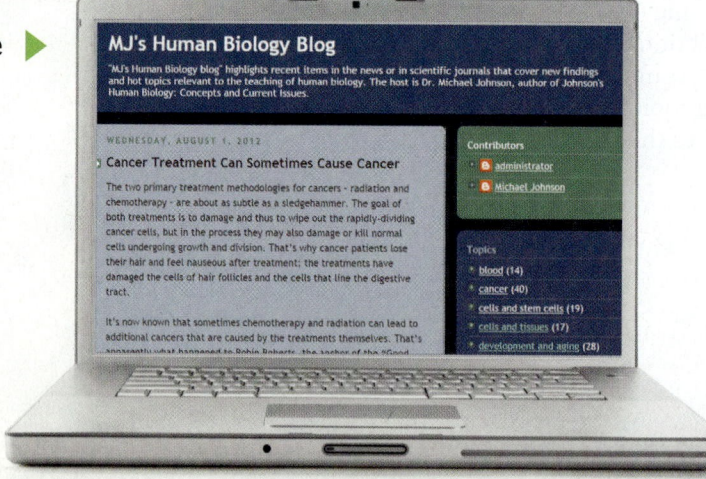

MJ's BlogInFocus Does radiation therapy for cancer treatment ever cause additional cancers? Visit MJ's blog in the Study Area in MasteringBiology and look under "Radiation and Cancer."

http://goo.gl/nAnIuL

◀ **BlogInFocus** MasteringBiology™ activities encourage students to read the blog and allow instructors to assess their understanding of the applied material.

ENGAGE WITH HIGH INTEREST ESSAYS

Each chapter opens with Michael Johnson's popular "Current Issue" essays, and BlogInFocus references within the chapter direct you to his frequently updated online blog for breaking human biology-related news.

Located at the start of each chapter, **Current Issue essays** draw you into the subject with interesting science and health news items, connecting human biology to real-world issues. Each essay provides contrasting views on the featured hot topic.

Many **NEW** Current Issue essays replace those from the previous edition, including:

- The 2013 Ebola outbreak (Chapter 9)
- Regulation of e-cigarettes (Chapter 10)
- Choosing between organic or conventional foods (Chapter 14)

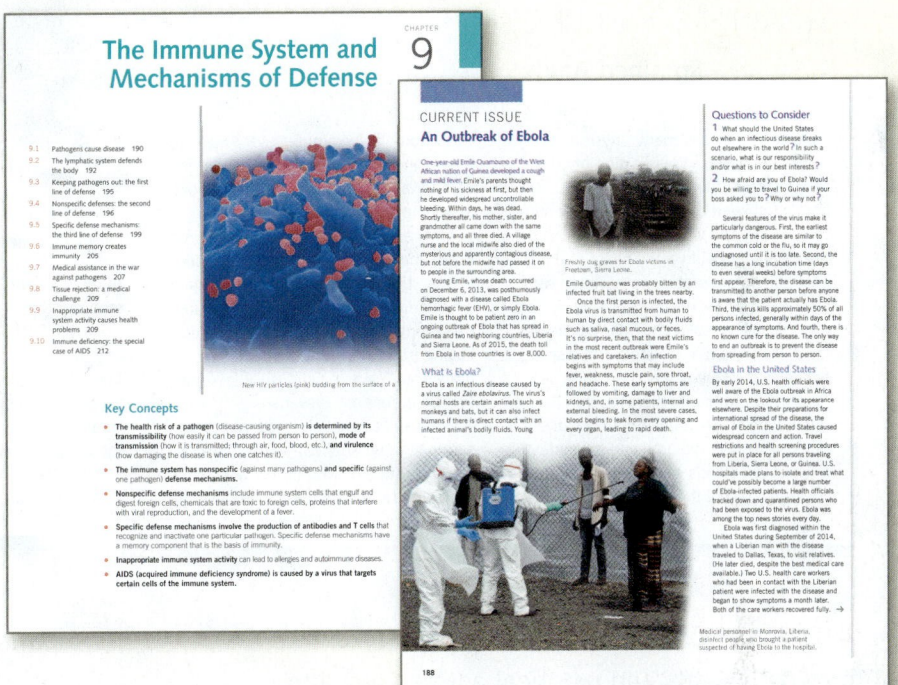

Questions to Consider at the end of each essay ask you to form your own opinions on the featured issue.

Key Concepts

- **The health risk of a pathogen** (disease-causing organism) **is determined by its transmissibility** (how easily it can be passed from person to person), **mode of transmission** (how it is transmitted; through air, food, blood, etc.), **and virulence** (how damaging the disease is when one catches it).
- **The immune system has nonspecific** (against many pathogens) **and specific** (against one pathogen) **defense mechanisms.**
- **Nonspecific defense mechanisms** include immune system cells that engulf and digest foreign cells, chemicals that are toxic to foreign cells, proteins that interfere with viral reproduction, and the development of a fever.
- **Specific defense mechanisms involve the production of antibodies and T cells** that recognize and inactivate one particular pathogen. Specific defense mechanisms have a memory component that is the basis of immunity.
- **Inappropriate immune system activity** can lead to allergies and autoimmune diseases.
- **AIDS (acquired immune deficiency syndrome) is caused by a virus that targets certain cells of the immune system.**

NEW! **Key Concepts** are now listed at the beginning of each chapter for a handy "big picture" overview of topics that will be discussed in greater detail in the pages that follow.

Questions to Consider

1 What should the United States do when an infectious disease breaks out elsewhere in the world? In such a scenario, what is our responsibility and/or what is in our best interests?

2 How afraid are you of Ebola? Would you be willing to travel to Guinea if your boss asked you to? Why or why not?

CONNECT CONCEPTS AND APPLICATIONS TO EVERYDAY LIFE

HEALTH & WELLNESS

Treating a Sprained Ankle

For a severe sprain, many physicians advise the frequent application of cold to the sprained area during the first 24 hours, followed by a switch to heat. Why the switch, and what is the logic behind the timing of cold versus heat? The biggest immediate problem associated with a sprain is damage to small blood vessels and subsequent bleeding into the tissues. Most of the pain associated with a sprain is due to the bleeding and swelling, not damage to ligaments themselves. The immediate application of cold constricts blood vessels in the area and prevents most of the bleeding. The prescription is generally to cool the sprain for 30 minutes every hour or 45 minutes every hour and a half. In other words, keep the sprain cold for about half the time, for as long as you can stand it. The in-between periods ensure adequate blood flow for tissue metabolism. It's also a good idea to keep the ankle wrapped in an elastic bandage and elevated between cooling treatments, to prevent swelling. If you're having trouble remembering all this, remember the acronym "RICE"—Rest, Ice, Compression, Elevation.

The key to a quick recovery from a sprain is rapid application of the RICE method. Athletes who try to "work through the pain" by continuing to compete while

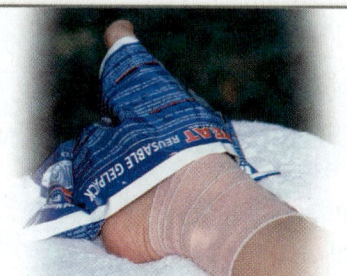

Treat sprains first with cold, then later with heat.

injured generally pay the price in a longer recovery time.

After 24 hours there shouldn't be any more bleeding from small vessels. The damage has been minimized, so now the goal is to speed the healing process. Heat dilates the blood vessels, improves the supply of nutrients to the area, and attracts blood cells that begin the process of tissue repair.

UPDATED! **Health & Wellness boxes** provide insights and practical advice on health topics, such as the causes and risks of carbon monoxide poisoning and the prevalence and consequences of Viagra abuse.

NEW! **Health & Wellness boxes include:**
- Donating Blood (Chapter 7)
- Water Intoxication (Chapter 15)
- What If You Could Save Someone's Life? (Chapter 18)

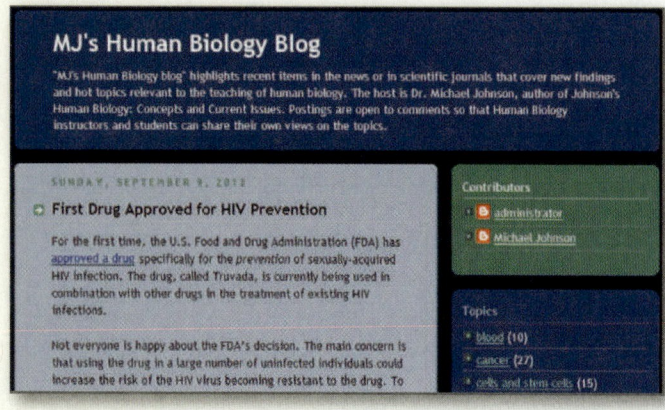

Michael Johnson's blog also features posts on recent health and wellness related news items.

NEW! "**Connections**" passage at the start of each chapter provides the student with a sense of how a chapter's specific topic interrelates with the overall subject of human biology, biology in general, and to the larger world.

Way back when life began, a single cell floating freely in the primordial sea received all its nutrients from the surrounding fluid and dumped all its wastes into it. Today, a single cell in the human body still does essentially the same thing; it receives its nutrients from (and dumps its waste into) the surrounding fluid, called the *interstitial* (between cells) *fluid*. In the human body, though, the cells are packed closely together, with very little fluid between them. A human cell would soon starve to death in a sea of waste if not for blood circulating through nearby blood vessels. Blood picks up nutrients from the digestive tract. It transports waste carbon dioxide gas to the lungs and picks up much-needed oxygen. It transports the waste products of metabolism to the liver for destruction or to the kidneys for removal from the body. It even transports waste heat to the skin, as part of the control mechanism for regulating body temperature. Last but not least, blood contains specialized cells of the immune system that are essential to our defense against invading microorganisms. Always and everywhere throughout the body, blood is seeing to it that each living cell is bathed in a fluid conducive to life. Blood is our internal primordial ocean. ∎

Steroid hormones enter target cells

Figure 13.2 depicts the mechanism of action for steroid hormones. Recall that the cell membrane is primarily composed of a bilayer of phospholipids. **①** Because they are lipid soluble, steroid hormones can easily diffuse right across both the cell membrane and the nuclear membrane. Once inside the cell, steroid hormones bind to specific hormone receptors, forming a hormone-receptor complex either within the nucleus or within the cytoplasm (not shown). If the hormone-receptor complex was formed in the cytoplasm, it too can diffuse into the nucleus.

② Once inside the nucleus, the hormone-receptor complex attaches to DNA, activating specific genes. **③** Gene activation causes the formation of messenger RNA, which then leaves the nucleus and directs the synthesis of certain proteins. **④** The proteins then carry out the cellular response to the hormone, whatever it might be.

Steroid hormones tend to be slower acting than nonsteroid hormones because the entire protein production process (starting with gene activation) begins only after arrival of the steroid hormone within the cell. Starting from the time the steroid hormone first enters the cell, it can take minutes or even hours to produce a new protein.

Nonsteroid hormones bind to receptors on target cell membranes

Nonsteroid hormones have an entirely different mechanism of action from steroid hormones (**Figure 13.3**). Nonsteroid hormones cannot enter the target cell because they are not lipid soluble. Instead, they bind to receptors located on the outer surface of the cell membrane. The receptors are generally associated with, or are part of, protein molecules floating in the phospholipid bilayer of the cell membrane (Chapter 3). **①** The binding of hormone to receptor causes a change in the shape of the membrane protein, which in turn initiates a change within the cell. It's like turning the lights

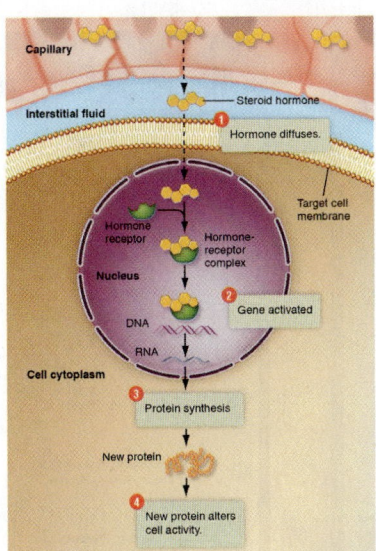

Figure 13.2 Mechanism of steroid hormone action on a target cell. Lipid-soluble steroid hormones diffuse across the cell and nuclear membranes into the nucleus, where they bind to hormone receptors that activate genes. Gene activation results in the production of a specific protein.

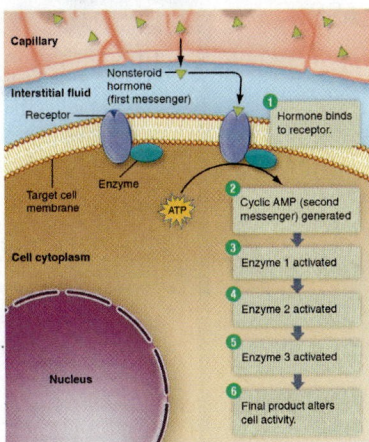

Figure 13.3 Mechanism of nonsteroid hormone action on a target cell. Nonsteroid hormones bind to receptors in the cell membrane, leading to the conversion of ATP to cyclic AMP (the second messenger) within the cell. Cyclic AMP initiates a cascade of enzyme activations, amplifying the original hormonal signal and generating a cellular response.

NEW! **The use of numbered steps.** Where complex processes are described, numbered step icons are included both in the text and in the accompanying figure. These correlating step icons help students follow the logical sequence of events as those events unfold within a complex process.

SUPPORT FOR STUDENTS
ANYTIME, ANYWHERE

MasteringBiology®

is an online homework, tutorial, and assessment program that helps you quickly master biology concepts and skills. Self-paced tutorials provide immediate wrong-answer feedback and hints to help keep you on track to succeed in the course.

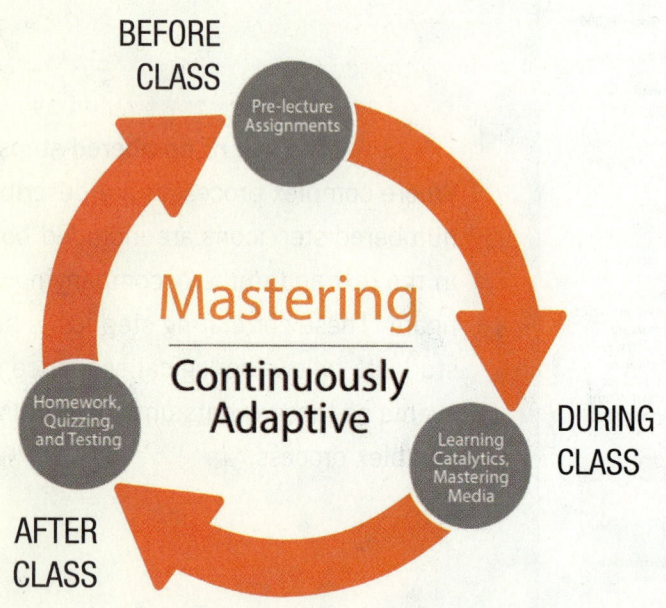

BEFORE CLASS

NEW! **eText 2.0** Allow your students to access their text anytime, anywhere.

- Now available on smartphones and tablets
- Seamlessly integrated digital and media resources
- Fully accessible (screen-reader ready)
- Configurable reading settings, including resizable type and night reading mode
- Instructor and student note-taking, highlighting, bookmarking and search

NEW! **Dynamic Study Modules** help students acquire, retain, and recall information faster and more efficiently than ever before. These convenient practice questions and detailed review explanations can be accessed using a smartphone, tablet, or computer.

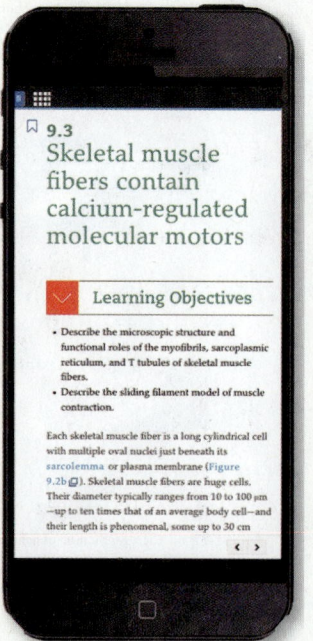

eText 2.0

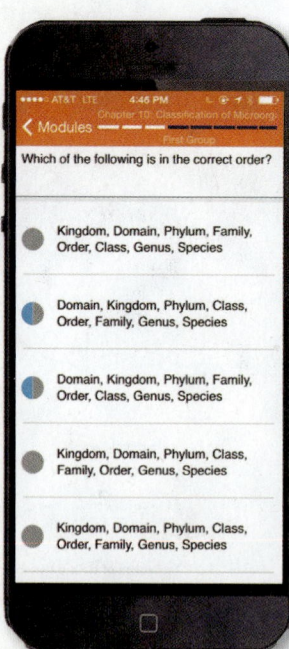

Dynamic Study Modules

DURING CLASS

Image copyright: Pearson Education, Inc.
What sequence of nucleotides would you expect to see IN THE BOX in the image?

(List sequence from the left to right.)

A. GTACA
B. GUACA
C. CAUGU
D. CATGT

NEW! **Learning Catalytics** is an assessment and classroom activity system that works with any web-enabled device and facilitates collaboration with your classmates. Your MasteringBiology subscription with eText includes access to Learning Catalytics.

◀ **NEW!** **Everyday Biology Videos** briefly explore interesting and relevant biology topics that relate to concepts in the course. These 20 videos, produced by the BBC, can be shown in class or assigned as homework in MasteringBiology.

AFTER CLASS

A wide range of question types and activities are available for homework assignments, including the following assignment options for the Eighth Edition:

- **NEW!** **Interactive Physiology 2.0 tutorials** help students advance beyond memorization to a genuine understanding of complex physiological processes. Full-color animations and videos demonstrate difficult concepts to reinforce the material. IP 2.0 features brand new graphics, quicker navigation, and more robust mobile-ready interactivities where students can explore, experiment, and predict.

- **Blog In Focus activities** ask students to read Michael Johnson's blog and answer questions.

- **NEW!** **Evaluating Science in the Media activities** challenge you to evaluate various types of information from web sites, articles, and videos.

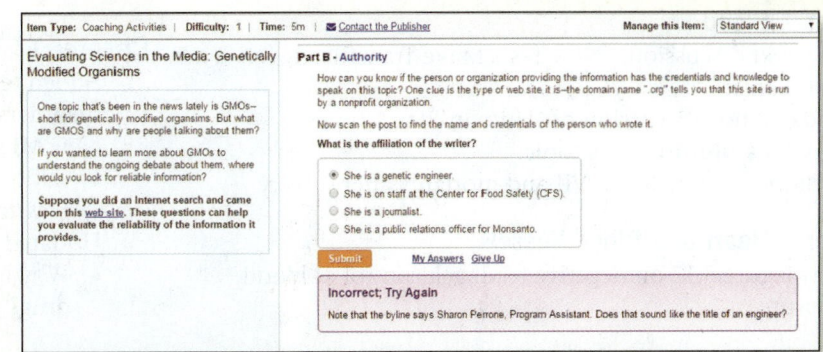

CHAPTER-SPECIFIC CHANGES

Chapter 1 Human Biology, Science, and Society
- Updated *Current Issue*, "Mandated Childhood Vaccinations"
- New **MJ's BlogInFocus** topics:
  - The human hand makes a good fist
  - Science and the popular press
- New **MJ's BlogInFocus** question:
  - Who should pay for very expensive drugs?

Chapter 2 The Chemistry of Living Things
- Updated Figure 2.1, the Periodic Table of the Elements
- New material added on free radicals
- New Figure 2.7, a polar molecule
- Redrawn Figure 2.13, dehydration synthesis and hydrolysis
- New **MJ's BlogInFocus** topic:
  - The purity of herbal supplements
- New **MJ's BlogInFocus** question:
  - Should the government approve a powdered alcohol product?

Chapter 3 Structure and Function of Cells
- New **MJ's BlogInFocus** topic:
  - An inexpensive microscope

Chapter 4 From Cells to Organ Systems
- Updated *Current Issue*, "Reshaping Your Body"
- New **MJ's BlogInFocus** topics:
  - Wearable skin patches
  - Severe sunburns and risk of melanoma

Chapter 5 The Skeletal System
- New Figure 5.14, the fontanels in a baby's head
- New **MJ's BlogInFocus** topics:
  - Bone density scans to measure bone mass
  - Why are knee and hip surgeries on the rise?
  - Smoking and bone deposition in young women

Chapter 6 The Muscular System
- Updated *Current Issue*, "Drug Abuse Among Athletes"
- New **MJ's BlogInFocus** topics:
  - Growth hormone to athletic performance
  - Xenon gas and athletic performance

Chapter 7 Blood
- New text discussion, "New Tests Make Transfused Blood Safer"
- Updated text discussion of "Human Blood Types"
- New **MJ's BlogInFocus** topic:
  - Blood clotting factor VII and mortality after surgery

Chapter 8 Heart and Blood Vessels
- New Figure 8.17 on negative feedback control of blood pressure

- New **MJ's BlogInFocus** topics:
  - A new class of cholesterol-lowering drugs
  - A next-generation artificial heart

Chapter 9 The Immune System and Mechanisms of Defense
- New *Current Issue*, "An Outbreak of Ebola"
- New Figure 9.24, persons living with HIV by sex and exposure categories
- New **MJ's BlogInFocus** topics:
  - Purchasing human breast milk
  - A home test for HIV
- New **MJ's BlogInFocus** question:
  - Why is a pill to prevent HIV infection not popular?

Chapter 10 The Respiratory System: Exchange of Gases
- New *Current Issue*, "The Fight over Regulation of E-Cigarettes"
- New **MJ's BlogInFocus** topics:
  - Smoking and shortened life expectancy
  - Who should be screened for lung cancer?
- New **MJ's BlogInFocus** question:
  - Why are rates of smoking continuing to decline?

Chapter 11 The Nervous System: Integration and Control
- Replaced Figures 11.15, parts of the brain, and 11.17, the limbic system, with new art
- Streamlined the discussions of sleep and wakefulness
- Revised the section on the limbic system
- Added a passage on changing societal views on marijuana
- New **MJ's BlogInFocus** topics:
  - New ways to diagnose a concussion
  - An outbreak of meningitis

Chapter 12 Sensory Mechanisms
- Added a new figure to the Health & Wellness on Lasik eye surgery
- Updated **MJ's BlogInFocus** topic:
  - State laws on texting while driving

Chapter 13 The Endocrine System
- Added text discussion on hypogonadism to endocrine disorders section
- New **MJ's BlogInFocus** topics:
  - Endocrine disruptor Bisphenol A (BPA) in food cans
  - Low testosterone; how common is it?
- Updated **MJ's BlogInFocus** question:
  - Why hasn't inhalable insulin been a blockbuster drug?

Chapter 14 The Digestive System and Nutrition

- New *Current Issue*, "Choosing Organic Versus Conventional Foods"
- New Health & Wellness, "Should You Drink Raw Milk?"
- Added new Figure 14.8, peptic ulcers
- Added new Figure 14.17 on saturated and unsaturated fats
- New **MJ's BlogInFocus** topics:
  - Antioxidants in organic foods
  - A human feces bank
  - Drinking bone broth for good health?
- New **MJ's BlogInFocus** question:
  - What do food "sell by" and "best if used by" dates mean?

Chapter 15 The Urinary System

- Revised *Current Issue*, "A Shortage of Kidneys" to include recent changes in the allocation procedure
- New Health & Wellness, "Water Intoxication"
- Revised Figure 15.10, urinary dilution and concentration
- Revised several pieces of nephron art for consistency and clarity
- Expanded the discussion of acute and chronic renal failure
- New **MJ's BlogInFocus** topics:
  - Economic theory and kidney donations
  - The connection between a kidney disease and African sleeping sickness

Chapter 16 Reproductive Systems

- Combined old Figures 16.7 and 16.8, both on the menstrual cycle, into new Figure 16.7
- Re-rendered Figure 16.11, pelvic inflammatory disease
- New **MJ's BlogInFocus** topics:
  - A state restricts the use of Mifeprex
  - What is an embryoscope?
  - Does vaccination against HPV change sexual behavior?

Chapter 17 Cell Reproduction and Differentiation

- New *Current Issue*, "Therapeutic Cloning"
- New **MJ's BlogInFocus** topics:
  - An anti-aging protein in blood
  - Cloning goes commercial
- New **MJ's BlogInFocus** question:
  - Could stem cells be used to produce edible meat?

Chapter 18 Cancer: Uncontrolled Cell Division and Differentiation

- Updated *Current Issue*, "Preventive Double Mastectomy to Reduce Breast Cancer Risk"
- Added a discussion of cancer stages
- New Health & Wellness, "What If You Could Save Someone's Life"
- New Figure 18.4 on proto-oncogenes and tumor suppressor genes

- Added the current recommendations regarding mammograms and self-examination for detecting breast cancer
- New discussion of pancreatic cancer
- New discussion and Figure 18.12, esophageal cancer
- New **MJ's BlogInFocus** topics:
  - Radiation therapy for cancer can sometimes cause cancer
  - Double mastectomies to prevent breast cancer
  - An alternative to the Pap test for cervical cancer

Chapter 19 Genetics and Inheritance

- Revised the *Current Issue* for greater emphasis on the risks and benefits of genetic testing.
- New **MJ's BlogInFocus** topics:
  - Marketing genetic tests and predicting risk of genetic disease
  - State laws on screening newborns for genetic diseases.
  - Accuracy of commercially available genome tests

Chapter 20 DNA Technology and Genetic Engineering

- Revised *Current Issue*, "Genetically Modified Plants" to include concerns about labeling GM foods
- Expanded the Health & Wellness feature on DNA-based vaccines
- Updated the text discussion of DNA fingerprinting
- New **MJ's BlogInFocus** topics:
  - Patenting human genes
  - The long-term effects of herbicide resistance in weeds
  - A genetically modified potato
- New **MJ's BlogInFocus** question:
  - Are alcoholic beverages made from non-GMO grains any safer to drink?

Chapter 21 Development, Maturation, Aging, and Death

- New chapter title includes "Maturation" and "Death"
- New *Current Issue*, "Death with Dignity (Britanny Maynard's Journey)"
- New **MJ's BlogInFocus** topics:
  - Taking acetaminophen during pregnancy
  - When should the umbilical cord be cut?
- New **MJ's BlogInFocus** question:
  - Why do older fathers pass on more genetic mutations to their offspring than mothers?

Chapter 22 Evolution and the Origins of Life

- New Figure 22.8 to illustrate genetic drift and gene flow
- New **MJ's BlogInFocus** topics:
  - Pinpointing the time of Earth's largest mass extinction
  - How many species of extinct humans are there?

Chapter 23 Ecosystems and Populations

- New *Current Issue*, "Overharvesting is Depleting the Oceans' Wildlife Populations"
- Updated Figure 23.15 and the text discussion, human population dynamics
- New **MJ's BlogInFocus** topics:
  - Why societies collapse

Chapter 24 Human Impacts, Biodiversity, and Environmental Issues

- Updated *Current Issue*, "Global Warming and Global Climate Change"

- Revised Figure 24.2 on solar radiation and the greenhouse effect
- New **MJ's BlogInFocus** topics:
  - Regional climate changes due to global warming
  - Depletion of a freshwater aquifer
  - Advanced biofuels
- New **MJ's BlogInFocus** question:
  - What is a carbon tax?

ACKNOWLEDGMENTS

The Eighth Edition of *Human Biology: Concepts and Current Issues* is once again the product of the continued hard work and dedication of the people at Pearson Education, led by VP, Editor-in-Chief Beth Wilbur, Executive Editorial Manager Ginnie Simione Jutson, and Senior Acquisitions Editor Star Mackenzie Burruto. Star directs a team that functions as smoothly and professionally as any in the business.

On a day-to-day basis, I depended on Developmental Editor Susan Teahan. Her experience, her insight, and above all, her dogged determination to get it exactly right have made this edition what it is. I am forever grateful for her support and counsel.

Changes to the art and photos in the Eighth Edition are the result of the hard work of artists at Imagineering and Rights and Permissions Project Managers Donna Kalal, at Pearson Education, and Candice Velez, at QBS Learning. Photo Researcher Pat Holl found the new photos you see in this edition.

Accuracy and clarity have been checked and rechecked by the hundreds of insightful faculty members around the country over the past 10 years. Reviewers specific to this edition are listed below.

Thanks go to the outstanding support team at Pearson Education. It includes Project Managers Mae Lum and Brett Coker (Pearson) and Andrea Stefanowicz (Lumina Datamatics, Inc.), Program Manager Anna Amato, Editorial Content Producer Joe Mochnick, Supervising Project Manager–Instructor Media Eddie Lee, and Editorial Assistant Maja Sidzinska.

Once again, the textbook is supported by a wonderful set of ancillary materials. Thanks go to Robert Sullivan of Marist College, Hyde Park, New York, who wrote the *Instructor Resource Manual*; Suzanne Long of Monroe Community College, Rochester, New York, who developed the PowerPoint Lecture Slides; Janette Gomos Klein of Hunter College of the City University of New York and Kristine Williams and Angela Cordle of the University of Iowa, who revised the Test Bank, checked it for accuracy, and supplied the MasteringBiology Reading Quizzes; Maria Cendon of Miami Dade College, Miami, who was responsible for the Instructor Quiz Shows; and Julie Posey of Columbus State Community College, who revised the Mastering Study Area quizzes. Finally, thanks to Bert Atsma of Union County College, whose Laboratory Manual continues to complement the best human biology courses.

Last but not least, I'd like to thank my wife, Pamela, for her wholehearted support and understanding over the years.

Reviewers of the eighth edition

Laurel Carney-Zelko,
Joliet Junior College

Renee Ehrenstrom,
Missouri State University

Michele Finn,
Monroe Community College

Sheldon Gordon,
Oakland University

Sarah Hanna,
University of Wisconsin, Green Bay

Janette Klein,
Hunter College, The City University of New York

Jacqueline Nesbit,
University of New Orleans

Joanne Oellers,
Yavapai College

David Opon,
Joliet Junior College

Tara Reed,
University of Wisconsin, Green Bay

Derek Sims,
Hopkinsville Community College

Larry Taylor,
Arapahoe Community College

Diane Wickham,
Trocaire College

Reviewers of the seventh edition

Andrea Abbas,
Washtenaw Community College

Wade Bell,
Virginia Military Institute

Samantha Butler,
University of Southern California

Anne Casper,
Eastern Michigan University

Chris Chabot,
Plymouth State University

Jennifer Ellie,
Wichita State University

Tom Kennedy,
College of New Mexico

Gary Lange,
Saginaw Valley State University

Suzanne Long,
Monroe Community College

Nancy O'Keefe,
Purdue University Calumet

Samiksha Raut,
Dalton State College

Lisa Runco,
New York Institute of Technology

Sean Senechal,
CSU Monterey Bay

Derek Sims,
Hopkinsville Community College

Corinne Ulbright,
Indiana University–Purdue University, Indianapolis

Peggy Wright,
Columbia College

Human Biology, Science, and Society

Crew of the space shuttle *Atlantis*, November 20, 2007.

Key Concepts

- **Living things have certain characteristics** that make them different from nonliving things. All living things are composed of cells that harness energy to create unique chemical compounds. Living things grow and reproduce.

- **Humans are just one of several million different life-forms** on Earth. Our closest relatives are the other primates (including monkeys and apes). Features that *taken together* define humans as unique are bipedalism, opposable thumbs, a large brain, and a capacity for complex language.

- **Science is a process for studying the natural world.** It is based on observable, quantifiable data obtained by repeatedly questioning, observing, and drawing conclusions.

- **Science helps us understand what *is,* not what *should* be.** It does not provide us with "right" answers or give meaning to our lives.

- **We make choices** about how to use scientific knowledge every day whether we are consciously aware of it or not. We owe it to ourselves to make informed choices.

Mandatory Childhood Vaccinations

Questions to Consider

1 What should medical professionals, politicians, or even just concerned citizens do, if anything, to help parents understand the risks and benefits of vaccines?

2 Will you vaccinate your children? Why or why not? What would you like to know in order to make an informed decision?

All 50 states now require that school-age children be properly vaccinated before they can attend school. The trend is toward requiring specific vaccinations even for preschoolers. In 2009, New Jersey became the first state to require a vaccination against the flu for children who attend licensed day care and preschool programs. Connecticut followed suit in 2010, as did New York in 2014.

At the same time, more and more parents are seeking exemptions from vaccinations for their children. (All 50 states permit an exemption for medical reasons and 48 states also allow for an exemption for religious or personal beliefs.) What is going on?

Childhood Vaccinations Save Lives

The states' rationale is clear: Childhood vaccines introduced since the 1950s have all but wiped out many communicable diseases in the United States, including measles, mumps, whooping cough (pertussis), polio, and diphtheria. In the 1940s and 50s, before vaccines against these diseases were available, the five diseases combined caused an estimated 900,000 cases of disease and 7,700 deaths per year. By 2004, there were only 27 deaths from all five diseases combined—a 99.6% reduction. The number of cases of measles dropped from more than 500,000 per year before the measles vaccine was available to about 60 cases per year between 2001 and 2010.

Vaccines Become Controversial

In 1998, the prestigious medical journal *The Lancet* published a paper in which the author concluded that the vaccine for measles, mumps, and rubella (or a preservative in the vaccine, called thimerosal) was a likely cause of autism. Autism spectrum disorder, as it is more properly called, is a baffling group of neurological disorders that lead to social, communication, and behavioral difficulties. It generally develops at about the same time that most children are vaccinated. Since the cause of autism was not known at the time (and still isn't known), the paper caused widespread concern. However, the

paper was later shown to be fraudulent and was retracted, and the author was found guilty of serious professional misconduct. Unfortunately, those facts barely made the news.

Since that paper appeared, scientists have searched for any connection between vaccinations and autism and have failed to find one. Nevertheless some parents, including actress and former *Playboy* model Jenny McCarthy and actress Alicia Silverstone, continue to promote their belief that vaccines may cause autism. Jenny McCarthy is on the board of Generation Rescue, a nonprofit organization that claims to be able to treat autism effectively with a special diet. Alicia Silverstone has written a book about parenting in which she dismisses the scientific evidence for vaccine safety. For parents unwilling to accept the research, emotions (fear and anxiety) are likely to continue to trump science until we know for certain what does cause autism. Science is just no match for an appearance by Ms. McCarthy on the *Oprah Winfrey Show* with an emotional story about an autistic child.

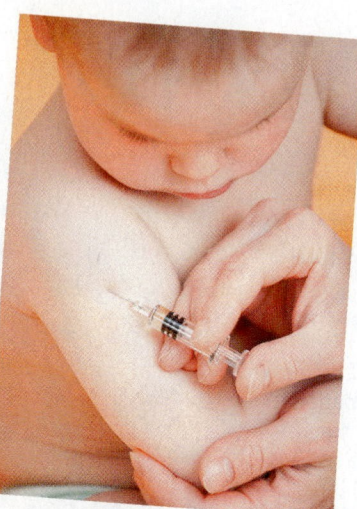

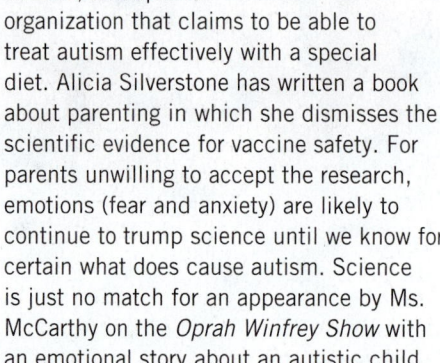

A child receiving a vaccination.

Vaccination Rates Decline, Preventable Diseases Return

In recent years, the number of exemptions from school immunization programs has increased. These exemptions, granted for philosophical or personal beliefs, coincide with a sharp uptick in the number of cases of measles and whooping cough. Because of their highly contagious nature, the two diseases are extremely sensitive to vaccination rates. In the first eight months of 2014, there were nearly 600 cases of measles in the United States, according to the Centers for Disease Control and Prevention (CDC). Nearly all of the measles victims had not been

vaccinated, even though they were old enough. The CDC recommends that children be vaccinated against measles at 1 year old.

Public health officials are watching the decline in vaccinations against measles with growing concern. They know that the success of any vaccine is based on a concept called *herd immunity*. When most people in a community, or herd, have been vaccinated, a disease has a much harder time spreading among unvaccinated people. So, in addition to protecting the person who has been vaccinated, high vaccination rates protect the community as a whole from widespread disease outbreaks, especially among young children. Although there are always some people who aren't vaccinated and therefore at risk of contracting vaccine-preventable diseases, herd immunity substantially undercuts that risk. People who aren't vaccinated include children under 1 year old, whose immune systems are not yet developed enough for vaccines to be effective, and patients receiving chemotherapy or immunosuppressive therapy, whose immune systems are compromised.

Mandatory Vaccinations Remain Controversial

Compared to parents who vaccinate their children, parents who choose not to vaccinate their children are more likely to believe that the risk of their child getting a contagious disease is low and that the disease itself is not severe. The latter view is understandable, because most parents today have not lived through a major outbreak of any communicable disease. Today's parents were born after the →

scourge of polio, for example. Polio killed nearly 10% of its victims and crippled countless others for life before the polio vaccine became available in 1955.

Some parents oppose mandatory childhood vaccinations because they are philosophically opposed to government intervention into what they see as a personal choice. Says Barbara Loe Fisher, a mother and the cofounder of the National Vaccine Information Center, representing parents against forced vaccinations, ". . . If the State can tag, track down and force citizens against their will to be injected with biologicals of unknown toxicity today, there will be no limit on which individual freedoms the State can take away in the name of the greater good tomorrow."[1]

Parents in favor of mandatory vaccinations are mounting lobbying campaigns as well. Their celebrity advocate is actress Amanda Peet, now a spokesperson for Every Child By Two, a vaccine-advocacy group founded by former first lady Rosalynn Carter.

Ms. Peet once called anti-vaccine parents "parasites" for relying on other children's immunity to protect their own. She later apologized for the word and suggested that parents should get their advice from doctors, not celebrities like herself (and presumably Ms. McCarthy and Ms. Silverstone).

Health officials continue to stress that vaccines don't cause autism. It would be a shame if misinformation and fear allowed preventable diseases such as polio to return. We need to find a way to address parents' concerns about vaccine safety and about the role of government in our lives, while at the same time protecting the public from preventable, communicable diseases. How we do that is up to all of us.

_____

[1] www.vaccineawakening.blogspot.com

SUMMARY

- **Childhood vaccination programs have been effective in all but eliminating certain communicable diseases.**
- **All 50 states have childhood vaccination (immunization) programs as a requirement for school attendance—all states also allow for certain exemptions.**
- **Exemptions from vaccination (and communicable diseases) are on the rise. Many parents object to mandatory vaccination programs out of concern that the vaccines may cause autism or certain other chronic childhood diseases.**
- **The available scientific evidence does not support the argument that vaccinations can cause childhood diseases, including autism.**

You were born into exciting times, when scientific discoveries are happening more rapidly than at any other time in human history. Like the Industrial Revolution of the nineteenth century and the discovery of DNA in the twentieth, today's scientific innovations will change the human condition forever.

In your lifetime, people may be able to select or modify their children's features before they are born. People may even be able to have clones (copies) made of themselves. At the very least, certain diseases that threaten us now will become curable. Perhaps your grandchildren will not even know what AIDS is because the disease will have disappeared.

What you are witnessing is the power of science. **Science** is the study of the *natural world*, which includes all matter and all energy. Because all living organisms are also made of matter and energy, they are part of the natural world. Biology is one of many branches of science. More specifically, **biology** (from the Greek words *bios*, meaning "life," and *logos*, meaning "word or thought") is the study of living organisms and life's processes. It is the study of life. Within biology, *anatomy* is the study of structure and *physiology* is the study of function. Other branches of science are chemistry, physics, geology, astronomy, and related fields such as medicine.

This text is specifically about *human* biology. We will explore what it means to be alive. We will see how

the molecules that make up our bodies are created from molecules in the air and in our food and drink. We will learn how the trillions of cells that comprise our bodies grow and divide. We will explore how our bodies function, why we get diseases, and how we manage to survive them. We will look at how we develop into adults, reproduce, and influence the destinies of other organisms on Earth.

A recurrent theme in all of biology is the theory of *evolution*: that over the billions of years of Earth's history, living organisms (including humans) have undergone slow change over time. Based on the evidence available to us, it is hard to escape the conclusions that all living organisms evolved from single-celled organisms and that single-celled organisms arose from nonliving chemical elements nearly 3.5 billion years ago. We'll explore evolution more thoroughly later in the book.

With the power of science comes an awesome responsibility. All of us, individually and collectively, must choose how to use the knowledge that science gives us. Will human cloning be acceptable? Can we prevent global warming? Should you be required to vaccinate your children against certain infectious childhood diseases? (See the Current Issue feature, *Mandatory Childhood Vaccinations*.)

We all have to make responsible decisions concerning not only our own health and well-being but also the long-term well-being of our species. This book considers many aspects of human connections with the natural

Photo taken from the Hubble Space Telescope showing a tiny portion of the universe. Studies by astronomers have shown that all matter on Earth originated inside stars or with the Big Bang.

Studying unusual species such as this deep sea glass squid allows biologists to understand the processes by which a species successfully survives. Many different environments exist in the world, but the same physical and chemical laws govern them all.

The natural world comprises all matter and energy. An erupting volcano spewing liquid rock and heat is the result of energy that still remains from the creation of Earth nearly 4.6 billion years ago.

Jane Goodall has dedicated her life to studying the needs and behaviors of chimpanzees. The DNA of humans and chimps is almost the same, yet important physical and behavioral differences are obvious. Evolution examines how these differences arose.

Figure 1.1 Studies of the natural world.

world (**Figure 1.1**). We'll contemplate how humans function within the environment, as well as the impact of humans on the environment. Along the way, we'll confront a variety of social and personal issues and discuss the choices we might make about them. Finally, we'll discuss our place in Earth's history when we explore the topic of evolution. Because biology is the study of life, we begin by defining life itself.

↺ **Recap** Science is the study of the natural world, which consists of all matter and energy. Biology is the study of living organisms. ∎

1.1 The characteristics of life

What is life? On the one hand, this question seems easy and on the other hand so abstract that it is more like a riddle. We all think we can recognize life even if we can't define it easily. Children learn early to distinguish between living and

nonliving things. Remember that childhood game "animal, vegetable, or mineral"? In it, children distinguish what is alive (animals and plants) from what is not (minerals). Most biologists accept the following criteria as signs of life:

- *Living things have a different molecular composition than nonliving things.* Everything in the natural world, both living and nonliving, is composed of the same set of approximately 100 different chemical *elements*. However, only a few elements are present in any abundance in living organisms. In addition, living organisms can combine elements in unique ways, creating certain *molecules* (combinations of elements) that nonliving things cannot create. These molecules of life (proteins, carbohydrates, lipids, and nucleic acids) are found in all living organisms and often persist in the remains of dead organisms. Variations in these molecules in different life-forms account for the diversity of life.

- *Living things require energy and raw materials.* The creation of the molecules of life doesn't happen by accident, at least under present conditions on Earth. The transformation of molecules from one form to another requires energy. The term **metabolism** refers to the physical and chemical processes involved in transforming energy and molecules so that life can be sustained. All living things take in raw materials and energy from the environment and metabolize them into the molecules and energy that they need to survive. Plants use the energy of sunlight and chemicals obtained from soil, water, and air. Animals and all other forms of life ultimately obtain their energy and raw materials from water, air, plants, or other animals.

- *Living things are composed of cells.* A **cell** is the smallest unit that exhibits all the characteristics of life. All cells originate only from existing cells. Living things always have at least one cell, and some organisms (called *unicellular* organisms) are *only* one cell (**Figure 1.2a**). *Multicellular* organisms are composed of many cells or many different types of cells (**Figure 1.2b**).

- *Living things maintain homeostasis.* All living organisms must maintain an internal environment compatible with life, and the range of chemical and physical conditions compatible with life is very narrow. The maintenance of a relatively constant internal environment is called **homeostasis.** Living things have developed remarkable ways of regulating their internal environment despite sometimes dramatic changes in the external environment. Single cells and unicellular organisms are surrounded by a membrane that allows the cell (or organism) to maintain internal homeostasis by providing a selective barrier to the entry and exit of various substances. In multicellular organisms, the tissues, organs, and organ systems work together to maintain homeostasis of the fluid that surrounds all cells. We discuss the importance of homeostasis further in Chapter 4.

- *Living things respond to their external environment.* Stay out in the cold too long and you are likely to respond by moving to a warm room. Plants respond to their environment by turning their leaves toward light or by growing roots toward sources of nutrients and water. Even bacteria respond to their environment by moving toward nutrients (and away from noxious stimuli) and by increasing their growth rate.

- *Living things grow and reproduce.* Living organisms have the capacity to grow and ultimately produce more living organisms like themselves (**Figure 1.3**).

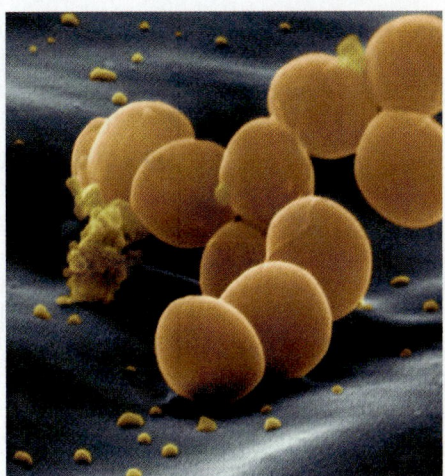

a) Several *Staphylococcus aureus*, one of several types of bacteria that causes food poisoning (SEM ×50,000).

b) Some of the many cells that line the inner surface of the human stomach (SEM ×500).

Figure 1.2 Cells are the smallest units of life. Some organisms consist of just one cell (unicellular), whereas others contain many cells (multicellular).

Mycobacterium dividing

Milkweed going to seed

Bison and calf

Figure 1.3 Living things grow and reproduce.

The ability to grow and reproduce is determined by the genetic material in cells, called d̲eoxyrib̲on̲ucleic a̲cid (DNA). Some nonliving things can get larger, of course; examples are glaciers and volcanic mountains. However, they cannot create copies of themselves.

- *Populations of living things evolve.* The various forms of life may change over many generations, a process known as evolution. Evolution explains why there are so many different forms of life on Earth today.

✔ Why do living things require energy? Where does that energy ultimately come from?

Although all these characteristics are necessary to describe life fully, not all of them apply to every living thing all the time. Individual organisms do not evolve nor do they necessarily reproduce or always respond to their surroundings. However, populations of similar organisms have the *capacity* to perform these functions.

↺ Recap All living things are composed of cells. Living things require energy and raw materials, maintain homeostasis, respond to their external environments, grow and reproduce, and evolve over many generations. ■

1.2 How humans fit into the natural world

Living things are grouped according to their characteristics

To find order in the diversity of life, biologists have long sought ways to categorize living things. Although methods of classification have changed over time and several schemes can be found in scientific literature, the most widely used classification system defines all organisms by hierarchical categories from largest to smallest, based on common features. The categories of the classification system (from largest to smallest) are **domain, kingdom, phylum, order, class, family, genus,** and **species.**

The classification system begins with three domains: Bacteria, Archaea, and Eukarya. **Figure 1.4** shows how the domains are thought to have evolved from the first primitive life-forms. Organisms in the domains Bacteria and Archaea are all single-celled bacteria without nuclei. (Cells without nuclei are called *prokaryotes;* organisms whose cells have nuclei are called *eukaryotes.*) Most of the bacteria with which you may be familiar belong to the domain Bacteria, kingdom Eubacteria. The bacteria in domain Archaea, kingdom Archaebacteria tend to live in unique environments under extreme conditions, such as in deep-sea thermal vents.

All organisms whose cells have nuclei belong to the domain Eukarya. This domain is further divided into four kingdoms: Protista (protozoans and algae), Animalia (animals), Fungi (fungi), and Plantae (plants).

The kingdom **Protista** comprises unicellular and relatively simple multicellular eukaryotes such as protozoa, algae, and slime molds. The remaining three kingdoms (Animalia, Fungi, and Plantae) all consist of multicellular organisms. The criteria for classifying animals, plants, and fungi are based largely on the organism's life cycle, structure, and mode of nutrition. Plants, for example, contain a green pigment called *chlorophyll* that allows them to capture the energy of sunlight, which they convert for their own use in a process called **photosynthesis.** Animals get the energy they need

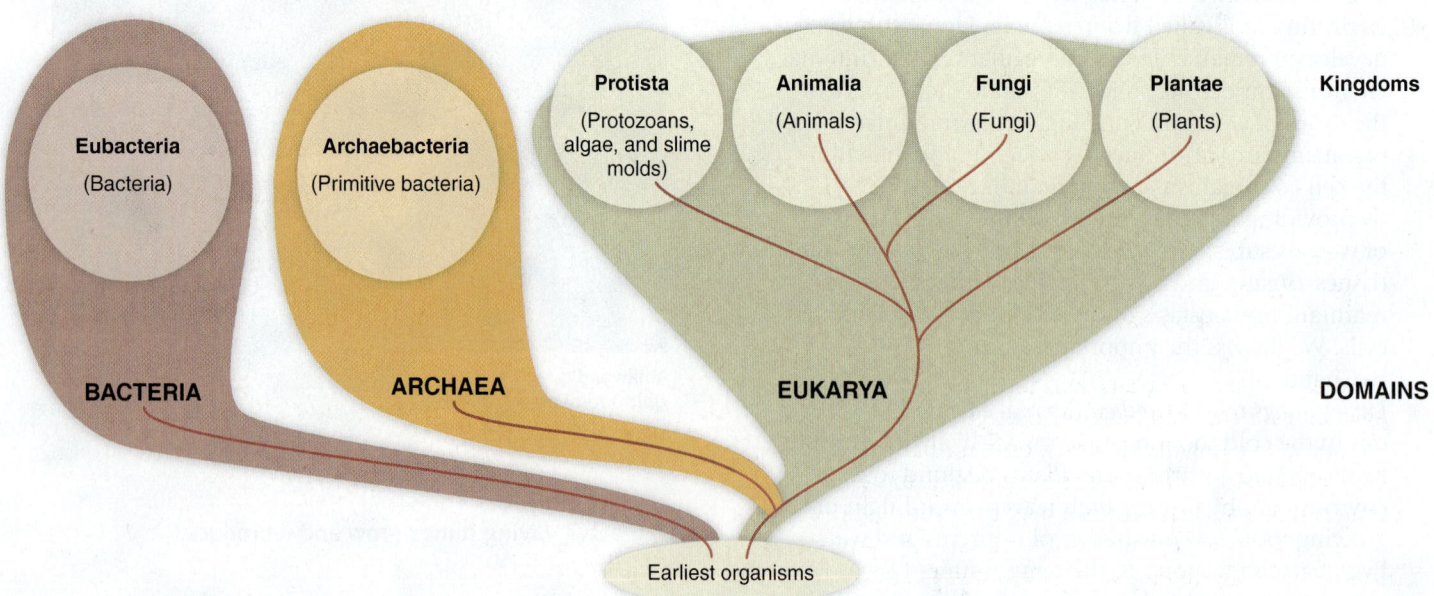

Figure 1.4 Domains and kingdoms. Classification of living organisms is based on common features. All organisms in domains Bacteria and Archaea are single-celled bacteria without nuclei (prokaryotes). Organisms in the domain Eukarya have nuclei; that is, they are eukaryotes. The domain Eukarya is subdivided into four kingdoms.

by eating plants or other animals, which requires structures specialized to digest and absorb food. Most animals can move about to obtain food. Fungi (yeasts, molds, and mushrooms) are *decomposers,* meaning that they obtain their energy from decaying material.

Humans belong to the domain Eukarya, kingdom Animalia. Based on their features in common with other animals, humans are further classified into the phylum Chordata (animals with neural cords), class Mammalia (chordates with mammary glands for nursing their young), order Primates (humans, lemurs, monkeys, and apes), family Hominidae (ancient and modern humans), and finally genus and species *(Homo sapiens).*

A species is one or more populations of organisms with similar physical and functional characteristics that interbreed and produce fertile offspring under natural conditions. All humans currently living on Earth belong to the genus and species *Homo sapiens.* We share common features that make us different from any other species on Earth, and we can interbreed.

No one knows how many species of living organisms exist on Earth. Estimates range from about 3 million to 30 million, but only about 2 million species have been identified so far.

☑ While studying a drop of pond water under a microscope, you notice two tiny single-celled organisms, one with a nucleus and one without. In what domain and kingdom is each organism? Which organism is more closely related to humans? Explain.

The defining features of humans

Humans are not the largest animal nor the fastest or strongest. Our eyesight and hearing are not the best. We cannot fly, we swim poorly, and we don't dig holes in the ground very well with our hands. Nevertheless we possess several features that, taken together, define how we are different from other organisms and explain how we have managed to survive for so long:

- *Bipedalism.* Humans are the only mammals that prefer to stand upright and walk on two legs. Bipedalism (from the Latin *bi-,* meaning "two," and *pes,* meaning "foot") frees our hands and forearms for carrying items ranging from weapons to infants. Birds walk upright, too, of course, but they do not have the advantage of being able to carry things with their forelimbs.
- *Opposable thumbs.* Humans and several other primates have thumbs that can be moved into position to oppose the tips of the fingers. However, only humans have the well-developed muscles that enable us to exert a certain type of precise control over the thumb and fingers. For instance, we tend to pick up and manipulate small objects between the tip of the thumb and the tip of either the index or second finger (**Figure 1.5**). In contrast, chimpanzees more naturally grasp objects between the thumb and the side of the index finger. Threading a needle or suturing a wound would be difficult for a chimpanzee.

MJ's **BlogInFocus** What other feature of the human hand may have conferred an evolutionary advantage? To find out, visit MJ's blog in the Study Area in MasteringBiology and look under "The Human Hand."

http://goo.gl/c2ZMbi

- *Large brain.* Humans have a large brain mass relative to body size. The evolution of a large brain seems to have coincided with the advent of stone tools, leading some scientists to suggest that a large brain was required for the complex motions associated with tool use. Other scientists believe that a large brain was necessary for language and that language developed as social interactions among humans became more important.
- *Capacity for complex language.* Many animals vocalize (produce sounds) to warn, threaten, or identify other members of their species, and a few (such as dolphins) have developed fairly complex forms of communication. However, humans have developed both complex vocal language and a system of signs, symbols, and gestures for communicating concepts and emotions. Throughout the world, every group of human beings has developed a complex spoken language. Humans have also placed their languages into written form, permitting communication over great distances and spans of time.

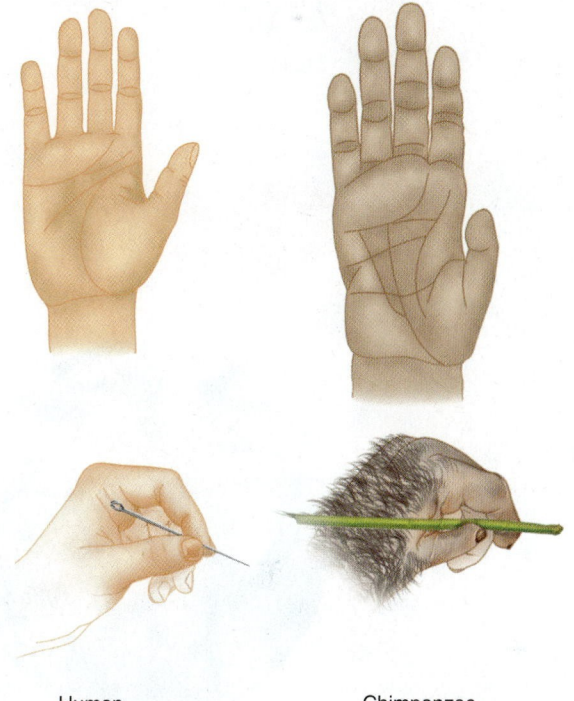

Human   Chimpanzee

Figure 1.5 How humans and chimpanzees hold small objects. Although the hands of humans and chimpanzees seem similar, only humans tend to hold objects between the tip of a thumb and the tips of the fingers.

It should be stressed that none of these features necessarily make us any better than any other species, only different.

☑ Suppose you are a scientist who has just uncovered a complete fossil skeleton of a human-like primate. Which areas of the skeleton could you study to determine whether the fossil is closely related to humans? Explain your reasoning.

Human biology can be studied on any level of biological organization

Figure 1.6 shows how humans fit into the grand scheme of things in the natural world. The figure also shows how humans—or any living thing, for that matter—can be studied on any level of biological organization, from the level of the atom to the level of the biosphere. This text examines human biology on progressively larger scales.

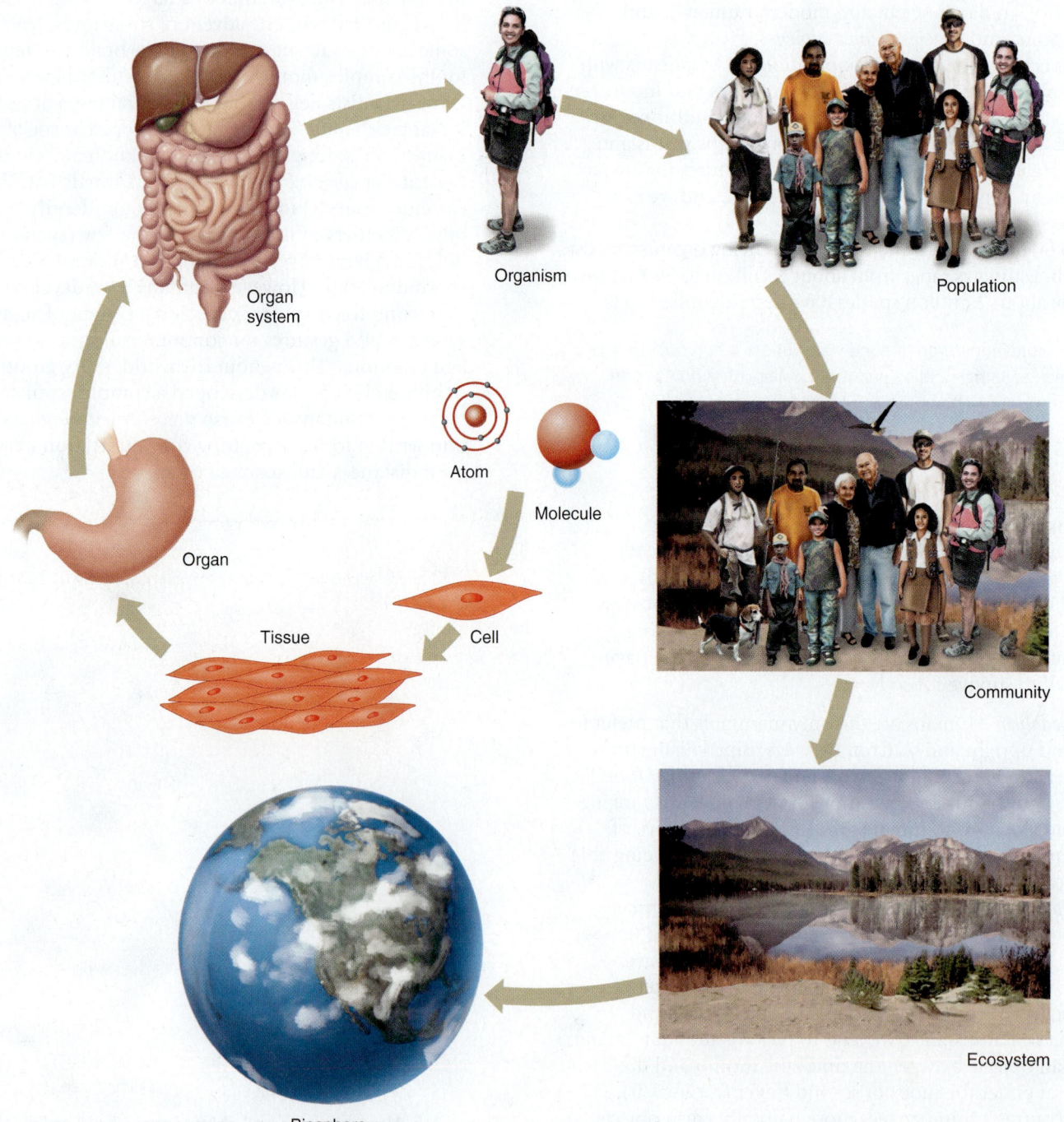

Organ system

Organism

Population

Atom

Molecule

Organ

Community

Tissue

Cell

Ecosystem

Biosphere

Figure 1.6 Levels of organization in human biology.

Our study of human biology begins with the smallest units of life. Like rocks, water, and air, humans are composed of small units of natural elements called *atoms* and *molecules*. The basic chemistry of living things will be introduced, and we'll look at how the atoms and molecules of living things are arranged into the smallest of living units, called *cells*. We'll then learn how groups of similar cells become *tissues*, how groups of tissues that carry out a specific function constitute an *organ*, and how organs work together in an *organ system* to carry out a more general function. The structures and functions of specific human organs and organ systems will be examined. For example, you'll learn why—and how—more blood flows through the lungs than through any other organ and what happens to your dinner as it makes its way through your digestive system. Humans as complete *organisms*, will be considered, including how cells reproduce, how we inherit traits from our parents, and how we develop, age, and die. Finally, we'll discuss how life began, how *communities* of living organisms evolved, and how humans fit into and alter the *ecosystems* in which we live and the entire *biosphere* of the natural world.

Table 1.1 lists some current issues, controversies, and "hot topics" that relate to human biology. Many of these issues and controversies also concern fields outside the sphere of science, such as economics, law, politics, and ethics. What you learn in this course will help you make informed decisions about these and future issues that will come up in your lifetime.

Your ability to make good judgments in the future and to feel comfortable with your decisions will depend on your critical thinking skills. We turn now to a discussion of what science is, how we can use the methodology of science to improve our critical thinking skills, and how science influences our lives.

↩ **Recap** Classification systems place living things into groups. The most inclusive group is a domain, and the smallest is a species. Humans belong to the kingdom Animalia within the domain Eukarya. Our genus and species is *Homo sapiens*. Humans walk on two legs (are bipedal) and can grasp small objects between the tips of the thumb and first finger (have opposable thumbs). Humans have large brains relative to body mass and the capacity for complex spoken and written languages. Biology can be studied at any level, from atom to biosphere. ∎

Table 1.1 Examples of issues and controversies associated with the levels of human biology

Level of organization	Definition	Issues and controversies
Atom and molecule	**Atom:** Smallest unit of an element of matter	Disposing of radioactive wastes
	Molecule: More than one atom in a stable association	Role of unstable molecules (free radicals) in cancer and aging
Cell	Smallest unit of life	Cloning adult animals, plants, and humans from a single cell
Tissue	An association of cells with the same general structure and function	Using human fetal tissues in research
Organ	An association of several tissue types that carry out a specific function	Increasing the supply of human organs for transplantation
		Transplanting animal organs into humans
Organ system	Two or more organs that work together to carry out a general function, such as digestion or movement	Enhancing human performance with drugs or by genetic engineering
Organism	An individual living being composed of several organs or organ systems	Testing for heritable diseases for which there are no cures
		Abortion
		Deciding who should pay for human behavior–related illnesses such as those caused by smoking
Population	A group of individuals of the same species living in the same area	Rationing medical care
		Determining who gets the scarce human organs available for transplantation
Community	Several populations of different species who inhabit the same area and interact with each other	Impact of humans on the well-being and survival of other species
		Genetic engineering of plants and animals for human purposes
		Using animals in medical research and cosmetics testing
Ecosystem	All of the organisms in a given area plus all of the nonliving matter and energy	Environmental pollution
		Destruction of ecosystems due to overuse by humans
Biosphere	All ecosystems combined. The portion of Earth occupied by living organisms, plus those organisms.	Global warming
		Destruction of the ozone layer

1.3 Science is both a body of knowledge and a process

Science is two things: *knowledge* (organized, reliable information) about the natural world and the *process* we use to get that knowledge. The process of science, or the way scientific knowledge is acquired, is generally called the **scientific method,** although in practice this term encompasses a variety of methods. Throughout this book, you will be presented with scientific knowledge, but it's worth remembering that this information was obtained slowly over time by the scientific method.

Scientific knowledge enables us to describe and predict the natural world. Through the scientific method, scientists strive to accumulate information that is as free as possible of bias, embellishment, or interpretation.

The scientific method is a process for testing ideas

Although there is more than one way to gather information about the natural world, the scientific method is a systematic process for developing and testing predictions (**Figure 1.7**).

You probably already use the scientific method, or at least elements of it, in your own everyday problem solving.

❶ Observe and generalize When we observe the world around us and make generalizations from what we learn, we are employing *inductive reasoning* (extrapolating from the specific to the general case). Usually, we don't even think about it and don't bother to put our observations and generalizations into any kind of formal language, but we do it just the same. For example, you are probably convinced that it will *always* be colder in winter than in summer (a generalization) because you have observed that every winter in the past was colder than the preceding summer (specific observation). The difference between common experience and good science is that science uses generalization to make a prediction that can be tested. Consider an example from biological research:

Observation 1: Rats given a particular drug (call it Drug X) have lower blood pressures than rats not fed the drug.

Observation 2: Independently, researchers in Canada showed that Drug X lowers blood pressures in dogs and cats.
Generalization: Drug X lowers blood pressure in all mammals.

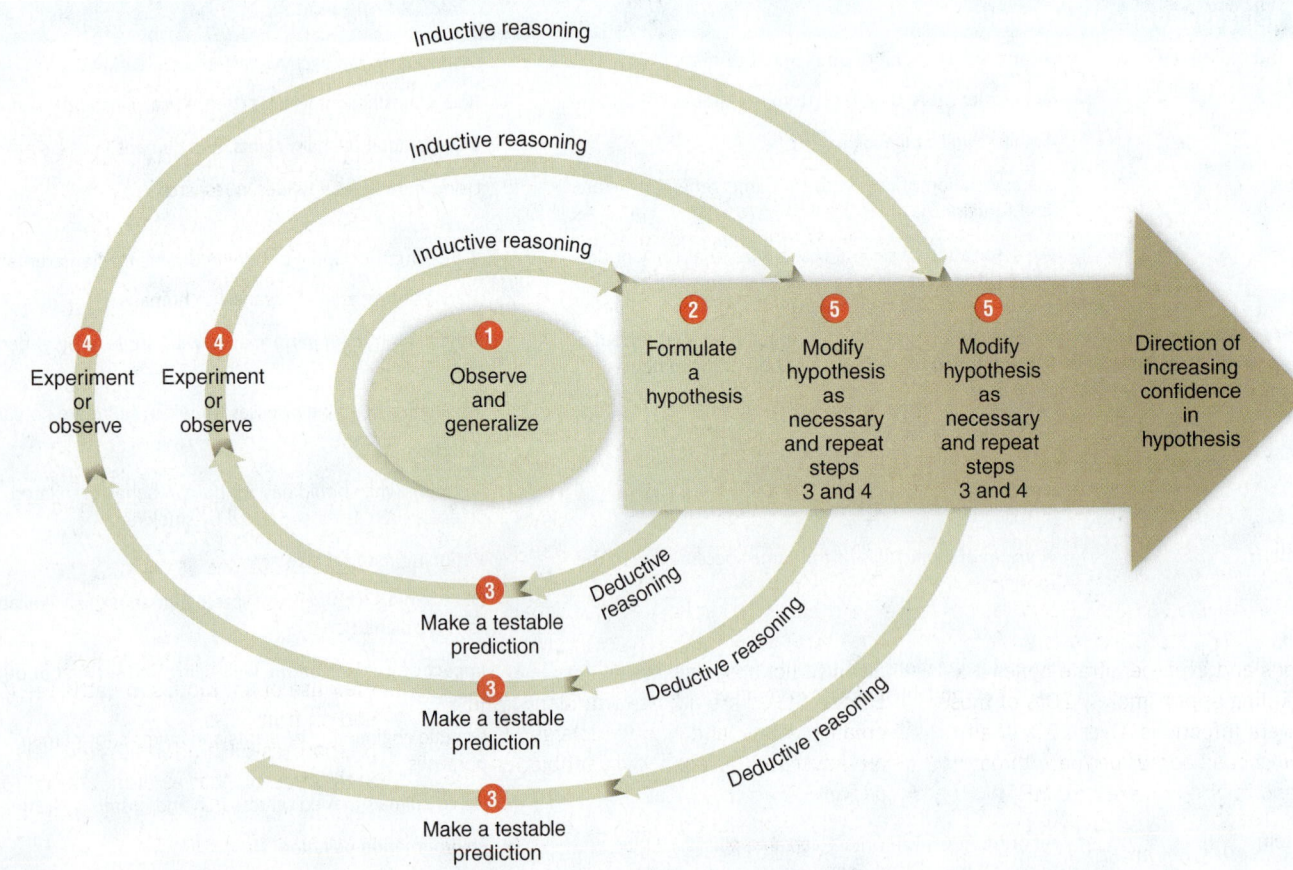

Figure 1.7 The scientific method. Observations and generalizations lead to the formulation of a hypothesis. From the hypothesis, specific predictions are made that can be tested by experimentation or observation. The results either support the hypothesis or require that it be modified to fit the new facts. The cycle is repeated. Ultimately, the scientific method moves in the direction of increased confidence in the modified hypothesis.

❷ Formulate a hypothesis Observations and generalizations are used to develop a *hypothesis*. A **hypothesis** is a tentative statement about the natural world. Importantly, it is a statement that can lead to testable deductions.

Hypothesis: Drug X would be a safe and effective treatment for high blood pressure in humans.

❸ Make a testable prediction Hypotheses that cannot be tested are idle speculation, so much hot air. But many hypotheses are so sweeping and comprehensive that ways must be found to test them under a variety of conditions. For example, you probably would not be convinced that Drug X is safe and effective for all people under all conditions until you had at least tested it in quite a few people under many different conditions. To have confidence in your hypothesis, you must make testable predictions (also called *working hypotheses*) based on the hypothesis and then test them one at a time. Predictions employ *deductive reasoning* (applying the general case to the specific). Often they are put in the form of an *if . . . then* statement, in which the *if* part of the statement is the hypothesis. For example,

Prediction: If Drug X is a safe and effective treatment for high blood pressure in humans, then 10 mg/day of Drug

X will lower blood pressure in people with high blood pressure within one month.

Notice that the prediction is very specific. In this example, the prediction specifies the dose of drug, the medical condition of the persons on whom it will be tested, the expected effect of the drug if the prediction is correct, and a specified time period for the test. Its specificity makes it testable—yes or no, true or false.

☑ Suppose you are designing a research study to test the hypothesis that regular exercise helps people sleep better. Develop a specific, testable prediction from this hypothesis.

❹ Experiment or observe The truth or falsehood of your prediction is determined by observation or by experimentation. An **experiment** is a carefully planned and executed manipulation of the natural world to test your prediction. The experiment that you conduct (or the observations you make) will depend on the specific nature of the prediction.

❺ Modify the hypothesis as necessary and repeat steps 3 and 4 If your prediction turns out to be false, for instance, the drug didn't lower blood pressure under the conditions of your experiment, you will have to modify your hypothesis to

HEALTH & WELLNESS
The Growing Threat of Antibiotic-Resistant Bacteria

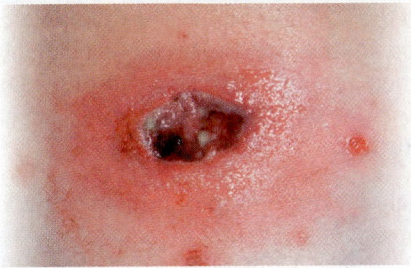
Skin lesions caused by MRSA.

When antibiotics—drugs that kill bacteria—became available in the 1940s, they were hailed as a breakthrough. Indeed, many people owe their lives to them. But there is a downside: Indiscriminate use of antibiotics leads to antibiotic resistance in some strains of bacteria. For example, in recent years a new strain of the *Staphylococcus aureus* bacterium has appeared that is resistant to all penicillin-type antibiotics. This frightening superbug, called methicillin-resistant *Staphylococcus aureus* (MRSA) can cause serious skin infections and even penetrate bones and lungs, killing approximately 20% of those with severe infections. Over 25% of all *Staphylococcus aureus* ear/nose/throat infections in children are now MRSA.

The rise of MRSA actually makes biological sense. Whenever an antibiotic kills most but not all of a population of bacteria, the surviving bacteria are naturally most resistant to the antibiotic.

With the total population of bacteria now decreased, these antibiotic-resistant bacteria multiply. The more we use antibiotics, the more we encourage the rise of resistant strains of bacteria.

Because antibiotics have been so effective in the past, we have grown to rely on them, and we tend to use them often and indiscriminately. Over 50 million pounds of antibiotics are produced each year in the United States, about half of which are fed to livestock or sprayed on fruit trees. Researchers estimate that fully one-third of all antibiotic prescriptions given to nonhospitalized patients in the United States are not needed. Many hand creams, soaps, and laundry detergents are advertised as containing antibiotics as well.

What can you do to ensure that we continue to have effective antibiotics?

- Don't ask your doctor for antibiotics for a viral illness such as a cold or flu.

Antibiotics kill bacteria but have no effect on viruses.
- Take antibiotics only when needed and as prescribed. Complete the full course of treatment so that even the most resistant bacteria are killed.
- Reduce your use of antibacterial hand creams, soaps, and laundry detergents.
- Support farmers' efforts to reduce their use of antibiotics in cattle feed and on fruit trees.
- Support research efforts to find new antibiotics, so that as older antibiotics begin to fail, we have newer antibiotics to take their place.

Used properly and judiciously, antibiotics will remain in our antibacterial arsenal for many years.

fit the new findings and repeat steps 3 and 4. For example, perhaps the drug would lower blood pressure if you increased the dose or administered the drug for a longer period of time.

✅ Consider the following statement: "Giant apes called Sasquatches are living in the forests of the Pacific Northwest, but they avoid people and leave no evidence of their existence." Does this statement qualify as a scientific hypothesis?

Designing and conducting the experiment

Once you've developed your working hypothesis (prediction) you'll need to test it with an experiment. In a **controlled experiment,** all possible **variables** (factors that might vary during the course of an experiment) are controlled so that they cannot affect the outcome.

An *independent variable* is one that stands alone and isn't changed by the variable you wish to measure. Time, distance, age, or drug treatment versus no drug treatment are generally independent variables. A *dependent variable* is one that depends on an independent variable, such as growth (may depend on time) or blood pressure (may depend on drug treatment). Wherever possible, scientists design experiments so that just one independent variable, called the *controlled variable,* is manipulated between groups. In this case, the controlled variable is drug treatment. Therefore, in the example of Drug X, you could follow these steps in your controlled experiment (**Figure 1.8**):

1. Select a large group of human subjects with high blood pressure.
2. *Randomly* divide the larger pool of subjects into two groups. Designate one the **experimental group** and the other the **control group.** The importance of random assignment to the two groups is that all other independent variables that might affect the outcome (such as age or gender or previous health problems) are automatically equalized between the two groups. In effect, the control group accounts for all unknown factors.
3. Treat subjects in the two groups *exactly* the same except that only the experimental group gets the drug. Treat the experimental group with 10 mg/day of Drug X for one month and the control group with a **placebo,** or "false treatment." If you deliver Drug X in a pill, the placebo should be an identical-looking pill with no drug in it. If you administer Drug X as an injection in a saline solution, the placebo should be an injection of the same volume of saline.

   When working with human subjects, scientists must take the power of suggestion into account. For example, if subjects in a study of the effectiveness of a pain medication were told they were in the medication group and not the placebo group, they might rate their pain lower. To eliminate the power

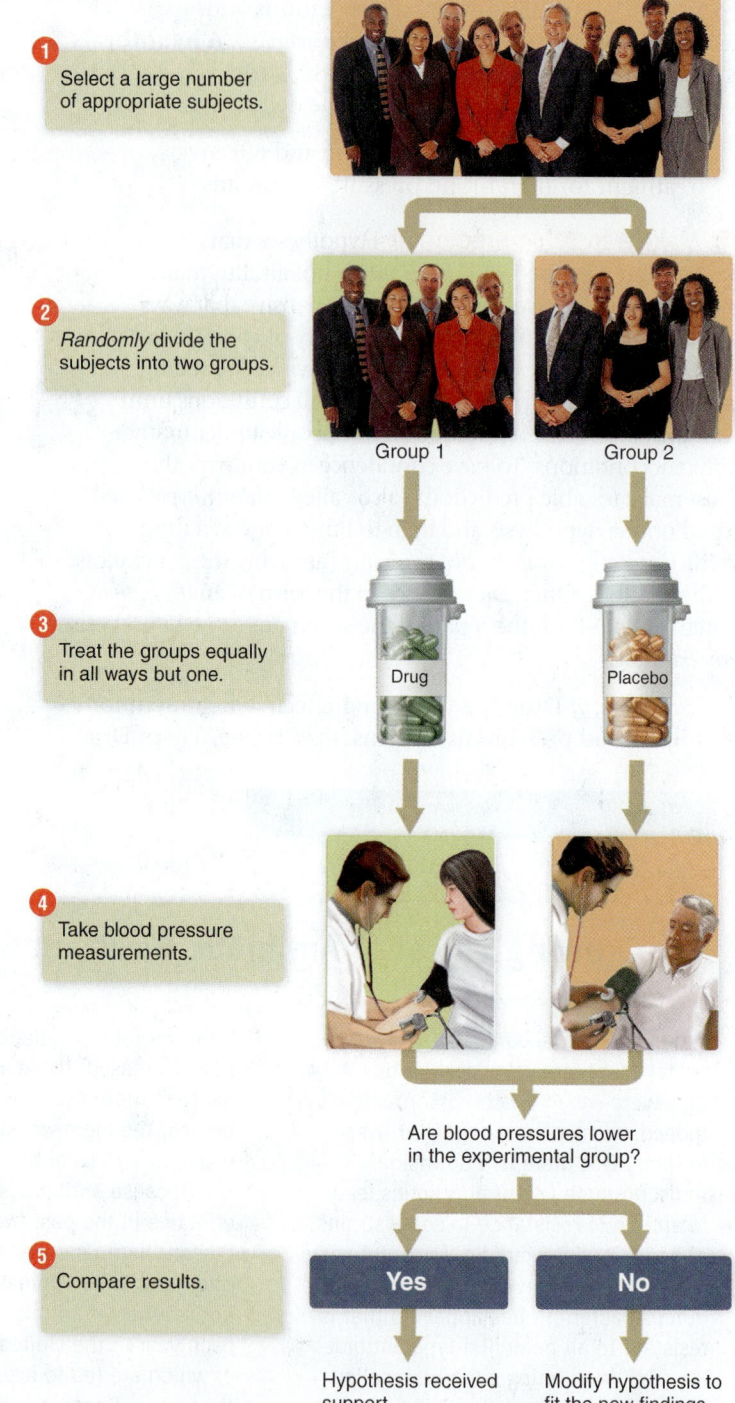

1 Select a large number of appropriate subjects.

2 *Randomly* divide the subjects into two groups.

Group 1   Group 2

3 Treat the groups equally in all ways but one.

Drug   Placebo

4 Take blood pressure measurements.

Are blood pressures lower in the experimental group?

5 Compare results.

Yes   No

Hypothesis received support.   Modify hypothesis to fit the new findings.

Figure 1.8 The steps in a controlled experiment. In this example, the experimental variable is blood pressure. Because the subjects were assigned randomly to the two groups, the groups will be identical except for the presence or absence of the experimental treatment (drug). Therefore, any differences in blood pressures between the two groups can be ascribed to the drug.

✅ Suppose that instead of randomly assigning subjects to the two groups, subjects with high blood pressure are more likely to be put in the experimental group. Would this be a better or a worse experiment, or would it make no difference? Explain.

of suggestion as a variable, researchers conduct experiments "blind," meaning that the subjects aren't told whether they are getting the placebo or the drug. Sometimes experiments are done "double-blind," so that even the person administering the drugs and placebos does not know which is which until the experiment is over.

4 Measure blood pressures in both groups at the end of one month.

5 Compare the results. If the experimental group's blood pressures are lower when compared using appropriate statistical (mathematical) tests, then your prediction is verified and your hypothesis receives support.

Even if your prediction turns out to be true, however, you're still not done. This is because you've tested only one small part of the hypothesis (its effectiveness under one specific set of conditions), not all the infinite possibilities. How do we know the drug is safe, and how safe is "safe," anyway? Does it cause a dangerous drop in blood pressure in people with normal blood pressures? Does it cause birth defects if taken by pregnant women? Does long-term use lead to kidney failure? Does the drug's effectiveness diminish over time? Specific predictions may have to be stated and tested for each of these questions before the drug is declared effective and safe and allowed on the market.

Only after many scientists have tried repeatedly (and failed) to disprove a hypothesis do they begin to have more confidence in it. A hypothesis cannot be proved true; it can only be supported or disproved. As this example shows, the scientific method is a process of elimination that may be limited by our approach and even by our preconceived notions about what to test. We move toward the best explanation for the moment, with the understanding that it may change in the future.

✅ A researcher tries to test the hypothesis that "exercise helps people sleep better" with the following study: He asks 50 regular joggers and 50 non-joggers how well they sleep at night. What are some problems with this experiment?

Making the findings known

New information is not of much use if hardly anybody knows about it. For that reason, scientists need to let others know of their findings. Often, they publish the details in scientific journals to announce their findings to the world. Articles in *peer-reviewed* journals are subjected to the scrutiny of several experts (the scientists' peers) who must approve the article before it can be published. Peer-reviewed journals often contain the most accurate scientific information.

An unspoken assumption in any conclusion is that the results are valid only for the conditions under which

the experiment was done. This is why scientific articles go into such detail about exactly how the experiment was performed. Complete documentation allows other scientists to repeat the experiments themselves or to develop and test their own predictions based on the findings of others.

Try to apply the scientific method to a hypothesis dealing with some aspect of evolution or to a global problem such as cancer or AIDS, and you begin to appreciate how scientists can spend a lifetime of discovery in science (and enjoy every minute). At times, the process seems like three steps forward and two steps back. You can bet that at least some of what you learn in this book will not be considered accurate 10 years from now. Nevertheless, even through our mistakes we make important new observations, some of which may lead to rapid advances in science and technology. Just in the past 100 years, we've developed antibiotics, sent people to the moon, and put computers on millions of desks.

 MJ's BlogInFocus How many deaths per year are caused by antibiotic-resistant organisms, and what is the government's current plan to combat antibiotic resistance? Visit MJ's blog in the Study Area in MasteringBiology and look under "Antibiotic Resistance."

http://goo.gl/hl4etY

Many people associate science with certainty, whereas in reality scientists are constantly dealing with uncertainty. That is why we find scientists who don't agree or who change their minds. Building and testing hypotheses is slow, messy work, requiring that scientists constantly question and verify each other.

A well-tested hypothesis becomes a theory

Many people think that a *theory* is a form of idle speculation or a guess. Scientists use the word quite differently. To scientists, a **theory** is a broad hypothesis that has been extensively tested and supported over time and that explains a broad range of scientific facts with a high degree of reliability.

A theory is the highest status that any hypothesis can achieve. Even theories, however, may be modified over time as new and better information emerges. Only a few hypotheses have been elevated to the status of theories in biology. Among them are the theory of evolution and the cell theory of life.

↩ **Recap** The scientific method is a systematic process of observation, hypothesis building, and hypothesis testing. A theory is a broad hypothesis that has withstood numerous tests. ■

1.4 Sources of scientific information vary in style and quality

We are constantly bombarded by scientific information, some of it accurate and some not. What can you believe when the facts seem to change so quickly? All of us need to know how to find good information and evaluate it critically. Different sources of scientific information may have very different goals, so look for those that can best inform you at your own level of understanding and interest.

Some scientific knowledge is highly technical. As a result, scientists have a tendency to speak on a technical level and primarily to each other. Scientists often communicate by means of articles in specialized peer-reviewed journals such as *Nature* and *Science*. Articles in peer-reviewed journals are concise, accurate, and documented so thoroughly that another scientist ought to be able to duplicate the work after reading the article. Articles generally include an extensive list of previous references on the subject. Articles in peer-reviewed journals make for laborious reading, and they are usually as dry as toast. But bear in mind that their purpose is primarily to inform other experts.

Other helpful print sources are science magazines and nonfiction books meant for the well-educated public. The goal is to inform the interested reader who may have only a limited background in science. The authors are usually science writers or experts who translate the finer scientific points into language that we can all understand. The information is generally accurate and readable, although the reader may not understand some of the details. Generally, these articles and books tell readers who want to delve more deeply into the subject where to find more information.

General interest news magazines and daily newspapers also report on selected "hot topics" in science. Their goal is to get the information out as quickly as possible to a wide audience. Coverage is timely but less in-depth than in science magazines and may not include the details you need to check the validity of the statements. A decided plus is that magazines and newspapers often discuss social, political, economic, legal, or ethical ramifications of the scientific findings, something generally lacking in the previous sources. Although the scientific information is usually accurate, the reporter may not understand the subject fully and may not provide adequate context. The best articles point readers to the original sources. Television (for instance, the Discovery Channel, *Nova*) also presents science-related topics to the public.

Since the 1980s, scientists and researchers have used the Internet to communicate and share ideas. The recent expansion of the World Wide Web has made the Internet accessible to the general public, opening up exciting new sources of scientific information. Nearly all universities now have Web pages; the site addresses end in ".edu" (for educational) rather than ".com" (for commercial). A number of scientific and professional organizations have created Web sites that offer helpful information for both scientists and consumers. Examples of organizations with Web sites include the National Institutes of Health, the American Cancer Society, and the American Heart Association. The Web addresses of government agencies and nonprofit organizations generally end in ".gov" and ".org," respectively.

Be aware that the Internet can also be a source of misinformation. At present, the Internet is less closely regulated than print and broadcast media, so it can be difficult to tell the difference between objective reports and advertisements. In addition, participants in online chat rooms and special-interest groups may promote their own opinions as proven truths. It pays to be skeptical.

Recap The best sources of scientific information translate difficult or complex information accurately into understandable terms and have enough references that you can check the information if you wish. ∎

1.5 Learning to be a critical thinker

Many scientists are motivated by strong curiosity or a sense of wonder and awe about how the natural world works. Exploring the frontiers of knowledge requires a great deal of creativity and imagination. Like many people, however, scientists may leap to conclusions or resist new ideas. A few may be driven by self-interest. To combat these natural human tendencies, good scientists try to use certain tools of critical thinking. You too can learn to use these tools, regardless of whether you choose a career in science. The following sections describe some of the simple tools that anyone can use to improve their critical thinking skills.

Become a skeptic

Good scientists combine creativity and imagination with skepticism, a questioning attitude. If you've ever bought something based on claims about how well it works and then been disappointed, you know the value of skepticism. Question everything and dig a little deeper before believing something you read and hear. Here are some questions you might ask yourself:

- Who says that a particular statement is true?
- What evidence is presented?
- Are the persons speaking on a subject qualified by training or skill to speak authoritatively about it?
- Are they being paid, and if so, how might that affect what they have to say?
- Where is the evidence to back up a claim?

Skepticism is particularly important for claims that are new, startling, and not yet verified by other scientists. Listen carefully to the debate between scientists in the public arena. A new scientific claim may take several years to be checked out adequately.

Learn how to read graphs

Just like a picture, a graph is worth a thousand words. Graphs display data (facts obtained from observations or experimental results) in a way that is economical and easy to grasp. Graphs can also be used to clarify the meaning of experimental results.

Most graphs are plotted on two lines, or axes (singular: axis). The horizontal axis at the bottom is called the *abscissa* (from math you may know this as the *x*-axis), and the vertical axis is called the *ordinate* (*y*-axis). By convention, the *independent variable* is generally plotted on the abscissa and the *dependent variable* is plotted on the ordinate.

Graphs can take a variety of forms, from plots of individual data points to lines or bars of average values (Figure 1.9). When reading a graph, first check the scales and the legends on the abscissa and the ordinate to determine what the graph is about. Be careful to look for a *split axis*, in which the scale changes. An example is shown in Figure 1.10. A split axis is sometimes a convenient way of representing data that cover a wide range on one axis, but it can also be used to deliberately mislead people unfamiliar with reading graphs.

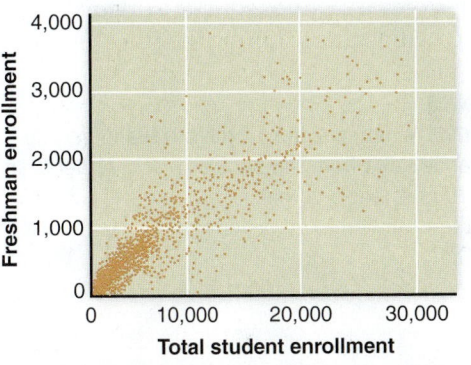

a) A scatter plot showing enrollment at each individual college. Each data point represents one U.S. college.

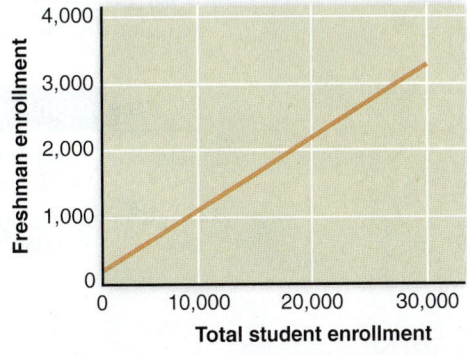

b) A line graph representing the best straight line fit of the data in a).

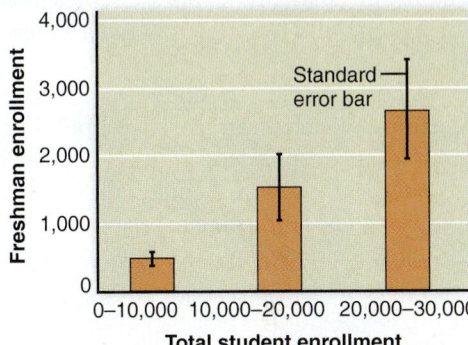

c) A bar graph in which university enrollments are lumped together in three class sizes and the freshman enrollments are then averaged. Standard error bars indicate that the data have been analyzed statistically.

Figure 1.9 Types of graphs. Each of these graphs reports the relationship between freshman enrollment and total student enrollment at approximately 1,500 U.S. colleges and universities.

☑ In the bar graph, why are the standard error bars in the third bar much higher than in the first bar? Put another way, what do standard error bars actually tell us? (Look back at the first graph for a hint.)

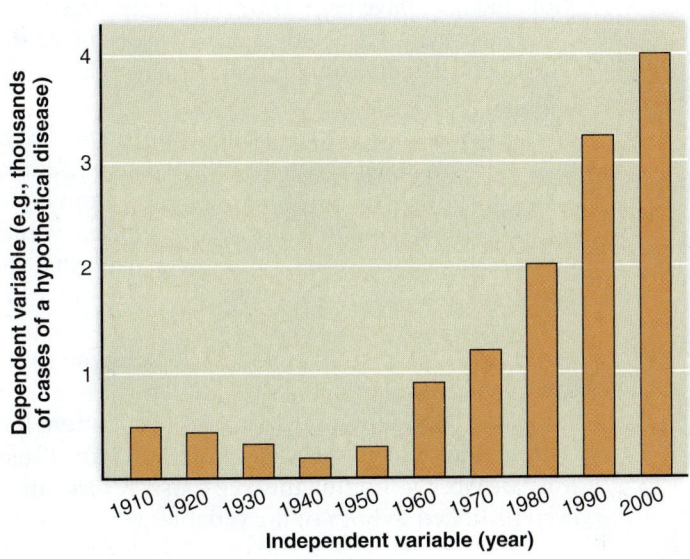

a) Graph with regular axis.

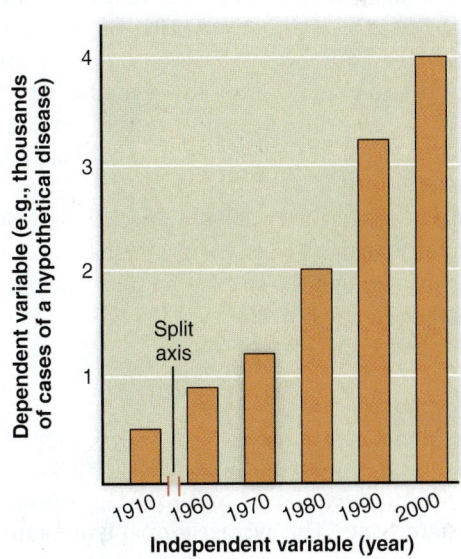

b) Graph with split axis.

Figure 1.10 How a split axis affects a graph. The graph in (b) is redrawn from the data in (a) by splitting the abscissa and omitting the data for the years 1920–1950. The effect is a consolidated graph that fits in less space, but it might mislead you into thinking that the number of cases of the disease has been rising steadily since 1910, instead of only since 1960.

Appreciate the value of statistics

Statistics is the mathematics of organizing and interpreting **data.** Scientists use statistics to determine how much confidence they should place in information. Most scientists would be willing to accept experimental results with confidence if (according to statistical tests) they would get the same outcome 19 of every 20 times they repeat the experiment, or 95% of the time. When you see numerical averages followed by a smaller "+/−" number, the smaller number represents an expression of confidence in the certainty of the results, called the *standard error*. In graphs, the standard errors are represented as small lines that extend above and below the average number (see Figure 1.9c).

Statistics are important in many disciplines. During elections, we may hear pollsters report, for example, that "52% of the respondents said that they will vote for the president. The poll has a margin for error of +/− 3%." This tells you the pollsters are relatively certain that the actual percentage who will vote for the president is somewhere between 49 and 55%, still too close to call.

Distinguish anecdotes from scientific evidence

Anecdotal evidence takes the form of a testimonial or short unverified report. Although an anecdote may be true as stated, it in no way implies scientific or statistical certainty. It cannot be generalized to the larger population because it is not based on empirical evidence. Advertising agencies sometimes use anecdotes to influence you. The actor on television who looks sincerely into the camera and says "Drug X worked for me" may be telling the truth—the drug may have worked for him. But this does not prove the drug will work for everyone, or even for 10% of the population.

Nonscientists (and even scientists) often say things like "My grandmother swears by this remedy." Again, the statement may be true, but it is not scientific evidence. Listen carefully to how the evidence for a statement is presented.

☑ One brand of cold medication has a personal testimonial on its Web site from a famous actor. Another brand of cold medication has testimonials from 10 people who are not actors. Do any of these testimonials make a difference to you in deciding which medication is better? Why or why not?

Separate facts from conclusions

A *fact* is a verifiable piece of information, whereas a *conclusion* is a judgment based on the facts. Consider the following statement: "The average global temperature went up 0.1°C last year, proving that global warming is occurring." The first part of the statement (temperature went up 0.1°C last year) may be a verifiable fact, but the second part of the statement is a conclusion. The conclusion may not be warranted if temperature fluctuations up and down of 0.3°C are normal from year to year.

Understand the difference between correlation and causation

A close pattern or relationship (a correlation) between two variables does not necessarily mean that one causes the other. The catch-phrase is "correlation does not imply causation." A good example of a correlation without causation is the close correlation between ice cream sales and drownings—when ice cream sales are up in the summer months, so are drownings. Does that mean that eating ice cream causes people to drown? Hardly. Ice cream sales and drownings also correlate with (and are most likely caused by) a third factor not mentioned in the original correlation—warmer temperatures during the summer.

 MJ's BlogInFocus What are some of the pitfalls in getting your science news from the popular press, rather than directly from science journals? To find out, visit MJ's blog in the Study Area in MasteringBiology and look under "Science Reporting."

http://goo.gl/BVA4pf

If this example seems too obvious a correlation without causation, try this one: In 1999, a study at a major university found that children who slept with a light on were more likely to develop nearsightedness (myopia) later in life. But does this mean that sleeping with a light on *causes* nearsightedness? In fact, a follow-up study in 2000 found no direct causal relationship between sleeping with a light on and the development of nearsightedness. The follow-up study showed that children who develop nearsightedness are more likely to have parents who are nearsighted, suggesting (but not proving) a genetic cause. It also showed that parents who are nearsighted are just more likely to leave the light on at night!

In the aforementioned example, the original scientific observation was stated correctly (lights on *correlates* with nearsightedness). But anyone who became convinced that sleeping with a light on *causes* nearsightedness would have been wrong. Be skeptical of causal statements that are based only on a good correlation, for the true cause may not be obvious at first.

Of course, a close correlation is likely whenever a true causal relationship does exist. In other words, although a correlation does not necessarily prove causation, it can be a strong hint that you *may* have found the true cause, or at least are close to finding the true cause. A correlation may even be linked to both of the variables you're observing.

↩ **Recap** Healthy skepticism, a basic understanding of statistics, and an ability to read graphs are important tools for critical thinking. Know anecdotal evidence when you see it, and appreciate the differences between fact and conclusion and between correlation and causation. ■

1.6 The role of science in society

How do we place science in its proper perspective in our society? Why do we bother spending billions of dollars on scientific research when there are people starving in the streets? These are vital questions for all of us, so let's look at why we study the natural world in the first place.

Science improves technology and the human physical condition

Science gives us information about the natural world upon which we can base our societal decisions. Throughout history, some of the greatest benefits of science have been derived from the *application* of science, called technology, for the betterment of humankind (**Figure 1.11**). Time and time again, scientific knowledge has led to technological advances that have increased the productivity and hence prosperity of both industries and nations. Science has given us larger crop yields, more consistent weather predictions, better construction materials, better health care, and more efficient and cleaner sources of power, to name just a few benefits. It has made global transportation and communication possible.

Many people are concerned that overuse of our technological capabilities may lead to problems in the future. Science can help here, too, by providing the technology and research that allow us to identify problems early on. We can see the early warning role of science in the Health & Wellness discussion of bacterial resistance to antibiotics. Only by understanding a problem can we learn how to solve it. Science helps us correct our mistakes.

Science has limits

A practical limitation of science is that some information, including data that may be useful in improving human health, cannot be obtained by observation or experimentation. Our society places a very high value on human life, and therefore we don't experiment on humans unless the experiment is likely to be of direct benefit to the subject (the use of experimental cancer drugs falls into this category). This is why it is hard to investigate the danger of street drugs like cocaine or anabolic steroids. No good scientist would ever deliberately give healthy humans a drug that might cause injury or death, even if the resulting information could save lives in the future.

Scientific knowledge is limited to physical explanations for observable events in the natural world. It cannot prove or disprove the existence, or importance to us, of things that fall outside the realm of the natural world, such as faith or spiritual experiences. Many scientists have a strong faith or belief that cannot be tested by science, because faith does not depend on logical proof or material evidence. They tend to believe that the search for meaning and the search for knowledge are complementary, not contradictory.

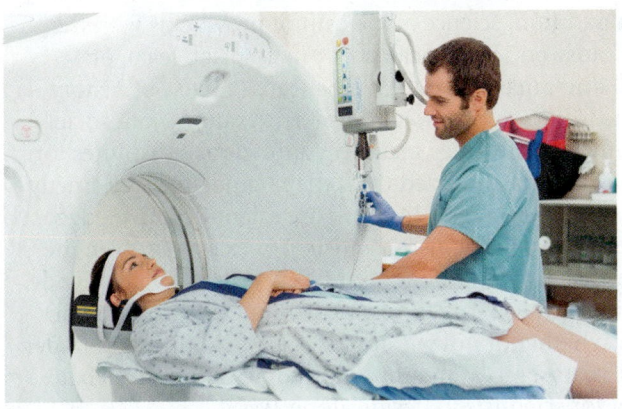

a) **A patient about to undergo a computed axial tomography (CAT) scan.** Modern non-invasive medical imaging techniques such as this have improved patient diagnoses.

b) **This scientist is collecting insects from a tree-top in a tropical rain forest.** Studies such as this improve our understanding of the interrelationships of organisms in an ecosystem and have yielded rare natural chemical compounds, many of which have proven useful in human and animal medicine.

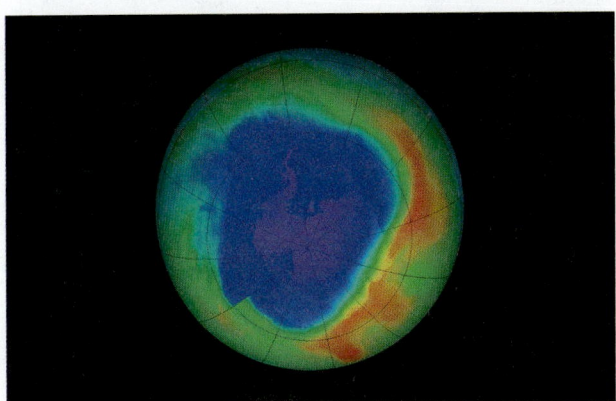

c) **A satellite map documenting depletion of the ozone layer over Antarctica in 2013.** The area of greatest depletion appears dark blue. Studies such as this one allow scientists to document when and where ozone depletion occurs so that they can better understand its causes and cures.

Figure 1.11 Benefits of science. These photos show typical scenes of scientists and the application of science (technology) for the betterment of the human condition. The range of scientific endeavor is as vast as the natural world itself.

In addition, science alone cannot provide us with the "right" answers to political, economic, social, legal, or ethical dilemmas. Humans have minds, a sense of history and the future, and a moral sense. For example, our society currently permits experiments on animals as substitutes for human subjects, provided that federal guidelines are strictly followed (Figure 1.12). How we use scientific knowledge is up to all of us, not just scientists. For example, given the current state of knowledge about how cells grow and divide, scientists may eventually be able to clone an adult human being. Whether or not we should permit cloning, and under what circumstances, are important topics of public debate. It's not for scientists to decide alone.

This does not mean scientists are without moral obligation. As experts in their fields, they are in a unique

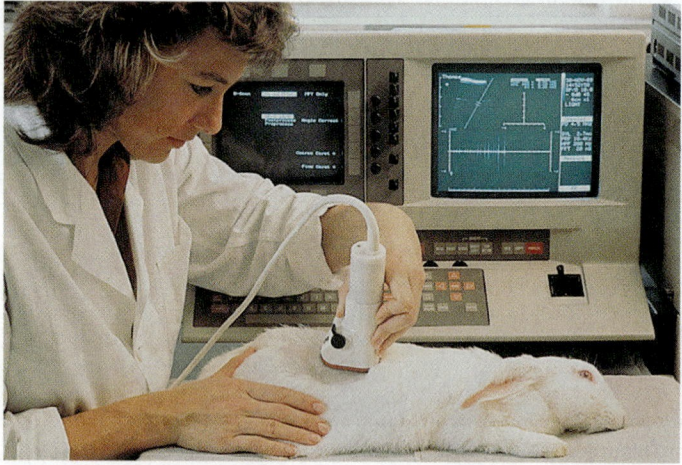

Figure 1.12 Animals in research. In this society, we allow the use of animals for research in certain circumstances. This researcher is using a noninvasive technique for measuring blood flow in the skin as part of a study of vascular diseases.

position to advise us about the application of scientific knowledge, even if the choices ultimately rest with all of us.

The importance of making informed choices

You live in a science-oriented society. Throughout this book, we present a common theme: Every day, you make decisions about how you and society choose to use the knowledge that science gives us. Whether or not you are conscious of it, whether or not you deliberately take action, you make these choices daily.

Should Olympic athletes be allowed to use bodybuilding drugs? Do you think the use of pesticides is justified in order to feed more people? How do you feel about the cloning of human beings? Are you willing to eat a proper diet to stay healthy—and by the way, what *is* a proper diet? Who should pay for health care for the poor?

Our knowledge has advanced rapidly. By the end of this book, you will know more about genetics and evolution than did the scientists who originally developed the theories about them. With knowledge comes the responsibility for making choices. From global warming to genetic engineering to personal health, each of us must deal with issues that concern our well-being and the future of the biological world in which we live. We owe it to ourselves, as individuals and as a society, to acquire the knowledge and skill we need to make intelligent decisions. Your choices can make a difference.

Recap Science and technology have improved the human condition. Science cannot, however, resolve moral dilemmas. Scientists can advise us on issues of science, but we as a society must decide how to put this scientific knowledge to use. ■

Chapter Summary

1.1 The characteristics of life *p. 4*

- All living things acquire and transform both matter and energy from their environment for their own purposes.
- The basic unit of life is a single cell.
- Living things maintain homeostasis, respond to their external environments, and reproduce.

1.2 How humans fit into the natural world *p. 6*

- The biological world can be organized. A common scheme classifies life into three domains (Bacteria, Archaea, and Eukarya) and six kingdoms (Eubacteria, Archaebacteria, Protista, Fungi, Plantae, and Animalia).
- Features that define humans are bipedalism, well-developed opposable thumbs, a large brain, and the capacity for complex language.
- Humans are part of communities of different organisms living together in various ecosystems.

1.3 Science is both a body of knowledge and a process *p. 10*

- Scientific knowledge allows us to describe and make predictions about the natural world.
- The *scientific method* is a way of thinking, a way of testing statements about the natural world (*hypotheses*) by trying to prove them false.
- A *theory* is a hypothesis that has been extensively tested and that explains a broad range of scientific facts with a high degree of reliability.

1.4 Sources of scientific information vary in style and quality *p. 14*

- Articles in peer-reviewed scientific journals are generally written for other scientists. They may be difficult to read but are very accurate.
- Journals, books, and television shows on popular science present scientific knowledge efficiently to the general public.
- Web sites vary widely in the quality and accuracy of the information they present.

1.5 Learning to be a critical thinker *p. 14*

- Skepticism is a questioning attitude ("prove it to me"). Critical thinking requires skepticism.
- Knowing how to read graphs and understand basic statistics can help you evaluate numerical data.
- Being able to recognize an anecdote, tell fact from conclusion, and distinguish between correlation and causation can help you evaluate the truth of a claim.

1.6 The role of science in society *p. 17*

- The application of science is called technology.
- Science is limited to physical explanations of observable events.
- How to use science is up to us.

Terms You Should Know

biology, *3*
cell, *5*
control group, *12*
data, *16*
experiment, *11*
experimental group, *12*
homeostasis, *5*
hypothesis, *11*

metabolism, *5*
photosynthesis, *6*
placebo, *12*
Protista, *6*
science, *3*
scientific method, *10*
theory, *13*
variables, *12*

Concept Review

Answers can be found in the Study Area in MasteringBiology.

1. Living things have a different molecular composition from non-living things. What makes this possible?
2. Explain the meaning of the term *homeostasis*.
3. Name four features that together contribute to our uniqueness and define us as human.
4. Describe the difference between a hypothesis and a prediction (or working hypothesis).
5. Discuss the role of scientists in helping us solve economic, social, and ethical dilemmas.

Test Yourself

Answers can be found in the Appendix.

1. To which of the following domains of life do humans belong?
   a. Prokarya
   b. Eukarya
   c. Animalia
   d. Mammalia
2. To which of the following domains do unicellular organisms which lack nuclei belong?
   a. Eukarya
   b. Archaea
   c. Bacteria
   d. both Archaea and Bacteria
3. New scientific knowledge is gained through a multistep process known as:
   a. the scientific method
   b. hypothesis development
   c. variable testing
   d. observation testing

4. An experiment designed and conducted under strictly managed conditions is a:
   a. replicated experiment
   b. controlled experiment
   c. "blind" experiment
   d. peer-reviewed experiment
5. A broad hypothesis that has been supported by repeated experimentation is known as:
   a. a proven hypothesis
   b. a supported hypothesis
   c. a theory
   d. dogma
6. Which of the following is used when developing a hypothesis?
   a. observations
   b. inductive reasoning
   c. controlled experiments
   d. both observations and inductive reasoning
7. The smallest unit of life that demonstrates all the properties of life is a(n):
   a. organism
   b. organ system
   c. molecule
   d. cell
8. Consider all of the organisms (human as well as nonhuman) that occupy your college campus. From a biological standpoint, this would be a/an:
   a. community
   b. ecosystem
   c. biome
   d. population
9. Which of the following lists the steps of the scientific method in order?
   a. observation—prediction—experimentation—hypothesis development
   b. hypothesis development—observation—experimentation—prediction
   c. prediction—hypothesis development—experimentation—observation
   d. observation—hypothesis development—prediction—experimentation
10. In graphs, which of the following is usually plotted on the abscissa (*x*-axis)?
    a. controlled variable
    b. independent variable
    c. dependent variable
    d. placebo
11. An acceptable scientific hypothesis:
    a. can be tested
    b. can be proven true
    c. can be proven false
    d. both (a) and (c)
12. Drug A is being tested for its effectiveness in shortening the duration and severity of influenza in humans. In designing an experiment to test Drug A, which of the following would be an important consideration?
    a. Participants can choose whether to be in the experimental or control group.
    b. The experimental group will contain only males and the control group will contain only females.
    c. The experimental group should contain 1,000 subjects, but the control group should include 100 subjects.
    d. The experimental group will receive Drug A and the control group will receive a placebo.
13. Jenna has been telling her friends about how successful she was at losing 10 pounds by using a dietary supplement she purchased at a health food store. This is an example of a/an:
    a. proven hypothesis
    b. anecdotal evidence
    c. controlled experiment
    d. scientific theory
14. The maintenance of a relatively stable internal environment is:
    a. metabolism
    b. evolution
    c. constancy
    d. homeostasis

15. All of the following are features that collectively distinguish humans from other animals except:
    a. bipedalism
    b. large brain
    c. ability to evolve as a species
    d. capacity for complex language

Apply What You Know

Answers can be found in the Study Area in MasteringBiology.

1. A magician has a coin that he says (hypothesizes) has heads on both sides, but he's unwilling to show you both sides. To convince you, he flips it three times and gets heads each time. Do you believe that the coin has two heads? What if he gets heads 10 times in a row? 100 times? What would it take (by coin flip) to prove that the coin does *not* have two heads? With this example, explain the difference between having relative confidence in the truth of a hypothesis, proving it to be true, and proving it to be false.

2. Your roommate is writing a paper on the subject of cocaine and birth defects in humans and wonders why there don't seem to be published reports of controlled experiments in humans on the subject; all the studies are on rats! Describe to her how such a controlled experiment would have to be designed and conducted, and convince her that it would never be permitted by any responsible regulatory agency.

3. You have a friend who truly believes in the existence of ghosts and says he has scientific evidence; he and his two roommates have all seen them. Explain to your friend what it means to have scientific evidence. Think about what data are, how they are gathered, and why personal experiences do not meet the criteria to be considered scientific evidence.

4. An episode of an old TV show was about doctors living in a tropical environment where the heat is unusually oppressive. An orderly comes to seek relief from the heat, and one doctor gives a supply of sugar pills to the orderly and tells him they are an experimental drug designed to keep humans cool in hot weather. The gullible orderly takes them, and while others are sweating, he claims to suffer no effects from the heat to the point of not even sweating. He later finds out the drug is fake and immediately complains of being overheated.

    This is a fictitious demonstration of the *placebo effect*. Explain how the placebo effect can be avoided when testing new drugs.

5. You are trying to convince your friend who smokes cigarettes that he should quit. You explain to him that smoking and the incidence of lung cancer are strongly correlated. Your friend says that that does not prove smoking causes lung cancer. Is your friend correct? If so, explain why he is correct. What would you say to him?

6. On the radio, you hear an interview with a climatologist discussing global warming. The interviewer asks the scientist what proof she has that humans are to blame for global warming and the resulting rise in sea levels. The scientist responds that while we cannot prove humans are to blame for global warming there is much evidence that human activities are responsible for the rising temperatures. The interviewer says, "Aha, so you have no proof that humans are to blame, and this is all nothing more than a theory, and isn't a theory nothing more than an opinion?" What differentiates a scientific theory from an opinion?

7. Explain why religious explanations cannot disprove a scientific theory, and conversely, why science cannot prove or disprove a religious belief.

MJ's BlogInFocus

Answers can be found in the Study Area in MasteringBiology.

1. Who should pay for (or should we even consider using) drugs that may cause hundreds of thousands of dollars a year? For an example of such a drug, visit MJ's blog in the Study Area in MasteringBiology and look under "An Expensive Drug." http://goo.gl/56bybA

2. Should new vaccines be tested in children (as is currently done in adults) before they are allowed on the market? Why or why not? For a current controversy involving the anthrax vaccine, visit MJ's blog in the Study Area in MasteringBiology and look under "Vaccine Testing in Children." http://goo.gl/p2ikTw

MasteringBiology®

Students Go to MasteringBiology for assignments, the eText, and the Study Area with animations, practice tests, and activities.

Professors Go to MasteringBiology for automatically graded tutorials and questions that you can assign to your students, plus Instructor Resources.

Answers to ✓ questions are available in the Appendix.

The Chemistry of Living Things

Polarized light micrograph of sucrose crystals (table sugar).

Key Concepts

- **The natural world consists of matter and energy.** The smallest functional unit of matter is an atom.

- **Chemical bonds link atoms together to form molecules.** These bonds form naturally because molecules are more stable than the atoms that comprise them. One of the most important naturally occurring stable molecules is water.

- **Water is the universal biological solvent.** Water comprises most of the fluid within cells and surrounds all cells in multicellular organisms. Most of life's chemical reactions take place in water.

- **Living things harness energy and use it to make complex molecules not otherwise found in nature.** The four classes of *organic* molecules made by living organisms are proteins, carbohydrates, lipids, and nucleic acids.

- **Carbon is the common building block of all four classes of organic molecules** because of the many ways it can form chemical bonds with other atoms.

21

Functional Foods and Dietary Supplements— Safe and Effective?

Red Bull energy drink to boost our energy; TripleFlex (containing glucosamine and chondroitin) for our aching joints; extracts of *Ginkgo biloba* to sharpen our memories: We seem to have an appetite for functional foods and dietary supplements that promise to improve our health or make us feel better. But do they work?

Functional foods, also sometimes called *nutraceuticals*, are food or drink products that are said to have benefits beyond basic nutrition. Some are natural products; others are fortified foods or completely artificially created products. Red Bull is a functional food because of the manufacturer's claim that it "boosts energy levels."

Dietary supplements are products that are not normally part of your diet but that you choose to take to improve your health or well-being. Your daily multivitamins are

Do you know what is in this can?

dietary supplements, as are any supplemental minerals, amino acids, hormones, bodybuilding products, and plant extracts (herbal remedies) you take by choice.

Traditional herbal remedies are by far the oldest of the dietary supplements. In some cultures, they have been in use for hundreds or even thousands of years, long before we thought to call them dietary supplements. Some of the ingredients in herbal remedies may indeed have specific health benefits. However, they have likely never been tested scientifically to *prove* that they are safe and effective. To understand why, you need to know how new pharmaceutical drugs are tested and approved and why those same testing and approval processes are not applied to functional foods and dietary supplements.

Questions to Consider

1 Who do you think should be responsible for ensuring that functional foods and dietary supplements are safe? Would you be willing to accept more regulation if it meant fewer products would be available? Explain your position.

2 What, if any, dietary supplements or functional foods do you use? Do you know what is in these products, and do you understand why you or other people use them?

Regulatory Issues

The U.S. Food and Drug Administration (FDA) is responsible for overseeing the safety and efficacy of pharmaceutical drugs (drugs created specifically for the treatment or prevention of disease). By law, a pharmaceutical company must prove beyond a reasonable doubt that a new drug is safe and effective in humans before it can be sold to the public. On average, it takes 12 to 15 years and costs hundreds of millions of dollars to bring a new drug to market. Companies can afford it only because they can patent the drug, giving them exclusive ownership and marketing rights for a certain number of years. The high prices of some prescription medications reflect the pharmaceutical companies' need to recover their steep development costs.

Because the ingredients in functional foods and dietary supplements occur in nature, they cannot be patented. Anyone can purify and package them. But without the assurance of patent protection, functional food and dietary supplement manufacturers cannot afford to spend what it would cost to test the safety and effectiveness of their products. Recognizing this, the dietary supplement manufacturers asked for (and were granted) an exemption from the FDA drug approval process. Under the Dietary Supplements Health and Education Act of 1994, dietary supplements and functional foods can be produced and sold until they are *proven unsafe*. But why would any manufacturer choose to spend time and money to prove its own product unsafe?

Functional food and dietary supplement manufacturers are not even required to notify the FDA of adverse events; the →

How will you determine if the supplement you are taking is safe and effective?

agency must rely on voluntary information supplied by consumers and health professionals. As a result, the FDA reported only about 500 adverse events per year associated with dietary supplements over a five-year period. In contrast, the American Association of Poison Control Centers received nearly 37,000 calls about dietary supplements in 2012 alone.

Producers and distributors of functional foods and dietary supplements also have considerable latitude in advertising their products; the only restriction is that they are not allowed to claim that their products prevent or treat specific medical conditions or diseases. For example, producers of cranberry juice products are free to say that cranberry juice "helps maintain urinary tract health" (a rather vague health claim), but they cannot claim that cranberry juice "prevents the recurrence of urinary tract infections" because that would represent a specific medical claim. Nevertheless, many consumers do use cranberry juice to treat urinary tract infections or to prevent their recurrence, simply because they believe that it works. And indeed it may; it is just that it has never been scientifically tested to the standards of a pharmaceutical drug.

With all that latitude in producing and marketing their products, it is not surprising that the functional foods and dietary supplements industries and the advertising industry that supports them have grown rapidly. U.S. sales of dietary supplements now top $25 billion a year. Annual sales of ginkgo alone top several hundred million dollars a year, despite the evidence that ginkgo does not prevent memory loss or dementia.

Questions of Safety and Efficacy

Proponents of functional foods and dietary supplements argue that because many of these products have been in use for a long time, any adverse effects should have shown up by now. Critics argue that many of the ingredients can now be synthesized chemically and thus used at much higher concentrations and in different combinations than ever occur in nature. Slam Energy drink, for example, contains over 100 mg of caffeine in only two ounces of liquid. Health officials worry that too much caffeine in a user unaccustomed to it could raise heart rate and blood pressure to potentially dangerous levels.

Other concerns include inaccurate product labeling and improper manufacturing processes. Manufacturers are not required to report quality control information to the FDA, so there is no assurance that the product actually contains what the manufacturer says it does. For example, independent tests found that products labeled as containing the same dosage of ginseng actually varied by a factor of 10 (some contained none at all). And California investigators found that nearly one-third of all imported Asian herbal remedies they tested contained lead, arsenic, mercury, or drugs not mentioned on the label.

Consumers want to be assured that the dietary supplements and natural and fortified food products they use are safe. They would like to know that the health claims about these products are true. How to achieve that goal and still ensure that the products remain available is an ongoing issue. In the meantime, it is up to you to know what is in the products you choose to put in your body.

SUMMARY

- Functional foods and dietary supplements are popular with consumers. Americans consume more than $18 billion worth of dietary supplements every year.
- Functional foods and dietary supplements don't pass through the same rigorous approval procedure that is required of all new pharmaceutical drugs.
- Some functional foods and dietary supplements may not be effective; others may not be safe. However, increased regulation to ensure efficacy and safety would mean fewer of these products would be available.

Chemistry is the study of matter and energy. Matter is the physical stuff; anything that has mass and occupies space. Energy is the power to do work.

The natural world, from rocks to humans, is comprised of the same building blocks of matter and the same kinds of energy. As part of the natural world, we are intimately connected to it in various ways. We take in oxygen gas from the air and nitrogen, hydrogen, and carbon from the food we eat. Our bodies use these molecules to provide the energy we need to grow, move about, and reproduce. At regular intervals, we eliminate molecules as waste into the biosphere. And when we die, all of the matter and stored energy composing our bodies returns again to the Earth. Everything in the natural world, including us, is governed by the laws of chemistry. ■

MJ's BlogInFocus How often do the herbal supplement products sold in the United States and Canada actually contain what they say they contain? Visit MJ's blog in the Study Area in MasteringBiology and look under "Herbal Supplements."

http://goo.gl/ucXpfQ

We begin this chapter by examining the basic principles of chemistry and then move to consider more specifically how chemistry serves life. Let's turn now to a discussion on matter and how it relates to one of the basic principles of chemistry.

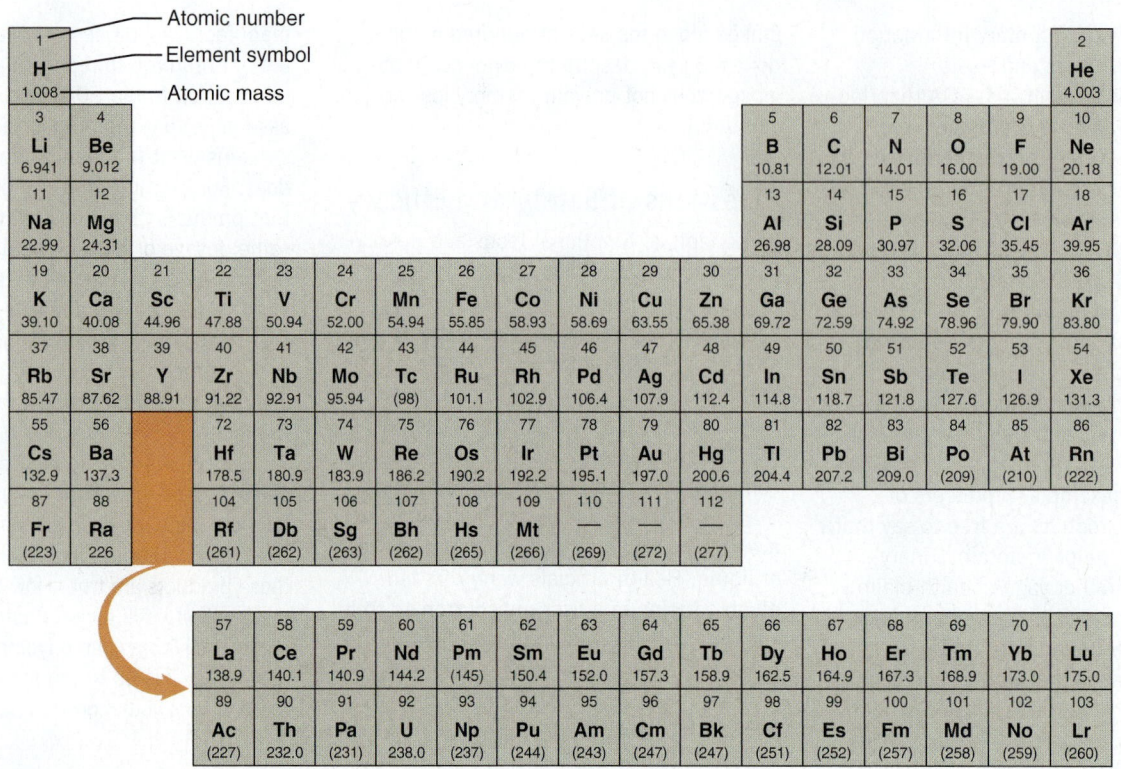

Figure 2.1 **The periodic table of the elements shows all known elements in order of increasing atomic number.** The atomic number represents the number of protons in an atom of the element; atomic mass is the total mass of the atom's protons and neutrons.

2.1 All matter consists of elements

All matter is composed of elements. An **element** is a fundamental (pure) form of matter that cannot be broken down to a simpler form. Aluminum and iron are elements, as are oxygen and hydrogen. The 118 known elements account for all matter. The *periodic table of elements* (**Figure 2.1**) arranges the elements into rows and columns according to their similar properties. Each named element is designated by a one- or two-letter symbol taken from English or Latin. For example, oxygen is designated by the letter O, nitrogen by N, sodium by Na (from the Latin word for sodium, *natrium*), and potassium by K (Latin *kalium*).

Atoms are the smallest functional units of an element

Elements are made up of particles called *atoms*. An **atom** is the smallest unit of any element that still retains the physical and chemical properties of that element. All matter, both living and nonliving, is composed of atoms (**Figure 2.2**). Although

we now know that atoms can be split apart under unusual circumstances (such as a nuclear reaction), atoms are the smallest units of matter that can take part in chemical reactions. So, for all practical purposes, atoms are the smallest functional units of matter.

The central core of an atom is called the *nucleus*. The nucleus is made of positively charged particles called

a) A magnified view (×15,000) of a portion of a skeletal muscle cell.

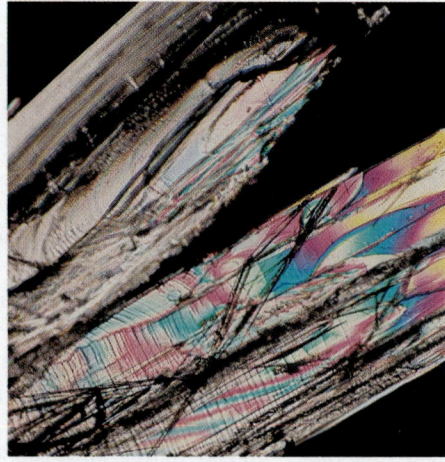

b) A magnified view (×30) of a crystal of aspirin.

Figure 2.2 **All matter is made of atoms.** The three most common atoms in both muscle and aspirin are the same; carbon, hydrogen, and oxygen.

protons and a nearly equal number of neutral particles called **neutrons,** all tightly bound together. An exception is the smallest atom, hydrogen, whose nucleus consists of only a single proton. Each element has an *atomic number*, representing the number of protons in an atom's nucleus, and an *atomic mass* (or mass number), which is generally fairly close to the total number of protons and neutrons (see Figure 2.1). Smaller negatively charged particles called **electrons** orbit the nucleus. Electrons have negligible mass. Because electrons are constantly moving, their precise position at any one time is unknown. You can think of electrons as occupying one or more spherical "clouds" of negative charge around the nucleus. These spheres are called *shells*. Each shell can accommodate only a certain number of electrons. The first shell, the one closest to the nucleus, can hold two electrons, the second can hold up to eight, and the third shell (if there is one) also can hold eight. Each type of atom has a unique number of electrons. Under most circumstances, the number of electrons equals the number of protons, and as a result, the entire atom is electrically neutral (**Figure 2.3**).

Protons and neutrons have about the same mass, and both have much more mass than electrons. (*Mass* is measured chemically and is not dependent on gravity. For the purpose of this text, however, mass and *weight* are about the same.) The protons and neutrons in the atom's nucleus account for over 99.9% of the atom's mass.

In chemical formulas for molecules composed of more than one atom, a subscript numeral following the element symbol indicates more than one atom of that element is present. For example, the chemical formula O_2 represents two atoms of oxygen linked together, the most stable form of elemental oxygen. The chemical formula for water, H_2O, indicates that a water molecule consists of two hydrogen atoms bonded to one oxygen atom.

☑ Nitrogen's atomic number is 7. Just from the atomic number, can you determine how many protons, neutrons, and electrons a nitrogen atom has and how many electrons are in its first and second electron shells? (Assume the atom is electrically neutral.)

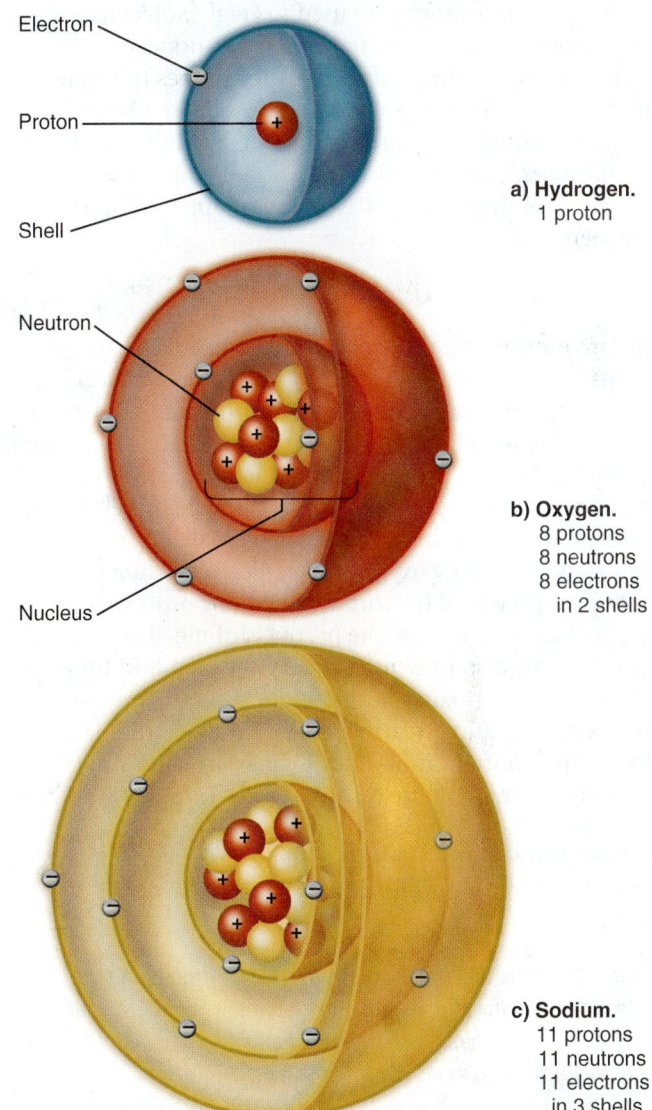

a) **Hydrogen.**
1 proton

b) **Oxygen.**
8 protons
8 neutrons
8 electrons
in 2 shells

c) **Sodium.**
11 protons
11 neutrons
11 electrons
in 3 shells

Electron
Proton
Shell
Neutron
Nucleus

Figure 2.3 The structure of atoms. Atoms consist of a nucleus, comprising positively charged protons and neutral neutrons, surrounded by spherical shells of negatively charged electrons.

Isotopes have a different number of neutrons

Although all the atoms of a particular element have the same number of protons, the number of neutrons can vary slightly. Atoms with either more or fewer neutrons than the usual number for that element are called **isotopes.** Isotopes of an element have the same atomic number as the more common atoms but a different atomic mass. For example, elemental carbon (C) typically consists of atoms with six protons and six neutrons, for an atomic mass of 12 (see Figure 2.1). The isotope of carbon known as carbon-14 has an atomic mass of 14 because it has two extra neutrons.

Isotopes are always identified by a superscript mass number preceding the symbol. For instance, the carbon-14 isotope is designated ^{14}C. The superscript mass number of

the more common elemental form of carbon is generally omitted because it is understood to be 12.

Many isotopes are unstable. Such isotopes are called *radioisotopes* because they tend to give off energy (in the form of radiation) and particles until they reach a more stable state. The radiation emitted by radioisotopes can be dangerous to living organisms because the energy can damage tissues.

Certain radioisotopes have a number of important scientific and medical uses. Because the rate of decay to more stable energy states is known for each radioisotope, scientists can determine when rocks and fossils were formed by measuring the amount of radioisotope still present. The carbon-14 isotope is commonly used for this purpose.

In medicine, radioisotopes are used to "tag" molecules so that radiation sensors can track their location in the body. For example, physicians use radioisotopes to locate areas of damaged tissue in a patient's heart after a heart attack. Radioisotopes are also used to target and kill certain kinds of cancer. Certain radioisotopes that emit energy for long periods of time are used as a power supply in heart pacemakers.

Free radicals have unpaired electrons

Atoms are most stable when each shell is filled and all electrons are paired with another electron. A *free radical* is an atom or molecule that has one or more unpaired electrons in its outer shell but still has an independent existence—hence the term *free*. Free radicals tend to be highly reactive, readily taking electrons from other molecules or giving up electrons to other molecules. In living organisms, small amounts of free radicals (especially oxygen free radicals, missing one electron) are produced by normal metabolic processes. They are essentially toxic waste products of metabolism. Free radicals are also present in the environment, in tobacco smoke and various toxins and pollutants.

Because they are so reactive, free radicals can damage biologically important molecules such as proteins and DNA (the genetic material of the cell). There is also some evidence that free radicals may speed up the cellular aging process. How cells deal with free radicals will be described when we discuss cell structure and function (Chapter 3).

↩ **Recap** Atoms are made up of protons, neutrons, and electrons. Radioisotopes, unstable atoms with an unusual number of neutrons, give off energy and particles as they decay (move to a more stable state). ∎

2.2 Atoms combine to form molecules

A **molecule** consists of a stable association between two or more atoms. For example, a molecule of water is two atoms of hydrogen plus one atom of oxygen (written H_2O). A molecule of ordinary table salt (written NaCl) is one atom of sodium (Na) plus one atom of chlorine (Cl). A molecule of hydrogen gas (written H_2) is two atoms of hydrogen. To understand *why* atoms join together to form molecules, we need to know more about energy.

Energy fuels life's activities

Energy is the capacity to do work, to cause some change in matter. Joining atoms is one type of work, and breaking up molecules is another—and both require energy. Stored energy, that which is not actually performing any work at the moment, is called **potential energy,** because it has the *potential* to make things happen (**Figure 2.4a**). Energy that

is actually *doing* work—that is, energy in motion—is called **kinetic energy** (**Figure 2.4b**).

You can visualize the difference between potential energy and kinetic energy in the water held behind a dam: There is tremendous potential energy in the water held behind the dam. When the water is released, potential energy is converted into kinetic energy: rushing water that can be put to work turning turbines. Similarly, the spark of a match converts the potential energy in firewood to kinetic energy in the form of heat and light.

Potential energy is stored in the bonds that hold atoms together in all matter, both living and nonliving. Living organisms take advantage of this general principle of chemistry by using certain molecules to store energy for their own use. When the chemical bonds of these energy-storage molecules are broken, potential energy becomes kinetic energy. We rely on this energy to do biological work, such as breathing, moving, and digesting food.

Recall that electrons carry a negative charge, whereas protons within the nucleus have a positive charge. Electrons are simultaneously attracted to the positively charged nucleus and repelled by each other. As a result of these opposing attractive and repulsive forces, each electron occupies a specific shell around the nucleus. Each shell corresponds to a specific level of electron potential energy, and each shell farther out represents a higher potential energy level than the preceding one closer to the nucleus.

a) Potential energy is locked up in the chemical bonds of energy-storage molecules in Michael Phelps' tissues.

b) Kinetic energy is energy in motion.

Figure 2.4 Energy.

Table 2.1 Summary of the three types of chemical bonds

Type	Strength	Description	Examples
Covalent bond	Strong	A bond in which the sharing of electrons between atoms results in each atom having a maximally filled outermost shell of electrons	The bonds between hydrogen and oxygen in a molecule of water
Ionic bond	Moderate	The bond between two oppositely charged ions (atoms or molecules that were formed by the permanent transfer of one or more electrons)	The bond between Na^+ and Cl^- in salt
Hydrogen bond	Weak	The bond between oppositely charged regions of molecules that contain covalently bonded hydrogen atoms	The bonds between molecules of water

When an electron moves to a shell closer to the nucleus, it loses energy. To move to a shell that is farther from the nucleus, the electron must absorb energy.

Chemical bonds link atoms to form molecules

A key concept in chemistry is that atoms are most stable when their outermost electron shell is completely filled with the maximum number of electrons that it can accommodate. An atom whose outermost electron shell is not normally completely filled tends to interact with one or more other atoms in a way that fills its outermost shell. Such interactions generally cause the atoms to be bound to each other by attractive forces called **chemical bonds.** The three principal types of chemical bonds are called covalent, ionic, and hydrogen bonds (Table 2.1).

Covalent bonds involve sharing electrons One way that an atom can fill its outermost shell is by sharing a pair of electrons with another atom. An electron-sharing bond between atoms is called a **covalent bond** (Figure 2.5). Covalent bonds between atoms are among the strongest chemical bonds in nature, so strong that they rarely break apart. In structural formulas, a covalent bond is depicted as a line drawn between two atoms. For example, the structural formula for hydrogen is H—H.

Hydrogen gas offers an example of how a covalent (electron-sharing) bond fills the outermost shells of two atoms. Each of the two hydrogen atoms has just one electron in the first shell, which could accommodate two electrons. When joined together by a covalent bond (forming H_2 a gas), each atom has, in effect, a "full" first shell of two electrons. As a result, H_2 gas is more stable than the same two hydrogen atoms by themselves. The sharing of one pair of electrons, as in H_2, is called a *single bond*.

Oxygen gas is another example of covalent bonding. An oxygen atom has eight electrons: two of these fill the first electron shell, and the remaining six occupy the second electron shell (which can accommodate eight). Two oxygen atoms may join to form a molecule of oxygen gas by sharing two pairs of electrons, thus completing the outer shells of both atoms. When two pairs of electrons are shared, the bond is called a *double bond*. In structural formulas, double bonds are indicated by two parallel lines. For example, the structural formula for oxygen is O=O.

A molecule of water forms from one oxygen and two hydrogen atoms because this combination completely fills the outermost shells of both hydrogen and oxygen. The prevalence of water on Earth follows a basic principle of chemistry: Matter is most stable when it contains the least potential energy. Both hydrogen and oxygen are more stable when bonded together (as H_2O) than as independent atoms.

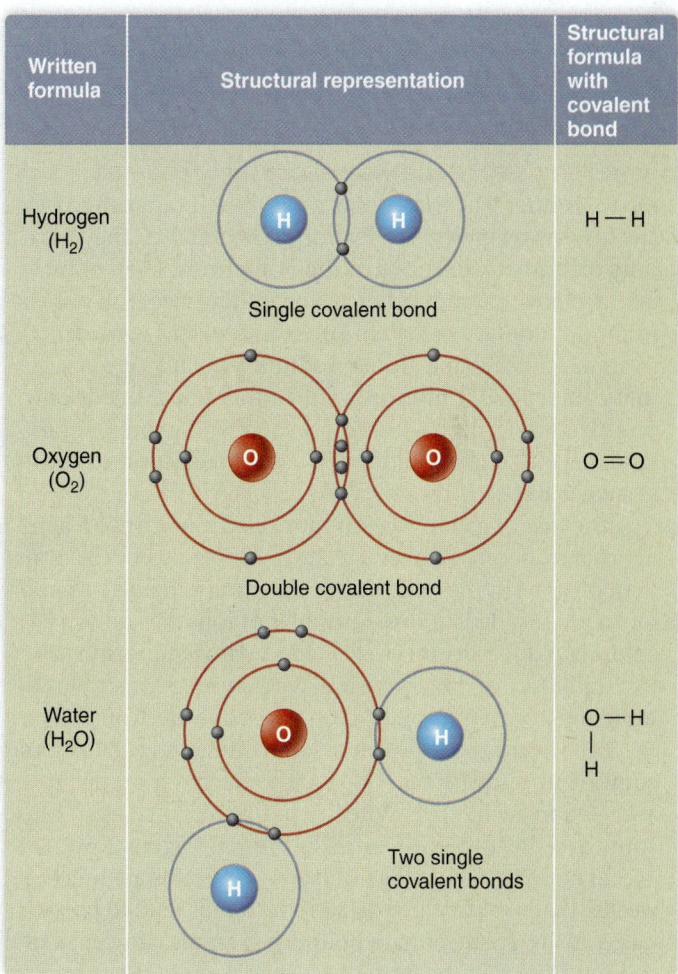

Figure 2.5 Covalent bonds. Sharing pairs of electrons is a way for atoms to fill their outermost shell.

✓ Draw a model, like those in Figure 2.5, for CH_4. Hint: Carbon has two electrons in its inner shell and four in its second shell.

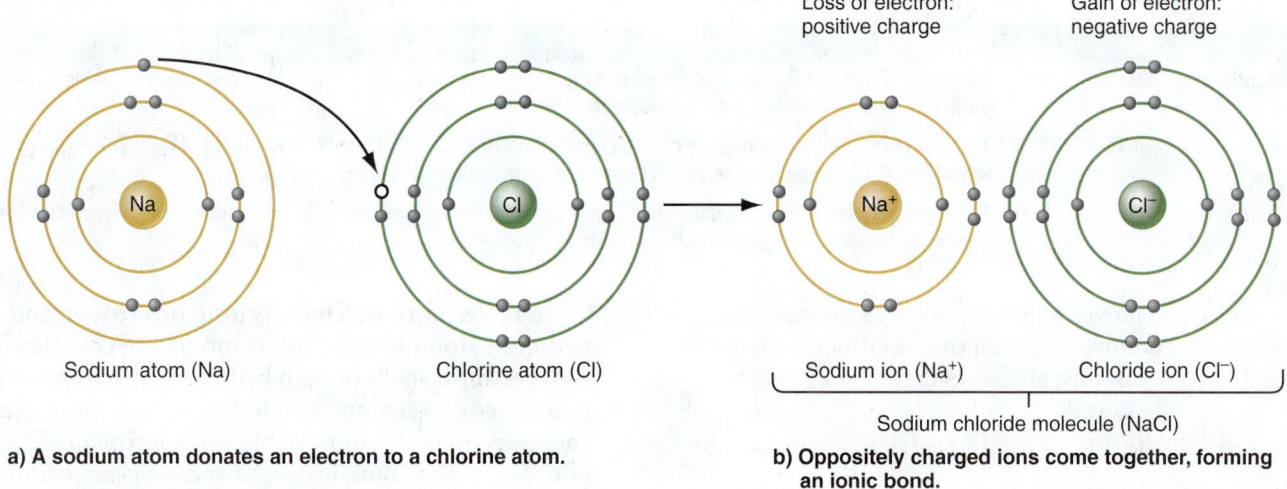

a) A sodium atom donates an electron to a chlorine atom.

b) Oppositely charged ions come together, forming an ionic bond.

Figure 2.6 Ionic bonds.

☑ Can you determine the number of bonds between atoms in a single molecule of carbon dioxide, CO_2? Hint: Oxygen has eight electrons and carbon has six electrons.

Ionic bonds occur between oppositely charged ions
Another way that atoms can fill their outer shell of electrons is by giving up electrons completely if they have only one or two electrons in their outermost shell or by taking electrons from other atoms if they need one or two to fill their outermost shell. For example, let's consider sodium and chlorine atoms. Sodium has only one electron in its outer shell. Chlorine, on the other hand, needs only one electron to fill its outermost shell. There is a natural tendency, then, for sodium to donate an electron to chlorine (**Figure 2.6a**).

The loss or gain of one or more electrons gives an atom a net charge, because now there are fewer (or more) electrons than protons in the nucleus. An electrically charged atom or molecule is called an **ion**. Examples of ions are sodium (Na^+), chloride (Cl^-), calcium (Ca^{2+}), and hydrogen phosphate (HPO_4^-). Notice that ions can have a shortage or surplus of more than one electron. (Ca^{2+} has lost two electrons.)

Ever heard the expression "opposites attract"? It should come as no surprise that oppositely charged ions are attracted to each other. When two oppositely charged ions come together, an **ionic bond** is formed (**Figure 2.6b**).

In aqueous (watery) solutions, ionic bonds are much weaker than covalent bonds, so ionic bonds tend to break rather easily. In the human body, for example, almost all of the sodium (Na) is in the form of Na^+ and most of the chlorine (Cl) is in its ionized form, called *chloride* (Cl^-). Very little exists as NaCl. Ions in aqueous solutions are sometimes called *electrolytes* because solutions of water containing ions are good conductors of electricity. As you will see, cells can control the movement of certain ions, creating electrical forces essential to the functioning of nerves, muscles, and other living tissues.

Weak hydrogen bonds form between polar molecules
A third type of attraction occurs between molecules that do not have a net charge. Glance back at the water molecule in Figure 2.5 and note that the two hydrogen atoms are found not at opposite ends of the water molecule but fairly close together. Although the oxygen and the two hydrogen atoms share electrons, the sharing is unequal. The shared electrons in a water molecule actually spend slightly more of their time near the oxygen atom than near the hydrogen atoms because the oxygen atom attracts electrons more strongly than do the hydrogen atoms. The uneven sharing gives the oxygen region of a water molecule a partial negative charge and the two hydrogen regions a partial positive charge, even though the water molecule as a whole is electrically neutral.

Molecules such as water that are electrically neutral overall but still have partially charged regions, or *poles*, are called **polar** molecules (**Figure 2.7**). According to the

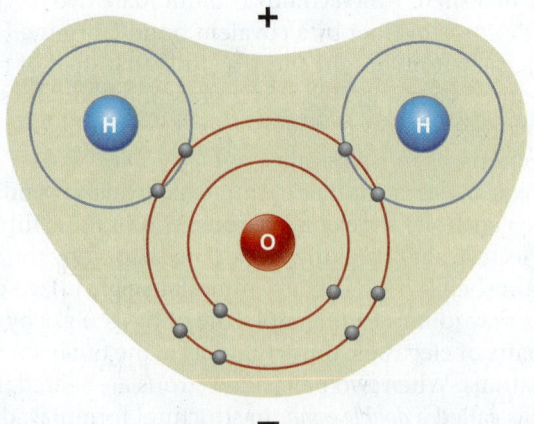

Figure 2.7 Water is a polar molecule. Although polar molecules are electrically neutral overall, they have equal but oppositely charged regions (poles) due to unequal sharing of electrons.

principle that opposites attract, polar molecules arrange themselves so that the partial negative pole of one molecule is oriented toward (attracted by) the partial positive pole of another molecule. The weak attractive force between oppositely charged regions of polar molecules that contain covalently bonded hydrogen is called a **hydrogen bond.**

Hydrogen bonds are important in biological molecules. They're what give proteins their three-dimensional shape, and they keep the two strands of the DNA molecule together. The structures of both proteins and DNA are described later in this chapter.

Living organisms contain only certain elements

Although there are nearly 100 different elements in nature, living organisms are constructed from a limited number of them (Table 2.2). In fact, about 99% of your body weight consists of just six elements: oxygen, carbon, hydrogen, nitrogen, calcium, and phosphorus. These particular elements are especially important because they are stable and because they readily form ions, making them able to combine with other elements to create the many molecules of life. However, some of the less common elements are important, too. Life on Earth would not be possible without them.

Even the largest atoms are small compared to the structures in living organisms. To appreciate the vast size differences between the atoms, cells, and organs in your body, imagine that a sodium atom is the size of a penny. On this scale, one of your red blood cells would be half a mile in diameter, and your heart would be larger than the entire Earth!

Next, let's look at some of the most important matter of living systems: water, hydrogen ions, and a host of molecules that contain a backbone of carbon atoms.

Recap Electrons farthest from the nucleus have more potential energy than electrons close to the nucleus. Strong covalent bonds form between atoms when they share pairs of electrons, ionic bonds form between oppositely charged ions, and weak hydrogen bonds occur between oppositely charged regions of polar molecules. ■

2.3 Life depends on water

No molecule is more essential to life than water. Indeed, it accounts for 60% of your body weight. The following properties of water are especially important to living organisms:

- Water is an excellent solvent.
- Water is a liquid at body temperature.
- Water can absorb and hold heat energy.
- Evaporation of water uses up heat energy.
- Water participates in essential chemical reactions.

Water is the biological solvent

A **solvent** is a liquid in which other substances dissolve, and a **solute** is any dissolved substance. Water is the ideal solvent in living organisms specifically because it is a polar liquid at body temperature. As the solvent of life, water is the substance in which the many chemical reactions of living organisms take place. Let's look at a simple example of a solute dissolving in water to better understand how the polar nature of water facilitates the reaction.

Table 2.2 The 12 most common and important elements in living organisms*

Element	Atomic symbol	Atomic number	Atomic mass	% of human weight	Functions in life
Oxygen	O	8	16.0	65	Part of water and most organic molecules; also molecular oxygen
Carbon	C	6	12.0	18	The backbone of all organic molecules
Hydrogen	H	1	1.0	10	Part of all organic molecules and of water
Nitrogen	N	7	14.0	3	Component of proteins and nucleic acids
Calcium	Ca	20	40.1	2	Constituent of bone; also essential for the action of nerves and muscles
Phosphorus	P	15	31.0	1	Part of cell membranes and of energy storage molecules; also a constituent of bone
Potassium	K	19	39.1	0.3	Important in nerve action
Sulfur	S	16	32.1	0.2	Structural component of most proteins
Sodium	Na	11	23.0	0.1	The primary ion in body fluids; also important for nerve action
Chlorine	Cl	17	35.5	0.1	Component of digestive acid; also a major ion in body fluids
Magnesium	Mg	12	24.3	Trace	Important for the action of certain enzymes and for muscle contraction
Iron	Fe	26	55.8	Trace	A constituent of hemoglobin, the oxygen-carrying molecule

*The elements are listed in descending order of their contribution to total body weight. Atomic number represents the number of protons in the nucleus. Atomic mass is roughly equivalent to the total number of protons and neutrons because electrons have very little mass. Note that 99% of your body weight is accounted for by just six elements.

Consider a common and important solid, crystals of sodium chloride (NaCl), or table salt. Crystals of table salt consist of a regular, repeating pattern of sodium and chloride ions held together by ionic bonds (**Figure 2.8**). When salt is placed in water, individual ions of Na^+ and Cl^- are pulled away from the crystal and are immediately surrounded by the polar water molecules. The water molecules form such a tight cluster around each ion that they are prevented from reassociating back into the crystalline form. In other words, water keeps the ions dissolved. Note that the water molecules are oriented around ions according to the principle that opposite charges attract.

Polar molecules that are attracted to water and interact with it easily are called **hydrophilic** molecules (Greek, meaning "water-loving"). Nonpolar, neutral molecules such as cooking oils do not interact easily with water and generally will not dissolve in it. They are said to be **hydrophobic** (Greek, meaning "water-fearing"). When water and oil are mixed, the water molecules tend to form hydrogen bonds with each other, excluding the oil from regions occupied by water. Over time, the oil is forced together into larger and larger drops until it is separated from the water completely.

✅ A red dye is placed in a container that contains an equal amount of cooking oil and water. After the mixture is shaken, you notice that the water is red but the oil is not. From what you know about the properties of oil and water, what can you say about the dye: Is it comprised of neutral, polar, or charged molecules? Explain your reasoning.

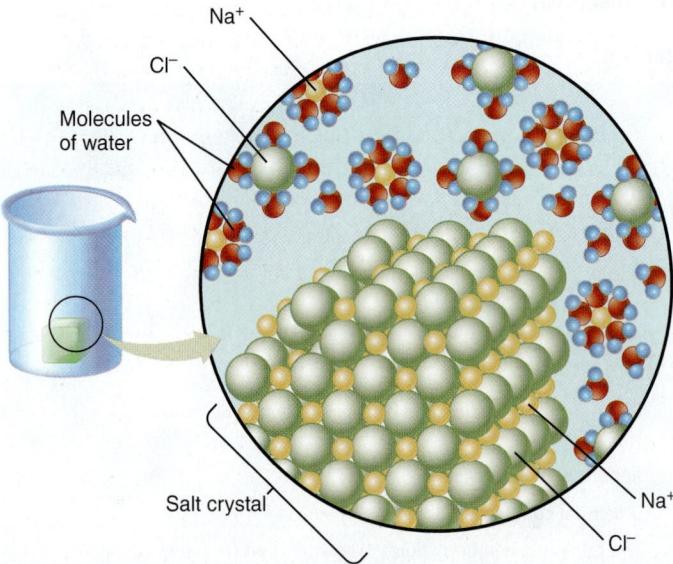

Na$^+$
Cl$^-$
Molecules of water
Salt crystal
Na$^+$
Cl$^-$

Figure 2.8 How water keeps ions in solution. The slightly negative ends of polar water molecules are attracted to positive ions, whereas the slightly positive ends of water molecules are attracted to negative ions. The water molecules pull the ions away from the crystal and prevent them from reassociating with each other.

Water is a liquid at body temperature

Water is a liquid at temperatures between 0 and 100 degrees. That is because within this temperature range, there is just enough heat energy in water to temporarily break some of the weak hydrogen bonds between water molecules. New hydrogen bonds will soon re-form with other nearby water molecules, but the pattern of bonding between adjacent water molecules is seemingly random (**Figure 2.9a**). However, when the temperature falls below zero degrees centigrade, there is no longer enough heat energy to break the hydrogen bonds between water molecules. Below zero degrees centigrade, the water molecules orient themselves into a stable, unchanging, rigid lattice structure (ice; **Figure 2.9b**). At the other end of the temperature scale (above 100 degrees centigrade), all hydrogen bonds between adjacent water molecules are broken completely, and water molecules escape into the atmosphere as a gas (water vapor; **Figure 2.9c**).

Because water is a liquid at body temperature, it is an excellent medium for transporting solutes from one place to another in our bodies. Indeed, transport is the primary function of the blood (which is about 90% water) in our cardiovascular system. Under pressure generated by the heart, blood transports oxygen and nutrients to all living cells and cellular wastes (including carbon dioxide) away from the cells.

Water is the main constituent of all of the fluid-filled spaces within our bodies. It fills our cells (the intracellular space); it occupies the spaces between cells (the intercellular space); it even fills the fluid-filled spaces not occupied by living cells, such as the urine in our bladders, the watery solutions in our digestive system, and the fluid in our eyes. Of course, these fluids contain many important dissolved solutes and solids as well. Overall, though, approximately 60% of our body weight is water.

Water helps regulate body temperature

An important property of water is that it can absorb and hold a large amount of heat energy with only a modest increase in temperature. In fact, it absorbs heat better than most other liquids. Water thus may prevent large increases in body temperature when excess heat is produced. Water also holds heat well when there is a danger of too much heat loss (for instance, when you go outdoors wearing shorts on a cool day). The ability of water to absorb and hold heat helps prevent rapid changes in body temperature when changes occur in metabolism or in the environment.

Our bodies generate heat during metabolism. We usually generate more heat than we need to maintain a constant body temperature of 98.6°F (37°C), so being able to lose heat is important to our survival. One way we lose heat is by evaporation of water (sweat) from the surface of our bodies. Remember that it takes a great deal of heat energy to break all of the hydrogen bonds between adjacent water molecules. As sweat evaporates, then, it cools the blood in the blood

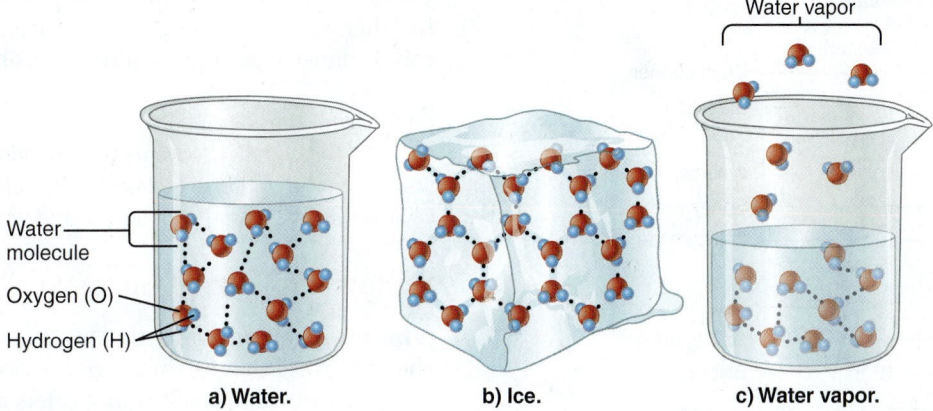

Figure 2.9 Hydrogen bonds in water and ice. In water, weak hydrogen bonds continually form, break, and re-form between adjacent water molecules. Ice is a solid because a repeating pattern of hydrogen bonds forms between adjacent water molecules.

capillaries near the surface of the skin. Evaporation of sweat is just one of several mechanisms involved in the overall maintenance of a constant body temperature.

You can demonstrate the cooling power of evaporation for yourself. The next time you perspire heavily, notice that even though you may feel hot, your exposed skin may actually feel cool to the touch.

Water participates in chemical reactions

Water is involved in many of the chemical reactions within living organisms. Examples are the synthesis and breakdown of carbohydrates, lipids, and proteins, all of which are essential to life. The equivalent of a water molecule is either used up or created anew whenever these molecules are synthesized or later deconstructed, as you will see later in this chapter.

> ⮌ **Recap** Most biological molecules dissolve readily in water because water is a polar molecule. The liquid nature of water facilitates the transport of biological molecules. Water absorbs and holds heat and can lower body temperature through evaporation. It also participates in essential chemical reactions. ▪

2.4 The importance of hydrogen ions

One of the most important ions in the body is the hydrogen ion (a single proton without an electron). In this section, we will see how hydrogen ions are created and why it is so important to maintain an appropriate concentration of them.

Acids donate hydrogen ions, bases accept them

Although the covalent bonds between hydrogen and oxygen in water are strong and thus rarely broken, it can happen.

When it does, the electron from one hydrogen atom is transferred to the oxygen atom completely, and the water molecule breaks into two ions—a *hydrogen ion* (H^+) and a *hydroxide ion* (OH^-).

In pure water, only a very few molecules of water are dissociated (broken apart) into H^+ and OH^- at any one time. However, there are other sources of hydrogen ions in aqueous solutions. An **acid** is any molecule that can donate (give up) an H^+ ion. When added to pure water, acids produce an *acidic* solution, one with a higher H^+ concentration than pure water. (By definition, an aqueous solution with the same concentration of H^+ as that of pure water is a *neutral* solution.) Common acidic solutions are vinegar, carbonated beverages, black coffee, and orange juice. Conversely, a **base** is any molecule that can accept (combine with) an H^+ ion. When added to pure water, bases produce a basic or *alkaline* solution, one with a lower H^+ concentration than that of pure water. Common alkaline solutions include baking soda in water, detergents, and drain cleaner.

Because acids and bases have opposite effects on the H^+ concentration of solutions, they are said to neutralize each other. You have probably heard that a spoonful of baking soda in water is a time-honored way to counteract an "acid stomach." Now you know that this home remedy is based on sound chemical principles.

MJ's BlogInFocus Do you drink energy drinks or alcoholic drinks containing caffeine? To learn more about such drinks and about the latest developments in their regulation, visit MJ's blog in the Study Area in MasteringBiology and look under "Caffeine."

http://goo.gl/mtmNXF

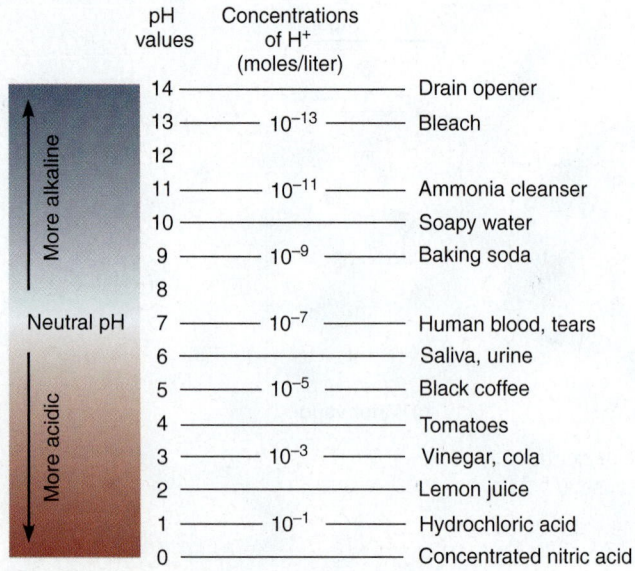

Figure 2.10 The pH scale. The pH scale is an indication of the H^+ concentration of a solution.

The pH scale expresses hydrogen ion concentration

Scientists use the pH scale to indicate the acidity or alkalinity of a solution. The **pH scale** is a measure of the hydrogen ion concentration of a solution. **Figure 2.10** shows the pH scale and indicates the pH values of some common substances and body fluids. The scale ranges from 0 to 14, with pure water having a pH of 7.0, the neutral point. A pH of 7 corresponds to a hydrogen ion concentration of 10^{-7} moles/liter (a *mole* is a term used by chemists to indicate a certain number of atoms, ions, or molecules). An *acidic* solution has a pH of *less* than 7, whereas a *basic* solution has a pH of *greater* than 7. Each whole number change in pH represents a 10-fold change in the hydrogen ion concentration in the opposite direction. For example, an acidic solution with a pH of 5 has an H^+ concentration of 10^{-5} moles/liter (100 times greater than pure water), whereas an alkaline solution with a pH of 9 has an H^+ concentration of 10^{-9} moles/liter (1/100 that of water).

The pH of blood is 7.4, just slightly more alkaline than neutral water. The hydrogen ion concentration of blood plasma is low relative to the concentration of other ions. (The hydrogen ion concentration of blood plasma is less than one-*millionth* that of sodium ions, for example.) It is important to maintain homeostasis of this low concentration of hydrogen ions in the body because hydrogen ions are small, mobile, positively charged, and highly reactive. Hydrogen ions tend to displace other positive ions in molecules, and when they do, they alter molecular structures and change the ability of the molecule to function properly.

Changes in the pH of body fluids can affect how molecules are transported across the cell membrane and how rapidly certain chemical reactions occur. They may even alter

the shapes of proteins that are structural elements of the cell. In other words, a change in the hydrogen ion concentration can be dangerous because it threatens homeostasis.

✅ A chemist has a solution that has a pH of 3. She adds a chemical to it, and shortly afterwards the solution has a pH of 5. What was the concentration of hydrogen ions before adding the chemical, what was it afterwards, and did she add an acid or a base? Explain.

Buffers minimize changes in pH

A **buffer** is any substance that tends to minimize the changes in pH that might otherwise occur when an acid or base is added to a solution. Buffers are essential to our ability to maintain homeostasis of pH in body fluids.

In biological solutions such as blood or urine, buffers are present as *pairs* of related molecules that have opposite effects. One of the pair is the acid form of the molecule (capable of donating an H^+ ion), and the other is the base form (capable of accepting an H^+ ion). When an acid is added and the number of H^+ ions increases, the base form of the buffer pair accepts some of the H^+ ions, minimizing the fall in pH that might otherwise occur. Conversely, when a base is added that might take up too many H^+ ions, the acid form of the buffer pair releases additional H^+ ions and thus minimizes the rise in pH. Buffer pairs are like absorbent sponges that can pick up excess water and then can be wrung out to release water when necessary.

One of the most important buffer pairs in body fluids such as blood is bicarbonate (HCO_3^- the base form) and carbonic acid (H_2CO_3 the acid form). When blood becomes too acidic, bicarbonate accepts excess H^+ according to the following reaction:

$$HCO_3^- + H^+ \rightarrow H_2CO_3$$

When blood becomes too alkaline, carbonic acid donates H^+ by the reverse reaction:

$$HCO_3^- + H^+ \leftarrow H_2CO_3$$

In a biological solution such as blood, bicarbonate and carbonic acid take up and release H^+ all the time. Ultimately, a chemical *equilibrium* is reached in which the rates of the two chemical reactions are the same, as represented by the following combined equation:

$$HCO_3^- + H^+ \leftrightarrow H_2CO_3$$

When excess acid is produced, the combined equation shifts to the right as the bicarbonate combines with H^+. The reverse is true for alkalinity.

The body has many other buffers as well. The more buffers that are present in a body fluid, the more stable the pH.

↩ **Recap** Acids can donate hydrogen ions to a solution, whereas bases can accept hydrogen ions from a solution. The pH scale indicates the hydrogen ion concentration of a solution. The normal pH of blood is 7.4. Buffers help maintain a stable pH in body fluids. ∎

2.5 The organic molecules of living organisms

Organic molecules are molecules that contain carbon and other elements held together by covalent bonds. The name *organic* came about at a time when scientists believed that all organic molecules were created only by living organisms and all *inorganic* molecules came from nonliving matter. Today, we know that organic molecules can be synthesized in the laboratory under the right conditions and that they probably existed on Earth before there was life.

Carbon is the common building block of organic molecules

Carbon (**Figure 2.11**) is relatively rare in the natural world, representing less than 0.03% of Earth's crust. However, living organisms actively accumulate it. Carbon accounts for about 18% of body weight in humans.

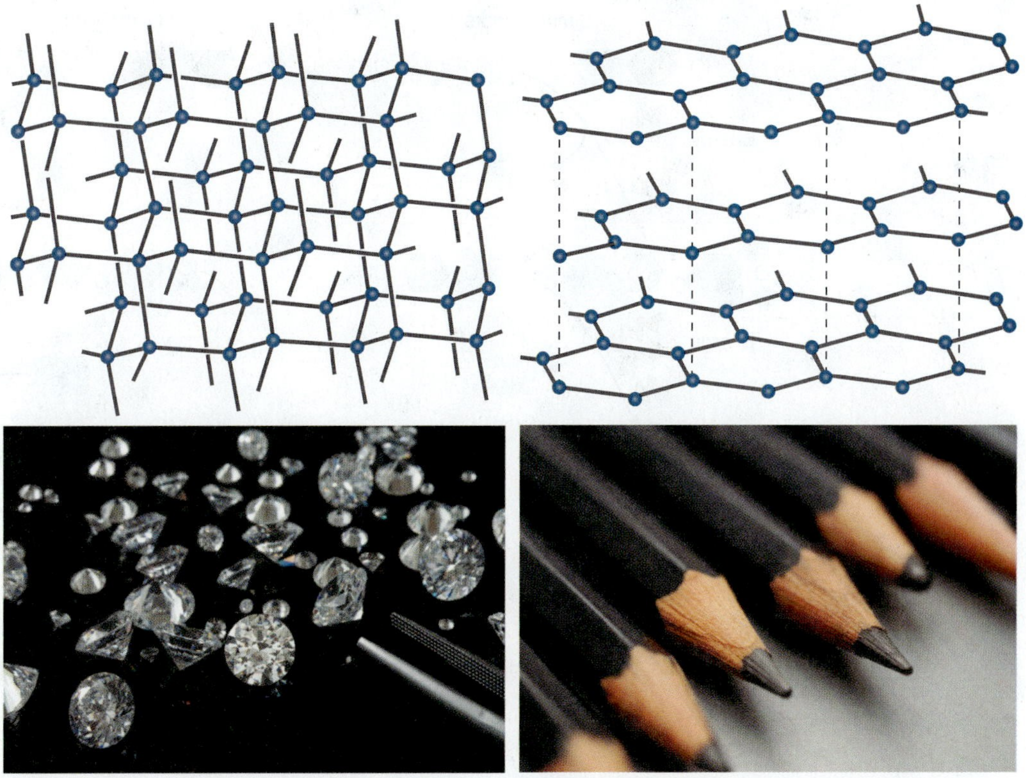

a) **Diamonds are formed only under conditions of extreme temperature and pressure.** The structure of diamond resembles the steel framework of a large building; each atom is covalently bonded to four neighboring carbon atoms. This explains the hardness of diamonds.

b) **Graphite is produced as a result of decay of older carbon-based substances.** Its structure consists of layers of hexagonal rings of carbon atoms. Graphite is fairly soft (hence its use in pencils) because these layers of carbon atoms can slide past one another.

Figure 2.11 Carbon. Graphite and diamond are both elemental forms of carbon.

Carbon is the common building block of all organic molecules because of the many ways it can form strong covalent bonds with other atoms. Carbon has six electrons; two in the first shell and four in the second. Because carbon is most stable when its second shell is filled with eight electrons, *its natural tendency is to form four covalent bonds with other molecules.* This makes carbon an ideal structural component, one that can branch in a multitude of directions.

Using the chemist's convention that a line between the chemical symbols of atoms represents a pair of shared electrons in a covalent bond, let's look at some of the many structural possibilities for carbon. Carbon can form covalent bonds with hydrogen, nitrogen, oxygen, or another carbon (**Figure 2.12**). It can form double covalent bonds with oxygen or another carbon. It can even form five- or six-membered carbon rings, with or without double bonds between carbons.

In addition to their complexity, there is almost no limit to the size of organic molecules derived from carbon. Some, called *macromolecules* (from the Greek *makros*, meaning "long"), consist of thousands or even millions of smaller molecules.

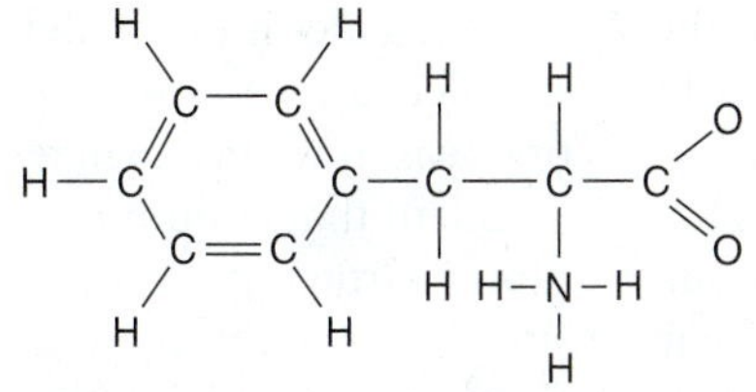

Macromolecules are synthesized and broken down within the cell

Macromolecules are built (synthesized) within the cell itself. In a process called **dehydration synthesis** (also called the

a) **Carbon is the backbone of amino acids, the building blocks of protein.** This amino acid is phenylalanine.

b) **In carbon dioxide, a carbon atom forms two covalent bonds with each oxygen atom.**

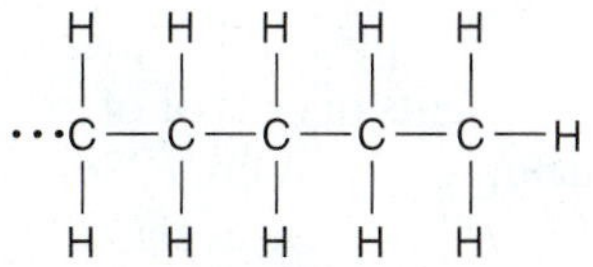

c) **Lipid molecules (a portion of one is shown here) contain long chains of carbon atoms covalently bound to hydrogen.**

Figure 2.12 Examples of the structural diversity of carbon.

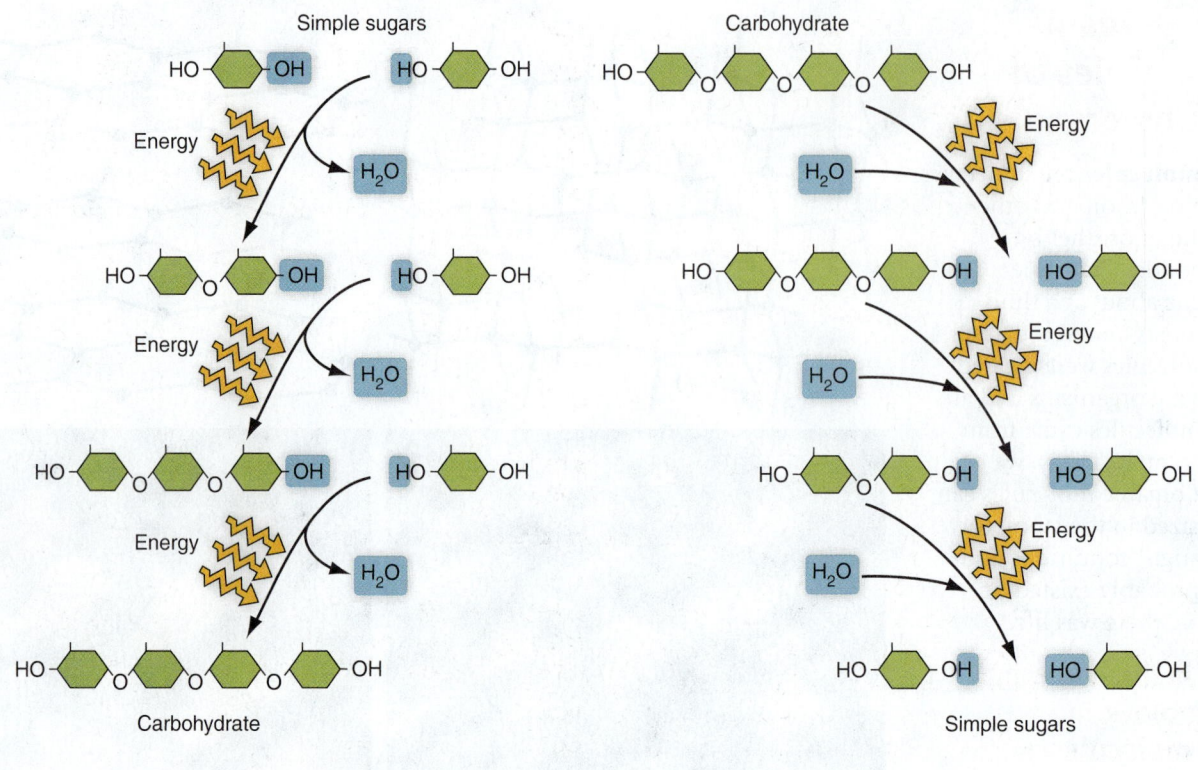

a) Dehydration synthesis.

b) Hydrolysis.

Figure 2.13 Dehydration synthesis and hydrolysis. a) Dehydration synthesis. The synthesis
of larger molecules by dehydration synthesis requires energy. In this example, the smallest units
are simple sugars and the macromolecule is a carbohydrate. **b) Hydrolysis.** The breakdown of large
molecules into smaller units requires a molecule of water and release of stored energy.

condensation reaction), smaller molecules called *subunits* are
joined by covalent bonds, like pearls on a string. The name
of the process accurately describes what is happening: Each
time a subunit is added, the equivalent of a water molecule
is removed ("dehydration"; **Figure 2.13a**). The subunits
needed to synthesize macromolecules come from the foods
you eat and from the biochemical reactions in your body
that break down other large molecules into smaller ones.

The synthesis of macromolecules from smaller molecules
requires energy, which is one reason we need energy to
survive and grow. It is no accident that children seem to
eat enormous amounts of food. Growing children require
energy to make the macromolecules necessary to create new
cell membranes, muscle fibers, and other body tissues.

Some macromolecules are synthesized specifically for
the purpose of storing energy within our cells. The ability to
store energy internally allows organisms to survive even when
food is not plentiful. Other macromolecules serve as structural
components of cells or of extracellular (outside the cell)
structures such as bone. Still others direct the many activities
of the cell or serve as signaling molecules between cells.

Organic macromolecules are broken down by a process
called **hydrolysis.** During hydrolysis, the equivalent of
a water molecule is added each time a covalent bond
between single subunits in the chain is broken. Hydrolysis
is essentially the reverse of dehydration synthesis.

Thus, it should not surprise you that the breakdown of
macromolecules releases energy (**Figure 2.13b**). The energy
was stored as potential energy in the covalent bonds between
atoms. The body obtains much of its energy through
hydrolysis of energy-storage molecules, such as glycogen.
Hydrolysis is also used to break down molecules of food
during digestion, to recycle materials for reuse, and to get rid
of substances that are no longer needed by the body.

Living organisms synthesize four classes of organic
molecules: *carbohydrates, lipids, proteins,* and *nucleic acids.* The
many different molecules within each class are constructed
of the same handful of chemical elements. However, there is
essentially no limit to the number of different molecules that
could be created. No one knows how many different organic
molecules there are. On a chemical level, the tremendous
diversity among the many species of organisms on Earth is
due to differences in their organic molecules, especially their
proteins and nucleic acids.

Recap Carbon is a key element of organic molecules because
of the multiple ways it can form strong covalent bonds with
other molecules. Synthesizing organic molecules requires
energy; breaking them down liberates energy. The four classes
of organic molecules are carbohydrates, lipids, proteins, and
nucleic acids. ∎

2.6 Carbohydrates: used for energy and structural support

A clue to the basic structure of carbohydrates is found in their name. Carbohydrates have a backbone of carbon atoms with hydrogen and oxygen attached in the same proportion as they appear in water (2-to-1); hence, the carbon is *hydrated*, or combined with water.

Most living organisms use carbohydrates for energy, and plants use at least one carbohydrate (cellulose) as structural support.

Monosaccharides are simple sugars

The simplest kind of carbohydrate is called a **monosaccharide** (meaning "one sugar"). Monosaccharides have relatively simple structures consisting of carbon, hydrogen, and oxygen in a 1-2-1 ratio. The most common monosaccharides contain five or six carbon atoms arranged in either a five- or six-membered ring.

Ribose, deoxyribose, glucose, and *fructose* are four of the most important monosaccharides in humans. Ribose and deoxyribose (**Figure 2.14a**) are both five-carbon monosaccharides that are components of nucleotide molecules. The only difference between the two is that deoxyribose has one less oxygen atom than ribose. Glucose, a six-carbon monosaccharide, is an important source of energy for cells. When more energy is available than can be used right away, glucose molecules can be linked together or with other molecules by dehydration synthesis to form larger carbohydrate molecules (**Figure 2.14b**).

Oligosaccharides: more than one monosaccharide linked together

Oligosaccharides are short strings of monosaccharides (*oligo* means "a few") linked together by dehydration synthesis. One common oligosaccharide is table sugar, or sucrose. Sucrose is also called a *disaccharide* because it consists of just two monosaccharides (glucose + fructose). Another is lactose (glucose + galactose), the most common disaccharide in human milk and an important source of energy for infants.

Some oligosaccharides are covalently bonded to certain cell-membrane proteins (called *glycoproteins*). Glycoproteins participate in linking adjacent cells together and in cell-cell recognition and communication.

☑ When simple monosaccharides are linked together into larger oligosaccharides by the process of dehydration synthesis, what other molecule is created?

Polysaccharides store energy

During dehydration synthesis, thousands of monosaccharides are joined together into straight or branched chains to form complex carbohydrates called **polysaccharides** (*poly* means "many"; **Figure 2.15**). Polysaccharides are a convenient way

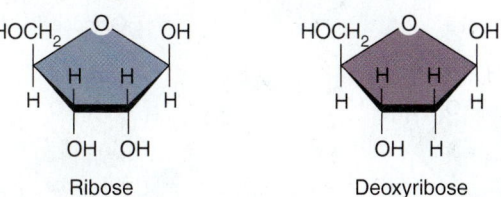

Ribose   Deoxyribose

a) The five-carbon monosaccharides ribose and deoxyribose.

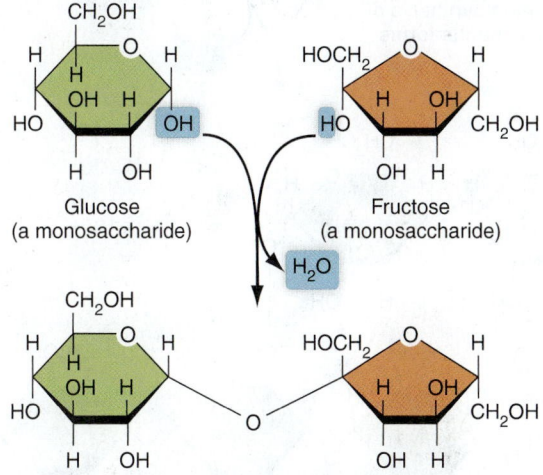

Glucose
(a monosaccharide)   Fructose
(a monosaccharide)

H_2O

Sucrose (a disaccharide)

b) Two 6-carbon monosaccharides (glucose and fructose) are joined together by dehydration synthesis, forming sucrose.

Figure 2.14 Monosaccharides. By convention, in a ringed structure the symbol C for carbon is often omitted because its presence is inferred by the union of two bond lines at an angle.

for cells to stockpile extra energy by locking it in the bonds of the polysaccharide molecule.

The most important polysaccharides in living organisms consist of long chains of glucose monosaccharides. In animals, the storage polysaccharide is **glycogen,** whereas in plants, it is **starch.** The flour we obtain by grinding plant grains is high in starch, which we then utilize for our own energy needs by breaking it down to glucose. Any glucose not consumed for energy in the short term can be used to create glycogen or lipids and stored within our cells for later use.

Cellulose is a slightly different form of glucose polysaccharide. Plants use it for structural support rather than for energy storage. The nature of the chemical bonds in cellulose is such that most animals, including humans, cannot break cellulose down to glucose units (which is why we cannot digest wood). But there is plenty of energy locked in the chemical bonds of cellulose, as demonstrated by the heat generated by a wood fire.

Undigested cellulose in the food we eat contributes to the fiber or *roughage* in our diet. A certain amount of fiber is thought to be beneficial because it increases the movement of wastes through the digestive tract. The more rapid excretion of wastes decreases the time of exposure to any carcinogens (cancer-causing agents) that may be in the waste material.

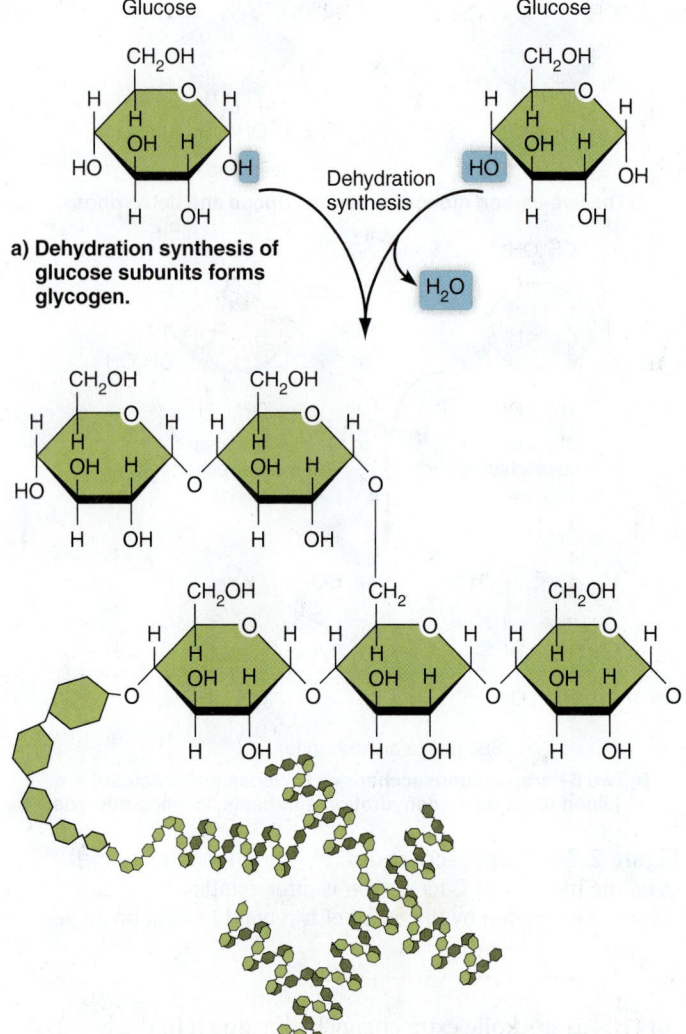

a) **Dehydration synthesis of glucose subunits forms glycogen.**

b) **A representation of the highly branched nature of glycogen.**

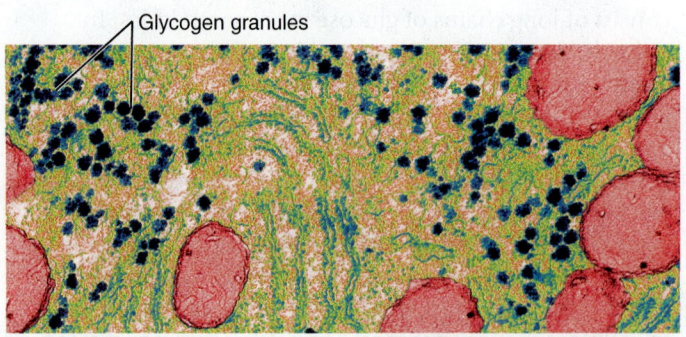

Glycogen granules

c) **A portion of an animal cell showing granules of stored glycogen (blue).** The large pink structures are mitochondria.

Figure 2.15 Glycogen is the storage carbohydrate in animals.

☑ Name two common polysaccharides made by plants. Are they also made of glucose? Can we digest both of them, and why or why not?

↩ **Recap** Carbohydrates contain carbon, hydrogen, and oxygen in a 1-2-1 ratio. Simple sugars such as glucose provide immediate energy for cells. Complex carbohydrates, called polysaccharides, store energy in animals and plants and provide structural support in plants. ■

2.7 Lipids: insoluble in water

For biology, the most important physical characteristic of the class of organic molecules called **lipids** is that they are relatively insoluble, meaning they do not dissolve in water. The most important subclasses of lipids in your body are *triglycerides*, *phospholipids*, and *steroids*.

Triglycerides are energy-storage molecules

Triglycerides, also called neutral fats or just fats, are synthesized from a molecule of glycerol and three fatty acids (**Figure 2.16a**). **Fatty acids** are chains of hydrocarbons (usually about 16 to 18 carbons long) that end in a group of atoms known as a carboxyl group (HO—C═O). Fats vary in the length of their fatty acid tails and the ratio of hydrogen atoms to carbon atoms in the tails.

Saturated fats have a full complement of two hydrogen atoms for each carbon in their tails (**Figure 2.16b**). In saturated fats, the tails are fairly straight, allowing them to pack closely together. As a result, saturated fats are generally solid at room temperature. Animal fats, such as butter and bacon grease, are saturated fats. A diet rich in saturated fats is thought to contribute to the development of cardiovascular disease.

Unsaturated fats, also called *oils*, have fewer than two hydrogen atoms on one or more of the carbon atoms in the tails (**Figure 2.16c**). As a result, double bonds form between adjacent carbons, putting kinks in the tails and preventing the fats from associating closely together. Consequently, unsaturated fats (oils) are generally liquid at room temperature.

Triglycerides are stored in adipose (fat) tissue and are an important source of stored energy in our bodies. Most of the energy is located in the bonds between carbon and hydrogen in the fatty acid tails.

☑ Cocoa butter is solid at room temperature; canola oil is liquid at room temperature. Which lipid has more double bonds between the carbon atoms (and fewer than two hydrogen atoms per carbon atom) in its fatty acid tails? Explain.

Phospholipids are the primary component of cell membranes

Phospholipids are a modified form of lipid. They are the primary structural component of cell membranes.

Like fats, phospholipids have a molecule of glycerol as the backbone, but they have only two fatty acid tails. Replacing the third fatty acid is a negatively charged phosphate group (PO_4^-) and another group that varies depending on the phospholipid but is generally positively charged. The presence of charged groups on one end gives the phospholipid a special property: One end of the molecule is polar and thus soluble (dissolves) in water, whereas the other end

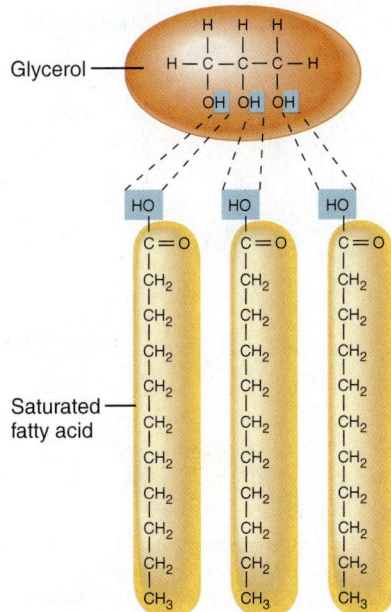

Glycerol

Saturated fatty acid

a) **Triglycerides (neutral fats) are synthesized from glycerol and three fatty acids by dehydration synthesis.**

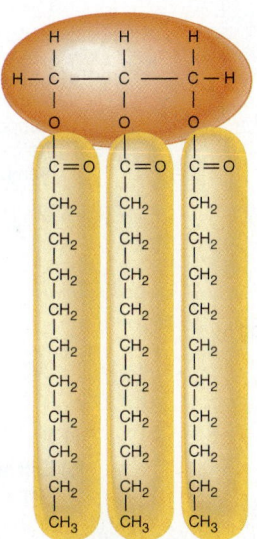

b) **Triglycerides with saturated fatty acids have straight tails, allowing them to pack closely together.**

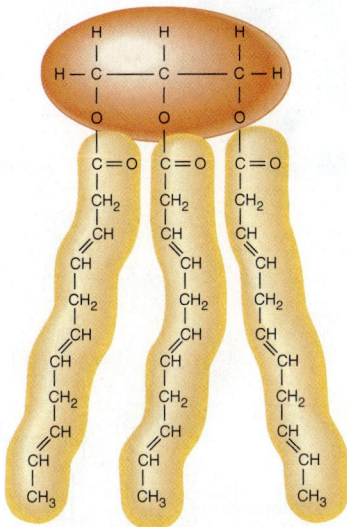

c) **Triglycerides with unsaturated fatty acids have kinked tails, preventing them from packing closely together.**

Figure 2.16 **Triglycerides.**

HEALTH & WELLNESS
Radon: A Known Cancer Risk

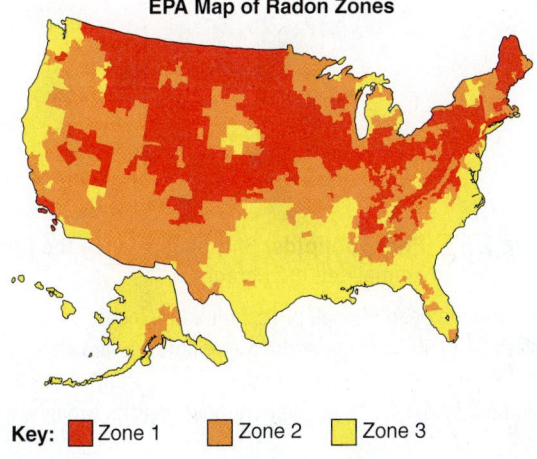

EPA Map of Radon Zones

Key: ■ Zone 1 ■ Zone 2 □ Zone 3

EPA Map of Radon Zones.

The element radon (Rn) is an odorless, colorless gas. We don't talk much about radon in biology because it is not an important molecule in biological systems. That's because radon is inert; unlike carbon and nitrogen, for example, it does not combine readily with other elements to create the kinds of complex molecules required by living systems. Radon exists in nature primarily as elemental radon gas. Most of it is trapped in the soil beneath rock layers.

However, radon gas is highly diffusible. Although it cannot diffuse through solid rock, it can pass right through semisolid objects such as cardboard, wood, sheet rock (wallboard), and even cement blocks typically used in home construction. As a result, in areas of the world with a high concentration of radon in the soil and where houses typically have basements, radon readily diffuses out of the soil and into the air in the basement.

People who spend a lot of time in their basements may be exposed to potentially dangerous levels of radon gas in the air they breathe.

Why is radon considered dangerous? A key property of radon is that it is a radioisotope, meaning it gives off radiation energy as it decays to other elements of lower energy. Radiation exposure from radon is a known risk for lung cancer. Radon is the number one cause of lung cancer in nonsmokers, and the second leading cause of lung cancer overall.

How much radon you may be exposed to depends on a number of factors, including where you live, whether or not you have a basement, how much time you spend in the basement, and how often the air in your basement exchanges with outside air. The Environmental Protection Agency (EPA) publishes maps of the average levels of radon found in basements

in different regions of the country. Levels typically found in Zone 1 are considered dangerous to human health. The EPA and the U.S. Surgeon General recommend that you test your basement for radon even if you live in Zones 2 and 3.

Fortunately, you can determine the level of radiation in your basement with a simple $25 test kit available from most megastores. If you do find high levels of radon, take steps to increase the rate of air exchange in the basement to flush out as much radon as possible.

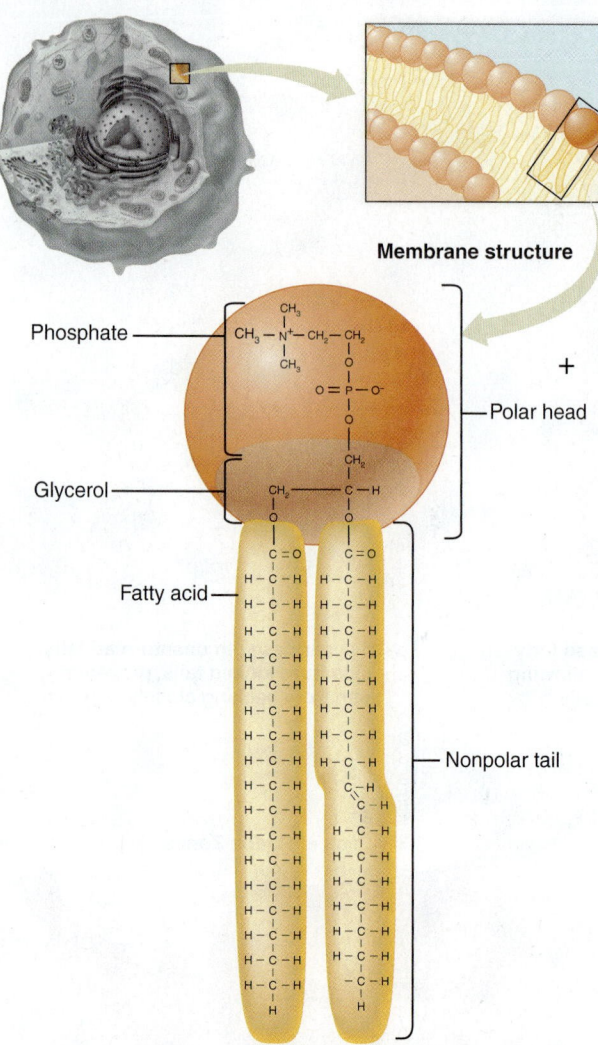

Membrane structure

Phosphate

Glycerol

Fatty acid

Polar head

Nonpolar tail

Figure 2.17 Phospholipids. Phospholipids are the primary constituent of animal cell membranes.

☑ Why is the head of each phospholipid molecule oriented toward the outer surface of each side of the membrane (toward water), instead of toward the middle of the membrane interior? In other words, what prevents a phospholipid molecule from orienting the wrong way?

(represented by the two fatty acid tails) is neutral and relatively insoluble in water (**Figure 2.17**).

Steroids are composed of four rings

Steroids do not look at all like the lipids described previously but are classified as lipids because they are relatively insoluble in water. Steroids consist of a backbone of three 6-membered carbon rings and one 5-membered carbon ring to which any number of different groups may be attached.

One steroid you may be familiar with is **cholesterol** (**Figure 2.18**). High levels of it in the blood are associated with cardiovascular disease. However, we need a certain

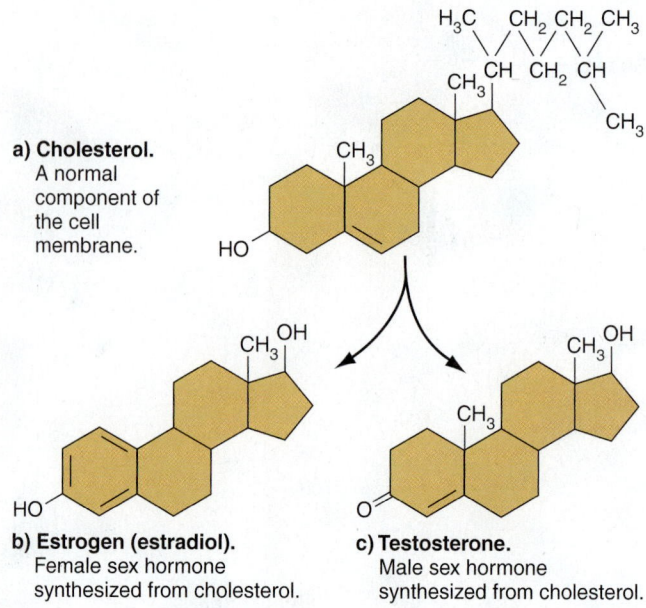

a) Cholesterol.
A normal component of the cell membrane.

b) Estrogen (estradiol).
Female sex hormone synthesized from cholesterol.

c) Testosterone.
Male sex hormone synthesized from cholesterol.

Figure 2.18 Steroids. Steroids consist of a backbone of three 6-membered carbon rings and one 5-membered ring.

amount of cholesterol. It is an essential structural component of animal cell membranes and the source of several important hormones, including the sex hormones estrogen and testosterone. Our bodies manufacture cholesterol even though we generally get more than we need from our diet.

↩ **Recap** Lipids (triglycerides, phospholipids, and steroids) are all relatively insoluble in water. Triglycerides are an important source of stored energy. Phospholipids, an important component of cell membranes, have a polar (water-soluble) head and two fatty acid (water-insoluble) tails. Steroids, such as cholesterol, have a four-ring structure. ∎

2.8 Proteins: complex structures constructed of amino acids

Proteins are macromolecules constructed from long strings of single units called **amino acids** (**Figure 2.19**). All human proteins are constructed from only 20 different amino acids. Each amino acid has an *amino group* (NH_3) on one end, a carboxyl group on the other, a COH group in the middle, and an additional group (designated R) that represents everything else. Some of the R groups are completely neutral, others are neutral but polar, and a few carry a net charge (either positive or negative). Differences in the charge and structure of the amino acids affect the shape and functions of the proteins constructed from them. Our bodies can synthesize (make) 11 of the amino acids if necessary. However, we generally get enough of most of them, including the nine we cannot synthesize, in the food we eat.

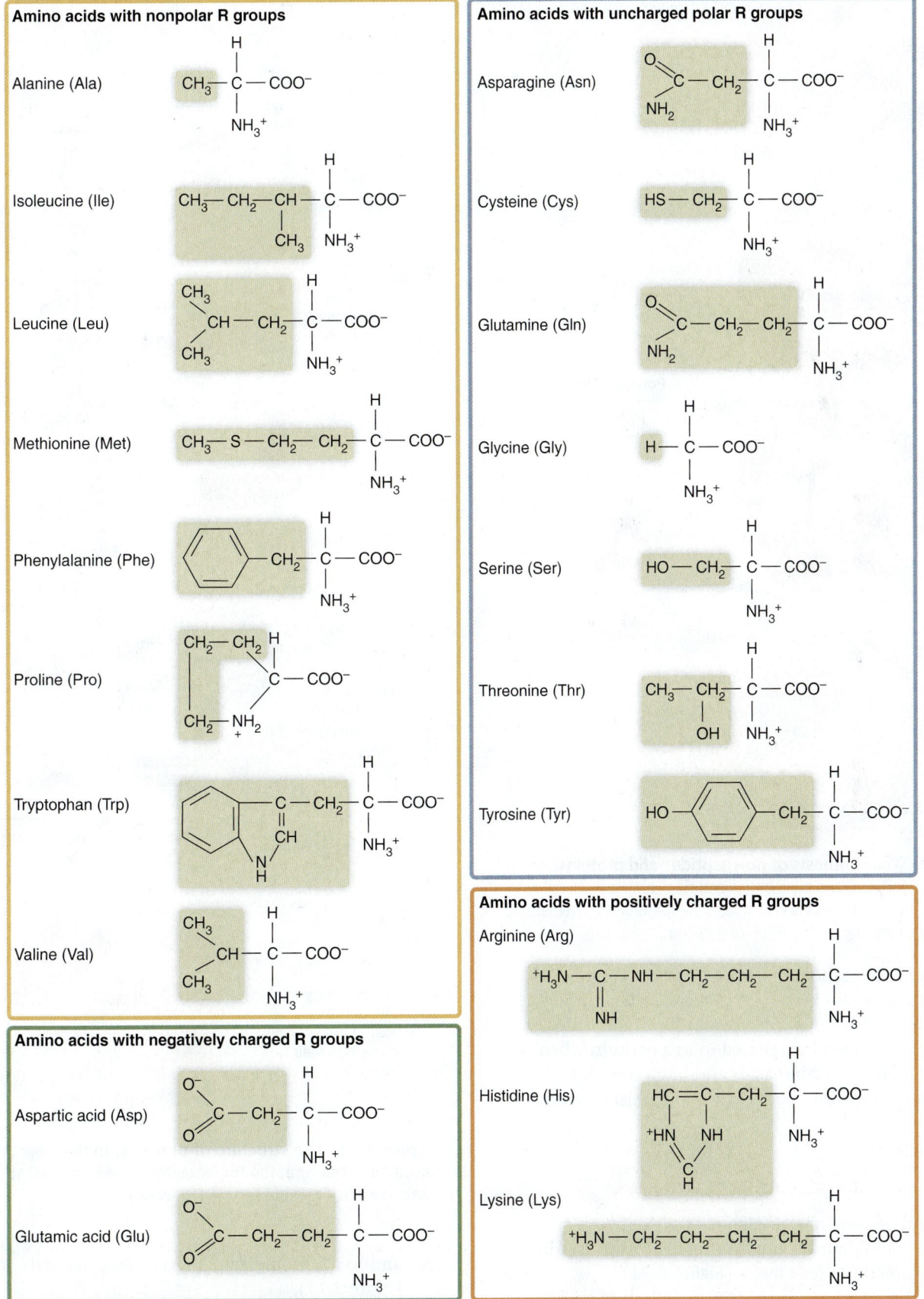

Figure 2.19 The 20 amino acid building blocks of proteins. The portions of amino acids that make them different from each other, called R groups, are colored. The three-letter codes in parentheses designate the amino acids in written formulas.

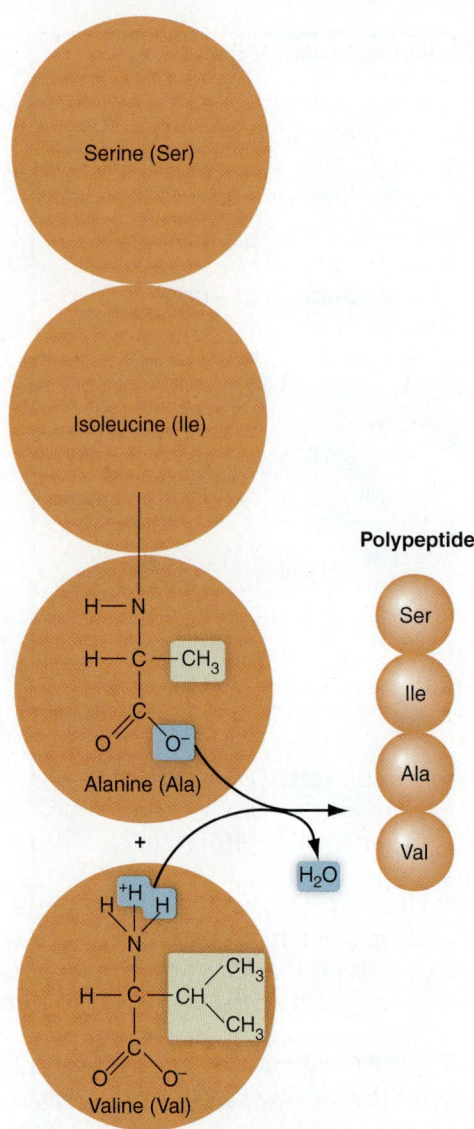

Figure 2.20 The synthesis of polypeptides and proteins. Polypeptides and proteins are synthesized from amino acids by dehydration synthesis. In this example, a polypeptide of three amino acids becomes four by the addition of a valine.

Like complex carbohydrates and fats, proteins are formed by dehydration synthesis (**Figure 2.20**). A single string of 3 to 100 amino acids is called a **polypeptide.** A polypeptide is generally referred to as a **protein** when it is longer than 100 amino acids and has a complex structure and a function. Some proteins consist of several polypeptides linked together.

Protein function depends on structure

The function of every protein depends critically on its structure. We can define protein structure on at least three levels and sometimes four levels (**Figure 2.21**):

- *Primary structure.* The primary structure of a protein is represented by its amino acid sequence. In writing, each

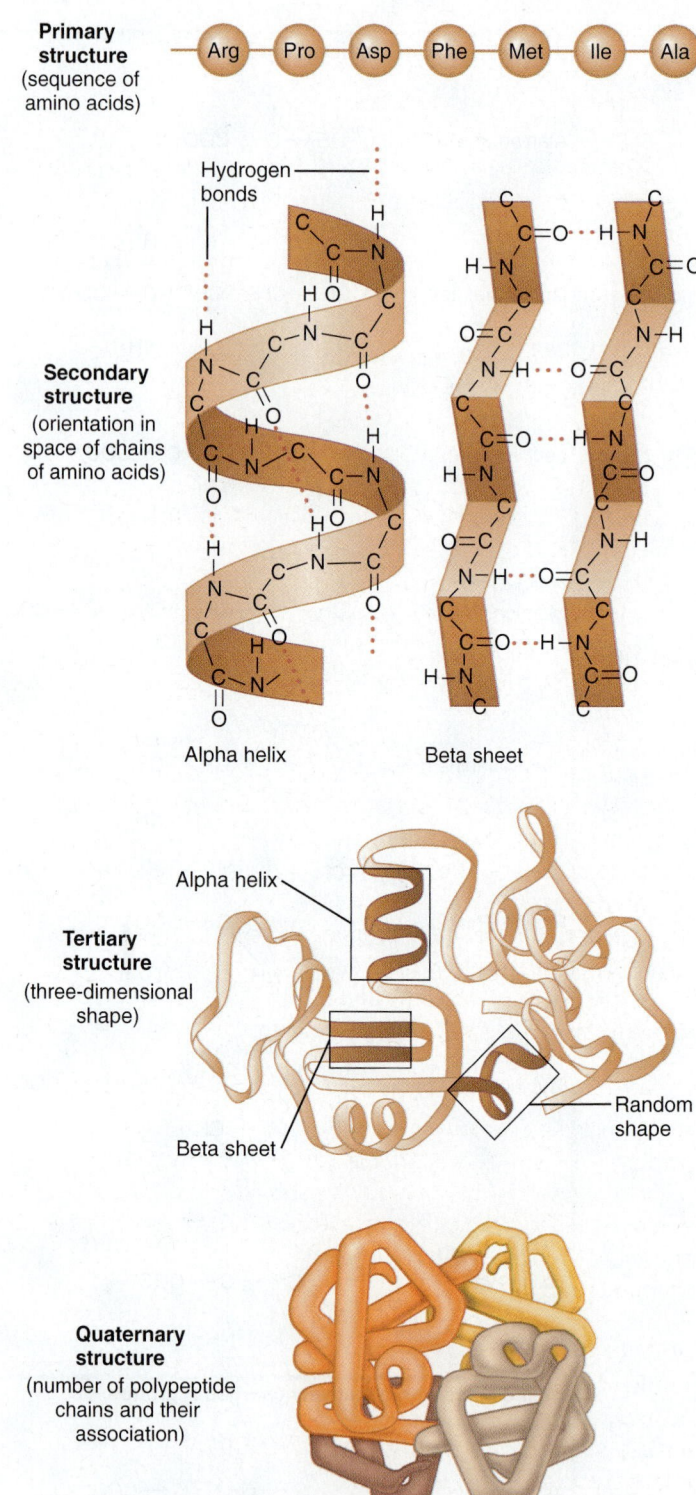

Figure 2.21 The structure of proteins. In the diagrams of secondary structure, the R groups have been omitted so that the basic backbone can be seen more easily.

amino acid is indicated by a three-letter code (review Figure 2.19).
- *Secondary structure.* The secondary structure describes how the chain of amino acids is oriented in space. A

common secondary structure of proteins is an *alpha helix*. An alpha helix is a right-hand spiral that is stabilized by hydrogen bonds between amino acids at regular intervals. Another common secondary structure also stabilized by hydrogen bonds is a flat ribbon called a *beta sheet*. A beta sheet is formed when hydrogen bonds join two primary sequences of amino acids side by side. In addition to forming these two structures, proteins can coil into an almost infinite variety of seemingly random shapes depending on which amino acids make up the sequence.

- *Tertiary structure.* Tertiary structure, the third level, refers to how the protein twists and folds to form a three-dimensional shape. The protein's three-dimensional structure depends in part on its sequence of amino acids, because the locations of the polar and charged groups within the chain determine the locations of hydrogen bonds that hold the whole sequence together. In addition, occasionally a covalent bond called a *disulfide* (S—S) bond forms between the sulfur molecules of two cysteine amino acids (see Figure 2.19). Finally, proteins tend to fold in such a way that neutral amino acids are more likely to end up in the interior, whereas charged and polar amino acids are more likely to face the outside (aqueous environment). Proteins acquire their characteristic tertiary structure by a folding process that occurs either during synthesis or shortly thereafter.

- *Quaternary structure.* The quaternary (fourth) structure of some proteins refers to how many polypeptide chains make up the protein (if there is more than one) and how they associate with each other.

The human body has thousands of different proteins, each serving a different function. Some proteins are primarily for structural support. Others are involved in muscle contraction. Others form part of the cell membrane, where they help transmit information and materials into and out of cells. Still others, called *enzymes*, regulate the rates of biochemical reactions within cells (see next section).

Because the links that determine the secondary and tertiary structures of protein are relatively weak hydrogen bonds, they may be broken by nearby charged molecules. This means that *the shape of proteins can change* in the presence of charged or polar molecules. The ability to change shape is essential to the functions of certain proteins.

Protein structure can also be damaged, sometimes permanently, by high temperatures or changes in pH. **Denaturation** refers to permanent disruption of protein structure, leading to a loss of biological function. For an example of denaturation, consider what happens to an egg when it is exposed to high temperatures: The soluble proteins in the egg become damaged and clump together as a solid mass, causing the egg to harden.

Most proteins are water soluble, meaning that they dissolve in water; however, there are exceptions. Many of the proteins that are part of our cell membranes either are insoluble in water or have water-insoluble regions. Water-insolubility allows them to associate with the water-insoluble regions of the phospholipids that comprise most of the cell membrane's structure.

Enzymes facilitate biochemical reactions

An **enzyme** is a protein that functions as a biological catalyst (**Figure 2.22**). A **catalyst** is a substance that speeds up the rate of a chemical reaction without being altered or consumed by the reaction. Enzymes help biochemical reactions to occur, but they do not change the final result of the reaction. That is, they can only speed reactions that would have happened anyway, although much more slowly. A chemical reaction that could take hours by itself might reach the same point in minutes or seconds in the presence of an enzyme.

Without help from thousands of enzymes, most biochemical reactions in our cells would occur too slowly to sustain life. Each enzyme facilitates, or catalyses, a particular chemical reaction or group of reactions. Enzymes serve as catalysts because as proteins they can change shape. Some enzymes break molecules apart; others join molecules together. In general, an enzyme takes one or more *reactants* (also called *substrates*) and turns them into one or more *products*. Figure 2.22 illustrates how an enzyme facilitates the formation of a product from two reactants.

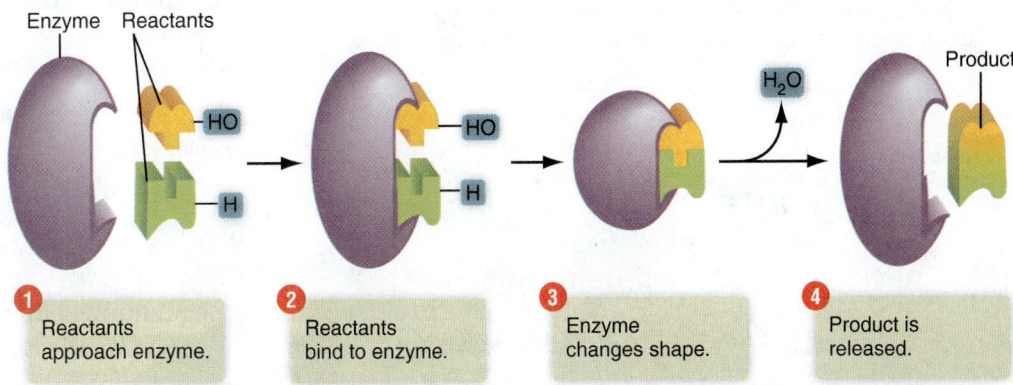

| ① Reactants approach enzyme. | ② Reactants bind to enzyme. | ③ Enzyme changes shape. | ④ Product is released. |

Figure 2.22 Enzymes function as catalysts. This particular enzyme facilitates a dehydration synthesis reaction in which two reactants join to create one larger product plus a molecule of water. Note that the enzyme is not used up during the reaction.

1 Two reactants, each with a shape that corresponds to a specific binding site on the enzyme, approach the enzyme.
2 The reactants bind to the enzyme at their respective binding sites.
3 The binding of both substrates to the enzyme triggers a change in the enzyme's shape. The reactants are brought together, and a dehydration synthesis reaction occurs.
4 The binding of the two substrates together triggers another change in the enzyme's shape, and the final product is released. Note that the enzyme is not used up during the reaction; once the product is released the enzyme is able to repeat the process again with new substrates.

Just how important are enzymes? As one example, the reason we can digest glycogen and starch is that we possess specific enzymes that break the chemical bonds between the glucose monosaccharides in these molecules. In contrast, we cannot digest cellulose because we lack the right enzyme to break it apart. Termites can utilize cellulose only because their digestive systems harbor bacteria that have a cellulose-digesting enzyme.

The changeable shape of an enzyme shows why homeostasis within our cells is so important. Protein shape is in part determined by the chemical and physical environment inside a cell, including temperature, pH, and the concentrations of certain ions. Any deviation from homeostasis can affect the shapes and biological activities of dozens of different enzymes and thus alter the course of biochemical reactions within the cell.

✔️ You've isolated an unknown macromolecule, and you are trying to identify it. So far, all you know is that it consists mostly of carbon and hydrogen, doesn't contain any nitrogen at all, and is insoluble in water. Is it most likely to be a protein, a carbohydrate, or a lipid? Explain.

↩ **Recap** Proteins consist of strings of amino acids. The function of a protein relates to its shape, which is determined by its amino acid sequence and the twisting and folding of its chain of amino acids. Enzymes are proteins that facilitate biochemical reactions in the body. Without enzymes, many biochemical reactions would occur too slowly to sustain life. ∎

2.9 Nucleic acids store genetic information

Another important class of organic molecules is the *nucleic acids*, **deoxyribonucleic acid (DNA)** and **ribonucleic acid (RNA).** You have probably heard of such subjects as cloning, genetic engineering, and DNA fingerprinting. These subjects relate to the nucleic acids: DNA and RNA.

DNA, the genetic material in living things, directs everything the cell does. It is both the organizational plan and the set of instructions for carrying the plan out. Because it directs and controls all of life's processes including growth, development, and reproduction, DNA is key to life itself. RNA, a closely related macromolecule, is responsible for carrying out the instructions of DNA and, in some cases, for regulating the activity of DNA itself. In some viruses, RNA (rather than DNA) serves as the genetic material.

To fully appreciate the importance of DNA and RNA, consider that

- DNA contains the instructions for producing RNA.
- RNA contains the instructions for producing proteins.
- Proteins direct most of life's processes.

Both DNA and RNA are composed of smaller molecular subunits called **nucleotides.** Nucleotides consist of (1) a five-carbon sugar, (2) a single- or double-ringed structure containing nitrogen called a *base*, and (3) one or more phosphate groups. Eight different nucleotides exist: four in DNA and four in RNA.

Figure 2.23 shows the structures of the four nucleotides that make up DNA. Each nucleotide is composed of a five-carbon sugar molecule called *deoxyribose* (like the five-carbon ribose but missing one oxygen atom), a phosphate group, and one of four different nitrogen-containing base molecules; *adenine* (A), *thymine* (T), *cytosine* (C), or *guanine* (G). In a single strand of DNA, these nucleotides are linked together by covalent bonds between the phosphate and sugar groups. The complete molecule of DNA is actually composed of two intertwined strands of nucleotides that are held together

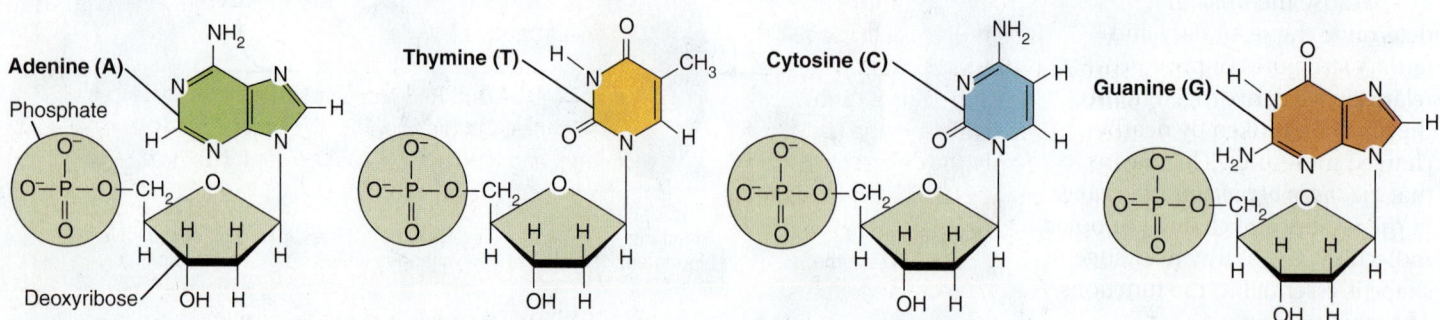

Figure 2.23 The four nucleotides that compose DNA. The phosphate and sugar groups are identical in all four nucleotides.

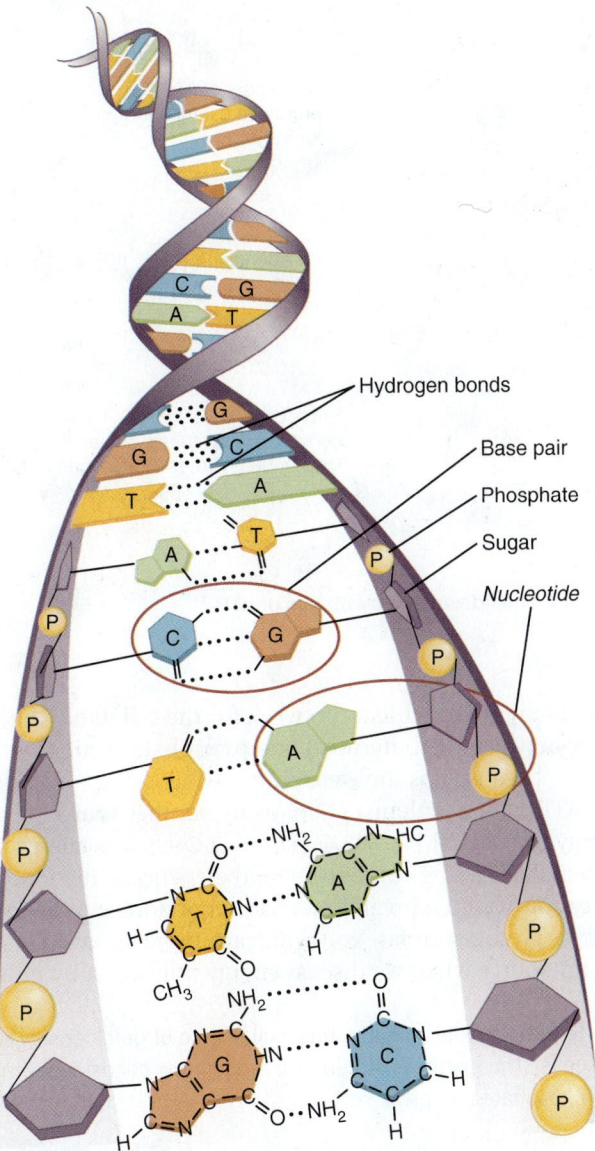

Figure 2.24 The double helical structure of DNA.

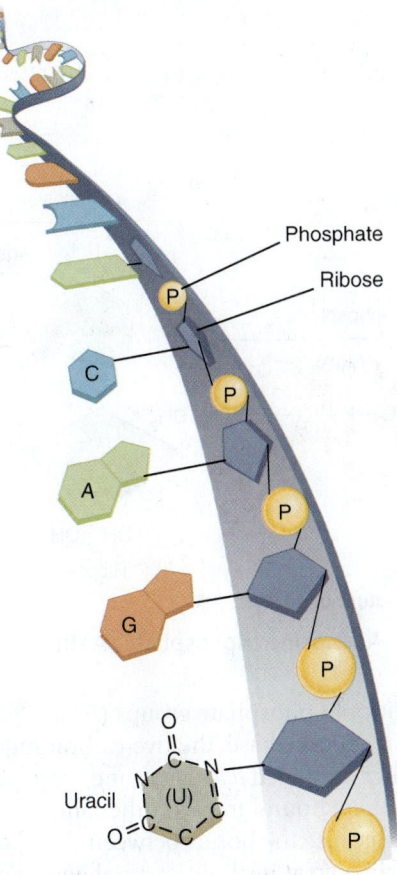

Figure 2.25 The structure of RNA.

by weak hydrogen bonds (**Figure 2.24**). The sequence of one strand determines the sequence of the other (they are complementary strands), because adenine bonds only with thymine (two hydrogen bonds are required) and cytosine bonds only with guanine (three hydrogen bonds are required).

The *code* for making a specific protein resides in the specific sequence of base pairs in one of the two strands of the DNA molecule. Notice that the entire genetic code is based entirely on the sequence of only four different molecular units (the four nucleotides). You will learn more about DNA and the genetic code when we discuss cell reproduction and inheritance.

A single molecule of DNA carries the code for making a lot of different proteins. It is like an entire bookshelf of information, too big to be read all at once. To carry out their function, portions of the DNA molecule are transcribed into

smaller fragments of RNA. RNA is structurally like DNA, with a few exceptions (**Figure 2.25**):

- The sugar unit in all four of the nucleotides in RNA is *ribose* rather than deoxyribose (hence the name ribonucleic acid).
- One of the four nitrogen-containing base molecules is different (uracil is substituted for thymine).
- RNA is a single-stranded molecule, representing a complementary copy of a portion of only *one* strand of DNA.
- RNA is shorter, representing only the segment of DNA that codes for one or more proteins.
- In Chapter 17, we discuss in more detail how RNA is used to make proteins.

Recap DNA and RNA are constructed of long strings of nucleotides. Double-stranded DNA represents the genetic code for life, and RNA, which is single-stranded, is responsible for carrying out those instructions. ∎

2.10 ATP carries energy

One additional related nucleotide with an important function is **adenosine triphosphate (ATP).** ATP is identical to the adenine-containing nucleotide in RNA except that ATP

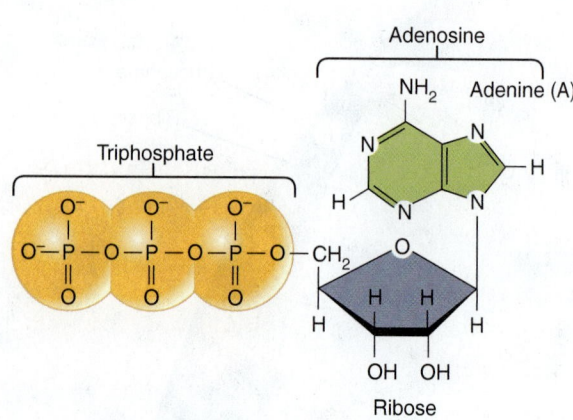

a) The structure of ATP.

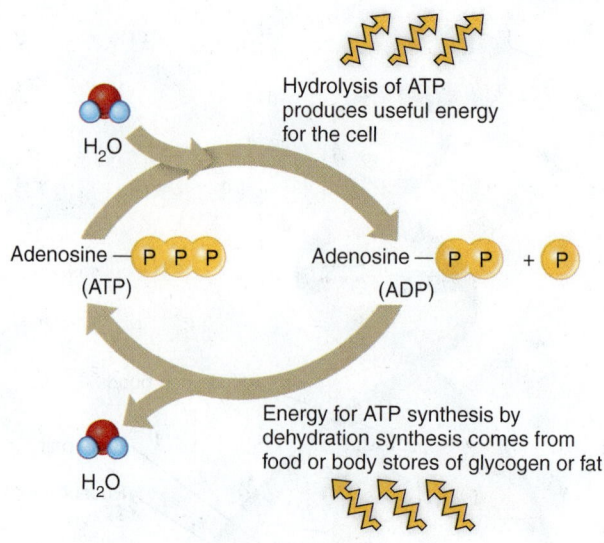

b) The breakdown and synthesis of ATP.

Figure 2.26 Adenosine triphosphate (ATP).

has two additional phosphate groups (**Figure 2.26a**). ATP consists of an adenine base, the five-carbon sugar ribose, which together are called *adenosine*, and three phosphate groups, which are bound to each other and therefore called *triphosphate*. Because the bonds between the phosphate groups contain a great deal of potential energy, ATP is a universal energy source for cells. ATP is like that energy bar in your backpack. Any time a cell needs energy for virtually any function, it can break the bond between the outermost two phosphate groups of an ATP molecule.

The breakdown of ATP produces **adenosine diphosphate (ADP)** plus an inorganic phosphate group (P_i), plus energy (**Figure 2.26b**):

$$ATP \rightarrow ADP + P_i + energy$$

The energy is available to do work for the cell. Only the bond between the outermost two phosphate groups is broken because it is the weakest bond.

ATP can be replenished by using another source of energy to reattach P_i to ADP (Figure 2.26b). The energy to replenish ATP may come from stored energy in the food we eat or from the breakdown of energy storage molecules such as glycogen or fat. You will learn more about ATP as an energy source when we discuss energy utilization by muscles.

Recap ATP is a nearly universal source of quick energy for cells. The energy is stored in the chemical bonds between phosphate groups. ∎

Chapter Summary

2.1 All matter consists of elements *p. 24*

- Atoms, the smallest functional unit of any element, contain a nucleus and a cloud of electrons.
- The protons and neutrons in an atom's nucleus account for most of its mass.
- Radioisotopes are unstable isotopes; an isotope has more or fewer neutrons than the usual number for that atom.

2.2 Atoms combine to form molecules *p. 26*

- Energy exists as either kinetic energy or potential energy.
- Three types of chemical bonds account for the structures of molecules: covalent, ionic, and hydrogen bonds. Covalent bonds are the strongest; hydrogen bonds are the weakest.
- Over 99% of your body weight consists of just six elements: oxygen, carbon, hydrogen, nitrogen, calcium, and phosphorus.

2.3 Life depends on water *p. 29*

- The polar nature of the water molecule accounts for its physical properties and for its unusually good qualities as a solvent for most other molecules and ions.
- Water is important in human temperature regulation.

2.4 The importance of hydrogen ions *p. 31*

- Molecules that can donate a hydrogen ion (H^+) are called acids. Molecules that can accept H^+ are called bases.
- The hydrogen ion concentration of a solution is expressed as pH.
- Buffers are pairs of molecules that tend to minimize changes in pH when an acid or base is added to a solution.

2.5 The organic molecules of living organisms *p. 33*

- The backbone of all organic molecules is carbon.
- Organic molecules are formed by a process called dehydration synthesis (requiring energy and releasing a water molecule).

- Organic molecules are broken down by a process called hydrolysis (releasing energy and using up a water molecule).

2.6 Carbohydrates: used for energy and structural support p. 35

- Monosaccharides, or simple sugars, are a source of quick energy for cells.
- Complex carbohydrates (polysaccharides) are formed by linking simple sugars (monosaccharides) together by dehydration synthesis.
- Carbohydrates are primarily energy-storage molecules. Plants use them for structural support as well.
- In animals, the storage molecule is glycogen; in plants, it is starch.

2.7 Lipids: insoluble in water p. 36

- Lipids include fats and oils, phospholipids, and steroids. Lipids are insoluble in water.
- Fats store energy. Phospholipids and cholesterol are important structural components of the cell membrane. The sex hormones are steroids synthesized from cholesterol.

2.8 Proteins: complex structures constructed of amino acids p. 38

- Proteins have unique three-dimensional structures that depend on their primary structure (their amino acid sequences). Living organisms construct a tremendous number of different proteins using just 20 different amino acids.
- The human body contains thousands of proteins, each with a different function.
- Enzymes are proteins that facilitate the rates of chemical reactions.

2.9 Nucleic acids store genetic information p. 42

- DNA is composed of two long strands of nucleotides intertwined into a double helix. DNA is constructed from just four different DNA nucleotides.
- RNA is a shorter single strand of RNA nucleotides, representing the code for one or more proteins.

2.10 ATP carries energy p. 43

- The nucleotide ATP is an energy source for cells. Energy is stored in the bonds between phosphate groups.
- Energy is released when the outermost phosphate-phosphate bond is broken.
- ATP can be replenished, to be used again.

Terms You Should Know

acid, 31
atom, 24
adenosine triphosphate (ATP), 43
base, 31
buffer, 32
catalyst, 41
covalent bond, 27
dehydration synthesis, 33
deoxyribonucleic acid (DNA), 42
electrons, 25
enzyme, 41
hydrogen bond, 29
hydrolysis, 34
ion, 28
ionic bond, 28
lipid, 36
molecule, 26
neutrons, 25
pH scale, 32
protein, 40
protons, 25
ribonucleic acid (RNA), 42
solvent, 29
solute, 29

Concept Review

Answers can be found in the Study Area in MasteringBiology.

1. Describe the electrical charges and relative masses of protons, neutrons, and electrons.
2. Explain why two atoms of hydrogen tend to combine into a molecule of hydrogen gas (H_2).
3. Explain why polar and charged molecules tend to be soluble in water.
4. How is a covalent bond different from an ionic bond?
5. Compare and contrast potential energy and kinetic energy.
6. Distinguish between saturated and unsaturated fats.
7. Describe the process known as dehydration synthesis.
8. Explain why proteins come in an almost unlimited variety of shapes.
9. Discuss the importance of enzymes in living organisms.
10. Describe the role of ATP in energy transfer within a cell.

Test Yourself

Answers can be found in the Appendix.

1. If a molecule of starch is repeatedly hydrolyzed, which of the following would be the final product?
   a. glucose   c. ribose   e. deoxyribose
   b. fructose   d. sucrose

2. Which of these molecules would be described as hydrophobic?
   a. glucose   c. cholesterol   e. RNA
   b. sodium chloride d. DNA

3. _____ bonds form between the oxygen and hydrogen within water molecules, whereas _____ bonds form between different water molecules.
   a. Hydrogen . . . covalent   d. Hydrogen . . . ionic
   b. Covalent . . . hydrogen   e. Covalent . . . ionic
   c. Ionic . . . hydrogen

4. ^{13}C and ^{14}C are _____ of carbon.
   a. isotopes   c. compounds   e. isomers
   b. ions   d. molecules

5. Which of the following substances has the lowest hydrogen ion concentration?
   a. water (pH 7)   d. vinegar (pH 3)
   b. bleach (pH 13)   e. black coffee (pH 5)
   c. baking soda (pH 9)

6. When sugar is dissolved in water, sugar is the _____ and water is the _____.
   a. acid . . . base   d. solute . . . solvent
   b. base . . . acid   e. inorganic molecule . . .
   c. solvent . . . solute   organic molecule

7. A monosaccharide is to a polysaccharide as an amino acid is to
   a _____.
   a. nucleic acid   c. protein   e. triglyceride
   b. carbohydrate   d. nucleotide

8. Which of these bonds is the easiest to disrupt, simply by raising
   the temperature?
   a. hydrogen bonds   d. nonpolar covalent bonds
   b. ionic bonds   e. peptide bonds
   c. polar covalent bonds

9. The primary structure of a protein is maintained by _____
   bonds.
   a. hydrophilic   c. hydrogen   e. ionic
   b. hydrophobic   d. covalent

10. DNA ultimately contains the instructions for the assembly of:
    a. proteins   c. triglycerides   e. steroids
    b. polysaccharides   d. nucleotides

11. If one strand of DNA has the sequence A-A-C-T-G-T-G, what
    will be the nucleotide sequence of the complementary strand?
    a. A-A-C-T-G-T-G   c. T-T-G-U-C-U-C   e. T-T-G-A-C-A-C
    b. U-U-G-A-C-A-C   d. G-G-C-A-G-A-G

12. Which of the following is true regarding the synthesis of a tri-
    glyceride?
    a. Three water molecules would be removed.
    b. The triglyceride would be hydrophilic.
    c. It would require three amino acids.
    d. It would require three monosaccharides.
    e. Hydrolysis reactions would be involved.

13. Which of the following is true regarding enzymes?
    a. The synthesis of an enzyme involves hydrolysis reactions.
    b. Enzymes provide energy for biochemical reactions.
    c. One enzyme can catalyze many different types of reactions.
    d. The instructions for the synthesis of an enzyme are found
       in DNA.
    e. Each enzyme molecule can be used only once.

14. Synthesis of proteins requires the input of energy, which can be
    provided by:
    a. enzymes   d. amino acids
    b. synthesis of ATP from   e. isotopes
       ADP and P_i
    c. hydrolysis of ATP to form
       ADP and P_i

15. The element phosphorus (P) has an atomic number of 15 and
    a mass number of 31. Which of the following represents the
    numbers of particles in phosphorus?
    a. 15 protons, 15 electrons, 31 neutrons
    b. 15 protons, 16 electrons, 15 neutrons
    c. 31 protons, 31 electrons, 15 neutrons
    d. 15 protons, 15 electrons, 16 neutrons
    e. 16 protons, 16 electrons, 15 neutrons

Apply What You Know

Answers can be found in the Study Area
in MasteringBiology.

1. Athletes are sometimes advised to eat large amounts of
   complex carbohydrates (such as whole-wheat pasta) for a day
   or two before a competitive event. Explain the reasoning
   behind this.

2. Physicians become concerned about the potential for
   irreversible brain damage when body temperatures approach
   $105°F$. Which of the four classes of macromolecules do you
   think is most likely affected by high temperatures? Explain.

3. Many people use cholesterol-lowering drugs to reduce
   their high cholesterol, because they know that a high
   cholesterol level is a risk factor for heart disease. Would it be
   advisable to take a little extra dose of these drugs to try to
   lower your cholesterol to below normal levels, just to be on the
   safe side?

4. In Miami, when it is 90 degrees outside and very humid,
   the heat feels stifling. Yet, most people report feeling fine in
   Arizona, where the humidity is generally low, even when it is
   100 degrees. Why does the humidity have such an effect on our
   perception of comfort in terms of temperature?

5. Coca-Cola is a very acidic drink; its pH is around 3. Blood has a
   pH of about 7. Yet when you drink a Coke the pH of the blood
   does not change measurably.
   Why is that?

MJ's BlogInFocus

Answers can be found in the Study Area in MasteringBiology.

1. Hormones are sometimes added to food products. Explain
   why this is permissible. Hint: Visit MJ's blog
   in the Study Area in MasteringBiology and
   look under "Melatonin in Brownies."
   http://goo.gl/XjS3OT

2. An enterprising company has managed to create a
   "powdered alcohol" product—just add water and you have
   an alcoholic drink. So far, state liquor control boards
   have failed to approve it for sale. Do you think they
   should approve it? Why do you think they are reluctant
   to approve it? Would *you* use it if
   it were available? For more on this
   subject, visit MJ's blog in the Study
   Area in MasteringBiology and look under
   "Powdered Alcohol." http://goo.gl/1Ljs9V

MasteringBiology®

Students Go to MasteringBiology for assignments, the
eText, and the Study Area with animations, practice tests,
and activities.

Professors Go to MasteringBiology for automatically graded
tutorials and questions that you can assign to your students,
plus Instructor Resources.

Answers to ✓ questions are available in the Appendix.

Structure and Function of Cells

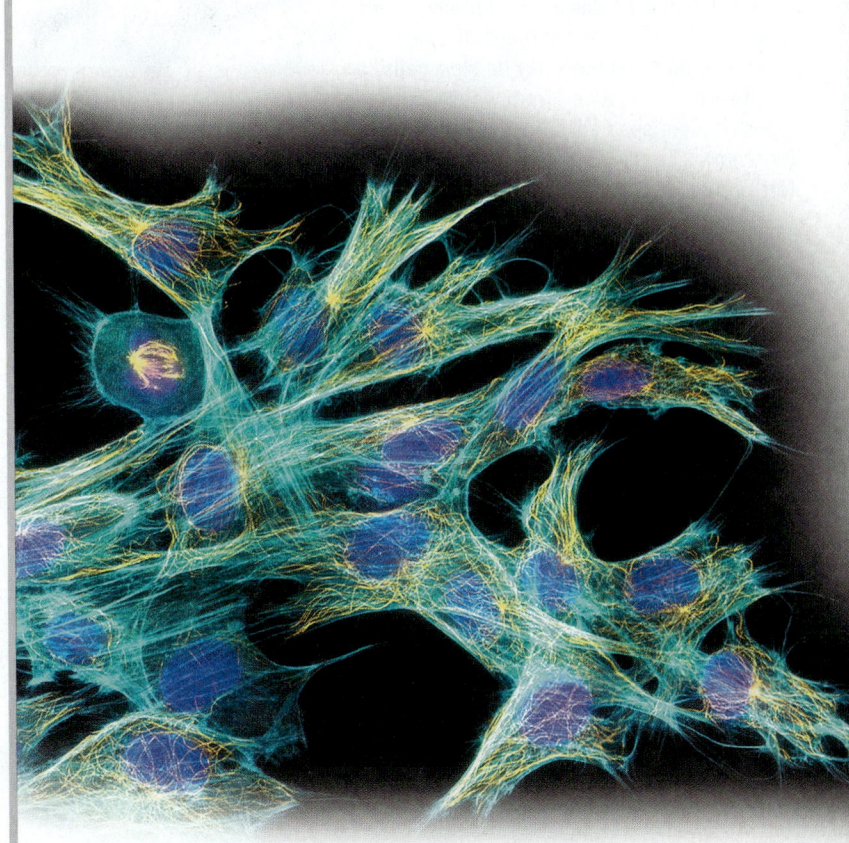

Immunofluorescence light micrograph of fibroblast cells from connective tissue.

Key Concepts

- **A single cell is the smallest unit of life.** All living things are comprised of one or more cells (plus cell products), and all cells are derived from preexisting cells.

- **Human cells are surrounded by a plasma membrane.** The plasma membrane serves to contain the cellular structures within the cell and to regulate the kinds and quantities of molecules that can enter and exit the cell.

- **The nucleus of a human cell contains the cell's DNA.** The genetic code of DNA specifies the amino acid sequences of the proteins produced by the cell.

- **Mitochondria, which are cell structures, produce usable energy in the form of adenosine triphosphate (ATP).** Most of the energy for making ATP comes from the complete breakdown of glucose, a mitochondrial process that requires oxygen and results in the production of carbon dioxide, a waste product.

- **Cells use ATP to carry out cellular functions,** including manufacturing and exporting biological molecules; providing for cell support, movement, growth, and division; defending against foreign cells and toxic chemicals; and getting rid of cellular waste.

CURRENT ISSUE

The Use of Human Stem Cells

What do boxing champion Muhammad Ali and actor Michael J. Fox have in common? They both suffer from Parkinson's disease, a debilitating neurological disorder. The key to curing Parkinson's disease and many other diseases, including Alzheimer's disease, leukemia, and diabetes may be **stem cells.**

A stem cell is a cell from which other types of more specialized cells originate (or stem). The ultimate stem cell is the fertilized egg, because *all* of the specialized cells of the body originate from it. The first eight cells of a human embryo are also stem cells, because they have not yet begun to differentiate (become different from each other). But shortly after the eight-cell stage, the cells begin to specialize. Some become muscle; others become skin; still others become nerve cells in the brain.

Stem cells have several properties that make them desirable for research and for the treatment of certain diseases. They are easier to work with in the laboratory than adult cells because they don't adhere tightly to each other, and they grow better in culture (in controlled conditions, such as in a laboratory). They generally don't provoke a *tissue rejection* immune response in the patient because they are undifferentiated

Muhammad Ali and Michael J. Fox, both sufferers of Parkinson's disease, support stem cell research.

and thus not recognized as foreign cells. They are also easier to administer to a patient—usually they can be injected and allowed to migrate to their target site rather than having to be surgically transplanted. And most important, they still have the capacity to become the type of specialized cell the patient needs, under the direction of the patient's own cell division/differentiation control mechanisms.

Traditionally most human embryonic stem cells have come from very early-stage embryos. Currently, the only available

Questions to Consider

1 What is your opinion on this controversy? What basic beliefs do you hold that cause you to feel as you do?

2 Suppose that you held frozen embryos at a private fertility clinic and that you knew you would never need them. Would you donate them for stem cell research? Why or why not?

embryos of this age are those created "in excess of clinical need" by *in vitro* fertilization at private fertility clinics. Only a few researchers have access to such embryos. However, cells used to treat specific diseases don't need to be completely undifferentiated. If nerve cells are needed, a good source of young cells is the very first embryonic neural tissue in the human fetus. Already, over 100 people with Parkinson's disease have received fetal nerve cell transplants worldwide, and some of these recipients have shown measurable improvements in brain function. Nonliving human fetuses are widely available as a consequence of the more than 1 million legal abortions performed in the United States each year.

Controversy and Compromise

Not surprisingly, the use of human embryonic cells derived from fertilized eggs or from cells harvested from aborted fetuses is highly controversial. On the one hand, patient advocacy groups recognize the potential benefits of human embryonic and fetal cells and promote efforts to harvest and use them. On the other hand, some human rights groups object strongly to harvesting or using human embryonic stem cells or fetal cells under any circumstances, calling such research a destruction of precious human life. Both sides believe strongly in their position and both sides are active politically, and as a result politicians have been forced to take a stand.

According to guidelines developed under the Bush administration in 2001, federal funds could only be used to study stem cell lines derived from embryos before that year. (A stem cell *line* is a group of identical cells grown from a single stem cell.) In effect, these federal guidelines prohibited the National Institutes →

Many people object strongly to the harvesting and use of human embryonic stem cells.

of Health, which funds most biomedical research, from financing any stem cell research that might require the *future* death of a human embryo. However, the guidelines stopped short of prohibiting human stem cell research altogether; privately funded human embryonic stem cell research was permitted under certain conditions.

The guidelines of 2001 were a political compromise at best. They allow selected federally funded research projects to go forward with the few stem cell lines that already existed at the time the law was passed, while respecting concerns about the sanctity of human life. The guidelines were opposed by stem cell researchers, who contended that the limited number of cell lines available were not enough for the United States to stay in the forefront of this important research area. Some stem cell researchers moved to Britain, which had developed a facility for storing thousands of cell lines, or to other places in the world where stem cell research was not only allowed but also encouraged.

The election of President Obama changed the political environment yet again. Shortly after taking office in 2009, President Obama signed an executive order lifting the restrictions on stem cell research laid down by President Bush. Federally funded researchers are now free to use the hundreds of stem cell lines in existence today, as well as new stem cell lines created by private funding in the future. But there is still a prohibition in place, called the Dickey-Wicker Amendment of 1996, which prohibits federal funding for research "in which human embryos are created, destroyed, discarded, or knowingly subjected to risk of injury or death." Although federally funded researchers will be able to use stem cell lines created by private funds (because the researchers themselves did not destroy any embryos), they will still be prohibited from creating their own new cell lines from human embryos. Obviously, the controversy over stem cells is not over.

What Is the Solution?

Both sides in this controversy actually do have something they can agree on; they both hope that someday we will not need to use stem cells from embryos or fetuses at all. The key may be the development of methods to create undifferentiated stem cells from fully differentiated adult cells—in essence, reversing the entire process of cellular differentiation and specialization. Several groups of scientists already claim to have done it successfully. But some scientists caution that new techniques for creating stem cell lines may not necessarily translate quickly into cures for specific diseases. Although the day may yet come when human embryonic stem cells truly aren't necessary, that day has not yet arrived, just ask Michael J. Fox or Muhammad Ali.

SUMMARY

- Stem cells obtained from human embryos and fetuses have the potential to treat or cure diseases.
- Research using cells obtained from human embryos or fetuses is controversial.
- In the United States, stem cell research has been affected by changes in the political environment.
- In the distant future, human embryos and fetuses may no longer be needed as a source of stem cells.

Scientists first observed living cells under a microscope in 1674. Since then, countless observations and experiments have confirmed the **cell doctrine,** which consists of three basic principles:

1. All living things are composed of cells and cell products.
2. A single cell is the smallest unit that exhibits all the characteristics of life.
3. All cells are derived from preexisting cells.

If we examine any part of the human body under a microscope, we find living cells and/or cell products. *Cell products* include materials composed of dead cells (such as the outer layer of your skin) and substances resulting from cellular activity (such as the hard crystalline elements of bone). There are no living units smaller than cells. All of our cells—all 100 trillion of them—are derived from earlier cells, going all the way back to our first cell, the fertilized egg. Even that original cell came from preexisting cells: the sperm and egg from our parents.

Aside from common functions such as cellular metabolism, growth, and division, most cells in the human body have a specialized function. Heart muscle cells contract rhythmically, pumping blood to all parts of the body. Blood cells carry oxygen and nutrients to all organs, including the heart and brain. Nerve cells in the brain receive information from and transmit information to all parts of the body, including the heart. In a complex organism such as a human being, essentially all cells are functionally interconnected and interdependent, even though they may not be located near each other. ■

We begin with a description of the common structures and functions of all cells. In subsequent chapters, we describe how cells are grouped together into specialized tissues and organs, each with its own specialized functions.

3.1 Cells are classified according to their internal organization

All cells are surrounded by an outer membrane called the **plasma membrane.** The plasma membrane encloses the material inside the cell, which is mostly water but also contains ions, enzymes, and other structures the cell requires to maintain life. All living cells are classified as either eukaryotes or prokaryotes, depending on their internal organization.

Eukaryotes have a nucleus, cytoplasm, and organelles

Human cells, like those of most species, are **eukaryotes** (*eu-* means "true" and *karyote* means "nucleus"). Nearly every eukaryotic cell has three basic structural components (Figure 3.1a):

1. *A plasma membrane.* The plasma membrane forms the outer covering of the cell.

2. *A nucleus. Nucleus* is a general term for *core.* In the last chapter, we saw that chemists define a nucleus as the core of an atom. In biology, the nucleus is a membrane-bound compartment that houses the cell's genetic material and functions as its "information center." Most eukaryotic cells have one nucleus; however, there are a few exceptions, to be discussed in later chapters.

3. *Cytoplasm ("cell material").* The cytoplasm includes everything inside the cell except the nucleus. It is composed of a soft, gel-like fluid called the *cytosol* ("cell solution"). The cytosol contains a variety of microscopic structures called **organelles** ("little organs") that carry out specialized functions, such as digesting nutrients or packaging cellular products.

Prokaryotes lack a nucleus and organelles

Prokaryotes (*pro-* means "before" and *karyote* means "nucleus") are the bacteria (kingdom Monera; Figure 3.1b). Prokaryotes have a plasma membrane that is surrounded by a rigid cell wall. Their genetic material is concentrated in a particular region, but it is not specifically enclosed within a membrane-bound nucleus. Prokaryotes also lack most of the organelles found in eukaryotes. Nevertheless, they are living organisms that fit the definition of a cell according to the cell doctrine.

In the rest of this chapter and throughout the book, we concentrate on the structure and function of eukaryotic cells. However, we discuss bacteria again in terms of how they can make us ill (Chapter 9) and in the context of evolution (Chapter 22).

3.2 Cell structure reflects cell function

Eukaryotic cells are remarkably alike in their structural features regardless of which organism they come from. This is because all cells carry out certain activities to maintain life, and there is a strong link between structure and function.

All cells must gather raw materials, excrete wastes, make macromolecules (the molecules of life), and grow and reproduce. These are not easy tasks. The specific activities carried out by a living cell (and the structures required

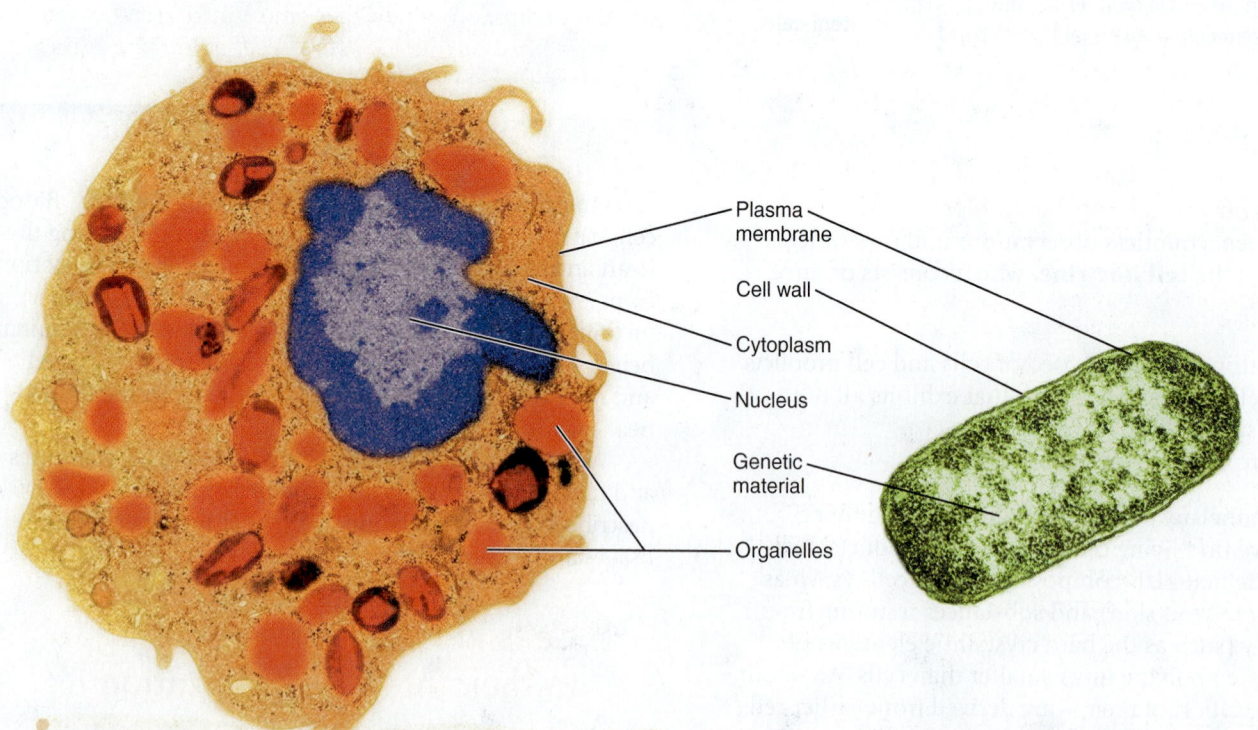

Plasma membrane

Cell wall

Cytoplasm

Nucleus

Genetic material

Organelles

a) **A eukaryotic animal cell has a large nucleus and numerous small organelles.** The cytoplasm is enclosed by a flexible plasma membrane.

b) **Prokaryotic cells such as this bacterium have a rigid cell wall surrounding the plasma membrane.** The genetic material is not surrounded by a membrane, and there are no organelles in the cell.

Figure 3.1 Eukaryotes versus prokaryotes.

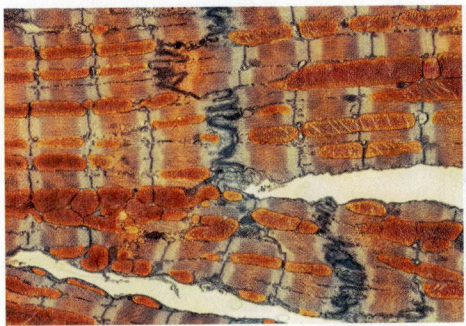

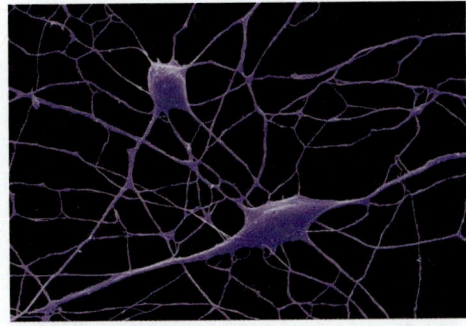

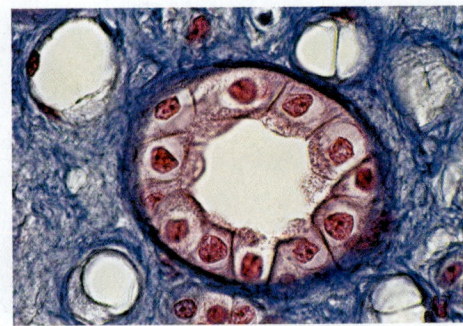

a) A portion of several muscle cells of the heart (×1,500).

b) Nerve cells of the central nervous system (×830).

c) Cells lining a tubule of a kidney (×250).

Figure 3.2 Human cells vary in shape.

to perform them) would rival those of any large city or even a country! There is an outer structure that defines its border; an infrastructure for support; an information center; manufacturing facilities; refining, packaging, and shipping centers; transportation systems for supplying raw materials and energy; stockpiles of energy; and mechanisms for recycling or removing toxic waste. Cells even possess sophisticated defense mechanisms to combat invaders.

Most of the structural differences between cells reflect differences in function. Muscle cells contain numerous mitochondria that produce the energy for muscle contraction (**Figure 3.2a**). Many nerve cells are long and thin; the longest nerve cells carry impulses all the way from your toes to your spinal cord (**Figure 3.2b**). The cells that line the kidney tubules are cube shaped and tightly bound together, reflecting their role in the transport of water and other molecules (**Figure 3.2c**). Essentially every cell has a specialized function of some sort or it would be of little use to the organism.

Cells that serve the same function are often remarkably similar between species. For example, a human nerve cell has more in common (structurally and functionally) with a nerve cell in a cockroach than it does with a human liver cell. Furthermore, cells in a mouse are not that much different in size from those in an elephant; an elephant just has more of them.

Cells remain small to stay efficient

Despite their structural differences and incredible complexity, all cells have at least one feature in common: They are small in one or more dimensions. Given the variety of life-forms on Earth, why are there so few giant cells? (An exception is the egg of some species.)

The answer shows that nature obeys certain simple and understandable principles:

- The total metabolic activities of a cell are proportional to its volume of cytoplasm, which is in effect its size. To support its activities, every cell needs raw materials in proportion to its size. Every cell also needs a way to get rid of its wastes.
- All raw materials, energy, and waste must cross the plasma membrane to enter or leave the cell.

- As cells get larger, their volume increases more than their surface area. For both spheres and cubes, for example, an eightfold increase in volume is accompanied by only a fourfold increase in surface area.

The larger a cell gets, then, the more likely that its growth and metabolism will be limited by its ability to supply itself across the plasma membrane. Put another way, the smaller the cell, the more effectively it can obtain raw materials and get rid of wastes (**Figure 3.3**).

Some cells have numerous microscopic projections of the plasma membrane called *microvilli*. Microvilli are an effective way to increase surface area relative to volume (**Figure 3.3c**). Plasma membranes with microvilli are especially common in cells that transport substances into and out of the body, such as the cells that line the digestive tract and the tubules of the kidneys.

Visualizing cells with microscopes

Because cells can't be seen effectively with the naked eye, the only way to visualize them is with the use of a microscope. The three main types of microscopes in use today are the standard light microscope (LM) and two newer types of microscopes that utilize beams of electrons, rather than visible light.

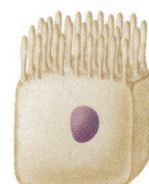

a) One large cell.

b) Eight small cells.

c) Cell with microvilli on one surface.

Figure 3.3 Cell size and plasma membrane shape affect surface area and volume. The volume is the same in these three examples: the cell in **(a)**, the eight small cells in **(b)**, and the cell with microvilli in **(c)**. However, the eight small cells in (b) have twice the surface area of the single cell in (a). As (c) illustrates, the surface area of any cell can be increased by the presence of microvilli.

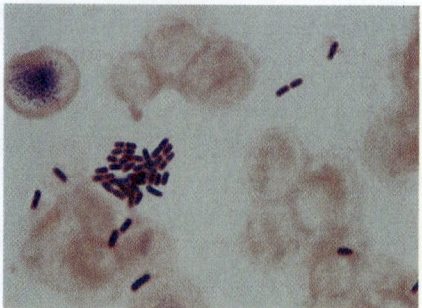

a) The light microscope (LM).

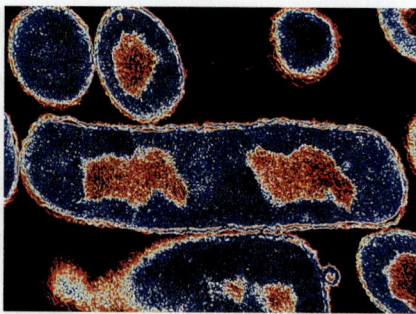

b) The transmission electron microscope (TEM).

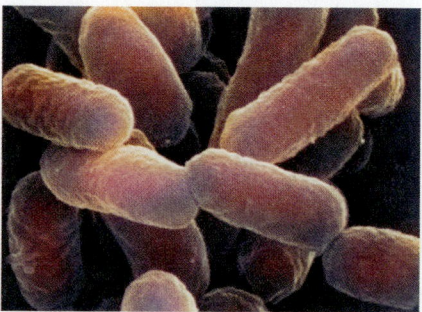

c) The scanning electron microscope (SEM).

Figure 3.4 **Visualizing cells with microscopes.** Photographs taken by the various methods of microscopy are called photomicrographs. All three of the photomicrographs shown here are of *Escherichia coli*, a normally harmless bacterium found in the digestive tract.

Light microscope (LM) The LM has been in use for over 300 years. If you've ever used a microscope at all, this is undoubtedly the type you used. As the name implies, an LM uses visible light to illuminate a small sample viewed through magnifying lenses. Small enough to sit on a laboratory bench, an LM can magnify a sample up to about 1,000-fold; not high by today's standards. However, the LM is still the only microscope that can be used on living samples (**Figure 3.4a**).

Transmission electron microscope (TEM) A TEM bombards a thin slice of an object (a thin slice through a single cell, for example) with a beam of electrons. Some of the electrons are transmitted through the sample to create a two-dimensional image that can be magnified up to 100,000-fold. A TEM is the only microscope that can reveal structural details *inside* a cell at high magnifications (**Figure 3.4b**).

Scanning electron microscope (SEM) An SEM uses a beam of electrons to scan the surface of an object, producing a three-dimensional image of the outer surface (**Figure 3.4c**). Like the TEM, the SEM produces magnifications of about 100,000-fold.

SEMs and TEMs are large and very expensive, so you won't see one in a student laboratory. The computer images they produce are only black and white, so they are usually colorized after the image is obtained. You may see the designation "false-color SEM" (or TEM) to indicate that they have been colorized, but nowadays it is generally just assumed that they have been.

MJ's **BlogInFocus** What is being done to increase the availability of microscopes in very poor countries? Visit MJ's blog in the Study Area in MasteringBiology and look under "An Inexpensive Microscope."

http://goo.gl/RRqHMf

Recap Common features of nearly all eukaryotic cells are a plasma membrane, a nucleus, organelles, and the cytoplasm. Cells exchange materials with their environment across their plasma membrane. Cells are small because this makes them more efficient at obtaining nutrients and expelling wastes. ▪

3.3 Internal structures carry out specific functions

So far we have discussed the plasma membrane that surrounds the cell. Now we move inside the cell, where we find a number of different membrane-bound and non-membrane-bound structures. The membrane-bound structures are called *organelles,* because they are like tiny organs in that they have a specific function to perform. **Figure 3.5** presents a cutaway view of an animal cell with its nucleus and organelles.

The nucleus controls the cell

The most conspicuous organelle of a living eukaryote is its *nucleus* (**Figure 3.6**). As the information center of a cell, the nucleus contains most of the cell's genetic material in the form of long molecules of DNA. (As you will learn, mitochondria have their own DNA.) Ultimately, DNA controls nearly all the activities of a cell. Details of how DNA controls cellular function are discussed when we review cell reproduction and genetic inheritance (Chapters 17 and 19).

The outer surface of the nucleus consists of a double-layered membrane, called the *nuclear membrane,* that keeps the DNA within the nucleus. The nuclear membrane is bridged by *nuclear pores* that are too small for DNA to pass through but that permit the passage of certain small proteins and RNA molecules.

Within the nucleus is a dense region called the **nucleolus,** where the components of ribosomes (RNA and ribosomal proteins) are synthesized. The components pass through the nuclear pores to be assembled into ribosomes in the cytoplasm.

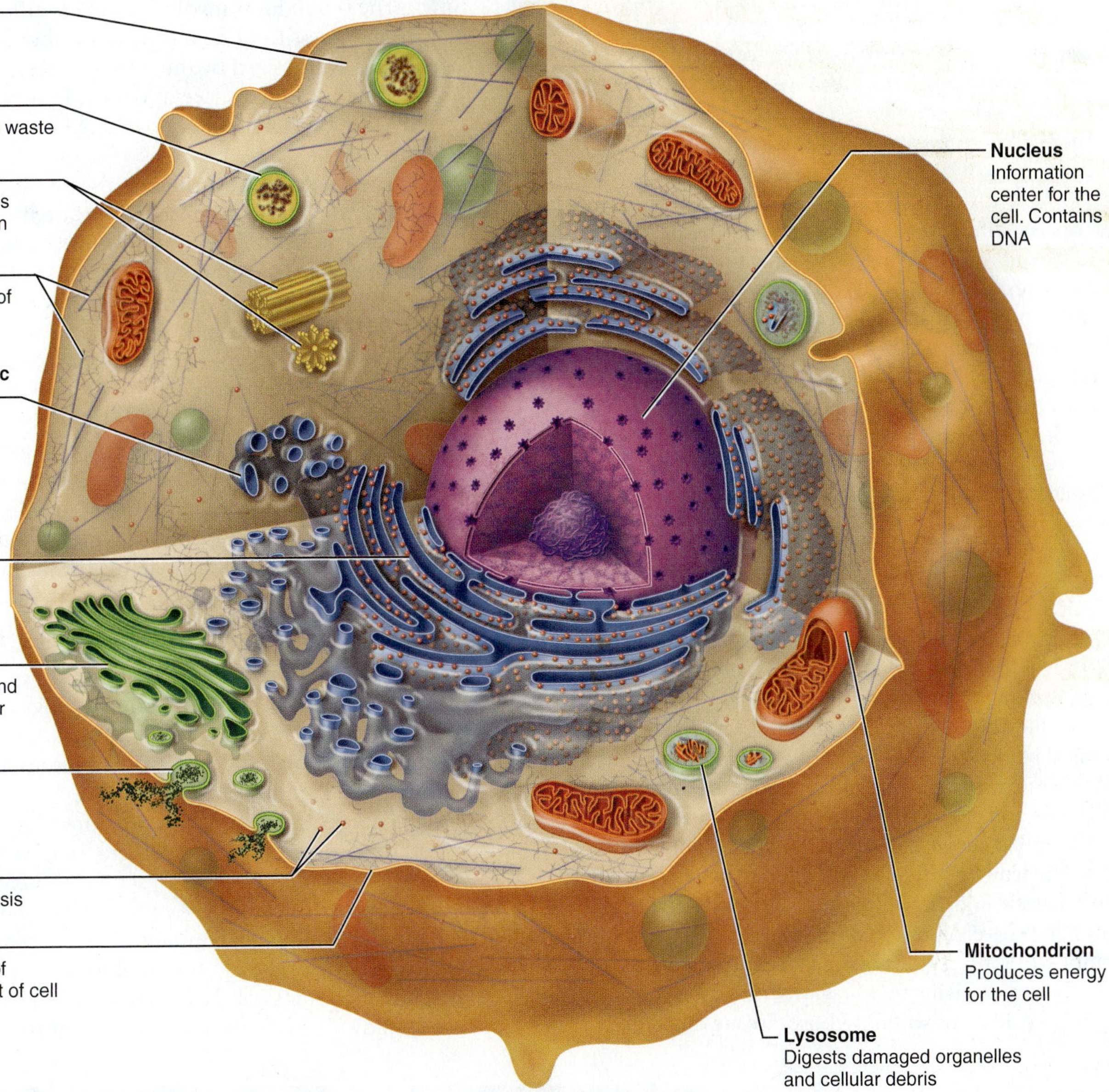

Cytosol
Semifluid gel material inside the cell

Peroxisome
Destroys cellular toxic waste

Centrioles
Microtubular structures involved in cell division

Cytoskeleton
Structural framework of the cell

Smooth endoplasmic reticulum
Primary site of macromolecule synthesis other than proteins

Rough endoplasmic reticulum
Primary site of protein synthesis by ribosomes

Golgi apparatus
Refines, packages, and ships macromolecular products

Secretory vesicle
Membrane-bound shipping container

Ribosomes
Site of protein synthesis

Plasma membrane
Controls movement of materials into and out of cell

Nucleus
Information center for the cell. Contains DNA

Mitochondrion
Produces energy for the cell

Lysosome
Digests damaged organelles and cellular debris

Figure 3.5 A typical animal cell. Not shown in this cell are cilia, a flagellum, glycogen granules, or fat deposits, as only certain cells have these features.

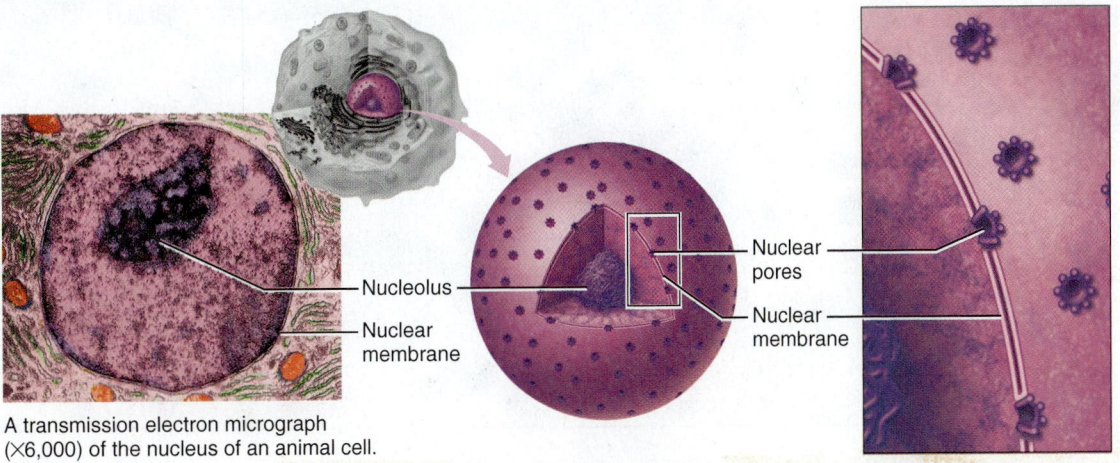

Nucleolus

Nuclear membrane

Nuclear pores

Nuclear membrane

A transmission electron micrograph (×6,000) of the nucleus of an animal cell.

Figure 3.6 The nucleus. The nucleus contains the cell's genetic material. The nucleolus produces the protein and RNA components of ribosomes. These components exit the nucleus through nuclear pores.

Ribosomes are responsible for protein synthesis

Ribosomes are small structures composed of RNA and certain proteins that are either floating freely in the cytosol or are attached to the endoplasmic reticulum, the cell organelle that synthesizes most biological molecules. Ribosomes are responsible for making specific proteins. They assemble amino acids into proteins by connecting the appropriate amino acids in the correct sequence according to an RNA template. We describe this process in more detail when we review how cells reproduce (Chapter 17).

Ribosomes that are attached to the endoplasmic reticulum release their proteins into the folds of the endoplasmic reticulum. Many of these proteins are packaged in membrane-bound vesicles, transported to the cell membrane, and secreted. Free-floating ribosomes generally produce proteins for immediate use by the cell, such as enzymes that serve as catalysts for chemical reactions within the cytoplasmic fluid.

The endoplasmic reticulum is the manufacturing center

The **endoplasmic reticulum (ER),** in conjunction with its attached ribosomes, synthesizes most of the chemical compounds made by the cell. If a cell were an industrial city, then the ER would be the city's steel mills, sawmills, and chemical plants. Like the output of a steel mill, the materials manufactured by the ER are often not in their final form. They are refined and packaged by the Golgi apparatus, discussed later.

Figure 3.7 shows the structure of the ER and its role in the manufacture of proteins and other materials. The ER is an extensively folded, membranous system surrounding a fluid-filled space. A portion of the ER connects to the nuclear membrane. Some regions of the ER's outer surface are dotted with ribosomes, giving those regions, called *rough ER*, a granular appearance. Regions without ribosomes are called *smooth ER*.

The rough ER is involved in the synthesis of proteins, as you may guess from the presence of ribosomes. Most of the proteins synthesized by the attached ribosomes are released into the fluid-filled space of the ER. Eventually they enter the smooth ER, where they are packaged for transfer to the Golgi apparatus.

The smooth ER synthesizes macromolecules other than protein. Most notable among these are the lipids, including some hormones. Numerous enzymes embedded in the inner surface of the ER membrane facilitate the chemical reactions necessary for macromolecule synthesis.

The smooth ER is also responsible for packaging the proteins and lipids for delivery to the Golgi apparatus. Newly synthesized proteins and lipids collect in the outermost layers of smooth ER. There, small portions of the fluid-filled space are surrounded by ER membrane and pinched off, forming vesicles that contain fluid, proteins, and lipids. The vesicles migrate to the Golgi apparatus, fuse with the Golgi apparatus membrane, and release their contents into the Golgi apparatus for further processing.

The Golgi apparatus refines, packages, and ships

The **Golgi apparatus** is the cell's refining, packaging, and shipping center. To continue our analogy of the cell as an industrial city, here is where steel bars are shaped into nails and screws, raw lumber assembled into doors and window frames, and grain turned into bread.

Figure 3.8 diagrams the structure of the Golgi apparatus and the processes that occur there. In cross section, the Golgi apparatus appears as a series of interconnected fluid-filled spaces surrounded by membrane, much like a stack of plates. Like the ER, it contains enzymes that further refine the products of the ER into final form.

The contents of the Golgi apparatus move outward by a slow but continuous process. At the outermost layer

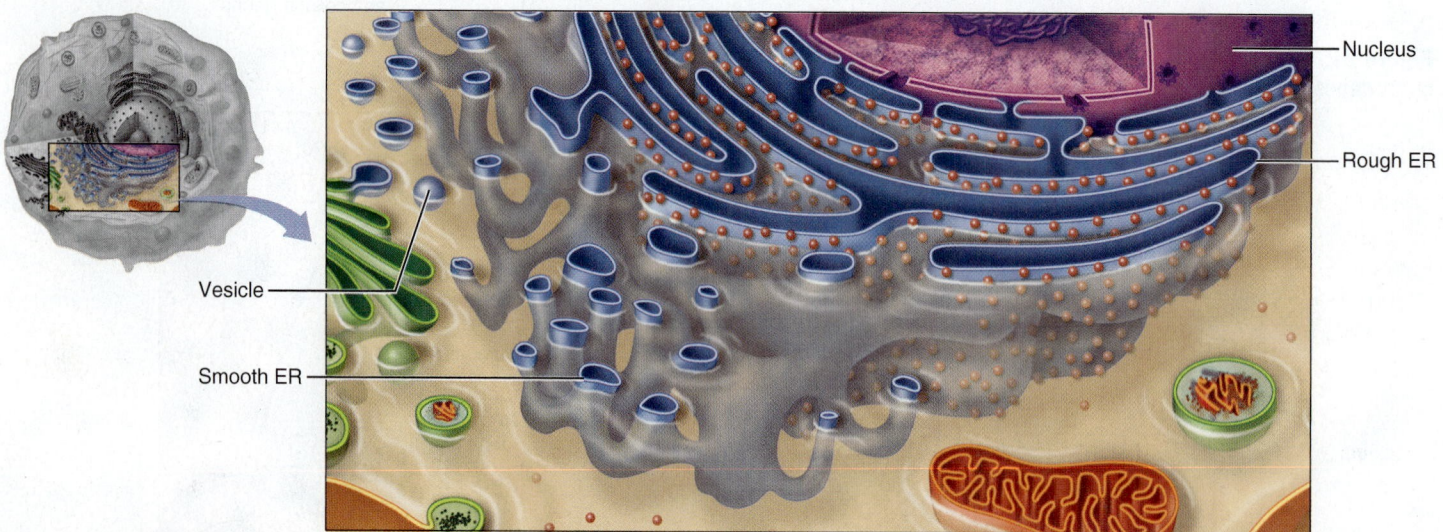

Vesicle

Smooth ER

Nucleus

Rough ER

Figure 3.7 The endoplasmic reticulum (ER). The rough ER is studded with ribosomes, where proteins are made. The smooth ER packages the proteins and other products of the ER and prepares them for shipment to the Golgi apparatus in vesicles.

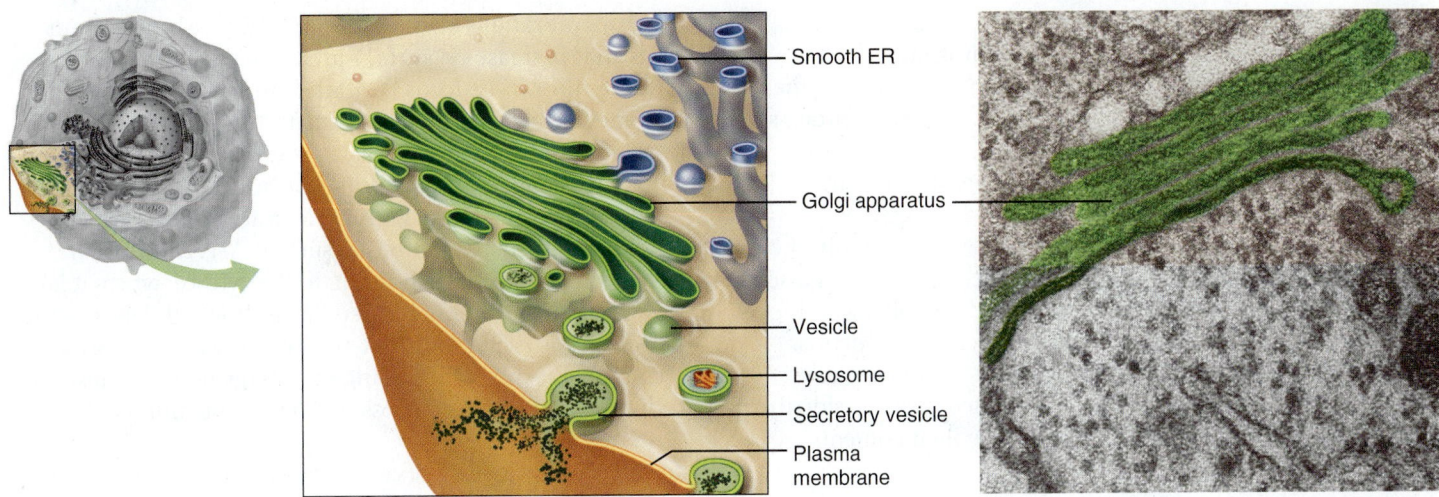

Figure 3.8 **The Golgi apparatus.** The Golgi apparatus receives substances from the ER, refines them into final products, and packages them into vesicles for their final destinations.

of the Golgi apparatus, the products are finally ready to be packaged into vesicles and shipped to their final destinations.

Vesicles: membrane-bound storage and shipping containers

Vesicles are membrane-bound spheres that enclose something within the cell. There are several types of vesicles, each with a different origin and purpose.

Vesicles that ship and store cellular products These vesicles enclose and transport the products of the ER and Golgi apparatus. Each vesicle contains only one product out of the thousands of substances made by the Golgi apparatus.

The contents of each vesicle depend on certain proteins in the vesicle membrane that act as "shipping labels." They determine which product is put into the vesicle and where the vesicle is sent. If a vesicle's products are not immediately needed it remains in the cell cytoplasm, like a box stored in a warehouse awaiting shipment.

Secretory vesicles Secretory vesicles contain products destined for export from the cell. They migrate to the plasma membrane and release their contents outside the cell. Because most secretory products are made in the Golgi apparatus, secretory vesicles generally derive from Golgi apparatus membrane.

Endocytotic vesicles These structures enclose bacteria and raw materials from the extracellular environment. They bring them into the cell by endocytosis.

Peroxisomes and lysosomes These vesicles contain enzymes so powerful that they must be kept within the vesicle to avoid damaging the rest of the cell. Both are produced by the Golgi apparatus. **Figure 3.9** shows their functions.

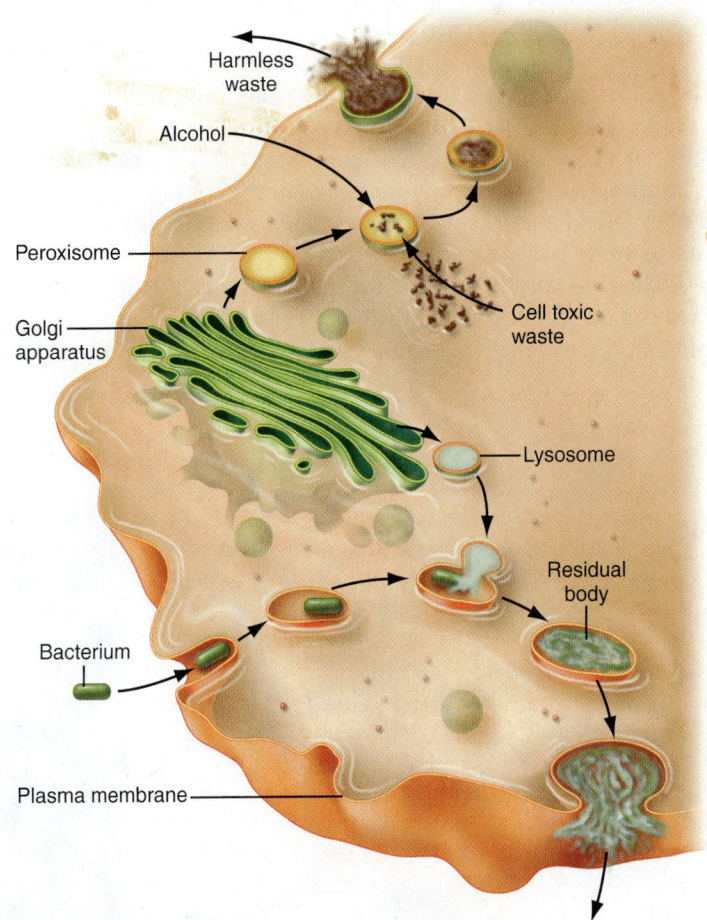

Figure 3.9 **Lysosomes and peroxisomes.** Lysosomes formed by the Golgi apparatus fuse with endocytotic vesicles containing a bacterium. Following digestion of the bacterium, the residual waste is excreted. Peroxisomes take up toxic wastes (including alcohol) and degrade them to harmless waste, which is also excreted.

The enzymes in **peroxisomes** destroy various toxic wastes produced in the cell, including hydrogen peroxide (H_2O_2). They also destroy compounds that have entered the cell from outside, such as alcohol. The detoxification process occurs entirely within the peroxisome.

Lysosomes (from the Greek *lysis*, meaning "dissolution," and *soma*, meaning "body") contain powerful digestive enzymes. Lysosomes fuse with endocytotic vesicles within the cell, digesting bacteria and other large objects. Lysosomes also perform certain housekeeping tasks, such as dissolving and removing damaged mitochondria and other cellular debris. When their digestive task is complete, they become *residual bodies*, analogous to small bags of compacted waste. Residual bodies can be stored in the cell, but usually their contents are eliminated from the cell.

Mitochondria provide energy

Nearly all of a cell's functions require energy. Energy is available in the chemical bonds of the food we eat, but cells cannot use it directly. Most energy in ingested nutrients must be converted to a more usable form before it can power the chemical and physical activities of living cells.

Mitochondria (singular: mitochondrion) are the organelles responsible for providing most of this usable energy; they are often called the cells' "power plants." Not surprisingly, their number within different cells varies widely according to the energy requirements of the cells. A cell with a high rate of energy consumption, such as a muscle cell, may contain over 1,000 mitochondria.

The structure of a mitochondrion can be seen in **Figure 3.10**. A smooth outer membrane, similar to the plasma membrane, covers the entire surface. Within the outer membrane is an inner membrane with numerous folds that increase its surface area. The inner membrane and the fluid in its folds contain hundreds of protein enzymes, which serve as catalysts to break down chemical bonds in our food and release the energy. This process consumes oxygen and produces carbon dioxide.

The energy liberated within the mitochondria is used to create high-energy molecules such as ATP. ATP is then exported from the mitochondria to the cytosol, where it is available as a quick source of energy for the cell. Like electric power, ATP is useful for a variety of purposes. We have already seen one of its uses—providing the energy used to transport sodium and potassium across the plasma membrane.

✅ While studying a human cell under a microscope, you spot a small, round organelle that seems to have a single membrane. Is it more likely to be a nucleus, ribosome, vesicle, or mitochondrion? What might it contain?

Fat and glycogen: sources of energy

The mitochondria generally manufacture ATP as it is needed. To avoid the possibility of running out of fuel, some cells store energy in raw form. These energy stores are not enclosed in any membrane-bound container. They are more like large piles of coal on the ground, awaiting delivery to the power plants (mitochondria) for conversion to electricity (ATP).

Some cells store raw energy as lipids (fat). Our so-called fat cells are so specialized for this purpose that most of their volume consists of large droplets of stored lipids. Dieting and exercise tend to reduce the amount of stored fat—that is, they make the fat cells leaner. However, dieting and exercise do not reduce their number. The cells are available to store fat again, which is why it is so hard to keep lost weight off.

Other cells store energy as glycogen granules (review Figure 2.16). Muscle cells rely on glycogen granules rather than on fat deposits because the energy stored in the chemical bonds of glycogen can be used to produce ATP more quickly than the energy derived from fat.

↩ **Recap** The nucleus contains most of the cell's genetic material. Ribosomes are responsible for protein assembly. The endoplasmic reticulum manufactures most other cellular products in rough form. The Golgi apparatus refines cellular products and packages them into membrane-bound vesicles. Some vesicles store, ship, and secrete cellular products; others digest and remove toxic waste and cellular debris. Mitochondria manufacture ATP for the cell. ∎

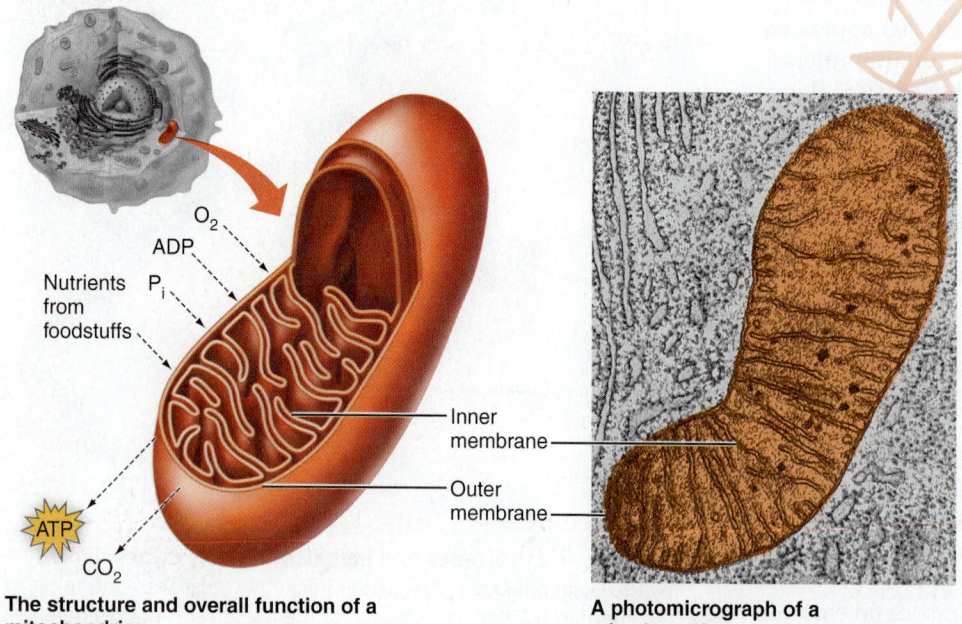

O_2
ADP
Nutrients from foodstuffs
P_i
Inner membrane
Outer membrane
ATP
CO_2

The structure and overall function of a mitochondrion.

A photomicrograph of a mitochondrion.

Figure 3.10 Mitochondria.

☑ The cells of the pancreas make large quantities of proteins that are constantly secreted outside the cell. Would you expect a pancreatic cell to have a relatively large or small number of ribosomes, and would you expect it to have a lot of smooth or rough ER (compared to a cell that does not secrete proteins)?

3.4 Cells have structures for support and movement

The soft plasma membrane is supported by an internal scaffolding that helps the cell maintain its shape. Some cells have specialized structures to help them move around, and all cells contain structures that are involved in moving cellular components during cell division. Structural elements for support and movement include the cytoskeleton, cilia and flagella, and centrioles.

The cytoskeleton supports the cell

The **cytoskeleton** (Figure 3.11) consists of a loosely structured network of fibers called *microtubules* and *microfilaments*. As their names imply, microtubules are tiny hollow tubes, and microfilaments are thin solid fibers. Both are composed of protein. They attach to each other and to proteins in the plasma membrane, called *glycoproteins*, which typically have carbohydrate group components.

The cytoskeleton forms a framework for the soft plasma membrane, much as tent poles support a nylon tent. The cytoskeleton also supports and anchors the other structures within the cell.

Cilia and flagella are specialized for movement

A few cells have hairlike **cilia** (singular: cilium) or longer **flagella** (singular: flagellum) that extend from the surface. Cilia are generally only 2 to 10 microns long (1 micron equals one-millionth of a meter). In cells that have them, cilia are numerous (Figure 3.12a). Cilia move materials

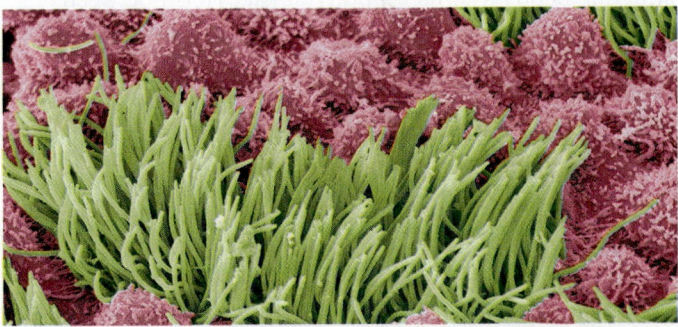

a) **The surface of an oviduct, showing cells with cilia (green).**

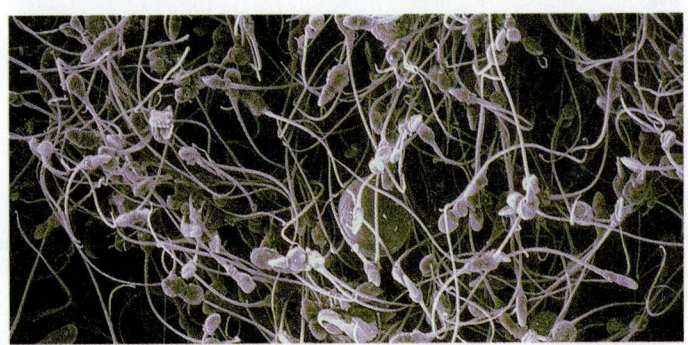

b) **Human sperm cells with flagella.**

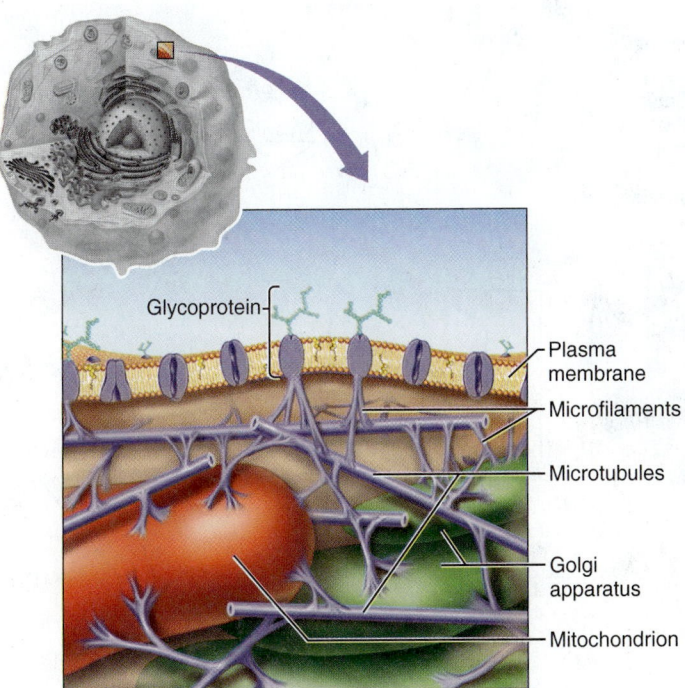

Glycoprotein
Plasma membrane
Microfilaments
Microtubules
Golgi apparatus
Mitochondrion

Figure 3.11 The cytoskeleton. The cytoskeleton consists of microtubules and microfilaments that attach to and support the cell's organelles and plasma membrane.

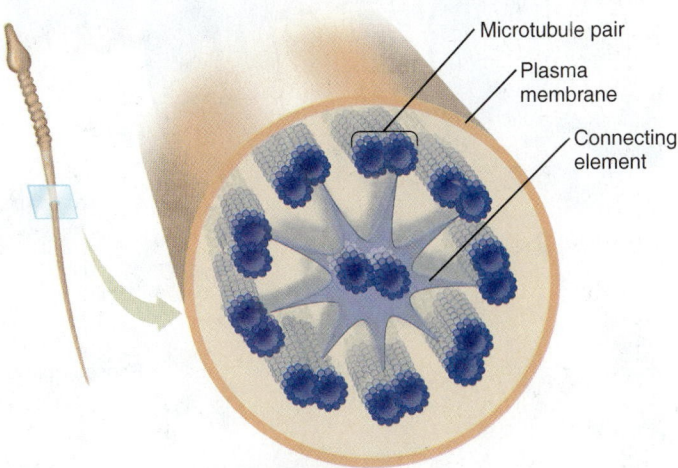

Microtubule pair
Plasma membrane
Connecting element

c) **Composition of cilia and flagella.**

Figure 3.12 Cilia and flagella. Cilia and flagella are composed of nine pairs of microtubules surrounding a central pair.

along the surface of a cell with a brushing motion. They are common on the surfaces of cells that line the airways and in certain ducts within the body.

In humans, flagella (approximately 200 microns long) are found only on sperm cells (**Figure 3.12b**). The whiplike movement of the flagellum moves the entire sperm cell from one place to another.

Cilia and flagella are similar in structure (**Figure 3.12c**). They are composed primarily of protein microtubules held together by connecting elements and surrounded by a plasma membrane. Nine pairs of fused microtubules surround two single microtubules in the center. The entire structure bends when temporary linkages form between adjacent pairs of microtubules, causing the pairs to slide past each other. The formation and release of these temporary bonds requires energy in the form of ATP.

Centrioles are involved in cell division

Centrioles are short, rodlike microtubular structures located near the nucleus. Centrioles are essential to the process of cell division because they participate in aligning and dividing the genetic material of the cell. We discuss centrioles when we describe how a cell divides in two (Chapter 17).

↩ **Recap** The cytoskeleton forms a supportive framework for the cell. Cilia and flagella are specialized for movement, and centrioles are essential to cell division. ∎

3.5 A plasma membrane surrounds the cell

Consider a house. Its walls and roof are composed of special materials that prevent rain and wind from entering. They also form a barrier that allows the temperature inside the house to stay warmer or cooler than the temperature outside. At the same time, the house interacts with its environment. Windows allow light in; doors open and close to allow entry and exit. Water and power lines permit the regulated entry of water and energy, and sewer lines remove wastes. The house exchanges information with the outside world through mail slots, telephone lines, and computer cables.

The exterior structure of a living cell is its plasma membrane. Like the roof and walls of a house, the plasma membrane must permit the movement of some substances into and out of the cell yet restrict the movement of others. It must also allow the transfer of information across the membrane.

The plasma membrane is a lipid bilayer

The plasma membrane is constructed of two layers of phospholipids, called a *lipid bilayer*, plus some cholesterol and various proteins (**Figure 3.13**). Each of the three

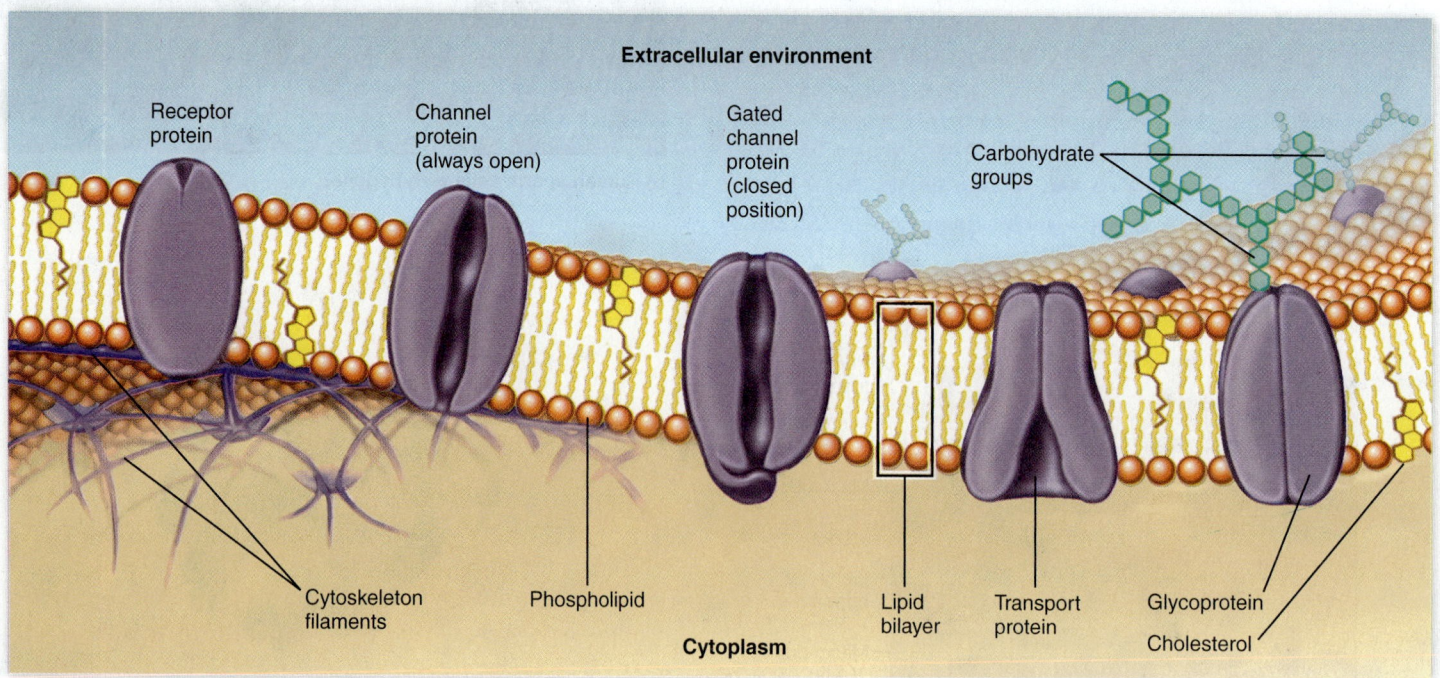

Figure 3.13 The plasma membrane. The plasma membrane is a phospholipid bilayer containing cholesterol and proteins. Cholesterol provides mechanical strength. The proteins transfer information, permit the passage of certain molecules, and provide structural support for the cytoskeleton.

components contributes to the membrane's overall structural and functional properties:

- *Phospholipids.* Recall that phospholipids are a particular type of lipid with a polar head and neutral nonpolar tails. In the plasma membrane, the two layers of phospholipids are arranged so that the nonpolar tails meet in the center of the membrane. One layer of polar (water-soluble) heads faces the watery solution on the outside of the cell, and the other layer of polar heads faces the watery solution of the cell's cytoplasm.
- *Cholesterol.* Cholesterol increases the mechanical strength of the membrane by preventing it from becoming either too rigid or too flexible. It also prevents the phospholipids from moving around too much and helps to anchor the proteins within the membrane.
- *Proteins.* Various proteins are embedded in the phospholipid bilayer of the plasma membrane. Like the doors, windows, and wires of a house, they provide the means for transporting molecules and information across the plasma membrane. A few membrane proteins anchor the cell's internal scaffoldlike support network. Some proteins span the entire membrane; others protrude from only one surface. Plasma membrane proteins generally have one region that is electrically neutral and another that is electrically charged (either + or −). The charged regions tend to extend out of the membrane and thus are in contact with water, whereas the neutral portions are often embedded within the phospholipid bilayer.

The phospholipid bilayer of the plasma membrane is only about 3.5 nanometers thick (a nanometer, abbreviated nm, is a billionth of a meter), too small to be seen in detail even with microscopes. To appreciate relative sizes, imagine (as we did in Chapter 2) that a single sodium ion is the size of a penny. At this scale, the phospholipid bilayer of a typical cell would be about 13 inches thick. It is no wonder that many substances are restricted from passing through the membrane unless there is some sort of channel or transport mechanism available.

Although we have likened the plasma membrane to the exterior of a house, there are two key differences. First, the plasma membrane of animal cells is not rigid. If you could touch a plasma membrane, it would probably feel like a wet sponge, giving way under your touch and springing back when you remove your hand. Most cells do maintain a certain shape, but it is mainly due to a supporting network of fibers inside the cell, the fluid within the cell, and the limitations imposed by contact with surrounding cells, not the stiffness of the plasma membrane itself.

Second, the phospholipids and proteins are not anchored to specific positions in the plasma membrane. Many proteins drift about in the lipid bilayer like icebergs floating on the surface of the sea. Imagine if you were to get up in the morning and find that the front door to your house had moved 3 feet to the left! The plasma membrane of animal cells is often described as a "fluid mosaic" to indicate that it is not a rigid structure and that the pattern of proteins within it constantly changes.

 **Recap** The plasma membrane is comprised of phospholipids, cholesterol, and proteins. The proteins transfer information and transport molecules across the membrane and provide structural support. ∎

3.6 Molecules cross the plasma membrane in several ways

The plasma membrane creates a barrier between the cell's external environment and the processes of life going on within. Life was impossible until these functions could be enclosed and concentrated in one place, keeping what was needed for growth and reproduction inside and limiting the entry of other materials. Molecules (and ions) cross the plasma membrane in three major ways: (1) passive transport (diffusion and osmosis), (2) active transport, and (3) endocytosis or exocytosis.

Passive transport: principles of diffusion and osmosis

Passive transport is "passive" because it transports a molecule without requiring the cell to expend any energy. Passive transport relies on the mechanism of diffusion.

Diffusion Molecules in a gas or a liquid move about randomly, colliding with other molecules and changing direction. The movement of molecules from one region to another as the result of this random motion is known as **diffusion.**

If there are more molecules in one region than in another, then strictly by chance more molecules will tend to diffuse away from the area of high concentration and toward the region of low concentration. In other words, the *net* diffusion of molecules requires that there be a difference in concentration, called a *concentration gradient*, between two points. Once the concentration of molecules is the same throughout the solution, a state of equilibrium exists in which molecules diffuse randomly but equally in all directions.

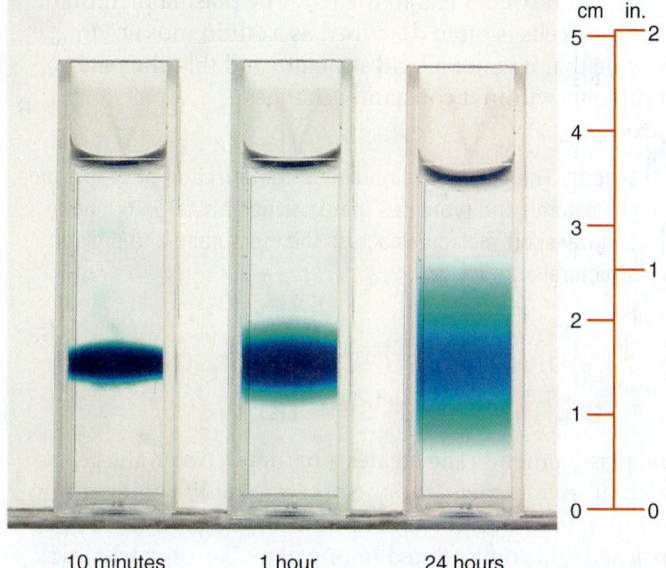

Figure 3.14 Diffusion. At time zero, a concentrated solution of a blue dye was placed in the middle of a tube of water. Over time, the random motion of molecules in solution has caused some molecules of dye to move from the region of highest dye concentration to the region of low dye concentration. Note that diffusion is effective only over short distances; it has taken 24 hours for the dye to move only 1 cm in any direction.

Figure 3.14 illustrates diffusion by showing what happens when a concentrated solution of blue dye is placed in the middle of a tube of water. Over time, the dissolved molecules diffuse away from their region of highest concentration and toward the regions of low concentration.

A hallmark of diffusion is that it is only effective over very short distances. In the human body, diffusion is effective only within cells, across cell membranes, or between adjacent cells. It is not an effective transport method between organs in the human body; for that we need a cardiovascular system.

Water also diffuses from the region of its highest concentration toward the region of its lowest concentration. However, the concentration of water (the liquid, or solvent) in a solution is *opposite* to that of the molecules other than water (the solutes). The higher the concentration of solutes, the lower the concentration of water. Pure water is the solution with the highest possible concentration of water. Net diffusion of water is always toward the solution with the higher concentration of solutes and away from the solution with a higher concentration of water.

Osmosis Not all substances diffuse readily into and out of living cells. The plasma membrane is **selectively permeable,** meaning that it allows some substances to cross by diffusion but not others. It is highly permeable to water but not to all ions or molecules. The net diffusion of water across a selectively permeable membrane is called **osmosis.**

Figure 3.15 demonstrates the process of osmosis.
❶ A selectively permeable membrane—in this case permeable only to water—separates pure water from a solution of glucose in water. Although the glucose cannot diffuse, the water diffuses toward its region of lower concentration, from right to left. ❷ As osmosis occurs, the volume in the left chamber rises, creating a fluid pressure that begins to oppose the continued osmosis of

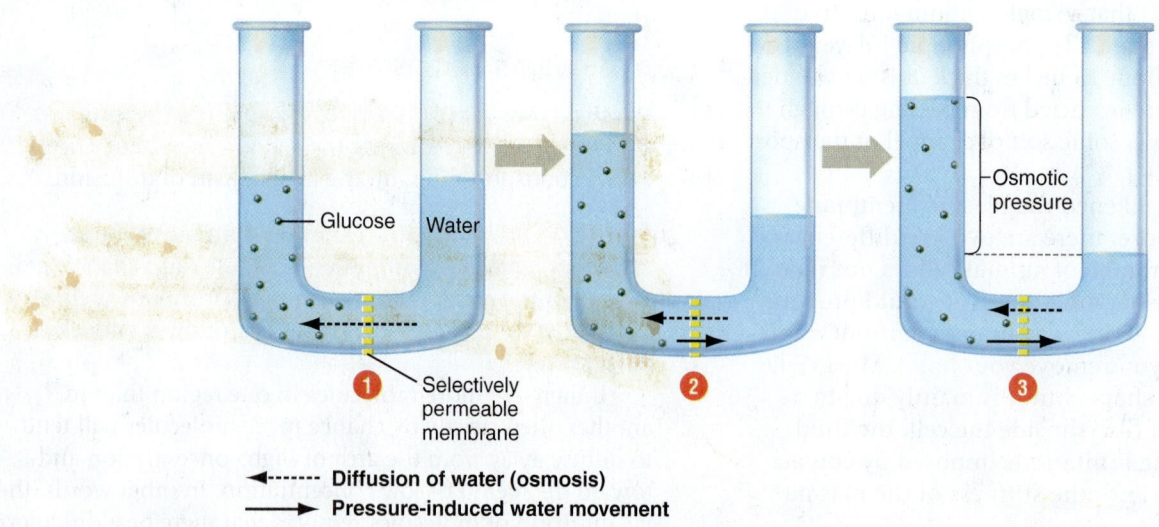

➤ Diffusion of water (osmosis)
➤ Pressure-induced water movement

Figure 3.15 Generation of osmotic pressure by osmosis. Starting in ❶, there is a net movement of water from right to left until the diffusion of water (osmosis) is opposed by an equal rate of movement of water in the opposite direction due to osmotic pressure (❸).

✔ In Figure 3.15 step 3, what would happen if you now add glucose to the right side—adding exactly the same total number of glucose molecules as are in the left side? Explain your answer.

water. **3** Eventually, the movement of water from left to right (because of differences in fluid pressure) equals the movement from right to left (by osmosis), and there is no further net change in the volume of water on each side of the membrane.

The fluid pressure required to exactly oppose osmosis is called *osmotic pressure*. Osmotic pressure is represented in step 3 of Figure 3.15 by the extra weight of the higher column of water on the left than on the right.

Passive transport moves with the concentration gradient

Most substances cross cell membranes by passive transport. Passive transport always proceeds "downhill" with respect to the concentration gradient, meaning that it relies on diffusion in some way. Three forms of passive transport across the cell membrane are (1) diffusion through the lipid bilayer, (2) diffusion through channels, and (3) facilitated transport.

Diffusion through the lipid bilayer The lipid bilayer structure of the plasma membrane allows the free passage of some molecules while restricting others (**Figure 3.16a**). For instance, small uncharged nonpolar molecules can diffuse right through the lipid bilayer as if it did not exist. Such molecules simply dissolve in the lipid bilayer, passing through it like a ghost through a wall. Polar or electrically charged molecules, in contrast, cannot cross the lipid bilayer because they are not soluble in lipids.

Two important lipid-soluble molecules are oxygen (O_2), which diffuses into cells and is used up in the process of metabolism, and carbon dioxide (CO_2), a waste product of metabolism, which diffuses out of cells and is removed from the body by the lungs. Another substance that crosses the lipid bilayer by diffusion is urea, a neutral waste product removed from the body by the kidneys.

✅ Do you think sodium (Na^+) or chloride (Cl^-) ions can diffuse directly through the lipid bilayer? How about water molecules? Explain.

Diffusion through channels Water and many ions diffuse through channels in the plasma membrane (**Figure 3.16b**). The channels are constructed of proteins that span the entire lipid bilayer. The sizes and shapes of these protein channels, as well as the electrical charges on the various amino acid groups that line the channel, determine which molecules can pass through.

Some channels are open all the time (typical of water channels). The diffusion of any molecule through the membrane is largely determined by the number of channels through which the molecule can fit. Other channels are "gated," meaning that they can open and close under certain conditions. Gated channels are particularly important in regulating the transport of ions (sodium, potassium, and calcium) in cells that are electrically excitable, such as nerve cells (Chapter 11). Look at Figure 3.13, which represents a number of the proteins instrumental in transport.

Facilitated transport In **facilitated transport,** also called *facilitated diffusion,* the molecule does not pass through a channel at all. Instead, it attaches to a membrane protein, triggering a change in the protein's shape or orientation that transfers the molecule to the other side of the membrane, where it is released (**Figure 3.16c**). Once the molecule is released, the protein returns to its original form. A protein that carries a molecule across the plasma membrane in this manner, rather than opening a channel through it, is called a **transport protein** (carrier protein).

Facilitated transport is highly selective for particular substances. The direction of movement is always from a region of higher concentration to one of lower

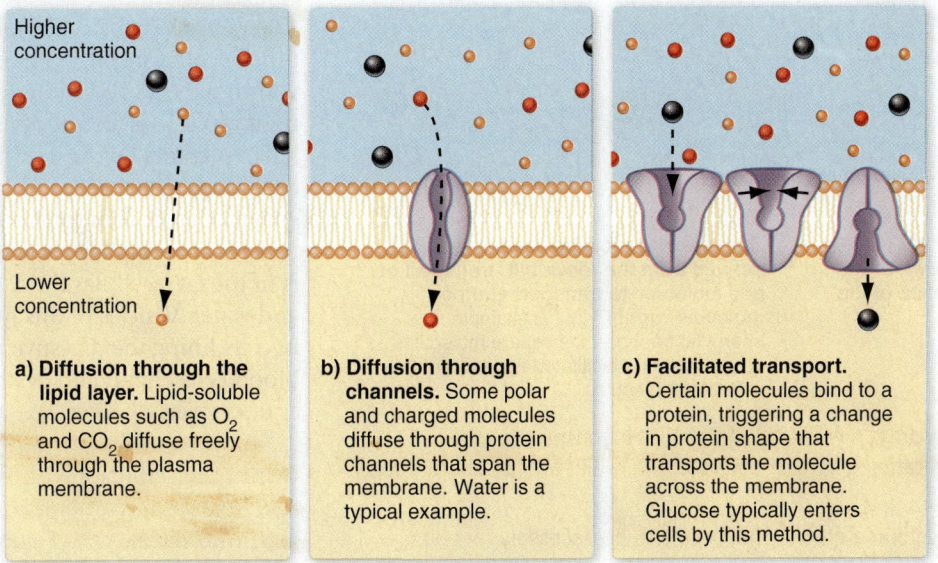

a) **Diffusion through the lipid layer.** Lipid-soluble molecules such as O_2 and CO_2 diffuse freely through the plasma membrane.

b) **Diffusion through channels.** Some polar and charged molecules diffuse through protein channels that span the membrane. Water is a typical example.

c) **Facilitated transport.** Certain molecules bind to a protein, triggering a change in protein shape that transports the molecule across the membrane. Glucose typically enters cells by this method.

Figure 3.16 The three forms of passive transport. All involve transport down a concentration gradient without the expenditure of additional energy.

concentration, and thus it does not require the cell to expend energy. The normal process of diffusion is simply being "facilitated" by the transport protein. Glucose and other simple sugars enter most cells by this method.

Active transport requires energy

All methods of passive transport allow substances to move only down their concentration gradients, in the direction they would normally diffuse if there were no barrier. However, **active transport** can move substances through the plasma membrane *against* their concentration gradient. Active transport allows a cell to accumulate essential molecules even when their concentration outside the cell is relatively low and to get rid of molecules that it does not need. Active transport requires the expenditure of energy.

Like facilitated transport, active transport is accomplished by proteins that span the plasma membrane. The difference is that active transport proteins must have some source of energy to transport certain molecules. Some active transport proteins use the high-energy molecule adenosine triphosphate (ATP) for this purpose (Figure 3.17a). They break ATP down to adenosine diphosphate (ADP) and a phosphate group (P_i) and use the released energy to transport one or more molecules across the plasma membrane against their concentration gradient.

Proteins that actively transport molecules across the plasma membrane are sometimes called "pumps." Some pumps can transport several different molecules at once and even in both directions at the same time. One of the most important plasma membrane pumps is the **sodium-potassium pump,** which uses energy derived from breaking down ATP to transport sodium out of the cell and potassium into the cell.

Not all active transport pumps use ATP as the energy source. Some derive energy from the "downhill" facilitated transport of one molecule and use it to transport another molecule "uphill," against its concentration gradient (Figure 3.17b). This type of transport is analogous to an old-fashioned mill that grinds grain into flour by using energy derived from the downhill movement of water.

Endocytosis and exocytosis move materials in bulk

Most ions and small molecules move across the cell membrane by one or more of the passive and active transport mechanisms just described. However, some molecules are too big to be transported by these methods.

To move large molecules or transport several kinds of molecules in bulk, some cells resort to endocytosis and/or exocytosis. These two processes are based on the same principle but have different directions of movement. **Endocytosis** moves materials into the cell, and **exocytosis** moves materials out of the cell.

In endocytosis, molecules dissolved in the extracellular fluid are surrounded by a pocket formed by an infolding of the plasma membrane (Figure 3.18a). Eventually the pocket pinches off, forming a membrane-bound **vesicle** within the cell. To facilitate the selection of the right molecules for endocytosis, some vesicles have receptors on their surface that bind only to certain specific molecules. Insulin and certain enzymes enter cells by this method. Other vesicles are nonselective, engulfing whatever is in the extracellular fluid, such as nutrients and water. Vesicles of this type are often found in cells lining the digestive tract. Some white blood cells engulf and destroy whole bacteria by endocytosis (Chapter 9).

In exocytosis, a vesicle already present within the cell fuses with the plasma membrane and releases its contents into the fluid surrounding the cell (Figure 3.18b). This is how certain cells release toxic waste products, get rid of indigestible material, or secrete their special products.

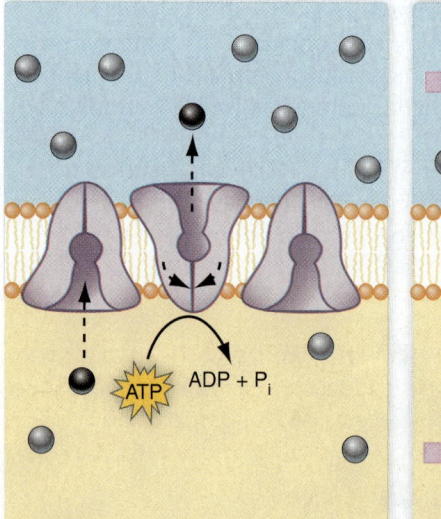
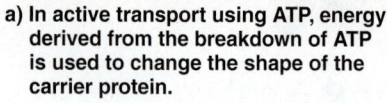
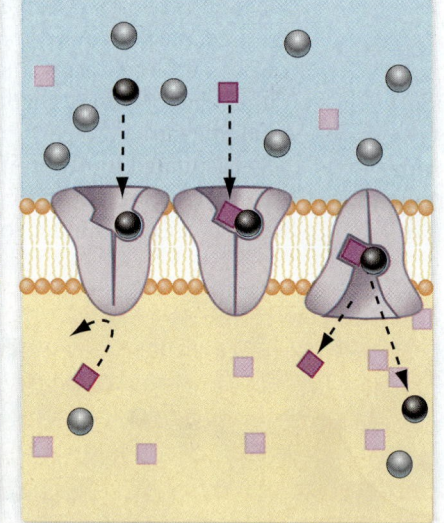

a) **In active transport using ATP, energy derived from the breakdown of ATP is used to change the shape of the carrier protein.**

b) **Some carrier proteins use energy derived from the "downhill" transport of one molecule to transport another molecule "uphill."** In this example, the energy to transport the square molecules comes from the facilitated transport of the spherical molecules.

Figure 3.17 Active transport. A cell can employ active transport to move a molecule against a concentration gradient. Because this is an "uphill" effort, energy is required.

☑ You've discovered a membrane protein that seems to be necessary for cells to transport fructose from higher to lower concentration. Predict whether this membrane protein requires ATP to function, and name the type of transport that it is most likely doing. Explain your reasoning.

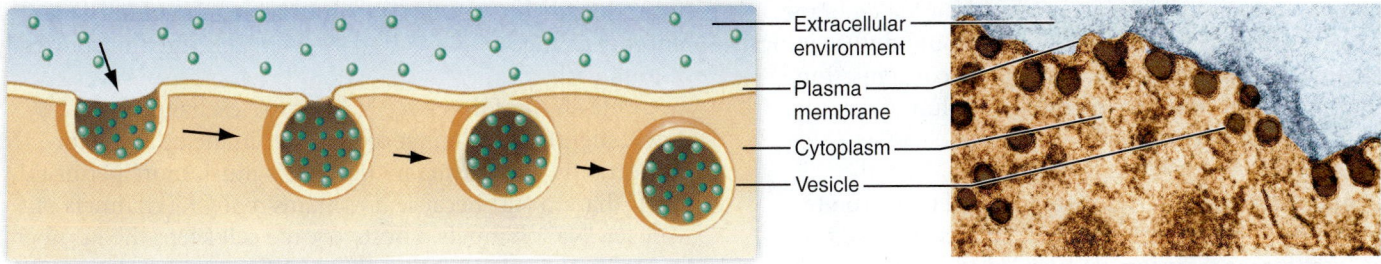

a) Endocytosis. In endocytosis, material is surrounded by the cell membrane and brought into the cell.

b) Exocytosis. In exocytosis, a membranous vesicle fuses with the plasma membrane, expelling its contents outside the cell.

Figure 3.18 **Endocytosis and exocytosis.**

Information can be transferred across the plasma membrane

Receptor proteins that span the plasma membrane can receive and transmit information across the membrane. The information received by receptor proteins generally causes something to happen within the cell even though no molecules cross the membrane.

Figure 3.19 illustrates how a receptor protein works. ❶ A *signaling molecule* approaches a receptor protein

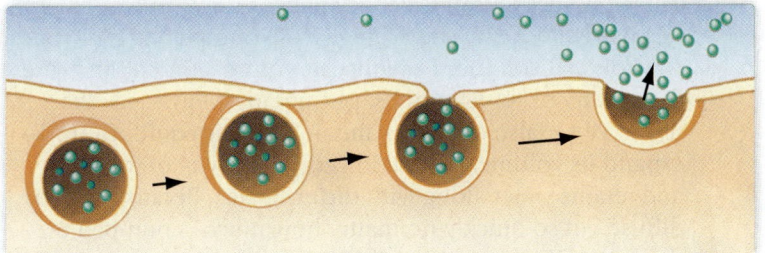

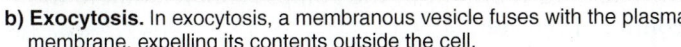

Figure 3.19 **Receptor protein action.** A specific signaling molecule approaches a receptor site on a receptor protein and binds to it. The binding of the molecule to the receptor protein causes a series of chemical reactions within the cell. In this example, a particular cellular product (the squares) is produced from substrate molecules (the circles).

embedded in the cell membrane and ❷ binds to a specific *receptor site* in a lock-and-key fashion. This binding triggers a series of biochemical events that ultimately cause changes within the cell; in this example, a *substrate* molecule is used to produce a *product*. ❸ Once the signaling molecule detaches from the receptor site, the reaction stops. Receptor proteins are highly specific for a particular molecule or a group of similar molecules. For example, the receptor protein for the hormone insulin responds only to insulin and not to any other hormone. Furthermore, different cells have different sets of receptor proteins, which explains why some cells and tissues respond to a particular hormone and others do not.

We discuss receptor proteins and hormones in more detail when we review the endocrine system. For now, remember that certain molecules can influence what happens inside a cell merely by coming in contact with the cell's outer surface.

The sodium-potassium pump helps maintain cell volume

Probably the most critical task facing a cell is maintaining its volume. Why? Recall that the plasma membrane is soft and flexible. It cannot withstand much stretching or high fluid pressures. Furthermore, cells tend to accumulate certain materials depending on what is available in their *extracellular* environment (the area outside the cell, beyond the plasma membrane). Cells already contain a nucleus and organelles. In addition, they produce or stockpile molecules, including amino acids, sugars, lipids, ions, and many others. These molecules are necessary for the cell to function normally, but they represent a lot of solute particles within the cell.

Because water can diffuse across the plasma membrane rather easily, you might expect that water would diffuse into the cell, toward the high cytoplasmic solute concentration. This inward diffusion would increase cell volume, eventually causing the cell to swell and even rupture.

The only way to avoid this is for the cell to keep the solute concentration in its cytoplasm identical to the solute concentration of the extracellular fluid. Then there is no net driving force for the diffusion of water. What the cell actually does is get rid of ions it *doesn't* need in large quantities (primarily sodium) in exchange for those it must stockpile. This is the primary function of a specialized protein embedded in the cell membrane called the sodium-potassium pump.

Figure 3.20a shows how the sodium-potassium (Na^+-K^+) pump works. The pump has three binding sites for sodium ions (Na^+) that are accessible from inside the cell. ❶ The binding of three cytoplasmic Na^+ ❷ triggers the breakdown of an ATP molecule to ADP and an inorganic phosphate (P_i). ❸ The energy released by ATP causes the pump to change shape, expelling the Na^+ and ❹ exposing two binding sites for potassium (K^+) that are accessible only from outside the cell. ❺ The binding of potassium triggers another change of shape, and ❻ the two K^+ are transported into the cell.

One effect of this three Na^+/two K^+ exchange is to lower slightly the number of ions within the cell. ❼ More importantly, the plasma membrane is much more permeable to K^+ than to Na^+ because it contains many K^+ channels but very few Na^+ channels. Effectively, the cell keeps the number of Na^+ in the cytoplasm low by pumping them out again just as soon as a few leak in. The inward active transport of K^+ does not increase the intracellular K^+ concentration very much because K^+ can leak back out so easily.

As **Figure 3.20b** shows, the N^+-K^+ pump effectively controls cell volume. To reduce its volume, the cell increases the activity of N^+-K^+ pumps, getting rid of more Na^+ than usual. Water also exits to maintain osmotic equilibrium. To expand its volume, the cell lowers the activity of the pumps and retains water along with the extra Na^+. Because K^+ can diffuse out so quickly no matter how much is pumped in, the rate of K^+ transport by the pumps is not relevant to the control of cell volume.

HEALTH & WELLNESS

Do Antioxidant Supplements Slow the Rate of Cellular Aging?

The 100 trillion cells in your body are slowly aging. Some will live just a couple of months before they die and are replaced by newer cells. Others will live as long as you do, though they may show signs of aging. What causes cells to age? And most importantly, is there anything you can do to maintain your cells' good health and delay the aging process?

All cells produce waste products during the normal course of their metabolic activities. Some of these wastes are toxic. A predominant theory of cell aging is that these toxic substances may accumulate over time, eventually leading to an inability of the cell to function properly. One such group of toxic substances is *oxygen free radicals*—oxygen-containing molecules with an unpaired electron in the oxygen's outer shell. Because of that unpaired electron, oxygen free radicals (or just free radicals for short) are unstable molecules. They have a strong tendency to oxidize (remove electrons from) other molecules, damaging or destroying them in the process. Eventually, an accumulation of molecular damage leads to a decline in cell function (aging).

Under normal circumstances, certain enzymes produced within the cell and several vitamins defend our cells against free radicals. Collectively these enzymes and vitamins are called *antioxidants*. Some of the naturally occurring antioxidant enzymes are so powerful that they must be stored within peroxisomes in the cell so that they do not damage the cell itself. As their name implies, antioxidants inactivate free radicals (think of them as mopping up cellular garbage). The antioxidants include vitamins E and C, beta-carotene, and the enzyme superoxide dismutase.

The importance of the body's natural antioxidants in maintaining cellular health has led to the popular belief that taking high doses of antioxidants as dietary supplements could slow the cellular aging process and even the course of certain diseases, such as cancer. That's why there has been so much hype lately about the alleged health benefits of taking antioxidant vitamins or eating antioxidant-rich foods. (Try searching the term *antioxidants* on the Internet and see how many hits you get.) But a word of caution should be interjected here. Animal studies have been encouraging but not definitive, and the few human studies have yielded mixed results. In its fact sheet on cancer and antioxidants, the National Cancer Institute describes in some detail how antioxidants *might* prevent cancer but stops short of recommending antioxidant supplements for cancer treatment.[1]

So should you be taking antioxidant supplements, just in case? It's an open question with no real answer. The evidence is not yet convincing that you will live a healthier or longer life, but on the other hand there is no real evidence that moderate doses of antioxidants do any harm, either. Perhaps time and more research will tell us more, but we're just not there yet.

---

[1]www.cancer.gov/cancertopics/factsheet/prevention/antioxidants

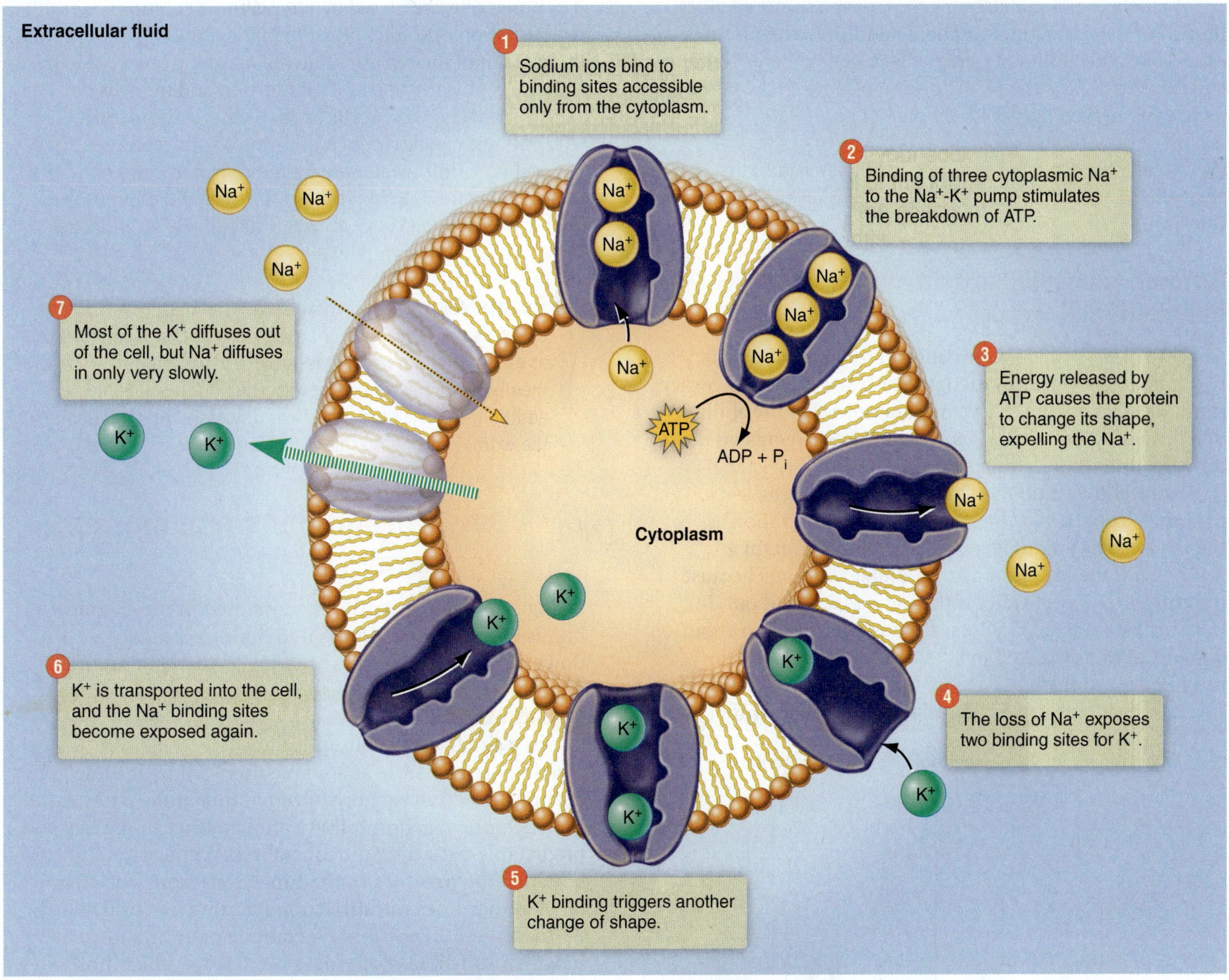

Extracellular fluid

1 Sodium ions bind to binding sites accessible only from the cytoplasm.

2 Binding of three cytoplasmic Na+ to the Na+-K+ pump stimulates the breakdown of ATP.

7 Most of the K+ diffuses out of the cell, but Na+ diffuses in only very slowly.

3 Energy released by ATP causes the protein to change its shape, expelling the Na+.

ATP

ADP + P_i

Cytoplasm

6 K+ is transported into the cell, and the Na+ binding sites become exposed again.

4 The loss of Na+ exposes two binding sites for K+.

5 K+ binding triggers another change of shape.

a) **The cell membrane contains Na+-K+ pumps, and also channels that permit the rapid outward diffusion of K+ but only a slow inward diffusion of Na+.**

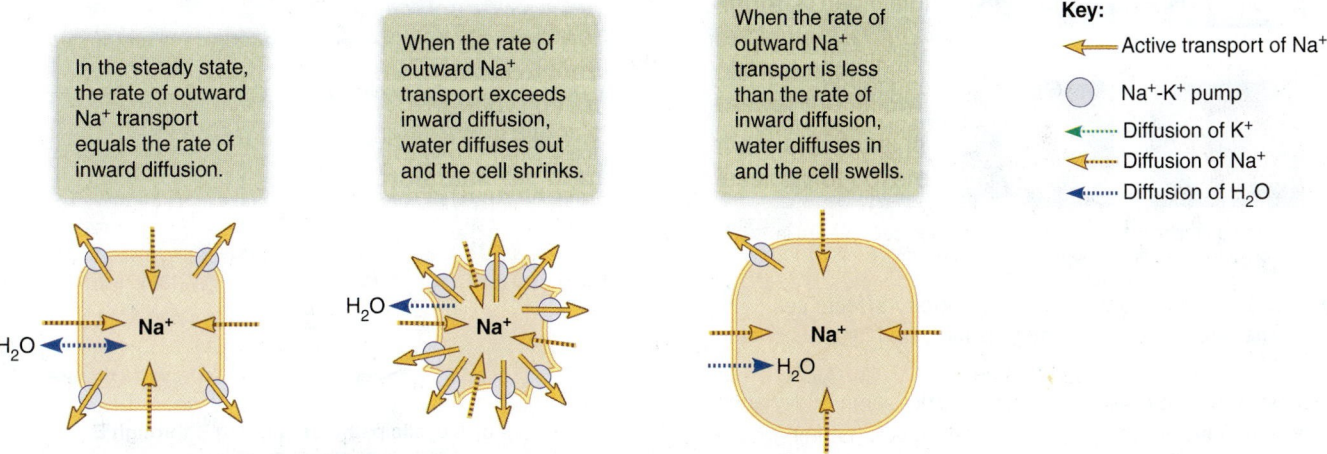

In the steady state, the rate of outward Na+ transport equals the rate of inward diffusion.

When the rate of outward Na+ transport exceeds inward diffusion, water diffuses out and the cell shrinks.

When the rate of outward Na+ transport is less than the rate of inward diffusion, water diffuses in and the cell swells.

Key:

↞ Active transport of Na+

◯ Na+-K+ pump

◁······ Diffusion of K+

◁······ Diffusion of Na+

◁······ Diffusion of H₂O

b) **The rate of transport by the Na+-K+ pumps determines cell volume.**

Figure 3.20 Control of cell volume by the sodium-potassium pump (Na+-K+ pump).

A single red blood cell may have over a hundred N^+-K^+ pumps in its plasma membrane. In addition, the pump is essential to the ability of nerve cells to generate an electric current. We discuss the N^+-K^+ pump in more detail when we review the nervous system (Chapter 11).

☑ If a cell's sodium-potassium pumps are poisoned so that they stop working, will the cell tend to swell, shrink, or stay the same? Explain.

Isotonic extracellular fluid also maintains cell volume

Tonicity refers to the relative concentrations of solutes in two fluids (*tonic* means "strength" or "tone"). Because water can diffuse across the cell membrane so easily, the ability of a human cell to control its volume also depends on the tonicity of the extracellular fluid.

Extracellular fluid that is **isotonic** (Greek *isos*, meaning "equal") has the same solute concentration as the *intracellular* fluid (**Figure 3.21a**). Cells maintain a normal volume in isotonic extracellular solutions because the concentration of water is the same inside and out. In humans, isotonic extracellular fluid is equivalent to about 9 grams of salt dissolved in a liter of solution. Regulatory mechanisms in the body ensure that the extracellular fluid solute concentration remains relatively constant at that level.

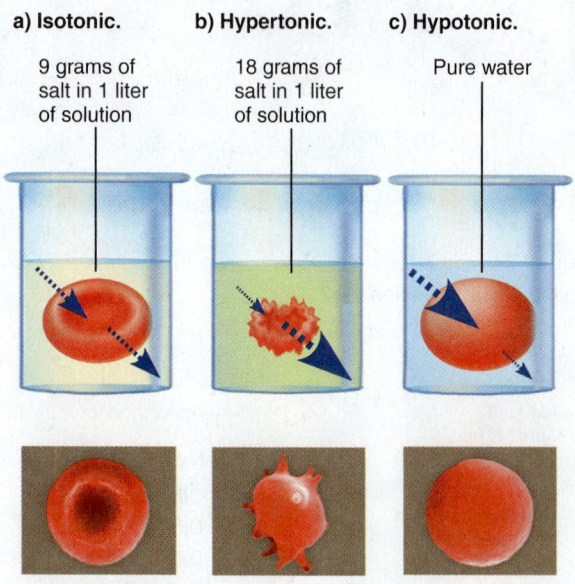

a) **Isotonic.**
9 grams of salt in 1 liter of solution

b) **Hypertonic.**
18 grams of salt in 1 liter of solution

c) **Hypotonic.**
Pure water

Scanning electron micrographs of red blood cells placed in isotonic, hypertonic, or hypotonic solutions.

Figure 3.21 How extracellular fluid tonicity affects cell volume. The rate of water movement is indicated by the sizes of the arrows.

☑ Suppose a woman runs a marathon on a hot summer day and becomes extremely dehydrated. Would you expect that her extracellular fluid would be isotonic, hypotonic, or hypertonic? Predict what might happen to her red blood cells.

When cells are placed in a **hypertonic** solution, one with a concentration of solutes *higher* than the intracellular fluid, water diffuses out of the cells, and the cells shrink (**Figure 3.21b**). Eventually, this impairs normal function and the cells die.

Conversely, when cells are placed in a **hypotonic** solution with a *lower* concentration of solutes than intracellular fluid, water enters the cells and causes them to swell (**Figure 3.21c**). Pure water is the most hypotonic solution possible. Most human cells quickly swell, burst, and die when placed in pure water.

↩ **Recap** Molecules may move across the plasma membrane by diffusion, by passive or active transport, or by endocytosis or exocytosis. Sodium-potassium exchange pumps in the cell membrane are essential for the regulation of cell volume. In addition, homeostatic regulatory processes keep the tonicity of the extracellular fluid relatively constant. ∎

3.7 Cells use and transform matter and energy

Living cells can release the energy stored in the chemical bonds of molecules and use it to build, store, and break down still other molecules as required to maintain life. **Metabolism** is the sum of all of the chemical reactions in the organism.

In a single cell, thousands of different chemical reactions are possible at any one time. Some of these chemical reactions are organized as metabolic pathways in which one reaction follows after another in orderly and predictable patterns. Some metabolic pathways are linear, in which the **product** (or end material) from one chemical reaction becomes the **substrate** (starting material) for the next (**Figure 3.22a**). Other metabolic pathways form a cycle in which substrate molecules enter and product molecules exit, but the basic chemical cycle repeats over and over again (**Figure 3.22b**).

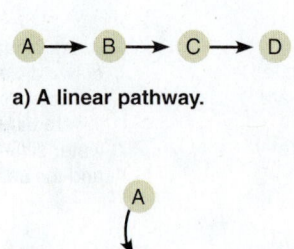

a) **A linear pathway.**

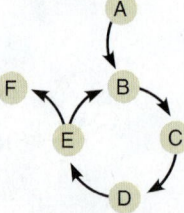

b) **A cyclic pathway in which B through E repeat over and over again.**

Figure 3.22 Types of metabolic pathways.

There are two basic types of metabolic pathways:

1. *Anabolism* (from the Greek *anabole*, meaning "a throwing up"). Molecules are assembled into larger molecules that contain more energy, a process that requires energy. The assembly of a protein from many amino acids is an example of an anabolic pathway.
2. *Catabolism* (from the Greek *katabole*, meaning "a throwing down"). Larger molecules are broken down, a process that releases energy. The breakdown of glucose into water, carbon dioxide, and energy is an example of a catabolic pathway.

Two facts are important about metabolic pathways. First, nearly every chemical reaction requires a specific enzyme. The cell regulates and controls the rates of chemical reactions through the specificity and availability of key enzymes. Second, the metabolic activities of a living cell require a lot of energy. Energy is required for building the complex macromolecules found only in living organisms, such as proteins, DNA, cholesterol, and so on. Energy is also used to power cellular activities such as active transport and muscle contraction.

Cells get their energy by catabolism of molecules that serve as chemical stores of energy. The most immediately useful source of energy, a sort of "energy cash" if you will, is ATP. The energy in ATP is locked in the chemical bond between the second and third phosphate group. Every time the third phosphate group is removed from an ATP molecule, energy is released that the cell can use to do work. The reaction is reversible, meaning that the application of energy to ADP in the presence of a phosphate group can *phosphorylate* (add a phosphate group to) ADP again, re-creating ATP. The equation is written as:

$$ATP \leftrightarrow ADP + P_i + energy$$

In this equation, P_i is used as the abbreviation for the inorganic phosphate (PO_4^{-2}) to distinguish it from the chemical symbol for pure phosphorus (P).

Glucose provides the cell with energy

Cells can use a variety of fuels to make the ATP energy "cash" they need. The most readily available fuel generally is glucose, derived either from food recently eaten or from stored glycogen. However, if glucose is not available, cells may turn to stored fats and even proteins for fuel. Regardless of the fuel used, most of the ATP is produced by very similar metabolic pathways. Let's look at the use of glucose as a fuel source first, and then we'll describe briefly how other fuels are used.

Recall that glucose is a six-carbon sugar molecule with the chemical formula $C_6H_{12}O_6$. This seemingly simple little molecule packs a lot of potential energy in its chemical bonds. Just as a gallon of gas provides enough energy to power your car for 25 to 30 miles, a single glucose molecule provides the

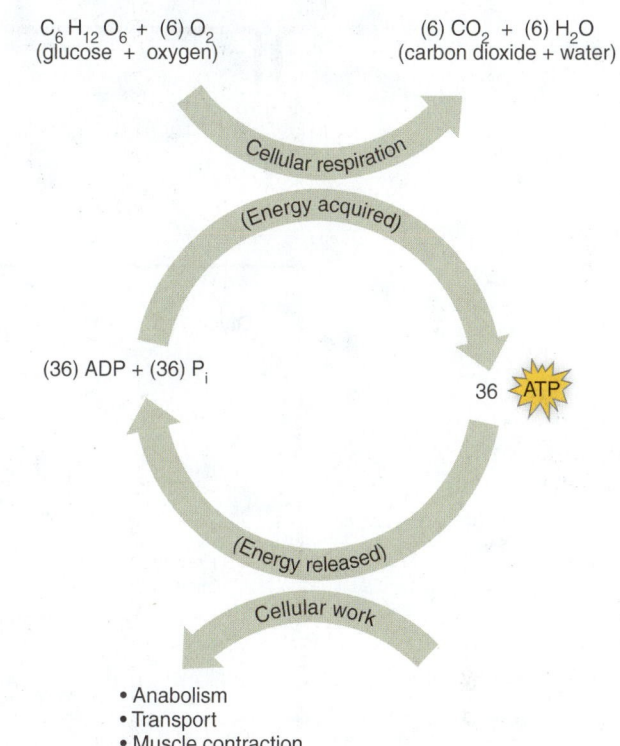

Figure 3.23 Glucose provides energy for the cell. The complete catabolism of glucose uses oxygen, produces carbon dioxide and water, and generates 36 molecules of ATP that can be used to do cellular work.

cell enough energy to produce approximately 36 molecules of ATP (**Figure 3.23**). The production of ATP from glucose occurs in four stages, collectively called cellular respiration:

1. *Glycolysis.* The six-carbon glucose molecule is split into two 3-carbon pyruvate molecules. Energy is required to get the process started.
2. *The preparatory step.* In preparation for the citric acid cycle, pyruvate enters a mitochondrion. A series of chemical reactions yields a two-carbon molecule called *acetyl CoA*, plus some energy.
3. *The citric acid cycle.* An acetyl CoA molecule is broken down completely by mitochondrial enzymes, and its energy is released. Most of the energy is captured by certain high-energy electron transport molecules.
4. *The electron transport system.* Most of the energy derived from the original glucose molecule is used to phosphorylate ADP, producing high-energy ATP.

Glycolysis: glucose is split into two pyruvate molecules

Glycolysis is the first stage in the complete breakdown of glucose (*glyco-* means "sweet," referring to sugar, and *-lysis* means "to break"). Glycolysis occurs within the cell's cytoplasm, not within mitochondria.

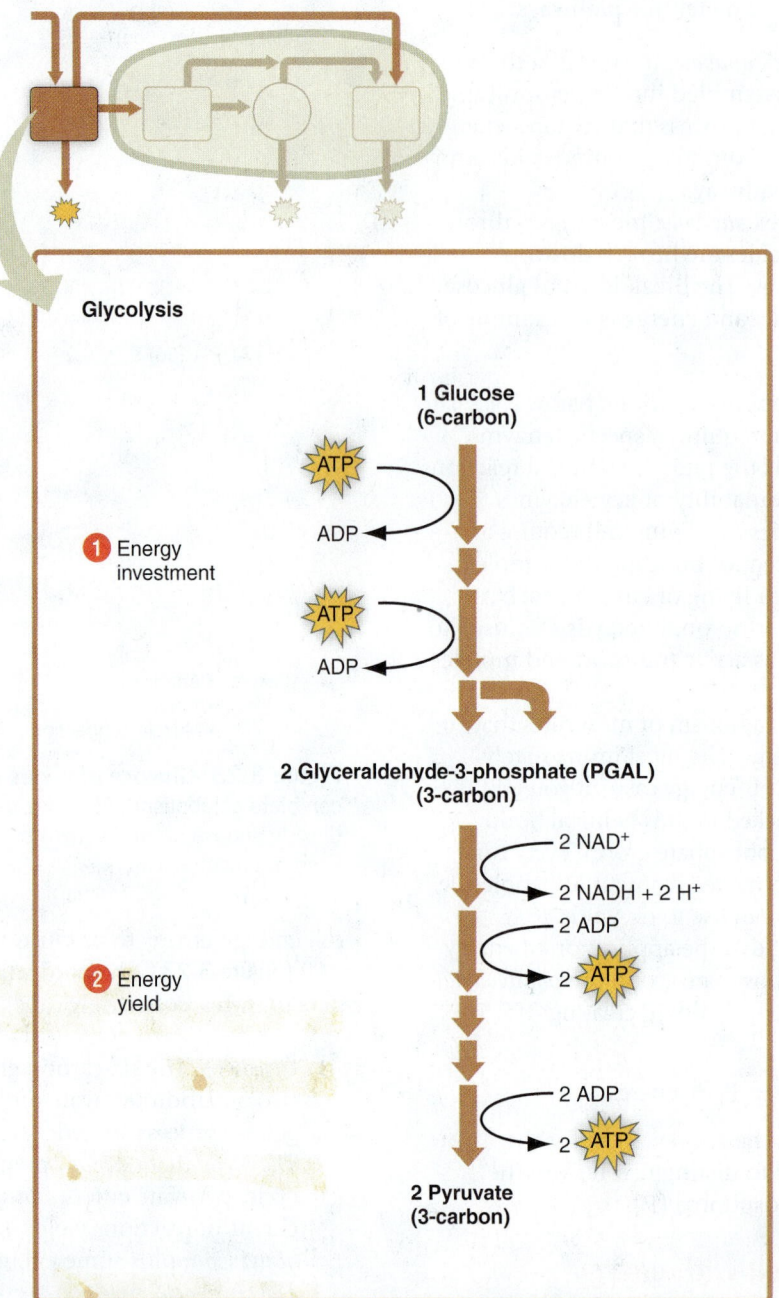

Figure 3.24 Glycolysis. This initial breakdown of glucose to two molecules of glyceraldehyde-3-phosphate (PGAL) requires energy. Thereafter, energy is generated as the two PGAL molecules are further degraded to two molecules of pyruvate.

Glycolysis occurs in two steps (Figure 3.24): ❶ In the *energy investment step*, energy derived from two molecules of ATP initiates a chemical reaction—like putting a match to a bonfire—in which glucose, a six-carbon molecule, is split into two molecules of *glyceraldehyde-3-phosphate* (PGAL), each of which receives a high-energy phosphate group from an ATP molecule. ❷ In the *energy yielding step*, the two molecules of PGAL are broken down to two molecules of *pyruvate* by enzymes within the cytoplasm. The process yields four molecules of ATP plus some additional energy in the form of

high-energy hydrogen ions (H^+) and electrons (e^-). These are picked up by a *coenzyme* (a coenzyme is a small molecule that assists an enzyme by transporting small molecules) called nicotinamide adenine dinucleotide (**NAD$^+$**). NAD$^+$ picks up one high-energy H^+ ion and two high-energy electrons to become the higher-energy molecule **NADH.**

So far, then, the net energy yield from glycolysis is only two molecules of ATP (four were formed, but two were used to get the process started) and two molecules of NADH. The energy carried by NADH is used to make ATP within the mitochondria.

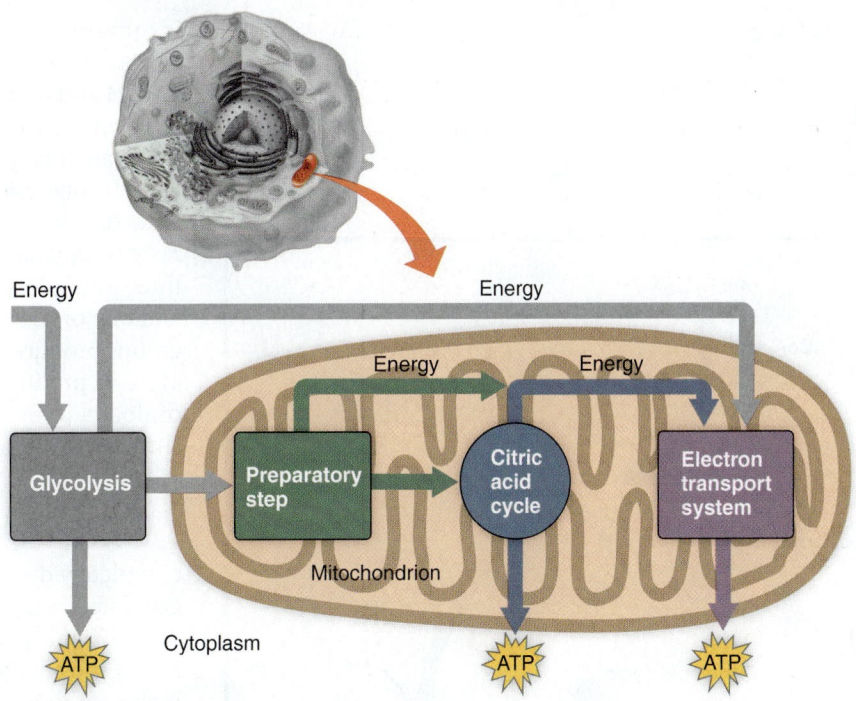

Figure 3.25 Aerobic respiration: an overview. The products of glycolysis, two molecules of pyruvate, enter mitochondria. The preparatory step and the citric acid cycle result in the complete breakdown of the two pyruvate molecules but only limited ATP production.

Aerobic respiration uses oxygen

Although some ATP is made in the cytoplasm during glycolysis, glycolysis does not require oxygen. Cellular metabolic processes that use oxygen and produce carbon dioxide in the process of making ATP are collectively termed **aerobic respiration.** Aerobic respiration takes place entirely within mitochondria and includes the preparatory step, the citric acid cycle, and electron transport system (Figure 3.25).

The preparatory step: pyruvate is converted to acetyl CoA
After glycolysis, the two pyruvate molecules enter the mitochondria, where all the rest of the ATP-generating reactions occur. In the preparatory step (Figure 3.26), ❶ each pyruvate molecule is converted to a two-carbon *acetyl group*. A carbon dioxide (CO_2) waste molecule and another high-energy NADH molecule is generated from each pyruvate. ❷ Each acetyl group is then picked up by a coenzyme called *coenzyme A* to form acetyl CoA for delivery to the *citric acid cycle*. The preparatory step results in a net gain of two NADH since two pyruvate molecules were generated from a single glucose molecule during glycolysis.

The citric acid cycle harvests energy The **citric acid cycle,** also called the *Krebs cycle* for its discoverer, Hans Krebs, is a series of steps in which each acetyl group is completely disassembled to CO_2 waste and various high-energy products. Each reaction is regulated by a different enzyme. The citric acid cycle begins when the two-carbon acetyl group from the

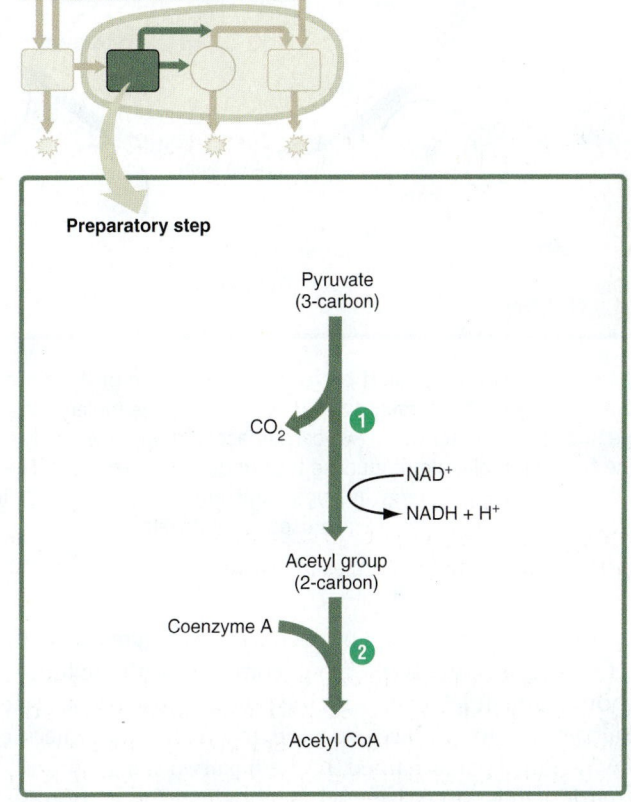

Figure 3.26 The preparatory step. Pyruvate is transported into a mitochondrion and catabolized to a two-carbon acetyl group. Energy is released, and the freed carbon is given off as carbon dioxide waste. The acetyl group combines with coenzyme A for delivery to the next step, the citric acid cycle.

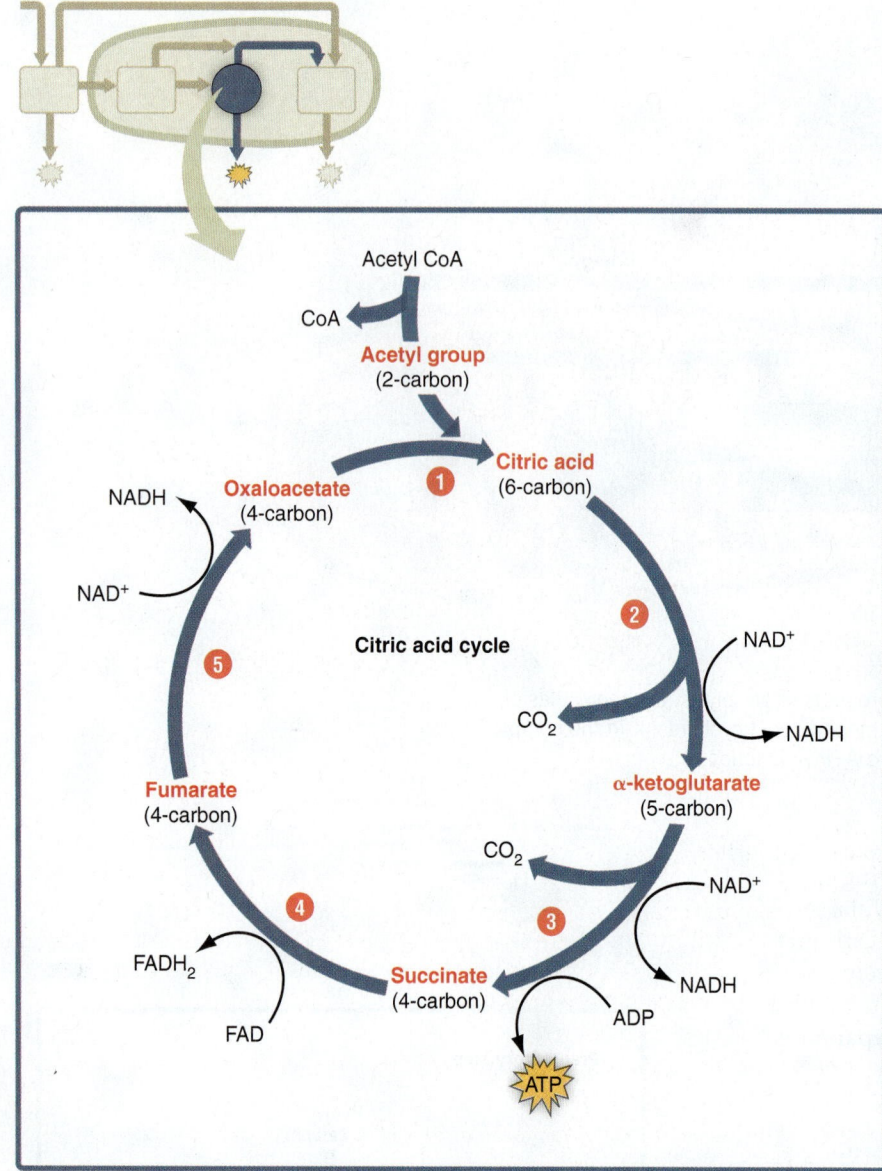

Figure 3.27 The citric acid cycle. A complete turn of the citric acid cycle produces two molecules of CO_2 waste, an ATP molecule, three molecules of NADH, and a molecule of $FADH_2$ for every two-carbon acetyl group used as fuel. The cycle occurs twice for each molecule of glucose that undergoes glycolysis. The NADH and $FADH_2$ molecules carry high-energy hydrogen ions and electrons to the electron transport system, where their energy is also used to synthesize ATP.

is converted first to *fumarate* and finally to *oxaloacetate*. Energy in the form of one NADH and one $FADH_2$ (the hydrogen- and electron-carrying form of another coenzyme, flavin adenine dinucleotide) are produced at each conversion, but there is no further reduction in the number of carbons. The net energy gain from the citric acid cycle is three molecules of NADH, one of $FADH_2$, and one of ATP for each of the two acetyl groups produced from one glucose molecule. The end product of the citric acid cycle, oxaloacetate, is available to participate in the citric acid cycle again.

The electron transport system produces ATP So far, the glucose molecule has been completely dismantled and CO_2 has been generated as a waste product, but only four new ATP molecules have been generated. The rest of the energy is still in the electrons and hydrogen ions that are part of NADH and $FADH_2$.

At this point, NADH and $FADH_2$ move to the inner membrane of the mitochondria and release their cargo to the **electron transport system** (Figure 3.28). Here, the energy-rich electrons are transferred sequentially from one protein carrier molecule to another. The sequential transfer is important because it allows the energy in the electrons to be released in manageable quantities. The sequence of events is as follows:

1. NADH and $FADH_2$ release the H^+ and high-energy electrons they acquired in the citric acid cycle to a carrier protein of the electron transport system.
2. The electrons pass from one protein carrier molecule to the next in the electron transport system.
3. Each time an electron is transferred, the carrier molecule acquires some of its energy and the electron loses energy. The carrier protein uses the energy to transport H^+ from the inner compartment of the mitochondria to the outer compartment.

The active transport of H^+ into the outer compartment of the mitochondria sets the stage for the actual production of ATP. Because the concentration of H^+ is now higher in the outer compartment than in the inner compartment, there is a concentration gradient that favors diffusion of H^+ back to the inner compartment. However, H^+ can diffuse only through special channels. These channels are actually an enzyme called **ATP synthase** that uses the

preparatory step is released by coenzyme A (Figure 3.27). In the first step, ①, the acetyl group combines with the four-carbon fragment left over from the previous turn of the cycle (oxaloacetate) to produce *citric acid*, the six-carbon molecule for which the cycle is named. In the next two steps, ② and ③, citric acid is successively broken down to five-carbon *α-ketoglutarate* and then four-carbon *succinate*. Energy released by these two steps is stored in the form of two molecules of NADH and one of ATP. Two atoms of carbons are released as CO_2 waste. In the last two steps, ④ and ⑤, succinate

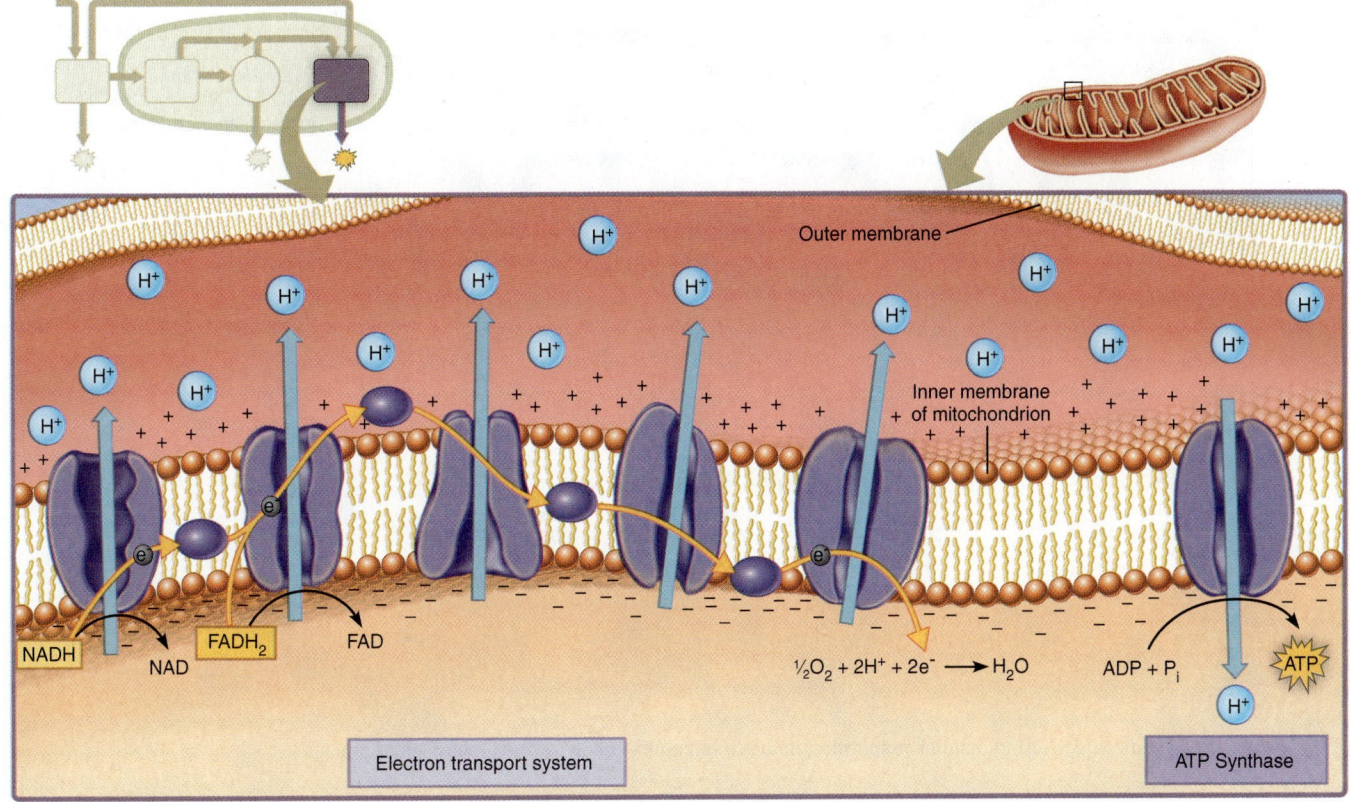

Figure 3.28 The electron transport system and oxidative phosphorylation. During electron transfer, the high-energy molecules of NADH (and $FADH_2$) give up electrons and hydrogen ions, releasing energy. The energy is used to transport hydrogen ions into the space between the two mitochondrial membranes. Diffusion of hydrogen ions back into the inner mitochondrial space provides the energy for the enzyme ATP synthase to synthesize ATP from ADP and inorganic phosphate (P_i). The process of using energy derived from the electron transport system to produce ATP by phosphorylation of ADP is called oxidative phosphorylation.

☑ Explain in your own words what exactly is causing the ATP synthase to make ATP. Do NADH or $FADH_2$ interact directly with the ATP synthase?

energy derived from the diffusion of H^+ to catalyze the synthesis of ATP from ADP and P_i. Once it is formed, ATP leaves the inner compartment of the mitochondria via a special channel protein, to be used by the cell as an energy source.

As mentioned earlier, *phosphorylation* is the addition of a phosphate group. The process of producing ATP from ADP plus P_i, using the energy obtained as electrons are transferred from one molecule to another in the electron transport system, is called **oxidative phosphorylation.** The term *oxidative* indicates that the process uses oxygen and that electrons have been removed.

By the time they have reached the end of the electron transport system, most of the energy of the electrons and H^+ has been spent. At this point, the spent hydrogen ions and electrons combine with oxygen to form water, a waste product. The low-energy NAD^+ and FAD molecules, now lacking hydrogen ions and electrons, are recycled and used

again. The ability to recycle NAD^+, FAD, and ADP increases cellular efficiency because it means that these do not need to be synthesized anew each time.

☑ If ATP synthase completely stopped working, could glycolysis or the citric acid cycle still produce any ATP? Explain.

Summary of energy production from glucose The most effective way to harvest the energy in any fuel, whether gasoline or glucose, is to release the energy *slowly*, under carefully controlled conditions. Holding a match to a gallon of gasoline results in a useless fire or even a potentially dangerous explosion, but burning it drop by drop in the pistons of your car allows the energy to be converted to mechanical work. Similarly, in a living cell, the whole point of glycolysis and aerobic respiration is to release the energy in the chemical bonds of glucose *slowly* so that the energy can be harvested effectively.

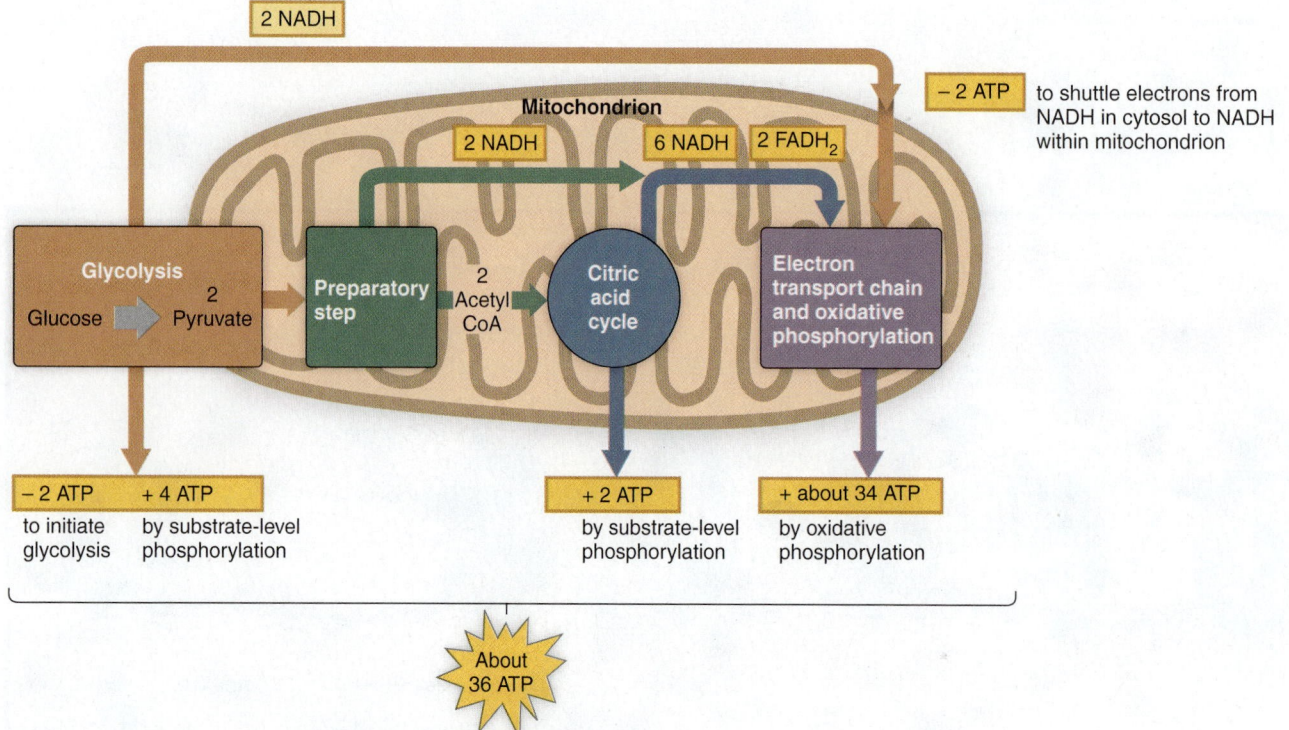

a) Most of the ATP generated during cellular respiration is synthesized in the electron transport system.

b) Cellular respiration powers the activities of humans and many other organisms.

Figure 3.29   Cellular respiration: a recap.

That is exactly what the cell does (**Figure 3.29**). The complete breakdown of glucose during glycolysis, the preparatory step, and the citric acid cycle are accomplished by a sequence of over 20 different chemical reactions, each controlled by an enzyme. The net yield of high-energy molecules (prior to the electron transport chain) is four molecules of ATP, 10 molecules of NADH, and two molecules of $FADH_2$. Each NADH carries sufficient energy to produce about three ATP molecules in the electron transport system. $FADH_2$ carries less energy, enters the electron transport chain at a lower energy point, and so produces only about two ATP

molecules per $FADH_2$ molecule. Add it all up and you should get about 38 ATP molecules per glucose molecule.

But there's a catch. Two of the NADH (the two produced during glycolysis) were produced in the cytoplasm. Yet the electron transport system utilizes NADH located in the inner compartment of the mitochondria. Because the inner mitochondrial membrane is impermeable to NADH, the cytosolic NADH molecules release their H^+ ions and electrons, which are then transported across by transport proteins. Once inside, they may be picked up again by other NAD^+ molecules. But every transfer and active transport

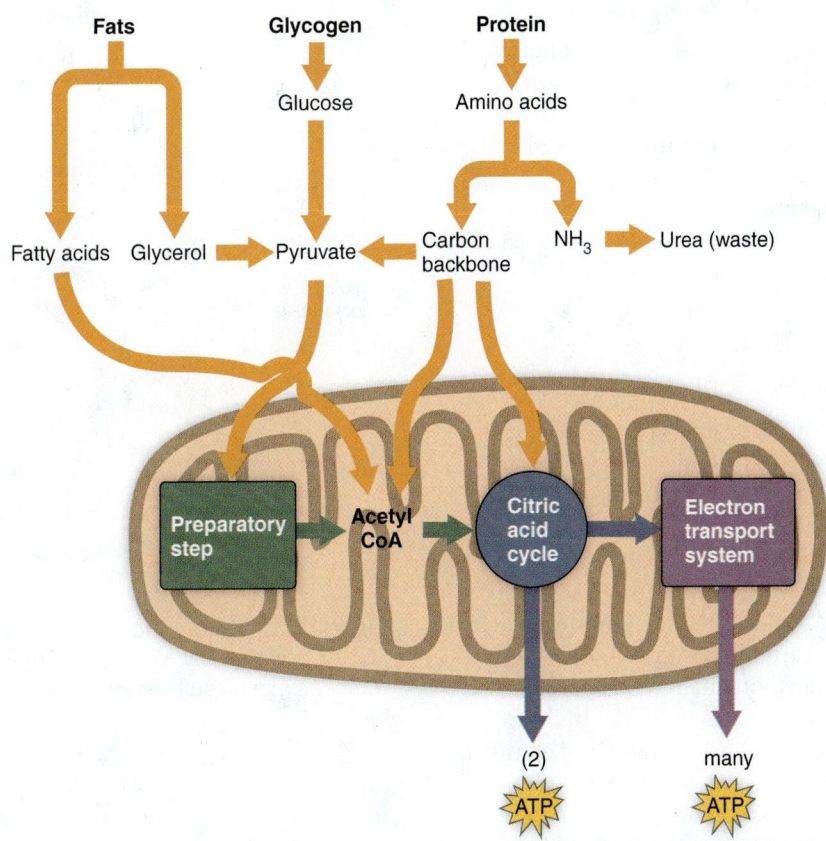

Figure 3.30 Metabolic pathways for fats, glycogen, and proteins as sources of cellular energy. All three sources produce pyruvate and acetyl groups. In addition, a few components of protein can enter the citric acid cycle directly.

results in a small energy loss—in this case, it's the equivalent of about two ATP molecules. Therefore, the net maximum yield of energy from a single glucose molecule is closer to 36 ATP molecules. In fact, it may be even lower than that because some of the energy may be used directly to do other work for the cell, but for the sake of simplicity we'll settle on a number of 36.

Fats and proteins are additional energy sources

So far, we have concentrated on the cellular catabolism of glucose. Normally, the blood glucose concentration remains fairly constant even between meals because glycogen (the storage form of glucose in humans) is constantly being catabolized to replenish the glucose that is used by cells.

However, most of the body's energy reserves do not take the form of glycogen. In fact, the body stores only about 1% of its total energy reserves as glycogen; about 78% are stored as fats and 21% as protein. After glycogen, our bodies may utilize fats and proteins. **Figure 3.30** illustrates the metabolic pathways for fat, glycogen, and protein.

Energy is constantly being transferred into and out of the body. Immediately after a meal, when plenty of glucose, lipids, and amino acids are readily available in

the blood, we tend to use primarily glucose as an energy source. When we eat more calories than we can use immediately, some of the excess energy goes to replenish the body's stores of glycogen and the rest is converted to fat and stored in fat tissue. After we have not eaten for hours, the body uses first glycogen, then fats, and eventually even proteins for energy.

Pound for pound, fats carry more than twice the energy of carbohydrates such as glycogen. During fat catabolism, triglycerides (the storage form of fat) are first broken down to glycerol and fatty acids. The glycerol can be converted to glucose in the liver or it can be converted to pyruvic acid, which then enters the citric acid cycle. Enzymes break down the fatty acid tails to two-carbon acetyl groups, which also enter the citric acid cycle. Each molecule of triglyceride yields a great deal of ATP because the fatty acid tails are generally 16 to 18 carbons long and so generate many acetyl groups.

Proteins carry about the same amount of energy per pound as glycogen. Proteins are first broken down to amino acids, and the amine group ($-NH_2$) of each amino acid is removed. The carbon backbones then enter the citric acid cycle at various points, depending on the specific amino acid. The amine groups are converted to urea by the liver and then excreted in urine as waste.

Proteins serve primarily as enzymes and structural components of the body, not as a stored form of energy. Nevertheless, protein catabolism increases significantly during starvation. Prolonged protein catabolism causes muscle wasting, but at least it keeps the individual alive.

☑ If a person wants to lose weight, why is it better to reduce caloric intake moderately over a long period of time, rather than to just stop eating until the target weight is achieved?

Anaerobic pathways make energy available without oxygen

As we have seen, cellular respiration requires oxygen to complete the chemical reactions of the citric acid cycle and the electron transport chain. However, a small amount of ATP *can* be made in humans by anaerobic metabolism (without oxygen) for at least brief periods of time. Glycolysis, for example, is an anaerobic metabolic pathway. In the absence of oxygen, glucose is broken down to pyruvate (glycolysis occurs), but then the pyruvate cannot proceed through the citric acid cycle and electron transport chain (**Figure 3.31**). Instead, the pyruvate is converted to *lactic acid*. The buildup of lactic acid is what causes the burning sensation and cramps associated with muscle fatigue when not enough oxygen is available to muscle tissue. When oxygen becomes available again, the lactic acid is metabolized by aerobic pathways.

Because glycolysis is the only step that can occur without oxygen, glucose (and glucose derived from glycogen) is the only fuel that can be used under anaerobic conditions. The amounts of ATP are very limited, however; only two molecules of ATP are produced per molecule of glucose instead of the usual 36.

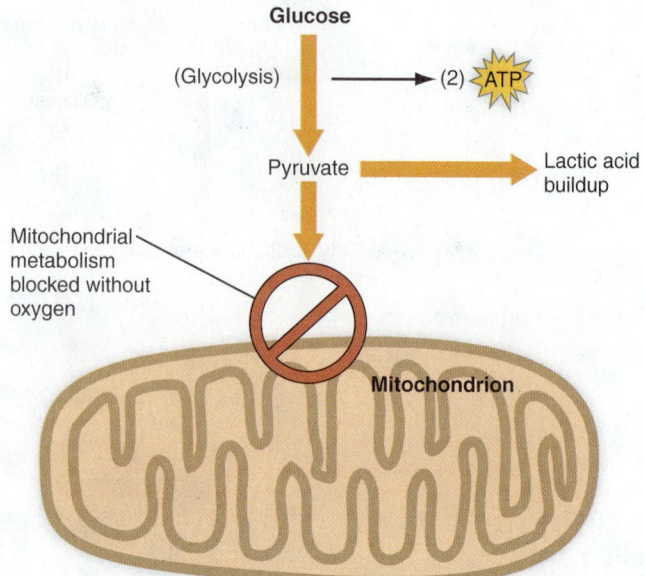

Figure 3.31 Anaerobic metabolism. In the absence of oxygen, glycolysis is the only ATP-producing step available. Glycolysis without oxygen results in lactic acid buildup.

↩ **Recap** Metabolism refers to all of a cell's chemical processes. Metabolic pathways either create molecules and use energy (anabolism) or break them down and liberate energy (catabolism). The primary source of energy for a cell is ATP, produced within mitochondria by the complete breakdown of glucose to CO_2 and water. The process requires oxygen. A single molecule of glucose yields about 36 molecules of ATP. Fats and proteins can also be used to produce energy if necessary. ∎

Chapter Summary

3.1 Cells are classified according to their internal organization *p. 49*

- All cells have a plasma membrane that surrounds and encloses the cytoplasm.
- Eukaryotic cells have a nucleus.

3.2 Cell structure reflects cell function *p. 50*

- Limits to cell size are imposed by the mathematical relationship between cell volume and cell surface area.
- Various types of microscopes with magnifications up to 100,000-fold enable us to visualize cells and their structures.

3.3 Internal structures carry out specific functions *p. 52*

- The nucleus directs all of the cell's activities.
- Ribosomes, the endoplasmic reticulum, and the Golgi apparatus participate in the synthesis of life's molecules.
- Vesicles are membrane-bound spheres that transport, store, and ship cellular products and toxic or dangerous materials.

- Mitochondria make energy available for the cell in the form of the high-energy molecule ATP.

3.4 Cells have structures for support and movement *p. 57*

- A cytoskeleton of microtubules and microfilaments serves as structural support and anchors the various organelles.
- Cilia and flagella provide for movement in certain types of cells. Both cilia and flagella are made of pairs of protein microtubules.

3.5 A plasma membrane surrounds the cell *p. 58*

- The plasma membrane is a bilayer of phospholipids that also contains cholesterol and various proteins.

3.6 Molecules cross the plasma membrane in several ways *p. 59*

- Some molecules are transported across the plasma membrane passively (by diffusion), whereas others are transported by active processes requiring the expenditure of energy.
- Receptor proteins transfer information across the plasma membrane.
- The sodium-potassium pump is a plasma membrane protein with a critical role in the maintenance of cell volume.

3.7 Cells use and transform matter and energy *p. 66*

- The creation and destruction of molecules either requires energy or liberates energy.
- The most readily useful form of energy for cells is ATP.
- The production of ATP from glucose requires four consecutive stages: glycolysis, a preparatory step, the citric acid cycle, and the electron transport system.
- Cells can utilize glycogen, fats, or proteins for energy.
- Only a small amount of ATP can be made in the absence of oxygen.

Terms You Should Know

active transport, *62*

ATP synthase, *70*

citric acid cycle, *69*

cytoskeleton, *57*

diffusion, *59*

electron transport system, *70*

endocytosis/exocytosis, *62*

endoplasmic reticulum (ER), *54*

glycolysis, *67*

Golgi apparatus, *54*

metabolism, *66*

mitochondria, *56*

NAD$^+$, *68*

NADH, *68*

osmosis, *60*

passive transport, *59*

plasma membrane, *49*

ribosomes, *54*

sodium-potassium pump, *62*

vesicle, *62*

Concept Review

Answers can be found in the Study Area in MasteringBiology.

1. Explain why being small is advantageous to a cell.
2. List the basic tenets of the *cell doctrine*.
3. Describe how phospholipids are oriented in the plasma membrane and why they orient naturally that way.
4. Define *passive transport* and name the three passive transport methods that are used to transport different molecules across the plasma membrane.
5. Compare and contrast endocytosis and exocytosis.
6. Describe the activity of the sodium-potassium pump and indicate its importance to the cell.
7. Explain what happens to cells placed in a high-salt or a low-salt environment and why.
8. Define *vesicles*, and name at least two different types of vesicles.
9. What are the four stages of ATP production from glucose, and which one yields the most ATP?
10. Describe what happens to a cell's ability to produce ATP when oxygen is not available.

Test Yourself

Answers can be found in the Appendix.

1. Which of the following adaptations would increase the surface area of a cell?
   a. increased number of mitochondria
   b. increased number of ribosomes
   c. presence of microvilli
   d. increased number of channel proteins
   e. changing a cell shape from cubical to spherical

2. Which of the following is always at the same concentration inside and outside the cell?
   a. potassium ions
   b. water
   c. hydrogen ions
   d. glucose
   e. sodium ions

3. Cells transport sodium ions out of the cell against the sodium concentration gradient. This is an example of:
   a. facilitated diffusion
   b. simple diffusion
   c. diffusion via channel proteins
   d. endocytosis
   e. active transport

4. Red blood cells placed in distilled water will:
   a. swell as water moves into the cells by osmosis
   b. shrink as sodium moves out of the cells by diffusion
   c. swell as water moves into the cells by active transport
   d. shrink as proteins move out of the cell by diffusion
   e. remain unchanged because of homeostatic mechanisms

5. Which organelles are most active during vigorous exercise?
   a. ribosomes
   b. endoplasmic reticulum
   c. mitochondria
   d. lysosomes
   e. cilia

6. Phagocytic white blood cells engulf and digest bacteria and cellular debris. Which organelles would be most involved in the digestion of the engulfed material?
   a. mitochondria
   b. lysosomes
   c. Golgi apparatus
   d. ribosomes
   e. endoplasmic reticulum

7. Some lymphocytes (white blood cells) synthesize and secrete defensive proteins known as antibodies. Which of the following represents the most likely path of these proteins from synthesis to secretion?
   a. endoplasmic reticulum—ribosomes—Golgi apparatus—vesicles—plasma membrane
   b. ribosomes—Golgi apparatus—vesicles—endoplasmic reticulum—plasma membrane
   c. ribosomes—Golgi apparatus—lysosome—endoplasmic reticulum—plasma membrane
   d. ribosomes—endoplasmic reticulum—Golgi apparatus—vesicles—plasma membrane
   e. ribosomes—lysosomes—endoplasmic reticulum—vesicles—Golgi apparatus

8. Which organelles would be active in liver cells that are detoxifying alcohol?
   a. Golgi apparatus
   b. lysosomes
   c. mitochondria
   d. endoplasmic reticulum
   e. peroxisomes

9. Cells lining the respiratory passages have numerous filamentous structures that sweep mucus and debris up and away from the lungs. These filamentous structures are:
   a. microtubules
   b. flagella
   c. cilia
   d. microfilaments
   e. microvilli

10. Which of the following is/are the most immediate source of energy for cellular work?
    a. glucose
    b. ATP
    c. glycogen
    d. triglycerides
    e. amino acid

11. All of the following cellular activities require energy except:
    a. facilitated transport
    b. active transport
    c. movement
    d. protein synthesis
    e. cell division

12. Which of the following can occur within a cell in the absence of oxygen?
    a. glycolysis
    b. lactate production
    c. citric acid cycle
    d. electron transport system
    e. both (a) and (b)

13. In which stage of cell respiration does oxygen play a role?
    a. glycolysis
    b. electron transport system
    c. citric acid cycle
    d. lactate production
    e. transport of pyruvate into the mitochondria

14. Which of the following is/are recycled during cellular respiration and do not appear in the net equation?
    a. NAD^+ and NADH
    b. ATP, ADP, and P_i
    c. glucose
    d. carbon dioxide
    e. oxygen

15. Most of the ATP produced during cell respiration is a by-product of:
    a. glycolysis
    b. citric acid cycle
    c. electron transport system
    d. lactate production
    e. anaerobic pathways

Apply What You Know

Answers can be found in the Study Area in MasteringBiology.

1. Imagine that you are shown two cells under the microscope. One is small, has lots of mitochondria, and contains numerous glycogen granules. The other is somewhat larger and has only a few mitochondria and no glycogen granules. Which cell do you think is more metabolically active? Explain your reasoning.

2. The sodium-potassium pump is a large protein molecule. Where do you think the sodium-potassium pumps are made in the cell, and how do you think they become inserted into the lipid bilayer of the plasma membrane?

3. Mitochondria resemble a bacterial cell in a number of ways. Some scientists hypothesize that mitochondria evolved from aerobic prokaryotes that were engulfed by anaerobic eukaryotes, and now both have evolved together in a mutually advantageous way. Can you think of an explanation for why it might have been advantageous for both cells to enter into such an arrangement?

4. You have decided that you need to lose a little weight. You have heard a lot about no-carbohydrate and low-carbohydrate diets, and you have decided to use one of these diet plans. Explain how a low-carbohydrate diet works. Can you think of any possible negative side effects of such a diet?

5. Recently, a young man from Derby in the United Kingdom entered a contest and drank 26 pints of water in a very short time. He later died of complications due to hypotonic hydration, also known as water intoxication. How were his body's cells affected by the excess water, and how might that have contributed to his death?

6. You have been selected to serve on a jury for a trial involving a young man accused of public intoxication. His defense attorney argues that the alcohol found in his system was the result of natural fermentation, that he had just finished a grueling one-hour workout during which his body could not meet the oxygen demand, and that the excess lactic acid that was produced during the exercise was then converted to alcohol by the process of lactic acid fermentation. Should you believe the defense attorney? Explain why or why not.

7. Although normal physiological processes produce small amounts of free radicals, it is possible that your behavior, lifestyle, or environment contribute to an increased production of free radicals. What behavioral or environmental risks do you think you have that might promote the formation of free radicals?

MJ's BlogInFocus

Answer can be found in the Study Area in MasteringBiology.

1. What is a pluripotent stem cell, and why are scientists so interested in learning how to induce adult human cells into becoming pluripotent again? For more on this topic, visit MJ's blog in the Study Area in MasteringBiology and look under "Pluripotent Stem Cells." http://goo.gl/RTddxk

MasteringBiology®

Students Go to MasteringBiology for assignments, the eText, and the Study Area with animations, practice tests, and activities.

Professors Go to MasteringBiology for automatically graded tutorials and questions that you can assign to your students, plus Instructor Resources.

Answers to ✓ questions are available in the Appendix.

From Cells to Organ Systems

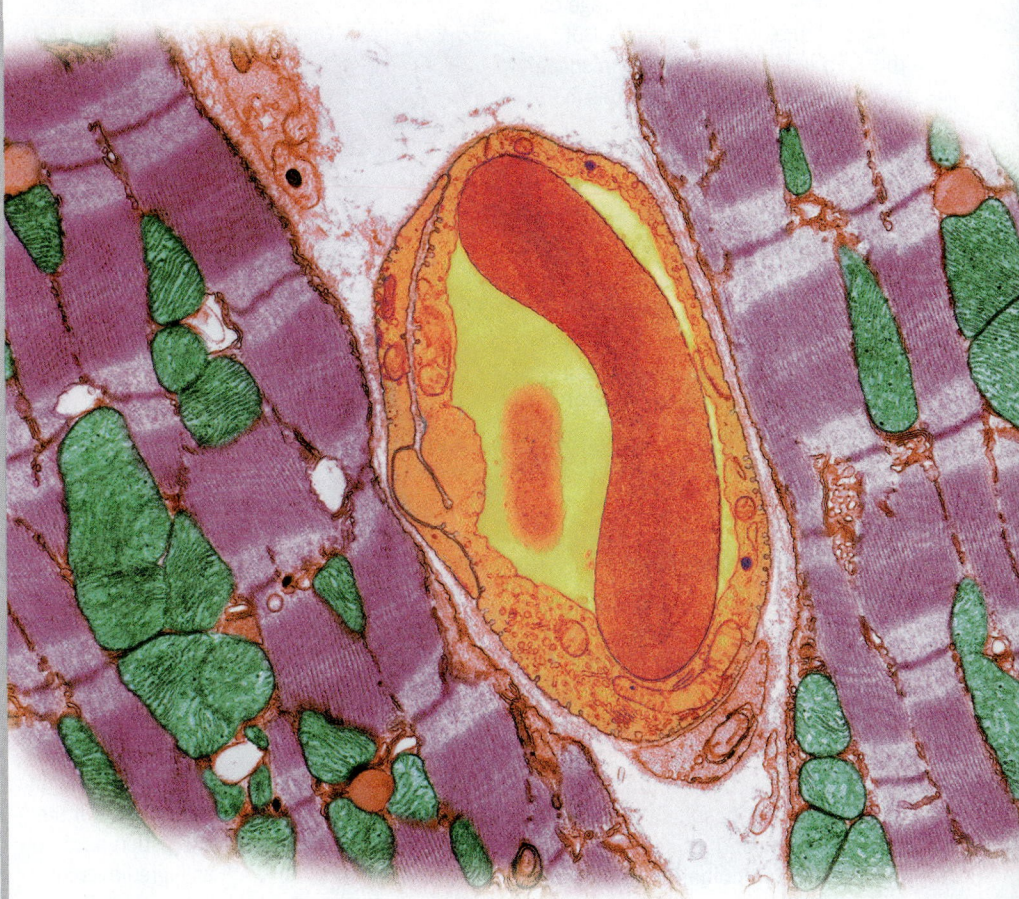

A red blood cell (red) within a capillary (orange) between two cardiac muscle cells (purple/green).

Key Concepts

- **In multicellular organisms, cells have specialized functions.** These functions evolved (along with multicellularity) because they benefit the entire organism.

- **Groups of cells with a common function are called tissues.** The four main tissue types are epithelial, connective, muscle, and nervous.

- **Organs and organ systems each perform one or more essential complex functions for the organism.** Humans have 11 different organ systems; examples are the male and female reproductive systems (reproduction), skeletal system (structural support), and muscular system (movement).

- **Multicellular organisms must maintain homeostasis** (constancy) of their internal environments. The maintenance of homeostasis allows cells to survive without direct contact with the external environment.

- **Homeostasis is maintained by negative feedback control systems.** In negative feedback systems, any deviation from a stable condition is detected and corrected.

CURRENT ISSUE
Reshaping Your Body

By now, you are familiar with the chemistry of lipids and the structure and function of fat cells. You understand that when your average daily caloric intake exceeds your average daily caloric expenditure, you gain weight. But does this knowledge help you keep in shape? Are you watching your diet and exercising regularly but still finding fat deposits in places where you don't want them?

What if you could just melt away unwanted fat? What if you could get rid of pesky fat once and for all? What if you could sculpt your body without having to diet or exercise? It's a fantasy many of us have indulged in, and it's one of the reasons for the meteoric rise of several cosmetic procedures, including lipodissolve, high-intensity focused ultrasound (HIFU), and fat transfer breast augmentation.

Is Lipodissolve Safe and Effective?

Lipodissolve, also called injection lipolysis, is described by its promoters as a safe, effective, nonsurgical way to sculpt the body into a desired shape. Usually the technique is performed as a series of six injections two weeks apart directly into subcutaneous fat deposits. The injections contain two active ingredients: a phospholipid called phosphatidylcholine and a bile acid called deoxycholate. The combination of these two drugs, phosphotidylcholine deoxycholate (PCDC), allegedly works by dissolving the bonds between the three fatty acids and the glycerol backbone that comprise a triglyceride molecule, the primary molecular form of stored fat in the body. PCDC is also thought to disrupt the cell membranes of fat cells in the vicinity of an injection, resulting in death of the fat cells themselves. Some lipodissolve cocktails include vitamins and plant extracts along with the PCDC.

So far there have been only anecdotal reports of minor unwanted side effects associated with the lipodissolve technique, such as swelling, skin blistering, pain, and blackened skin in some patients. No deaths have been reported. A clinical trial of the safety and efficacy of lipodissolve

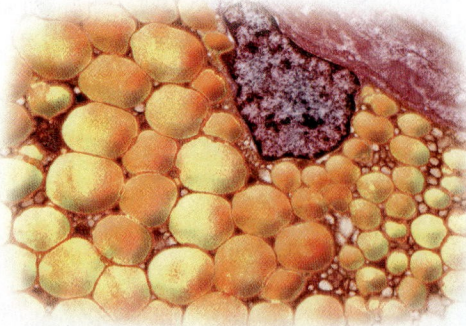

An SEM of a human fat cell (×3,000) filled with lipid droplets (yellow).

is underway, but it may be several more years before we know the outcome.

You might be wondering why lipodissolve is available if it has not been proven safe and effective. Lipodissolve clinics can legally administer PCDC because of a loophole in our drug regulatory laws. According to the law, a licensed doctor can legally prescribe compounded drugs (drugs made from more than one ingredient) specifically for individual patients, as long as all of the *ingredients* are approved by the FDA. The intent of the drug regulatory law is to protect the sanctity of the doctor–patient relationship, allowing the doctor to choose what drug combination might be best for that particular patient. Both of the compounds in a lipodissolve injection have been approved by the FDA as ingredients in other drugs, though neither was approved as a fat-dissolving drug. Health clinics and spas have been quick to capitalize on the loophole. At prices ranging from $375 to $1,500 per treatment for up to six treatments, they stand to make a lot of money.

Lipodissolve is injected directly into areas of excess fat.

Questions to Consider

1 Health clinics sometimes cite retrospective studies to support their claim that the lipodissolve technique is safe. What is a "retrospective study"? How is it different from a controlled study?

2 How important to you is FDA approval and/or scientific evidence on the safety of lipodissolve? If someone were to pay for it, would you undergo a lipodissolve or Liposonix procedure?

Critics of the lipodissolve technique argue that the law never intended to allow mass marketing by chains of clinics or spas. The American Society of Plastic Surgeons, the American Society of Aesthetic Plastic Surgery, and the American Society of Dermatologic Surgery have all issued cautionary warnings against the technique.

What About HIFU?

High-intensity focused ultrasound (HIFU) is a technique in which a high-intensity beam of ultrasound is focused on a particular tissue below the surface of the body. The high energy produced by the ultrasound waves generates heat, which destroys cells in the targeted area. HIFU was first used in the treatment of uterine fibroids and slow-growing prostate cancers. But lately, aesthetic medicine specialists have been adapting the technique for removal of fat deposits beneath the skin. Under the proprietary names Ultrashape, Liposonix, and Vaser, the HIFU technique is promising patients noninvasive body sculpting without anesthesia or pain and with minimal recovery time. It sounds great, but again, the technique has not been approved by the FDA for all of its current uses.

Fat Transfer Breast Augmentation

As an alternative to breast implants, which can leak or break, some plastic surgeons are now performing a technique called "fat transfer breast augmentation"—using fat cells harvested by liposuction for breast enhancement. Liposuction is an established technique that removes fat cells from a local area of the body using a suction device. A distinct advantage of using the patient's own fat cells for breast enhancement is that tissue incompatibility is not an issue; a body won't reject its own fat cells. However, the procedure's success depends critically on the skill and experience of the surgeon. Invariably, not all fat cells survive the procedure. If too many fat cells die, the result may be scarring, lumpiness, and calcium deposits in the breast, which are hard to distinguish from cancers on an X-ray. In addition, unlike breast implants the patient's breast size may decline if she loses weight after the surgery.

The regulatory environment for fat transfer breast augmentation is murky at best. Because it's not a technique that involves drugs or specialized medical equipment, the FDA has no jurisdiction over its regulation. At the moment, patients must rely on the opinion of professional societies such as the American Society of Plastic Surgeons and the American Society for Aesthetic Plastic Surgery, which could have a vested interest in the cosmetic procedure.

Anyone contemplating reshaping their body with any of these procedures should do their research carefully.

SUMMARY

- Several cosmetic procedures for removing unwanted body fat or augmenting breasts are being heavily promoted by health clinics and spas. None has been fully approved by the FDA.
- The lipodissolve technique involves a series of injections of fat-dissolving drugs directly into local fat deposits to dissolve fat cells.
- High-intensity focused ultrasound uses targeted high-intensity ultrasound to disrupt fat cells.
- Fat transfer breast augmentation is a technique for removing unwanted fat cells from one area of the body (by liposuction) and using the fat cells for breast enhancement.

For nearly two-thirds of the history of life, or more than 2 billion years, all organisms consisted of just one cell. Plenty of single-celled organisms exist today; in fact they far outnumber multicellular organisms. Theirs is a simple, uncomplicated life. They get their raw materials and energy from the fluid in which they are bathed, they dump their wastes into that same fluid, and they reproduce by dividing in two.

However, there are disadvantages to being a single cell. The single-celled organism is completely at the mercy of its immediate external environment for every requirement of life. If the pond in which it lives dries up, it will die. If salt levels in the water rise, if the temperature gets too hot, or if its food runs out, it dies. There must be another way!

There is another way, and that is for cells to join together. Collectively they can perform functions that no one cell could perform alone. A *multicellular organism* consists of many cells that share the functions of life. Advantages to multicellularity include greater size (better to eat, rather than be eaten) and the ability to seek out or maintain an external environment conducive to life. Multicellularity comes with its own challenges, however. Each cell in a multicellular organism must specialize so that it can contribute to the well-being of other cells, and in turn it must rely on the specialized functions of other cells for its own well-being. The carefree, simple, uncomplicated life of a single cell is replaced by an interconnected, interdependent life.

In this chapter, we look at how cells are organized into the tissues, organs, and organ systems that comprise your body. We consider the structure and function of your skin as an example of an organ system. And we discuss how tissues, organs, and organ systems work together to maintain the health and stability of your body.

4.1 Tissues are groups of cells with a common function

To accomplish a complex activity such as pumping blood or digesting food, cells are grouped together into tissues. **Tissues** are groups of specialized cells that are similar in structure and that perform common functions. An example would be heart muscle tissue, comprised of hundreds of thousands of heart muscle cells connected together. By coordinating their contractions, the cells in heart muscle tissue can accomplish what no one cell could do alone—pump blood. There are four major types of tissues in the human body: epithelial, connective, muscle, and nervous.

At higher levels of organization, several types of tissues may be grouped together to form an *organ* (all four types of tissues are present in the heart) and one or more organs may comprise an *organ system* (the heart is part of the cardiovascular system, which also includes the blood vessels). Organs and organ systems will be discussed later in the chapter. We turn now to the four types of tissues.

4.2 Epithelial tissues cover body surfaces and cavities

Most **epithelial tissues** consist of sheets of cells that line or cover various surfaces and body cavities. Two epithelial tissues that you can easily see are your skin and the lining of your mouth. Other epithelial tissues line the inner surfaces of your digestive tract, lungs, bladder, blood vessels, and the tubules of your kidneys.

Epithelial tissues are more than just linings. They protect underlying tissues. Often they are smooth to reduce friction;

the smooth epithelial tissue lining your blood vessels reduces resistance to blood flow in those vessels, for instance. Some epithelial tissues are highly specialized for transporting materials. Epithelial tissues (and cells) absorb water and nutrients across your intestines into your blood. They also secrete waste products across the tubules of your kidneys so that you can eliminate unneeded or toxic substances in urine.

A few epithelial tissues are *glandular epithelia* that form the body's glands. **Glands** are epithelial tissues that are specialized to synthesize and secrete a product. **Exocrine glands** (*exo-* means "outside" or "outward") secrete their products into a hollow organ or duct. Examples of exocrine glands are the glands in your mouth that secrete saliva, glands in your skin that excrete sweat, and glands in your

stomach that produce digestive acid. **Endocrine glands** (*endo-* means "within") secrete substances called *hormones* into the bloodstream. One endocrine gland is the thyroid gland, which secretes several hormones that help regulate your body's growth and metabolism. We describe various glands throughout the book where appropriate.

Epithelial tissues are classified according to cell shape

Biologists classify epithelial tissues into three types according to the shapes of the cells (**Figure 4.1**):

- *Squamous epithelium* consists of one or more layers of flattened cells. (*Squama* means "platelike." Think of squamous epithelium as "squashed flat.") Squamous

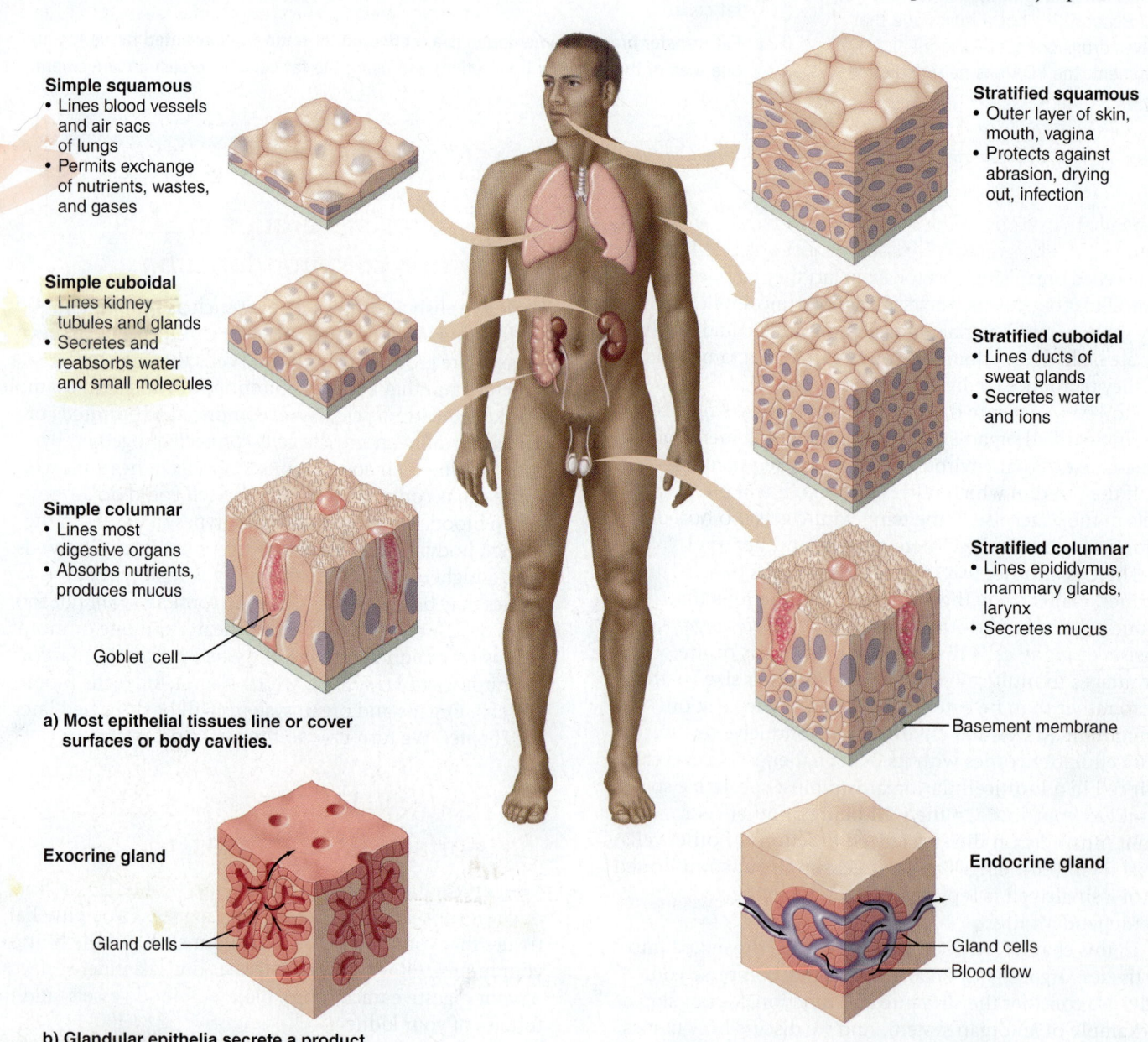

Simple squamous
- Lines blood vessels and air sacs of lungs
- Permits exchange of nutrients, wastes, and gases

Simple cuboidal
- Lines kidney tubules and glands
- Secretes and reabsorbs water and small molecules

Simple columnar
- Lines most digestive organs
- Absorbs nutrients, produces mucus

Goblet cell

a) **Most epithelial tissues line or cover surfaces or body cavities.**

Stratified squamous
- Outer layer of skin, mouth, vagina
- Protects against abrasion, drying out, infection

Stratified cuboidal
- Lines ducts of sweat glands
- Secretes water and ions

Stratified columnar
- Lines epididymus, mammary glands, larynx
- Secretes mucus

Basement membrane

Exocrine gland

Gland cells

b) **Glandular epithelia secrete a product.**

Endocrine gland

Gland cells
Blood flow

Figure 4.1 Types of epithelial tissues.

epithelium forms the outer surface of the skin and lines the inner surfaces of the blood vessels, lungs, mouth and throat, and vagina.

- *Cuboidal epithelium* is composed of cube-shaped cells. Cuboidal epithelium forms the kidney tubules and also covers the surfaces of the ovaries.
- *Columnar epithelium* is composed of tall, rectangular (column-shaped) cells. Columnar epithelium lines parts of the digestive tract, certain reproductive organs, and the larynx. Certain cells within columnar epithelium, called *goblet cells*, secrete mucus, a thick fluid that lubricates the tissues and traps bacteria, viruses, and irritating particles.

Epithelial tissues are classified not only by shape but also by the number of cell layers in the tissue. A *simple epithelium* is a single layer of cells, whereas a *stratified epithelium* consists of multiple layers (or strata). Simple epithelium is so thin that molecules can pass through it easily. Stratified epithelium is thicker and provides protection for underlying cells.

☑ What sort of epithelium would you expect to find lining a part of the digestive tract that absorbs food molecules and also secretes mucus? Explain your answer.

The basement membrane provides structural support

Directly beneath the cells of an epithelial tissue is a supporting noncellular layer called the **basement membrane** (see Figure 4.1a), and beneath that is generally a layer of connective tissue (described later). You can think of the basement membrane as the mortar that anchors the epithelial cells to the stronger connective tissue underneath. The basement membrane is composed primarily of proteins secreted by the epithelial cells and their underlying connective tissue layer. It should not be confused with the plasma membrane that is a part of every living cell.

In addition to being attached to a basement membrane, epithelial cells may be connected to each other by several types of **cell junctions** made up of various

proteins. Three different types of junctions may hold the cells together, depending on the type of epithelial tissue (**Figure 4.2**):

- *Tight junctions* seal the plasma membranes of adjacent cells so tightly together that nothing can pass between the cells. Tight junctions are particularly important in epithelial layers that must control the movement of substances into or out of the body. Examples include the cells that line the digestive tract (which bring in nutrients) and the bladder (which stores urine), and the cells that form the tubules of the kidneys (which remove waste products from the body).
- *Adhesion junctions*, sometimes called "spot desmosomes," are looser in structure. The protein filaments of adhesion junctions allow for some movement between cells so that the tissues can stretch and bend. Adhesion junctions in the epithelium of your skin, for instance, allow you to move freely.
- *Gap junctions* represent connecting channels made of proteins that permit the movement of ions or water between two adjacent cells. They are commonly found in the epithelial cells in the liver, heart, and some muscle tissues.

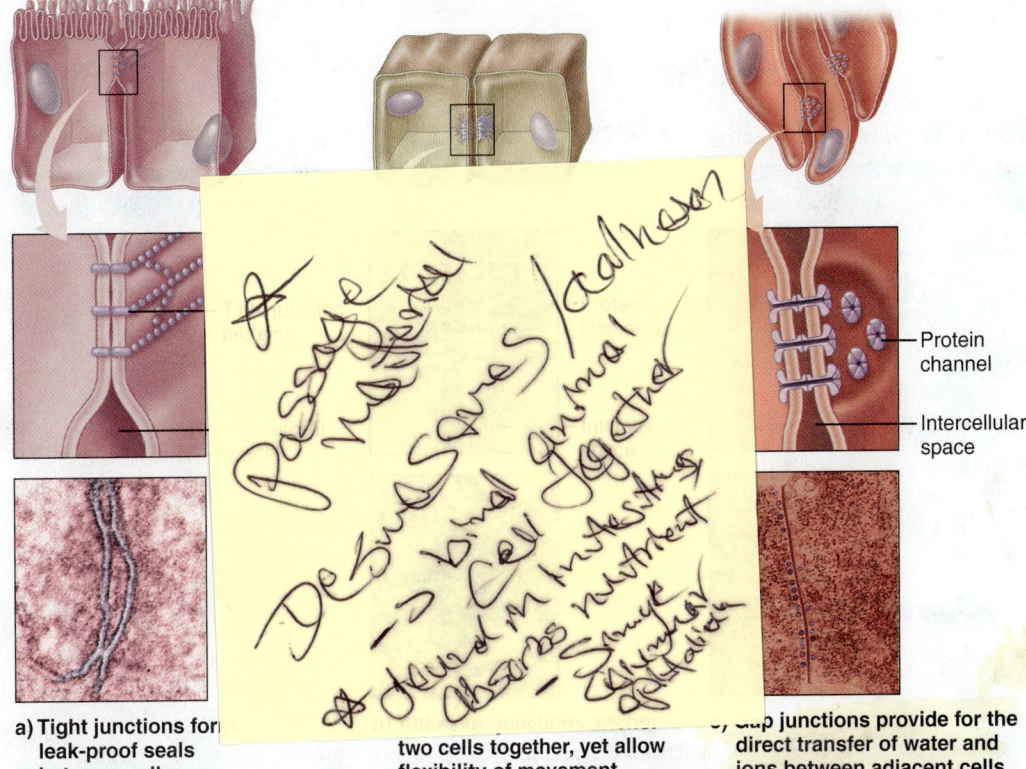

a) Tight junctions for[m] leak-proof seals between cells.

[...] two cells together, yet allow flexibility of movement.

[...] Gap junctions provide for the direct transfer of water and ions between adjacent cells.

— Protein channel

— Intercellular space

Figure 4.2 Examples of junctions between cells. Only one type of junction is generally present in any given tissue.

☑ Which type of junction would you expect to find between two cells that share ions or raw materials? Why?

Recap Epithelial tissues line body surfaces and cavities and form glands. They are classified according to cell shape (squamous, cuboidal, or columnar) and the number of cell layers (simple or stratified). Epithelial cells are supported by a basement membrane and may be attached to each other by cell junctions. ■

4.3 Connective tissue supports and connects body parts

Connective tissue supports the softer organs of the body against gravity and connects the parts of the body together. It also stores fat and produces the cells of blood.

Unlike epithelial tissue, most connective tissues have few living cells. Most of their structure consists of nonliving extracellular material, the matrix that is synthesized by connective tissue cells and released into the space between them. The strength of connective tissue comes from the matrix, not from the living cells themselves. The few living cells rarely make contact with each other; direct cell-to-cell junctions are not present.

Connective tissues are so diverse that any classification system is really a matter of convenience (Table 4.1). Broadly, we can divide them into fibrous and specialized connective tissues.

Fibrous connective tissues provide strength and elasticity

Fibrous connective tissues connect various body parts, providing strength, support, and flexibility. As indicated by their name, fibrous connective tissues consist of several types of fibers and cells embedded in a *ground substance*, consisting of water, polysaccharides, and proteins, that ranges in consistency from gel-like to almost rubbery. Figure 4.3 shows the structural elements of fibrous connective tissue.

Collagen fibers, made of protein, confer strength and are slightly flexible. Most fibrous connective tissues also contain thinner coiled **elastic fibers,** made primarily of the *protein elastin*, which can stretch without breaking. Some fibrous connective tissue also contains thinner fibers of collagen, called **reticular fibers,** that interconnect with each other. The reticular fibers often serve as an internal structural framework for some of the soft organs such as the liver, spleen, and lymph nodes.

The ground substance contains fat cells, mast cells (specialized connective tissue cells that release histamine), white blood cells (macrophages, neutrophils, lymphocytes, and plasma cells), and, most importantly, **fibroblasts.** The fibroblasts are the cells responsible for producing and secreting the proteins that compose the collagen, elastic, and reticular fibers. The fat cells, of course, store fat, and both the mast cells and white blood cells are involved in the body's immune system.

Table 4.1 Types of connective tissues

Type	Structure	Attributes	Locations
Fibrous Connective Tissue			
Loose	Mostly collagen and elastin fibers in no particular pattern; more ground substance	Flexible but only moderately strong	Surrounds internal organs, muscles, blood vessels
Dense	Mostly collagen in a parallel arrangement of fibers; less ground substance	Strong	In tendons, ligaments, and the lower layers of skin
Elastic	High proportion of flexible fibers	Stretches and recoils easily	Surrounds hollow organs that change shape or size regularly
Reticular (lymphoid)	Mostly thin, interconnecting fibers of collagen	Serves as a flexible internal framework	In soft organs such as liver, spleen, tonsils, and lymph glands
Special Connective Tissues			
Cartilage	Primarily collagen fibers in a ground substance containing a lot of water	Maintains shape and resists compression	Embryonic tissue that becomes bone. Also the nose, vertebral disks, and the lining of joint cavities
Bone	Primarily hard mineral deposits of calcium and phosphate	Very strong	Forms the skeleton
Blood	Red blood cells, white blood cells, platelets, and plasma	Transports materials and assists in defense mechanisms	Within cardiovascular system
Adipose tissue	Primarily *adipocytes* cells filled with fat deposits	Stores energy in the form of fat	Under the skin, around some internal organs

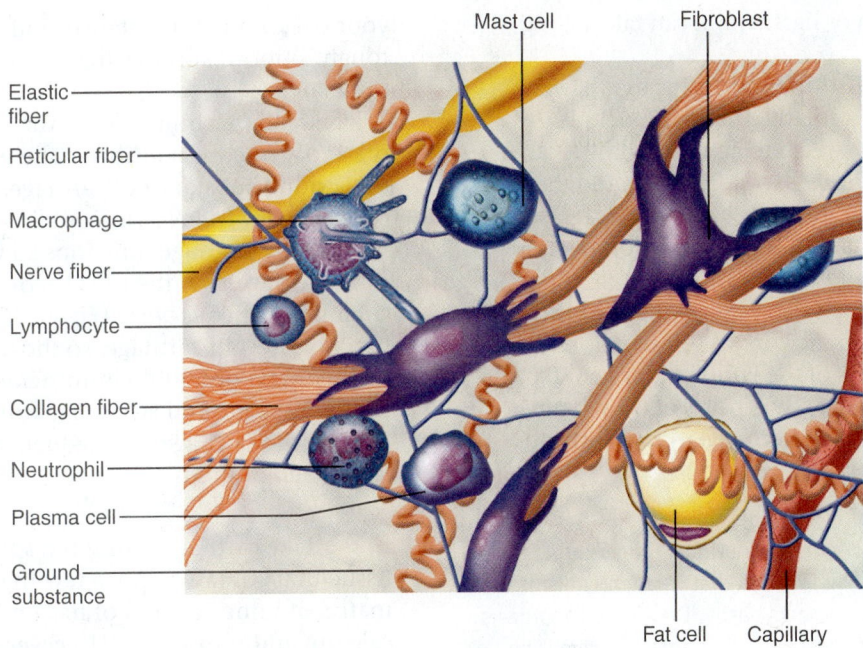

Figure 4.3 Fibrous connective tissue. The main elements are three types of fibers (collagen, elastic, and reticular) and a variety of cells (fibroblasts, fat cells, mast cells, and several types of white blood cells) in a ground substance of polysaccharides, proteins, and water. Blood vessels and nerves pass through or are associated with connective tissue. Fibrous connective tissues vary in their relative proportions of cells and fibers and in fiber orientation.

Fibrous connective tissues are subclassified according to the density and arrangement of their fiber types:

- *Loose connective tissue* (**Figure 4.4a**), also called *areolar connective tissue,* is the most common type. It surrounds many internal organs, muscles, and blood vessels. Loose connective tissue contains a few collagen fibers and elastic fibers in no particular pattern, giving it a great deal of flexibility but only a modest amount of strength.

- *Dense connective tissue* (**Figure 4.4b**), found in tendons, ligaments, and lower layers of skin, has more collagen fibers. The fibers are oriented primarily in one direction, especially in the tendons and ligaments in and around our joints. Dense connective tissue is the strongest connective tissue when pulled in the same direction as the orientation of the fibers, but it can tear if the stress comes from the side. Very few blood vessels in dense connective tissue supply the few living cells. This is why,

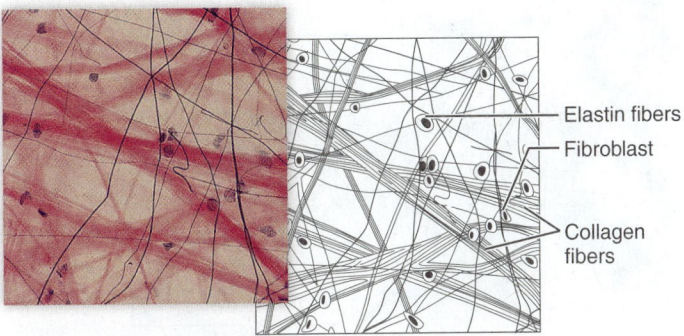

a) **Loose areolar connective tissue (×160).** In loose connective tissue, the collagen and elastin fibers are arrayed in a random pattern.

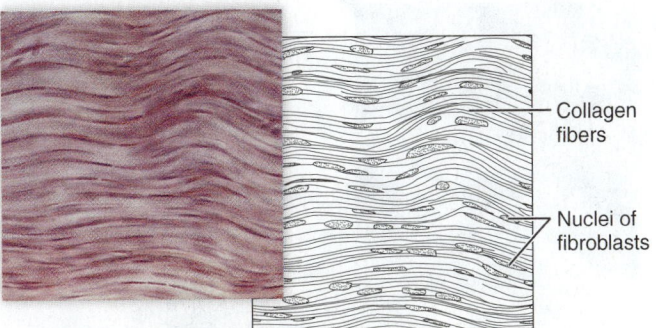

b) **Dense connective tissue (×160).** In dense connective tissue, the fibers are primarily collagen fibers. In tendons and ligaments, the fibers are oriented all in the same direction, with fibroblasts occupying narrow spaces between adjacent fibers.

Figure 4.4 Examples of fibrous connective tissues.

if you strain a tendon or ligament, it can take a long time to heal.

- *Elastic connective tissue* surrounds organs that have to change shape or size regularly. Examples include the stomach, which must stretch to accommodate food; the bladder, which stretches to store urine; and the vocal cords, which vibrate to produce sounds. Elastic connective tissue contains a high proportion of elastic fibers, which stretch and recoil easily.

- *Reticular connective tissue* (also called lymphoid tissue) serves as the internal framework of soft organs such as the liver and the tissues of the lymphatic system (spleen, tonsils, and lymph nodes). It consists of thin, branched reticular fibers (composed of collagen) that form an interconnected network.

☑ People with a hereditary condition known as Ehlers-Danlos syndrome (EDS) have hyperextensible joints that frequently dislocate and extremely stretchy skin that tears easily. Develop a hypothesis for what type of tissue, and what protein in particular, might be the cause of EDS.

Specialized connective tissues serve special functions

The so-called *specialized connective tissues* are a diverse group that includes cartilage, bone, blood, and adipose tissue. Each is specialized to perform particular functions in the body.

Cartilage The transitional tissue from which bone develops is called **cartilage.** The shape of certain body parts (such as the soft tip of your nose) is maintained by cartilage, and cartilage protects and cushions joints. Disks of cartilage separate and cushion the vertebrae in your backbone, for instance, and cartilage forms the tough, smooth surfaces that reduce friction in some body joints.

Like dense connective tissue, cartilage consists primarily of collagen fibers. The two tissues differ in that the ground substance of cartilage, which is produced by cells called *chondroblasts*, contains a great deal more water. This is why cartilage functions so well as a cushion. As cartilage develops, the cells become enclosed in small chambers called *lacunae* (**Figure 4.5a**). There are no blood vessels in cartilage, so the mature cells (called *chondrocytes*) obtain their nutrients only by diffusion through the ground substance from blood vessels located outside the cartilage. Consequently, cartilage is slow to heal when injured.

Bone Like cartilage, **bone** is a specialized connective tissue that contains only a few living cells. Most of the matrix of bone consists of hard mineral deposits of calcium and phosphate. However, unlike cartilage, bone contains numerous blood vessels, and for this reason it can heal within four to six weeks after being injured. We discuss bone in more detail when we discuss the skeletal system.

Blood Consisting of cells suspended in a fluid matrix called *plasma*, **blood** is considered a connective tissue because its cells derive from stem cells located within bone. Red blood cells transport oxygen and nutrients to body cells and carry away the waste products of the cells' metabolism. White blood cells function in the immune system that defends the body. Platelets participate in the mechanisms that cause blood to clot following an injury. The composition and functions of blood are covered in more detail in a separate chapter.

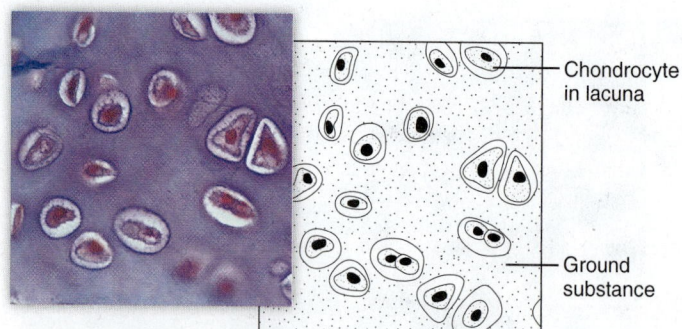

a) **Cartilage from the trachea (×300).** Mature cartilage cells, called chondrocytes, become trapped in chambers called lacunae within the hard, rubbery ground substance. Ground substance is composed of collagen fibers, polysaccharides, proteins, and water.

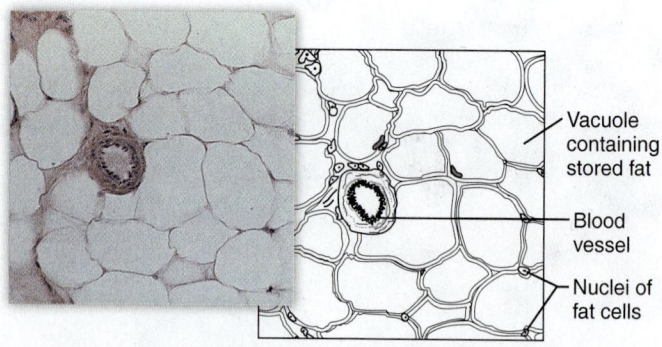

b) **Adipose tissue from the subcutaneous layer under the skin (×140).** Adipose tissue consists almost entirely of fat cells. The fat deposit within a fat cell can become so large that the nucleus is pushed to the side.

Figure 4.5 Examples of special connective tissues.

Heavy smokers may develop premature wrinkles.

HEALTH & WELLNESS

Suntans, Smoking, and Your Skin

Many of us think suntans are attractive. A suntan indicates we have leisure time to spend basking in the sun, and it gives the skin a healthy looking "glow." But are suntans really healthy?

The answer is clearly "no," according to medical experts. Strong light rays can age your skin prematurely and increase your risk for skin cancer. The rays penetrate to the dermis and damage its collagen and elastin fibers. Elastin fibers clump together, leading at first to fine wrinkles and later to a wrinkled, leathery skin texture.

Prolonged exposure to light rays can also damage small blood vessels. Sometimes the vessels remain permanently dilated, leading to a condition called *telangiectasis,* or spider veins. Sunlight also damages the keratinocytes and melanocytes in the epidermis. The keratinocytes become rough and thickened and no longer fit together as a smooth interlocking layer. The melanocytes begin to produce melanin unevenly, leading to patches of darker pigmentation known as freckles, age spots, or liver spots.

What about tanning beds—are they okay? The tanning salon industry has been quite aggressive in trying to convince the public that tanning beds are not only safe but that a healthy tan is good for you. Experts disagree. According to skin cancer researchers, repeated exposure to light rays, whether from sunlight or from tanning beds, is a risk factor for melanoma.

In addition, your skin is likely to age more quickly. Although a minimal amount of sunlight is necessary to activate vitamin D, the amount is far below what is necessary to cause a tan.

Perhaps you have heard that ultraviolet (UV) rays consist of "good" UVA rays that tan your skin and "bad" UVB rays that cause sunburn. Tanning lotions with high sun protection factor (SPF) numbers are designed to block the UVB rays, which at least prevents the acute damage and pain of a sunburn. However, UVA rays are also bad for you. They penetrate more deeply than the UVB rays and in fact cause most of the long-term changes that age skin prematurely.

What about smoking? Heavy smokers are nearly five times more likely to develop premature wrinkles. Smoking damages and thickens the elastin fibers in the dermis. It also dehydrates keratinocytes in the epidermis, causing the epidermis to develop a rough texture. Finally, smoking narrows blood vessels, reducing blood flow to the skin. As a result, the skin of smokers heals more slowly from injury than the skin of nonsmokers.

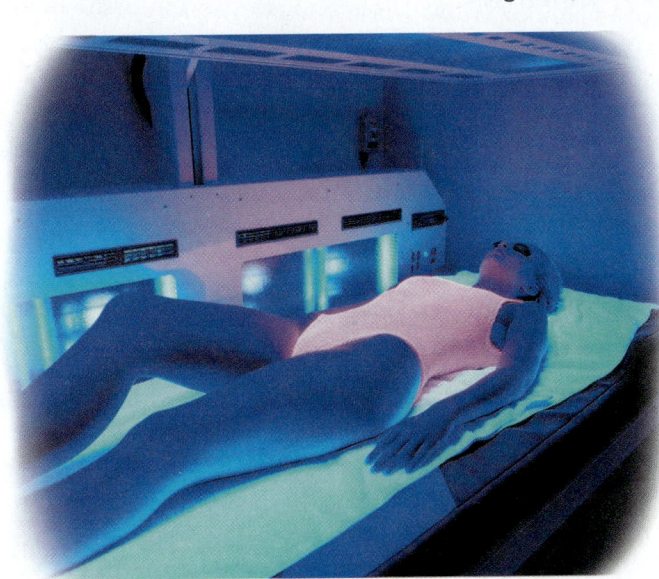

Repeated exposure to light rays is a risk factor for melanoma.

Adipose tissue   A loose connective tissue called **adipose tissue** is highly specialized for fat storage (**Figure 4.5b**). Adipose tissue has few connective tissue fibers and almost no ground substance. Most of its volume is occupied by **adipocytes** (fat cells). Adipose tissue is located primarily under the skin, where it serves as a layer of insulation. It also forms a protective layer around internal organs such as

MJ's BlogInFocus   Do severe sunburns increase the risk of melanoma in teens? In older women? Visit MJ's blog in the Study Area in MasteringBiology and look under "Sunburns and Melanoma."

http://goo.gl/3idkXK

the kidneys. The number of adipocytes you have is partly determined by your genetic inheritance. When you eat more food than your body can use, some of the excess energy is stored in your adipocytes as fat (the fat cells get "fatter"). When you lose weight, the fat cells slim down, too. In other words, weight loss reduces the volume of each fat cell but it does not necessarily reduce the number of fat cells. *Lipodissolve* is a controversial technique that disrupts and destroys fat cells chemically (review the Current Issue at the beginning of this chapter).

↩ **Recap** Fibrous connective tissues provide strength and elasticity and hold body parts together. Among the specialized connective tissues, cartilage and bone provide support, blood transports materials throughout the body, and adipose tissue stores energy in the form of fat. ∎

4.4 Muscle tissues contract to produce movement

Muscle tissue consists of cells that are specialized to shorten, or contract, resulting in movement of some kind. Muscle tissue is composed of tightly packed cells called muscle *fibers*. The fibers are generally long and thin and aligned parallel to each other. The cytoplasm of a muscle fiber contains proteins, which interact to make the cell contract.

There are three types of muscle tissue: *skeletal*, *cardiac*, and *smooth*. They vary somewhat in body location, structure, and function, but they all do essentially the same thing— when stimulated, they contract. We devote an entire chapter to muscles as an organ system. For now, we'll focus solely on differences between the three types of muscle tissue.

Skeletal muscles move body parts

Skeletal muscle tissue connects to tendons, which attach to bones. When skeletal muscles contract, they cause body parts to move. The individual cells, also called fibers, are too small to be seen with the naked eye, but they may be as long as the entire muscle (Figure 4.6a). Each muscle cell has many nuclei, a phenomenon that happens because many young cells fuse end-to-end during development, producing one long cell.

A skeletal muscle may contain thousands of individual cells, all aligned parallel to each other. This parallel arrangement enables the cells to all pull in the same direction, shortening the muscle between its two points of attachment. Skeletal muscle is called *voluntary* muscle because we can exert conscious control over its activity. Skeletal muscle cells are activated only by nerves.

Cardiac muscle cells activate each other

Cardiac muscle tissue (Greek *kardia* means "the heart") is found only in the heart. The individual cells are much shorter than skeletal muscle fibers, and they have only one

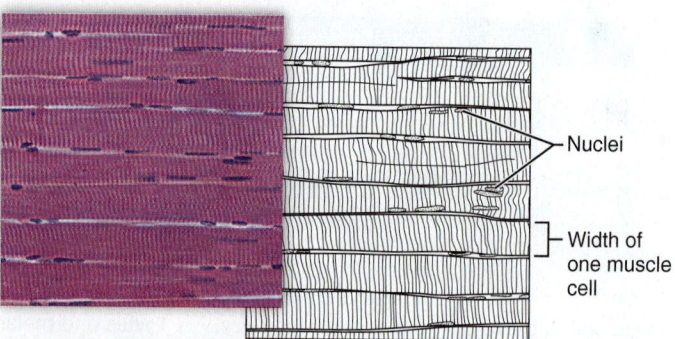

a) **Skeletal muscle (×100).** Skeletal muscle cells are very long and have many nuclei.

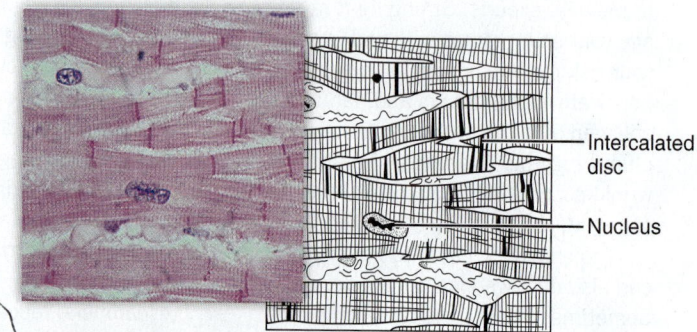

b) **Cardiac muscle (×225).** Cardiac muscle cells interconnect with each other.

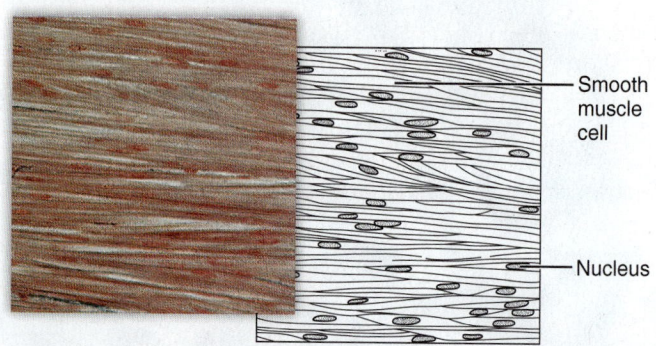

c) **Sheet of smooth muscle (×250).** Smooth muscle cells are thin and tapered.

Figure 4.6 Muscle tissue.

nucleus (Figure 4.6b). Like skeletal muscle, the cells are arranged parallel to each other. Cardiac muscle cells are short and blunt-ended, with gap junctions between the ends of adjoining cells. The gap junctions represent direct electrical connections between adjoining cells, so when one cell is activated it activates its neighbors down the line. Because of these gap junctions, the entire heart contracts in a coordinated fashion.

Cardiac muscle is considered involuntary because the heart can contract rhythmically entirely on its own, without any conscious thought on our part and without any stimulation by nerves.

☑ Does skeletal muscle, like cardiac muscle, have gap junctions between adjacent cells? Why or why not?

Smooth muscle surrounds hollow structures

Smooth muscle tissue surrounds hollow organs and tubes, including blood vessels, digestive tract, uterus, and bladder. These slim cells are much smaller than skeletal muscle cells and have only one nucleus, like cardiac muscle (**Figure 4.6c**). The cells are aligned roughly parallel to each other. In blood vessels, they are generally aligned in a circular fashion around the vessel. When smooth muscle cells shorten, the diameter of the blood vessel is reduced.

Smooth muscle cells taper at both ends, and there are gap junctions between adjacent cells so that when one contracts, nearby cells also contract. Like cardiac muscle, smooth muscle is involuntary in that we cannot control its contractions consciously.

🔄 **Recap** Skeletal muscle tissue connects to tendons and bones. Smooth muscle tissue surrounds organs and tubes. Cardiac muscle tissue comprises most of the heart. The common feature of all muscle tissues is that they contract, producing movement. ■

4.5 Nervous tissue transmits impulses

Nervous tissue consists primarily of cells that are specialized for generating and transmitting electrical impulses throughout the body. It forms a rapid communication network for the body. Nervous tissue is located in the brain, the spinal cord, and the nerves that transmit information to and from various organs. An entire chapter is devoted to the nervous system, so we describe nervous tissue only briefly here.

Nervous tissue cells that generate and transmit electrical impulses are called **neurons** (**Figure 4.7**). A neuron can be as long as the distance from your spinal cord to the tip of your toe. Typically, a neuron has three basic parts: (1) the *cell body*, where the nucleus is located; (2) *dendrites*, numerous cytoplasmic extensions that extend from the cell body and receive signals from other neurons; and (3) a long extension called an *axon* that transmits electrical impulses over long distances.

Nervous tissue also includes another type of cell called a **glial cell** that does not transmit electrical impulses. Glial cells play a supporting role by surrounding and protecting neurons and supplying them with nutrients.

🔄 **Recap** Nervous tissues serve as a communication network. Within nervous tissue, neurons transmit information by generating and transmitting electrical impulses. ■

4.6 Organs and organ systems perform complex functions

Many of the more complex functions of multicellular organisms (such as pumping blood or digesting food) cannot be carried out by one tissue type alone. **Organs** are structures composed of two or more tissue types joined together that perform a specific function or functions.

Your heart is an organ because the tissues that comprise it share one essential function: They pump blood. Most of the heart consists of cardiac muscle, but there is also smooth muscle in the blood vessels that supply the cardiac muscle. The heart contains nervous tissue that affects the rate at which the organ beats. Some connective tissue, primarily in the heart valves, open and close to control blood flow within the heart, and even a thin layer of epithelial tissue lines the heart chambers.

Some organs have several functions. For example, the kidneys remove wastes and help control blood pressure.

The human body is organized by organ systems

Organ systems are groups of organs that together serve a broad function that is important to survival either of the individual organism or a species. For example, survival of an individual organism depends on respiration, movement, or excretion of wastes, and survival of a species relies on reproduction. A good example is the organ system responsible for the digestion of food. Your *digestive system* includes your mouth, throat, esophagus, stomach, intestines, and even your liver, pancreas, and gallbladder. All of these organs must interact and be controlled and coordinated to accomplish their overall function.

Figure 4.8 depicts the 11 organ systems of the human body. Some organ systems perform several functions. For example, the lymphatic system has important functions related to defense against disease, circulation of certain body fluids, and digestion.

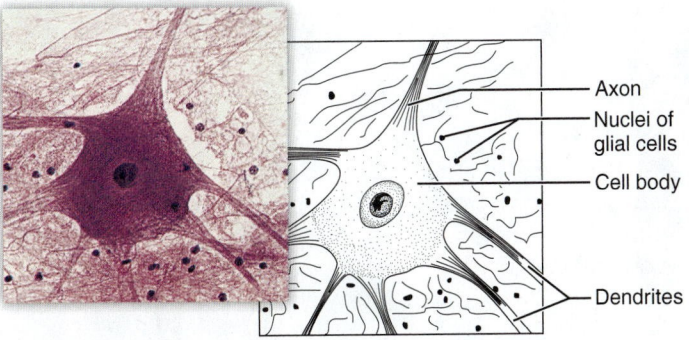

Figure 4.7 Nervous tissue: a neuron (×170). The neuron is the functional unit of nervous tissue. The single neuron shown here is surrounded by numerous supporting cells called *glial cells.* The cell bodies of the glial cells do not stain well, but their nuclei are clearly visible.

Labels: Axon · Nuclei of glial cells · Cell body · Dendrites

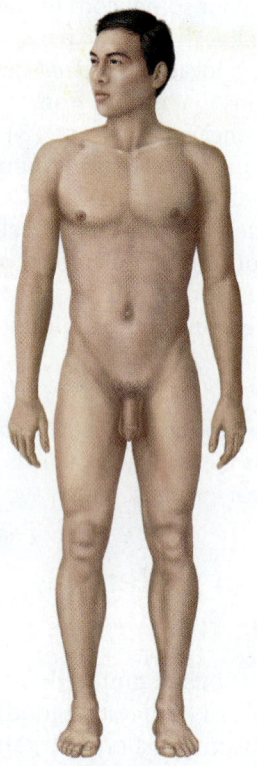

Integumentary system
- Protects from injury, infection, and dehydration
- Participates in temperature control
- Receives sensory input from the external environment

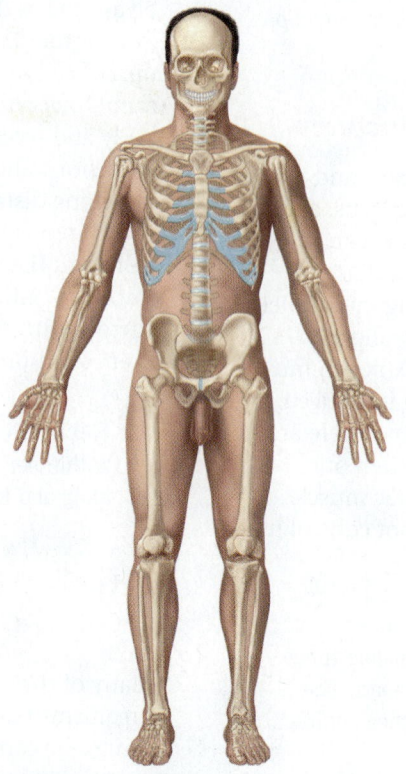

Skeletal system
- Protects, supports, and anchors body parts
- Provides the structural framework for movement
- Produces blood cells
- Stores minerals

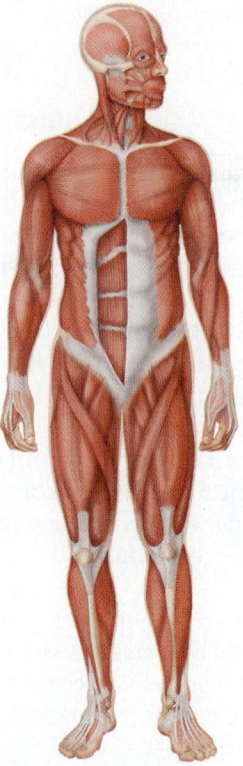

Muscular system
- Produces movement or resists movement
- Generates heat

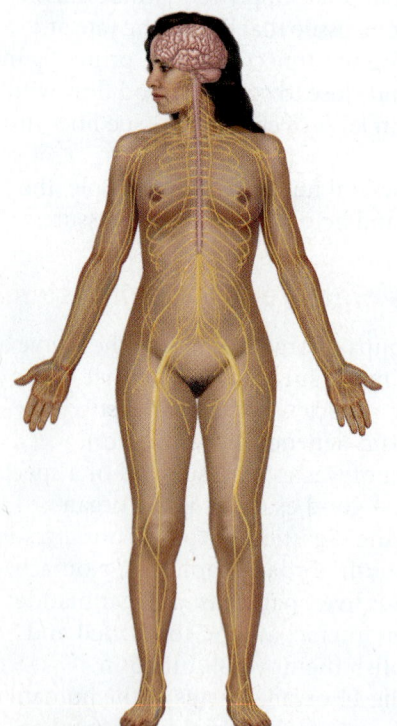

Nervous system
- Detects both external and internal stimuli
- Controls and coordinates rapid responses to these stimuli
- Integrates the activities of other organ systems

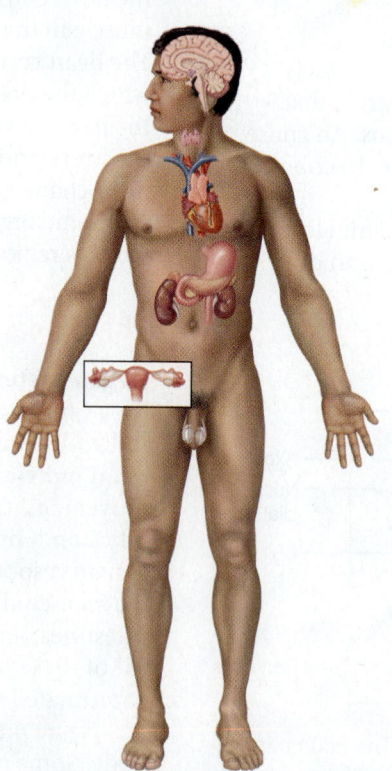

Endocrine system
- Produces hormones that regulate many body functions
- Participates with the nervous system in integrative functions

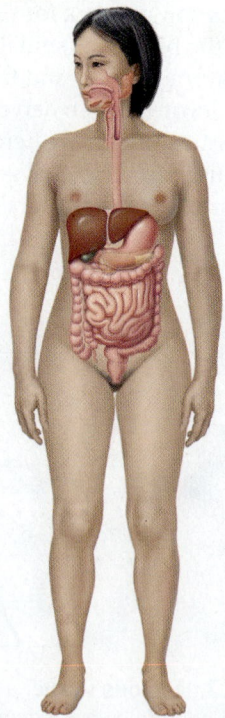

Digestive system
- Provides the body with water and nutrients
- (The liver) synthesizes certain proteins and lipids for the body
- (The liver) inactivates many chemicals, including hormones, drugs, and poisons

Figure 4.8 The 11 organ systems of the human body.

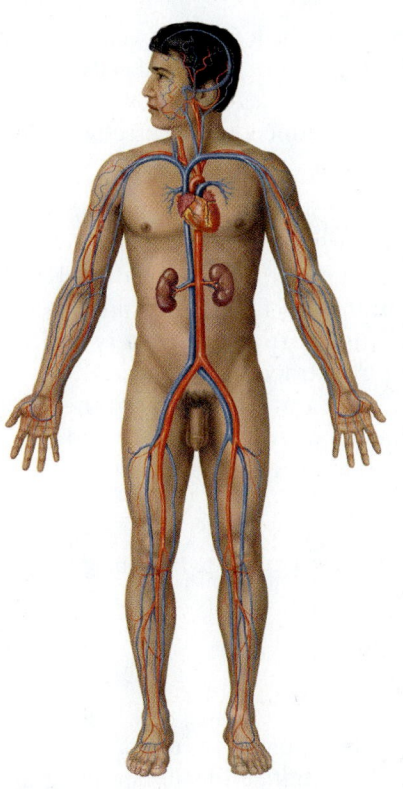

Circulatory system
- Transports materials to and from all cells
- Participates in the maintenance of body temperature
- Participates in mechanisms of defense against disease and injury

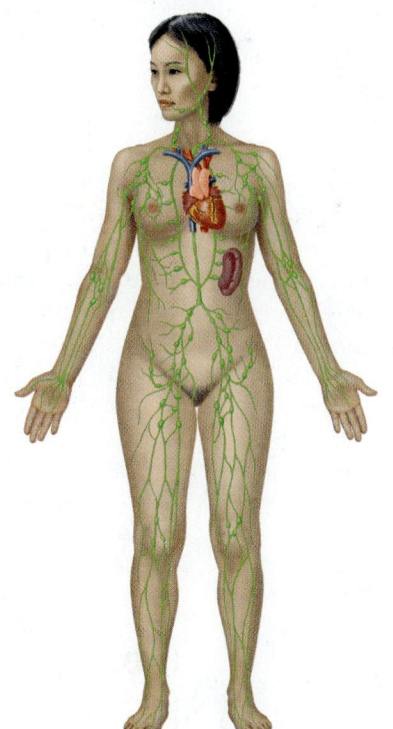

Lymphatic system
- Returns excess tissue fluid to the circulatory system
- Participates in both general and specific (immune) defense responses

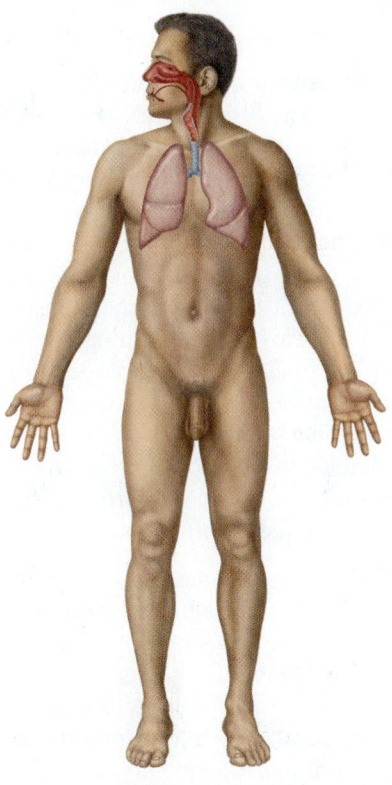

Respiratory system
- Exchanges gases (oxygen and carbon dioxide) between air and blood
- Participates in the production of sound (vocalization)

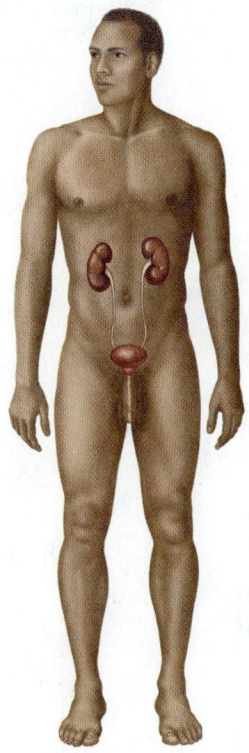

Urinary system
- Maintains the volume and composition of body fluids
- Excretes some waste products

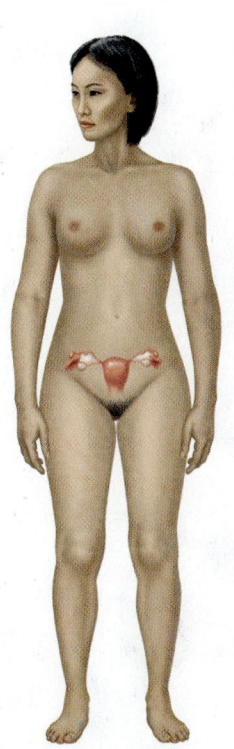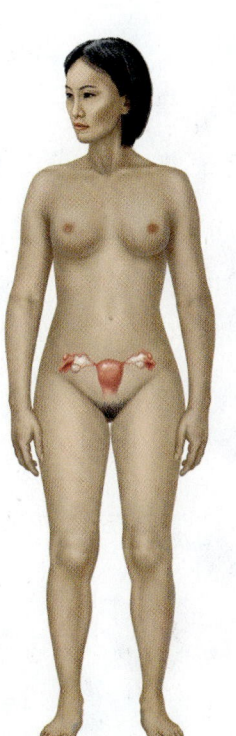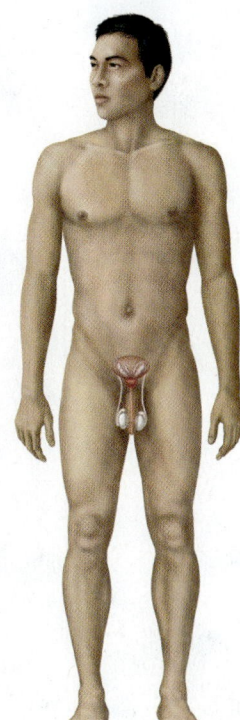

Reproductive system
- Female: Produces eggs
- Female: Nurtures the fertilized egg, developing embryo, and fetus until birth
- Male: Produces sperm
- Male: Participates in the delivery of sperm to the female

Tissue membranes line body cavities

Some of the organs and organ systems are located in hollow cavities within the body (**Figure 4.9**). The large anterior cavity is divided into the thoracic cavity and abdominal cavity by the diaphragm between them. The thoracic cavity is in turn divided into two pleural cavities, each containing a lung, and the pericardial cavity, which encloses the heart. The lower part of the abdominal cavity is sometimes called the *pelvic cavity*. The smaller posterior cavity consists of the cranial cavity and the spinal cavity (vertebral canal). Many other smaller cavities exist as well, such as the synovial cavities in movable joints.

Tissue membranes consisting of a layer of connective tissue and (with one exception) a layer of epithelial cells line each body cavity and form our skin. There are four major types of tissue membranes:

- *Serous membranes.* Line and lubricate internal body cavities to reduce friction between internal organs.
- *Mucous membranes.* Line the airways, digestive tract, and reproductive passages that are open to the outside of the body. Goblet cells within the epithelial layer secrete *mucus*, which lubricates the membrane's surface and entraps foreign particles.
- *Synovial membranes.* Line the very thin cavities between bones in movable joints. These membranes secrete

a watery fluid that lubricates the joint. Synovial membranes do not have a layer of epithelial cells.
- *Cutaneous membrane.* Our outer covering. You know it as skin, and it serves several functions discussed later in this chapter.

By now, you may have noticed that *membrane* is a general term for a thin layer that covers or surrounds a structure in the body. You have been introduced to three different membranes so far: the *plasma* membrane of phospholipids surrounding every cell, the *basement* membrane of extracellular material on which epithelial tissue rests, and *tissue* membranes that consist of several layers of tissue sandwiched together that cover or surround cavities, organs, and entire organ systems.

☑ What kind of membrane would you expect to find lining a pleural cavity? Explain.

Describing body position or direction

When describing parts of the body, biologists use precise terms to define position and direction. Generally speaking, an organ or even the entire body can be described by three planes known as the *midsagittal, frontal,* and *transverse*

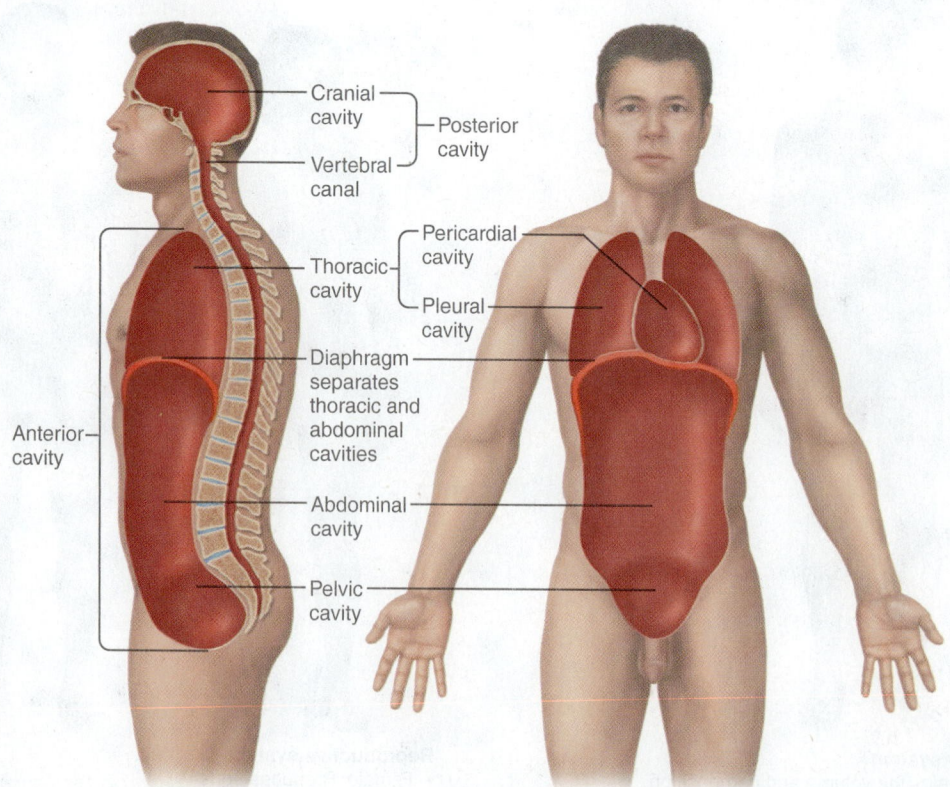

Figure 4.9 **The main body cavities.** The pelvic cavity and the abdominal cavity are continuous (not separated by a membrane).

(**Figure 4.10**). These planes divide the body into left and right, front and back, and top and bottom, respectively. *Anterior* means "at or near the front" and *posterior* means "at or near the back." *Proximal* means "nearer (in closer proximity) to" any point of reference, usually the body trunk, and *distal* means "farther away." For example, your wrist is distal to your elbow. *Superior* means "situated above" or "directed upward," and *inferior* means "situated below" or "directed downward."

 Recap An organ consists of several tissue types that join together to perform a specific function. An organ system is a group of organs that share a broad function important for survival. The body's hollow cavities are lined by tissue membranes that support, protect, and lubricate cavity surfaces. ∎

4.7 The skin as an organ system

The proper name for the skin and its accessory structures such as hair, nails, and glands is the **integumentary system** (from the Latin *integere*, meaning "to cover"). We describe the skin here as a representative organ system; other organ systems are covered later in the book.

Skin has many functions

The skin has several different functions related to its role as the outer covering of the body:

- Protection from dehydration (helps prevent our bodies from drying out)
- Protection from injury (such as abrasion)
- Defense against invasion by bacteria and viruses
- Regulation of body temperature
- Synthesis of an inactive form of vitamin D
- Sensation: provides information about the external world via receptors for touch, vibration, pain, and temperature

MJ's BlogInFocus How might wearable skin patches be used to improve human health? Visit MJ's blog in the Study Area in MasteringBiology and look under "Wearable Skin Patches."

http://goo.gl/8Q8yyN

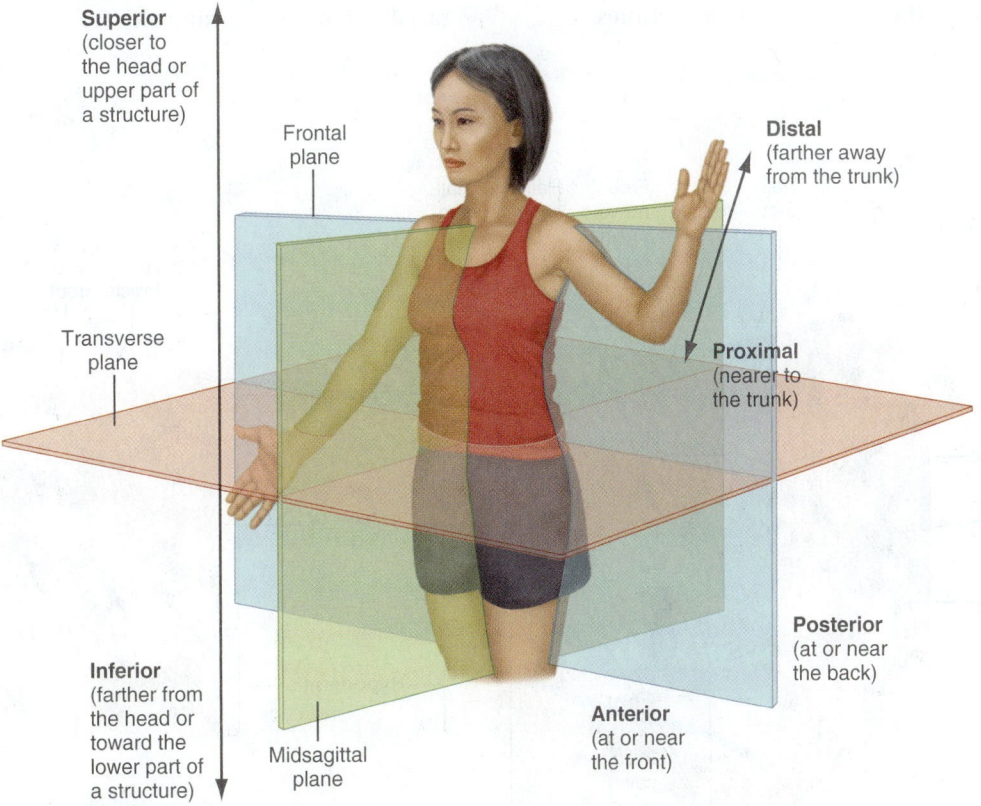

Superior
(closer to the head or upper part of a structure)

Frontal plane

Distal
(farther away from the trunk)

Transverse plane

Proximal
(nearer to the trunk)

Inferior
(farther from the head or toward the lower part of a structure)

Midsagittal plane

Posterior
(at or near the back)

Anterior
(at or near the front)

Figure 4.10 Planes of symmetry and terms used to describe position or direction in the human body. The frontal plane divides the body into anterior and posterior sections, the midsagittal plane divides it into left and right, and the transverse plane divides it into superior and inferior sections. Proximal and distal refer to points closer to or farther away from a point of reference, usually the trunk.

✔️ A friend who is an anatomist tells you that he has a blister on the inferior surface of his foot. He adds that it is posterior, not anterior. Where is the blister?

Skin consists of epidermis and dermis

Recall that skin is a tissue membrane and that tissue membranes contain layers of epithelial and connective tissue. The outer layer of the skin's epithelial tissue is the **epidermis,** and the inner layer of connective tissue is the **dermis** (Figure 4.11).

The skin rests on a supportive layer called the *hypodermis* (*hypo-* means "under"), consisting of loose connective tissue containing fat cells. The hypodermis is flexible enough to allow the skin to move and bend. The fat cells in the hypodermis insulate against excessive heat loss and cushion against injury.

Epidermal cells are replaced constantly The epidermis consists of multiple layers of squamous epithelial cells. A key feature of the epidermis is that it is constantly being replaced as cells near the base of the epidermis divide repeatedly, pushing older cells toward the surface.

Two types of cell make up the epidermis: keratinocytes and melanocytes. The more numerous of the two cell types are **keratinocytes,** which produce a tough, waterproof protein called *keratin*. Actively dividing keratinocytes located near the base of the epidermis are sometimes called *basal cells*. As keratinocytes derived from the basal cells move toward the skin surface, they flatten and become squamous. Eventually, they die and dry out, creating a nearly waterproof barrier that covers and protects the living cells below (Figure 4.12). The rapid replacement of keratinocytes allows the skin to heal quickly after injury.

One reason the outer layers of epidermal cells die is that the epidermis lacks blood vessels, so as mature cells are pushed farther from the dermis they can no longer obtain nutrients. The dead cells of the outer layers are shed over time, accounting for the white flakes you sometimes find on your skin or on dark clothing, especially when your skin is dry.

Less numerous cells, called **melanocytes,** are also located near the base of the epidermis. Melanocytes produce a dark-brown pigment called *melanin,* which is then picked up and stored by the nearby keratinocytes. The presence of melanin in the keratinocytes protects us against the sun's ultraviolet radiation. Exposure to sunlight increases the activity of melanocytes, accounting for the ability of some people to develop a suntan. (For more on tanning, see Health & Wellness, *Suntans, Smoking, and Your Skin.*) Because all humans have about the same number of melanocytes, racial differences in skin color reflect either differences in

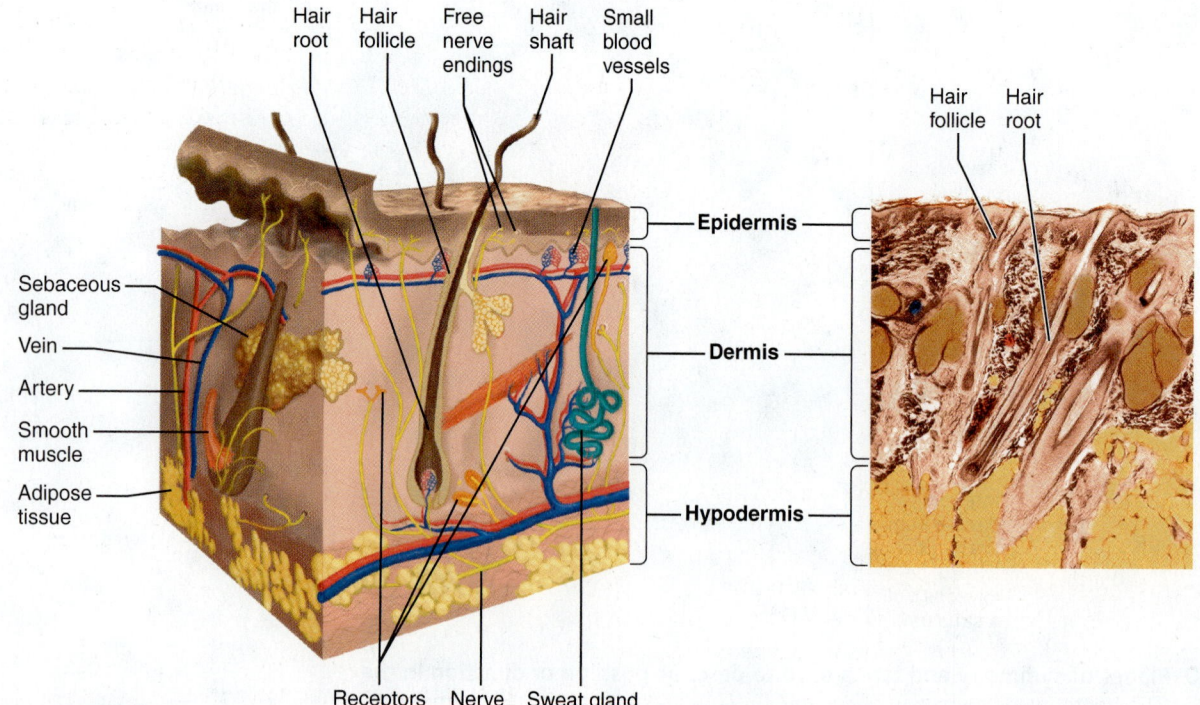

Figure 4.11 The skin. The two layers of skin (epidermis and dermis) rest on a supportive layer (hypodermis). Although not part of the skin, the hypodermis serves important functions of cushioning and insulation.

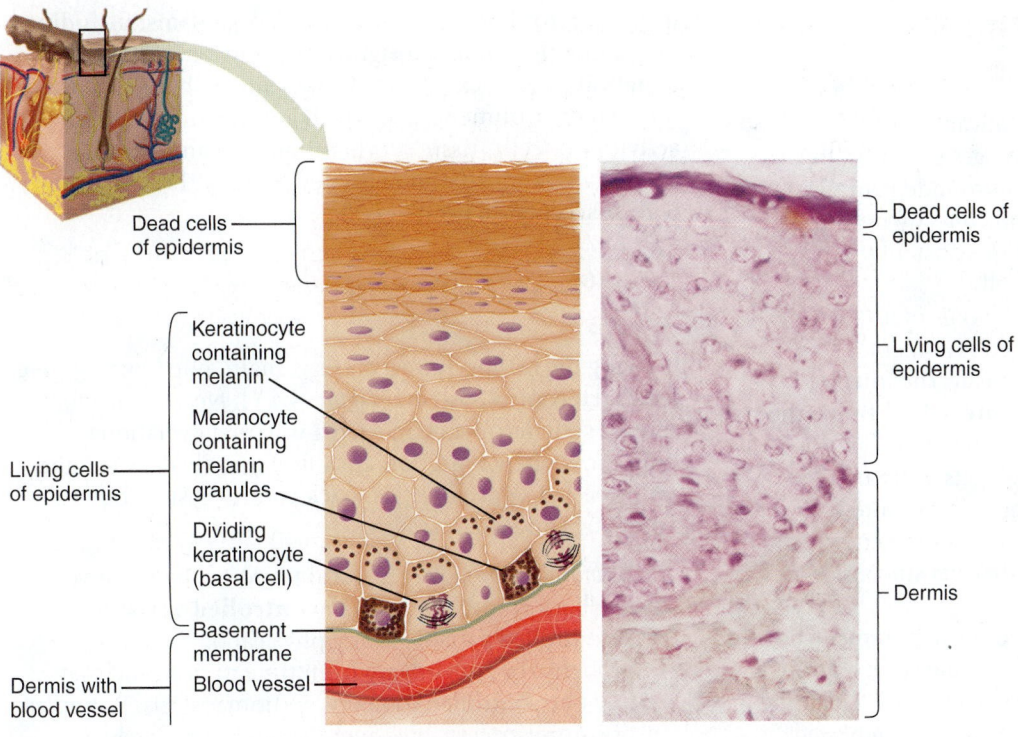

Figure 4.12 The epidermis. Living cells near the base of the epidermis divide, pushing more mature cells toward the surface. As cells migrate toward the surface they die and dry out, forming a tough, waterproof barrier. The cells of the epidermis are supplied only by blood vessels located in the dermis.

melanocyte activity or differences in the rate of breakdown of melanin once it is produced.

☑ What would happen to your skin if your keratinocytes were to start dividing slightly more rapidly than usual?

Fibers in dermis provide strength and elasticity The dermis is primarily dense connective tissue, consisting of collagen, elastic, and reticular fibers embedded in a ground substance of water, polysaccharides, and proteins. The fibers allow the skin to stretch when we move and give it strength to resist abrasion and tearing. Our skin becomes less flexible and more wrinkled as we age because the number of fibers in the dermis decreases.

The surface of the dermis has many small projections called *papillae* that contain sensory nerve endings and small blood vessels. When the skin is rubbed excessively—such as when your shoes are too tight—the epidermis and dermis separate from each other and a fluid-filled blister develops between them.

The most abundant living cells in the dermis are the fibroblasts that produce the various fibers, but there are also mast cells, white blood cells, and occasional fat cells. Other structures in the dermis include the following:

- *Hair.* Each hair has a *shaft* above the skin's surface and a *root* below the surface. Hair is actually composed of

several layers of cells enclosed in an outer layer of overlapping, dead, flattened keratinocytes. The root of a hair is surrounded by a sheath of several layers of cells called the *follicle.* The cells at the very base of the follicle are constantly dividing to form the hair root. As new hair cells are formed at the base, the hair root is pushed upward toward the skin's surface.

- *Smooth muscle.* Attached to the base of the hair follicle, smooth muscle contracts when you are frightened or cold, causing your hair to become more erect.
- *Sebaceous glands.* Also known as oil glands, these secrete an oily fluid that moistens and softens hair and skin.
- *Sweat glands.* These produce sweat, a watery fluid containing dissolved ions, small amounts of metabolic wastes, and an antibiotic peptide called *dermicidin.* Sweat helps regulate body temperature and protects against bacteria.
- *Blood vessels.* These supply the cells of the dermis and epidermis with nutrients and remove their wastes. The blood vessels also help regulate body temperature. They dilate to facilitate heat loss when we are too hot and constrict to prevent heat loss when we are too cool. The dermis also contains lymph vessels, which drain fluids and play a role in the immune system.
- *Sensory nerve endings.* These provide information about the outside environment. Separate receptors on nerve endings exist to detect heat, cold, light touch, deep pressure, and vibration.

As mentioned earlier, the skin synthesizes an inactive form of vitamin D. It is not known which cell type in the skin is responsible. But we do know that a cholesterol-like molecule in the skin becomes an inactive form of vitamin D when it is exposed to the ultraviolet rays of sunlight. The inactive form must then be modified in the liver and kidneys before it becomes active.

↩ **Recap** The skin is an organ because it consists of different tissues serving common functions. Functions of skin include protection, temperature regulation, vitamin D synthesis, and sensory reception. ▪

4.8 Multicellular organisms must maintain homeostasis

Although multicellularity offers many advantages to organisms, it presents certain disadvantages that must be overcome. For example, cells that are surrounded entirely by other cells can't obtain their nutrients directly from the organism's external environment and are constantly exposed to the waste products of neighboring cells.

The environment that surrounds the cells of a multicellular organism (their external environment) is the **internal environment** of the organism. The internal environment is a clear fluid called the **interstitial fluid** (the Latin noun *interstitium* means "the space between," in this case, the space between cells). Every cell gets nutrients from the interstitial fluid around it and dumps wastes into it. In a multicellular organism, the interstitial fluid is the equivalent of the ocean, lake, or tiniest drop of fluid that surrounds and nourishes single-celled organisms.

Because every cell must receive all of its requirements for life from the surrounding interstitial fluid, the composition of this fluid must be kept fairly constant to sustain life. In the long run, nutrients consumed by the cell must be replaced and wastes must be removed or the cell will die.

Relative constancy of the conditions within the internal environment is called **homeostasis** (*homeo-* means "unchanging" or "the same," and *-stasis* means "standing"). The maintenance of homeostasis is so important for life that multicellular organisms, including human beings, devote a significant portion of their total metabolic activities to it. Although small changes in the internal environment do occur from time to time, the activities of cells, tissues, organs, and organ systems are carefully integrated and regulated to keep these changes within acceptable limits.

Homeostasis is maintained by negative feedback

In living organisms, homeostasis is maintained by **negative feedback** control systems (**Figure 4.13**). Negative feedback control systems operate in such a way that deviations from the desired condition are automatically detected and counteracted. A negative feedback control system has the following components:

- *A controlled variable.* The focal point of any negative feedback control loop is the **controlled variable.** A controlled variable is any physical or chemical property that might vary from time to time and that must be controlled to maintain homeostasis. Examples of controlled variables are blood pressure, body temperature, and the concentration of glucose in blood.
- *A sensor (or receptor).* The **sensor** monitors the current value of the controlled variable and sends the information (via either nerves or hormones) to the control center.

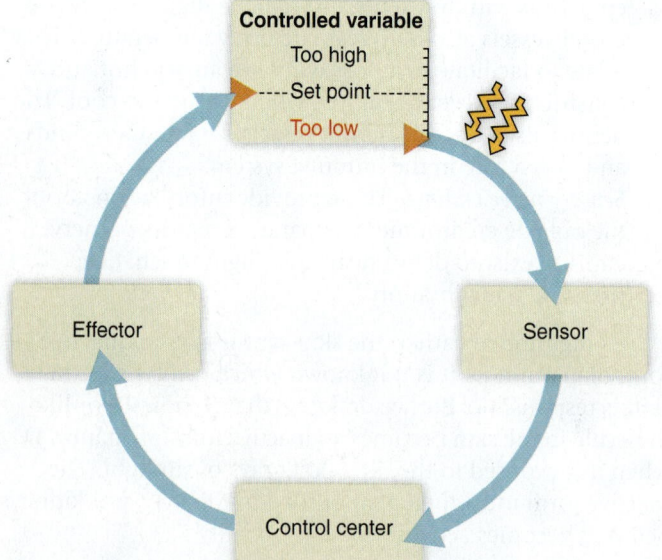

a) A *decrease* in the controlled variable causes events that *raise* the controlled variable toward its set point again.

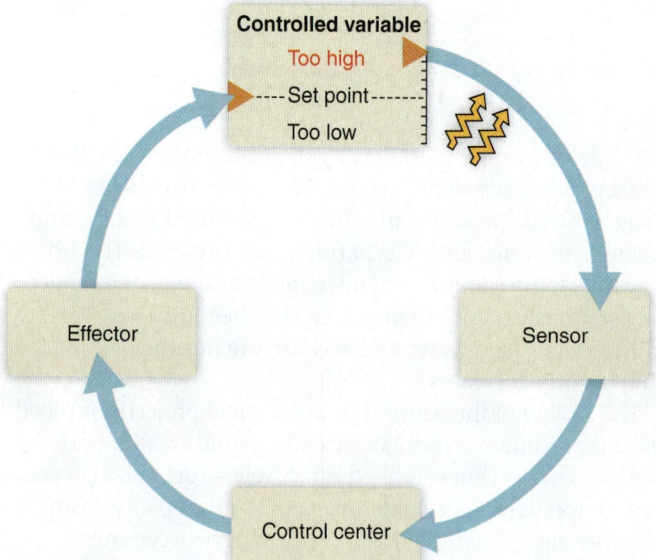

b) An *increase* in the controlled variable causes events that *lower* the controlled variable toward its set point again.

Figure 4.13 Components of a negative feedback control system. The focal point of the control system is the controlled variable. A sensor monitors the controlled variable and sends signals to a control center, which compares the current value of the controlled variable with its set point. If the controlled variable and set point do not match, the control center sends signals to effectors that take action to reverse the difference between the controlled variable and its set point.

- *A control center.* The **control center** receives input from the sensor and compares it to the correct, internally set value of the controlled variable, sometimes called the **set point.** When the current value and the set point are not in agreement, the control center sends signals (again, via either nerves or hormones) to an effector.
- *An effector.* The **effector** takes the necessary action to correct the imbalance, in accordance with the signals it receives from the control center.

The cycle is called *negative feedback* because any change in the controlled variable triggers a series of events that ultimately opposes (negates) the initial change, returning the variable to its set point. In other words, homeostasis is maintained.

☑ Your refrigerator's ability to maintain a relatively steady cool temperature is another example of a feedback system. Is it a negative or a positive feedback system? Identify the major components of the system.

Negative feedback helps maintain core body temperature

A prime example of negative feedback is the maintenance of homeostasis of your body temperature. In this case, multiple organ systems participate in maintaining homeostasis (**Figure 4.14**).

The controlled variable is your core temperature, meaning the temperature near the center of your body. Temperature sensors in your skin and internal organs monitor core temperature. These sensors transmit signals via nerves to the control center, located in a region of your brain called the *hypothalamus.* The control center uses different combinations of effector mechanisms to raise or lower core temperature as needed.

When your core temperature falls *below* its set point, the hypothalamus:

- Sends more nerve impulses to blood vessels in the skin, causing the blood vessels to constrict. This restricts blood flow to your skin and reduces heat loss.
- Stimulates your skeletal muscles, causing brief bursts of muscular contraction known as shivering. Shivering generates heat.

When your core temperature rises *above* its set point, the hypothalamus:

- Sends fewer nerve impulses to blood vessels in the skin, causing the blood vessels to dilate. This increases blood flow to your skin and promotes heat loss.
- Activates your sweat glands. As perspiration evaporates from your skin, you lose heat.

Even when your core temperature is normal, your hypothalamus is transmitting some nerve impulses to the

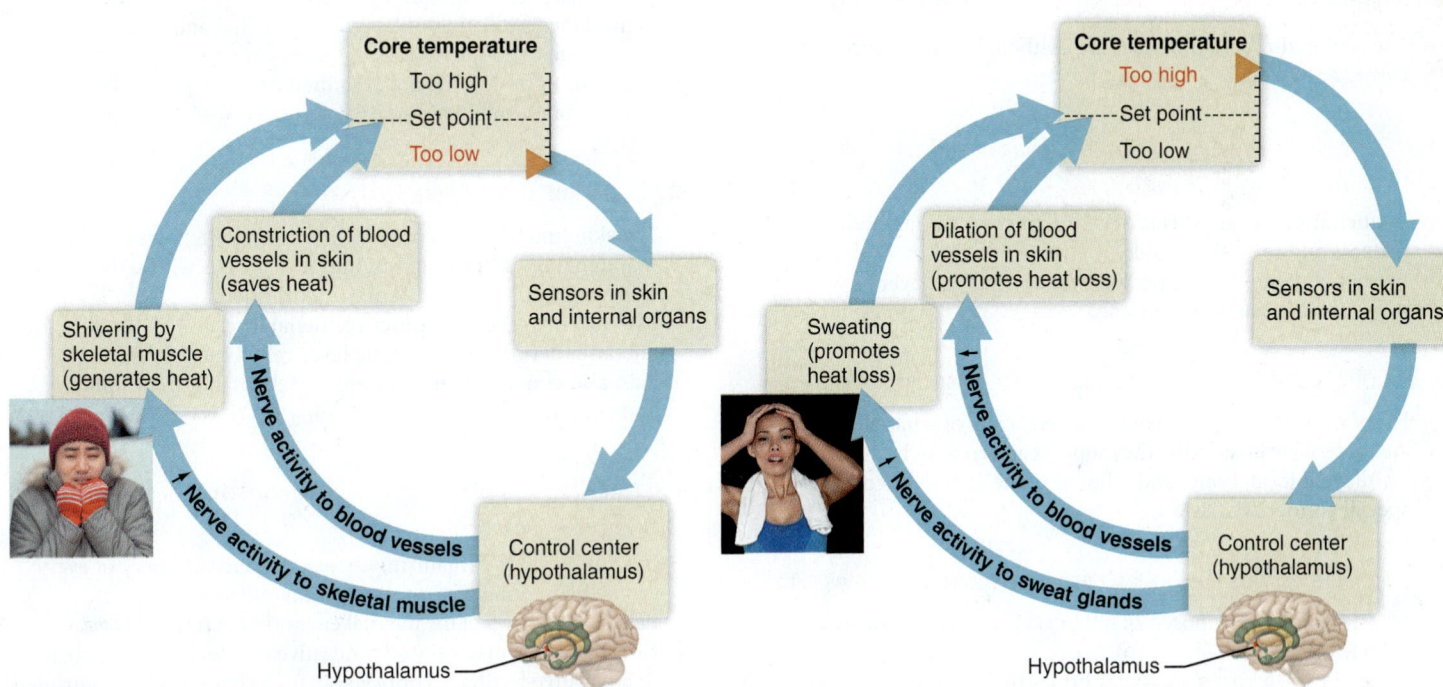

Figure 4.14 Negative feedback control of core temperature. Note that different combinations of effector mechanisms may be activated, depending on the direction of the initial change in core temperature.

☑ In most people, hunger (and food intake) appears to be regulated in such a way that the amount of body fat stays surprisingly constant over time (despite people's efforts to lose weight). Draw a diagram showing how hunger and food intake might be regulated in a negative feedback system to keep body fat stores at a certain set point.

blood vessels in your skin. Small changes in temperature, then, can be handled effectively just by increasing or decreasing the normal number of signals. Only when the variations in temperature are large is sweating or shivering called into play.

From this example, we can make the following points about negative feedback control:

- Many sensors may be active at once. In this case, sensors throughout the body monitor body temperature.
- The control center integrates all of this incoming information and comes up with an appropriate response.
- There can be multiple effectors as well as multiple sensors that may belong to different organ systems.

Both skin and muscles function in returning body temperature to its set point.

☑ People who have fevers will often get "chills"—feeling as if they are cold and shivering—even though body temperature is actually above normal. Later, when the fever "breaks," they will suddenly start sweating. Explain why this happens.

Positive feedback amplifies events

Positive feedback control systems are relatively uncommon in living organisms. In positive feedback, a change in the controlled variable sets in motion a series of events that *amplify* the original change, rather than returning the controlled variable to a set point. The process of childbirth once labor has started is governed by positive feedback mechanisms.

Obviously something must terminate positive feedback events. The contractions of childbirth end when the child is born. The important point is that positive feedback is not a mechanism for maintaining homeostasis.

↺ **Recap** All multicellular organisms must maintain homeostasis of their internal environment. In a negative feedback control system, any change in a controlled variable sets in motion a series of events that reverse the change, maintaining homeostasis. ■

Chapter Summary

4.1 Tissues are groups of cells with a common function *p. 79*

- The four main types of tissues are epithelial tissue, connective tissue, muscle tissue, and neural tissue.

4.2 Epithelial tissues cover body surfaces and cavities *p. 79*

- Epithelial tissues are sheets of cells that cover or line body surfaces and form the glands.
- Epithelial tissues are supported by a noncellular layer called the basement membrane.

4.3 Connective tissue supports and connects body parts *p. 82*

- Fibrous connective tissues contain several types of extracellular fibers and only a few living cells. They support and connect body parts.
- Cartilage, blood, bone, and adipose tissue are classified as special connective tissues.

4.4 Muscle tissues contract to produce movement *p. 86*

- Muscle tissue is composed of either skeletal, cardiac, or smooth muscle cells.
- Skeletal muscles are attached to bones by tendons.
- Skeletal muscle contraction is voluntary; contractions of smooth and cardiac muscle are involuntary.

4.5 Nervous tissue transmits impulses *p. 87*

- Neurons are specialized for conduction of electrical impulses.
- Glial cells surround and protect neurons and supply them with nutrients.

4.6 Organs and organ systems perform complex functions *p. 87*

- The human body is composed of 11 organ systems, each of which has at least one broad function.
- Membranes consisting of layers of epithelial and connective tissues line the body cavities.
- Positions of body parts are described on three planes: midsagittal, frontal, and transverse.

4.7 The skin as an organ system *p. 91*

- The skin functions as a protective barrier, participates in the maintenance of homeostasis, and provides us with sensory information about the external environment.
- Skin has two layers: an outer epithelial layer called the *epidermis* and an inner connective tissue layer called the *dermis*.
- Skin also contains nerves, blood vessels, glands, hair follicles, and smooth muscle.

4.8 Multicellular organisms must maintain homeostasis *p. 94*

- In a multicellular organism, the external environment of every cell is the internal environment of the organism.
- Relative constancy of the internal environment is called *homeostasis*.
- Homeostasis is maintained by negative feedback control systems.
- In a negative feedback control system, a change in the controlled variable sets in motion a sequence of events that tends to reverse (or negate) the initial change.
- In the regulation of body temperature, sensors located throughout the body send information about temperature to the control center, located in the hypothalamus of the brain.
- Possible responses to a change in body temperature include dilating or constricting the blood vessels to the skin, shivering (if temperature is too low), and sweating (if temperature is too high).

Terms You Should Know

basement membrane, *81*
cell junction, *81*
connective tissue, *82*
controlled variable, *94*
dermis, *92*
endocrine gland, *80*
epidermis, *92*
epithelial tissue, *79*
exocrine gland, *80*

homeostasis, *94*
internal environment, *94*
muscle tissue, *86*
negative feedback, *94*
nervous tissue, *87*
neuron, *87*
organ systems, *87*
set point, *95*

Concept Review

Answers can be found in the Study Area in MasteringBiology.

1. Describe some advantages and disadvantages of multicellularity.
2. Name the four main types of tissues in the human body, and list their main functions.
3. Describe the functions of the three types of cell junctions.
4. Distinguish between an organ and an organ system.
5. List the 11 organ systems of the body and give at least one function of each.
6. Define *interstitial fluid*.
7. Name the two cavities of the anterior body cavity that are separated from each other by the diaphragm.
8. Compare/contrast positive and negative feedback.
9. Discuss the purpose of homeostasis in the body.
10. Describe the function of a control center in a negative feedback control system.

Test Yourself

Answers can be found in the Appendix.

1. Collagen and elastin fibers are typically found in:
   a. connective tissue
   b. epithelial tissue
   c. intercellular junctions
   d. muscle tissue
   e. the epidermis

2. Cells in cardiac muscle are able to contract in a coordinated fashion because of communication made possible through:
   a. gap junctions
   b. spot desmosomes
   c. adhesion junctions
   d. tight junctions
   e. synapses

3. Which of the following membranes is not composed of cells?
   a. serous membrane
   b. synovial membrane
   c. cutaneous membrane
   d. basement membrane
   e. mucous membrane

4. Exocrine and endocrine glands are types of:
   a. loose connective tissue
   b. epithelial tissue
   c. squamous tissue
   d. fibrous connective tissue
   e. specialized connective tissue

_____ to the
_____ entral
_____ istal

_____ e in common?
_____ ments
_____ work

_____ ry to cartilage because:

_____ prenatal
   c. the polysaccharides in cartilage ground substance can't be replaced
   d. there is a richer blood supply to bone
   e. bone has a higher mineral content

8. Which of the following tissues may be found in the skin?
   a. smooth muscle
   b. fibrous connective tissue
   c. nervous tissue
   d. epithelial tissue
   e. all of these tissues

9. Which of the following is responsible for skin pigmentation?
   a. melanin
   b. keratin
   c. collagen
   d. sebum
   e. dermicidin

10. When a decrease in blood pressure is detected by the central nervous system, the central nervous system triggers several changes that will return the blood pressure to its set point. This is an example of:
    a. positive feedback
    b. thermoregulation
    c. negative feedback
    d. reverse feedback
    e. set point feedback

11. The presence of a full bladder triggers the bladder to contract. As a little urine is released, this causes more contractions, which will completely empty the bladder. This is an example of:
    a. homeostatic regulation
    b. uroregulation
    c. negative feedback
    d. positive feedback
    e. reverse feedback

12. Which type of tissue stores triglycerides?
    a. muscle tissue
    b. loose connective tissue
    c. fibrous connective tissue
    d. columnar epithelial tissue
    e. adipose tissue

13. A substantial amount of nonliving extracellular material, also known as the matrix, characterizes all:
    a. muscle tissue
    b. epithelial tissue
    c. connective tissue
    d. nervous tissue
    e. membranes

14. All of the following organ systems may be involved in thermoregulation except the:
    a. integumentary system
    b. muscular system
    c. circulatory system
    d. skeletal system
    e. nervous system

15. Reducing food intake may lead to weight loss by decreasing the:
    a. number of muscle cells
    b. amount of fibrous connective tissue
    c. volume of adipocytes
    d. volume of muscle cells
    e. number of adipocytes

Apply What You Know

Answers can be found in the Study Area in MasteringBiology.

1. Your roommate says that the concept of homeostasis is being violated when the rate of respiration goes up during exercise, because the rate of respiration clearly is not being held constant. Explain to him where his thinking is faulty.
2. What do you think would be some of the problems associated with severe third-degree burns, in which both the epidermis and the dermis are severely damaged or destroyed?
3. Sherlock Holmes, the greatest fictional detective of all time, is talking to a woman in her late 40s, when he suddenly says, "I see, my dear madam, you must have enjoyed your cigarettes and your suntans." The woman is amazed, because she mentioned nothing about these two former favorite activities. What physical characteristics might Mr. Holmes have seen in this woman to indicate she was an avid sun-worshiper and cigarette smoker? How do these characteristics develop?
4. Dieting is difficult. People who do manage to lose weight can gain it back if they're not careful. Are techniques such as liposuction or lipodissolve a good way to keep the weight off permanently? Why or why not?
5. Fibrous connective tissue consists of ground substance and fibers that provide strength, support, and flexibility. Concrete is used to make tough, durable structures in construction projects. How is a concrete structure like or unlike fibrous connective tissue?

6. By definition, an organ is a structure composed of two or more tissue types that perform a specific function. Performance of that function often requires coordination among many cells. Why is it so important that cardiac muscle cells of the heart be synchronized (coordinated) so that they beat nearly all at once?

MJ's BlogInFocus

Answers can be found in the Study Area in MasteringBiology.

1. What does the SPF on a suntan lotion product stand for? How high does the SPF number have to be for the product to be advertised as protective against cancer? For more on this topic, visit MJ's blog in the Study Area in MasteringBiology and look under "SPF." http://goo.gl/pzeb3s

MasteringBiology®

Students Go to MasteringBiology for assignments, the eText, and the Study Area with animations, practice tests, and activities.

Professors Go to MasteringBiology for automatically graded tutorials and questions that you can assign to your students, plus Instructor Resources.

Answers to ✓ questions are available in the Appendix.

The Skeletal System

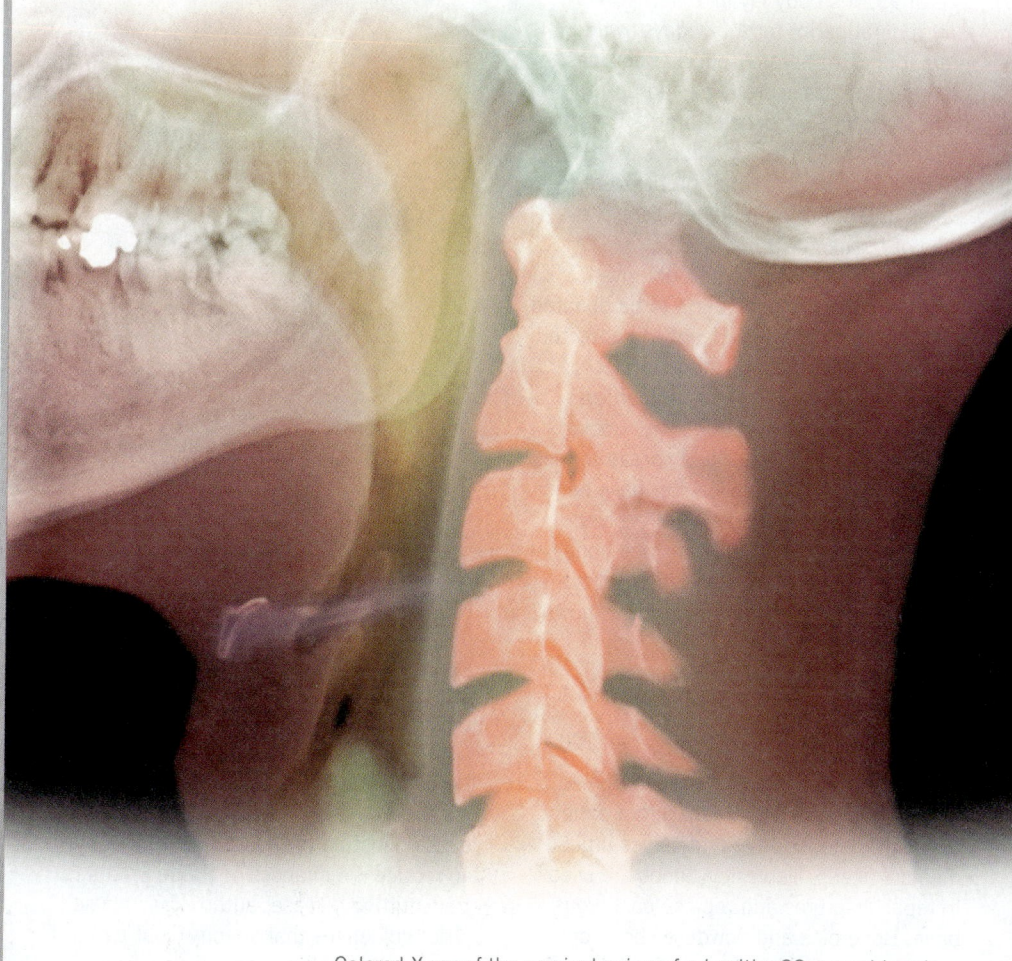

Colored X-ray of the cervical spine of a healthy 29-year-old male.

Key Concepts

- **The skeletal system is composed of bones, ligaments, and cartilage.** The skeletal system supports and protects the other organ systems of the body and provides a structure that enables movement.

- **Bones consist of living cells surrounded by extracellular deposits of calcium minerals.** Bone tissue undergoes constant replacement, remodeling, and repair.

- **Bones store minerals and produce the cellular components of blood** (red blood cells, white blood cells, and platelets).

- **Ligaments comprised of connective tissue hold bones together.** When damaged, ligaments are slow to heal because they have very few living cells and a poor blood supply.

- **Joints are the points of contact between bones.** In a movable joint, bone surfaces are covered by a layer of smooth cartilage and lubricated with fluid to reduce friction and wear.

CURRENT ISSUE

A Black Market in Human Bones?

Questions to Consider

1 Do you approve of harvesting bone from cadavers for the purpose of processing the tissue into bone-based products for patients, provided the tissue is legally obtained?

2 Do you think it should be legal to *sell* a cadaver or its parts?

Alistair Cooke, famed host of the PBS series *Masterpiece Theatre,* died in 2004 at the age of 95. His body lay in a New York City funeral parlor for a few days awaiting cremation. But before Cooke's body was cremated, it was secretly carved up in a back room and his bones were removed. Authorities allege that his bones were then sold for a substantial profit, to be transplanted into patients in desperate need of tissue grafts.

Cooke's family, who had not given permission for his body parts to be donated, knew nothing of this until police contacted them after the funeral. Understandably, they were appalled. But there is an even more horrifying side to this story: Cooke died of lung cancer that had spread to his bones. Could his deadly cancer have been transmitted to the people who received his bone tissue? It's unlikely since bone products generally are sterilized, but the answer to this question may not be known for decades.

Recycling Body Parts: A Legitimate Industry

The process of harvesting tissues from a human corpse for transplant into a patient is a legitimate industry that serves urgent medical needs. Harvested bones are used to repair fractures and replace cancerous bone. Bone pins and powdered bone created from donated bone is used in dental surgery; bone paste plugs holes. Tendons and ligaments taken from donors are used to repair joints and tissues damaged by sports

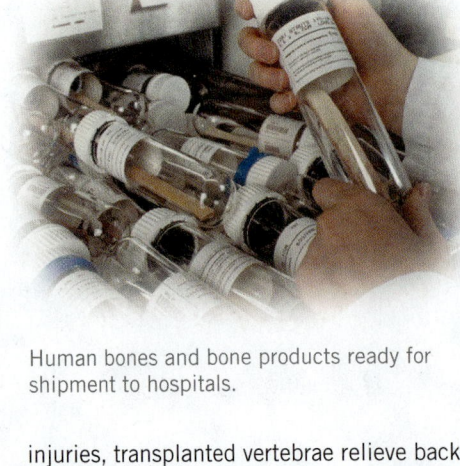

Human bones and bone products ready for shipment to hospitals.

injuries, transplanted vertebrae relieve back pain, and veins and heart valves are used in heart surgeries. The bones, tendons, veins, and heart valves from just one corpse can be worth over $200,000 to surgeons, hospitals, and recipients.

Under federal law, it is illegal to sell human body parts for a profit—they can only be donated, either by the patient while he or she is still alive or by the family after death. Several hundred licensed nonprofit tissue banks in the United States receive donated tissues and test them for infectious diseases such as HIV, syphilis, and the viruses that cause hepatitis (inflammation of the liver). To reduce the chances of tissue products transmitting disease, authorities impose strict guidelines that specify what types of tissues may be harvested, from whom they may be harvested, and how they must be processed. For instance, to prevent any risk of transmitting cancer, federal guidelines prohibit the use of bones from cancer patients for tissue implants.

After donation, bone tissue is shaped into usable forms, such as pins, plates, and powders. The final products are sterilized and shipped to hospitals and surgeons all over the country, where they

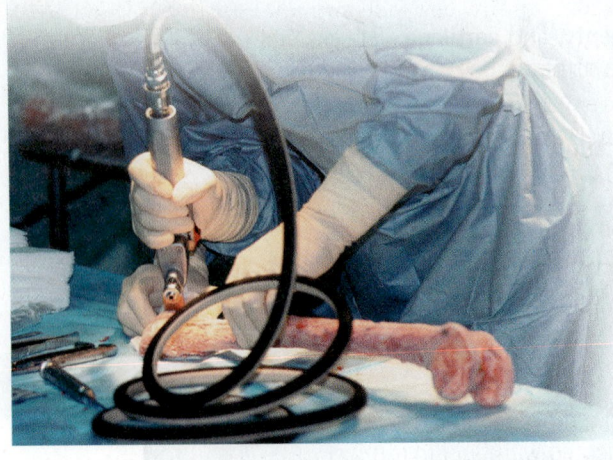

Donated bones are cleaned, sterilized, and shaped into bone products.

are used in more than 600,000 surgical procedures every year. The patient pays all fees incurred in the handling, processing, testing, and shipping of the products, but the tissue banks themselves do not make a profit.

Illegal Body Parts Enter the Supply Chain

The imbalance between supply and demand for human body parts leaves the entire system vulnerable to exploitation. In the Alistair Cooke case, prosecutors alleged that Michael Mastromarino, an oral surgeon who had lost his license, arranged for a Brooklyn funeral parlor to deliver bodies to a secret operating room. There, Mastromarino and his accomplices removed body parts before the bodies were buried or cremated. Authorities say that the men paid the funeral parlor up to $1,000 per body and then sold the harvested tissues for up to $7,000 per body to a legitimate but unsuspecting tissue-processing company.

In some cases, Mastromarino and his accomplices falsified records indicating the deceased's age and cause of death. Mr. Cooke died of cancer at the age of 95, but his records were falsified to indicate that he died at age 85 of a heart attack. They also allegedly looted body parts from a 43-year-old woman who had died of ovarian cancer; they then forged a signature on a consent form and listed the cause of death as a head injury. Prosecutors eventually identified over a thousand corpses from which body parts were taken without permission between 2001 and 2005.

In 2008, Mr. Mastromarino pled guilty in a plea bargain in exchange for providing information about others who were involved. He was expected to spend at least 18 years in prison, but he died of liver and bone cancer in 2013. An accomplice and seven funeral home directors received lesser sentences.

The Alistair Cooke case is not the first such incident, and it is not likely to be the last. In 1999, the University of California →

at Irvine discovered that the director of its Willed Body Program was selling human spines to a Phoenix hospital for $5,000 apiece. And in 2008, the director of UCLA's Willed Body Program was sentenced to four years in prison for selling more than a million dollars worth of body parts. Regulators say that abuses such as these are most likely to occur when relatively poorly paid directors (including funeral home directors) have access to valuable body parts and when oversight is lax. UC Irvine and UCLA have both tightened their oversight procedures as a result of the scandals.

The donation of a single cadaver to a nonprofit tissue-processing company can benefit several dozen patients. Patients should only have to pay the legitimate costs associated with the body parts processing industry—not the added fees paid to traffickers in illegal body parts.

Safeguards need to be put in place to prevent abuses so that we can be assured of the legitimacy and the safety of the supply of human body parts.

SUMMARY

- It is illegal to buy or sell human body parts for a profit. Patients or their families can donate body parts only to nonprofit tissue banks.
- Legitimately donated tissues are tested, sterilized, processed, and transplanted into patients who urgently need them.
- Only about 20,000 cadavers are donated annually for body parts—not enough to meet the growing legitimate demand for human tissues and tissue products.
- The supply/demand imbalance may be contributing to a black market in body parts harvested illegally from cadavers.

The human body is capable of an awesome array of physical activities. With training, some individuals can run a mile in less than four minutes or lift more than their own weight. Exquisitely sensitive motor skills allow us to thread a needle, turn our head to focus on a single star, and throw a baseball into the strike zone. Considered individually, any one of these activities may not seem amazing, but for a single structure (the human body) to be capable of all of them is remarkable indeed. From an engineering standpoint it would be like designing a bulldozer that is strong enough to flatten a building, yet delicate enough to pick up a dime.

The one organ system that makes all this possible is the skeletal system. Without the skeletal system, we'd be about as capable of movement as a jellyfish or a slug. In partnership with our muscles, the bones of our skeletal system permit us to stand upright and to move about. The skeletal system supports and encloses our internal organs and provides a structure for our external covering, the skin. And last but not least, bones of the skeletal system surround and protect the brain, our body's central control system. ■

In this chapter, we describe the structure and development of the bones that comprise the skeletal system, how bones develop, and how they remodel and repair themselves. We'll review how bones fit together to form the skeleton, look at how joints enable bones and muscles to work in unison, and consider what can go wrong with the skeletal system.

5.1 The skeletal system consists of connective tissue

The skeletal system comprises three types of connective tissue—bones, ligaments, and cartilage. *Bones* are the hard elements of the skeleton with which we are most familiar.

Ligaments consist of dense fibrous connective tissue—they bind the bones to each other. *Cartilage* is a specialized connective tissue consisting primarily of fibers of collagen and elastin in a gel-like fluid called *ground substance.* Cartilage has several functions, including reducing friction in joints.

Bones are the hard elements of the skeleton

Most of the mass of **bones** consists of nonliving extracellular crystals of calcium minerals that give bones their hard, rigid appearance and feel. But bone is actually a living tissue that contains several types of living cells involved in bone formation and remodeling, plus nerves and blood vessels. Indeed, bones bleed when cut during orthopedic surgery or when they break.

Bones perform five important functions. The first three—*support, protection,* and *movement*—are the same as the functions of the skeleton overall, which is, after all, primarily bone. The rigid support structure of bones allows us to sit and to stand upright. The bones of the skeleton also support, surround, and protect many of our soft internal organs, such as the lungs, liver, and spleen. The attachment of bones to muscles makes it possible for our bodies to move.

The fourth and fifth functions of bones—*blood cell formation* and *mineral storage*—are harder to remember, but they are just as important. Cells in certain bones are the only source of new red and white blood cells and platelets for blood. Without this ability to produce new blood cells, we would die within months. Bones also serve as an important long-term storage depot for two minerals, calcium and phosphate, that can be drawn from bone when necessary (for example, during pregnancy to support the growth of a fetus).

Bone contains living cells

A typical long bone, so called because it is longer than it is wide, consists of a cylindrical shaft (called the *diaphysis*) with an enlarged knob called an *epiphysis* at each end (**Figure 5.1a**). Dense **compact bone** forms the shaft and covers each end, and less dense **spongy bone** fills the inner regions of the epiphyses.

A central cavity in the diaphysis is filled with *yellow bone marrow*. Yellow bone marrow is primarily fat that can be utilized for energy. In certain long bones, most notably the long bones of the upper arms and legs (humerus and femur, respectively), the spaces within spongy bone are filled with *red bone marrow*. Special cells called *stem cells* in the red bone marrow produce red and white blood cells and platelets.

The outer surface of the bone is covered by a tough layer of connective tissue, the *periosteum*, which contains specialized bone-forming cells. If an epiphysis of a long bone forms a movable joint with another bone, the joint surface is covered by a smooth layer of cartilage that reduces friction.

A closer look at a section of bone taken from an epiphysis (**Figure 5.1b**) reveals that compact bone is a nearly solid structure, with central canals containing nerves and blood vessels. In contrast, spongy bone is a latticework of

hard, relatively strong *trabeculae* (from Latin, meaning "little beams"). Although spongy bone is as hard as compact bone (it is spongy only in appearance), it is less dense, allowing bones to be light but strong.

Taking an even closer look at compact bone (**Figure 5.1c**), we see that it is made up largely of extracellular deposits of calcium phosphate enclosing and surrounding living cells called **osteocytes** (from the Greek words for "bone" and "cells"). Osteocytes are arranged in rings in cylindrical structures called **osteons** (sometimes called *Haversian systems*). Osteocytes nearest the center of an osteon receive nutrients by diffusion from blood vessels that pass through a **central canal** (Haversian canal).

As bone develops and becomes hard, the osteocytes become trapped in hollow chambers called *lacunae* (**Figure 5.1d**). However, the osteocytes remain in direct contact with each other via thin canals called *canaliculi*. Within the canaliculi, extensions of the cell cytoplasm in adjacent osteocytes are joined together by gap junctions (channels that permit the movement of ions, water, and other molecules between two adjacent cells). By exchanging nutrients across gap junctions, osteocytes can be supplied with nutrients even though most osteocytes are not located near a blood vessel. Waste products produced by the osteocytes are exchanged in the

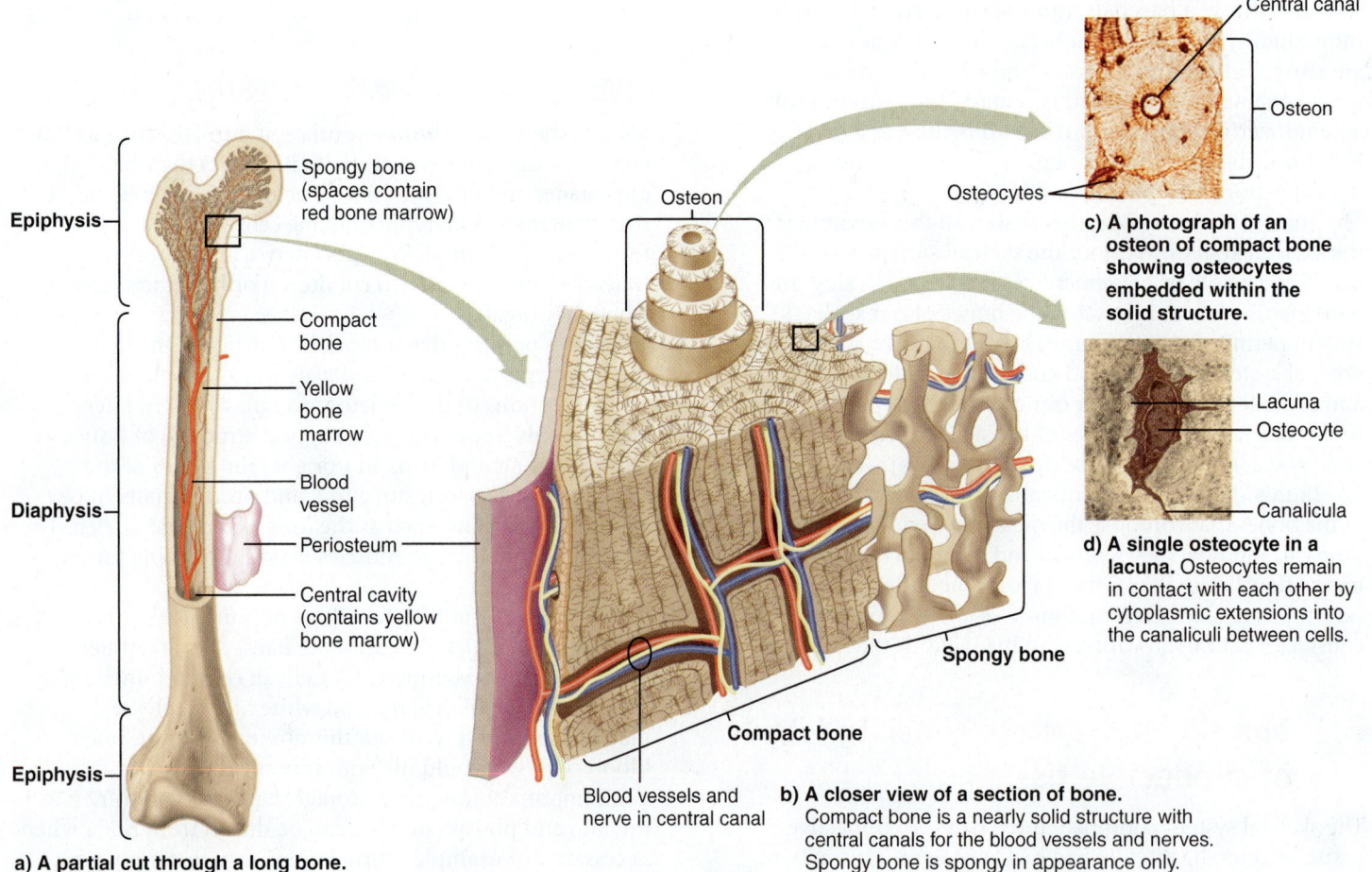

Epiphysis

Spongy bone (spaces contain red bone marrow)

Osteon

Central canal

Osteon

Osteocytes

c) A photograph of an osteon of compact bone showing osteocytes embedded within the solid structure.

Compact bone

Yellow bone marrow

Blood vessel

Periosteum

Diaphysis

Central cavity (contains yellow bone marrow)

Lacuna

Osteocyte

Canalicula

d) A single osteocyte in a lacuna. Osteocytes remain in contact with each other by cytoplasmic extensions into the canaliculi between cells.

Spongy bone

Epiphysis

Compact bone

Blood vessels and nerve in central canal

b) A closer view of a section of bone. Compact bone is a nearly solid structure with central canals for the blood vessels and nerves. Spongy bone is spongy in appearance only.

a) A partial cut through a long bone.

Figure 5.1 Structure of bone.

opposite direction and are removed from the bone by blood vessels. In spongy bone, osteocytes do not need to rely on central canals for nutrients and waste removal. The slender trabecular structure of spongy bone gives each osteocyte access to nearby blood vessels in red bone marrow.

☑ If osteocytes did not have gap junctions in their cell membranes, would they be able to survive? Explain.

Ligaments hold bones together

Ligaments attach bone to bone. Ligaments consist of dense fibrous connective tissue, meaning that they are a regular array of closely packed collagen fibers all oriented in the same direction, with just a few fibroblasts in between. (Recall that fibroblasts are cells that produce and secrete the proteins that compose collagen, elastic, and reticular fibers.) Ligaments confer strength to certain joints while still permitting movement of the bones in relation to each other.

Cartilage lends support

Cartilage contains fibers of collagen and/or elastin in a ground substance of water and other materials. Cartilage, smoother and more flexible than bone, is found where support under pressure is needed and movement is necessary.

There are three types of cartilage in the human skeleton. *Fibrocartilage* consists primarily of collagen fibers arranged in thick bundles. It withstands both pressure and tension well. The intervertebral disks between the vertebrae, and also certain disklike supportive structures in the knee joint called *menisci*, are made of fibrocartilage. *Hyaline cartilage* is a smooth, almost glassy cartilage of thin collagen fibers. Hyaline cartilage forms the embryonic structures that later become the bones. It also covers the ends of mature bones in joints, creating a smooth, low-friction surface. *Elastic cartilage* is mostly elastin fibers, so it is highly flexible. It lends structure to the outer ear and to the epiglottis, a flap of tissue that covers the larynx during swallowing.

↻ **Recap** Bones contribute to support, movement, and protection. Bones also produce the blood cells and store minerals. Ligaments hold bones together, and cartilage provides support. ▪

5.2 Bones develop from cartilage

In the earliest stages of fetal development, the rudimentary models of future bones are created out of hyaline cartilage (**Figure 5.2a**) by cartilage-forming cells called **chondroblasts.** Most chondroblasts are short-lived, however; within 2–3 months they begin to die, and as they do the cartilage

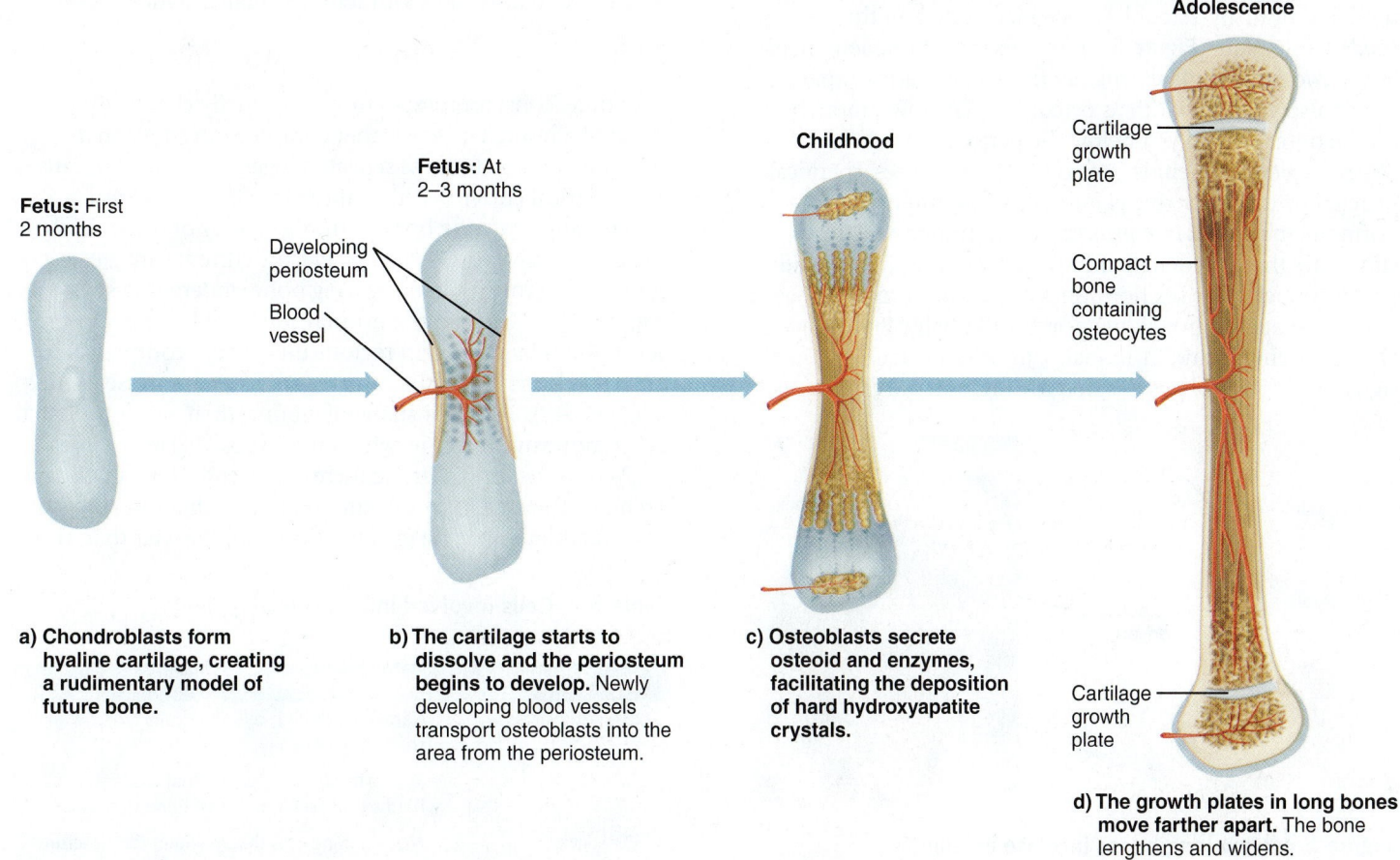

a) Chondroblasts form hyaline cartilage, creating a rudimentary model of future bone.

b) The cartilage starts to dissolve and the periosteum begins to develop. Newly developing blood vessels transport osteoblasts into the area from the periosteum.

c) Osteoblasts secrete osteoid and enzymes, facilitating the deposition of hard hydroxyapatite crystals.

d) The growth plates in long bones move farther apart. The bone lengthens and widens.

Figure 5.2 How bone develops. The first two phases of bone development occur in the fetus. Bones continue to grow longer throughout childhood and adolescence because of growth at the growth plates.

dissolves, making room for blood vessels (Figure 5.2b). At the same time, the periosteum begins to form on the outer surface of the developing bone. The blood vessels carry bone-forming cells called **osteoblasts** (from the Greek words for "bone" and "to build") into the area from the developing periosteum. Osteoblasts secrete a mixture of proteins (including collagen) called *osteoid*, which forms a matrix that provides internal structure and strength to bone (Figure 5.2c). Osteoblasts also secrete enzymes that facilitate the crystallization of hard mineral salts of calcium phosphate around and between the osteoid matrix. As more and more calcium phosphate is deposited, the osteoblasts become embedded in the hardening bone tissue. In mature compact bone, approximately one-third of the structure is osteoid and two-thirds is crystals of calcium phosphate.

Eventually, the rate at which osteoblasts produce the osteoid matrix and stimulate the mineral deposits declines, and osteoblasts become mature osteocytes embedded in their individual lacunae. Mature osteocytes continue to maintain the bone matrix, however. Without them, the matrix would slowly disintegrate.

Some bones, called long bones, continue to lengthen throughout childhood and adolescence. This is because chondroblasts and osteoblasts remain active in a narrow strip of cartilage called the **growth plate** (or *epiphyseal plate*) (Figure 5.2d). Chondroblast activity (and hence the deposition of new cartilage) is concentrated on the *outside* of the plate, whereas the conversion of the cartilage model to bone by osteoblasts is concentrated on the *inside* of the plate (Figure 5.3). In effect, the bone lengthens as the two growth plates migrate farther and farther apart. Bones also grow in width as osteoblasts lay down more bone on the outer surface just below the periosteum.

Bone development is controlled by hormones, chemicals secreted by the endocrine glands. The most important hormone in preadolescents is growth hormone, which stimulates the bone-lengthening activity of the growth plate. During puberty, the sex hormones (testosterone and estrogen) also stimulate the growth plate, at least initially. But at about age 18 in females and 21 in males, these same sex hormones signal the growth plates to stop growing, and the cartilage is

replaced by bone tissue. At this point, the bones can no longer lengthen, though they can continue to grow in width.

> **↺ Recap** Bone-forming cells called *osteoblasts* produce a protein mixture (including collagen) that becomes bone's structural framework. Osteoblasts also secrete an enzyme that facilitates mineral deposition. ▪

5.3 Mature bone undergoes remodeling and repair

Even though bones stop growing longer, they do not remain the same throughout life. Bone is a dynamic tissue that undergoes constant replacement, remodeling, and repair. Remodeling may be so extensive that there is a noticeable change in bone shape over time, even in adults.

Bone remodeling and repair are in part due to a third type of bone cell called an **osteoclast** (from the Greek words for "bone" and "to break"). Osteoclasts cut through mature bone tissue, dissolving the hard calcium phosphate crystals and digesting the osteoid matrix in their path. The released calcium and phosphate ions enter the blood. The areas from which bone has been removed attract new osteoblasts, which lay down new osteoid matrixes and stimulate the deposition of new calcium phosphate.

Table 5.1 summarizes the four types of cells that contribute to bone development and maintenance.

Bones can change in shape, size, and strength

Over time, constant remodeling can actually change the shape of a bone. The key is that compression stress on a bone, such as the force of repeated jogging on the legs, causes tiny electrical currents within the bone. These electrical currents stimulate the bone-forming activity of osteoblasts. The compressive forces and the electric currents are greatest at the inside curvature of the long bone undergoing stress (Figure 5.4a). Thus, in the normal course of bone turnover, new bone is laid down in regions under high compressive stress and bone is resorbed in areas of low compressive stress (Figure 5.4b). The final shape of a bone, then, tends to match the compressive forces to which it is exposed (Figure 5.4c).

Weight-bearing exercise increases overall bone mass and strength. The effect is pronounced enough that the bones of trained athletes may be visibly thicker and heavier than those

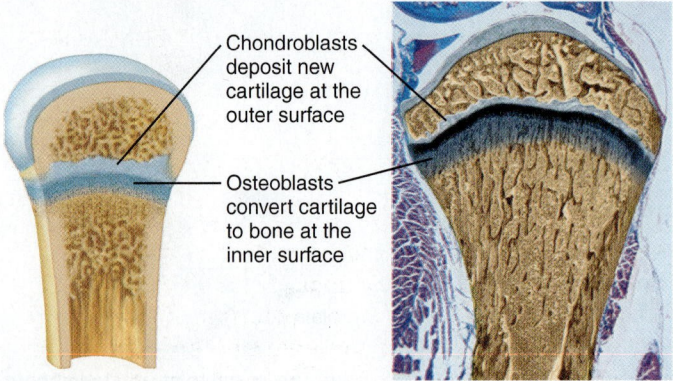

Chondroblasts deposit new cartilage at the outer surface

Osteoblasts convert cartilage to bone at the inner surface

Figure 5.3 How long bones increase in length.

✅ What two substances do osteoblasts produce that lead to the formation of hard bone?

Table 5.1 Cells involved in bone development and maintenance

Type of cell	Function
Chondroblasts	Cartilage-forming cells that build a model of the future bone
Osteoblasts	Young bone-forming cells that cause the hard extracellular matrix of bone to develop
Osteocytes	Mature bone cells that maintain the structure of bone
Osteoclasts	Bone-dissolving cells

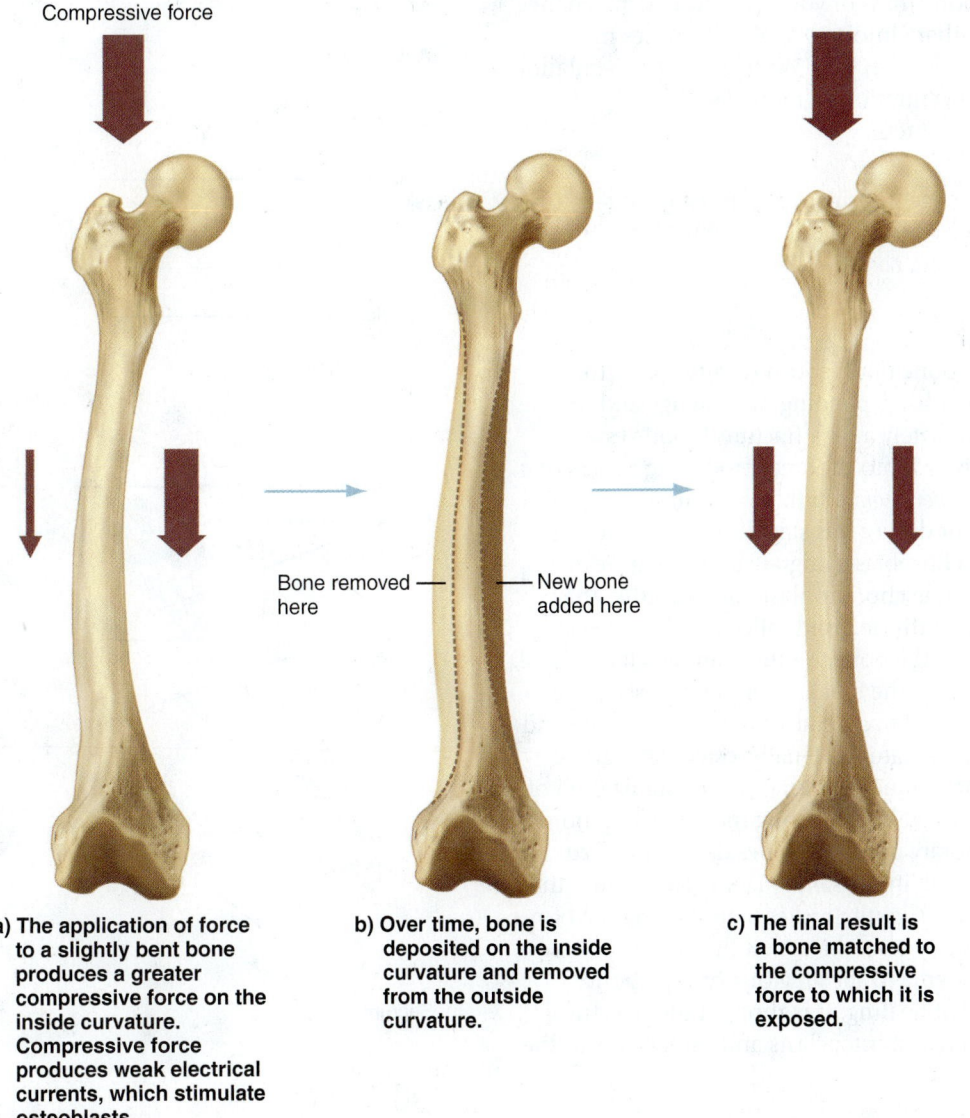

Compressive force

Bone removed here — New bone added here

a) **The application of force to a slightly bent bone produces a greater compressive force on the inside curvature. Compressive force produces weak electrical currents, which stimulate osteoblasts.**

b) **Over time, bone is deposited on the inside curvature and removed from the outside curvature.**

c) **The final result is a bone matched to the compressive force to which it is exposed.**

Figure 5.4 Bone remodeling.

of nonathletes. You don't have to be a professional athlete to get this benefit, however. If you begin a regular program of any weight-bearing exercise, such as jogging or weight lifting, your bones will become denser and stronger as your osteoblasts produce more bone tissue.

MJ's BlogInFocus Should you have a bone density scan to determine whether you're losing bone mass? Is it currently recommended? To find out, visit MJ's blog in the Study Area in MasteringBiology and look under "Bone Density Scans."

http://goo.gl/RRyMxp

Homeostasis of bone structure depends on the precise balance of the activities of osteoclasts and osteoblasts. **Osteoporosis** is a common condition in which bones lose a great deal of mass (seemingly becoming "porous") because

of an imbalance over many years in the rates of activities of these two types of bone cells.

☑ Swimming is considered good exercise for muscle mass and overall cardiorespiratory fitness. How would you expect swimming to affect bone mass and density?

Bone cells are regulated by hormones

Like bone growth, the rates of activities of osteoblasts and osteoclasts in adulthood are regulated by hormones that function to maintain calcium homeostasis. When blood levels of calcium fall below a given point, *parathyroid hormone* (*PTH*) stimulates osteoclasts to secrete more bone-dissolving enzymes. The increased activity of osteoclasts causes more bone to be dissolved, releasing calcium and phosphate into the bloodstream. If calcium levels rise, then another hormone called *calcitonin* stimulates osteoblast activity, causing calcium and phosphate to be removed from blood and deposited in bone.

Although the total bone mass of young adults doesn't change much, it's estimated that almost 10% of their bones may be remodeled and replaced each year. We'll review the regulation of blood calcium concentration in more detail when we discuss the endocrine system.

☑ Suppose a man is not getting sufficient calcium in his diet, such that his blood calcium level is chronically low. Would his PTH levels and calcitonin levels be low, normal, or high? Explain.

Bones undergo repair

A broken (fractured) bone undergoes a repair process that can take weeks to months, depending on your age and the bone involved. Immediately after a fracture, blood vessels supplying the bone bleed into the area, producing a mass of clotted blood called a *hematoma*. Inflammation, swelling, and pain are likely to occur during this stage. The repair process begins within days as fibroblasts migrate to the area. Some of the fibroblasts become chondroblasts, and together they produce a tough fibrocartilage bond called a *callus* between the two broken ends of the bone. A callus can be felt as a hard, raised ring at the point of the break. Then osteoclasts arrive and begin to remove dead fragments of the original bone and the blood cells of the hematoma. Finally, osteoblasts arrive to deposit osteoid matrix and encourage the crystallization of calcium phosphate minerals, converting the callus into bone. Eventually, the temporary union becomes dense and hard again. Bones rarely break in the same place twice because the repaired union remains slightly thicker than the original bone.

Recently, it has been discovered that the application of weak electrical currents to the area of a broken bone can increase the rate of healing. It is thought that electrical current works by attracting osteoclasts and osteoblasts to the area under repair.

↩ **Recap** Healthy bone replacement and remodeling depend on the balance of activities of bone-resorbing osteoclasts and bone-forming osteoblasts. When a bone breaks, a fibrocartilage callus forms between the broken ends and is later replaced with bone. ■

5.4 Bones fit together to form the skeleton

Now that we have reviewed the dynamic nature of bone tissue, we turn to how bones are classified and organized in the human **skeleton** (Figure 5.5). Bones can be classified into four types based on shape: long, short, flat, and irregular. So far we have discussed *long bones,* which include the bones of the limbs and fingers. *Short bones* (the bones of the wrists) are approximately as wide as they are long. *Flat bones* (including the cranial bones, the sternum, and the ribs) are thin, flattened, and sometimes curved, with only a small amount of spongy bone sandwiched between two layers of compact bone. *Irregular bones* such as the coxal (hip) bones and the vertebrae include a variety of shapes that don't fit into the other categories. A few flat and irregular bones, including the

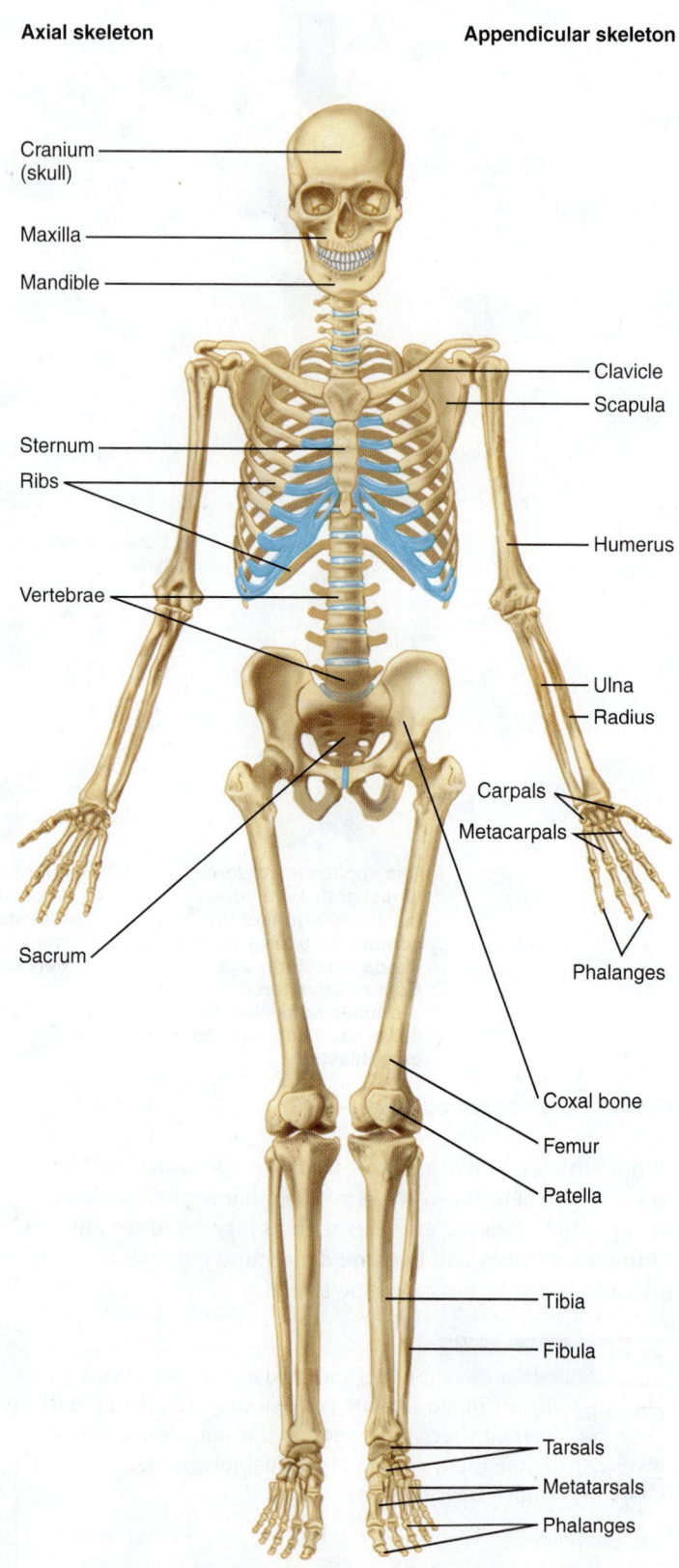

Axial skeleton   Appendicular skeleton

Cranium (skull)
Maxilla
Mandible
Clavicle
Scapula
Sternum
Ribs
Humerus
Vertebrae
Ulna
Radius
Carpals
Metacarpals
Sacrum
Phalanges
Coxal bone
Femur
Patella
Tibia
Fibula
Tarsals
Metatarsals
Phalanges

Figure 5.5 The human skeleton.

☑ On this and subsequent figures, find the anatomical terms corresponding to the following common names: breastbone, collarbone, shoulder blade, hip bone, thighbone, shinbone.

sternum and the hip bones, contain red bone marrow that produces blood cells.

The human skeleton comprises 206 bones and the various connective tissues that hold them together. The skeleton has three important functions. First, it serves as a structural framework for support of the soft organs. Second, it protects certain organs from physical injury. The brain, for example, is enclosed within the bones of the skull, and the heart and lungs are protected by a bony cage consisting of ribs, the sternum, and vertebrae. Third, because of the way that the bony elements of the skeleton are joined together at joints, the presence of the skeleton permits flexible movement of most parts of the body. This is particularly true of the hands, feet, legs, and arms.

The skeleton is organized into the *axial skeleton* and the *appendicular skeleton.*

The axial skeleton forms the midline of the body

The **axial skeleton** consists of the skull (including the maxilla and mandible), sternum, ribs, and vertebral column (including the sacrum; see Figure 5.5).

The skull: cranial and facial bones
The human **skull** (cranium) comprises over two dozen bones that protect the brain and form the structure of the face. **Figure 5.6** illustrates some of the more important bones of the skull.

The *cranial bones* are flat bones in the skull that enclose and protect the brain. Starting at the front of the skull, the *frontal bone* comprises the forehead and the upper ridges of the eye sockets. At the upper left and right sides of the skull are the two *parietal bones,* and forming the lower left and right sides are the two *temporal bones.* Each temporal bone is pierced by an opening into the ear canal that allows sounds to travel to the eardrum. Between the frontal bone and the temporal bones is the *sphenoid bone,* which forms the back of both eye sockets. The *ethmoid bone* contributes to the eye sockets and also helps support the nose.

The *facial bones* compose the front of the skull. On either side of the nose are the two *maxilla* (maxillary) bones, which form part of the eye sockets and contain the sockets that anchor the upper row of teeth. The hard palate (the "roof" of the mouth) is formed by the maxilla bones and the two *palatine bones.* Behind the palatine bones is the *vomer bone,* which is part of the nasal septum that divides the nose into left and right halves. The two *zygomatic bones* form the cheekbones and the outer portion of the eye sockets.

The two small, narrow *nasal bones* underlie only the upper bridge of the nose; the rest of the fleshy protuberance called the nose is made up of cartilage and other connective tissue. Part of the space formed by the maxillary and nasal bones is the nasal cavity. The small *lacrimal bones,* at the inner eye sockets, are pierced by a tiny opening through which the tear ducts drain tears from the eye sockets into the nasal cavity.

The *mandible,* or lower jaw, contains the sockets that house the lower row of teeth. All the bones of the skull

are joined tightly together except for the mandible, which attaches to the temporal bone by a joint that, because it permits a substantial range of motion, allows us to speak and chew.

Curving underneath to form the back and base of the skull is the *occipital bone.* Near the base of the occipital bone is a large opening called the *foramen magnum* (Latin for "great opening"). This is where the vertebral column connects to the skull and the spinal cord enters the skull to communicate with the brain.

Several of the cranial and facial bones contain air spaces called **sinuses,** which make the skull lighter and give the human voice its characteristic tone and resonance. Each sinus is lined with tissue that secretes mucus, a thick, sticky fluid that helps trap foreign particles in incoming air. The sinuses connect to the nasal cavity via small passageways through which the mucus normally drains. However, if you develop a cold or respiratory infection, the tissue lining your sinuses can become inflamed and block these passages. Sinus inflammation is called *sinusitis.* If fluid accumulates inside the sinuses, the resulting sensation of pressure may give you a "sinus headache."

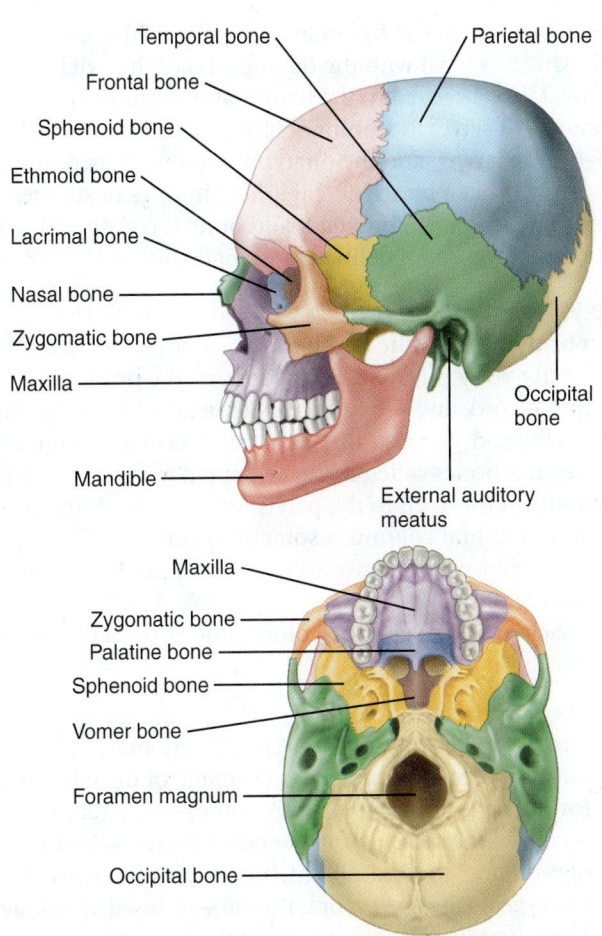

Figure 5.6 The human skull. Except for the mandible, which has a hinged joint with the temporal bone, the bones of the skull are joined tightly together. Their function is protection, not movement.

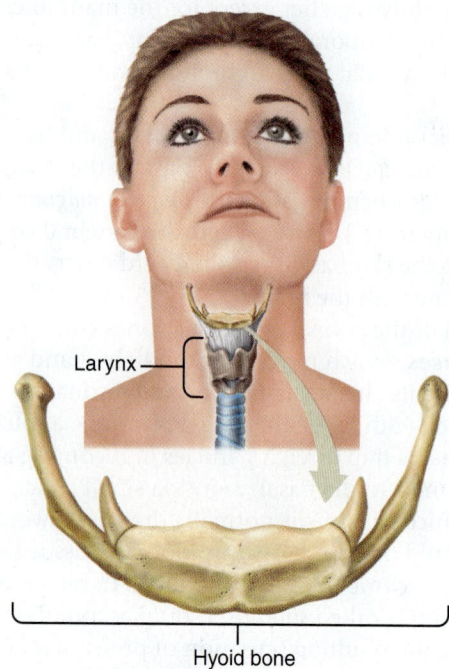

Larynx

Hyoid bone

Figure 5.7 The hyoid bone.

The hyoid bone The **hyoid** bone (**Figure 5.7**) does not make direct contact with the other bones of the axial skeleton; it is attached to the temporal bone only by ligaments. It serves as a point of attachment for muscles of the tongue, the larynx, and the pharynx. Because of its position, it is rarely broken by accidental injury. In cases of suspected homicide, however, a broken hyoid bone is considered to be a strong indication of deliberate strangulation.

The vertebral column: the body's main axis The **vertebral column** (the backbone or spine) is the main axis of the body (**Figure 5.8**). It supports the head, protects the spinal cord, and serves as the site of attachment for the four limbs and various muscles. It consists of a column of 33 irregular bones called *vertebrae* (singular: vertebra) that extend from the skull to the pelvis. When viewed from the side, the vertebral column is somewhat curved, reflecting slight differences in structure and size of vertebrae in the various regions.

We classify the vertebral column into five anatomical regions:

- Cervical (neck)—7 vertebrae.
- Thoracic (the chest or thorax)—12 vertebrae.
- Lumbar (the lower portion or "small" of the back, which forms the lumbar curve of the spine)—5 vertebrae.
- Sacral (in the sacrum or upper pelvic region)—In the course of evolution, the 5 sacral vertebrae have become fused.
- Coccygeal (the coccyx or tailbone)—4 fused vertebrae. The coccyx is all that remains of the tails of our ancient ancestors. It is an example of a *vestigial* structure, meaning one that no longer has any function.

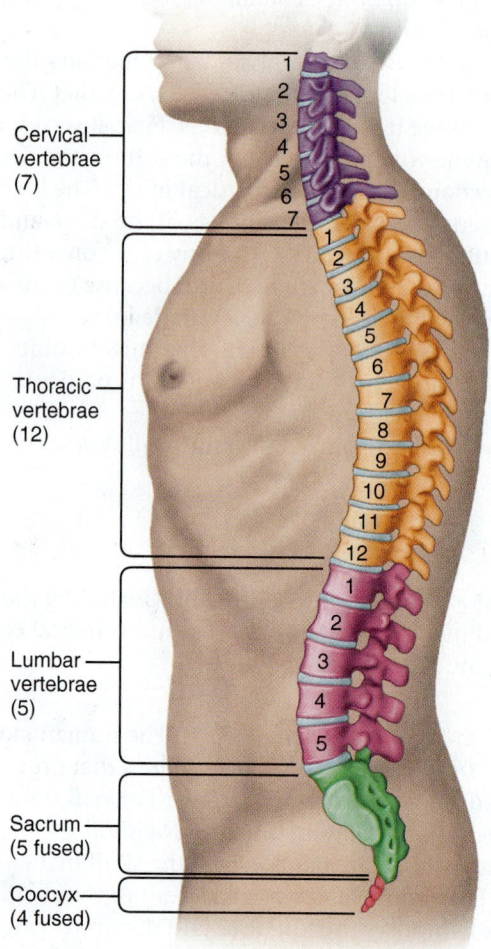

Cervical vertebrae (7)

Thoracic vertebrae (12)

Lumbar vertebrae (5)

Sacrum (5 fused)

Coccyx (4 fused)

Figure 5.8 The vertebral column. Vertebrae are named and numbered according to their location. The vertebral column is moderately flexible because of the presence of joints and intervertebral disks.

A closer look at vertebrae (**Figure 5.9a**) shows how they are stacked on each other and how they are joined. Vertebrae share two points of contact, called *articulations*, located behind their main body. There are also articulations with the ribs. The spinal cord passes through a hollow cavity between the articulations and the main body. Neighboring vertebrae are separated from each other by a flat, elastic, compressible **intervertebral disk** composed of a soft gelatinous center and a tough outer layer of fibrocartilage. Intervertebral disks serve as shock absorbers, protecting the delicate vertebrae from the impact of walking, jumping, and other movements. In conjunction with the vertebral joints, vertebral disks also permit a limited degree of movement. This lends the vertebral column greater flexibility, allowing us to bend forward, lean backward, and rotate the upper body.

An especially strong impact or sudden movement can compress an intervertebral disk, forcing the softer

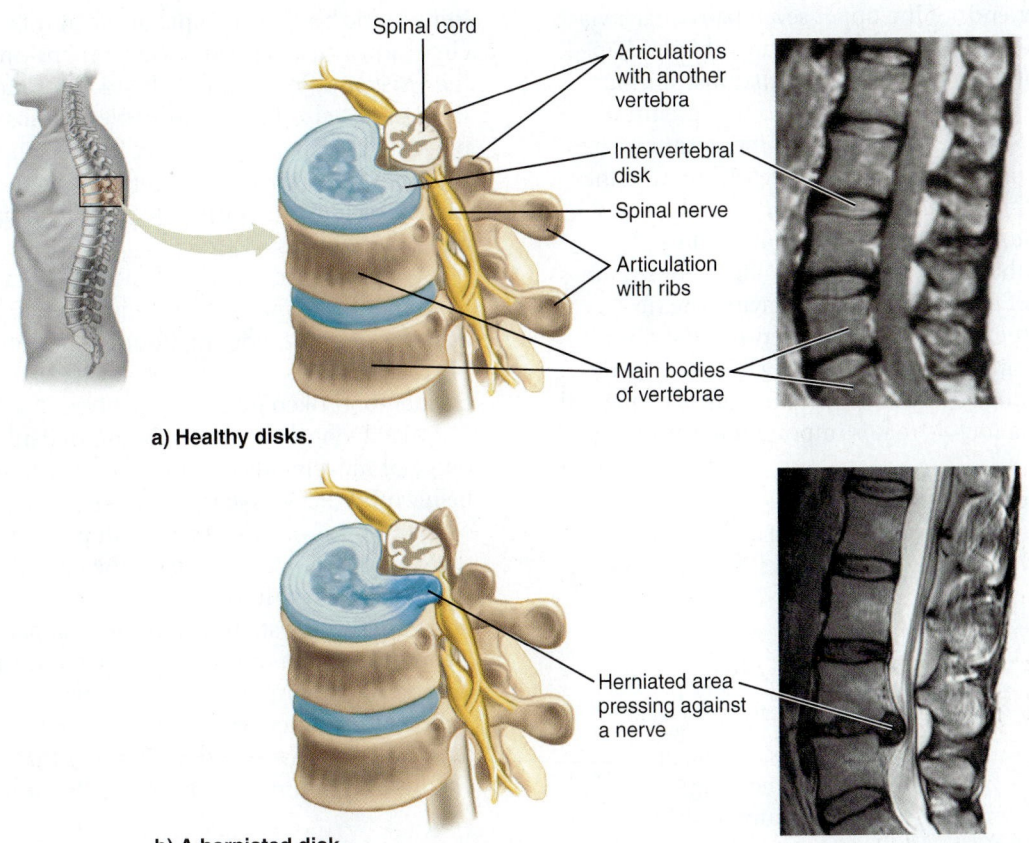

Figure 5.9 **Vertebrae.**

center to balloon outward, press against spinal nerves, and cause intense back pain. This condition is referred to as a "herniated" or "slipped disk" (**Figure 5.9b**), and it occurs most often in the lumbar vertebrae. Occasionally the disk may rupture, releasing its soft, pulpy contents. The pain that accompanies a herniated disk can be alleviated by surgery to remove the damaged disk, relieving the pressure against the nerve. However, surgical correction of a herniated disk reduces spinal flexibility somewhat because the two adjacent vertebrae must be fused together with bone grafts.

Generally, the bony vertebral column does an effective job of shielding the softer spinal cord, which consists of nervous tissue that connects the brain to the rest of the body. However, injury to the vertebral column can damage the spinal cord or even sever it, resulting in partial or complete paralysis of the body below that point. Persons with suspected vertebral injuries should not be moved until a physician can assess the situation, because any twisting or bending could cause additional, perhaps permanent, damage to the spinal cord. You may have noticed that when athletes are injured on the field, they are instructed to lie absolutely still until a trainer and physician have examined them thoroughly.

The ribs and sternum: protecting the chest cavity

Humans have 12 pairs of **ribs** (**Figure 5.10**). One end of each rib branches from the thoracic region of the vertebral

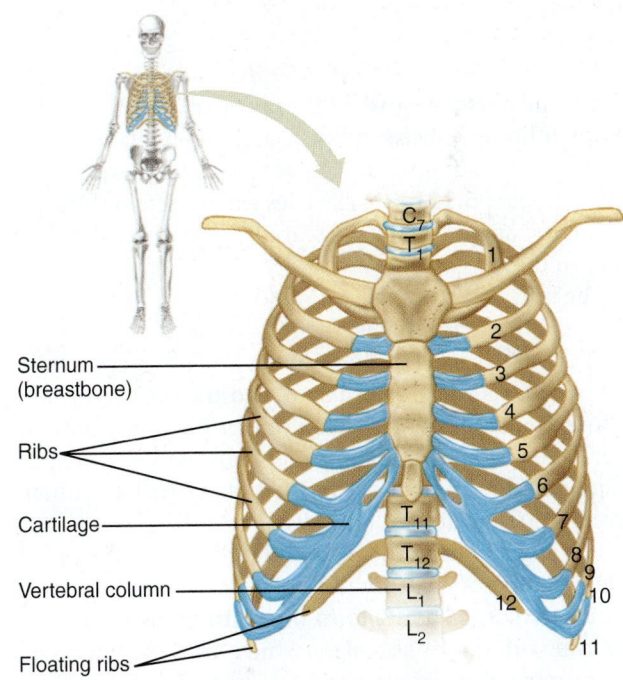

Figure 5.10 **Ribs.** The 12 pairs of ribs are numbered according to their attachment to the thoracic vertebrae. Only the first seven pairs attach directly to the sternum.

☑ What function do the ribs and sternum have that other parts of the skeleton do not have? How might this explain the fact that the ribs and sternum are connected by flexible cartilage rather than by bone?

column. The other ends of the upper seven pairs attach via cartilage to the **sternum,** or breastbone, a flat blade-shaped bone composed of three separate bones that fuse during development. Rib pairs 8–10 are joined to the seventh rib by cartilage, and thus attach indirectly to the sternum. The bottom two pairs of ribs are called *floating ribs* because they do not attach to the sternum at all.

The ribs, sternum, and vertebral column form a protective *rib cage* that surrounds and shields the heart, lungs, and other organs of the chest (thoracic) cavity. The rib cage also helps us breathe, because muscles between the ribs lift them slightly during breathing, expanding the chest cavity and inflating the lungs. The base of the sternum is connected to the diaphragm, a muscle that is important to breathing.

✔ Humans have more sacral vertebrae than most mammals do, and these sacral bones are fused into an unusually strong structure. Given what you know about the functions of the vertebral column, why do you think that is?

The appendicular skeleton: pectoral girdle, pelvic girdle, and limbs

Those parts of the body that attach to the axial skeleton are called *appendages,* from the Latin word meaning "to hang upon." The second division of the human skeleton, the **appendicular skeleton,** includes the arms, legs, and their attachments to the trunk, which are the pectoral and pelvic girdles.

The pectoral girdle lends flexibility to the upper limbs
The **pectoral girdle,** a supportive frame for the upper limbs, consists of the right and left **clavicles** (collarbones) and right and left **scapulas** (shoulder blades) (**Figure 5.11**). The clavicles extend across the top of the chest and attach to the scapulas, the triangular bones in the upper back.

The arm and hand consist of 30 different bones. The upper end of the **humerus,** the long bone of the upper arm, fits into a socket in the scapula. The other end of the humerus meets with the **ulna** and **radius,** the two bones of the forearm, at the elbow. If you've ever hit your elbow and experienced a painful tingling, you know why this area is nicknamed the "funny bone," you've just struck the ulnar nerve that travels along the elbow.

The lower ends of the ulna and radius meet the *carpal* bones, a group of eight small bones that make up the wrist. The five *metacarpal* bones form the palm of the hand, and they join with the 14 *phalanges,* which form the fingers and thumb.

The pectoral girdle and arms are particularly well adapted to permit a wide range of motion. They connect to the rest of the body via muscles and tendons—a relatively loose method of attachment. This structure gives the upper body of humans a degree of dexterity unsurpassed among large animals. We can rotate our upper arms almost 360 degrees—a greater range of movement than with any other

joint in the body. The upper arm can rotate in roughly a circle, the arm can bend in one dimension and rotate, and the wrist and fingers can all bend and rotate to varying degrees. We also have "opposable thumbs," meaning we can place them opposite our other fingers. The opposable thumb has played an important role in our evolutionary history, as it makes it easier to grasp and manipulate tools and other objects.

We pay a price for this flexibility, because freedom of movement also means relative instability. If you fall on your arm, for example, you might dislocate your shoulder joint or crack a clavicle. In fact, the clavicle is one of the most frequently broken bones in the body.

Although our upper limbs are well adapted to a wide range of movements, too much of one kind of motion can be harmful. Repetitive motions—performing the same task over and over—can lead to health problems called *repetitive stress syndromes.* Depending on the part of the body that is overused, these injuries can take many forms. A well-known repetitive stress syndrome is *carpal tunnel syndrome,* a condition often due to repetitive typing at a computer keyboard. The carpal bones of the wrist are held together by a sheath of connective tissue. The blood vessels, nerves, and tendons to the hand and fingers pass through the sheath via the "carpal tunnel." Overuse of the fingers and hands

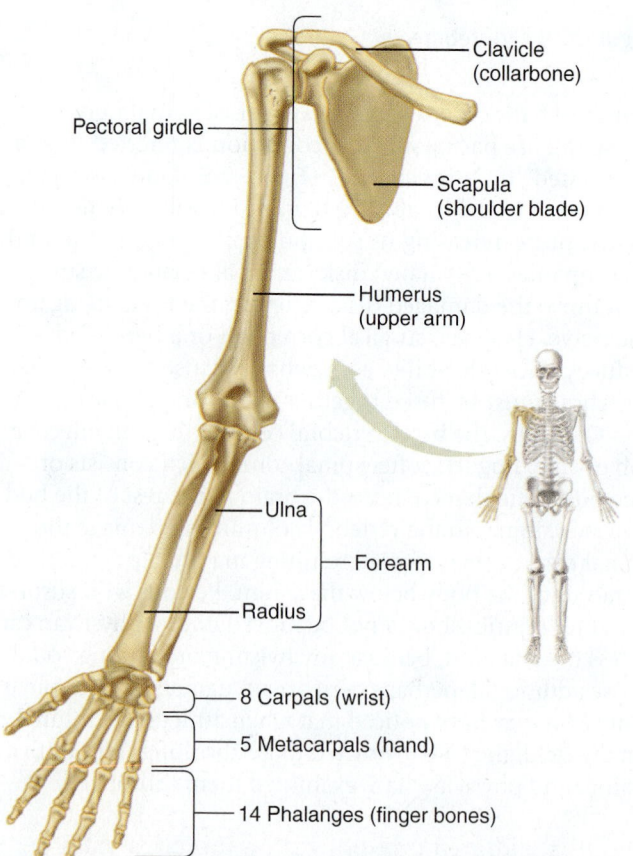

Figure 5.11 Bones of the right side of the pectoral girdle and the right arm and hand.

produces swelling and inflammation of the tendons, which causes them to press against the nerve supplying the hand. The result may be pain, tingling, or numbness in the wrist and hand. Mild episodes of carpal tunnel syndrome respond to rest and pain relievers. Severe cases can be treated with surgery to relieve the pressure.

The pelvic girdle supports the body. The **pelvic girdle** consists of the two **coxal bones** and the sacrum and coccyx of the vertebral column (**Figure 5.12**). The coxal bones

attach to the sacral region of the vertebral column in back, then curve forward to meet in front at the *pubic symphysis,* where they are joined by cartilage. You can feel the upper curves of the coxal bones (the iliac region) as your hip bones. Together, these structures form the pelvis.

The primary function of the pelvic girdle is to support the weight of the upper body against the force of gravity. It also protects the organs inside the pelvic cavity and serves as a site of attachment for the legs. The structure of the pelvic girdle reflects a trade-off between dexterity and stability. Partly because the pelvic girdle and lower limbs are larger and more firmly connected to the rest of the body than the pectoral girdle and upper limbs, the lower limbs are less dexterous than the upper limbs.

The **femur** (thighbone) is the longest and strongest bone in the body. When you jog or jump, your femurs are exposed to forces of impact of several tons per square inch. The rounded upper end of each femur fits securely into a socket in a coxal bone, creating a stable joint that effectively supports the body while permitting movement. The lower end of the femur intersects at the knee joint with the larger of the two bones of the lower leg, the **tibia,** which, in turn, makes contact with the thinner **fibula.** The *patella,* or kneecap, is a triangle-shaped bone that protects and stabilizes the knee joint.

At the ankle, the tibia and fibula join with the seven *tarsal* bones that make up the ankle and heel. Five long bones, the *metatarsals,* form the foot. The 14 bones of the toes, like those of the fingers, are called *phalanges.*

In adult women, the pelvic girdle is broader and shallower than it is in men, and the pelvic opening is wider (**Figure 5.13**). The wider pelvic opening allows for safe passage of a baby's head during labor and delivery. Differences in pelvic structures between men and women also account for the different degrees of hip sway between men and women as they walk. These characteristic differences appear during puberty when a woman's body begins to produce sex hormones. The sex hormones trigger a process of bone remodeling that shapes the female pelvic girdle to adapt for pregnancy and birth.

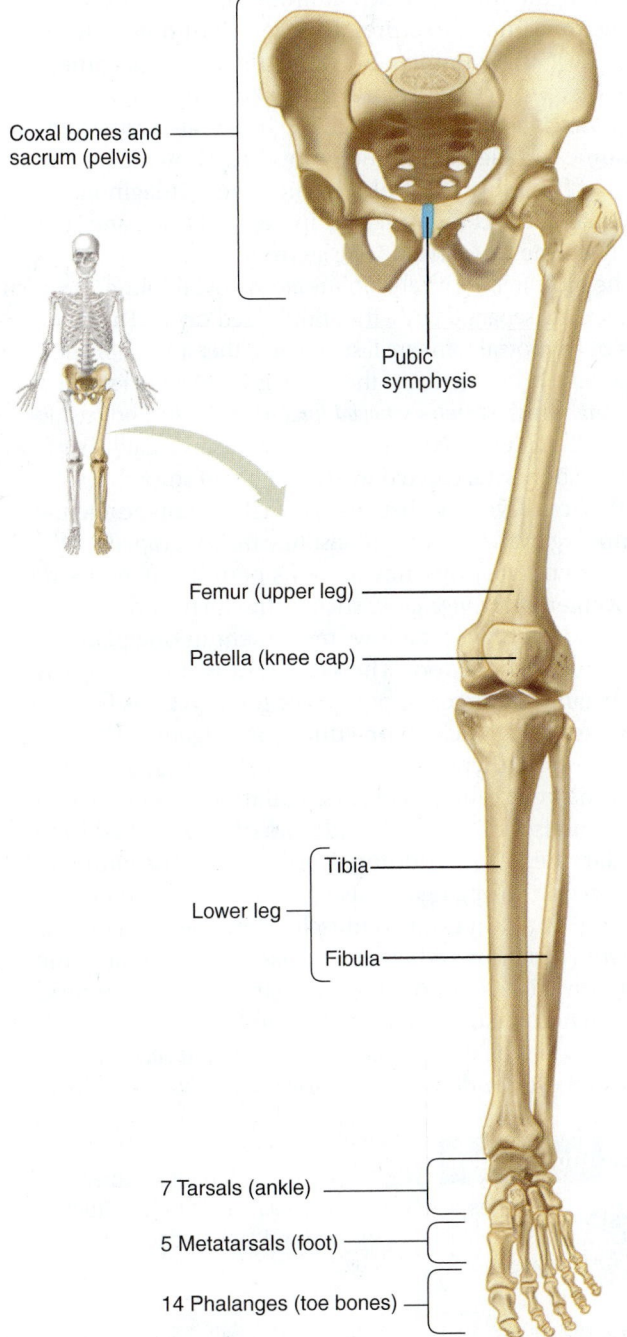

Coxal bones and sacrum (pelvis)

Pubic symphysis

Femur (upper leg)

Patella (knee cap)

Tibia

Lower leg

Fibula

7 Tarsals (ankle)

5 Metatarsals (foot)

14 Phalanges (toe bones)

Figure 5.12 Bones of the pelvic girdle and the left leg and foot.

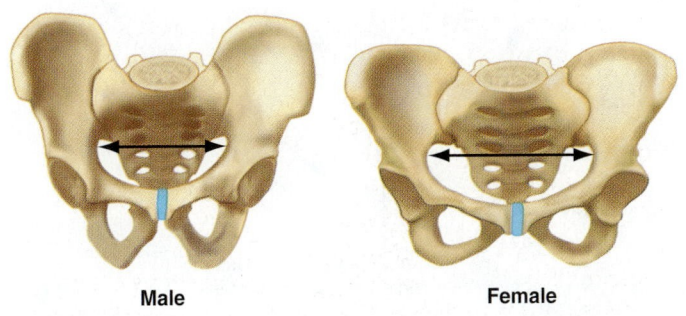

Male

Female

Figure 5.13 The pelvis. Note the wider pelvic opening (arrows) in the female.

↩ **Recap** The skull and vertebral column protect the brain and spinal cord, the rib cage protects the organs of the chest cavity, and the pelvic girdle supports the body's weight and protects the pelvic organs. The upper limbs are capable of a wide range of motions (dexterous movement). The lower limbs are stronger but less dexterous than the upper limbs. ■

5.5 Joints form connections between bones

We now turn to the structures and tissues that hold the skeleton together while still permitting us to move about freely: joints, ligaments, and tendons. **Joints,** also called articulations, are the points of contact between bones. Ligaments and tendons are connective tissues that stabilize many joints.

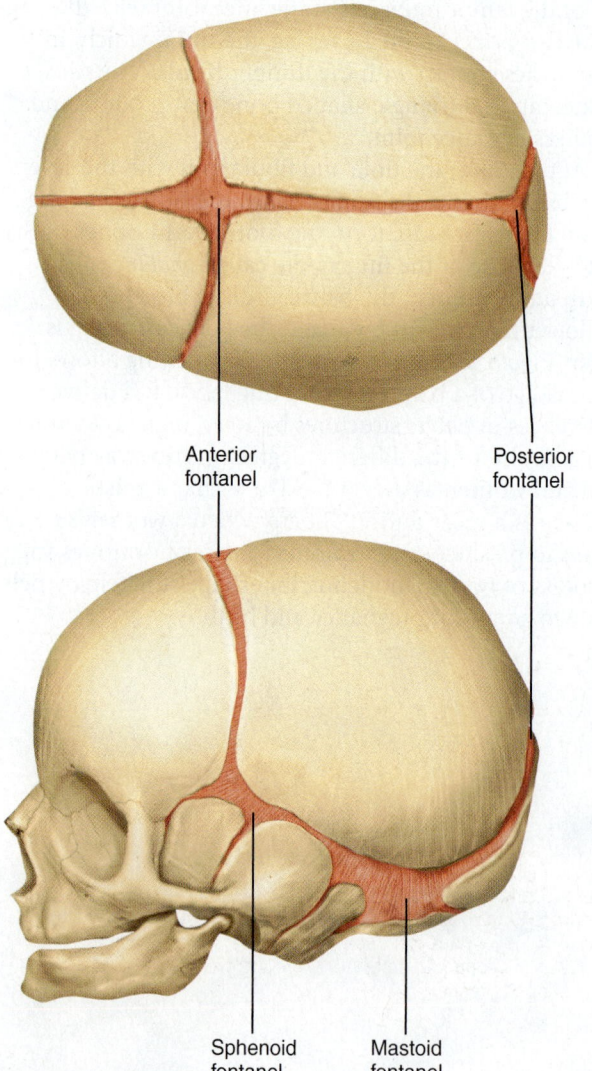

Anterior fontanel

Posterior fontanel

Sphenoid fontanel

Mastoid fontanel

Figure 5.14   Fontanels. The four fontanels allow a baby's head to change shape slightly during childbirth.

Joints vary from immovable to freely movable

Joints vary considerably from basically immovable to freely movable. Types of joints include fibrous, cartilaginous, and synovial joints.

Fibrous joints are immovable. At birth, the flat bones in a baby's skull are separated by relatively large spaces filled with fibrous connective tissue. These "soft spots," called *fontanels,* enable the baby's head to change shape slightly so that it can squeeze safely through the mother's pelvic opening during childbirth (Figure 5.14). The presence of joints also allows for brain growth and development after birth. During childhood, these fibrous joints gradually harden. By the time we reach adulthood, the joints have become thin lines, or sutures, between skull bones (review Figure 5.6). These immovable joints firmly connect the bones that protect and stabilize the skull and brain.

Cartilaginous joints, in which the bones are connected by hyaline cartilage, are slightly movable, allowing for some degree of flexibility. Examples include the cartilaginous joints that connect the vertebrae in the backbone and those that attach the lower ribs to the sternum.

The most freely movable joints are **synovial joints,** in which the bones are separated by a thin fluid-filled cavity. The two bones of a synovial joint are fastened together and stabilized by ligaments. The interior of the cavity is lined with a *synovial membrane,* which secretes *synovial fluid* to lubricate and cushion the joint. To reduce friction even further, the articulating surfaces of the two bones are covered with a tough but smooth layer of hyaline cartilage. Together, the synovial membrane and the surrounding hyaline cartilage constitute the *joint capsule.*

Different types of synovial joints permit different kinds of movements. A *hinge joint,* such as the knee and elbow, gets its name because it allows movement in one plane like the hinges on a door. The knee joint is strong enough to withstand hundreds of pounds of force, yet it is flexible enough to swing freely in one direction (Figure 5.15). To reduce friction, there are small disks of cartilage on either side of the knee called *menisci* (singular: meniscus). The knee joint also includes 13 small sacs of fluid, called *bursae* (singular: bursa), for additional cushioning. The entire joint is wrapped in strong ligaments that attach bone to bone and tendons that attach bone to muscle. Note the two *cruciate ligaments* (posterior and anterior) that join the tibia to the femur bone. The anterior cruciate ligament is sometimes injured when the knee is hit with great force from the side.

A second type of synovial joint, a *ball-and-socket joint,* permits an even wider range of movement. Examples include

---

MJ's BlogInFocus   Why are knee and hip replacement surgeries on the rise? Visit MJ's blog in the Study Area in MasteringBiology and look under "Knee and Hip Surgeries."

http://goo.gl/AarPvE

HEALTH & WELLNESS
Treating a Sprained Ankle

For a severe sprain, many physicians advise the frequent application of cold to the sprained area during the first 24 hours, followed by a switch to heat. Why the switch, and what is the logic behind the timing of cold versus heat? The biggest immediate problem associated with a sprain is damage to small blood vessels and subsequent bleeding into the tissues. Most of the pain associated with a sprain is due to the bleeding and swelling, not damage to ligaments themselves. The immediate application of cold constricts blood vessels in the area and prevents most of the bleeding. The prescription is generally to cool the sprain for 30 minutes every hour or 45 minutes every hour and a half. In other words, keep the sprain cold for about half the time, for as long as you can stand it. The in-between periods ensure adequate blood flow for tissue metabolism. It's also a good idea to keep the ankle wrapped in an elastic bandage and elevated between cooling treatments, to prevent swelling. If you're having trouble remembering all this, remember the acronym "RICE"—Rest, Ice, Compression, Elevation.

The key to a quick recovery from a sprain is rapid application of the RICE method. Athletes who try to "work through the pain" by continuing to compete while

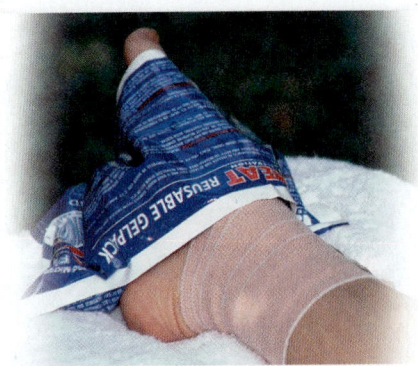

Treat sprains first with cold, then later with heat.

injured generally pay the price in a longer recovery time.

After 24 hours there shouldn't be any more bleeding from small vessels. The damage has been minimized, so now the goal is to speed the healing process. Heat dilates the blood vessels, improves the supply of nutrients to the area, and attracts blood cells that begin the process of tissue repair.

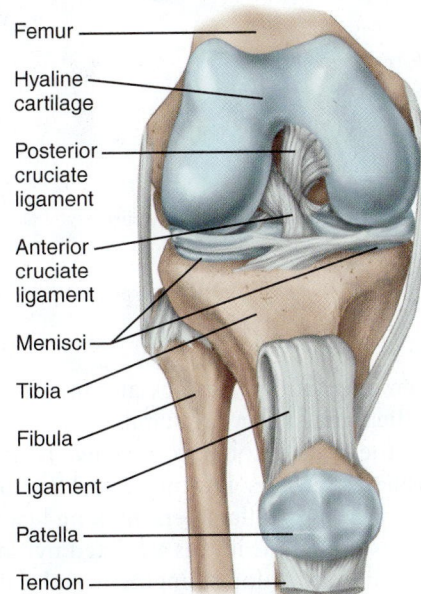

Femur
Hyaline cartilage
Posterior cruciate ligament
Anterior cruciate ligament
Menisci
Tibia
Fibula
Ligament
Patella
Tendon

a) A cutaway anterior view of the right knee with muscles, tendons, and the joint capsule removed and the bones pulled slightly apart so that the two menisci are visible.

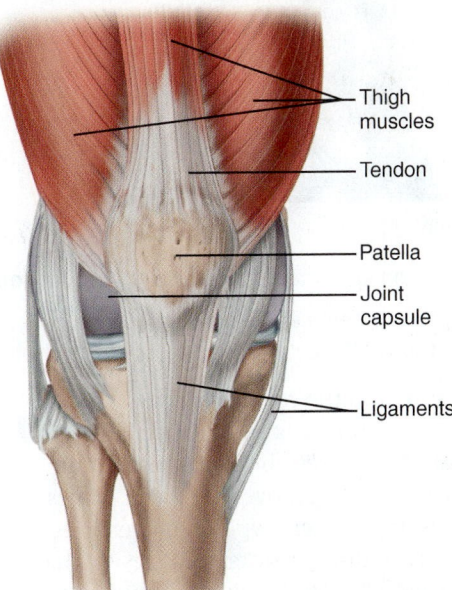

Thigh muscles
Tendon
Patella
Joint capsule
Ligaments

b) A view of the knee with muscles, tendons, and ligaments in their normal position surrounding the intact joint capsule. The combination of ligaments, tendons, and muscles holds the knee tightly together.

Figure 5.15   The knee joint is a hinged synovial joint.

☑ What is the difference between a ligament and a tendon?
Hint: Find all the ligaments and tendons in this figure and notice what they are attached to.

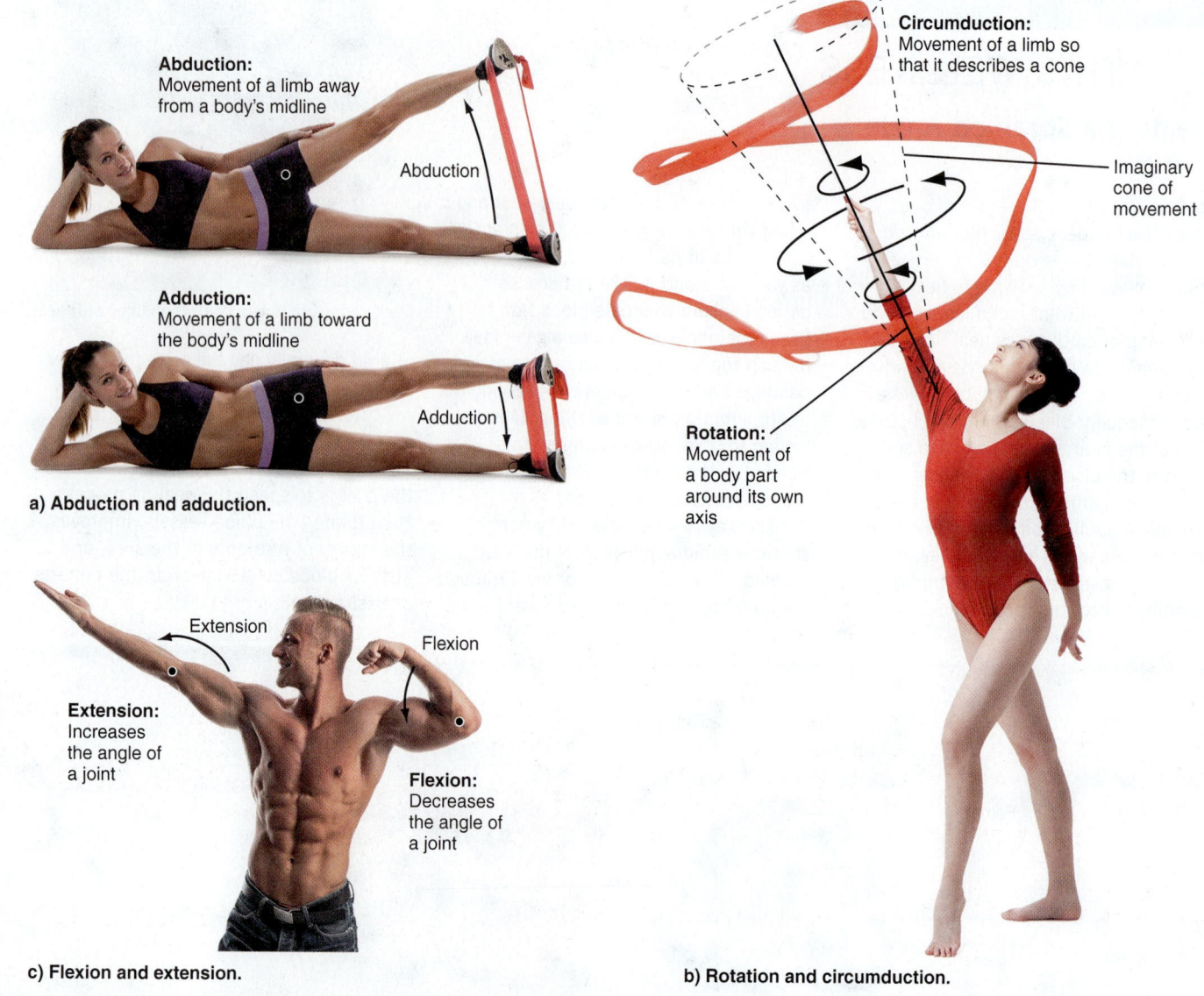

Abduction:
Movement of a limb away from a body's midline

Abduction

Adduction:
Movement of a limb toward the body's midline

Adduction

a) Abduction and adduction.

Circumduction:
Movement of a limb so that it describes a cone

Imaginary cone of movement

Rotation:
Movement of a body part around its own axis

Extension

Flexion

Extension:
Increases the angle of a joint

Flexion:
Decreases the angle of a joint

c) Flexion and extension.

b) Rotation and circumduction.

Figure 5.16   Types of movements made possible by synovial joints.

☑ Which of these types of movement can be produced by a hinge joint? Which by a ball-and-socket joint? Which by a fibrous joint?

the joint between the femur and the coxal bone (see Figure 5.12) and between the humerus and the pectoral girdle (see Figure 5.11). In both cases, the rounded head of the bone fits into a socket, allowing movement in all planes.

Figure 5.16 illustrates the different types of movements made possible by hinge and ball-and-socket joints. Note that you can rotate your arm and your leg because the shoulder and hip are ball-and-socket joints, but you cannot rotate the hinge joint in your knee.

Ligaments, tendons, and muscles strengthen and stabilize joints

Thanks to its design, a synovial joint can withstand tremendous pounding day after day, year after year without wearing out. But where does it get its strength? For that, we turn to ligaments, tendons, and muscles. As we have

seen, the bones of a synovial joint are held tightly together by ligaments. They are stabilized even more by **tendons,** another type of tough connective tissue, which join the bones to muscles. Ligaments and tendons contain collagen arranged in parallel fibers, making ligaments and tendons as strong and as flexible as a twisted nylon rope. In addition, muscle contraction strengthens and stabilizes certain joints at the very moment they need it the most.

To appreciate the role of muscle contraction in stabilizing a joint, try this simple experiment. Sit in a low chair, stretch one of your legs straight out in front of you with your heel resting on the floor, and relax your muscles. Move your kneecap (patella) from side to side gently with your hand. Notice how easily you can shift it out of position. Now, without changing position, tense the muscle of your thigh and again try to move your kneecap with your hand. See the difference? The patella is attached to the tibia by a ligament and to the muscles of the thigh by a tendon (review Figure 5.14b). Contraction of the thigh muscle (as when you take a step while walking) puts tension on the tendon and the ligament. The increased tension holds the patella and the rest of the joint firmly in place. If you move your hand to just below the kneecap, you can feel the tightening of the patellar ligament as you alternately contract and relax your thigh muscle.

Recap Joints are the points of contact between bones. Fibrous joints are immovable in adults, cartilaginous joints permit some movement, and synovial joints are highly movable. Synovial joints are held together by ligaments and lubricated by synovial fluid. ∎

5.6 Diseases and disorders of the skeletal system

In this chapter, we have already discussed several health conditions related to the skeletal system, including fractures and carpal tunnel syndrome. Now we look at several more.

Osteoporosis is caused by excessive bone loss

Osteoporosis is a condition caused by excessive bone loss over time (Figure 5.17), leading to brittle, easily broken bones. Symptoms include hunched posture (Figure 5.18), difficulty walking, and an increased likelihood of bone fractures, especially of the spine and hip. Osteoporosis is a major health problem in the United States. Over 10 million Americans have the condition, and it accounts for more than 1.5 million debilitating fractures every year.

A very slow progressive bone loss occurs in both men and women after age 35 because of a slight imbalance between the rates of bone breakdown by osteoclasts and new bone formation by osteoblasts. Overall, the rate of bone loss in men (and in women before menopause) is only about 0.4% per year. That means that on average, a man will lose only about 20% of his bone mass by age 85—not enough to cause disability in most cases.

For women, it's a different story, because a decline in estrogen after menopause leads to a more rapid rate of bone loss in the decade immediately after menopause—as high as 2 to 3% per year. After that, the rate of loss begins to decline slowly toward 0.4% again. Nevertheless, women tend to lose considerably more bone mass over a lifetime than men, which is why women are more prone to osteoporosis. Other risk factors include smoking, a sedentary lifestyle, low calcium intake, and being underweight.

MJ's BlogInFocus What is the effect of smoking on bone deposition in young women? To find out, visit MJ's blog in the Study Area in MasteringBiology and look under "Smoking and Bone Deposition."

http://goo.gl/LvvPLV

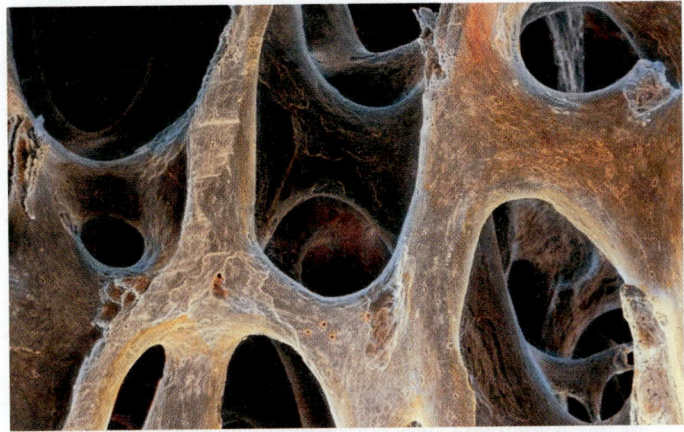

a) A scanning electron micrograph (SEM) of normal bone.

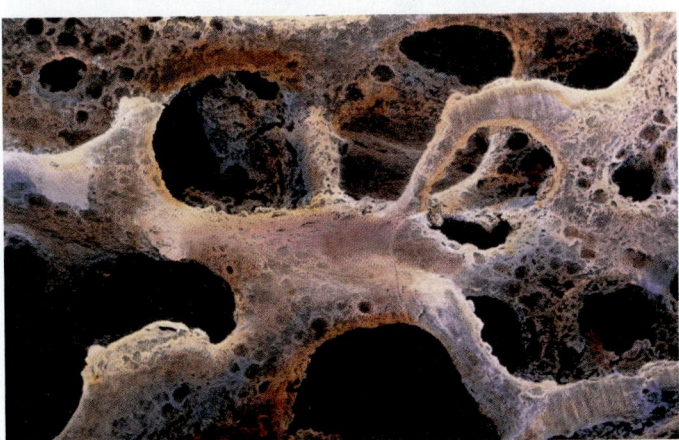

b) SEM of a bone showing osteoporosis.

Figure 5.17 Bone Loss in osteoporosis.

Figure 5.18 Osteoporosis. Osteoporosis can lead to repeated compression fractures of the spine and a permanent change in spine curvature.

The good news is that osteoporosis can be prevented. Two important strategies: get enough calcium and vitamin D, and maintain a consistent exercise program throughout your life. Calcium is crucial for the formation of new bone tissue. Current recommendations call for a daily intake of about 1,000 to 1,500 mg per day for adults, but women who have gone through menopause may benefit from even higher intakes. Both men and women can benefit from weight-bearing exercise (such as walking) and strength training (such as lifting weights), because these activities increase bone mass. For women especially, estrogen replacement therapy after menopause can slow the rate of bone loss.

Several medications are available to treat osteoporosis. A class of drugs called biphosphonates (alendronate and risedronate) act by inhibiting the bone-resorbing function of osteoclasts. The FDA recently approved a new biphosphonate medication, Boniva Injection, which can be administered intravenously every three months. Teriparatide, a medication that is a fragment of the normal parathyroid hormone molecule, is the first osteoporosis medication that can actually stimulate the activity of the bone-forming osteoblasts.

Sprains mean damage to ligaments

A sprain is due to stretched or torn ligaments. Often, it is accompanied by internal bleeding with subsequent bruising, swelling, and pain. The most common example is a sprained ankle. Sprains take a long time to heal because the ligaments have few cells and a poor blood supply. Minor sprains, in which the ligaments are only stretched, usually mend themselves with time. If a large ligament is torn completely, it generally does not heal by itself, and surgery may be necessary to remove it. Sometimes the joint can be stabilized with a piece of tendon or by repositioning other ligaments. Torn ligaments in the knee are particularly troublesome because they often leave the knee joint permanently unstable and prone to future injuries.

☑ Which is likely to heal faster—a broken bone or a sprained ligament? Why?

Bursitis and tendinitis are caused by inflammation

Bursitis and tendinitis refer to inflammation of the bursae or tendons. (The suffix *-itis* denotes that the disease is characterized by inflammation; perhaps the best-known inflammatory disease is appendicitis.) Inflammation is a response to injury characterized by redness, warmth, swelling, and pain.

Causes of bursitis and tendinitis may include tearing injuries to tendons, physical damage caused by blows to the joint, and some bacterial infections. Like ligaments, tendons and the tissues lining the bursae are not well supplied with blood vessels, so they do not heal quickly. Treatment usually involves applying cold during the first 24 hours and heat after that, resting the injured area, and taking pain-relieving medications. "Tennis elbow" is a painful condition caused by either bursitis or tendinitis. Other common locations for pain include the knee, shoulder, and the Achilles tendon that pulls up the back of the heel.

Arthritis is inflammation of joints

By their nature, joints are exposed to high compressive forces and are prone to excessive wear caused by friction. *Arthritis* is a general term for joint inflammation. The most common type of arthritis is *osteoarthritis*, a degenerative ("wear-and-tear") condition that affects about 20 million Americans, most over age 45. In osteoarthritis, the cartilage covering the ends of the bones wears out. With time the bone thickens and may form bony spurs, which further restrict joint movement. The result is increased friction between the bony surfaces, and the joint becomes inflamed and painful. Over-the-counter medications can reduce the inflammation and pain, and surgical joint replacements for severe osteoarthritis are fairly routine today. Injections of hyaluronic acid, a component of hyaline cartilage, can also reduce arthritic knee pain. Many physicians advise people with osteoarthritis to exercise regularly, which helps preserve the joints' healthy range of motion. Several promising new treatments to reduce joint inflammation are still in the experimental stage.

Osteoarthritis should not be confused with *rheumatoid arthritis*. Rheumatoid arthritis also involves joint inflammation, but it is caused by the body's own immune system, which mistakenly attacks the joint tissues. We take a closer look at rheumatoid arthritis when we discuss the immune system.

☑ A medical researcher is trying to develop a new drug that will help patients with osteoarthritis. Which is likely to be most helpful: a drug that increases osteoclast activity, one that increases osteoblast activity, or one that increases chondroblast activity? Explain.

Chapter Summary

5.1 The skeletal system consists of connective tissue *p. 101*

- Connective tissues of the skeletal system are bones, ligaments, and cartilage.
- Bone is a living tissue composed of cells and extracellular material.
- Ligaments, composed of dense fibrous connective tissue, attach bones to each other.
- Cartilage forms the intervertebral disks and lines the points of contact between bones.

5.2 Bones develop from cartilage *p. 103*

- After about two months of fetal development, rudimentary models of bones have been formed from cartilage.

- Throughout the rest of fetal development and on into childhood, bone-forming cells called *osteoblasts* replace the cartilage model with bone.
- Growth in the length of long bones centers on growth plates in each epiphysis.

5.3 Mature bone undergoes remodeling and repair *p. 104*

- Bone undergoes replacement throughout life.
- Bones can change shape over time, depending on the forces to which they are exposed.

- The process of bone repair includes (1) the formation of a hematoma, (2) the formation of a fibrocartilage callus that binds the broken ends together, and (3) the eventual replacement of the callus with new bone.

5.4 Bones fit together to form the skeleton p. 106

- The axial skeleton is represented by the skull, the vertebral column, the sternum, and the ribs.
- In the vertebral column, intervertebral disks of fibrocartilage absorb shock and permit limited movement.
- The appendicular skeleton includes the pectoral girdle, the pelvic girdle, and the upper and lower limbs.

5.5 Joints form connections between bones p. 112

- Three types of joints connect bones: fibrous, cartilaginous, and synovial.
- Synovial joints are designed for movement without friction. They are lined with a synovial membrane and lubricated by synovial fluid.

5.6 Diseases and disorders of the skeletal system p. 115

- Sprains are the result of stretched or torn ligaments. Bursitis and tendinitis are caused by injuries to the bursae and tendons.
- *Arthritis* is a general term for joint inflammation.
- Osteoarthritis is a condition in which the cartilage covering the ends of the bones wears out and joint friction increases.
- Osteoporosis is a condition caused by progressive bone loss over time.

Terms You Should Know

appendicular skeleton, *110*
axial skeleton, *107*
bone, *101*
cartilage, *103*
central canal, *102*
chondroblast, *103*
compact bone, *102*
growth plate, *104*
intervertebral disk, *108*
joint, *112*

ligament, *103*
osteoblast, *104*
osteoclast, *104*
osteocyte, *102*
osteon, *102*
osteoporosis, *105*
skeleton, *106*
spongy bone, *102*
tendon, *114*

Concept Review

Answers can be found in the Study Area in MasteringBiology.

1. List the five functions of bone.
2. Describe the functions of red and yellow bone marrow.
3. Explain how the two growth plates in a long bone account for the ability of a long bone to lengthen.
4. Explain what might cause a long bone to slowly change shape over many years.
5. Describe the process of bone remodeling and how it can reshape bones to make them stronger.
6. Name the three anatomical regions of the vertebral column that are above the *sacral* and *coccygeal* regions.

7. Explain why it is important not to move someone who may have suffered an injury to the vertebral column until a medical assessment can be made.
8. Describe the features of synovial joints that reduce friction and prevent the joint from wearing out prematurely.
9. Distinguish between *flexion* and *extension*.
10. Define the differences between *osteoarthritis* and *rheumatoid arthritis*.

Test Yourself

Answers can be found in the Appendix.

1. Which of the following might result from a parathyroid tumor that causes oversecretion of parathyroid hormone?
   a. joint inflammation leading to osteoarthritis
   b. bone loss due to stimulation of osteoclasts
   c. bone growth due to stimulation of osteoblasts
   d. conversion of cartilage to bone
2. Steps in the repair of a bone fracture include (1) bone deposition by osteoblasts, (2) bone and debris removal by osteoclasts, (3) hematoma, and (4) formation of a fibrocartilage callus. In what order do these steps occur?
   a. 1-2-3-4   c. 3-4-2-1
   b. 3-4-1-2   d. 4-3-2-1
3. All of the following bones form part of the eye socket except the:
   a. occipital bone   c. zygomatic bone
   b. lacrimal bone   d. ethmoid bone
4. All of the following bones of the skull are stationary except the:
   a. frontal bone   c. maxilla
   b. mandible   d. zygomatic bone
5. Which bones are found in both the hands and feet?
   a. carpals   c. tarsals
   b. metacarpals d. phalanges
6. The movement of the thumb to trace a circle might best be described as:
   a. abduction   c. circumduction
   b. rotation   d. pronation
7. Synovial joints may include cartilage, ligaments, tendons, and synovial fluid. Which of these attach bones to other bones within the joint?
   a. synovial membrane   c. tendons
   b. ligaments   d. cartilage
8. Which of the following is an example of a cartilaginous joint?
   a. knee joint   c. pubic symphysis
   b. skull sutures   d. hip joint
9. All of the following are bones of the axial skeleton except:
   a. vertebrae   c. skull
   b. ribs   d. clavicle
10. Which of the following would be likely to prevent or slow the bone loss of osteoporosis?
    a. stimulate the activity of fibroblasts
    b. stimulate the activity of osteoblasts
    c. inhibit the activity of osteoclasts
    d. both (b) and (c)

11. Which of the following contains the richest population of the stem-cell precursors for red and white blood cells?
    a. red bone marrow
    b. yellow bone marrow
    c. osteoid
    d. crystals of calcium phosphate

12. In the formation and development of bones within the fetus, which of these cell types functions earliest?
    a. osteocyte
    b. osteoblasts
    c. osteoclasts
    d. chondroblasts

13. Which of the following might be most helpful in determining whether an adolescent is no longer growing?
    a. measuring the length of the femur and humerus
    b. examining the growth plates near the ends of long bones
    c. examining bone density
    d. examining the fontanels in the skull

14. All of the following processes continue in the skeletal system throughout the life span except:
    a. bones continue to lengthen
    b. stem cells continue to form new blood cells
    c. bones continue to be remodeled
    d. bones continue to store minerals (calcium and phosphorus)

15. Which kind of joint is essentially immovable?
    a. hinge joint
    b. fibrous joint
    c. cartilaginous joint
    d. ball-and-socket joint

Apply What You Know

Answers can be found in the Study Area in MasteringBiology.

1. Compare and contrast swimming and running as forms of exercise training in terms of how they might affect muscle mass, bone mass, and the possibility of injuries to joints.
2. The administration of growth hormone is sometimes used clinically to stimulate growth in unusually short children who are deficient in growth hormone. However, growth hormone is ineffective in unusually short but otherwise normal adults. What accounts for the difference?
3. Although sports are getting more and more competitive at younger and younger ages, in baseball it is not recommended that children learn to throw curveballs at too young an age. What is the problem with throwing a curveball?
4. You and a friend decide to volunteer to help build houses with Habitat for Humanity over your spring break. Although you have only rarely used a hammer, you take on the task helping to construct the frame of the house. After your break is over and you have returned to campus, you notice a sharp pain in your elbow every time you bend your arm. You seek medical advice, and the doctor tells you that you have tendinitis. What is tendinitis? What might have caused your condition?

5. You just graduated and got your first job as a forensic investigator. Your first case is a skeleton that was discovered in the desert. The pathologist examines the bones and tells you that the skeleton belonged to an adult man. How can the pathologist be certain by examining only the bones? What else might the pathologist be able to tell you by examining the bones of this skeleton?
6. Obesity is a common problem in this country, even among children. What changes would you expect to see in the skeletal system of a person who has been obese for a long time?

MJ's BlogInFocus

Answers can be found in the Study Area in MasteringBiology.

1. Based on what you learned about synovial joints, do you think runners are at increased risk of degenerative hip and knee diseases later in life, compared to non-runners? To learn more about this issue, find MJ's blog in the Study Area in MasteringBiology and look under "Is Running Hard on Hips and Knees?" http://goo.gl/DYSqFI

MasteringBiology®

Students Go to MasteringBiology for assignments, the eText, and the Study Area with animations, practice tests, and activities.

Professors Go to MasteringBiology for automatically graded tutorials and questions that you can assign to your students, plus Instructor Resources.

Answers to ✓ questions are available in the Appendix.

The Muscular System

Climber in Banff National Park, Canada.

Key Concepts

- **Your body has three types of muscle:** skeletal muscle, cardiac muscle, and smooth muscle.

- **The fundamental activity of all muscles is contraction.** Depending on muscle type and location, muscle contraction can either *cause* or *resist* movement.

- **Skeletal muscle contraction is initiated by nerve activity.** Contraction requires energy that ultimately comes from stored carbohydrates and fats.

- **Skeletal muscle mass, strength, and endurance can be increased by exercise training.** The type of exercise training determines whether primarily strength or endurance is increased.

- **Cardiac and smooth muscle cells can be activated either by nerve activity or by other nearby cells.** As a result, cardiac and smooth muscles tend to contract in a coordinated fashion; when one cell contracts, nearby cells contract too.

CURRENT ISSUE
Drug Abuse Among Athletes

Baseball slugger Barry Bonds was convicted for lying under oath about using performance-enhancing drugs. Sprinter Marion Jones confessed to using performance-enhancing drugs and offered a tearful apology before relinquishing her five Olympic gold medals and being banned from participation in the 2008 Olympics in Beijing. Cyclist Lance Armstrong, accused of using performance-enhancing drugs, was stripped of his record of seven consecutive Tour de France victories. He later admitted on the Oprah Winfrey Show that he had been using performance-enhancing drugs for years. What is going on here?

The short answer: In sports where just a hundredth of a second can make the difference between a gold medal and relative obscurity, the temptation is high to use drugs that enhance athletic performance. These drugs work in ways that are predictable and understood, based on human physiology.

Anabolic Steroids

Anabolic steroids and related compounds such as dehydroepiandrosterone (DHEA) and androstenedione ("Andro") are the most widely abused drugs in athletics today. Although anabolic steroids are banned by sports federations and school systems, many of them are available over the counter as the result of a 1994 federal law that was written to ensure access to herbal remedies. In general, they are structurally and functionally related to the male sex steroid testosterone. And, like testosterone, they make it easy for the user to increase his or her muscle mass. Muscle strength improves as well, leading to improved athletic performance in sports that require short bursts of energy.

How common is anabolic steroid use? The National Institute on Drug Abuse reports that 2.1% of 12th-graders, 1.3% of 10th-graders, and 1.1% of 8th-graders have tried them. Information on steroid use by college and professional athletes is unreliable because the athletes are reluctant to talk about it. A few well-known athletes, including Arnold Schwarzenegger, sprinter Ben Johnson, and wrestler Hulk Hogan, have admitted to using them at one time or another.

Sophisticated "designer drugs" such as THG (tetrahydrogestrinone) are popular among athletes because they are formulated specifically

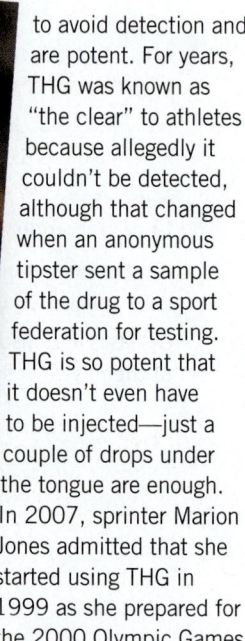

Cyclist Lance Armstrong, winner of seven consecutive Tour de France races.

to avoid detection and are potent. For years, THG was known as "the clear" to athletes because allegedly it couldn't be detected, although that changed when an anonymous tipster sent a sample of the drug to a sport federation for testing. THG is so potent that it doesn't even have to be injected—just a couple of drops under the tongue are enough. In 2007, sprinter Marion Jones admitted that she started using THG in 1999 as she prepared for the 2000 Olympic Games.

Aside from the obvious issue of conferring unfair advantage in athletic competition, anabolic steroids are banned by sports federations because of their side effects and possible health risks. Androgens have masculinizing effects in both sexes. Men may experience gynecomastia (enlargement of the breasts), shrinkage of the testicles, reduced sperm production, and impotence. In women, breast size and body fat decrease and the voice deepens. Women may lose scalp hair but gain body hair. Some of these changes are not reversible. Anabolic steroid abuse is also associated with irritability, hostility, and aggressive behavior ("roid rage"). Prolonged anabolic steroid abuse is associated with an increased risk of heart attack, stroke, and severe liver disease, →

Questions to Consider

1 Do you think we should continue to try to prevent the use of performance-enhancing drugs and block the development of genetic engineering techniques that enhance athletic performance? Why or why not?

2 A friend who uses anabolic steroids says that there is no convincing scientific evidence that anabolic steroid use will lead to health problems such as heart disease or cancer later in life. Is he right? What would you say to him?

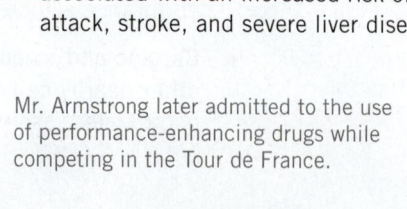

Mr. Armstrong later admitted to the use of performance-enhancing drugs while competing in the Tour de France.

including liver cancer. Although the number of cases of these diseases is fairly low (so far), the effects of steroid use/abuse may be underestimated because these diseases tend to come later in life. We just don't know what will happen to steroid abusers 30 years after they have taken the drug.

Blood Enhancers

Marathoners and cyclists aren't as interested in speed and strength; they're more interested in maintaining a high level of sustained performance over a long period of time. For that, they need increased aerobic capacity. Their drug of choice is erythropoietin (EPO), a hormone produced by the kidneys that increases the production of red blood cells. EPO is available by prescription only, for patients with *anemia* (too few red blood cells in the blood). Nevertheless, cyclists and marathon runners use it to improve their performance. Its potency is all a matter of normal human physiology: EPO produces more red blood cells, which leads to a higher oxygen-carrying capacity, which in turn leads to a higher level of sustainable muscle activity and faster times.

But a health risk is associated with EPO abuse. Excessive production of red blood cells can raise the hematocrit (the percentage of red blood cells in the blood) to dangerous levels. The blood becomes sludgelike, increasing the risk of high blood pressure, blood clots, and heart attacks. Statistically, one of the most common causes of death among professional cyclists is heart attack, although no deaths have ever officially been listed as having been caused by EPO.

It's hard to test for EPO abuse because it disappears·from the blood within days; however, an artificial boost of EPO leaves behind an increased hematocrit and an improved muscle endurance that lasts for a month or more. The cycling organizations are only able to curb EPO abuse by setting an upper limit for hematocrit of 50%; above that, EPO abuse is just assumed and the athlete is banned from competition. It is widely suspected that cyclists who choose to abuse EPO measure their hematocrit shortly before a race and then remove blood cells to just meet the 50% rule!

Next Up: Gene Doping

What if you could tinker with the genes that lead to the production of natural erythropoietin or testosterone, so that an athlete would just naturally produce more of these hormones? What if you could alter muscle biochemistry so that muscles used energy more efficiently or more rapidly? What if you could insert genes that caused muscle cells to store up more ATP? These ideas are not so far-fetched. Within decades, it will probably be possible to use genetic engineering techniques to modify an athlete's genes for improved athletic performance. It's called *gene doping*. Nearly all experts on the subject are convinced that if gene doping hasn't been tried already it soon will be. Gene doping will be extremely hard to detect or to prevent.

Have we lost our perspective on the role that sports should play in our lives?

SUMMARY

- **Performance-enhancing drugs such as anabolic steroids and erythropoietin (EPO) are used by some athletes because they improve certain types of athletic performance.**
- **Abuse of performance-enhancing drugs can lead to unwanted side effects, an increased risk of certain chronic diseases, and perhaps even premature death.**
- **Although most sports federations have banned the use of performance-enhancing drugs, enforcement has proven difficult.**
- **Soon it may be possible to use genetic engineering techniques to enhance athletic performance.**

Muscle cells are found in every organ in the body. There are three distinct types of muscle (skeletal, cardiac, and smooth), each with unique structural and functional characteristics. Together, they constitute nearly half of our body mass. *Skeletal muscles* attach to the skeleton, providing us with strength and mobility. Skeletal muscles also sculpt the body and contribute to our sense of attractiveness and well-being. One very important skeletal muscle is the diaphragm, without which the respiratory system would be unable to function. Some of the smallest skeletal muscles control the focus of our eyes; some of the largest are responsible for the shivering that helps keep us warm when it is cold. Nearly 40% of body weight in males and about 32% in females is skeletal muscle.

The other two types of muscle are less obvious but equally important. Rhythmic contractions of *cardiac muscle* in the heart provide the force needed to pump blood throughout the body. In the female reproductive system, powerful intermittent contractions of *smooth muscle* in the walls of the uterus propel the fetus through the birth canal at the time of birth. In the digestive and urinary systems, contractions of smooth muscle push food through the digestive tract and urine from the kidney into the bladder. And in every organ system in the body, steady, sustained contractions of smooth muscle in the walls of blood vessels regulate the delivery of nutrient-rich blood to every living cell. ∎

How do muscles accomplish all this? You'll find out in this chapter. We'll examine the structure of muscles, learn how they function, and see how their activity is controlled. Then we'll look at how muscles can become damaged. Finally, we'll familiarize ourselves with some of the diseases and disorders that affect muscles.

6.1 Muscles produce movement or generate tension

Figure 6.1 summarizes some of the major skeletal muscles of the body and their actions. Some muscle movements are *voluntary*, meaning that we have conscious control over the movements they produce. An example is deliberately picking up an object. Other muscle movements, such as the pumping action of the heart or the maintenance of

muscle tone in blood vessels, are *involuntary* in that they are generally beyond our conscious control. You cannot will your heart to stop beating.

We tend to think of muscles as producing movement, but another very important function of many muscles is to *resist* movement. When resisting movement, muscles generate a force that exactly opposes an equal but opposite force being applied to a body part. The maintenance of posture while standing is a good example. If you faint, you collapse

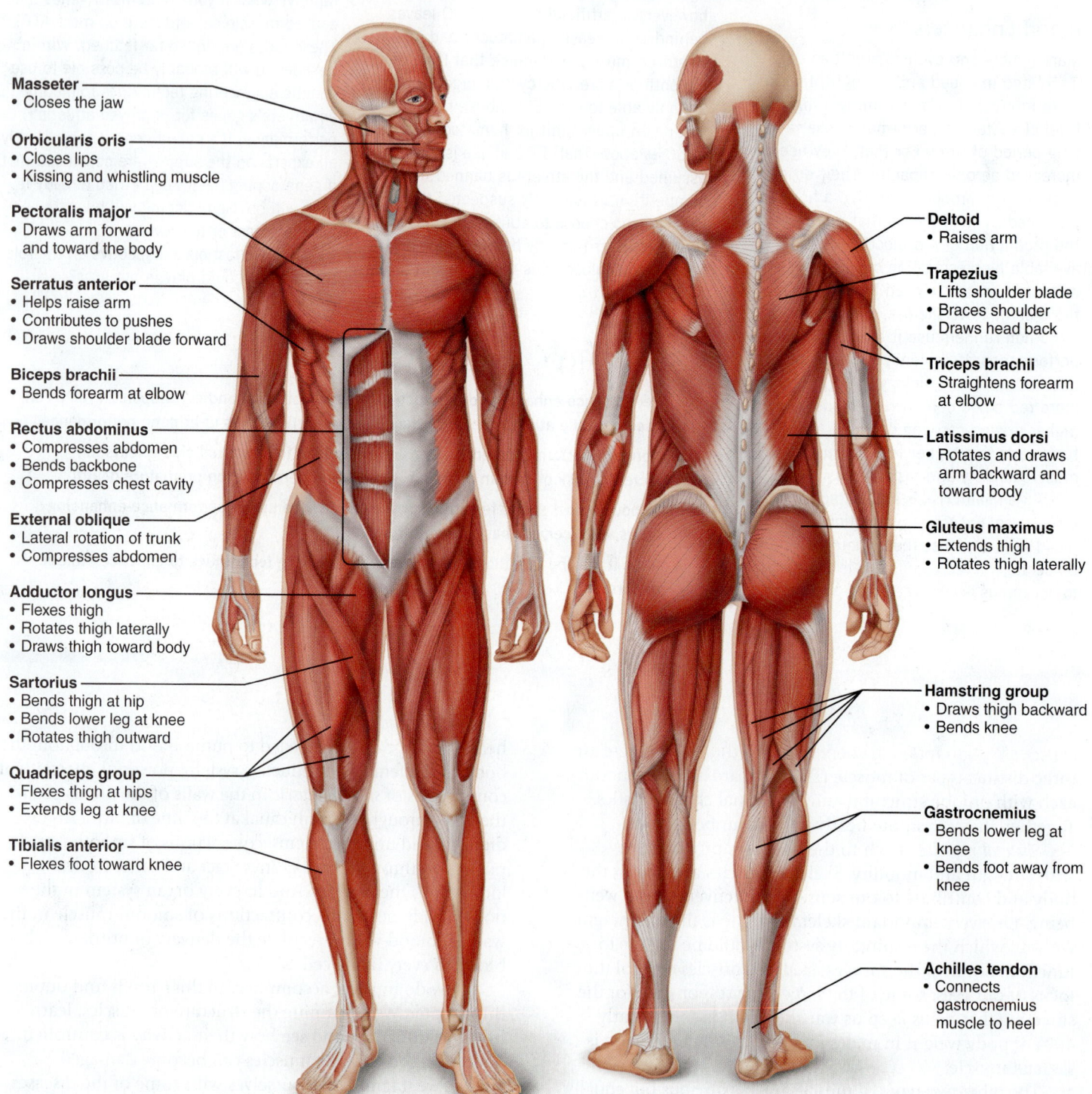

Masseter
• Closes the jaw

Orbicularis oris
• Closes lips
• Kissing and whistling muscle

Pectoralis major
• Draws arm forward and toward the body

Serratus anterior
• Helps raise arm
• Contributes to pushes
• Draws shoulder blade forward

Biceps brachii
• Bends forearm at elbow

Rectus abdominus
• Compresses abdomen
• Bends backbone
• Compresses chest cavity

External oblique
• Lateral rotation of trunk
• Compresses abdomen

Adductor longus
• Flexes thigh
• Rotates thigh laterally
• Draws thigh toward body

Sartorius
• Bends thigh at hip
• Bends lower leg at knee
• Rotates thigh outward

Quadriceps group
• Flexes thigh at hips
• Extends leg at knee

Tibialis anterior
• Flexes foot toward knee

Deltoid
• Raises arm

Trapezius
• Lifts shoulder blade
• Braces shoulder
• Draws head back

Triceps brachii
• Straightens forearm at elbow

Latissimus dorsi
• Rotates and draws arm backward and toward body

Gluteus maximus
• Extends thigh
• Rotates thigh laterally

Hamstring group
• Draws thigh backward
• Bends knee

Gastrocnemius
• Bends lower leg at knee
• Bends foot away from knee

Achilles tendon
• Connects gastrocnemius muscle to heel

Figure 6.1 Major skeletal muscle groups and their functions.

because you lose control over the muscles that support your upright posture. The maintenance of a constant blood vessel diameter even when blood pressure within the vessel changes is another example of how muscles resist movement.

Besides producing or resisting movement, muscles also generate heat. Under normal circumstances, contraction of our skeletal muscles accounts for over three-quarters of all the heat generated by the body. When our muscles generate too much heat (as they might during exercise), temperature-control mechanisms such as sweating allow the body to get rid of the excess heat. At other times, heat generated by muscles contributes to homeostasis of our body temperature. For example, if you spend time outdoors on a cold day, you may notice that you start to shiver. Shivering occurs because your temperature-control mechanisms cause skeletal muscles to alternately contract and relax so as to generate more heat.

The fundamental activity of muscle is contraction

All three types of muscle have certain fundamental features in common. First, muscles are *excitable*, meaning that they contract in response to chemical and/ or electrical signals from other organ systems. Second, all muscles have only one basic mechanism of action: they *contract* (shorten). In the absence of contraction, muscles relax, returning to their original length. The movements of your limbs, the beating of your heart, and the regulation of the diameters of your blood vessels all depend on muscle contraction.

The muscle type with which you are probably most familiar is skeletal muscle. In this chapter, we concentrate on skeletal muscle because skeletal muscle, in conjunction with the skeleton, is responsible for voluntary movement. At the end of this chapter, we briefly discuss cardiac and smooth muscle.

Skeletal muscles cause bones to move

Most **skeletal muscles** interact with the skeleton and cause bones to move (or prevent them from moving) relative to each other. All of the tasks accomplished by our skeletal muscles, whether shivering, threading a needle, lifting heavy weights, or even just standing completely still, are performed by skeletal muscles that are either contracting or relaxing. We have more than 600 skeletal muscles, often organized into pairs or groups. Hundreds of muscles, each controlled by

nerves and acting either individually or in groups, produce all possible human motions. Muscle groups that work together to create the same movement are called *synergistic muscles*. Muscles that oppose each other are called *antagonistic muscles*.

Most skeletal muscles attach to bones via tendons (Figure 6.2). There are exceptions, however; a few skeletal muscles attach only to other muscles or to skin (such as the muscles that permit you to smile). Muscles join to the skeleton in such a way that each individual muscle produces a very specific movement of one bone relative to

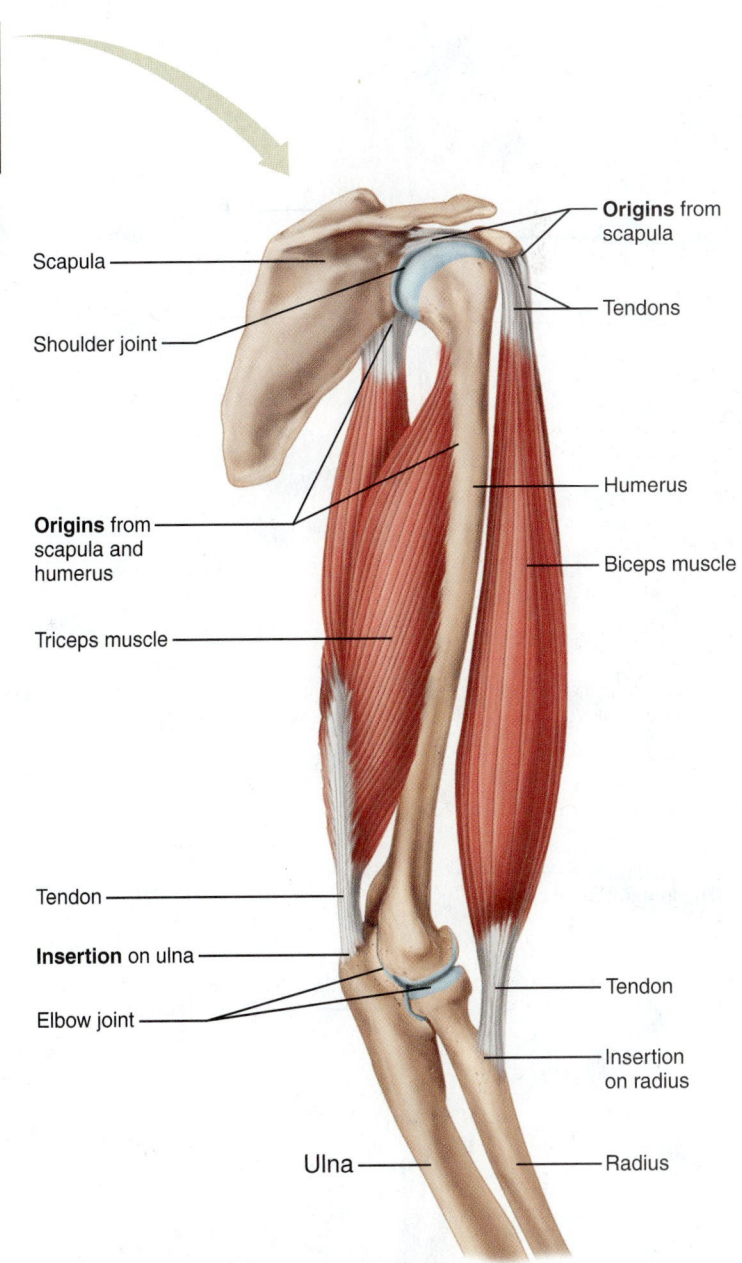

Figure 6.2 Muscles attach to bones via tendons. The point of attachment of a muscle to the stationary bone is its origin; the point of attachment to the moveable bone is its insertion.

another. The skeleton is a complex set of levers that can be pulled in many directions by contracting or relaxing skeletal muscles. One end of a skeletal muscle, called its **origin,** joins to a bone that remains relatively stationary. The other end of the muscle, called its **insertion,** attaches to another bone across a joint. When the muscle contracts, the insertion is pulled toward the origin. The origin is generally closer to the midline of the body and the insertion is farther away.

Figure 6.3 shows how the two antagonistic muscles of the upper arm, the biceps and triceps, oppose each other to bend (flex) and straighten (extend) the forearm.

When the triceps muscle relaxes and the biceps muscle contracts, the combined action pulls on the forearm and flexes it. When the biceps relaxes and the triceps contracts, the combined action pulls the forearm down, extending it again.

A muscle is composed of many muscle cells

A single *muscle* (sometimes referred to as a "whole muscle") is a group of individual muscle cells, all with the same origin and insertion and all with the same function. A cross section of muscle (Figure 6.4) reveals that it is arranged in bundles called *fascicles*, each enclosed in a sheath of a type of fibrous connective tissue called *fascia*. Each fascicle contains anywhere from a few dozen to thousands of individual muscle cells, or *muscle fibers*. The outer surface of the whole muscle is covered by several more layers of fascia. At the ends of the muscle, all of the fasciae (plural) come together, forming the tendons that attach the muscle to bone.

Individual muscle cells are tube shaped, larger, and usually longer than most other human cells. Some muscle cells are only a millimeter in length, whereas others may

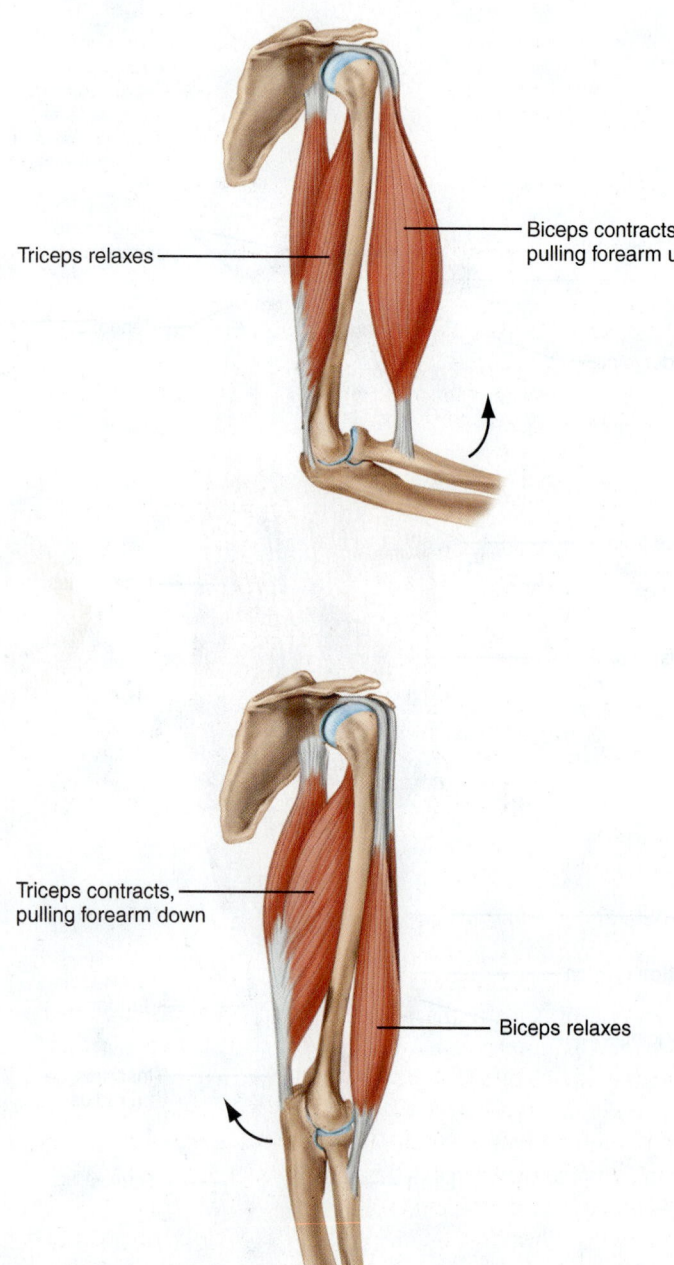

Triceps relaxes

Biceps contracts, pulling forearm up

Triceps contracts, pulling forearm down

Biceps relaxes

Figure 6.3 **Movement of bones.**

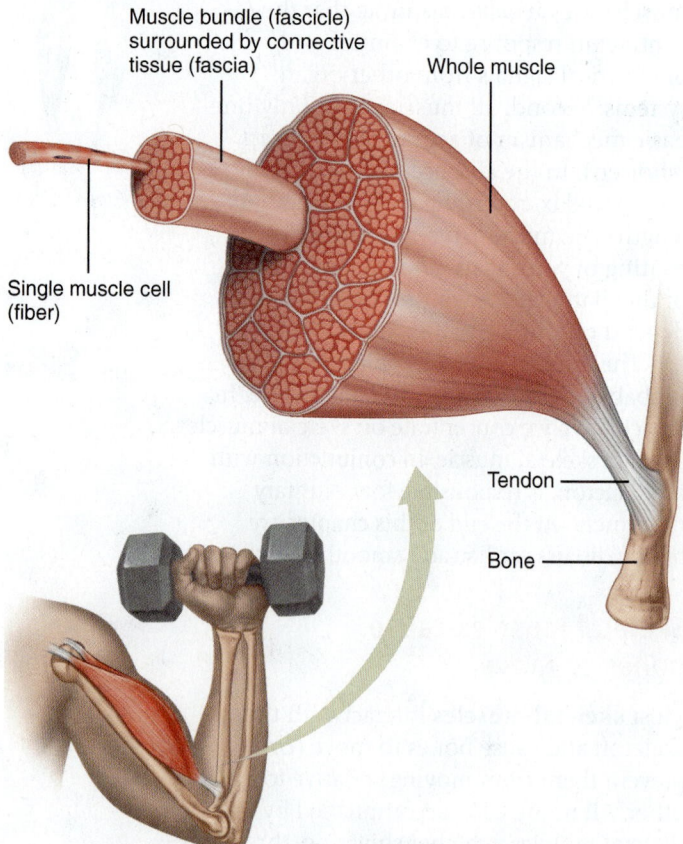

Muscle bundle (fascicle) surrounded by connective tissue (fascia)

Whole muscle

Single muscle cell (fiber)

Tendon

Bone

Figure 6.4 **Muscle structure.** A muscle is arranged in bundles called fascicles, each composed of many muscle cells and each surrounded by a sheath of fascia. Surrounding the entire muscle are several more layers of fascia. The fascia join together to become the tendon, which attaches the muscle to bone.

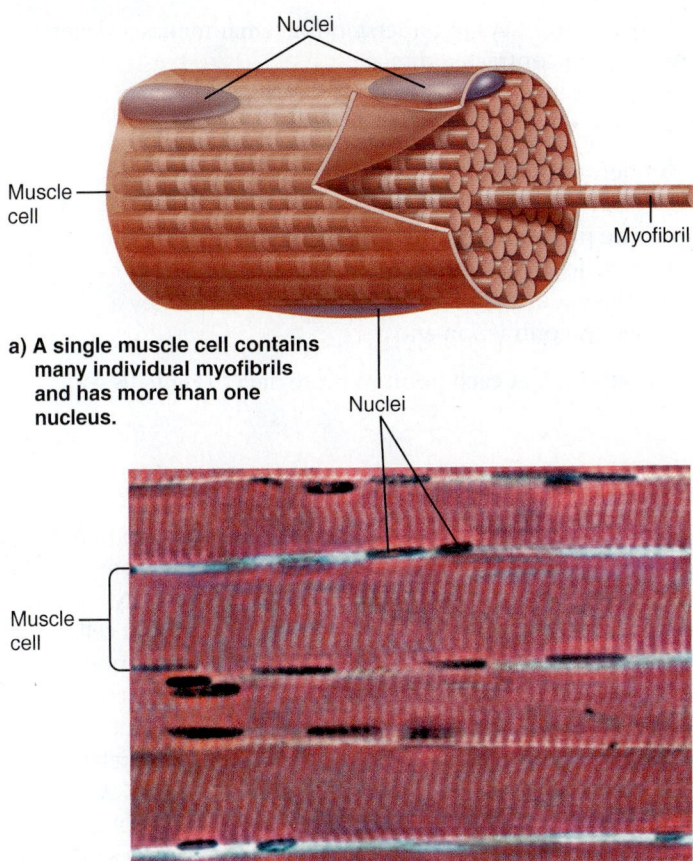

a) **A single muscle cell contains many individual myofibrils and has more than one nucleus.**

b) **A photograph of portions of several skeletal muscle cells.**

Figure 6.5 Skeletal muscle. a) A single muscle cell contains many individual myofibrils and has more than one nucleus. b) A photograph of portions of several skeletal muscle cells.

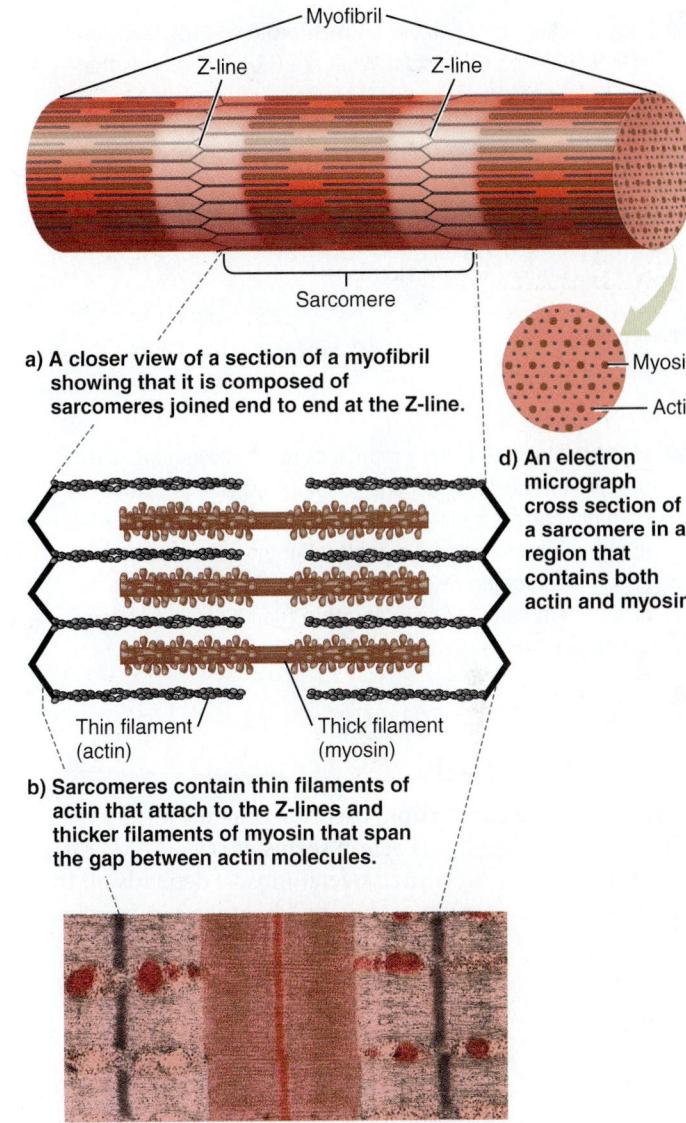

a) **A closer view of a section of a myofibril showing that it is composed of sarcomeres joined end to end at the Z-line.**

b) **Sarcomeres contain thin filaments of actin that attach to the Z-lines and thicker filaments of myosin that span the gap between actin molecules.**

c) **A transmission electron micrograph (×11,300) of a longitudinal section of a sarcomere.** The rounded red objects are mitochondria.

d) **An electron micrograph cross section of a sarcomere in a region that contains both actin and myosin.**

Figure 6.6 Structure of a myofibril.

be as long as 30 centimeters—roughly the length of your thigh muscle. Taking a closer look at a single muscle cell (Figure 6.5), we see that each cell contains more than one nucleus. The nuclei are located just under the cell membrane because nearly the entire interior of the cell is packed with long cylindrical structures arranged in parallel, called **myofibrils.** The myofibrils are packed with contractile proteins called *actin* and *myosin*, discussed in the following text. When myofibrils contract (shorten), the muscle cell also shortens.

The contractile unit is a sarcomere

Looking still closer at a single myofibril, we see a striated (or banded) appearance that repeats at regular intervals (Figure 6.6a). Various elements of the pattern stand out, but the one that is important for our discussion is a dark line called the *Z-line*. A segment of a myofibril from one Z-line to the next is called a **sarcomere** (Figure 6.6b and c). A single myofibril within one muscle cell in your biceps muscle may contain over 100,000 sarcomeres arranged end to end. The microscopic shortening of these 100,000 sarcomeres all at once is what produces contraction (shortening) of the

muscle cell and of the whole muscle. Understanding muscle shortening, then, is simply a matter of understanding how a single sarcomere works.

A sarcomere consists of two kinds of protein filaments (Figure 6.6d). Thick filaments composed of a protein called **myosin** are interspersed at regular intervals between paired filaments of a different protein called **actin.** Notice that the actin filaments are structurally linked to the Z-line and that myosin filaments are located near the middle of sarcomeres, stretching between two different actin filaments. Muscle contraction occurs when the myosin and actin filaments slide past each other, drawing the Z-lines closer together. The sliding process, called the *sliding filament mechanism* of contraction, will be described in more detail, but first we need to understand what triggers a muscle contraction in the first place.

☑ Beef, chicken, and fish are all high-protein foods that are primarily composed of muscle. What are the two proteins that comprise most of muscle tissue?

↩ **Recap** Muscles either produce or resist movement. Their fundamental activity is contraction. A muscle is composed of many muscle cells arranged in parallel, each containing numerous myofibrils. The contractile unit in a myofibril is called a *sarcomere*. A sarcomere contains thick filaments of a protein called myosin and thin filaments of a protein called actin. ∎

6.2 Individual muscle cells contract and relax

During a muscle contraction, each sarcomere shortens just a little. Subtle though this action seems, it is also powerful. The contraction of an entire skeletal muscle depends on the simultaneous shortening of the tiny sarcomeres in its cells.

There are four keys to understanding what makes a skeletal muscle cell contract and relax:

- A skeletal muscle cell must be activated by a nerve. It does not contract on its own.
- Nerve activation increases the concentration of calcium (Ca) in the vicinity of the contractile proteins.
- The presence of calcium permits contraction. The absence of calcium prevents contraction.
- When a muscle cell is no longer stimulated by a nerve, contraction ends.

Let's look at each point in more detail (**Figure 6.7**).

Nerves activate skeletal muscles

Skeletal muscle cells are stimulated to contract by certain nerve cells called **motor neurons.** The motor neurons secrete a chemical substance called *acetylcholine (ACh)*. Acetylcholine is a **neurotransmitter,** a chemical released by nerve cells that has either an excitatory or inhibitory effect on another excitable cell (another nerve cell or a muscle cell). In the case of skeletal muscle, acetylcholine excites (activates) the cells.

The junction between a motor neuron and a skeletal muscle cell is called the **neuromuscular junction.** As indicated in Figure 6.7 step ❶, when an electrical impulse traveling in a motor neuron arrives at the neuromuscular junction acetylcholine is released from the nerve terminal. The acetylcholine diffuses across the narrow space between

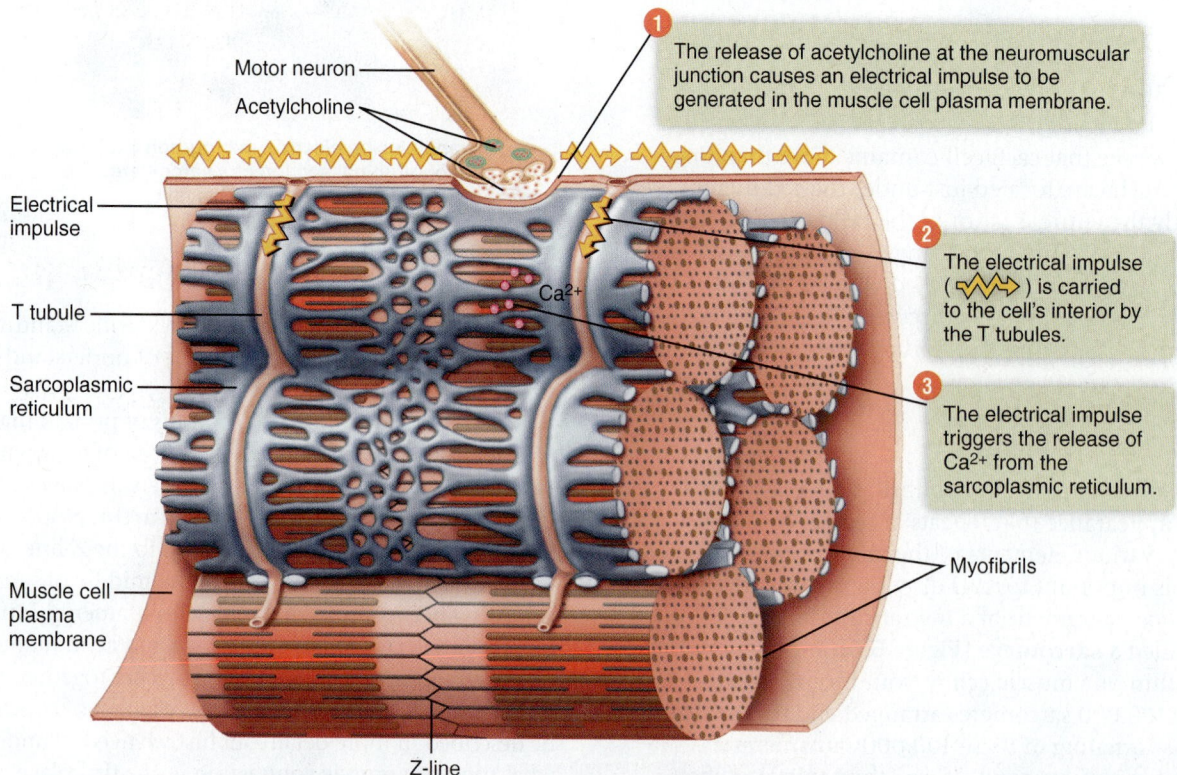

Motor neuron
Acetylcholine
Electrical impulse
T tubule
Sarcoplasmic reticulum
Muscle cell plasma membrane
Z-line
Ca²⁺
Myofibrils

❶ The release of acetylcholine at the neuromuscular junction causes an electrical impulse to be generated in the muscle cell plasma membrane.

❷ The electrical impulse (〰️) is carried to the cell's interior by the T tubules.

❸ The electrical impulse triggers the release of Ca²⁺ from the sarcoplasmic reticulum.

Figure 6.7 How nerve activation leads to calcium release within a muscle cell.

the neuron and the muscle cell and binds to receptor sites on the muscle cell membrane. In step ❷, the binding of acetylcholine to the receptors causes the muscle cell membrane to generate an electrical impulse of its own that travels rapidly along the cell membrane in all directions. In addition, tubelike extensions of the cell membrane called *T tubules* (the *T* stands for transverse) transmit the electrical impulse deep into the interior of the cell. The function of the T tubules is to get the electrical impulse to all parts of the cell as quickly as possible.

Activation releases calcium

As shown in Figure 6.7, T tubules are in close contact with a series of membrane-bound chambers called the **sarcoplasmic reticulum** (*sarco* is derived from a Greek word for "flesh" or "muscle"). The sarcoplasmic reticulum is similar to every other cell's smooth endoplasmic reticulum except that its shape is different, in part because it must fit into the small amount of space in the cell not occupied by myofibrils. The primary function of the sarcoplasmic reticulum is to store ionic calcium (Ca^{2+}).

Inside the muscle cell, an electrical impulse races down the T tubules to the sarcoplasmic reticulum. Figure 6.7 step ❸ illustrates how the arrival of an electrical impulse triggers the release of calcium ions from the sarcoplasmic reticulum. The calcium diffuses into the cell cytoplasm and then comes in contact with the myofibrils, where it sets in motion a chain of events that leads to contraction.

Calcium initiates the sliding filament mechanism

Muscles contract when sarcomeres shorten, and sarcomeres shorten when the thick and thin filaments slide past each other, a process described previously as the **sliding filament mechanism** of contraction. Taking a closer look at the arrangement of thick and thin filaments in a single sarcomere (**Figure 6.8a**), we see that every thin filament consists of two strands of actin molecules spiraling around each other and that thick filaments are composed of many individual molecules of myosin. Myosin molecules are shaped somewhat like a golf club, with a long shaft and a rounded head. Myosin shafts form the main part of the thick filaments. The heads stick out to the side, nearly touching the thin filaments of actin. When a muscle is relaxed, the myosin heads do not quite make contact with the thin filaments, however. Muscle contraction (**Figure 6.8b**) occurs when the myosin heads make contact with the thin filaments, forming a *cross-bridge* between the two filaments. The formation of a cross-bridge causes the head to bend relative to the shaft, pulling the actin molecules toward the center of the sarcomere. The processes of cross-bridge formation and bending (the molecular events of contraction) require energy.

But what initiates the process of contraction? Put another way, what prevents contraction from occurring all the time? The answer is calcium. A closer look at a section of myosin and actin shows why. Closely associated with the actin filaments are two other protein molecules called *troponin* and *tropomyosin* that together form the *troponin-tropomyosin protein complex*. In the absence of

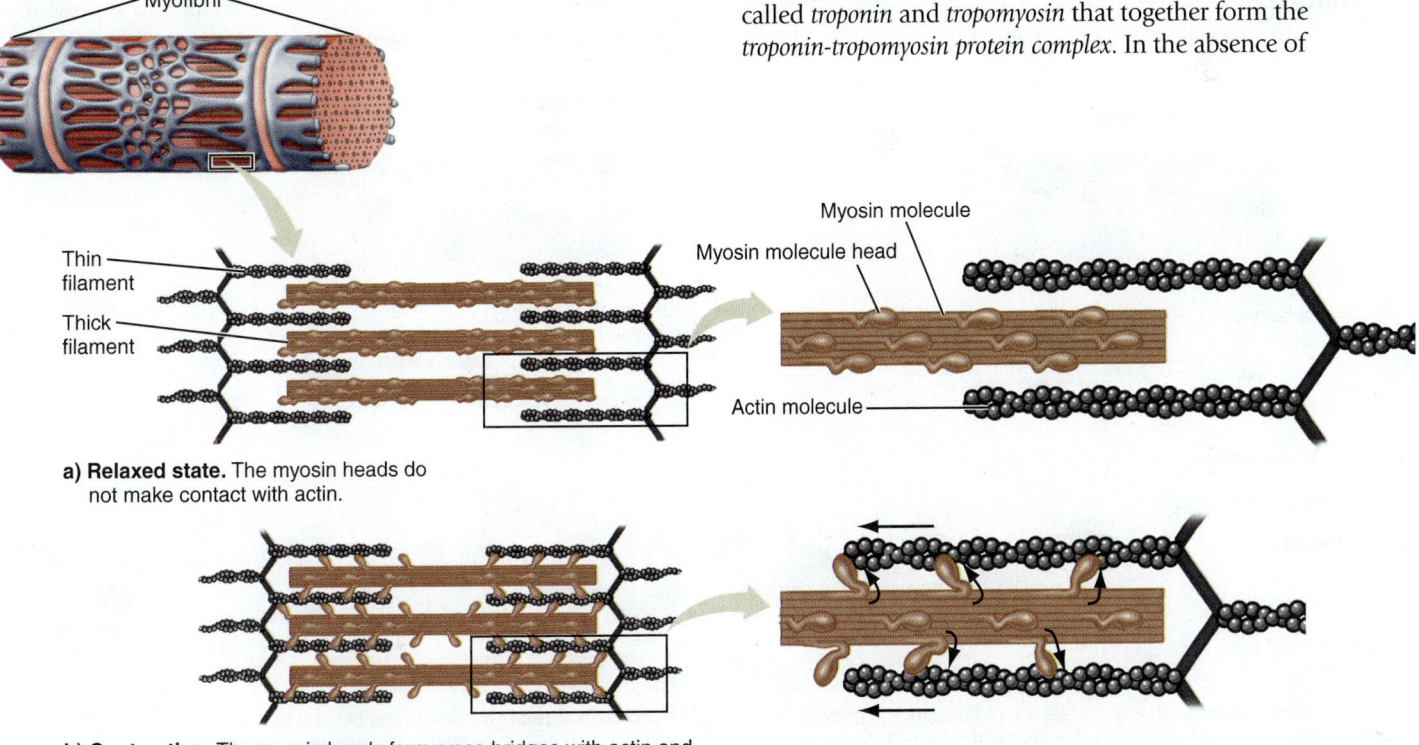

— Myofibril —

Thin filament

Thick filament

Myosin molecule

Myosin molecule head

Actin molecule

a) **Relaxed state.** The myosin heads do not make contact with actin.

b) **Contraction.** The myosin heads form cross-bridges with actin and then bend, pulling the actin filaments toward the center of the sarcomere.

Figure 6.8 Sliding filament mechanism of contraction.

calcium (Figure 6.9a), the troponin-tropomyosin protein complex interferes with the myosin binding sites on the actin molecule. Following an electrical impulse, Ca^{2+} released from the sarcoplasmic reticulum binds to troponin (Figure 6.9b), resulting in a shift in the position of the troponin-tropomyosin protein complex that exposes the myosin binding sites and permits the formation of cross-bridges. The myosin heads form cross-bridges with actin, undergo a bending process, and physically pull the actin filaments toward the center of the sarcomere from each end. With thousands of myosin cross-bridges doing this simultaneously, the result is a sliding movement of the thin filaments relative to the thick ones and a shortening of the sarcomere. As hundreds of thousands of sarcomeres shorten, individual muscle cells, and ultimately the whole muscle, shorten as well.

☑ Suppose a person had an unusual mutation in the troponin protein of his skeletal muscles, such that the troponin could not bind to calcium. Would this person's muscles be constantly contracted, constantly relaxed, or able to function normally? Explain.

When nerve activation ends, contraction ends

Relaxation of a muscle cell occurs when nerve activity ends. In the absence of nerve activity, no more calcium is released from the sarcoplasmic reticulum. The calcium released as a result of prior electrical impulses is transported back into the sarcoplasmic reticulum by active transport, which requires energy in the form of ATP. As the calcium concentration in the myofibrils falls, the troponin-tropomyosin protein complex shifts back into its original position, preventing the binding of the myosin cross-bridges to actin. The sarcomere stretches passively to its original resting length, and the muscle cell relaxes.

Any factor that interferes with the process of nerve activation can disrupt muscle function. In the disorder *myasthenia gravis*, the body's immune system attacks and destroys acetylcholine receptors on the cell membrane of muscle cells. Affected muscles respond only weakly to nerve impulses or fail to respond at all. Most commonly impaired are the eye muscles, and many people with myasthenia gravis experience drooping eyelids and double vision. Muscles in the face and neck may also weaken, leading to problems with chewing, swallowing, and talking. Medications that facilitate the transmission of nerve impulses can help people with this condition.

☑ Caffeine prolongs the life span of acetylcholine molecules in the motor junctions. Explain how this fact is related to caffeine's tendency to cause jitters, such as hand tremors and other small involuntary contractions.

Muscles require energy to contract and to relax

Muscle contraction requires a great deal of energy. Like most cells, muscle cells use ATP as the energy source. At rest, when the myosin head is detached from the actin filament (see Figure 6.8a), there's an ATP molecule attached to the myosin head. In the presence of calcium, myosin acts as an enzyme, splitting ATP into ADP and inorganic phosphate and releasing energy to do work. The energy is used to *energize* the myosin head so that it can form a cross-bridge with the thin filament and undergo bending (see Figure 6.8b). Once the bending has occurred, a fresh molecule of ATP binds to the myosin, which causes the myosin head to detach from actin and return to the relaxed position. As long as calcium is present, the cycle of ATP breakdown, myosin attachment, bending,

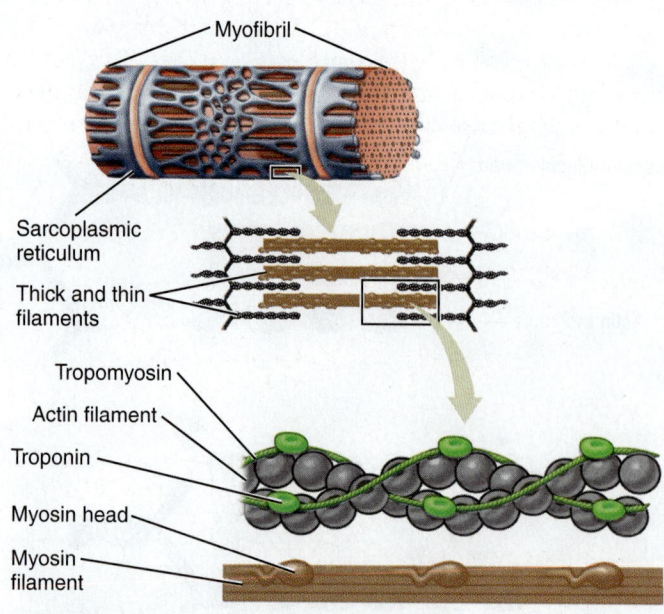

a) **Resting sarcomere.** In the absence of calcium, the muscle is relaxed because the myosin heads cannot form cross-bridges with actin.

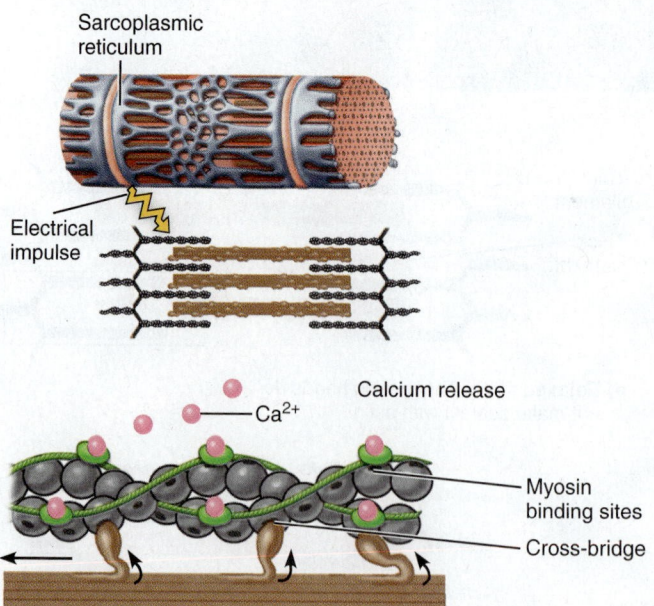

b) **Cross-bridge attachment.** The binding of calcium to troponin causes a shift in the troponin-tropomyosin protein complex, allowing cross-bridges to form.

Figure 6.9 Role of calcium in contraction.

and detachment is repeated over and over again in rapid succession. The result is a shortening of the sarcomere.

At the end of the contractile period (when nerve impulses end), energy from the breakdown of ATP is used to transport calcium back into the sarcoplasmic reticulum so that relaxation can occur. However, a second requirement for relaxation is that an intact molecule of ATP must bind to myosin before myosin can finally detach from actin. This last role of ATP is the explanation for *rigor mortis* (Latin, meaning "rigid death"), in which a body becomes stiff during the time period from about four hours to several days after death. Shortly after death, calcium begins to leak out of the sarcoplasmic reticulum, causing muscle contraction. The contractions use up the available ATP, but after death the ATP cannot be replenished. In the absence of ATP, the myosin heads cannot detach from actin, and so the muscles remain "locked" in the contracted state. Eventually, the stiffness of rigor mortis decreases as the muscle cells degenerate.

☑️ Name two things that must happen for the myosin to stop binding to the actin, that is, for the muscle to relax, and explain why each is necessary.

Producing and storing energy within muscle

Muscle cells store only enough ATP for about 10 seconds' worth of maximal activity. Once this is used up, the cells must produce more ATP from other energy sources, including creatine phosphate, glycogen, glucose, and fatty acids.

An important pathway for producing ATP involves creatine phosphate (creatine-P), a high-energy molecule with an attached phosphate group. Creatine phosphate can transfer a phosphate group and energy to ADP and therefore create a new ATP molecule quickly. This reaction is reversible: If ATP is not needed to power muscle contractions, the excess ATP can be used to build a fresh supply of creatine phosphate, which is stored until needed.

In recent years, creatine phosphate loading has become common among bodybuilders and athletes, particularly those who need short-term bursts of power. Unfortunately, muscles cannot store much more creatine phosphate than they usually have, even without creatine phosphate loading. Creatine phosphate also seems to improve muscle performance in certain neuromuscular diseases, such as amyotrophic lateral sclerosis (ALS), also known as Lou Gehrig's disease.

The combination of available ATP plus stored creatine phosphate produces only enough energy for up to 30–40 seconds of heavy activity. Beyond that, muscles must rely on stored glycogen, a complex sugar (polysaccharide) composed of many smaller molecules of glucose. For the first three to five minutes of sustained activity, a muscle cell draws on its internal supply of stored glycogen. Glucose molecules are removed from the glycogen, and their energy is used to synthesize ATP. Part of the process of the breakdown of glucose can happen without oxygen (called *anaerobic metabolism*) fairly quickly, but anaerobic metabolism yields only two ATP molecules per glucose molecule. It also has the unfortunate side effect of producing lactic acid, which causes the burning sensation one feels right at the end of a heavy weight-lifting session, just as exhaustion sets in.

The most efficient long-term source of energy is the aerobic metabolism of glucose in the blood, fatty acids derived from stored fat in fat cells, and other high-energy molecules such as lactic acid. Aerobic metabolism, as you already know, takes place in mitochondria and requires oxygen. The next time you engage in strenuous exercise, notice that it may take a minute or two for your respiratory rate to increase dramatically. Your heart rate also increases because there is an increase in blood flow to the exercising muscle. The increased respiration and heart rate are signs that aerobic metabolism is now taking place. Until aerobic metabolism kicks in, however, your cells are relying on stored ATP, creatine phosphate, and anaerobic metabolism of glycogen.

Weight lifters can rely almost exclusively on stored energy because their muscles perform for relatively short periods of time. Long-distance runners start out by depending on stored energy, but within several minutes, they are relying almost exclusively on aerobic metabolism. If they could not, they would collapse in exhaustion. Table 6.1 summarizes energy utilization by muscle.

Table 6.1 Energy sources for muscle

Energy source	Quantity	Time of use	Comments
Stored ATP	Stored only in small quantities	About 10 seconds	ATP is the only direct energy source. It must be replenished by the other energy sources.
Stored creatine phosphate	Three to five times amount of stored ATP	About 30 seconds	Creatine phosphate is converted quickly to ATP.
Stored glycogen	Variable; some muscles store large quantities	Primarily used during heavy exercise within the first 3–5 minutes	ATP yield depends on whether oxygen is available. One glucose molecule (derived from stored glycogen) yields only two ATP molecules in the absence of oxygen but 36 ATP molecules in the presence of oxygen.
Aerobic metabolism	Steady high yield; oxygen and nutrients (glucose and fatty acids derived from fat) are constantly supplied by the blood	Always present; increases dramatically within several minutes of onset of exercise, when blood flow and respiration increase	Complete metabolism of one glucose molecule yields 36 ATP molecules.

HEALTH & WELLNESS
Delayed Onset Muscle Soreness

Most of us are familiar with the feeling of stiffness and soreness after an unfamiliar form of exertion. The soreness is actually due to microscopic tears in myofibrils throughout the muscle. It is thought that the damage occurs because in the absence of regular use, some sarcomeres become like old, stiff rubber bands, unable to contract as well as they should. During exercise, these old or injured sarcomeres can stretch to the point that the thick and thin filaments no longer overlap, damaging the sarcomeres permanently. Exercises that cause muscles to be stretched (lengthened) while they are actively contracting (trying to shorten) are the most likely to cause damage and

soreness. Examples include running downhill, lowering very heavy weights, and the downward motion of push-ups and squats.

The feeling of soreness after exercise begins a day or so after exercise and usually reaches a peak 1–3 days later. The soreness is caused by chemicals released during the repair process, which involves inflammation, swelling, and the release of chemical substances such as prostaglandins. With time, the damaged sarcomeres are removed completely and new sarcomeres take their place.

Once muscles become accustomed to a particular exercise, damage (and the accompanying soreness) no longer

Delayed onset muscle soreness may occur after a strenuous unfamiliar exercise.

occurs. To minimize muscle injury and soreness, undertake any new exercise activity in moderation for the first few days.

After you finish exercising, note that you continue to breathe heavily for a period of time. These rapid, deep breaths help reverse your body's **oxygen debt,** incurred because your muscles used more ATP early on than was provided by aerobic metabolism. The additional ATP was produced by anaerobic metabolism, with the subsequent buildup of lactic acid. After exercise, you still need oxygen to metabolize the lactic acid by aerobic pathways and to restore the muscle's stores of ATP and creatine phosphate to their resting levels. The ability of muscle tissue to accumulate an oxygen debt and then repay it later allows muscles to perform at a near-maximal rate even before aerobic metabolism has increased.

Muscle **fatigue** is defined as a decline in muscle performance during exercise. The most common cause of fatigue is insufficient energy to meet metabolic demands, due to depletion of ATP, creatine phosphate, and glycogen stores within the muscle. However, fatigue can also be caused by psychological factors, including discomfort or the boredom of repetitive tasks.

↩ **Recap** Skeletal muscle contraction is initiated by nerves, which trigger the release of calcium within the cell. The calcium allows cross-bridges to form between myosin and actin, leading to contraction. Energy from the breakdown of ATP is required for contraction and for calcium transport. ATP is produced from metabolism of creatine phosphate and glycogen stores within the muscle and from glucose and fatty acids obtained from the blood. ∎

6.3 Muscles vary in movement, force, and endurance

The general functions of muscles are to move body parts or maintain a certain body position. How well they carry out their functions depends on a number of factors, including whether bones move or not, the degree of nerve stimulation, the type of muscle fiber, and the degree to which exercise has improved muscle mass and aerobic capacity.

Isotonic versus isometric contractions: movement versus static position

Most types of exercise include a combination of two different types of muscle contractions, called isotonic and isometric contractions.

Isotonic ("same" + "strength" or "tone") contractions occur whenever a muscle shortens while maintaining a constant force. An example of an isotonic contraction is the generation of enough muscle force to move an object or part of the skeleton. How heavy the object is doesn't matter, as long as parts of the skeleton actually move. It could be just your empty hand, a pencil, a book, or a 100-pound barbell.

In *isometric* ("same" + "length") contractions, force is generated, muscle tension increases, and the muscle may even shorten a little as tendons are stretched slightly, but bones and objects do not move. As a result, isometric contractions do not cause body movement. Examples are tightening your abdominal muscles while sitting still or straining to lift a weight too heavy to lift. Isometric contractions help to stabilize the skeleton. In fact, you contract your muscles isometrically whenever you stand, just to maintain an upright position. If you doubt it, think about how quickly you would fall down if you were to faint and lose control over your skeletal muscles. Isometric contractions are a useful way to strengthen muscles.

The degree of nerve activation influences force

A single muscle may consist of thousands of individual muscle cells. The individual cells in any muscle are organized into groups of cells that all work together. Each group of cells is controlled by a single nerve cell called a *motor neuron* (because it affects movement). The motor neuron and all of the muscle cells it controls are called a **motor unit** (Figure 6.10). A motor unit is the smallest functional unit of muscle contraction, because as the motor neuron is activated all the muscle cells in that motor unit are activated together.

Our strength and ability to move effectively depend on how forcefully our muscles contract. The mechanical force that muscles generate when they contract is called **muscle tension.** How much tension is generated by a muscle depends on three factors:

- The number of muscle cells in each motor unit (motor unit size)
- The number of motor units active at any one time
- The frequency of stimulation of individual motor units

Motor unit size can vary widely from one muscle to the next. The number of muscle cells per motor unit is a trade-off between brute strength and fine control. Larger motor units generate more force but offer less control. In the thigh muscle, where strength is more important than fine control, a single motor unit may consist of as many as a thousand muscle cells. In the eye muscles, where fine control is essential, a motor unit may consist of only 10 muscle cells.

According to the **all-or-none principle,** muscle cells are completely under the control of their motor neuron. Muscle cells never contract on their own. For an individual muscle cell, there is no such thing as a half-hearted contraction, and there is no such thing as disobeying an order. Muscle cells always respond with a complete cycle of contraction and relaxation (called a **twitch**) every time they are stimulated by an electrical impulse, called an *action potential*, from their motor neuron. You will learn more about action potentials when you study the nervous system. For now, you need only understand that they are the stimuli for muscle contraction.

Although individual motor units either are contracting or are relaxed, whole muscles generally maintain an intermediate level of force known as *muscle tone*. Muscle

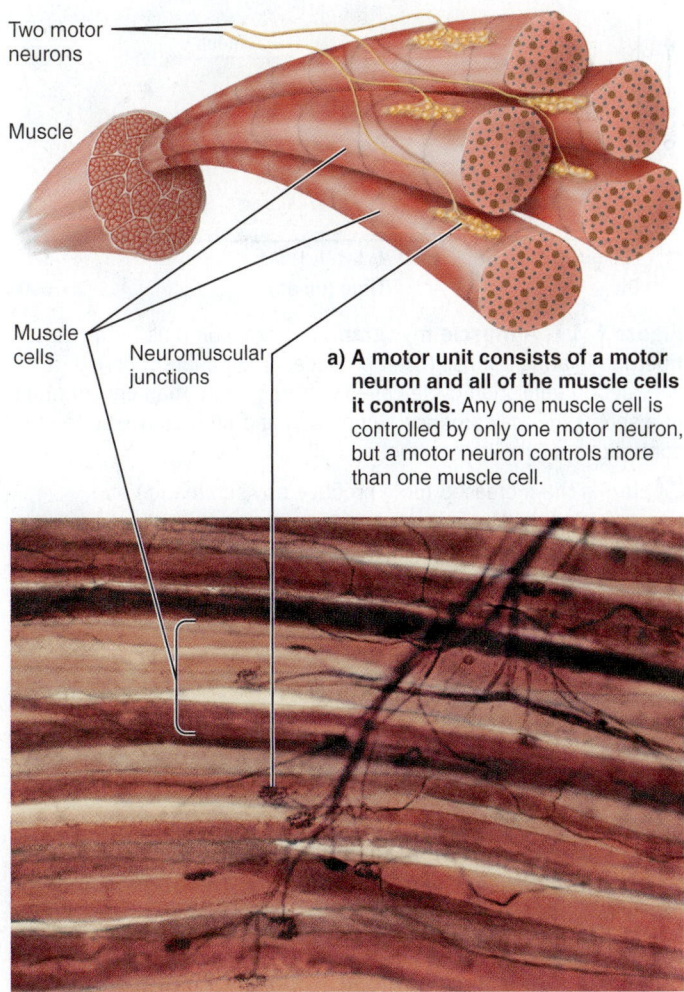

a) A motor unit consists of a motor neuron and all of the muscle cells it controls. Any one muscle cell is controlled by only one motor neuron, but a motor neuron controls more than one muscle cell.

b) Photograph of the muscle cells in a motor unit, showing branches of the motor neuron and neuromuscular junctions.

Figure 6.10 Motor units. a) A motor unit consists of a motor neuron and all the muscle cells that it controls. Any one muscle cell is controlled by only one motor neuron, but a motor neuron controls more than one muscle cell. b) A photograph of the muscle cells in a motor unit, showing branches of the motor neuron and neuromuscular junctions.

tone exists because, at any one time, some of the muscle's motor units are contracting while others are relaxed. The second factor that affects overall muscle force is the number of motor units active at any one time. Increasing tone (or force) by activating more motor units is called **recruitment.** The maintenance of muscle tone depends on the nervous system.

The third factor that affects force generation by a muscle is the frequency of stimulation of individual motor units. To understand how frequency of stimulation influences force, we need to take a closer look at what happens when a muscle cell is stimulated by its motor neuron. Although we cannot easily study the contraction of single muscle cells in the laboratory, a recording of the contractile activity of a whole muscle (called a *myogram*) reveals some important

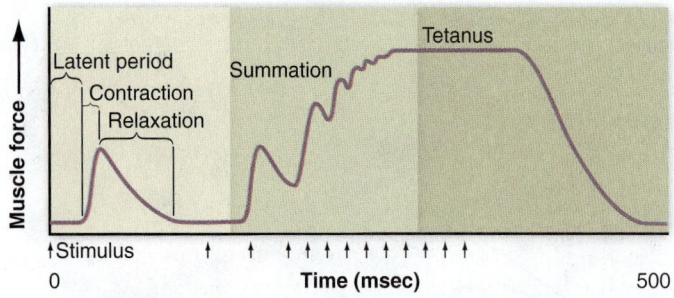

Figure 6.11 A muscle myogram. A single stimulus to the nerve fiber innervating a muscle cell produces, after a latent period, a contraction/relaxation cycle called a *twitch*. More than one stimulus in a short time may produce summation and ultimately a state of continuous maximum contraction called tetanus.

☑ How is the increased force produced by summation and tetanus different from recruitment? In what way are they similar?

components of the relationship between neural stimuli and muscle force (**Figure 6.11**):

1. *Latent period.* There is a time delay between neural stimulation and the start of contraction. This is the time it takes for the nerve impulse to travel to the sarcoplasmic reticulum, for calcium to be released, and for the myosin heads to bind to the actin filaments.
2. *Contraction.* Actin filaments are pulled toward the center of the sarcomere by the sliding filament mechanism. The muscle cell shortens.
3. *Relaxation.* Calcium is transported back into the sarcoplasmic reticulum. The troponin-tropomyosin protein complex shifts back into its original position,

and the sarcomere stretches passively to its original length. Provided that there is sufficient time between stimuli, each twitch (cycle of contraction and relaxation) looks exactly the same.

4. *Summation.* If additional stimuli arrive at the muscle cell before it has had a chance to transport calcium back into the sarcoplasmic reticulum and relax completely, the total force produced becomes greater than the force produced by one twitch alone. In effect, the force becomes greater because more calcium is present. Increasing muscle cell force by increasing the rate of stimulation of motor units is called **summation.**
5. *Tetanus.* If stimulation becomes so frequent that the muscle cell cannot relax at all, it will remain in a state of maximum contraction called *tetanus* or a *tetanic contraction.* On a myogram, tetanus appears as a straight horizontal line representing the fusion of the peaks and valleys of individual twitches. Tetanusic contraction may lead eventually to muscle fatigue.

Table 6.2 summarizes the mechanism of muscle cell activation and contraction.

☑ A friend tells you that the all-or-none principle means that all the motor units in a muscle always contract simultaneously. Explain what is wrong with his reasoning.

Slow-twitch versus fast-twitch fibers: endurance versus strength

As we have seen, all muscle cells can obtain ATP through both aerobic and anaerobic pathways. Humans have two types of skeletal muscle fibers, called *slow-twitch* and

Table 6.2 Summary of activation and contraction of skeletal muscle

Action	Description	Additional facts
Motor neuron activation	A brief electrical impulse known as an *action potential* travels down the motor neuron from the central nervous system.	One motor neuron innervates more than one muscle cell.
Neurotransmitter release	At the neuromuscular junctions between a motor neuron and each of its muscle cells, a chemical neurotransmitter called *acetylcholine* is released.	The acetylcholine is removed quickly, so that stimulation of the muscle cell is short-lived.
Muscle cell activation	The release of acetylcholine causes an electrical impulse in the cell membrane of each muscle cell.	The electrical impulse in the muscle cell is similar to the electrical impulse in a nerve.
Calcium release	An impulse in the muscle cell membrane causes calcium to be released into the muscle cell cytoplasm from the sarcoplasmic reticulum.	The sarcoplasmic reticulum is a network of membrane-bound storage sacs in the muscle cell.
Muscle cell contraction	The presence of calcium allows the thick and thin filaments to attach to each other and to slide past each other. Energy in the form of ATP is required.	Muscle cell contraction lasts longer than neuron activation.
Muscle cell relaxation	Calcium is pumped back into the sarcoplasmic reticulum. The thick and thin filaments detach from each other, and the muscle relaxes.	Calcium transport back into the sarcoplasmic reticulum also requires ATP.

fast-twitch fibers. The distinction is based on how quickly they can utilize ATP to produce a contraction and whether they use primarily aerobic or anaerobic metabolic pathways. Most muscles contain a mixture of both slow-twitch and fast-twitch fibers. The ratio of fiber types in any one muscle depends primarily on the function of the muscle.

Slow-twitch fibers break down ATP slowly, and so they contract slowly. They tend to make ATP as they need it by aerobic metabolism. Slow-twitch fibers contain many mitochondria and are well supplied with blood vessels, so they draw more blood and oxygen than fast-twitch fibers. They store very little glycogen because they can obtain glucose and fatty acids quickly from the blood. They store oxygen, however, in a molecule called *myoglobin*. The ability to maintain a temporary store of oxygen reduces the slow-twitch fiber's need for oxygen from the bloodstream. This is especially important during the early phases of an increase in activity, before blood flow to the muscle has increased. Myoglobin and the presence of numerous blood vessels make slow-twitch fibers reddish in color, so they are sometimes called "red" muscle.

Fast-twitch fibers can contract more quickly than slow-twitch fibers because they break down ATP more quickly. They have fewer mitochondria, fewer blood vessels, and little or no myoglobin compared to slow-twitch fibers, so they're called "white" muscle. Fast-twitch fibers store large amounts of glycogen and tend to rely heavily on creatine phosphate and anaerobic metabolism for quick bursts of high energy. Their contractions are rapid and powerful but cannot be sustained for long. Fast-twitch fibers depend on aerobic mechanisms for any activity that is sustained, but they have the capability of using anaerobic mechanisms for brief periods when bursts of power are needed. During periods of anaerobic activity, they tend to accumulate lactic acid, which causes them to become fatigued quickly.

Which type of fiber is better? It depends on the activity. Because slow-twitch fibers offer more endurance, they are most useful for steady activities such as jogging, swimming, and biking. Slow-twitch fibers are also important for maintaining body posture. Many of the muscles of the leg and back, for example, contain a high percentage of slow-twitch fibers because they must contract for long periods to support us when we stand. Fast-twitch fibers are more often used to power brief, high-intensity activities such as sprinting for short distances, lifting weights, or swinging a tennis racquet. Muscles in our hands, for example, contain a high proportion of fast-twitch fibers, allowing the muscles to contract quickly and strongly when necessary.

The percentage of slow- and fast-twitch fibers varies not only from muscle to muscle but from person to person. The percentages are determined in part by inheritance, and they can influence athletic ability. For example, most world-class marathoners have a higher-than-average percentage of slow-twitch fibers in their legs.

✅ Suppose a muscle biopsy done on an aspiring athlete shows that her leg muscles have an unusually large ratio of red muscle fibers to white muscle fibers. Would you recommend that she train for sprints or for marathons, and why?

Exercise training improves muscle mass, strength, and endurance

Although part of your athletic potential might be influenced by inheritance, a consistent, planned program of physical exercise (sometimes called *exercise training*) can improve your strength, endurance, and skill at any athletic endeavor. Whether primarily strength or endurance is improved by exercise training depends on the type and intensity of training. The two primary types of exercise training are strength (resistance) training and aerobic (endurance) training.

Strength training (**Figure 6.12a**) involves doing exercises that strengthen specific muscles, usually by providing some type of resistance that makes them work harder. Strength training is generally short, intense exercises such as weight lifting using free weights or weight machines. It builds more myofibrils, particularly in fast-twitch fibers, and causes the fast-twitch fibers to store more glycogen and creatine phosphate as quick energy sources. This increases the size of individual muscle cells and builds muscle mass and muscle strength, but it does not increase the number of muscle cells.

In general, the heavier the weight used, the more visible the increase in muscle size. However, this does not mean that strength training will necessarily build bulging biceps. The extent of muscle development depends on many factors, including the amount of resistance used, the duration and frequency of exercise, and your own genetic predisposition. However, even low to moderate weights can lead to noticeable improvements in muscle strength.

Aerobic training (**Figure 6.12b**) involves activities in which the body increases its oxygen intake to meet

a) Weight lifter (strength training).

b) Distance runner (aerobic training).

Figure 6.12 Strength training versus aerobic training.

the increased demands for oxygen by muscles. Whereas resistance training strengthens muscles, aerobic training builds endurance. With aerobic training, the number of blood capillaries supplying muscle increases. In addition, the number of mitochondria in muscle cells and the amount of myoglobin available to store oxygen both increase. The muscle fibers themselves do not increase much in mass nor do they increase in number. Aerobic exercise also improves the performance of the cardiovascular and respiratory systems. Less intense than strength training but carried out for prolonged periods, aerobic exercises include jogging, walking, biking, and swimming.

MJ's BlogInFocus How does breathing xenon gas affect athletic performance? Visit MJ's blog in the Study Area in MasteringBiology and look under "Xenon Gas."

http://goo.gl/yxCkcb

Gentle stretching before exercise increases your heart rate gradually, pumping additional blood to your muscles and preparing you for more strenuous exertion. After exercising, let your heart rate and breathing return gradually to normal. Walk slowly and do more stretching. Regular stretching improves joint mobility and range of motion. Whenever you stretch, do it gradually and hold each position for 30 seconds. You should feel a gentle pull in your muscles but not pain. Try not to bounce, because abrupt stretches could cause your muscles to contract quickly in response, increasing the risk of injury.

Recap A motor unit consists of a motor neuron and all of the muscle cells it controls. Greater muscle force is produced by activation of more motor units and/or increased frequency of stimulation of motor units. Most muscles contain a combination of slow-twitch and fast-twitch fibers. Slow-twitch fibers rely on aerobic metabolism and are most useful for endurance. Fast-twitch fibers are most useful where strength is required. Exercise increases aerobic capacity, muscle mass, and muscle strength but does not increase the number of muscle cells. ■

6.4 Cardiac and smooth muscles have special features

Most of the overall muscle mass of the body is skeletal muscle. Nevertheless, both cardiac and smooth muscle have unique features that set them apart from skeletal muscle. The primary difference between cardiac muscle and smooth muscle is in how they are activated. **Table 6.3** summarizes defining characteristics of the three types of muscle.

How cardiac and smooth muscles are activated

Cardiac and smooth muscle are called *involuntary muscle* because we generally do not have voluntary control over them. Both cardiac and smooth muscles can contract entirely on their own, in the absence of stimulation by nerves.

Like skeletal muscle, cardiac muscle has a regular array of thick and thin filaments arranged in sarcomeres, so it too is called *striated muscle* (**Figure 6.13a**). Cardiac muscle cells are joined at their blunt ends by structures called *intercalated*

Table 6.3 Defining characteristics of skeletal, cardiac, and smooth muscle

Defining characteristics	Skeletal muscle	Cardiac muscle	Smooth muscle
Location	Attached to bones (skeleton)	Found only in the heart	Found in the walls of blood vessels and in the walls of organs of the digestive, respiratory, urinary, and reproductive tracts
Function	Movement of the body. Prevention of movement of the body	Pumping of blood	Control of blood vessel diameter. Movement of contents in hollow organs
Anatomical description	Very large, cylindrical, multinucleated cells arranged in parallel bundles	Short cells with blunt, branched ends. Cells joined to others by intercalated discs and gap junctions	Small, spindle-shaped cells joined to each other by gap junctions
Initiation of contraction	Only by a nerve cell	Spontaneous (pacemaker cells), modifiable by nerves	Some contraction always maintained. Modifiable by nerves
Voluntary?	Yes	No	No
Gap junctions?	No	Yes	Yes
Speed and sustainability of contraction	Fast—50 milliseconds (0.05 second). Not sustainable	Moderate—150 milliseconds (0.15 second). Not sustainable	Slow—1–3 seconds. Sustainable indefinitely
Likelihood of fatigue	Varies widely depending on type of skeletal muscle and workload	Low. Relaxation between contractions reduces the likelihood	Generally does not fatigue
Striated?	Yes	Yes	No

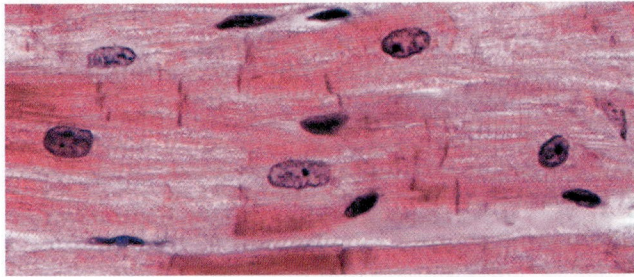

a) A photomicrograph of cardiac muscle tissue.

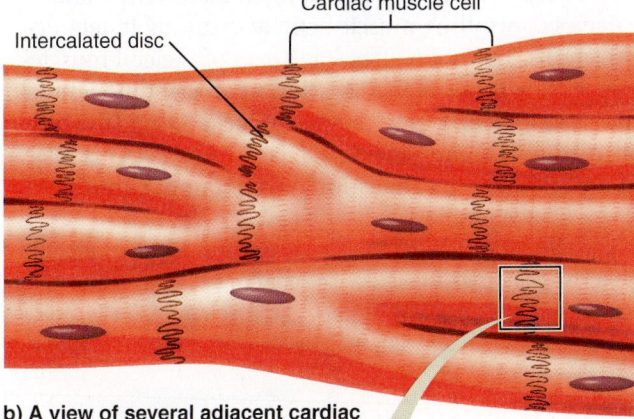

Intercalated disc

Cardiac muscle cell

b) A view of several adjacent cardiac muscle cells showing their blunt shape and the intercalated discs that join them together.

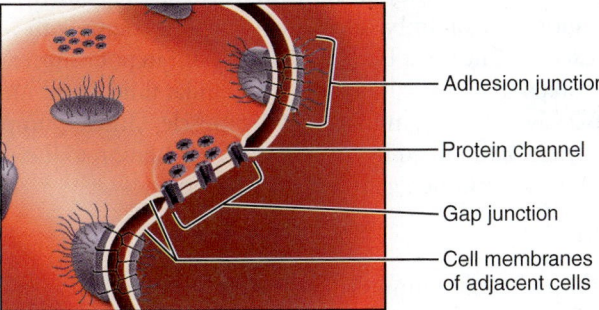

Adhesion junction

Protein channel

Gap junction

Cell membranes of adjacent cells

c) A closer view showing that intercalated discs are bridged by gap junctions that permit direct electrical connections between cells.

Figure 6.13 Cardiac muscle.

discs (Figure 6.13b). The intercalated discs contain gap junctions that permit one cell to electrically stimulate the next one (Figure 6.13c).

Although all cardiac muscle cells are capable of beating spontaneously and establishing their own cycle of contraction and relaxation, those with the fastest rhythm are called *pacemaker cells* because the rest of the cells follow their faster pace. The intercalated discs contain gap junctions that permit one cell to electrically stimulate the next one. In effect, the pacemaker cells dictate the rate of contraction of the whole heart, because their faster pace activates the slower cells before the slower cells would be activated by their own inherent rhythm.

Arrangement of myosin and actin filaments

In contrast to skeletal and cardiac muscles, the thick and thin filaments in smooth muscle are arranged in bundles that attach to cell membrane proteins in a crossing pattern (Figure 6.14a). During contraction, the points of attachment of the filaments are pulled toward each other and the cell gets shorter and fatter (Figure 6.14b). Because its filaments are not arranged in sarcomeres, smooth muscle lacks the striated appearance of skeletal and cardiac muscle. It is called *smooth* for this reason. Smooth muscle cells are also joined by gap junctions that permit the cells to activate each other, so that the whole tissue contracts together in a coordinated fashion.

Even though cardiac and smooth muscle cells can contract without signals from the nerves, they do respond to nerve activity as well. The nerves that activate cardiac and smooth muscle belong to the autonomic nervous system, meaning that you do not have conscious control over them, like you do of the nerves that activate skeletal muscle. The effect of nerve activity may be either inhibitory or stimulatory. Changes in both inhibitory and stimulatory nerve activity to the heart are responsible for the increase in your heart rate

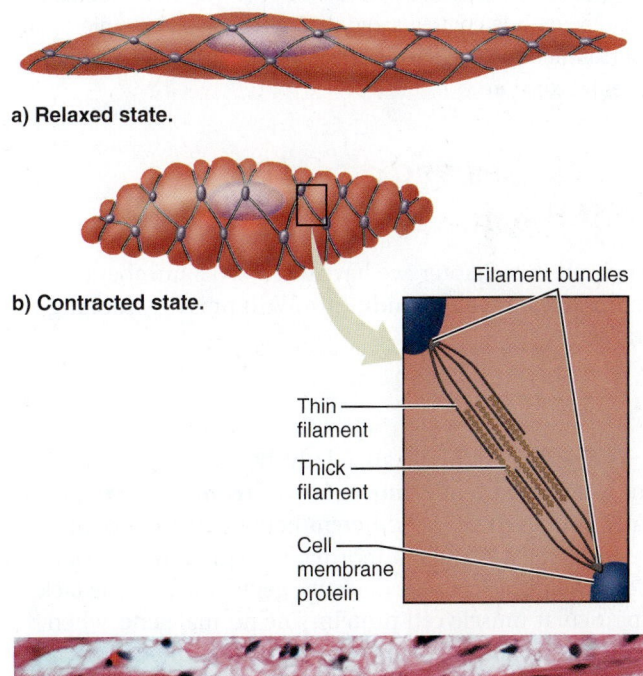

a) Relaxed state.

b) Contracted state.

Filament bundles

Thin filament

Thick filament

Cell membrane protein

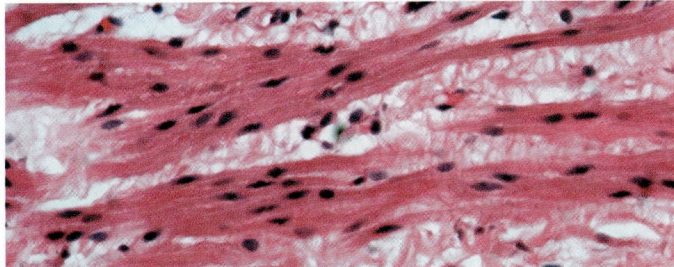

c) A photomicrograph of several smooth muscle cells.

Figure 6.14 Smooth muscle.

when you exercise, for example. Nerve stimulation can also change the contractile force of smooth muscle.

Speed and sustainability of contraction

In terms of speed and sustainability of contraction, skeletal muscle is the fastest, cardiac muscle is of moderate speed, and smooth muscle is very slow.

Cardiac muscle cells go through rhythmic cycles of contraction and relaxation. The relaxation periods are necessary periods of rest so that the muscle doesn't fatigue.

Smooth muscle generally is partially contracted all the time. This makes it ideally suited for situations in which contractions need to be sustained. Nevertheless, it almost never fatigues because it contracts so slowly that its ATP usage is always less than its production capability. Smooth muscle is a key player in the homeostatic regulation of blood pressure because it can maintain the diameter of blood vessels indefinitely, adjusting them slightly as necessary.

☑ If the gap junctions in heart muscle cells were eliminated, could the pacemaker cells still beat? Could they still set the pace of the entire heart? Explain.

↺ **Recap** Unlike skeletal muscle, both cardiac and smooth muscle can contract in the absence of any nerve stimulation. Cardiac muscle contracts and then relaxes in a rhythmic cycle. Smooth muscle can sustain a contraction indefinitely without ever relaxing. ∎

6.5 Diseases and disorders of the muscular system

Throughout this chapter, we have discussed a number of musculoskeletal health conditions. We'll finish by looking at several more.

Muscular dystrophy

Serious diseases of muscle are relatively uncommon, but foremost among them is **muscular dystrophy.** The term actually applies to several different hereditary diseases of muscle (*dystrophy* means "abnormal growth"). In *Duchenne muscular dystrophy,* a single defective gene results in the lack of a particular muscle cell protein. The normal gene, when present, directs the cell to produce a protein called *dystrophin* that is part of the muscle cell membrane. The function of dystrophin is to limit the inflow of calcium into muscle cells through calcium *leak* channels. People with muscular dystrophy lack dystrophin, and as a result too much calcium leaks into the muscle cell through the leak channels. The high intracellular calcium concentration activates enzymes that damage muscle proteins and ultimately may kill the cell. The result is a loss of muscle fibers and muscle wasting. Eventually, much of the muscle mass is replaced with fibrous connective tissue. Many people with muscular dystrophy die

before age 30, usually because of failure of the heart muscle or the skeletal muscles used for breathing. At the moment, there is no cure; however, it is an area of intense research interest, and progress is being made on several fronts.

Tetanus

Tetanus is caused by a bacterial infection. The disorder is called *tetanus* because this is the technical term for a maximal (tetanic) muscle contraction (see Figure 6.10). Generally, the infection is acquired by a puncture wound to a muscle. The bacteria produce a toxin that overstimulates the nerves controlling muscle activity, resulting in tetanic contractions. The toxin affects a variety of skeletal muscles, but especially those of the jaws and neck. Jaw muscles may contract so forcefully that they seem locked shut (the origin of its common name, "lockjaw"). Untreated, tetanus may lead to death due to exhaustion or respiratory failure.

☑ Suppose a doctor is trying to treat a patient who has tetanus, and he has two drugs available: one that mimics the action of acetylcholine and another that blocks the action of acetylcholine. Which would most likely be effective, and what might be a possibly dangerous side effect?

Muscle cramps

Muscle cramps are painful, uncontrollable, reflex-mediated muscle contractions. They are thought to be caused by the dehydration and ion imbalances that sometimes occur with heavy exercise. The most likely culprit is a shift in potassium ions between the intracellular and extracellular fluid, combined with excessive neural stimulation of the muscle. Muscle cramps generally can be soothed by increasing the circulation to the affected muscle through gentle stretching and massage.

Pulled muscles

Pulled muscles, sometimes called torn muscles, result from stretching a muscle too far, causing some of the fibers to tear apart. Internal bleeding, swelling, and pain often accompany a pulled muscle.

Fasciitis

Fasciitis involves inflammation of the connective tissue sheath, or fascia, that surrounds a muscle (see Figure 6.3). It is usually caused by straining or tearing the fascia. Most often, it affects the sole of the foot (plantar fasciitis), where it is a common cause of heel pain. Like tendons and ligaments, fascia mends slowly. Treatment includes resting the area and protecting it from pressure. Injections of corticosteroid drugs can relieve severe pain.

↺ **Recap** Muscular dystrophy is an inherited disease in which the absence of a single protein causes an abnormal leak of calcium into muscle cells. Ultimately, the leak of calcium damages muscle cell proteins and kills muscle cells. Tetanus is caused by a bacterial infection that overstimulates nerves to muscles. ∎

Chapter Summary

6.1 Muscles produce movement or generate tension *p. 122*

- All muscles produce movement or maintain position by contracting (shortening in length).
- The ways in which skeletal muscles attach to the skeleton determine what particular motion they cause.
- Within a single myofibril in a single muscle cell (fiber), thousands of contractile units called *sarcomeres* are arranged end to end.

6.2 Individual muscle cells contract and relax *p. 126*

- Skeletal muscle cells contract only when activated by their motor nerve.
- Motor nerve activation causes calcium to be released from the sarcoplasmic reticulum of the muscle cell.
- In the presence of calcium, the thick (myosin) and thin (actin) filaments slide past each other, and the sarcomere shortens.
- ATP supplies the energy for the entire muscle contraction/relaxation process.

6.3 Muscles vary in movement, force, and endurance *p. 130*

- An isotonic contraction occurs when a muscle shortens while maintaining a constant force; an isometric contraction occurs when tension is generated but the bones do not move.
- A motor unit is a single motor neuron and all of the skeletal muscle cells that it controls.
- The force generated by a muscle depends on the number of muscle cells in each motor unit, the number of motor units active at any one moment, and the frequency of stimulation of motor units.
- Muscle strength and endurance depend on the ratio of slow-twitch to fast-twitch fibers in the muscle and on the type and amount of exercise training.

6.4 Cardiac and smooth muscles have special features *p. 134*

- Cardiac and smooth muscles do not attach to bones.
- Both cardiac and smooth muscles can contract spontaneously, and both can be influenced by nerves of the autonomic nervous system.
- Cardiac muscle contracts rhythmically, with a period of relaxation between each contraction. Smooth muscle can maintain at least some contractile force indefinitely.

6.5 Diseases and disorders of the muscular system *p. 136*

- Muscular dystrophy is caused by inheritance of an abnormal gene.
- Pulled (torn) muscles occur when a muscle is stretched too far.

Terms You Should Know

actin, *125*	oxygen debt, *130*
all-or-none principle, *131*	recruitment, *131*
fatigue, *130*	sarcomere, *125*
motor neuron, *126*	sarcoplasmic reticulum, *127*
motor unit, *131*	sliding filament mechanism, *127*
myosin, *125*	summation, *132*
neurotransmitter, *126*	twitch, *131*

Concept Review

Answers can be found in the Study Area in MasteringBiology.

1. Describe how muscle contraction can resist movement rather than cause movement.
2. Describe how a muscle's origin and insertion determine the specific body movement that will result from muscle contraction.
3. Describe the roles of calcium in muscle contraction.
4. Explain what causes *rigor mortis*.
5. Discuss some possible reasons for muscle fatigue.
6. Define *summation*, and explain why it occurs when a muscle is stimulated rapidly and repetitively.
7. Explain why a spinal cord injury in the neck completely paralyzes the skeletal muscles of the limbs, whereas the cardiac muscle of the heart still beats rhythmically.
8. Compare and contrast how a constant degree of moderate tension, or tone, is maintained by a skeletal muscle that maintains posture versus a smooth muscle that maintains blood vessel diameter.
9. Define a *motor unit*, and describe how the size and the number of motor units in a muscle affect muscle strength and fine motor control.
10. List the sources of energy that a muscle cell may use to make more ATP, both from within and from outside the cell.

Test Yourself

Answers can be found in the Appendix.

1. Muscles that oppose each other and produce opposite movements are described as:
   a. synergistic
   b. antagonistic
   c. cooperative
   d. oppositional

2. Which of the following choices arranges the structures from the largest (most inclusive) to smallest? (1) muscle fiber, (2) fascicle, (3) myofibril, and (4) muscle
   a. 1-2-3-4
   b. 2-3-1-4
   c. 4-2-1-3
   d. 4-2-3-1

3. All of the following are functions of the muscular system except:
   a. maintenance of body calcium stores
   b. resisting movement
   c. maintenance of body temperature
   d. movement

4. Which of the following happens during muscle contraction?
   a. actin filaments shorten
   b. myosin filaments shorten
   c. sarcomeres shorten
   d. both (a) and (b)

5. Botulism toxin inhibits the release of acetylcholine at the neuromuscular junctions. What effect does this have on the muscle activity?
   a. Muscles will contract continuously.
   b. Muscles will contract sporadically, without conscious control.
   c. Muscles will not contract because they will not receive nerve stimulation.
   d. There will be no effect on muscle activity.

6. The sliding filament mechanism describes the process during which:
   a. actin and myosin slide relative to each other
   b. sarcomeres slide relative to each other
   c. troponin and tropomyosin slide relative to each other
   d. muscle fibers slide past each other

7. What is the first and most direct energy source for muscle contraction?
   a. glucose
   b. ATP
   c. creatine phosphate
   d. glycogen

8. As you clasp your hands in front of you and push them toward each other, this is an example of:
   a. an isotonic contraction
   b. an isometric contraction
   c. a tetanic contraction
   d. aerobic training

9. All of the following may happen in response to exercise training except:
   a. increase in the number of myofibrils
   b. increase in the storage of glycogen and creatine phosphate
   c. increase in the number of muscle fibers
   d. increase in the number of mitochondria

10. Which of the following is/are characteristic of slow-twitch fibers?
    a. large amounts of glycogen storage
    b. myoglobin content enables oxygen storage
    c. numerous mitochondria
    d. both (b) and (c)

11. Which of the following is the site of calcium ion storage within muscles?
    a. T tubules
    b. sarcoplasmic reticulum
    c. actin filaments
    d. myosin filaments

12. What is the role of ATP in muscle function?
    a. ATP provides energy that enables myosin to form cross-bridges with actin.
    b. ATP enables myosin to detach from actin.
    c. ATP provides energy to transport calcium back into storage.
    d. all of the above

13. Which of the following would have motor units with the smallest number of muscle cells?
    a. thigh muscle
    b. muscles in fingers
    c. abdominal muscles
    d. muscles of the back

14. Which type(s) of muscle cells can contract the fastest?
    a. smooth muscle cells
    b. cardiac muscle cells
    c. skeletal muscle cells
    d. All muscle cells can exhibit the same speed of contraction.

15. Which type(s) of muscle cells can contract spontaneously?
    a. smooth muscle cells
    b. cardiac muscle cells
    c. skeletal muscle cells
    d. both (a) and (b)

Apply What You Know

Answers can be found in the Study Area in MasteringBiology.

1. If a muscle cell's sarcoplasmic reticulum had little to no Ca^{2+}, could the muscle cell still produce an electrical impulse, and would the muscle cell still contract? Explain.

2. Why do you think it is generally accepted medical practice to get bedridden patients up and walking as soon as possible?

3. In what ways would you expect the training regimen for a sprinter to be different from that of a marathon runner, and why?

4. What would happen to a muscle if one of its tendons were torn? Would the muscle still be able to contract?

5. You and your friend are doing leg presses in the gym one day. As you extend your legs, the stack of weights goes up, and when you bend your legs, the stack goes down. Your friend says your muscles are actively pushing the weights up. Explain to him where he is wrong in his thinking.

6. You are outside on a cool fall day. You feel cool, but you think little of it until you notice yourself shivering. What is happening at the muscular level, and why is it at least partially effective in helping to maintain body temperature?

7. You have just joined an aerobics exercise class for the first time, and you have calculated your target heart rate. After class, you notice that your heart rate remains high for a while and only slowly returns to normal. Explain why this occurs and what is happening at the physiological level. How will this response change over time if you maintain your exercise program consistently?

8. Some weight lifters like to consume various products containing creatine phosphate. Why would this be useful? Why would weight lifters benefit more than marathon runners from creatine phosphate?

MJ's BlogInFocus

Answers can be found in the Study Area in MasteringBiology.

1. Do you think that stretching before exercise would reduce the muscle soreness that sometimes occurs a day or so after exercise or perhaps reduce the chances of injury? To review the available evidence, visit MJ's blog in the Study Area in MasteringBiology and look under "Pre-Exercise Stretching."
   http://goo.gl/kvwt4M

MasteringBiology®

Students Go to MasteringBiology for assignments, the eText, and the Study Area with animations, practice tests, and activities.

Professors Go to MasteringBiology for automatically graded tutorials and questions that you can assign to your students, plus Instructor Resources.

Answers to ✓ questions are available in the Appendix.

Blood

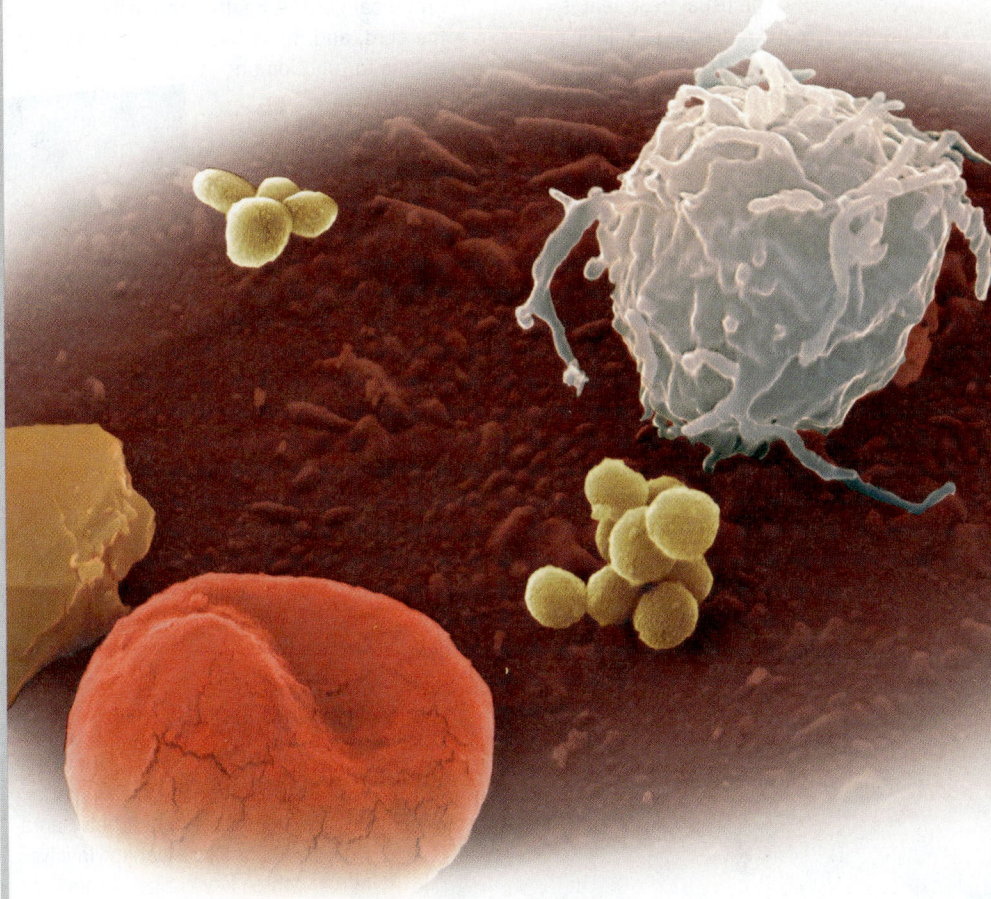

Scanning electron micrograph of a red blood cell, a white blood cell, and several *Staphylococcus* bacteria.

Key Concepts

- **Blood transports the essential requirements of life to all living cells.** Most blood consists of a watery fluid called plasma that contains ions, proteins, hormones, nutrients, and metabolic waste products.

- **Blood cells originate from stem cells located in bone marrow.** Blood cells have a short life span, so stem cells continue to divide throughout life to produce new blood cells.

- **Red blood cells are highly specialized for transporting oxygen and carbon dioxide.** Red blood cells contain a protein called hemoglobin that binds oxygen and carbon dioxide.

- **White blood cells defend the body against injury and disease.** White blood cells are part of the body's immune system.

- **Human blood types are A, B, AB, and O.** Blood type is determined by specific proteins called *antigens* on the surface of red blood cells.

CURRENT ISSUE
Should You Bank Your Baby's Cord Blood?

Questions to Consider

1 Do you agree with the federal government's decision to allocate $79 million (about 40 cents per adult) for a public cord blood collection and storage network?

2 When (or if) you have a child, what will you do with his or her cord blood? Explain your decision.

When she was 15 and a sophomore in high school, Jaclyn Albanese was diagnosed with acute leukemia—a type of stem cell cancer in bone marrow. The usual treatment is chemotherapy and radiation to kill the cancer cells (and normal stem cells), and then a bone marrow transplant to repopulate the bone marrow with stem cells. Traditionally these stem cells have come from bone marrow donated by a family member or an unrelated volunteer whose marrow is compatible. Compatibility is crucial because, as you will learn in this chapter, the body's immune cells recognize and attack foreign cells. Jaclyn had hoped to get a bone marrow transplant from one of her relatives, but none of them was a close enough match. Fortunately for Jaclyn, she found compatible units of cord blood from an unrelated donor. They saved her life. But, if her parents had banked the cord blood from her delivery when she was born, she wouldn't have needed to search for donors at all.

What Is Cord Blood?

During pregnancy, the fetus is connected to the mother via the umbilical cord and the placenta. Blood vessels in the umbilical cord and placenta filter out toxic substances, deliver nutrients from the mother, and remove waste products from the fetus. After the baby is delivered, the mother's body expels the placenta and umbilical cord. A health professional cuts the cord, and the baby's circulatory system begins to function on its own.

Until recently, the placenta and cord were discarded after birth. However, these structures still contain about 50 ml of cord blood. In addition to containing the usual components of blood, cord blood is rich in relatively immature stem cells from the fetus. These stem cells can be coaxed to divide repeatedly to produce immature blood cells, which in turn develop into platelets, red cells, and white cells.

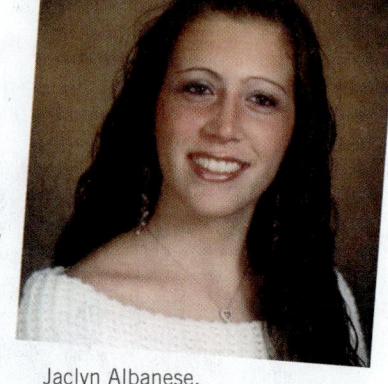

Jaclyn Albanese.

Bone Marrow, Cord Blood, and Compatibility Issues

A good bone marrow transplant match between a donor and a patient involves three key antigens known as HLA-A, HLA-B, and HLA-DR, each of which comes in two forms. The ideal match would be for the patient to have the same six forms as the donor (a 6/6 match). For unrelated donors and patients, the odds of that happening are only one in 20,000. Only about 10% of all patients who need a bone marrow transplant are able to find a compatible bone marrow match from among unrelated potential donors.

This is why cord blood has become such a precious commodity. The immune cells in cord blood are less mature than those in bone marrow, so cord blood transplants are less likely to cause transfusion reactions, wherein the recipient's immune system rejects the donor's blood. Even when transfusion reactions do occur, they tend to be less severe. As a result, the match between donor and recipient does not need to be a perfect 6/6— matches of 5/6 or even 4/6 are sufficient. This opens a much wider field of possibilities. Jaclyn Albanese could not find a single compatible bone marrow donor, but she was able to locate two units of compatible cord blood. She had a cord blood transplant in 1999 just prior to her junior year in high school, and today she is a college graduate. Jaclyn is one of approximately 6,000 patients who have benefited from cord blood transplants to date.

Banking Privately or Publicly

Should you bank your baby's cord blood privately or donate it to a public cord blood bank?

Private blood banks assert that banking your baby's cord blood privately is like taking out a medical insurance policy. They argue that you might want to use your baby's cord blood stem cells to treat a future disease (such as leukemia) in your child or a close family member. By banking your baby's cord blood, you ensure that your child always has access to the "perfect match"—his or her own stem cells. They also point out that scientists are working on stem-cell therapies for a variety of other conditions such as diabetes and heart disease. The implication is that in 30 or 40 years, by the time your newborn is at greater risk of developing these chronic ailments, he or she may be →

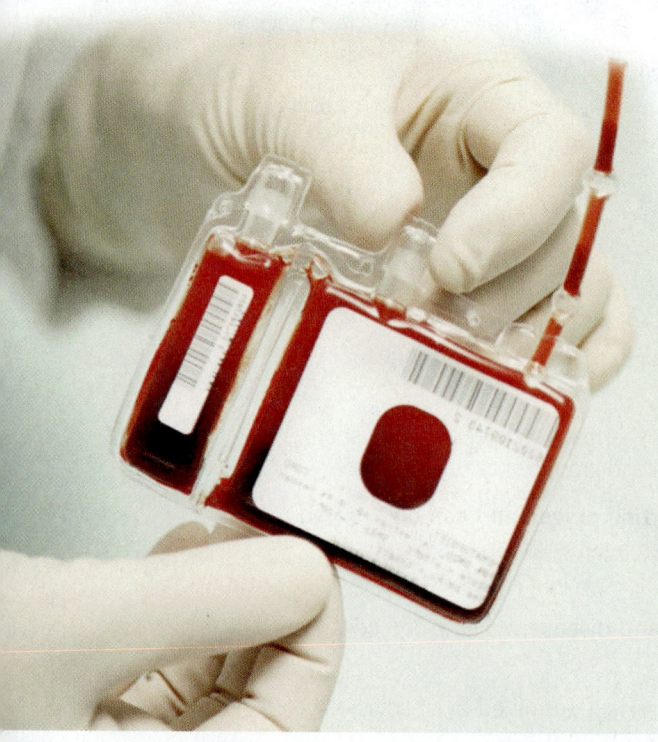

A cord blood collection unit.

able to use these stem cells for treatments as yet undreamed of. The initial charges for collecting, processing, and testing cord blood privately range from $1,400 to $2,300; then there's an annual storage fee of $115 to $150.

Proponents of public blood banking argue that the likelihood of a baby born to a healthy family ever needing his or her stem cells is about 1 in 2,500. Is such an unlikely event worth banking your baby's blood privately? If you and your family are healthy, you might want to consider helping others by donating your child's cord blood to a public cord blood bank. To support public donations, in 2005, the federal government authorized $79 million in federal funds to collect and store cord blood from ethnically diverse donors. To date, the

national Cord Blood Registry has banked over 500,000 units of cord blood and cord tissue, more than any other stem cell bank. As a result, the National Marrow Donor Program now lists a national inventory of nearly 165,000 cord blood units—enough to be able to provide more than half of all patients with a 5/6 antigen match.

Jaclyn Albanese's parents did not have the choice of banking her cord blood privately when she was born. Fortunately, she was able to find the cord blood she needed through one of the few public cord blood banks in existence at the time.

SUMMARY

- Cord blood—blood remaining in a newborn baby's placenta and umbilical cord—is rich in stem cells, similar to those found in bone marrow.
- Cord blood can benefit many patients who cannot find a suitable bone marrow donor.
- Private cord blood banks urge prospective parents to bank their baby's cord blood solely for their own child's future use.
- The federal government has established a public cord blood network for all Americans.

Way back when life began, a single cell floating freely in the primordial sea received all its nutrients from the surrounding fluid and dumped all its wastes into it. Today, a single cell in the human body still does essentially the same thing; it receives its nutrients from (and dumps its waste into) the surrounding fluid, called the *interstitial* (between cells) *fluid*. In the human body, though, the cells are packed closely together, with very little fluid between them. A human cell would soon starve to death in a sea of waste if not for blood circulating through nearby blood vessels. Blood picks up nutrients from the digestive tract. It transports waste carbon dioxide gas to the lungs and picks up much-needed oxygen. It transports the waste products of metabolism to the liver for destruction or to the kidneys for removal from the body. It even transports waste heat to the skin, as part of the control mechanism for regulating body temperature. Last but not least, blood contains specialized cells of the immune system that are essential to our defense against invading microorganisms. Always and everywhere throughout the body, blood is seeing to it that each living cell is bathed in a fluid conducive to life. Blood is our internal primordial ocean. ■

Blood is so effective at performing its functions that so far scientists' efforts to develop an artificial blood substitute have not been very successful. If someone needs blood, a transfusion of human blood is often the only solution.

Blood is just one component of the **circulatory system;** the heart and the blood vessels are the other components. The primary role of the cardiovascular system is *transport,* with the overall task of the maintenance of homeostasis of the internal environment (**Figure 7.1**). In this chapter, we concentrate on the composition and crucial functions of

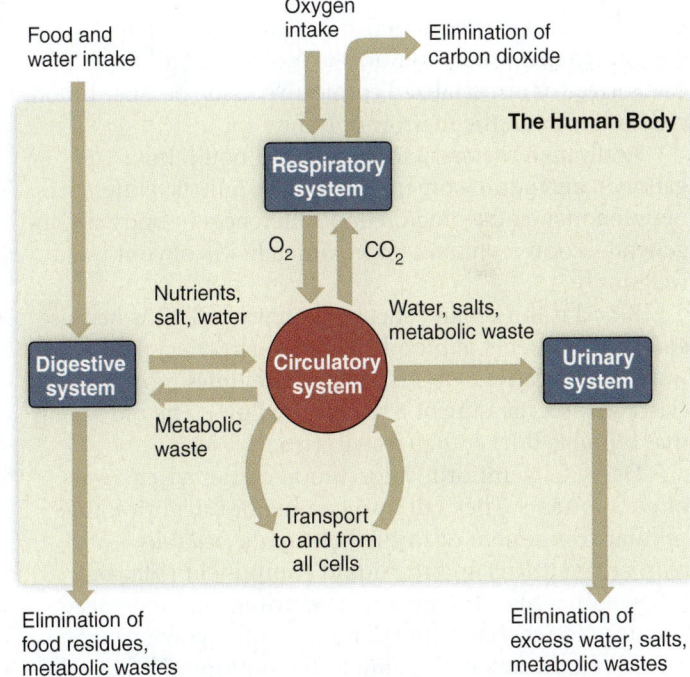

Figure 7.1 The transport role of the circulatory system. The circulatory system serves homeostasis by transporting nutrients and carrying off wastes from all parts of the body. Blood is the fluid within the circulatory system.

blood, the fluid that circulates within the heart and blood vessels. How the heart and blood vessels function together to transport blood throughout the body will be considered in a separate chapter.

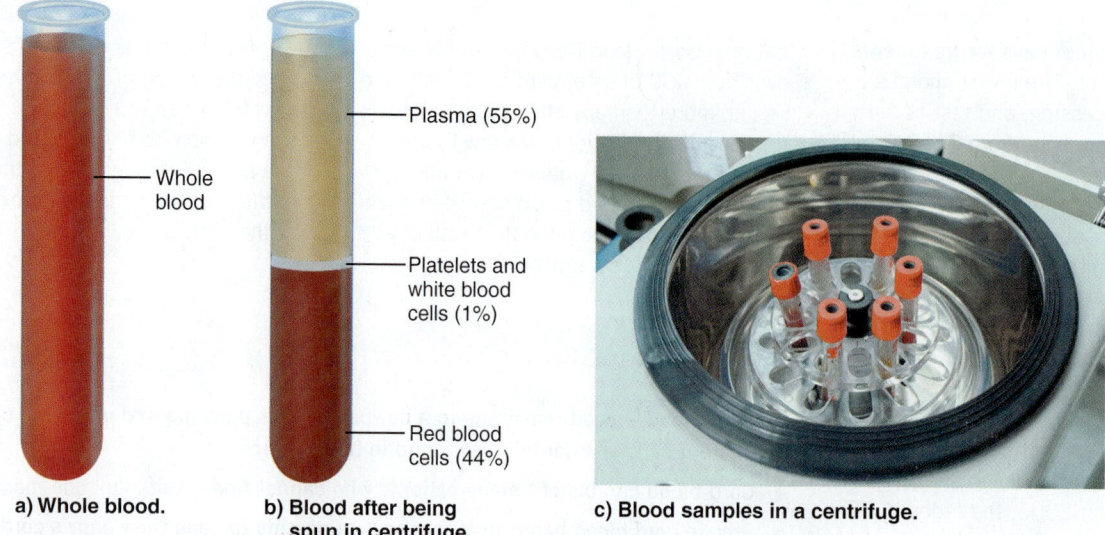

a) Whole blood.
b) Blood after being spun in centrifuge.
c) Blood samples in a centrifuge.

Whole blood

Plasma (55%)

Platelets and white blood cells (1%)

Red blood cells (44%)

Figure 7.2 Blood. The formed elements sink to the bottom during centrifugation. The percentage of the blood that is red blood cells is called the hematocrit; in this sample the hematocrit is 44%.

7.1 The composition and functions of blood

Blood consists of specialized cells and cell fragments suspended in a watery solution of molecules and ions. It is categorized as a specialized connective tissue because blood cells originate in the marrow of bone.

Adult men average 5 to 6 liters of blood (about 1.5 gallons), and adult women average 4 to 5 liters. Differences between men and women reflect differences in body size. In general, blood represents approximately 8% of your body weight.

Blood is thicker and stickier than water. This is because some components of blood are denser (heavier) than water and because blood is roughly five times more viscous (viscosity is a measure of resistance to flow). The old saying that blood is thicker than water is true.

Despite its uniform color, blood carries a rich array of components. They fall into two major categories: the cellular component or formed elements (*red cells, white cells,* and *platelets*) and the liquid component (plasma). If you spin a blood sample in a centrifuge (a high-speed rotation device that mimics and magnifies gravitational forces), formed elements sink to the bottom of a test tube because they are denser than plasma (**Figure 7.2**). Red blood cells (RBCs), representing the bulk of the formed elements, settle to the bottom. White blood cells (WBCs) and platelets appear just above red blood cells in a thin, grayish-white layer.

Blood plasma (the liquid component) contains electrolytes (ions), dissolved gases, proteins, hormones, and nutrients, along with the waste products of metabolism. **Table 7.1** summarizes the components of blood and their functions.

Plasma consists of water and dissolved solutes

The top layer of a centrifuged blood sample, representing about 55% of the total volume, consists of a pale yellow liquid called **plasma** (see Figure 7.2b). Plasma is the transport medium for blood cells and platelets. About 90% of plasma is water. The rest is dissolved proteins, hormones, more than 100 different small molecules (including amino acids, fats, small carbohydrates, vitamins, and various waste products of metabolism), and ions.

The largest group of solutes in plasma consists of **plasma proteins,** which serve a variety of functions. Important plasma proteins include albumins, globulins, and clotting proteins.

Nearly two-thirds of plasma proteins are **albumins,** which primarily serve to maintain the proper water balance between blood and the interstitial fluid. Manufactured in the liver, albumins also bind to certain molecules (such as bilirubin and fatty acids) and drugs (such as penicillin) and assist in their transport in blood.

Globulins (designated alpha, beta, and gamma) are a diverse group of proteins that transport various substances in the blood. Many beta globulins bind to lipid (fat) molecules, such as cholesterol. When a protein attaches to one of these molecules, it creates a complex called a *lipoprotein*. Two medically important lipoproteins are the low-density lipoproteins (LDLs) and high-density lipoproteins (HDLs), and medical exams often include taking a blood sample to measure LDL and HDL relative proportions. The LDLs are sometimes called "bad cholesterol" because high blood levels of these lipoproteins are associated with increased risk of cardiovascular health problems. High levels of HDLs often indicate a lower risk of cardiovascular disease.

Table 7.1 Composition of blood

Blood component	Examples and functions
Formed Elements (45%)	
Red blood cells	Transport oxygen to body tissues; transport carbon dioxide away from tissues
White blood cells	Defend the body against invading organisms, abnormal cells
Platelets	Take part in blood clotting as part of the body's defense mechanisms
Plasma (55%)	
Water	The primary constituent of blood plasma
Electrolytes (ions)	Sodium, potassium, chloride, bicarbonate, calcium, hydrogen, magnesium, others. Ions contribute to the control of cell function and volume, to the electrical charge across cells, and to the function of excitable cells (nerve and muscle). All ions must be kept at their normal concentrations for homeostasis to occur.
Proteins	Albumins maintain blood volume and transport electrolytes, hormones, and wastes. Globulins serve as antibodies and transport substances. Clotting proteins contribute to blood clotting.
Hormones	Insulin, growth hormones, testosterone, estrogen, others. Hormones are chemical messenger molecules that provide information needed to regulate specific body functions.
Gases	Oxygen is needed for metabolism; carbon dioxide is a waste product of metabolism. Both are dissolved in plasma as well as carried by RBCs.
Nutrients and wastes	Glucose, urea, many others. Nutrients, raw materials, and wastes (including heat) are transported by blood throughout the body.

We discuss lipoproteins and the health implications of high blood cholesterol levels when we discuss the heart and blood vessels.

Gamma globulins function as part of the body's defense system, helping to protect against infections and illness. We take a closer look at them when we discuss the immune system.

Clotting proteins, a third group of plasma proteins, play an important role in the process of blood clotting. As we see later in this chapter, blood clotting minimizes blood loss and helps maintain homeostasis after injury.

In addition to plasma proteins, plasma transports a variety of other molecules, including ions (also called electrolytes), hormones, nutrients, waste products, and gases. Electrolytes such as sodium and potassium contribute to the control of cell function and cell volume. Hormones, which are chemical "messengers" from the endocrine system, transport information throughout the body. Nutrients such as carbohydrates, amino acids, vitamins, and other substances are absorbed from the digestive tract or produced by cells' metabolic reactions. Waste products in plasma include carbon dioxide, urea, and lactic acid. Gases dissolved in plasma include oxygen, which is necessary for metabolism, and carbon dioxide, a waste product of metabolism.

Red blood cells transport oxygen and carbon dioxide

Just under half of the volume of whole blood consists of its formed elements (refer to Figure 7.2b). The most abundant are *red blood cells (RBCs)*, also called **erythrocytes** ("red cells" in Greek). Red blood cells function primarily as carriers of oxygen and carbon dioxide. Each cubic millimeter of blood contains approximately 5 million red blood cells. They give blood its color and are the major reason why it is viscous.

Red blood cells offer a great example of how structure serves function. Red blood cells are small, flattened, doughnut-shaped disks whose centers are thinner than their edges (Figure 7.3). This is an unusual shape among human cells, but it has several advantages for RBCs. It makes them flexible, so they can bend and flex to squeeze through tiny blood vessels. It also means that no point within an RBC's

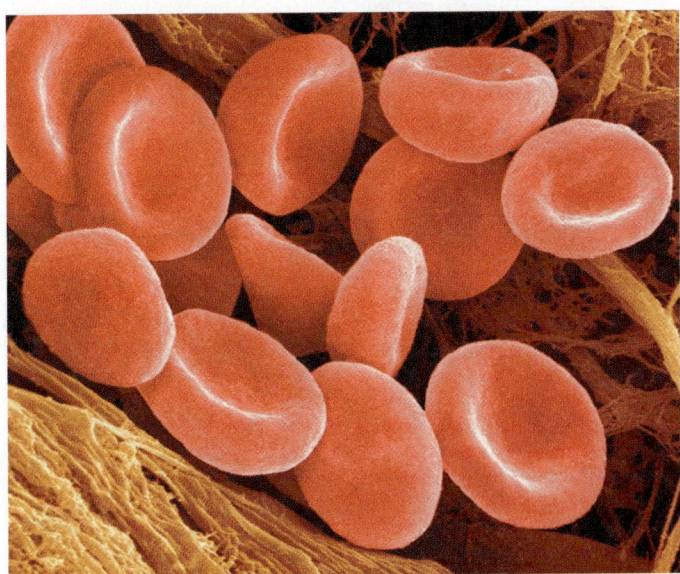

Figure 7.3 Red blood cells. Note that their flattened, biconcave shape gives them a sunken appearance.

cytoplasm is ever far from the cell surface, which facilitates the process of gas exchange.

Red blood cells are highly specialized to transport oxygen. Mature RBCs have no nucleus and essentially no organelles. Because RBCs lack mitochondria, they generate ATP by anaerobic pathways. This means RBCs don't consume any of the oxygen they carry. They are essentially fluid-filled bags made of plasma membrane, crammed with nearly 300 million molecules of an oxygen-binding protein called **hemoglobin.** Hemoglobin consists of four polypeptide chains, each containing a heme group (**Figure 7.4**). At the center of each heme group is an iron atom, which can readily form a bond with an oxygen molecule (O_2). In total, a single red blood cell can carry up to 1.2 billion molecules of oxygen.

Several factors influence the binding of hemoglobin to oxygen. Hemoglobin binds oxygen most efficiently when the concentration of oxygen is relatively high and the pH is fairly neutral. These are precisely the conditions that prevail in the lungs. In the lungs, oxygen diffuses into blood plasma and then into red blood cells, where it attaches readily to the iron atoms in hemoglobin. The binding of O_2 by hemoglobin removes some of the O_2 from the plasma, making room for more O_2 to diffuse from the lungs into the plasma. Hemoglobin, with four oxygen molecules attached, called *oxyhemoglobin,* has a characteristic bright red color.

The bond hemoglobin forms with oxygen must be temporary so that the oxygen can be released to the cells that need it. In body tissues that use oxygen in the course of their metabolic activities, the concentration of dissolved oxygen and the pH are both lower. Under these conditions, hemoglobin readily releases oxygen into body tissues, making it available to cells. Increased body heat also increases the rate at which hemoglobin releases oxygen. Hemoglobin that has given up its oxygen is called *deoxyhemoglobin.* Deoxyhemoglobin is characteristically dark purple, but because venous blood returning from the cells contains a mixture of oxyhemoglobin and deoxyhemoglobin, venous blood generally has a dark red or maroon color that is between red and purple.

Hemoglobin also transports some carbon dioxide (CO_2), a waste product of cellular metabolism. In tissues, where carbon dioxide levels are high, about 25% of the CO_2 binds to hemoglobin (at different sites than O_2). In the lungs, CO_2 detaches from hemoglobin and is eliminated through respiration. Gas transport and exchange is covered in more detail in the chapter on the respiratory system.

☑ Suppose a patient has an unusually low body temperature and his blood pH is unusually basic (at a higher pH value than normal). How might this affect oxygen delivery to the body tissues?

Hematocrit and hemoglobin reflect oxygen-carrying capacity

The percentage of blood that consists of red blood cells is called the **hematocrit** (review Figure 7.2b). The hematocrit is a relative measure of the oxygen-carrying capacity of blood, and thus it is often of interest to the health care professional. The normal hematocrit range is 43–49% in men and 37–43% in women. A related number is the amount of hemoglobin in the blood, expressed in units of grams per 100 ml of blood (abbreviated Hb gm%). Normal values for hemoglobin are 14–18 gm% in men and 12–14 gm% in women.

An unusual hematocrit (or Hb gm%) may be cause for concern. A low hematocrit may signal *anemia* or other disorders of inadequate red blood cell production (see section 7.4). A high hematocrit can also be risky because excessive red blood cells thicken blood and increase the risk of blood clots. In rare cases, a high hematocrit could signal *polycythemia,* a disorder of the bone marrow characterized by an overproduction of red blood cells. Polycythemia increases blood volume and blood viscosity, sometimes leading to headaches, blurred vision, and high blood pressure.

Some shifts in hematocrit (and hemoglobin) are normal and temporary. For example, if you visit the mountains on your next vacation and stay for at least several weeks, your hematocrit rises to compensate for lower levels of oxygen in the air you breathe. This is part of the normal homeostatic regulation of the oxygen-carrying capacity of the blood. After you return to your usual altitude, your hematocrit returns to normal.

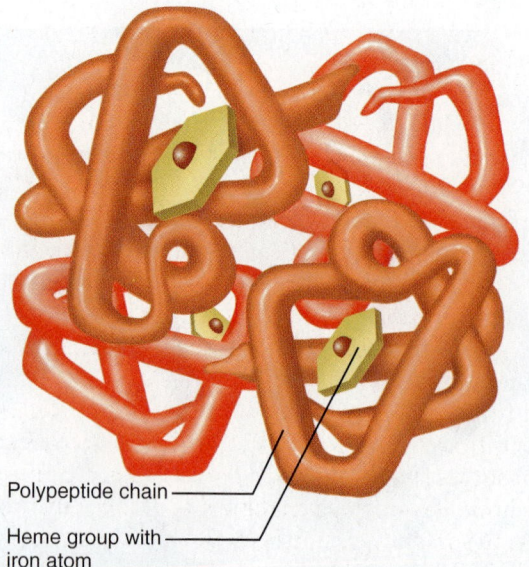

Polypeptide chain

Heme group with iron atom

Figure 7.4 A hemoglobin molecule. Hemoglobin consists of four polypeptide chains folded together, each with a heme group containing a single iron atom. There are nearly 300 million of these molecules in every red blood cell.

☑ Is the blood sample in Figure 7.2 more likely to be from a man or a woman? Why?

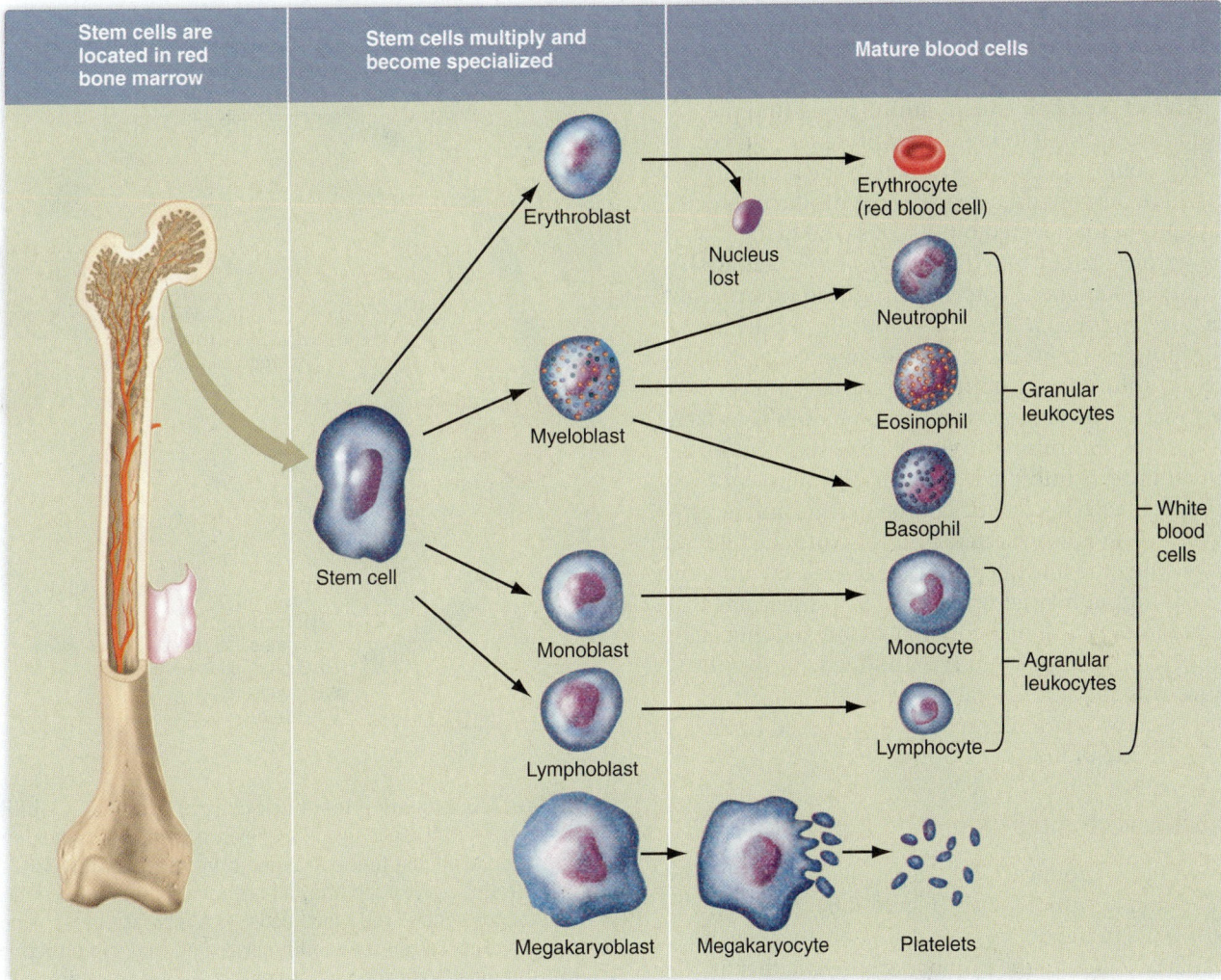

| Stem cells are located in red bone marrow | Stem cells multiply and become specialized | Mature blood cells |

Figure 7.5 The production of blood cells and platelets. Blood cells have short life spans and must be continually replaced. Stem cells in the red marrow of bones continually divide and give rise to a variety of types of blood cells. The production of erythrocytes (RBCs) is stimulated by a hormone, erythropoietin. The production of specific white blood cells is stimulated by specific colony-stimulating factors.

All blood cells and platelets originate from stem cells

All blood cells and platelets originate from cells in the red marrow of certain bones. These cells, called **stem cells,** divide repeatedly throughout our lives, continually producing immature blood cells. These immature cells develop into platelets and the various types of mature red and white blood cells described in **Figure 7.5**.

✔️ A medical researcher infuses a liter of blood plasma (whole blood minus the formed elements) into a patient. What do you think would be the effect of the infusion? What if he replaced *all* of the patient's blood with blood plasma?

RBCs have a short life span

Some stem cells develop into immature cells called *erythroblasts* ("red" + "immature"). Erythroblasts become

filled with hemoglobin and develop into mature RBCs, or erythrocytes, in about a week. As they mature, these cells lose their nucleus and organelles, so they cannot reproduce. Thus, all new RBCs must originate from dividing stem cells. Because they lack a nucleus and therefore cannot perform many standard cell activities (such as producing new proteins and phospholipids to renew their cell membranes), they wear out rather quickly. Red blood cells live for only about 120 days, but during that time, they make nearly 3,000 round-trips a day, ferrying O_2 from the lungs to the tissues and CO_2 from the tissues back to the lungs. Because they live for such a short time, red blood cells must be produced throughout life—at the incredible rate of more than 2 million per second—just to keep the hematocrit constant.

Old and damaged RBCs are removed from the circulating blood and destroyed in the liver and spleen by large cells called **macrophages.** Macrophages are derived from *monocytes,* the largest of the white blood cells.

Macrophages surround, engulf, and digest the red blood cell. The process is called **phagocytosis.** The four peptide chains of the hemoglobin molecules are then dismantled into their constituent amino acids, and the amino acids are recycled to make new proteins. The iron atoms of the heme groups are returned to the red bone marrow, where they are used again in the production of new hemoglobin for new red blood cells. The heme groups (minus the iron) are converted by the liver to a yellowish pigment called *bilirubin*. If you've ever noticed how a bruise slowly changes color as it heals, from purple to blue to green to yellow, you have observed the chemical breakdown of the heme groups to bilirubin at the site of damage. Under normal circumstances, when hemoglobin is broken down in the liver, bilirubin mixes with bile secreted during digestion and passes into the intestines. This pigment contributes to the characteristic colors of urine and feces.

When the liver fails to secrete bilirubin into the bile properly or when the bile duct from the liver to the intestines is blocked, bilirubin may accumulate in blood plasma. High circulating levels of bilirubin make skin and mucous membranes look yellowish and can turn the whites of the eyes yellow. This condition is called *jaundice* (from *jaune,* French for "yellow"). Jaundice may also be caused by an increase in the rate of RBC breakdown.

RBC production is regulated by a hormone

Regulation of RBC production is a negative feedback control loop that maintains homeostasis (**Figure 7.6**). The total number of RBCs in the body is not regulated; there are no cells capable of counting the number of RBCs. Rather, it is the effectiveness of RBCs in transporting oxygen that is actually regulated. Certain cells in the kidneys monitor the availability of oxygen. If oxygen availability falls for any reason, these cells cause the kidneys to secrete a hormone called **erythropoietin.** Erythropoietin is transported in the blood to the red bone marrow, where it stimulates stem cells to produce more red blood cells. When the oxygen-carrying capacity of blood returns to an appropriate level as monitored by kidney cells, the cells cut back on their production of erythropoietin, and RBC production returns to normal. Thus, the body maintains homeostasis of oxygen availability by adjusting the production rate of the RBCs that transport it.

Some people with kidney disease do not produce enough erythropoietin to regulate their RBC production properly. Fortunately, erythropoietin is now available commercially and can be administered to stimulate red cell production.

Some athletes have abused erythropoietin by injecting it to increase their RBC production and thus their blood oxygen-carrying capacity, a practice called **blood doping.** Three gold medalists at the 2002 Winter Olympics, Spain's Johann Mühlegg and Russia's Larissa Lazutina and Olga Danilova, were disqualified and stripped of their medals because of blood doping. All three skiers tested positive for

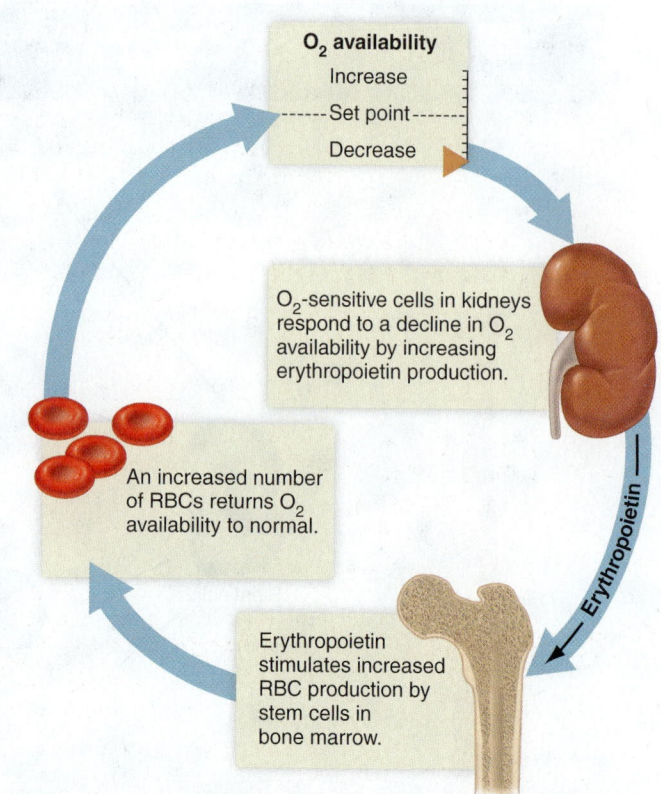

Figure 7.6 Negative feedback control of the availability of oxygen. Certain cells in the kidney are sensitive to the amount of oxygen available to them. When oxygen availability falls, these cells produce erythropoietin, a hormone that stimulates the bone marrow to produce red blood cells. The increase in red blood cells returns oxygen availability toward normal, which reduces the stimulus for further erythropoietin secretion. Ultimately, homeostasis of oxygen availability is achieved.

☑ What might happen in this feedback loop if bone marrow is diseased and is unable to make enough red blood cells?

darbepoetin, an erythropoietin-like drug that is 10 times more powerful than the natural hormone.

Blood doping can have serious health consequences. Excess red blood cells make blood more viscous, and so the heart must work harder to pump blood through the body. The dehydration that follows strenuous exercise can concentrate the blood even more, increasing the risk of blood clots, high blood pressure, heart attack, and stroke.

White blood cells defend the body

Approximately 1% of whole blood consists of **white blood cells** (WBCs or **leukocytes**). Larger than red blood cells, they are also more diverse in structure and function. They have a nucleus but no hemoglobin. Because they are translucent, they are difficult to identify under the microscope unless they have been stained. Each cubic millimeter of blood contains only about 7,000 of them, and there is only one WBC for every 700 RBCs. White blood cells play a number of crucial roles in defending against disease and injury.

Like red blood cells, white blood cells arise from stem cells in the red bone marrow. As shown in Figure 7.5, stem cells produce immature blood cells that develop into the various WBCs. There are two major categories of white blood cells: *granular leukocytes* (granulocytes) and *agranular leukocytes* (agranulocytes). Both types contain granules (actually vesicles) in their cytoplasm that are filled with proteins and enzymes to assist their defensive work. However, the granules of the agranular leukocytes are not visible when the cells are stained for viewing (*a–* means "without").

Most WBCs have a short life span. Many granular leukocytes die within a few hours to nine days, probably because of injuries sustained while fighting invading microorganisms. Monocytes may survive for several months; lymphocytes for several days to many years. Dead and injured WBCs are continually removed from the blood by the liver and spleen.

Circulating levels of white blood cells rise quickly whenever the body is threatened by viruses or bacteria. When activated by tissue injury or microbes, WBCs produce factors called *colony-stimulating factors* (because they stimulate a specific colony of WBC precursor cells). These colony-stimulating factors increase the rate of development of new WBCs from the bone marrow and also stimulate the release of stored WBCs from the spleen.

Red blood cells remain entirely within the vascular system except in cases of tissue injury, but some white blood cells leave the vascular system and circulate in the tissue fluid between cells, or in the fluid in the lymphatic system. Because WBCs can change their shape, they can squeeze between the cells that form the capillary walls. White blood cells, part of the body's defense system, are discussed in more detail in the chapter on the immune system. In this chapter, we describe the important characteristics of each type (refer to Figure 7.5).

Granular leukocytes: neutrophils, eosinophils, and basophils

The granular leukocytes include neutrophils, eosinophils, and basophils. These names are based on their staining properties:

- **Neutrophils,** the most abundant type of granulocyte, account for about 60% of WBCs. (Their name—which means "neutral-loving"—reflects the fact that their granules do not significantly absorb either a red or blue stain). The first white blood cells to combat infection, neutrophils surround and engulf foreign cells by phagocytosis (**Figure 7.7**). They especially target bacteria and some fungi, and their numbers can rise dramatically during acute bacterial infections such as appendicitis or meningitis.

- **Eosinophils** make up a relatively small percentage (2–4%) of circulating white blood cells. (Their name comes from their tendency to stain readily with an

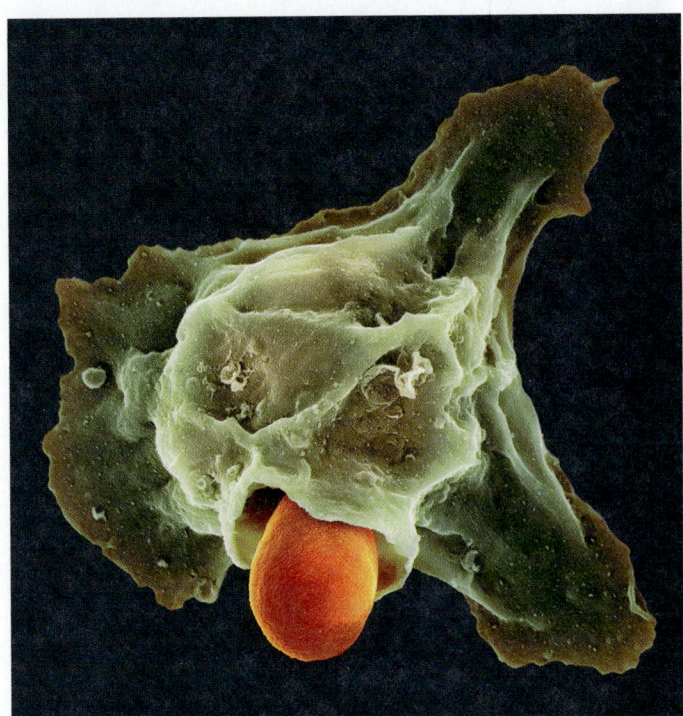

Figure 7.7 A neutrophil engulfing a fungal cell. In the first stage of phagocytosis, the neutrophil has approached the fungal cell using the protuberances on its surface. Next, it will engulf and destroy the fungal cell.

acidic red stain called *eosin*.) Eosinophils have two important functions. The first is to defend the body against large parasites such as worms (hookworms, tapeworms, flukes, and pinworms, among others). These parasites are too big to be surrounded and engulfed through phagocytosis. Instead, clusters of eosinophils surround each parasite and bombard it with digestive enzymes. The second function of eosinophils involves releasing chemicals that moderate the severity of allergic reactions.

- **Basophils,** the rarest white blood cells, account for only 0.5% of leukocytes. (They are named for their tendency to stain readily with basic blue stains.) The granules in the cytoplasm of basophils contain histamine, a chemical that initiates the inflammatory response. When body tissues are injured, basophils secrete histamine, causing adjacent blood vessels to release blood plasma into the injured area. The plasma brings in nutrients, various cells, and chemicals to begin the process of tissue repair. The swelling, itching, and redness associated with histamine release by basophils may not feel pleasant, but they are part of the immune system's defenses against molecules that are perceived as threatening.

Agranular leukocytes: monocytes and lymphocytes

The agranular leukocytes include monocytes and lymphocytes. The largest WBCs, **monocytes**, make up about

5% of circulating white blood cells. They can filter out of the bloodstream and take up residence in body tissues, where they differentiate into the macrophages that engulf invaders and dead cellular debris by phagocytosis. They also stimulate lymphocytes to defend the body. Monocytes seem especially active during chronic infections, such as tuberculosis, and against viruses and certain bacterial parasites.

Lymphocytes total about 30% of circulating white blood cells. They are found in the bloodstream, tonsils, spleen, lymph nodes, and thymus gland. They are classified into two types, *B lymphocytes* and *T lymphocytes* (or B cells and T cells). B lymphocytes give rise to *plasma cells* that produce *antibodies*, specialized proteins that defend against microorganisms and other foreign invaders. T lymphocytes target and destroy specific threats such as bacteria, viruses, and cancer cells. Both play a crucial role in the body's immune system.

☑ A friend sprains her ankle, and it soon becomes red, swollen, and sore. Which type of white blood cell is probably responsible for these symptoms, and what is its major function?

Platelets are essential for blood clotting

Less than 1% of whole blood consists of **platelets.** Platelets are derived from megakaryocytes, which are large cells derived from stem cells in the bone marrow (review Figure 7.5). Megakaryocytes never circulate—they remain in the bone marrow. Platelets are just small pieces of megakaryocyte cytoplasm and cell membrane. Because platelets are not living cells, they last only about five to nine days in circulation.

When a blood vessel is injured and leaks blood, platelets participate in the clotting process, thereby limiting the vascular and tissue damage. We examine the clotting process in the next section. Once the bleeding is stopped, platelets also participate in the repair process by releasing proteins that promote blood vessel growth and repair.

↺ **Recap** Blood consists of a watery fluid containing cells, proteins, nutrients, cellular waste products, and ions. Red blood cells are specialized for transporting oxygen and carbon dioxide; white blood cells protect against disease. Blood cells arise from stem cells in bone marrow. Platelets, important in blood clotting, are small pieces of bone marrow cells called *megakaryocytes.* ∎

7.2 Hemostasis: stopping blood loss

One of the most important properties of the circulatory system is its ability to limit blood loss following injury. **Hemostasis** (**Figure 7.8**), the natural process of stopping the flow or loss of blood, proceeds in three stages: ❶ vascular spasm, or intense contraction of blood vessels in the area, ❷ formation of a platelet plug, and ❸ blood clotting, also called *coagulation*. Once blood loss has stopped, tissue repair can begin.

Vascular spasms constrict blood vessels to reduce blood flow

As illustrated in Figure 7.8, ❶ when a blood vessel is damaged, smooth muscle in its wall undergoes spasms—intense contractions that constrict the vessels. If the vessels are medium-sized to large, the spasms reduce immediate outflow of blood, minimizing the damage in preparation for later steps in hemostasis. If the vessels are small, the spasms press the inner walls together and may even stop the bleeding

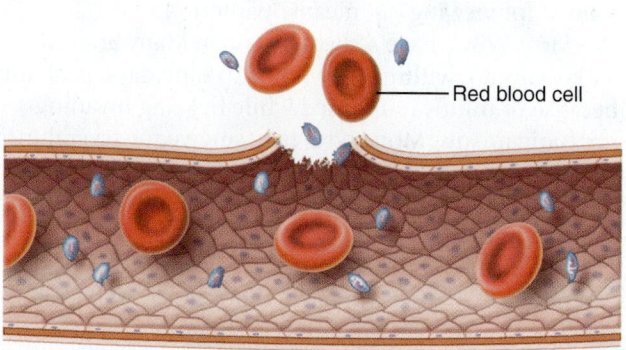

Damage to a blood vessel exposes the vessel muscle layers and the tissues to blood.

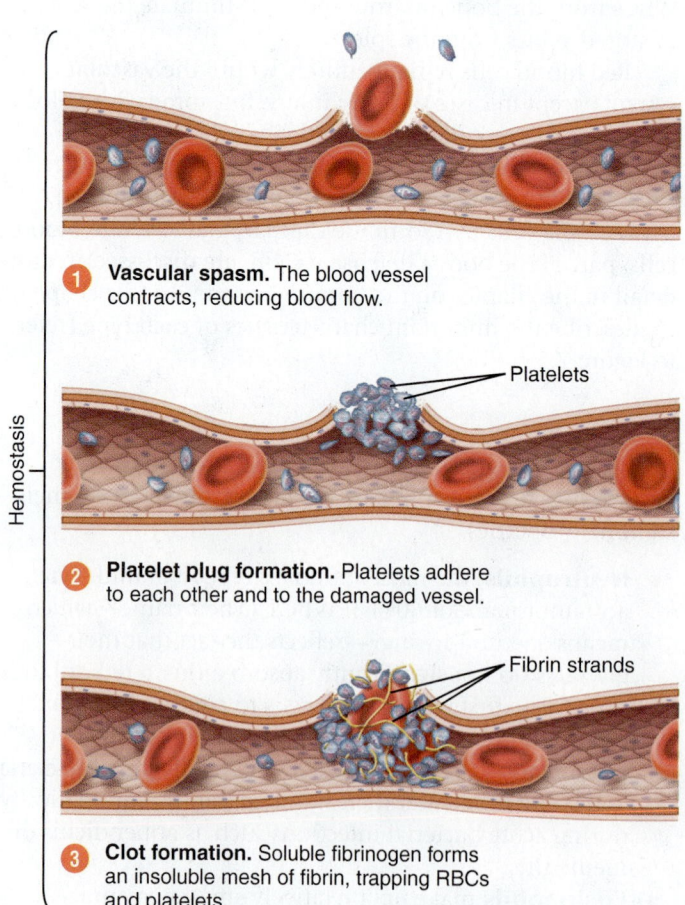

❶ **Vascular spasm.** The blood vessel contracts, reducing blood flow.

❷ **Platelet plug formation.** Platelets adhere to each other and to the damaged vessel.

❸ **Clot formation.** Soluble fibrinogen forms an insoluble mesh of fibrin, trapping RBCs and platelets.

Figure 7.8 The stages of hemostasis.

entirely. Vascular spasms generally last for about half an hour, long enough for the next two stages of hemostasis to occur.

Platelets stick together to seal a ruptured vessel

Normally, platelets circulate freely in blood. However, as seen in Figure 7.8, ❷ when the lining of a blood vessel breaks, exposing underlying proteins in the vessel wall, platelets swell, develop spiky extensions, and begin to clump together. They also become sticky and start adhering to the walls of the vessel and to each other. More platelets congregate and undergo these same changes. The result is a platelet plug that seals the injured area. If the rupture is fairly small, a platelet plug may be able to close it within several seconds. This may be enough to stop the bleeding. If damage is more severe, blood clotting occurs.

> **MJ's BlogInFocus**   How effective is a treatment called platelet-rich plasma therapy (PRPT) in improving the healing time of connective tissue injuries? Visit MJ's blog in the Study Area in MasteringBiology and look under "Platelet-Rich Plasma Therapy."
>
> http://goo.gl/G6Jk6k

A blood clot forms around the platelet plug

The third stage in hemostasis is ❸ the formation of a blood clot, during which the blood changes from a liquid to a gel. This involves a series of chemical reactions that ultimately produce a meshwork of protein fibers within the blood. At least 12 substances, known as clotting factors, participate in these reactions. We will focus on three clotting factors: prothrombin activator, thrombin, and fibrinogen.

Damage to blood vessels stimulates the vessels and nearby platelets to synthesize *prothrombin activator*. This activates the conversion of *prothrombin*, a plasma protein, into an enzyme called **thrombin.** The reaction requires the presence of calcium ions (Ca^{2+}). Thrombin in turn facilitates the conversion of a soluble plasma protein, *fibrinogen*, into long insoluble threads of a protein called **fibrin.** The fibrin threads wind around the platelet plug at the wound site, forming an interlocking net of fibers that traps and holds platelets, blood cells, and various molecules against the opening (refer to Figure 7.8).

The mass of fibrin, platelets, and trapped red blood cells coalesces into an initial *clot* that reduces the flow of blood at the site of injury (**Figure 7.9**). This initial fibrin clot can form in less than a minute. Shortly thereafter, platelets in the clot start to contract, tightening the clot and pulling the vessel walls together. Generally, the entire process of blood clot formation and tightening takes less than an hour.

If any step in this process is blocked, even a minor cut or bruise can become life threatening. Consider *hemophilia*, an

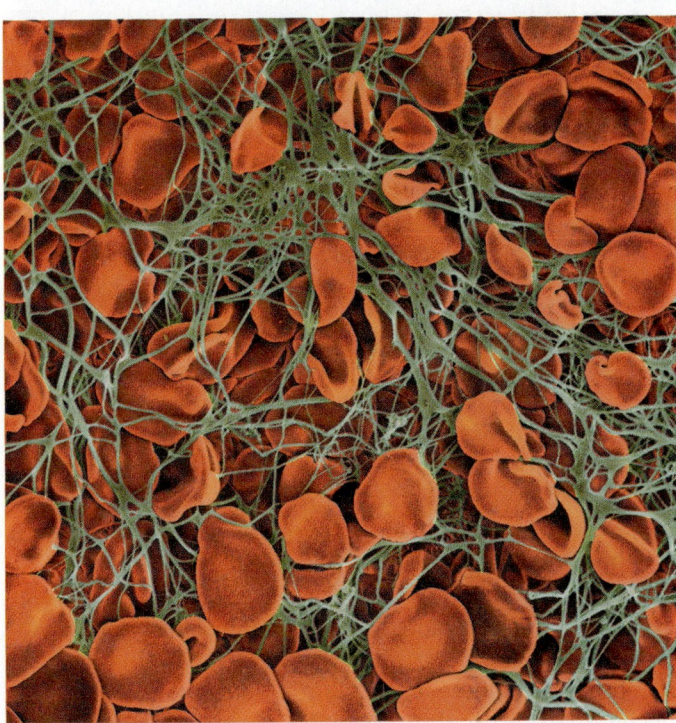

Figure 7.9   Magnified view of a developing clot, showing red blood cells trapped in a network of fibrin fibers.

inherited condition caused by a deficiency of one or more clotting factors. People with the most common form of the condition, hemophilia A, lack a protein known as *clotting factor VIII*. When a vessel is breached, blood clots slowly or not at all. Even if the skin is not broken, severe bruising can spread into joints and muscles. Fifty years ago, most people with hemophilia did not survive to adulthood. Today, many bleeding episodes can be controlled by administering another of the clotting factors, factor VIIa. Clotting factor VIIa was initially in short supply because it had to be purified from donor blood, but with genetic engineering techniques it is now possible to produce factor VIIa in the laboratory and in large quantities.

> **MJ's BlogInFocus**   Is factor VIIa effective in reducing mortality from excessive bleeding after traumatic injury or major surgery? Visit MJ's blog in the Study Area in MasteringBiology and look under "Recombinant Factor VIIa."
>
> http://goo.gl/qV0tQL

Certain medications can also interfere with hemostasis. If you cut yourself after taking aspirin, for example, you may notice that you bleed more than usual. This is because aspirin blocks platelet clumping and slows the formation of a platelet plug. If you plan to have surgery, your doctor will probably advise you to avoid taking aspirin for at least 7–10 days before the surgery.

Recap Damage to blood vessels causes the vessels to spasm (contract). Nearby platelets become sticky and adhere to each other, limiting blood loss. In addition, a series of chemical events causes the blood in the area to clot or coagulate (form a gel). ∎

7.3 Human blood types

Blood transfusions—the administration of blood directly into the bloodstream of another person—may seem like a miracle of modern medicine, but the concept is not new. For over a century, physicians have tried to counteract severe blood loss by transfusing blood from one living person into another. Sometimes these early attempts were successful. More often they were not, resulting in severe illness or even death for the recipient. Why did the first attempts at transfusing blood save some lives but not others?

Today, we know the success of blood transfusions depends largely on **blood type,** based primarily on the ABO blood group system. If you ever donate or receive blood, you will undergo testing to determine your blood type. This is necessary because if you receive blood from someone who does not belong to a compatible blood type you could suffer a severe reaction.

To understand the concept of blood typing, we must first be familiar with antigens and antibodies. Our cells have certain surface proteins that the immune system can recognize and identify as "self"—in other words, belonging to us. These are like passwords that cause our immune system to ignore our own cells. Foreign cells carry different surface proteins, which the immune system recognizes as "nonself."

An **antigen** (*anti* means "against," and the Greek word *gennan* means "to generate") is a "nonself" cell protein that stimulates the immune system of an organism to defend the organism. As part of this defense, the immune system produces an opposing protein called an **antibody** ("against" + "body").

Produced by lymphocytes, antibodies belong to the class of plasma proteins called gamma globulins, mentioned earlier. Antibodies mount a counterattack on antigens they recognize as "nonself" (**Figure 7.10a**). There are many antibodies, each one specialized to attack one particular antigen. This response has been compared to a lock and key: Only a specific antibody key can fit a specific antigen lock. Antibodies float freely in the blood and lymph until they encounter an invader with the matching antigen. They bind to the antigen molecule to form an antigen-antibody complex that marks the foreigner for destruction. The formation of an antigen-antibody complex often causes the foreign cells to clump together, effectively inactivating them (**Figure 7.10b**).

Antigens and antibodies are discussed in more detail in the chapter on the immune system. Let's look now at how their interactions relate to blood type and blood transfusions.

ABO blood typing is based on A and B antigens

Like other cells, red blood cells have proteins on the outer surface of their cell membranes that allow the body to identify them as "self." The interactions between these antigens, and the development of antibodies against the antigens of foreign red blood cells, underlie the reactions that can occur after blood transfusions.

Red blood cells are classified according to the ABO blood group system, in which nearly all individuals belong to one of four types: A, B, AB, or O. Type A blood has A antigens, type B blood has B antigens, type AB blood has *both* A and B antigens, and type O blood has neither (think of the O as a "zero"). In

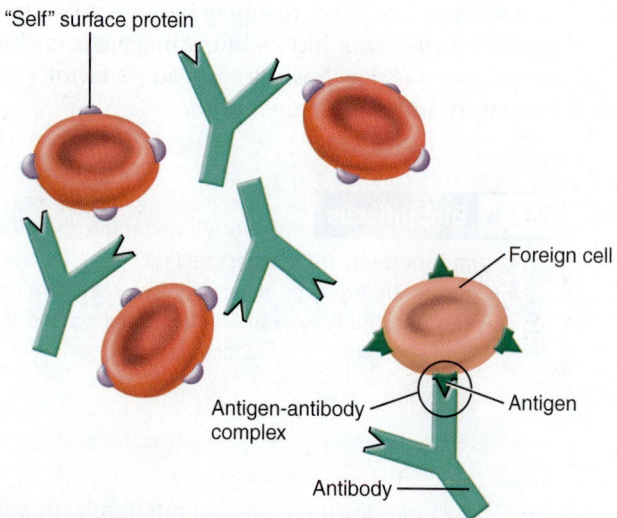

"Self" surface protein

Foreign cell

Antigen-antibody complex

Antigen

Antibody

a) Antibody binds to antigen. Antibodies ignore the "self" surface proteins but bind to the antigen of the foreign cell, forming an antigen-antibody complex.

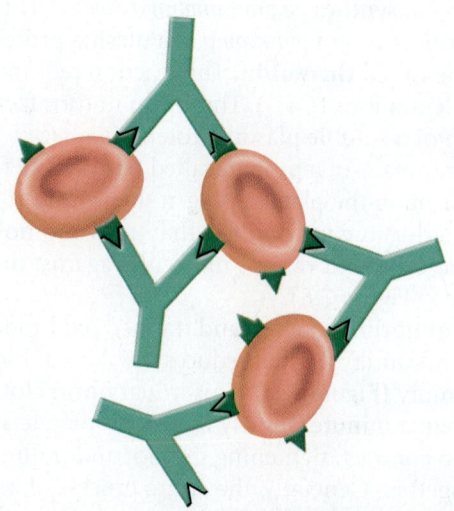

b) Antigen-antibody complexes clump together. Clumping effectively inactivates the foreign cells.

Figure 7.10 How antibodies recognize and inactivate foreign cells.

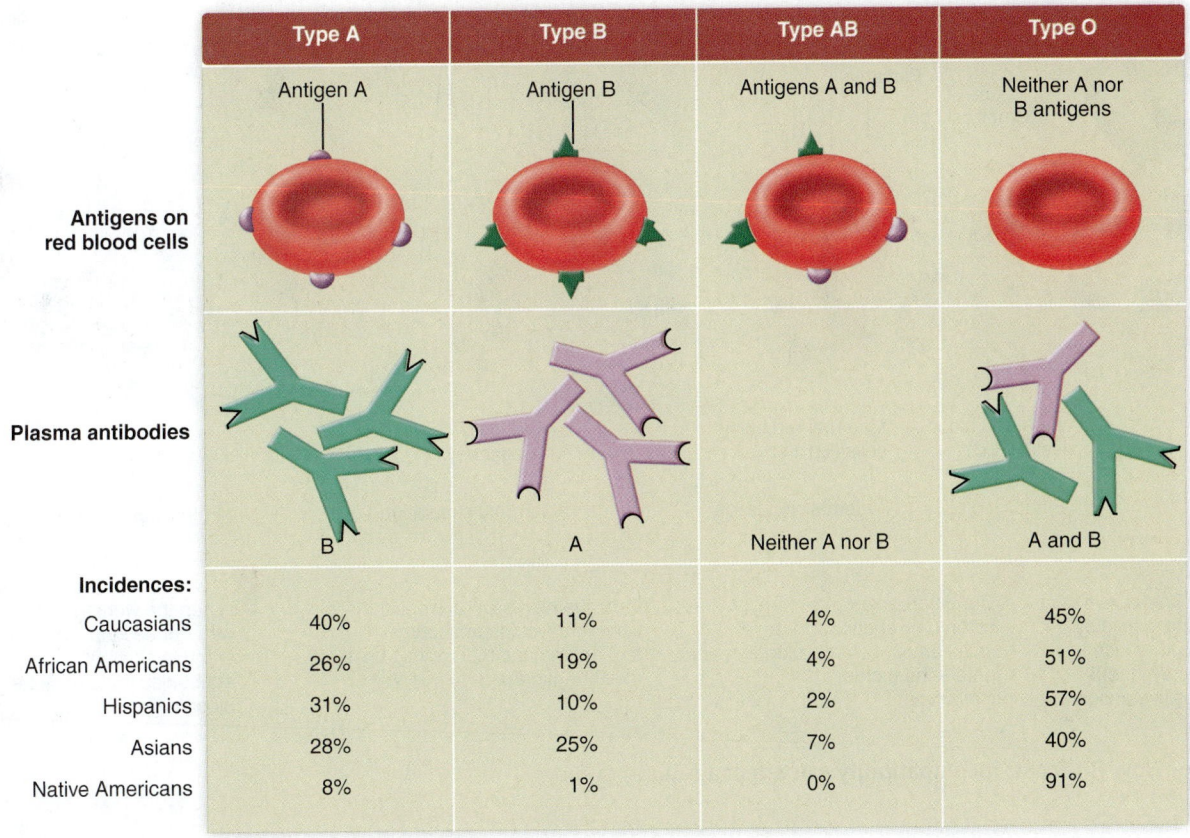

	Type A	Type B	Type AB	Type O
Antigens on red blood cells	Antigen A	Antigen B	Antigens A and B	Neither A nor B antigens
Plasma antibodies	B	A	Neither A nor B	A and B
Incidences:				
Caucasians	40%	11%	4%	45%
African Americans	26%	19%	4%	51%
Hispanics	31%	10%	2%	57%
Asians	28%	25%	7%	40%
Native Americans	8%	1%	0%	91%

Figure 7.11 **Characteristics of the four major blood types of the ABO typing system, showing their RBC surface antigens, antibodies, and relative incidences among various populations.**

✔ In which population—Caucasian, African American, Hispanic, Asian, or Native American—would it be least risky to do an emergency blood transfusion without blood typing either the donor or the recipient?

addition, all individuals have circulating antibodies (and the ability to make more antibodies) against any surface antigens different from their own; type A blood has type B antibodies, type B blood has type A antibodies, type O blood has both type A and B antibodies, and type AB blood has neither antibody. **Figure 7.11** shows these various blood types and also indicates the relative incidences of each type in various populations. The antibodies appear early in life, regardless of whether a person has ever received a blood transfusion. These antibodies attack red blood cells with foreign antigens, damaging them and causing them to *agglutinate*, or clump together. If agglutination is extreme, the clumps may block blood vessels, causing organ damage or even death. In addition, hemoglobin released by damaged red blood cells can block the kidneys, leading to kidney failure. Any adverse effect of a blood transfusion is called a *transfusion reaction*.

If you have type A blood, you are restricted to receiving transfusions of either type A or type O blood because neither of them has a foreign (type B) antigen. A transfusion of type B or type AB blood would provoke your antibodies to mount an attack against the B antigen of the donated RBCs, causing them to agglutinate. Similarly, if you're type B, you cannot receive any blood with type A antigens (A or AB). People with type AB blood can generally receive transfusions not only

from other AB individuals but from all three of the other blood types as well. People with type AB blood, however, can donate only to other type AB individuals. Type O persons can give blood to persons of A, B, or AB type, but they can receive blood only from type O. Notice that it is the antibodies of the recipient that generally cause the transfusion reaction. Though the donor blood may have antibodies against the recipient's RBCs, they rarely cause transfusion reactions because the volume of blood given is generally small compared to the volume of the recipient's blood.

✔ Suppose a man has a rare mutation in his blood cell antigens, such that he has only a single unique blood antigen, C. Nobody else in the world has the type C antigen and nobody else has antibodies that will react to it. Can he donate blood safely to anybody else? Explain.

Rh blood typing is based on Rh factor

Another red blood cell surface antigen, called **Rh factor** because it was first discovered in rhesus monkeys, is also important in blood transfusions. Approximately 85% of Americans are *Rh-positive*, meaning they carry the Rh antigen on their red blood cells. About 15% are *Rh-negative*—they

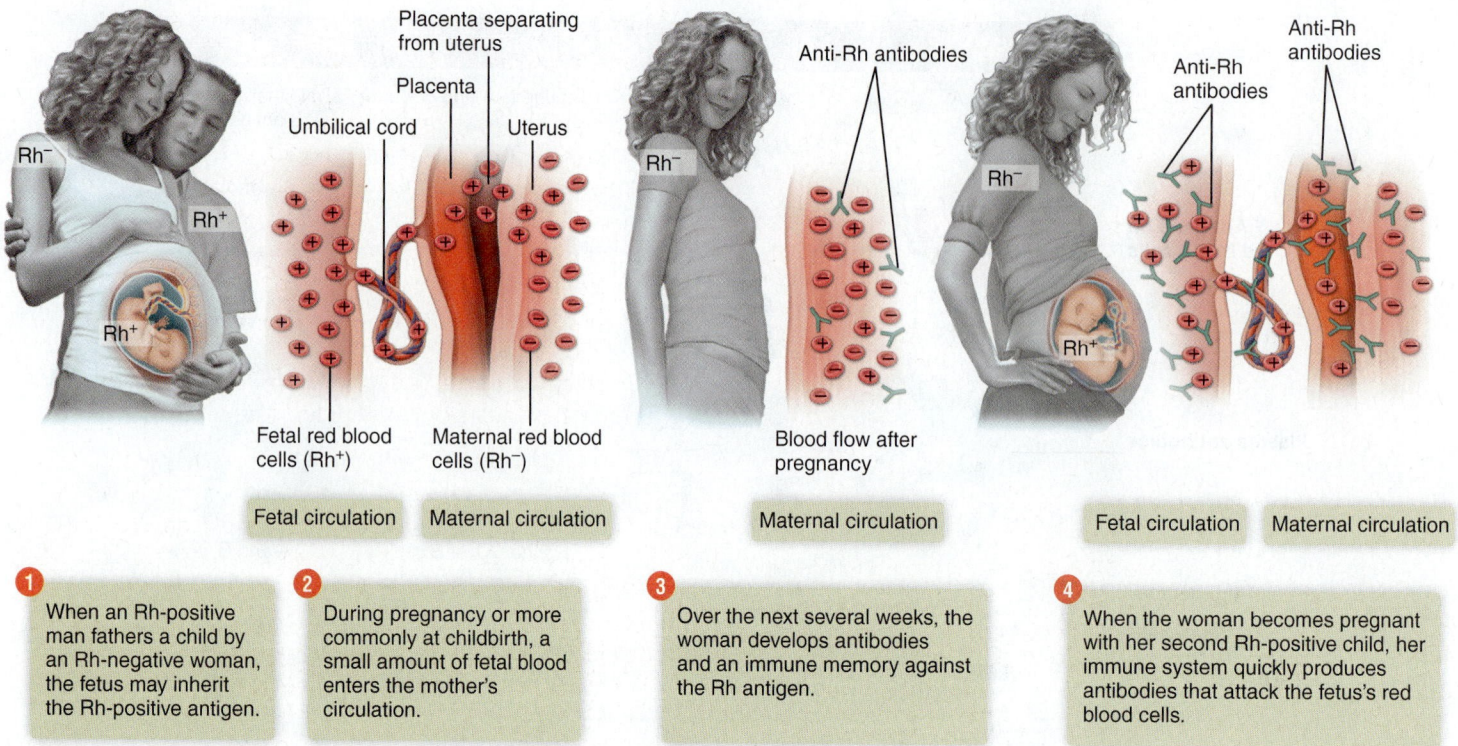

Placenta separating from uterus

Placenta

Umbilical cord

Uterus

Rh⁻

Rh⁺

Rh⁺

Fetal red blood cells (Rh⁺)

Maternal red blood cells (Rh⁻)

Fetal circulation | Maternal circulation

Anti-Rh antibodies

Rh⁻

Blood flow after pregnancy

Maternal circulation

Anti-Rh antibodies

Anti-Rh antibodies

Rh⁻

Rh⁺

Fetal circulation | Maternal circulation

1 When an Rh-positive man fathers a child by an Rh-negative woman, the fetus may inherit the Rh-positive antigen.

2 During pregnancy or more commonly at childbirth, a small amount of fetal blood enters the mother's circulation.

3 Over the next several weeks, the woman develops antibodies and an immune memory against the Rh antigen.

4 When the woman becomes pregnant with her second Rh-positive child, her immune system quickly produces antibodies that attack the fetus's red blood cells.

Figure 7.12 **How Rh factor incompatibility can affect a fetus.**

do not have the Rh antigen, and consequently their immune systems respond to any foreign Rh antigen by making antibodies against it.

The Rh factor is a particular concern for Rh-negative women who wish to have children (**Figure 7.12**). **1** If an Rh-negative woman becomes pregnant by an Rh-positive man, the fetus may be Rh-positive. If some of the fetus's Rh-positive blood cells leak into the mother's blood, **2** the mother starts producing anti-Rh antibodies. These maternal antibodies can cross the placenta and attack the fetus's red blood cells. The result may be *hemolytic disease of the newborn (HDN)*, a disorder characterized by a reduced number of red blood cells and toxic levels of hemoglobin breakdown products in the newborn. HDN can lead to mental retardation or even death.

The risk of HDN is much higher for the second and all subsequent Rh-positive fetuses than for the first. This is because it takes days or even weeks for antibodies to be produced after the first exposure to an antigen. **3** Although a few fetal cells may leak across the placenta during a normal pregnancy, presenting a slight chance that the first fetus could be affected, the greatest chance of maternal exposure to fetal blood generally occurs right at childbirth, when the placenta detaches from the uterus. The antibodies that develop from maternal exposure during the woman's first delivery come too late to affect the first fetus. **4** But the maternal immune system has learned its lesson, and it is ready and waiting to attack the blood of any subsequent Rh-positive fetus (see section 9.7, Immune memory creates immunity).

To prevent this reaction, an Rh-negative mother who *may* be carrying an Rh-positive child is given an injection of anti-Rh antibodies (RhoGAM) at 28 weeks of gestation, just in case. Then, if the newborn *is* Rh-positive, the mother is given a second injection no later than three days after childbirth. The injected antibodies quickly destroy any of the newborn's red blood cells that may have entered the woman's circulation during childbirth, before her immune system has time to react to them. The injected antibodies disappear in a short time.

In addition to its important medical applications, blood typing has many other uses. Because blood types are inherited, anthropologists can track early population migrations by tracing inheritance patterns. Blood typing is also used in criminal investigations to compare the blood of victims and perpetrators and to eliminate or identify suspects on the basis of matching antigens. DNA tests can be done on blood samples to help determine paternity.

☑ Will the immune system of an Rh-positive woman attack blood cells from an Rh-negative baby? Why or why not?

Blood typing and cross-matching ensure blood compatibility

Blood typing involves determining your ABO type and the presence or absence of the Rh factor. For example, if your blood type is "B-pos" (B+), you are type B and positive for the Rh factor. If you are "O-neg" (O−), you are type O and negative for the Rh factor.

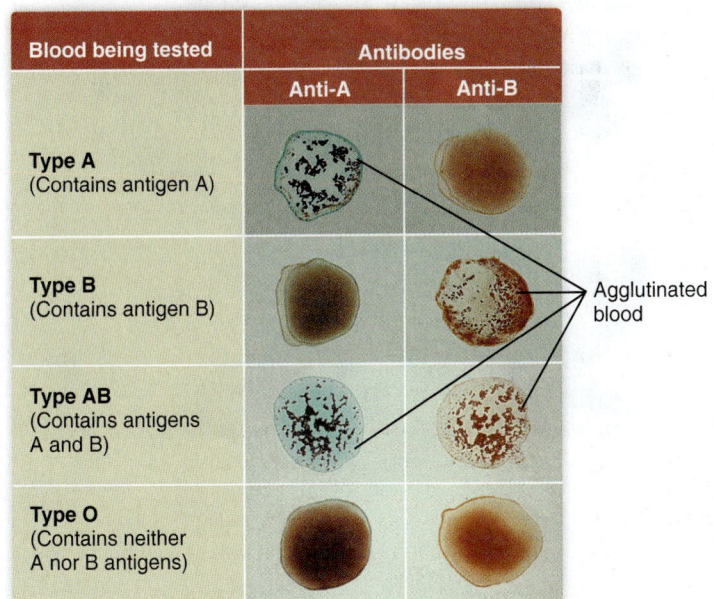

Blood being tested	Antibodies	
	Anti-A	Anti-B
Type A (Contains antigen A)		
Type B (Contains antigen B)		→ Agglutinated blood
Type AB (Contains antigens A and B)		
Type O (Contains neither A nor B antigens)		

Figure 7.13 Blood typing for ABO blood types. Anti-A and Anti-B antibodies are added to a drop of diluted blood. If the red blood cells have the surface antigen that matches the antibody, the blood agglutinates (the red cells clump together). Type O blood does not agglutinate in response to either antibody.

✅ Draw two more pictures illustrating what Rh-positive and Rh-negative blood will look like if they are each mixed with Rh antibodies.

ABO blood typing is done by adding plasma containing small amounts of anti-A and anti-B antibodies to diluted blood, and then placing a drop of the blood on a glass slide. If the blood agglutinates, then it must contain the antigens that match the antibodies (**Figure 7.13**).

AB+ individuals were once called *universal recipients* because they can generally receive blood from any other type. Type O− individuals were formerly called *universal donors* because their blood can usually be donated to any other type. However, because transfusion reactions can occur unexpectedly, the terms are now considered outdated. Why do transfusion reactions occur occasionally even when blood has been adequately typed for ABO blood type and the Rh factor? The reason is that there are over 100 other less common blood antigens in the human population, in addition to the very common A, B, and Rh antigens. Fortunately, most of them are fairly rare. To ensure that blood transfusions are absolutely safe, however, medical laboratories generally do blood typing and *cross-matching*. Cross-matching involves mixing small samples of donor blood with recipient plasma, and recipient blood with donor plasma, and examining both combinations for agglutination. If agglutination does not occur in either combination, the bloods are assumed to be a good match.

New tests make transfused blood safer

Aside from the issue of blood typing and cross-matching to ensure blood compatibility, there is also the issue of contracting a transmissible disease when receiving human blood from an unknown donor. In that regard, the human blood supply is safer than ever. Until 2002, donated blood was tested for HIV and hepatitis C viruses by looking for the presence of antibodies against the viruses in the blood. In rare cases, infections escaped detection, most likely because an antibody response to the presence of the virus had not yet taken place. Now, highly sensitive tests for specific nucleic acids present in the HIV and hepatitis C viruses have all but eliminated the risk from contracting these diseases via a blood transfusion.

↺ **Recap** Blood types A, B, AB, and O are defined by the presence (or absence) of type A and/or type B surface antigens on red blood cells. In addition to blood type, all persons are classified according to the presence or absence of another red blood cell surface antigen called the Rh factor. Antibodies to the Rh factor can cause a serious immune reaction of a mother to her own fetus under certain circumstances. ▪

7.4 Blood substitutes

Although it might seem that whole blood would always be the best choice for a patient who needs blood, there are a number of reasons why this might not be so: 1) Human blood can carry transmissible diseases, though these days the risk is small. 2) Blood must be typed and cross-matched to the recipient in order to avoid transfusion reactions. The right blood may not always be available for every patient. 3) Whole blood can only be stored for several months and must be stored under refrigerated conditions. Refrigeration during storage and transport is not always possible in some countries. 4) Demand for blood is rising more quickly than the supply.

For all of these reasons, the development of a good blood substitute has been an area of active research for over 70 years. Two types of blood substitutes are currently under investigation. Both are more properly called *oxygen-carrying* blood substitutes because they focus on only the oxygen-carrying function of blood. One type is based on various forms of modified hemoglobin in aqueous (watery) solutions. Generally, the hemoglobin comes from animals or is produced by genetic engineering in bacteria, so there are still the issues of limited supply and of possible disease transmission. In addition, hemoglobin protein in aqueous solution (not enclosed in a red blood cell membrane) is toxic to kidneys. Therefore, the hemoglobin must first be "packaged" in some way (encapsulated, cross-linked, or polymerized) before it can be used as an oxygen-carrying alternative to whole red blood cells.

HEALTH & WELLNESS
Donating Blood

A rapid loss of 30% or more of blood volume strains the body's ability to maintain blood pressure and deliver oxygen to cells throughout the body. When this happens, survival may depend on receiving a gift of donated blood. In addition, blood is often needed for certain planned surgical procedures. Approximately 15 million units of blood are donated every year, and almost 5 million people receive donated blood. Most people who donate blood get nothing more (and nothing less) than the satisfaction of knowing they have helped someone in need.

To donate blood, you must be at least 17 years old (16 in some states) and weigh at least 110 pounds. You'll be given a physical examination and asked for your health history, including a confidential questionnaire about your sexual history and recent international travel. This is not done to embarrass you but to ensure that it is safe for you to give blood and that your blood will be safe for others. You will not be allowed to donate if you are pregnant,

for example, and you may not be allowed to donate if you have traveled recently to a malaria-infected area or if you have ever received a blood transfusion in (or were born in) certain countries. For more specific information, consult the American Red Cross.

The blood withdrawal procedure itself is relatively painless (a needle is inserted into an arm vein) and takes about 10–20 minutes. All needles used are brand-new and sterile—you cannot catch AIDS or any other bloodborne disease by donating blood. Afterward, you'll be advised to drink and eat something and avoid rigorous physical exercise for the rest of the day. This is not the best day to go mountain climbing, but just about anything less strenuous is OK.

Most donors are allowed to give only one unit of blood (1 pint, about 10% of

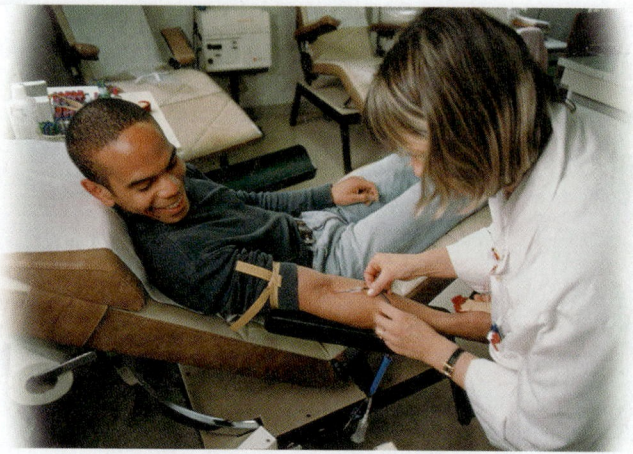

Donating blood. The procedure is relatively painless, takes only a short time, and can save lives.

your blood volume). This is not enough to affect you adversely. Usually the donated blood volume is replaced within several hours by any fluids that you drink. The liver replaces the lost plasma proteins within two days, and stem cells in bone marrow replace the lost RBCs in about a month.

What happens to the blood you give? Sometimes, it is stored as whole blood, but more often it is separated into three components: packed cells, platelets, and plasma. Each component may be given to different recipients, meaning that your single "gift of life" can benefit several people.

The other type of oxygen-carrying blood substitute seeks to take advantage of the high oxygen-carrying capacity of *perfluorocarbons* (PFCs), a group of colorless liquids containing only carbon and fluorine. PFCs are not soluble in water, so they must be dispersed (emulsified) as small droplets in water. Each droplet is about 1/40th the size of a red blood cell, making them small enough to travel through the smallest blood vessels. An aqueous solution containing PFC can carry several times more oxygen than whole blood, and because PFCs are completely man-made and can be heat-sterilized, they carry virtually no risk of transmitting disease. They could also be manufactured in unlimited quantities and stored for prolonged periods without refrigeration. However, a distinct disadvantage of the PFCs is that they are removed from the bloodstream within two days by exhalation and evaporation via the lungs. PFCs currently are used only as a temporary substitute for blood.

Several commercial hemoglobin-based and PFC-based products have been developed and tested in clinical trials, only to be withdrawn due to safety concerns. The bottom line is that there is still no ideal substitute for whole blood.

Research into blood substitutes continues, however, so don't be surprised if there is a viable blood substitute in commercial use in your lifetime.

Recap Blood substitutes currently under investigation include modified hemoglobin products and perfluorocarbons.

7.5 Blood disorders

Blood disorders include infections, several types of cancers, and disorders that affect the ability of the blood to deliver oxygen to the tissues or to clot properly when injury occurs. The effects of blood disorders are often widespread because blood passes through every organ in the body.

Mononucleosis: contagious viral infection of lymphocytes

Mononucleosis is a contagious infection of lymphocytes in blood and lymph tissues caused by the Epstein–Barr virus, a relative of the virus that causes herpes. Most common during adolescence, "mono" is nicknamed the "kissing disease" because it's frequently spread through physical contact.

Symptoms of mononucleosis can mimic those of the flu: fever, headache, sore throat, fatigue, and swollen tonsils and lymph nodes. A blood test reveals increased numbers of monocytes and lymphocytes. The disease is called mononucleosis because many of the lymphocytes enlarge and begin to resemble monocytes. There is no known cure for mononucleosis, but almost all patients recover on their own within four to six weeks. Extra rest and good nutrition help the body overcome the virus.

Blood poisoning: bacterial infection of blood

Although blood normally is well defended by the immune system, occasionally bacteria may invade the blood, overwhelm its defenses, and multiply rapidly in blood plasma. The bacteria may be toxic themselves, or they may secrete toxic chemicals as by-products of their metabolism. A bacterial infection of blood is called *blood poisoning*, or *septicemia*.

Blood poisoning may develop from infected wounds (especially deep puncture wounds), severe burns, urinary system infections, or major dental procedures. To help prevent it, wash wounds and burns thoroughly with soap and water. Consult your doctor immediately when an infection is accompanied by flushed skin, chills and fever, rapid heartbeat, or shallow breathing. An early sign of some blood poisonings is the sudden appearance of red streaks on healthy skin near the site of an infection (**Figure 7.14**). The red streaks are due to inflammation of veins or lymph vessels in the area, indicating that the infection is spreading toward the systemic circulation. Although blood poisonings can be very dangerous (even fatal) if left untreated, in most cases they can be treated effectively with antibiotics.

Anemia: reduction in blood's oxygen-carrying capacity

Anemia is a general term for reduction in the oxygen-carrying capacity of blood. All causes of anemia produce similar symptoms: pale skin, headaches, fatigue, dizziness, difficulty breathing, and heart palpitations—the uncomfortable feeling that one's heart is beating too fast as

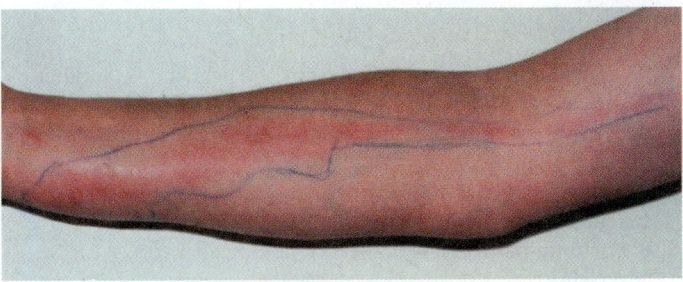

Figure 7.14 Blood poisoning.

it tries to compensate for the lack of oxygen delivery. Major types of anemia include the following:

- *Iron-deficiency anemia.* Recall that every hemoglobin molecule contains four molecules of iron. When the body is deficient in iron, hemoglobin cannot be synthesized properly. The result is fewer hemoglobin molecules per red blood cell, and thus a decreased ability to transport oxygen. Iron-deficiency anemia is the most common form of anemia worldwide. Usually, it is due to too little iron in the diet, but it can also be caused by an inability of the digestive tract to absorb iron properly. Generally, it can be treated by taking pills that contain iron or by eating foods rich in iron such as leafy green vegetables and meat.
- *Hemorrhagic anemia.* Anemia due to blood loss (hemorrhage) may be caused by injuries, bleeding ulcers, excessive menstrual flow, and even certain parasites. Treatment includes finding and treating the underlying cause of blood loss, if possible, and making sure one has enough iron in the diet to replenish the lost red blood cells.
- *Pernicious anemia.* Pernicious anemia is caused by a deficiency of vitamin B_{12} absorption by the digestive tract. Vitamin B_{12} is important for the production of normal red blood cells. Pernicious anemia can be treated by injections of B_{12}.
- *Hemolytic anemia.* Hemolytic anemia is the result of rupture (lysis) or early destruction of red blood cells. One cause is sickle-cell disease, an inherited disorder in which the red blood cells take on an abnormal sickle shape when the oxygen concentration is low. Because of their abnormal shape, sickled red blood cells become damaged as they travel through small blood vessels. Once damaged, they are destroyed by the body. Sickle-cell anemia is most prevalent in Africans who live near the equator and in African Americans. Another common cause of hemolytic anemia is the parasite that causes malaria.
- *Anemia due to renal failure.* When the kidneys fail, they do not produce enough erythropoietin to maintain normal red blood cell production. In this case, the anemia is secondary to the renal failure and the attendant decline in erythropoietin, not the primary problem. This type of anemia is easily corrected by treatment with exogenous erythropoietin, called EPO, which maintains red blood cell production within normal limits.

✓ Why does exogenous erythropoietin (EPO) effectively treat anemia due to renal failure but is ineffective in treating the other four types?

Leukemia: uncontrolled production of white blood cells

Leukemia refers to any of several types of blood cancer. Their common characteristic is uncontrolled proliferation of abnormal or immature white blood cells in the bone marrow. Overproduction of abnormal WBCs crowds out the production of normal white blood cells, red cells, and platelets. Huge numbers of leukemia cells enter and circulate in the blood, interfering with normal organ function.

There are two major categories of leukemia: acute, which develops rapidly; and chronic, which develops slowly. Both are thought to originate in the mutation of a white blood cell (a change in genetic structure) that results in uncontrolled cell division, producing billions of copies of the abnormal cell. Possible causes for the original mutation include viral infection or exposure to radiation or harmful chemicals. Genetic factors may also play a role.

Leukemia can produce a wide range of symptoms. Tissues may bruise easily because of insufficient production of platelets. Anemia may develop if the blood does not contain enough red blood cells. Bones may feel tender because the marrow is packed with immature white blood cells. Some people experience headaches or enlarged lymph nodes.

Treatment can cure leukemia in some cases and prolong life in others. Treatment generally involves radiation therapy and chemotherapy to destroy the rapidly proliferating cancer cells. This kills the normal stem cells as well, so transplants of bone marrow tissue are required to provide new stem cells. Cord blood transplants may be another option (see Current Issue, *Should You Bank Your Baby's Cord Blood?*). As with blood transfusions, all tissue must undergo testing to make sure the donor's antigens are compatible with those of the patient.

Multiple myeloma: uncontrolled production of plasma cells

Like leukemia, *multiple myeloma* is a type of cancer. In this case, abnormal plasma cells in the bone marrow undergo uncontrolled division. Plasma cells are a type of lymphocyte responsible for making a specific antibody. The proliferating plasma cells manufacture too much of an abnormal, frequently incomplete antibody, impairing production of other antibodies and leaving the body vulnerable to infections. Bones become tender as healthy bone marrow is crowded out by malignant plasma cells. Levels of calcium in the blood soar as bone tissue is destroyed. Treatment includes anticancer drugs and radiation therapy.

Thrombocytopenia: reduction in platelet number

Thrombocytopenia is a reduction in the number of platelets in the blood. Thrombocytopenia can occur for a number of reasons, such as viral infection, anemia, leukemia, other blood disorders, exposure to X-rays or radiation, and even as a reaction to certain drugs. Sometimes platelet levels decline for no apparent reason, in which case, they often rise again after several weeks.

Symptoms include easy bruising or bleeding, nosebleeds, bleeding in the mouth, blood in urine, and heavy menstrual periods. Treatment of the underlying cause generally improves the condition. If it persists, surgical removal of the spleen often helps.

Recap Blood poisoning and mononucleosis are types of blood infection. Several factors, including iron deficiency or hemorrhage, can lead to a reduction in oxygen-carrying capacity of blood. Leukemia and multiple myeloma are blood cell cancers that arise when abnormal cells in the bone marrow divide uncontrollably. Thrombocytopenia, a disease of too few platelets, is characterized by easy bleeding or bruising. ∎

Chapter Summary

7.1 The composition and functions of blood *p. 142*

- Blood consists of formed elements and plasma. Blood has transport, regulatory, and protective functions.
- Plasma contains numerous plasma proteins involved in transport, regulation of water balance, and protection. It also contains ions, hormones, nutrients, wastes, and gases.
- Erythrocytes (RBCs) are highly specialized for the transport of oxygen, but they also transport some carbon dioxide.
- Hemoglobin is the primary protein in red blood cells and gives blood its oxygen-carrying capacity.
- The formed elements of blood all originate from stem cells in red bone marrow.
- Leukocytes (WBCs) defend the body against disease and the effects of injury.
- RBCs and WBCs have short life spans and must continually be replaced. RBC production is stimulated when the body detects low oxygen levels in the blood.
- Platelets are cell products that participate in blood hemostasis.

7.2 Hemostasis: stopping blood loss *p. 148*

- Hemostasis is a three-phase process that prevents blood loss through damaged vessels. The phases are (1) vascular spasm, (2) the formation of a platelet plug, and (3) blood clotting.
- During the formation of a blood clot, substances released by damaged blood vessels cause soluble proteins called *fibrinogen* to become insoluble protein threads called *fibrin*. The threads form an interlocking mesh of fibers, trapping blood cells and sealing ruptured vessels.

7.3 Human blood types *p. 150*

- Successful transfusion of blood from one person into another depends on compatibility of their blood types, which is determined by antibodies in plasma and surface antigens on red blood cells.
- Blood types are classified primarily on the basis of the ABO system and the presence or absence of the Rh factor.
- Rh factor in particular can affect certain pregnancies adversely.

7.4 **Blood substitutes** *p. 153*

- Hemoglobin-based blood substitutes are made from modified hemoglobin derived from animals or genetically engineered bacteria.
- Perfluorocarbon-based blood substitutes are emulsions of color-less liquids with a high oxygen-carrying capacity, called perfluo-rocarbons (PFCs).

7.5 **Blood disorders** *p. 154*

- Mononucleosis is a contagious viral disease of lymphocytes and lymphatic tissue.
- *Blood poisoning* is a general term for infection of blood plasma by various microorganisms.
- Anemia is a reduction in blood oxygen-carrying capacity for any number of reasons, including insufficient red blood cell or hemoglobin production and excessive blood loss.
- Leukemia is a cancer characterized by uncontrolled production of abnormal leukocytes (white blood cells).

Terms You Should Know

anemia, *155*
blood type, *150*
erythrocyte (RBC), *143*
erythropoietin, *146*
fibrin, *149*
hematocrit, *144*
hemoglobin, *144*
hemostasis, *148*

leukocytes (WBC), *146*
phagocytosis, *146*
plasma, *142*
plasma proteins, *142*
platelet, *148*
Rh factor, *151*
stem cell, *145*

Concept Review

Answers can be found in the Study Area in MasteringBiology.

1. Describe the functions of blood.
2. Describe the role of hemoglobin in the transport of oxygen and carbon dioxide.
3. Explain how the production of red blood cells is regulated to maintain homeostasis of the oxygen-carrying capacity of blood.
4. Define *hematocrit,* and explain why it is important.
5. Describe how damaged or dead RBCs and the hemoglobin they contain are removed from the blood.
6. Describe the difference between the actions of neutrophils and eosinophils.
7. Describe the mechanism of hemostasis.
8. List the four ABO blood types. For each one, list its red blood cell surface antigen(s) and plasma antibody (antibodies).
9. Describe the Rh factor and its implications for pregnancy.
10. Compare and contrast the various causes of anemia.

Test Yourself

Answers can be found in the Appendix.

1. All of the following proteins are associated with blood. Which of these is found specifically inside red blood cells?
   a. prothrombin
   b. fibrinogen
   c. albumin
   d. hemoglobin

2. Which of the following blood components protects the individual from a variety of infectious agents such as bacteria and viruses?
   a. white blood cells
   b. platelets
   c. albumin
   d. red blood cells

3. Which of the following make(s) up the greatest volume of whole blood?
   a. platelets
   b. red blood cells
   c. plasma
   d. white blood cells

4. Which of the following influence(s) the bonding of oxygen to hemoglobin?
   a. pH
   b. oxygen concentration
   c. temperature
   d. all of the above

5. Jason has just spent four weeks in Rocky Mountain National Park, studying plants that grow above 10,000 feet elevation. Which of the following would be a likely change in his blood because of time spent at high elevation?
   a. increased number of red blood cells
   b. increased number of white blood cells
   c. increased number of platelets
   d. increased amount of globulins in the plasma

6. A person with Type A− (A-negative) blood will have:
   a. type A plasma antibodies
   b. type A antigens on the red blood cells
   c. Rh antigens on the red blood cells
   d. all of the above

7. A deficiency of platelets would result in:
   a. fatigue and dizziness
   b. bleeding and bruising
   c. increased susceptibility to infections
   d. all of the above

8. Which donor blood type would be most appropriate for trans-fusing an O− recipient?
   a. A−
   b. B−
   c. O−
   d. Any of these blood types could be successfully used for this recipient.

9. What do erythroblasts, myeloblasts, lymphoblasts, and mega-karyoblasts have in common?
   a. They are immature cells that develop into white blood cells.
   b. They are immature cells that develop into red blood cells.
   c. They are found in the circulating blood.
   d. They are immature cells found in the bone marrow.

10. Jaundice is caused by the presence of _____ in the blood plasma, which is a breakdown product of _____.
   a. hemoglobin . . . red blood cells
   b. bilirubin . . . hemoglobin
   c. albumin . . . white blood cells
   d. prothrombin . . . platelets

11. Which white blood cells are present in the greatest number in the blood and are the body's first responders to infection?
    a. neutrophils
    c. platelets
    b. lymphocytes
    d. monocytes

12. The steps in the hemostasis process are (1) platelets become sticky and form a platelet plug, (2) walls of a damaged blood vessel undergo spasms, (3) a clot forms from fibrin, platelets, and trapped red blood cells. Which of the following choices represents the correct order of these steps?
    a. 1-3-2
    c. 2-3-1
    b. 3-1-2
    d. 2-1-3

13. Hemophilia results from a(n):
    a. insufficient number of red blood cells
    c. lack of one or more plasma proteins involved in blood clotting
    b. insufficient number of platelets
    d. abnormal type of hemoglobin

14. Which of the following can lead to anemia?
    a. insufficient iron in the diet
    c. spending several weeks at a high altitude
    b. insufficient Vitamin B$_{12}$ absorption from the digestive tract
    d. both (a) and (b)

15. Which property do red blood cells and platelets have in common?
    a. Both lack a nucleus.
    b. Both transport oxygen.
    c. They are found in approximately equal numbers in the circulating blood.
    d. Both are derived from erythroblasts.

Apply What You Know

Answers can be found in the Study Area in MasteringBiology.

1. A 35-year-old white male is sent by his physician for a blood test. The lab results indicate his white blood cell count (number of WBCs per milliliter of blood) is 18,000. The typical WBC count for a man his age is 6,000–9,000, meaning his WBC count is considerably higher than normal. What may this mean?

2. One treatment for certain types of leukemia is to try to kill all of the stem cells in bone marrow through radiation and chemotherapy and then to give a bone marrow transplant from another person (a donor). Can just anyone be the donor? Who is most likely to be a good donor? Explain.

3. In the not too distant past, people with type O-negative blood were considered to be universal blood donors, and their blood was sought out during times of need. Explain what was meant by the term *universal donor*, why O-negative persons were considered to be universal donors, and why *universal donor* is now considered an outdated term.

4. The text states that when red blood cells reach actively metabolizing tissues, they release their cargo of oxygen because both the oxygen concentration and the pH are lower in metabolically active tissues than in the general circulation. The oxygen concentration is lower because actively metabolizing tissues are using oxygen at a rapid rate. But what causes the pH to fall? And how might a fall in pH cause the hemoglobin to release oxygen? Can you think of any other variables that might also lead to the release of oxygen by hemoglobin?

5. The term has just ended. Over the past three weeks, you wrote three term papers, studied for finals, and went to work at your part-time job. You passed all your courses and never missed a day of work. Of course, you spent the past three weeks living on soda, frozen pizza, and Ramen noodles (who has time for real food with all this work).

    Now that the term is over you notice that you are very tired, your skin is pale, you are experiencing headaches and dizziness, and even your breathing is difficult. You go to your doctor who diagnoses you with anemia. What type of anemia do you most likely have, how is it different from other forms of anemia, and what treatment will your doctor most likely suggest?

6. Coumadin is an anticoagulant drug that is sometimes given to patients who have just suffered a deep vein thrombosis, a pulmonary embolism, a heart attack, or to patients with artificial heart valves. It helps reduce the chance of future clots and the further risk of embolism. The active ingredient in Coumadin is warfarin, a rat poison. How do you think the same compound can be used for these two very different purposes?

MJ's BlogInFocus

Answers can be found in the Study Area in MasteringBiology.

1. A drug has been discovered that prevents newly formed blood clots from breaking down. What sorts of patients or medical conditions might be helped with such a drug? Visit MJ's blog in the Study Area in MasteringBiology and look under "Drug Reduces Bleeding."
   http://goo.gl/hHvbo2

MasteringBiology®

Students Go to MasteringBiology for assignments, the eText, and the Study Area with animations, practice tests, and activities.

Professors Go to MasteringBiology for automatically graded tutorials and questions that you can assign to your students, plus Instructor Resources.

Answers to 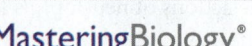 questions are available in the Appendix.

Heart and Blood Vessels

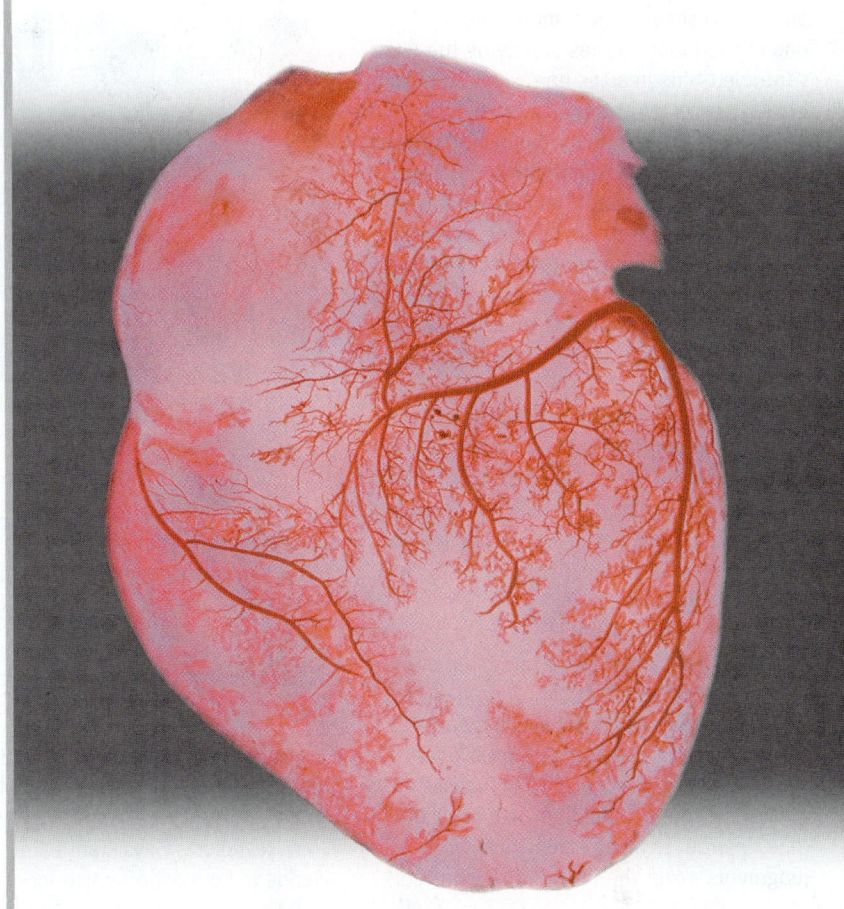

A human heart, showing the blood vessels that lie just beneath the surface.

Key Concepts

- **The structure of blood vessels reflects their function.** Thick-walled arteries and arterioles transport blood to the tissues under high pressure; capillaries allow fluid exchange between blood and interstitial fluid; large thin-walled veins store most of the blood and return it to the heart.

- **The heart is a pump composed primarily of muscle.** Its ability to pump blood depends on one-way valves and coordination of muscle contraction.

- **Arterial blood pressure is held fairly constant by homeostatic control mechanisms.** With arterial blood pressure held constant, blood flow to each tissue can be regulated by local control mechanisms.

- **Cardiovascular disorders are the number one cause of death in the United States.** Cardiovascular disorders include heart attack, heart failure, stroke, and cardiac arrhythmias.

- **Your risk of developing cardiovascular disease is affected by your lifestyle choices.** Risk factors include smoking, a lack of exercise, obesity, and chronic stress.

CURRENT ISSUE

How Should Comparative Effectiveness Research Be Used?

Mr. Reynolds has a heart problem. An angiogram shows that a short section of one of the main arteries supplying the left ventricle of his heart is narrowed, restricting blood flow to his heart muscle. His doctor tells him that he is at serious risk of a heart attack. The doctor explains that there are at least three techniques that could be used to restore blood flow to his heart: (1) balloon angioplasty, (2) a coronary artery stent, or (3) a coronary artery bypass graft (CABG). Which technique would be best for Mr. Reynolds? The doctor and Mr. Reynolds go over the options together, but the differences between the techniques are difficult for Mr. Reynolds to understand. He leaves the decision to his physician, whom he has known for 25 years. The physician chooses a coronary artery stent because it has worked best for his previous patients with heart disease.

The body of medical literature is now so vast and expanding so rapidly that even the best physicians can't know it all. This is where a relatively new field of medical science called Comparative Effectiveness Research (CER) comes in. CER focuses on analysis of the medical literature available to date, in order to reach scientifically sound judgments about the value (or lack thereof)

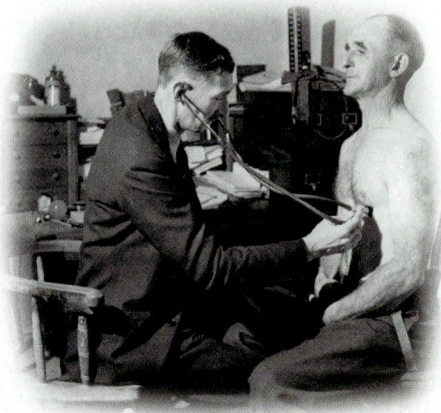

Doctoring, 1948.

of specific medical tests, treatments, and disease prevention strategies. In essence, CER seeks to determine the best practices in medicine based on our current knowledge.

Changing How Medicine Is Practiced

Consider how CER might benefit Mr. Reynolds's physician (and Mr. Reynolds, of course). By reviewing CER data, Mr. Reynolds's physician learns that a stent tends to be most effective for middle-aged

Questions to Consider

1 Who do you want to help you decide which treatment options would be best for you? If not a specific professional (doctor, patient representative, health insurance specialist), what other information would you like to have available to you?

2 Do you think cost-effectiveness should be a part of any comparative effectiveness analysis of treatment or diagnostic options? Why or why not?

white males with heart disease. However, the research shows that there's an age-related tipping point on the effectiveness of the technique and that the best treatment option depends on the severity of the narrowing: If the patient is over 55, the data indicates that balloon angioplasty is the best option. (Hmmm, how old *is* Mr. Reynolds this year?) If the degree of narrowing of a coronary artery is greater than 80%, for example, then the best option (again, for a middle-aged white male) would be a coronary artery bypass graft. (What is the degree of narrowing in Mr. Reynolds, anyway?) Toss in other factors like gender, race, physical condition, body weight, smoker-versus-nonsmoker, and you can begin to see the full power of CER. In theory, CER could analyze multiple factors at once to arrive at the best treatment option for patients who are described by a particular combination of factors. Even the most experienced physicians can't carry *that* much information around in their heads!

Some politicians believe that the federal government should invest in CER, because any money spent on the research now would be offset by reduced health care expenditures in the future. To jump-start a national CER program, Congress passed the Comparative Effectiveness Research Act of 2009 and funded it with $1.1 billion as part of the economic stimulus package. To keep the program free of bias, the prestigious Institute of Medicine of the National Academies of Science was asked to come up with a list of 100 top priority topics for CER funding. Among the topics are comparisons of the most effective practices to treat or prevent a number of cardiovascular diseases and risk factors, including high blood pressure, coronary artery disease, heart failure, ➔

Doctoring in the twenty-first century.

and abnormalities of heart electrical rhythm. This is not surprising because cardiovascular diseases are the number one cause of death in the United States (cancer is second).

Who Will Make Health Care Decisions?

CER could become a powerful tool for improving health care quality and lowering costs. Nevertheless, the CER Act of 2009 has stirred strong feelings among physicians, patients, politicians, and the health care industry because of the ways it could change how medicine is practiced. Physicians and patient advocacy groups worry that if "best practices" become defined by CER, doctors and patients could begin to lose the right to make decisions regarding treatment options. They fear that health care decisions may be dictated primarily by bureaucrats and insurance companies. In recognition of this concern, the CER Act includes language to the effect that the findings of CER research shall "not be construed as mandates for practice guidelines, coverage recommendations, payment, or policy recommendations." In other words, physicians and patients can still use their judgment in deciding the appropriate treatment option. But by the same token, neither private insurers nor our primary public health care system (Medicare) are required by law to pay for it.

Therein lie the big questions: Will physicians and patients continue to be the decision makers in medical treatment decisions? Or is it inevitable that the old way of practicing medicine is going to change? Do we really believe that health insurance companies, group health plans, and even Medicare/Medicaid will *not* find a way to use CER data to influence reimbursement policies and hence treatment decisions? Would it be a good thing or a bad thing if they did?

Flash forward 25 years. You're in the doctor's office, and the doctor is telling you that a scan of your heart shows a 63% narrowing of a section of your left-anterior descending coronary artery. She swings around to her computer, taps a few keys, and turns back to you to report that according to the latest data from the Comparative Effectiveness Research Institute, the most effective method for repair of your coronary artery is Robotic Artificial Vessel Extension (RAVE). A few more taps on her computer keyboard informs her that your government-supported health insurance will pay for the procedure and that there is an opening on the hospital's surgical schedule on Tuesday. Data further reveals that 99.7% of Dr. Sloan's RAVE surgeries have been successful and that 94% of all patients with your condition who undergo RAVE are discharged from the hospital on the same day as their surgery. You go ahead and book that vacation to London next month.

SUMMARY

- The medical literature is expanding so rapidly that even the best physicians can no longer keep up with it.
- Recognizing this, the government will spend $1.1 billion on Comparative Effective Research (CER) to determine the best practices in medicine based on our current knowledge and make that information available to everyone.
- CER could slow down rising health-care costs.
- A concern is that CER recommendations will eventually influence third-party payer reimbursement policies, so that patients and doctors will lose the ability to make treatment choices.

The heart and blood vessels play a critical role in the maintenance of homeostasis. Collectively known as the **cardiovascular system** (from the Greek *kardia*, heart, and the Latin *vasculum*, small vessel), the heart and blood vessels function as a centrally controlled blood distribution network. The heart provides the power to move the blood, and the vascular system represents the network of branching conduit vessels through which the blood flows. Central control (oversight and management, if you will) is provided by the nervous system. By controlling the rate at which the heart pumps and the resistance to flow through blood vessels, the nervous system apportions blood flow to the various tissues and organs according to need.

Homeostasis is not maintained by just the cardiovascular and nervous systems, however; other organ systems are also involved. The digestive system delivers nutrients to the blood passing through its blood vessels. The kidneys of the urinary system remove excess salt, water, and the waste products of cellular metabolism. Glands of the endocrine system secrete hormones into the blood. The respiratory system provides life-sustaining oxygen and removes the waste by-product carbon dioxide. Even the skin is involved in homeostasis, by adjusting the amount of heat lost from the body. ▪

We'll start the chapter by considering the blood vessels and the structure and function of the heart. Then we'll describe how the cardiovascular system is regulated. Finally, we'll take a look at some major cardiovascular disorders.

8.1 Blood vessels transport blood

A branching network of blood vessels transports blood to all parts of the body. The network is so extensive that if our blood vessels were laid end to end, they would stretch 60,000 miles!

We classify the body's blood vessels into three major types: *arteries, capillaries,* and *veins*. Thick-walled arteries transport blood to body tissues under high pressure. Microscopic capillaries exchange solutes and water with the cells of the body. Thin-walled veins store blood and return it to the heart. **Figure 8.1** illustrates the structures of each type of blood vessel, described in more detail below.

Arteries transport blood away from the heart

As blood leaves the heart, it is pumped into large, muscular, thick-walled **arteries.** Arteries transport blood away from the heart. The larger arteries have a thick layer of muscle because they must be able to withstand the high pressures generated by the heart. Arteries branch again and again, so the farther blood moves from the heart, the smaller in diameter the arteries become.

Large- and medium-sized arteries are like thick garden hoses, stiff yet somewhat elastic (distensible). Arteries stretch a little in response to high pressure but are strong enough to withstand high pressures year after year. The ability to stretch under pressure is important because a function of arteries is to store the blood that is pumped into them with each beat of the heart and then provide it to the capillaries (at high pressure) even between heartbeats. The elastic recoil of arteries is the force that maintains the blood pressure between beats. Think of the arteries as analogous to a city's water system of branching, iron or steel pipes that provide nearly constant water pressure to nearly every building in the vicinity.

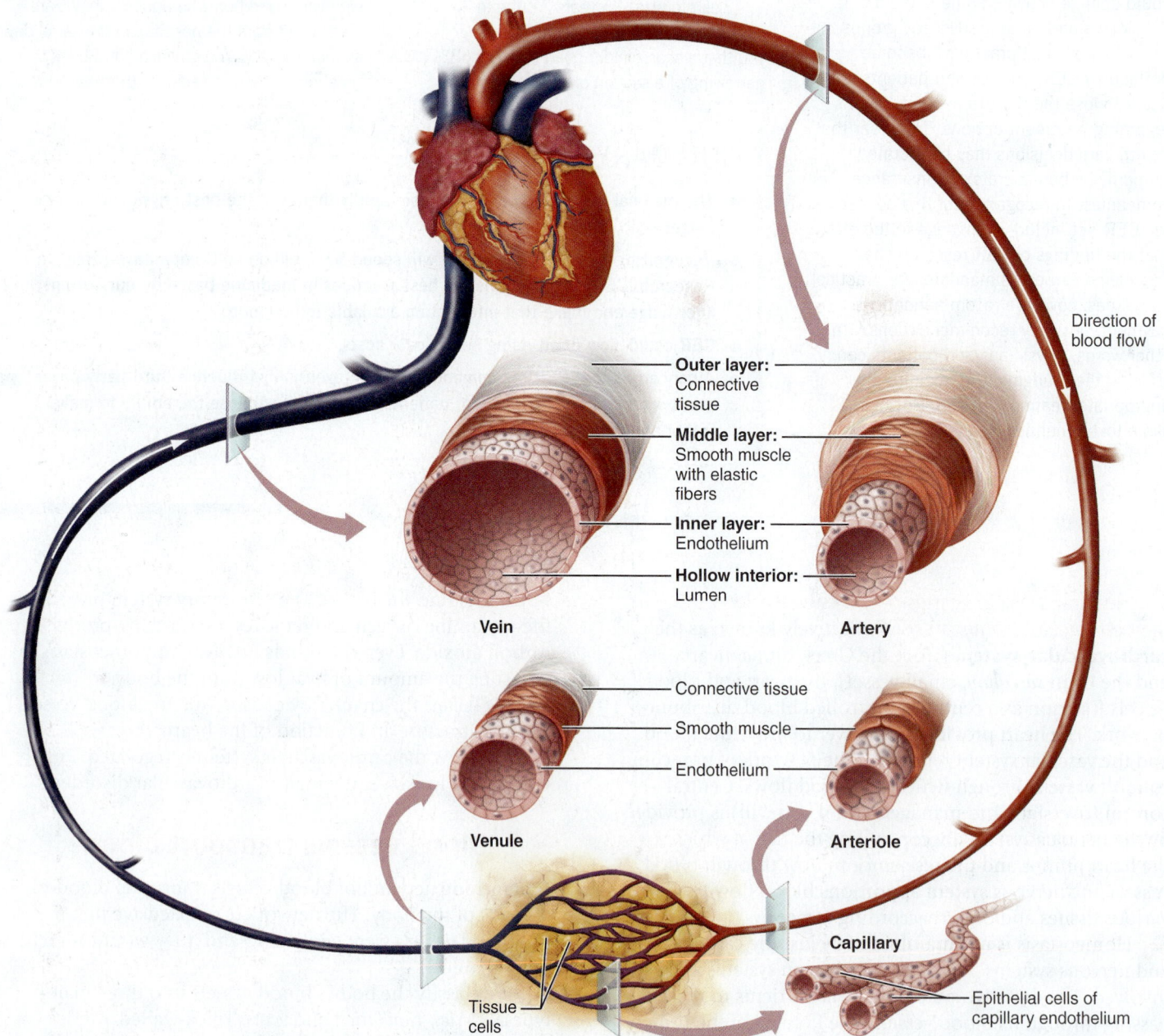

Figure 8.1 **The structures of blood vessels in the human body.**

The structure of the walls of large and medium-sized arteries is ideally suited to their functions. The vessel wall is a sandwich of three distinct layers surrounding the *lumen*, or hollow interior of the vessel:

1. The thin inner layer, the *endothelium*, is a layer of flat-tened, *squamous* epithelial cells. It is a continuation of the lining of the heart. The flattened cells fit closely together, creating a slick surface that keeps friction to a minimum and promotes smooth blood flow.

2. Just outside the endothelium is a layer composed primarily of smooth muscle with interwoven elastic connective tissue. In most arteries this is the thickest of the three layers. Steady partial contraction of the smooth muscle of large and medium-sized arteries stiffens the arteries and helps them resist the high pressures within, but it does not constrict them enough to alter blood flow. The elastic tissue makes large and medium-sized arteries slightly distensible so they can stretch passively to accommodate the blood that enters with each heartbeat.

3. The outermost layer of large and medium-sized arteries consists of a tough supportive layer of connective tissue, primarily collagen. This sturdy casing anchors vessels to surrounding tissues and helps protect them from injury.

The fact that arteries are constantly under high pressure places them at risk of injury. If the endothelium becomes damaged, blood may seep through the injured area and work its way between the two outer layers, splitting them apart. The result is an *aneurysm*, or ballooning of the artery wall. Some aneurysms cause the smooth muscle and endothelial layers to bulge inward as they develop, narrowing the lumen enough to reduce blood flow to an organ or region of the body. Others force the outer connective tissue layer to bulge outward. Sometimes aneurysms cause severe chest pain, but in other cases they are completely symptomless until they rupture or "blow out," causing massive internal bleeding and often death. If you've ever seen a water line burst, you know how quickly it can be devastating. Aneurysms of the aorta (see section 8.2) kill an estimated 25,000 Americans every year. Actor John Ritter's sudden death in 2003 was caused by a ruptured aneurysm.

Aneurysms often take years to develop. During this time, many can be detected and repaired surgically. Some physicians recommend that anyone with a family history of aneurysm should be examined, even if there are no symptoms. Doctors can sometimes detect inward-bulging aneurysms with a stethoscope (an instrument for listening to sounds inside the body) because flowing blood produces characteristic sounds as it passes through a narrowed arterial lumen. A computerized tomography (CT) scan may also locate aneurysms before they rupture.

Arterioles and precapillary sphincters regulate blood flow

Eventually blood reaches the smallest arteries, called **arterioles** (literally, "little arteries"). The largest artery in the body, the *aorta*, is about 2.5 centimeters (roughly 1 inch)

wide. In contrast, arterioles have a diameter of 0.3 millimeter or less, about the width of a piece of thread.

By the time blood flows through the arterioles, blood pressure has fallen considerably. Consequently, arterioles can be simpler in structure. Generally, they lack the outermost layer of connective tissue, and their smooth muscle layer is not as thick. In addition to blood transport and storage, arterioles have a third function not shared by the larger arteries: They help regulate the amount of blood that flows to each capillary. They do this by contracting or relaxing the smooth muscle layer, altering the diameter of the arteriole lumen.

Right where an arteriole joins a capillary is a band of smooth muscle called the **precapillary sphincter** (**Figure 8.2**). The precapillary sphincters serve as gates that control blood flow into individual capillaries.

Relaxation of vascular smooth muscle is called *vasodilation*. Vasodilation of arterioles and precapillary sphincters increases their diameter and thus increases blood flow to the capillaries. Conversely, contraction of vascular smooth muscle is called *vasoconstriction*. Vasoconstriction of arterioles and precapillary sphincters reduces their diameter and thus reduces blood flow to the capillaries.

A wide variety of external and internal factors can produce vasodilation or vasoconstriction, including nerves, hormones, and conditions in the local environment of the arterioles and precapillary sphincters. If you go outside on a cold day, you may notice that your fingers start to look pale. This is because vasoconstriction produced by nerves is narrowing your vessels

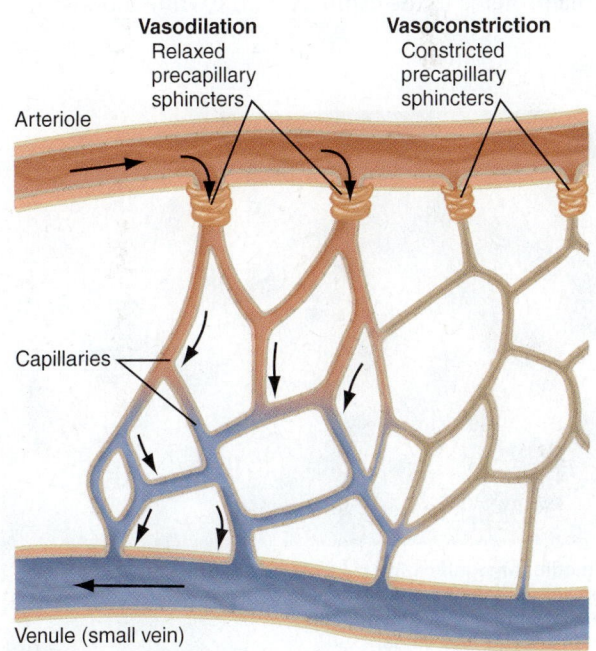

Figure 8.2 Precapillary sphincters control the flow of blood into individual capillaries. In this diagram, the two precapillary sphincters on the right are vasoconstricted, reducing flow in that region. Arrows indicate direction of blood flow.

to reduce heat loss from your body. On the other hand, hot weather will make your skin appear flushed as vasodilation occurs to speed up heat loss and cool you off. Emotions can also have an impact: Vasodilation is partly responsible for the surge in blood flow that causes the penis or clitoris to become erect when we are sexually aroused. Later in this chapter, we will talk more about how the cardiovascular system is regulated to maintain homeostasis.

Capillaries: where blood exchanges substances with tissues

Arterioles connect to the smallest blood vessels, called **capillaries** (Figure 8.3a). Capillaries are thin-walled vessels that average only about one-hundredth of a millimeter in diameter—not much wider than the red blood cells that travel through them. In fact, they are so narrow that red blood cells (RBCs) often have to pass through them in single file or even bend to squeeze through (Figure 8.3b).

Extensive networks of capillaries, called *capillary beds*, can be found in all areas of the body, which is why you are likely to bleed no matter where you cut yourself. The branching design of capillaries and their thin, porous walls allow blood to exchange oxygen, carbon dioxide, nutrients, and waste products with tissue cells. Capillary walls consist of a single layer of squamous epithelial cells (Figure 8.3c). Microscopic pores pierce this layer, and the cells are separated by narrow slits. These openings are large enough to allow the exchange of fluid and other materials between blood and the interstitial fluid (the fluid that surrounds every living cell), yet small enough to retain RBCs and most plasma proteins in the capillary. Some white blood cells

(WBCs) can also squeeze between the cells in capillary walls and enter the tissue spaces.

In effect, capillaries function as biological strainers that permit selective exchange of substances with the interstitial fluid. In fact, capillaries are the *only* blood vessels that can exchange materials with the interstitial fluid.

Figure 8.4 illustrates the general pattern of how water and substances move across a capillary. At the beginning of a capillary, fluid is filtered out of the vessel into the interstitial fluid, accompanied by oxygen, nutrients, and raw materials needed by the cell. The filtered fluid is essentially like plasma except that it contains very little protein because most protein molecules are too large to be filtered. Filtration of fluid is driven by the blood pressure generated by the heart. Waste materials such as carbon dioxide and urea (a nitrogen-containing substance) diffuse out of the cells and back into the blood.

Most of the filtered fluid is reabsorbed by diffusion back into the last half of the capillary before it joins a vein. The force for this reabsorption is the presence of protein in the blood but not in the interstitial fluid. In other words, fluid diffuses from an area of high water concentration (interstitial fluid) to an area of lower water concentration (blood plasma). However, the diffusional reabsorption of fluid does not quite match the pressure-induced filtration of fluid, so a small amount of filtered fluid remains in the interstitial space as excess interstitial fluid.

✅ Why doesn't exchange of gases and nutrients with the interstitial fluid occur in arteries and arterioles too, instead of just in capillaries? Put another way, what about the structure of an artery or an arteriole prevents such exchange from occurring?

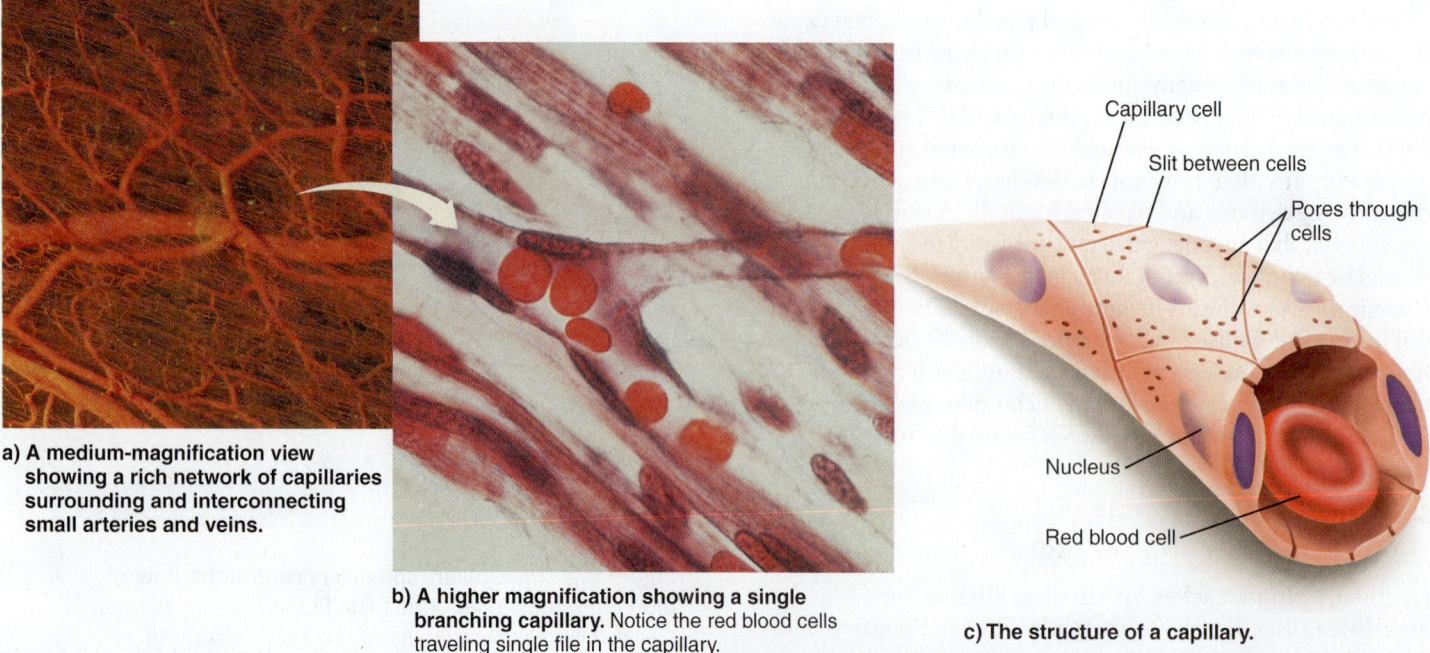

a) **A medium-magnification view showing a rich network of capillaries surrounding and interconnecting small arteries and veins.**

b) **A higher magnification showing a single branching capillary.** Notice the red blood cells traveling single file in the capillary.

Capillary cell

Slit between cells

Pores through cells

Nucleus

Red blood cell

c) **The structure of a capillary.**

Figure 8.3 Capillaries.

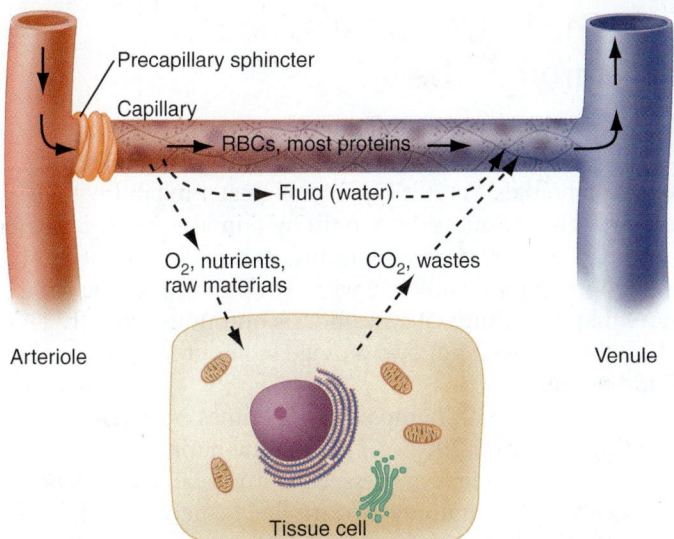

Figure 8.4 **The general pattern of fluid movement between capillaries, the interstitial fluid, and cells.** For simplicity, only a single tissue cell is shown, but a single capillary may supply many nearby cells.

✅ Why does most of the fluid that was filtered out of the proximal (beginning) end of the capillary move back into the distal (far) end?

The lymphatic system helps maintain blood volume

Although the imbalance between the amount of plasma fluid filtered by the capillaries and the amount reabsorbed is not large, over the course of a day it would amount to about 2 or 3 liters. This excess plasma fluid must be returned to the cardiovascular system somehow, or all the plasma would end up in the interstitial fluid.

The excess plasma fluid is absorbed by blind-ended capillaries that join together into a collection system of larger vessels, collectively called the *lymphatic system.* The lymphatic system is somewhat similar to (and nearly parallel to) the venous system of blood vessels, except that the fluid of the lymphatic system (called *lymph*) does not contain plasma proteins or RBCs. In addition to absorbing excess interstitial fluid, the capillaries of the lymphatic system also pick up objects in the interstitial fluid that are too large to diffuse into the blood capillaries, such as lipid droplets absorbed during digestion and invading microorganisms. The lymphatic system transports the lymph back to veins near the heart, where the lymph rejoins the venous blood.

Aside from its role in returning excess fluid and large objects to the blood, the lymphatic system also plays a major role in our immune defenses. You'll hear more about the lymphatic system when we discuss the immune system. For now, just be aware that the lymphatic system, though technically not part of the cardiovascular system, plays a vital role in maintaining the proper volumes of blood and interstitial fluid.

✅ There are certain parasitic worms that can enter the lymphatic system and completely block the lymphatic capillaries draining an arm or a leg. Predict what would happen to the arm or leg if this occurs.

Veins return blood to the heart

From the capillaries, blood flows back to the heart through *venules* (small veins) and **veins** (see Figures 8.1 and 8.4). Like the walls of arteries, the walls of veins consist of three layers of tissue. However, the outer two layers of the walls of veins are much thinner than those of arteries. Veins also have a larger diameter lumen than arteries.

The anatomical differences between arteries and veins reflect their functional differences. As blood moves through the cardiovascular system, blood pressure becomes lower and lower. The pressure in veins is only a small fraction of the pressure in arteries, so veins do not need nearly as much wall strength as arteries. The larger diameter and high distensibility of veins allows them to stretch like thin balloons to accommodate large volumes of blood at low pressures.

In addition to their transport function, then, veins serve as a blood volume reservoir for the entire cardiovascular system. Nearly two-thirds of all the blood in your body is in your veins. Thanks to their blood reservoir function, even if you become dehydrated or lose a little blood, your heart will still be able to pump enough blood to keep your blood pressure fairly constant.

The distensibility of veins, however, can lead to problems in returning blood to the heart against the force of gravity. When you stand upright, blood tends to collect in the veins of your legs and feet. People who spend a lot of time on their feet may develop *varicose veins,* permanently swollen veins that look twisted and bumpy from pooled blood. Varicose veins can appear anywhere, but they are most common in the legs and feet. In severe cases, the skin surrounding veins becomes dry and hard because the tissues are not receiving enough blood. Often, varicose veins can be treated by injecting an irritating solution that shrivels the vessels and makes them less visible. This should not affect blood flow because surrounding undamaged veins take over and return blood to the heart.

Fortunately, three mechanisms assist the veins in returning blood to the heart: (1) contractions of skeletal muscles, (2) one-way valves inside the veins, and (3) movements associated with breathing. Let's look at each in turn.

Skeletal muscles squeeze veins
On their path back to the heart, veins pass between many skeletal muscles. As we move and these muscles contract and relax, they press against veins and collapse them, pushing blood toward the heart. You may have noticed that you tire more easily when you stand still than when you walk around. This is because walking improves the return of blood to your heart and prevents fluid accumulation in your legs. It also increases blood flow and the supply of energy to your leg muscles.

One-way valves permit only one-way blood flow
Most veins contain valves consisting of small folds of the inner layer that protrude into the lumen. The structure of these valves allows blood to flow in one direction only: toward the heart. They open passively to permit blood to move toward the heart and then close whenever blood begins to flow backward. Together, skeletal muscles and valves form

what is called the "skeletal muscle pump" (Figure 8.5). Once blood has been pushed toward the heart by skeletal muscles or drained in that direction by gravity, it cannot drain back again because of these one-way valves. The opening and closing of venous valves is strictly dependent on differences in blood pressure on either side.

Pressures associated with breathing push blood toward the heart The third mechanism that assists blood flow involves pressure changes in the thoracic (chest) and abdominal cavities during breathing. When we inhale, abdominal pressure increases and squeezes abdominal veins. At the same time, pressure within the thoracic cavity decreases, dilating thoracic veins. The result is to push blood from the abdomen into the chest and toward the heart. This effect is sometimes called the "respiratory pump."

> ↩ **Recap** A branching system of thick-walled arteries distributes blood to every area of the body. Arterioles regulate blood flow to local regions, and precapillary sphincters regulate flow into individual capillaries. Capillaries consisting of a single layer of cells exchange materials with the interstitial fluid. The lymphatic system removes excess fluid. The thin-walled veins return blood to the heart and serve as a volume reservoir for blood. ■

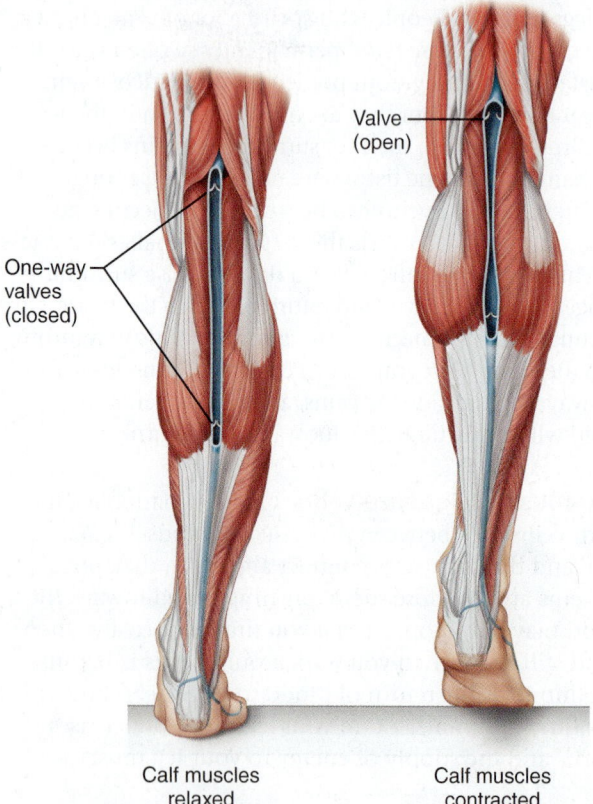

Figure 8.5 The skeletal muscle pump. With the calf muscle relaxed, blood accumulates in the vein, and backflow is prevented by one-way valves. When the calf muscle contracts, skeletal muscles press on the vein, forcing blood toward the heart through the upper one-way valve, while the lower one-way valve remains closed, preventing backflow.

8.2 The heart pumps blood through the vessels

The human heart is a pump, but it's a very special pump indeed. The heart is constructed entirely of living cells and cellular materials, yet it is capable of greater reliability than some of the best pumps ever built by humans. It can easily withstand 80–100 years of continuous service without ever stopping for repairs and without ever even resting for more than two-thirds of a second. Its output is also fully adjustable on demand, over a range of about 5–25 liters of blood per minute.

At rest your heart pumps about 75 times every minute, speeding up to over 200 beats per minute during periods of exertion or in response to stress. Under normal circumstances, the heart's rate of pumping is controlled by nerve signals originating in the brain. But unlike skeletal muscle, the heart can beat on its own without any nerve signals at all.

The heart is mostly muscle

The human heart is a muscular, cone-shaped organ slightly larger than your fist (Figure 8.6). It consists mostly of a special type of muscle called *cardiac muscle*. Unlike skeletal muscle, which attaches to bone and requires nerve signals for contraction, cardiac muscle contracts spontaneously and is not connected to bone. The heart contracts in a cyclic, coordinated, squeezing motion that propels blood through the blood vessels.

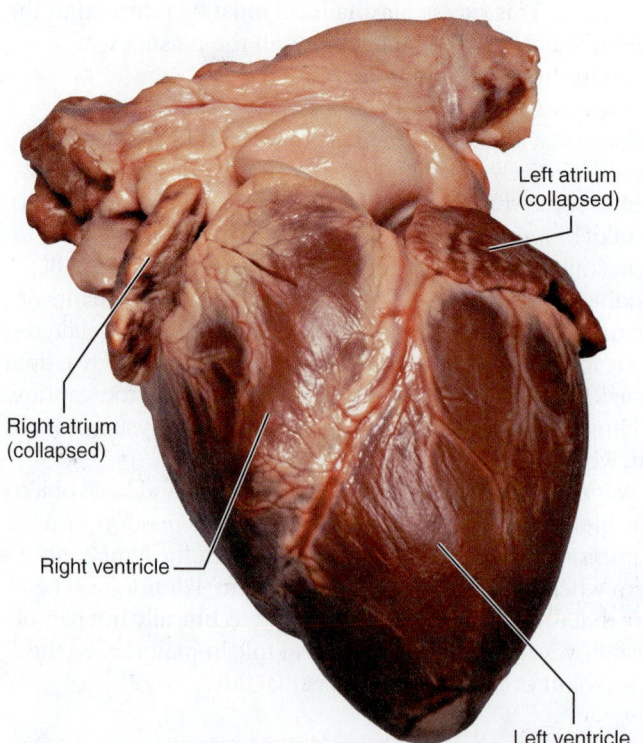

Figure 8.6 A human heart.

In its natural position within the chest cavity, the heart is closely surrounded by a tough fibrous sac called the *pericardium* (not shown in Figure 8.6). The pericardium protects the heart and anchors it to surrounding structures. The pericardium also prevents the heart from overfilling with blood, because although the heart is flexible, it is not very stretchable. Between the pericardium and the heart is a space called the *pericardial cavity.* The pericardial cavity contains a film of lubricating fluid that reduces friction and allows the heart and the pericardium to glide smoothly against each other when the heart contracts.

In cross section, we see that the walls of the heart consist of three layers: the epicardium, myocardium, and endocardium (**Figure 8.7**). The outermost layer, the *epicardium,* is a thin layer of epithelial and connective tissue. The middle layer is the **myocardium.** This is a thick layer consisting mainly of cardiac muscle that forms the bulk of the heart. The myocardium is the layer that contracts every time the heart beats. The structure of cardiac muscle cells allows electrical signals to flow directly from cell to cell. An electrical signal in one cardiac muscle cell can spread to adjacent cells, enabling large numbers of cells to contract as a coordinated unit. Every time the myocardium contracts, it squeezes the chambers inside the heart, pushing blood outward into the arteries. The innermost layer of the heart, the *endocardium,* is a thin endothelial layer resting on a layer of connective tissue. The endocardium is continuous with the endothelium that lines the blood vessels.

The heart has four chambers and four valves

Taking a closer look at the details of the structure of the heart, we see that it consists of four separate chambers (see Figure 8.7). The two chambers on the top are the **atria** (singular: *atrium*), and the two more-muscular bottom chambers are the **ventricles.** A muscular partition called the **septum** separates the right and left sides of the heart.

Blood returning to the heart from the body's tissues enters the heart at the *right atrium.* From the right atrium, the blood passes through a valve into the *right ventricle.* The right ventricle is more muscular than the right atrium because it pumps blood at considerable pressure through a second valve and into the artery leading to the lungs.

Blood returning from the lungs to the heart enters the *left atrium* and then passes through a third valve into the *left ventricle.* The very muscular left ventricle pumps blood through a fourth valve into the body's largest artery, the aorta. From the aorta, blood travels through the arteries and arterioles to the systemic capillaries, venules, and veins and then back to the right atrium again.

The left ventricle is the most muscular of the heart's four chambers because it must do more work than any other chamber. The left ventricle must generate pressures higher than aortic blood pressure in order to pump blood into the aorta. (We'll see how high aortic pressure is in a minute.) The right ventricle has a thinner wall and does less work because the blood pressure in the arteries leading to the lungs is only about one-sixth that of the aorta.

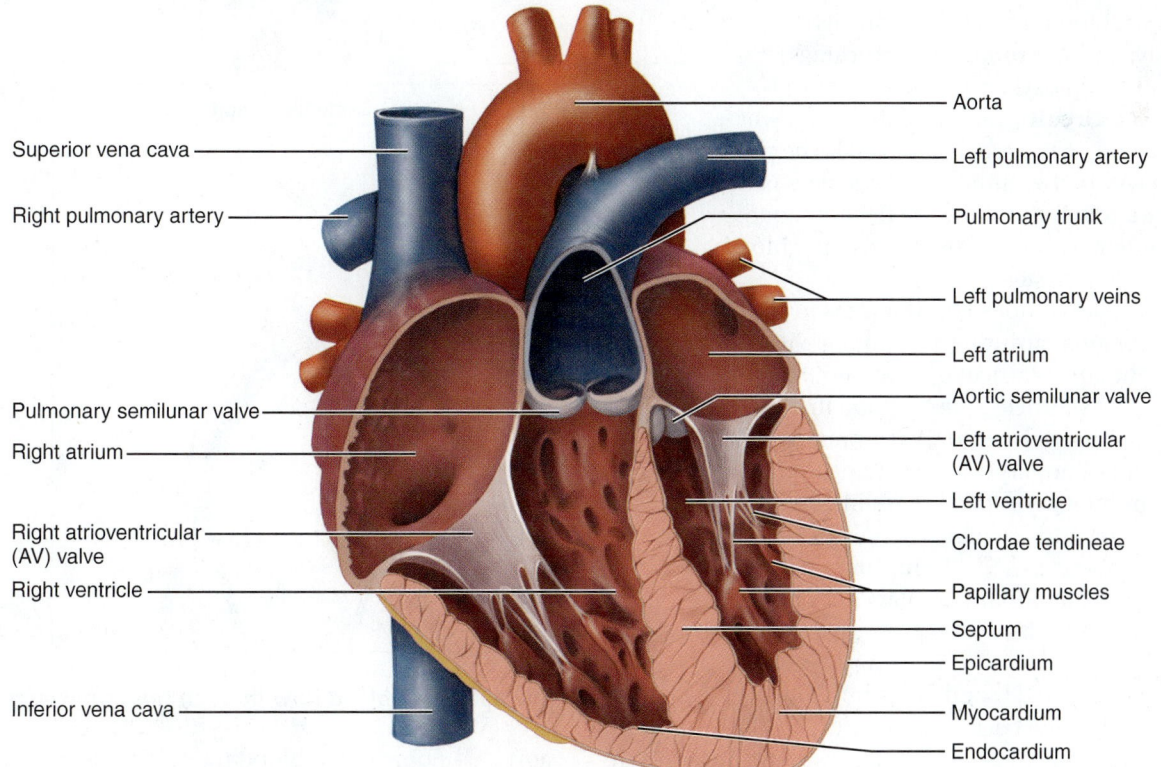

Figure 8.7 **A view of the heart showing major blood vessels, chambers, and valves.**

Four heart valves enforce the heart's one-way flow pattern and prevent blood from flowing backward. The valves open and shut passively in response to changes in the pressure of blood on each side of the valve. The right and left **atrioventricular (AV) valves** located between the atria and their corresponding ventricle prevent blood from flowing back into the atria when the ventricles contract. The AV valves consist of thin connective tissue flaps (cusps) that project into the ventricles. The right AV valve is called the *tricuspid valve* because it has three flexible flaps. The left AV valve has two flaps, so it is referred to as the *bicuspid* or *mitral valve*. These valves are supported by strands of connective tissue called *chordae tendineae* that connect to muscular extensions of the ventricle walls called *papillary muscles*. Together, the chordae tendineae and papillary muscles prevent the valves from everting (opening backward) into the atria when the ventricles contract.

Two **semilunar valves** (the pulmonary and the aortic) prevent backflow into the ventricles from the main arteries leaving the heart when the heart relaxes. Each semilunar valve consists of three pocketlike flaps. The valve name reflects the half-moon shape of these flaps (*semi* means "one-half"; *luna* comes from the Latin word for "moon").

The pattern of blood flow through the cardiovascular system

The heart pumps blood through two circuits simultaneously; the **pulmonary circuit** (lungs), where blood picks up oxygen and gets rid of CO_2, and the **systemic circuit** (the rest of the body) where oxygen is used and CO_2 waste is produced. The pattern of blood flow within the cardiovascular system is shown in **Figure 8.8**. Let's follow the flow of blood through the system, starting with the return of blood to the heart from the systemic circuit.

1 Deoxygenated venous blood returns to the heart and **2** enters the right atrium. **3** From there it passes through the right atrioventricular valve into the right ventricle. The right ventricle pumps blood through the pulmonary semilunar valve into **4** the pulmonary trunk (the main pulmonary artery) leading to the lungs. **5** The pulmonary trunk divides into the right and left pulmonary arteries, which supply the right and left lungs, respectively. **6** Blood entering the lungs passes through the pulmonary capillaries. This is where gas exchange occurs; blood gives up CO_2 and receives a fresh supply of O_2 from the air we inhale. **7** The freshly oxygenated blood flows into the pulmonary veins leading back to the heart.

8 On returning to the heart after its trip through the lungs, the now-oxygenated blood flows into the left atrium and **9** passes through the left atrioventricular valve to enter the ventricle. **10** The left ventricle pumps the blood into the

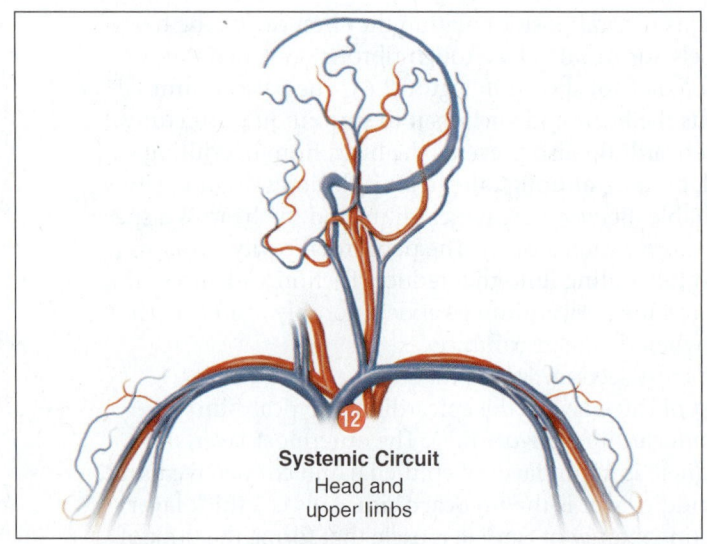

Systemic Circuit
Head and
upper limbs

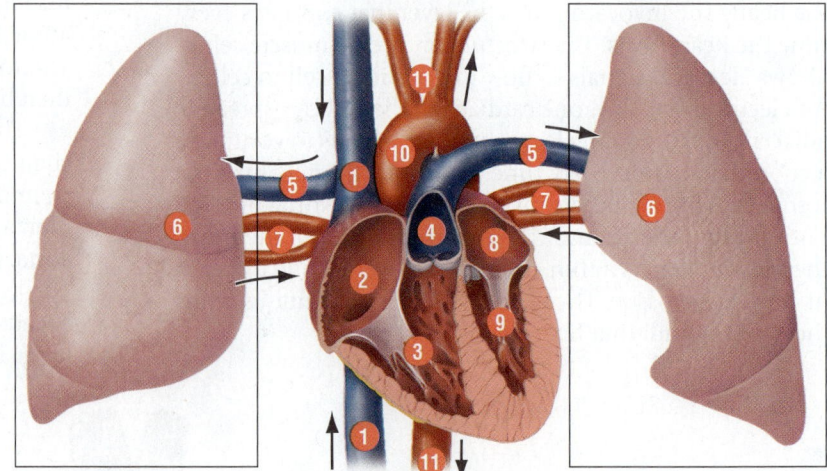

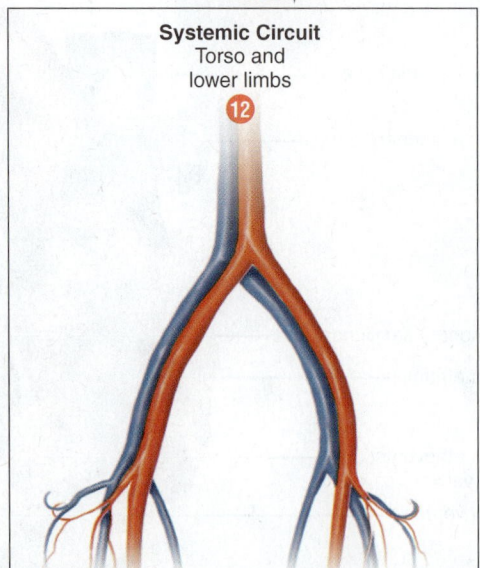

Systemic Circuit
Torso and
lower limbs

Figure 8.8 **The pattern of blood flow through the cardiovascular system.**

aorta. **11** Some of the blood travels up the main arteries to the head and upper body, and the rest passes down the aorta to the torso and lower limbs. **12** Upon arrival at the capillaries in the systemic circuit, blood delivers O_2 to the tissues and picks up waste CO_2. Then it returns to the venous system, eventually making its way back to the great veins that enter the heart.

Note that for every one trip around the body the blood passes through the heart twice; once as deoxygenated blood destined for the lungs, and once as oxygenated blood to be delivered to the systemic circuit. Deoxygenated blood passing through the right side of the heart never mixes with oxygenated blood passing through the left.

☑ Do the pulmonary arteries carry oxygenated blood or deoxygenated blood? What features of the pulmonary arteries make them arteries rather than veins?

Arteries and veins of the human body

Figure 8.9 shows some of the major arteries and veins of the human body. Arteries and veins serving the same vascular region often (but not always) have the same

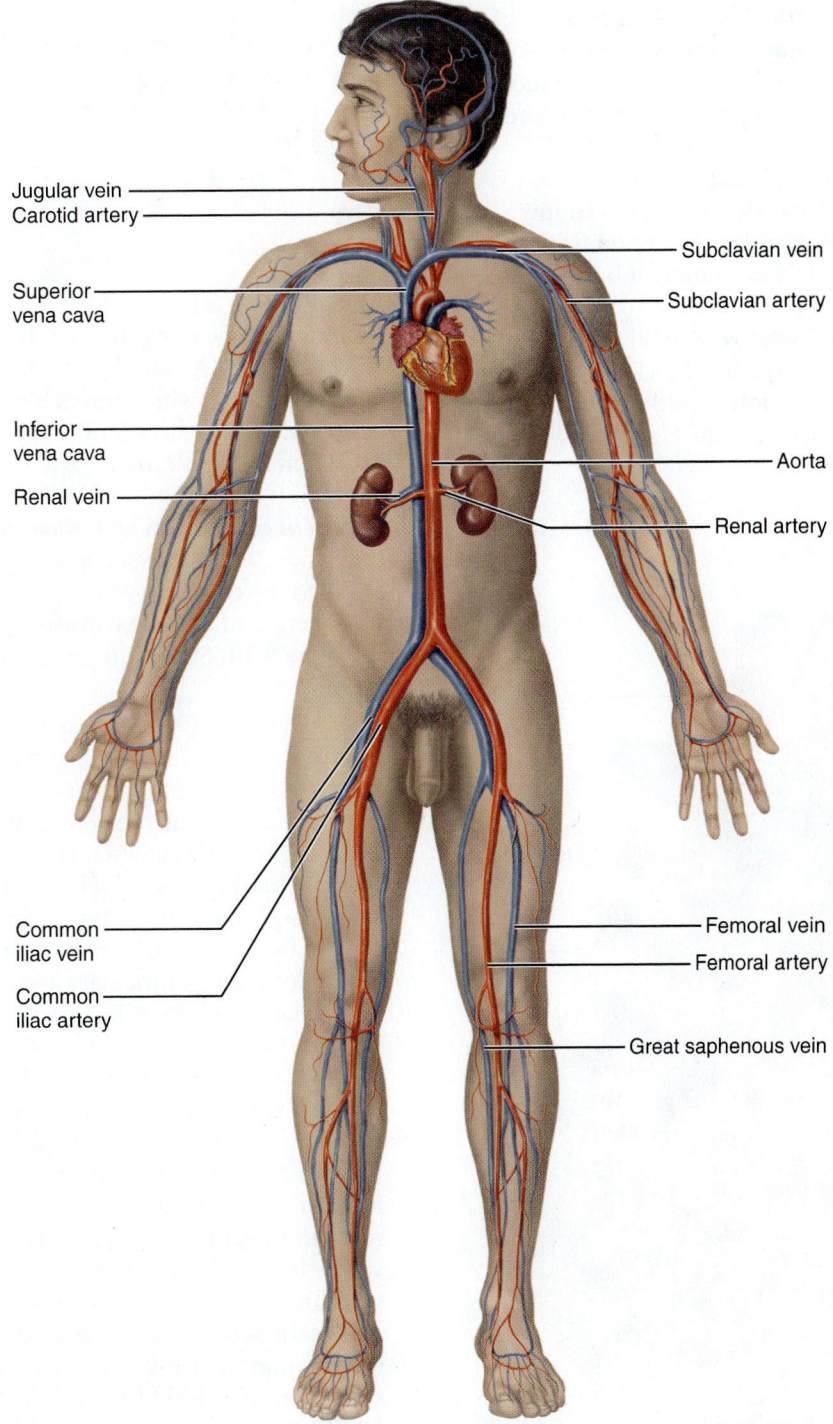

Jugular vein
Carotid artery
Superior vena cava
Inferior vena cava
Renal vein
Common iliac vein
Common iliac artery

Subclavian vein
Subclavian artery
Aorta
Renal artery
Femoral vein
Femoral artery
Great saphenous vein

Figure 8.9 Some of the major arteries and veins in the human body. For simplicity, the lungs and most of the internal organs have been omitted.

name and generally are located very near to each other. For example, a common iliac artery supplies blood to each leg, and a common iliac vein returns blood from the leg to the heart. However, carotid arteries supply the head, but jugular veins return the blood from the head.

As you might expect of such a hardworking muscle, the heart requires a great deal of oxygen and nutrients to fuel its own operations. The myocardium is too thick to be served by diffusion of oxygen and nutrients from the blood passing through. Thus, the heart has its own set of blood vessels called the **coronary arteries** that supply the heart muscle (Figure 8.10). The coronary arteries branch from the aorta just above the aortic semilunar valve and encircle the heart's surface (the word coronary comes from the Latin *corona*, meaning "encircling like a crown"). From the surface, they send branches inward to supply the myocardium. *Cardiac veins* collect the blood from the capillaries in the heart muscle and channel it back to the right atrium.

The coronary arteries are relatively small in diameter. If they become partially or completely blocked, perhaps as a result of atherosclerosis, serious health problems can result. For more on this subject, see the Health & Wellness feature.

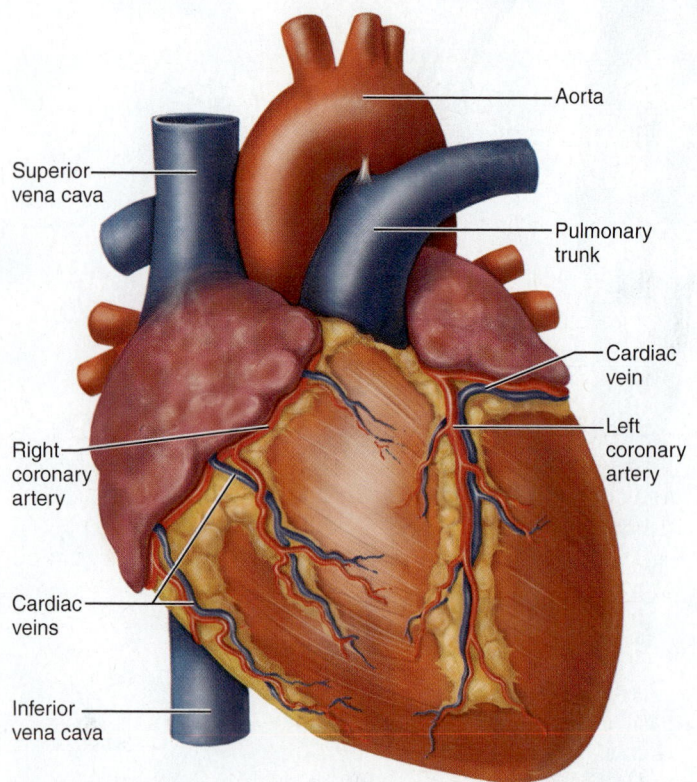

Figure 8.10 Blood vessels of the heart.

Labels: Aorta; Superior vena cava; Pulmonary trunk; Cardiac vein; Left coronary artery; Right coronary artery; Cardiac veins; Inferior vena cava

MJ's **BlogInFocus** What next-generation cholesterol-lowering drugs are pharmaceutical companies currently developing? Visit MJ's blog in the Study Area in MasteringBiology and look under "Cholesterol-Lowering Drugs."

http://goo.gl/qHbBEb

☑ Some babies are born with a heart defect in which the ventricles are connected to the wrong arteries—that is, the right ventricle sends blood to the aorta, and the left ventricle sends blood to the pulmonary trunk. What is the problem with this arrangement?

The cardiac cycle: the heart contracts and relaxes

The pumping action of the heart is pulsatile rather than continuous, meaning that it delivers blood in separate and distinct pulses. A complete cardiac cycle involves contraction of the two atria, which forces blood into the ventricles, followed by contraction of the two ventricles, which pumps blood into the pulmonary artery and the aorta, followed by relaxation of the entire heart. The term **systole** refers to the period of contraction and **diastole** refers to the period of relaxation. The entire sequence of contraction and relaxation is called the **cardiac cycle.**

Every cardiac cycle consists of three steps, as shown in Figure 8.11. Start with the heart as it first begins to contract:

❶ **Atrial systole** As contraction starts, the heart is already nearly filled with blood that entered the ventricles and atria passively during the previous diastole. Contraction of the heart begins with the atria. During atrial systole, both atria contract, raising blood pressure in the atria and giving the final "kick" that fills the two ventricles to capacity. Atrial systole also momentarily stops further inflow from the veins. Both atrioventricular valves are still open, and both semilunar valves are still closed.

❷ **Ventricular systole** The contraction that began in the atria spreads to the ventricles, and both ventricles contract simultaneously. The rapidly rising ventricular pressure produced by contraction of the ventricles causes the two AV valves to close, preventing blood from flowing backward into the atria and veins. At this time, the atria relax and begin filling again. The pressure within the ventricles continues to rise until it is greater than the pressure in the arteries, at which point, the pulmonary and aortic semilunar valves open and blood is ejected into the pulmonary trunk and the aorta. With each ventricular systole, about 60% of the blood in each ventricle is forcibly ejected.

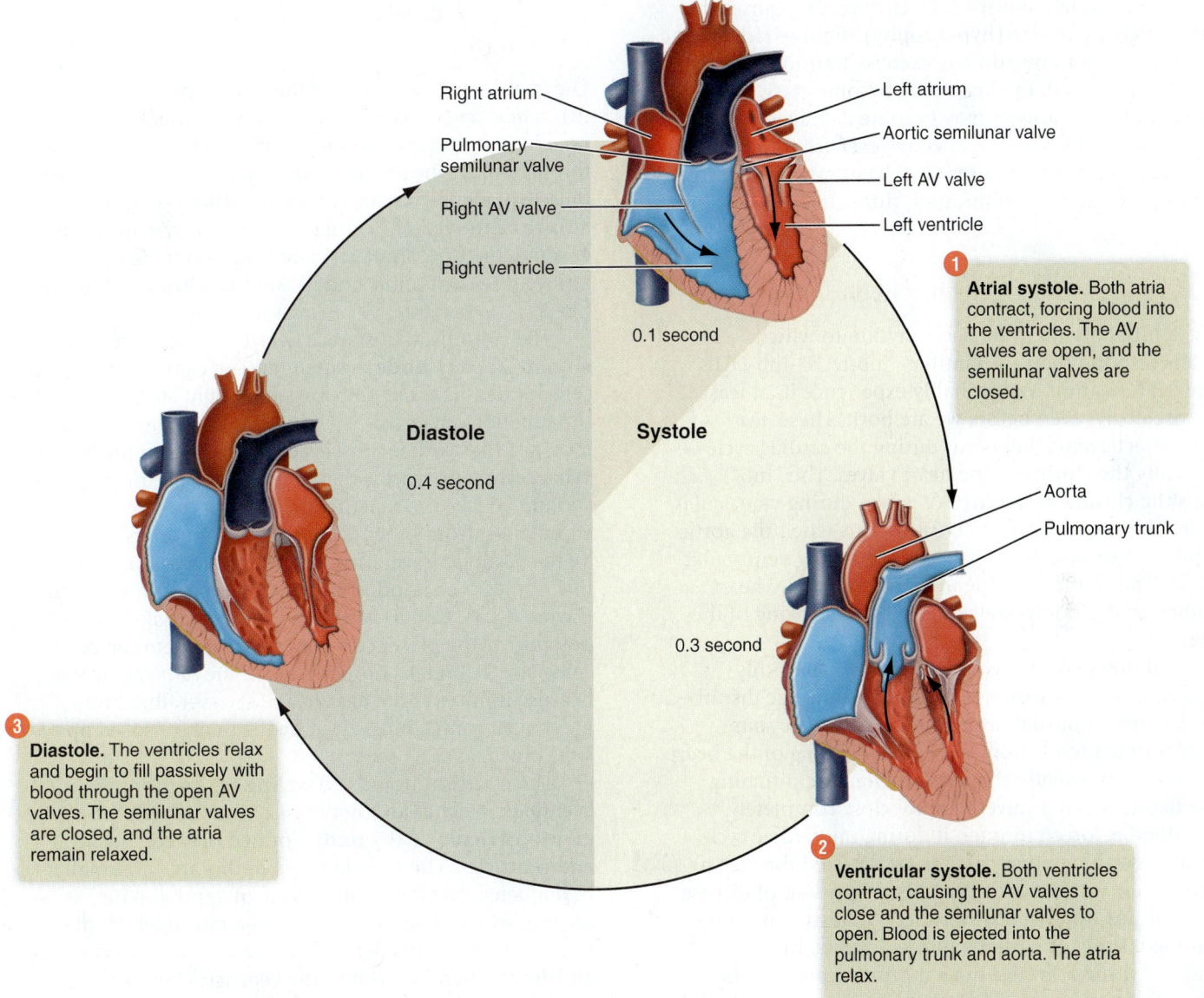

Right atrium
Pulmonary semilunar valve
Right AV valve
Right ventricle

Left atrium
Aortic semilunar valve
Left AV valve
Left ventricle

0.1 second

1 **Atrial systole.** Both atria contract, forcing blood into the ventricles. The AV valves are open, and the semilunar valves are closed.

Diastole
0.4 second

Systole

Aorta
Pulmonary trunk

0.3 second

3 **Diastole.** The ventricles relax and begin to fill passively with blood through the open AV valves. The semilunar valves are closed, and the atria remain relaxed.

2 **Ventricular systole.** Both ventricles contract, causing the AV valves to close and the semilunar valves to open. Blood is ejected into the pulmonary trunk and aorta. The atria relax.

Figure 8.11 **The cardiac cycle.**

✔️ What makes the AV and semilunar valves open and close?

3 **Diastole** Both atria and both ventricles are relaxed throughout diastole. At this point pressure within the ventricles begins to fall. As soon as ventricular pressures fall below arterial pressures during early diastole, the pulmonary and aortic semilunar valves close, preventing backflow of arterial blood. Once ventricular pressure falls below blood pressure in the veins, the AV valves open and blood begins to flow passively into the heart.

A complete cardiac cycle occurs every 0.8 second or so. These cycles repeat, from birth to death, without ever stopping. Atrial systole lasts about 0.1 second; ventricular systole about 0.3 second. During the remaining 0.4 second, the heart relaxes in diastole.

As each surge of blood enters the arteries during systole, the artery walls are stretched to accommodate the extra volume, and arterial pressure rises. Arteries recoil passively during diastole as blood continues to flow out of them through the capillaries. You can feel this cycle of rapid expansion and recoil in the wall of an artery if it's located close to the skin's surface. This is called a *pulse*. A good place to detect a pulse is the radial artery (inside your wrist, just below the base of the thumb).

The heart is composed primarily of cardiac muscle, and like our skeletal muscles it benefits from regular exercise. During sustained (aerobic) exercise, the heart beats more rapidly and more powerfully to sustain blood pressure in the face of increased blood flow to hardworking skeletal muscles. Over time, this causes the heart to increase in size (hypertrophy) slightly. However, it is important not to overdo any exercise training regimen. If it is overexercised, the heart may become starved for oxygen, and heart muscle may become damaged. The usual guideline for safe but effective exercise training is to perform an activity that raises your heart rate to its "target heart rate" for at least 20 minutes, three times a week or more.

Heart sounds reflect closing heart valves

There is probably no more basic rhythm to which humans respond than the familiar "lub-DUB–lub-DUB" of the heart beating. We probably experience it, at least subconsciously, even before we are born. These *heart sounds* reflect events that occur during the cardiac cycle—specifically the closing of the heart valves. The "lub" signals the closure of the two AV valves during ventricular systole. The slightly louder "DUB" occurs when the aortic and pulmonary semilunar valves close during ventricular diastole. The sounds are due to vibrations in the heart chambers and blood vessels caused by the closing of the valves.

Blood flows silently as long as it flows smoothly. However, if blood encounters an obstruction, the disturbed flow can create unusual heart sounds called *murmurs*. Many murmurs result from incomplete closing of the heart valves due to unusually shaped valve flaps or stiffening of the flap tissue. If a valve does not close completely, some blood is forced through it during the cardiac cycle, creating a swishing noise that can be detected through a stethoscope. Murmurs are not necessarily a sign of disease, but physicians can diagnose a variety of heart conditions, including leaking or partially blocked valves, from their sound and timing. Even serious murmurs can often be treated with surgery to replace the defective valve with an artificial valve (**Figure 8.12**).

✅ Do any sounds occur at the moment when the atria contract? When the ventricles contract? Explain.

The cardiac conduction system coordinates contraction

The coordinated sequence of the cardiac cycle is due to the *cardiac conduction system*, a group of specialized cardiac muscle cells that initiate and distribute electrical impulses throughout the heart. These impulses stimulate the heart muscle to contract in an orderly sequence that spreads from atria to ventricles. The cardiac conduction system consists of four structures: sinoatrial node, atrioventricular node, atrioventricular bundle and its two branches, and Purkinje fibers.

The stimulus that starts a heartbeat begins in the **sinoatrial (SA) node,** a small mass of cardiac muscle cells located near the junction of the right atrium and the superior vena cava. The muscle cells of the SA node look just like cardiac muscle cells elsewhere in the heart. What sets them apart is that they initiate electrical signals spontaneously and repetitively, without the need for stimulation from other cells. Electrical impulses initiated by the SA node travel from cell to cell across both atria like ripples on a pond, stimulating waves of contraction (**Figure 8.13**). Cell-to-cell electrical transmission is made possible by the presence of gap junctions between adjacent cells. The SA node is properly called the *cardiac pacemaker* because it initiates the heartbeat. However, the cardiac pacemaker can be influenced by the brain to speed up or slow down, as we'll see.

The electrical impulse traveling across the atria eventually reaches another mass of muscle cells called the **atrioventricular (AV) node,** located between the atria and ventricles. The muscle fibers in this area are smaller in diameter, causing a slight delay of approximately 0.1 second, which temporarily slows the rate at which the impulse travels. This delay gives the atria time to contract and empty their blood into the ventricles before the ventricles contract.

From the AV node, the electrical signal sweeps to a group of conducting fibers in the septum between the two ventricles called the **atrioventricular (AV) bundle.** These fibers branch and extend into **Purkinje fibers,** smaller

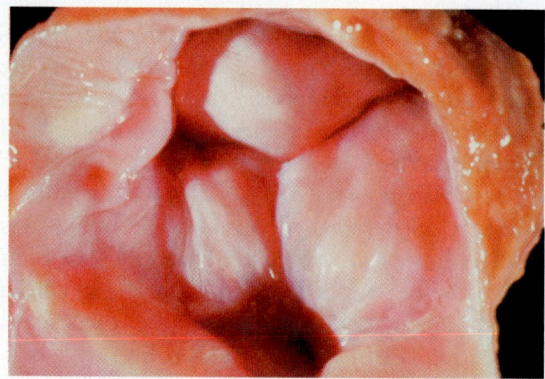

a) A pulmonary semilunar valve.

b) An artificial heart valve.

Figure 8.12 Heart valves.

complex, representing the spread of the electrical impulse down the septum and around the two ventricles in the Purkinje fibers, causing ventricular systole. Finally, the T wave occurs when the electrical activity in the ventricles ends and the ventricular muscle cells rest in preparation for the next cycle.

If something goes wrong with the cardiac conduction system or if the heart muscle becomes damaged, abnormal heart electrical impulses and contractions may occur. An abnormality of the rhythm or rate of the heartbeat is called

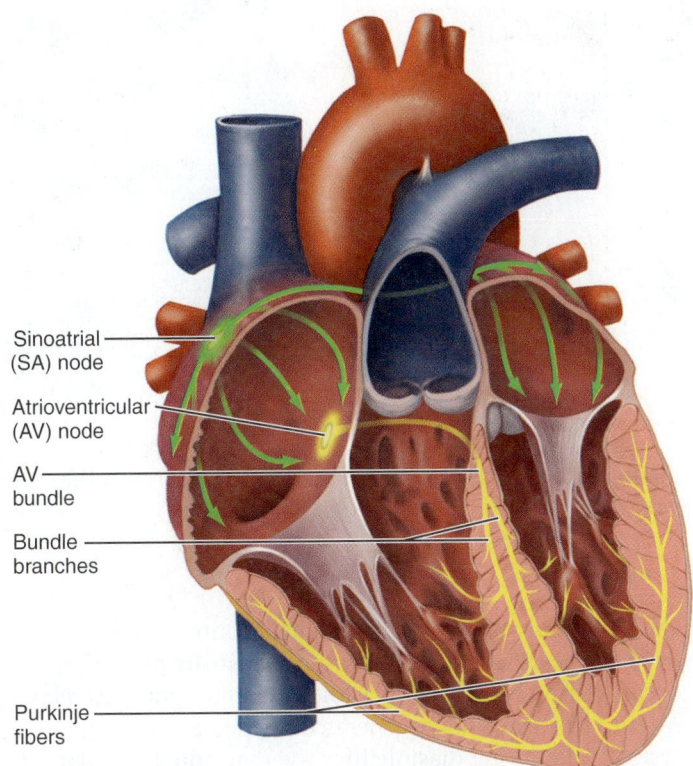

Sinoatrial (SA) node
Atrioventricular (AV) node
AV bundle
Bundle branches
Purkinje fibers

Figure 8.13 **The cardiac conduction system.** Electrical activity of the heart normally starts at the SA node, spreads across the atria to the AV node, and then progresses down the AV bundle and its branches to the Purkinje fibers.

fibers that carry the impulse to all cells in the myocardium of the ventricles. Because the electrical impulse travels down the septum to the lower portion of the ventricles and then spreads rapidly upward through the Purkinje fibers, the lower part of the ventricles contract before the upper part. This lower-to-upper squeezing motion pushes blood into the pulmonary trunk and aorta.

Electrocardiogram records the heart's electrical activity

Because the body is largely water and water conducts electrical activity well, we can track the electrical activity of the heart as weak differences in voltage at the surface of the body. An **electrocardiogram (ECG or EKG)** is a record of the electrical impulses in the cardiac conduction system (**Figure 8.14a**). An ECG involves placing electrodes on the skin at the chest, wrists, and ankles. The electrodes transmit the heart's electrical impulses, which are recorded as a continuous line on a screen or moving graph.

A healthy heart produces a characteristic pattern of voltage changes. A typical ECG tracks these changes as a series of three formations: P wave, QRS complex, and T wave (**Figure 8.14b**). First is the small P wave, representing the electrical impulse traveling across the atria (see Figure 8.13), causing atrial systole. Second is the QRS

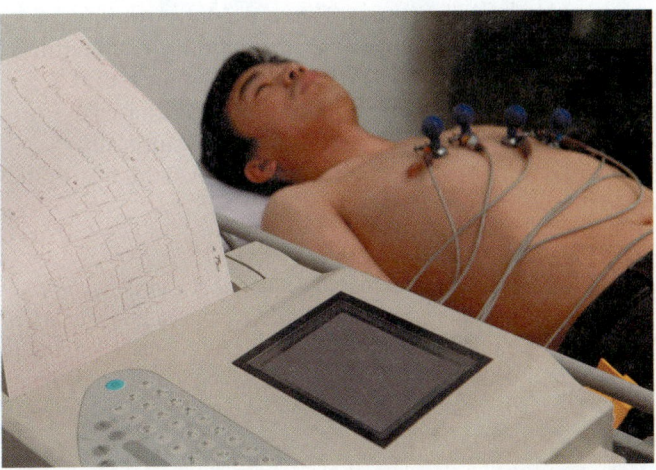

a) An ECG being recorded.

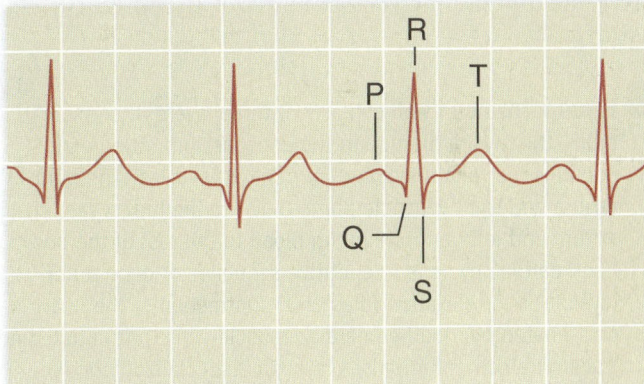

b) A normal ECG recording.

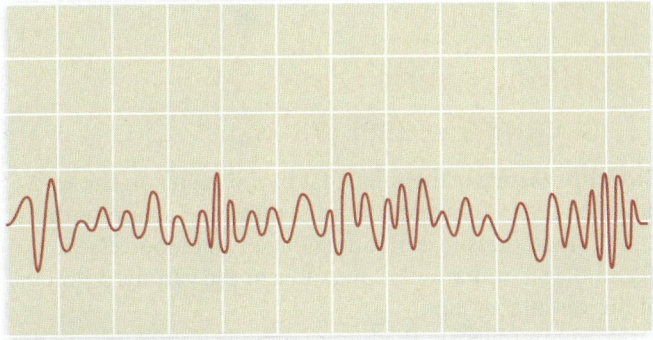

c) Ventricular fibrillation.

Figure 8.14 **The ECG is a tool for diagnosing heart arrhythmias.**

an *arrhythmia.* Arrhythmias take many forms. Occasional skipped heartbeats, for example, are fairly common and usually of no consequence. However, a type of rapid irregular ventricular contraction known as *ventricular fibrillation* (or "V-fib") is very quickly fatal unless treated immediately. Ventricular fibrillation is the leading cause of cardiac death in otherwise healthy people. In a hospital, ventricular fibrillation is treated by *cardioversion,* in which a strong electrical current is applied to the chest to eliminate the abnormal fibrillating pattern and restore the normal rhythm.

Arrhythmias produce characteristic ECG tracings, and the ECG is a valuable tool for identifying the cause, type, and location of arrhythmias (**Figure 8.14c**). Arrhythmias less life threatening than ventricular fibrillation can sometimes be treated with medications. In some cases, an artificial pacemaker (a small generating unit that automatically stimulates the heart at set intervals) can be surgically implanted under the chest skin to normalize the heart rate.

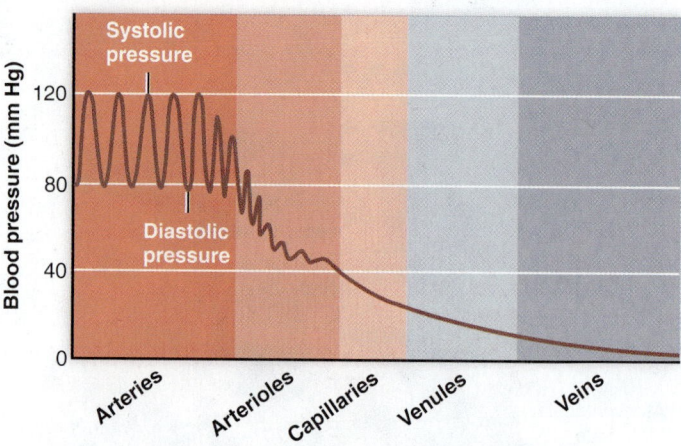

Figure 8.15 Blood pressure in different segments of the vascular system.

> **MJ's BlogInFocus**
>
> Is there anything that can be done for patients who need a new heart valve but who are too old or sick to withstand the extensive surgery usually required? Visit MJ's blog in the Study Area in MasteringBiology and look under "Heart Valve Replacement."
>
> http://goo.gl/qxnlxK

 **Recap** The heart wall consists of three layers: the epicardium, the myocardium, and the endocardium. The heart contains four chambers and four one-way valves. The right atrium and right ventricle pump blood to the lungs; the left atrium and left ventricle pump blood to the rest of the body. Each cardiac cycle is a repetitive sequence of contraction (systole) and relaxation (diastole). Contraction of the heart is coordinated by modified cardiac muscle cells that initiate and transmit electrical impulses through a specialized conduction system. An electrocardiogram (ECG) is a recording of the heart's electrical activity taken from the surface of the body. ∎

8.3 Blood exerts pressure against vessel walls

Blood pressure is the force that blood exerts on the wall of a blood vessel as a result of the pumping action of the heart. Blood pressure is not the same in all blood vessels. **Figure 8.15** compares the pressures in the various segments of the vascular system.

You can see from the highs and lows shown in Figure 8.15 that pressure is pulsatile in the arteries; that is, it varies with each beat of the heart. The highest pressure of the cycle, **systolic pressure,** is the pressure reached during ventricular systole when the ventricles contract to eject blood from the heart. The lowest pressure, **diastolic pressure,** occurs during ventricular diastole when the ventricles relax. Arteries store the energy generated by the heart during systole, and during diastole they use that stored energy to supply blood to the tissues.

Maintenance of arterial blood pressure is crucial to drive the flow of blood throughout the body and all the way back to the heart. Recall that fluid always flows from a region of high pressure toward a region of lower pressure. By the time it reaches the capillaries, blood flow is steady rather than pulsatile, and pressure continues to fall as blood flows through venules and veins. The differences in the blood pressure of arteries, capillaries, and veins keep blood moving through the body.

Measuring blood pressure

When health professionals measure your blood pressure, they are assessing the pressure in your main arteries. From a clinical standpoint, blood pressure gives valuable clues about the relative volume of blood in the vessels, the condition or stiffness of the arteries, and the overall efficiency of the cardiovascular system. Trends in blood pressure over time are a useful indicator of cardiovascular changes.

Blood pressure is recorded as mm Hg (millimeters of mercury, because early equipment used a glass column filled with mercury to measure the pressure). In young, healthy individuals, systolic pressures of less than 120 mm Hg and diastolic pressures of less than 80 mm Hg are considered desirable. With advancing age, there is a slight tendency for systolic blood pressure in particular to increase slightly, as a consequence of age-related stiffening of the arteries.

Blood pressure is measured with a *sphygmomanometer* (*sphygmo* comes from the Greek word for "pulse"; a manometer is a device for measuring fluid pressures). An inflatable cuff is placed over the brachial artery in your

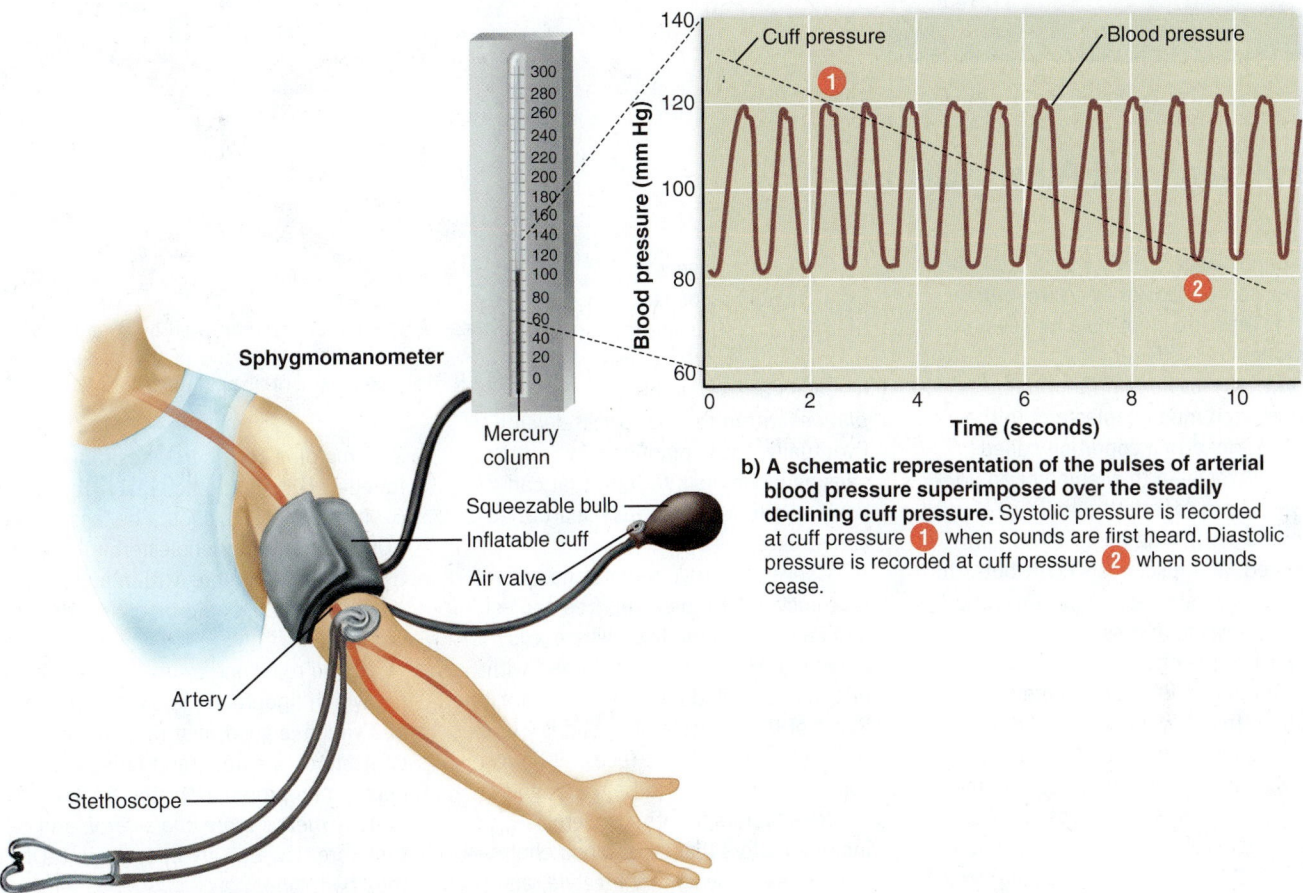

b) **A schematic representation of the pulses of arterial blood pressure superimposed over the steadily declining cuff pressure.** Systolic pressure is recorded at cuff pressure ① when sounds are first heard. Diastolic pressure is recorded at cuff pressure ② when sounds cease.

a) **A clinician inflates the cuff with air and then allows the pressure in the cuff to fall gradually while using a stethoscope to listen for the sounds of blood movement through the artery.**

Figure 8.16   How blood pressure is measured.

✅ Explain in your own words why the tapping noise is only heard when the cuff pressure is between systolic and diastolic pressure. In other words, what's making the noise?

upper arm and connected to the manometer, which in most cases is a column of mercury (**Figure 8.16a**). When the cuff is inflated to a pressure above systolic pressure, blood flow through the brachial artery stops because the high cuff pressure collapses the artery. The cuff is then deflated slowly while a health professional listens with a stethoscope for the sounds of blood flowing in your artery. As soon as pressure in the cuff falls below the peak of systolic pressure, some blood spurts briefly through the artery during the high point of the pressure pulse, making a characteristic light tapping sound that is audible through the stethoscope. The cuff pressure at which this happens is recorded as systolic pressure (**Figure 8.16b**). As the cuff continues to deflate, eventually blood flow through the artery becomes continuous and the tapping sound ceases. The point where the sound disappears is your diastolic pressure. This procedure yields two numbers, corresponding to your systolic and diastolic pressures. These represent the high and low points of blood pressure during the cardiac cycle.

Hypertension: high blood pressure can be dangerous

Blood pressure higher than normal is called **hypertension** (*hyper* comes from the Greek for "excess"). **Table 8.1** presents the systolic and diastolic readings health professionals use to classify blood pressure as normal or hypertensive.

Table 8.1 Systolic and diastolic blood pressure

Blood pressure category	Systolic (mm Hg)		Diastolic (mm Hg)
Normal	Less than 120	and	Less than 80
Prehypertension	120–139	or	80–89
Hypertension, Stage 1	140–159	or	90–99
Hypertension, Stage 2	160 or higher	or	100 or higher

Source: National Institutes of Health, National Heart, Lung, and Blood Institute, March 2003.

HEALTH & WELLNESS
Cholesterol and Atherosclerosis

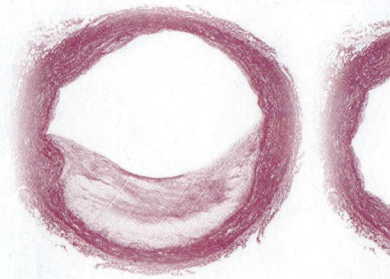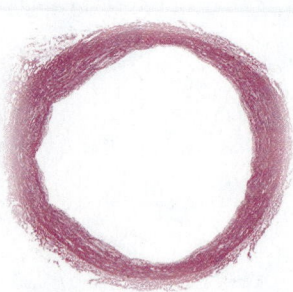

Left: Cross section of an artery narrowed by atherosclerotic plaque. Right: The same photo with the atherosclerotic plaque removed, showing how a normal artery would look.

As you probably already know, cholesterol is a key component of all cell membranes and the precursor molecule for several hormones. All cells require a certain amount of it for normal functioning. However, too much cholesterol in the blood can lead to a condition called *atherosclerosis*—a thickening of an arterial vessel wall due to the buildup of fatty materials containing cholesterol. Left untreated, atherosclerosis contributes to heart attacks, strokes, aneurysms, and peripheral vascular disease.

Most of the cholesterol in the blood is bound to certain carrier proteins. Together, the cholesterol and the protein are called a *lipoprotein*. There are two types of lipoproteins, based on their densities: In terms of atherosclerosis, *low-density lipoprotein (LDL)* is considered "bad" and *high-density lipoprotein (HDL)* is considered "good."

When present in normal amounts, LDL transports cholesterol throughout the body and makes it available to cells. However, when there is too much LDL, it begins to attach to the cells lining the arterial blood vessel wall and then makes its way into the cells. Once inside the cell, LDL triggers an inflammatory response that ultimately results in the buildup of fatty deposits called atherosclerotic *plaques* within the blood vessel wall. Eventually, these plaques may rupture, causing blood clots to form that can occlude arteries and cause heart attacks and strokes.

Conversely, HDL targets cholesterol for removal. HDL picks up free cholesterol and carries it to the liver, where it is detached from the protein, mixed with bile, and secreted into the small intestine. Some of the cholesterol in bile is excreted from the body with the feces, although some is reabsorbed, to be used again.

Risk factors for atherosclerosis include factors that raise blood cholesterol (obesity, sedentary lifestyle, and a high-fat diet), smoking, diabetes, hypertension, and a family history of atherosclerosis. Before age 45, men have a 10 times greater risk of atherosclerosis than women; however, women's risk rises after menopause.

According to the American Heart Association, a total cholesterol of under 200 mg/dl is considered desirable. Ideally, HDL should be greater than 60 mg/dl and LDL should be less than 100 mg/dl. A total cholesterol of greater than 240 mg/dl along with a high LDL and/or low HDL would be cause for concern.

If you're having trouble remembering which lipoprotein is bad for you and which is good, just remember that cholesterol is a lipid, and lipids are less dense than protein or water. So *low-density means more cholesterol*, and, therefore, low-density lipoprotein (LDL) is the "bad" one.

Some degree of atherosclerosis is common with advancing age. However, lifestyle can make a big difference in how rapidly atherosclerosis develops and whether it becomes severe. At the end of this chapter, we look at what you can do to lower your risk of atherosclerosis and other cardiovascular conditions.

Hypertension is a significant risk factor for cardiovascular disease, because the greater the pressure, the greater the strain on the cardiovascular system. Blood vessels react to the pounding by becoming hardened and scarred, which makes them less able to stretch during systole. Hypertension also places a greater strain on the heart, because the work it must do is directly proportional to the arterial pressure against which it must pump.

Hypertension is called "the silent killer" because usually it has no symptoms. The American Heart Association estimates that approximately 50 million Americans have hypertension and a third of them don't even realize it. If left untreated, hypertension increases the risk of serious health problems such as heart attack, heart failure, stroke, kidney damage, even damage to the tissues inside the eyes.

What causes hypertension? Many times it happens because blood vessels become narrowed from atherosclerosis. Certain other factors also increase the risk, as summarized in Table 8.2. The only sure way to diagnose it is to have your blood pressure measured.

Blood pressure varies from minute to minute even in healthy individuals. Simply getting up in the morning raises it, as does exercise, emotions, smoking cigarettes, eating, drinking, and many other factors. Even having your blood pressure measured can make you nervous enough for your blood pressure to rise—a situation that health professionals call "white coat hypertension." This is why physicians generally have you sit quietly while measuring your blood pressure.

If hypertension is suspected, your physician will probably measure your blood pressure on at least three different occasions before making a firm diagnosis. True hypertension is a *sustained* elevation in blood pressure above normal levels—a systolic pressure of 140 mm Hg or greater or a diastolic pressure of 90 mm Hg or greater. Even blood

Table 8.2 Risk factors for hypertension

Risk factor	Comments
Heredity	Family history of hypertension raises risk.
Age	Blood pressure tends to rise throughout life.
Race	African Americans have twice the incidence found in Caucasian Americans and Asian Americans.
Sex	Males are more likely than females to develop hypertension.
Obesity	The heart must pump harder to push blood through vessels.
High salt intake	In some individuals (but not others), a high salt intake raises blood pressure slightly.
Smoking	Smoking raises the blood concentration of epinephrine, a hormone that stimulates the heart.
Sedentary lifestyle	Not well understood. May be due to higher blood lipids or weight gain.
Persistent emotional stress	Emotional stress activates portions of the nervous system that elevate pressure.
Diabetes mellitus	Diabetics have a higher incidence of hypertension, for reasons not yet known.
Heavy alcohol consumption	Mechanisms unknown.
Oral contraceptives and certain medications	Mechanisms vary by medication.

pressure that is consistently just below the hypertensive level (prehypertension) may carry a slightly higher risk of health complications.

Generally, if systolic pressure is high, diastolic pressure will be, too. However, sometimes systolic pressure can register at above-normal levels while diastolic pressure remains normal, a condition called *isolated systolic hypertension*. Most common in older adults, it is diagnosed as a systolic pressure of 160 mm Hg or higher with a diastolic reading of less than 90 mm Hg. Like the more common form of hypertension, isolated systolic hypertension is associated with increased health problems.

At the end of this chapter, we discuss what you can do to lower your risk of hypertension as well as other cardiovascular problems. If hypertension does develop, however, a number of medical treatments are available to lower blood pressure to a healthy level. It is important for people on antihypertensive drugs to take their medication consistently; even though hypertension has no symptoms, it increases the risk of related health problems.

☑ A nurse taking blood pressure in a patient hears a tapping sound begin when the cuff pressure is 141 mm Hg and the sound ends when cuff pressure is 95 mm Hg. What are the systolic and diastolic pressures of this patient, and is this enough to tell you if this patient has normal blood pressure?

Hypotension: when blood pressure is too low

It is also possible for blood pressure to fall below normal levels, a condition called *hypotension* (*hypo* comes from the Greek for "under"). Generally, hypotension is a problem only if blood pressure falls enough to reduce blood flow to the brain, causing dizziness and fainting. Drops in blood pressure can follow abrupt changes in position, such as standing up suddenly. Other causes include severe burns or injuries involving heavy blood loss.

↺ **Recap** Blood pressure is the force that blood exerts on the wall of a blood vessel. It is measured as two numbers corresponding to systolic and diastolic pressures. Hypertension (high blood pressure) is a serious risk factor for cardiovascular disease and other health problems. ■

8.4 How the cardiovascular system is regulated

The overall function of the cardiovascular system is to provide every cell with precisely the right blood flow to meet its needs at all times. This might seem like a complicated process because different types of tissues have different needs, which change according to circumstances. In principle, however, the system is rather simple.

Consider for a minute how your community provides water to every house according to the needs of the occupants. Basically, municipal water systems provide constant water *pressure* to every house. With steady water pressure available, each household simply adjusts the *flow* of water into the house by turning faucets on and off.

The human cardiovascular system is based on similar principles. Consider the following key points:

- Homeostatic regulation of the cardiovascular system centers on *maintaining a constant arterial blood pressure*.
- A constant arterial blood pressure is *achieved* by regulating the **cardiac output** (the amount of blood the heart pumps into the aorta each minute) and by regulating the diameters of the arterioles (adjusting the overall resistance to flow through the blood vessels).
- With arterial blood pressure held relatively constant, local blood flows are adjusted to meet local requirements.

As we will see, this overall regulation is achieved with nerves, hormones, and local factors coupled with metabolism.

Baroreceptors maintain arterial blood pressure

There is probably no more important regulated variable in the entire body than blood pressure. Without a relatively constant blood pressure as the driving force for supplying blood to the capillaries, homeostasis simply would not be possible. Blood pressure does vary, but always within a range consistent with being able to provide blood to all cells. Normally, blood pressure is regulated within very narrow limits by a negative feedback control loop (Figure 8.17).

To regulate blood pressure, the body must have some way of measuring it. In fact, several of the large arteries, including the aorta and the two carotid arteries, have certain regions called **baroreceptors** (the Greek *baro* denotes "pressure"). The baroreceptors regulate arterial blood pressure in the following manner:

1 An increase in blood pressure stretches the baroreceptors passively, just as it stretches all blood vessels. **2** The baroreceptors are innervated by sensory neurons, which fire at an increased rate when the baroreceptors are stretched. **3** These neuron signals travel to a region of the brain called the *cardiovascular center.*

The cardiovascular correctly interprets increased sensory neuron activity coming from the baroreceptors as an arterial pressure that is too high. **4** It responds by changing its normal rate of nerve stimulation of the heart and blood vessels. **5** The effect on the heart is to lower heart rate and the force of contraction, thereby lowering cardiac output.

6 The effect on blood vessels is to reduce vascular resistance, leading to increased blood flow through the blood vessels. Together, decreases in cardiac output and vascular resistance lower blood pressure back toward normal.

Exactly the opposite sequence of events occurs when arterial pressure falls *below* normal. When pressure falls and the arteries are stretched less than normal, the baroreceptors send fewer sensory neuron signals to the brain. The brain correctly interprets this as a fall in pressure and sends signals to the heart and blood vessels that increase cardiac output and vascular resistance, raising arterial blood pressure again.

All day long—every time you stand up, sit down, get excited, run for the bus—your blood pressure fluctuates up or down. But it is quickly brought back within normal range by a negative feedback loop initiated by baroreceptors.

☑ A young girl with an allergy to peanuts suffers a severe allergic reaction that includes a sudden drop in blood pressure. How will her body attempt to restore the blood pressure over the next few seconds and minutes?

Local requirements dictate local blood flows

With arterial blood pressure held relatively constant, the flow of blood through each precapillary sphincter (and hence each capillary) can be adjusted according to need, just like turning a faucet on or off. How is the flow adjusted? When a particular tissue is metabolically active, such as when a muscle is contracting, it consumes more oxygen and nutrients. Increased metabolism also raises the production of carbon dioxide and other waste products. One or more of these changes associated with increased metabolism cause precapillary sphincters within the tissue to vasodilate, increasing flow (Figure 8.18). Scientists do not yet know the precise mechanisms of this vasodilation or the identity of all the chemical substances that influence the vasodilation, but we do know that it occurs.

Look at the overall process and you can see how together these control mechanisms deliver blood efficiently to all tissues. First, a negative feedback control loop works to maintain a relatively constant arterial blood pressure. Second, with pressure held constant, cells and tissues can get exactly the amount of blood they need by adjusting the vascular resistance of the small vessels supplying them.

Some tissues and organs need a more consistent blood supply than others. This is taken into account by our control systems. Consider what would happen if you lost a substantial volume of blood due to injury, producing a precipitous fall in blood pressure. The negative feedback control of arterial blood pressure would stimulate your heart and constrict your arterioles, reducing flow to most organs in an effort to raise blood pressure. However, organs whose survival depends critically on a constant blood supply (your brain and heart, for example) can

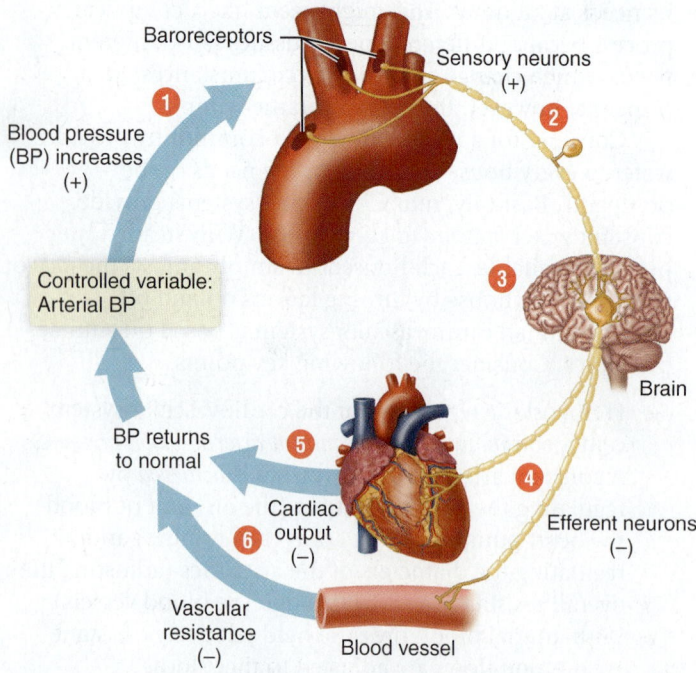

Figure 8.17 Negative feedback control of arterial blood pressure by baroreceptors. Increases or decreases are indicated by + or − signs, respectively, with blood pressure, cardiac output, and vascular resistance.

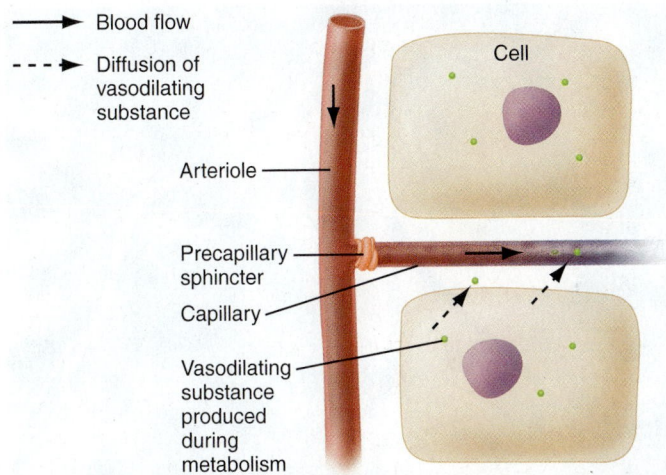

Blood flow

Diffusion of vasodilating substance

Cell

Arteriole

Precapillary sphincter

Capillary

Vasodilating substance produced during metabolism

a) At rest, very little of the vasodilating substance would be produced, and flow would be minimal.

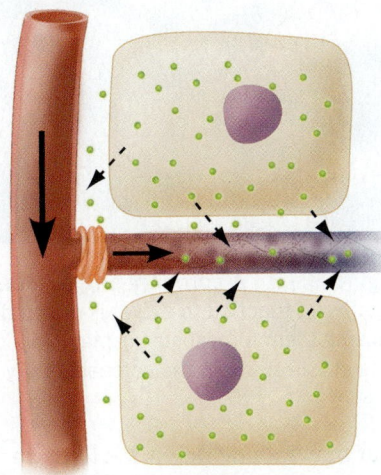

b) With increased metabolic activity, the presence of more of the vasodilating substance in the interstitial space would cause the arteriole and precapillary sphincter to vasodilate, increasing flow.

Figure 8.18 How an increase in metabolism increases local blood flow.

override the generalized vasoconstriction with their local control mechanisms. Organs whose metabolic activities are not required for their immediate survival (such as your kidneys and digestive tract) just remain vasoconstricted for a while. In effect, the limited available blood supply is shunted to essential organs.

You may wonder why, if these homeostatic controls are working properly, people ever develop hypertension. Every time blood pressure went up a little, wouldn't feedback mechanisms bring it back to normal? This is the great mystery of hypertension. Although scientists don't fully understand it yet, for some reason the body seems to adjust to high pressure once it has been sustained for a long time. Apparently, the feedback system slowly resets itself to

recognize the higher pressure as normal. The reason may be that hypertension tends to develop slowly over decades, so there is never a single defining moment when pressure registers as too high.

Exercise: increased blood flow and cardiac output

During exercise, the metabolic activity of the active skeletal muscles goes up dramatically. As a result, the production of vasodilator waste products increases, and the local concentration of oxygen falls. Both of these contribute to dilation of the blood vessels. Consequently, blood flow through the active skeletal muscles increases.

To sustain blood pressure in the face of increased blood flow, the heart must increase its output. From what you know, you might guess that an increase in blood flow to muscle would cause a fall in blood pressure, which in turn would cause a baroreceptor-mediated reflex increase in cardiac output. However, blood pressure doesn't fall very much (if at all) during exercise—if anything, it rises a little. During exercise, the primary cause of increased cardiac output is sensory input from moving muscles and joints. This sensory input signals the cardiovascular center to stimulate the heart and increase cardiac output even before blood pressure can fall very much. In other words, during exercise the body anticipates the need for increased cardiac output and prevents blood pressure from falling in the first place.

In nonathletic people, cardiac output reaches a maximum of about 20–25 liters per minute during heavy exercise. However, trained athletes whose heart muscles have hypertrophied as a result of exercise training can reach cardiac outputs of up to 35 liters per minute, or almost seven times their resting cardiac output.

↺ **Recap** Homeostatic regulation of the cardiovascular system centers on maintaining a relatively constant arterial blood pressure. Arterial pressure is sensed by baroreceptors located in the carotid arteries and aorta. Two opposing sets of nerves (sympathetic and parasympathetic) and a hormone (epinephrine) adjust cardiac output and arteriole diameters to maintain arterial blood pressure fairly constant. Local factors regulate blood flow into individual capillaries by altering the diameters of precapillary sphincters. ■

8.5 Cardiovascular disorders: a major health issue

Disorders affecting the cardiovascular system are a major health problem in Western countries. Cardiovascular disorders cause more than 740,000 deaths per year in the United States alone. They are the number one killer in the United States, far ahead of the number two killer, cancer (approximately 585,000 deaths per year).

We have already discussed hypertension, aneurysms, and atherosclerosis. In the rest of this chapter we look at several other conditions, including arrhythmias, angina, heart attack, heart failure, embolism, and stroke. Finally, we examine what you can do to reduce your own risk of cardiovascular health problems.

Angina: chest pain warns of impaired blood flow

As a hardworking muscle, the heart requires a constant source of blood. Normally the coronary arteries and their branches provide all the blood the heart needs, even during sustained exercise. However, if these arteries become narrowed, blood flow to the heart may not be sufficient for the heart's demands. This may lead to *angina*, a sensation of pain and tightness in the chest. Often, angina is accompanied by shortness of breath and a sensation of choking or suffocating (*angina* comes from the Latin word for "strangling"). Many angina episodes are triggered by physical exertion, emotional stress, cold weather, or eating heavy meals, because the heart requires more blood and oxygen at these times.

Angina is uncomfortable but usually temporary. Stopping to rest and taking several deep breaths can often relieve the discomfort. However, angina should never be ignored, because it is a sign of insufficient circulation to the heart. *Angiography* is a procedure that enables blood vessels to be visualized after they are filled with a contrast medium (a substance that is opaque to X-rays). Angiography allows health professionals to take X-ray pictures of blood vessels (called *angiograms*) and assess their condition (**Figure 8.19**).

Certain medications can increase blood flow to the heart muscle. Another treatment for narrowed coronary arteries is *balloon angioplasty*, which involves threading into the blocked artery a slender flexible tube with a small balloon attached. When the balloon reaches the narrowest point of the vessel, it is inflated briefly so that it presses against the fatty plaques that narrow the vessel lumen, flattening them and widening the lumen. Balloon angioplasty has a high success rate, although in some cases the vessel narrows again over time, requiring repeat treatments. A coronary artery bypass graft (see the next section) can yield longer-lasting benefits.

Heart attack: permanent damage to heart tissue

If blood flow to an area of the heart is impaired long enough, the result is a **heart attack**—sudden death of an area of heart tissue due to oxygen starvation. (The clinical term is *myocardial infarction*; infarction refers to tissue death from inadequate blood supply.) Many people who suffer a heart attack have a previous history of angina. The classic symptoms of a major heart attack, especially in men but also in women, include intense chest pain, a sense of tightness or pressure on the chest that makes it hard to breathe, and pain

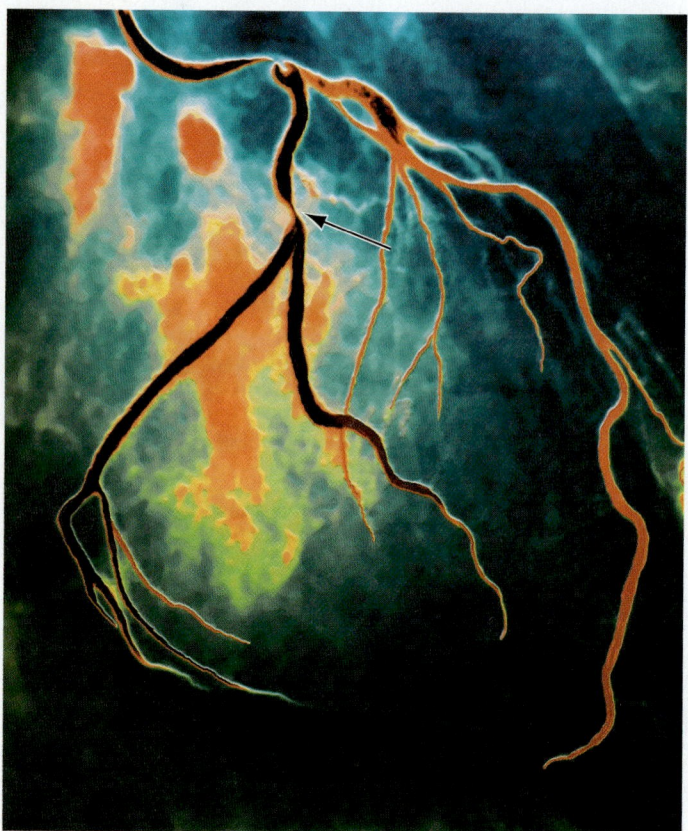

Figure 8.19 **A coronary angiogram.** A contrast medium (similar in function to a dye) is injected into the coronary arteries, causing them to become visible on an X-ray photograph. Note the area of narrowing in the coronary artery indicated by the arrow.

that may radiate down the left arm. Women tend to experience nausea and jaw and back pain more frequently than men, but as these symptoms do not seem as immediately life threatening, women's heart attacks are not diagnosed as soon as men's based on symptoms alone. The symptoms may come and go for several minutes at a time, leading many people to delay seeking medical attention.

A heart attack causes permanent damage to the heart. Because the body cannot replace cardiac muscle cells, damage to the heart impairs its ability to function. Most heart attack fatalities occur because of ventricular fibrillation, the serious heart arrhythmia brought about by damage to heart muscle.

Prompt treatment is crucial for recovery from a heart attack. If a heart attack is even suspected, the person should be rushed immediately to a hospital. The diagnosis of a heart attack is generally made on the basis of the ECG and the presence in the blood of certain enzymes that are released from dead and damaged heart cells. Health professionals can often control cardiac arrhythmias and other complications and administer clot-dissolving drugs to unblock vessels. The sooner treatment begins, the more successful it is likely to be.

Later (or even before a heart attack has occurred, if the narrowing of coronary vessels has been diagnosed by angiography), a *coronary artery bypass graft (CABG)* can be performed to improve coronary blood flow. In this procedure, a piece of blood vessel is removed from somewhere else in the body (often a leg vein is used) and grafted onto the blocked artery to bypass the damaged region (**Figure 8.20**). Over time, the grafted vein thickens and takes on the characteristics of an artery.

Thanks to these and other treatments, the heart attack survival rate has risen dramatically. Eighty percent of heart attack survivors are back at work within three months.

Heart failure: the heart becomes less efficient

Generally, our bodies maintain adequate arterial pressure because of the tight control mechanisms described earlier. However, if the heart muscle becomes damaged for any reason, the heart may become weaker and less efficient at pumping blood, a condition called **heart failure.**

When the heart begins to pump less blood, blood backs up in the veins and pressure in the veins and capillaries rises. The high capillary blood pressure causes more fluid than usual to filter out of the capillaries and into the interstitial space, causing fluid congestion. When that happens a person is said to have **congestive heart failure.** (*Congestive* refers to the buildup of interstitial fluid.)

People with congestive heart failure get out of breath when walking or climbing even a short flight of stairs. They may even have trouble breathing while lying down because the horizontal position results in even higher venous pressure and fluid accumulation in the lungs. Other symptoms include swollen ankles and legs, swollen neck veins, and weight gain from the extra fluid.

There are several reasons why a heart might begin to fail. Aging is one factor, but certainly not the only one because older people do not necessarily develop heart failure. Other possible causes include past heart attacks, leaking heart valves, heart valves that fail to open normally, uncontrolled hypertension, or serious arrhythmias. Lung conditions such as emphysema can raise blood pressure in the lungs and strain the heart.

Treatments for congestive heart failure focus on improving cardiac performance and efficiency while preventing the accumulation of interstitial fluid. Regular mild exercise promotes more efficient blood flow, and frequent resting with the feet elevated helps fluid drain from leg veins. Physicians may prescribe diuretics (drugs that help the body get rid of excess fluid), vasodilating drugs to expand blood vessels, or medications to help the heart muscle beat more forcefully.

Embolism: blockage of a blood vessel

Embolism refers to the sudden blockage of a blood vessel by material floating in the bloodstream. Most often the obstacle (an *embolus*) is a blood clot that has broken away from a larger clot elsewhere in the body (often a vein) and lodged in an artery at a point where arterial vessels branch and get smaller in diameter. Other possible emboli include cholesterol deposits, tissue fragments, cancer cells, clumps of bacteria, or bubbles of air.

Embolism conditions are named according to the area of the body affected. A *pulmonary embolism* blocks an artery supplying blood to the lungs, causing sudden chest pain and shortness of breath. A *cerebral embolism* impairs circulation to the brain, possibly causing a stroke. A *cardiac embolism* can cause a heart attack.

Stroke: damage to blood vessels in the brain

To function normally, the brain requires a steady blood supply—about 15% of the heart's output at rest. Any impairment of blood flow to the brain rapidly damages brain cells. A **stroke** (*cerebrovascular accident*) represents damage to part of the brain caused by an interruption to its blood supply. In effect, it is the brain equivalent of a heart attack. Strokes are the most common cause of brain injury and a leading cause of death in Western nations. The two most common causes are an embolism blocking a vessel and rupture of a cerebral artery.

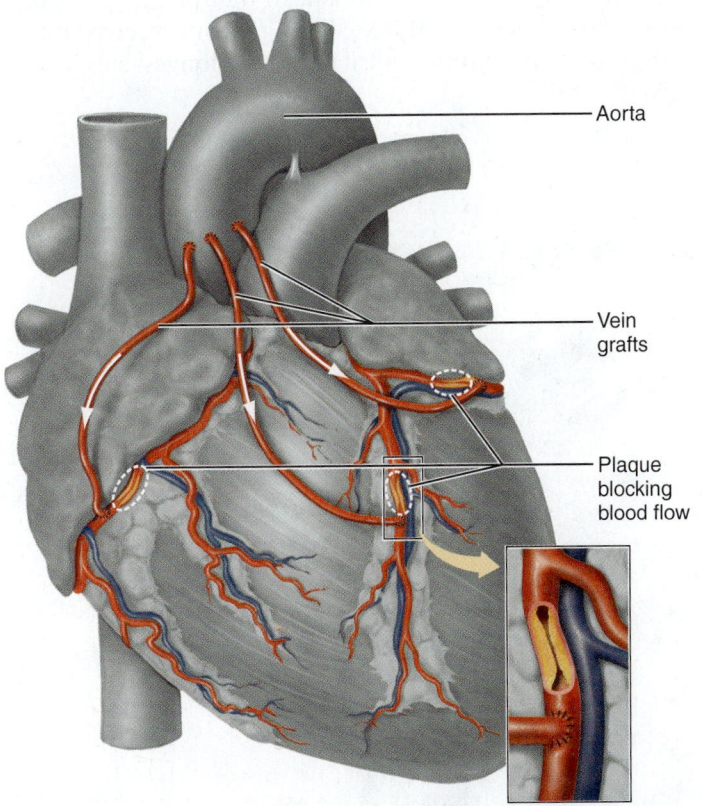

Aorta

Vein grafts

Plaque blocking blood flow

Figure 8.20 Coronary artery bypass grafts. This is an example of a "triple CABG," in which three vein grafts have been placed to bypass three areas of atherosclerotic plaque. Arrows indicate additional blood flow to the area beyond the restricted artery.

Symptoms of strokes appear suddenly and vary according to the area of the brain affected. They may include weakness or paralysis on one side of the body, fainting, inability to speak or slurred speech, difficulty in understanding speech, impaired vision, nausea, or a sudden loss of coordination.

Immediate medical care is crucial. If the stroke resulted from a cerebral embolism, clot-dissolving drugs or an embolectomy (surgery to remove the embolus) can save a life. Eliminating the clot as quickly as possible (within minutes or hours) can also limit the area of damage and reduce the severity of the permanent injury. If a ruptured artery is responsible, health providers may be able to surgically drain the excess blood.

Although some people recover well from a stroke or suffer only minor permanent effects, others do not recover much lost function at all despite intensive physical therapy over many months. The reason for generally poor recovery rates is that the body does not grow new nerve cells to replace damaged ones. Nevertheless, rehabilitation with skilled health professionals generally offers the best chance of at least a partial recovery. Recovery involves retraining nerve pathways that already exist so they can take over the functions of damaged nerve cells.

↻ Recap Cardiovascular disorders are the number one killer in the United States. Most disorders are caused either by conditions that result in failure of the heart as a pump or by conditions in which damage to blood vessels restricts flow or ruptures vessels. ∎

8.6 Replacing a failing heart

Sometimes heart failure is so complete that it becomes necessary to replace the heart entirely. For most patients, the only option is a heart transplant, using a heart from a human cadaver. Heart transplants are expensive (nearly $700,000), but the chances of success are fairly good; currently the average length of survival after a heart transplant is about 15 years. The main problem is a shortage of healthy hearts from recently deceased human donors. About one third of all patients on heart transplant lists die before they get a suitable heart.

When a human heart is not available soon enough or the patient is not eligible for one, the only short-term solution may be an artificial heart. Several different artificial hearts are currently under development, but only two have been approved by the FDA so far. Each is approved only for specific purposes. The SynCardia heart (**Figure 8.21a**) is approved only for temporary use, that is, until a suitable donor can be found. So far, nearly a thousand patients have received the SynCardia heart. About 80% of them survived until they received a human heart transplant. The other artificial heart, called the AbioCor Replacement Heart (**Figure 8.21b**), is only approved to extend the lives of patients who by virtue of their medical condition are not eligible for a heart transplant. According to the manufacturer (AbioMed), 14 patients received the AbioCor heart during clinical trials. The longest survivor lived 512 days.

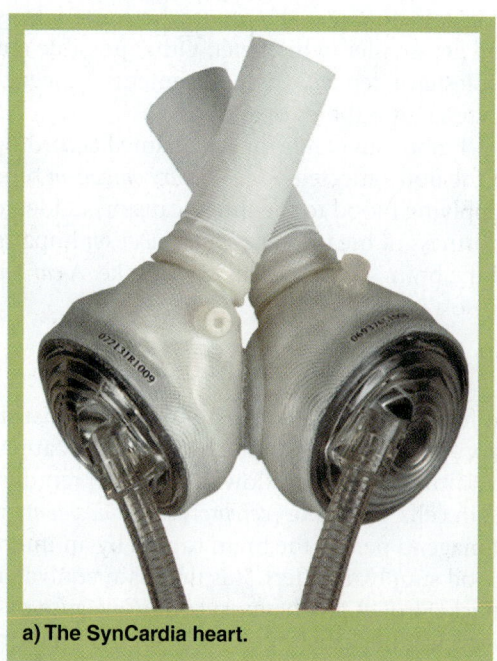

a) The SynCardia heart.

b) The AbioCor heart.

Figure 8.21 Artificial hearts.

Another option being explored is a *xenotransplant* (an organ from another species), most likely from pigs or baboons. These are less than ideal because tissue incompatibility can lead to transplant rejection by the immune system and also because these species do not live as long as humans, raising the question of whether the hearts would last long enough. If these issues could be resolved, xenotransplants could save thousands of lives a year.

There's work to be done before either artificial hearts or xenotransplants are considered viable alternatives to human hearts from cadavers. From the simple perspective of pumping blood adequately, artificial hearts actually perform rather well. If we could overcome the occasional mechanical failures, infections, and the twin problems of blood clotting or excessive bleeding, then an artificial heart might become a permanent replacement for a failed heart, not just a temporary fix.

MJ's BlogInFocus What is being done to improve artificial hearts so that they might replace a human heart permanently, not just temporarily. Visit MJ's blog in the Study Area in MasteringBiology and look under "Artificial Hearts."

http://goo.gl/09Rq9T

Recap The best solution for a failed heart is a heart transplant from a human cadaver. The current generation of artificial hearts is only a temporary solution. Xenotransplants are still in the research stage, not approved for human use. ∎

8.7 Reducing your risk of cardiovascular disease

Cardiovascular disorders are among the most preventable of chronic health conditions. Although some factors are beyond your control—such as sex, race, age, and genetic inheritance—your lifestyle choices can also affect your risk. Things you can do to reduce your risk include:

- Don't smoke, or if you do, quit. Smokers have more than twice the risk of heart attack that nonsmokers do, and smokers who suffer a heart attack are nearly four times as likely to die from it. Some researchers think secondhand smoke poses a risk as well.
- Watch your cholesterol levels. Cardiovascular risk rises with the blood cholesterol level. There is also evidence that high cholesterol increases risk even more when it is combined with other factors such as hypertension and tobacco smoke.

Figure 8.22   Moderate, regular exercise improves cardiovascular performance and lowers the risk of cardiovascular disease.

- Keep moving (Figure 8.22). Regular, moderate exercise lowers the risk of cardiovascular disease. This is not surprising because the heart is, after all, a muscle. Most physicians recommend exercising for at least 20 to 30 minutes, at least three times per week. Physical activity tends to lower blood pressure and cholesterol and makes it easier to maintain a healthy body weight. Always consult your physician before starting an exercise program.
- If your blood pressure is on the high side, seek treatment. As discussed earlier, untreated hypertension damages blood vessels and increases the workload on the heart.

In addition to these major risks, at least three other factors are associated with cardiovascular disease, although the precise link has not yet been determined. This is why doctors recommend the following:

- Maintain a healthy weight. It's not clear how obesity contributes to cardiovascular problems, but overweight people have a higher rate of heart disease and stroke even if they do not have other risk factors. One hypothesis is that increased weight strains the heart. Increased weight also has adverse effects on other risk factors such as blood cholesterol and hypertension.
- Keep diabetes under control. Diabetes mellitus is a disorder of blood sugar levels. Untreated diabetes damages blood vessels, but effective treatments reduce cardiovascular damage significantly. (Diabetes will be described in more detail when we discuss the endocrine system.)
- Avoid chronic stress. Again the mechanism is unclear, but there is an association between a person's perceived stress and behavior patterns and the development of cardiovascular disease. Stress may also affect other risk factors, for example, how much a smoker smokes or whether a person starts smoking.

Recap You can reduce your risk of developing cardiovascular disease by not smoking, exercising regularly, watching your weight and cholesterol, and avoiding prolonged stress. If you have diabetes and/or hypertension, try to keep these conditions under control. ∎

Chapter Summary

8.1 Blood vessels transport blood *p. 161*

- The primary function of blood vessels is to bring blood into close proximity with all living cells.
- Thick-walled arteries transport blood to the capillaries at high pressure.
- Small arterioles and precapillary sphincters regulate the flow of blood into each capillary.
- Thin-walled capillaries are the only vessels that exchange fluids and solutes with the interstitial fluid.
- Distensible venules and veins store blood at low pressure and return it to the heart.

8.2 The heart pumps blood through the vessels *p. 166*

- The heart is composed primarily of cardiac muscle. Structurally, it consists of four separate chambers and four one-way valves. Its primary function is to pump blood.
- The heart pumps blood simultaneously through two separate circuits: the pulmonary circuit, where blood picks up oxygen and gets rid of carbon dioxide, and the systemic circuit, which supplies the rest of the body's cells.
- The heart contracts and relaxes rhythmically. Contraction is called *systole*, and relaxation is called *diastole*.
- The coordinated contraction of the heart is produced by a system of specialized cardiac muscle cells that initiate and distribute electrical impulses throughout the heart muscle.
- An electrocardiogram, or ECG, records electrical activity of the heart from the surface of the body. An ECG can be used to diagnose certain cardiac arrhythmias and disorders.

8.3 Blood exerts pressure against vessel walls *p. 174*

- The heart generates blood pressure, and the arteries store the blood under pressure during diastole.
- Systolic and diastolic arterial blood pressures can be measured with a sphygmomanometer.
- High blood pressure, called *hypertension*, is a major risk factor for cardiovascular disease.

8.4 How the cardiovascular system is regulated *p. 177*

- The most important controlled variable in the cardiovascular system is arterial blood pressure.
- Arterial blood pressure is monitored by stretch receptors located in certain large arteries.
- Cardiac output and the diameters of the arterioles are regulated (controlled) to keep arterial blood pressure relatively constant.
- With pressure held constant, local blood flows can be adjusted according to the metabolic needs of the tissues and cells in that area of the body.

8.5 Cardiovascular disorders: a major health issue *p. 179*

- The heart muscle is always working. Impairment of blood flow to the heart can lead to a sense of pain and tightness in the chest (angina) and/or permanent damage to heart tissue (myocardial infarction, or heart attack).

- Slowly developing, chronic failure of the heart as a pump can lead to excessive interstitial fluid, a condition known as *congestive heart failure*.
- An embolism is the sudden blockage of a blood vessel by any object.
- Strokes, also called *cerebrovascular accidents*, can be caused by either embolisms or rupture of blood vessels. The result is damage to a part of the brain when its blood supply is interrupted.

8.6 Replacing a failing heart *p. 182*

- The best solution for a patient with a failed heart is a heart transplant from a human cadaver.
- Currently, artificial hearts are only intended as temporary solutions for heart failure.

8.7 Reducing your risk of cardiovascular disease *p. 183*

- Your chances of developing a cardiovascular disorder depend on certain risk factors. Some risk factors you cannot change, whereas others depend on the choices you make in life.

Terms You Should Know

arteries, *162*	pulmonary circuit, *168*
atrioventricular (AV) valves, *168*	semilunar valves, *168*
atria, *167*	sinoatrial (SA) node, *172*
capillaries, *164*	systemic circuit, *168*
diastole, *170*	systole, *170*
myocardium, *167*	veins, *165*
precapillary sphincter, *163*	ventricles, *167*

Concept Review

Answers can be found in the Study Area in MasteringBiology.

1. Why is it so important that the heart always get a consistent and adequate blood supply?
2. Compare and contrast the structures and functions of the three types of blood vessels.
3. Describe how arterial blood pressure is measured by the body.
4. Describe the function of the heart valves.
5. Compare the functions of the pulmonary circuit and the systemic circuit.
6. Describe the cardiac cycle of relaxation and contraction, and explain what causes each of the two heart sounds during the cycle.
7. Describe the function of the SA and AV nodes as they control contraction of the heart.
8. Of what functional value is the distensibility of veins? In other words, why not have veins just as thick and stiff as arteries?
9. Name the one hormone that has a stimulatory effect on the heart (increases heart rate).
10. List ways to reduce the risk of cardiovascular disease.

Test Yourself

Answers can be found in the Appendix.

1. _____ carry blood away from the heart and _____ carry blood toward the heart.
   a. Veins . . . arteries
   b. Arteries . . . arterioles
   c. Veins . . . capillaries
   d. Arteries . . . veins
   e. Arteries . . . capillaries

2. Which blood vessel is best suited for exchange of gases and nutrients with the surrounding tissue?
   a. artery
   b. vein
   c. capillary
   d. arteriole

3. Which of the choices represents the order of vessels through which blood passes after leaving the heart?
   a. artery—arteriole—capillary—venule—vein
   b. artery—capillary—arteriole—venule—vein
   c. vein—venule—capillary—arteriole—artery
   d. artery—vein—capillary—arteriole—venule

4. All of the following mechanisms assist in returning venous blood to the heart except:
   a. an increase in heart rate
   b. pressure changes in the abdominal and thoracic cavities due to breathing
   c. contraction of skeletal muscles in the legs
   d. one-way valves located inside veins

5. Which of the following represents the order of structures beginning inside the ventricle and traveling outward?
   a. pericardium—epicardium—myocardium—endocardium
   b. epicardium—myocardium—endocardium—pericardium
   c. endocardium—myocardium—epicardium—pericardium
   d. endocardium—pericardium—myocardium—epicardium

6. Which vein(s) carry oxygenated blood?
   a. superior vena cava and inferior vena cava
   b. right and left pulmonary veins
   c. aorta
   d. both (a) and (b)

7. Which of the following statements regarding the cardiac cycle is false?
   a. When the ventricles contract, the atrioventricular valves close.
   b. When the ventricles relax, the semilunar valves close.
   c. When the atria contract, the semilunar valves open.
   d. When the atria contract, the atrioventricular valves open.

8. A pacemaker is used to correct:
   a. coronary artery disease
   b. cardiac arrhythmias
   c. heart murmurs
   d. hypertension

9. As the blood travels through the circulatory system, the greatest drop in pressure occurs in:
   a. arteries
   b. arterioles
   c. capillaries
   d. venules

10. All of the following are part of the cardiac conduction system except the:
    a. chordae tendineae
    b. Purkinje fibers
    c. sinoatrial node
    d. atrioventricular bundle

11. Which of the following is/are involved in regulating blood pressure?
    a. heart
    b. baroreceptors
    c. cardiovascular center in medulla oblongata
    d. all of the above

12. Which of the following would be an appropriate homeostatic response to a drop in blood pressure below what is normal?
    a. heart rate decreases
    b. vasoconstriction of arterioles
    c. force of cardiac contraction decreases
    d. both (a) and (b)

13. Which of the following might be treated appropriately by administering clot-dissolving drugs?
    a. hypertension
    b. hypotension
    c. pulmonary embolism
    d. a hemorrhagic stroke

14. Which of the following is a true statement regarding strokes?
    a. Most of them are caused by a drop in blood pressure.
    b. The symptoms of a stroke develop slowly, sometimes over years.
    c. Immediate medical care is crucial to recovery.
    d. Once symptoms develop, no recovery is possible.

15. A heart attack occurs as a result of:
    a. prolonged hypotension
    b. narrowing or blockage of the coronary arteries
    c. improper closure of semilunar valves
    d. disruption of the cardiac conduction system

Apply What You Know

Answers can be found in the Study Area in MasteringBiology.

1. During exercise, arterial blood pressure changes very little. However, cardiac output may double and blood flow to exercising muscle may go up 10-fold, while at the same time the blood flow to kidneys may decline by nearly 50%. Explain possible mechanisms that might account for these very different changes.

2. When a coronary artery bypass graft (CABG) is performed, the vessels used for the bypass grafts are usually veins taken from the patient's legs. Over time the grafted veins take on many of the characteristics of arteries; that is, they become thicker and stiffer. What might this suggest about the possible cause(s) of the structural differences between arteries and veins? Hypothesize what might happen if you took a section of artery and implanted it into a vein.

3. Soldiers have to stand in formation at full attention for long periods of time. Sometimes this can cause otherwise very fit and healthy young people to pass out during long inspections. What would cause this?

4. Workers who spend hours per day standing can develop circulation problems in their legs. A recommended solution is to wear graduated compression stockings. These stockings are tighter around the ankles and less tight higher up on the legs. Why does standing for long periods sometimes lead to circulatory problems, and how can wearing something tight on the legs help prevent this?

5. Inflammation of the pericardium can lead to a condition called *pericardial effusion*, in which fluids collect in the space surrounding the heart. What effect would this have on the functioning of the heart? How might the condition be treated?

6. Ventricular fibrillation is a potentially fatal condition where the cells of the heart are no longer coordinated in their contractions. Why is it important that contraction of the heart muscle occurs nearly all at once?

7. VSD, or ventricular septal defect, is a condition that accounts for half of all congenital cardiovascular anomalies. In this condition, a hole exists in the septum, the muscular wall between the ventricles, which allows blood from the right and left side of the heart to mix. What sort of problems might this condition cause?

8. Last night, you and your roommate were sound asleep when the phone rang. Your roommate, startled awake, jumped from her bed, and rushed for the phone across the room. Before she reached the phone, she suddenly felt dizzy and had to sit down to avoid fainting. What happened?

MJ's BlogInFocus

Answers can be found in the Study Area in MasteringBiology.

1. Based on what you know about the heart, do you think that lowering *prehypertensive* blood pressures to *normal* might be cost-effective in preventing cardiovascular disease? To view a research paper that relates tangentially to this question, visit MJ's blog in the Study Area in MasteringBiology and look under "Treating Prehypertension."

   http://goo.gl/6SAKL2

MasteringBiology®

Students Go to MasteringBiology for assignments, the eText, and the Study Area with animations, practice tests, and activities.

Professors Go to MasteringBiology for automatically graded tutorials and questions that you can assign to your students, plus Instructor Resources.

Answers to ✓ questions are available in the Appendix.

The Immune System and Mechanisms of Defense

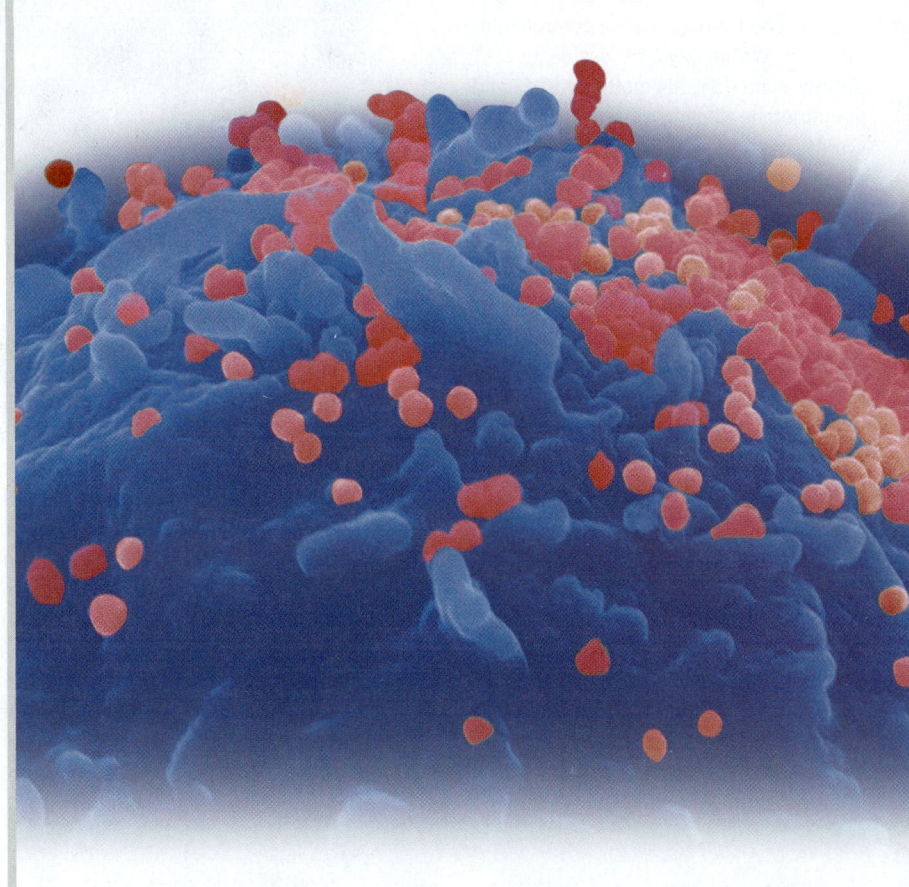

New HIV particles (pink) budding from the surface of a T lymphocyte (blue).

Key Concepts

- **The health risk of a pathogen** (disease-causing organism) **is determined by its transmissibility** (how easily it can be passed from person to person), **mode of transmission** (how it is transmitted; through air, food, blood, etc.), **and virulence** (how damaging the disease is when one catches it).

- **The immune system has nonspecific** (against many pathogens) **and specific** (against one pathogen) **defense mechanisms.**

- **Nonspecific defense mechanisms** include immune system cells that engulf and digest foreign cells, chemicals that are toxic to foreign cells, proteins that interfere with viral reproduction, and the development of a fever.

- **Specific defense mechanisms involve the production of antibodies and T cells** that recognize and inactivate one particular pathogen. Specific defense mechanisms have a memory component that is the basis of immunity.

- **Inappropriate immune system activity** can lead to allergies and autoimmune diseases.

- **AIDS (acquired immune deficiency syndrome) is caused by a virus that targets certain cells of the immune system.**

CURRENT ISSUE
An Outbreak of Ebola

One-year-old Emile Ouamouno of the West African nation of Guinea developed a cough and mild fever. Emile's parents thought nothing of his sickness at first, but then he developed widespread uncontrollable bleeding. Within days, he was dead. Shortly thereafter, his mother, sister, and grandmother all came down with the same symptoms, and all three died. A village nurse and the local midwife also died of the mysterious and apparently contagious disease, but not before the midwife had passed it on to people in the surrounding area.

Young Emile, whose death occurred on December 6, 2013, was posthumously diagnosed with a disease called Ebola hemorrhagic fever (EHV), or simply Ebola. Emile is thought to be patient zero in an ongoing outbreak of Ebola that has spread in Guinea and two neighboring countries, Liberia and Sierra Leone. As of 2015, the death toll from Ebola in those countries is over 8,000.

What Is Ebola?

Ebola is an infectious disease caused by a virus called *Zaire ebolavirus*. The virus's normal hosts are certain animals such as monkeys and bats, but it can also infect humans if there is direct contact with an infected animal's bodily fluids. Young

Freshly dug graves for Ebola victims in Freetown, Sierra Leone.

Emile Ouamouno was probably bitten by an infected fruit bat living in the trees nearby.

Once the first person is infected, the Ebola virus is transmitted from human to human by direct contact with bodily fluids such as saliva, nasal mucous, or feces. It's no surprise, then, that the next victims in the most recent outbreak were Emile's relatives and caretakers. An infection begins with symptoms that may include fever, weakness, muscle pain, sore throat, and headache. These early symptoms are followed by vomiting, damage to liver and kidneys, and, in some patients, internal and external bleeding. In the most severe cases, blood begins to leak from every opening and every organ, leading to rapid death.

Questions to Consider

1 What should the United States do when an infectious disease breaks out elsewhere in the world? In such a scenario, what is our responsibility and/or what is in our best interests?

2 How afraid are you of Ebola? Would you be willing to travel to Guinea if your boss asked you to? Why or why not?

Several features of the virus make it particularly dangerous. First, the earliest symptoms of the disease are similar to the common cold or the flu, so it may go undiagnosed until it is too late. Second, the disease has a long incubation time (days to even several weeks) before symptoms first appear. Therefore, the disease can be transmitted to another person before anyone is aware that the patient actually has Ebola. Third, the virus kills approximately 50% of all persons infected, generally within days of the appearance of symptoms. And fourth, there is no known cure for the disease. The only way to end an outbreak is to prevent the disease from spreading from person to person.

Ebola in the United States

By early 2014, U.S. health officials were well aware of the Ebola outbreak in Africa and were on the lookout for its appearance elsewhere. Despite their preparations for international spread of the disease, the arrival of Ebola in the United States caused widespread concern and action. Travel restrictions and health screening procedures were put in place for all persons traveling from Liberia, Sierra Leone, or Guinea. U.S. hospitals made plans to isolate and treat what could've possibly become a large number of Ebola-infected patients. Health officials tracked down and quarantined persons who had been exposed to the virus. Ebola was among the top news stories every day.

Ebola was first diagnosed within the United States during September of 2014, when a Liberian man with the disease traveled to Dallas, Texas, to visit relatives. (He later died, despite the best medical care available.) Two U.S. health care workers who had been in contact with the Liberian patient were infected with the disease and began to show symptoms a month later. Both of the care workers recovered fully. →

Medical personnel in Monrovia, Liberia, disinfect people who brought a patient suspected of having Ebola to the hospital.

In addition, a medical aid worker for Doctors Without Borders in Guinea returned to the United States with an Ebola infection. He was quickly isolated and treated, and he ultimately survived without infecting others.

By early 2015 and with no new cases of Ebola reported in the United States, the media and the public began to relax. The country had dodged a bullet (at least for a time).

Why Was West Africa Worst Hit?

Over 98% of the deaths from Ebola worldwide in 2014 were in Guinea, Liberia, and Sierra Leone. It's as if a "perfect storm" of factors came together in just those three countries.

First, Ebola is common in certain animal populations in Africa. Cultural practices on the continent of butchering and eating "bush meat," such as monkeys, increases the risk of contact with the virus.

Second, the first appearance of a rare disease is always a surprise, especially if the symptoms look like those of so many other diseases. People and health officials are therefore unprepared for the disease, giving it time to spread before containment efforts are mounted. If patient zero had been someone who became ill in a country where a new disease would be more readily identified and health care more accessible, the outcome might have been quite different than it was; the spread of the disease would likely be much more contained than it is currently in Africa.

Third, social norms and customs may have exacerbated the spread of the outbreak. In West Africa, it is customary for the family of a deceased person to participate in washing the body and preparing it for burial—a practice that may have contributed to the disease's spread before its mode of transmission was known.

Fourth, the three countries most affected do not have prosperous economies and stable governments, both of which are invaluable in containing disease. It's expensive to build, staff, and maintain quarantine facilities and to track and monitor patients exposed to a disease. In addition, both Liberia and Sierra Leone had suffered through devastating civil wars in the 1990s, and consequently, there is general distrust of the military, the institution in charge of containing disease outbreaks. Early on in the Ebola outbreak, efforts by these countries' militaries (not doctors) to locate and quarantine infected persons were often stymied because relatives hid their afflicted at home or refused to seek help in caring for them. In south Guinea, eight people, including three journalists, were killed by villagers who mistakenly believed that the strangers were deliberately sent to contaminate their village with the virus.

Fifth and perhaps most importantly, the existing medical infrastructure was simply not up to the task of detection, treatment, and containment. In Liberia, half of all health centers closed at the most critical time of need in the Ebola outbreak because medical personnel were not available. Some medical workers refused to come to work because they feared (perhaps rightly) for their *own* lives.

Lessons Learned

An intensive, coordinated medical response is perhaps the only way to stop an outbreak of an infectious disease quickly. Hospitals, local medical clinics, and health care workers have to be ready in advance. Surveillance and communication are required to identify the earliest cases of the disease and to track down and quarantine those exposed to the disease. None of these essential elements are present in West Africa.

Concerted efforts by other countries and international medical agencies are now beginning to slow the spread of the disease, but the outbreak is by no means over. But even if this outbreak of Ebola can be stopped soon, the three most affected countries could still drift toward economic collapse and political chaos. This story is not over.

SUMMARY

- In 2014, an outbreak of Ebola killed over 8,000 people in the West African nations of Guinea, Liberia, and Sierra Leone.
- Only four patients were diagnosed with Ebola in the United States, and only one of them died.
- There is no cure for an Ebola infection. About half of all persons infected with Ebola die within weeks.
- Stopping the spread of an Ebola outbreak requires an intensive, coordinated medical response that is generally not available in less developed countries.

The world is swarming with living organisms (bacteria) and even some nonliving entities (viruses and prions) too small to be seen with the naked eye. They're found on doorknobs, the money we handle, and our clothes. They're in the food we eat and the air we breathe. Most are harmless—indeed, some are highly beneficial. The ones that cause disease are called **pathogens** (from the Greek *pathos,* meaning "disease," plus *gennan,* meaning "to produce"). As a group, pathogens account for a large fraction of all human disease and suffering. Challenges to our health occasionally come from some of our own cells as well: Mutations (alterations to a cell's DNA) may cause cells to become abnormal and may even lead to cancer, and damaged cells must be recognized and removed. The key to our defenses against pathogens and our own abnormal cells is the **immune system,** a complex group of cells, proteins, and structures of the lymphatic and circulatory systems. Most organ systems have at least one important function that contributes to the maintenance of homeostasis of the internal environment. The immune system's primary function is to protect all other organ systems from attack. Without the immune system standing guard, multiple organ system failures soon follow—consider what happens when the immune system itself is rendered inoperable by a pathogen, such as human immunodeficiency virus (HIV). ▪

Our body's immune system and other general defense mechanisms include:

- *Barriers to entry or ways of expelling or neutralizing pathogens before they can do harm.* These include skin, stomach acid, tears, and such actions as vomiting and defecation.

- *Nonspecific defense mechanisms.* Nonspecific defenses help the body respond to generalized tissue damage and many of the more common or obvious pathogens, including most bacteria and some viruses.
- *Specific defense mechanisms.* These enable the body to recognize and kill specific bacteria and other foreign cells and to neutralize viruses. Our specific defense mechanisms employ sophisticated weaponry indeed. The specific defense mechanisms are also the basis of immunity from future disease.

All three mechanisms are operating night and day to protect us. The latter two (nonspecific and specific defenses) are carried out by the immune system. However, even the immune system is not perfect; it can only kill or neutralize pathogens or abnormal cells that it can recognize. This has implications for how the human body deals with certain pathogens.

What is the immune system on alert for? We start by looking at some of the kinds of pathogens that can invade our bodies.

9.1 Pathogens cause disease

Pathogens include bacteria, viruses, fungi, a few protozoa, and possibly prions. Larger parasites, including various worms, can also be pathogens, although they are relatively rare in industrialized countries. Because bacteria and viruses are by far the most numerous and problematic pathogens in Western nations, we focus on them as we describe how we protect and defend ourselves. We'll also discuss prions as a newly discovered class of pathogens.

Bacteria: single-celled living organisms

Bacteria (singular: bacterium) are single-celled organisms that do not have a nucleus or membrane-bound organelles. All the DNA in most bacteria is contained in just one chromosome, which usually forms a continuous loop that is anchored to the plasma membrane. Bacterial ribosomes

are smaller than ours and float freely in the cytoplasm. The outer surface of bacteria is covered by a rigid cell wall that gives bacteria their distinctive shapes, including spheres, rods, and spirals (**Figure 9.1**).

Judging by their variety and numbers, bacteria are among the most successful organisms on Earth. Although they are smaller than the typical human cell, their small size is actually an advantage. Like all living organisms, bacteria need energy and raw materials to maintain life and to grow and divide. Their small size means that bacteria have a high surface-to-volume ratio, a decided advantage when it comes to diffusion, the means by which they obtain raw materials and get rid of wastes.

Like our own cells, bacteria use ATP as a direct energy source and amino acids for making proteins. They store energy as carbohydrates and fats. Where do they obtain those raw materials? Anywhere they can. Some bacteria break down raw sewage and cause the decomposition of dead animals and plants, thereby playing an essential role in the recycling of energy and raw materials. Others obtain nutrients from the soil and air.

Humans have learned to harness bacteria to produce commercial products, including antibiotic drugs, hormones, vaccines, and foods ranging from sauerkraut to soy sauce. Some bacteria even live within our digestive tract, drawing energy from the food we eat in exchange for manufacturing vitamins or controlling the populations of other, more harmful bacteria. Life, as we know it, would not be possible without these little organisms.

A few bacteria are pathogens, however. Pathogens rely on living human cells for their energy supply, and in the process they damage or kill the human cells. They cause pneumonia, tonsillitis, tuberculosis, botulism, toxic shock syndrome, syphilis, Lyme disease, and many other diseases. Although we concentrate on pathogens in this chapter, do keep in mind that most bacteria are harmless and many are even beneficial.

Bacterial infections are generally treated with **antibiotics**—chemotherapeutic agents that inhibit or abolish the growth of bacteria, fungi, and protozoa.

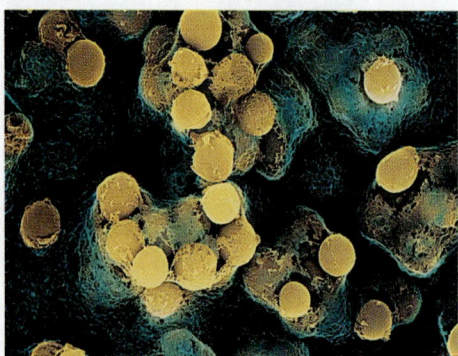
a) SEM (×2,000) of *Streptococcus*, a spherical bacterium that causes sore throats.

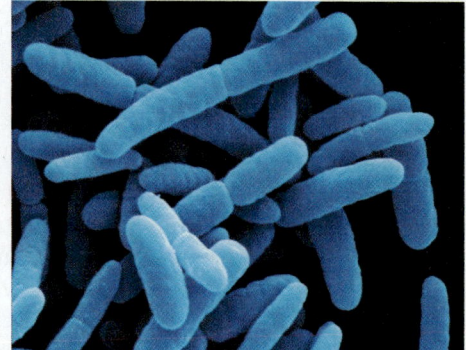
b) SEM (×5,600) of *Escherichia coli*, a common intestinal bacteria that is usually harmless.

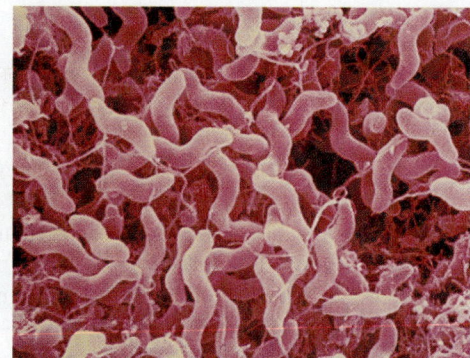
c) SEM (×12,000) of *Campylobacter jejuni*, a spiral-shaped bacterium that causes food poisoning.

Figure 9.1 Electron micrographs of the three common shapes of bacteria.

Viruses: tiny infectious agents

Viruses are extremely small infectious agents, perhaps one-hundredth the size of a bacterium and one-thousandth the size of a typical eukaryotic cell (Figure 9.2). Structurally, a virus is very simple, consisting solely of a small piece of genetic material (either RNA or DNA) surrounded by a protein coat. Viruses have no organelles of their own, so they can't grow and reproduce without access to the organelles of eukaryotic cells.

Are viruses alive? Biologists are divided on the answer to this question. Most would say that viruses are not alive because they cannot reproduce on their own. Viruses have no observable activity associated with life when they are not in contact with another living cell. However, when they enter a living cell, they take it over and use the cell's organelles to replicate.

Viruses have several ways of gaining entry into living cells. Most viruses that infect human cells are taken into the cell cytoplasm by endocytosis; once inside the cell, the protein coats are dissolved and the viral genetic material is released for incorporation into the cell's genetic material. Other viruses merge their outer coat with the cell membrane and release their genetic contents into the cell's cytoplasm. Still other viruses attach to the outer surface of the cell membrane and inject just their genetic material into the cell, much as a needle and syringe inject drugs into the body. Regardless of the method of entry, the presence of the viral genetic material causes the cell to begin producing thousands of copies of the virus instead of carrying out its own metabolic activities. Sometimes the newly formed viruses are released by a type of budding from the cell membrane while the cell is still alive. In other cases, the cell becomes so packed with viruses that it dies and bursts, releasing a huge number of viruses all at once.

Diseases caused by different types of viruses range from serious—AIDS, hepatitis, encephalitis, rabies—to annoying—colds, warts, or chicken pox. Viral infections can be minor for some people but serious for others. An otherwise healthy person may be ill for only a few days with a viral infection, whereas someone who is very young, very old, or in poor health may die.

Antibiotics generally don't work against viral infections. The best ways to cure a viral infection are either to prevent the viruses from entering living cells or to stop an infected host cell from producing more viruses.

Prions: infectious proteins

In 1986, scientists identified a disease in British cattle that destroyed nerve cells in the animals' brains and spinal cords, causing the animals to stagger, jerk, tremble, and exhibit other bizarre behaviors. The press nicknamed the condition "mad cow disease." Then between 1994 and 1995, 10 Britons aged 19 to 39 developed signs of a new human disease called *variant Creutzfeldt-Jakob disease (vCJD)*. Eight of them died. Alarmingly, researchers found that all of the vCJD patients had eaten beef from animals suspected of having mad cow disease. In 1996, scientists confirmed that a *prion* was responsible for both the mad cow disease and the first 10 cases of vCJD.

A **prion** is a misfolded form of a normal brain cell protein. But it is not just a misfolded protein—it is misfolded protein that can trigger the misfolding of nearby *normal* forms of the protein as well. Once prions enter a nerve cell, the misfolding process becomes self-propagating—one prion produces another, which produces another, and so on. Eventually, so many prions accumulate within infected

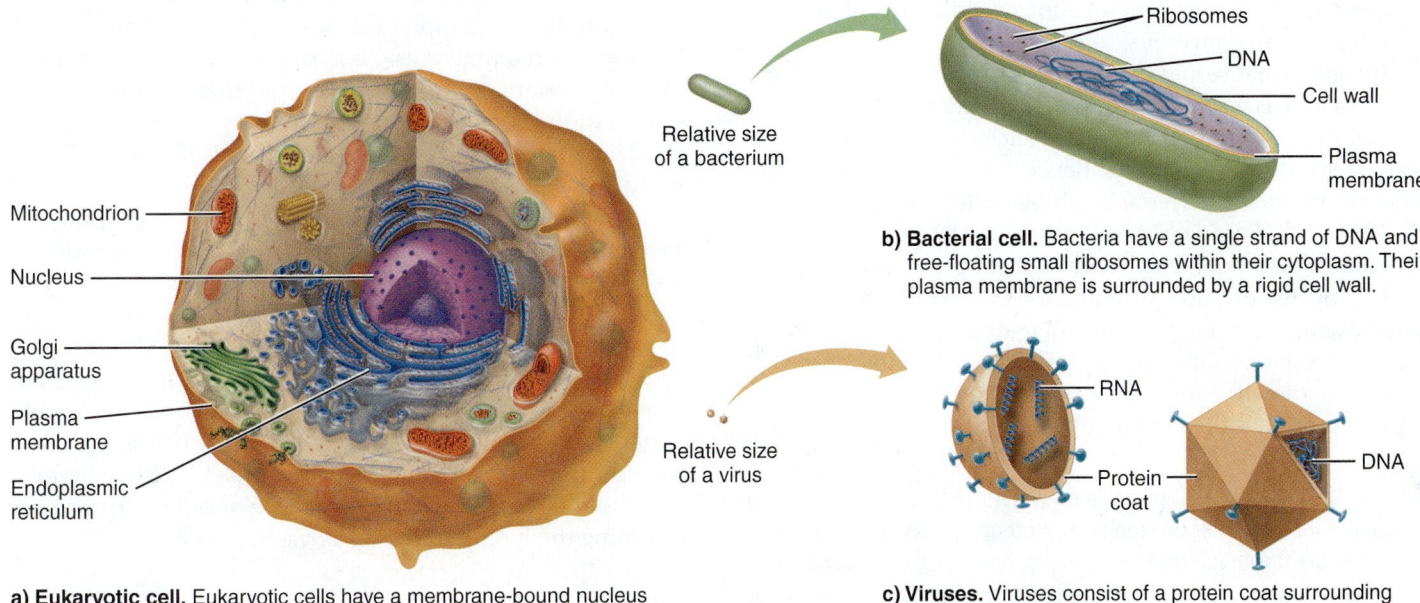

a) Eukaryotic cell. Eukaryotic cells have a membrane-bound nucleus and well-defined membrane-bound organelles.

Mitochondrion
Nucleus
Golgi apparatus
Plasma membrane
Endoplasmic reticulum

Relative size of a bacterium

Relative size of a virus

Ribosomes
DNA
Cell wall
Plasma membrane

b) Bacterial cell. Bacteria have a single strand of DNA and free-floating small ribosomes within their cytoplasm. Their plasma membrane is surrounded by a rigid cell wall.

RNA
Protein coat
DNA

c) Viruses. Viruses consist of a protein coat surrounding either RNA or DNA.

Figure 9.2 Size and structural differences between a eukaryotic cell, a bacterium (prokaryotic cell), and viruses.

brain cells that the cells die and burst, releasing prions to infect other brain cells. The death of nerve cells accounts for the debilitating neurological symptoms and progressive degeneration seen in both mad cow disease and human vCJD.

Prions are resistant to cooking, freezing, and even drying. There is no known cure for prion infection. Because infection occurs when humans (or cattle) eat prion-infected cattle tissues, the best way to prevent vCJD in humans is to limit the spread of mad cow disease in cattle. Global cooperation is making this possible. In 1994, the European Union banned the use of mammalian meat and bone meal products as cattle feed, and since that time the number of cases of mad cow disease has fallen dramatically.

✅ Suppose you are studying a mysterious disease, and you discover that it is caused by a tiny pathogen that contains some nucleic acid and some protein, but does not have a plasma membrane. Is it most likely a bacterium, a virus, or a prion? Explain your answer.

Transmissibility, mode of transmission, and virulence determine health risk

Some pathogens are clearly more risky to human health than others. Factors that determine the danger of a particular pathogen include *transmissibility* (how easily it is transferred from one person to another), *mode of transmission* (the method of transfer), and *virulence* (how severe the resulting disease is).

For instance, the viruses that cause the common cold are easily transmitted from hands to mucous membranes, as well as in the fluid particles spread by a sneeze. However, these viruses tend not to be very virulent. The HIV virus that causes AIDS, by contrast, is only moderately transmissible. Its mode of transmission is limited to exchange of body fluids (blood, semen, breast milk, or vaginal secretions). However, the HIV virus is tremendously virulent, and this is what makes it so dangerous.

Imagine what would happen if a disease as transmissible as the flu and as virulent as AIDS were to arise in the human population. In fact, there have been such diseases, and they have caused deadly epidemics. Between 1348 and 1350, the **bubonic plague,** a bacterial infection, killed an estimated 25–40% of the European population. A 1918 outbreak of influenza killed more than 20 million people worldwide.

Pathogens continue to challenge human defenses. A prime example is the Ebola virus that arose in Africa in 1976 and still presents a threat today. It is one of the most virulent pathogens known, killing more than 80% of an exposed population in less than two weeks.

↩ **Recap** Like all cells, bacteria draw their energy and raw materials from their environment. Pathogenic bacteria get the materials they need from living cells, damaging or killing the cells in the process. A virus consists of a single strand of DNA or RNA surrounded by protein. Viruses use their DNA or RNA to force a living cell to make more copies of the virus. Prions are infectious proteins that cause normal proteins to misfold. ∎

9.2 The lymphatic system defends the body

As noted in the chapter on blood, the **lymphatic system** is closely associated with the cardiovascular system. The lymphatic system performs three important functions:

- It helps maintain the volume of blood in the cardiovascular system.
- It transports fats and fat-soluble vitamins absorbed from the digestive system to the cardiovascular system.
- It defends the body against infection.

We briefly described how the lymphatic system helps to maintain blood volume and interstitial fluid volume when we discussed the cardiovascular system. We'll discuss its role in transporting fats and vitamins when we describe the digestive system. In this chapter, we turn to the third function: the role of the lymphatic system in protecting us from disease. Most of the cells of the immune system are housed in the lymphatic system, although they can also circulate in blood and enter the interstitial fluid. Here, we describe the structural components of the system; in later sections, we discuss how specific immune system cells carry out their function.

The basic components of the lymphatic system are a network of lymph vessels throughout the body, the lymph nodes, the spleen, the thymus gland, the tonsils, and the adenoids (**Figure 9.3**).

Lymphatic vessels transport lymph

The lymphatic system begins as a network of small, blind-ended *lymphatic capillaries* in the vicinity of the cells and blood capillaries. Lymph capillaries have wide spaces between overlapping cells. Their structure allows them to take up substances (including bacteria) that are too large to enter a blood capillary.

The fluid in the lymphatic capillaries is *lymph*, a milky body fluid that contains white blood cells, proteins, fats, and the occasional bacterium and virus. Lymphatic capillaries merge to form the *lymphatic vessels*. Like veins, lymphatic vessels have walls consisting of three thin layers, and they contain one-way valves to prevent backflow of lymph. Also like veins, flow in lymphatic vessels is aided by skeletal muscle contractions and pressure changes in the chest during respiration. The lymphatic vessels merge to form larger and larger vessels, eventually creating two major lymphatic ducts: the *right lymphatic duct* and the *thoracic duct*. The two lymph ducts join the subclavian veins near the shoulders, thereby returning the lymph to the cardiovascular system.

Lymph nodes cleanse the lymph

Located at intervals along the lymphatic vessels are small organs called **lymph nodes.** Lymph nodes remove

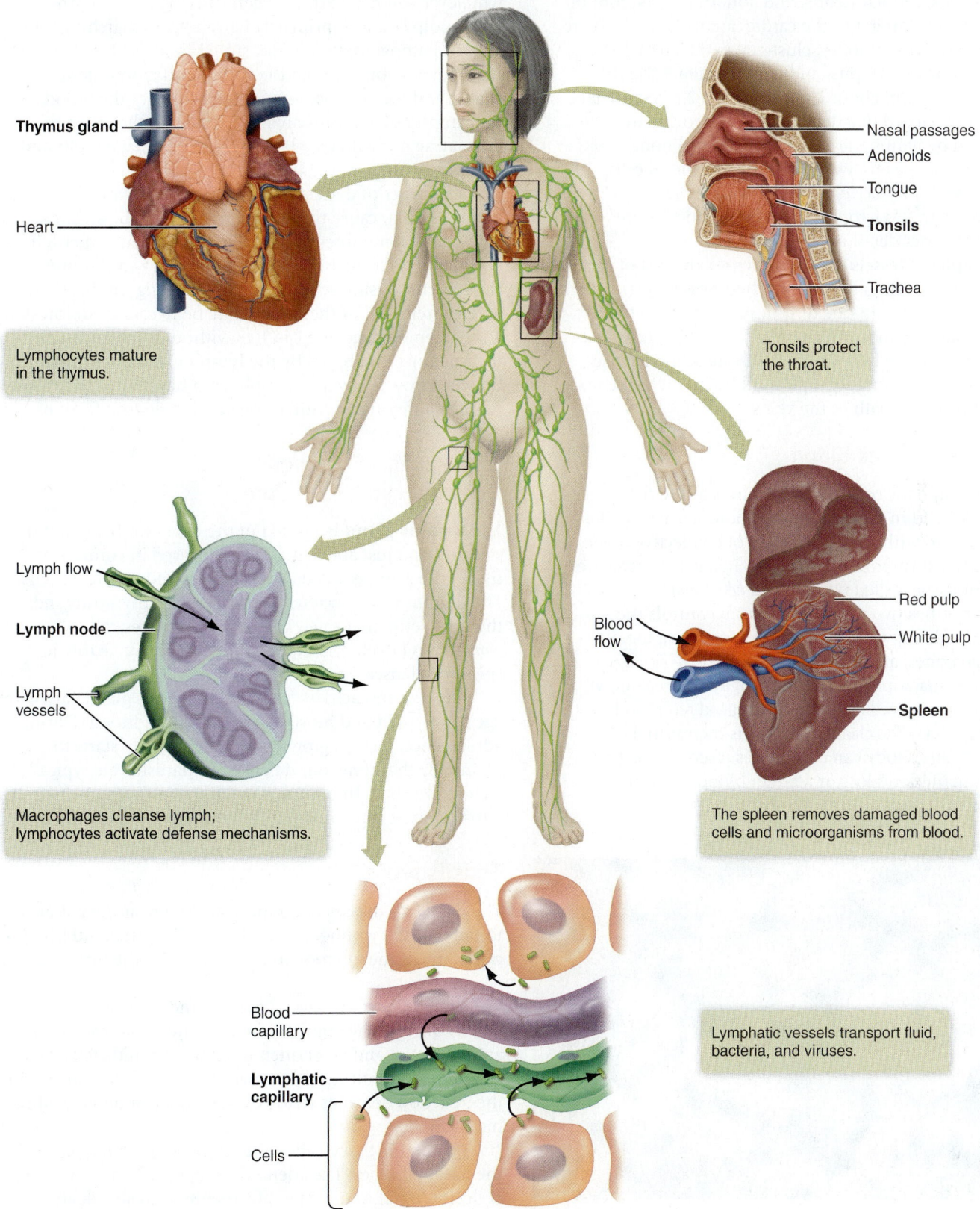

Thymus gland

Heart

Lymphocytes mature
in the thymus.

Nasal passages
Adenoids
Tongue
Tonsils
Trachea

Tonsils protect
the throat.

Lymph flow
Lymph node
Lymph
vessels

Macrophages cleanse lymph;
lymphocytes activate defense mechanisms.

Red pulp
White pulp
Spleen

Blood
flow

The spleen removes damaged blood
cells and microorganisms from blood.

Blood
capillary

**Lymphatic
capillary**

Cells

Lymphatic vessels transport fluid,
bacteria, and viruses.

Figure 9.3 The lymphatic system. The lymphatic system consists of a network of lymphatic
vessels throughout the body, lymph nodes, the thymus gland, tonsils, adenoids, and the spleen.

microorganisms, cellular debris, and abnormal cells from the lymph before returning it to the cardiovascular system. There are hundreds of lymph nodes, clustered in the areas of the digestive tract, neck, armpits, and groin (**Figure 9.4**). They vary in diameter from about 1 millimeter to 2.5 centimeters. Each node is enclosed in a dense capsule of connective tissue pierced by lymphatic vessels. Inside each node are connective tissue and two types of white blood cells, *macrophages* and *lymphocytes*, which identify microorganisms and remove them. (Macrophages and lymphocytes are discussed in greater detail in later sections.)

The lymphatic vessels carry lymph into and out of each node (see Figure 9.3). Valves within these vessels ensure that lymph flows only in one direction. As the fluid flows through a node, the macrophages destroy foreign cells by phagocytosis, and the lymphocytes activate other defense mechanisms. The cleansed lymph fluid flows out of the node and continues on its path to the veins.

The spleen cleanses blood

The largest lymphatic organ, the **spleen,** is a soft, fist-sized mass located in the upper-left abdominal cavity. The spleen is covered with a dense capsule of connective tissue interspersed with smooth muscle cells. Inside the organ are two types of tissue, called *red pulp* and *white pulp.*

The spleen has two main functions: it controls the quality of circulating red blood cells by removing the old and damaged ones, and it helps fight infection. The red pulp contains macrophages that scavenge and break down microorganisms as well as old and damaged red blood cells and platelets. The cleansed blood is then stored in the red pulp. Your body can call on this reserve for extra blood in case of blood loss or a fall in blood pressure, or

whenever you need extra oxygen-carrying capacity. The white pulp contains primarily lymphocytes searching for foreign pathogens; it does not store blood. Notice that the main distinction between the spleen and lymph nodes is *which* fluid they cleanse—the spleen cleanses the blood, and the lymph nodes cleanse lymph. Together, they keep the circulating body fluids relatively free of damaged cells and microorganisms.

A number of diseases, such as infectious mononucleosis and leukemia, cause the spleen to enlarge. The swollen spleen can sometimes be felt as a lump in the upper-left abdomen. A strong blow to the abdomen can rupture the spleen, causing severe internal bleeding. In this case, surgical removal of the spleen may be necessary to forestall a fatal hemorrhage. We can live without a spleen because its functions are shared by the lymph glands, liver, and red bone marrow. However, people who have had their spleen removed are often a little more vulnerable to infections.

Thymus gland hormones cause T lymphocytes to mature

The **thymus gland** is located in the lower neck, behind the sternum and just above the heart. Encased in connective tissue, the gland contains lymphocytes and epithelial cells. The thymus gland secretes two hormones, thymosin and thymopoietin, that cause certain lymphocytes called *T lymphocytes* (T cells) to mature and take an active role in specific defenses.

The size and activity level of the thymus gland vary with age. It is largest and most active during childhood. During adolescence, it stops growing and then slowly starts to shrink. By that time, our defense mechanisms are typically well established. In old age, the thymus gland may disappear entirely, to be replaced by fibrous and fatty tissue.

Tonsils protect the throat

The *tonsils* are masses of lymphatic tissue near the entrance to the throat. Lymphocytes in the tonsils gather and filter out many of the microorganisms that enter the throat in food or air.

We actually have several tonsils, and some are not readily visible. The familiar tonsils at the back of the throat are the largest and most often infected. When they become infected, the resulting inflammation is called *tonsillitis.* If the infection becomes serious, the tissues can be surgically removed.

Lymphatic tissue called the *adenoids* lies at the back of the nasal passages. The adenoids tend to enlarge during early childhood, but in most people, they start to shrink after age 5 and usually disappear by puberty. In some cases, they continue to enlarge and obstruct airflow from nose to throat. This can cause mouth breathing, a nasal voice, and snoring. Like the tonsils, the adenoids can be surgically removed if they grow large enough to cause problems.

Figure 9.4 Lymph node. Color scanning electron micrograph (×1,000) of a section through a lymph node, showing macrophages (pink) and lymphocytes (yellow) lying in wait to attack and destroy foreign and damaged cells. A few red blood cells (red) are also present.

↺ **Recap** The lymphatic system helps protect us from disease. Macrophages and lymphocytes within the lymph nodes identify microorganisms and remove them. The spleen removes damaged red blood cells and foreign cells from blood. The thymus gland secretes hormones that help T lymphocytes mature. Cells in the tonsils gather and remove microorganisms that enter the throat. ■

9.3 Keeping pathogens out: the first line of defense

Our bodies have several ways to prevent entry of pathogens. We also have ways to neutralize them, inhibit their growth, or expel them when they do get in. Here are some of the body's defense strategies.

Skin: an effective deterrent

The most important barrier against entry of any pathogen into our bodies is the skin. Skin has four key attributes that make it such an effective barrier: (1) its structure, (2) the fact that it is constantly being replaced, (3) its acidic pH, and (4) the production of an antibiotic by sweat glands.

Regarding structure, the outermost layers of the skin's epidermis consist of dead, dried-out epithelial cells. These cells contain a fibrous protein called *keratin,* which is also a primary component of fingernails and hair. When skin cells die and their water content evaporates, the keratin forms a dry, tough, somewhat elastic barrier to entry by microorganisms.

Skin is continually being renewed throughout life. Dead cells shed from the surface are replaced by new cells at the base of the epidermis. Any pathogens deposited on the surface are shed along with the dead cells.

Healthy skin has a pH of about 5 to 6, primarily because of the sweat produced by sweat glands. This relatively low (acidic) pH makes skin a hostile environment for many microorganisms.

Sweat glands produce and secrete *dermicidin,* a natural antimicrobial peptide. Dermicidin is effective against a range of harmful bacteria, as well as some fungi.

For further proof of intact skin's effectiveness as a barrier to infection, look at what happens when the skin is damaged by a cut or scratch. If the damage reaches the moist layers of living cells underneath the skin, you may see signs of infection in the area within a few days. One of the most critical problems in treating patients with extensive burns is the infections that often result from the loss of the barrier function of skin.

Impeding pathogen entry in areas not covered by skin

Most successful pathogens enter the body at places where we do not have skin. They enter through the mucous membranes that line the digestive, urinary, respiratory, and reproductive tracts, where they can take advantage of moist surfaces in direct contact with living cells. They enter around the eyes or in the ears. However, even these areas have ways to impede pathogen entry. Defenses include mucus, tears, saliva, earwax, digestive and vaginal acids, all of which impede entry; the ability to remove pathogens by vomiting, urination, and defecation; and even competition created by nonpathogenic bacteria that normally live in (and on) the body.

Tears, saliva, and earwax Although we may not think of tears as a defense mechanism, they perform a valuable service by lubricating the eyes and washing away particles. Tears and saliva both contain **lysozyme,** an enzyme that kills many bacteria. In addition, saliva lubricates the delicate tissues inside the mouth so that they do not dry out and crack. It also rinses microorganisms safely from the mouth into the stomach, where most of them are killed by stomach acid. Earwax traps small particles and microorganisms.

Mucus Mucus is a thick, gel-like material secreted by cells at various surfaces of the body, including the lining of the digestive tract and the branching airways of the respiratory system. Microorganisms that come into contact with the sticky mucus become mired and cannot gain access to the cells beneath. In addition, the cells of the airways have tiny hairlike projections called *cilia* that beat constantly in a wavelike motion to sweep mucus upward into the throat. There we get rid of the mucus by coughing or swallowing it. Sometimes, we remove mucus and microorganisms by sneezing, which is also one of the primary ways we pass microorganisms to other people (**Figure 9.5**).

Digestive and vaginal acids Undiluted digestive acid is strong enough to kill nearly all pathogens that enter the digestive tract on an empty stomach. Only one strain of bacteria, *Helicobacter pylori,* has actually evolved to thrive in the highly acidic environment of the stomach. *H. pylori* is now known to contribute to many cases of stomach ulcers (see Chapter 14). Vaginal secretions are slightly acidic, too, though not nearly as acidic as stomach secretions.

Figure 9.5 Sneezing removes mucus and microorganisms from the body. Sneezing also promotes the transmission of certain respiratory infections from person to person.

Vomiting, urination, and defecation Vomiting, though unpleasant, is certainly an effective way of ridding the body of toxic or infected stomach contents.

Generally speaking, the urinary system does not have a resident population of bacteria. Urine is usually slightly acidic, and in addition the constant flushing action of urination tends to keep bacterial populations low. Urine pH can vary from fairly acidic to slightly basic, depending on diet. Some physicians advise patients with bladder or urethral infections to drink cranberry juice, which is acidic. The increased acidity of the urine inhibits bacterial growth, and the increased urine volume flushes the bacteria out.

The movement of feces and the act of defecation also help remove microorganisms from the digestive tract. When we become ill, the muscles in the intestinal wall may start to contract more vigorously, and the intestine may secrete additional fluid into the feces. The result is *diarrhea*—increased fluidity, frequency, or volume of bowel movements. Unpleasant though diarrhea may be, mild cases serve a useful function by speeding the removal of pathogens.

Resident bacteria Certain strains of beneficial bacteria normally live in the mucous membranes lining the vagina and the digestive tract. They help control population levels of more harmful organisms by competing successfully against them for food. They may also make the body less vulnerable to pathogens. For example, *Lactobacillus* bacteria in the vagina produce a substance that lowers vaginal pH to levels that many fungi and bacteria cannot tolerate.

One might ask how any beneficial bacteria ever get to the small and large intestine if they have to pass through the stomach first. The answer is that following a meal the stomach contents are not so acidic because food both dilutes and buffers the stomach acid, so some bacteria pass through the stomach with the food we eat.

↩ **Recap** Various mechanisms create an inhospitable environment for pathogenic microorganisms. Skin is a dry outer barrier. Tears, saliva, earwax, and mucus trap pathogens or wash them away. Acidic conditions kill pathogens or inhibit their growth; urination, defecation, and vomiting forcibly expel them; and resident bacteria compete with pathogens for food. ∎

9.4 Nonspecific defenses: the second line of defense

If pathogens manage to breach our physical and chemical barriers and start to kill or damage cells, we have a problem of a different sort. Now the body must actively seek out the pathogens and get rid of them. It must also clean up the injured area and repair the damage.

Table 9.1 The second line of defense: nonspecific defense

Defense	Functions
Complement system	A group of proteins that assists other defense mechanisms; enhances inflammation and phagocytosis; kills pathogens.
Phagocytes	Neutrophils and macrophages engulf and digest foreign cells; eosinophils bombard large parasites with digestive enzymes and phagocytize foreign proteins.
Inflammatory response	Four components include redness, warmth, swelling, and pain; attracts phagocytes and promotes tissue healing.
Natural killer cells	Release chemicals that disintegrate cell membranes of tumor cells and virus-infected cells.
Interferons	Stimulate the production of proteins that interfere with viral reproduction.
Fever	Modest fever makes internal environment less hospitable to pathogens; fosters ability to fight infections.

Our second line of defense includes a varied group of defense mechanisms. We refer to them as *nonspecific* because they do not target specific pathogens. Instead, they appear in response to all types of health challenges without discriminating between them. **Table 9.1** summarizes nonspecific defenses, which include the complement system, phagocytes, the inflammatory response, natural killer cells, interferons, and fever.

The complement system assists other defense mechanisms

The **complement system,** or *complement*, comprises at least 20 plasma proteins that circulate in the blood and complement, or assist, other defense mechanisms. Normally, these proteins circulate in an inactive state. When activated by the presence of an infection, however, they become a potent defense force. Once one protein is activated, it activates another, leading to a cascade of reactions. Each protein in the complement system can activate many others, creating a powerful "domino effect."

Figure 9.6 shows how some activated complement proteins attack and destroy bacteria. ❶ Activated complement proteins link together, forming protein complexes that create large holes through the bacterial cell wall. ❷ Water and salts leak into the bacterium through the holes. ❸ Eventually, the bacterium swells and bursts (lyses).

Other activated complement proteins bind to bacterial cell membranes, marking them for destruction by phagocytes. Still, others stimulate mast cells to release histamine or serve as chemical attractants to draw additional phagocytes to the infection.

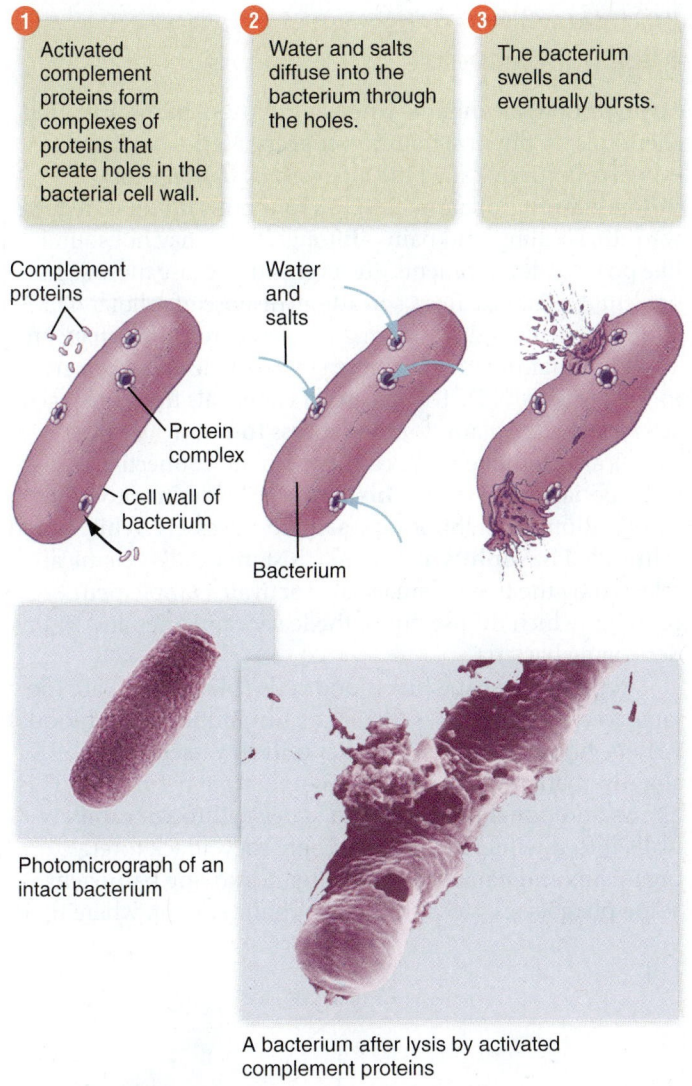

1 Activated complement proteins form complexes of proteins that create holes in the bacterial cell wall.

2 Water and salts diffuse into the bacterium through the holes.

3 The bacterium swells and eventually bursts.

Complement proteins

Water and salts

Protein complex

Cell wall of bacterium

Bacterium

Photomicrograph of an intact bacterium

A bacterium after lysis by activated complement proteins

Figure 9.6 How activated complement proteins kill bacteria.

Phagocytes engulf foreign cells

As noted in the chapter on blood, *phagocytes* are white blood cells that destroy foreign cells through the process of **phagocytosis** (Figure 9.7). **1** A phagocyte first captures a bacterium with its cytoplasmic extensions. **2** Then it draws the bacterium in, eventually engulfing it (endocytosis) and **3** enclosing it in a membrane-bound vesicle. **4** Inside the cell, the vesicle containing the bacterium fuses with lysosomes, and **5** the powerful enzymes in the lysosomes dissolve the bacterial membranes. Once digestion is complete, **6** the phagocyte jettisons the bacterial wastes by exocytosis.

Some white blood cells can slip through the walls of blood vessels into tissue spaces, attracted by substances released by injured cells at the site of infection. Other phagocytes reside permanently in connective tissues of the lymph nodes, spleen, liver, lungs, and brain.

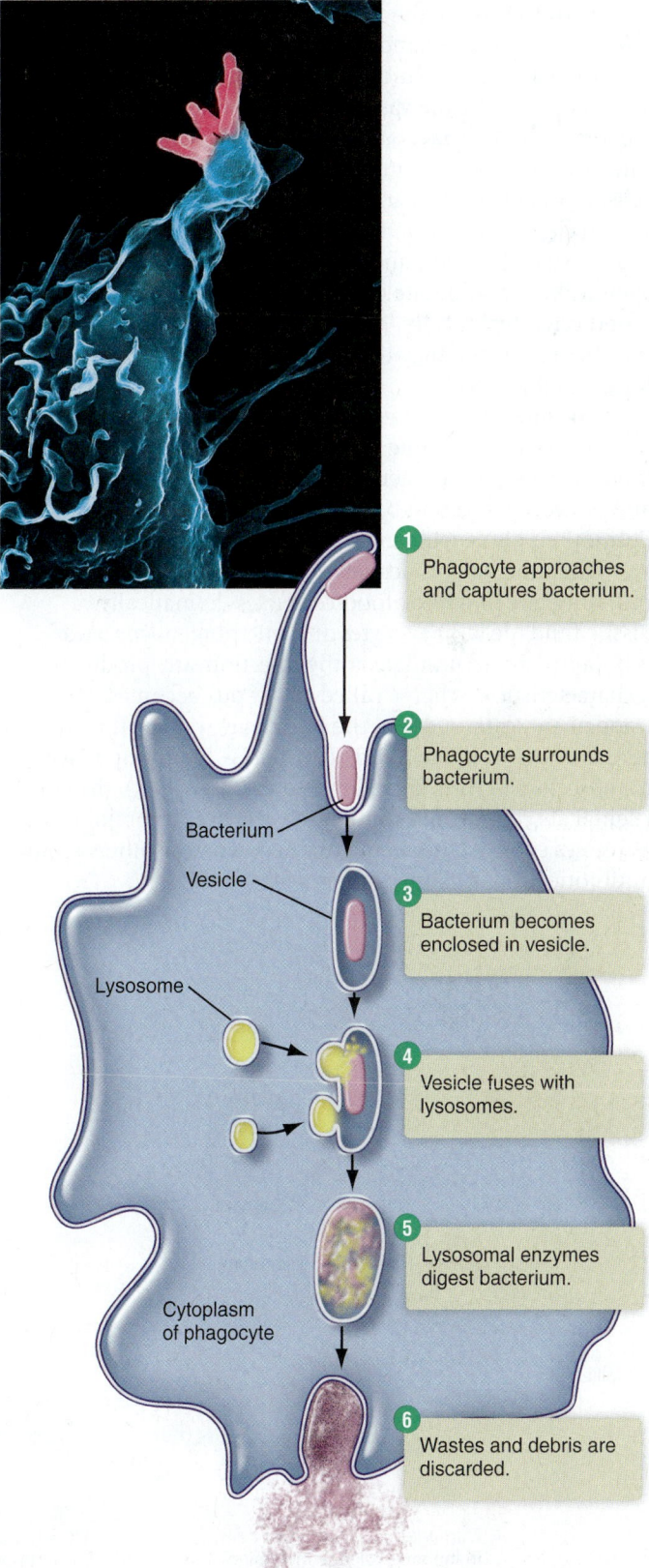

1 Phagocyte approaches and captures bacterium.

2 Phagocyte surrounds bacterium.

Bacterium

Vesicle

3 Bacterium becomes enclosed in vesicle.

Lysosome

4 Vesicle fuses with lysosomes.

5 Lysosomal enzymes digest bacterium.

Cytoplasm of phagocyte

6 Wastes and debris are discarded.

Figure 9.7 Phagocytosis. (Top) An electron micrograph of a macrophage (blue) capturing several bacteria (pink). (Bottom) Steps in the process of phagocytosis.

Neutrophils are the first white blood cells to respond to infection. They digest and destroy bacteria and some fungi in the blood and tissue fluids. Other white blood cells, known as *monocytes*, leave the vascular system, enter the tissue fluids, and develop into **macrophages** (from the Greek for "large eater") that can engulf and digest large numbers of foreign cells, especially viruses and bacterial parasites. Macrophages serve a cleanup function by scavenging old blood cells, dead tissue fragments, and cellular debris. They also release chemicals that stimulate the production of more white blood cells. Technically, macrophages are not white blood cells because when they develop from monocytes they are no longer in the blood.

When invaders are too big to be engulfed and digested by phagocytosis, other white blood cells called **eosinophils** take action. Eosinophils cluster around large parasites such as flukes and pinworms and bombard them with digestive enzymes. Eosinophils also engulf and digest certain foreign proteins.

When the body is actively fighting an infection, the mortality rate of white blood cells rises dramatically. Tissue fluid, dead phagocytes and microorganisms, and cellular debris accumulate at the infection site, producing a characteristic discharge called *pus*. If pus becomes trapped and cannot drain, the body may wall it off with connective tissue. The result is an *abscess*. Common places for abscesses to form include the breast (mastitis), the gums (dental abscesses), and more rarely the liver or brain. Many abscesses subside after being drained, whereas others require antibiotic drugs or surgical removal.

Inflammation: redness, warmth, swelling, and pain

Any type of tissue injury—whether infection, burns, irritating chemicals, or physical trauma—triggers a series of related events collectively called the *inflammatory response*, or **inflammation.** Inflammation has four outward signs: redness, warmth, swelling, and pain. Although these may not sound like positive developments, the events that cause these signs prevent the damage from spreading, dispose of cellular debris and pathogens, and set the stage for tissue-repair mechanisms.

The inflammatory response starts whenever tissues are injured (**Figure 9.8**). The release of chemicals from damaged cells sounds the alarm for the process to begin. **1** These chemicals stimulate **mast cells,** which are connective tissue cells specialized to release **histamine.** Histamine promotes vasodilation of neighboring small blood vessels. White blood cells called **basophils** also secrete histamine. **2** Chemicals released by the tissue damage also activate complement proteins, which diffuse out of the leaky capillaries and begin destroying bacteria.

Recall that most white blood cells (phagocytes) are too large to cross capillary walls. **3** As histamine dilates blood vessels, however, the endothelial cells in vessel walls pull slightly apart, and the vessels become more permeable. This allows additional phagocytes to squeeze through capillary walls into the interstitial fluid. There they attack foreign organisms and damaged cells. After destroying pathogens, some phagocytes travel to the lymphatic system, where their

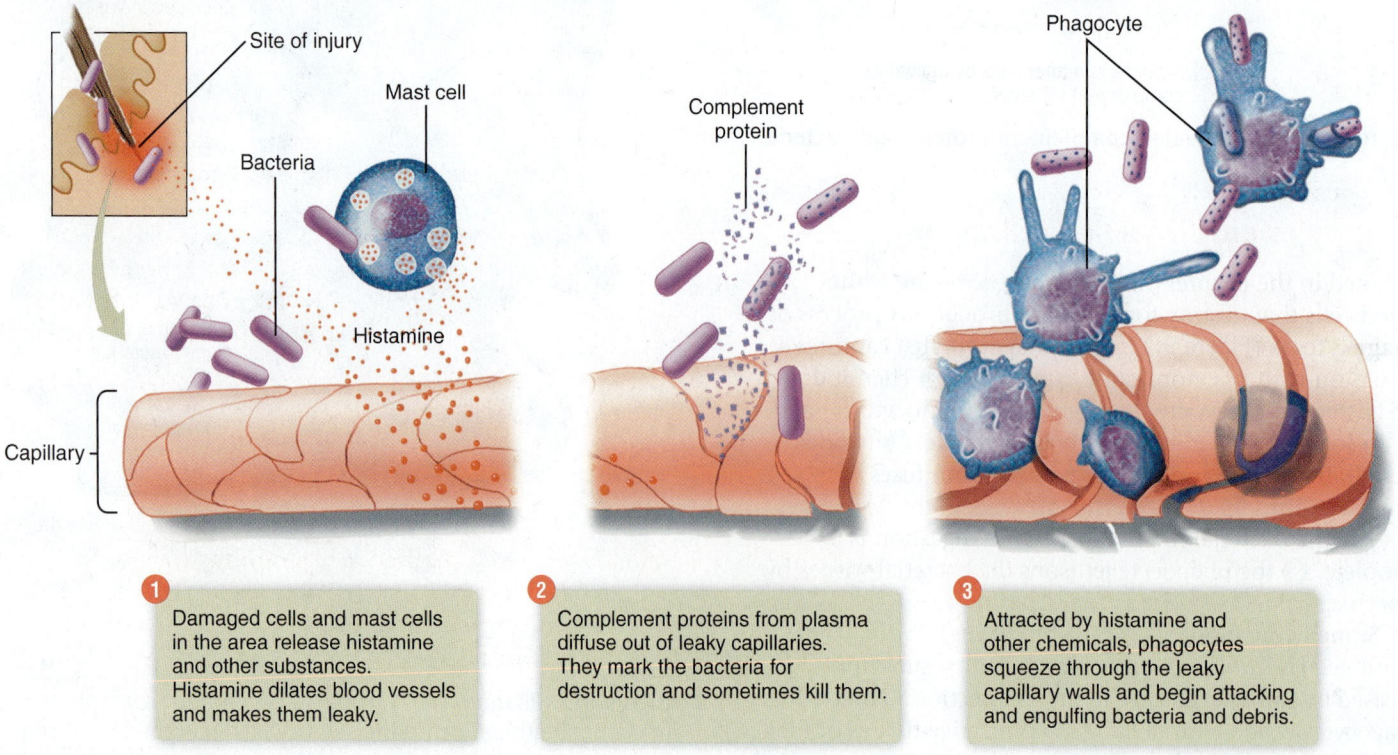

1 Damaged cells and mast cells in the area release histamine and other substances. Histamine dilates blood vessels and makes them leaky.

2 Complement proteins from plasma diffuse out of leaky capillaries. They mark the bacteria for destruction and sometimes kill them.

3 Attracted by histamine and other chemicals, phagocytes squeeze through the leaky capillary walls and begin attacking and engulfing bacteria and debris.

Figure 9.8 The inflammatory response.

presence activates lymphocytes to initiate specific defense mechanisms (discussed later).

Vasodilation brings more blood into the injured area, making it red and warm. The rising temperature increases phagocyte activity. The increased leakiness of capillary walls allows more fluid to seep into tissue spaces, causing swelling. The extra fluid dilutes pathogens and toxins and brings in clotting proteins that form a fibrin mesh to wall off the damaged area from healthy tissue. As a bonus, the fluid carries in extra oxygen and nutrients to promote tissue healing and carries away dead cells, microorganisms, and other debris from the area.

Swollen tissues press against nearby nerve endings. This swelling, plus the sensitizing effects of inflammatory chemicals, creates the sensation of pain that accompanies inflammation. However, even pain can be positive. The discomfort hinders active movement and forces the injured person to rest, facilitating the healing process.

☑ Antihistamines are drugs that block the effect of histamine. Why do antihistamines help alleviate the "stuffy nose" and nasal congestion of a cold?

Natural killer cells target tumors and virus-infected cells

Natural killer (NK) cells are a group of white blood cells (lymphocytes) that destroy tumor cells and cells infected by viruses. NK cells are able to recognize certain changes that take place in the plasma membranes of tumor cells and virus-infected cells. The name *natural killer* reflects the fact that NK cells are nonspecific killers, unlike other killer cells discussed later in this chapter that target only specific enemies.

NK cells are not phagocytes. Instead, they release chemicals that break down their targets' cell membranes. Soon after an NK attack, the target cell's membrane develops holes, and its nucleus rapidly disintegrates. NK cells also secrete substances that enhance the inflammatory response.

Interferons interfere with viral reproduction

One of the most interesting defense mechanisms is an early warning system between virus-infected and still-healthy cells. As mentioned earlier, viruses cannot reproduce on their own. Instead, they invade body cells and use the cells' machinery to make more viruses.

Cells that become infected by viruses secrete a group of proteins called **interferons.** Interferons diffuse to nearby healthy cells, bind to their cell membranes, and stimulate the healthy cells to produce proteins that interfere with the synthesis of viral proteins, making it harder for the viruses to infect the protected cells.

Interferons are now being produced in pharmaceutical laboratories. At least one interferon protein (*alpha interferon*)

has shown promise against certain viral diseases, including genital warts, hepatitis B, and one form of leukemia.

Fever raises body temperature

A final weapon in our second line of defense is **fever,** an abnormally high body temperature. Your body's "thermostat" is set to approximately 98.6°F (37°C), with a normal range of about 97–99°F (36–37.2°C). When macrophages detect and attack bacteria, viruses, or other foreign substances, they release certain chemicals into the bloodstream. These chemicals, called *pyrogens,* cause the brain to reset your thermostat to a higher temperature.

There is a tendency to treat all fevers as if they were a problem. But a modest fever may be beneficial because it makes our internal environment less hospitable to pathogens and enhances the body's ability to fight infections. Fever increases the metabolic rate of body cells, speeding up both defense mechanisms and tissue-repair processes. When the infection is gone, the process reverses: Macrophages stop releasing pyrogens, the thermostat setting returns to normal, and your fever "breaks."

Although a moderate fever tends to be beneficial, a high fever can be dangerous. Because the chemical bonds that give a protein its shape are relatively weak, they may be broken by high temperatures. If that happens, the shape of the protein will change and the protein may not function normally. It's a good idea to monitor the course of any fever, particularly in children and older adults. Health professionals recommend seeking medical advice for any fever that lasts longer than two days or rises above 100°F.

🔄 **Recap** Nonspecific defense mechanisms involve a general attack against all foreign and damaged cells. Neutrophils and macrophages engulf and digest bacteria and damaged cells, and eosinophils bombard larger organisms (too large to be engulfed) with digestive enzymes. The inflammatory response attracts phagocytes and promotes tissue healing. Interferons interfere with viral reproduction. A modest fever enhances our ability to fight infections. ▪

9.5 Specific defense mechanisms: the third line of defense

Even if foreign cells manage to bypass physical and chemical barriers and overcome nonspecific defenses, they must still cope with the body's third line of defense, the most sophisticated weapon of all. The *immune system* comprises cells, proteins, and the lymphatic system, all working together to detect and kill particular pathogens and abnormal body cells. The activities of the immune system are collectively called the **immune response.** Because the immune system targets specific enemies, we refer to these operations as *specific defense mechanisms*.

The immune response has three important characteristics:

- It recognizes and targets specific pathogens or foreign substances.
- It has a "memory," the capability to store information from past exposures to a pathogen so that it can respond more quickly to later invasions by the same pathogen.
- It protects the entire body; the resulting immunity is not limited to the site of infection.

The key to specific defenses is the body's ability to distinguish between its own cells and those of foreign invaders. Among its own cells, it must also be able to distinguish between those that are healthy, those that are abnormal (such as cancer cells), and those that are dead or dying.

The immune system targets antigens

An **antigen** is any substance that mobilizes the immune system and provokes an immune response. Generally, antigens are large protein or polysaccharide molecules. In much the same way that a key fits a lock, each antigen has a unique shape, and every bacterium or virus has a different one. The immune system responds to each uniquely shaped antigen by producing specific *antibodies* to attack and inactivate the antigen (and the bacterium or virus carrying it).

All antigens are located only on the outer surface of a cell or virus. Hence, the immune system cannot detect viruses (or viral DNA/RNA) once they are safely inside a living human cell.

Like any cell, human cells also have surface proteins that can act as antigens under the right circumstances. Your cells have a unique set of proteins on their surfaces that your immune system uses to recognize that those cells belong to you. These *self markers* are known as **major histocompatibility complex (MHC) proteins.** Your MHC proteins are unique to you by virtue of your unique set of genes. Normally, they signal your immune system to bypass your own cells. They are a sort of password, the equivalent of a cellular fingerprint. Your immune system "reads" the password and leaves your cells alone.

However, the same MHC proteins that define your cells as belonging to you would be read as *nonself markers* in another person. In other words, your MHC proteins would be antigens in another person. Abnormal and cancerous cells in your own body also have MHC proteins that are not recognized as *self*. The immune system targets all antigens, including those on pathogens and foreign and damaged human cells, for destruction.

✔ Why does pregnancy pose a problem for the mother's immune system?

Lymphocytes are central to specific defenses

Lymphocytes play crucial roles in our specific defense mechanisms. As described in the chapter on blood, lymphocytes are white blood cells originating from stem cells in bone marrow. They are fairly small white blood cells with a single nucleus that fills nearly the entire cell. They total about 30% of circulating white blood cells. Lymphocytes are found in the bloodstream, tonsils, spleen, lymph nodes, and thymus gland.

There are two types of these white blood cells: B lymphocytes and T lymphocytes, also called **B cells** and **T cells.** (Their names are based on where they mature: B cells mature in bone marrow; T cells in the thymus gland.) Although both types of lymphocytes can recognize and target antigen-bearing cells, they go about this task in different ways.

B cells are responsible for **antibody-mediated immunity.** B cells produce **antibodies**—proteins that bind with and neutralize specific antigens. They release antibodies into the lymph, bloodstream, and tissue fluid, where they circulate throughout the body. Antibody-mediated immunity works best against viruses, bacteria, and foreign molecules that are soluble in blood and lymph.

T cells are responsible for **cell-mediated immunity,** which depends on the actions of several types of T cells. Unlike B cells, T cells do not produce antibodies. Instead, some T cells directly attack foreign cells that carry antigens. Other T cells release proteins that help coordinate other aspects of the immune response, including the actions of T cells, B cells, and macrophages. Cell-mediated immunity protects us against parasites, bacteria, viruses, fungi, cancerous cells, and cells perceived as foreign (including, unfortunately, transplanted tissue—see section 9.8, Tissue rejection: a medical challenge). T cells can identify and kill infected human cells even before the cells have a chance to release new bacteria or viruses into the blood.

Both B cells and T cells are activated by specific antigens, and both store information about their first exposure to a specific antigen in the form of memory cells. The presence of memory cells enables B cells and T cells to undergo clonal expansion (replication) more quickly the second and subsequent times they are exposed to the same antigen.

✔ Suppose a baby was born without a functional thymus gland. Could this child still produce antibodies? Explain your answer.

B cells: antibody-mediated immunity

In adults, B cells mature in the bone marrow. As they mature, they develop unique surface receptors (with the same structure as an antibody) that allow them to recognize specific antigens (**Figure 9.9**). Then they travel in the bloodstream to the lymph nodes, spleen, and tonsils, where they remain inactive until they encounter a foreign cell with that particular antigen. ❶ When a B cell with just the right surface receptor encounters the appropriate antigen, its surface receptors bind to the antigen. ❷ This activates the B cell to grow and then multiply rapidly, producing more B cells exactly like the original and bearing the same surface receptors. The resulting identical cells, all descended from the same cell, are called *clones*.

❸ Although the B cells themselves tend to remain in the lymphatic system, most of the cells of the clone are called **plasma cells** because they begin to secrete their antibodies into the lymph fluid and ultimately into the blood plasma. A typical plasma cell can make antibody molecules at a staggering rate—about 2,000 molecules per second. A plasma cell maintains this frantic pace for a few days and then dies, but its antibodies continue to circulate in blood and lymph.

❹ Some of the clone cells become memory cells, long-lived cells that remain inactive until that same antigen reappears in the body at some future date. Memory cells store information about the pathogen; if there is a second exposure, the immune response is even faster than the first time. Upon exposure, these memory cells quickly become plasma cells and start to secrete antibodies. Memory cells are the basis for long-term immunity.

When the antibodies encounter matching antigens (**Figure 9.10**), ❶ they bind to them and create an *antigen-antibody complex*. Antibodies specialize in recognizing certain proteins; thus one particular antibody can bind to one particular antigen. ❷ Some antibodies inactivate pathogens by causing the cells to agglutinate (clump together), preventing them from entering human cells and causing disease. ❸ More commonly, the formation of an antigen-antibody complex marks the antigen (and the foreign cell that carries it) for destruction either by phagocytes or by activated complement proteins.

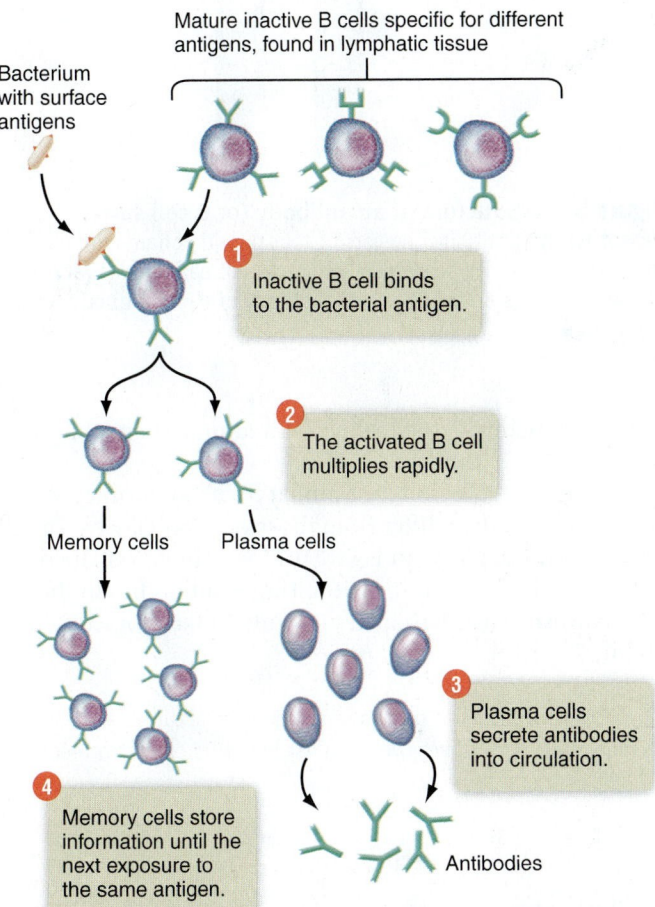

Figure 9.9 The production of antibodies by B cells. The surface antigen of a pathogen binds to the matching receptor on a mature, inactive B cell in lymphatic tissue. The B cell becomes activated and grows larger. The enlarged cell begins to divide rapidly, forming a clone. Some of the clone cells become memory cells; others become plasma cells. Memory cells lie in wait for the next exposure to the antigen. Plasma cells secrete antibodies into the lymph fluid.

Within figure 9.9:
Bacterium with surface antigens
Mature inactive B cells specific for different antigens, found in lymphatic tissue
❶ Inactive B cell binds to the bacterial antigen.
❷ The activated B cell multiplies rapidly.
Memory cells   Plasma cells
❸ Plasma cells secrete antibodies into circulation.
Antibodies
❹ Memory cells store information until the next exposure to the same antigen.

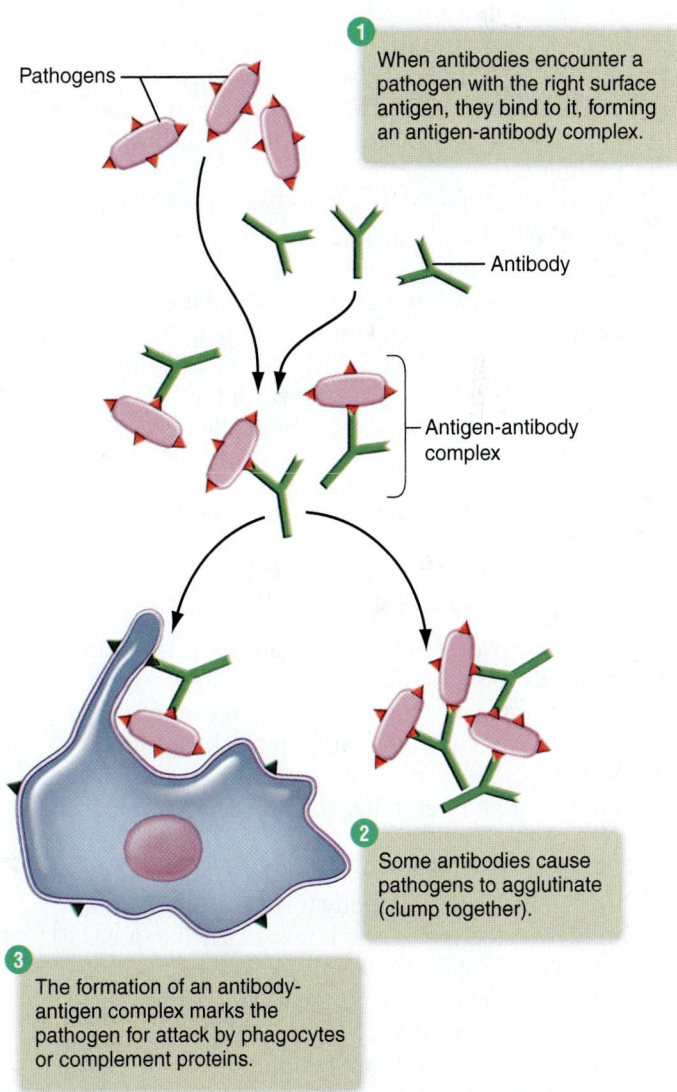

Figure 9.10 How antibodies inactivate pathogens.

Within figure 9.10:
Pathogens
❶ When antibodies encounter a pathogen with the right surface antigen, they bind to it, forming an antigen-antibody complex.
Antibody
Antigen-antibody complex
❷ Some antibodies cause pathogens to agglutinate (clump together).
❸ The formation of an antibody-antigen complex marks the pathogen for attack by phagocytes or complement proteins.

The five classes of antibodies

Antibodies belong to the class of blood plasma proteins called *gamma globulins.* Because they play such a crucial role in immunity, the word **immunoglobulin (Ig)** is often used. There are five classes of immunoglobulins, each designated by a different letter: IgG, IgM, IgA, IgD, and IgE. Each type has a different size, location in the body, and function:

- *IgG (75% of immunoglobulins).* This is the most common class. Found in blood, lymph, intestines, and tissue fluid, the long-lived IgG antibodies activate the complement system and neutralize many toxins. They are the only antibodies that cross the placenta during pregnancy and pass on the mother's acquired immunities to the fetus.
- *IgM (5–10%).* IgM antibodies are the first to be released during immune responses. Found in blood and lymph, they activate the complement system and cause foreign cells to agglutinate. ABO blood cell antibodies belong to this class.
- *IgA (15%).* IgA antibodies enter areas of the body covered by mucous membranes, such as the digestive, reproductive, and respiratory tracts. There they neutralize infectious pathogens. They are also present in a mother's milk and are transmitted to the infant during breast-feeding.
- *IgD (less than 1%).* IgD antibodies are in blood, lymph, and B cells. Their function is not clear, but they may play a role in activating B cells.
- *IgE (approximately 0.1%).* The rarest of the immunoglobulins, IgE antibodies are in B cells, mast cells, and basophils. They activate the inflammatory response by triggering the release of histamine. They are also the troublemakers behind allergic responses (covered in section 9.9).

An antibody's structure enables it to bind to a specific antigen

An antigen provides all the information the immune system needs to know about a foreign substance. Essentially, antigens are the locks on an enemy's doors. The immune system can identify the lock, produce the antibody key, and then send in immune system cells to open the door and neutralize the invader. How does this happen?

All antibodies share the same basic structure, represented by an IgG antibody (**Figure 9.11**). Each IgG antibody (or surface receptor, if it is attached to a B cell) consists of four linked polypeptide chains arranged in a Y shape. The two larger chains are called "heavy" chains, and the two smaller ones are called "light" chains. Each of the four chains has a constant region that forms the trunk and two branches and a variable region that represents the antigen-binding site. Because it has a unique amino acid

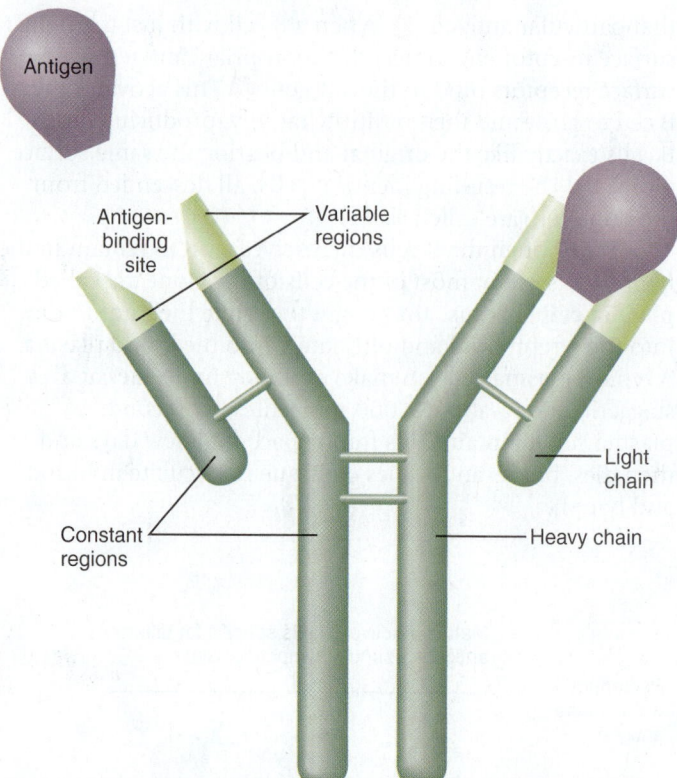

Figure 9.11 Structure of an antibody (or B cell surface receptor). An antibody consists of four peptide chains linked together to form a Y shape. Only the ends of each chain vary. These variations determine the specificity of each antibody for only one antigen.

sequence, each variable region has a unique shape that fits only one specific antigen.

The constant regions are similar for all antibodies in one class, although they differ from those of other classes. The IgG antibody depicted in Figure 9.11 is a single Y-shaped molecule with two binding sites. Larger antibodies in the IgM class consist of five Y-shaped molecules linked together, with 10 binding sites.

✔ Explain, in terms of details of antibody structure, why a memory cell that protects you against hepatitis cannot protect you against the common cold.

T cells: cell-mediated immunity

T cells develop from stem cells in bone marrow but migrate to the thymus gland, where they become mature but remain inactive. During maturation, they also develop one of two sets of surface proteins, CD4 or CD8. These proteins determine what type of T cell they will become. CD4 T cells will become helper and memory cells, and CD8 T cells will become cytotoxic cells and suppressor cells. Let's look at how inactive T cells become activated and then delve into the roles of CD4 and CD8 proteins.

Macrophages and B cells activate T cells

Inactive T cells cannot recognize and attack whole foreign cells. Before T cells can begin to do their jobs, they need to have the foreign cell's antigens presented to them in a way that they can recognize. Certain macrophages and activated B cells—called **antigen-presenting cells (APCs)**—fulfill this role (**Figure 9.12**). **1** First, an APC encloses and then engulfs a foreign cell or pathogen in a vesicle and **2** partially digests it. **3** Then the vesicle fuses with another vesicle containing MHC molecules, **4** forming antigen-MHCs. **5** Finally, the vesicle migrates to the surface of the cell and fuses with the cell membrane, displaying the antigen-MHCs on the cell's surface. In essence, the cell "presents" the antigen for T cells to recognize, along with its own cell-surface self marker.

Helper T cells stimulate other immune cells

When a T cell with CD4 receptors encounters an APC displaying an antigen (**Figure 9.13**), **1** it is activated to become a **helper T cell**. **2** The new helper T cell undergoes clonal expansion, quickly producing a clone of identical helper T cells. Because all the cells in the clone carry the same receptors, they all recognize the same antigen.

3 Most of the cells in the helper T cell clone begin secreting a class of signaling molecules called **cytokines.** Cytokines are proteins that stimulate other immune cells such as phagocytes, natural killer cells, and T cells with CD8 receptors. They also attract other types of white blood cells to the area, enhancing nonspecific defenses and activate B cells, creating an important link between antibody-mediated and cell-mediated immunity. **4** A few helper cloned cells become memory T cells. Like memory B cells, memory T cells become inactive, to be reactivated when an antigen encountered in the past appears again.

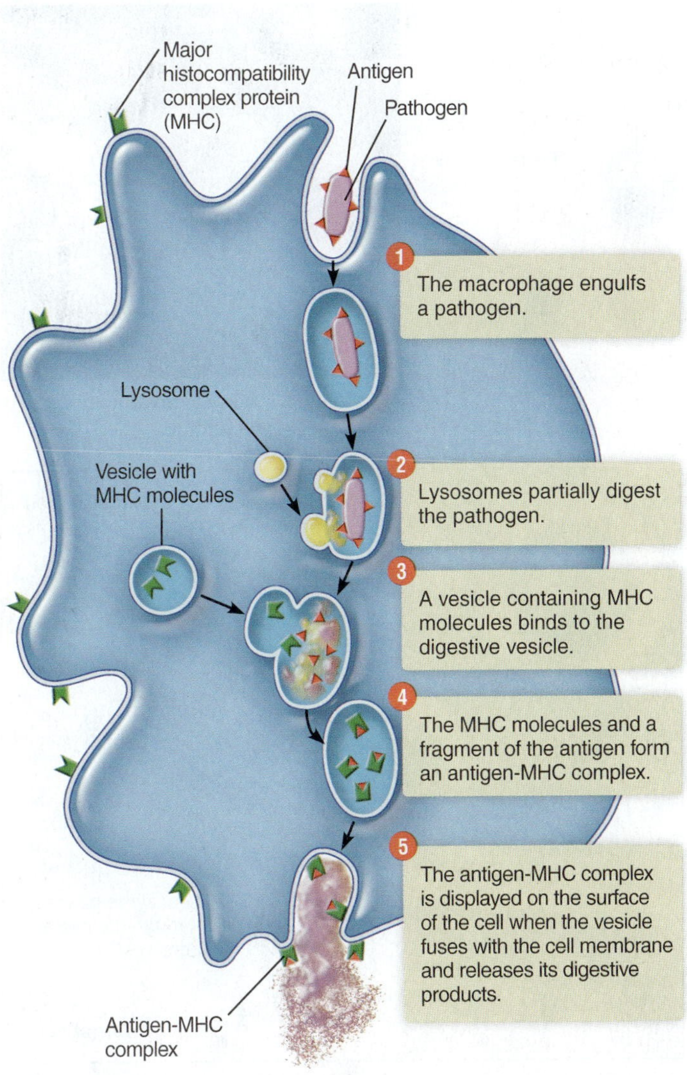

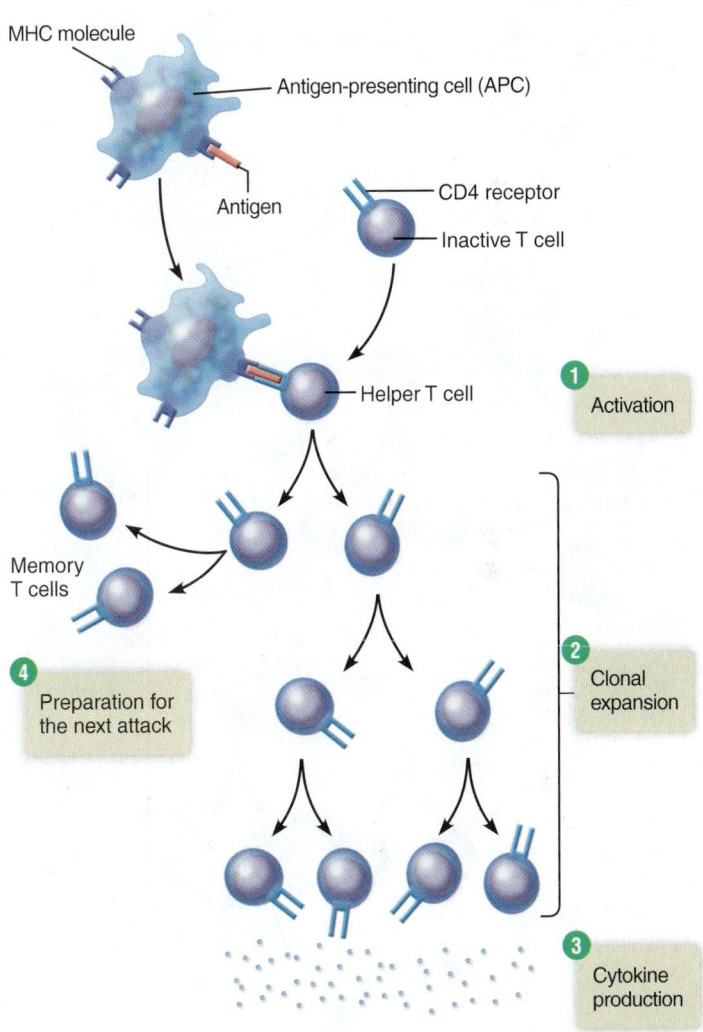

Figure 9.13 The activation and clonal expansion of helper T cells. Activation occurs when a mature but inactive helper T cell comes in contact with an antigen-presenting cell displaying the appropriate antigen. After clonal expansion, the clones produce cytokines. A few activated clones become memory T cells.

✔ Why does clonal expansion occur? That is, why can't the activated T cells simply start producing cytokines without undergoing clonal expansion?

Major histocompatibility complex protein (MHC)

Antigen

Pathogen

1 The macrophage engulfs a pathogen.

Lysosome

Vesicle with MHC molecules

2 Lysosomes partially digest the pathogen.

3 A vesicle containing MHC molecules binds to the digestive vesicle.

4 The MHC molecules and a fragment of the antigen form an antigen-MHC complex.

5 The antigen-MHC complex is displayed on the surface of the cell when the vesicle fuses with the cell membrane and releases its digestive products.

Antigen-MHC complex

Figure 9.12 How a macrophage acts as an antigen-presenting cell (APC).

The role of helper T cells and their cytokines in directing the activities of other immune cells is crucial to an effective immune response—without them the immune response would be severely impaired or nonexistent. The reason AIDS is so devastating is that HIV destroys helper T cells, and thus weakens the body's ability to mount a cell-mediated immune response to a wide variety of other diseases. (See section 9.10, Immune deficiency: the special case of AIDS.)

Cytotoxic T cells kill abnormal and foreign cells
When a mature T cell with CD8 receptors meets an APC that displays an antigen (**Figure 9.14**), ❶ the T cell is activated and

❷ begins to produce a clone of **cytotoxic T cells,** also called *killer T cells.* ❸ These are the only T cells that directly attack and destroy other cells. ❹ Like helper T cells, some cytotoxic T cells become memory T cells.

Once activated, cytotoxic T cells roam throughout the body. They circulate through blood, lymph, and lymphatic tissues in search of cells that display recognizable antigens. Or cytotoxic T cells may migrate to a tumor or site of infection, where they release chemicals that are toxic to abnormal cells.

Figure 9.15 illustrates cytotoxic T cells in action. ❶ When a cytotoxic T cell locates and binds to a target cell, secretory vesicles migrate to the cell surface and fuse with it, ❷ releasing proteins called *perforins* into the space between the two cells. The perforin molecules assemble themselves into a pore in the target cell, allowing water and salts to enter. That alone should eventually kill the cell in much the same way that activated complement protein does (review Figure 9.8). ❸ But just to make sure, the cytotoxic T cell also releases granzyme, a toxic enzyme that is small enough

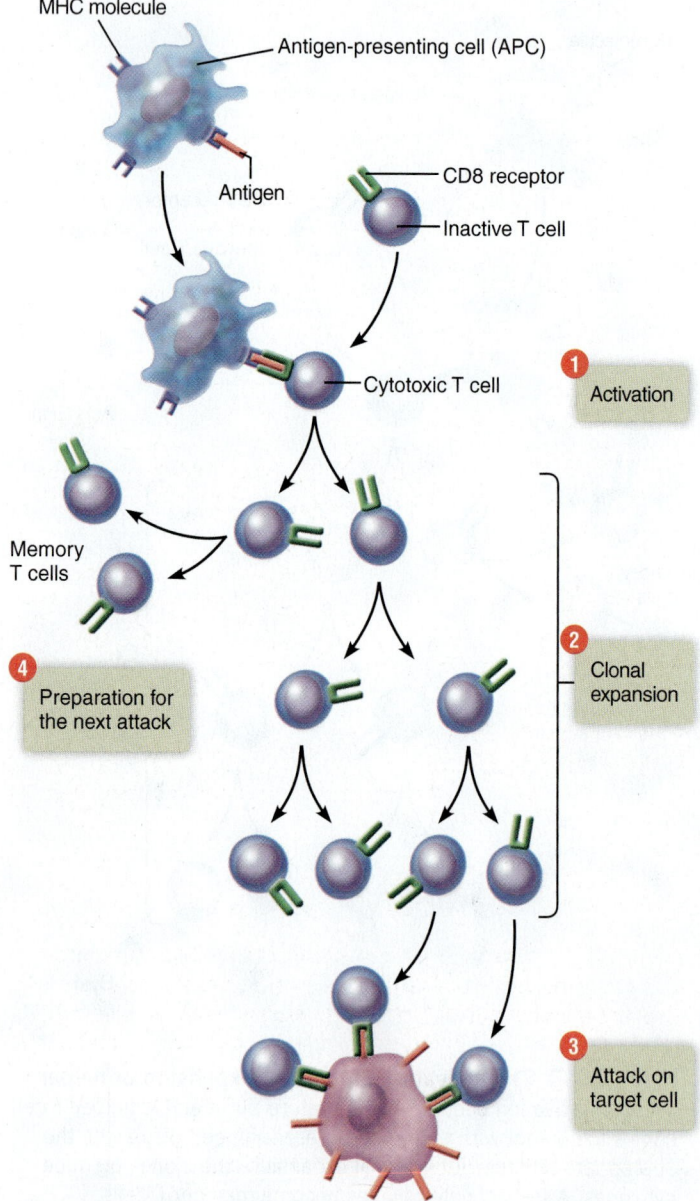

Figure 9.14 The activation and clonal expansion of cytotoxic T cells. Following activation and clonal expansion, the clones directly attack and kill cells carrying the antigen they recognize.

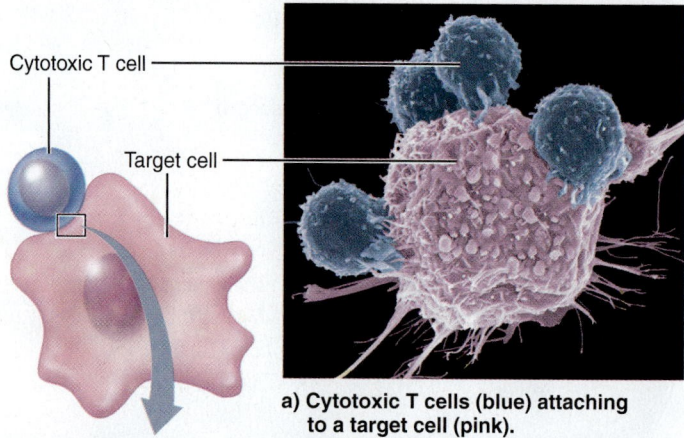

a) Cytotoxic T cells (blue) attaching to a target cell (pink).

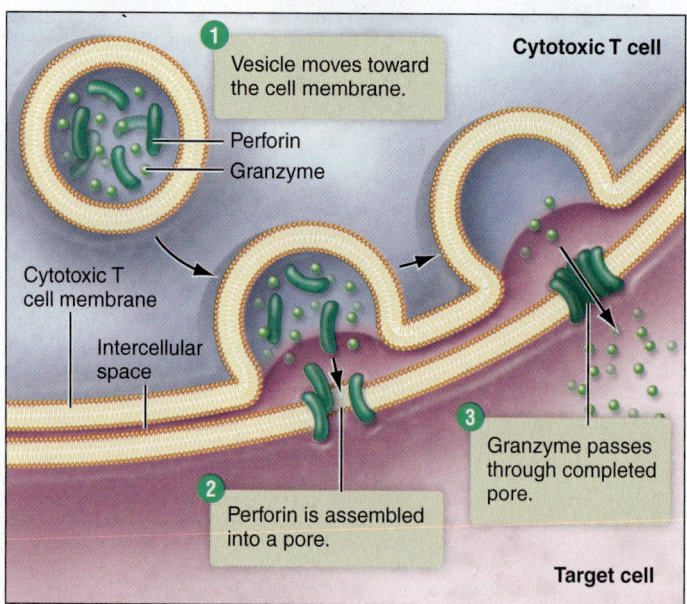

b) How cytotoxic T cells kill a target cell.

Figure 9.15 Cell-mediated immunity in action.

to pass through the pore. The cytotoxic T cell then detaches from the target cell and goes off in search of other prey.

Several promising medical treatments involve harnessing the defensive capabilities of cytokines, specifically interferons. Genetically engineered gamma interferon is used to treat the chronic viral disorder hepatitis C. Another type of interferon has been moderately successful for treating multiple sclerosis, and a third is being used to treat certain types of cancer.

Memory T cells reactivate during later exposures

Some activated T cells become inactive memory cells, retaining receptors for the antigen that originally stimulated their production. If that antigen is presented to them again, the memory cells are reactivated. Some form new helper T cells that multiply quickly to marshal an immune response. Others form a new army of cytotoxic T cells to attack and destroy. Like memory B cells, memory T cells are an important factor that distinguishes specific defenses from nonspecific defense mechanisms.

Table 9.2 summarizes the various cells and proteins involved in specific defense mechanisms.

↩ **Recap** An antigen is any substance that provokes an immune response. When activated by first exposure to a specific antigen, lymphocytes called B cells quickly produce antibodies against the antigen. They also produce a few long-lived memory cells that remain inactive until the next exposure to the same antigen. Other lymphocytes called T cells mature in the thymus gland. Helper T cells stimulate other immune cells, cytotoxic T cells attack abnormal and foreign cells, and memory T cells store information about an encountered antigen until the next exposure to the same antigen. ∎

Table 9.2   Cells and proteins involved in specific defenses

Cell/protein	Function
B cells	Mature in bone marrow. Responsible for antibody-mediated immunity.
Plasma cells	Produce and secrete specific antibodies.
Memory B cells	Store information. Upon subsequent exposure to a specific antigen become plasma cells and secrete antibodies.
Immunoglobulins	Five classes of antibodies. Every antibody has a unique shape that fits one specific antigen.
T cells	Mature in thymus. Responsible for cell-mediated immunity.
Helper T cells	Produce cytokines. Enhance immune responses by stimulating other immune cells.
Cytotoxic T cells	Attack and destroy abnormal cells.
Memory T cells	Store information. Upon subsequent exposure to a specific antigen become helper and cytotoxic T cells.
Cytokines	A class of signaling molecules that stimulate various immune system activities.

☑ What causes memory T cells (both helper and cytotoxic type) to begin clonal expansion?

9.6 Immune memory creates immunity

When you are first exposed to an antigen, your immune system protects you with the wealth of defense mechanisms described so far. Your first exposure to a particular antigen generates a *primary immune response*. As we have seen, this involves recognition of the antigen and production and proliferation of B and T cells.

Typically, the primary immune response has a lag time of three to six days after the antigen first appears. During this period, B cells specific to that antigen multiply and develop into plasma cells. Antibody concentrations rise, typically reaching their peak about 10–12 days after first exposure. Then they start to level off (**Figure 9.16**).

However, as you have learned, B and T cells create a population of memory cells. The presence of these memory cells is the basis for **immunity** from disease. (The Latin word *immunis* means "safe" or "free from.") Subsequent exposure to the pathogen elicits a *secondary immune response*

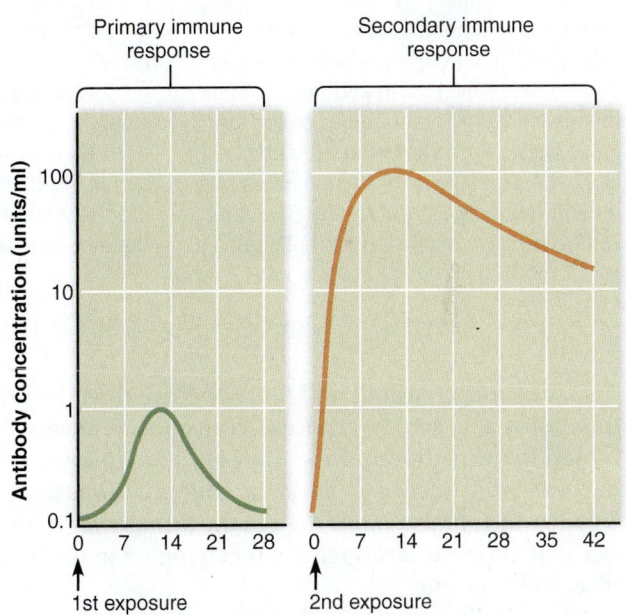

Figure 9.16 The basis of immunity. The antibody response to a first exposure to antigen (the primary immune response) declines in about a month. A second exposure to the same antigen results in a response that is more rapid in onset, larger, and longer lasting. On this graph, the abscissa scale is broken to indicate that it may be months or even years before second exposure. Note that the concentration of antibody during a secondary response may be 100 times higher than the primary response.

☑ Suppose a person lacks the capability to produce any memory cells, but otherwise has a normal immune system. Sketch new lines on each of the graphs showing what this person's antibody responses will look like during a first exposure and a second exposure.

HEALTH & WELLNESS

The Case for Breast Milk

When human infants are born, their immune systems are still very immature. At birth, they don't have a sufficient number of phagocytes (to engulf foreign cells) or lymphocytes (that produce antibodies). Normally, the populations of these cells begin to increase shortly after birth in response to colonization of the infant's gut by bacteria. But it takes time, so newborns are more vulnerable than adults to relatively minor ailments such as middle ear infections, diarrhea, and respiratory and urinary tract infections.

Human milk has several benefits at this early stage beyond its nutritional value. Human milk contains both phagocytes and lymphocytes, which boost the infant's population of these immune cells until the infant's immune system is more fully developed. In addition, human milk is loaded with antibodies of the IgA type. These antibodies offer protection against essentially all of the microbes that inhabit the mother's gut, many of which will eventually colonize the infant's gut

as well. Furthermore, human milk has a high concentration of a protein called *lactoferrin*, which stimulates the immune system and kills bacteria while preventing an inflammatory response. Together, human IgA and lactoferrin account for about 30% of the proteins in human milk, compared to just 5% in cow's milk. Finally, some of the oligosaccharides (long sugar molecules) in human milk also serve a protective function. By virtue of their shape, the oligosaccharides prevent certain harmful bacteria from binding to the infant's gut, thereby preventing an infection from developing.

Human milk has some long-term benefits as well. Children that were breastfed are better protected against certain infections than non-breastfed children, even for several years after breastfeeding has stopped. Breastfed children also seem to respond to some vaccinations with better T cell and antibody responses. Precisely how human milk encourages the development and maturation of the

infant's immune system is still under investigation.

A word of caution: Substances detrimental to an infant, including heroin, cocaine, and alcohol, can be transferred via breastfeeding. For mothers using any of these substances, the benefits to their infants of breastfeeding may not outweigh the risks

---

that is faster, longer lasting, and more effective than the first. Within hours after second exposure to an antigen, memory cells bind to the pathogen. New armies of T and plasma cells form, and within a few days antibody concentrations rise rapidly to much higher levels than in the primary response. Notice that antibody levels remain much higher in the body after second exposure.

Memory cells are long-lived, and many retain their ability to generate a secondary immune response over a lifetime. The secondary immune response can be so effective that you don't even realize you've been exposed to the pathogen a second time. At worst, you may experience only a fleeting sensation of feeling unwell. Some memory cells, such as the ones for the bacterial infection that causes tetanus, need to be reactivated (exposed to the pathogen again to undergo clonal expansion) every 10 years or so.

Given this immunity, though, why is it possible to get a cold or the flu over and over, sometimes several times a year? One reason is that there are more than 100 different viruses that can cause colds and flu. Even if your latest respiratory ailment feels like the previous one, it may actually be due to an entirely

different pathogen. Furthermore, the viruses that cause colds and flu evolve so rapidly that they are essentially different each year. Their antigens change enough that each one requires a different antibody, and each exposure triggers a primary response. Rapid evolution is their survival mechanism. Our survival mechanism is a healthy immune system.

Recap First exposure to a specific antigen generates a primary immune response. Subsequent exposure to the same antigen elicits a secondary immune response that is faster, longer lasting, and more effective than the primary immune response. ■

MJ's BlogInFocus Where can one purchase human breast milk? Visit MJ's blog in the Study Area in MasteringBiology and look under "Purchasing Human Breast Milk."

http://goo.gl/kD7jnx

9.7 Medical assistance in the war against pathogens

Our natural defenses against pathogens are remarkable. Nevertheless, we humans have taken matters into our own hands by developing the science of medicine. We have been able to conceive of and produce other sophisticated weaponry to help us combat pathogens. Important milestones in human health include the development of active and passive methods of immunization, which help the body resist specific pathogens; the production of monoclonal antibodies; and the discovery of antibiotics.

Active immunization: an effective weapon against pathogens

It is said that an ounce of prevention is worth a pound of cure, and this is certainly true when dealing with pathogens. The best weapon against a known pathogen is to give the body a laboratory-prepared dose of that particular pathogen's antigen in advance so that the immune system will mount a primary immune response against it. Then, if exposed to the pathogen in the environment, the body is already primed with the appropriate antibodies and memory cells. The immune system can react swiftly with a secondary response, effectively shielding you from the danger of the disease and discomfort of its symptoms.

The process of activating the body's immune system in advance is called **active immunization.** This involves administering an antigen-containing preparation called a **vaccine.** Most vaccines are produced from dead or weakened pathogens. An example is the oral polio vaccine (the Sabin vaccine), made from weakened poliovirus. Other vaccines are made from organisms that have been genetically altered to produce a particular antigen.

Of course, vaccines created from dead or weakened pathogens have their limitations. First, there are issues of safety, time, and expense. Living but weakened pathogens generally make better vaccines because they elicit a greater immune response. However, a vaccine that contains weakened pathogens has a slight potential to cause disease itself. This has happened, although rarely, with the polio vaccine. It takes a great deal of time, money, and research to verify the safety and effectiveness of a vaccine. Second, a vaccine confers immunity against only one pathogen, so a different vaccine is needed for every virus. This is why doctors may recommend getting a flu vaccine each time a new flu strain appears (nearly every year). Third, vaccines are not particularly effective after a pathogen has struck; that is, they do not cure an already existing disease.

Nonetheless, vaccines are an effective supplement to our natural defense mechanisms. Active immunization generally produces long-lived immunity that can protect us for many years. The widespread practice of vaccination has greatly reduced many diseases such as polio, measles, and whooping cough.

In the United States, immunization of adults has lagged behind that of children. It's estimated that more than 50,000 Americans die each year from infections, including pneumonia, hepatitis, and influenza, that could have been prevented with timely vaccines.

In many countries, vaccines are too costly and difficult to administer: Generally, they must be injected by a health care worker with some basic level of training. To get around this problem, some researchers are developing potatoes or bananas that are genetically modified to produce vaccines against diseases such as hepatitis B, measles, and the diarrhea-causing Norwalk virus. An oral vaccine against Norwalk virus has already undergone limited testing in humans.

Passive immunization can help against existing or anticipated infections

To fight an existing or even anticipated infection, a person can be given antibodies prepared in advance from a human or animal donor with immunity to that illness. Usually this takes the form of a *gamma globulin* shot (serum containing primarily IgG antibodies). The procedure is called **passive immunization.** In essence, the patient is given the antibodies that his/her own immune system might produce if there were enough time.

Passive immunization has the advantage of being somewhat effective against an existing infection. It can be administered to prevent illness in someone who has been unexpectedly exposed to a pathogen, and it confers at least some short-term immunity. However, protection is not as long-lasting as active immunization following vaccine administration because the administered antibodies disappear from the circulation quickly. Passive immunization also can't confer long-term immunity against a second exposure, because the person's own B cells aren't activated and so memory cells for the pathogen do not develop.

Passive immunization has been used effectively against certain common viral infections, including those that cause hepatitis B and measles, bacterial infections such as tetanus, and Rh incompatibility. Passive immunization of the fetus and newborn also occurs naturally across the placenta and through breast-feeding.

✔ In New York City in the late 1800s, children ill with diphtheria were often given an extract from the blood of horses that had previously had diphtheria. Was this an example of active or passive immunization, and did it provide permanent or temporary protection?

Monoclonal antibodies: laboratory-created for commercial use

An antibody preparation used to confer passive immunity in a patient is actually a mixture of many different antibody molecules, because a single pathogen can have many different antigens on its surface. **Monoclonal antibodies,** on the other hand, are antibodies produced in the laboratory from cloned descendants of a single hybrid B cell. As such, monoclonal antibodies are relatively pure preparations of antibodies specific for a single antigen. Monoclonal antibodies are proving useful in research, testing, and cancer treatments because they are pure and they can be produced cheaply in large quantities.

Figure 9.17 summarizes a technique for preparing monoclonal antibodies using mice. ❶ Typically, after a mouse has been immunized with a specific antigen to stimulate B cell production, ❷ B cells are removed from the mouse's spleen. ❸ The B cells are fused with myeloma (cancer) cells to create *hybridoma* (hybrid cancer) cells that have desirable traits of both parent cells: They each produce a specific antibody, and they proliferate with cancer-like

rapidity. As these hybridoma cells grow in culture, ❹ those that produce the desired antibody are separated out and ❺ cloned, ❻ producing millions of copies. ❼ The antibodies they produce are harvested and processed to create preparations of pure monoclonal antibodies. (The term *monoclonal* means these antibody molecules derive from a group of cells cloned from a single cell.)

Monoclonal antibodies have a number of commercial applications, including home pregnancy tests, screening for prostate cancer, and diagnostic testing for hepatitis, influenza, and HIV/AIDS. Monoclonal antibody tests tend to be more specific and more rapid than conventional diagnostic tests. In the future, it may be possible to use monoclonal antibodies to deliver anticancer drugs directly to cancer cells. The first step would be to bond an anticancer drug to monoclonal

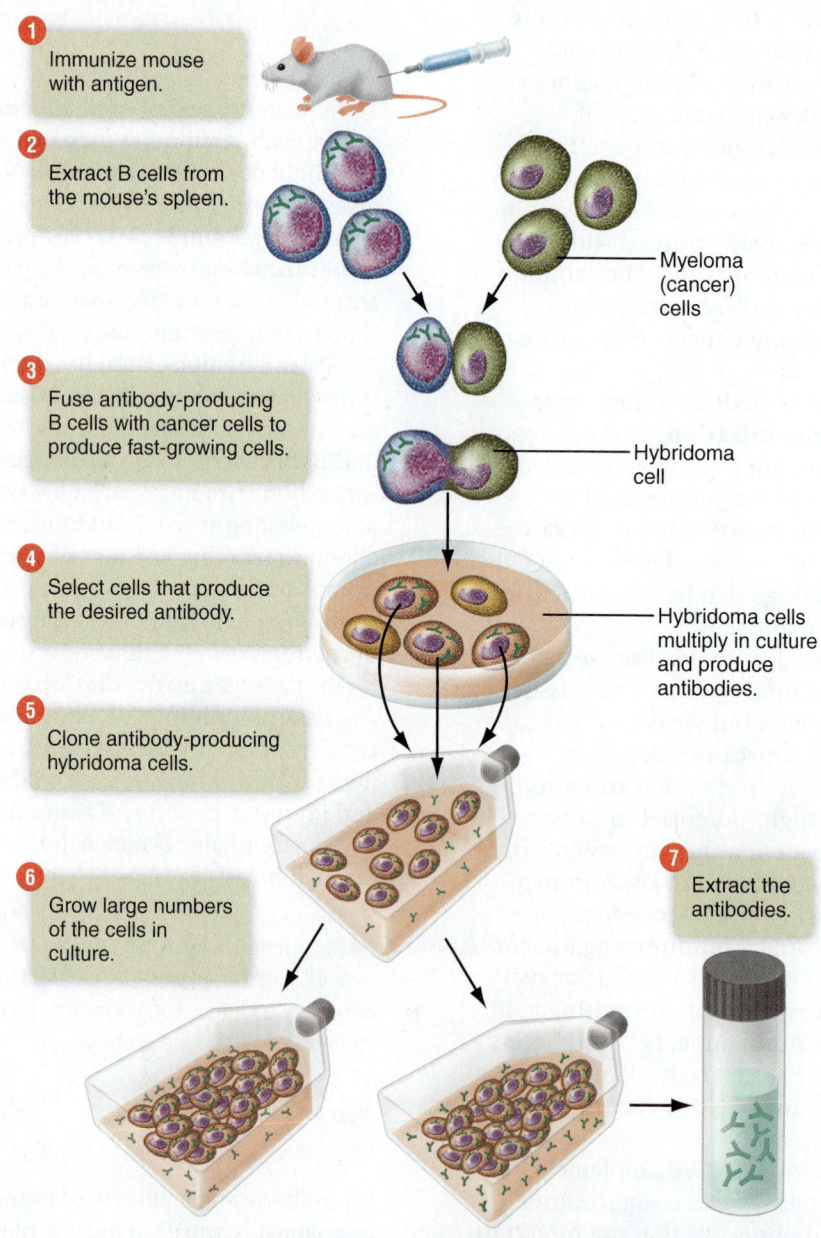

1 Immunize mouse with antigen.

2 Extract B cells from the mouse's spleen.

Myeloma (cancer) cells

3 Fuse antibody-producing B cells with cancer cells to produce fast-growing cells.

Hybridoma cell

4 Select cells that produce the desired antibody.

Hybridoma cells multiply in culture and produce antibodies.

5 Clone antibody-producing hybridoma cells.

6 Grow large numbers of the cells in culture.

7 Extract the antibodies.

Figure 9.17 **How monoclonal antibodies are produced.**

antibodies prepared against the cancer. Upon injection into the patient, the antibodies would deliver the drug directly to the cancer cells, sparing nearby healthy tissue.

Antibiotics combat bacteria

Literally, *antibiotic* means "against life." Antibiotics kill bacteria or inhibit their growth. The first antibiotics were derived from extracts of molds and fungi, but today most antibiotics are synthesized by pharmaceutical companies.

There are hundreds of antibiotics in use today, and they work in dozens of ways. In general, they take advantage of the following differences between bacteria and human cells: (1) bacteria have a thick cell wall, human cells do not; (2) bacterial DNA is not safely enclosed in a nucleus, human DNA is; (3) bacterial ribosomes are smaller than human ribosomes; and (4) bacterial rate of protein synthesis is very rapid as they grow and divide. Consider two examples of how antibiotics work: Penicillin focuses on the difference in cell walls and blocks the synthesis of bacterial cell walls, and streptomycin inhibits bacterial protein synthesis by altering the shape of the smaller bacterial ribosomes.

MJ's BlogInFocus An innovative approach to fighting bacterial infections is to use "good" bacteria to fight the "bad" ones. To learn more, visit MJ's blog in the Study Area in MasteringBiology and look under "Using Bacteria to Fight Bacteria."

http://goo.gl/K8wv22

Some antibiotics combat only certain types of bacteria. Others, called *broad-spectrum antibiotics,* are effective against several groups of bacteria. By definition, however, antibiotics are ineffective against viruses. Recall that viruses do not reproduce on their own; they replicate only when they are inside living cells. Using antibiotics to fight viral infections such as colds or the flu is wasted effort and contributes to the growing health problem of bacterial resistance to antibiotics.

Recap Active immunization with a vaccine produces a primary immune response and readies the immune system for a secondary immune response. Administration of prepared antibodies (passive immunization) can be effective against existing infections but does not confer long-term immunity. Most antibiotics kill bacteria by interfering with bacterial protein synthesis or bacterial cell wall synthesis. ▪

9.8 Tissue rejection: a medical challenge

Because the goal of the immune system is to protect the body from invasion by nonself cells, it is not surprising that the immune system attacks foreign human tissues with vigor.

This phenomenon is called *tissue rejection.* A very common type of tissue rejection occurs with blood transfusion, when a donor's blood is rejected by the recipient patient. It's called a *transfusion reaction,* and it can be fatal.

Surgical techniques for performing many organ transplants are really not that difficult. Historically, the major stumbling block to widespread transplantation of most organs has been the effectiveness of the immune system in rejecting foreign tissue. In the normal immune response, cytotoxic T cells swiftly attack and destroy any and all foreign cells. Before a transplant is even attempted, then, the donor's and recipient's ABO and other blood group antigens must first be determined. Next, donor and recipient tissues are tested to compare MHC antigens, because cytotoxic T cells target foreign MHC proteins. The closer the relationship between donor and recipient the better, because their MHC antigens are likely to be similar. Although successful transplants can be done between unrelated people, at least a 75% match between tissues is essential. After surgery, the patient must take *immunosuppressive drugs* that block the immune response, such as corticosteroid drugs to suppress inflammation or cytotoxic medications that kill rapidly dividing cells (to block activated lymphocytes).

Immunosuppressive therapy can dramatically prolong the lives of transplant patients, but it brings other risks. An impaired immune system cannot protect the body as effectively against pathogens and abnormal cells, so patients who are taking immunosuppressive drugs are more vulnerable to infections and certain cancers. The key to a successful transplant is to suppress the immune system enough to prevent rejection, while preserving as much immune function as possible. Antibiotic drugs can help control infections as they arise.

In recent years, three factors have made organ transplants a viable option for many people: (1) improvements in immunosuppressive drugs, (2) better techniques for cross-matching (or "typing") tissue, and (3) national sharing of information and donor organs through organ-bank systems. The organ-bank system allows patients to receive the best matches possible regardless of where they live.

Recap The major obstacle to organ transplantation is the recipient's immune response, as cytotoxic T cells usually attack all foreign cells. ▪

9.9 Inappropriate immune system activity causes health problems

Conditions characterized by inappropriate immune system activity include allergies and autoimmune disorders such as lupus erythematosus and rheumatoid arthritis.

Allergies: a hypersensitive immune system

Many of us—perhaps 10% of North Americans—suffer from allergies. Some allergies are relatively minor; others are quite severe. Examples include hay fever, poison ivy rashes, and severe reactions to specific foods or drugs. Some allergic reactions even require hospitalization.

An **allergy** is an inappropriate response of the immune system to an **allergen** (any substance that causes an allergic reaction). The key word is *inappropriate:* The allergen is not a dangerous pathogen, but the body reacts as if it were.

Recall that there are five classes of immunoglobulins. As shown in Figure 9.18, the culprits involved in allergic reactions are those in the IgE group. ❶ At some point, exposure to an allergen triggers a primary immune response, causing B cells to produce specific IgE antibodies. ❷ The IgE antibodies bind to mast cells (found primarily in connective tissue) and to circulating basophils.

❸ When the same allergen enters the body a second time, it binds to the IgE antibodies on mast cells and basophils, causing them to release histamine. ❹ The result is an allergic reaction, a typical inflammatory response that includes warmth, redness, swelling, and pain in the area of contact with the allergen. Histamine also increases secretion of mucus in the region. Every time the body is exposed to this allergen, the body reacts as if it has an injury or infection, even though it doesn't.

Some allergens affect only the areas exposed. Other allergens, including food allergens and bee sting venom, are absorbed or injected into the bloodstream. These substances are carried quickly to mast cells throughout the body, including connective tissue in the respiratory, digestive, and circulatory systems. Such allergens often elicit a *systemic response,* meaning they affect several organ systems. Systemic responses include constriction of smooth muscle in the lungs and digestive system and dilation of blood vessels.

Symptoms of a severe systemic allergic reaction can include difficulty breathing (caused by constricted airways), severe stomach cramps (muscle contractions), swelling throughout the body (increased capillary permeability), and circulatory collapse with a life-threatening fall in blood pressure (dilated arterioles). This is known as *anaphylactic shock.* Anyone who appears to be suffering from anaphylactic shock should be rushed to a hospital, because the reaction can be fatal. The symptoms often begin suddenly, and doctors advise people with a history of strong allergic reactions to carry an emergency kit with them at all times. The kit contains a self-injected hypodermic syringe of epinephrine, a hormone that dilates the airway and constricts peripheral blood vessels, preventing shock.

Most allergies, however, are more annoying than dangerous. Antihistamines—drugs that block the effects of histamine—are often effective treatments for mild to moderate reactions. Allergy shots can help by causing the body to produce large numbers of IgG antibodies, which combine with the allergen and block its attachment to IgE.

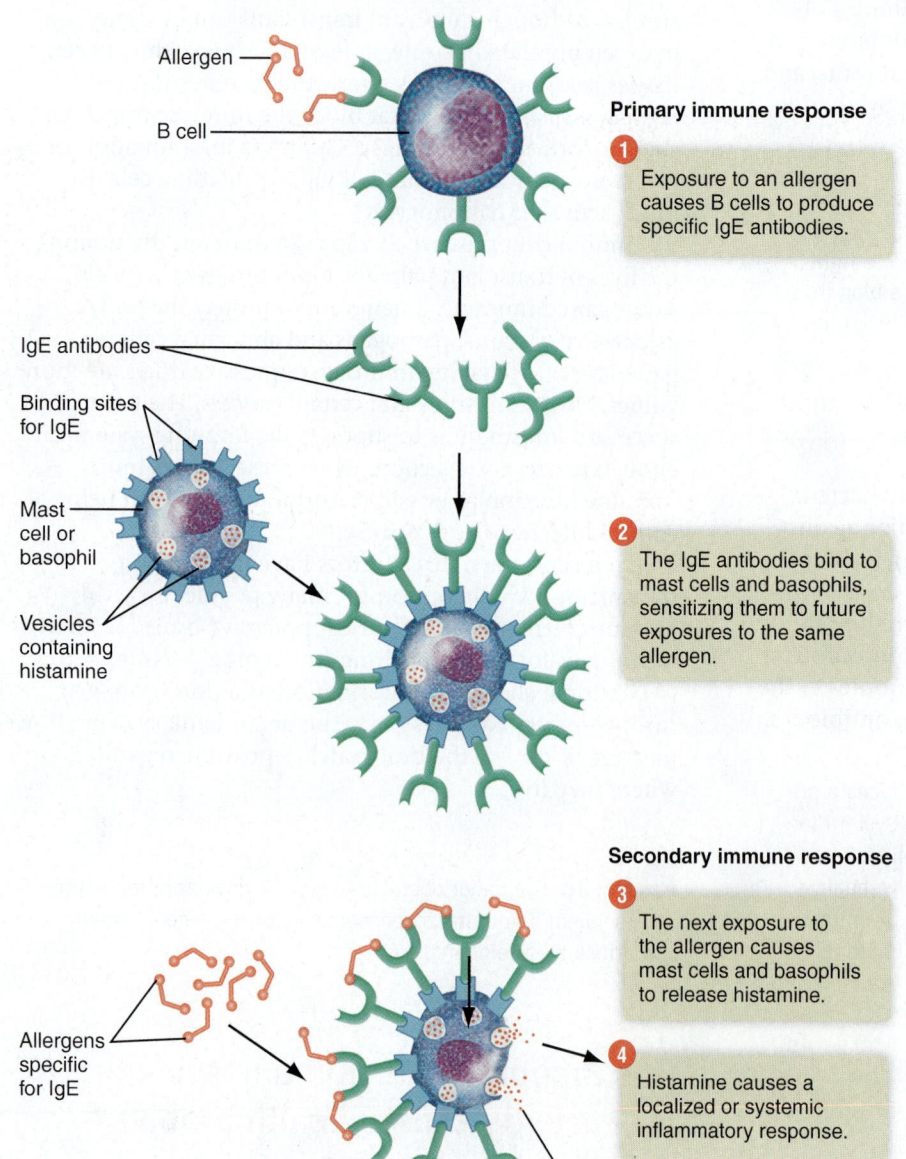

Allergen

B cell

Primary immune response

❶ Exposure to an allergen causes B cells to produce specific IgE antibodies.

IgE antibodies

Binding sites for IgE

Mast cell or basophil

Vesicles containing histamine

❷ The IgE antibodies bind to mast cells and basophils, sensitizing them to future exposures to the same allergen.

Secondary immune response

❸ The next exposure to the allergen causes mast cells and basophils to release histamine.

Allergens specific for IgE

❹ Histamine causes a localized or systemic inflammatory response.

Histamine

Figure 9.18 How an allergic reaction develops.

✔ Why do allergic reactions often get worse with each successive exposure to the antigen?

Autoimmune disorders: defective recognition of self

On rare occasions, the immune system's remarkable ability to distinguish self from nonself fails. When that happens, the immune system may produce antibodies and cytotoxic T cells that target its own cells. Conditions in which this happens are called **autoimmune disorders.**

Approximately 5% of adults in North America, two-thirds of them women, have some type of autoimmune disorder. We don't yet know all the details of how these diseases arise. In some cases, certain antigens simply are never exposed to the immune system as it undergoes fetal development. These antigens were never programmed into the system as self, so when tissue damage exposes them, the mature immune system responds as if they are foreign. In other cases, antibodies produced against a foreign antigen may cross-react with the person's own tissues.

At the moment, there are no cures for autoimmune disorders. Treatments include therapies that depress the body's defense mechanisms and relieve the symptoms. Autoimmune conditions include a wide range of diseases, including multiple sclerosis, a progressive disorder of the central nervous system (see Chapter 11), and Type 1 diabetes mellitus, which targets cells in the pancreas (see Chapter 13). Next, we will look at two other autoimmune disorders: lupus erythematosus and rheumatoid arthritis.

Lupus erythematosus: inflamed connective tissue

Lupus erythematosus (or *lupus*) is an autoimmune disorder in which the body attacks its own connective tissue. One type of lupus, called *discoid lupus erythematosus*, primarily affects areas of the skin exposed to sunlight. More serious is systemic lupus erythematosus, which may affect various tissues and organs including the heart, blood vessels, lungs, kidneys, joints, and brain.

Lupus often starts as a red skin rash on the face or head. Other symptoms include fever, fatigue, joint pain, and weight loss. Spreading inflammation can lead to osteoarthritis (see Chapter 5), pericarditis (see Chapter 8), or pleurisy (inflammation of the lining of the lungs).

Lupus affects nine times as many women as men. Typically, it occurs during childbearing age and is more common in certain racial groups such as African Americans, West Indians, and Chinese. Medications can reduce the inflammation and alleviate the symptoms.

Rheumatoid arthritis: inflamed synovial membranes

Rheumatoid arthritis is a type of arthritis involving inflammation of the synovial membrane that lines certain joints (see Chapter 5). In rheumatoid arthritis, B cells produce antibodies against a protein in the cartilage of synovial membranes. The resulting immune response releases inflammatory chemicals that cause further tissue damage. At first, fingers, wrists, toes, or other joints become painful and stiff. Over time, the inflammation destroys joint

cartilage and the neighboring bone. Eventually, bony tissue begins to break down and fuse, resulting in deformities (Figure 9.19) and reduced range of motion. The disease is intermittent, but with each recurrence the damage is progressively worse.

Pain-relieving medications can help many people with rheumatoid arthritis, as can regular mild exercise and physical therapy to improve range of motion. Powerful drugs that neutralize chemicals in the inflammatory response can prevent joints from becoming deformed. Surgery to replace damaged joints with artificial joints can restore the ability to move and prevent painful disabilities.

Recap An allergy is an inappropriate inflammatory response. An autoimmune disorder occurs when the immune system fails to distinguish self from nonself cells and begins to attack the body's own cells. Examples of autoimmune disorders are lupus erythematosus (inflammation of connective tissue) and rheumatoid arthritis (inflammation of synovial membranes). ∎

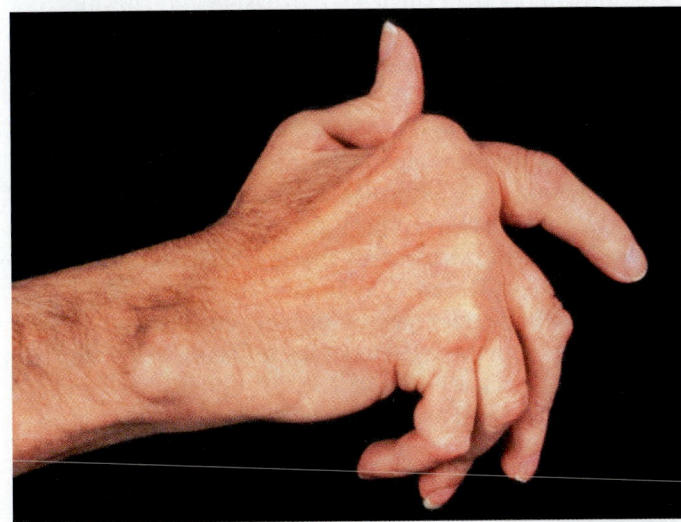

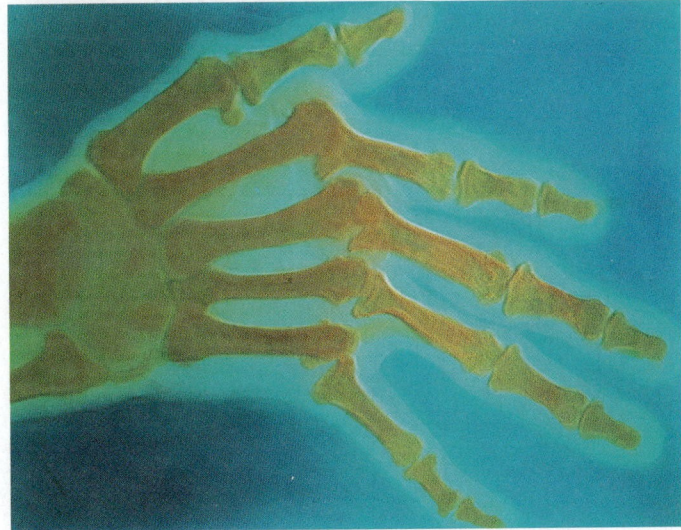

Figure 9.19 **Rheumatoid arthritis.**

9.10 Immune deficiency: the special case of AIDS

Immune deficiency is a general term for an immune system that is not functioning properly. One immune deficiency disease is *severe combined immunodeficiency disease (SCID)*. For the rare person who inherits SCID, even a minor infection can become life-threatening. People with SCID have too few functional lymphocytes to defend the body against infections. The most common and best-known severe immune deficiency condition is **AIDS (acquired immune deficiency syndrome).** A *syndrome* is a medical term for a group of symptoms that occur together, and *acquired* means that a person catches a disease—in this case by becoming infected with **HIV (human immunodeficiency virus).**

HIV targets helper T cells of the immune system

Figure 9.20 shows the structure of HIV. The virus consists of nothing more than single-stranded RNA and enzymes, wrapped in two protein coats and a phospholipid membrane with protein spikes. It has no nucleus and no organelles. Like other viruses, HIV infects by entering a cell and using the cell's machinery to reproduce. HIV targets helper T cells, gaining entry by attaching to CD4 receptors.

HIV belongs to a particular class of viruses, called *retroviruses*, that have a unique way of replicating (**Figure 9.21**). Retroviruses first attach to the CD4 receptor of a helper T cell. The attachment fuses the retrovirus's envelope with the cell's membrane, releasing the viral RNA and enzymes into the cell. Under the influence of the viral enzymes and using the viral RNA as a template (a pattern), the host cell is forced to make a single strand of DNA complementary to the viral RNA, and from it a second strand of DNA complementary to the first. The new double-stranded DNA fragment is then inserted into the cell's DNA. The cell, not recognizing the DNA as foreign, uses it to produce more viral RNA and proteins, which are then assembled into thousands of new viruses within the cell. The sheer magnitude of viral replication so saps the T cell's energy that eventually it dies and ruptures, releasing the viral copies. The new viruses move on to infect other helper T cells.

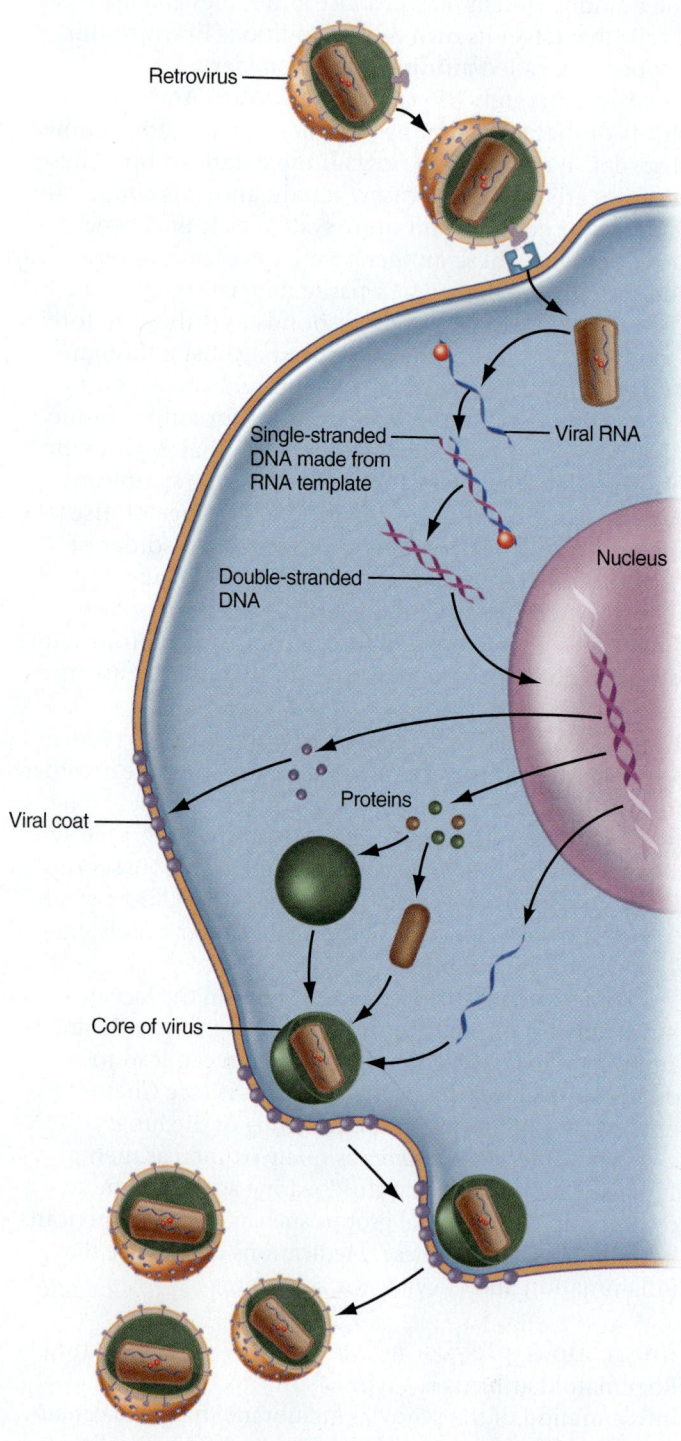

Figure 9.21 labels: Retrovirus; Single-stranded DNA made from RNA template; Viral RNA; Double-stranded DNA; Nucleus; Viral coat; Proteins; Core of virus

Figure 9.21 How HIV replicates. After binding to a helper T cell's CD4 receptor, the virus injects its RNA and enzymes into the cell and uses them to force the cell to make single- and then double-stranded DNA. The DNA is inserted into the cell's own DNA within the nucleus, where it directs the cell to make more protein coat and viral RNA and protein core. The coats and cores join, exiting the cell as new viruses.

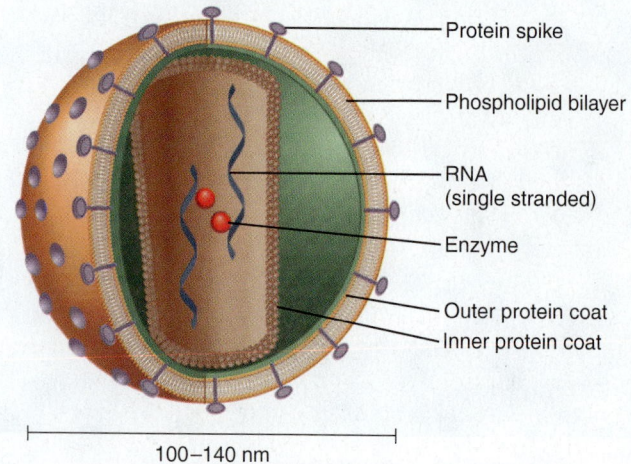

Figure 9.20 labels: Protein spike; Phospholipid bilayer; RNA (single stranded); Enzyme; Outer protein coat; Inner protein coat; 100–140 nm

Figure 9.20 The structure of HIV.

HIV is transmitted in body fluids

HIV is a fragile virus that cannot survive dry conditions for even a short time. This means that it cannot be transmitted through the air, by casual contact, or by doorknobs or toilet seats. However, HIV can be transmitted in at least four body fluids—blood, semen, breast milk, and vaginal secretions. It is not known to be transmitted by contact with urine, feces, saliva, perspiration, tears, or nasal secretions unless they contain blood.

Most commonly, HIV is transmitted by sexual contact or contaminated hypodermic needles. In addition, infected mothers can pass it to their children during pregnancy, labor and delivery, or when breast-feeding. It has been estimated, for example, that a healthy newborn who is breastfed by an infected mother has about a 20% chance of contracting HIV from its mother's milk. A significant number of cases of HIV infection resulted from blood transfusions until the mid-1980s, when it became routine to test blood for HIV antibodies.

Because transmission of HIV generally requires direct contact with body fluids, HIV is considered relatively hard to transmit from person to person (at least when compared to such common diseases as the flu). However, unlike the flu, HIV is extremely virulent unless it is aggressively treated with an expensive regimen of drugs. Over 90% of all U.S. citizens diagnosed with AIDS by 1988 (before effective drugs were available) are now dead.

AIDS develops slowly

The symptoms of HIV infection and the subsequent development of full-blown AIDS progress slowly. The course of the disease falls into three phases (**Figure 9.22**).

Phase I Phase I lasts anywhere from a few weeks to a few years after initial exposure to the virus. There is a brief spike in HIV in the blood, followed by typical flulike symptoms of swollen lymph nodes, chills and fever, fatigue, and body aches. The immune system's T cell population may decline briefly then rebound as the body begins to produce more cells and antibodies against the virus. The presence of antibodies against the virus is the basis for a diagnosis of HIV infection, and a person having these antibodies is said to be HIV-positive. However, he or she will not yet have the disease syndrome called *AIDS*.

Unless they suspect they have been exposed to HIV, most people would not associate Phase I symptoms with HIV infection and so would not think to be tested. The antibodies do not destroy the virus entirely because many of the virus particles remain inside cells, where antibodies and immune cells cannot reach them.

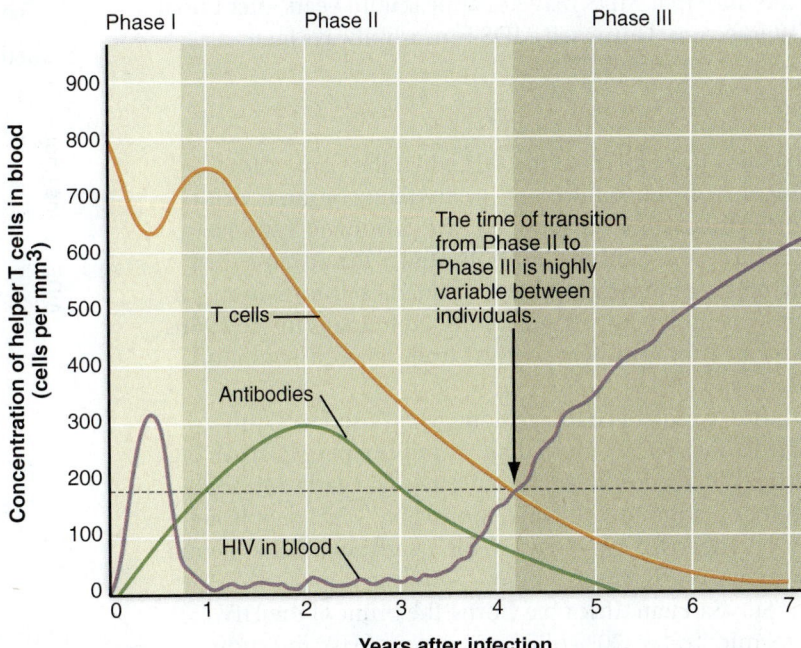

Figure 9.22 **A representative time course of the progression toward AIDS after HIV infection.** Phase I is characterized by flulike symptoms followed by apparent recovery. There is a brief spike in HIV in the blood. In Phase II, there is a slow decline in the number of T cells. Most of the HIV is harbored inside cells. Phase III (AIDS) is characterized by a T cell count of less than 200 per cubic millimeter of blood, a decline in HIV-specific antibodies, a rise in HIV in blood, opportunistic infections, and certain cancers. The patient represented by this graph progressed to AIDS in just over four years.

Phase II In Phase II the virus begins to do its damage, wiping out more and more of the helper T cells. The loss of T cells makes the person more vulnerable to *opportunistic infections*—infections that take advantage of the weakened immune system to establish themselves in the body. During Phase II, people *may* have persistent or recurrent flulike symptoms but may have no symptoms at all if they do not have an opportunistic infection. If they have not been tested for HIV antibodies they may not know they are infected.

Two-thirds to three-quarters of all people who test positive for antibodies to HIV do not exhibit symptoms associated with AIDS. During Phases I and II, many people pass the virus on to others without realizing it. Those they infect may transmit the virus to still others. Phase II can progress to Phase III in as little as 6 months, but on average it takes about 10 years to progress to Phase III. Left untreated, 95% of people in Phase II progress to Phase III.

Phase III Once the number of helper T cells of an HIV-positive person falls below 200 per cubic millimeter of blood *and* the person has an opportunistic infection or type of cancer associated with HIV infection, the person is said to have AIDS. Infections and cancers associated with AIDS include pneumonia, meningitis, tuberculosis, encephalitis, Kaposi's sarcoma, and non-Hodgkin's lymphoma, among others.

Notice that AIDS may not appear until years after initial HIV infection. Untreated AIDS is nearly always fatal.

The AIDS epidemic: a global health issue

AIDS was first described in 1981 when the Centers for Disease Control and Prevention in Atlanta began to notice a disturbing similarity between cases involving a strange collection of symptoms. (This illustrates the advantage of having a central clearinghouse for medical information.) It is now believed that HIV first infected humans in the 1960s in Africa after "jumping species" from other primates to humans.

The worldwide damage done by HIV so far is truly astonishing. Today, more than 35 million people are living with HIV infection or AIDS, representing nearly 1% of the adult population worldwide. More than 5,700 people are newly diagnosed every day. So far, nearly 39 million people have died of AIDS.

Sub-Saharan Africa has borne the brunt of the HIV epidemic. Today, 70% of all new cases of HIV infection occur in sub-Saharan Africa. Notably, the patterns of AIDS infection and transmission in Africa differ from those of industrialized countries. In the United States, more men than women are infected with HIV because the virus is generally transmitted via homosexual sex. In Africa, more than half of all HIV-infected persons are women who have contracted the virus heterosexually. Studies in several African nations have found that females aged 15–19 are four to five times more likely to be infected than males their age. According to one report, in Africa older HIV-infected men coerce or pay impoverished girls to have sex in the mistaken belief that sex with a virgin will cure AIDS.

The problem of AIDS in sub-Saharan Africa is made worse by political, economic, and social instability. Some sub-Saharan African nations are at war or are suffering intense internal conflict. In many countries, the economies are weak, transportation is difficult, sanitation is poor, and there are too few hospitals and medical personnel. Only 37% of sub-Saharan Africans with HIV currently are being treated for their infections.

And yet, there is hope on the horizon. Efforts to improve health care delivery by international agencies such as the World Health Organization (WHO) and various nongovernmental organizations such as Doctors Without Borders are beginning to have an effect. The number of new cases of HIV infection each year is now declining, even in sub-Saharan Africa. Worldwide, new HIV infections and AIDS-related deaths are down more than 35% from their peaks in the previous decade.

The picture is even better in the United States, thanks to a well-developed AIDS reporting system and the availability of good health care. In the United States, the number of deaths has declined to under 14,000 per year after peaking sharply in 1995 at more than 50,000 per year (**Figure 9.23**). Most of the decline after 1995 is due to the availability

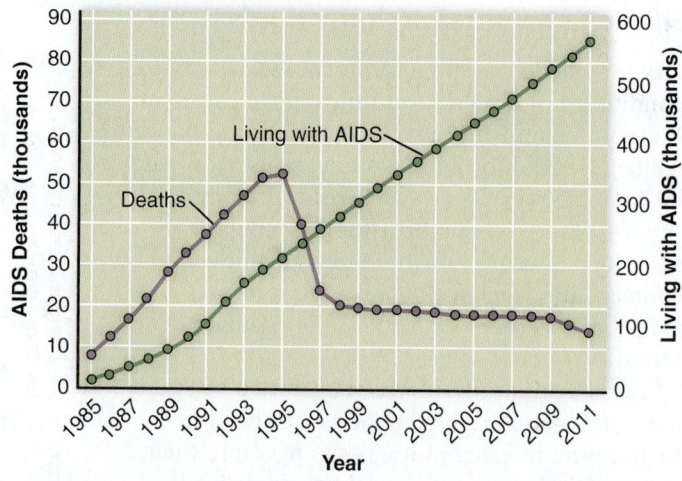

Figure 9.23 HIV/AIDS in the United States. AIDS deaths and the number of people living with AIDS each year since 1985.
Source: Centers for Disease Control and Prevention (CDC).

of drugs that suppress the active component of an HIV infection and keep infected people alive much longer. However, to be completely effective the drugs need to be taken for the rest of the patient's lifetime.

Although the death rate from AIDS has stabilized in the United States, the number of persons living with HIV/AIDS continues to climb. More than 500,000 persons are now living with AIDS in the United States. An additional 700,000 persons are currently infected with HIV but are not yet showing symptoms of AIDS. About 20% of all HIV-infected persons do not even know they are infected. Treatment of the rising number of persons living with HIV/AIDS is likely to place a heavy financial burden on our health care system in the future. The current cost of treating an HIV-infected person with HIV suppressive drugs is about $14,000–$20,000 per year.

Risky behaviors increase your chances of getting AIDS

Figure 9.24 shows how HIV is contracted in the United States among adults and adolescents (people above 12 years of age). In the United States, nearly three-quarters of all HIV-infected persons in 2009 were men. The riskiest behaviors for both men and women are engaging in sex with HIV-infected men and sharing needles during injected drug use.

Making sex safer

The only sure way to eliminate the risk of contracting HIV/AIDS completely is to eliminate sexual contact altogether. Some people do choose abstinence for ethical, practical, or professional reasons or at certain times in their lives. However, most persons do not remain abstinent throughout their lives. Given the risks of contracting HIV/AIDS, anyone who is sexually active should think seriously about how to make their sexual contacts as safe as possible. Below are several *safer*

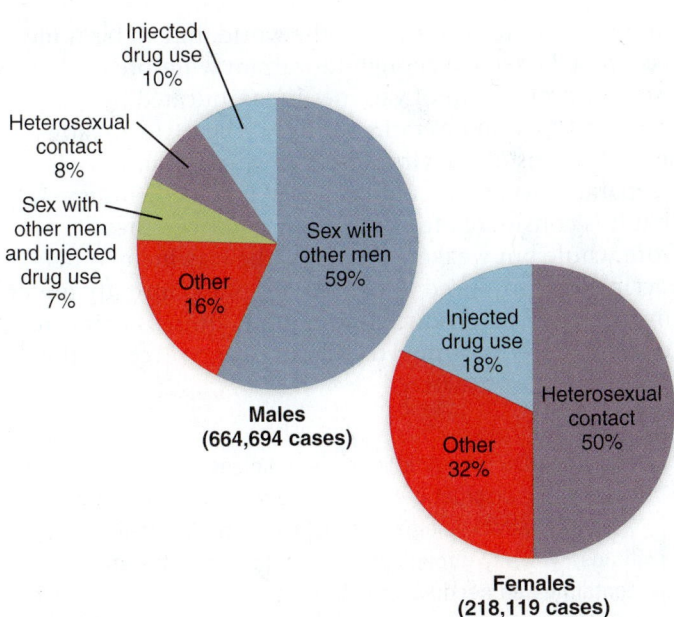

Other includes hemophilia, blood transfusion, perinatal exposure, and risk factors not reported or not identified.
Source: Centers for Disease Control and Prevention, HIV/AIDS Surveillance Report, 2012.

Figure 9.24 Persons living with HIV infection in the United States in 2009, by sex and exposure category.

sex guidelines and some evidence for them. The suggestions do not guarantee complete safety, which is why they are called recommendations for "safer" rather than "safe" sex.

- *Reduce the number of sexual partners.* Evidence shows that this is particularly effective among homosexual men. The evidence is not yet conclusive regarding the relationship between HIV infection and number of partners in heterosexual transmission, perhaps because the median number of partners is generally lower. Nevertheless, it makes sense that fewer partners translates to lower risk.

- *Choose a sexual partner with low-risk behavior.* This is especially important. A partner who has a history of injecting drugs or a man who has sex with other men is a high-risk partner. One study reported that choosing a partner who is not in any high-risk group lowers the risk of AIDS almost 5,000-fold.

- *Avoid certain high-risk sexual practices.* Any sexual behavior that increases the risk of direct blood contact, such as anal-genital sex, should be considered risky. Oral-genital sex is less risky but not risk-free.

- *Use latex or polyurethane condoms or other barriers.* In laboratory tests, latex or polyurethane condoms prevent the passage of HIV. How effective they are in actual use is not certain, but because they are about 90% effective as a birth control method, a reasonable hypothesis (not yet tested) is that they reduce the risk of HIV by about the same amount. Condoms are effective only if used consistently and correctly. Natural skin condoms are

not as safe as latex condoms. The use of dental dams is recommended for oral-vaginal sex.

- *Use nonoxynol-9, a spermicidal agent.* In laboratory tests, nonoxynol-9 has been shown to inhibit the growth of HIV. Be aware, however, that the evidence is conflicting regarding whether nonoxynol-9 offers any protection during sexual contact.

- *Get tested (and have your partner tested).* Notice, however, that the HIV tests currently available are designed to detect the human *antibodies* to HIV, not the virus itself. It can take up to six months after exposure to produce enough antibodies for the antibody test to become positive. For this reason, retesting at least six months after the last possible exposure is a good idea. Evidence shows that close to 90% of HIV-infected persons will have a positive HIV antibody test result within six months. A negative test at six months indicates substantially lower risk, but it does not provide complete assurance. A few individuals may not test positive for three or more years.

 **MJ's BlogInFocus** Do insurance companies pay for HIV tests? Are there any home tests for HIV? Visit MJ's blog in the Study Area in MasteringBiology and look under "HIV Tests."

http://goo.gl/GmurhZ

New treatments offer hope

As yet, there is no cure for AIDS, but researchers are investigating more than a hundred drugs to treat the condition. In general, AIDS drugs work by interfering with HIV replication at one of the steps in the virus's life cycle:

- *Interference with viral entry into the cell.* If the virus can't enter the cell, it can't take over the cell's metabolic machinery and force it to make more viruses. These drugs, called *entry inhibitors,* work by blocking the ability of the virus to lock onto receptors on the cell's surface, a necessary first step to entry.

- *Inhibition of enzymes required for replication of viral RNA.* Called *reverse transcriptase inhibitors,* these drugs prevent the virus from taking over the cell's metabolic pathways that lead to replication of viral RNA after the virus has gained entry to the cell.

- *Prevention of insertion of the viral genome into the cell's DNA.* These drugs, called *integrase inhibitors,* block viral enzyme required to insert the viral RNA into the host cell's DNA.

- *Inhibition of enzymes required to assemble viral proteins.* Called *protease inhibitors,* these drugs prevent the virus from being assembled within an infected cell, even though the virus may already have forced the cell to replicate its RNA and proteins.

Because each class of drugs works by an entirely different mechanism, more effective treatment can sometimes be achieved by taking several different drugs in combination. However, no combination of drugs has yet been found that completely cures an HIV infection. People who are HIV-positive have to be prepared to take AIDS drugs throughout their lives.

With an ever-expanding arsenal of AIDS drugs at their disposal, doctors are now starting to treat people who are infected with HIV but who don't yet show the symptoms of AIDS. The hope is that early treatment of an HIV infection will prevent the widespread destruction of the immune system that precedes the onset of AIDS symptoms. People who are at-risk for HIV infections are advised to be tested so that asymptomatic cases can be detected and treated.

Some researchers believe that safe and effective vaccines offer the only real hope of conquering HIV. Several dozen potential vaccines are already being tested on human volunteers around the world, but so far none has proved effective enough to warrant widespread use. The production of vaccines is complicated by the fact that HIV mutates rather quickly. There are already several strains of the virus, each of which would need a separate vaccine. In addition, HIV is so dangerous that it is considered too risky to produce vaccines from whole but weakened viruses, the way many other vaccines are produced. Most HIV vaccines currently under investigation are produced from other viruses engineered to contain pieces of the HIV virus or from pieces of the HIV viral genetic material.

Recap AIDS (acquired immune deficiency syndrome) is a devastating disorder of the immune system caused by a virus (HIV) that attacks helper T cells. HIV is transmitted in body fluids, typically through sexual contact, blood transfusions, contaminated needles, or breast-feeding. ∎

Chapter Summary

9.1 Pathogens cause disease p. 190

- Pathogens include bacteria, viruses, fungi, protozoa, worms, and possibly prions.
- Bacteria have very little internal structure and are covered by a rigid outer cell wall.
- Viruses cannot reproduce on their own. Viral reproduction requires a living host cell.
- Prions are misfolded proteins that replicate by causing a normal protein to misfold.
- The danger from a particular pathogen depends on *how* it is transmitted, how *easily* it is transmitted, and how *damaging* the resulting disease is.

9.2 The lymphatic system defends the body p. 192

- The lymphatic system consists of vessels, lymph nodes, the spleen, the thymus gland, and the tonsils.
- The lymphatic system helps protect us against disease.
- Phagocytic cells in the spleen, lymph nodes, and tonsils engulf and kill microorganisms.
- The thymus gland secretes hormones that help T lymphocytes mature.

9.3 Keeping pathogens out: the first line of defense p. 195

- Skin is an effective barrier to the entry of microorganisms.
- Tears, saliva, mucus, and earwax trap organisms and/or wash them away.
- Digestive acid in the stomach kills many microorganisms.
- Vomiting, defecation, and urination physically remove microorganisms after entry.

9.4 Nonspecific defenses: the second line of defense p. 196

- Phagocytes surround and engulf microorganisms and damaged cells.
- Inflammation has four outward signs: redness, warmth, swelling, and pain.
- Natural killer cells kill their targets by releasing damaging chemicals.
- Circulating proteins of the complement system either kill microorganisms directly or mark them for destruction.
- Interferons are proteins that interfere with viral reproduction.
- Fever raises body temperature, creating a hostile environment for some microorganisms.

9.5 Specific defense mechanisms: the third line of defense p. 199

- Cells of the immune system can distinguish foreign or damaged cells from our own healthy cells.
- All cells have cell-surface markers called MHC proteins that identify the cells as self.
- An antigen is a substance that stimulates the immune system and provokes an immune response.
- B cells produce antibodies against foreign antigens.
- T cells of several types release chemicals that enhance the immune response and kill foreign cells directly.

9.6 Immune memory creates immunity p. 205

- Information about an antigen is stored in memory cells after first exposure.
- The second exposure to the antigen produces a much greater immune response than the first.
- The rapidity of the second response is the basis of immunity from disease.

9.7 Medical assistance in the war against pathogens *p. 207*

- Vaccines immunize the body in advance against a particular disease.
- Injected antibodies provide temporary immunity and are of some benefit against an existing infection.
- Monoclonal antibodies are used primarily in medical tests.
- Antibiotics are effective against bacteria but not against viruses.

9.8 Tissue rejection: a medical challenge *p. 209*

- The phenomenon of tissue rejection is a normal consequence of the body's ability to recognize self from nonself.
- Immunosuppressive drugs, the ability to test for various antigens, and organ-donor matching programs have increased the success rate of organ transplantation in humans.

9.9 Inappropriate immune system activity causes health problems *p. 209*

- Allergies occur when the immune system responds excessively to foreign particles that are not otherwise harmful.
- Autoimmune disorders develop when a person's immune system attacks the person's own cells as if they were foreign.

9.10 Immune deficiency: the special case of AIDS *p. 212*

- AIDS is caused by a virus (HIV).
- The disease can take years to develop after initial HIV infection, but it is nearly always fatal.
- Worldwide, the number of cases of HIV infection and of AIDS is still rising rapidly.
- The chances of contracting AIDS can be reduced (but never eliminated completely) by practicing "safer" sex.

Terms You Should Know

active immunization, *207*
AIDS (acquired immune deficiency syndrome), *212*
antibiotics, *190*
antibody, *200*
antibody-mediated immunity, *200*
antigen, 200
bacteria, *190*
B cell, *200*
cell-mediated immunity, *200*
cytokines, *203*
HIV (human immunodeficiency virus), *212*
immune response, *199*
inflammation, *198*
interferons, *199*
macrophage, *198*
monoclonal antibodies, *208*
passive immunization, *207*
pathogen, *189*
phagocytosis, *197*
prion, *191*
T cell, *200*
vaccine, *207*
viruses, *191*

Concept Review

Answers can be found in the Study Area in MasteringBiology.

1. Define the term *pathogen* and give some examples.
2. Describe what is unusual about viruses that cause some people to question whether they are living things.
3. Explain how prions affect human health.
4. Describe the functions of organs of the lymphatic system.
5. List the four components of an inflammatory response.
6. Describe in general terms the distinction between *nonspecific* and *specific* defense mechanisms.
7. Describe the concept of an antigen and how it relates to self and nonself markers.
8. Describe how cells that belong to a particular individual are identified so that the individual's immune system doesn't attack them.
9. Explain how cytotoxic T cells kill target cells.
10. Describe how a vaccine produces immunity from a specific disease.

Test Yourself

Answers can be found in the Appendix.

1. In which of the following ways are bacterial cells similar to human cells?
   a. Bacterial cells have cell walls.
   b. Bacterial cells have a single, circular chromosome.
   c. Bacterial cells use ATP to fuel cellular activities.
   d. Bacterial cells lack mitochondria.

2. Which of the following statements about viruses is true?
   a. Viruses require a host cell in which to reproduce.
   b. Viruses are very small bacteria.
   c. Viral infections can generally be controlled with antibiotics.
   d. Viruses are composed of protein only.

3. Which of the following pathogenic agents causes a self-propagating misfolding of proteins in nerve cells?
   a. bacteria   c. viruses
   b. prions   d. helminths (worms)

4. Consider the following group of diseases: hepatitis, chicken pox, warts, and measles. What do these diseases have in common?
   a. They are all caused by bacteria.
   b. They are readily treated with antibiotics.
   c. They are all caused by viruses.
   d. They are very common in patients infected with HIV.

5. Which of the following is true regarding prion diseases?
   a. They are caused by a deadly type of virus.
   b. They can readily be treated with antibiotics.
   c. They can be prevented by vaccinations.
   d. They cause accumulation of misfolded proteins in brain cells.

6. Which of the following is a benefit of resident bacteria?
   a. Resident bacteria cause the stomach to be acidic.
   b. Resident bacteria produce antiviral compounds that prevent viral infections.
   c. Resident bacteria can out-compete harmful bacteria and lower the incidence of infection.
   d. Resident bacteria digest cellulose within the human digestive tract.

7. DiGeorge syndrome is a congenital disease that results in a poorly developed, nonfunctioning thymus gland. Which of the following would be a likely problem experienced by a baby with DiGeorge syndrome?
   a. lack of B cells
   b. lack of antibodies
   c. lack of T cells
   d. lack of macrophages

8. The following are steps in phagocytosis: (1) Bacterium is digested by lysosomal enzymes, (2) phagocyte approaches bacterium, (3) phagocytic vesicle fuses with lysosome, and (4) phagocyte engulfs bacterium, forming a phagocytic vesicle. In which order do these steps occur?
   a. 4-2-3-1
   b. 2-4-3-1
   c. 2-3-4-1
   d. 4-1-3-2

9. In which of the following choices is the cell correctly matched with its function?
   a. eosinophil: produces antibodies
   b. B lymphocyte: directly attacks foreign cells
   c. basophil: secretes histamine
   d. T lymphocyte: phagocytizes bacteria

10. Each of the following processes helps combat infection except:
    a. inflammation
    b. fever
    c. autoimmunity
    d. antibody production

11. The primary immune response is:
    a. faster than the secondary immune response
    b. longer lasting than the secondary immune response
    c. less effective than the secondary immune response
    d. due to the presence of memory cells

12. Compared to active immunization, passive immunization:
    a. provides immediate protection
    b. is longer lasting
    c. creates a large number of memory cells
    d. may occasionally cause the disease it is intended to prevent

13. Which of the following increases the likelihood of successful organ transplant?
    a. matching the ABO blood group antigens
    b. matching the MHC tissue antigens
    c. administration of immunosuppressive drugs
    d. all of these choices

14. Which of the following does not belong with the others?
    a. lupus erythematosus
    b. rheumatoid arthritis
    c. anaphylactic shock
    d. Type I diabetes mellitus

15. Which of the following statements about HIV is true?
    a. Latex condoms are 100% effective in blocking transmission of HIV.
    b. Most individuals who progress to Phase II of HIV infection remain in Phase II and never progress to Phase III (AIDS).
    c. HIV specifically impairs the cell-mediated immune response.
    d. Anti-HIV medications such as AZT and maraviroc can cure HIV infection.

Apply What You Know

Answers can be found in the Study Area in MasteringBiology.

1. When you get a minor infection in a small cut in the skin, sometimes soaking it in hot water speeds the healing process. How might the heat help?

2. Explain why antibiotics don't work against viruses.

3. Everyone knows that bacteria can cause disease. Suppose that we could actually remove all bacteria from our bodies all at once. Would that be a good thing or a bad thing?

4. In 1918, a pandemic of a deadly flu strain killed upward of 30 million people. If that same flu strain were to come around again, would we be better off or worse off than those who were alive in 1918?

5. Researchers have been working on an effective vaccine for gonorrheal infections for some years. One promising vaccine is delivered via a spray into the nasal cavity. This is clearly not the site of gonococcal infection. Why would one administer a vaccine meant to protect the reproductive tract into the nasal cavity?

6. Why is it that most people get the chicken pox only once, but they can get a cold or the flu over and over again throughout a lifetime?

7. The immune system is supposed to defend us from harmful microorganisms. Why doesn't it always work? In other words, why do some people still get sick and die?

MJ's BlogInFocus

Answers can be found in the Study Area in MasteringBiology.

1. There is now an antivenom for the treatment of scorpion stings. The antivenom is given after a sting has occurred. Based on what you know, is it a form of active immunization or is it passive immunization? Visit MJ's blog in the Study Area in MasteringBiology and look under "Antivenom Against Scorpion Stings." http://goo.gl/Nvil1L

2. A pill has been available since 2012 that can prevent HIV infection in healthy people who are at risk of becoming infected (for example, partners of HIV-infected persons). But the pill is not popular among its potential users. Why do you suppose that is? Visit MJ's blog in the Study Area in MasteringBiology and look under "A Pill to Prevent HIV Infection." http://goo.gl/vwaeKv

MasteringBiology®

Students Go to MasteringBiology for assignments, the eText, and the Study Area with animations, practice tests, and activities.

Professors Go to MasteringBiology for automatically graded tutorials and questions that you can assign to your students, plus Instructor Resources.

Answers to ✓ questions are available in the Appendix.

The Respiratory System: Exchange of Gases

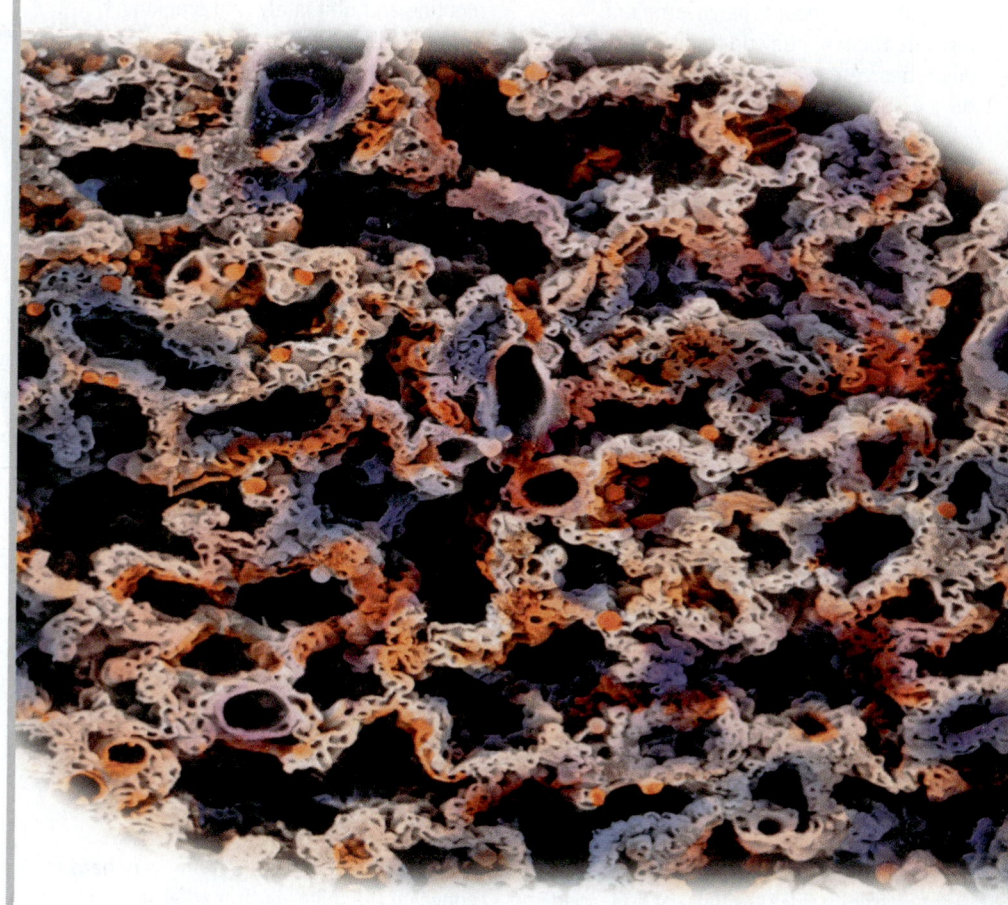

False color SEM (×150) of a section of lung tissue.

Key Concepts

- **The function of the respiratory system is to facilitate the diffusion of gases (O_2 and CO_2) between the atmosphere and blood**.

- **The actual structure across which gases are exchanged is only two living cell layers thick.** It is essential that this delicate structure be kept moist, clean, and free of microorganisms.

- **Breathing requires physical effort by skeletal muscles.** The muscles of breathing include the dome-shaped *diaphragm* and the *intercostal muscles* between the ribs.

- **Breathing is controlled by powerful feedback control mechanisms** that seek to maintain homeostasis of blood O_2 and CO_2. That's why your breathing increases automatically during exercise and why you cannot hold your breath indefinitely even if you try.

- **Red blood cells are essential for transporting O_2 from the lungs to all living cells.** Blood plasma alone could not carry enough O_2 for our survival.

The Fight over Regulation of E-Cigarettes

There's a fight smoldering over the regulation and proper use of electronic cigarettes (e-cigarettes). Should the government regulate them as they do traditional tobacco cigarettes? Are e-cigarettes a drug delivery device? Should they be regulated at all?

What Is an E-Cigarette?

An e-cigarette is a device that looks vaguely like a cigarette. It is designed to deliver a vapor-containing nicotine without the usual tar and other cancer-causing agents present in a traditional tobacco cigarette. An e-cigarette is a hand-held device consisting of a replaceable cartridge that contains nicotine and flavors in a liquid solution, a heating element, and a small battery. When the user draws on an e-cigarette, the element heats up, vaporizing some of the solution and producing an aerosol that looks like smoke but isn't smoke. To complete the illusion of smoking, some e-cigarettes even have a red LED light that lights up when the user draws on the cigarette!

The FDA Versus E-Cigarette Marketers

When e-cigarettes first came out, no government regulation on the devices was yet in force. The Food and Drug Administration (FDA) initially had planned to regulate e-cigarettes as drug delivery devices, approving them only to help tobacco smokers wean themselves off nicotine and ultimately quit smoking. In the FDA's view, e-cigarettes would fall into the same regulatory category as nicotine gums and patches, products proven to be effective in smoking cessation. FDA approval of e-cigarettes as drug delivery devices hinged on data becoming available that would prove the health, safety, and smoking-cessation efficacy of the devices. The FDA would then approve e-cigarettes as drug delivery devices, smokers would use them to stop smoking, and countless lives would be saved. And, like tobacco cigarettes and nicotine gum and patches, if e-cigarettes were regulated as drug delivery devices the FDA would designate them as controlled substances, making them unavailable to minors.

E-cigarette marketers, it seems, had an entirely different agenda: sales, sales, sales! It's no accident that e-cigarettes come in many shapes and lots of colors and flavors. E-cigarette marketers are trying very hard to convince the huge youth market (possibly you!) that e-cigarettes are trendy and

Questions to Consider

1 Do you think e-cigarettes should be regulated, and if so, how? Should they be taxed?

2 What safety concerns do you have, if any, about e-cigarettes?

safer than smoking tobacco. Of course, the marketers also hoped that e-cigarettes would be used by smokers who want to quit, but that demographic comprises just a small percentage of the potential market: Compare the number of existing tobacco smokers to an entire generation of young people who could become e-cigarette users.

Are E-Cigarettes Safe? Do They Help Tobacco Smokers Quit?

E-cigarettes are considered to be much safer than traditional cigarettes because they do not contain tobacco. The real question is not whether they are an alternative for tobacco smokers but whether they are safe enough to be used by nonsmokers. Most e-cigarettes contain nicotine, an addictive substance with no known health benefit. Health officials worry that e-cigarettes may spawn a whole generation of nicotine addicts, and for no good reason. The aerosol emitted by e-cigarettes contains propylene glycol, glycerol, and trace amounts of flavor particles. So far, there is no evidence that the amounts of these substances reaches a level that would lead to a public health concern. On the other hand, it is too early to know if e-cigarettes have any long-term health effects.

As for whether e-cigarettes help smokers quit, the evidence is encouraging but inconclusive. Several reviews of the scientific literature on the subject conclude that e-cigarettes reduce a smoker's desire to smoke and lead to fewer traditional cigarettes smoked per day; however, statistics are not yet available on the percentage of smokers who have given up tobacco cigarettes and switched entirely to e-cigarettes.

E-Cigarette Regulation— A Timeline

2006—E-cigarettes from China were imported into the United States. →

Will e-cigarettes lead to a whole new generation of nicotine addicts?

2009—The FDA directed the U.S. Customs and Border Protection agency to reject the entry of e-cigarettes into the United States. Nevertheless, sales of e-cigarettes reached $40 million.

2010—The FDA sent letters to five marketers of e-cigarettes warning them that claims that their products were an aid to stopping smoking were "unsubstantiated medical claims" not permitted under the Food, Drug, and Cosmetic Act. The FDA also signaled its intent to regulate e-cigarettes as drug delivery devices, a move which would essentially block the importation and sale of e-cigarettes.

Also in 2010, a leading marketer sued the FDA for seizing shipments of e-cigarettes. A federal district court judge and later a federal appeals court both ruled that the FDA could *not* regulate e-cigarettes as drug-delivery devices, paving the way for the importation and sale of e-cigarettes. However, the court left the door open for e-cigarettes to be regulated in the same way as tobacco cigarettes. Sales of e-cigarettes reach $82 million.

2013—Three years after federal court rulings, the FDA still had not finalized regulations on e-cigarettes. Concerned over the rising use of e-cigarettes by high school students, the city council of Los Angeles decided to take matters into its own hands. In December, the L.A. city council voted unanimously to regulate e-cigarettes with all the same restrictions on sales and use as tobacco cigarettes. Other municipalities and states begin to consider similar regulation. E-cigarette sales surpass $1 billion.

2014—The FDA proposes to regulate e-cigarettes in the same way it regulates tobacco products. Under the proposal, makers of products deemed to "meet the statutory definition of a tobacco product" would be required to register with the FDA and report product ingredients. E-cigarette manufacturers could only market new products after FDA review and would be prohibited both from making claims of reduced risk, unless substantiated by scientific evidence, and from distributing free samples. Furthermore, the FDA would impose a minimum age restriction on e-cigarette sales, a requirement for health warnings, and a ban on vending machine sales in places where the underage have access.

2015 and beyond—E-cigarettes don't actually contain tobacco, so it will be interesting to see whether the FDA's proposal will ever be finalized, and if it is, whether it will stand up in court. Other questions include whether e-cigarettes will be taxed like cigarettes. This story is a long way from over.

SUMMARY

- Electronic cigarettes (e-cigarettes) are devices that deliver nicotine to the user without the tar and other cancer-causing agents in a traditional tobacco cigarette. E-cigarettes were developed primarily to help smokers quit smoking.
- The FDA initially attempted to regulate e-cigarettes as a drug delivery device, but regulation was stymied by the courts.
- Sales of e-cigarettes continue to climb as manufacturers design their products for a younger generation of nonsmokers.
- Municipalities, states, and the FDA are moving toward regulating e-cigarettes as tobacco products, even though the devices contain no tobacco.

Your dinner companion is turning blue. One minute ago, he was laughing and drinking and spearing his steak with enthusiasm, and now, he can't breathe or talk and is frantically pulling at his collar.

Choking is an emergency, but one that is seldom handled by emergency medical personnel. The victim's fate is almost always decided before professional help can arrive. Unless someone in the immediate vicinity can quickly intervene to relieve the choking, there's a good chance the victim won't survive.

What is the physical basis of this urgency? Our cells require energy and certain raw materials (sugars, fats, and proteins) in order to grow, reproduce, and carry out their special functions. To create energy in a form they can use, cells rely on a process called aerobic metabolism, in which oxygen (O_2) is consumed and carbon dioxide (CO_2) waste is produced. Without a consistent supply of oxygen and a way to get rid of carbon dioxide, essential cellular processes cannot continue.

The primary function of the **respiratory system** is to facilitate the exchange of oxygen and carbon dioxide between the air and blood. When we inhale (breathe in), we extract oxygen from the air, and when we exhale (breathe out) we get rid of carbon dioxide. The exchange of these two gases is perhaps the most urgent of all the exchanges of materials between an animal and its environment. We can live for days, perhaps even weeks without food or water, but we'll die within minutes if we cannot breathe. The respiratory system thus plays an essential role in the maintenance of homeostasis of the internal environment, that is, the maintenance of just the right concentrations of oxygen and carbon dioxide in the blood. ∎

Plants do just the opposite for their primary metabolic processes; they absorb carbon dioxide through small pores in their leaves and use it in their own energy-producing process called *photosynthesis*. In the process, they produce oxygen for their (and coincidentally our) use. We consider photosynthesis and other examples of the delicate balance of life on Earth when we discuss ecosystems and environmental issues. In this chapter, we focus on the human respiratory system.

10.1 Respiration takes place throughout the body

The term *respiration* encompasses four processes:

- *Breathing* (also called **ventilation**). The movement of air into and out of the lungs.
- *External respiration*. The exchange of gases between inhaled air and blood.
- *Internal respiration*. The exchange of gases between the blood and tissue fluids.
- *Cellular respiration*. The process of using oxygen to produce ATP within cells.

Cellular respiration generates carbon dioxide as a waste product. Breathing is facilitated by the respiratory system and its associated bones, muscles, and nerves. External respiration takes place within the lungs, and internal respiration and cellular respiration take place in the tissues throughout the body.

The respiratory system in humans and most animals has another function in addition to gas exchange, and that is the production of sound (vocalization). The production of sound is an important mechanism that infants use to signal their adult caregivers. Sound production has, of course, also proved valuable in information exchange and cultural development, and thus contributes to the long-term survival of our species.

10.2 The respiratory system consists of upper and lower respiratory tracts

The respiratory system (Figure 10.1) consists of a system of passageways for getting air to and from the lungs and the lungs themselves, where gas exchange actually occurs. Also important to respiration are the bones, muscles, and components of the nervous system that cause air to move into and out of the lungs.

For convenience, the respiratory system can be divided into the upper and lower respiratory tracts. The *upper respiratory tract* comprises the nose (including the nasal cavity)

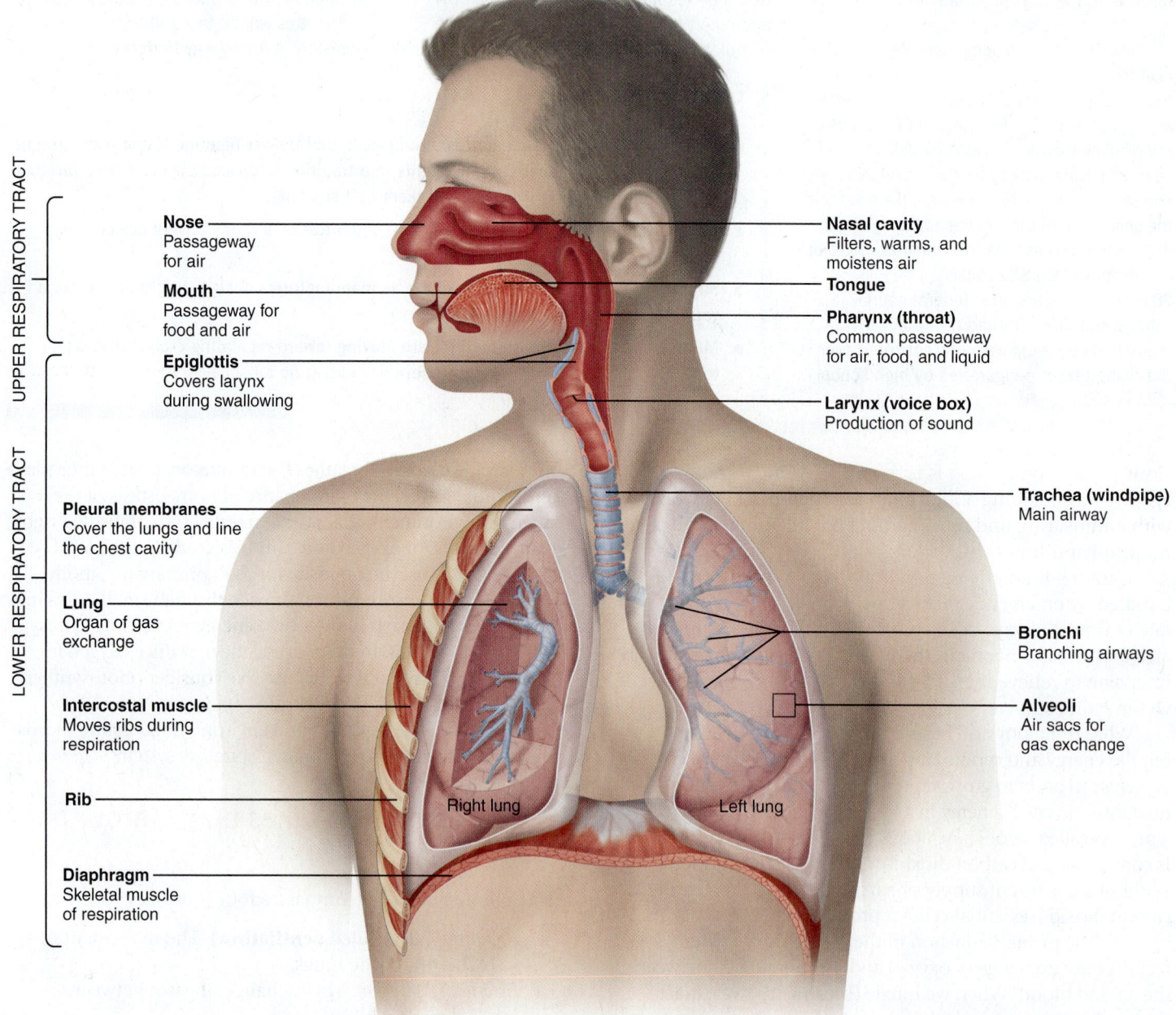

UPPER RESPIRATORY TRACT

Nose
Passageway for air

Mouth
Passageway for food and air

Epiglottis
Covers larynx during swallowing

Nasal cavity
Filters, warms, and moistens air

Tongue

Pharynx (throat)
Common passageway for air, food, and liquid

Larynx (voice box)
Production of sound

LOWER RESPIRATORY TRACT

Pleural membranes
Cover the lungs and line the chest cavity

Lung
Organ of gas exchange

Intercostal muscle
Moves ribs during respiration

Rib

Diaphragm
Skeletal muscle of respiration

Trachea (windpipe)
Main airway

Bronchi
Branching airways

Alveoli
Air sacs for gas exchange

Right lung   Left lung

Figure 10.1 The human respiratory system. The functions of each of the anatomical structures are included.

and pharynx—structures above the Adam's apple. The *lower respiratory tract* starts with the larynx and includes the trachea, the two bronchi that branch from the trachea, and the lungs.

The upper respiratory tract filters, warms, and humidifies air

When you inhale, air enters through your nose or your mouth (Figure 10.2). Your *nose* is to be appreciated, as it does more than serve as a passageway for respiration. The nose also

- Contains receptors for the sense of smell
- Filters inhaled air and screens out some foreign particles
- Moistens and warms incoming air
- Provides a resonating chamber that helps give your voice its characteristic tone

The visible portion of the nose is known as the **external nose.** The internal portion of the nose is called the **nasal cavity.** The external nose consists of cartilage in the front and two nasal bones behind the cartilage. The nose varies in size and shape from person to person, primarily as a result of individual differences in the cartilage tissue.

The nose is divided into chambers by the nasal septum. Air enters the nostrils, the two openings at the base of the external nose, where it is partially filtered by nose hairs, and then flows into the nasal cavity. This cavity is lined with moist epithelial tissue that is supplied with blood vessels. The blood vessels warm incoming air, and the epithelial tissue secretes mucus, which humidifies the air. The epithelium is covered with hairlike projections called *cilia.*

The mucus in the nasal cavity traps dust, pathogens, and other particles in the air before they get any further into the respiratory tract. The cilia beat in a coordinated motion, creating a gentle current that moves the particle-loaded mucus toward the back of the nasal cavity and pharynx. There, we cough it out, or swallow it to be digested by powerful digestive acids in the stomach. Ordinarily, we are unaware of our nasal cilia. However, exposure to cold temperatures can slow down their activity, allowing mucus to pool in the nasal cavity and drip from the nostrils. This is why your nose "runs" in cold weather.

Air spaces called *sinuses* inside the skull are also lined with tissue that secretes mucus and helps trap foreign particles. The sinuses drain into the nasal cavity via small passageways. Two tear ducts, carrying fluid away from the eyes, drain into the nasal cavity as well. This is why excess production of tears, perhaps due to strong emotions or irritating particles in your eyes, also makes your nose "runny."

Incoming air enters the **pharynx** (throat), which connects the mouth and nasal cavity to the larynx (voice box). The upper pharynx extends from the nasal cavity to the roof of the mouth. Into it open the two *auditory tubes* (eustachian tubes) that drain the middle ear cavities and equalize air pressure between the middle ear and outside air. The lower pharynx is a common passageway for food and air. Food passes through on its way to the esophagus, and air flows through to the lower respiratory tract.

✓ Why can upper respiratory tract infections sometimes cause ear infections?

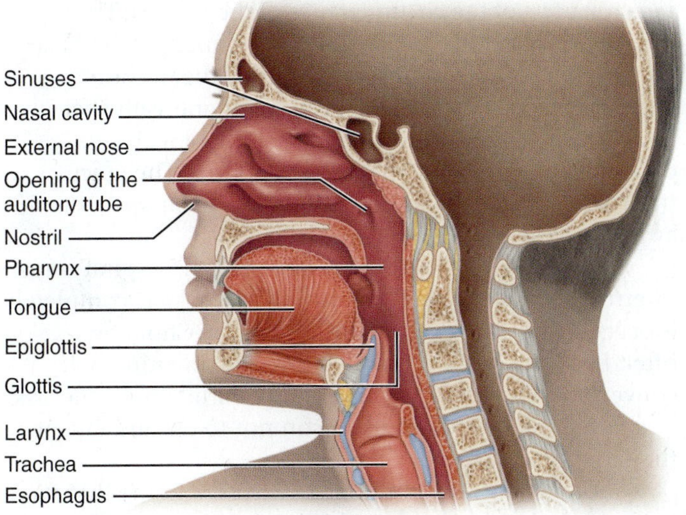

The lower respiratory tract exchanges gases

The lower respiratory tract includes the larynx, the trachea, the bronchi, and the lungs with their bronchioles and alveoli (Figure 10.3).

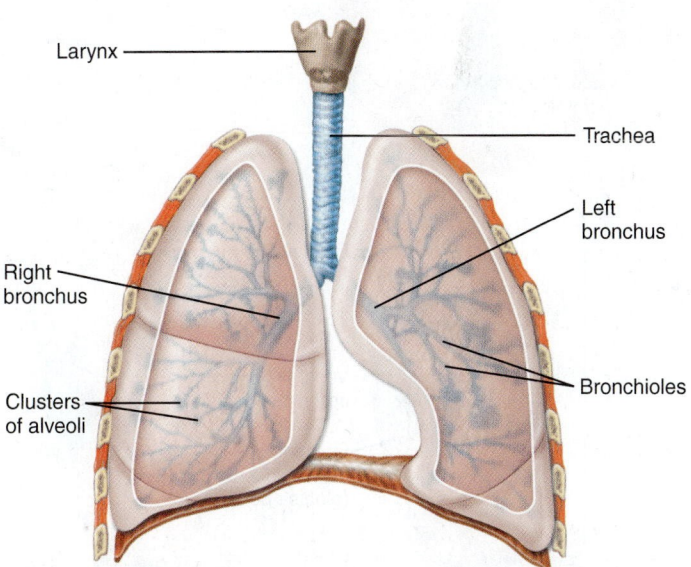

Figure 10.3 Components of the lower respiratory tract.
Individual alveoli are too small to be seen clearly in this figure.

✓ List, in order, all of the anatomical structures that a molecule of oxygen passes through as it moves from the nose to a pulmonary capillary. Where does diffusional gas exchange occur?

Figure 10.2 Components of the upper respiratory tract.

The larynx produces sound The **larynx,** or voice box, extends for about 5 cm (2 inches) below the pharynx. The larynx serves to

- Maintain an open airway.
- Route food and air into the appropriate channels.
- Assist in the production of sound.

The larynx contains two important structures: the epiglottis and the vocal cords. The **epiglottis** is a flexible flap of cartilage located at the opening to the larynx. When air is flowing into the larynx, the epiglottis remains open, but when we swallow food or liquids, the epiglottis tips to block the opening temporarily. This "switching mechanism" routes food and beverages into the esophagus and digestive system, rather than into the trachea. This is why it is impossible to talk while you are swallowing.

The **vocal cords** consist of two folds of connective tissue that extend across the airway. They surround the opening to the airway, called the **glottis.** The vocal cords are supported by ligaments and enclosed within a cartilaginous structure nicknamed the Adam's apple.

We produce most sounds by vibration of the vocal cords, although we can also make a few sounds by moving our tongue and teeth. The tone of the sounds produced by the vocal cords depends on how tightly the vocal cords are stretched, which is controlled by skeletal muscle. When we are not talking, the vocal cords are relaxed and open (Figure 10.4a). When we start to talk, they stretch tightly across the tracheal opening, and the flow of air past them causes them to vibrate (Figure 10.4b).

Like any string instrument, cords that are relatively short yield higher-pitched tones than longer cords do. Also, the tighter the vocal cords stretch, the higher the tone they produce. Men tend to have deeper voices than women and a more prominent Adam's apple, due to testosterone that causes the larynx to enlarge at puberty. We can exert some control over the volume and pitch of the voice by adjusting the tension on our vocal cords. (You may have noticed that when you're nervous, your voice becomes higher pitched.) The resulting vibrations travel through the air as sound waves.

Most of us can be recognized by the distinctive quality of our voices. In addition to the shape and size of vocal cords, individual differences in voice are determined by many components of the respiratory tract and mouth, including the pharynx, nose and nasal cavity, tongue, and teeth. Muscles in the pharynx, tongue, soft palate, and lips cooperate to create recognizable sounds. The pharynx, nose, and nasal sinuses serve as resonating chambers to amplify and enhance vocal tone.

Sound production does not contribute to homeostasis of respiratory gases. Nevertheless, humans take advantage of the structures of the respiratory system to accomplish speech. Speech has played an important role in our evolutionary history and the development of human culture.

The trachea transports air As air continues down the respiratory tract, it passes to the **trachea,** the windpipe that extends from the larynx to the left and right bronchi. The trachea consists of a series of C-shaped, incomplete rings of cartilage held together by connective tissue and muscle. As shown in Figure 10.5a, each cartilage ring extends only three-quarters of the circumference of the trachea. The rings of cartilage keep the trachea open at all times, but because they are not complete circles they permit the trachea to change diameter slightly when we cough or breathe heavily.

Like the nasal cavity, the trachea is lined with cilia-covered epithelial tissue that secretes mucus. The mucus traps foreign particles and the cilia move them upward, away from the lungs.

If a foreign object lodges in the trachea, respiration is interrupted and choking occurs. If the airway is completely blocked, death can occur within minutes. Choking often happens when a person carries on an animated conversation while eating. Beyond good manners, the risk of choking provides a good reason not to eat and talk at the same time.

Choking typically stimulates receptors in the throat that trigger the *cough reflex*. This is a sudden expulsion of air from the lungs in an attempt to dislodge foreign

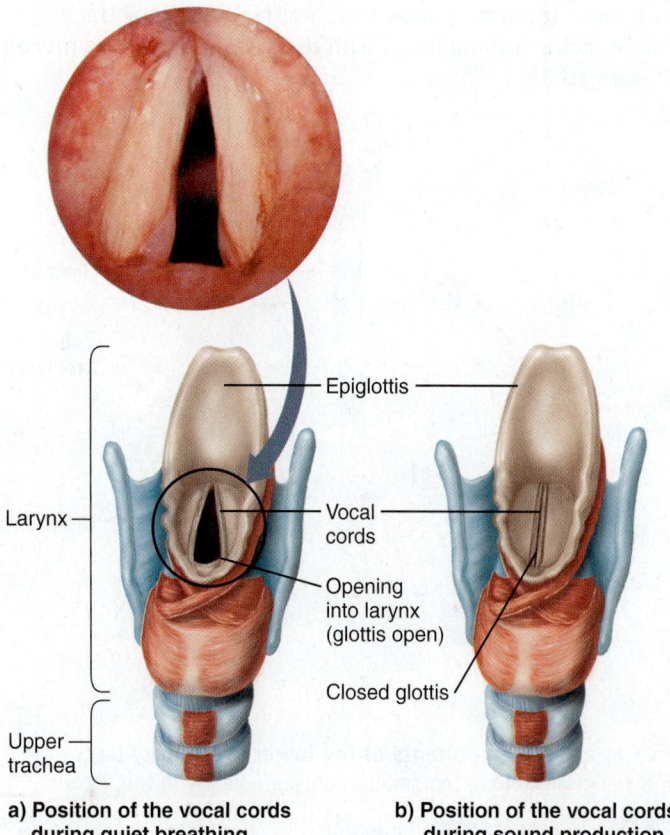

Larynx

Upper trachea

Epiglottis

Vocal cords

Opening into larynx (glottis open)

Closed glottis

a) Position of the vocal cords during quiet breathing.

b) Position of the vocal cords during sound production.

Figure 10.4 Structures associated with the production of sound.

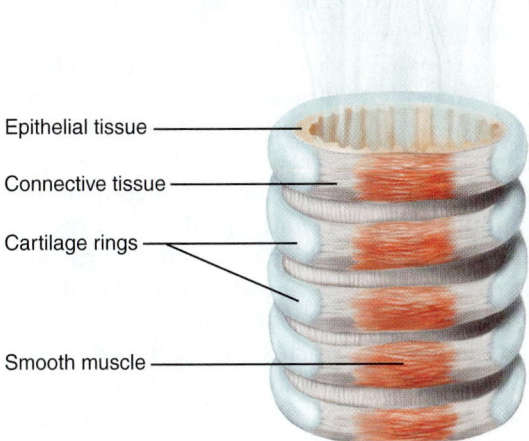

Epithelial tissue

Connective tissue

Cartilage rings

Smooth muscle

a) Relaxed state. The maximum diameter facilitates air movement in and out.

b) During the cough reflex, the smooth muscle contracts briefly, reducing the diameter of the trachea. Combined with contraction of the abdominal muscles, this increases the velocity of air movement, forcibly expelling irritants or mucus from the trachea.

Figure 10.5 The trachea. The trachea consists of smooth muscle and layers of epithelial and connective tissue held open by tough, flexible C-shaped bands of cartilage.

material (Figure 10.5b). If the object blocks the airway completely before the person has finished inhaling, there may not be much air in the lungs. This will make the obstacle more difficult to remove. If the object blocks air flow only partially, it may be possible to dislodge it by inhaling slowly, then coughing.

Bronchi branch into the lungs The trachea branches into two airways called the right and left **bronchi** (singular: bronchus) as it enters the lung cavity (refer to Figure 10.3). Like the branches of a tree, the two bronchi divide into a network of smaller and smaller bronchi. The bronchi walls contain fibrous connective tissue and smooth muscle reinforced with cartilage. As the airways branch, the amount of cartilage declines. By definition, the smaller airways that lack cartilage are called **bronchioles.** The smallest bronchioles are 1 mm or less in diameter and consist primarily of a thin layer of smooth muscle surrounded by a tiny amount of elastic connective tissue.

The bronchi and bronchioles have several other functions in addition to air transport. They also clean the air, warm it to body temperature, and saturate it with water vapor before it reaches the delicate gas-exchange surfaces of the lungs. The air is warmed and humidified by contact with the moist surfaces of the cells lining the bronchi and bronchioles. With the exception of

the very smallest bronchioles, the bronchi and bronchioles are lined with ciliated epithelial cells and occasional mucus-secreting cells. The thin, watery mucus produced by the mucus-secreting cells traps dust, bacteria, and other small particles. The ciliated cells then sweep the accumulated mucus and trapped material upward toward the pharynx so that it can be swallowed.

Tobacco smoke contains chemicals and particles that irritate the respiratory tract. Mucus production increases in response, but the smoke impairs the activity of the cilia (Figure 10.6). Continued smoking destroys the cilia, allowing mucus and debris from the smoke to accumulate in the airway. "Smoker's cough" refers to the

a) Healthy airway.

b) Smoker's airway.

Figure 10.6 Effects of smoking on the cilia of the airways.

violent coughing necessary to dislodge the mucus from the airway. Mucus pooling leads to frequent infections because pathogens and irritants remain in the respiratory tract. It also increases the risk of bronchitis, emphysema, and lung cancer (see section 10.6, Disorders of the respiratory system).

From the nose and mouth to the tiniest bronchioles in the lungs, none of the airways we have described so far participate in gas exchange. Essentially, they are all tubes for getting air to the lungs, where gas exchange actually occurs.

MJ's BlogInFocus How much does smoking shorten one's life expectancy? If you've smoked and then quit, do you stay at higher risk of a smoking-related death or do you "recover"? Visit MJ's blog in the Study Area in MasteringBiology and look under "The Benefits of Quitting Smoking."

http://goo.gl/tlKZGJ

The lungs are organs of gas exchange The lungs are organs consisting of supportive tissue enclosing the bronchi, bronchioles, blood vessels, and the areas where gas exchange occurs. They occupy most of the thoracic cavity. There are two lungs, one on the right side and one on the left, separated from each other by the heart (**Figure 10.7**). The shape of the lungs follows the contours of the rib cage and the thoracic cavity. The base of each lung is broad and shaped to fit against the convex surface of the diaphragm.

Each lung is enclosed in two layers of thin epithelial membranes called the **pleural membranes.** One of these layers represents the outer lung surface and the other lines the thoracic cavity. The pleural membranes are separated by a small space, called the *pleural cavity*, that contains a very small amount of watery fluid. The fluid reduces friction between the pleural membranes as the lungs and chest wall move during breathing. Inflammation of the pleural membranes, a condition called *pleurisy*, can reduce the secretion of pleural fluid, increase friction, and cause pain during breathing. Pleurisy can be a symptom of pneumonia (see section 10.6).

Lungs consist of several lobes, three in the right lung and two in the left. Each lobe contains a branching tree of bronchioles and blood vessels. The lobes can function fairly independently of each other, so it is possible to surgically remove a lobe or two without totally eliminating lung function.

Gas exchange occurs in alveoli If you could touch a living lung, you would find that it is very soft and frothy. In fact, most of it is air. The lungs are basically a system of branching airways that end in 300 million tiny air-filled sacs called **alveoli** (singular: alveolus). It is here that gas exchange takes place. Alveoli are arranged in clusters at the end of every terminal bronchiole, like grapes clustered on

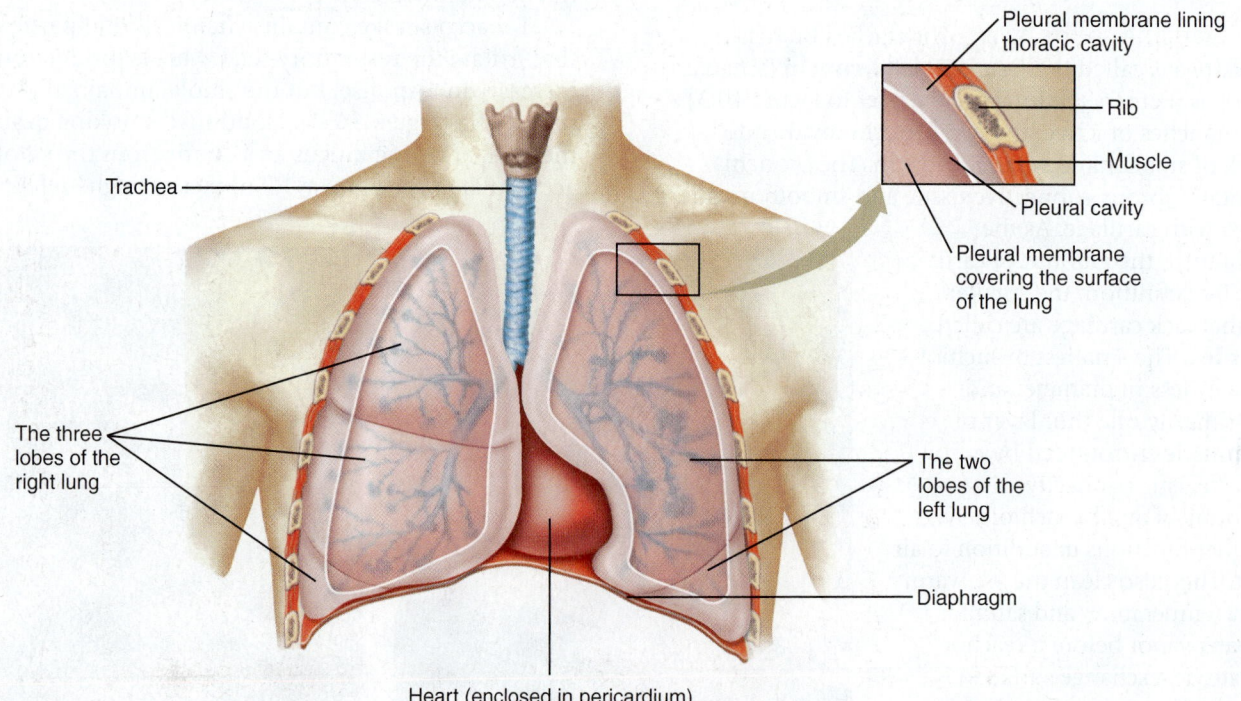

Trachea

Pleural membrane lining thoracic cavity

Rib

Muscle

Pleural cavity

Pleural membrane covering the surface of the lung

The three lobes of the right lung

The two lobes of the left lung

Diaphragm

Heart (enclosed in pericardium)

Figure 10.7 The lungs, pleural membranes, and pleural cavity. In reality, the pleural membrane is no more than a very thin, watery space that reduces friction between the two pleural membranes.

a stem (**Figure 10.8a**). A single alveolus is a thin bubble of living squamous epithelial cells only one cell layer thick. Their combined surface area is nearly 800 square feet, approximately 40 times the area of our skin. This tremendous surface area and the thinness of the squamous type of epithelium facilitate gas exchange with nearby capillaries (**Figure 10.8b**).

Within each alveolus, certain epithelial cells secrete a lipoprotein called *surfactant* that coats the interior of the alveoli and reduces surface tension. Surface tension is due to the attraction of water molecules toward each other. Without surfactant, the force of surface tension could collapse the alveoli. This can occur in infants who are born prematurely, because the surfactant-secreting cells in their lungs are underdeveloped. Called *infant respiratory distress syndrome*, the condition is treated with surfactant replacement therapy.

Pulmonary capillaries bring blood and air into close contact The right ventricle of the heart pumps deoxygenated blood into the pulmonary trunk, which splits into the left and right pulmonary arteries. The pulmonary arteries divide into smaller and smaller arteries and arterioles, eventually terminating in a capillary bed called the *pulmonary capillaries*. In the pulmonary capillaries, blood comes into very close proximity to the air in the alveoli. Only two living cells (the squamous epithelial cell of the alveolus and the cell of the capillary wall) separate blood from air at this point. A series of venules and veins collects the oxygenated blood from the pulmonary capillaries and returns the blood to the left side of the heart, from whence it is transported to all parts of the body.

The close contact between air and blood and the large air surface area of the lungs suggests that the lungs might be

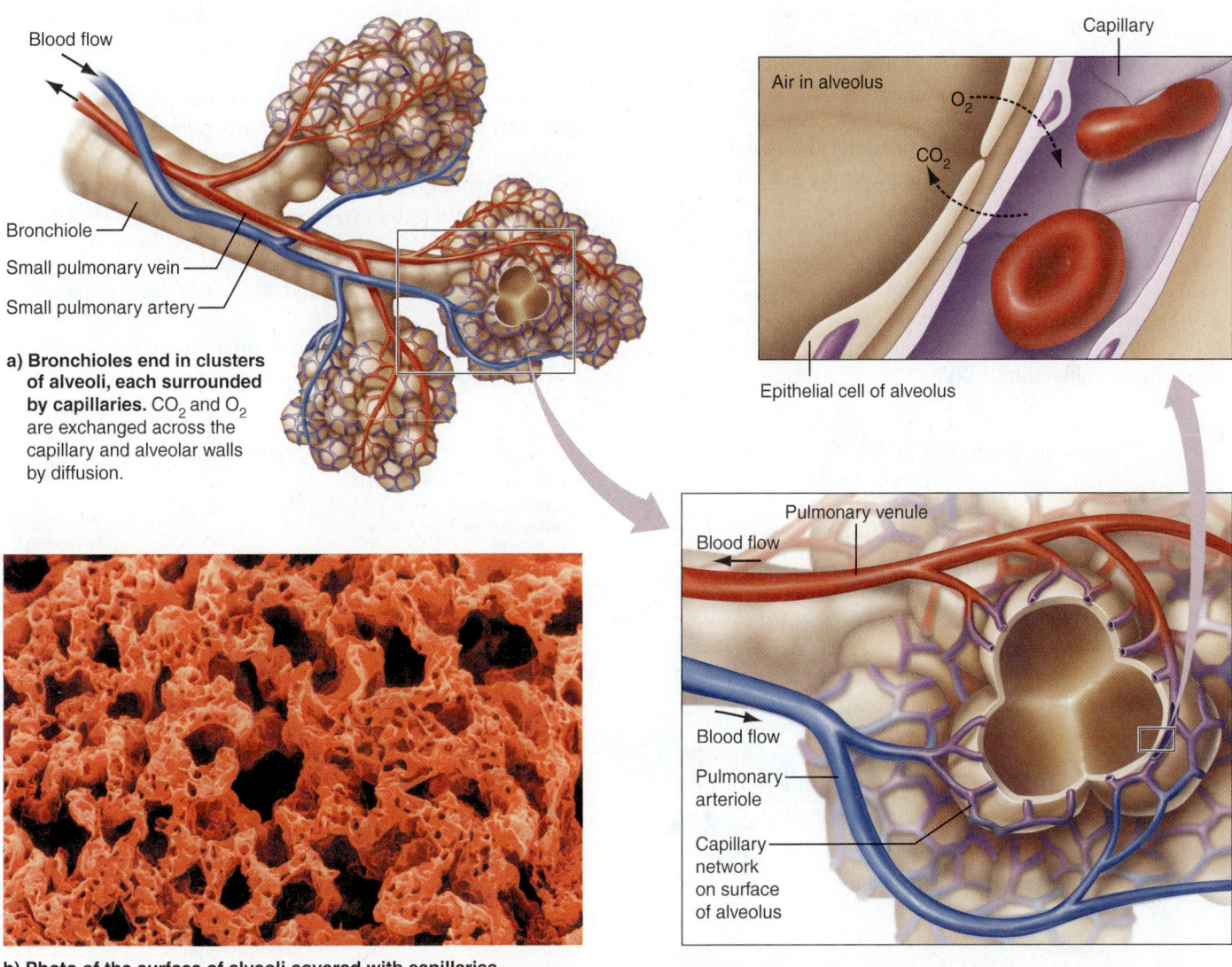

a) **Bronchioles end in clusters of alveoli, each surrounded by capillaries.** CO_2 and O_2 are exchanged across the capillary and alveolar walls by diffusion.

b) **Photo of the surface of alveoli covered with capillaries.**

Figure 10.8 Gas exchange between the blood and alveoli.

useful as an alternative method for delivering medications to the bloodstream. Several pharmaceutical companies have developed fine mists or powders that can be inhaled deep into the lungs. The latest, an inhalable insulin product called Afrezza, was approved by the FDA in 2014.

↩ **Recap** The respiratory system is specialized for the exchange of oxygen and carbon dioxide with the air. Sound is produced by vibration of the vocal cords of the larynx as air passes through the glottis. The trachea, or windpipe, branches into the right and left bronchi. The bronchi and bronchioles filter, warm, and humidify the incoming air. The lungs are organs containing a branching system of bronchi and bronchioles, blood vessels, and 300 million alveoli. Gas exchange occurs between the alveoli and pulmonary capillaries. ∎

10.3 The process of breathing involves a pressure gradient

Breathing involves getting air into and out of the lungs in a cyclic manner and that requires muscular effort. However, the lungs themselves don't have any skeletal muscle tissue. The lungs expand passively because the surrounding bones and muscles expand the size of the chest cavity.

The bones and muscles of respiration include the ribs, the intercostal muscles between the ribs, and the main muscle of respiration, called the **diaphragm,** a broad sheet of muscle that separates the thoracic cavity from the abdominal cavity. The intercostal muscles and the diaphragm are skeletal muscles.

Inspiration brings in air, expiration expels it

To understand why air moves into and out of the lungs in a cyclic manner, we need to understand the following general principles of gas pressure and of how gases move:

- Gas pressure is caused by colliding molecules of gas.
- When the volume of a closed space *increases,* the molecules of gas in that space are farther away from each other, and the pressure inside the space *decreases.* Conversely, when the volume in a closed space *decreases,* the gas pressure *increases.*
- Gases flow from areas of *higher* pressure to areas of *lower* pressure.

As we have seen, the lungs are air-filled structures consisting almost entirely of bronchioles, alveoli, and blood vessels. Lacking skeletal muscle, they cannot expand (increase in volume) or contract (decrease in volume) on their own. The lungs expand and contract only because they are compliant (stretchable) and because they are surrounded by the pleural cavity, which is airtight and sealed. If the volume of the pleural cavity expands, the lungs will expand with it.

Inspiration (inhalation) pulls air into the respiratory system as lung volume expands, and *expiration* (exhalation) pushes air out as lung volume declines again. Let's look at a cycle of inspiration and expiration, starting from the relaxed state at the end of a previous expiration (**Figure 10.9**):

❶ *Relaxed state.* At rest, both the diaphragm and the intercostal muscles are relaxed. The relaxed diaphragm appears dome shaped.

❷ *Inspiration.* As inspiration begins, the diaphragm contracts, flattening it and pulling its center downward.

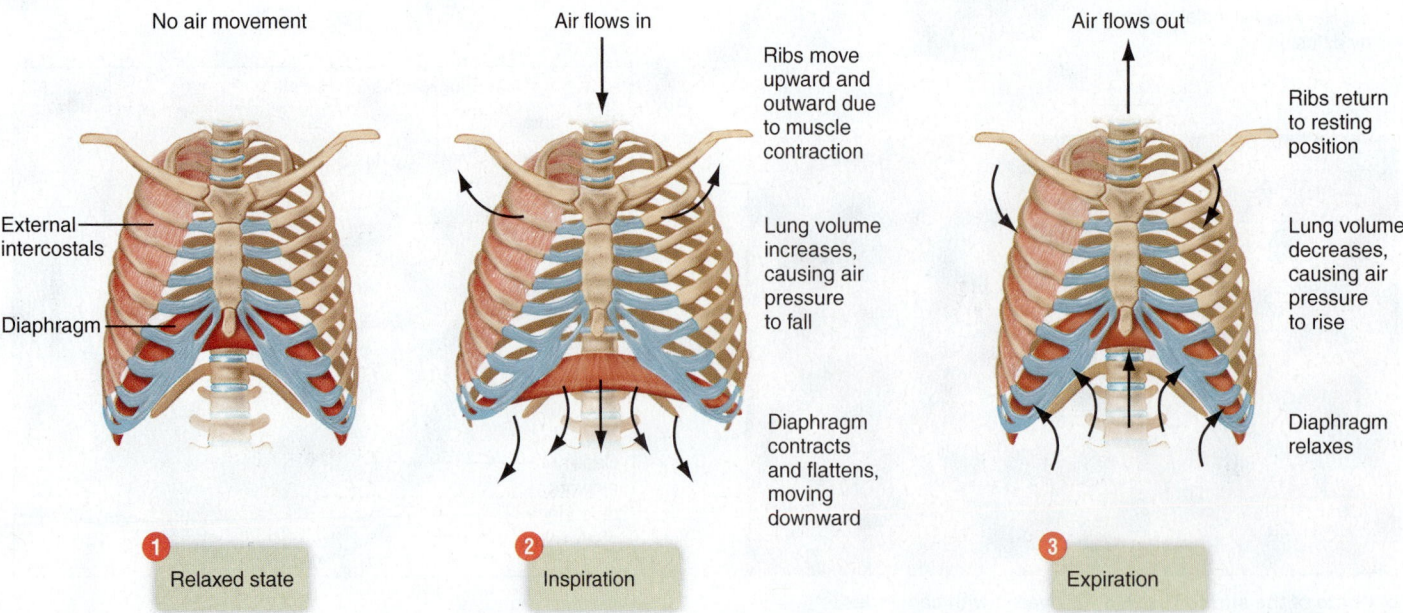

Figure 10.9 The respiratory cycle.

At the same time, the intercostal muscles contract, pulling the ribs upward and outward. These two actions of skeletal muscle increase the volume of the pleural cavity and lower the pressure within the pleural space. Because the lungs are elastic and the pressure around them has just fallen relative to the atmosphere, they expand with the pleural cavity. Expansion of the lungs reduces air pressure within the lungs relative to the atmosphere, allowing air to rush in.

3 *Expiration.* Eventually, the muscle contractions end. As the muscles relax the diaphragm returns to its domed shape, the ribs move downward and inward, and the pleural cavity becomes smaller. The rest of the process reverses as well. The lungs become smaller, so pressure within the lungs rises relative to the atmosphere and air flows out.

During quiet breathing, inspiration is active (requiring muscular effort) and expiration is passive. When we are under physical or emotional stress, however, we need to breathe more frequently and more deeply. At this point, both inspiration and expiration may become active. We can take bigger breaths because additional rib cage muscles raise the rib cage higher. As we exhale deeply, abdominal muscles contract and push the diaphragm even higher into the thoracic cavity, and the inner intercostal muscles contract to pull the rib cage downward. These events combine to increase the speed and force of respiration.

We also exhale forcibly when we sneeze or cough. During sneezing and coughing, the abdominal muscles contract suddenly, raising abdominal pressure. The rapid increase in abdominal pressure pushes the relaxed diaphragm upward against the pleural cavity, forcing air out of the lungs.

Lung volumes and vital capacity measure lung function

Lung volumes and vital capacity are measured with a device called a *spirometer* (Figure 10.10a). The measurements are made by having a person breathe normally into the device and then take a maximum breath and exhale it forcibly and as completely as possible.

A recording of lung volumes and vital capacity are shown in Figure 10.10b. At rest, you take about 12 breaths every minute. Each breath represents a **tidal volume** of air of approximately 500 milliliters (ml), or about 1 pint. The maximal volume that you can exhale after a maximal inhalation is called your **vital capacity.** Your vital capacity is about 4,800 ml, almost 10 times your normal tidal volume at rest. The amount of additional air that can be inhaled beyond the tidal volume (about 3,100 ml) is called the *inspiratory reserve volume,* and the amount of the air that we can forcibly exhale beyond the tidal volume (about 1,200 ml) is the *expiratory reserve volume.* No matter how forcefully you exhale, some air always remains in your lungs. This is your *residual volume,* approximately 1,200 ml.

On average, only about 350 ml of each normal breath actually reach the alveoli and become involved in gas exchange. The other 150 ml remain in the airways, and because that amount does not participate in gas exchange, the air is referred to as *dead space volume.* Dead space volume is not measured by a spirometer.

Lung volumes and rates of change of volume are useful in diagnosing various lung diseases. For example, emphysema is a condition in which the smaller airways lose elasticity, causing them to collapse during expiration

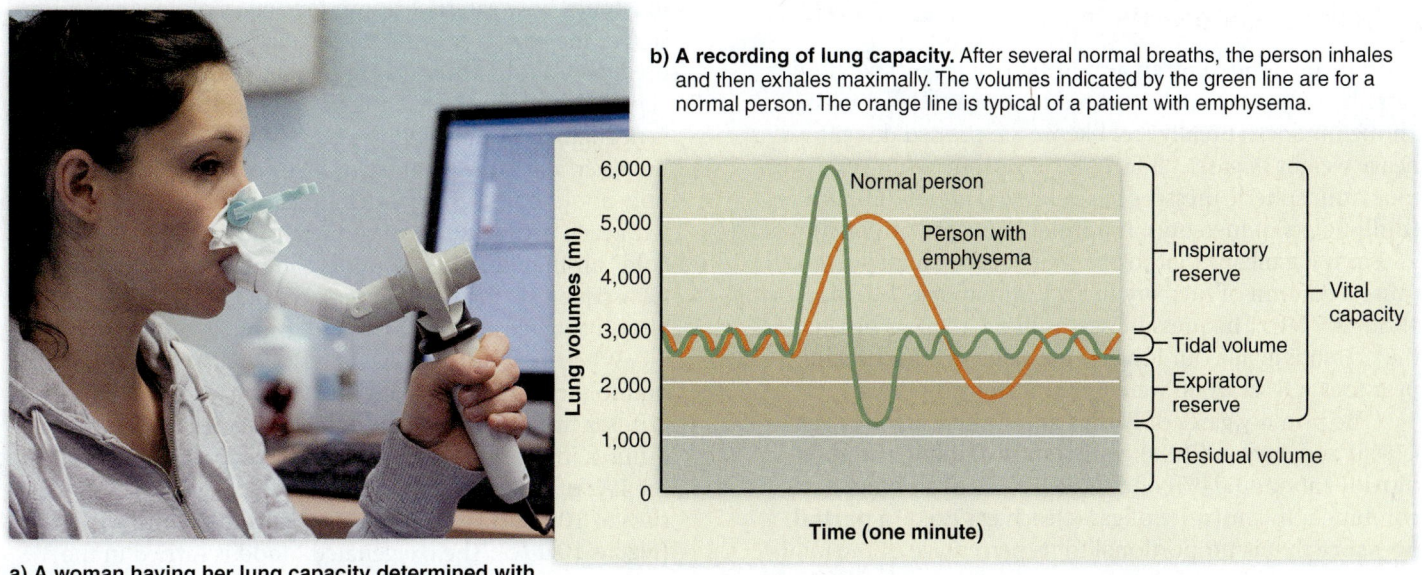

b) A recording of lung capacity. After several normal breaths, the person inhales and then exhales maximally. The volumes indicated by the green line are for a normal person. The orange line is typical of a patient with emphysema.

a) A woman having her lung capacity determined with a spirometer.

Figure 10.10 **Measurement of lung capacity.**

and impairing the ability to exhale naturally. A spirometer recording of someone with emphysema might show a decreased inspiratory reserve and a prolonged period of expiration after a maximum inspiration because of resistance to air outflow.

↩ **Recap** Inspiration is an active process (requiring energy) that occurs when the diaphragm and intercostal muscles contract. Normally, expiration is passive, but it can become active when we forcibly exhale, cough, or sneeze. Although we normally take breaths of about 500 ml, the maximum breath we can inhale and then forcibly exhale is about 4,800 ml. Some air, called the *residual volume*, remains in the lungs even at the end of expiration. ■

☑ When a person who is initially at rest begins exercising, do you think tidal volume will increase, decrease, or stay the same? How about vital capacity? Explain.

10.4 Gas exchange and transport occur passively

So far, we have focused on the first process in respiration: breathing. Once air enters the alveoli, gas exchange and transport occur. In this section, we review some basic principles governing the diffusion of gases to set the stage for our discussion of external and internal respiration (the second and third processes of respiration). We also describe how the gases are transported by blood. The fourth process of respiration, called cellular respiration, is the use of oxygen by cells in the production of energy. Cellular respiration was described in the chapter on the structure and function of cells.

Gases diffuse according to their partial pressures

Earth is surrounded by an atmosphere of gases. Like liquids, gases have mass and are attracted to the earth by gravity. Though it doesn't really feel like we are pressed down by a heavy weight of gases, in fact, the atmosphere (air) exerts a total atmospheric pressure at sea level of about 760 mm Hg (millimeters of mercury). A normal atmospheric pressure of 760 mm Hg means that the pressure of the atmosphere will cause a column of mercury in a vacuum to rise 760 mm, or about 2.5 feet. The pressure seems like zero to us because the pressure inside our lungs is the same as atmospheric pressure, at least when we are resting between breaths.

The primary gases of Earth's atmosphere are nitrogen (78%) and oxygen (21%), with a trace amount of carbon dioxide (about 0.04%) and less than 1% of all other gases combined. In a mixture of gases, each gas exerts a **partial pressure** that is proportional to its percentage of the total gas composition. Partial pressure is represented by P and, like atmospheric pressure, it is measured in mm Hg. The pressure of the atmosphere is thus the sum of the partial

pressures of each of the gases found in the atmosphere at sea level. Because we know the percentages of each gas in our atmosphere, we can find the partial pressure of each. For example, the partial pressure of oxygen (P_{O_2}) in air is about 160 mm Hg (760 mm Hg $\times$ 0.21).

 **MJ's BlogInFocus** What if a person's breath could be used to diagnose disease? To learn more about breath as a diagnostic tool of the future, visit MJ's blog in the Study Area in MasteringBiology and look under "Breath Analysis."

http://goo.gl/iT9QHv

Because partial pressures in a mixture of gases are directly proportional to concentrations, a gas will always diffuse down its *partial pressure gradient*, from a region of higher partial pressure to a region of lower partial pressure. As we shall see, the exchanges of O_2 and CO_2, both between the alveoli and the blood and between the blood and the tissues, are purely passive. No consumption of ATP is involved; changes in partial pressures are entirely responsible for the exchange and transport of these gases.

☑ At the top of Mount Everest, the atmospheric pressure is usually around 260 mm Hg, and the percentages of nitrogen, oxygen, and other gases are the same as they are at sea level. What is the partial pressure of oxygen at the summit of Everest?

External respiration: the exchange of gases between air and blood

A comparison of the partial pressures of O_2 and CO_2 in inhaled air, in the alveoli, and in the blood in the lungs illustrates how external respiration takes place. As previously stated, the partial pressure of the oxygen (P_{O_2}) in the air we breathe is about 160 mm Hg (**Figure 10.11a**). As there is very little CO_2 in the air, the partial pressure of inspired CO_2 is negligible. The partial pressures of alveolar air are not, however, the same as those of inspired air. This is because only about one-eighth of the air is actually exchanged with each breath, so most of the air in the lungs is actually "old" air that has already undergone some gas exchange. Consequently, the partial pressures of O_2 and CO_2 in the alveoli average about 104 and 40 mm Hg, respectively.

When venous (deoxygenated) blood arrives at the pulmonary capillaries from the pulmonary arteries, O_2 diffuses from the alveoli into the capillaries, and CO_2 diffuses in the opposite direction (**Figure 10.11b**). As a result, the P_{O_2} of oxygenated (arterial) blood leaving the lungs rises to 100 mm Hg and the P_{CO_2} falls to 40 mm Hg (**Figure 10.11c**). The oxygenated blood is carried in the pulmonary veins to the heart and then throughout the body in the arterial blood vessels. The CO_2 that diffuses into the alveoli is exhaled, along with some water vapor.

Notice that the definition of venous blood is that it is deoxygenated, not that it happens to be in a vein. The pulmonary arteries transport venous blood to the lungs, and the pulmonary veins transport arterial blood to the heart.

☑ Suppose an airplane flying at high altitude loses cabin pressure, such that the partial pressure of oxygen in the air in the passengers' alveoli drops to 35 mm Hg (while the partial pressure of oxygen of their venous blood remains normal). Which way will oxygen diffuse—from the venous blood to air or from air to the venous blood—and why?

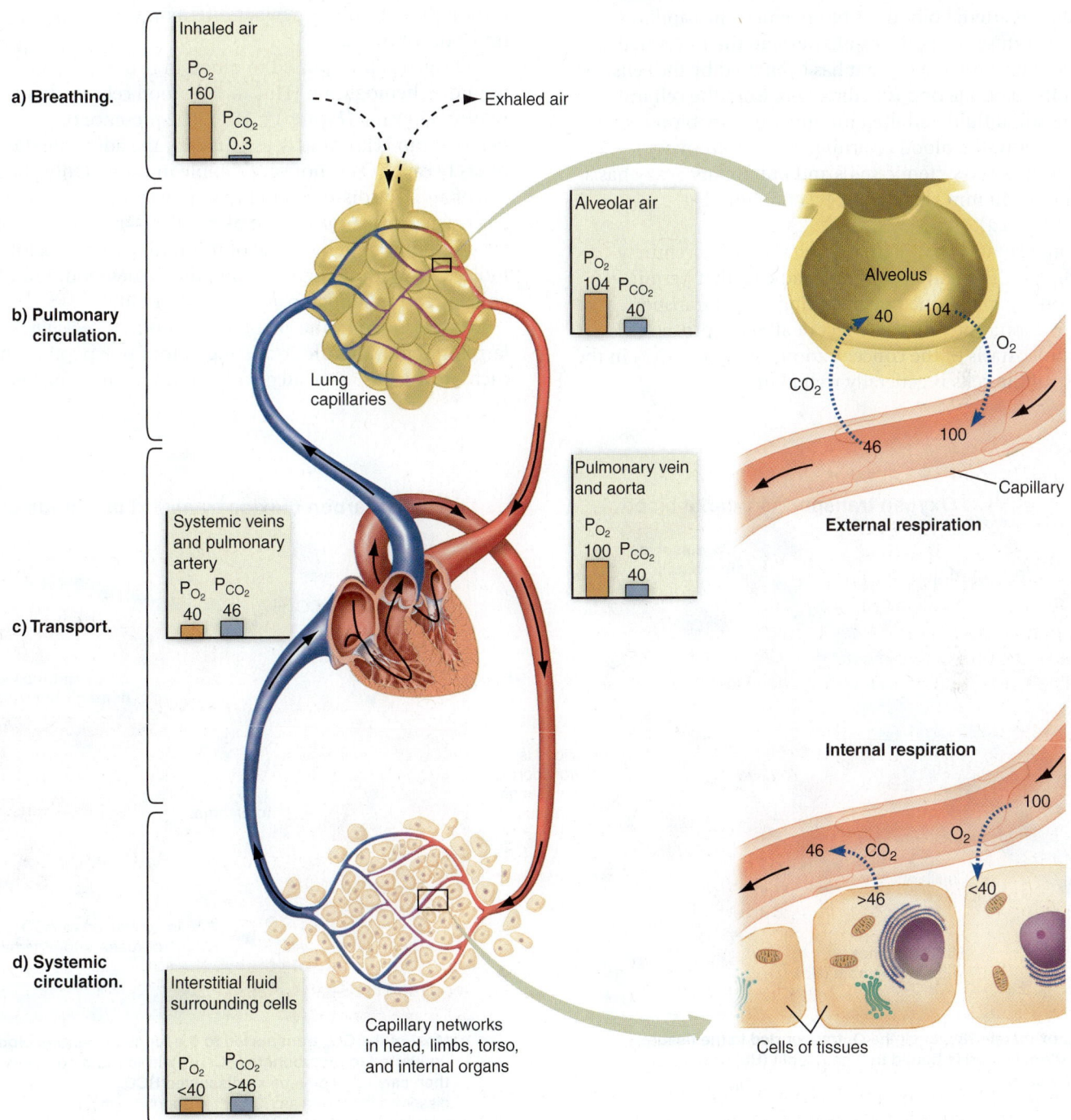

Figure 10.11 Partial pressures. All partial pressures are expressed in units of mm Hg. Differences in partial pressures account for the diffusion of O_2 and CO_2 between the lungs and blood, and between blood and the body's tissues.

☑ Why is the P_{O_2} lower (less than 40 mm Hg) in the interstitial fluid surrounding cells and the cells themselves than it is in the aorta (approximately 100 mm Hg)?

Internal respiration: the exchange of gases with tissue fluids

The body's cells get their supply of O_2 for cellular respiration from the interstitial fluid that surrounds them. Because the cells are constantly drawing oxygen from the interstitial fluid, the interstitial fluid P_{O_2} is usually quite a bit lower than that of arterial blood. As blood enters the capillaries, then, O_2 diffuses from the capillaries into the interstitial fluid, replenishing the O_2 that has been used by the cells. CO_2 diffuses in the opposite direction, from the cell into the interstitial fluid and then into the capillary blood. Consequently, the blood returning to the pulmonary circulation in the systemic veins and pulmonary artery has a P_{O_2} of only 40 mm Hg and a P_{CO_2} of 46 mm Hg (Figure 10.11d).

Both external and internal respiration occur entirely by diffusion. The partial pressure gradients that permit diffusion are maintained by breathing, blood transport, and cellular respiration. The net effect of all these processes is that homeostasis of the concentrations of O_2 and CO_2 in the vicinity of the cells is generally well maintained.

Hemoglobin transports most oxygen molecules

Our discussion of external and internal respiration brings us to an important aspect of the overall subject of gas exchange, and that is how the respiratory gases (O_2 and CO_2) are transported between the lungs and tissues in the blood. The mechanisms of transport of oxygen and carbon dioxide are somewhat different; we'll start with the transport of oxygen.

Oxygen is transported in blood in two ways: either bound to hemoglobin (Hb) in red blood cells or dissolved in blood plasma (Figure 10.12a). The presence of hemoglobin is absolutely essential for the adequate transport of O_2 because O_2 is not very soluble in water. Only about 2% of all O_2 is dissolved in the watery component of blood known as *blood plasma*. Most of it—98%—is taken out of the watery component by virtue of its binding to hemoglobin molecules. Without hemoglobin, the tissues would not be able to receive enough oxygen to sustain life.

As described in the chapter on blood, *hemoglobin* is a large protein molecule consisting of four polypeptide chains, each of which is associated with an iron-containing heme

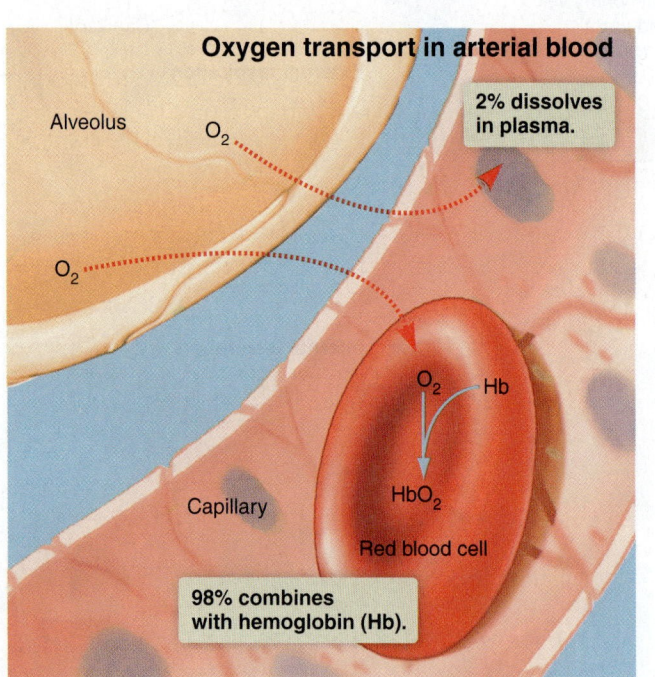

a) Approximately 98% of all the O_2 transported to the tissues by arterial blood is bound to hemoglobin (Hb) within red blood cells.

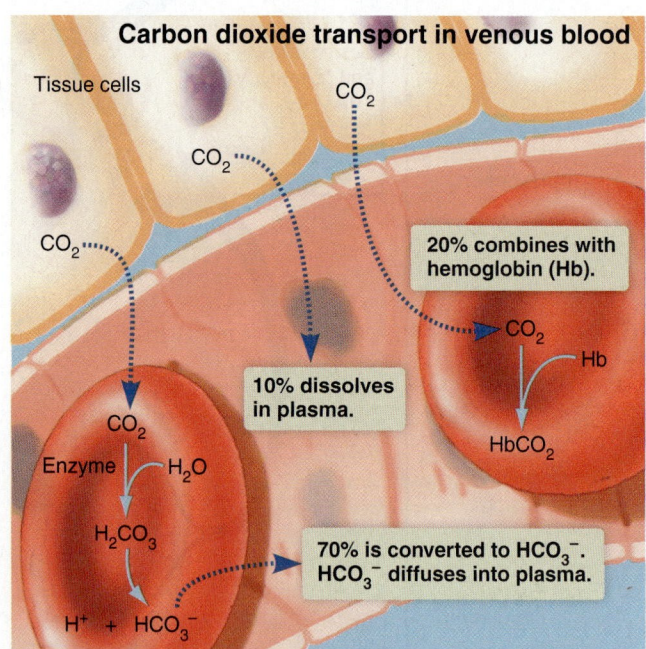

b) Most of the CO_2 transported to the lungs by venous blood is converted to bicarbonate (HCO_3^-) within red blood cells and then carried in plasma as dissolved HCO_3^-. The rest is either dissolved in plasma as CO_2 or transported within the red blood cells bound to hemoglobin.

Figure 10.12 How oxygen and carbon dioxide are transported in blood.

✓ People who have insufficient iron in their diets will often become chronically oxygen-deprived, even to the point of passing out, despite having normal levels of O_2 dissolved in the blood plasma. Explain how they can be oxygen-deprived if they have a normal O_2 concentration in the blood plasma.

group that can bind oxygen. Because there are four heme groups, each hemoglobin molecule can bind four oxygen molecules at a time, forming **oxyhemoglobin** (HbO_2). We can represent this reaction as:

$$Hb \quad + \quad O_2 \quad \rightarrow \quad HbO_2$$
hemoglobin $\quad\quad$ oxygen $\quad\quad\quad$ oxyhemoglobin

This reaction is reversible and highly dependent on the partial pressures of O_2 in plasma. When the P_{O_2} rises (in the lungs), oxygen attaches to hemoglobin and is transported in arterial blood. When the P_{O_2} falls (at the tissues), oxygen detaches from hemoglobin.

Several other factors affect O_2 attachment to hemoglobin as well. Hemoglobin binds O_2 most efficiently in conditions of fairly neutral pH and relatively cool temperatures—similar to conditions existing in the lungs. Body regions with warmer temperatures and lowered pH—such as in body tissues—reduce hemoglobin's affinity for binding O_2. Consequently, O_2 and hemoglobin tend to combine in the lungs, facilitating the transport of oxygen to the tissues, and to detach in body tissues, making O_2 available to cells. Hemoglobin's affinity for oxygen is also greatly reduced by carbon monoxide, a colorless, odorless, highly poisonous gas (see the Health & Wellness feature).

Most CO_2 is transported in plasma as bicarbonate

Cellular metabolism in body tissues continuously produces carbon dioxide as a waste product. One of the most important functions of blood (other than the transport of oxygen) is to transport CO_2 away from tissues and back to the lungs, where it can be removed from the body. Because the partial pressure of CO_2 is higher in the tissues than it is in blood, CO_2 readily diffuses from tissues into the bloodstream. Once in the blood, CO_2 is transported in three ways: dissolved in blood plasma, bound to hemoglobin, or in the form of bicarbonate (**Figure 10.12b**).

Only about 10% of the CO_2 remains dissolved in blood plasma. Another 20% binds with hemoglobin to form **carbaminohemoglobin** ($HbCO_2$). This reaction is represented as:

$$Hb \quad + \quad CO_2 \quad \rightarrow \quad HbCO_2$$
hemoglobin $\quad$ carbon $\quad\quad\quad$ carbaminohemoglobin
$\quad\quad\quad\quad\quad$ dioxide

Hemoglobin can transport O_2 and CO_2 molecules simultaneously because the two gases attach to different sites on the hemoglobin molecule: O_2 combines with heme, CO_2 with globin.

About 70% of all the CO_2 produced by the tissues is converted to bicarbonate (HCO_3^-) prior to transport. When bicarbonate is produced, CO_2 combines with water (H_2O) to become carbonic acid (H_2CO_3). This first reaction is catalyzed by an enzyme called *carbonic anhydrase*. The carbonic acid immediately breaks apart into bicarbonate and hydrogen (H) ions, as follows:

$$CO_2 \quad + \quad H_2O \quad \rightarrow \quad H_2CO_3 \quad \rightarrow \quad HCO_3^- \quad + \quad H^+$$
carbon $\quad$ water $\quad\quad$ carbonic $\quad\quad$ bicarbonate hydrogen
dioxide $\quad\quad\quad\quad\quad$ acid

The formation of bicarbonate from CO_2 occurs primarily inside red blood cells because this is where the carbonic anhydrase enzyme is located. However, most of the bicarbonate quickly diffuses out of red blood cells and is transported back to the lungs dissolved in plasma.

Some of the hydrogen ions formed along with bicarbonate stay inside the red blood cells and bind to hemoglobin. Their attachment to hemoglobin weakens the attachment between hemoglobin and oxygen molecules and causes hemoglobin to release more O_2. This is the chemistry behind the previously mentioned effect of pH on oxygen binding. The overall effect is that the presence of CO_2 (an indication that cellular metabolism has taken place) actually enhances the delivery of O_2 to the very sites where it is most likely to be needed.

At the lungs, dissolved CO_2 diffuses out of the blood and into the alveolar air. The loss of CO_2 from the blood plasma causes the P_{CO_2} to fall, which in turn causes the chemical reaction that formed bicarbonate in the first place to reverse, as follows:

$$HCO_3^- \quad + \quad H^+ \rightarrow \quad H_2CO_3 \quad \rightarrow \quad CO_2 \quad + \quad H_2O$$
bicarbonate $\quad$ hydrogen $\quad$ carbonic $\quad\quad$ carbon $\quad$ water
$\quad\quad\quad\quad\quad\quad\quad\quad$ acid $\quad\quad\quad$ dioxide

Thus, as CO_2 is removed by breathing, the bicarbonate and hydrogen ions formed in the peripheral tissues to transport CO_2 are removed as well.

Recap In a mixture of gases such as air, each gas exerts a partial pressure. Gases diffuse according to differences in their partial pressures. Diffusion accounts for both external and internal respiration. Nearly all of the oxygen in blood is bound to hemoglobin in red blood cells. Most carbon dioxide produced in the tissues is transported in blood plasma as bicarbonate. ■

10.5 The nervous system regulates breathing

As mentioned earlier, breathing depends on the contractions of skeletal muscles, which as you know are activated only by motor neurons. In other words, respiration is controlled by the nervous system. The nervous system regulates the rate and depth of breathing in order to maintain homeostasis of the concentrations of certain key variables, most notably the concentrations of CO_2, H^+, and O_2. In addition, we can exert a certain amount of conscious control over breathing if we wish.

A respiratory center establishes rhythm of breathing

The basic cyclic pattern of inspiration and expiration and the rate at which we breathe are established in an area near the base of the brain called the *medulla oblongata*. Within this area, called the **respiratory center** (Figure 10.13a), groups of nerve cells automatically generate a cyclic pattern of electrical impulses every 4–5 seconds. The impulses travel along nerves to the diaphragm and the intercostal muscles and stimulate those muscles to contract. As these respiratory muscles contract, the rib cage expands, the diaphragm is pulled downward, and we inhale. As inhalation proceeds, the respiratory center receives sensory input from stretch

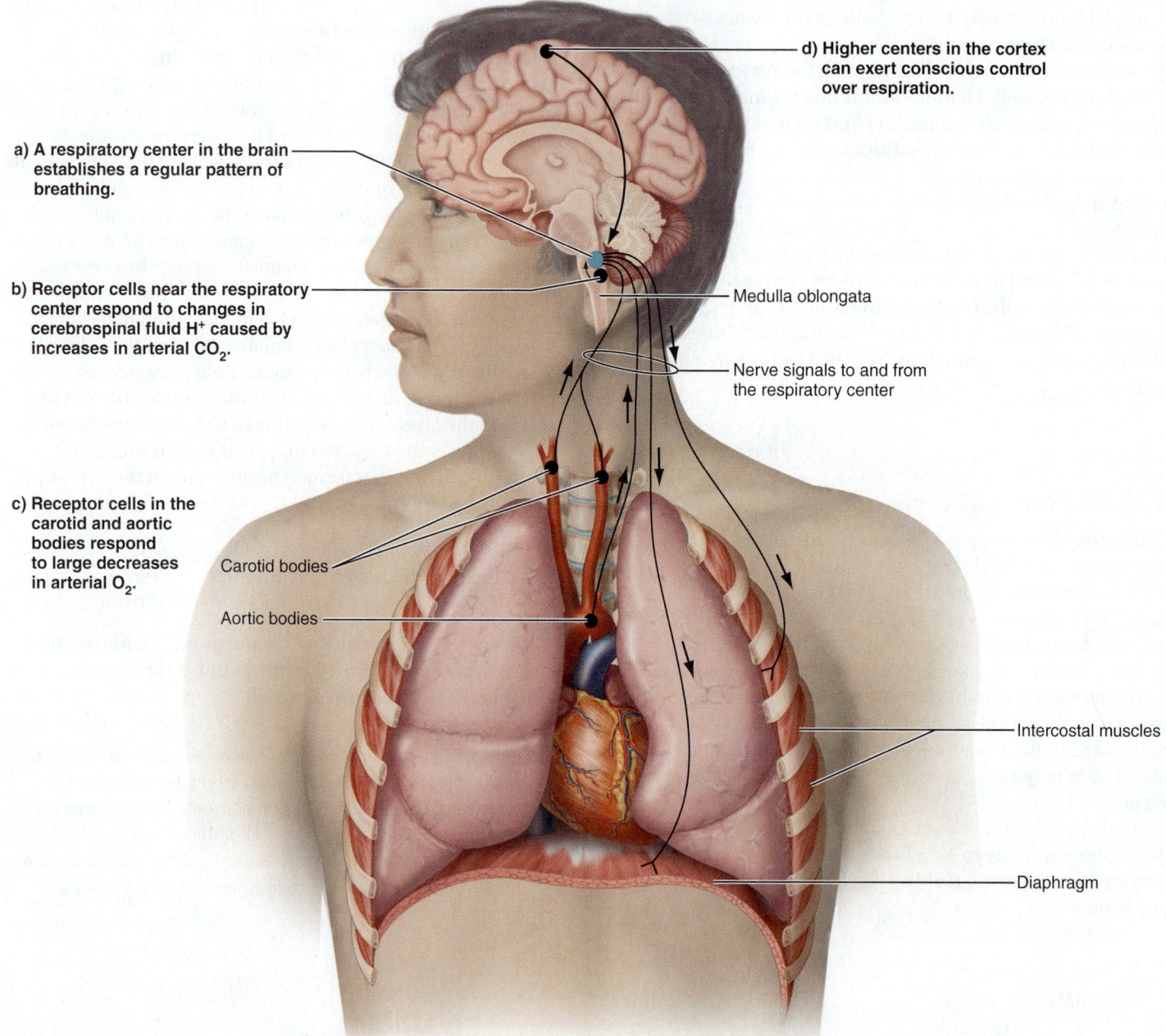

a) A respiratory center in the brain establishes a regular pattern of breathing.

b) Receptor cells near the respiratory center respond to changes in cerebrospinal fluid H^+ caused by increases in arterial CO_2.

c) Receptor cells in the carotid and aortic bodies respond to large decreases in arterial O_2.

d) Higher centers in the cortex can exert conscious control over respiration.

Medulla oblongata

Nerve signals to and from the respiratory center

Carotid bodies

Aortic bodies

Intercostal muscles

Diaphragm

Figure 10.13 Regulation of breathing. The basic pattern of breathing is established by a respiratory center within the medulla oblongata. Normally, the rate and depth of respiration are set primarily to regulate blood CO_2 levels, but respiration can also be stimulated by a substantial drop in blood O_2 levels. In addition, respiration can be controlled (at least for short periods of time) by conscious control.

receptors in the lungs. These receptors monitor the degree of inflation of the lungs and serve to limit inhalation and initiate exhalation. When nerve impulses from the respiratory center to the muscles end, the respiratory muscles relax, the rib cage returns to its original size, the diaphragm moves upward again, and we exhale.

Any disorder that interferes with the transmission of these nerve impulses can affect breathing. Consider amyotrophic lateral sclerosis (ALS, also known as Lou Gehrig's disease, for the famous baseball player who succumbed to it). In ALS, the nerves to skeletal muscle become damaged and no longer conduct impulses properly. Over time, the skeletal muscles, including the diaphragm and intercostal muscles, weaken and waste away from lack of use. Most ALS patients die within five years of initial diagnosis. Although ALS is not a respiratory condition per se, the immediate cause of death is usually respiratory failure.

Chemical receptors monitor CO_2, H^+, and O_2 levels

The body modifies, or regulates, the rate and depth of the breathing pattern to maintain homeostasis. Under normal circumstances, the regulation of breathing centers on maintaining homeostasis of CO_2, H^+, and O_2, with the main emphasis on CO_2.

The sensory mechanisms for detecting changes in CO_2 levels are actually indirect. Rather than detecting changes in CO_2 levels, certain cells in the medulla oblongata near the respiratory center (**Figure 10.13b**) detect changes in the H^+ concentration of the cerebrospinal fluid (the interstitial fluid around the cells in the brain). This is pertinent to control of CO_2 because, as you may recall, a rise in CO_2 concentration is accompanied by a rise in the hydrogen ion concentration according to the following reaction:

$$CO_2 + H_2O \rightarrow H_2CO_3 \rightarrow HCO_3^- + H^+$$

Thus, when the P_{CO_2} of arterial blood rises, the concentration of H^+ in cerebrospinal fluid also rises. Receptor cells detecting an elevated H^+ concentration transmit signals to the respiratory center, causing it to increase the rate of the cyclic pattern of impulses and the number of impulses per cycle. As a result, we breathe more frequently and more deeply, exhaling more CO_2 and lowering blood levels of the gas back to normal. Our normal arterial P_{CO_2} is maintained at about 40 mm Hg because the normal regulation of respiration keeps it there. Any rise above 40 mm Hg stimulates breathing, and any fall below 40 mm Hg inhibits it.

Certain other receptor cells respond to blood P_{O_2} rather than P_{CO_2}. These receptors are located in small structures associated with the carotid arteries (the large arteries to the head) and the aorta, called the *carotid* and *aortic bodies* (**Figure 10.13c**). Normally, P_{O_2} in arterial blood is about 100 mm Hg. If P_{O_2} drops below about 80 mm Hg, these receptors signal the respiratory center to increase the rate and

depth of respiration in order to raise blood O_2 levels back toward normal. The carotid and aortic bodies also can be stimulated by an increase in the blood concentrations of H^+ and CO_2 if they are great enough.

Note that the receptors for O_2 in the carotid and aortic bodies become activated only when arterial P_{O_2} falls by at least 20%. In contrast, even a 2–3% change in P_{CO_2} stimulates respiration. Under normal circumstances, then, respiration is controlled entirely by receptors that respond to CO_2, rather than O_2. In other words, the rate and depth of normal breathing is set by the need to get rid of CO_2, not to obtain O_2. It just so happens that the regulation of CO_2 also keeps O_2 within the normal range. There are circumstances, however, when the O_2 receptors do come into play. These include certain disease states, drug overdoses, and breathing at high altitudes where the partial pressure of O_2 is much lower than normal.

☑ Some medical conditions can cause an abnormally low concentration of H^+ ions in the blood. Explain what effect this will have on respiratory rate, and why.

We can exert some conscious control

Finally, there is another way to regulate breathing, and that is by conscious control. Conscious control resides in higher brain centers, most notably the cortex (**Figure 10.13d**). The ability to modify our breathing allows us to speak and sing. You can choose to hold your breath for a minute or even to hyperventilate (breathe rapidly) for a short while.

But if you don't think automatic regulatory mechanisms are powerful, just try to hold your breath indefinitely. You will find your conscious control overpowered by the automatic regulatory mechanisms described previously.

↩ **Recap** The respiratory center in the brain establishes a regular pattern of cyclic breathing. The rate and depth of breathing are then adjusted by regulatory mechanisms that monitor arterial concentrations of CO_2, H^+, and O_2. Conscious control can modify regulatory control but cannot override it completely. ■

10.6 Disorders of the respiratory system

Many factors can lead to disorders of respiration. Among them are conditions that reduce air flow or gas exchange, infections by microorganisms, cancer, diseases of other organs, and genetic diseases.

Reduced air flow or gas exchange impedes respiratory function

Respiration depends on the flow of air between the atmosphere and the alveoli and on the diffusional exchange of gases across the alveolar and capillary walls. Any factor that impairs these activities impedes respiratory function.

Asthma is a condition characterized by spasmodic contraction of bronchial muscle, bronchial swelling, and increased production of mucus. An asthma attack causes partial closure of the bronchi, making breathing difficult. It is a recurrent, chronic lung disorder that affects 17 million people in North America. Its incidence is on the rise both in the United States and around the world.

Symptoms of an asthma attack include coughing while exercising, shortness of breath, wheezing, and a sense of tightness in the chest. People with asthma often wheeze when they exhale. The symptoms can be triggered by any number of causes, including viruses, air particles, allergies (such as allergies to pollen, house dust, and animal fur), exercise (especially in cold temperatures), tobacco smoke, and air pollution. The symptoms may come and go.

Most asthma attacks are caused by a hyperactive immune system. When a person with asthma breathes in allergens such as pollen or tobacco smoke, the body reacts with excessive production of immunoglobulin E. The IgE stimulates mast cells in the lungs to release chemical weapons such as histamine, leading to excessive inflammation and constriction of bronchiolar smooth muscle.

Drugs are available to dilate the bronchi (bronchodilators), reduce the inflammation (corticosteroids), and restore normal breathing. Treatments also focus on preventing attacks by isolating the cause and avoiding it when possible.

Emphysema is a chronic disorder in which the alveoli become permanently damaged. It begins with destruction of connective tissue in the smaller airways. As a result, the airways become less elastic, do not stay open properly, and tend to collapse during expiration. The high pressures in the lungs caused by the inability to exhale naturally through the collapsed airways eventually damage the fragile alveoli. The result is a permanent reduction in the surface area available for diffusion, and eventually breathlessness and reduced capacity to exchange gases across the lung.

At least one form of emphysema is inherited, but most cases are associated with smoking or long-term exposure to air pollutants. The difference between asthma and emphysema is that asthma is an episodic, recurrent condition of increased airway resistance that largely goes away between episodes, whereas emphysema involves permanent damage to airways that eventually destroys alveoli.

Bronchitis refers to inflammation of the bronchi, resulting in a persistent cough that produces large quantities of phlegm. Bronchitis may be acute (comes on suddenly and clears up within a couple of weeks) or chronic (persists over a long period and recurs over several years). Both forms are more common in smokers and in people who live in highly polluted areas.

Symptoms include wheezing, breathlessness, and a persistent cough that yields yellowish or greenish phlegm. Sometimes there is fever and a feeling of discomfort behind the sternum. Acute bronchitis can be treated by humidifying the lungs (using a home humidifier or

inhaling steam), drinking plenty of fluids, and taking antibiotics if the infection is caused by bacteria. People with chronic bronchitis may need further testing to rule out other health conditions. Bronchodilators can widen the bronchi; oxygen may be prescribed to raise blood oxygen levels.

Emphysema and chronic bronchitis are both characterized by poor airflow and difficulty breathing. Hence both conditions are generally referred to by the more general term, *chronic obstructive pulmonary disease* (COPD).

Cystic fibrosis is an inherited condition in which a single defective gene causes the mucus-producing cells in the lungs to produce a thick, sticky mucus. The disease affects other organ systems as well. In the lungs, the abnormally thick mucus impedes air flow and also provides a site for the growth of bacteria. People with cystic fibrosis tend to get frequent infections of the airways. Treatment of the disease includes consistent physical therapy to try to dislodge the mucus and keep the airways open. Several promising new drugs are now on the market for this disease. For more on cystic fibrosis, see the Health & Wellness feature in the chapter on genetics and inheritance.

☑ For the past three years, a friend of yours has had symptoms of breathlessness, a tendency to wheeze, and a chronic cough that produces large amounts of yellowish phlegm. He is a smoker. Name the most likely cause of his symptoms, and explain your reasoning.

Microorganisms can cause respiratory disorders

The lungs are particularly prone to infections. One of the reasons is that they are moist, warm, and covered in a thin layer of fluid, exactly the conditions favored by microorganisms. Besides the common cold and the flu (both caused by viruses), more serious diseases caused by microorganisms include pneumonia, tuberculosis, and botulism.

Colds and the flu are common respiratory diseases; nearly everyone has had them at some time in their life. Both are caused by viruses. Colds (sometimes called an *upper respiratory infection* or *URI*) are generally caused by viruses of the rhinovirus or coronavirus families; both are highly contagious but not very virulent. The primary symptoms are coughing, runny nose, nasal congestion, and sneezing. The flu is caused by viruses of the influenza family, and although the symptoms are generally more severe, the flu is also not very virulent. Symptoms of the flu include sore throat, fever, and a cough, sometimes accompanied by aches and chills, muscle pains, and headache. The flu and a cold are easily mistaken for each other, except that colds are generally not accompanied by a fever. In any case, it hardly matters because there is no medical treatment for either of them. Rest and plenty of fluids are the best prescription. Only rarely, colds or the flu lead to inflammation of the lungs and pneumonia, accompanied by a bacterial infection that requires antibiotics (see the next section).

Colds and the flu can be caught over and over again throughout a person's lifetime. The reason is that these viral infections evolve rapidly, so that each year, they are just a little bit different from the previous year so are not recognized by the immune system.

Pneumonia is an inflammatory condition of the lungs. It is usually caused by a viral or bacterial infection. In pneumonia, the alveoli secrete excess fluid, impairing the exchange of oxygen and carbon dioxide. Symptoms typically include fever, chills, shortness of breath, and a cough that produces yellowish-green phlegm and sometimes blood. Some people experience chest pain when breathing due to inflammation of the membranes that line the chest cavity and cover the lungs.

In North America, pneumonia ranks among the top 10 causes of death, primarily because it is a frequent complication of many serious illnesses. Treatment depends on the microorganism involved; if it is bacterial, antibiotics may be effective. In severe cases, oxygen therapy and artificial ventilation may be necessary. However, most people who develop pneumonia recover completely within a few weeks.

Tuberculosis (TB) is an infectious disease caused by the bacterium *Mycobacterium tuberculosis*. People pass the infection in airborne droplets by coughing or sneezing. The bacteria enter the lungs and multiply to form an infected "focus." In most cases, the immune system fights off the infection, although it may leave a scar on the lungs. In perhaps 5% of cases, however, the infection spreads via lymphatic vessels to the lymph nodes and may enter the bloodstream. Sometimes the bacteria become dormant for many years, then reactivate later to cause more lung damage.

Major symptoms include coughing (sometimes bringing up blood), chest pain, shortness of breath, fever, night sweats, loss of appetite, and weight loss. A chest X-ray usually reveals lung damage, such as cavities in the lungs or old infections that have healed, leaving scarred lung tissue. A skin test called the tuberculin test can indicate whether someone has been exposed to the infection.

HEALTH & WELLNESS

Carbon Monoxide: An Invisible, Odorless Killer

On January 2, 2006, an explosion roared through the Sago Mine in West Virginia, trapping 13 coal miners 260 feet underground. The men retreated deeper into the mine and, following standard safety guidelines, hung up a fabric barrier that was supposed to block poisonous gases. Rescue workers labored for a day and a half to reach them. Tragically, all but one of them were found dead.

What killed the miners? Not the explosion itself but a by-product of the combustion, carbon monoxide (CO). It is a colorless, odorless, highly poisonous gas produced by burning carbon-containing substances such as coal, natural gas, gasoline, wood, and oil.

Carbon monoxide has a high affinity for hemoglobin and competes with oxygen for hemoglobin binding sites. CO binds to hemoglobin almost 200 times more tightly than oxygen and, once bound, remains attached for many hours. Even trace amounts significantly reduce the oxygen-transporting capacity of blood. High concentrations reduce the oxygen-carrying capacity of hemoglobin so much that body tissues, including those of the brain, literally starve.

Early symptoms of CO poisoning include flushed skin (especially the lips), dizziness, headache, nausea, and feeling faint. Continued exposure leads to unconsciousness, brain damage, and death. The leading cause of accidental poisoning in the United States, CO kills about 500 people and contributes to 40,000 emergency room visits every year. About 2,000 people per year in the United States commit suicide by inhaling it. Common sources of CO include industrial pollution, automobile exhaust, furnaces and space heaters, and cigarette smoke.

If you are exposed to CO, it's vital that you get away from the gas and into fresh air as soon as possible. People with severe cases of CO poisoning should be taken to a hospital, where pressurized oxygen or transfusions of red blood cells can be administered to increase blood oxygen levels. Depending on the degree of exposure, survivors can show permanent

Randall McCloy, Jr., sole survivor of the Sago Mine explosion.

heart or brain impairment. In one study, almost 40% of CO survivors with heart damage died within seven years, compared to 15% of those without heart damage.

Fortunately, taking a few precautions can prevent CO poisoning. Keep your gas or oil furnace and your car's exhaust system in good repair, and consider installing a CO detector in your home. Never run a car's engine in an enclosed space. Don't smoke—cigarette smoke contains measurable amounts of CO that are delivered directly to your lungs and blood. As much as possible, avoid spending a lot of time in smoke-filled rooms and high-traffic urban areas.

A century ago, tuberculosis was a major cause of death worldwide. With the development of antibiotics the incidence of tuberculosis declined precipitously, and most patients in industrialized countries now recover fully. The disease remains a major health problem in undeveloped nations, however, and recently the incidence of tuberculosis has increased in industrialized countries as well. Many authorities attribute this increased prevalence to immigration of people from developing nations. Worldwide, the increased prevalence of AIDS may also be a factor. Furthermore, some strains of tuberculosis are becoming resistant to antibiotics.

Botulism is a form of poisoning caused by a bacterium, *Clostridium botulinum*, occasionally found in improperly cooked or preserved foods. The bacterium produces a powerful toxin that blocks the transmission of nerve signals to skeletal muscles, including the diaphragm and intercostal muscles.

Symptoms of botulism poisoning usually appear 8 to 36 hours after eating the contaminated food. They can include difficulty swallowing and speaking, double vision, nausea, and vomiting. If not treated, botulism can be fatal because it paralyzes the respiratory muscles.

☑ A patient comes into a hospital with a bloody cough, shortness of breath, chest pain, and a fever. Name two possible causes for these symptoms, and describe at least two tests that might be helpful in arriving at the correct diagnosis.

Lung cancer is caused by proliferation of abnormal cells

Cancer is the uncontrolled growth of abnormal cells. In the lung, cancerous cells crowd out normal cells and eventually impede normal function. Lung cancer can impair not only the movement of air in the airways but also the exchange of gases in the alveoli and the flow of blood in pulmonary blood vessels (Figure 10.14). It accounts for one-third of all cancer deaths in the United States.

Lung cancer takes years to develop and is strongly associated with smoking. More than 90% of lung cancer patients are current or former smokers, and some of those who don't smoke were exposed to secondhand smoke from other people. Two other important causes are radon gas (which forms in rock as a breakdown product of uranium and seeps into uncirculated air in a building's foundation or basement) and chemicals in the workplace, the best known of which is asbestos.

Symptoms of lung cancer include chronic cough, wheezing, chest pain, and coughing up blood. Most people don't go to the doctor for every trivial cough, but any cough that is persistent or accompanied by chest pain—especially coughs that bring up blood—should be checked out immediately. If cancer is caught early, it may be possible to remove the cancerous lobe or region of the lung or to eradicate the tumor with radiation therapy and chemotherapy.

Lung cancer is highly preventable. It is a good idea to not smoke, to know the conditions of your work environment, and, depending on where you live, to have your home inspected for radon gas.

MJ's **BlogInFocus** Who should be screened for the early signs of lung cancer? Visit MJ's blog in the Study Area in MasteringBiology and look under "Screening for Lung Cancer."

http://goo.gl/rcwk2n

Exposure to asbestos can lead to mesothelioma

Mesothelioma is a particularly deadly cancer of the lining of the lungs, heart, and abdomen. It now appears that most cases of mesothelioma are caused by prior exposure to asbestos. Unfortunately, asbestos was widely used as a fire retardant and insulating material for decades before it was recognized as a cancer-causing agent. Today, efforts are under way to contain the asbestos within the buildings where it is still found or to remove it altogether.

Symptoms of mesothelioma generally appear 20–50 years after exposure to asbestos. The earliest symptoms of mesothelioma (shortness of breath, coughing) are similar to many other respiratory conditions, and consequently the disease may go undiagnosed until the cancer is well established. On average, patients given a diagnosis of mesothelioma live only 8–14 months.

Pneumothorax and atelectasis: a failure of gas exchange

A pneumothorax is collapse of one or more lobes of the lungs. The most common cause is a penetrating wound of the chest that allows air into the pleural cavity around the lungs, but it can also occur spontaneously as the result of disease or injury to a lung. Pneumothorax can be a life-threatening event, because the inability to inflate the lung results in reduced exchange of oxygen and carbon dioxide. Treatment requires repairing the damage to the chest wall or lung and removing the air from the pleural cavity.

Atelectasis refers to a lack of gas exchange within the lung as a result of alveolar collapse or a buildup of fluid within alveoli. In either case, there is no exchange of gases between the atmosphere and the blood in regions affected. Atelectasis is sometimes a complication of surgery, but it can also occur when the amount of surfactant is deficient. Treatment involves finding and reversing the underlying cause. Post-surgical patients in general are encouraged to take deep breaths, cough, and get up and start walking as soon as possible to avoid atelectasis. Sometimes positive pressure ventilation helps to force alveoli open again.

Congestive heart failure impairs lung function

In the chapter on the heart, we discussed *congestive heart failure* as a cardiovascular condition in which the heart

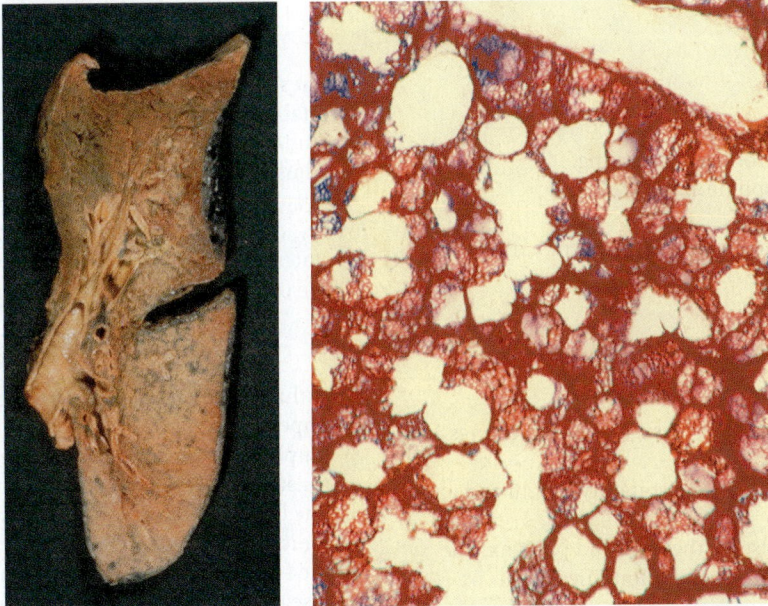

a) **Normal lung.** Alveoli (clear spaces) surrounded by numerous intact blood vessels.

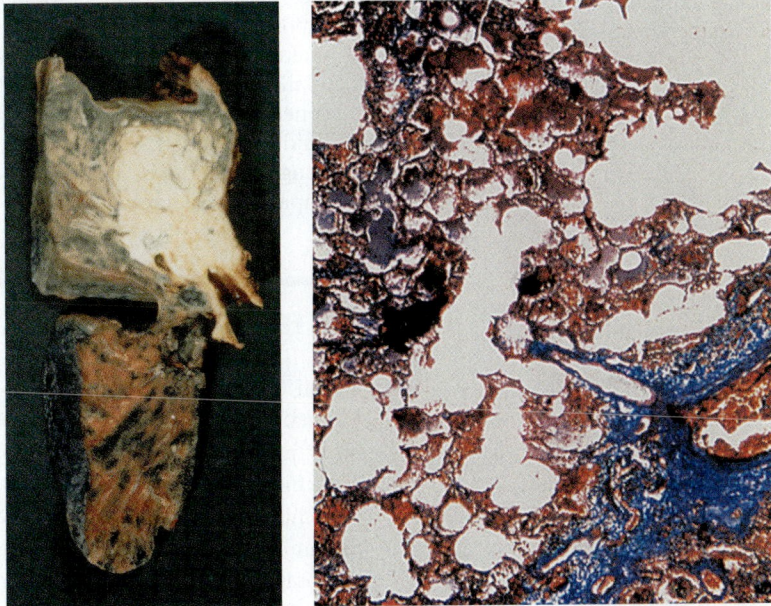

b) **Cancerous lung of a smoker.** Destruction of alveoli and blood vessels. Some alveoli are filled with blood.

Figure 10.14 Lung cancer.

gradually becomes less efficient. Even though it starts as a heart disorder, heart failure eventually causes a severe impairment of lung function as well.

Recall that in congestive heart failure, the heart begins to fail as a pump. When the left side of the heart fails, blood backs up in the pulmonary blood vessels behind that side of the heart. The result is a rise in blood pressure in the pulmonary vessels. When pulmonary capillary pressure increases, the balance of physical pressure and osmotic forces across the capillary wall favors fluid loss from the capillary. As a result, fluid builds up in the interstitial spaces between capillaries and alveoli and sometimes within alveoli themselves. This increases the diffusional distance and reduces diffusion of gases. Treatments focus on reducing this fluid buildup by helping the body get rid of fluid and improving the heart's pumping action.

Recap The lungs are prone to damage by environmental pollutants, tobacco smoke, and infections by microorganisms. Cases of both asthma and tuberculosis are on the rise. ∎

Chapter Summary

10.1 Respiration takes place throughout the body p. 221

- Respiration encompasses four processes: breathing, external respiration, internal respiration, and cellular respiration.
- External respiration occurs in the lungs; internal respiration and cellular respiration take place in the tissues.

10.2 The respiratory system consists of upper and lower respiratory tracts p. 222

- The respiratory system includes the air passageways to the lungs and the lungs themselves.
- Bones and skeletal muscles support the respiratory system and participate in breathing.
- The upper respiratory tract consists of the nose and pharynx. The upper respiratory tract filters, warms, and humidifies the air we breathe.
- The lower respiratory tract consists of the larynx, the trachea, the right and left bronchi, and the lungs.
- Within the lungs, the bronchi branch many times, becoming smaller airways called bronchioles that end in air-filled sacs called *alveoli*.
- The tremendous surface area of the alveoli, coupled with the thinness of the respiratory membrane, facilitates gas exchange with the pulmonary capillaries.

10.3 The process of breathing involves a pressure gradient p. 228

- Inspiration occurs as the lungs expand due to the action of the diaphragm and the intercostal muscles; expiration occurs when these muscles relax.
- When the lungs expand, the pressure within them falls relative to atmospheric pressure and air rushes in; during expiration, the lungs become smaller and increasing pressure within them forces air out.
- During normal breathing, inspiration is active (requiring energy) and expiration is passive.
- Normally, we breathe at about 12 breaths per minute with a tidal volume of 500 ml per breath.
- Vital capacity is the maximum amount of air a person can exhale after a maximal inhalation.

10.4 Gas exchange and transport occur passively p. 230

- The diffusion of a gas is dependent on a partial pressure gradient, which is equivalent to a concentration gradient.
- External and internal respiration are both processes that occur entirely by diffusion.
- Nearly all (98%) of the oxygen transported by blood is bound to hemoglobin in red blood cells.
- Although some carbon dioxide is transported as dissolved CO_2 or is bound to hemoglobin, most CO_2 (70%) is converted to bicarbonate and then transported in plasma.

10.5 The nervous system regulates breathing p. 233

- A respiratory center in the medulla oblongata of the brain establishes a regular cycle of inhalation and exhalation.
- Under normal conditions, the rate and depth of breathing is adjusted primarily to maintain homeostasis of arterial blood P_{CO_2}.
- Regulation of respiration by O_2 comes into play only when the P_{O_2} concentration falls by more than 20%, such as in disease states or at high altitude.
- We can exert some conscious control over breathing.

10.6 Disorders of the respiratory system p. 235

- Asthma is episodic, spasmodic contractions of the bronchi that impede air flow.
- Emphysema is a chronic disorder characterized by high resistance to air flow and destruction of alveoli.
- The lungs are prone to infections because their surface is kept moist and warm in order to facilitate gas exchange. Colds and the flu can occur nearly every year. Tuberculosis is an infectious disease caused by a bacterium. In pneumonia, infected alveoli secrete excess fluid, impairing gas exchange.
- Cancers of the respiratory system include lung cancer (a risk factor is smoking) and mesothelioma (a risk factor is prior exposure to asbestos).
- Pneumothorax and atelectasis are conditions characterized by a failure of gas exchange, either due to collapse of a lung (pneumothorax) or alveolar collapse or filling of alveoli with fluid (atelectasis).
- Lung disease can be a secondary condition resulting from impairment of another organ, as in congestive heart failure.

Terms You Should Know

alveoli, *226*	oxyhemoglobin, *233*
asthma, *236*	partial pressure, *230*
bronchi, *225*	pharynx, *223*
bronchioles, *225*	pleural membranes, *226*
bronchitis, *236*	respiratory center, *234*
carbaminohemoglobin, *233*	tidal volume, *229*
diaphragm, *228*	trachea, *224*
emphysema, *236*	tuberculosis (TB), *237*
epiglottis, *224*	ventilation, *221*
glottis, *224*	vital capacity, *229*
larynx, *224*	vocal cords, *224*

Concept Review

Answers can be found in the Study Area in MasteringBiology.

1. List at least three functions of your nose.
2. Explain why men's voices are lower than women's.
3. Explain why smokers sometimes have a chronic cough.
4. Distinguish between *bronchi* and *bronchioles*.

5. Explain what happens to gas pressure in a closed container when the size of the container becomes larger and smaller.

6. Describe the sequence of pressure changes that occurs during inspiration and expiration.

7. Define *partial pressure* and explain its importance to gas diffusion.

8. Explain why the partial pressure of oxygen in the alveoli is always lower than the partial pressure of oxygen in the air we breathe.

9. Explain (or use a diagram to show) how bicarbonate is formed from dissolved CO_2 in the region of metabolizing tissues.

10. Describe how heart failure can lead to a decrease in the ability of O_2 and CO_2 to diffuse between the blood and the alveoli.

Test Yourself

Answers are available in the Appendix.

1. Which of the following lists the order of structures through which air will pass during inspiration?
   a. nose, pharynx, larynx, trachea, bronchi, bronchioles, alveoli
   b. nose, pharynx, trachea, larynx, bronchioles, bronchi, alveoli
   c. alveoli, bronchioles, bronchi, trachea, larynx, pharynx, nose
   d. nose, pharynx, larynx, bronchi, bronchioles, trachea, alveoli

2. During acute epiglottitis, the epiglottis swells to several times its normal size. This would interfere directly with:
   a. internal respiration
   b. external respiration
   c. breathing
   d. gas exchange in the alveoli

3. Much of the respiratory tract is lined with ciliated epithelium. The exception is the:
   a. alveoli
   b. bronchi
   c. nasal cavity
   d. trachea

4. The bronchioles are:
   a. kept open by cartilage rings
   b. larger than the bronchi
   c. located between the bronchi and the alveoli
   d. an important site of gas exchange in the lungs

5. Why do smokers experience a higher incidence of respiratory infection than nonsmokers?
   a. Smoking introduces disease-causing bacteria.
   b. The nicotine and tar in smoke promote bacterial growth.
   c. Smoking inhibits mucus production.
   d. Smoking diminishes the number and action of cilia.

6. Unlike capillaries in the systemic circuit, in the pulmonary circuit, _____ blood enters the capillaries from the arterioles and _____ blood leaves the capillaries for the venules.
   a. oxygenated . . . oxygenated
   b. oxygenated . . . deoxygenated
   c. deoxygenated . . . oxygenated
   d. deoxygenated . . . deoxygenated

7. Botulism toxin interferes with the transmission of nerve impulses (action potentials) to skeletal muscles. This interferes directly with:
   a. external respiration
   b. gas exchange
   c. cellular respiration
   d. breathing

8. During a breathing cycle, as the diaphragm and intercostal muscles _____ , the volume of the pleural cavity _____ and air moves _____ the lungs.
   a. contract . . . decreases . . . into
   b. contract . . . increases . . . into
   c. relax . . . increases . . . out of
   d. relax . . . decreases . . . into

9. Which of the following has the highest P_{O_2}?
   a. alveolar air
   b. pulmonary artery
   c. interstitial fluid
   d. pulmonary vein

10. All of the following conditions favor the attachment of O_2 to hemoglobin except:
    a. high P_{O_2}
    b. cooler temperature
    c. neutral pH
    d. presence of carbon monoxide

11. In the blood, O_2 is transported primarily _____ , whereas CO_2 is transported primarily _____ .
    a. as oxyhemoglobin . . . as carbaminohemoglobin
    b. dissolved in plasma . . . as carbaminohemoglobin
    c. as oxyhemoglobin . . . as bicarbonate in plasma
    d. as oxyhemoglobin . . . dissolved in plasma

12. Which of the following is characterized by loss of elasticity in the bronchioles and permanently damaged alveoli?
    a. bronchitis
    b. emphysema
    c. asthma
    d. pneumonia

13. Which of the following is an inherited condition characterized by the production of abnormally thick mucus that can interfere with air flow within the lungs?
    a. asthma
    b. emphysema
    c. cystic fibrosis
    d. congestive heart failure

14. Which of the following will happen when someone holds their breath?
    a. The P_{CO_2} will increase.
    b. The P_{O_2} will increase.
    c. The H^+ concentration will decrease.
    d. all of the above

15. All of the following respiratory disorders are due to infectious microorganisms except:
    a. tuberculosis
    b. emphysema
    c. influenza
    d. the common cold

Apply What You Know

Answers can be found in the Study Area in MasteringBiology.

1. Why would someone administering a Breathalyzer test for alcohol ask the person being tested to give one deep breath rather than many shallow ones?

2. At a murder trial, a pathologist is asked whether the dead baby he examined had been alive at the time of birth or was stillborn. How can the pathologist tell?

3. A lifetime of heavy smoking often leads to destruction of the cilia of the cells that line the bronchi and bronchioles. How does this contribute to an increased chance of lung infections?

4. Laryngitis is an inflammation of the larynx. Why does someone with laryngitis have a hard time speaking?

5. If you take a sharp blow to the stomach you can be said to have the "wind knocked out of you." Why would an unexpected blow to the stomach leave someone unable to breathe for a moment?

6. When a hole is created in the chest wall, the lung collapses, even if the lung is not damaged. A collapsed lung is called a pneumothorax. Why does the lung collapse if it is not damaged? Will the patient be able to breathe?

7. A person who gets overly excited or has an anxiety attack may begin to hyperventilate. This can lead to a feeling of dizziness, light-headedness, and muscle weakness. One remedy is to have the person breathe into a paper bag. Why would this help?

MJ's BlogInFocus

Answers can be found in the Study Area in MasteringBiology.

1. A group of scientists who were studying breast sagging (ptosis) were surprised to find a correlation between smoking and breast sagging, that is, women who smoked were more likely to have a greater degree of breast sagging. For a reference to the research paper, visit MJ's blog in the Study Area in MasteringBiology and look under "Smoking and Breast Sagging." However, a correlation does not prove causation. Based on what you know about how scientific research is conducted, why is it unlikely that it will ever be proven definitively that smoking actually causes breast sagging? http://goo.gl/Ax9f3X

2. The rate of smoking among U.S. adults continues to decline, from nearly 25% in 1997 to less than 18% today. What do you think are some of the reasons for the decline? Visit MJ's blog in the Study Area in MasteringBiology and look under "Smoking Rates Decline." http://goo.gl/ORrS10

MasteringBiology®

Students Go to MasteringBiology for assignments, the eText, and the Study Area with animations, practice tests, and activities.

Professors Go to MasteringBiology for automatically graded tutorials and questions that you can assign to your students, plus Instructor Resources.

Answers to ✔ questions are available in the Appendix.

The Nervous System: Integration and Control

Scanning electron micrograph of nerve cells (green) with their supporting neuroglial cells (orange).

Key Concepts

- **The nervous system is the body's main control system.** It receives input from a variety of sources to control our body's physical movements, maintain homeostasis of internal variables, and initiate our higher thought processes and emotions.

- **The unit of function in the nervous system is a neuron** (nerve cell). Neurons generate and transmit electrical impulses called *action potentials*.

- **Neurons communicate with each other chemically.** Information is passed from one neuron to another via the release of chemical substances called *neurotransmitters*.

- **The brain has the capacity for long-term memory.** Long-term memory involves permanent chemical and physical changes within the brain.

- **Psychoactive drugs** (drugs that affect states of consciousness, emotions, or behavior) all act by altering communication between neurons in the brain. They do so by influencing the concentrations or actions of neurotransmitters.

Head Trauma in Young Athletes

Late in the first half of a middle school football game in 2006, Zackery Lystedt went down hard and banged his head against the ground. He struggled to his feet holding his head in pain. In the second half, he returned to the game, forcing a fumble by the opposing team on the goal line and preserving the victory for his team. But in the process, he banged his head again. After the game, he said to his father, "Dad, my head hurts," and then "I can't see." Then Zackery lapsed into a coma.

A Single Blow to the Head Can Cause a Concussion

Zackery most likely suffered a mild *concussion* after the first impact. A concussion is a trauma-induced change in mental status, often accompanied by headache, confusion, dizziness, double vision, lack of coordination,

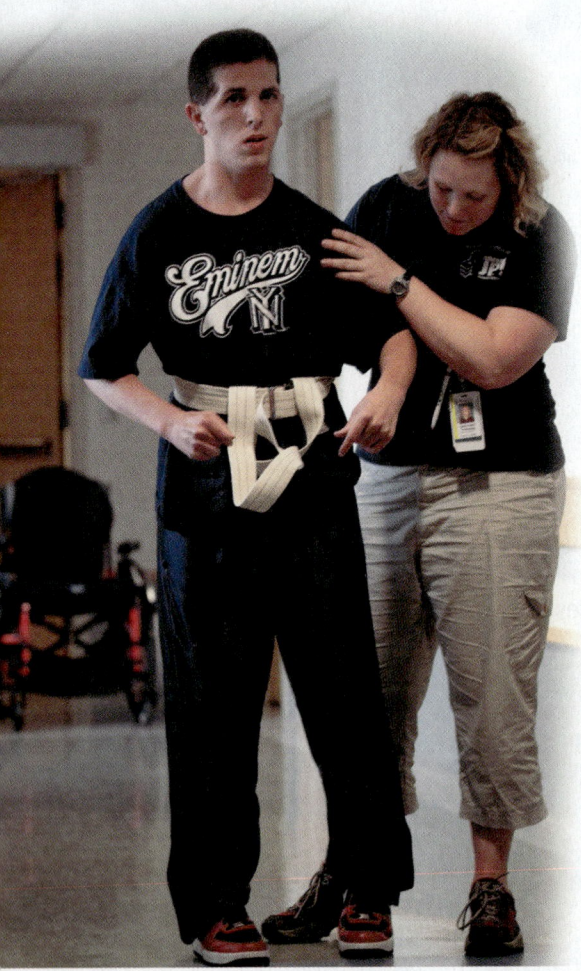

Zackery Lystedt in physical therapy after his accident.

confused speech, and (occasionally) loss of consciousness. Concussion is a description of symptoms. The medical term for the underlying cause is *mild traumatic brain injury (MTBI)*. The brain is actually a fairly soft and fragile tissue, which is precisely why it is encased in a bony skull for protection. A strong blow to the head jars the brain against the inside of the hard skull, causing temporary disruption of function of higher brain centers (a concussion), followed by mild brain swelling. In most cases, the swelling and the concussive symptoms go away with time, and the person recovers completely.

Zackery might have recovered from the first impact, except that he suffered a second impact soon after the first. That set the stage for a more serious condition known as *second impact syndrome,* in which the brain swelling is more severe. Zackery experienced the most extreme condition of all, an acute intracranial hemorrhage. He was in a coma for 30 days, didn't speak for nearly nine months, and even today can only walk short distances with a cane.

In response to his sports-related head trauma, in 2009, Zackery's home state of Washington passed one of the toughest laws in the country at the time for protecting young athletes from brain injuries. Under Washington law, athletes under the age of 18 who are suspected of having sustained a concussion *must* be removed from the game (or practice) and are not allowed to return until they have written authorization-to-play from a medical professional trained to diagnose and manage head injuries. Other states quickly followed suit, and by 2014, all 50 states had passed what have become known as Lystedt Laws.

Concussions and other acute brain injuries are more common among young athletes than you might think. Approximately one in five high school football players suffer a concussion or a more serious brain injury in just four years. The

Questions to Consider

1 Do you see any change in the popularity of high school football in your former high school or in your state? Do you think the recent lawsuit against the NFL affect the popularity of professional football?

2 What do you think could feasibly be done to find out if repetitive head trauma leads to brain damage and dementia later in life?

second most dangerous high school sport in terms of head injuries is girls' soccer. It used to be thought that younger athletes bounced back quickly from "getting their bell rung" (a blow to the head), but neuroscientists now have evidence that it's just the opposite. The effects of acute concussions are longer lasting in adolescents than in adults, and the cognitive deficits are more severe. Fortunately, increased awareness is leading to improvements in safety in contact sports, including better helmets, changes in game rules, increased parental and coach staff awareness of the symptoms of a concussion, and more medical supervision.

The Long-Term Effects of Repetitive Head Trauma

Until recently, the potential long-term risks of repetitive head trauma were completely unknown. New evidence raises the possibility that athletes who compete in sports with high rates of head impact may be at risk for permanent degenerative brain disease later in life, even if they never suffer an obvious concussion. Medical doctors call it chronic traumatic encephalopathy (CTE), permanent pathological changes in the brain brought about by repeated blows to the head. Impact sensors inside football helmets show that when two professional football linemen collide helmet-to-helmet, the head impact force is the equivalent of hitting your head against the windshield in a car crash at 25 miles per hour. Do these repetitive "minor" impacts have an additive effect?

What little information we have so far is cause for concern. In 2007, the National Football League (NFL) commissioned a telephone survey of more than 1,000 retired professional football players who had played at least three seasons in the NFL. The University of Michigan researchers →

who conducted the survey found that 6.1% of the players over 50 reported that they had been diagnosed with a dementia-related condition, more than five times higher than the national average for their age group. For players aged 30 to 49, it was *nineteen times* the age-matched national average. (*Dementia* is the term for the loss of function—including memory, language, judgment, behavior, and thinking—associated with *permanent* brain damage, as opposed to acute injury.)

These and other findings led more than 4,500 former NFL players to sue the league for damages due to concussion-related brain injuries. In 2014, the NFL and the players reached a final agreement in which the league will pay for medical exams, compensate victims of concussion-related brain injuries, and pay for research on how to prevent such

injuries in the future. The total cost to the NFL could be more than $870 million over the next 60 years. The highest award that any one player could receive is $5 million.

It would be a shame if an inability to protect our athletes from injuries ultimately led to a decline in contact sports. More information is needed about acute and

chronic head injuries. We need to know how and why the injuries are occurring and what can be done to prevent them. We owe it to the athletes to make contact sports as safe as possible.

Zackery Lystedt graduated from high school in 2011. He is now taking college courses while continuing his rehabilitation.

SUMMARY

- A high-impact blow to the head can cause a concussion. The medical consequences of concussions are greater in young athletes than in adults.
- All 50 states now have laws requiring approval of a medical professional before an injured athlete may return to a game or practice.
- Preliminary data suggests (but does not prove) that repetitive head trauma over many years leads to permanent brain damage and dementia later in life.
- In 2014, The National Football League agreed to compensate former athletes for concussion-related injuries.

You're on your way to class. It is cold, but you are not wearing your coat, and you are hungry. Your mind is elsewhere as you think about your psychology exam coming up. As you cross the street you see a doughnut shop and you smell doughnuts and coffee in the air. Suddenly, you hear sounds—the screech of a skidding truck behind you and the blaring of the truck's horn. Quickly, you whirl around, glance at the approaching truck, and leap to safety on the sidewalk. The doughnuts and the psychology exam are forgotten momentarily.

Your nervous system is constantly receiving input from all kinds of sources. It's also sorting through a huge amount of stored information in your brain's large memory banks, looking for the probable meaning of all sensory inputs, which in this scenario includes the meaning of screeching truck tires. The nervous system adds up and makes sense of this seemingly unrelated information quickly and enables you to react. In addition to helping you avoid an accident such as the one described above, the nervous system also monitors all sorts of internal conditions, including body temperature, blood pressure, and the concentrations of oxygen and carbon dioxide in the blood. It is involved in controlling how quickly you digest the food you eat and even (via receptors for taste and smell) what you eat. It controls the timing of urination and defecation. With the possible exception of the endocrine system, no other organ system is more involved in the maintenance of homeostasis than the nervous system. ■

Here are four characteristics of the nervous system:

- The nervous system receives information from many different senses simultaneously.
- The nervous system integrates information. Integration is the process of taking different pieces of information from

many different sources and assembling the pieces into a whole that makes sense. This process entails sifting through mountains of data and coming up with a plan or course of action.
- The nervous system is very fast. It can receive information, integrate it, and produce a response within tenths of a second. In this chapter, we learn about the special properties of the nervous system that give it such speed.
- The nervous system can initiate specific responses, including muscle contraction, glandular secretion, and even conscious thought and emotions.

Most of the functions of the nervous system are completely automatic and can be carried on simultaneously without requiring our attention or conscious decision making. However, the nervous system can also bring selected information to the level of conscious awareness. Working with the endocrine system (Chapter 13), the nervous system maintains homeostasis. It also allows us to feel emotions, to be aware of ourselves, and to exert conscious control over the incredible diversity of our physical movements.

11.1 The nervous system has two principal parts

The nervous system includes the **central nervous system (CNS)** and the **peripheral nervous system (PNS).** The CNS consists of the brain and the spinal cord. It receives, processes, stores, and transfers information. The PNS includes the components of the nervous system that lie outside the CNS. The PNS has two functional

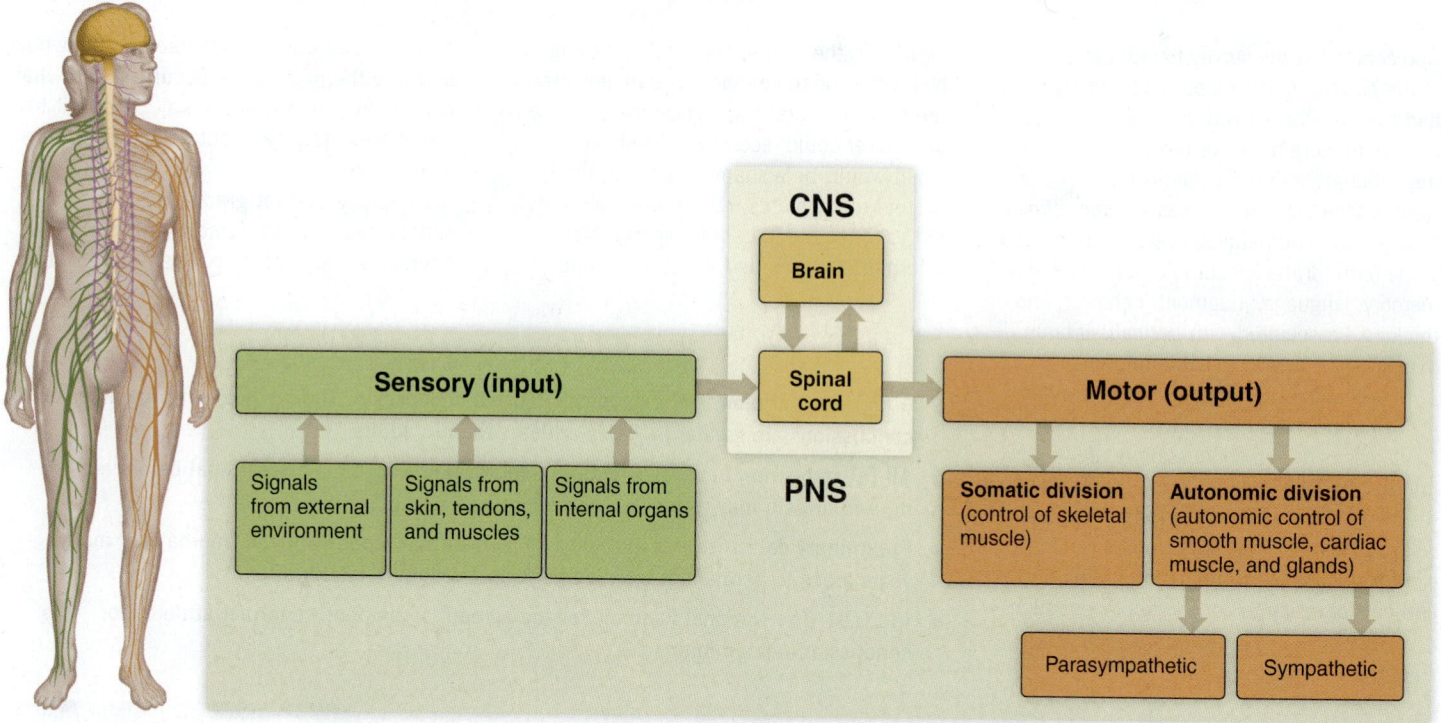

Figure 11.1 Components of the nervous system. The CNS receives input from the sensory component of the PNS, integrates and organizes the information, and then sends output to the periphery via the motor components of the PNS.

subdivisions: the sensory division of the PNS carries information to the brain and spinal cord, and the motor division of the PNS carries information from the CNS to other parts of the body (**Figure 11.1**).

The motor division of the peripheral nervous system is further subdivided along functional lines. The *somatic division* of the PNS controls skeletal muscles, and the *autonomic division* of the PNS controls smooth muscles, cardiac muscles, and glands. In turn, the autonomic division has two subdivisions called the *sympathetic* and *parasympathetic* divisions. In general, the actions of the sympathetic and parasympathetic divisions oppose each other. They work antagonistically to accomplish the automatic, subconscious maintenance of homeostasis.

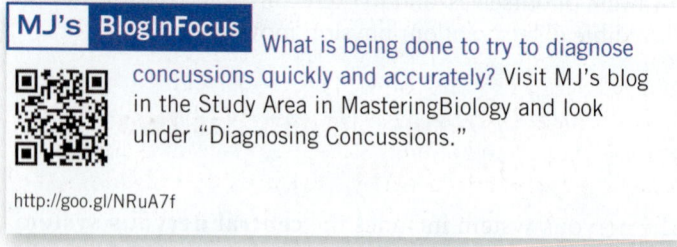

MJ's | **BlogInFocus** What is being done to try to diagnose concussions quickly and accurately? Visit MJ's blog in the Study Area in MasteringBiology and look under "Diagnosing Concussions."

http://goo.gl/NRuA7f

We examine each of these components in more detail later in this chapter. But first, we discuss the structure and function of neurons and of the neuroglial cells that accompany and support them.

↻ **Recap** The nervous system has two major subdivisions: the central nervous system (CNS), consisting of the brain and spinal cord, and the peripheral nervous system (PNS), which includes all parts of the nervous system that lie outside the CNS. The motor division of the PNS has a somatic division, which controls skeletal muscles, and an autonomic division, which controls smooth muscles, cardiac muscles, and glands. ■

11.2 Neurons are the communication cells of the nervous system

Neurons are cells specialized for communication. They generate and conduct electrical impulses, also called *action potentials*, from one part of the body to another. The longest neurons extend all the way from your toes to your spinal cord. These are single cells, remember.

There are three types of neurons in the nervous system, as shown in **Figure 11.2**.

- **Sensory neurons** of the PNS are specialized to respond to a certain type of stimulus, such as pressure or light. They transmit information about this stimulus to the CNS in the form of electrical impulses. In other words, sensory neurons provide input to the CNS.

All neurons consist of a cell body, one or more dendrites, and an axon. The main body of a neuron is called the **cell body.** The nucleus, with its content of DNA, is located in the cell body, as are the mitochondria and other cell organelles.

Slender extensions of the cell body, called **dendrites,** receive information from receptors or incoming impulses from other neurons (see Figure 11.2). Interneurons and motor neurons have numerous dendrites that are fairly short and extend in many directions from the cell body. Sensory neurons are an exception, because their dendrites connect directly to an axon.

An **axon** is a long, slender tube of cell membrane containing a small amount of cytoplasm. Axons are specialized to conduct electrical impulses. Axons of sensory neurons originate from a dendrite, whereas the axons of interneurons and motor neurons originate from the point of union with the cell body, called the *axon hillock* (see Figure 11.2). At its other end, the axon branches into slender extensions called *axon terminals*. Each axon terminal ends in a small, rounded tip called an *axon bulb*.

Typically, an interneuron or motor neuron receives incoming information from other neurons at its dendrites or cell body. If the incoming information is of the right kind and is strong enough, the neuron responds by generating an electrical impulse of its own at its axon hillock. In contrast, in a sensory neuron the impulse is initiated where the dendrite joins the axon. The impulse is then transmitted from one end of the axon to the other, bypassing the cell body entirely. We talk more about different types of receptors in the chapter on sensory mechanisms (Chapter 12).

Recap Neurons generate and transmit electrical impulses from one part of the body to another. Sensory neurons transmit impulses to the CNS. Interneurons transmit impulses between components of the CNS. Motor neurons transmit impulses away from the CNS to muscles and glands. ∎

11.3 Neurons initiate action potentials

The function of a neuron is to transmit information from one part of the body to another in the form of electrical impulses. In all cells, including neurons, the sodium-potassium pump transports three sodium ions (Na^+) out of the cell for every two potassium ions (K^+) transported in. Each time the sodium-potassium pump goes through a cycle, the net effect is the removal of one osmotic particle (a Na^+ ion) and also one positive charge (because the Na^+ ion is positively charged). In other words, the sodium-potassium pump removes both an osmotic particle and a positive charge from the cell at the same time. As a result of the activity of the sodium-potassium pumps and also the presence of negatively charged protein molecules trapped within the cell, the cell cytoplasm has a slight excess of negative charge compared to the interstitial fluid. The difference in charge results in a small but measurable difference in voltage across the cell membrane, called the *membrane potential*.

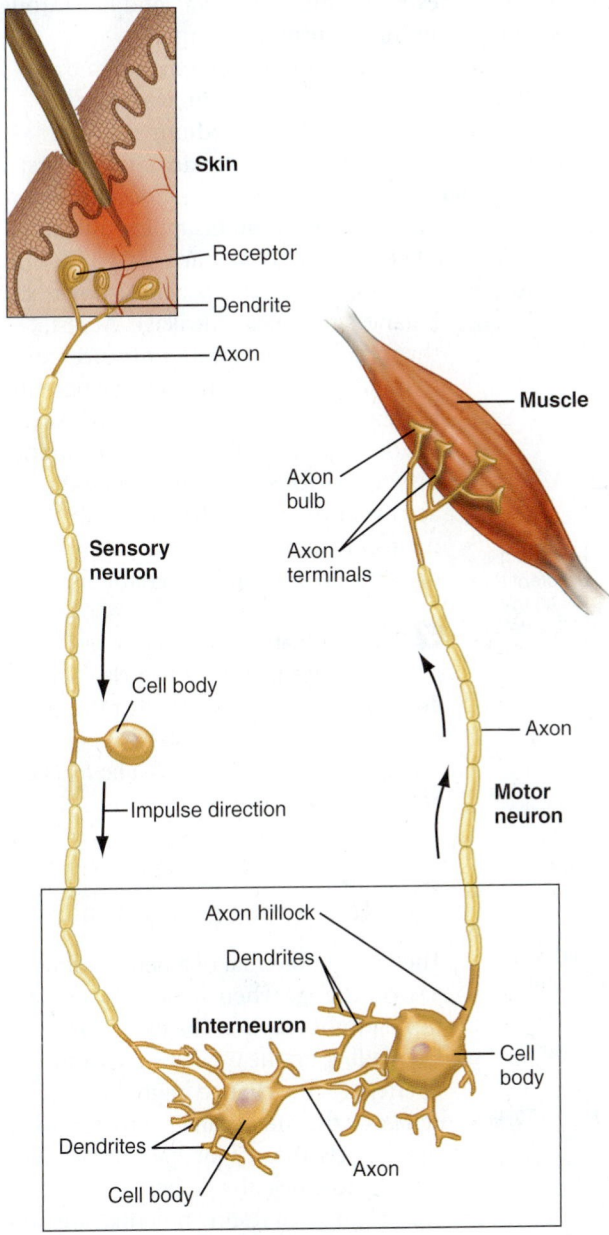

Figure 11.2 Types of neurons in the nervous system. The flow of information begins at a receptor near a dendritic ending of a sensory neuron and ends at the axon bulbs of a motor neuron. In this example, the receptors are in skin. Interneurons are located entirely within the central nervous system and typically have short axons.

- **Interneurons** within the CNS transmit impulses between components of the CNS. Interneurons receive input from sensory neurons, integrate this information, and influence the functioning of other neurons.
- **Motor neurons** of the PNS transmit impulses away from the CNS. They carry the nervous system's output, still in the form of electrical impulses, to all of the tissues and organs of the body.

Sodium-potassium pump maintains resting potential

In a neuron capable of producing an electrical impulse but not generating one at the moment, the membrane potential at rest, called its **resting potential**, is the same as the membrane potential in all cells. The resting potential of a neuron is about −70 millivolts (mV), meaning that the inside of the neuron is negatively charged compared to the outside. Figure 11.3 illustrates the role of the sodium-potassium pump and the processes of diffusion of Na⁺ and K⁺ (through channels) in maintaining the resting potential.

The concentration of sodium is much higher in the interstitial fluid than it is in the cytoplasm, whereas the inverse is true for potassium. Because sodium is always leaking into the cell and potassium is leaking out by passive diffusion, you might think that the concentration differences between the inside and outside would disappear over time. This doesn't happen because the active transport of sodium and potassium by the pump in the opposite directions exactly balances the rates of leakage. At rest, the ionic concentration differences are maintained and the membrane potential stays at −70 mV, inside negative. As we'll see in a moment, the large concentration differences for sodium and potassium across the axon membrane enable the neuron to generate impulses simply by opening and closing ion channels.

✅ Suppose that a neuron has unusual Na⁺/K⁺ pumps that move four Na⁺ out for every two K⁺ in. Predict whether the resting potential of this neuron will be more negative, less negative, or the same as a normal cell, and explain your reasoning.

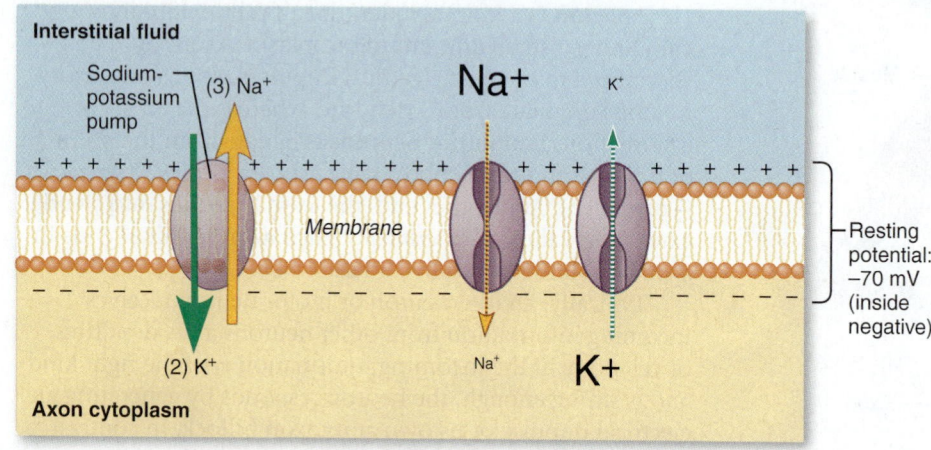

Figure 11.3 Maintenance of the resting potential. The sodium-potassium pump actively transports three Na⁺ ions out of the cell in exchange for two K⁺ ions. Most of the K⁺ leaks out again. However, the inward leak of sodium is so slow that the pump effectively is able to exclude most Na⁺ from the cell. The exclusion of positive ions (Na⁺) from the cell creates a voltage difference (an electrical potential) across the cell membrane of about −70 mV (inside negative). In this figure, the relative concentrations of Na⁺ and K⁺ in the interstitial fluid and the cytoplasm are indicated by the sizes of the letters associated with the diffusion channels.

Graded potentials can initiate an action potential

The resting potential of a neuron undergoes a slight change when an impulse arrives from another neuron (Figure 11.4). Depending on the type of signal and its strength, the change might be to ❶ *depolarize* the membrane (move the voltage closer to zero) or ❷ *hyperpolarize* it (make it even more negative). These transient local changes in resting potential are called **graded potentials** because they can vary in size. A key feature of graded potentials is that they can add up in space and time, ❸ meaning that many incoming signals from other neurons produce a bigger change in membrane potential than does one impulse alone. This is known as **summation.**

❹ If the sum of all graded potentials is sufficiently strong to reach a certain triggering membrane voltage called the **threshold,** an action potential results. ❺ An **action potential** is a sudden, temporary reversal of the voltage difference across the cell membrane. Once an action potential is initiated, it sweeps rapidly down the axon at a constant amplitude and rate of speed until it reaches the axon terminals.

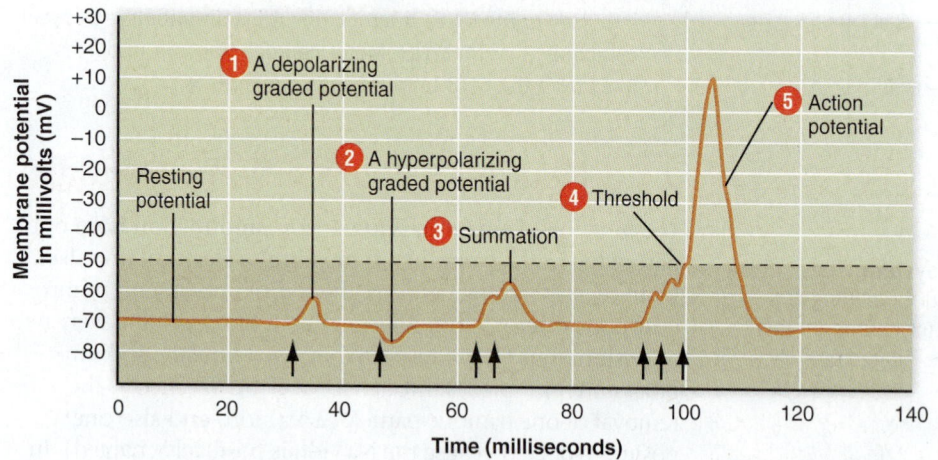

Figure 11.4 The resting potential, graded potentials, and an action potential in a neuron. Stimulation by other neurons is indicated by the arrows. Note that graded potentials can either depolarize or hyperpolarize the membrane and that they can summate. If the sum of all graded potentials is sufficient to depolarize the membrane to threshold, an action potential occurs.

✅ Extend the graph to draw what you think will happen if this neuron now receives an even stronger depolarizing stimulus (say, a series of four depolarizing graded potentials) that results in another action potential. Will this new action potential be higher and/or longer lasting than the first one?

Functionally, an action potential (also known as an *electrical impulse*) is the only form in which information is transmitted long distance by the nervous system. An action potential is like a single monotonic "beep" in a touch-tone phone or a single on-off pulse of current. Its meaning is provided only by where it came from and where it goes—that is, by how the nervous system is wired.

Action potentials occur because the axon membrane contains voltage-sensitive ion channels that open and close sequentially once threshold has been reached. An action potential occurs as a sequence of three events: depolarization, repolarization, and reestablishment of the resting potential (**Figure 11.5**).

❶ Depolarization: sodium moves into the axon When threshold is exceeded, voltage-sensitive Na$^+$ channels in the axon's membrane open briefly and Na$^+$ ions from interstitial fluid diffuse rapidly into the cytoplasm of the axon. This influx of positive ions causes *depolarization*, meaning that the membrane potential shifts from negative (-70 mV) to positive (about $+30$ mV).

❷ Repolarization: potassium moves out of the axon After a short delay, the Na$^+$ channels close automatically. But the reversal of the membrane polarity triggers the opening of K$^+$ channels. This allows more K$^+$ ions than usual to diffuse rapidly out of the cytoplasm of the axon into surrounding interstitial fluid. The loss of positive ions from the cell leads to *repolarization*, meaning that the interior of the axon becomes negative again.

❸ Reestablishment of the resting potential Because the K$^+$ channels are slow to close, there is a brief overshoot of membrane voltage during which the interior of the axon is

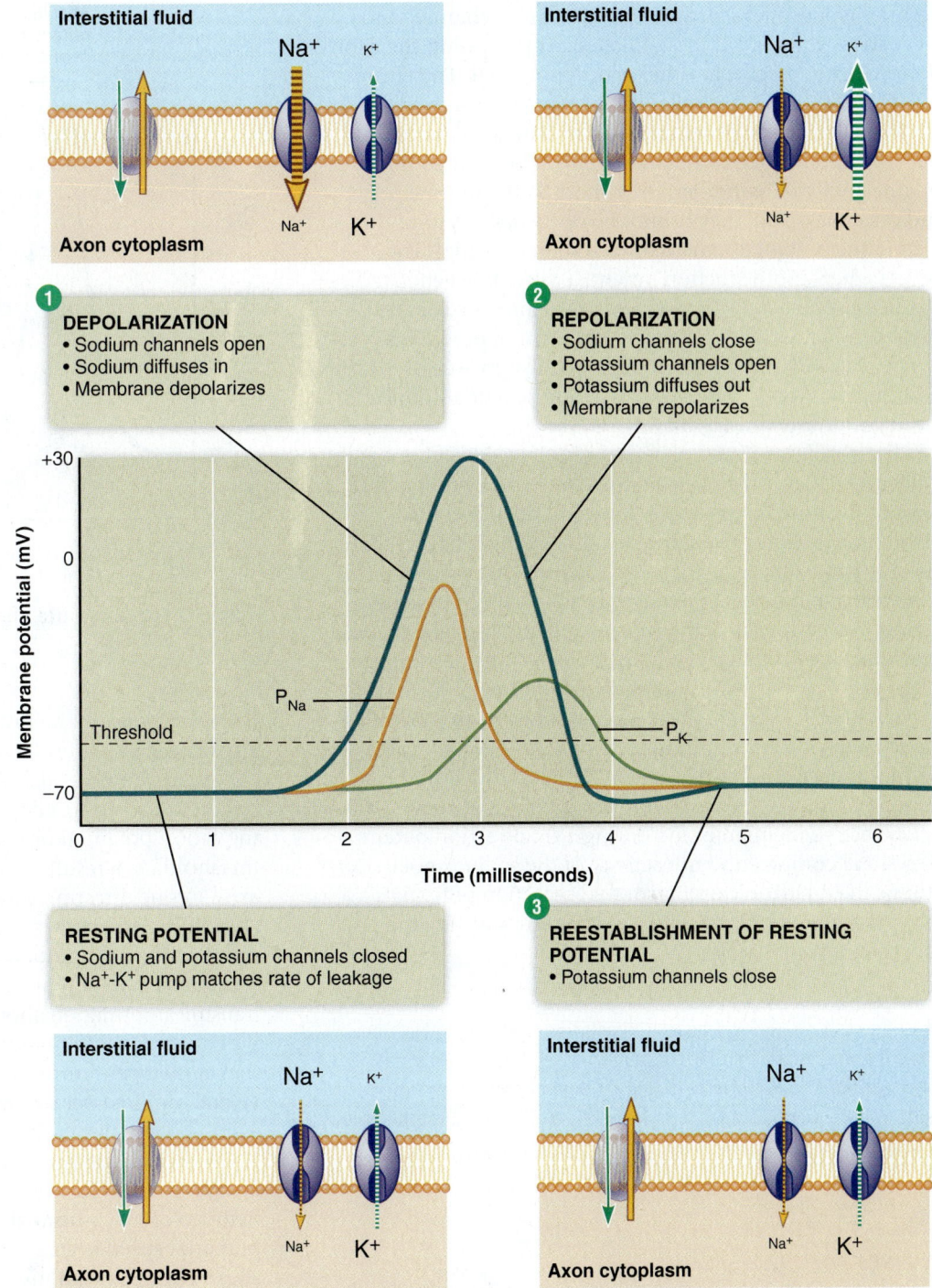

Figure 11.5 Propagation of an action potential. This is a graph of changes in membrane potential over time recorded from a single point on the axon. At rest, the sodium-potassium pump transports sodium and potassium, and the sodium and potassium channels are essentially closed.
❶ Depolarization: Once threshold is exceeded, voltage-sensitive sodium channels open briefly and the membrane permeability to sodium (P$_{Na}$) increases. Sodium diffuses rapidly into the axon, depolarizing the membrane. ❷ Repolarization: Sodium channels close. Potassium channels open and the membrane permeability to potassium (P$_K$) increases. Potassium diffuses outward into interstitial fluid, repolarizing the membrane. ❸ Reestablishment of resting potential: After a brief hyperpolarization caused by a delay in the full closure of potassium channels, the membrane potential returns to its normal resting value.

✓ Explain in your own words why the blue line goes down when the green line goes up.

slightly hyperpolarized. Shortly after the K$^+$ channels close, the resting potential is reestablished. At this point, the axon is prepared to receive another action potential. The entire sequence of three steps takes about three milliseconds.

While an action potential is under way, an axon can't generate another action potential. This is because sodium channels only remain open for a short period of time, and once they close they cannot be opened again until the membrane potential returns to its resting potential. The period when another action potential cannot be generated is called the *absolute refractory period*. The presence of an absolute refractory period ensures that action potentials always travel in one direction only, because an area of the axon that has just experienced an action potential cannot produce another one right away.

The absolute refractory period is followed by a brief *relative refractory period*. During the relative refractory period, most of the potassium channels are still open and the membrane is slightly hyperpolarized, making it harder than usual to generate the next action potential. In practical terms, this means that by encroaching into the relative refractory period, very strong stimuli can generate more frequent action potentials than relatively weak stimuli can. The absolute and relative refractory periods are shown in **Figure 11.6**.

Whether or not a neuron is generating action potentials, the sodium-potassium pumps continue to maintain the normal concentrations of Na$^+$ and K$^+$ and the normal resting potential. The actual number of Na$^+$ and K$^+$ ions that diffuse across the membrane during a single action potential is so small compared to the activity of the sodium-potassium pumps that an axon can transmit many action potentials in a short period of time without adversely affecting the subsequent resting potential.

✅ Suppose a neuron is placed in a solution that has no Na$^+$ ions but is a normal extracellular solution in other respects (normal concentrations of osmotic particles, normal charge). If the membrane voltage were brought to threshold, would the neuron be able to generate an action potential? Explain your answer.

Action potentials are all-or-none and self-propagating

As we have seen, an action potential does not occur unless a certain threshold membrane voltage is reached. Once it is triggered, however, it always looks exactly the same in form and voltage. Even if the graded stimulus that causes it is significantly higher than the threshold level, the form and amplitude of the action potential are always the same. Consequently, an action potential is an *all-or-none* phenomenon: either it occurs or it does not. It's like firing a gun. A certain amount of pressure—the threshold level—is required to make the gun fire: If you pull the trigger too lightly, the gun won't fire. If you pull the trigger harder

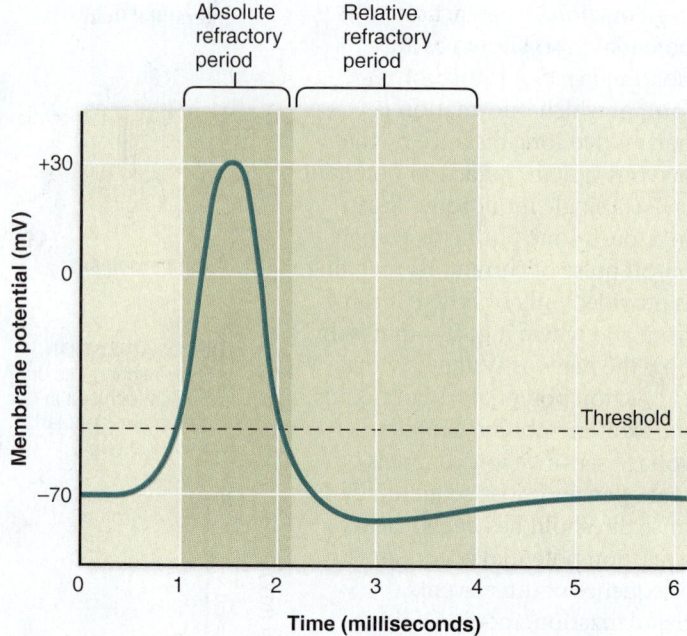

Figure 11.6 Absolute and relative refractory periods.

than is necessary, the bullet won't shoot from the gun any faster.

An action potential is also a self-propagating event, because the leading edge of the depolarizing portion of the action potential brings the next region of the axon to threshold. As a result, an action potential moves down the axon in one direction only, at a constant rate of speed and amplitude. An action potential is like an ocean wave, always moving forward and never turning back.

If all action potentials are alike, how does a neuron transmit information about stimulus intensity? Simple, really: It's encoded by the number of action potentials generated and transmitted per time unit, ranging from near zero to perhaps several hundred per second. Stronger stimuli generate *more* action potentials per time unit, not faster or bigger ones.

Although the speed of travel of an action potential is always constant for a given neuron, the speed in different neurons can vary from about 5 mph up to 250 mph. Action potential speed is greater in larger-diameter axons and in axons covered by an insulating sheath, as we'll see in a minute.

↩ **Recap** A neuron's resting potential of about −70 millivolts is maintained by the constant action of the sodium-potassium pump. Impulses arriving from other neurons can cause small, local changes in the neuron's membrane potential called *graded potentials*. The sum of all graded potentials may initiate a self-propagating, all-or-none action potential in a neuron. An action potential involves three events: depolarization, repolarization, and reestablishment of the resting potential. ∎

11.4 Neuroglial cells support and protect neurons

Only about 20% of cells in the human nervous system are neurons. The rest are **neuroglial cells,** which provide physical support and protection to neurons and help maintain healthy concentrations of important chemicals in the fluid surrounding them. (Neuroglia derives from the Greek words for "neuron" and "glue.") Neuroglial cells do not generate or transmit impulses.

In the peripheral nervous system, many neuron axons are enclosed and protected by specialized neuroglial cells called *Schwann cells* (**Figure 11.7a**). Schwann cells produce a fatty insulating material called *myelin.* During development, individual Schwann cells wrap themselves around a short segment of an axon many times as a sort of insulating blanket, creating a shiny white protective layer around the axon called a **myelin sheath** (**Figure 11.7b**). Between adjacent Schwann cells are short uninsulated gaps called **nodes of Ranvier,** where the surface of the axon is still exposed. Neurons that have axons wrapped in a sheath of myelin are called **myelinated neurons.**

The myelin sheath around the axon serves three important functions:

- *It saves the neuron energy.* The insulating layer of myelin prevents some of the slow inward leak of sodium and outward leak of potassium that would otherwise occur. These leaks normally have to be replaced by active transport processes requiring energy (see section 11.3).
- *It speeds up the transmission of impulses.* The myelin sheath causes action potentials to jump from node to node at a very fast rate. This "leaping" pattern of conduction along myelinated neurons is called **saltatory conduction** (from *saltare,* Latin for "dance") (**Figure 11.7c**). Continuously propagated action potentials in unmyelinated neurons travel at a speed of only about 5 mph (2.3 meters per second). In contrast, saltatory conduction in myelinated neurons can reach speeds of up to 250 mph (about 110 meters per second).
- *It helps damaged or severed axons of the peripheral nervous system regenerate.* If a neuron axon is severed, the portion of the axon distal to the cell body may degenerate. However, the cut end of the axon still attached to the cell body can regrow through the channel formed by the sheath, eventually reconnecting with the cell that it serves. Depending on the length of the axon, the regeneration process can take anywhere from a few weeks to more than a year.

In the central nervous system, protective sheaths of myelin are produced by another type of neuroglial cell called an *oligodendrocyte.* Two very debilitating diseases

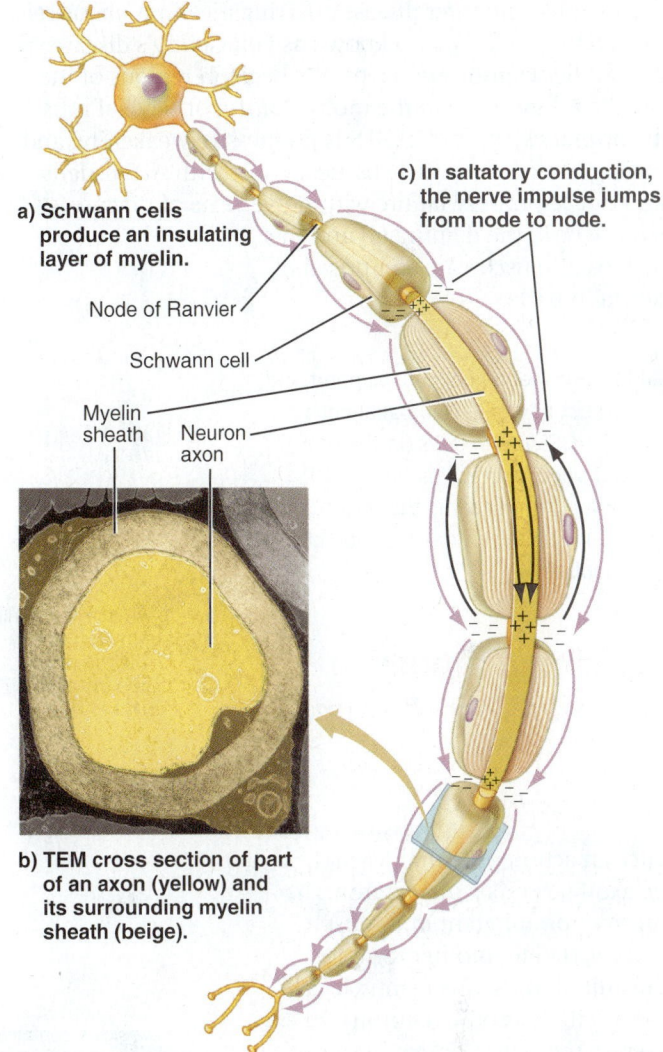

a) **Schwann cells produce an insulating layer of myelin.**

Node of Ranvier

Schwann cell

Myelin sheath

Neuron axon

c) **In saltatory conduction, the nerve impulse jumps from node to node.**

b) **TEM cross section of part of an axon (yellow) and its surrounding myelin sheath (beige).**

Figure 11.7 A myelinated motor neuron of the peripheral nervous system. The axon of the motor neuron is wrapped by individual Schwann cells, which produce an insulating substance called *myelin.* Between adjacent Schwann cells are uninsulated sections called *nodes of Ranvier.* In myelinated neurons, electrical impulses appear to jump rapidly from node to node as a result of local current flows both within and outside the neuron (shown by thin, black arrows).

are caused by oligodendrocyte dysfunction. In the autoimmune disorder *multiple sclerosis* (MS), the protective sheaths produced by oligodendrocytes in the CNS become progressively damaged. Eventually, they form hardened (sclerotic) scar tissue that no longer insulates the neurons effectively. As a result, transmission of impulses in CNS neurons is disrupted. Common symptoms include muscle weakness, visual impairment, and urinary incontinence. The disease usually appears in people between 20 and 40 years of age, and although it can be fatal in the most severe cases, the probability of an almost normal life span (with physical limitations) is fairly high.

A similar but rarer disease affecting myelin is *amyotrophic lateral sclerosis* (ALS), also known as Lou Gehrig's disease. In ALS, the sclerotic areas typically begin in regions of the spinal cord involved in the motor control of skeletal muscle. The primary symptom of ALS is progressive weakening and wasting of skeletal muscle tissue. People with ALS usually die from respiratory failure within five years of diagnosis, because both the diaphragm and the intercostal muscles are comprised of skeletal muscle.

↺ Recap Neuroglial cells support and protect neurons. Neuroglial cells called *Schwann cells* (in the PNS) and *oligodendrocytes* (in the CNS) form myelin sheaths that protect axons and speed transmission of impulses. ∎

11.5 Information is transferred from a neuron to its target

Once an action potential reaches the axon terminals of a neuron, the information inherent in it must be converted to another form for transmittal to its target (muscle cell, gland cell, or another neuron). In essence, the action potential causes the release of a chemical, called a **neurotransmitter,** that crosses a specialized junction, called a **synapse,** between the two cells. The neurotransmitter represents a signal from a neuron to its target cell. The entire signaling process—from a neuron to a target cell—is called *synaptic transmission.* Synaptic transmission also occurs between a neuron and a muscle cell and between neurons and some glands. Now we'll see how synaptic transmission occurs between neurons.

Neurotransmitter is released

Figure 11.8 illustrates the structure of a typical synapse and the events that occur during synaptic transmission. Each axon terminates with an axon bulb that contains neurotransmitter

stored in membrane-bound vesicles. At a synapse, the *presynaptic membrane* is the cell membrane of the neuron that is sending the information. The *postsynaptic membrane* is the membrane of the cell that is about to receive the information. The small fluid-filled gap that separates the

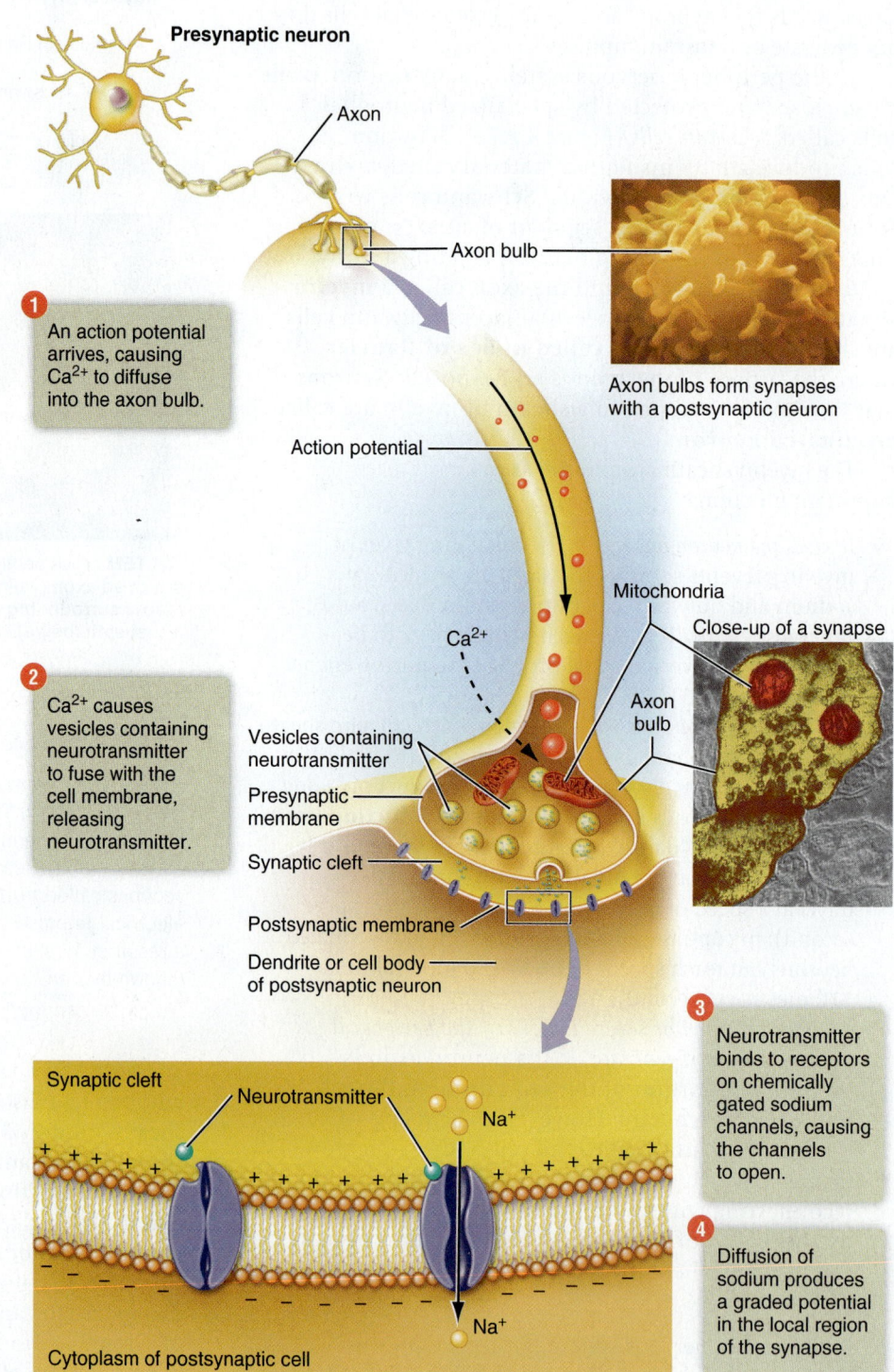

Axon bulbs form synapses with a postsynaptic neuron

Close-up of a synapse

Presynaptic neuron

Axon

Axon bulb

1 An action potential arrives, causing Ca²⁺ to diffuse into the axon bulb.

Action potential

Mitochondria

Ca^{2+}

2 Ca²⁺ causes vesicles containing neurotransmitter to fuse with the cell membrane, releasing neurotransmitter.

Axon bulb

Vesicles containing neurotransmitter

Presynaptic membrane

Synaptic cleft

Postsynaptic membrane

Dendrite or cell body of postsynaptic neuron

Synaptic cleft

Neurotransmitter

Na^+

Cytoplasm of postsynaptic cell

Na^+

3 Neurotransmitter binds to receptors on chemically gated sodium channels, causing the channels to open.

4 Diffusion of sodium produces a graded potential in the local region of the synapse.

Figure 11.8 Summary of synaptic transmission.

presynaptic and postsynaptic membranes is the *synaptic cleft*.

The events that occur during synaptic transmission follow a particular pattern:

❶ An action potential arrives at the axon bulb, causing calcium (Ca^{2+}) channels in the presynaptic membrane to open. Ca^{2+} diffuses into the axon bulb.

❷ The presence of Ca^{2+} causes the vesicles to fuse with the presynaptic membrane and release their contents of neurotransmitter directly into the synaptic cleft. Because the synaptic cleft is so narrow (about 0.02 micrometer, or one-millionth of an inch), the neurotransmitter molecules reach the postsynaptic membrane by diffusion.

❸ Molecules of neurotransmitter bind to receptors on the postsynaptic membrane. This causes certain chemically gated channels, such as those for sodium, to open.

❹ Sodium ions diffuse inward, producing a graded depolarization of the postsynaptic membrane in the area of the synapse.

Note that graded potentials are caused by the opening of *chemically* sensitive ion channels (the chemical being a neurotransmitter), rather than voltage-sensitive ones.

☑ If there were no available Ca^{2+} in the synaptic cleft, would action potentials still travel down the presynaptic neuron's axon in the usual fashion? Would the presynaptic neuron be able to release neurotransmitters into the synaptic cleft? Explain your answer.

Neurotransmitters exert excitatory or inhibitory effects

How the postsynaptic cell responds to neurotransmitter depends on several factors, including the type of neurotransmitter, its concentration in the synaptic cleft, and the types of receptors and chemically sensitive ion channels in the postsynaptic membrane.

To date, scientists have identified more than 50 chemicals that can function as neurotransmitters. All are stored in vesicles within the axon bulb and released into the synaptic cleft in response to an action potential. All produce a graded potential in the postsynaptic cell, though the effects may be quite different because they can affect different ion channels.

Neurotransmitters are classified as excitatory, inhibitory, or both. Excitatory neurotransmitters depolarize the postsynaptic cell, causing threshold to be approached or exceeded. Therefore, excitatory neurotransmitters encourage the generation of new impulses in the postsynaptic neuron. Inhibitory neurotransmitters cause the postsynaptic cell to hyperpolarize, meaning that the cell's interior becomes even more negative than before. Hyperpolarization makes it harder for threshold to be reached, so inhibitory neurotransmitters tend to prevent the generation of action potentials in the postsynaptic neuron. Some neurotransmitters can be excitatory or inhibitory, depending on the type of receptor to which they bind on the postsynaptic membrane.

Table 11.1 summarizes the actions of several common neurotransmitters.

The effects of neurotransmitters are relatively short-lived because the neurotransmitter remains in the synaptic cleft for only a short time. Prompt removal of neurotransmitter causes neural signals to be terminated almost as rapidly as they are initiated. Only then can the next message be received and recognized. The neurotransmitter may be removed from the synaptic cleft in any or all of three ways: (1) It may be taken back up again by the presynaptic neuron and repackaged into membrane-bound vesicles, to be used again; (2) it may be destroyed by enzymes in the synaptic cleft; or (3) it may diffuse away from the synaptic cleft into the general circulation, where it will ultimately be destroyed.

Table 11.1 Actions of common neurotransmitters

Neurotransmitter	Sites where released	Principal actions
Acetylcholine	Neuromuscular junctions, autonomic nervous systems, brain	Excitatory on skeletal muscles; excitatory or inhibitory at other sites, depending on receptors
Norepinephrine	Areas of brain and spinal cord, autonomic nervous system	Excitatory or inhibitory, depending on receptors; plays a role in emotions
Serotonin	Areas of brain, spinal cord	Usually inhibitory; involved in moods, sleep cycle, appetite
Dopamine	Areas of brain, parts of peripheral nervous system	Excitatory or inhibitory, depending on receptors; plays a role in emotions
Glutamate	Areas of brain, spinal cord	Usually excitatory; major excitatory neurotransmitter in brain
Endorphins	Many areas in brain, spinal cord	Natural opiates that inhibit pain; usually inhibitory
Gamma-aminobutyric acid	Areas of brain, spinal cord	Usually inhibitory; principal inhibitory neurotransmitter in brain
Somatostatin	Areas of brain, pancreas	Usually inhibitory; inhibits pancreatic release of growth hormone

Postsynaptic neurons integrate and process information

The conversion of the incoming signal from electrical (action potential) to chemical (neurotransmitter) allows the postsynaptic neuron to do a lot of integration and information processing. This is because a single all-or-none action potential in the presynaptic neuron does not always cause a corresponding all-or-none action potential in the postsynaptic neuron. A single action potential produces only a small graded potential in the postsynaptic neuron, and a lot of these graded potentials generally are required for threshold to be reached.

One way for threshold to be reached in a postsynaptic neuron is for the presynaptic neuron (or a few neurons) to increase their frequency of stimulation, sending lots of action potentials in a short time (hundreds per minute). In addition to stimulus intensity, other factors affecting whether threshold is reached in the postsynaptic neuron include (1) how many other neurons form synapses with it and (2) whether those synaptic connections are stimulatory or inhibitory. Typically, one neuron receives input from many neurons, a phenomenon known as *convergence*. In turn, its action potential goes to many other neurons, a phenomenon known as *divergence* (Figure 11.9).

Though it would be incorrect to say that neurons "interpret" information—that would imply that a neuron has a brain—the effect of convergence is that most neurons may integrate and process thousands of simultaneous incoming stimulatory and inhibitory signals before generating and transmitting their own action potentials. They also reroute the information to many different destinations. Individual neurons do not see, smell, or hear, yet their combined actions allow us to experience those complex sensations.

Although muscle cells are also target cells of presynaptic neurons, they do not process and integrate information like postsynaptic neurons do. There are two reasons for this. First, a muscle cell receives input from only one neuron. There are no converging neural inputs to process. Second, the neuromuscular junction is quite large, with many points of contact between neuron and muscle cell. Consequently, threshold is reached in a skeletal muscle cell every time the motor neuron sends even a single action potential. Functionally, this means that the nervous system has absolute and final control over the activity of skeletal muscle cells.

↩ **Recap** A neuron transmits information to another cell across a synapse. At a synapse, neurotransmitter released from a presynaptic neuron diffuses across the synaptic cleft and binds to receptors on the postsynaptic membrane, opening ion channels and causing a graded potential in the postsynaptic cell. Postsynaptic neurons may integrate incoming signals from many different presynaptic neurons. ∎

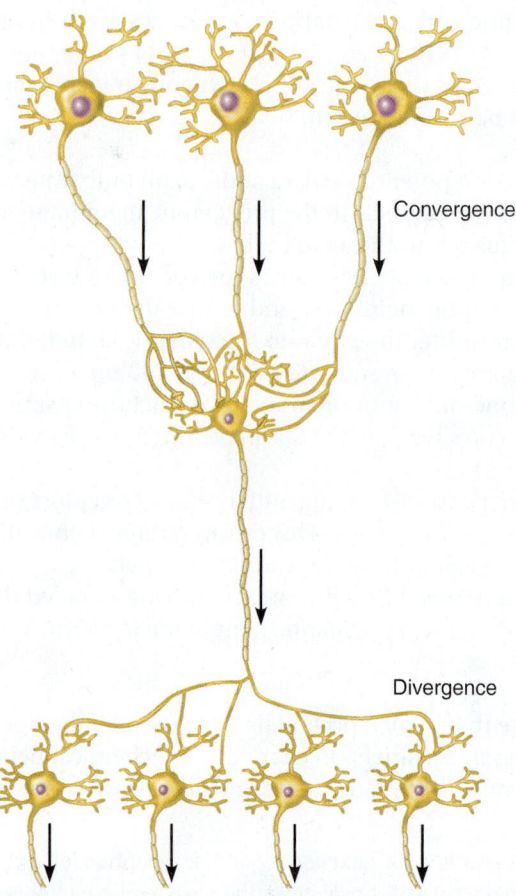

Figure 11.9 Neural information processing by convergence and divergence. Convergence occurs when several presynaptic cells form synapses with a single postsynaptic cell. Divergence occurs when one nerve cell provides information to a variety of postsynaptic cells.

11.6 The PNS relays information between tissues and the CNS

Now that we have seen how neurons generate and transmit action potentials, we look at how this information travels throughout the nervous system. As stated earlier, the nervous system has two major subdivisions: the central nervous system (CNS), which includes the spinal cord and brain, and the peripheral nervous system (PNS), which consists of nerves that transmit information between body tissues and the CNS.

Nerves carry signals to and from the CNS

A single nerve axon is like a single fiber-optic fiber in that both transmit specific kinds of information from one location to another. By the same token, a nerve is like a large fiber-optic cable strung between two cities. A **nerve** consists of the axons of many neurons, all wrapped together in a protective sheath of connective tissue and all coming from and going to the same place. The function of each nerve depends on where it originates and the organs or tissues to which it travels.

The peripheral nervous system consists of cranial nerves and spinal nerves. The 12 pairs of **cranial nerves** connect directly with the brain. Cranial nerves carry action potentials between the brain and the muscles, glands, and receptors of the head, neck, and thoracic and abdominal cavities.

The 31 pairs of **spinal nerves** connect with the spinal cord. Each spinal nerve attaches to the spinal cord via two short branches of the spinal cord called the *dorsal root* and the *ventral root*. The dorsal root contains the sensory neurons that transmit incoming action potentials from body tissues to the spinal cord. The ventral root contains motor neurons that carry action potentials away from the spinal cord to the rest of the body. Thus, each spinal nerve carries both sensory and motor (incoming and outgoing) information.

Sensory neurons provide information to the CNS

The nervous system must have good information if it is to take appropriate action. That information arrives at the CNS as action potentials traveling in sensory neurons located throughout the body. Because information provided by sensory neurons to the CNS may affect the motor output of both the somatic and autonomic divisions of the PNS, the sensory (input) component of the PNS cannot be classified as belonging specifically to either the somatic or autonomic motor (output) divisions.

The somatic division controls skeletal muscles

Recall that the motor (or output) neurons of the peripheral nervous system are organized into the somatic division and the autonomic division. The **somatic division** controls voluntary and

involuntary skeletal muscle movement. Somatic motor neurons transmit information from the spinal cord to skeletal muscles. Somatic motor neurons are activated either by conscious control from the brain or by an involuntary response called a *reflex*.

Spinal reflexes are involuntary responses that are mediated primarily by the spinal cord and spinal nerves. The brain receives sensory input and can modify spinal reflexes slightly, but it cannot prevent them from occurring altogether. **Figure 11.10** illustrates spinal reflexes. ❶ Stepping on a sharp object activates pain receptors, ❷ producing action potentials in sensory neurons traveling to your spinal cord. ❸ The sensory neurons stimulate interneurons within the spinal cord, which in turn transmit signals to your brain and then on to motor neurons in your leg, ❹ causing you to flex the appropriate muscles to lift your foot. This is called the *flexor (withdrawal) reflex.*

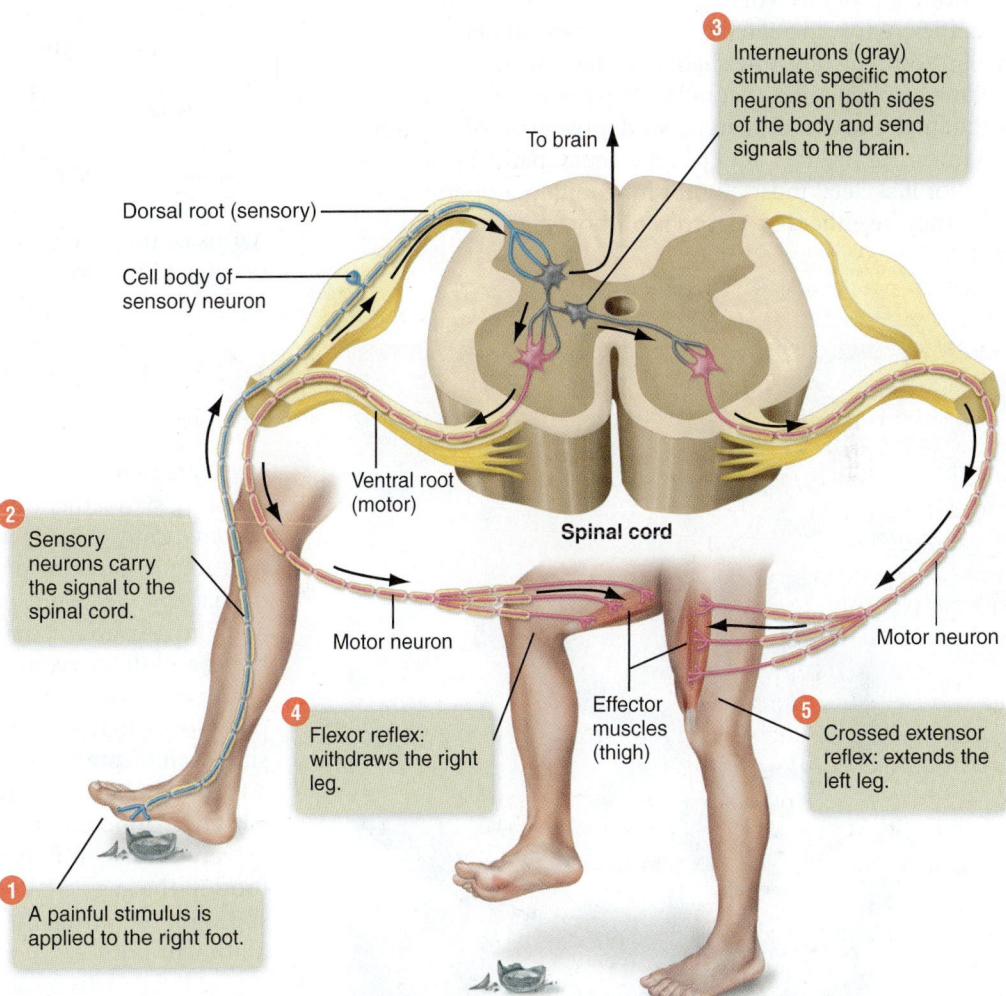

❸ Interneurons (gray) stimulate specific motor neurons on both sides of the body and send signals to the brain.

To brain

Dorsal root (sensory)

Cell body of sensory neuron

Ventral root (motor)

Spinal cord

❷ Sensory neurons carry the signal to the spinal cord.

Motor neuron

Motor neuron

❹ Flexor reflex: withdraws the right leg.

Effector muscles (thigh)

❺ Crossed extensor reflex: extends the left leg.

❶ A painful stimulus is applied to the right foot.

Figure 11.10 Spinal reflexes. In this example, a painful stimulus on one side of the body initiates different reflex actions on the two sides of the body: a flexor reflex (flexion of the leg) on the stimulated side and an extension reflex (extension of the leg) on the opposite side. Sensory information is also relayed to the brain to make you aware of the painful stimulus.

Meanwhile, other interneurons that cross over to the opposite side of the spinal cord are also stimulated by the sensory neuron. These interneurons stimulate motor neurons to extend, rather than flex, certain skeletal muscles in your other leg. After all, it wouldn't do to flex both legs and fall on the sharp object! ❺ This second reflex is called a *crossed extensor reflex*—crossed because the signal initiating it crossed over from the other side of the body. Notice that these events did not require conscious thought. A simple spinal reflex is processed primarily at the level of the spinal cord, allowing a rather complex series of actions to occur quickly.

Another type of spinal reflex is a *stretch reflex*. Stretch reflexes play an important role in maintaining upright posture and movement, as they allow us to stand and move without having to concentrate on our actions. Physicians often test stretch reflexes to determine whether the spinal cord and areas of the brain involved are intact and functioning properly. An example is the well-known knee-jerk reflex. When the patellar tendon is tapped lightly below the kneecap, the resulting slight stretch of the thigh muscles initiates a spinal reflex that ultimately results in contraction of the thigh muscles and an upward movement of the foot and lower leg (Figure 11.11). If the neural pathways are not intact or if skeletal muscle is not functioning properly, the reflex may be either less vigorous or more vigorous than normal.

Figure 11.11 The patellar (knee-jerk) reflex. Integrity of the patellar reflex verifies that certain sensory and motor nerves, the spinal cord, areas of the brain, and skeletal muscle are all functioning normally.

☑ Compare and contrast the patellar (knee-jerk) reflex and the reflex responses to a painful stimulus applied to the foot. What are the major similarities and differences?

The autonomic division controls automatic body functions

Motor neurons carrying information to smooth or cardiac muscle and to all other tissues and organs except skeletal muscle belong to the **autonomic division** (*autonomic* derives from the Greek words for "self" and "law"). The autonomic division of the peripheral nervous system carries signals from the CNS to the periphery that control "automatic" functions of the body's internal organs. Many of these activities occur without our ever being aware of them.

Unlike the somatic division in which one neuron extends from the CNS to the target cell, the autonomic division requires two neurons to transmit information from the CNS to the target cell. The cell bodies of the first neurons (called *preganglionic neurons*) lie within the central nervous system. The axons of many of these neurons go to structures called *ganglia* (singular: ganglion) that lie outside the CNS. Ganglia are clusters of neuron cell bodies of the second neurons (called *postganglionic neurons*). Axons of the postganglionic neurons extend to the far reaches of the body to communicate with internal glands and organs.

The sympathetic and parasympathetic divisions oppose each other

There are two types of motor neurons in the autonomic division: sympathetic and parasympathetic. Both function automatically, usually without our conscious awareness, to regulate body functions. They innervate smooth muscle, cardiac muscle, glands, the epithelial cells of the kidneys, and virtually every internal organ except skeletal muscle, bone, and the nervous system itself.

In most organs, the actions of the sympathetic and parasympathetic divisions have very opposite effects, as shown in Figure 11.12. Which division predominates at any one time determines what happens in your body. And with two opposing systems active at all times, control can be very precise. It is as if you were driving a car with one foot on the gas pedal and one on the brake—you could speed up by hitting the gas pedal, releasing the brake pedal a little, or both.

The sympathetic division arouses the body Preganglionic motor neurons of the **sympathetic division** originate in the thoracic and lumbar regions of the spinal cord. Many of their axons extend to a chain of sympathetic ganglia that lie alongside the spinal cord, where they

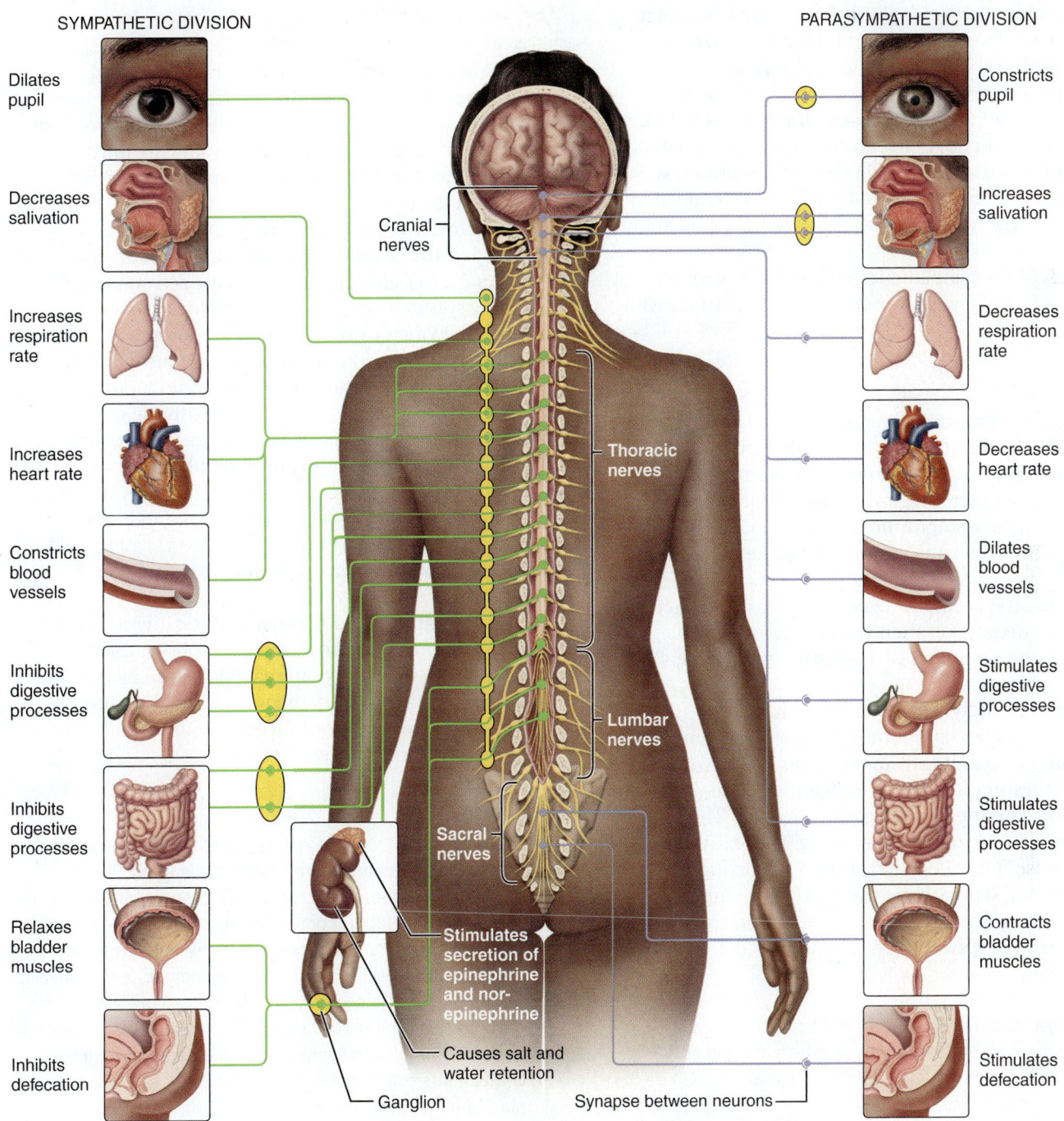

SYMPATHETIC DIVISION

PARASYMPATHETIC DIVISION

Dilates pupil

Decreases salivation

Increases respiration rate

Increases heart rate

Constricts blood vessels

Inhibits digestive processes

Inhibits digestive processes

Relaxes bladder muscles

Inhibits defecation

Cranial nerves

Thoracic nerves

Lumbar nerves

Sacral nerves

Stimulates secretion of epinephrine and nor-epinephrine

Causes salt and water retention

Ganglion

Synapse between neurons

Constricts pupil

Increases salivation

Decreases respiration rate

Decreases heart rate

Dilates blood vessels

Stimulates digestive processes

Stimulates digestive processes

Contracts bladder muscles

Stimulates defecation

Figure 11.12 The sympathetic and parasympathetic divisions of the autonomic nervous system. The parasympathetic division predominates in the body's relaxed state. The sympathetic division is associated with fight-or-flight responses. The nerves of the parasympathetic division arise from cranial and sacral segments, whereas the nerves of the sympathetic division arise in the thoracic and lumbar segments of the spinal cord. Ganglia (groups of nerve cell bodies) are shown in yellow.

form synapses with the postganglionic cells. Because these ganglia are connected to each other, the sympathetic division tends to produce a unified response in all organs at once. Axons of the postganglionic neurons branch out to innervate all the organs. The neurotransmitter released by sympathetic postganglionic neurons in the target organs is *norepinephrine.*

The sympathetic division transmits signals that prepare the body for emergencies and situations requiring high levels of mental alertness or physical activity, such as fighting or running away from danger ("fight or flight"), and even play and sexual activity. It increases heart rate and respiration, raises blood pressure, and dilates the pupils—effects that help you detect and respond quickly to changes in your environment. The sympathetic division also reduces blood flow to organs that do not help you cope with an immediate crisis, such as the intestines and kidneys, and inhibits less important body functions, such as digestion and production of saliva (which is why your mouth feels dry when you are anxious).

The effects of the sympathetic division are sometimes called the *fight-or-flight response,* reflecting its association with emergency situations.

The parasympathetic division predominates during relaxation

Preganglionic neurons of the **parasympathetic division** originate either in the brain, becoming part of the outflow of certain cranial nerves, or from the sacral region of the spinal cord. The ganglia where the preganglionic neurons synapse with the postganglionic neurons is generally some distance from the CNS and may even be in the target organ itself. The neurotransmitter released by parasympathetic postganglionic neurons at their target organs is acetylcholine, the same neurotransmitter used at the neuromuscular junction.

In contrast to the sympathetic division, the parasympathetic division predominates in situations in which you are relaxed. It transmits signals that lower heart rate and respiration, increase digestion, and permit defecation and urination, among other actions. The parasympathetic division exerts calming, restorative effects ("rest and repose") that counteract the fight-or-flight stimulation of the sympathetic division. Parasympathetic nerves are also responsible for the vasodilation that causes erection in males and swelling of the labia and erection of the clitoris in females.

☑ Suppose a person took a drug that blocked the release of norepinephrine from all peripheral nerves. Which division of the peripheral nervous system would be affected, and what would be some of the likely symptoms?

The sympathetic and parasympathetic divisions work antagonistically to maintain homeostasis

As we have seen, the actions of the sympathetic and parasympathetic divisions oppose each other. They work antagonistically to accomplish the automatic, subconscious maintenance of homeostasis. Both systems are intimately involved in feedback loops that control homeostasis throughout the body.

An example is the control of blood pressure, described in the chapter on the heart and blood vessels (Chapter 8). Certain sensory nerves are activated by stretch receptors (called *baroreceptors*) located in various places in the vascular system. When blood pressure rises, the baroreceptors are activated, and the sensory nerves send more frequent action potentials to the brain. The brain integrates this information with the other information it is receiving. Then it decreases the number of action potentials traveling in sympathetic nerves and increases the number of action potentials traveling in parasympathetic nerves serving the heart and blood vessels. As a result, heart rate slows, blood vessels dilate, and blood pressure falls until it is back to normal.

The sympathetic and parasympathetic divisions can also cause a widespread response, such as arousing the entire body. They can even carry out many localized homeostatic control mechanisms simultaneously and independently. Table 11.2 summarizes the functions of the somatic and autonomic divisions of the PNS.

↩ **Recap** Twelve pairs of cranial nerves transmit action potentials between the brain and the head, neck, and thoracic and abdominal cavities. Thirty-one pairs of spinal nerves transmit action potentials to and from the spinal cord to all regions of the body. Sensory neurons provide information (input) to the CNS. The somatic output division of the PNS controls voluntary (conscious) and involuntary (reflex) skeletal muscle movement. The autonomic division of the PNS includes sympathetic and parasympathetic components that regulate automatic body functions to maintain homeostasis. ∎

Table 11.2 The somatic and autonomic divisions of the PNS

	Somatic division	Autonomic division	
		Sympathetic	Parasympathetic
Functions	Serves skeletal muscles	Fight or flight; arouses body to deal with situations involving physical activity, mental alertness	Relaxes body; promotes digestion and other basic functions
Neurotransmitter	Acetylcholine	Norepinephrine	Acetylcholine
Number of neurons required to reach the target cell	One	Two (preganglionic and postganglionic)	Two (preganglionic and postganglionic)

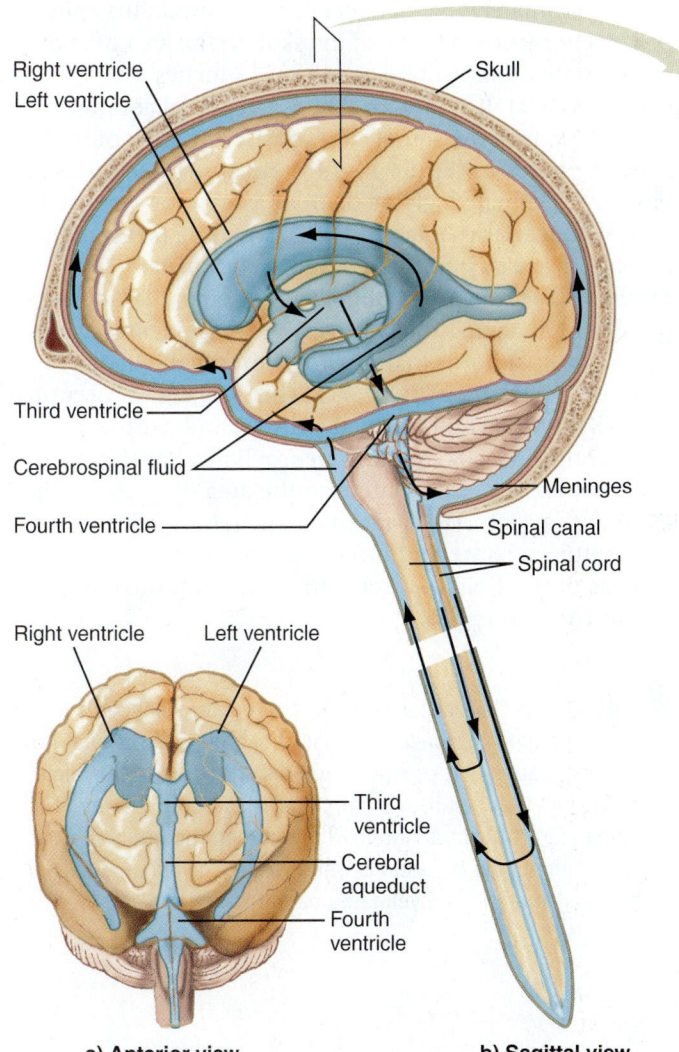

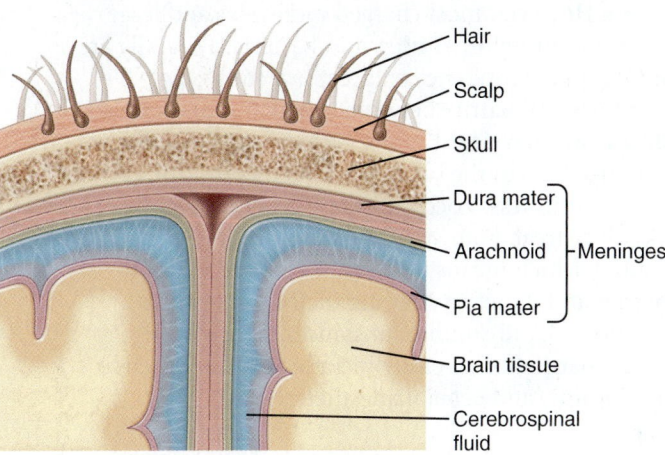

Figure 11.13 **The ventricles of the brain and circulation of cerebrospinal fluid.** Cerebrospinal fluid is secreted into the four brain ventricles (left, right, third, and fourth). From there, it flows through a series of cavities, bathing the brain and spinal cord.

a) Anterior view.   b) Sagittal view.

11.7 The brain and spinal cord constitute the CNS

A neuron does not think or love, yet humans do. Scientists are still trying to unravel the biology of emotion and thought, but we do know that they originate in the central nervous system, specifically in the brain.

The central nervous system, consisting of the brain and spinal cord, is "central" both in location and action. It is where integration, or processing, of information occurs. It is where information is received and where complex outputs are initiated. It's where billions of action potentials traveling in millions of neurons all come together as a conscious thought. That is not a very satisfactory explanation for "thinking," but it may be the best we can do.

Bone, meninges, and the blood-brain barrier protect the CNS

The CNS is vital to proper functioning of the entire body, so it is no surprise that the CNS is well protected against physical injury and disease.

First, the CNS is protected by bone. The brain is encased in the skull, and the spinal cord is enclosed in a hollow channel within the vertebrae. This shell of bone around the central nervous system helps shield it from physical trauma.

Second, the CNS is enclosed by three membranes of connective tissue, called **meninges,** named (from outermost to innermost layers) the *dura mater,* the *arachnoid,* and the *pia mater.* These three meninges protect the neurons of the CNS and the blood vessels that service them.

Third, the CNS is bathed in its own special liquid, called **cerebrospinal fluid,** which fills the space between the arachnoid and the pia mater. In addition to serving as a liquid shock absorber around the brain and spinal cord, cerebrospinal fluid tends to isolate the central nervous system from infections. Inflammation of the meninges due to a bacterial or viral infection is called *meningitis.* Meningitis can be serious because it may spread to the nervous tissue of the CNS.

Cerebrospinal fluid is somewhat similar to the interstitial fluid that bathes all cells, but it does not exchange substances as freely with blood. It is secreted from specialized capillaries into four hollow cavities (called *ventricles*) in the brain. From the ventricles, it flows through a system of fluid-filled cavities and canals, bathing and surrounding the neurons in the brain and spinal cord (**Figure 11.13**).

Most capillaries in the body are fairly leaky because there are narrow slits between adjacent capillary cells. However, the cells of the capillaries that produce cerebrospinal fluid are fused tightly together. Consequently, substances must pass through rather than between the capillary cells in order to get from blood to cerebrospinal fluid. Lipid-soluble substances such as oxygen and carbon dioxide can diffuse easily through the lipid bilayer of the cell membrane. Certain important molecules such as glucose are actively transported. However, most charged molecules and large objects, such as proteins, viruses, and bacteria, generally are prevented from entering the cerebrospinal fluid.

This functional barrier between blood and brain is called the **blood-brain barrier.** Thanks to this isolation, bacterial and viral infections of the brain and spinal cord are rare, but when these illnesses occur they are generally serious and difficult to treat. Many of our best antibiotics are not lipid soluble, which means they cannot cross the blood-brain barrier and therefore they are ineffective against CNS infections. Lipid-soluble substances that cross the blood-brain barrier rather easily include alcohol, caffeine, nicotine, cocaine, and general anesthetics.

☑ What are the major differences in fluid composition between cerebrospinal fluid and interstitial fluid elsewhere in the body? Also, do you think antibiotics would be effective in fighting an infection in the central nervous system? Why or why not?

The spinal cord relays information

The **spinal cord** serves as the superhighway for action potentials traveling between the brain and the rest of the body. As we have seen, the spinal cord can also process and act on certain information on its own via spinal reflexes, without necessarily involving the brain. However, its power to act independently is limited, and such actions never reach the level of conscious awareness.

Approximately the diameter of your thumb, the spinal cord extends from the base of the skull to the area of about the second lumbar vertebra, or about 17 inches (Figure 11.14a). It is protected by the vertebral column. The outer portions of the spinal cord consist primarily of bundles of axons, or nerves (Figure 11.14b). Because these axons are generally myelinated, giving them a whitish appearance, the areas of the cord occupied by these ascending (sensory) and descending (motor) nerves are called *white matter* (Figure 11.14c). Neurons of the PNS enter and leave the spinal cord at regular intervals via the dorsal (sensory) and ventral (motor) horns that fuse to form spinal nerves.

Near the center of the spinal cord is a region occupied primarily by the cell bodies, dendrites, and axons of neurons of the CNS, and also neuroglial cells. These structures are not myelinated, so the area they occupy is referred to as *gray matter* (see Figure 11.14c). Within the gray matter, sensory and motor neurons synapse with neurons of the CNS that transmit signals up the spinal cord to the brain.

↺ **Recap** The CNS consists of the brain and spinal cord. It is well protected by bone, the meninges, and cerebrospinal fluid. The capillaries that produce cerebrospinal fluid have special properties that result in a functional blood-brain barrier. The spinal cord transmits information between the brain and the rest of the body. It contains white matter (myelinated axons) and gray matter (unmyelinated neural structures). ∎

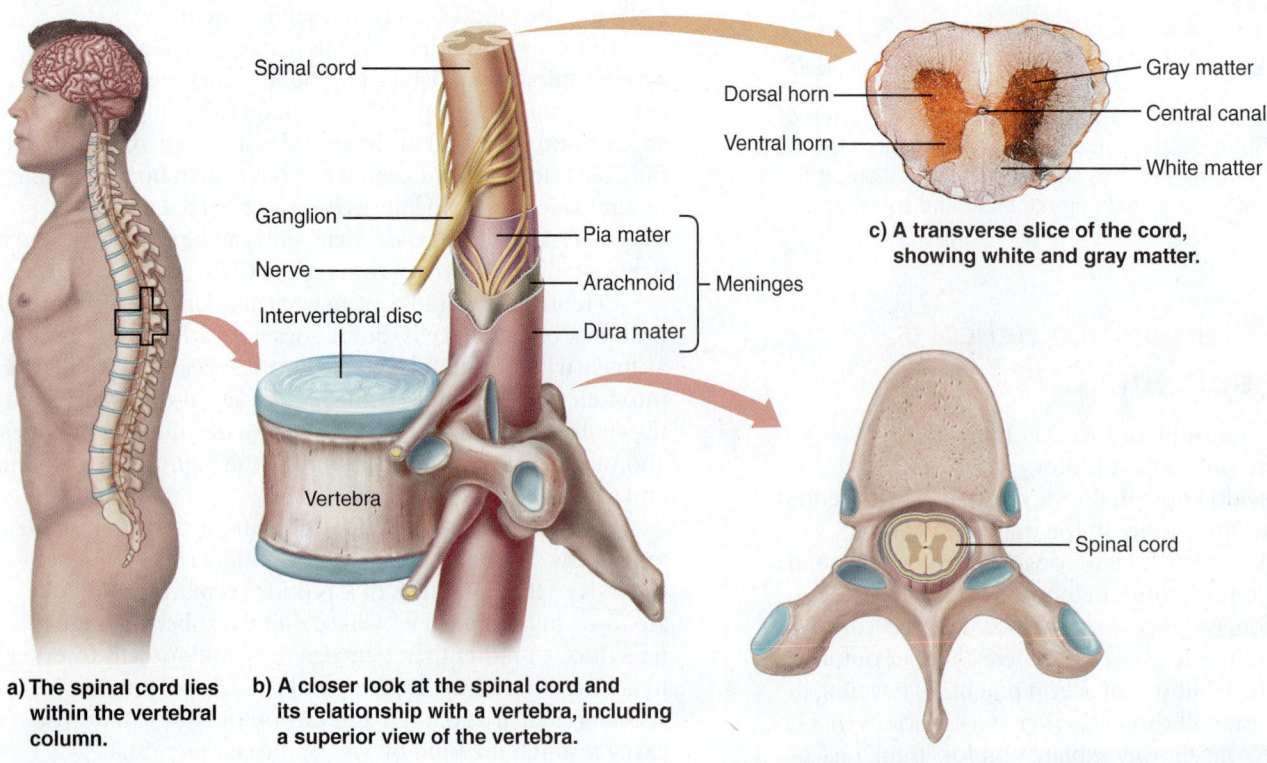

Dorsal horn
Ventral horn
Gray matter
Central canal
White matter

c) A transverse slice of the cord, showing white and gray matter.

Spinal cord
Ganglion
Nerve
Intervertebral disc
Pia mater
Arachnoid   } Meninges
Dura mater
Vertebra
Spinal cord

a) The spinal cord lies within the vertebral column.

b) A closer look at the spinal cord and its relationship with a vertebra, including a superior view of the vertebra.

Figure 11.14 The human spinal cord.

11.8 The brain processes and acts on information

The **brain** is the command center of the body. The brain receives information in the form of action potentials from various nerves and the spinal cord, integrates that information, and generates the appropriate response.

Within the brain, certain areas specialize in integrating different types of information or performing certain motor tasks. Action potentials derived from various receptors, such as light and sound, arrive at the brain in different sensory neurons that specialize in processing those signals. For example, one area of the brain responds to visual stimuli and another area responds to stimuli that help us maintain our balance.

Three major anatomical and functional divisions of the brain have been identified:

- The *hindbrain* coordinates basic, automatic, and vital tasks.
- The *midbrain* helps coordinate muscle groups and responses to sights and sounds.
- The *forebrain* receives and integrates sensory input from the external environment and determines most of our more complex behaviors.

Figure 11.15 illustrates these major divisions and their components.

The hindbrain: movement and automatic functions

The **hindbrain** is connected to the spinal cord. From an evolutionary standpoint, the hindbrain is the oldest, most primitive brain division and the one most similar among animals. Important components of the hindbrain include the medulla oblongata, the cerebellum, and the pons.

The medulla oblongata controls automatic functions

The **medulla oblongata** connects to the spinal cord. It contains areas that control vital automatic functions of internal organs. Within the medulla oblongata, the *cardiovascular center* regulates heart rate and blood pressure, and the *respiratory center* integrates information about blood levels of oxygen and carbon dioxide, and adjusts respiration accordingly. Other areas coordinate reflexes such as coughing, vomiting, sneezing, and swallowing.

All information passing between the higher areas of the brain and the spinal cord must pass through the medulla oblongata. An interesting feature of the way the body is "wired" is that the motor nerves from one side of the forebrain cross over to the other side of the body in the medulla oblongata. As a result, the left side of the brain controls the right side of the body, and vice versa.

The cerebellum coordinates basic movements

Located just behind the medulla oblongata, the **cerebellum** coordinates basic body movements that are below the level of conscious control. For example, the cerebellum ensures that antagonistic muscles do not contract at the same time. The cerebellum also stores and replicates (without our conscious thought) whole sequences of skilled movements, such as tying a shoe, shifting a car, or hitting a home run (hitters sometimes say they just "pull the trigger" and the rest happens automatically). To coordinate these skilled activities, the cerebellum receives sensory input from many sources, including joint and muscle receptors, balance and position receptors in the ear, and visual receptors.

Excessive consumption of alcohol acts on the cerebellum and disrupts its normal functions. This is why people who are drunk can't coordinate their movements or walk a straight line.

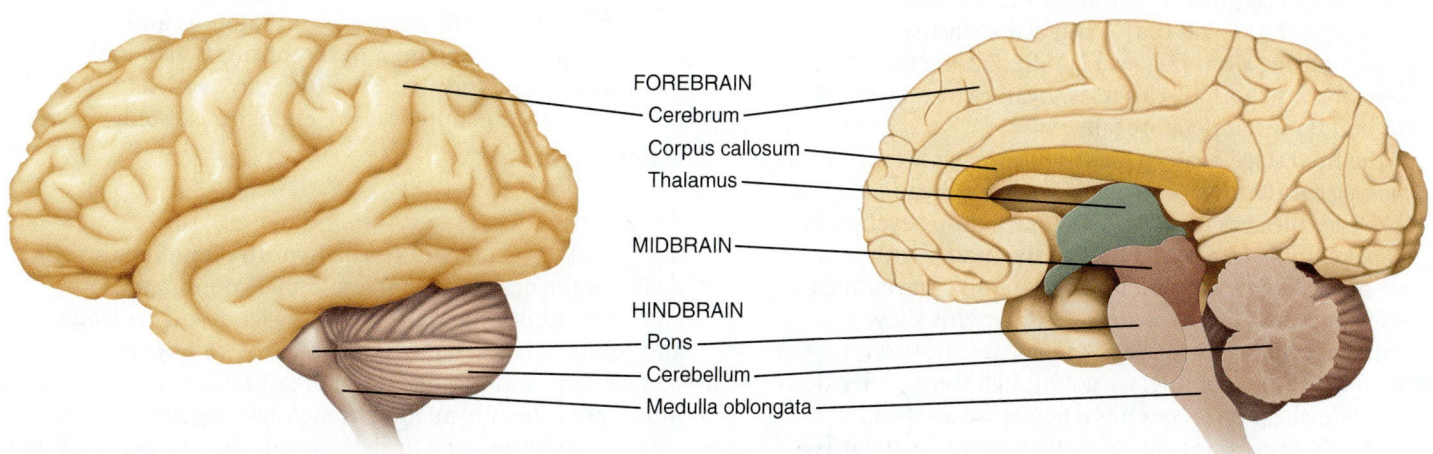

FOREBRAIN
Cerebrum
Corpus callosum
Thalamus

MIDBRAIN

HINDBRAIN
Pons
Cerebellum
Medulla oblongata

Figure 11.15 The human brain. a) A surface view. b) A midline sagittal section (front to back on a vertical plane).

The pons aids information flow Just above and partially surrounding the medulla oblongata, the pons (from the Latin for "bridge") connects higher brain centers and the spinal cord. The pons contains groups of axons that extend from the cerebellum to the rest of the CNS and is especially important in coordinating the flow of information between the cerebellum and higher brain centers. In addition, the pons aids the medulla oblongata in regulating respiration.

✓ If a person's medulla oblongata were temporarily rendered nonfunctional for just one hour, do you think he could survive? How about if the cerebellum were rendered nonfunctional for an hour? Explain.

The midbrain: vision, hearing, and sleep/wakefulness

Visual and auditory sensory inputs from the eyes and ears pass through the **midbrain** before being relayed to higher brain centers. In turn, the midbrain coordinates movements of the head related to vision and hearing, such as turning toward a sudden sound or flash of light, and it controls movement of the eyes and size of the pupils. The midbrain also monitors the unconscious movement of our skeletal muscles so that their actions are smooth and coordinated.

Several groups of neurons in the midbrain play a pivotal role in controlling our daily cycles of sleep/wakefulness. One group of neurons transmits a steady stream of action potentials to the cerebrum, keeping us awake and alert. Another group of neurons comprise what is called the "sleep center"; they release the neurotransmitter serotonin, which induces sleep by inhibiting the neurons that arouse the brain.

Why do we sleep? This may seem like a simple question, but scientists are still looking for the answer. The obvious hypothesis is that we sleep to get rest. However, measurements of brain electrical activity, called an *electroencephalogram* (EEG), show that compared to wakefulness (Figure 11.16a), brain activity actually increases during sleep (Figure 11.16b). During the deepest phases of sleep, our muscles relax, heart rate and respiration slow, and body temperature falls.

The final stage of sleep before we awaken is called *rapid eye movement* (REM) sleep. This is when we dream. During REM sleep, the eyes move rapidly under closed eyelids. Heart rate, respiration, and blood flow to the brain increase but most of the body's skeletal muscles go limp, preventing us from responding to our dreams. EEG patterns show a level of brain activity comparable to that of wakefulness (Figure 11.16c). Typically, we go through three to five deep sleep/REM sleep cycles per night before we awaken.

Although it is not clear exactly why we sleep, it is clear that we need a certain amount of sleep. Persons

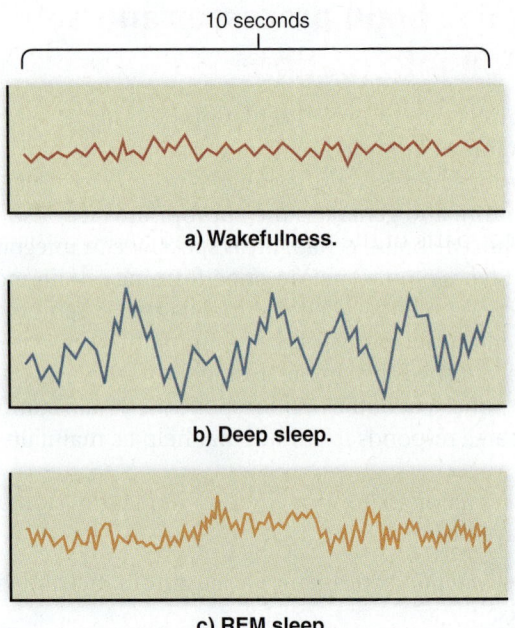

a) Wakefulness.

b) Deep sleep.

c) REM sleep.

Figure 11.16 Brain patterns of electrical activity during wakefulness and sleep.

deprived of REM sleep become moody, irritable, and depressed and may even suffer hallucinations if deprived long enough.

↺ **Recap** The brain remains active even during sleep. Stages of sleep have been correlated with brain wave patterns. The reticular activating system controls levels of sleep and wakefulness. ∎

The forebrain: emotions and conscious thought

The **forebrain** determines our most complex behavior, including emotions and conscious thought. Important areas in the forebrain are the hypothalamus and thalamus, limbic system, and cerebrum. It also includes two glands, the pineal and the pituitary, which are described in the chapter on the endocrine system (Chapter 13).

The hypothalamus and thalamus maintain homeostasis and process information The **hypothalamus** is a small region at the base of the forebrain that coordinates some automatic functions of the pituitary gland, such as water and solute balance, temperature control, carbohydrate metabolism, and production of breast milk. It plays an important role in regulating homeostasis because it monitors sensory signals relating to sight, smell, taste, noise, and body temperature. A "hunger center" located in the hypothalamus alerts us when we need to eat, and a "thirst center" signals us to drink something when body fluids become too concentrated.

The larger **thalamus**, located just above the hypothalamus, is primarily a receiving, processing, and transfer center. It integrates and relays some outgoing motor activity as well. The thalamus accepts sensory signals from other parts of the body and brain and channels them to the cerebrum to be interpreted. As sensory information—for example, pain, pressure, and temperature changes—arrives at the thalamus, we are conscious of the sensation but cannot identify where it originates. After the information reaches the cerebrum, we become aware of which part of the body is experiencing the sensation.

The limbic system is involved in emotions The name *limbic system* was originally used to describe those structures that bordered the basal regions of the cerebrum. Functionally, the term describes all of the neuronal structures that, together, control emotional behavior and motivational drives (**Figure 11.17**). When different areas of the limbic system are stimulated, we experience strong feelings and emotions such as fear, anger, sorrow, pleasure, or love. These feelings and emotions influence our behavior, particularly self-gratifying behaviors such as satisfying hunger, thirst, and sexual desire.

The intensity with which we feel emotions and the extent to which we act on them depends on modification by the cerebrum. To appreciate the importance of cerebral modification, just imagine what would happen if our most basic urges were no longer influenced by social rules or codes of conduct!

The cerebrum deals with higher functions In humans, the large **cerebrum** (**Figure 11.18**) is the most highly developed of all brain regions. It deals with many of the

functions that we associate with being human: language, decision making, and conscious thought.

The outer layer of the cerebrum is called the **cerebral cortex** (see Figure 11.18a). The cerebral cortex is primarily

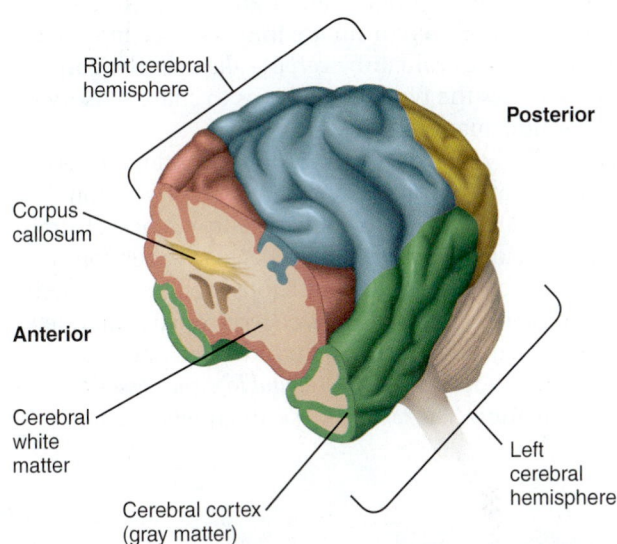

a) **The cerebral cortex (gray matter; outer layer, shown in four colors) consists of interneurons that integrate and process information.** White matter (inner core) consists of ascending and descending nerve tracts. The two separate hemispheres are joined by the corpus callosum.

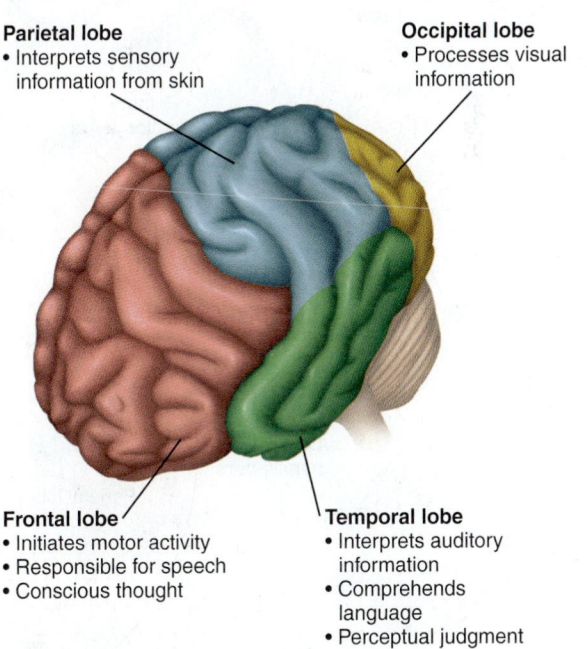

Parietal lobe
• Interprets sensory information from skin

Occipital lobe
• Processes visual information

Frontal lobe
• Initiates motor activity
• Responsible for speech
• Conscious thought

Temporal lobe
• Interprets auditory information
• Comprehends language
• Perceptual judgment

b) **The functions of the four lobes of the cerebral cortex are location-specific.**

Figure 11.18 The cerebrum.

✔ Explain how a blow to the back of the head can sometimes cause blindness, even if the eyes are undamaged.

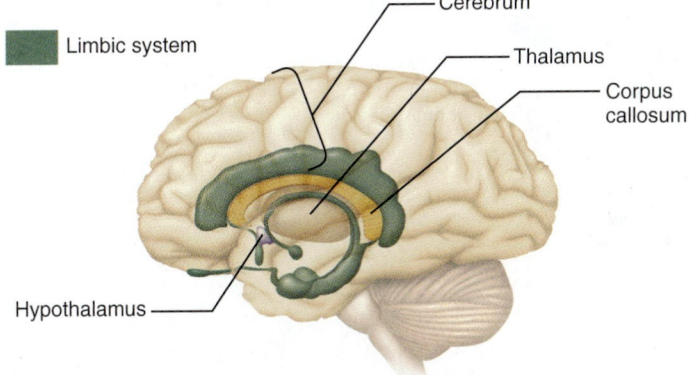

Figure 11.17 The limbic system. The limbic system is a collection of neuronal areas near the base of the cerebrum that are involved in emotions and basic behaviors.

gray matter, consisting of neurons with unmyelinated axons and their associated neuroglial cells. It has numerous folds on its surface, like a walnut. The inner portion of the cerebrum is mostly white matter, containing myelinated nerve axons that connect lower brain areas to the cerebral cortex. The left and right halves (called hemispheres) of the cerebrum are joined in the middle by a network of nerves called the *corpus callosum*. The corpus callosum enables the two hemispheres to share sensory and motor information.

The cerebral cortex receives, integrates, and interprets sensory information from every organ and every region of the body. It is also responsible for memory storage, abstract thought, conscious awareness, and even conscious control of skeletal muscle. All of these functions are location-specific (see Figure 11.18b). The *parietal lobe* receives and interprets sensory information from the skin, such as temperature, touch, pain, and pressure. The *occipital lobe* processes visual information. The *temporal lobe* interprets auditory information, comprehends spoken and written language, and is responsible for perceptual judgment. Finally, toward the front of the cortex is the *frontal lobe*, perhaps the most highly developed of all brain regions. The frontal lobe initiates motor activity and is also responsible for speech and conscious thought. All four lobes are involved in memory storage.

Two areas of the brain that have been studied extensively are the region in the parietal lobe that receives sensory input from the skin (called *the primary somatosensory area*) and the corresponding region of the frontal lobe that initiates motor activity (called the *primary motor area*). Researchers have found very specific regions within each of these areas corresponding to specific parts of the body (**Figure 11.19**). Body parts that are extremely sensitive, such as the lips and genitals, involve more neurons and consequently larger portions of the primary somatosensory area, whereas less sensitive body parts, such as a thigh, involve fewer neurons and smaller portions of the primary somatosensory area.

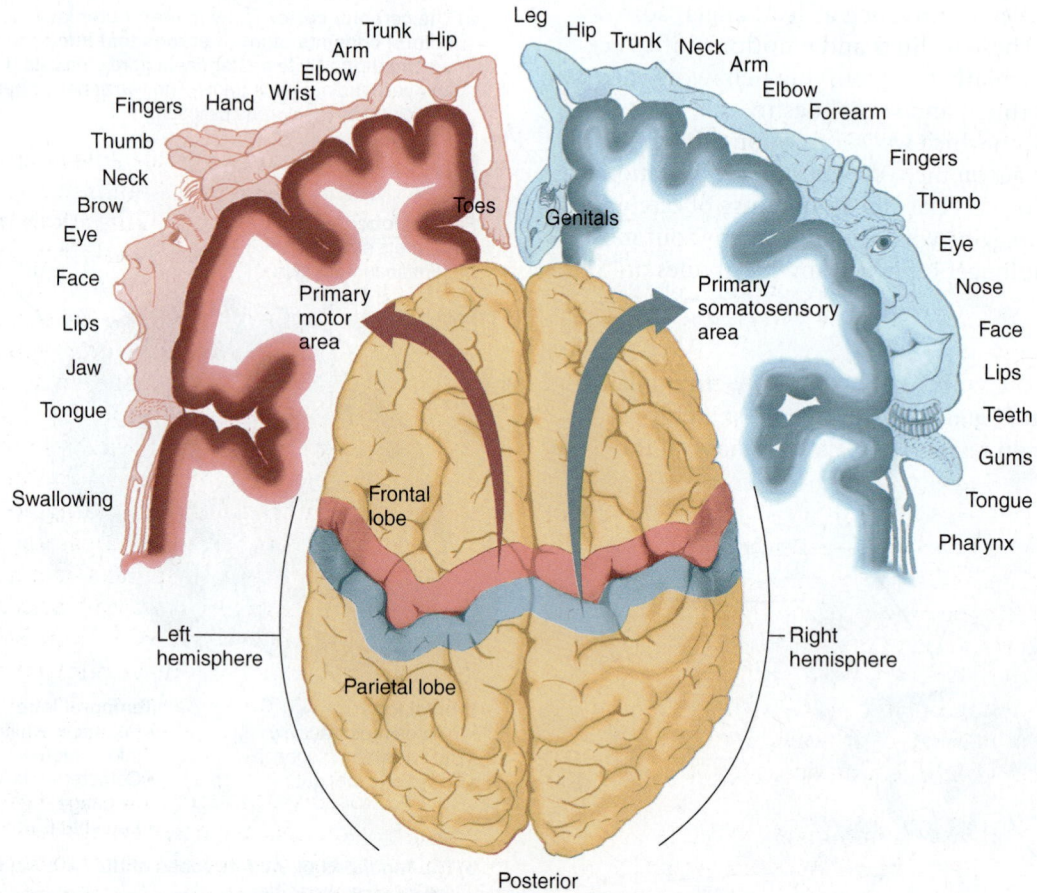

Figure 11.19 Primary somatosensory and motor areas of the cerebral cortex. Note the close similarity of the maps of the two areas. The larger the anatomical representation on the diagram, the more precisely controlled or more sensitive the area. The lips, face, and tongue are particularly well represented.

Similarly, body parts that can perform intricate and precise movements, such as the fingers, involve more neurons in the primary motor area than do less dexterous regions, such as an elbow.

The location-specific nature of information processing explains why damage to different brain regions can cause such varied, seemingly contradictory, effects. This is often apparent in stroke victims. One stroke patient might be unable to speak yet still be fully capable of understanding speech. Another could have trouble controlling basic skeletal muscle movements but retain the capacity for abstract thought. Physicians use their understanding of these known location-specific functions to diagnose which areas of the brain have been affected.

Recap The brain has three divisions: hindbrain, midbrain, and forebrain. The evolutionarily oldest, more primitive hindbrain controls most automatic activities. The midbrain functions in vision, hearing, and movement. Structures in the forebrain help maintain homeostasis, process information, control emotions and memory, and are responsible for language, decision making, and conscious thought. ∎

11.9 Memory involves storing and retrieving information

Memory involves storing information and retrieving it later, as needed. Memory has two stages: Short-term ("working") memory involves retrieval of information that was stored within the past few hours and long-term memory is the ability to retrieve information days or years later.

There is a clear difference between how the brain manages the two stages of memory. Short-term memory occurs in the limbic system. When you receive new sensory information (such as reading an unfamiliar telephone number), the stimulus triggers a quick burst of action potentials in the limbic system. You can remember the number for a few seconds or minutes, because neurons that have just fired are easier to fire again in the short term. The pathway becomes entrained, if only for a brief time. If the number is not important, you quickly forget it because the enhancement of the limbic pathway is not long lasting. That particular piece of information does not move into long-term memory.

However, if the number is important to you, as evidenced by the fact that you say it out loud several times, write it down and then read it to yourself, or use it repeatedly over several weeks, the information may be transmitted to your cerebral cortex for storage in long-term memory centers. During this process, neurons undergo a permanent chemical or physical change. Evidence from experimental animals suggests that long-term memory storage creates additional synapses between connecting neurons, enabling those circuits to be activated more easily in the future.

Recap Short-term memory involves quick bursts of action potentials in the limbic system. Long-term memory resides in the cerebral cortex and involves permanent changes in neurons and synapses. ∎

11.10 Psychoactive drugs affect higher brain functions

A drug is any substance introduced into the body for the purpose of causing a physiological change. Psychoactive drugs are those that affect states of consciousness, emotions, or behavior—in other words, the higher brain functions. Some psychoactive drugs are legal, such as alcohol and nicotine. Others are not legal but have found their way into the fabric of our society.

Most drugs, including psychoactive substances, were originally discovered or designed to treat disease. The nonmedical use of drugs is sometimes euphemistically called "recreational" use, but the term *illicit* is certainly a more accurate reflection of the truth.

All psychoactive drugs are able to cross the blood-brain barrier (not all drugs can do so). They work by influencing the concentrations or actions of brain neurotransmitters. Recall that normally the effects of neurotransmitters are short-lived because the neurotransmitter remains in the synaptic cleft for only a brief time. Psychoactive drugs either bind to neurotransmitters or affect their release, action, or reuptake (reabsorption) in some way. This changes normal patterns of electrical activity in the brain.

Depending on which neurotransmitter they influence and where they act, psychoactive drugs can exert a variety of effects. For example, dopamine is one of the most important neurotransmitters in areas of the brain associated with pleasure. Cocaine blocks the reuptake of dopamine back into the presynaptic bulb, leading to too much dopamine in the synaptic cleft and therefore excessive stimulation of postsynaptic neurons. Cocaine initially allows the user to experience pleasurable effects longer than normal.

Psychoactive drugs can lead to *psychological dependence*, meaning that users begin to crave the feelings associated with a drug and alter their behavior to obtain it. Eventually, tolerance develops—they need to use more of it, or use it more frequently, to achieve the same effect. Tolerance occurs in part because the liver produces enzymes that detoxify many drugs, and repeated drug use increases circulating levels of these detoxifying enzymes. Soon, the original dose is no longer enough to create the desired effect, or sometimes any effect at all. A second reason for tolerance, in the case of cocaine, is that if too much dopamine remains in the synaptic cleft for prolonged periods, the body starts to produce less of it. So with repeated cocaine use, a person needs more and more drug to maintain even a normal amount of dopamine neurotransmitter in the synaptic cleft. Eventually, cocaine addicts become incapable of experiencing pleasure without the drug and cannot even maintain a normal emotional state. At this point, severe anxiety and depression set in.

Tolerance leads to **addiction,** the need to continue obtaining and using a substance despite one's better judgment and good intentions. Essentially, choice is lost. Addiction has serious negative consequences for both emotional and physical health. Eventually, most addicts find it impossible to concentrate and hold a steady job. At the same time, they must spend large amounts of money to support their habit. Personal relationships suffer, and many addicts lose their families, their employment, and their physical health. In addition, they face the probability of **withdrawal,** the symptoms that occur if the drug is removed suddenly.

Cocaine and nicotine are both highly addictive. Alcohol is addictive in some people but not in others, for reasons we do not yet fully understand. One of the most addictive street drugs is crystal methamphetamine (also known as "crank," "ice," "tina," or "glass").

The federal government keeps track of illicit drug use through an annual survey. **Figure 11.20** shows reported illicit drug use during the month prior to the survey by persons

aged 12 and older. Illicit drug use peaks in 18- to 20-year-olds and then declines with age. The most commonly used illicit drugs in 18- to 25-year-olds are marijuana and the nonmedical use of psychotherapeutic drugs (narcotic and opioid pain relievers, sedatives, tranquilizers, stimulants, antidepressants, and antianxiety drugs). Less commonly used are cocaine and hallucinogens (primarily ecstasy, or MDMA).

Societal views about marijuana seem to be undergoing a shift. The federal government classifies marijuana as a Schedule 1 drug right along with heroin and ecstasy, meaning that it is considered to have a high potential for abuse and no currently accepted medical use. Not everyone agrees, apparently. Twenty-three states now have laws that permit the medical use of marijuana (that is, by prescription). By 2015, four states (Colorado, Washington, Alaska, and Oregon) had legalized the recreational use of marijuana as well. This presents a dilemma for the federal government, which for now is taking a wait-and-see enforcement approach in states in which marijuana is legal. It is entirely possible that someday marijuana will be taken off the list of illicit drugs.

Recap Psychoactive drugs work by altering either the concentration or action of brain neurotransmitters. Some lead to psychological dependence, tolerance, and addiction. ■

11.11 Disorders of the nervous system

Because the nervous system is so complex and controls so many body functions, disorders of the nervous system can be particularly debilitating. We have already discussed stroke in the context of cardiovascular disease. Here, we discuss several classes of disorders that affect the nervous system, including trauma, infections, disorders of neural or synaptic transmission, and brain tumors.

Trauma

Trauma (physical injury) to either the brain or spinal cord can be particularly dangerous because the brain controls so many body functions and the spinal cord is the information pathway to virtually all the organs.

Concussion disrupts electrical activity A violent blow to the head or neck can cause *concussion,* a brief period of unconsciousness in which electrical activity of brain neurons is briefly disrupted. After the person "comes to," he or she may experience blurred vision, confusion, and nausea and vomiting. Although most concussions are not associated with permanent injury, one of the biggest dangers is

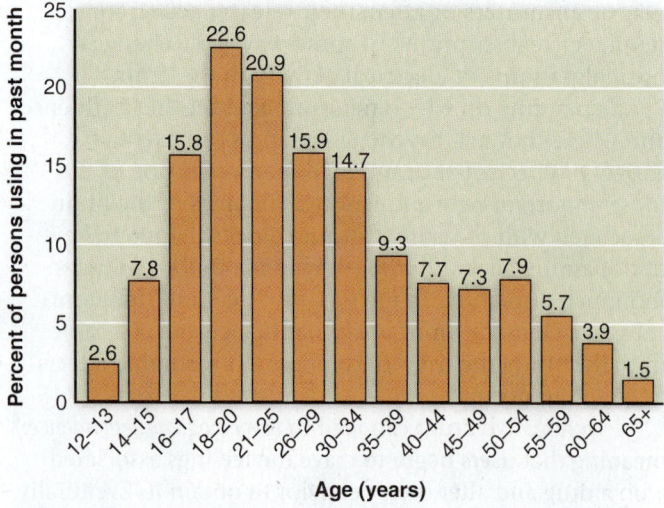

Figure 11.20 Past-month illicit drug use among persons aged 12 and older, 2012. *Source:* Substance Abuse and Mental Health Services Administration, *Results from the 2013 National Survey on Drug Use and Health: Summary of National Findings.*

a *subdural hematoma*, or hemorrhage (bleeding) into the space between the meninges.

Because the brain is encased in a rigid skull, any bleeding increases pressure within the skull. Eventually, the trapped blood presses against healthy brain tissue and disrupts function. Symptoms can include drowsiness, headache, and growing weakness of one side of the body. Depending on how rapidly bleeding occurs, symptoms may take weeks or even months to appear. Sometimes surgeons drill through the skull to drain the hematoma, relieve the pressure, and repair damaged blood vessels.

Spinal cord injuries impair sensation and function

Violent trauma to vertebrae may press against the spinal cord, tear it, or even sever it completely. Damage to the spinal cord usually impairs sensation and body function below the injured area, so the extent of injury depends on the site of damage. Injuries below the neck can cause paraplegia (weakness or paralysis of the legs and sometimes part of the trunk). Injuries to the neck may lead to quadriplegia (weakness or paralysis of the legs, arms, and trunk). The higher the spinal cord injury, the more likely it is to be fatal. Spinal cord injuries may also cause problems with bladder or bowel control because the autonomic division no longer functions properly.

Researchers are investigating many exciting treatments for spinal cord injuries. For more information, see the Health & Wellness feature, *Repairing Spinal Cord Injuries*.

Infections

The brain and spinal cord are spared most of the infections that affect the rest of the body because of the blood-brain barrier. Nevertheless, certain viruses and bacteria do occasionally infect the brain or spinal cord.

Encephalitis: inflammation of the brain
Encephalitis refers to inflammation of the brain, often caused by a viral infection. Frequent sources of encephalitis include the herpes simplex virus (which causes cold sores) and a virus transmitted through mosquito bites. HIV infection is responsible for a growing number of encephalitis cases.

Symptoms of encephalitis can vary, but often include headache, fever, and fatigue, progressing to hallucinations, confusion, and disturbances in speech, memory, and behavior. Some patients also experience epileptic seizures. Treatment varies depending on the cause, but antiviral drugs may help.

Meningitis: inflammation of the meninges
Meningitis involves inflammation of the meninges, usually due to viral or bacterial infection. Symptoms include headache, fever, nausea and vomiting, sensitivity to light, and a stiff neck.

Mild cases may resemble the flu. Viral meningitis tends to be relatively mild and usually improves within a few weeks. Bacterial meningitis can be life-threatening and requires antibiotics.

Rabies: infectious viral disease
Rabies is an infectious viral brain disease of mammals, especially dogs, wolves, raccoons, bats, skunks, foxes, and cats. It is transmitted to humans by direct contact, usually through a bite or a lick over broken skin. The virus makes its way through the sensory neurons in the region of the bite to the spinal cord and then to the brain, where it multiplies and kills cells.

Signs of rabies include swollen lymph glands, painful swallowing, vomiting, choking, spasms of throat and chest muscles, fever, and mental derangement. The symptoms generally appear one to six months after initial infection, followed by coma and death within 2 to 20 days.

If an animal has bitten you, wash the wound thoroughly and consult a physician. If possible, have the animal tested for rabies. If there is any possibility it could be rabid, your doctor will advise you to be immunized, because untreated rabies may be fatal. The rabies vaccine is most effective if administered within a few days after the bite, but it may still work when given weeks or months later.

Brain tumors: abnormal growths

A brain tumor is an abnormal growth in or on the brain. Brain tumors are not necessarily cancerous, but even noncancerous growths can cause problems if they press against neighboring tissue and increase pressure within the skull. Over time, the rising pressure disrupts normal brain function. Depending on the location and size of the tumor, symptoms include headache, vomiting, visual impairment, and confusion. Some people also experience muscle weakness, difficulty speaking, or epileptic seizures.

Some brain tumors originate from the brain itself, but most are seeded by cancer that spreads from other parts of the body. For more on cancer, see the chapter devoted to that subject (Chapter 18).

Some brain tumors can be surgically removed. If tumors are inaccessible or too large to be removed, radiation and chemotherapy treatments may shrink them or limit their growth.

MJ's BlogInFocus Whatever happened to the compounding pharmacy that was responsible for an outbreak of meningitis that killed 64 people? Visit MJ's blog in the Study Area in MasteringBiology and look under "An Outbreak of Meningitis."

http://goo.gl/Pv9Pg1

HEALTH & WELLNESS
Repairing Spinal Cord Injuries

On September 7, 2007, Buffalo Bills football player Kevin Everett suffered a fracture and dislocation of the cervical spine while making a tackle during a game against the Denver Broncos. As he was taken off the field, Kevin moved only his eyes. Two days after the accident, Kevin still could barely feel or move his arms and legs. Doctors gave him a "statistically very small" chance of ever walking again.

Injuries like Kevin Everett's generally crush the spinal cord axons rather than actually severing the cord. Although the nerves are initially intact, over the next few weeks many of the damaged neurons die or lose the ability to function.

Until recently, it was believed that damaged neurons could not regenerate. However, researchers are investigating a number of exciting possibilities for helping spinal cord patients recover lost function. Current areas of research or treatment include:

- *Limiting cell injury and death.* At least three mechanisms of cell injury and death are involved. Researchers are exploring ways to block these injury and death pathways to preserve as many neurons and oligodendrocytes as possible.
- *Controlling inflammation.* Ironically, the inflammatory process, which normally promotes healing, makes matters worse by causing secondary damage to cells that were not injured initially. Anti-inflammatory drugs such as methylprednisolone can help, especially if given within eight hours of the injury.

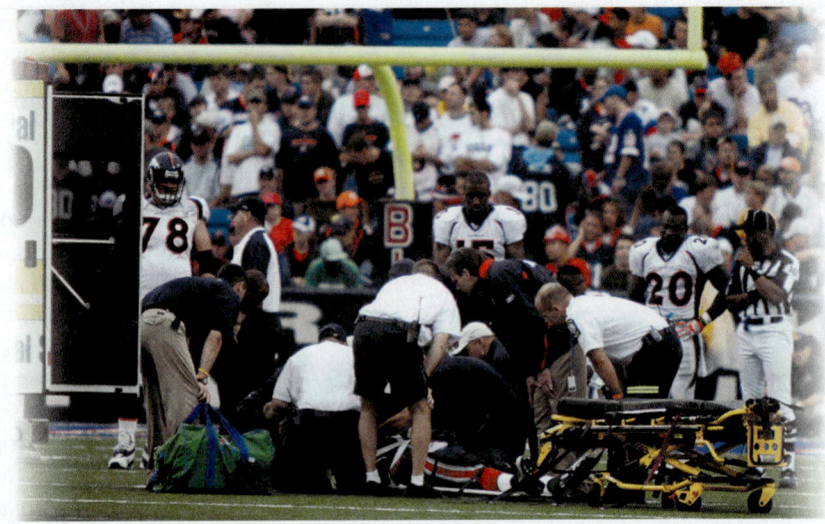

Kevin Everett suffered a fracture of the spine on September 7, 2007. He was walking again just five months later.

- *Blocking certain molecules that prevent regeneration.* The CNS of adult mammals contains inhibitory molecules that block neurons from continually growing and dividing inappropriately in healthy adults. After an injury, however, they become an obstacle to tissue repair. Scientists are working on ways to neutralize these molecules.
- *Stimulating axons to grow.* Some researchers are working with growth factors and guidance molecules to stimulate damaged axons to "sprout" new extensions across the injured area.
- *Replacing lost cells with new tissue.* Stem cells, which have the potential to differentiate into many types of cells, might be stimulated to develop into neurons and neuroglial cells for transplant into a damaged area of the spinal cord.
- *Bridging the injury site or removing barriers to regrowth.* Some researchers are using artificial materials to support and guide neurons

as they regenerate across an injured area. Other groups are experimenting with grafts of neural tissue, or even proteins that can assemble into nanofibers smaller than a human hair. Still others are experimenting with enzymes that dissolve scar tissue and allow neurons to grow through the affected area.
- *Stimulating muscles electronically.* Computer-controlled functional electrical stimulation (FES) systems help patients move by activating their muscles with electrodes.

Some of this research is beginning to pay off. Kevin Everett attended the Super Bowl on January 31, 2008, and was able to walk under his own power during an appearance on the *Oprah* show that same day. Today, 90% of people with spinal cord injuries survive and live a near-normal life span. Unfortunately, fewer than half of spinal cord injury victims are able to return to the workforce. Perhaps with further advances that number will rise.

Disorders of neural and synaptic transmission

The nervous system depends on action potentials traveling in neurons and the chemical transfer of signals from one neuron to another across a synapse. It should come as no surprise, then, that problems involving neural or synaptic transmission can cause disease. The symptoms of these disorders depend on which nerves and/or neurotransmitters are affected.

Epilepsy: recurring episodes of abnormal electrical activity
Epilepsy is a disorder characterized by transient, recurring bouts of abnormal brain cell activity. Fatigue,

stress, or patterns of flashing lights may trigger epileptic attacks, known as seizures. Seizures vary widely, because virtually any activity controlled by the brain may be affected. Some epileptic seizures are almost unnoticeable: perhaps a short episode of staring or nonpurposeful movements, a shaking limb, or repeated hand rubbing. Some seizures result in changes in the way things smell, taste, or sound to the patient or cause temporary speech disturbances. These seizures end quickly without causing any loss of consciousness. Other seizures affect larger areas of the brain and thus have more noticeable symptoms, such as sudden collapse, convulsions (stiffening and shaking of the extremities or the entire body), or temporary loss of consciousness, from which the person recovers fully after a short time. Epilepsy usually appears during childhood or adolescence. Some cases result from infections (such as meningitis or encephalitis), head injuries, brain tumors, or stroke, but often the cause is not known.

Alzheimer's disease: a buildup of abnormal proteins

Alzheimer's disease is the most common cause of *dementia*, the general term for loss of memory and other intellectual abilities serious enough to interfere with daily living. One in every 10 people over age 65 has Alzheimer's disease. There appears to be a genetic component, as a family history of the disease increases the risk. Recent studies suggest that an abnormal gene on chromosome 21 may be involved.

Alzheimer's disease begins with memory lapses and progresses to severe memory loss, particularly of recent events. Long-term memory is affected more slowly. Patients suffering from Alzheimer's disease become progressively more disoriented and confused until they may not even recognize their spouses or relatives. For relatives, the hardest part of caring for an Alzheimer patient is their change in personality. Rapid mood changes—from anger and rage to tears—may occur for no reason and at any time. The disease progresses until eventually the patient can no longer function independently.

The study of affected brains at autopsy reveal structural changes in the brain, including tangles of abnormal neurons and the development of unusually large deposits of an abnormal protein called *beta amyloid* around neurons. As the disease progresses, these abnormal proteins choke off neurons until the number of neurons and their supply of neurotransmitter (acetylcholine) becomes insufficient for normal neurotransmission. Metabolic activity in the frontal and temporal lobes declines (Figure 11.21).

Unfortunately, currently, the diagnosis of Alzheimer's disease is made only when symptoms appear, and by that time it is too late to do anything about it. Drugs that enhance the brain's production of acetylcholine can slow the progression of the disease once it is identified, but as yet there is no cure. Nevertheless, there may be hope on the horizon. Researchers are studying the possibility of identifying biomarkers (changes in the concentration of

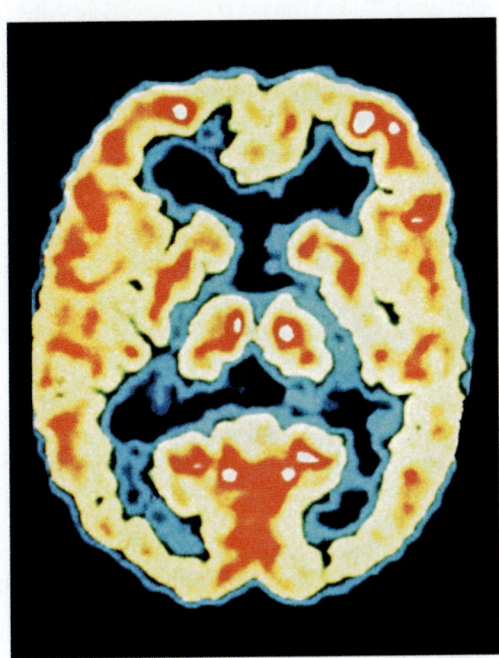

a) **A PET scan of a healthy brain.** Areas of highest brain activity appear red and yellow.

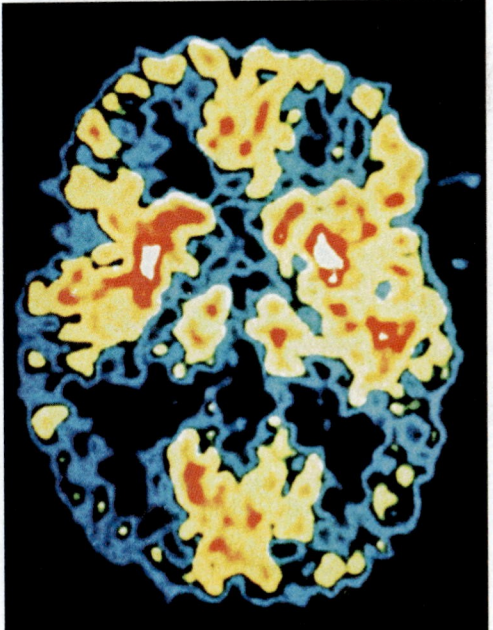

b) **A PET scan of a patient with Alzheimer's disease, showing decreased activity and an irregular pattern.**

Figure 11.21 Changes in brain metabolic activity in Alzheimer's disease. A positron emission tomography (PET) scan reveals differences in metabolic activity of tissue. Reds and yellows represent high metabolic activity, blues and greens represent lower activity.

a particular protein, for example) that would identify the very earliest stages of a buildup of beta-amyloid before the damage is sufficient for symptoms to appear. They also hope to identify drugs that could halt the progressive buildup of beta-amyloid in potential Alzheimer's sufferers while they are still asymptomatic, once it is known who is at risk or is on a path to developing the disease.

Parkinson's disease: loss of dopamine-releasing neurons Parkinson's disease is a progressive, degenerative disorder that strikes nearly 50,000 people a year in North America, most over age 55. In contrast to Alzheimer's disease, which affects primarily memory and intellectual abilities, Parkinson's disease affects primarily physical activity. Symptoms include stiff muscles and muscle tremors in the hands and feet. Eventually, people with Parkinson's lose mobility; they may also become mentally impaired and depressed.

We now know that Parkinson's disease is caused by degeneration of dopamine-releasing neurons in the area of the midbrain that coordinates muscle movement.

The shortage of dopamine impairs the ability to perform smooth, coordinated motions. The symptoms can be improved by taking L-dopa, a drug that the body converts into dopamine, but this does not slow the loss of neurons. As the disease progresses, even L-dopa becomes ineffective.

However, there is an experimental and highly controversial treatment for patients with Parkinson's disease that might replace lost cells: receiving transplants of fetal neurons directly into the brain. Experiments with animals have shown that transplanted fetal neurons will survive and grow in the recipient animal. In humans, the fetal cells would come from aborted fetuses, which is why the technique generates such controversy.

Recap Disorders of the nervous system are often serious because of the central role of the nervous system in regulating other body systems. Common disorders include physical trauma and infections, as well as acquired or genetic diseases that disrupt neural transmission in some way, such as epilepsy, Parkinson's disease, and Alzheimer's disease. ∎

Chapter Summary

11.1 The nervous system has two principal parts p. 245

- The central nervous system receives, processes, and stores information, and the peripheral nervous system provides sensory input and motor output.
- The PNS has somatic and autonomic divisions.
- The autonomic division of the PNS has sympathetic and parasympathetic divisions.

11.2 Neurons are the communication cells of the nervous system p. 246

- Neurons generally consist of a cell body, several short extensions called dendrites, and a long, thin extension called an axon.

11.3 Neurons initiate action potentials p. 247

- Incoming signals from other neurons produce short-lived graded potentials in a neuron that alter its resting potential slightly.
- If the sum of all graded potentials exceeds threshold, voltage-sensitive sodium channels open briefly and an action potential is initiated.
- The three components of an action potential are (1) depolarization, in which sodium channels open briefly and then close; (2) repolarization, in which potassium channels open briefly and then close; and (3) reestablishment of the resting potential by the continuous action of the sodium-potassium pump.

11.4 Neuroglial cells support and protect neurons p. 251

- Schwann cells produce myelin, an insulating material that increases conduction velocities in neurons.

- Multiple sclerosis is a disease in which the myelin sheath of neurons becomes damaged.

11.5 Information is transferred from a neuron to its target p. 252

- An action potential in a presynaptic neuron causes ion channels for Ca^{2+} to open in the membrane of the axon bulb.
- Ca^{2+} diffuses into the axon bulb and causes a chemical neurotransmitter to be released into the synaptic cleft.
- The neurotransmitter binds to receptors on the postsynaptic cell membrane, causing a graded potential in the postsynaptic cell.
- Information processing depends on differences in stimulus intensity and on converging and diverging neural connections.

11.6 The PNS relays information between tissues and the CNS p. 254

- The peripheral nervous system consists of 12 pairs of cranial nerves and 31 pairs of spinal nerves. Each nerve may contain a mixture of sensory, somatic motor, and autonomic motor nerve axons.
- The somatic motor division of the peripheral nervous system controls skeletal muscle.
- The autonomic motor division of the peripheral nervous system controls automatic functions and helps to maintain homeostasis.

11.7 The brain and spinal cord constitute the CNS p. 259

- The brain and spinal cord are protected by bone, by three layers of connective tissue called meninges, and by cerebrospinal fluid.
- The spinal cord consists of ascending (going to the brain) and descending (coming from the brain) nerves, plus neurons and neuroglial cells of the CNS.

11.8 The brain processes and acts on information *p. 261*

- The brain receives sensory information, integrates and stores it, and generates the appropriate response.
- The medulla oblongata contains areas that control blood pressure and respiration.
- The cerebellum coordinates basic and certain skilled body movements.
- The brain exhibits characteristic cyclic patterns of activity during sleep.
- The functional importance of sleep is unknown.
- The hypothalamus coordinates certain automatic functions and participates in the maintenance of homeostasis.
- The limbic system is the center of emotions such as fear, anger, sorrow, love, and basic behaviors such as seeking food or sexual gratification.
- The cerebrum is involved in language, decision making, and conscious thought. It is also the site of long-term memory storage.

11.9 Memory involves storing and retrieving information *p. 265*

- Short-term memory resides in the limbic system. Long-term memory involves permanent changes to neurons in the cerebral cortex.

11.10 Psychoactive drugs affect higher brain functions *p. 265*

- Psychoactive drugs act either by altering the concentrations of brain neurotransmitters or by altering the responses of postsynaptic neurons to them.
- Addiction to psychoactive drugs is characterized by the compulsion to use the drug, often at increasing doses, despite good intentions not to use and the likelihood of negative consequences.

11.11 Disorders of the nervous system *p. 266*

- Head trauma can be dangerous because any bruising or bleeding within the confined space of the skull presses against the brain.
- Spinal cord injury can interrupt neural signals traveling up and down the spinal cord, resulting in a loss of sensation and paralysis of muscle.
- Rabies is an infectious viral disease that is transmitted by the bite of an infected animal. The virus travels to the spinal cord and brain via the axon of a sensory nerve.
- Alzheimer's disease is characterized by a decline in the amount of the neurotransmitter acetylcholine in the limbic system and frontal lobe. Although the cause of Alzheimer's is unknown, the disease is associated with deposits of an abnormal protein in the brain.

Terms You Should Know

action potential, *248*
autonomic division, *256*
axon, *247*
blood-brain barrier, *260*
central nervous system (CNS), *245*
cerebellum, *261*
cerebral cortex, *263*
cerebrospinal fluid, *259*
cerebrum, *263*
dendrites, *247*
forebrain, *262*
graded potential, *248*
hindbrain, *261*
interneurons, *247*
midbrain, *262*
motor neurons, *247*
myelin sheath, *251*
nerve, *254*
neuroglial cells, *251*
neurons, *246*
neurotransmitter, *252*
parasympathetic division, *258*
peripheral nervous system (PNS), *245*
resting potential, *248*
sensory neurons, *246*
somatic division, *255*
synapse, *252*
sympathetic division, *256*
threshold, *248*

Concept Review

Answers can be found in the Study Area in MasteringBiology.

1. Distinguish between the central nervous system and the peripheral nervous system.
2. Explain what is meant by summation of graded potentials, and explain how summation may lead to an action potential.
3. Describe, in terms of the opening and closing of ion channels and the subsequent diffusion of ions, how the depolarization and repolarization phases of an action potential take place.
4. Describe the functions of neuroglial cells.
5. Describe the role of neurotransmitters in the function of neurons.
6. List the three major divisions of the brain and identify their general functional roles.
7. Explain what is meant by the statement that the functions of the cerebral cortex are "location-specific."
8. Define *REM sleep* and indicate the physiological changes that might be observed during REM sleep.
9. Describe why a subdural hematoma is so much more dangerous than a hematoma in a muscle (known commonly as a bruise).
10. Explain why an injury to the spinal cord may impair sensations and motor movement below, but not above, the point of injury.

Test Yourself

Answers can be found in the Appendix.

1. Which of the following statements correctly characterizes an action potential?
   a. Action potentials are initiated by depolarization of the membrane to threshold.
   b. Action potentials reverse the membrane potential so that the interior is negatively charged and the exterior is positively charged.
   c. A stronger action potential will travel faster than a weaker action potential.
   d. Action potentials are more likely to result when the membrane is hyperpolarized.

2. Which of the following influence(s) the speed of an action potential?
   a. the presence of a myelin sheath
   b. the extent of depolarization that initiates the action potential
   c. the diameter of the axon
   d. both the presence of a myelin sheath and the diameter of the axon

3. Which of the following cell types makes up the majority of cells in the nervous system?
   a. sensory neurons
   b. motor neurons
   c. neuroglial cells
   d. interneurons

4. What do multiple sclerosis and ALS (amyotropic lateral sclerosis) have in common?
   a. They both are more common in young men than in young women.
   b. They both impair the initiation of action potentials.
   c. They both damage the myelin sheath, disrupting action potential transmission.
   d. They both impair nerve function in the peripheral nervous system.

5. All of the following are directly involved in synaptic transmission except
   a. voltage-sensitive ion channels on the postsynaptic membrane
   b. chemical-sensitive ion channels on the postsynaptic membrane
   c. Ca^{2+}
   d. neurotransmitters

6. An _____ neurotransmitter causes _____ of the postsynaptic membrane.
   a. inhibitory . . . depolarization
   b. excitatory . . . depolarization
   c. inhibitory . . . hyperpolarization
   d. Both (b) and (c) are correct.

7. Information coming into the central nervous system arrives via _____ while information going from the central nervous system to the muscles, glands, and organs travels via _____.
   a. motor neurons . . . sensory neurons
   b. neuroglial cells . . . motor neurons
   c. sensory neurons . . . motor neurons
   d. sensory neurons . . . interneurons

8. The brain and spinal cord make up the:
   a. sensory nervous system
   b. central nervous system
   c. sympathetic nervous system
   d. parasympathetic nervous system

9. A spinal reflex requires the participation of each of the following except:
   a. cerebral cortex
   b. sensory neuron
   c. interneuron
   d. motor neuron

10. Which of the following would cause an acceleration of heart rate, an increase in blood pressure, and a slowing of digestive processes?
    a. motor division of the somatic nervous system
    b. sensory branch of the central nervous system
    c. parasympathetic division of the autonomic nervous system
    d. sympathetic division of the autonomic nervous system

11. Which part of the brain is responsible for regulating the heart and respiratory rates and blood pressure?
    a. cerebral cortex
    b. medulla oblongata
    c. cerebellum
    d. thalamus

12. The coordination necessary for a musician to play the piano depends on control exerted by the:
    a. thalamus
    b. cerebral cortex
    c. cerebellum
    d. medulla oblongata

13. All of the following are associated with storage of long-term memory except:
    a. limbic system
    b. cerebral cortex
    c. permanent changes in neurons
    d. formation of new synapses

14. Which of the following is characteristic of psychoactive drugs?
    a. They do not readily cross the blood-brain barrier.
    b. They alter levels of neurotransmitters.
    c. They can alter the amplitude of action potentials.
    d. They alter the speed with which action potentials travel.

15. Which of the following statements regarding infections of the nervous system is accurate?
    a. Most cases of encephalitis are caused by bacteria.
    b. Rabies can be treated with antibiotics.
    c. Bacterial meningitis is generally more severe than viral meningitis.
    d. The blood-brain barrier is not effective at preventing microorganisms from entering the central nervous system.

Apply What You Know

Answers can be found in the Study Area in MasteringBiology.

1. Explain in terms of brain anatomy and function how it would be possible for people to remember whole events in the distant past yet not be able to recall what they had for breakfast.

2. What do you suppose would happen to a person's behavior and emotional expression if the neural connections between the limbic system and the cerebral cortex were severed?

3. A friend is washing dishes one evening and gets a deep cut from a sharp knife hidden in the soap suds. Several weeks after the visit to the ER, the stitches are removed. Your friend mentions that one part of his finger is numb; he cannot feel anything when he touches one spot. Will this numbness ever go away?

4. You're feeding a semi-wild cat outside the campus cafeteria when it accidentally nips you, drawing blood. Your friends advise you to report to the campus medical center immediately for a rabies vaccination. Should you go for the vaccination?

5. Overnight, your roommate develops a high fever. She says she has a severe headache and complains of neck stiffness. One hour later, she seems confused. You take her to the hospital

and the physician recommends a spinal tap—a test in which a sample of cerebrospinal fluid is removed and analyzed. What might the physician be looking for?

6. Some states still administer the death penalty by lethal injection. The injection may contain an anesthetic (for pain) combined with a high dose of potassium to stop the heart. Why would a large dose of potassium stop the heart?

7. At a party, someone offers you an ecstasy pill. Aside from the legal issues involved, what are the short-term and long-term effects of trying it just one time?

8. Certain bacteria, such as those that cause botulism or tetanus, produce toxins that interfere with neuromuscular transmission. From what you know of neuromuscular transmission, how might such toxins work?

MJ's BlogInFocus

Answers can be found in the Study Area in MasteringBiology.

1. According to current learning theory, anxious thoughts compete for short-term memory space in the limbic system. As a result, people with severe test anxieties may perform well below their abilities on exams; the more important the exam the worse they perform. Knowing this, can you think of any strategies that might help people with test anxiety perform better on exams? Visit MJ's blog in the Study Area in MasteringBiology and look under "Anxiety and Memory."
http://goo.gl/CZ9OGU

MasteringBiology®

Students Go to MasteringBiology for assignments, the eText, and the Study Area with animations, practice tests, and activities.

Professors Go to MasteringBiology for automatically graded tutorials and questions that you can assign to your students, plus Instructor Resources.

Answers to ✓ questions are available in the Appendix.

Sensory Mechanisms

Colorized SEM of the surface of the tongue, showing papillae (pointing toward the back of the tongue).

Key Concepts

- **Sensory mechanisms provide information** about the world around us and about conditions within our own bodies.

- **Various types of receptors convert different types of sensory stimuli (physical touch, odors, light, etc.) into action potentials.** How the brain interprets incoming action potentials depends on where in the brain they are received.

- **Some receptors adapt quickly,** which is why you don't notice some stimuli and strong stimuli seem less intense after a while.

- **Receptors are located throughout the body.** They provide us with information about body position, touch, temperature, vibration, pressure, and especially pain.

- **The five special senses (taste, smell, hearing, balance, and vision) originate from special areas of the body.** Four of them (taste, smell, hearing, and vision) provide us with detailed information about the external world.

CURRENT ISSUE
DWD: Driving While Distracted

On August 5, 2010, a 19-year-old driving a pickup truck slammed into the back of a tractor truck on an interstate highway near Gray Summit, Missouri. Two school buses carrying high school students to an outing at the Six Flags Missouri amusement park collided with each other and with the two trucks, smashing the pickup truck into a nearly unrecognizable pile of metal. The driver of the pickup truck and a 15-year-old student on one of the buses were killed, and 38 students were injured. The driver of the pickup truck had been texting on his cell phone at the time of the accident.

It's not hard to find examples of the disastrous consequences of driving while distracted, and once you understand how the brain works it's easy to understand why. Your brain is encased in the skull, entirely in the dark. It assesses what is happening in the outside world (and then reacts to it) by processing and organizing a steady stream of sensory input from various sources. It copes with all this sensory input by focusing on the input it considers most important at the moment, sometimes at the expense of other input. It should be no surprise, then, that anything that distracts you while driving is likely to lower your ability to drive safely. "Every single time someone takes their eyes or their focus off the road—even for just a few seconds—they put their lives and the lives

of others in danger," said Transportation Secretary Ray LaHood at a national summit on distracted driving. "Distracted driving is unsafe, irresponsible and in a split second, its consequences can be devastating." Distractions may be visual (taking your eyes off the road), cognitive (thinking about something else), or manual (taking your hands off the wheel). Typical distractions include texting or talking on the phone, putting on makeup, eating, reading, or using a navigation system. Texting while driving is considered especially risky because it involves all three types of distractions at once (visual, cognitive, and manual).

According to the National Highway Traffic Safety Administration (NHTSA), in 2012 there were 421,000 injuries and 3,328 deaths in motor vehicle crashes involving a distracted driver. When broken down by age, the highest proportion of distracted drivers involved in fatal crashes is under 20 years old.

Driving while on the phone and texting while driving are the two most common

Texting while driving is now illegal in 44 states.

distractions. The NHTSA recommends that drivers not use their cell phones while driving except in an emergency, but it doesn't appear that anyone is listening. Seventy-one percent of teens and young adults say they have composed/sent text messages while driving, according to the NHTSA. At any given moment during daylight hours, approximately 660,000 drivers are either on their cell phones or manipulating electronic devices while driving.

In the United States, driving is a privilege granted by the states, not the federal government. Although the National Transportation Safety Board (NTSB) has issued a strongly worded statement to the states urging them to ban texting while driving, it is up to each state to develop its own laws pertaining to driving and to enforce the penalties for breaking them.

Reacting to the statistics and to high-profile deaths like those in Missouri and elsewhere, the states are beginning to take action. Although no states currently ban all cell phone use (including hands-free phones), 44 states now prohibit texting while driving. Many of these states have special restrictions on cell phone use by drivers with learner's →

Questions to Consider

1 Is it fair that in some states, a ban on texting while driving applies to drivers only under a certain age?

2 Should the use of all hand-held devices, including just talking on the phone, be banned completely while driving?

3 Do you approve of the law in five states that a driver can only be cited for a hand-held phone violation if the driver commits another moving violation while on the phone?

The driver of the pickup truck, between the tractor truck bed and the bottom of the bus, was texting while driving.

permits or drivers under a certain age. Fourteen states now prohibit all drivers from using hand-held cell phones while driving, even for making telephone calls. Other states are likely to follow suit.

Although all this sounds like progress, enforcement is still an issue. In five of the 44 states that currently ban texting while driving (Florida, Iowa, Ohio, Nebraska, South Dakota), texting while driving is not a primary offense, but merely a secondary one. In those five states, a driver cannot be cited merely for texting while driving unless he/she has an accident or violates some other driving law while texting. The penalty for a texting-while-driving offense also varies from state to state.

In Missouri, where the high-profile accident described earlier took place, only drivers under 21 years old are prohibited from texting while driving. Older drivers are still free to do as they wish.

SUMMARY

- More than 3,200 deaths and 420,000 injuries per year are attributed to driving while distracted.
- At any given moment during daylight hours, 660,000 drivers are talking on their cell phones.
- The highest proportion of distracted drivers is under 20 years old.
- Laws against driving while distracted (texting, talking on a hand-held phone) vary from state to state, from complete bans to no ban at all. There is no national standard.

Look around you at the complex and colorful world you inhabit. Then, after reading this paragraph, close your eyes and concentrate for a moment on what you can hear. Identify the sources of any sounds you hear, and try to judge from what direction the sounds are coming and their distance from you. Describe any odors that you smell, and make a guess at the temperature in the room. Now concentrate on your own body. Move your arms and legs about, and then identify what type of clothing you are wearing. Finally, visualize in your mind the precise position of every part of your body, right down to your fingers and toes.

Your brain is your body's central command center. It is completely encased within the skull, so it is in the dark both literally and figuratively. It gets information about both the external and internal environments only in the form of nerve impulses originating from sensory nerves connected to sensory receptors. Sensory mechanisms, therefore, are critical to the brain's ability to gather and integrate information and to respond accordingly. Sensory mechanisms allow us to see and hear what is going on in the external environment. They are responsible for our senses of taste, smell, and touch. They enable us to experience pain, to monitor temperature, and to respond accordingly. They even contribute to our ability to maintain our balance and to control our body movements. ▪

In this chapter, we learn how sensory information is received by the body, how that information is converted to nerve impulses (action potentials), and how the nerve impulses are transmitted to the brain in such a way that the brain can make sense of them. Then we'll examine the various sensory mechanisms in more detail.

12.1 Receptors receive and convert stimuli

Sensory input that causes some change within or outside the body is called a **stimulus** (plural: stimuli). The stimulus is often a form of physical energy such as heat, pressure, or sound waves, but it can also be a chemical. A **receptor**—a structure specialized to receive certain stimuli—accepts the stimulus and converts its energy into another form.

Some receptors are evolved from the dendritic structures of a sensory neuron. They convert the stimulus into a graded potential that, if it is powerful enough, initiates an impulse within the sensory neuron. Other receptors are parts of cells that produce graded potentials and release a neurotransmitter, stimulating a nearby sensory neuron. In the end, the effect is the same—generation of an impulse in a sensory neuron.

When the central nervous system (CNS) receives these impulses, we often experience a **sensation,** meaning that we become consciously aware of the stimulus. A sensation is different from **perception,** which means understanding what the sensation means. As an example, hearing the sound of thunder is a sensation, whereas the belief that a storm is approaching is a perception.

Receptors are classified according to stimulus

Receptors are classified according to the type of stimulus energy they convert. The classification of receptors is as follows:

- **Mechanoreceptors** respond to forms of mechanical energy, such as waves of sound, changes in fluid pressure, physical touch or pressure, stretching, or forces generated by gravity and acceleration.
- **Thermoreceptors** respond to heat or cold.
- **Pain receptors** respond to tissue damage or excessive pressure or temperature.
- **Chemoreceptors** respond to the presence of chemicals in the nearby area.
- **Photoreceptors** respond to light.

Many of these receptors contribute to sensations, although some (such as joint receptors) contribute no more than a general sense of where our limbs are located.

Table 12.1 lists examples of each type of receptor and the sensations they give us.

A few receptors are "silent" in the sense that we are not consciously aware of their actions. These receptors function in negative feedback loops that maintain homeostasis inside the body. They include stretch receptors for monitoring and regulating blood pressure and fluid volumes, and chemoreceptors for regulating the chemical composition of our internal environment. They aren't discussed in detail here, as we discuss most of them throughout the book in the context of the homeostatic controls of various organ systems.

The CNS interprets nerve impulses based on origin and frequency

How can cells in the CNS interpret some incoming impulses as images and others as sounds? How do the cells distinguish a loud sound from a soft one and bright lights from dim?

As described in the chapter on the nervous system (Chapter 11), nerve impulses (action potentials) are transmitted from receptors to specific brain areas. For example, impulses generated by visual stimuli travel in sensory neurons whose axons go directly to brain regions associated with vision. All incoming impulses traveling in these neurons are interpreted as light, regardless of how they were initiated. This is why, when you are hit in the eye, you may "see stars." The blow to your eye triggers impulses

in visual sensory neurons. When these signals arrive in the visual area of the brain, they are interpreted as light.

Stronger stimuli activate more receptors and trigger a greater frequency of impulses in sensory neurons. In effect, the CNS gets all the information it needs by monitoring where impulses originate and how frequently they arrive.

☑ Suppose the axon of a sound-detecting (auditory) neuron in a person's ear was somehow redirected so that it connected to the visual area of the brain. How will this person likely perceive the action potentials produced in this neuron by the auditory stimulus? Explain.

Some receptors adapt to continuing stimuli

The CNS can ignore one sensation to concentrate on others. For example, you ignore the feel of your clothing when you are more interested in other things. Some sensory inputs are ignored after a while because of **receptor adaptation,** in which the sensory neuron stops sending impulses even though the original stimulus is still present. To demonstrate, try touching a few hairs on your forearm. Notice that the initial feeling of light pressure goes away within a second or so, even though you maintain the same amount of pressure.

Receptors in the skin for light touch and pressure adapt rather quickly. This confers a survival advantage, because they can keep the CNS informed of *changes* in these stimuli, without constantly bombarding it with relatively

Table 12.1 Types of receptors

Type of receptor	Examples	Sensation
Mechanoreceptors		
Touch, pressure	Unencapsulated dendritic endings, Meissner's corpuscles, Pacinian corpuscles, Ruffini endings, Merkel disks	Touch and pressure on skin or body hair
Vibration	Pacinian corpuscles	Vibration on skin
Stretch	Muscle spindles	Muscle length
	Tendon receptors	Tendon tension
	Joint position	Joint receptors
Hearing	Hair cells in cochlea of the inner ear	Vibration of fluid originating from sound waves
Balance	Hair cells in semicircular canals of the inner ear, hair cells in the utricle and saccule of the inner ear	Rotational movement of the head, static position and linear acceleration and deceleration of the head
Thermoreceptors		
Temperature	Receptors in unencapsulated dendritic endings in skin	Heat or lack of heat (cold)
Pain Receptors		
Pain	Receptors in unencapsulated dendritic endings in skin and internal organs	Pain caused by excessive pressure, heat, light, or chemical injury
Chemoreceptors		
Taste	Taste receptors in taste buds	Salty, sweet, sour, and bitter tastes
Smell	Olfactory receptors of nasal passages	Over a thousand different chemical smells
Photoreceptors		
Vision	Rods of the retina	Light (highly sensitive black-and-white vision)
	Cones of the retina	Light (color vision)

unimportant stimuli. Olfactory (smell) receptors also adapt rapidly. Olfactory adaptation can be hazardous to people if they are continually exposed to low levels of hazardous chemicals that they can no longer smell.

Other receptors—pain receptors, joint and muscle receptors that monitor the position of our limbs, and essentially all of the silent receptors involved in homeostatic feedback control loops—adapt slowly or not at all. Lack of adaptation in these receptors is important to survival. Persistent sensations such as pain alert us to possible tissue damage from illness or injury and prompt us to take appropriate action. Persistent activity of silent receptors is essential to our ability to maintain homeostasis.

MJ's BlogInFocus What are the laws regarding cell phone use while driving in your state? Visit MJ's blog in the Study Area in MasteringBiology and look under "Distracted Driving."

http://goo.gl/n9ZIJa

Somatic sensations and special senses provide sensory information

The sensations (or senses) provided by receptors are categorized as either somatic or special. The **somatic sensations** originate from receptors present at more than one location in the body (*soma* is the Greek word for "body"). The somatic sensations include temperature, touch, vibration, pressure, pain, and awareness of body movements and position. The five **special senses** (taste, smell, hearing, balance, and vision) originate from receptors that are restricted to particular areas of the body, such as the ears and eyes. The special senses deliver highly specialized information about the external world.

Next, we look at the somatic sensations first and then discuss the five special senses in detail.

↩ **Recap** Receptors receive a physical or chemical stimulus, causing nerve impulses to be generated in sensory neurons. Receptors are classified according to the type of stimulus they convert. The five classes of receptors are mechanoreceptors, thermoreceptors, pain receptors, chemoreceptors, and photoreceptors. Stronger stimuli activate more receptors and trigger a greater frequency of action potentials in sensory neurons. Some receptors adapt rapidly; others adapt slowly or not at all. ▪

12.2 Somatic sensations arise from receptors throughout the body

The somatic sensations of temperature, touch, pressure, vibration, pain, and awareness of body movements and position are essential to help us coordinate muscle

movements, avoid danger, and maintain body temperature. Some somatic sensations contribute to pleasurable feelings as well, such as our response to the gentle touch of a loved one.

Receptors that detect the somatic sensations are located throughout the body in skin, joints, skeletal muscles, tendons, and internal organs. Sensory neurons linked to these receptors send their impulses to the brain, specifically to the primary somatosensory area of the parietal lobe of the cerebral cortex (review Figure 11.17). As you already know, body parts with the greatest sensory sensitivity, such as the mouth and fingers, involve more neurons in the somatosensory area, whereas less sensitive body parts involve fewer neurons.

The somatosensory area processes the information and sends it to the nearby primary motor area in the frontal lobe. If necessary, impulses are then generated in motor (output) neurons of the peripheral nervous system to cause body movement.

The skin contains a variety of sensory receptors

The skin is ideally positioned to receive information about the external environment. Skin contains *mechanoreceptors* for sensing touch, pressure, and vibration; *thermoreceptors* for monitoring temperature; and pain receptors for helping us avoid danger. All of these receptor types are the dendritic endings of sensory neurons. Any force that deforms the plasma membrane of the dendritic ending produces a typical graded potential (Chapter 11). If the graded potential (or the sum of multiple graded potentials) is large enough to exceed threshold, the sensory neuron initiates an impulse.

Some dendritic endings are *unencapsulated*, meaning that they are just the free endings of sensory nerves. Most unencapsulated dendritic endings respond to pain or temperature, though a few are mechanoreceptors that respond to light touch. Other dendritic endings are *encapsulated*, that is, they are enclosed in layers of connective tissue. Nearly all encapsulated dendritic endings are mechanoreceptors.

Mechanoreceptors for touch, pressure, and vibration differ in their locations, the intensity of the stimulus that must be applied in order to generate an impulse (pressure is a stronger mechanical force than light touch), and the degree to which they adapt. Vibration-sensitive receptors adapt so quickly that it takes a rapidly changing physical deformation (many on-and-off cycles per second) to keep them stimulated. Vibration-sensitive receptors are particularly useful for providing information about potentially harmful flying insects because those receptors respond to the kinds of vibrations generated by insects' wings.

Figure 12.1 depicts the skin with several types of receptors for detecting somatic sensations. They are:

- *Unencapsulated dendritic endings.* Naked dendritic endings of sensory neurons around hairs and near the skin surface signal pain, light pressure, and changes in temperature.
- *Merkel disks.* Merkel disks are modified unencapsulated dendritic endings that detect light touch and pressure.

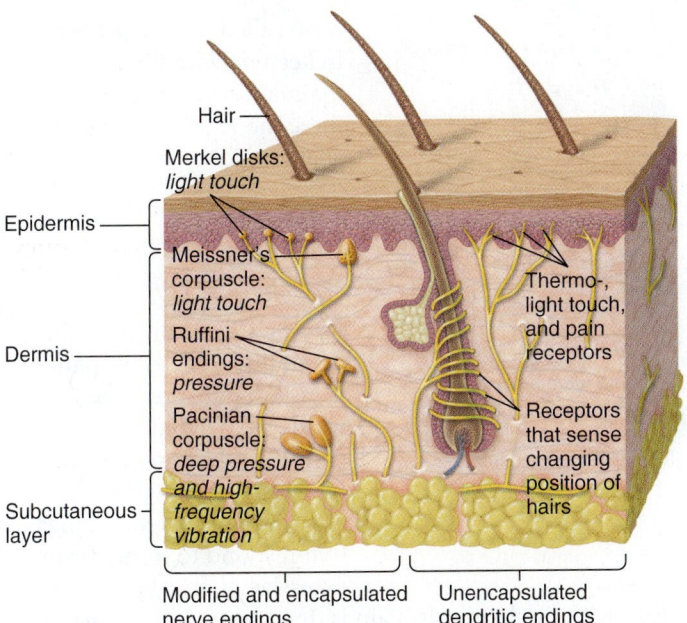

Figure 12.1 Sensory receptors in skin. Thermoreceptors, some of the receptors for light touch, and the receptors for pain are the free endings of sensory neurons. Other receptors for touch, pressure, and vibration are modified or encapsulated endings of sensory neurons.

- *Meissner's corpuscles.* Encapsulated touch receptors located close to the skin surface detect the beginning and the end of light pressure and touch.
- *Ruffini endings.* These encapsulated receptors respond continually to ongoing pressure.
- *Pacinian corpuscles.* Pacinian corpuscles are the best-understood of all skin sensory receptors. Pacinian corpuscles are encapsulated nerve endings located in the dermis that respond to either deep pressure or high-frequency vibration. The nerve ending of a Pacinian corpuscle is covered by loose layers of connective tissue (**Figure 12.2a**). When a stimulus is first applied, the layers of connective tissue are pushed inward and deformed, bending the nerve ending and causing action potentials to be generated in the sensory nerve (**Figure 12.2b**). However, even if the stimulus remains, the layers of encapsulation quickly slide past each other, allowing the nerve ending to return to its initial position (**Figure 12c**). When that happens, the nerve firing rate returns to baseline as well.

Mechanoreceptors indicate limb position, muscle length, and tension

With your eyes closed, try to describe the positions of your arms and legs. You can probably describe their position rather well. You may even be able to tell which muscle groups are contracting. You do this with a variety of mechanoreceptors in joints (for joint position), in skeletal

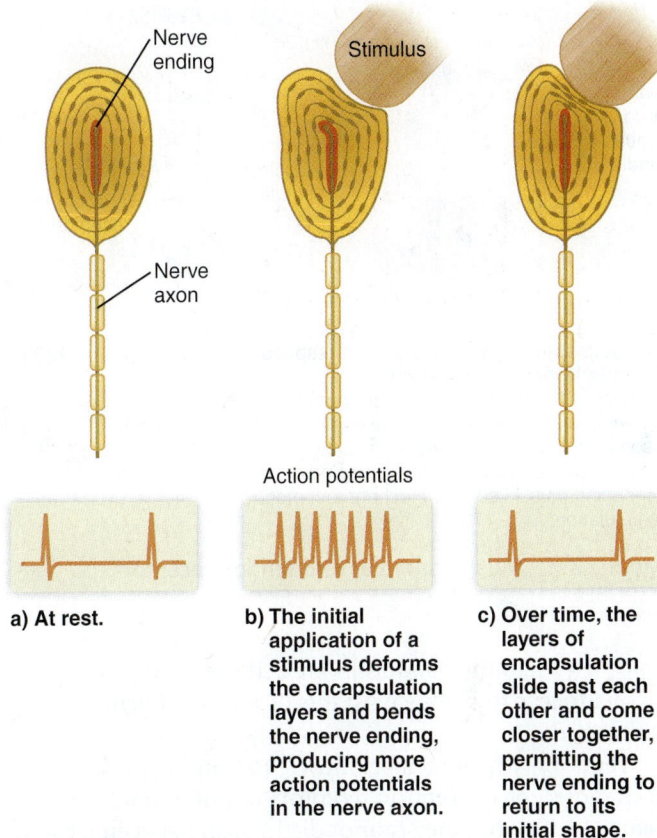

a) **At rest.**

b) **The initial application of a stimulus deforms the encapsulation layers and bends the nerve ending, producing more action potentials in the nerve axon.**

c) **Over time, the layers of encapsulation slide past each other and come closer together, permitting the nerve ending to return to its initial shape.**

Figure 12.2 The Pacinian corpuscle.

muscles (for muscle length), and in tendons (for tension). Perhaps the best known of these mechanoreceptors are the specialized structures for monitoring muscle length, called **muscle spindles.** For most joints, muscle length determines joint position because of the way the muscle is attached to the bones.

A muscle spindle is a small bundle of modified skeletal muscle cells located within a skeletal muscle (**Figure 12.3a**). Muscle spindles are innervated by sensory nerves whose dendritic endings are mechanoreceptors that respond to stretch (lengthening). One type of dendritic ending wraps around the middle of a muscle spindle cell. Another has several branches that attach nearer the end of the muscle spindle cell. When the whole muscle is stretched, the receptors attached to the muscle spindle cells are stretched as well. Mechanical distortion of the mechanoreceptors produces local graded potentials in the dendritic endings and (if threshold is passed) an action potential.

This is exactly what happens in the classic patellar stretch reflex, described in the chapter on the nervous system. When a physician taps your patellar tendon (at the front of your knee) with a hammer, the patellar tendon and quadriceps (thigh) muscle are stretched, stretching the muscle spindle as well. Stretch of the muscle causes sensory neurons innervating the muscle spindle to transmit action potentials to the spinal cord. Within the cord,

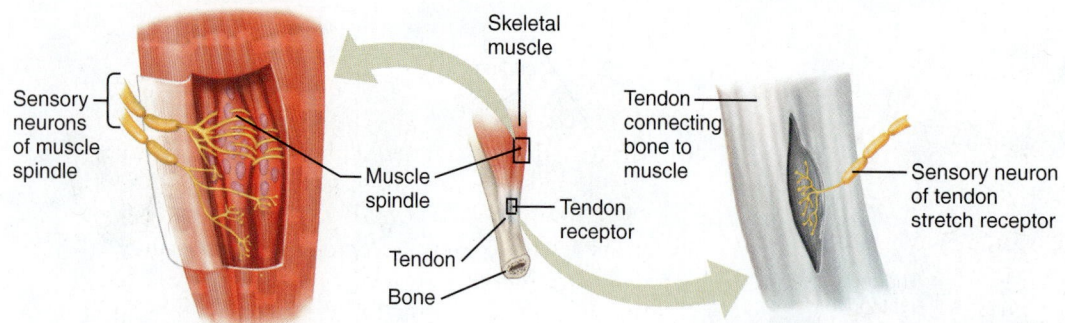

a) **Muscle spindle.** A muscle spindle responds to muscle length. Passive stretch of a muscle stretches the muscle spindle, stimulating mechanoreceptors in the nerve endings of the sensory neurons. Conversely, muscle contraction shortens the muscle spindle, reducing muscle spindle mechanoreceptor stimulation.

b) **Tendon receptor.** A tendon receptor responds to tension in tendons. When a muscle contracts and also when it is stretched passively, tension on the tendon increases, activating tendon receptors.

Figure 12.3 **Muscle spindles and tendon receptors.**

motor nerves to the quadriceps are activated, causing the quadriceps to contract. All this occurs in a fraction of a second.

Tendons, which connect muscle to bone, also have mechanoreceptors. Tendon receptors respond to tension (Figure 12.3b), but they cannot distinguish between tension produced by passive stretch and tension produced by active muscle contraction. Tendon receptors may play a protective role in preventing injury by extremely high tensions. For example, if you were to fall out of a tree but land on your feet, the impact when you hit the ground would produce high levels of tension, activating your tendon stretch receptors. The sensory information they provide inhibits muscle contraction, so you would crumple to the ground rather than stay upright. This reflex-induced collapse protects your muscles from tearing and your bones from breaking.

✅ A friend is trying to touch his toes (with his legs straight) by bouncing his upper body up and down. This sort of bouncing typically stimulates muscle spindles, but not tendon receptors. Explain whether this is a good method of stretching the hamstring muscles in the back of the legs.

Thermoreceptors detect temperature

Near the skin's surface, thermoreceptors for heat and cold provide useful information about the external environment. These receptors adapt quickly, allowing us to monitor changes in temperature accurately and yet adjust sensory input so it becomes more bearable. For example, when you step into a hot shower the water may seem uncomfortably warm at first, but you soon become used to it.

Other thermoreceptors located in abdominal and thoracic organs throughout the body monitor internal temperature

(also called *core temperature*). In keeping with their role in maintaining homeostasis, these thermoreceptors do not adapt rapidly.

Pain receptors signal discomfort

As much as we dislike pain, the ability to perceive pain is essential for survival. Pain warns us to avoid certain stimuli and informs us of injuries. Pain receptors are unencapsulated endings that respond to injury from excessive physical pressure, heat, light, or chemicals. Pain is described as either fast or slow, depending on its characteristics.

Fast pain, also called *sharp* or *acute pain*, occurs as soon as a tenth of a second after the stimulus. Receptors for fast pain generally respond to physical pressure or heat and usually are located near the surface of the body. They inform us of stimuli to be avoided—for example, stepping on a nail or touching a hot burner on the stove. The reflex withdrawal response to fast pain is rapid and strong.

In contrast, *slow pain* generally arises from muscles or internal organs. Slow pain, which may not appear until seconds or even minutes after injury, is due to activation of chemically sensitive pain receptors by chemicals released from damaged tissue. Slow pain from internal organs is often perceived as originating from an area of the body completely removed from the actual source. This phenomenon, called **referred pain,** happens because action potentials from internal pain receptors are transmitted to the brain by the same spinal neurons that transmit action potentials from pain receptors in the skin and skeletal muscles. The brain has no way of knowing the exact source of the pain, so it assigns the pain to another location.

Referred pain is so common that physicians use it to diagnose certain disorders of internal organs. For example, a heart attack usually manifests itself as pain in the left shoulder and down the left arm. Figure 12.4 shows where pain from internal organs may be felt. The close association between areas of referred pain from the heart and esophagus is why it can be difficult, by sensations alone, to distinguish a heart attack from the less serious problem of esophageal spasm.

Pain receptors generally do not adapt. From a survival standpoint, this is beneficial, because the continued presence of pain reminds us to avoid doing whatever causes it. However, it also means that people with chronic diseases or disabilities can end up experiencing constant discomfort.

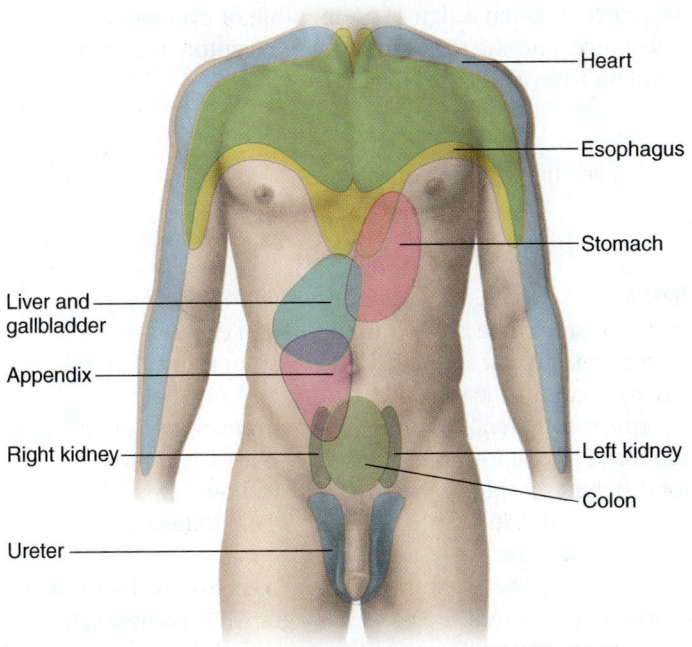

Figure 12.4 **Locations of referred pain from various internal organs.** Note that the areas of overlap for esophageal and heart pain are so extensive that it is often difficult to tell them apart by location.

↩ **Recap** Somatic sensations include temperature, touch, pressure, vibration, pain, and awareness of body movements and position. Mechanoreceptors in skeletal muscle, tendons, and joints provide information about the position of our limbs and the amount of tension on muscles. Thermoreceptors near the skin surface monitor the temperature of the external environment, and thermoreceptors in internal organs maintain homeostasis of body temperature. Fast pain arises from receptors near the body surface, whereas slow pain originates in muscles or internal organs. Pain is important to survival because it warns us of injury. ∎

12.3 Taste and smell depend on chemoreceptors

We turn now to the five special senses: taste, smell, hearing, balance, and vision. As mentioned earlier, these senses originate from receptors located in special areas of the body.

The first senses we consider, taste and smell, are both due to the action of chemoreceptors. The rich fragrance of a rose, the less pleasing odor of a skunk, the tart taste of an orange—all depend on our ability to convert a chemical stimulus into the only language the CNS can understand: action potentials.

Taste: chemoreceptors bind with dissolved substances

The tongue (Figure 12.5a) has a rough texture because its surface is composed of numerous small projections, called *papillae*, surrounded by deep folds (Figure 12.5b). About 10,000 **taste buds** are located at the surface of the papillae within the folds (Figure 12.5c). A taste bud is a cluster of about 25 taste cells and 25 supporting cells that separate the taste cells from each other.

The exposed tip of a taste cell has hairs that project into the mouth (Figure 12.5d). The hairs contain chemoreceptors that are specific for certain chemicals (called *tastants*). When we eat or drink, tastants dissolve in our saliva and bind to the chemoreceptor. The binding of tastant to chemoreceptor causes the taste cell to release a neurotransmitter, which stimulates a nearby sensory neuron to produce action potentials. Note that the taste cell itself is not a sensory neuron. It takes two cells (the taste cell and a sensory neuron) to convert the chemical stimulus into a nerve impulse, but the ultimate effect is the same as if it were one cell.

The five commonly accepted primary taste qualities are sweet, sour, salty, bitter, and umami. Taste receptors for umami, discovered by Japanese researchers, allow us to

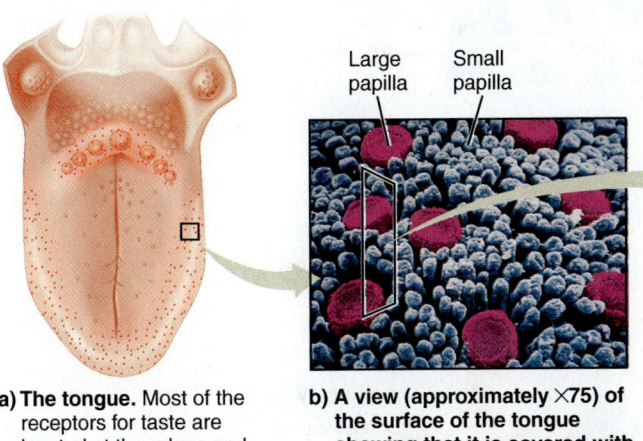

a) The tongue. Most of the receptors for taste are located at the edges and at the front and back of the tongue.

b) A view (approximately ×75) of the surface of the tongue showing that it is covered with large and small papillae with many deep folds in-between.

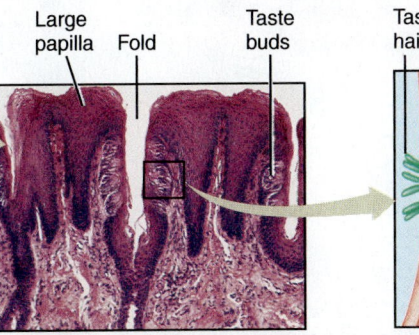

c) A sagittal section (approximately ×200) through several large papillae showing the location of the taste buds along their sides, just beneath the tongue's surface.

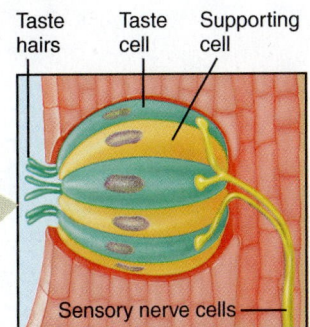

d) A taste bud is composed of a group of taste cells and supporting cells. Each taste cell connects to a sensory neuron.

Figure 12.5 **The structure of the receptors for taste.**

appreciate foods like parmesan cheese, mushrooms, beef, soy sauce, and Worcestershire sauce. Although an individual taste receptor and its associated sensory neuron seem to be "tuned" to respond most strongly to only one of the taste qualities, they do respond weakly to the other taste qualities as well.

MJ's **BlogInFocus** Why do some people like Brussels sprouts, and other people do not? Visit MJ's blog in the Study Area in MasteringBiology and look under "Tasting Bitter Foods."

http://goo.gl/4EGJu5

Most of the taste receptors are located around the edges and at the front and back of the tongue. Although all receptor types can be found in each of these regions, there is some evidence that their distribution is not entirely even. The tongue is slightly more sensitive to sweet tastes on the tip, sour on the sides, salty on the tip and sides, and bitter toward the back (the distribution of umami receptors is not yet well documented). Therefore, to gain the maximum taste sensation, we move food around in our mouth before swallowing. The combined sensitivities of our taste receptors allow us to distinguish hundreds of different flavors, far beyond the basic five qualities. On the other hand, if we don't want to taste something, we pass it over the center of the tongue and swallow quickly.

It is probably no accident that bitter taste receptors are heavily concentrated at the very back of the tongue.

Bitterness is often a sign of an inedible or even poisonous substance, and the location of these receptors makes it difficult to bypass them.

✅ If you blot your tongue completely dry, you will find that you temporarily lose the ability to taste dry foods such as sugar or salt. Why?

Smell: chemoreceptors bind with odorants

The sense of smell also relies on a chemoreceptor mechanism. There are two fundamental differences between taste and smell: (1) The receptors for smell are located on true sensory neurons and (2) there are receptors for over 1,000 different odorant chemicals, as opposed to just four tastant classes. With an abundance of different receptor types for smell and our sense of taste, we can distinguish as many as 10,000 taste and smell sensations.

Odors are detected by **olfactory receptor cells** located in the upper part of the nasal passages. The receptor cells are sensory neurons, each with a modified dendritic ending that branches to become several olfactory hairs (Figure 12.6). The olfactory hairs extend into a layer of mucus covering the surface of the nasal passages. The mucus, produced by nearby olfactory glands, keeps these hairs from drying out.

Gaseous and airborne odorants enter the nasal passages, become dissolved in the mucus, bind to chemoreceptors located on the olfactory hairs, and cause the olfactory receptor cell to generate impulses. Olfactory receptor cells synapse with olfactory neurons in a nearby area

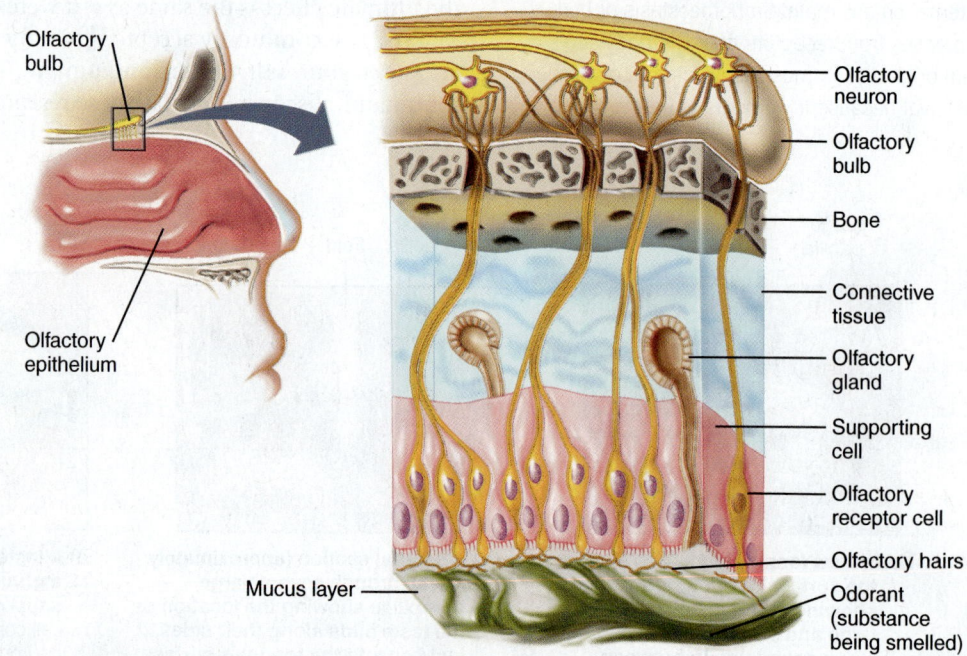

Figure 12.6 Olfactory receptors and the mucus-producing olfactory glands. Odorants become trapped in mucus and then bind to the olfactory hairs of receptor cells.

Figure 12.7 Smell. Over 1,000 different olfactory receptors contribute to our sense of smell and taste.

of the brain called the *olfactory bulb*, where information is partially integrated and then passed to higher brain centers.

Our sense of smell complements our sense of taste because when food is chewed it releases chemicals that come into contact with olfactory receptors. The combined sensations create an effect that is interpreted at higher brain levels (Figure 12.7). When we have a cold, most foods taste less appetizing. In part this is because increased production of mucus prevents odorant molecules from reaching olfactory receptors, impairing our sense of smell.

↩ **Recap** Taste buds contain chemoreceptor cells that bind dissolved chemicals. The five currently accepted basic taste qualities are sweet, salty, sour, bitter, and umami. Our sense of smell augments and complements our sense of taste, allowing us to experience over 10,000 combinations of taste and smell. ∎

12.4 Hearing: mechanoreceptors detect sound waves

Sounds are waves of compressed air. Sound intensity is a physical property of sound that refers to the energy of the sound, whereas loudness refers to our subjective interpretation of sound. Both intensity and perceived loudness are related to the amplitude of the sound waves, and both are measured in units called decibels (dB). The threshold of human hearing (barely audible) is set at 0 dB. A 10-dB increase represents a 10-fold increase in sound intensity (sound energy), but only a doubling of perceived loudness. In other words, a sound of 30 dB has 100 times more energy than a 10-dB sound, but humans would perceive it as only four times as loud. Hearing loss can result from exposure to sounds louder than 85 dB, which can easily be exceeded by attending a rock concert or listening to music with headphones or earbuds at near maximum volumes. Table 12.2 lists the decibels associated with some common sounds.

The tone (or pitch) of a sound is determined by its frequency, the number of wave cycles that pass a given point per second (cycles/sec). Frequency is expressed in *hertz* (Hz). Higher tones are those of a higher frequency (Figure 12.8). The frequency range for normal human hearing is from 20 to 20,000 Hz.

When waves of sound strike the ear, they cause vibration-sensitive mechanoreceptors deep within the ear to vibrate. The vibration causes the production of impulses that travel to the brain for interpretation.

Table 12.2 Common noises and your hearing

Sound	Decibels	How it affects your hearing
Rustling leaves	20	Does not cause hearing loss
Normal conversation	50–60	Does not cause hearing loss
Vacuum cleaner	70	Does not cause hearing loss
Power lawn mower	85–100	Cumulative exposure causes hearing loss; newer mowers may be quieter
MP3 player max, ear buds	105–120	Definite risk of permanent hearing loss
Live rock concert	110–130	Definite risk of permanent hearing loss
Shotgun	140	Repeated exposure likely to cause permanent hearing loss
Jet engine at 100 ft	150	Exposure likely to cause permanent hearing loss

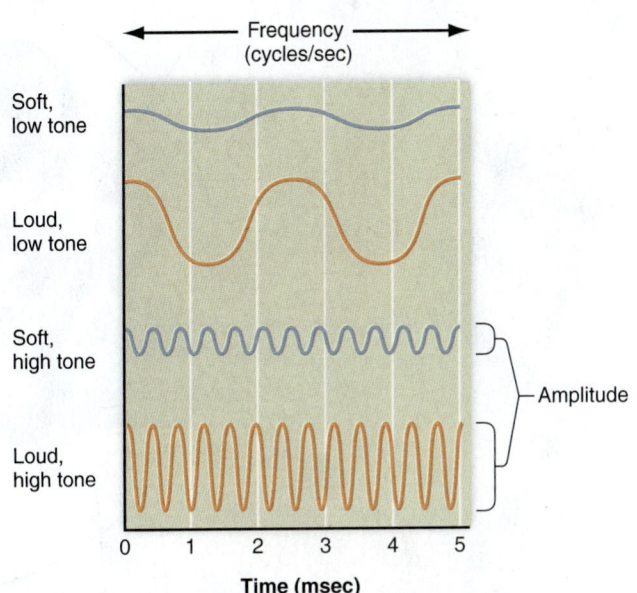

Figure 12.8 Properties of sound waves. Loudness (intensity) is determined by sound wave amplitude. Tone (pitch) is determined by sound wave frequency.

☑ Draw one more graph, to the same scale, illustrating a sound wave of extremely high tone and intermediate loudness, compared to the sounds illustrated in this figure.

The outer ear channels sound waves

The human ear consists of three functionally different regions: the outer, middle, and inner ear (Figure 12.9).

The **outer ear** consists of the pinna, or visible portion of the ear, and the auditory canal. Sound waves arrive at the pinna and are directed into the auditory canal, which channels them to the **tympanic membrane (eardrum)**. The tympanic membrane serves as a partition between the outer and middle ears.

The middle ear amplifies sound

The **middle ear** consists of an air-filled chamber within the temporal bone of the skull, bridged by three small bones called the *malleus* ("hammer"), *incus* ("anvil"), and *stapes* ("stirrup"). When sound waves strike the tympanic membrane, it moves back and forth slightly (vibrates). Vibration of the tympanic membrane makes the malleus, incus, and stapes vibrate. The stapes touches a smaller membrane, called the **oval window,** causing it to vibrate, too.

All the vibrating force of the much larger tympanic membrane is now concentrated on the smaller oval window. In effect, the middle ear amplifies the sound waves severalfold. Amplification is essential because the oval window must vibrate with sufficient force to produce pressure waves in the watery fluid of the inner ear.

The air-filled middle ear is kept at atmospheric pressure by the **auditory tube** (eustachian tube), a narrow tube that runs from the middle ear chamber to the throat. Although the tube is normally held closed by the muscles along its sides, it opens briefly when you swallow or yawn to equalize air pressure in the middle ear with that of the surrounding atmosphere. When you change altitude quickly, your ears "pop" as the auditory tube opens briefly and air pressures are suddenly equalized. When you have a cold or allergy, inflammation and swelling may prevent the auditory tube from equalizing pressures normally, which is why airplane trips can be painful when you have a cold or allergies. Children tend to get a lot of earaches (middle ear infections) via the auditory tube until the skull elongates, changing the angle of the tube to the ear.

☑ Suppose human ears had no outer ear and no middle ear and just had the inner ear—that is, with the oval window in direct contact with the external air. In what way would our sense of hearing probably be different?

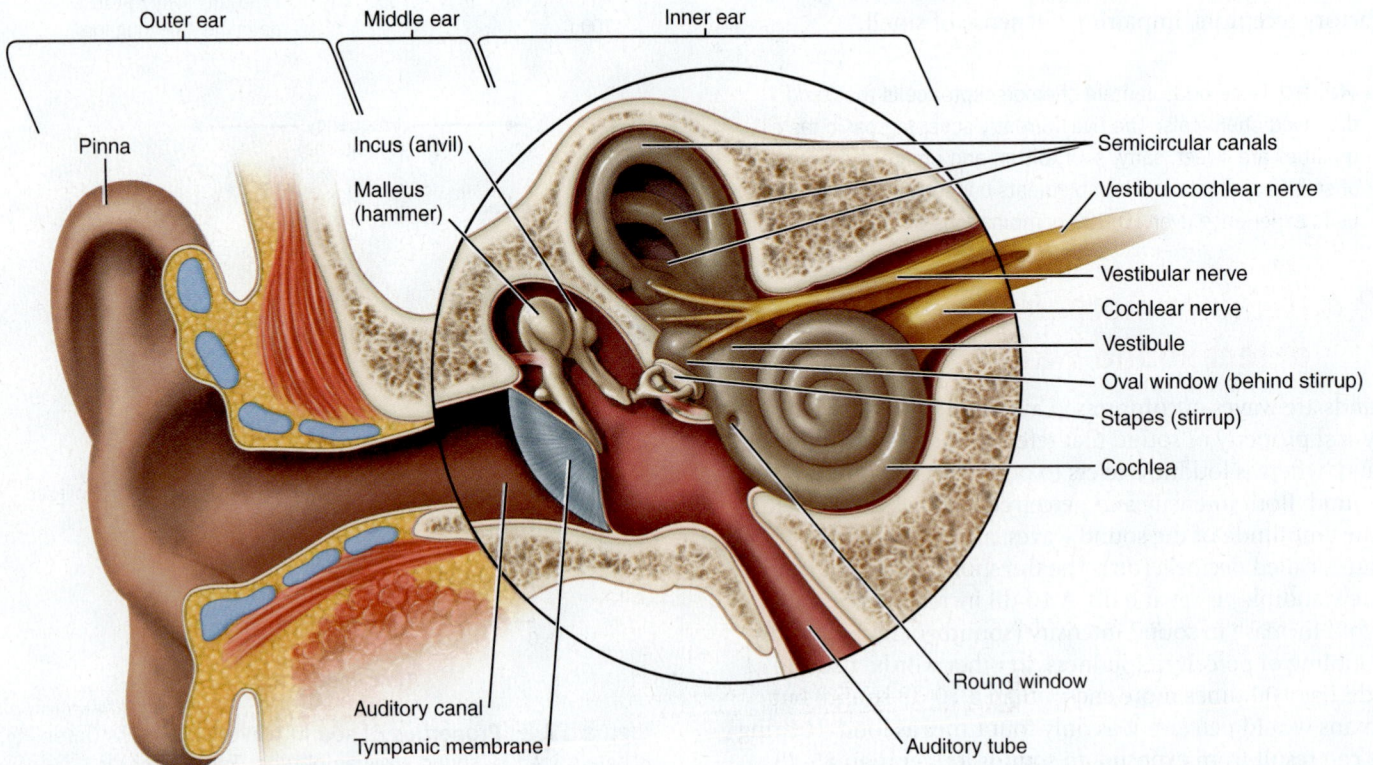

Figure 12.9 Structure of the human ear. The outer ear receives and channels sound, the middle ear amplifies sound, and the cochlea of the inner ear converts the sound to action potentials. The vestibule and the three semicircular canals are responsible for balance.

The inner ear sorts and converts sounds

The **inner ear** sorts sounds by tone and converts them into impulses. It consists of the **cochlea,** where sound is converted, and the *vestibular apparatus*, consisting of the vestibule and the three semicircular canals, which does not contribute to hearing at all. We will discuss the vestibular apparatus later in connection with the sense of balance.

The cochlea is a coiled structure shaped a bit like a snail. **Figure 12.10a** shows what the cochlea would look like if it were uncoiled. It is a tapered tube containing two interconnected outer canals called the *vestibular canal* and the *tympanic canal*, surrounding a third, closed fluid-filled space called the *cochlear duct* (**Figure 12.10b**). The base of the cochlear duct is formed by the **basilar membrane** (**Figure 12.10c**). The basilar membrane supports a population of about 15,000 **hair cells,** the mechanoreceptor cells of the ear. Hair cells have hairlike projections that are embedded in an overhanging structure called the **tectorial membrane** (the Latin word for "roof" is *tectum*), which is composed of a firm, gelatinous, noncellular material. Together, the hair cells and the tectorial membrane are called the *organ of Corti*, the organ that converts pressure waves to action potentials.

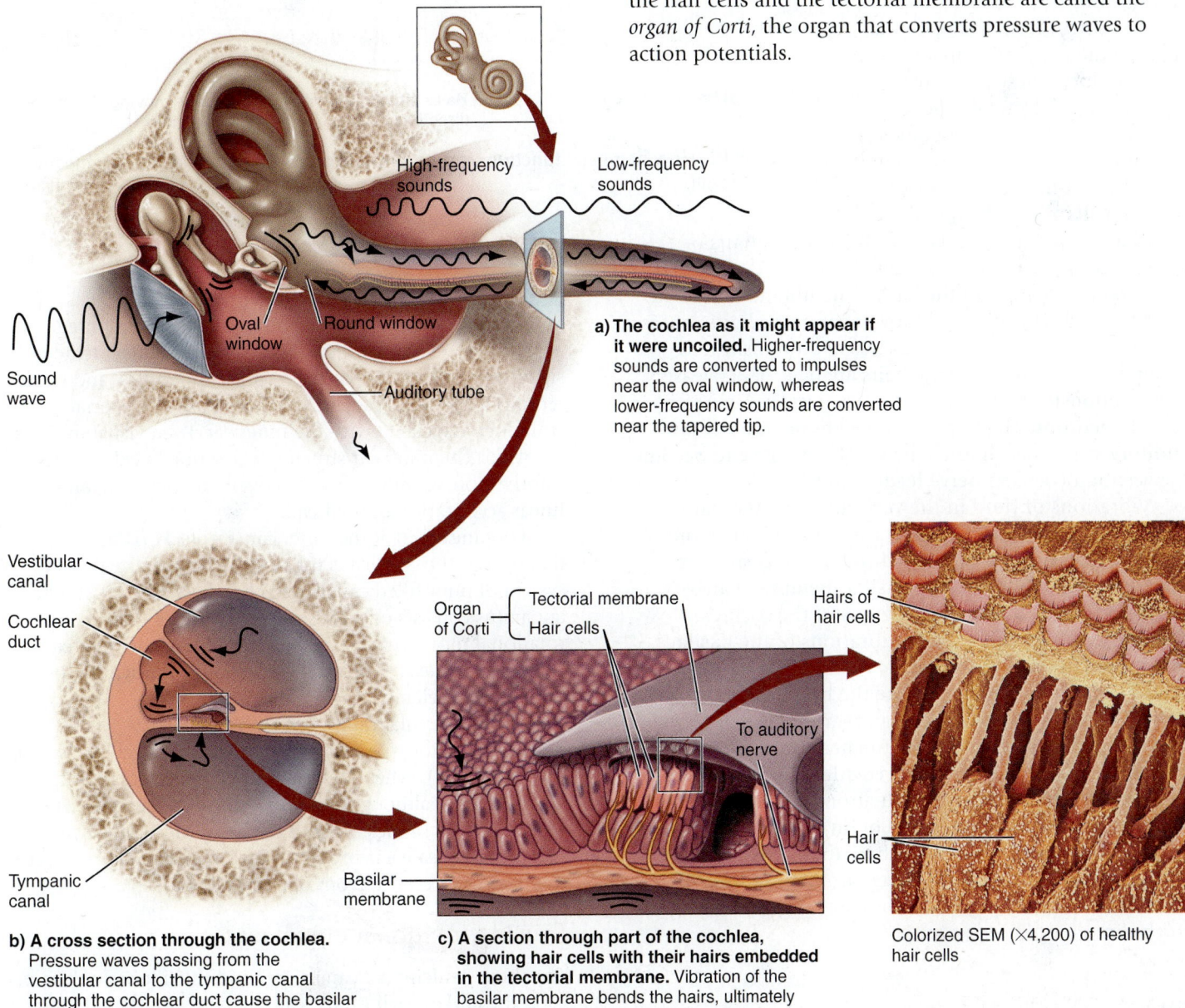

a) **The cochlea as it might appear if it were uncoiled.** Higher-frequency sounds are converted to impulses near the oval window, whereas lower-frequency sounds are converted near the tapered tip.

b) **A cross section through the cochlea.** Pressure waves passing from the vestibular canal to the tympanic canal through the cochlear duct cause the basilar membrane to vibrate.

c) **A section through part of the cochlea, showing hair cells with their hairs embedded in the tectorial membrane.** Vibration of the basilar membrane bends the hairs, ultimately generating impulses in sensory neurons.

Colorized SEM (×4,200) of healthy hair cells

Figure 12.10 Structure and function of the cochlea.

✅ Suppose the hair cells at the very beginning of the cochlea began malfunctioning, such that they began sending constant action potentials (even if no vibrations were occurring anywhere in the cochlea). What sort of perception would probably result, and why?

When sound waves strike the oval window, it generates pressure waves in the watery fluid in the vestibular canal (see Figure 12.10). These waves travel around the cochlea to the tympanic canal. Eventually, the waves strike another membrane called the *round window*, which bulges inward and outward in synchrony with the oval window, dissipating some of the pressure.

Some of the pressure waves take a shortcut from the vestibular canal to the tympanic canal via the cochlear duct. These pressure waves cause the basilar membrane to vibrate like the strings of a musical instrument. The slight movement of the basilar membrane causes the hair cells above the membrane to vibrate as well (**Figure 12.11**). Because the hairs of the hair cells are embedded in the less movable tectorial membrane, vibration of the basilar membrane causes the hairs to bend. This physical movement of the hairs causes the hair cells to release more or less neurotransmitter, depending on which way the hairs are bent. The release of neurotransmitter in turn generates impulses in nearby sensory neurons. The impulses travel in neurons of the auditory nerve, which joins the vestibular nerve to become the vestibulocochlear nerve leading to the brain.

Vibrations of fluid in the vestibular and tympanic canals are of the same frequency as the original sound waves, which may be of many different tones at once. But because the fibers of the basilar membrane are of different lengths at different regions of the cochlea, sound waves are converted to vibrations of the basilar membrane in a location-specific manner according to tone. High-pitched tones vibrate the basilar membrane closer to the base of the cochlea (near the oval and round windows), whereas low-pitched tones vibrate the basilar membrane closer to the cochlear tip. The identical impulses generated in sensory neurons at these different locations are interpreted by the brain as sounds of specific tones on the basis of where they originated on the basilar membrane. This is yet another example of the location-specific nature of information processing by the brain.

> ↺ **Recap** The outer ear receives and directs sound to the middle ear. The three bones of the middle ear amplify the sound and pass it to the cochlea. Vibrations of fluid in the cochlea cause the hairs of receptor hair cells to bend, generating action potentials in nearby neurons. ∎

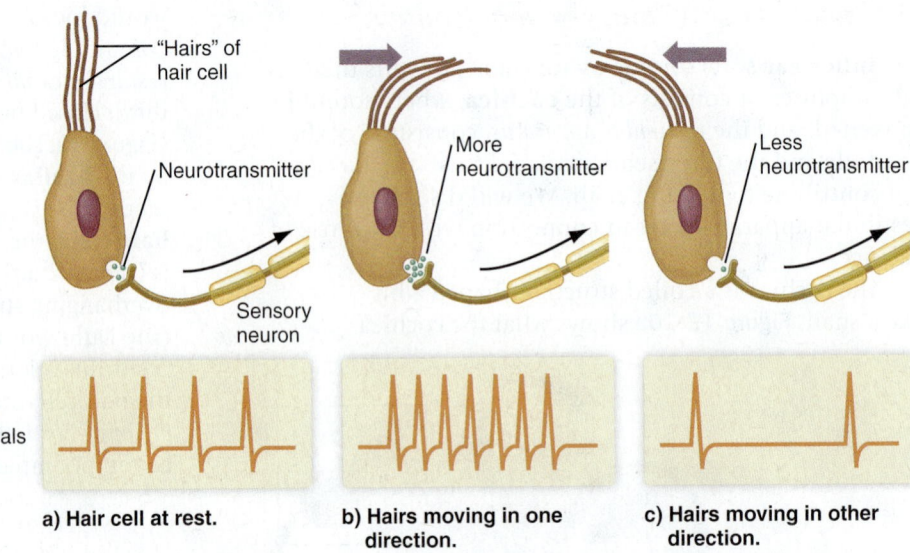

a) **Hair cell at rest.**   b) **Hairs moving in one direction.**   c) **Hairs moving in other direction.**

Figure 12.11 How hair cells function as mechanoreceptors. The hairs are actually modified cilia or microvilli.

12.5 The inner ear plays an essential role in balance

Maintaining your balance against gravity requires integration of multiple sensory inputs. Those sensory inputs include signals from the joint receptors, muscle spindles, and tendon receptors described previously, as well as from special structures associated with the inner ear. Even visual input is involved. The inner ear structures, described next, provide information about rotational movement, position, and linear acceleration of the head.

Looking again at the inner ear (**Figure 12.12a**), we see that next to the cochlea is the **vestibular apparatus,** a system of fluid-filled canals and chambers. The vestibular apparatus consists of three *semicircular canals* for sensing rotational movement of the head and body and an area called the *vestibule*, which senses static (nonmoving) position and linear acceleration and deceleration.

The semicircular canals and the vestibule have mechanoreceptor-type hair cells embedded in a gel-like material (**Figure 12.12b**). When the head moves or changes position, the gel also moves, although a bit more slowly because of inertia. The movement of the gel bends the hairs, which causes the hair cells to release a neurotransmitter that ultimately generates impulses in sensory neurons of the vestibular nerve.

Sensing rotational movement

The three semicircular canals are fluid-filled tubes of bone. Near the end of each canal is a bulging region called the *ampulla*, containing a soft, gel-like dome called the *cupula* (Figure 12.12b). Hairs of mechanoreceptor sensory neurons are embedded in the cupula.

When you turn your head, the fluid in your semicircular canal moves more slowly than your head because of inertia.

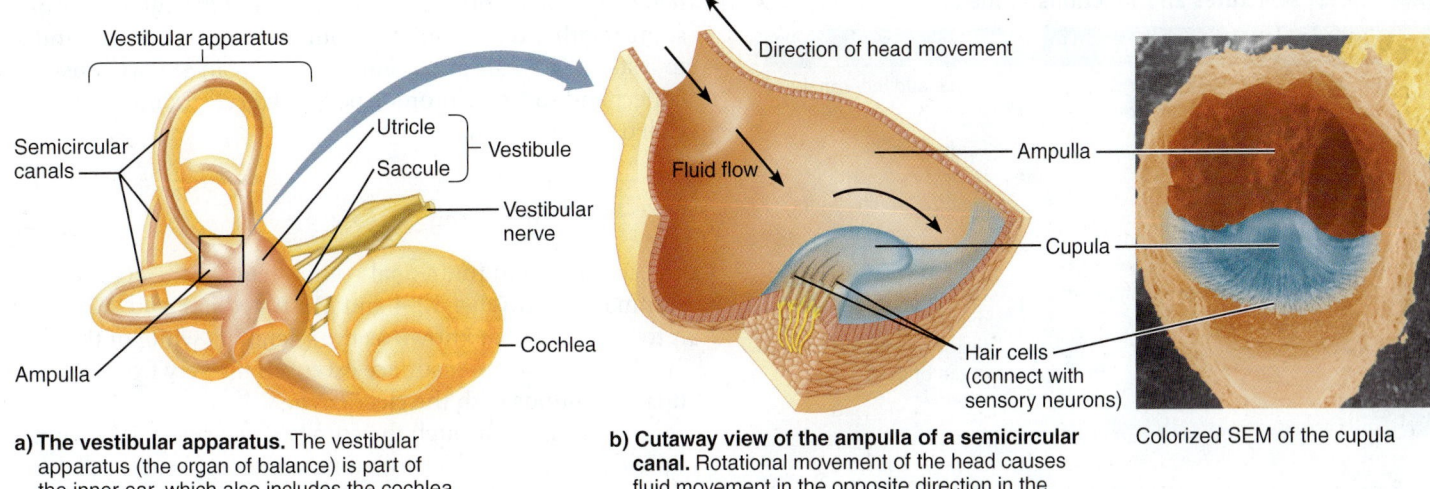

a) **The vestibular apparatus.** The vestibular apparatus (the organ of balance) is part of the inner ear, which also includes the cochlea.

b) **Cutaway view of the ampulla of a semicircular canal.** Rotational movement of the head causes fluid movement in the opposite direction in the semicircular canal. The fluid pushes against the cupula, bending it and activating the mechanoreceptor hairs of the hair cells.

Colorized SEM of the cupula

Figure 12.12 The vestibular apparatus.

The shifting position of the fluid bends the soft cupula and the hairs within it, producing impulses. The structure of the hairs enables them to detect the direction of the movement, in addition to movement in general. Because there are three semicircular canals and each is oriented on a different plane, together they can detect any and every possible rotational movement of the head.

☑ If you spin in place many times and then suddenly stop, you will experience a sensation as if you are spinning in the other direction (even though you are now standing still). Propose an explanation for this phenomenon.

Sensing head position and acceleration

To sense the position of the head relative to the Earth, we need a mechanism that is sensitive to gravity. The ear is such a mechanism. The two fluid-filled chambers of the vestibule, called the **utricle** and the **saccule,** each contain mechanoreceptor hair cells, gel, and hard crystals of bonelike material called **otoliths** (literally, "ear stones") (**Figure 12.13a**).

The otoliths are embedded in the gel near the gel's surface, unattached to the bone wall of the vestibule. The otoliths are indeed like little rocks, and they are considerably heavier than the gel itself. When the head tips forward, for instance, the otoliths slide toward the center of the gravitational pull, causing the gel to slip and bending the hairs of hair cells (**Figure 12.13b**). Again, this produces impulses in nearby sensory neurons that are transmitted to the brain for interpretation.

As well as static position, the utricle and saccule respond to linear acceleration and deceleration (**Figure 12.13c**). When you are in a car that is accelerating rapidly, you feel like you are being pushed backward in your seat. This is the force of acceleration acting against your own inertia. Likewise,

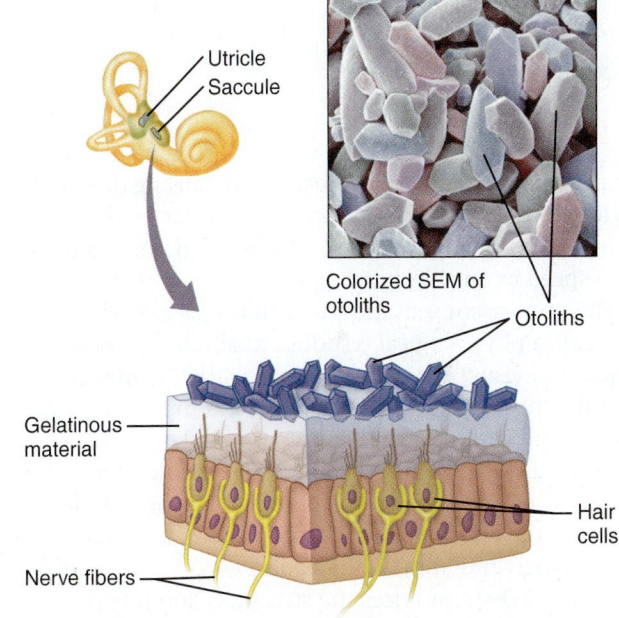

Colorized SEM of otoliths

a) **A cutaway view of the vestibule with the head at rest.**

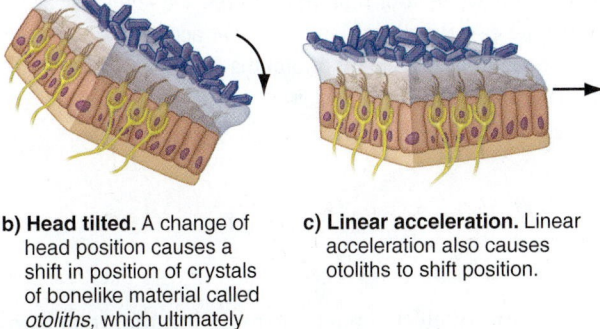

b) **Head tilted.** A change of head position causes a shift in position of crystals of bonelike material called *otoliths*, which ultimately stimulates hair cells.

c) **Linear acceleration.** Linear acceleration also causes otoliths to shift position.

Figure 12.13 How the utricle and saccule sense static head position and linear acceleration.

Table 12.3 **Structures and functions of the ear**

Region of ear	Functions	Structures
Outer ear	Receives and channels sound	Pinna, auditory canal
Middle ear	Amplifies sound	Eardrum, malleus, incus, stapes, oval window, auditory tube
Inner ear	Sorts sounds by tone and converts them into impulses	Cochlea, vestibular canal, tympanic canal, cochlear duct, basilar membrane, hair cells, tectorial membrane, organ of Corti, round window
	Senses rotational movements and converts them into impulses	Semicircular canals, ampulla, cupula, hair cells
	Senses static (nonmoving) position and linear acceleration and deceleration; converts them into impulses	Vestibule, utricle, saccule, otoliths, hair cells

your otoliths are being pushed backward, causing the hair cells to bend. The otoliths respond only to acceleration and deceleration (a change in linear velocity), and not to a constant rate of speed in one direction.

The unpleasant sensation we call *motion sickness* represents a physiological response to abnormal body motion and visual experience. Motion sickness may be due to conflicting sensory inputs from the vestibular apparatus, the eyes, and various receptors in muscles, tendons, and joints throughout the body. The CNS is unable to integrate these conflicting signals into an understandable whole, producing the characteristic symptoms of nausea, cold sweats, hyperventilation, and headache.

Table 12.3 summarizes the structures and functions of the ear.

↩ **Recap** Rotational movement of the head is sensed as the result of movement of fluid in the three semicircular canals of the inner ear. Head position and linear acceleration are sensed as the result of movement of otoliths (ear stones) in the utricle and saccule of the inner ear. ▪

12.6 **Vision: detecting and interpreting visual stimuli**

Light is a form of electromagnetic radiation that travels in waves at a speed of 186,000 miles per second. Our eyes allow us to receive and process light. They enable us to detect light from objects both nearby and distant and from sources either dim or bright, from the candle in your hand to a star light-years away. But before we can see, we must collect and focus light onto specialized cells in our eyes called *photoreceptors.*

Structure of the eye

The structure and functional components of the eye are summarized in Table 12.4 and illustrated in Figure 12.14. A tough outer coat known as the sclera, or "white of the eye," covers the outer surface except in the very front, where it is continuous with the clear **cornea.**

Light passes through the cornea and a space filled with fluid called *aqueous humor* that nourishes and cushions the cornea and lens. Light then either strikes the **iris,** a colored, disk-shaped muscle that determines how much light enters the eye, or passes through the **pupil,** the adjustable opening in the center of the iris. Light that passes through the pupil strikes the **lens,** a transparent, flexible structure attached by connective tissue fibers to a ring of circularly arranged smooth muscle called the *ciliary muscle.*

After passing through the main chamber of the eye filled with *vitreous humor,* light encounters the layers at the back and sides of the eye. This is the **retina,** comprising primarily photoreceptor cells, neurons, and a few blood vessels. Between the retina and the sclera at the back of the eye lies the choroid, consisting of pigmented cells and blood vessels. The pigmented cells absorb light not sensed by photoreceptors so that the image does not become distorted by reflected light, and the blood vessels nourish the retina. At the back of the eyeball is the **optic nerve,** which carries

Table 12.4 **Parts of the eye and their functions**

Eye structure	Functions
Sclera	Covers and protects eyeball
Cornea	Bends incoming light (focuses light)
Aqueous humor	Nourishes and cushions cornea and lens
Iris	Adjusts amount of incoming light
Lens	Regulates focus
Ciliary muscle	Adjusts curvature of the lens
Vitreous humor	Transmits light to retina
Retina	Absorbs light and converts it into impulses
Rods	Photoreceptors responsible for black-and-white vision in dim light
Cones	Photoreceptors responsible for color vision and high visual acuity
Macula	Central region of the retina with the highest density of photoreceptors
Optic disk	"Blind spot" where optic nerve exits eye
Optic nerve	Transmits impulses to the brain
Choroid	Nourishes retina and absorbs light not absorbed by retinal photoreceptors

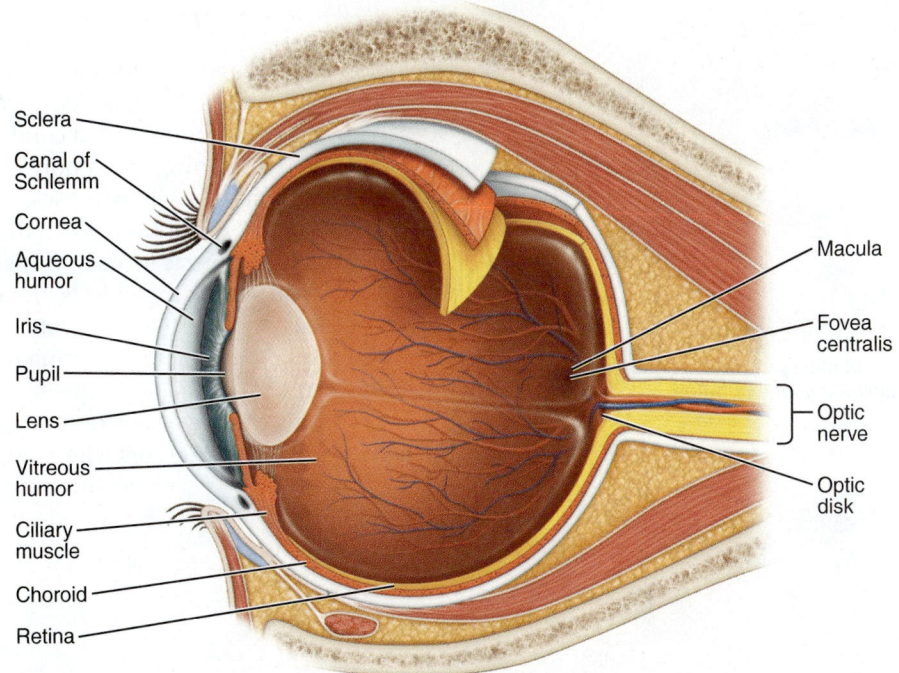

Figure 12.14 Structure of the eye. Light enters through the transparent cornea and passes through aqueous humor, the pupil, the lens, and vitreous humor before striking the retina at the back of the eye. The macula is the central area of the retina with the highest visual acuity. The nerves and blood vessels exit the eye through the optic disk.

information to the thalamus, to be forwarded to the visual cortex for interpretation. Finally, skeletal muscles surround the eye and control its movements, so we can choose exactly where to look.

Several sites on the retina deserve special mention. The **macula** is the central region of the retina, where photoreceptor density is the highest. When we want high visual acuity, we look directly at an object, focusing the image on the macula. At the very center of the macula is a small pit called the *fovea centralis* that is lined with highly packed photoreceptors.

The **optic disk** is the area where the axons of the optic nerve and associated blood vessels exit the eye, so there are no photoreceptors there at all. The optic disk leaves us with a "blind spot" in each eye. To find the blind spots, try the exercise in Figure 12.15.

Regulating the amount of light and focusing the image

The iris adjusts the amount of light entering the eye with two sets of smooth muscle. When bright light strikes the eye, contraction of muscles arranged circularly around the pupil causes the pupil to contract. Otherwise, the intensity of daylight would overwhelm our photoreceptors and temporarily blind us. In dim light, contraction of smooth muscles arranged radially around the pupil causes the pupil to dilate.

Figure 12.15 Find your blind spot. To find the blind spot in your right eye, position the figure about 4 inches directly in front of you. Cover your left eye. While focusing your right eye on the square, slowly move your head away from the figure. Eventually, the circle will disappear, because at a certain distance from your right eye, its image will strike the blind spot. You can do the same demonstration with your left eye by covering your right eye and focusing on the circle; the square will disappear.

Nerves control each set of muscles. When examining an unconscious or injured person, a physician may shine a light in each eye. The pupils should contract in response to light and dilate when the light is removed. Pupils that are *fixed and dilated* (unresponsive to light) are a bad sign, possibly indicating more widespread failure of the nervous system.

Light entering the eye is focused by the cornea and the lens. The cornea, which is curved, is actually responsible for bending most of the incoming light. However, the curvature of the cornea is not adjustable. Our ability to regulate the

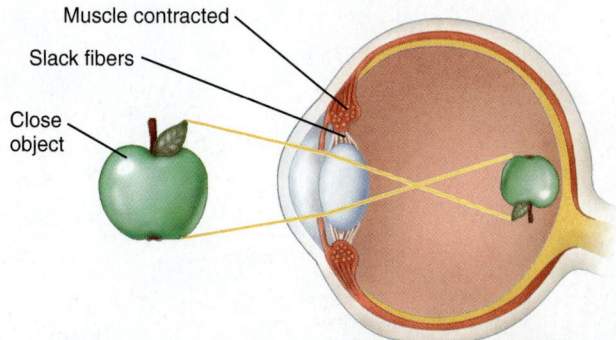

a) For near objects, contraction of ciliary muscles reduces the tension on the fibers and allows the lens to bulge.

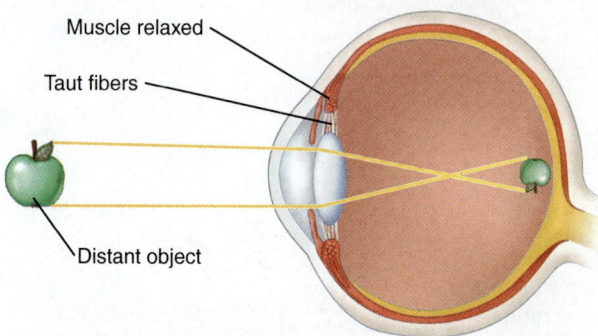

b) For focus on distant objects, relaxation of ciliary muscles increases tension on the fibers, which pull on the lens and flatten it.

Figure 12.16 Accommodation. Notice that the images created on the retina are upside down.

degree to which incoming light is bent, and therefore our ability to change focus between near and far objects, is accomplished solely by adjusting the curvature of the lens. This is done by the ciliary muscle.

When the ciliary muscle contracts, the inner radius of the muscle shrinks, reducing the tension on the fibers attached to the lens (Figure 12.16a). This allows the lens to bulge, and we focus on a near object. When the ciliary muscle relaxes, the ring of muscle increases the tension on the lens, stretching and flattening it and bringing more distant objects into focus (Figure 12.16b). **Accommodation** refers to the adjustment of lens curvature so we can focus on either near or far objects.

The light rays from each point of an object are bent and focused so that the image created on the retina is inverted (upside down). However, the brain interprets the image as right side up.

As we get older, the lens tends to stiffen and cannot resume a bulging shape even when the ciliary muscles are maximally contracted. The result is *presbyopia*, the inability to focus on nearby objects, which generally appears after age 40. People with presbyopia must hold reading material farther away to see the print clearly. Reading glasses can easily correct presbyopia.

☑ Suppose a person's lenses were removed entirely. Would the person become entirely blind, able to see some light but no images, able to see blurry images, or able to see sharp images? Explain.

Eyeball shape affects focus

Differences in the shape of the eyeball can affect the ability to focus properly. Glasses and contact lenses correct improper focus.

Myopia is a common inherited condition in which the eyeball is slightly longer than normal (or more rarely, myopia occurs when the ciliary muscle contracts too strongly) (Figure 12.17a). Even when the lens is flattened maximally, distant objects focus in front of the retina. People with myopia can see nearby objects, but distant objects are out of focus, which is why it is called *nearsightedness*. Concave lenses, which bend incoming light so it focuses on the retina, can correct nearsightedness.

Less common is the opposite condition, *hyperopia* (farsightedness), which occurs when the eyeball is too

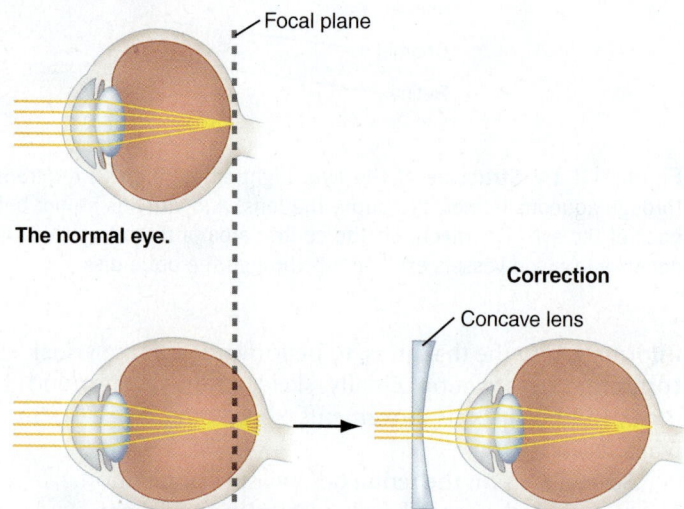

The normal eye.

Correction

a) Nearsightedness (myopia). Nearsighted persons can see near objects clearly, but distant objects are out of focus because the focal point is in front of the retina.

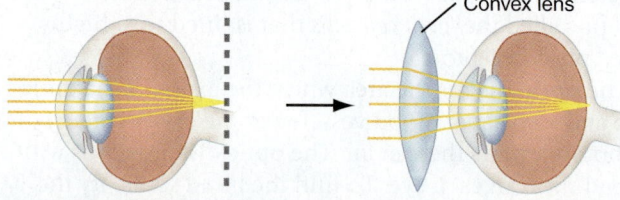

b) Farsightedness (hyperopia). Farsighted persons can see distant objects clearly, but near objects are out of focus.

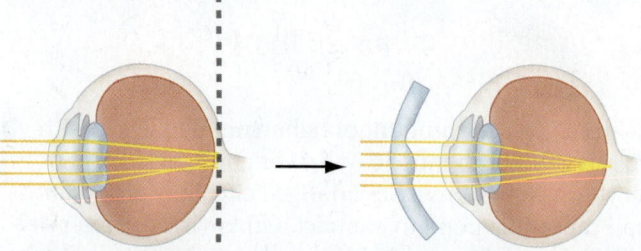

c) Astigmatism. Astigmatism is due to an abnormal curvature of either the cornea or the lens.

Figure 12.17 Examples of abnormal vision. All three types of abnormal vision are correctable.

short and nearby objects focus behind the retina (Figure 12.17b). Farsighted people can see distant objects clearly, but nearby objects are out of focus. Convex lenses correct farsightedness.

Astigmatism refers to blurred vision caused by irregularities of the shape of the cornea or lens (Figure 12.17c). The result is that light is scattered and may not focus evenly on the retina. Astigmatism can be corrected with specially ground lenses that exactly compensate for the irregularities of the cornea and lens.

Light is converted into action potentials

Like other sensory organs, the eyes convert a stimulus (light) into impulses. In the eye, this process occurs in the retina. The unique structure of the retina allows us to see in color, adapt to varying light intensities, and perceive images of the world around us.

Figure 12.18 shows a close-up of the retina, which consists of four layers:

1. The outermost layer consists of pigmented cells that, along with the choroid, absorb light not captured by the photoreceptor cells.
2. Next is a layer of photoreceptor cells called **rods** and **cones** because of their shape.
3. The rods and cones synapse with the third layer, a layer of neurons called *bipolar cells*. Bipolar cells partially process and integrate information and then pass it on to the fourth layer.
4. The innermost layer consists of ganglion cells. Ganglion cells are also neurons. Their long axons become the optic nerve going to the brain.

Notice that light must pass through the third and fourth layers to reach the photoreceptor cells, but this does not significantly hinder the ability of the photoreceptor cells to receive and convert light.

Rods and cones respond to light

Figure 12.19 takes a closer look at the photoreceptor cells, rods, and cones. One end of the cell consists of a series of flattened disks arranged to form either a rod- or cone-shaped structure, as the names imply. The flattened disks contain numerous molecules of a particular light-sensitive protein called a *photopigment*.

A photopigment protein undergoes a change of shape when it is exposed to energy in the form of light. The change in photopigment molecule shape causes the photoreceptor

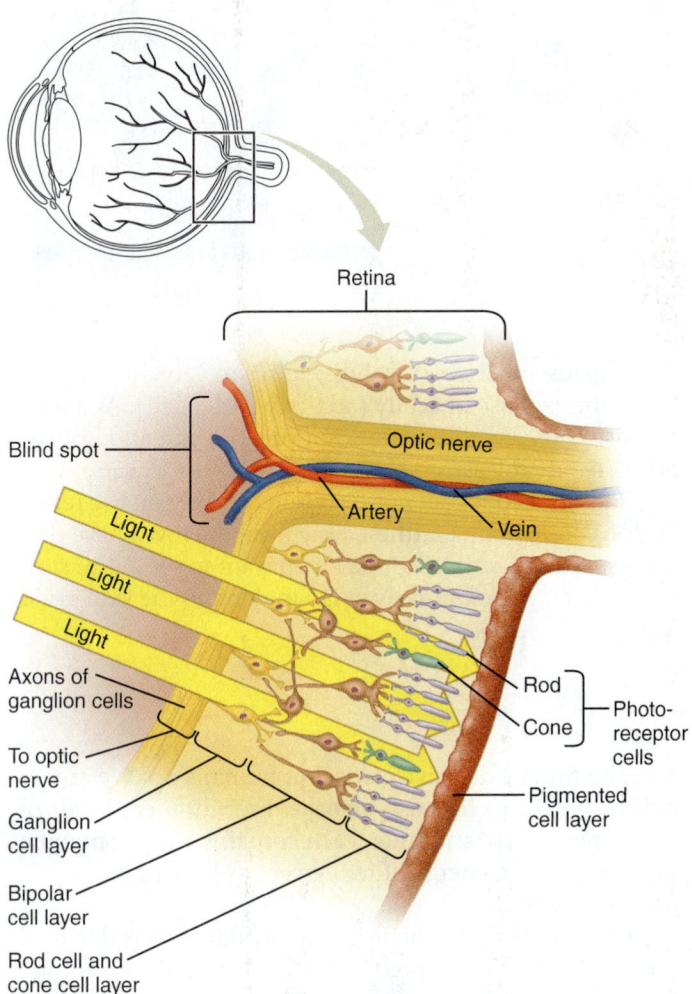

Figure 12.18 Structure of the retina. Light must pass between the axons of the ganglion cells and through the ganglion and bipolar cell layers to reach the photoreceptor cells, the rods and cones.

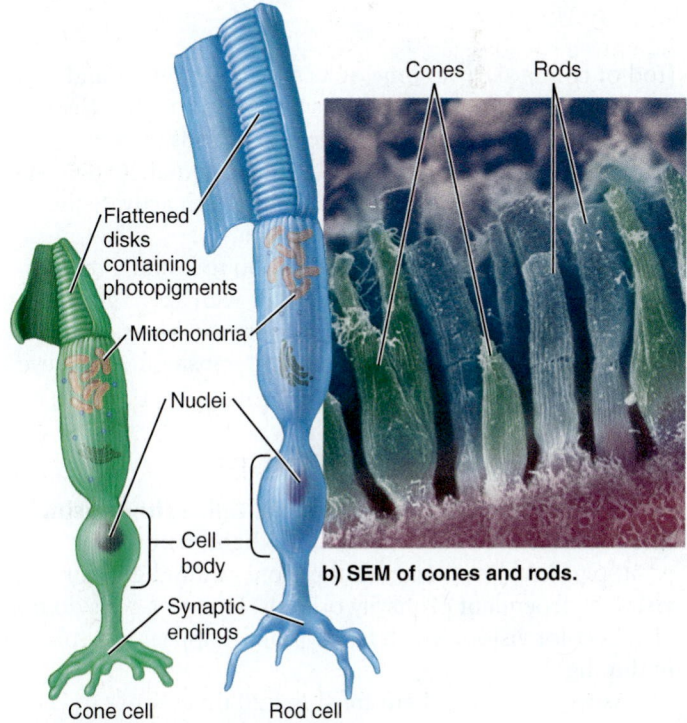

a) The structure of cones and rods.

Figure 12.19 The photoreceptor cells, cones and rods. The photopigments in the flattened disks are sensitive to light.

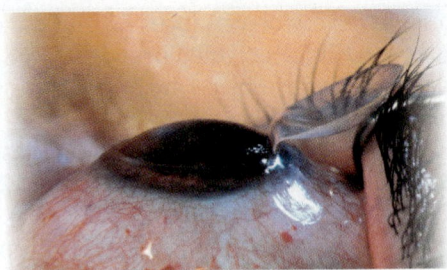

HEALTH & WELLNESS
LASIK to Correct Vision Problems

Nearsightedness, farsightedness, and astigmatism are all visual problems that can be corrected with eyeglasses or contact lenses of one sort or another. But if you want a more permanent solution, you might consider LASIK surgery.

LASIK stands for laser in-situ keratomileusis, a generally safe and effective surgical technique in which lasers are used to permanently change the shape of the cornea. The technique can be done under local anesthesia in the ophthalmologist's office. **1** The surgeon creates a thin flap of corneal tissue and folds it back. **2** A laser beam is then used to remove a small amount of corneal tissue from the cornea to reshape the cornea and correct the visual problem. **3** The flap of tissue is returned to its original position, and the eye is allowed to heal.

Most patients who undergo LASIK surgery will have vision close to 20/20 (normal) once the eye has healed completely. Normal vision is not assured, but most patients come pretty close to it. Some people experience halos of light around objects or are sensitive to glare after surgery, but these side effects generally are temporary. Every patient is different, of course, but complications from the surgery are rare. Questions about *your* chances of attaining normal vision with LASIK should be addressed to your surgeon.

Not everyone is a good candidate for LASIK. People whose eyesight is still changing are generally advised to wait until their prescription for eyeglasses or contacts has been stable for at least two years. In addition, LASIK does not correct presbyopia, the condition for which one uses reading glasses. Nevertheless, many people who wear glasses or contacts all the time for nearsightedness, farsightedness, and astigmatism experience the freedom of being able to get rid of their corrective lenses entirely.

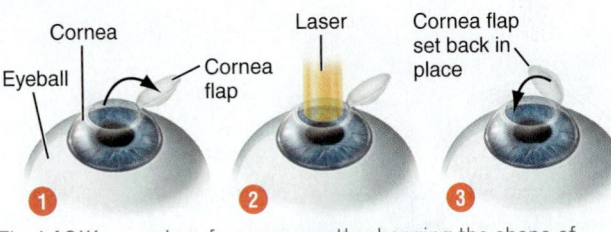

The LASIK procedure for permanently changing the shape of the cornea.

(rod or cone) to close some of its sodium channels and reduce the amount of neurotransmitter it normally releases. Because the neurotransmitter released by rods or cones normally inhibits the bipolar cells, light ultimately increases the activity of the bipolar cells, which in turn activate the ganglion cells (see Figure 12.18).

There are approximately 120 million rods and 6 million cones, but only 1 million ganglion cells with axons going to the brain. Clearly, a great deal of convergence and summation of signals occurs at each level of neuron transfer from the eye to the brain.

Rods provide vision in dim light

Rods all have the same photopigment, called **rhodopsin.** Rhodopsin is much more sensitive to light than the photopigments in cones, and therefore in dim light our vision is dependent primarily on rods. However, rods do not give us color vision, which is why objects appear less colorful in dim light.

As we have seen, there are about 20 times more rods than cones. If we imagine that all 120 million rods converge on only half of the ganglion cells (one-half million), there would be 240 rods converging on a single ganglion cell. This

convergence increases our ability to see in very dim light, but at the expense of acuity (accuracy and detail). As a result, our vision in very dim light is not very colorful and not very detailed, but at least we can see.

Rods and cones are not distributed evenly on the retina. Regions of the retina farthest away from the fovea have the highest ratio of rods to cones. If you want to see a dim star at night, don't look directly at it—look off to the side just a little.

Cones provide color vision and accurate images

Not all animals perceive color, but humans do. We are able to see colors because we have three different kinds of cones: red, green, and blue. Each contains a photopigment that absorbs the energy of red, green, or blue light particularly well.

Our ability to distinguish a variety of colors is due to the way the brain interprets the ratios of impulses coming from the ganglion cells connected to the three types of cones. When all three types are activated by all different wavelengths, we perceive white light. The perception of black is no light at all.

Cones also are responsible for visual acuity. However, cones require stronger light to be activated because the cone photopigments are much less sensitive to light than the rhodopsin in rods. This is why your ability to distinguish between colors declines in dim light and it becomes difficult to tell whether a dark-colored car is green or red. In dim light, you are seeing primarily with your rods, which can detect only black and white.

✅ Do you think that animals active only at night, such as most owls, can see in color? Explain your answer.

Visual receptors adapt

From your own experience, you know that vision adapts to changing light conditions over several seconds or minutes. Adaptation seems to take longer when going from bright to dim light than it does in the opposite direction. Adaptation depends on rapid adjustment of the pupil by the iris and on adaptation by the rods.

The absorption of light by rhodopsin "uses up" the photopigment temporarily. Light energy actually breaks the rhodopsin into two molecules. These molecules can be resynthesized into rhodopsin again, but only over a period of minutes.

When you have been out in bright light, most of the rhodopsin in the light-sensitive rods has been broken down. Then, when you enter a dim room and the cones no longer are functioning, there is not enough intact rhodopsin for good vision in dim light. With time (5 to 15 minutes), the rhodopsin is resynthesized and available again.

Conversely, when you go out into sunlight from a dimly lit room, the light seems very bright because you have the maximal amount of photopigment available in both rods and cones. Most of the rhodopsin in the rods is quickly used up, however, and you revert to using primarily cones in bright light.

↩ **Recap** Light passes through the cornea, the lens, and the aqueous fluid of the eye before striking the photoreceptors in the retina. The iris regulates the amount of light entering the eye, and the cornea and the lens adjust the focus. Light energy alters the structure of photopigments—rhodopsin in rods; and red, blue, or green photopigments in cones. Rods provide vision in dim light. Cones are responsible for color vision and detailed images. ▪

12.7 Disorders of sensory mechanisms

Deafness: loss of hearing Deafness can have many causes. Deafness caused by damage to hair cells (Figure 12.20) is called **nerve deafness** because sounds cannot be converted into impulses in sensory nerves. Nerve deafness is generally the result of frequent exposure to loud sounds. Damage to the tympanic membrane or the bones of the middle ear, on the other hand, is called **conduction deafness.** In conduction deafness, sound waves simply are not transferred to the inner ear at all. Conduction deafness is frequently due to arthritis of the middle ear bones.

Some types of hearing loss may be partially corrected with a hearing aid, which amplifies incoming sounds, but even amplification won't be very effective if hair cells are missing. Some people with nerve deafness benefit from a cochlear implant, a tiny implanted microprocessor that converts sound waves into electrical signals.

Otitis media: inflammation of the middle ear A common cause of earache is *otitis media,* or inflammation of the middle ear. Usually, this results from an upper respiratory tract infection that extends up the auditory tube. The tube may become blocked and trap fluid in the middle ear. Otitis media usually responds to antibiotics.

Ménière's syndrome: inner ear condition impairs hearing and balance Ménière's syndrome is a chronic condition of the inner ear. The cause is unknown, but it may be due to excess fluid in the cochlea and semicircular canals. Symptoms include repeated episodes of dizziness and nausea accompanied by progressive hearing loss. Balance is usually affected to the point that patients find it difficult to stay upright. Mild cases can be treated with motion sickness medications. Severe cases may benefit from surgery to drain excess fluid from the inner ear.

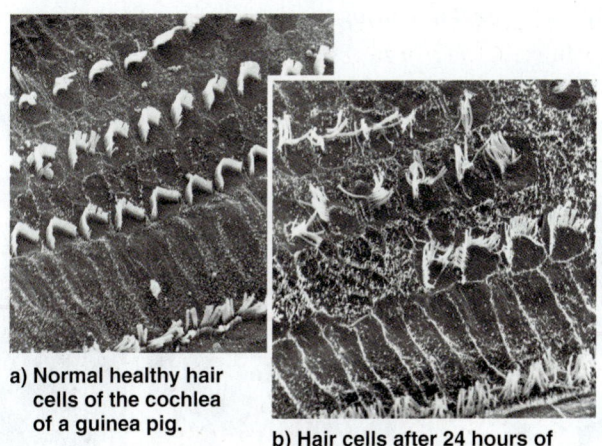

a) Normal healthy hair cells of the cochlea of a guinea pig.

b) Hair cells after 24 hours of exposure to loud sounds.

Figure 12.20 Damage to hair cells in the ears.

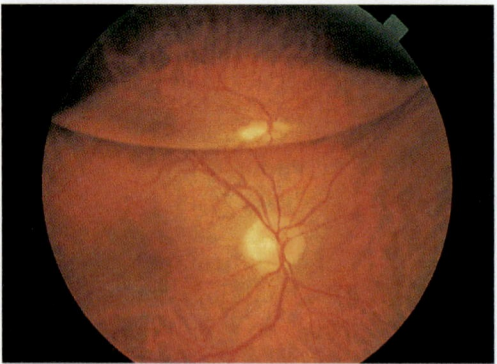

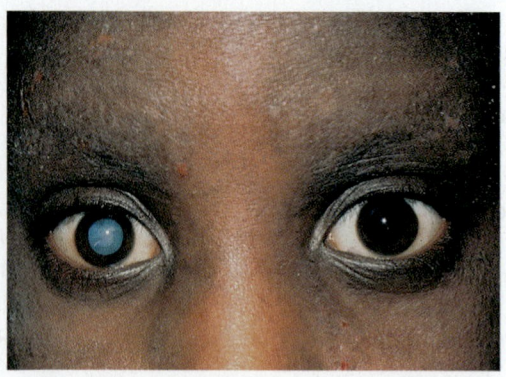

a) Retinal detachment. In this ophthalmoscope view of a patient's eye, the retina has peeled away from the choroid (upper third of the eye). At lower center is the yellow optic disk, through which the blood vessels travel.

b) Cataracts. The cataract in this person's right eye was caused by a chronic inflammatory disease process.

Figure 12.21 Common eye disorders.

Retinal detachment: retina separates from choroid

If the thin retina is torn, the vitreous humor may leak through the tear and peel the retina away from the choroid, like peeling wallpaper from a wall (**Figure 12.21a**). The most common cause of retinal detachment in healthy persons is a blow to the head. The detached region of the retina loses most of its blood supply and its ability to focus an image properly. Symptoms include flashes of light, blurred vision, or loss of peripheral vision. Retinal detachment is an emergency, but prompt surgery can usually repair the damage.

Cataracts: the lens becomes opaque

Cataracts are a decrease in the normal transparency of a lens (**Figure 12.21b**). Normally, the lens contains a particular kind of protein folded in such a way that it is transparent. If the delivery of nutrients to the lens is not sufficient, these proteins may denature, or clump, making the lens opaque. Some cataracts are congenital, but most are age-related or secondary to diabetes or other diseases.

About 5.5 million people in the United States (1 in 50) have a cataract impairing their vision. The treatment, which is usually successful, involves removing the lens surgically and replacing it with an artificial lens.

Glaucoma: pressure inside the eye rises

As mentioned previously, aqueous humor fills the chamber between the cornea and the lens. Capillaries located near the ciliary muscles of the eye produce aqueous humor, which brings nutrients and oxygen to the lens and cornea and carries away wastes. Normally, aqueous humor is drained from the eye and returned to the venous blood by a vessel called the *canal of Schlemm* (review Figure 12.14). Glaucoma is a condition in which this drainage vessel becomes blocked. The excess fluid increases pressure inside the eye and compresses blood vessels supplying the retina. Cells of the retina or optic nerve may eventually die, impairing vision and ultimately leading to blindness. The impairment of vision often starts as a loss of peripheral vision (**Figure 12.22a**).

Glaucoma generally develops slowly and painlessly over several years, without any outward symptoms. About 2.2 million people over the age of 40 in the United States (1 in 70) are visually impaired by glaucoma. Most patients do not realize they have it until they have already lost some vision. Less commonly, glaucoma develops acutely (within hours) with symptoms of blurred vision and red, swollen, sometimes painful eyes. Acute glaucoma is a medical emergency that requires immediate attention by an eye doctor.

Chronic (slow-onset) glaucoma can be detected by a peripheral vision test. If it is detected early enough, the progression of glaucoma can be slowed or stopped with drugs or surgery to lower the fluid pressure within the eye. However, any vision that has already been lost cannot be restored. Because it is such a risk factor for glaucoma, test of the fluid pressure in the eye is usually included in any comprehensive eye examination.

Normal vision.

a) Glaucoma.

b) Age-related macular degeneration.

Figure 12.22 Visual impairments caused by glaucoma and age-related macular degeneration.

Age-related macular degeneration Age-related macular degeneration (AMD) is a disease of visual impairment caused by detachment of the retina and degeneration of photoreceptor cells in the macular region of the retina. The most common cause of AMD is accumulation of cellular debris between the choroid and the retina. Less commonly, it may be caused by abnormal growth of blood vessels in the region. In both cases, the result is a loss of visual acuity in the center of the visual field, making it difficult to recognize faces or to read (Figure 12.22b). AMD affects about 1.6 million people over the age of 50 in the United States (1 in 65). The prevalence of AMD is expected to rise dramatically as the U.S. population continues to age, unless a cure is found. There is no effective cure yet for most cases of AMD, although various vascular growth factors and vitamin treatments show promise at slowing or delaying the progression of the disease.

Color blindness: inability to distinguish the full range of colors *Color blindness* is a general term for the inability to distinguish the full range of colors. Most forms of color blindness are caused by deficient numbers of particular types of cones. Less common are conditions in which one of the three cones is completely missing.

People with red-green color blindness, the most common form, are deficient in either red cones or green cones. They have trouble distinguishing between red and green, or they perceive them as the same color. The color they see is red or green, depending on which type of cone is missing.

The inability to perceive any color at all is quite rare in humans because it occurs only when two of the three types of cones are completely missing.

Color blindness is often inherited. Red-green color blindness is an X-linked recessive trait that is more common in men than in women. It affects approximately 8% of Caucasian males but only 1% of Caucasian females and is less common in other races. Traffic lights always have the red ("stop") light at the top and the green ("go") light at the bottom so that color-blind people can tell them apart by position. Color blindness is easily tested with a series of colored test plates (Figure 12.23).

🔄 **Recap** A detached retina is a retina that has separated from the underlying choroid. Glaucoma is a condition in which fluid pressure inside the eye rises, constricting the blood vessels that normally supply the living cells. Age-related macular degeneration is the most common cause of loss of vision in older persons. ▪

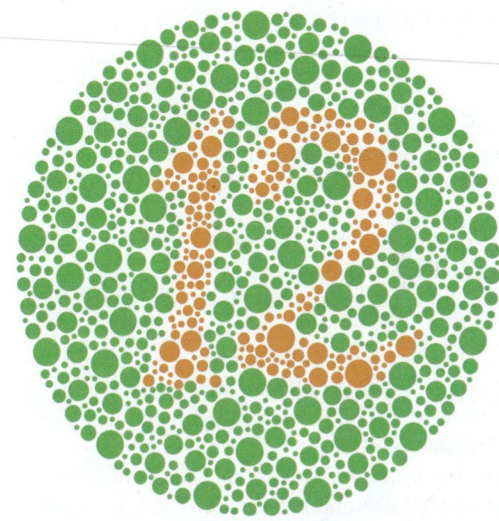

Figure 12.23 Test plate for diagnosing color blindness. A person with red-green color blindness would not be able to see the number 12 in a test plate such as this. An original plate should be used for accurate testing of color blindness.

✅ Suppose a woman had no blue cones at all in her retinas but had normal numbers of the other types of rods and cones. Could she see the "12" in this picture? Explain.

Chapter Summary

12.1 Receptors receive and convert stimuli *p. 276*

- Receptors convert the energy of a stimulus into impulses in sensory neurons.
- The brain interprets incoming impulses on the basis of where the sensory neurons originate.
- Some types of receptors adapt to continuous stimulation, whereas others (most notably pain receptors) do not.
- The somatic senses (temperature, touch, vibration, pressure, pain, and awareness of body position and movements) arise from receptors throughout the body.
- The five special senses (taste, smell, hearing, balance, and vision) arise from only certain specialized areas of the body.

12.2 Somatic sensations arise from receptors throughout the body *p. 278*

- The skin contains receptors for heat, cold, vibration, pressure, light touch, and pain. Receptors for temperature and for pain are also located in some internal organs.
- Receptors in joints sense joint position.
- Muscle spindles sense muscle length, and tendon receptors sense tendon (and muscle) tension.

12.3 Taste and smell depend on chemoreceptors *p. 281*

- Chemoreceptors for taste are located in structures called *taste buds*, most of which are on the tongue.
- The receptors for smell (called *olfactory receptors*) are located in the nasal passages. Humans have over 1,000 different olfactory receptor types.

12.4 Hearing: mechanoreceptors detect sound waves *p. 283*

- The outer ear channels sound waves to the tympanic membrane, the middle ear amplifies the sound waves, and the cochlea of the inner ear converts sound waves to impulses.
- Sound waves of different tones are converted to impulses at different locations in the cochlea.

12.5 The inner ear plays an essential role in balance *p. 286*

- The vestibular apparatus of the inner ear consists of three semicircular canals and the vestibule.
- The semicircular canals are structures that sense rotational movement of the head on three planes.
- Structures of the vestibule sense static (nonmoving) head position and linear acceleration.

12.6 Vision: detecting and interpreting visual stimuli *p. 288*

- The eye is the organ of vision, with structures for moving the eye (muscles), regulating the amount of light entering the eye (iris), focusing the visual image (lens), and converting light to impulses (retina).
- The photoreceptor cells in the retina are called *rods* and *cones*.
- Rods are 20 times more numerous than cones. They give us vision in dim light, but not in color and not with the highest detail.
- The three types of cones have different photopigments, so each type responds best to a different wavelength (color) of light. Cones are most effective in bright light, and they give us color vision and the sharpest image.
- Photoreceptors adapt to conditions of varying light intensities (brightness). It takes about 15 minutes to fully adapt to very dim light.

12.7 Disorders of sensory mechanisms *p. 293*

- Inflammation of the middle ear is a common cause of earache.
- Common eye disorders include retinal detachment, cataracts, glaucoma, age-related macular degeneration, and color blindness.

Terms You Should Know

accommodation, *290*
auditory tube, *284*
basilar membrane, *285*
chemoreceptors, *276*
cochlea, *285*
cones, *291*
cornea, *288*
hair cells, *285*
iris, *288*
lens, *288*
macula, *289*
mechanoreceptors, *276*
muscle spindle, *279*
olfactory receptor cell, *282*

optic disk, *289*
photoreceptors, *276*
receptor adaptation, *277*
retina, *288*
rhodopsin, *292*
rods, *291*
somatic sensations, *278*
special senses, *278*
taste bud, *281*
thermoreceptors, *276*
tympanic membrane
   (eardrum), *284*
vestibular
   apparatus, *286*

Concept Review

Answers can be found in the Study Area in MasteringBiology.

1. List the five classifications of receptors in terms of types of stimulus they convert.
2. Describe how receptor adaptation works.
3. Describe what information the somatic sensations provide.
4. Name the five special senses.
5. Compare and contrast fast pain, slow pain, and referred pain.
6. Describe the overall function of the three bones (the malleus, the incus, and the stapes) of the middle ear.
7. The eye contains both skeletal muscle and smooth muscle. Describe where each is located and indicate their functions.
8. Compare and contrast rods and cones.
9. Explain why it may take up to 15 minutes to become fully adapted to very dim light conditions.
10. Explain what causes glaucoma.

Test Yourself

Answers can be found in the Appendix.

1. What do the receptors in the skin for light touch and the receptors in the nose for smell have in common?
   a. They both respond to pressure.
   b. They both respond to chemicals.
   c. They send action potentials to the same area of the brain.
   d. They both adapt very quickly.

2. Which of the following are NOT found among the somatosensory receptors?
   a. mechanoreceptors   c. chemoreceptors
   b. thermoreceptors   d. pain receptors

3. A more powerful stimulus will result in:
   a. action potentials that travel more quickly
   b. activation of more receptors
   c. more frequent action potentials
   d. both (b) and (c)

4. Receptor adaptation results when:
   a. sensory neurons stop sending action potentials
   b. a stimulus is discontinued
   c. action potentials from different neurons cancel one another out
   d. the action potentials do not reach the brain

5. In what order do the structures of the ear come into play when hearing?
   a. malleus—tympanic membrane—cochlea—stapes—cochlear nerve
   b. auditory tube—malleus—cochlea—incus—cochlear nerve
   c. auditory canal— tympanic membrane—malleus—cochlea—cochlear nerve
   d. auditory canal—cochlea—tympanic membrane—cochlear nerve—malleus

6. Different pitched sounds will result in stimulation of receptors located in different regions of:
   a. hair cells   c. the vestibular apparatus
   b. round and oval windows   d. the cochlea

7. Which of the following correctly traces the path of light through the eye?
   a. cornea—aqueous humor—pupil—lens—vitreous humor—retina
   b. aqueous humor—pupil—cornea—lens—vitreous humor—retina
   c. pupil—cornea—vitreous humor—lens—aqueous humor—retina
   d. cornea—vitreous humor—lens—pupil—aqueous humor—retina

8. When we adjust our head and eye position to carefully look at something, we are directing the image to focus on the:
   a. optic disk
   b. macula
   c. periphery of the retina
   d. optic nerve

9. Which of the following will regulate the amount of light entering the eye?
   a. cornea
   b. lens
   c. iris
   d. retina

10. Which of the following is/are involved in focusing light?
    a. lens
    b. pupil
    c. cornea
    d. both (a) and (c)

11. Which of the following correctly describes the distribution of rods and cones in the retina?
    a. Rods and cones are evenly distributed throughout the retina.
    b. There are more cones on the periphery of the retina.
    c. Cones are more highly concentrated in the fovea centralis.
    d. Rods are more highly concentrated in the optic disk.

12. Which of the following statements explains why vision is less sharp in dim light than in bright light?
    a. Cones are more responsive in dim light.
    b. There are more rods than cones in the eye.
    c. Many rods converge on a single ganglion.
    d. The dilated pupil lets in more light than the lens can focus.

13. Conduction deafness would most likely be caused by damage to the:
    a. cochlear nerve
    b. malleus, incus, and/or stapes
    c. hair cells in the cochlea
    d. vestibular apparatus

14. Which of the following would most readily be treated with antibiotics?
    a. cataracts
    b. otitis media
    c. glaucoma
    d. Ménière's syndrome

15. Which of the following can be corrected with corrective lenses?
    a. myopia
    b. color blindness
    c. hyperopia
    d. both (a) and (c)

Apply What You Know

Answers can be found in the Study Area in MasteringBiology.

1. With the eyes closed, would an astronaut in outer space be able to detect lateral movement of the head (would the semicircular canals be functioning normally)? Would the astronaut be able to detect static head position (would the utricle and saccule be functioning normally)? Explain.

2. Why do you suppose that you are not normally aware of the blind spot in each eye?

3. Our sense of smell is closely linked to our feelings and emotions. Bad odors usually evoke negative feelings, whereas pleasant odors give us a general sense of well-being or may be associated with good memories (perfume makers know this). The house of a pet owner can have an odor that a guest finds objectionable but the owner may not even be aware of. With a long enough visit even the guest may not notice the pet odors. Why is this?

4. Walk into any drugstore and you'll find dozens of over-the-counter products for reducing pain. Why is the experience of pain so common and so problematic?

5. An amateur boxer reported on an online forum that he saw lights flash before his eyes after being hit on the top of the head. This was unusual since he's normally hit in the chin or ear and knows what it's like to brownout during a knockout blow. Once the lights faded, he was able to continue the fight. What happened when he was hit on the top of the head?

6. Try taking a really, really deep breath. If done correctly, there should be a moment of sharp pain in your chest. What's causing this?

7. What do you think might cause motion sickness, the combination of nausea and dizziness that some people get after a roller-coaster ride or when riding in a boat or a car? And why don't gymnasts get motion sick when they're performing?

MJ's BlogInFocus

Answers can be found in the Study Area in MasteringBiology.

1. Humans and other mammals respond to chemicals (called *pheromones*) released into the air by other members of their species, even though they can't consciously smell them. Based on what you have learned in this chapter about sensory mechanisms, where do you think the receptors for pheromones would most likely be located? For more on pheromones, visit MJ's blog in the Study Area in MasteringBiology and look under "Pheromones." http://goo.gl/vqiU0u

MasteringBiology®

Students Go to MasteringBiology for assignments, the eText, and the Study Area with animations, practice tests, and activities.

Professors Go to MasteringBiology for automatically graded tutorials and questions that you can assign to your students, plus Instructor Resources.

Answers to ✓ questions are available in the Appendix.

13 The Endocrine System

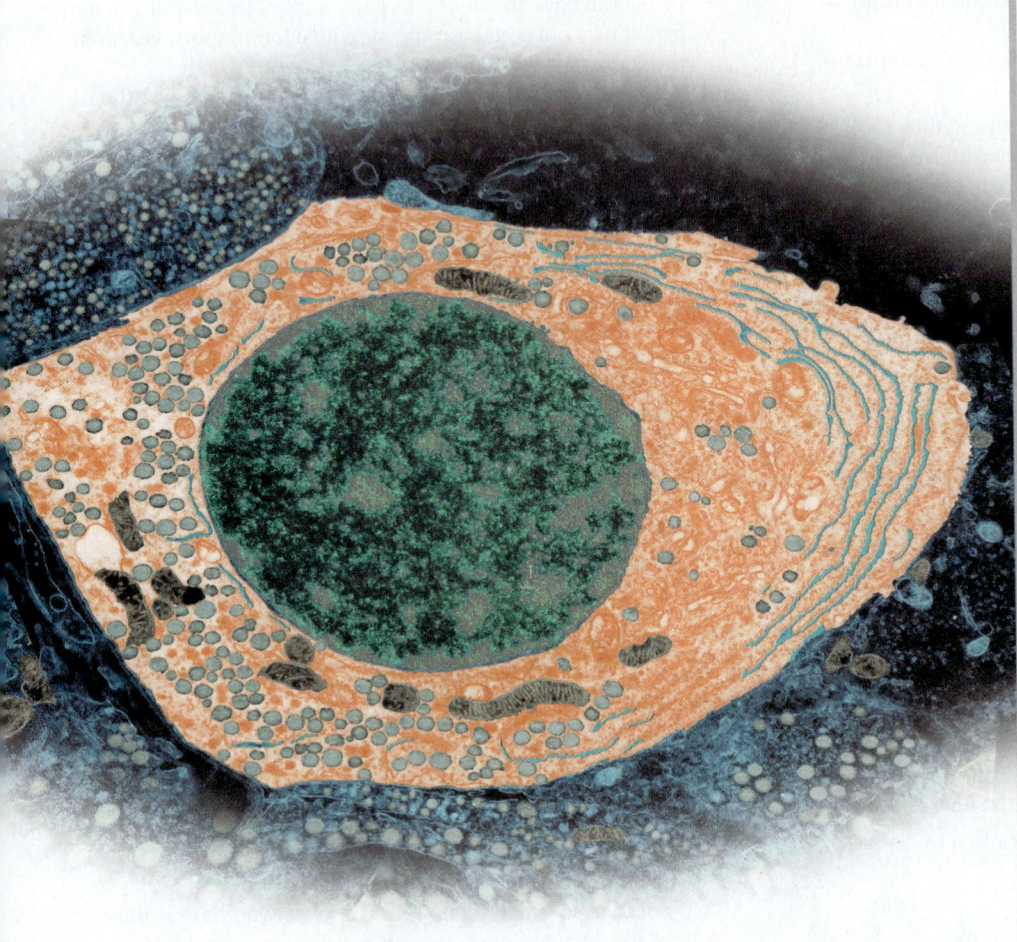

A growth hormone–producing cell in the pituitary gland. The hormone is stored in secretory vesicles (light blue) within the cytoplasm (orange), awaiting secretion.

Key Concepts

- The endocrine system produces chemical messenger molecules called hormones and secretes them into the blood.

- Hormones only act on certain cells, called the hormone's *target cells,* because only a hormone's target cells have specific receptors for the hormone.

- Steroid and nonsteroid hormones have different mechanisms of action. Steroid hormones enter the target cell before binding to a receptor, whereas nonsteroid hormones bind to receptors at the outer surface of the cell membrane.

- Some hormones participate in negative feedback loops that maintain homeostasis. Other hormones regulate major life-changing processes such as growth and sexual maturation.

- The pituitary gland is sometimes called *the master gland.* It secretes eight different hormones that in turn regulate the production of many other hormones.

CURRENT ISSUE

Endocrine Disruptors in the Environment

In the fall of 1997, right before the Christmas shopping season, the well-known European environmental group Greenpeace launched a highly visible, hazardous chemical awareness campaign called "Play Safe" simultaneously in London and New York. Its target was a little-known group of chemicals called **phthalates,** added to polyvinyl chloride (PVC) plastics to make them soft and pliable. PVC plastics are used to make many consumer goods, including children's toys and other items.

Underway at the same time as the Play Safe campaign was a grassroots movement to ban another potentially hazardous chemical, **Bisphenol A (BPA).** BPA is present in the plastic lining of food cans, hard plastic food and beverage containers, and dental sealants. The chemical is also used to coat the paper for heat-sensitive printers, so it is even found in many cash register receipts. BPA is widespread in our environment due to runoff at sites where products containing the chemical are manufactured and leaching from landfills where the chemical and products containing it have been dumped.

Phthalates and BPA are suspected of disrupting the function of the body's endocrine system. Playing a critical role in the maintenance of homeostasis, the endocrine system triggers or supports key physiological events, such as sexual maturation and puberty, childbirth, and lactation, and certain aspects of metabolism. The endocrine system relies on hormones as signalling molecules, and chemicals that interfere with hormonal signals are known collectively as *endocrine disruptors*. The Environmental Protection Agency (EPA) defines endocrine disruptors as "exogenous agents that interfere with the production, release, transport, metabolism, binding, action, or elimination of the natural hormones in the body responsible for the maintenance of homeostasis and the regulation of developmental processes."

Endocrine disruptors are found nearly everywhere in trace amounts: in the environment, in the consumer products we

Children's toys, such as these plastic ducks, once contained phthalates.

use, and even in our bodies. Should we be concerned? The answer to that question is largely unknown and is a hot topic of debate.

Phthalates: The Plastic Duck Scare of 1997

When Greenpeace targeted phthalates as dangerous endocrine disruptors that should be banned from children's toys, the public became concerned. Certain toys were shunned that Christmas season, and there were calls for a ban on phthalates in children's toys. The public outcry had the desired effect, though it took 10 years before action was taken. In 2008, the

Questions to Consider

1 Why do you think that endocrine disruptors might cause more harm to children than to adults?

2 Do you think that phthalates and/or BPA should be further restricted from additional consumer products? Why or why not? If you're not sure, what additional information would you like to have before making a decision?

Consumer Product Safety Commission banned several of the phthalates from children's toys and child care articles, even though there was no clear scientific evidence to support a ban. To ensure compliance, since 2012, all children's toys and child care articles that could be placed in a child's mouth have been tested for phthalates by an approved independent testing laboratory.

Phthalates are still being widely used today. They're found in nail polishes, hair sprays, vinyl floors, wall coverings, and soft plastic consumer products such as shower curtains. They're even found in blood collection bags, medical tubing, and some pharmaceutical pills. People have been exposed to phthalates for nearly 50 years, and still no convincing evidence exists of a human disease having been caused by phthalates.

Should we be worried? Perhaps yes, perhaps no. The evidence against phthalates comes primarily from toxicology studies in animals, in which →

Most plastic reusable water bottles are now BPA-free.

animals are deliberately fed ever-larger amounts of phthalates to determine at what point they cause a negative health effect. In animal toxicology studies, phthalates are endocrine disruptors, interfering with the action of testosterone on the development of the male reproductive system. They can cause low sperm counts, liver failure, and even cancer. No human has ever been subjected to these kinds of doses. The U.S. Centers for Disease Control and Prevention (CDC) says that the human health effects of exposure to low levels of phthalates are unknown and that more research is needed.

Of Greater Concern: Bisphenol A (BPA)

According to a report from the National Health and Nutrition Examination Survey, BPA can be detected in the urine of 92% of all persons over the age of six. BPA has been called an "environmental estrogen" because it interferes with (or at times may mimic) the natural hormone, estrogen. There is concern (but no hard evidence yet) that BPA may contribute to a number of trends in human health over the past several decades, including an increase in obesity and Type 2 diabetes, decreases in sperm count in males, increases in hormonally mediated cancers (prostate

and breast), and early sexual maturation in females.

BPA represents a difficult problem for policymakers because it is a high-volume production chemical used in a wide variety of manufacturing processes and consumer products. Proposals to Congress to restrict or ban the use of BPA have consistently failed due to intense lobbying by the chemical industry. Industry lobbyists argue that BPA is safe, that there are not good alternatives to BPA, and that jobs are at stake.

The greatest concern with BPA has always been for the health of children, whose development is heavily influenced by hormones. By 2012, 12 states had passed legislation banning the use of BPA in children's bottles and sippy cups. It almost didn't matter, because by then plastics manufacturers had abandoned the use of BPA-based materials in baby bottles and sippy cups.

BPA is still being used in industrial processes and in products used by adults, however. Over 8 billion pounds of it are produced each year—a pound for every person on earth. The long-term health effects are still unknown.

As these two examples show, the issue of endocrine disruptors is likely to remain controversial for a while. We are just beginning to learn how widespread endocrine disruptors already are in our environment and the many ways they might affect body systems. More research is needed, of course. The big question is what to do about them now, if anything.

When something is suspected of causing harm but there is no proof of harm, perhaps it makes sense to err on the side of caution. But where is the proper balance between caution and jobs? You decide.

SUMMARY

- Our environment contains many chemicals, collectively called endocrine disruptors, that have been proven (in animal studies) to disrupt the function of endocrine systems.

- The effects of endocrine disruptors on human health and development are largely unknown.

- Regulators face the difficult task of deciding whether to ban or restrict the use of endocrine disruptors, many of which are important chemicals in manufacturing processes.

At about ages 11–13, something strange and wonderful happens to the human body and mind. The body grows rapidly, and secondary sex characteristics develop. Many boys get interested in girls, and many girls get interested in boys with an intensity that sometimes seems to border on obsession, when only months before they couldn't care less.

Welcome to the endocrine system, an internal system of powerful chemicals that controls a host of body functions. The endocrine system triggers sexual maturation, sexual desire, uterine contractions during childbirth, and milk letdown. It is involved in responses to stress, digestion, cellular metabolism, and overall organ growth and development. And it is wholly or partly responsible for the maintenance of homeostasis of many of the most important variables in the body, including salt and water balance, blood pressure, the production of red blood cells, and blood

calcium concentration. Essentially all body systems are impacted in some way by the endocrine system. ■

13.1 The endocrine system produces hormones

The **endocrine system** is a collection of specialized cells, tissues, and glands that produce and secrete circulating chemical messenger molecules called **hormones**. Figure 13.1 shows the various tissues and organs that make up the endocrine system and the most important hormones they secrete. Hormones are secreted by **endocrine glands**— ductless organs that secrete their products into interstitial fluid, lymph, and blood (*endocrine* means "secreted internally"). In contrast, **exocrine glands** secrete products such as mucus, sweat, tears, and digestive fluids into ducts

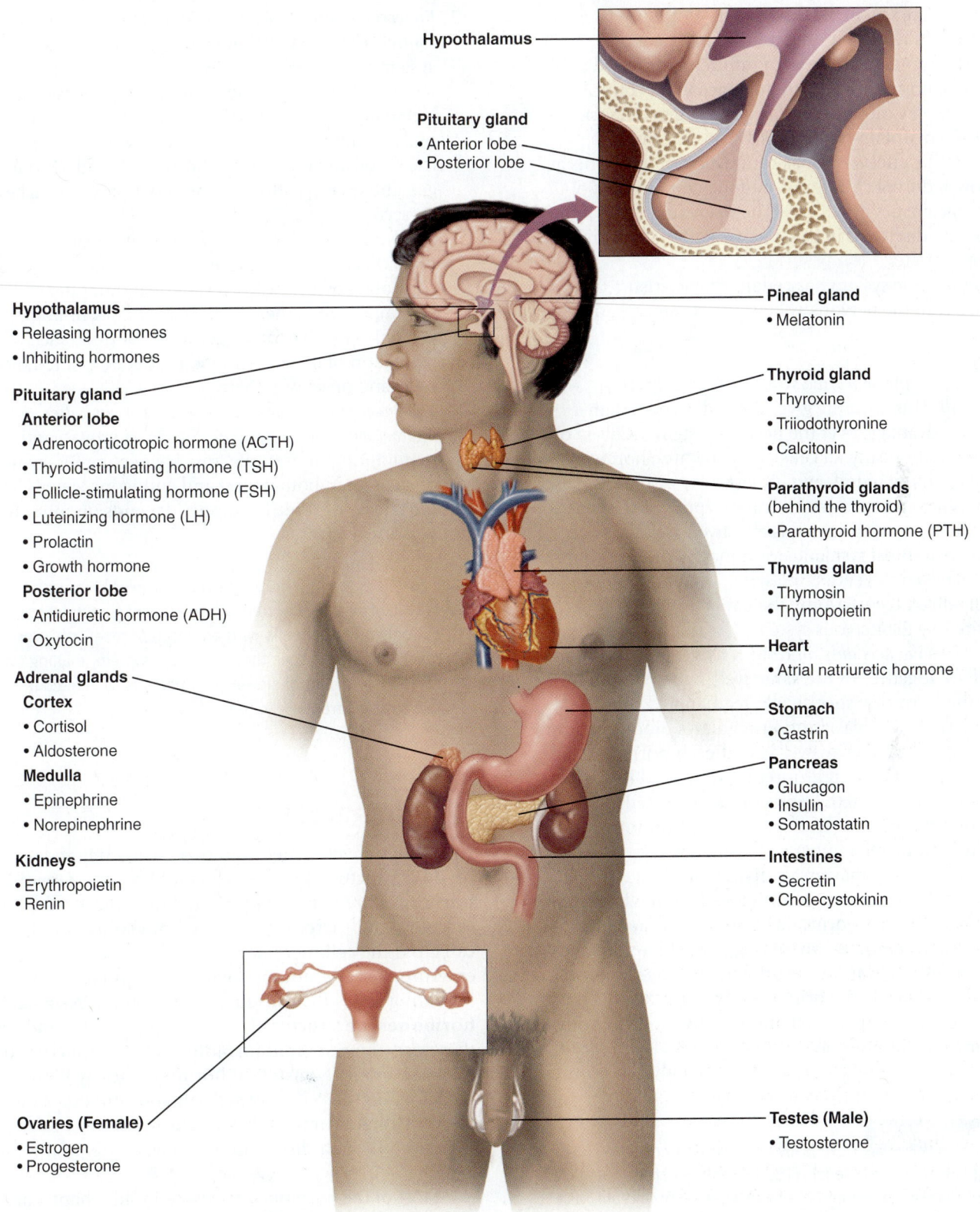

Hypothalamus

Pituitary gland
• Anterior lobe
• Posterior lobe

Hypothalamus
• Releasing hormones
• Inhibiting hormones

Pituitary gland
Anterior lobe
• Adrenocorticotropic hormone (ACTH)
• Thyroid-stimulating hormone (TSH)
• Follicle-stimulating hormone (FSH)
• Luteinizing hormone (LH)
• Prolactin
• Growth hormone
Posterior lobe
• Antidiuretic hormone (ADH)
• Oxytocin

Adrenal glands
Cortex
• Cortisol
• Aldosterone
Medulla
• Epinephrine
• Norepinephrine

Kidneys
• Erythropoietin
• Renin

Ovaries (Female)
• Estrogen
• Progesterone

Pineal gland
• Melatonin

Thyroid gland
• Thyroxine
• Triiodothyronine
• Calcitonin

Parathyroid glands
(behind the thyroid)
• Parathyroid hormone (PTH)

Thymus gland
• Thymosin
• Thymopoietin

Heart
• Atrial natriuretic hormone

Stomach
• Gastrin

Pancreas
• Glucagon
• Insulin
• Somatostatin

Intestines
• Secretin
• Cholecystokinin

Testes (Male)
• Testosterone

Figure 13.1 Components of the human endocrine system. Note that some organs with non-endocrine primary functions (the heart, stomach, intestines, and kidneys) produce and secrete hormones as secondary functions.

that empty into the appropriate sites. Approximately 50 known hormones circulate in the human bloodstream, and new ones are still being discovered.

Hormones are bloodborne units of information, just as nerve impulses are units of information carried in nerves. Some hormones participate in feedback control loops regulating various body functions. We need these hormones to maintain homeostasis. Other hormones produce specific effects, such as contraction of the uterus at birth, growth during childhood, and the development of sexual characteristics at puberty. The many functions of these hormones are discussed in this chapter and in other chapters where appropriate.

The endocrine system has certain characteristics that set it apart from the nervous system as a communications system.

1. *Hormones of the endocrine system reach nearly every living cell*. This attribute gives the endocrine system a distinct advantage over the nervous system. Delivery of hormones to nearly all cells occurs because hormones circulate in the blood, and almost every cell in the body is near a blood vessel. (An exception is the central nervous system, which is effectively isolated from exposure to most systemic hormones by the blood-brain barrier.) With the endocrine system, there is no need for the direct connections (the hard wiring, so to speak) required by the nervous system.

2. *Each hormone acts only on certain cells.* As all hormones circulate together in the same well-mixed blood, how can one hormone specifically regulate red blood cell production, whereas another regulates blood calcium concentration? The answer is that each hormone acts only on a certain group of cells, called its **target cells,** because only the hormone's target cells have the appropriate receptor to fit it. When a hormone binds to a receptor on its target cell, a change occurs within the cell. The cell may grow, divide, or change its metabolism in some way. All other cells of the body fail to respond to the hormone because they lack the appropriate receptor. As an analogy, consider that your car is a cell, the car's ignition switch is the receptor, and the car key is the hormone. Your car key fits only your car (the "target" car) and no other. Fitting your key into the ignition and turning it causes the car to

start. Gas is consumed, engine parts rotate, and heat is generated.

3. *Endocrine control tends to be slower than nervous system control.* There is a price for using the cardiovascular system as the message delivery system, and that is a reduction in speed—a lengthening of the response time. Nerve impulses can travel from head to toe in a fraction of a second. It takes longer to secrete a hormone into the bloodstream, have the hormone circulate to its target, and exert its effect. At the very fastest, this takes 20 seconds or so.

Given this difference in speed, it is not surprising that reflexes that prompt us to avoid a hot flame are controlled by the nervous system and not the endocrine system. On the other hand, endocrine communication is highly effective for longer-term controls, such as regulation of blood pressure, production of red blood cells, and onset of puberty.

4. *The endocrine and nervous systems can (and often do) interact with each other.* The timing of growth and sexual maturation, for example, involves a complex sequence of changes in both neural and endocrine signals, and the release of some hormones depends on input from sensory neurons.

⤺ Recap Hormones secreted by glands of the endocrine system act only on target cells with appropriate receptors. Hormones reach their targets via the circulatory system, making endocrine system control slower than nervous system control. The two systems frequently interact. ∎

13.2 Hormones are classified as steroid or nonsteroid

Hormones are classified into two basic categories based on their structure and mechanism of action. **Steroid hormones** are structurally related to cholesterol (a lipid), and therefore, they are lipid soluble and can cross the cell membrane. Steroid hormones act by entering the cell, binding to an intracellular receptor, and activating genes that produce new proteins. **Nonsteroid hormones** are structurally related to proteins, and therefore, they are lipid insoluble and cannot cross the cell membrane. Nonsteroid hormones act by binding to receptors on the cell's surface, which initiates a series of events that ultimately alters cellular activity in some way, even though the hormones never entered the cell at all.

Most of the hormones discussed in this chapter are nonsteroid. The only steroid hormones are those produced by the cortex of the adrenal gland (cortisol and aldosterone) and the sex hormones produced by the testes (testosterone) and ovaries (estrogen and progesterone).

MJ's BlogInFocus How easily does the endocrine disruptor bisphenol A (BPA) leach out of the lining of food cans and into the foods we eat? To find out, visit MJ's blog in the Study Area in MasteringBiology and look under "Endocrine Disruptors."

http://goo.gl/5LR8Wb

Steroid hormones enter target cells

Figure 13.2 depicts the mechanism of action for steroid hormones. Recall that the cell membrane is primarily composed of a bilayer of phospholipids. ❶ Because they are lipid soluble, steroid hormones can easily diffuse right across both the cell membrane and the nuclear membrane. Once inside the cell, steroid hormones bind to specific hormone receptors, forming a hormone-receptor complex either within the nucleus or within the cytoplasm (not shown). If the hormone-receptor complex was formed in the cytoplasm, it too can diffuse into the nucleus.

❷ Once inside the nucleus, the hormone-receptor complex attaches to DNA, activating specific genes. ❸ Gene activation causes the formation of messenger RNA, which then leaves the nucleus and directs the synthesis of certain proteins. ❹ The proteins then carry out the cellular response to the hormone, whatever it might be.

Steroid hormones tend to be slower acting than nonsteroid hormones because the entire protein production process (starting with gene activation) begins only after arrival of the steroid hormone within the cell. Starting from the time the steroid hormone first enters the cell, it can take minutes or even hours to produce a new protein.

Nonsteroid hormones bind to receptors on target cell membranes

Nonsteroid hormones have an entirely different mechanism of action from steroid hormones (**Figure 13.3**). Nonsteroid hormones cannot enter the target cell because they are not lipid soluble. Instead, they bind to receptors located on the outer surface of the cell membrane. The receptors are generally associated with, or are part of, protein molecules floating in the phospholipid bilayer of the cell membrane (Chapter 3). ❶ The binding of hormone to receptor causes a change in the shape of the membrane protein, which in turn initiates a change within the cell. It's like turning the lights

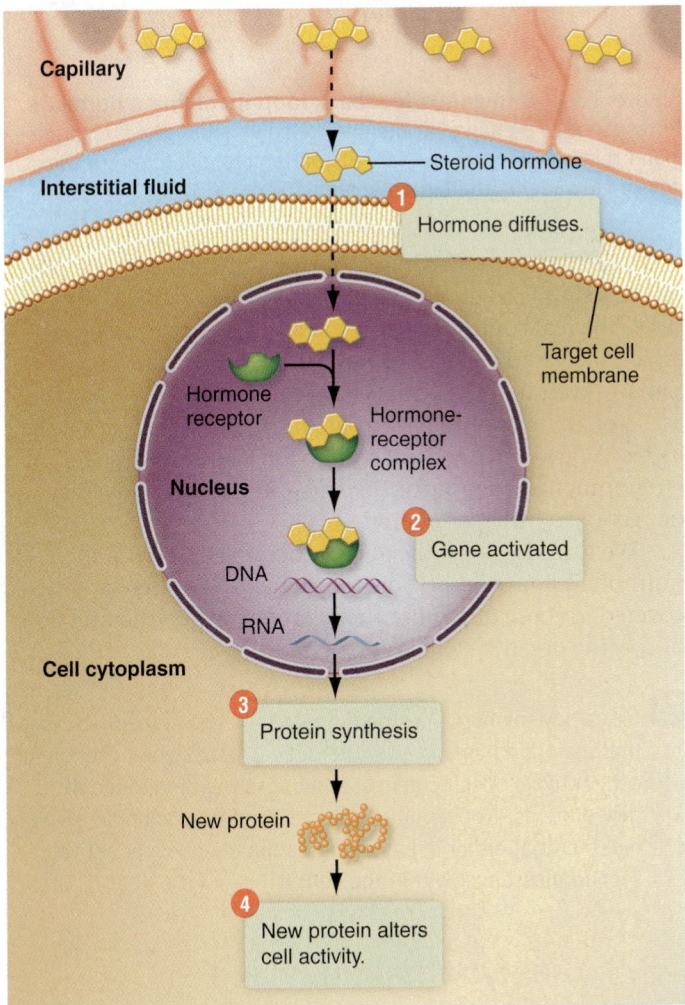

Figure 13.2 Mechanism of steroid hormone action on a target cell. Lipid-soluble steroid hormones diffuse across the cell and nuclear membranes into the nucleus, where they bind to hormone receptors that activate genes. Gene activation results in the production of a specific protein.

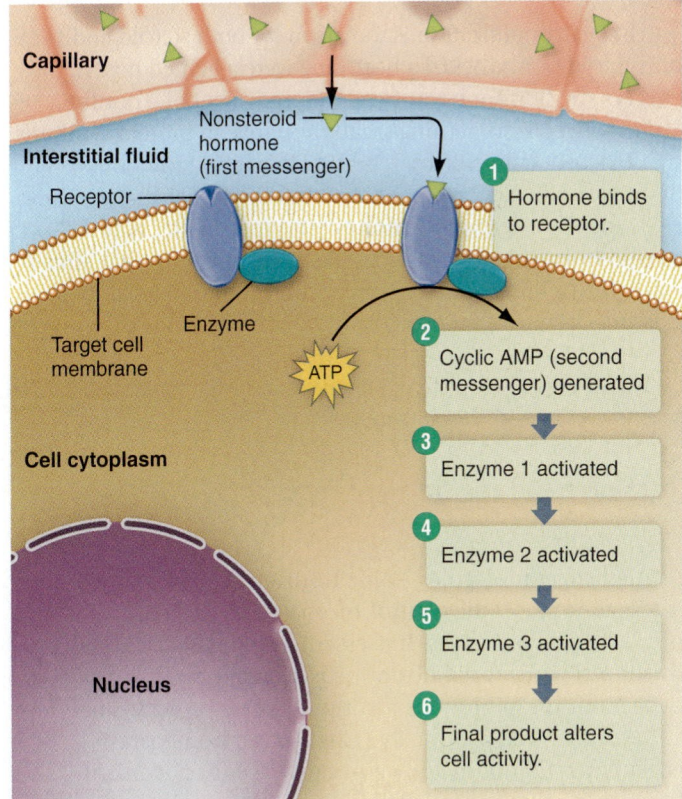

Figure 13.3 Mechanism of nonsteroid hormone action on a target cell. Nonsteroid hormones bind to receptors in the cell membrane, leading to the conversion of ATP to cyclic AMP (the second messenger) within the cell. Cyclic AMP initiates a cascade of enzyme activations, amplifying the original hormonal signal and generating a cellular response.

on in a room by flipping a switch located in the hall outside, without ever having to enter the room.

Some nonsteroid hormones cause ion channels in the cell membrane to open or close, similar to the action of a neurotransmitter. More commonly, hormone-receptor binding converts an inactive molecule within the cell into an active molecule. ❷ The activated molecule generated within the cell is called a **second messenger** because it carries the message provided by the hormone (the "first messenger") without the hormone ever entering the cell. A common second messenger is cyclic AMP, produced from ATP. ❸ Cyclic AMP then activates an enzyme already present within the cell, ❹ which activates another, ❺ which in turn activates another, and so on. ❻ The last enzyme catalyzes the conversion of an inactive molecule to the final product, which goes on to alter cell activity. At each successive activation step, the effect is amplified, so that even a small amount of hormone can produce a large amount of final product.

☑ If the binding of one hormone molecule to its receptor on the cell's surface results in production of 10 cyclic AMPs and then there is an additional 10-fold amplification at each subsequent step, how many molecules of final product will be produced?

In fact, amplification is the norm for both steroid and nonsteroid hormones. If a hormone were a lighted match, the effect of the hormone would be the equivalent of a bonfire. Hormones start something small (a change in protein synthesis or activation of an enzyme), but the final effect can be as dramatic as the development of secondary sexual characteristics.

Most nonsteroid hormones tend to be faster acting than steroid hormones because they activate enzymes that already exist within the cell in inactive form. An entire cascade of existing enzymes can be activated within seconds or minutes.

☑ Would you expect a steroid hormone to result in production of a second messenger? Why or why not?

Some hormones participate in negative feedback loops

As messenger molecules, some hormones participate in internal homeostatic control mechanisms and control vital physiological processes. It is essential to regulate carefully the rate at which each hormone is secreted, so that its concentration in blood is just right to carry out its function.

As described previously (Chapter 4), homeostasis is generally maintained by a negative feedback control loop. In a negative feedback loop involving a hormone, the endocrine gland is the control center, the hormone represents the pathway between the control center and the effectors, and the effectors are the hormone's target cells, tissues, or organs (**Figure 13.4**). A negative feedback loop involving an endocrine gland and a hormone is a stable, self-adjusting mechanism for maintaining homeostasis of the controlled variable, because any change in the controlled variable sets in motion a response that reverses that change.

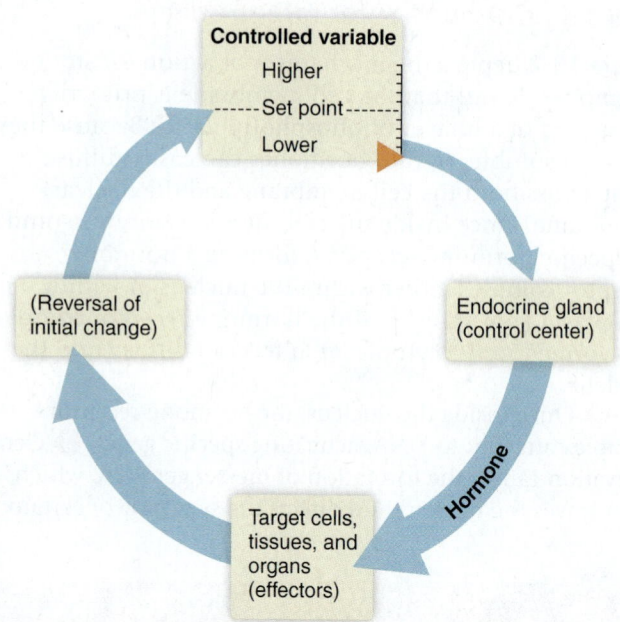

Figure 13.4 A negative feedback loop involving a hormone. In response to a change in the controlled variable, the endocrine gland releases a hormone that acts on target cells to return the controlled variable to its normal state.

Not all negative feedback loops involving hormones are quite as simple as the one depicted in Figure 13.4. In some cases, the real control center is the brain, which activates an endocrine gland via nerves. Nevertheless, the effect is the same—reversal of the initial change in the controlled variable.

Not all hormones participate in negative feedback loops. A few hormones are secreted in response to specific environmental cues or for specific purposes, such as to avoid danger or to initiate puberty.

We turn now to some key endocrine glands and their hormones. Some of the hormones mentioned here are covered in more depth in other chapters, when the function of specific organ systems is discussed.

↩ **Recap** Many hormones participate in negative feedback loops that maintain homeostasis. Steroid hormones enter the target cell, activate specific genes, and cause the production of new proteins. Nonsteroid hormones bind to a cell membrane receptor that either opens or closes ion channels or activates a second messenger within the cell. ▪

13.3 The hypothalamus and the pituitary gland

The **hypothalamus** is a small region in the brain that serves as a homeostatic control center. It is an important link between the nervous system and the endocrine system. The hypothalamus receives neural input about certain internal environmental conditions such as water and solute balance, temperature, and carbohydrate metabolism. It produces

Table 13.1 Hormones of the pituitary gland

Hormone	Abbreviation	Main targets	Primary actions
Posterior Pituitary			
Antidiuretic hormone	ADH	Kidneys	Reduces amount of water lost in urine
Oxytocin	—	Uterus	Induces uterine contractions
		Mammary glands	Induces ejection of milk from mammary glands
Anterior Pituitary			
Adrenocorticotropic hormone	ACTH	Adrenal cortex	Stimulates adrenal cortex to release glucocorticoids
Thyroid-stimulating hormone	TSH	Thyroid gland	Stimulates synthesis and secretion of thyroid hormones
Follicle-stimulating hormone	FSH	Ovaries	In females, stimulates egg maturation and the secretion of estrogen
		Testes	In males, stimulates the formation of sperm
Luteinizing hormone	LH	Ovaries	In females, stimulates ovulation (egg release) and the secretion of progesterone
		Testes	In males, stimulates testosterone secretion
Prolactin	PRL	Mammary glands	Stimulates the development of mammary gland cells and production of milk
Growth hormone	GH	Most cells	Stimulates growth in young individuals; plays multiple roles in cell division, protein synthesis, and metabolism in adults

two hormones of its own, and it monitors and controls the hormone secretions of the pituitary gland. The pituitary gland is located directly beneath the hypothalamus and connected to it by a thin stalk of tissue.

The **pituitary gland** is sometimes called the "master gland" because it secretes eight different hormones that in turn regulate many of the other endocrine glands (Table 13.1). It consists of two lobes: a posterior lobe that appears to be a bulging extension of the connecting stalk and a larger, more distinct anterior lobe (Figure 13.5).

The posterior pituitary stores ADH and oxytocin

The interaction between the hypothalamus and posterior lobe of the pituitary gland is an example of the close working relationship between the nervous and endocrine systems. Cells in the hypothalamus called **neuroendocrine cells** function as both nerve cells and endocrine cells, because the same cells generate nerve impulses and secrete hormones into blood vessels.

Individual neuroendocrine cells of the hypothalamus and the posterior pituitary secrete either antidiuretic hormone (ADH) or oxytocin. The hormones enter the blood supply of the posterior pituitary and make their way to the general circulation. ADH acts on the kidneys to regulate water balance in the body. Oxytocin primarily acts on the mammary glands and uterus muscles of lactating and pregnant women, respectively.

Figure 13.5 illustrates the posterior pituitary lobe and its connections to the hypothalamus. The cell bodies of the neuroendocrine cells are located in the hypothalamus. Their axons extend down the stalk between the hypothalamus to

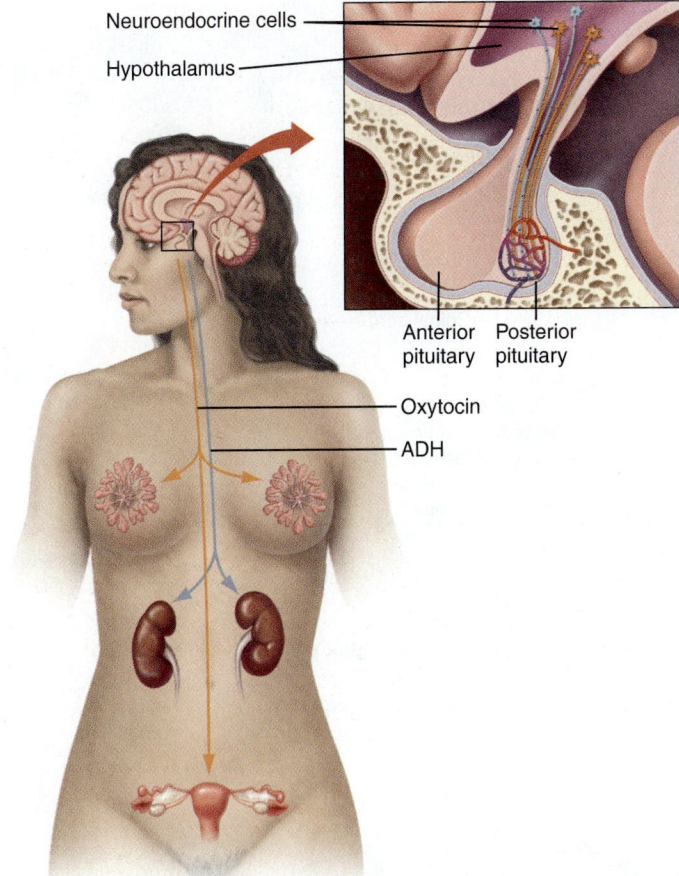

Figure 13.5 Pituitary gland and hypothalamus.

the posterior pituitary and into the posterior pituitary itself. Neurons make either ADH or oxytocin in the hypothalamus and then transport the hormones down the axon to the posterior pituitary, where they are stored prior to release.

When the hypothalamus is stimulated to release the hormones, the neuroendocrine cells send impulses down the axon, causing hormone to be secreted into nearby capillaries. The hormones are carried to their target cells by the blood.

ADH regulates water balance You can think of ADH as the "water-conserving hormone." ADH reduces the amount of water lost in urine, helping to maintain the body's overall water balance. The primary target cells for ADH are in the kidneys, where it causes changes in cell permeability to water in the kidney's tubules. The action of ADH is discussed in more detail in the chapter on the urinary system (Chapter 15).

The control of ADH secretion is an example of a negative feedback loop, in this case, one that maintains homeostasis of water volume and concentration. When the hypothalamus detects low concentrations of water (equivalent to a high concentration of solute particles) in the blood, neuroendocrine cells release ADH from the posterior pituitary. ADH circulates in blood to the kidneys, where it stimulates

the reabsorption of water. When water concentrations rise to normal levels, the hypothalamus is no longer stimulated to release ADH, and ADH secretion declines.

☑ What would happen to your ADH levels if you drank a large amount of water, and what effect would this have on the kidneys?

Oxytocin causes uterine contractions and milk ejection The primary target cells for oxytocin are in the uterus and the mammary glands (breasts) of pregnant and lactating females, respectively. Oxytocin is secreted at high concentrations during childbirth and also when an infant nurses at its mother's breast (**Figure 13.6**). It stimulates contractions of the uterus during childbirth and ejection of milk into the ducts of the mammary glands during suckling.

The stimulation of oxytocin secretion by nursing is an example of a *neuroendocrine reflex*, in which a nervous system stimulus (activation of sensory mechanoreceptors in the mother's breast) is responsible for secretion of a hormone. ➊ When a baby suckles, sensory receptors in the nipple are stimulated. ➋ Nerve impulses generated by the sensory receptors travel in sensory and spinal nerves to neuroendocrine cells in the hypothalamus. ➌ The neuroendocrine cells release

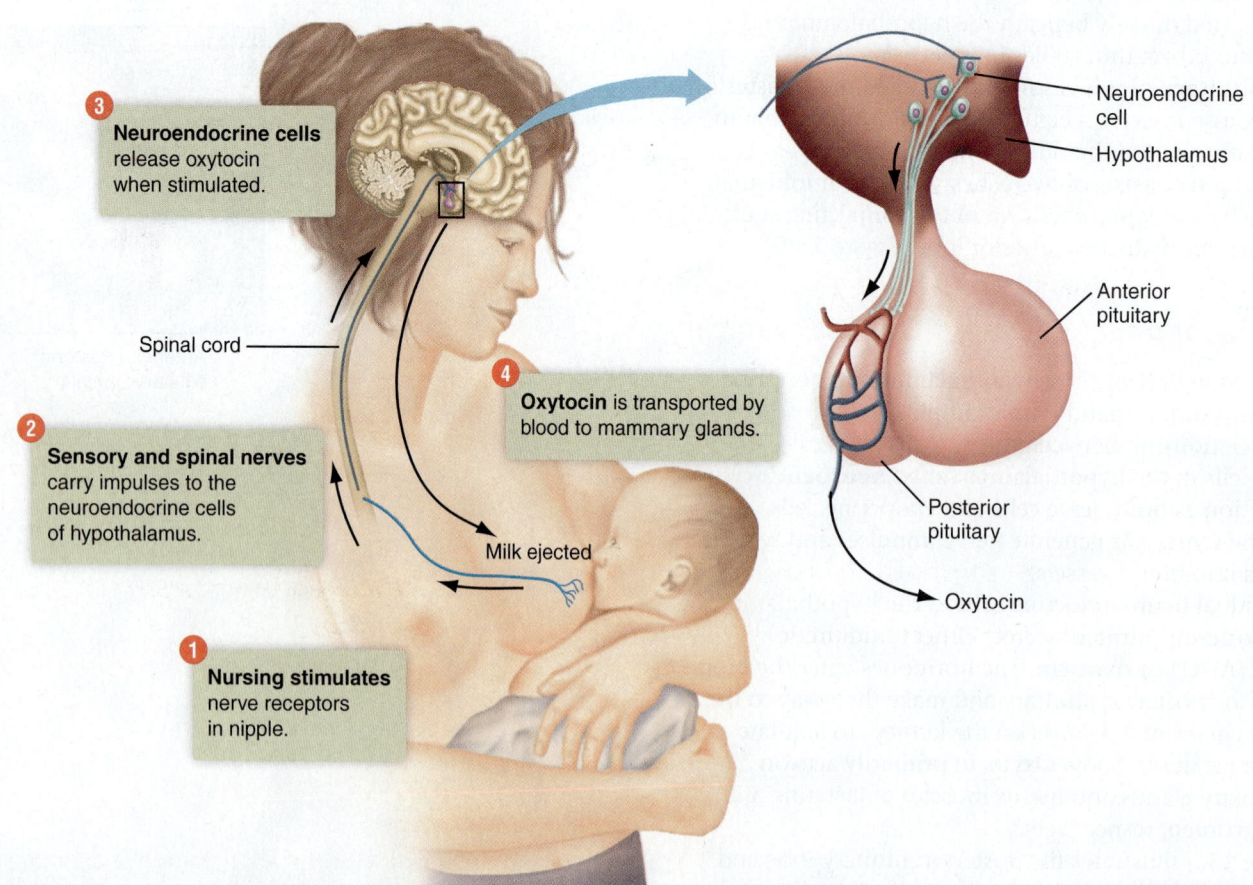

➌ **Neuroendocrine cells** release oxytocin when stimulated.

Spinal cord

➋ **Sensory and spinal nerves** carry impulses to the neuroendocrine cells of hypothalamus.

➊ **Nursing stimulates** nerve receptors in nipple.

➍ **Oxytocin** is transported by blood to mammary glands.

Milk ejected

Neuroendocrine cell

Hypothalamus

Anterior pituitary

Posterior pituitary

Oxytocin

Figure 13.6 The control of oxytocin secretion by nursing. The act of suckling triggers a series of events that ultimately eject milk from the nipple.

oxytocin from their neural endings located in the posterior pituitary. ④ Oxytocin circulates in the blood to breast tissue, where it causes contractions of smooth muscle that eject milk.

Oxytocin is also present in males. In both sexes, it is thought to play a role in sexual arousal and the feeling of sexual satisfaction after orgasm.

The anterior pituitary produces six key hormones

The anterior lobe of the pituitary gland produces the following six key anterior pituitary hormones:

- Adrenocorticotropic hormone (ACTH), also called corticotropin
- Thyroid-stimulating hormone (TSH), also called thyrotropin
- Follicle-stimulating hormone (FSH)
- Luteinizing hormone (LH)
- Prolactin (PRL)
- Growth hormone (GH)

Each anterior pituitary hormone is produced and secreted by a separate cell type within the anterior pituitary.

The secretion of anterior pituitary hormones depends partly on the hypothalamus, but the mechanism differs from that of the posterior pituitary (**Figure 13.7**). The connection between the hypothalamus and anterior pituitary is endocrine, not neural. ① When stimulated, different groups of neuroendocrine cells in the hypothalamus secrete one of five specific releasing hormones. Releasing hormones have very unimaginative names; the releasing hormone for ACTH is ACTH-RH, that for TSH is TSH-RH, and so on. The exception is gonadotropin releasing hormone (GnRH), which is responsible for stimulating the release of both FSH and LH. ② A releasing hormone is secreted directly into a special blood supply that runs directly between the hypothalamus and anterior pituitary (called the *pituitary portal system*), so even though the releasing hormone is produced in miniscule quantities its concentration is sufficient to stimulate the secretion of an anterior pituitary hormone. ③ Upon arrival in the anterior pituitary, a releasing hormone increases the secretion of one of the six anterior pituitary hormones (or two, in the case of GnRH).

For each releasing hormone, there is also a corresponding inhibiting hormone, each produced by neurons in the hypothalamus. The actions of an inhibiting hormone are opposite to those of the releasing hormone.

ACTH stimulates the adrenal cortex The names of hormones generally offer a clue to their function or site of action. The target cells for ACTH are located in the outer layer, or cortex, of the adrenal gland (in *adrenocorticotropic* the suffix *-tropic* means "acting upon").

ACTH stimulates the adrenal cortex to release another group of hormones called *glucocorticoids,* which are steroid hormones involved in stress-related conditions and the control of glucose metabolism (see section 13.5). ACTH secretion is regulated by a negative feedback loop involving a releasing hormone from the hypothalamus and the concentration of glucocorticoids in blood.

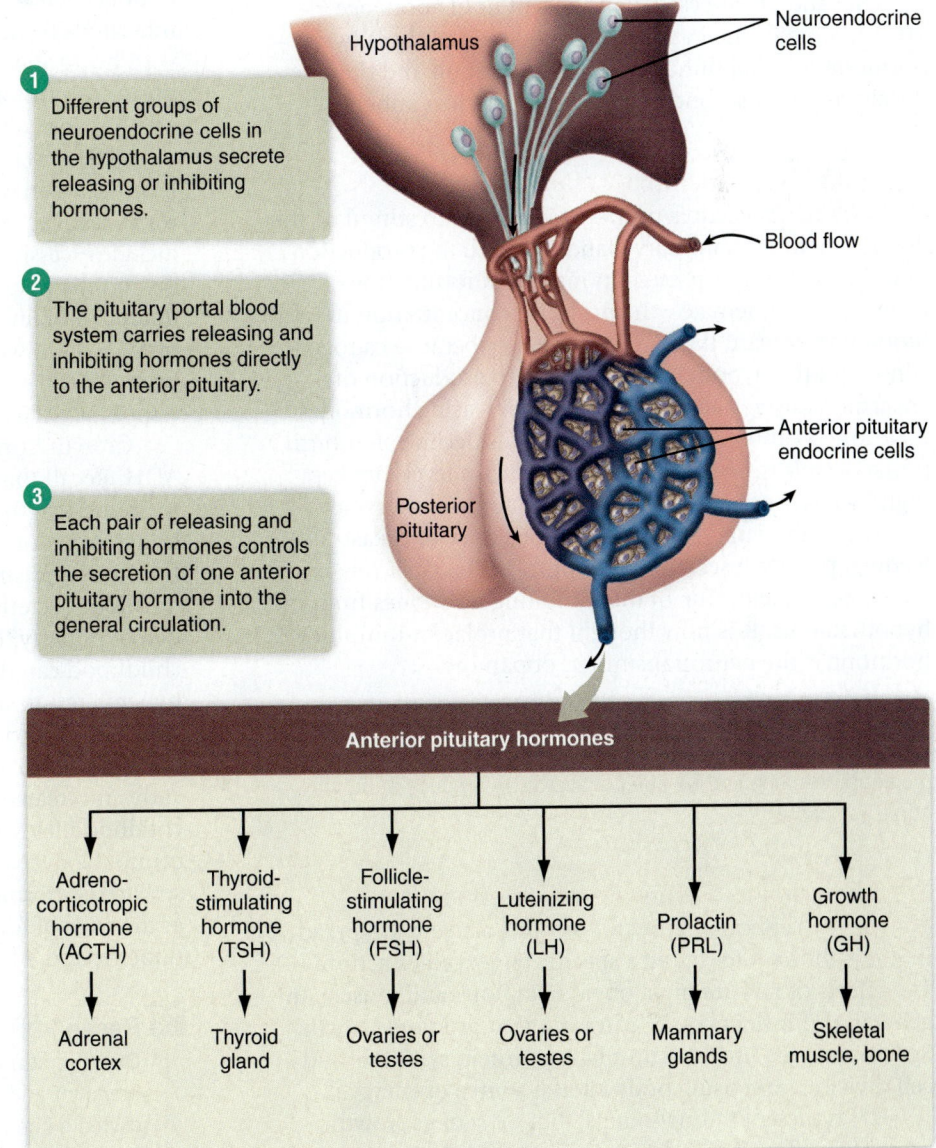

① Different groups of neuroendocrine cells in the hypothalamus secrete releasing or inhibiting hormones.

② The pituitary portal blood system carries releasing and inhibiting hormones directly to the anterior pituitary.

③ Each pair of releasing and inhibiting hormones controls the secretion of one anterior pituitary hormone into the general circulation.

Figure 13.7 The relationship between the hypothalamus and the anterior pituitary gland. Neuroendocrine cells in the hypothalamus secrete releasing and inhibiting hormones into a system of capillaries that goes directly to the anterior pituitary. Specific releasing and inhibiting hormones control the secretion of each anterior pituitary hormone.

TSH acts on the thyroid gland   TSH stimulates the thyroid gland to synthesize and release thyroid hormones. The secretion of TSH is controlled by both releasing and inhibiting hormones from the hypothalamus, as mentioned earlier.

FSH and LH stimulate the reproductive organs   FSH and LH are called *gonadotropins* because they stimulate the growth, development, and function of the reproductive organs (gonads) in both males and females. In females, FSH induces egg development, LH promotes ovulation (egg release), and both hormones stimulate the secretion of the ovarian hormone estrogen. In addition, LH stimulates the secretion of the ovarian hormone progesterone after ovulation. In males, FSH induces sperm development and LH stimulates the production of the hormone testosterone by the testes.

FSH and LH are essentially absent until about age 10–13. A major increase in the production of these two hormones is what initiates sexual maturation and the development of secondary sex characteristics (puberty).

Prolactin (PRL): mammary glands and milk production   Prolactin's primary function is to stimulate the development of mammary gland cells and the production of milk. Prolactin is present in males but its function is unknown. In females, the prolactin concentration in blood rises toward the end of pregnancy because estrogen stimulates the hypothalamus to step up production of *prolactin-releasing hormone*, one of the releasing hormones mentioned earlier. When estrogen levels decline after birth, prolactin-releasing hormone and prolactin levels are kept high by a sensory neural reflex associated with nursing.

When a woman is not in late pregnancy or breast-feeding, prolactin secretion is suppressed by *prolactin-inhibiting hormone*, one of the inhibiting hormones from the hypothalamus. It is now thought that prolactin-inhibiting hormone is the neurotransmitter, dopamine.

✅ Suppose a pregnant woman has a mutant version of oxytocin that cannot bind to oxytocin receptors, but she produces normal prolactin. Will she still be able to give birth, produce milk, and nurse her baby?

Growth hormone (GH): widespread effects on growth   The effects of growth hormone are so widespread that it is difficult to define a specific target cell or action. The effects of GH are most obvious in bone and muscle. In general, GH influences cells in ways that promote growth. Specific actions include stimulating protein synthesis and cell division, and using body fat as a source of energy.

GH is present throughout life, but most of its growth-promoting effects occur during childhood and adolescence. At these times, it causes a dramatic increase in the mass and length of muscles and bones. Abnormal GH concentrations during development can lead to permanent changes in the final height

of an adult. In adults, it continues to help regulate metabolism by promoting the utilization of fats and amino acids as cellular fuel and by helping to maintain blood glucose levels.

Pituitary disorders: hypersecretion or hyposecretion

As you might guess, maintaining just the right quantities of hormones in the blood is a delicate balance that can sometimes go awry. Many diseases and disorders are chronic conditions of too much hormone (hypersecretion) or too little hormone (hyposecretion). Some are inherited. As a general rule, an endocrine feedback control loop can be disrupted either by lack of a hormone or by lack of target cell receptors for that hormone. Either way, the system does not function effectively. Endocrine disorders due to defects in target cell receptors are quite common and are often inherited. To illustrate the types of things that can go wrong with hormone secretion, we focus on hyposecretion of one posterior pituitary hormone (ADH) and both hyposecretion and hypersecretion of an anterior pituitary hormone (GH).

ADH affects maintenance of water balance by the kidneys. Hyposecretion of ADH causes the kidneys to lose too much water, leading to a disorder called *diabetes insipidus*. Symptoms include excessive urination, dehydration, thirst, headache, and dry mouth. Some cases of diabetes insipidus are caused by head trauma or brain surgery that disrupts the normal production of ADH. Even when ADH hormone levels are normal, diabetes insipidus can still occur if the kidney cells that normally respond to ADH are lacking the specific ADH receptors.

Growth hormone has much longer-acting effects than ADH. Recall that in children, growth hormone stimulates the bone-lengthening activity of the growth plate, so that bones grow longer along with the child's overall growth. It is not surprising, therefore, that abnormalities of growth hormone secretion are most damaging when they occur before puberty. Hyposecretion of growth hormone during childhood can result in *pituitary dwarfism*. Conversely, hypersecretion of growth hormone during childhood and adolescence produces *gigantism* (Figure 13.8).

Dwarfism can be prevented if children are diagnosed early and treated with a growth hormone throughout childhood. However, once adulthood is reached, dwarfism cannot be overcome by administering a growth hormone. The sex steroid hormones cause the cartilaginous growth plates at the ends of long bones to be replaced by bone during puberty, and after that time, bones cannot grow longer.

↩ **Recap**   ADH acts on the kidneys to regulate water balance. Oxytocin stimulates uterine contractions and milk ejection in pregnant and lactating females, respectively. In both sexes, oxytocin contributes to feelings of sexual satisfaction. The six hormones produced by the anterior pituitary are ACTH, TSH, FSH, LH, PRL, and GH. The release of each hormone is controlled in part by a releasing or inhibiting hormone from the hypothalamus. ∎

2. Beta cells secrete **insulin,** which lowers blood sugar. After a meal, blood glucose levels rise as sugars are absorbed from the digestive tract. The high glucose concentration stimulates the beta cells to secrete insulin into the blood, where it does the opposite of glucagon. Insulin promotes the uptake of glucose by cells of the liver, muscle, and fat tissue. It also promotes the conversion of glucose into glycogen in the liver, both glycogen and proteins in muscle, and fats in adipose tissue.

3. Delta cells secrete somatostatin. The physiological significance of pancreatic somatostatin is not well understood, though it appears to inhibit the secretion of both glucagon and insulin. Somatostatin is also secreted by the hypothalamus, where it is the inhibitory hormone for growth hormone secretion, and the digestive tract, where it helps regulate several digestive secretions. The example of somatostatin reinforces the principle that the function of a hormone in any particular tissue or organ depends on the types of target cells present in those locations.

The normal range for blood glucose concentration is 80–120 mg/100 ml. Figure 13.9 illustrates how the pancreas might respond when blood glucose concentration rises after a meal. ❶ As blood glucose concentration approaches 120 mg/100 ml, the pancreas responds by increasing the secretion of insulin and decreasing the secretion of glucagon. Together, these two actions serve to reduce the blood glucose concentration. ❷ If blood glucose concentration later falls to near 80 mg/100 ml, the opposite happens; insulin

Figure 13.8 **Pituitary gigantism.** The world's tallest man, Sultan Kosen, poses with his doctors at the University of Virginia.

13.4 The pancreas secretes glucagon, insulin, and somatostatin

The **pancreas** is both an endocrine gland (secreting hormones into the blood) and an exocrine gland (secreting enzymes, fluids, and ions into the digestive tract to aid in digestion). In this chapter, we deal with its endocrine function.

The endocrine cells of the pancreas are located in small clusters scattered throughout the pancreas called the *islets of Langerhans* (an islet is a little island), or just the pancreatic islets. The pancreatic islets contain three types of cells and produce three hormones, all having to do with the regulation of blood sugar (glucose). One hormone raises blood sugar, the second lowers it, and the third inhibits the secretion of the other two hormones.

1. Alpha cells secrete **glucagon,** which raises blood sugar. As blood glucose levels decline between meals, glucagon is secreted into the bloodstream. In the liver, glucagon causes the breakdown of glycogen (a stored form of sugar) to glucose, raising blood glucose levels. Glucagon is particularly important between meals for increasing the supply of glucose to cells that need it.

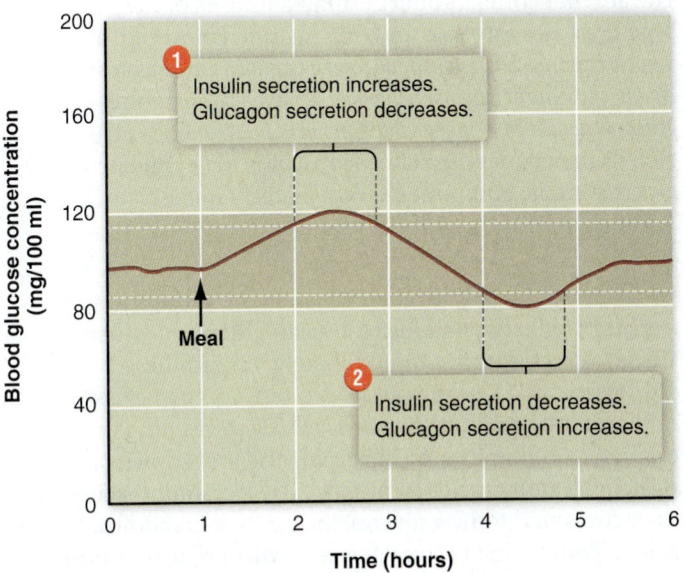

Figure 13.9 **How the pancreas responds to a meal.** As blood glucose concentration nears its normal upper limit after a meal (dashed line), the pancreas secretes more insulin and less glucagon, promoting the uptake and storage of glucose. If glucose concentration falls to near its normal lower limit, the pancreas secretes less insulin and more glucagon. In the steady state, both hormones are secreted in modest amounts.

secretion decreases and glucagon secretion increases. In actual fact, insulin and glucagon are both secreted in modest amounts all of the time. Adjusting the blood sugar is just a matter of altering the relative secretion rates of these two hormones—more insulin and less glucagon when blood sugar is high, and the inverse when blood sugar is low.

↩ **Recap** Hormones secreted by the pancreas include glucagon, insulin, and somatostatin. Glucagon raises blood glucose levels by increasing production of glucose from glycogen. Insulin lowers glucose levels by facilitating glucose uptake and storage. Somatostatin inhibits the secretion of both glucagon and insulin. ▪

13.5 The adrenal glands comprise the cortex and medulla

The **adrenal glands** are two small endocrine organs located just above the kidneys. Each gland has an outer layer called the **adrenal cortex** and an inner core called the *adrenal medulla*. The cortex and medulla function as two separate endocrine glands. (**Cortex** and **medulla** are general terms for outer and inner regions; you'll hear the terms again in the context of the kidneys, and you're already aware of the cerebral cortex of the brain.)

The adrenal cortex: glucocorticoids and mineralocorticoids

The adrenal cortex produces small amounts of the sex hormones estrogen and testosterone (most of the sex hormones are produced by the ovaries and testes, discussed later) and two classes of steroid hormones called *glucocorticoids* and *mineralocorticoids*. Their names give a clue to their functions: Glucocorticoids help regulate blood glucose levels, and mineralocorticoids help regulate the minerals sodium and potassium.

Glucocorticoids: cortisol The adrenal cortex produces a group of glucocorticoids with nearly identical structures. **Cortisol** accounts for approximately 95% of these glucocorticoids. During prolonged fasting, cortisol assists in maintaining blood glucose levels by promoting the utilization of fats and by increasing the breakdown of protein to amino acids in muscle. The free amino acids are then converted to new glucose by the liver. In addition to its role in controlling blood glucose, cortisol is also important in suppressing inflammatory responses that occur following infection or injury.

Cortisol secretion is controlled by a typical feedback loop. A releasing factor from the hypothalamus stimulates the secretion of ACTH from the anterior pituitary gland, which in turn stimulates the adrenal cortex to secrete cortisol. When cortisol reaches its upper limit of normal blood level,

it inhibits further secretion of the releasing hormone from the hypothalamus and of ACTH from the anterior pituitary.

A fall in blood glucose concentration can override the normal feedback control mechanism and increase cortisol secretion significantly. So, too, can physical injury and emotional stress. All of these stimuli act at the level of the hypothalamus, which secretes more releasing hormone, ultimately raising both ACTH and cortisol. The elevated cortisol concentration helps to restore the blood glucose concentration to normal and makes more amino acids available to repair damaged tissues. It also reduces inflammation by decreasing fluid loss from capillaries.

Cortisol and cortisol-like drugs can be administered topically to treat inflammation or injury. When used for prolonged periods, however, these drugs suppress the immune system.

☑ How does an increase in cortisol in response to stressful experiences such as physical injury and emotional stress help your body adjust to the stressful situation?

Mineralocorticoids: aldosterone The other hormones produced by the adrenal cortex are the mineralocorticoids, the most abundant of which is **aldosterone**. Aldosterone is the hormone primarily responsible for regulating the amounts of sodium and potassium in the body. With ADH, it helps maintain body water balance.

Adrenal secretion of aldosterone increases whenever the total amount of sodium and water in the body is too low or when too much potassium is present (see Chapter 15 for details). In turn, aldosterone acts on cells in the kidneys to promote sodium reabsorption and potassium excretion, precisely what you would expect it to do if it were part of a homeostatic feedback control loop for these ions.

The adrenal medulla: epinephrine and norepinephrine

The **adrenal medulla** produces the nonsteroid hormones **epinephrine** (adrenaline) and **norepinephrine** (noradrenaline). These hormones affect cellular metabolism as well as blood pressure, respiration, and heart activity.

Epinephrine and norepinephrine are considered both hormones and neurotransmitters, depending on the situation. When they are released into the blood from the adrenal medulla and act on distant target cells, they are functioning as hormones. When they are released by a presynaptic neuron into a synaptic cleft and then bind to receptors on a postsynaptic neuron (Chapter 11), they are functioning as neurotransmitters.

Like the hypothalamus and the posterior pituitary, the adrenal medulla is really a neuroendocrine organ. Epinephrine and norepinephrine are synthesized and stored in specialized cells called *chromaffin cells* in the adrenal medulla that are innervated by sympathetic nerves. When

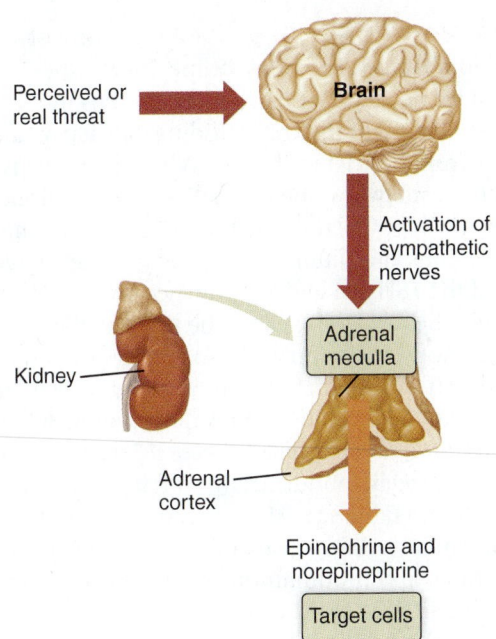

Figure 13.10 Secretion of epinephrine and norepinephrine by the adrenal medulla. The secretion of these two hormones is closely linked with the activity of the sympathetic division of the autonomic nervous system (ANS). Neural pathways are noted in dark brown; hormonal pathways in light brown.

sympathetic nerves are activated, the chromaffin cells secrete epinephrine and norepinephrine (**Figure 13.10**).

Once released, the two hormones participate in a wide range of actions that are essentially like the fight-or-flight responses of the sympathetic division. They raise blood glucose levels, increase heart rate and force of contraction, increase respiration rate, and constrict or dilate blood vessels in many organs, all actions that prepare the body for emergency activity.

You may have noticed that approximately 30 seconds to a minute after you are frightened by a dangerous situation such as avoiding an auto accident, your heart starts to pound, you breathe faster, and you feel "pumped up," alert, and a little shaky. These changes may occur even though the danger has already passed. You are experiencing an "adrenal rush" due to intense stimulation of your adrenal glands by the sympathetic nervous system. The adrenal hormonal response occurs about half a minute after the event because it takes that long for the two hormones to be released, circulate to all tissues, bind to receptors on their target cells, and initiate the appropriate intracellular cascade of events. The response is also short-lived, lasting only a couple of minutes at most. Epinephrine and norepinephrine from

the adrenal glands enhance the neural action of the sympathetic nervous system because the hormones reach more cells and are released in large amounts.

Recap The main secretory products of the adrenal cortex are cortisol and aldosterone. Cortisol raises blood glucose levels and suppresses inflammatory responses, and aldosterone regulates water balance by promoting sodium reabsorption. The hormones secreted by the adrenal medulla (epinephrine and norepinephrine) contribute to the fight-or-flight response initiated by the sympathetic nervous system. ∎

13.6 Thyroid and parathyroid glands

The thyroid and parathyroid glands are anatomically linked. The **thyroid gland** is situated just below the larynx at the front of the trachea, and the two lobes of the thyroid gland wrap part of the way around the trachea (**Figure 13.11**). The four small **parathyroid glands** are embedded in the back of the thyroid.

The thyroid and parathyroid glands are also functionally linked; both help to regulate calcium balance. In addition, the thyroid has a separate and important role in controlling metabolism.

The two main hormones produced by the thyroid gland are thyroxine and calcitonin. The parathyroid glands produce parathyroid hormone (PTH).

The thyroid gland: thyroxine speeds cellular metabolism

The thyroid gland produces two very similar hormones called **thyroxine** (T_4) and *triiodothyronine* (T_3). Thyroxine and triiodothyronine are identical except that thyroxine contains four molecules of iodine (hence the abbreviation T_4), whereas triiodothyronine contains only three. Because they are so similar, we consider them together as thyroxine.

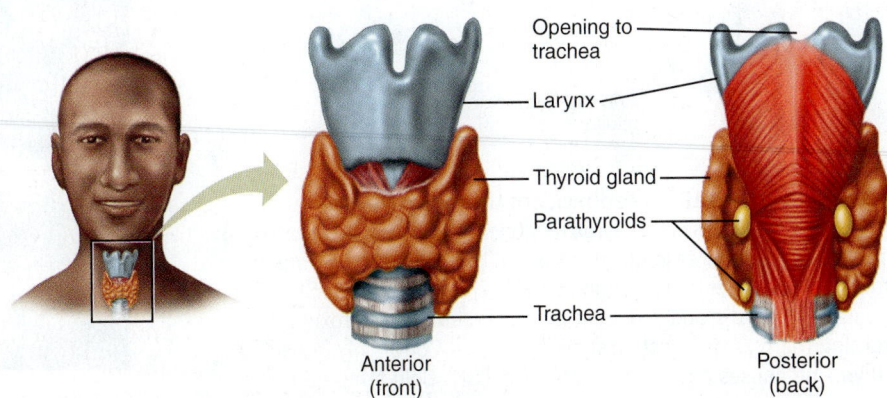

Figure 13.11 The thyroid and parathyroid glands.

Structurally, thyroxine is not a steroid hormone. Nevertheless, it acts like a steroid because it is lipid soluble. It crosses the cell and nuclear membranes, binds to nuclear receptors, and activates genes as its mode of action. Those genes carry the code for the various enzymes that regulate our metabolic rate.

Thyroxine increases the production and use of ATP from glucose in nearly all body cells. When thyroxine concentration increases, the basal metabolic rate (BMR) also increases (see section 14.11). Conversely, decreasing the blood concentration of thyroxine reduces energy utilization and BMR.

The basal rate of thyroxine secretion is regulated by thyroid-stimulating hormone (TSH) in a typical negative feedback loop (**Figure 13.12**). Any fall in thyroxine concentration ultimately leads to an increase in the secretion of the releasing hormone for TSH, called TSH-RH, from the hypothalamus. TSH-RH stimulates the secretion of TSH from the pituitary, which in turn stimulates the secretion of thyroxine by the thyroid gland, driving thyroxine concentration back toward normal. In addition, conditions that increase the body's energy requirements (such as exposure to cold temperatures) can stimulate the hypothalamus and raise the steady-state concentrations of *all* hormones in the loop.

✅ Do you think thyroxine levels would rise or fall in people who are starving (or who are dieting)? Why?

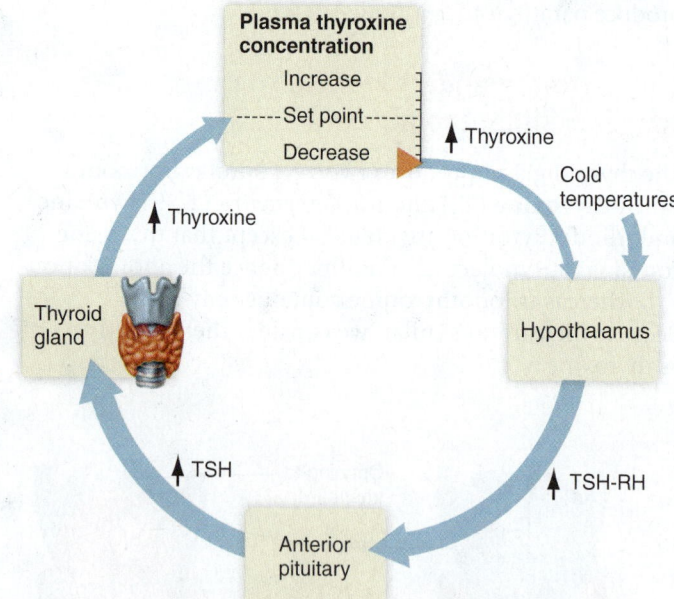

Figure 13.12 Negative feedback control of thyroxine secretion. A fall in plasma thyroxine concentration causes the hypothalamus to increase its secretion of TSH-RH and the anterior pituitary to increase its secretion of TSH. The increase in TSH stimulates thyroid gland production and secretion of thyroxine, returning the plasma concentration to normal. Conversely, a rise in thyroxine causes a fall in TSH-RH and TSH. Cold temperatures alter the feedback loop (raise plasma thyroxine concentration) by increasing hypothalamic secretion of the TSH-RH.

Iodine deficiency can cause goiter

Production of active thyroid hormones requires iodine. In fact, the main reason we need iodine in our diet is to ensure adequate production of thyroxine. Iodine deficiency can result in a thyroid-deficiency disease. When thyroxine is absent or abnormally low, the normal feedback inhibitory controls on the hypothalamus and pituitary are missing, so the hypothalamus and pituitary continue to secrete large quantities of the TSH-RH and TSH unchecked.

The high TSH levels stimulate the thyroid gland to grow to enormous size in an effort to get the thyroid to make more hormone, which it cannot do because it lacks iodine. An enlarged thyroid, as a result of iodine deficiency or other factors, is called a *goiter* (**Figure 13.13**). Goiter is less common today in industrialized nations because iodine is added to table salt. However, it is still a public health problem in certain areas of the world, most notably in Africa, China, and mountainous regions where rain has leached iodine from the soil.

Calcitonin promotes bone growth

The other main hormone of the thyroid gland, **calcitonin**, is produced by a separate group of thyroid cells. Calcitonin decreases the rate of bone resorption by inhibiting the activity of osteoclasts. It also increases the rate of calcium uptake by bones, leading to bone deposition. These two actions tip the normal balance of bone resorption/deposition toward deposition, which over time increases bone mass. Calcitonin is part of a negative feedback control mechanism for calcium, in which high blood calcium levels stimulate calcitonin release and low blood calcium levels inhibit its release.

Calcitonin is particularly important for the growth and development of bones in children. Once adulthood

Figure 13.13 Goiter. A goiter is an enlarged thyroid gland caused by dietary iodine deficiency.

is reached, bone responsiveness to calcitonin declines. In adults, calcitonin is not essential for the normal regulation of blood calcium concentration. However, calcitonin may play a role in protecting the skeleton when calcium demand is high, such as during pregnancy.

Parathyroid hormone (PTH) controls blood calcium levels

The parathyroid glands produce only one hormone, **parathyroid hormone (PTH)**, which

1. Removes calcium and phosphate from bone,
2. Increases absorption of calcium by the digestive tract, and
3. Causes the kidneys to retain calcium and excrete phosphate.

Together, these actions tend to raise the blood calcium concentration (and coincidentally lower blood phosphate concentration). As you might expect for a negative feedback control system, the secretion of PTH is stimulated by low blood calcium levels and inhibited by high blood calcium levels. **Figure 13.14** shows the overall regulation of blood calcium concentration.

The effect of PTH on calcium absorption by the digestive tract is actually indirect, through vitamin D. Vitamin D is required for calcium absorption by the intestines. However, vitamin D circulates in an inactive form until it is activated by a biochemical alteration that occurs in the kidneys. PTH is required for the renal (kidney) activation step. An increase in PTH concentration increases the renal activation of vitamin D, which in turn enhances calcium absorption from the digestive tract.

As mentioned previously, calcitonin plays little or no role in the normal maintenance of blood calcium concentration once adulthood is reached. In adults, the primary regulator of blood calcium concentration is PTH.

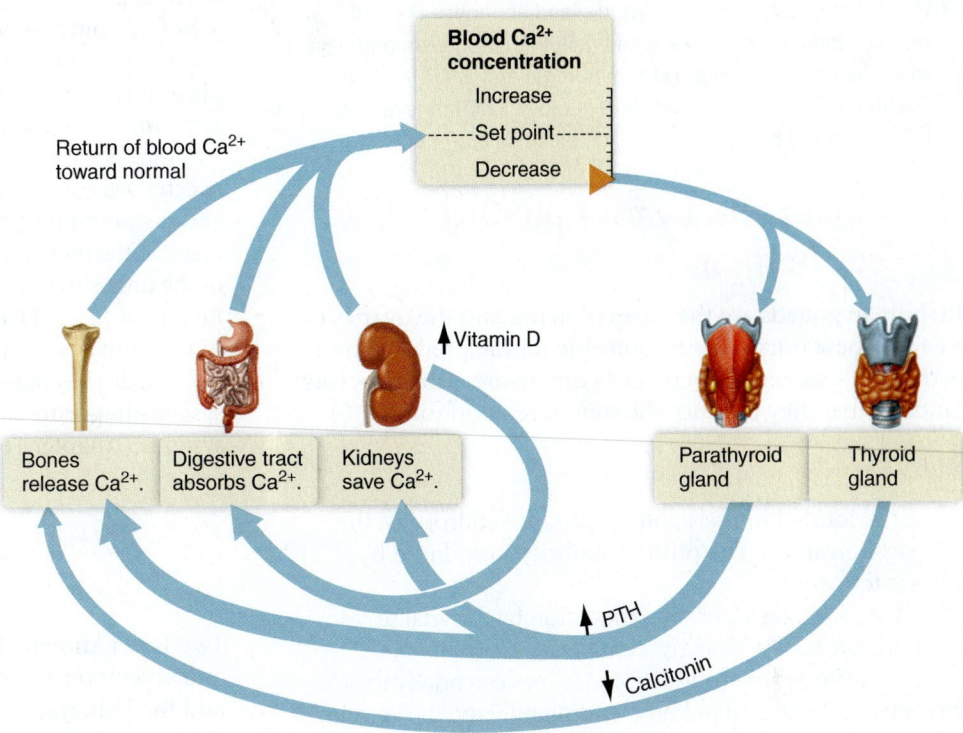

a) **Responses to a fall in blood calcium.**

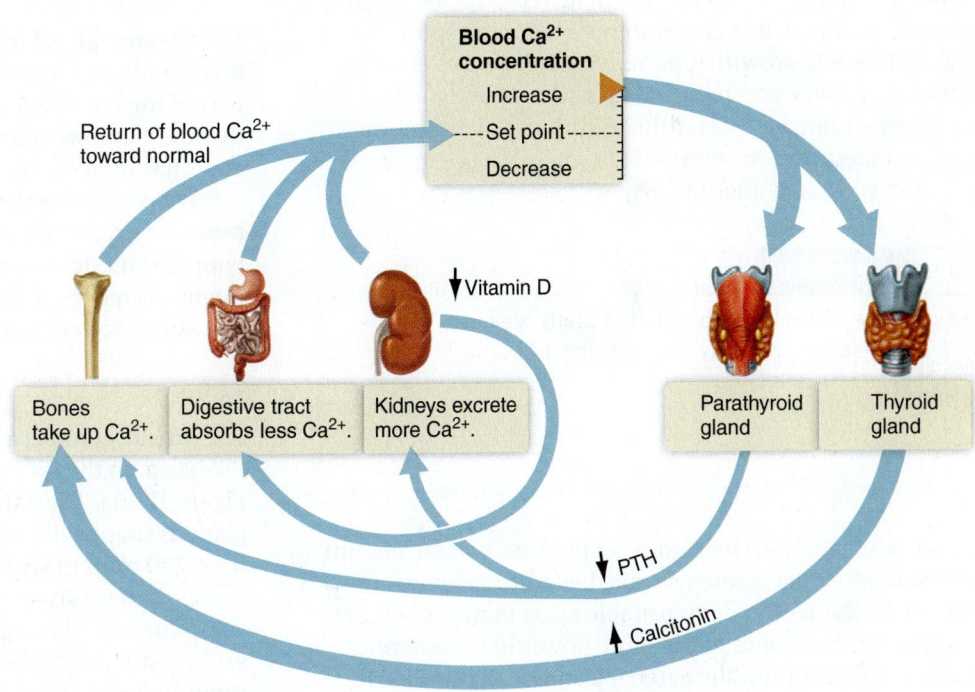

b) **Responses to a rise in blood calcium.** The importance of calcitonin is greatest in children, declining once adulthood is reached.

Figure 13.14 The homeostatic regulation of blood calcium concentration.

☑ Draw a diagram similar to part b illustrating how blood glucose is brought back to normal after it rises following a meal. (Refer to Figure 13.9 for a reminder of the principles involved.)

⮌ **Recap** Production of the thyroid gland hormones (T_4 and T_3) requires iodine. T_3 and T_4 speed cellular metabolism and have widespread effects on growth in children and BMR in adults. Calcitonin (from the thyroid) lowers blood calcium levels, and PTH (from the parathyroids) raises blood calcium. ∎

13.7 Testes and ovaries produce sex hormones

The human gonads are the testes of males and the ovaries of females. These organs are responsible for the production of sperm and eggs, respectively. Both organs are also endocrine glands in that they produce the steroid sex hormones.

Testes produce testosterone

The testes, located in the scrotum, produce androgens, the male sex hormones. The primary androgen produced by males is testosterone.

Before birth, testosterone production by the fetal testes is responsible for the development of the external male genitalia. Testosterone production declines essentially to zero between birth and puberty. During puberty, the anterior pituitary gland begins to release luteinizing hormone (LH), which stimulates the testes to resume testosterone production.

In males, testosterone regulates the development and normal functioning of sperm, the male reproductive organs, and male sex drive. It is also responsible for the spurt of bone and muscle growth at puberty and the development of male secondary sex characteristics such as body hair and a deepening voice. LH, follicle-stimulating hormone (FSH), and testosterone continue to maintain the male reproductive system after puberty.

MJ's **BlogInFocus**

What is "low T," and how common is it? Visit MJ's blog in the Study Area in MasteringBiology and look under "Low T."

http://goo.gl/FZIUZS

The adrenal glands in both sexes produce a small amount of a testosterone-like androgen called *dihydroepiandrosterone* (DHEA). DHEA has no demonstrable effect in males because they have an abundance of the more powerful testosterone. In females, DHEA from the adrenal glands is responsible for many of the same actions as testosterone in males, including enhancement of female pubertal growth, the development of axillary (armpit) and pubic hair, and the development and maintenance of the female sex drive.

Ovaries produce estrogen and progesterone

The ovaries, located in the abdomen, produce the female sex hormones known collectively as *estrogens* (17b-estradiol, estrone, and estriol), and also progesterone. The estrogens are often commonly referred to just as estrogen.

During puberty, the anterior pituitary starts to release LH and FSH. These hormones stimulate the ovaries to begin secreting estrogen and progesterone.

Estrogen initiates the development of female secondary sex characteristics such as breast development, widening pelvis, and distribution of body fat. Both estrogen and progesterone regulate the menstrual cycle, the monthly changes in the tissue of the uterus that prepare the female body for pregnancy. Release of LH, FSH, estrogen, and progesterone continues throughout a woman's reproductive years.

We discuss male and female hormones and reproductive systems in greater detail in the chapter on reproductive systems (Chapter 16).

13.8 Other glands and organs also secrete hormones

Two lesser known glands (the thymus and pineal glands) and several organs (the heart, those of the digestive system, and the kidneys) also produce hormones. The hormones of the heart, digestive system, and kidneys are described briefly here and then discussed in more detail in later chapters.

Thymus gland hormones aid the immune system

The **thymus gland** is located between the lungs, behind the breastbone and near the heart. Sometimes it is considered part of the lymphatic system. For years, the function of the thymus gland went unrecognized because its removal in an adult has little effect.

We now know that the thymus gland secretes peptide hormones called *thymosin* and *thymopoietin*, which help lymphocytes develop into mature T cells (Chapter 9). The thymus is relatively important in children, but it shrinks and reduces its secretion in adults, especially after age 30.

The pineal gland secretes melatonin

The **pineal gland** is a pea-sized gland located deep within the brain, in the roof of the third ventricle (refer to Figure 11.13). The name *pineal* derives from the fact that the gland is shaped like a small pine cone (*pineas* in Latin). More than 200 million years ago the pineal gland of our ancestors was a photosensitive area, or "third eye," located near the skin's surface. Although it is now shielded from light by the skull, it still retains its photosensitivity because it receives input indirectly from the eyes via the optic nerve and nerve pathways in the brain.

The pineal secretes the hormone **melatonin** (not to be confused with the skin pigment, melanin) in a cyclic manner, coupled to the daily cycle of light and dark. Melatonin is sometimes called the "hormone of darkness" because its rate of secretion rises nearly 10-fold at night and then falls again during daylight. Its secretion appears to be regulated by the absence or presence of visual cues. During

Table 13.2 Hormones of endocrine glands other than the hypothalamus and pituitary

Gland	Hormone(s)	Main targets	Primary actions
Pancreas	Glucagon	Liver	Raises blood sugar levels
	Insulin	Liver, muscle, adipose tissue	Lowers blood sugar levels
	Somatostatin	Pancreas, hypothalamus, digestive tract	Inhibits release of glucagon, insulin, growth hormone, and digestive secretions
Adrenal cortex	Glucocorticoids (including cortisol)	Most cells	Raises blood glucose levels and promote breakdown of fats and proteins; suppresses inflammation
	Mineralocorticoids (including aldosterone)	Kidneys	Regulates body water balance by promoting sodium reabsorption, potassium excretion
Adrenal medulla	Epinephrine and norepinephrine	Liver, muscle, adipose tissue	Raises blood sugar; increases heart rate and force of contraction; dilates or constricts blood vessels; increases respiration
Thyroid	Thyroxine (T_4) and triiodothyronine (T_3)	Most cells	Regulates the rate of cellular metabolism; affects growth and development
	Calcitonin	Bone	Lowers blood calcium levels
Parathyroid glands	Parathyroid hormone (PTH)	Bone, digestive tract, kidneys	Raises blood calcium levels
Testes	Testosterone	Most cells	Involved in development of sperm, male secondary sex characteristics, and reproductive structures
Ovaries	Estrogen, progesterone	Most cells	Involved in development of female secondary sex characteristics and reproductive structures; regulation of menstrual cycle
Thymus	Thymosin, thymopoietin	Lymphocytes	Helps lymphocytes mature, especially in children
Pineal	Melatonin	Many cells	Synchronizes body rhythms; may be involved in onset of puberty

the day when a person is exposed to strong light, nerve impulses from the retina inhibit its release.

We are just beginning to understand what melatonin does. Researchers believe that it may be important in synchronizing the body's rhythms to the daily light/dark cycle (the *circadian cycle*). Some scientists suggest that declining sensitivity to melatonin may contribute to the onset of puberty; evidence in favor of this hypothesis is that destruction of the pineal gland by disease is associated with early puberty.

Melatonin is readily available over the counter as a pill. Some people take it as a natural sleep aid, although its effectiveness is controversial.

Table 13.2 lists all of the primary endocrine organs other than the hypothalamus and the pituitary, and summarizes their targets and hormonal actions.

☑ During winter, when days are short, many office workers never see the sun at all because they spend their days indoors working under relatively dim office lights. How might this affect their melatonin levels over a 24-hour period and what effect, if any, might that have on their mood or behavior?

Endocrine functions of the heart, the digestive system, and the kidneys

The heart, the digestive system, and the kidneys all have functions unrelated to the endocrine system. Nevertheless, all three of these organs secrete at least one hormone. The hormones they secrete and the functions of the hormones are described in the following list and summarized in Table 13.3.

Table 13.3 Hormones of the digestive tract, kidneys, and heart

System or organ	Hormone(s)	Main targets	Primary actions
Digestive tract	Gastrin, secretin, cholecystokinin	Stomach, pancreas, gallbladder	Stimulates activities of the stomach, pancreas, and gallbladder
Kidneys	Erythropoietin	Bone marrow	Stimulates red blood cell production
Kidneys and liver	Renin-angiotensin	Adrenal cortex, blood vessels	Stimulates aldosterone secretion, constricts blood vessels
Heart	Atrial natriuretic hormone (ANH)	Kidney, blood vessels	Increases sodium excretion, dilates blood vessels

- *Atrial natriuretic hormone (ANH)* is a peptide (nonsteroid) hormone secreted by the atria of the heart that helps regulate blood pressure. When blood pressure rises, ANH increases the rate at which sodium and water are excreted in urine. This decreases blood volume and lowers blood pressure. We discuss how ANH acts on the kidneys in the chapter on the urinary system (Chapter 15).
- *Gastrin, secretin,* and *cholecystokinin* are all hormones secreted by the digestive system. They have effects on the stomach, pancreas, and gallbladder (Chapter 14).
- *Erythropoietin* and *renin* are secreted by the kidneys. Erythropoietin is a hormone that stimulates the production of red blood cells in bone marrow (Chapter 7). Renin is actually an enzyme, but it is generally included in any discussion of hormones because it is a critical component of the *renin-angiotensin system*, which functions overall as a hormonal negative feedback control system. The other critical component of the renin-angiotensin system comes from the liver. The renin-angiotensin system is discussed in the chapter on the urinary system (Chapter 15).

13.9 Other chemical messengers

Some chemical messengers can function in ways similar to hormones but are not considered hormones because they are not secreted into the blood. Their actions are primarily local (near their site of release). You already know about neurotransmitters, one group of chemical messengers that act on nearby cells. Other nonhormonal chemical messengers include histamine, prostaglandins, nitric oxide, and various growth factors and growth-limiting factors, plus dozens of other molecules about which we still know very little (Table 13.4). Most of these substances are short-lived because they are either quickly destroyed or taken back up by the cells that produced them.

Table 13.4 **Other chemical messengers**

Messenger(s)	Primary actions
Histamine	Important in inflammatory response; increases secretion of mucus; dilates blood vessels; increases leakiness of capillaries
Prostaglandins	Controls local blood flow; constricts blood vessels; dilates blood vessels; contributes to inflammatory response and blood clotting
Nitric oxide (NO)	Regulates local blood flow; controls penile erection; regulates smooth muscle contraction in digestive tract; acts as a "chemical warfare" agent against bacteria; interferes with clotting mechanisms; acts as a neurotransmitter in brain
Growth factors	Modulates growth of tissues at local level

Histamine is important in inflammation

Histamine was mentioned in the chapter on the immune system (Chapter 9). Mast cells release histamine into the local interstitial fluid in response to tissue injury or the presence of an allergen. In turn, the histamine increases local secretion of mucus by mucus-secreting cells, dilates blood vessels, and increases the leakiness of capillaries, all part of a constellation of symptoms called *inflammation*.

Histamine is largely responsible for the runny nose of a cold and the nasal congestion of allergies. This is why over-the-counter drugs for colds and allergies generally contain compounds that block the action of histamine.

Prostaglandins: local control of blood flow

Prostaglandins are a group of chemicals derived from a fatty acid precursor. Their name comes from the fact that they were first discovered in semen and were thought to originate from the prostate gland, but they have since been found in nearly every tissue in the body.

Prostaglandins have a variety of functions, including local control of blood flow. Some prostaglandins constrict blood vessels; others dilate blood vessels. They also contribute to the inflammatory response and are involved in blood clotting at the site of an injury.

Nitric oxide has multiple functions

Nitric oxide (NO) is a gas that has only recently received attention as a chemical messenger. NO was originally called "endothelial-derived relaxing factor" because an extract of the endothelial cells lining blood vessels caused blood vessels to relax, or dilate. We now know that NO is produced in many other tissues as well.

Nitric oxide functions include regulating local blood flow in many tissues of the body, controlling penile erection, and regulating smooth muscle contraction in the digestive tract. Macrophages use it as a "chemical warfare" agent against bacteria, and it also interferes with clotting mechanisms. It even functions as a neurotransmitter in the brain.

Growth factors regulate tissue growth

Chemical messenger molecules called *growth factors* modulate development of specific tissues at the local level. The list includes nerve growth factor, epidermal growth factor, fibroblast growth factor, platelet-derived growth factor, tumor angiogenesis growth factors, insulin-like growth factor, and many more. Growth factors influence when a cell will divide and even in what direction it will grow. For example, nerve growth factors help to determine the direction of neuron axon growth so that nerves develop properly.

Recap *Various chemical messengers (histamine, prostaglandins, nitric oxide, and growth factors) act primarily locally. Among their actions are participating in local defense responses, regulating local blood flow, and regulating growth of specific types of tissue.* ■

13.10 Disorders of the endocrine system

Because the endocrine system is one of the two primary systems for controlling body functions, any disruption to it can have dramatic and widespread effects on the body. We have already discussed several endocrine disorders earlier in this chapter. Here, we look at several more.

Diabetes mellitus: inadequate control of blood sugar

Diabetes mellitus means "a high flow of urine with a sweet taste." Called diabetes for short, it is a disease of sugar regulation. The common feature of the several types of diabetes is an inability to get glucose into cells where it can be used. The result is an abnormally high glucose concentration in the blood—so high that it overwhelms the kidney's ability to reabsorb all that is filtered. Glucose and excessive water appear in the urine (hence the name of the disease), causing dehydration and thirst. In addition to these symptoms, many diabetics experience fatigue, frequent infections, blurred vision, cuts that are slow to heal, and tingling in the feet or hands.

Without the ability to metabolize glucose properly, the body turns to the metabolism of fat and proteins. The abnormal metabolism of glucose, fats, and proteins causes most of the other medical problems associated with diabetes, including cardiovascular and neural diseases, renal failure, blindness, and damage to small blood vessels that may lead to the need for amputations. There are two types of diabetes.

- *Type 1 diabetes* (5–10% of all cases) is caused by the failure of the pancreas to produce enough insulin. It is called *insulin-dependent diabetes* because the person depends on daily injections of insulin to allow glucose to enter the cells. Typically, it is not governed by lifestyle; rather, it is an autoimmune disorder in which the person's own immune system attacks the insulin-producing cells in the pancreas. It is thought to have a strong genetic component (close relatives of a person with Type 1 diabetes are at increased risk), but it may also require an environmental trigger, such as a viral infection. Typically, it appears in childhood or adolescence.
- *Type 2 diabetes* (*non-insulin-dependent diabetes*) represents 90–95% of all cases. Its hallmark is *insulin resistance*—cells fail to respond adequately to insulin even when it is present. Type 2 diabetes usually occurs in adults beyond the age of 40; however, in recent years it has been occurring more frequently at younger ages. Type 2 diabetes may have a genetic component, but lifestyle factors are thought to be the major determinants.

How does lifestyle affect the development of Type 2 diabetes? Apparently, persistent exposure of a cell to too much glucose causes cells to become resistant to the effects of insulin. As a result, blood glucose rises even higher. That, in turn, may cause the pancreas to secrete even more insulin in an attempt to overcome the resistance, which leads to a vicious cycle of further increases in insulin resistance, blood glucose, and insulin. Eventually, the pancreas may wear out and insulin secretion may decline. Type 2 diabetics may or may not have enough insulin, but what they all have in common is resistance to insulin and blood sugar concentrations out of control.

Some Type 2 diabetics and "prediabetics" whose blood sugars are just approaching the diabetic range can sometimes control the disease entirely through lifestyle changes—maintaining a healthy weight, eating a nutritious diet, and exercising regularly. Some need to take insulin, especially to supplement their lifestyle changes when necessary. Other treatments include oral drugs that stimulate the pancreas to secrete more insulin, drugs that improve the action of the body's natural insulin, and drugs that increase the uptake of glucose by liver and muscle cells. More permanent solutions, such as implanting an artificial pancreas or transplanting human pancreatic cells, are under investigation.

☑ In general, do you think it would be easier to treat disorders that result from failure to produce a hormone (such as Type 1 diabetes) or disorders that result from failure to respond to a hormone (such as Type 2 diabetes)? Why?

Hypothyroidism: underactive thyroid gland

Hypothyroidism refers to underactivity of the thyroid gland (hyposecretion of thyroid hormones). Mild cases of hypothyroidism may not have any signs, but more severe deficiencies can cause a variety of symptoms.

In children, insufficient thyroxine production for any reason can slow body growth, alter brain development, and delay the onset of puberty. Left untreated, this deficiency can result in cretinism, a condition of mental retardation and stunted growth.

In adults, insufficient thyroxine can lead to myxedema, a condition characterized by edema (swelling) under the skin, lethargy, weight gain, low BMR, and low body temperature. Hyposecretion of the hormone can be treated with thyroxine pills.

Hyperthyroidism: overactive thyroid gland

Hyperthyroidism involves an overactive thyroid gland and hypersecretion of thyroid hormones. Too much thyroxine increases BMR and causes hyperactivity, nervousness, agitation, and weight loss.

The most common form of hyperthyroidism is Graves' disease, an autoimmune disorder in which a person's antibodies stimulate the thyroid to produce too much thyroxine. Graves' disease is often accompanied by protruding eyes caused by fluid accumulation behind the eyes.

HEALTH & WELLNESS

Dealing with Diabetes: Prevention or Treatment?

The number of people in the United States with diabetes has nearly quadrupled in the past 30 years. Worldwide, the diabetes problem is even worse: According to the World Health Organization (WHO), the number of people worldwide living with diabetes has risen from 30 million in 1985 to 347 million in 2015—a 10-fold increase. What are the health care costs associated with diabetes, and how are we going to deal with them?

The true cost of diabetes includes the many medical problems associated with the disease if it is left untreated. For example, in the United States, approximately 40,000 diabetics develop kidney disease each year. Many of them progress to kidney failure, at which point they require treatment at a cost of over $50,000 per year (assuming the treatment, called *dialysis,* is available to them). Their lives and the lives of their families begin to revolve around managing their disease as they spend several half-days a week in a hospital or dialysis unit. Eventually, they may need a kidney transplant, for which they will be billed nearly $250,000 if they are uninsured. Currently, the total cost of treating the

long-term complications of diabetes in the United States is estimated to be $245 billion/year—the equivalent of $1,000 for every U.S. adult.

A concerted effort by health professionals and governments aimed at promoting healthy lifestyles so that fewer people develop diabetes in the first place, combined with early detection and effective management of those who do develop diabetes, would reduce the total cost of diabetes in the long run. Unfortunately, the U.S. health care reimbursement system does not focus on maintaining wellness; it focuses on treating disease once it occurs. Hospitals, doctors, and health care centers make their money through payments from government sources (such as Medicare and Medicaid) and private insurers. At current reimbursement rates, they generally lose money on preventive care and early disease management, but make money every time a diabetic is hospitalized for a major complication such as a kidney transplant or an amputation. One study estimated that New York health care centers lost $40–$55 every time a diabetic patient met with a diabetes

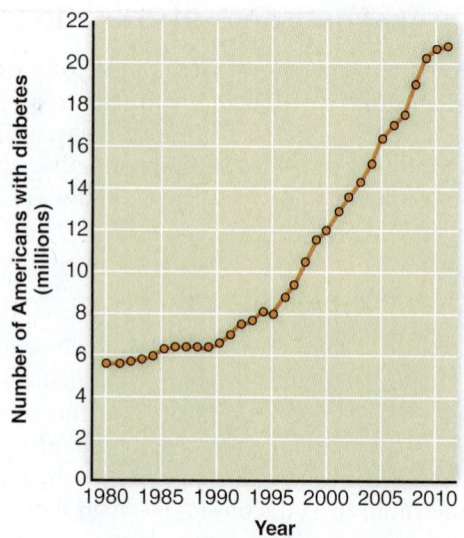

The number of Americans with diabetes since 1980.

educator to learn how to monitor blood sugar properly or administer insulin or other treatment, and another $40–$55 when the patient met with a nutritionist to discuss controlling blood sugar through weight loss and a proper diet. In contrast, hospitals earned $1,500 to over $10,000 for every amputation, depending on the type of amputation and its complications.

Is it any wonder that hospitals are more likely to hire a surgeon than to add nutritionists or health educators to their staff? Until something changes, the health care costs related to diabetes will continue to rise.

Addison's disease: too little cortisol and aldosterone

Addison's disease is caused by failure of the adrenal cortex to secrete sufficient cortisol and aldosterone. Lack of cortisol lowers blood glucose levels, and lack of aldosterone lowers blood sodium. Addison's disease tends to develop slowly, with chronic symptoms of fatigue, weakness, abdominal pain, weight loss, and a characteristic "bronzed" skin color. It can be successfully treated with medications that replace the missing hormones.

Cushing's syndrome: too much cortisol

The symptoms of Cushing's syndrome are due to the exaggerated effects of too much cortisol, including

(1) excessive production of glucose from glycogen and protein and (2) retention of too much salt and water. Blood glucose concentration rises, and muscle mass decreases because of the utilization of protein to make glucose. Some of the extra glucose is converted to body fat, but only in certain areas of the body, including the face, abdomen, and the back of the neck (Figure 13.15). Symptoms of Cushing's syndrome include muscle weakness and fatigue, edema (swelling due to too much fluid), and high blood pressure. The disease can be caused by tumors of the adrenal gland or the ACTH-secreting cells of the pituitary. It can also be due to excessive use of cortisol and cortisol-like drugs (cortisone, prednisone, dexamethasone, and others) to control chronic inflammatory conditions such as allergies and arthritis.

a) Before the disease.

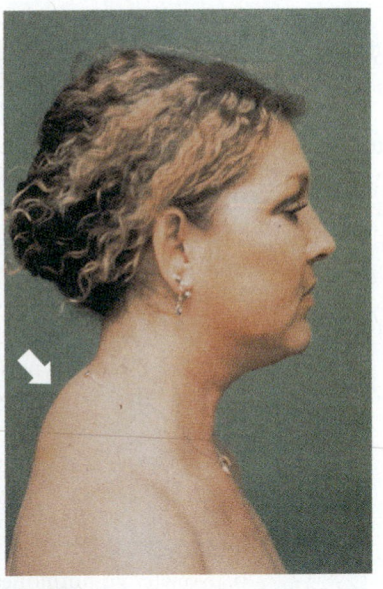

b) During the disease, a characteristic hump of fat develops at the back of the neck.

Figure 13.15 Cushing's syndrome. Cushing's syndrome is caused by too much cortisol.

Hypogonadism: too little testosterone

Hypogonadism is a condition caused by abnormally low testosterone levels. If it is present before birth, a male child may be born with ambiguous, underdeveloped, or female genitals. After birth but before puberty, hypogonadism may result in delayed or incomplete male sexual maturation. In adult males, a diagnosis of hypogonadism requires a low testosterone level and the presence of one or more clinically relevant symptoms, including a low sperm count, erectile dysfunction, unusual fatigue, decreased sex drive, reduced strength, and depressed mood. The condition is effectively treated by testosterone replacement therapy, using gels or creams that are absorbed through the skin.

Only 3% of men under the age of 70 have hypogonadism. It is normal for testosterone levels to decline with age, so a low testosterone level in an 80-year-old man is not necessarily defined and treated as hypogonadism. It is also normal for men over 40 to feel more tired and to begin to experience a reduced sex drive. A diagnosis of hypogonadism should be made before beginning a testosterone replacement regimen.

Chapter Summary

13.1 The endocrine system produces hormones *p. 300*

- The endocrine system is a collection of glands that secrete hormones into the blood.
- Hormones are chemical messengers secreted by the endocrine system that circulate in the blood and act on specific target cells.

13.2 Hormones are classified as steroid or nonsteroid *p. 302*

- Steroid hormones are lipid soluble; they enter the target cell in order to function.
- Nonsteroid hormones generally are not lipid soluble. They bind to receptors on the outer surface of the cell membrane of the target cell.

13.3 The hypothalamus and the pituitary gland *p. 304*

- The hypothalamus of the brain and the pituitary gland of the endocrine system link the nervous and endocrine systems.
- The posterior lobe secretes two hormones (ADH and oxytocin) that are produced by the hypothalamus.
- The anterior lobe of the pituitary produces and secretes six hormones (ACTH, TSH, FSH, LH, PRL, and GH). The first four stimulate the release of other hormones from other endocrine glands.
- Abnormalities of ADH secretion lead to disorders of water balance.
- Hypersecretion of growth hormone in a child leads to gigantism; hyposecretion causes dwarfism.

13.4 The pancreas secretes glucagon, insulin, and somatostatin *p. 309*

- Within the pancreas, clusters of cells called the *pancreatic islets* produce and secrete three hormones (glucagon, insulin, and somatostatin).

- Glucagon raises blood glucose levels, insulin lowers blood glucose levels, and somatostatin appears to inhibit the secretion of glucagon and insulin.

13.5 The adrenal glands comprise the cortex and medulla *p. 310*

- The outer zone of the adrenal gland, called the *cortex*, secretes steroid hormones that affect glucose metabolism (glucocorticoids) and salt balance (mineralocorticoids).
- The inner zone of the adrenal gland, called the *medulla*, is actually a neuroendocrine organ. The adrenal medulla secretes epinephrine and norepinephrine into the blood.

13.6 Thyroid and parathyroid glands *p. 311*

- The thyroid gland produces thyroxine (T_4), triiodothyronine (T_3), and calcitonin.
- The four small parathyroid glands produce PTH, which, along with calcitonin, is responsible for the regulation of blood calcium concentration.

13.7 Testes and ovaries produce sex hormones *p. 314*

- The primary male sex hormone is testosterone, produced by the testes.
- The ovaries produce the two female sex hormones, estrogen and progesterone.

13.8 Other glands and organs also secrete hormones *p. 314*

- The thymus gland produces two hormones (thymosin and thymopoietin) that help lymphocytes mature into T cells.
- Melatonin from the pineal gland synchronizes the body's daily rhythms to the daily light/dark cycle.
- The heart secretes atrial natriuretic hormone; the kidneys secrete renin and erythropoietin; and the digestive tract secretes gastrin, secretin, and cholecystokinin.

13.9 Other chemical messengers p. 316

- Some substances not defined as hormones (because they are not specifically secreted into the bloodstream) nevertheless serve as chemical messengers.
- Most of these chemical messengers tend to act locally, near their site of secretion. They include histamine, prostaglandins, nitric oxide, and growth factors.

13.10 Disorders of the endocrine system p. 317

- Because the endocrine system controls so many body functions, disruptions can have widespread effects.
- Hypothyroidism can lead to cretinism (stunted growth and mental retardation) in children and myxedema in adults. Graves' disease is the most common form of hyperthyroidism.
- Insufficient cortisol and aldosterone can cause Addison's disease, and hypersecretion of cortisol results in Cushing's syndrome.

Terms You Should Know

adrenal cortex, 310
adrenal medulla, 310
aldosterone, 310
calcitonin, 312
cortisol, 310
endocrine glands, 300
epinephrine, 310
glucagon, 309
hormones, 300
hypothalamus, 304
insulin, 309
neuroendocrine cells, 305
nonsteroid hormones, 302
norepinephrine, 310
parathyroid gland, 311
parathyroid hormone (PTH), 313
pineal gland, 314
pituitary gland, 305
second messenger, 304
steroid hormones, 302
target cells, 302
thymus gland, 314
thyroid gland, 311
thyroxine, 311

Concept Review

Answers can be found in the Study Area in MasteringBiology.

1. List the functions of the six hormones secreted by the anterior lobe of the pituitary gland.
2. Explain the characteristics of the endocrine system that make its reactions different from those of the nervous system.
3. Compare and contrast the mechanisms by which steroid and nonsteroid hormones act.
4. Define when you would call a norepinephrine a neurotransmitter and when you would call it a hormone.
5. Explain why a 30-year-old short person cannot be made taller by taking a growth hormone.
6. Describe the function of the two hormones from the thymus gland. Why is this function likely to be more important in children than in adults?
7. Explain how the pineal gland can be sensitive to light, even though it is located deep within the brain.
8. Describe why substances such as histamine and nitric oxide, although they are chemical messengers, are not considered true hormones.

9. Name the hormone that is released from the posterior pituitary of lactating females when the infant nurses.
10. Describe the role of the pineal gland in the circadian cycle.

Test Yourself

Answers can be found in the Appendix.

1. _____ hormones enter target cells and bind to intracellular receptors, whereas _____ hormones bind to cell membrane receptors and never enter the target cells.
   a. Nonsteroid . . . steroid
   b. Protein . . . peptide
   c. Steroid . . . nonsteroid
   d. Steroid . . . carbohydrate
2. Which of the following statements comparing steroid and nonsteroid hormones is true?
   a. Steroid hormones typically act through second messengers such as cAMP.
   b. Nonsteroid hormones usually are faster acting than steroid hormones.
   c. Steroid hormones are secreted by the anterior pituitary gland, and nonsteroid hormones are secreted by the posterior pituitary gland.
   d. Steroid hormones are produced only after puberty, whereas nonsteroid hormones are produced throughout the lifespan.
3. Secretion of hormones by the anterior pituitary gland is closely regulated by the:
   a. hypothalamus
   b. posterior pituitary gland
   c. cerebellum
   d. adrenal cortex
4. Which two hormones most directly regulate lactation?
   a. estrogen and progesterone
   b. estrogen and prolactin
   c. progesterone and oxytocin
   d. oxytocin and prolactin
5. All of the following represent hormone pairs with antagonistic activities except:
   a. estrogen . . . progesterone
   b. insulin . . . glucagon
   c. parathyroid hormone . . . calcitonin
   d. aldosterone . . . ANH
6. Which of the following hormones/chemical messengers is produced exclusively in males?
   a. prostaglandins
   b. melatonin
   c. testosterone
   d. none of these choices
7. Hormones secreted by the anterior pituitary gland control the activity of all of the following glands except:
   a. adrenal medulla
   b. thyroid gland
   c. testes
   d. ovaries
8. Which of the following hormones is/are involved in normal functioning of the male reproductive system?
   a. testosterone
   b. LH
   c. FSH
   d. all of these choices
9. Which of these endocrine disorders is associated with the inability of the pancreas to produce insulin?
   a. diabetes insipidus
   b. Addison's disease
   c. Type 1 diabetes mellitus
   d. Type 2 diabetes mellitus

10. Which of these statements comparing the nervous system and the endocrine system is true?
    a. The endocrine system exclusively controls development, whereas the nervous system controls homeostasis.
    b. The nervous system can access and communicate with all cells in the body, whereas the access of the endocrine system to different cells and tissues is more limited.
    c. The nervous system enables more rapid communication and signaling than does the endocrine system.
    d. The nervous system is fully functioning at birth, whereas the endocrine system doesn't begin to function until puberty.

11. Which of the following statements about the adrenal cortex or its hormones is true?
    a. The adrenal cortex secretes a mixture of steroid and nonsteroid hormones.
    b. The adrenal cortex secretes hormones that may decrease the inflammatory response and suppress immune responses.
    c. The secretion of hormones by the adrenal cortex is controlled by the sympathetic division of the nervous system.
    d. Hypersecretion of cortisol and aldosterone is seen in Addison's disease.

12. Which of the following statements about glucagon is true?
    a. Glucagon causes glycogen to break down to form glucose.
    b. Glucagon secretion is highest following a heavy meal.
    c. Glucagon promotes the uptake of glucose by liver and muscle cells.
    d. Glucagon works with cortisol to lower blood glucose levels.

13. Which of the following would be least likely to be affected by a tumor of the pituitary gland?
    a. cortisol secretion
    b. thyroid hormone secretion
    c. lactation
    d. secretion of epinephrine by the adrenal medulla

14. Which of the following is incorrectly paired?
    a. insulin . . . diabetes insipidus
    b. thyroid . . . Grave's disease
    c. cortisol . . . Cushing's syndrome
    d. ADH . . . diabetes insipidus

15. Which of the following would be triggered by a drop in blood Ca^{2+} levels?
    a. increase in PTH secretion
    b. decrease in calcitonin secretion
    c. increase in ADH secretion
    d. both (a) and (b)

Apply What You Know

Answers can be found in the Study Area in MasteringBiology.

1. If the biological effectiveness of a hormone were greatly reduced because its target cells lacked the appropriate receptors, would you expect the hormone concentration to be high, normal, or low? Explain. (*Hint:* It may help to draw the components of a negative feedback loop.)
2. Explain why an injection of epinephrine to combat an acute immune reaction (such as an allergic response to a bee sting) has an almost immediate effect, whereas an injection of a steroid hormone can take hours to have an effect.

3. From what you read in section 13.5, draw the negative feedback control loop for cortisol secretion. Then indicate on your diagram how stress or injury would affect the feedback loop. Check your diagram with a classmate or your teacher—if you can draw it and explain it, you understand it!
4. When a pregnant woman goes into labor, a series of strong muscle contractions force the baby from the uterus through the birth canal. Labor can be painful. Normally, such pain would initiate a negative feedback loop that would inhibit muscle contractions to prevent such pain, but that doesn't happen during labor. Why not?
5. Diabetics have to be particularly careful about sugar intake. Some are able to control the condition through careful diet. What sorts of foods should diabetics avoid, what foods should they try to get more of, and what can they do if they make a mistake?
6. A student finishes up the fall semester with several term papers and a number of big exams. After the semester, she goes home for the holidays and comes down with a bad cold. The following year the same thing happens; she's sick for the holidays. During her senior year, the fall semester isn't too bad, but applications to graduate school are due and she spends most of her semester break filling out forms and nervously worrying about what she will do next year. She's again sick for the beginning of the next term. Is this student just unlucky, or is something else at work?

MJ's BlogInFocus

Answers can be found in the Study Area in MasteringBiology.

1. Several pharmaceutical companies developed inhalable insulin products as far back as 2007. They expected the products to be blockbuster drugs, but the companies gave up on the products in 2008. Now a third company has announced that it will try to market an inhaled insulin product. Why do you think the first inhalable insulin product was not a success, and do you think it will be any different this time around? Visit MJ's blog in the Study Area in MasteringBiology and look under "Inhalable Insulin." http://goo.gl/TUa9yB

MasteringBiology®

Students Go to MasteringBiology for assignments, the eText, and the Study Area with animations, practice tests, and activities.

Professors Go to MasteringBiology for automatically graded tutorials and questions that you can assign to your students, plus Instructor Resources.

Answers to ✓ questions are available in the Appendix.

The Digestive System and Nutrition

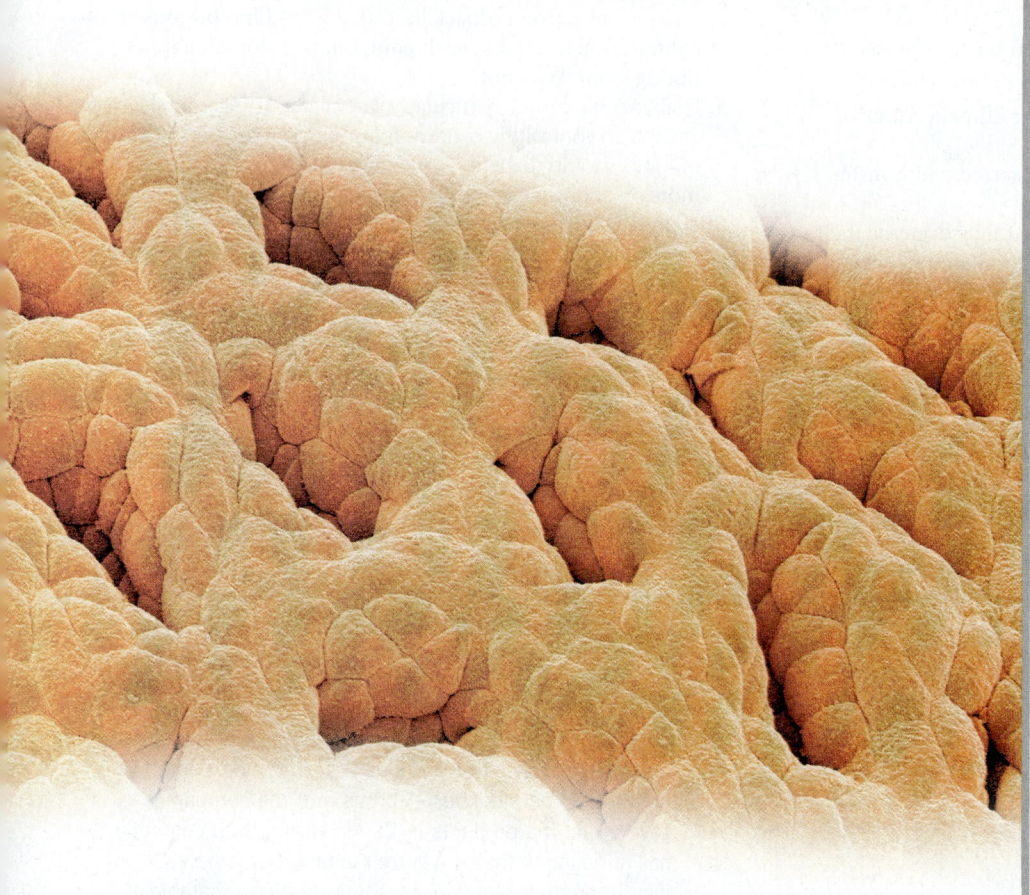

Colored scanning electron micrograph (SEM) of the lining of the stomach. The indents are gastric pits.

Key Concepts

- **The digestive system and its organs digest and absorb nearly everything we eat and drink**. Nonabsorbed products and bacteria are eliminated as feces.

- **The stomach stores ingested food and water** and secretes a strong acid—hydrochloric acid (HCl)—that breaks down proteins and kills most bacteria.

- **Nutrients and water are absorbed in the small intestine** and, to a lesser extent, the large intestine. Enzymes from the pancreas and the small intestine break down carbohydrates and fats so that they can be absorbed.

- **Nutritional requirements for health include vitamins and minerals.** Some of these are not stored within the body for long periods, so we need a consistent supply.

- **To maintain a constant body weight, energy intake must equal energy expenditure.** An imbalance in energy intake versus expenditure over time leads to weight gain or loss. Exercise can change how many calories we expend.

- **A healthy diet contains grains, fruits, vegetables, and low-fat milk products.** Saturated fats, cholesterol, and refined sugar should be consumed in moderation.

CURRENT ISSUE
Choosing Organic Versus Conventional Foods

The organic foods industry is growing at a rate of about 9% per year, according to industry experts. When people are asked why they choose organic foods over conventional foods, their answers, by and large, are that (1) organic foods are more nutritious and contain fewer harmful synthetic pesticides and (2) organic farming techniques are better for the environment. Let's look at these two beliefs.

ORGANIC FARM PLEASE DO NOT SPRAY WATCH WIND SPEED AND DIRECTION

Are Organic Foods More Nutritious?

The nutritional values of both organic and conventionally grown foods have been extensively tested. A few studies have shown that some organic foods may be slightly more nutritious in one nutrient or another, but most studies have found no differences. Where different nutritional values have been found, size is the main factor: Conventionally grown fruits and vegetables tend to contain slightly fewer of certain nutrients per pound because they are often larger than their organic counterparts. However, these size-related discrepancies in nutrients are generally slight, which is why they are not universally found. Anyone eating, say, conventionally grown strawberries might have to eat one more to attain the same nutritional value as someone eating organically grown strawberries. But anyone eating strawberries at all (either organic or conventional) is probably getting enough of whatever nutrition is in strawberries.

What About Pesticides in Our Foods?

In nearly all studies in which they have been measured, the concentrations of synthetic pesticides are higher in conventional foods. But that alone is not proof that consuming conventional foods is dangerous. Before any synthetic pesticide is approved for sale and use, it is tested extensively at different doses in animals to determine what concentrations of the

Questions to Consider

1 Do you buy organic foods or conventional foods when both are available? On what do you base your decision?

2 If you were concerned about the possible harm caused by a particular chemical used in conventional farming, where would you go for information? What would you do if you found conflicting information?

pesticide are toxic and cause harm. Such toxicology studies have shown that it takes concentrations thousands of times higher than in our foods to do any harm. If studies revealed otherwise, the pesticide would not be approved. However, there is one important *caveat:* It is not possible in short-term toxicology studies to duplicate the potential effect of ingesting low levels of synthetic pesticide over decades. Mice are the usual test subjects, and mice only live about two years. Evidence for the long-term safety of low concentrations of pesticides in our foods would either have to be based on testing in animals with life-expectancies as long as our own or on *post hoc* (after the fact) studies of humans exposed to low levels of pesticides for most of their lives. So while there isn't any proof that conventional foods *aren't* safe, neither is there proof that they *are.* A person's food choice, therefore, must be based on incomplete information, at least for now.

By the way, organic farmers use pesticides and fertilizers, too. They just use certain approved "natural" (biological) ones, such as soaps and manures. There aren't necessarily fewer pesticides in organic foods, just allegedly safer ones.

Are Organic Farming Techniques Better for the Environment?

The answer is almost surely yes. One cannot deny that mistakes were made with the use of synthetic pesticides in the past. The classic example is the use of dichlorodiphenyltrichloroethane (DDT) as an agricultural insecticide, starting in the 1940s. By the 1960s, it had become clear that DDT was having major negative effects on the health of wildlife, particularly birds, and there was a suspicion that it might contribute to cancer in humans. DDT was banned →

A plane spraying an alfalfa field with DDT, prior to 1970.

as an agricultural pesticide in the United States in the 1970s, but not before it decimated the population of birds of prey in this country. The question today is: Could widespread pesticide poisoning happen again?

Although the guidelines for approval of agricultural chemicals have become stricter over the years, there is always the chance of an oversight. How do we feel about that risk, however small it might be? It's hard to argue against the notion that it would be better to avoid any risk entirely by banning the use of all potentially damaging chemicals on our crops.

Are There Any Reasons Not to Choose Organic?

For someone who wishes to eat organic foods, the primary drawbacks are availability and price. Organic foods are not available everywhere, and where they are available they generally are more expensive. The higher price of organic foods reflects higher production costs due to the more labor-intensive pest-control tactics required by organic farming techniques. Price may not be a limiting factor to some, but for people living in poor regions, who are just trying to get enough to eat, expense may be a major factor.

Moreover, how we choose to farm, conventional versus organic, has economic consequences for the population as a whole. The one big advantage of conventional

farming, and the reason that its techniques rely on synthetic pesticides in the first place, is efficiency. Efficient farming equates to larger yields, and for many people around the world (some of whom often go hungry), this is a big plus indeed. The importance of farming efficiency is highly debated: Some scientists claim that a switch to all-organic farming worldwide would result in food shortages, whereas other experts dispute that claim.

Efficiency aside, the steady growth of the organic foods movement could get a boost from increased consumer demand. For those who can afford it and whose risk tolerance is low, choosing organic may be a wise personal choice. That choice could improve the

market for organically grown foods, which would in turn increase organic farming and improve its economic feasibility. And that certainly wouldn't harm the environment.

In the long run, lessons learned from organic farming techniques may partially transform conventional farming techniques. Someday, instead of "organic" the buzzword could be "ecoagriculture"—environmentally friendly farming practices in which all farmers use far fewer synthetic pesticides than they do today. If ecoagriculture were to take hold, then perhaps we could stop debating the merits of organic versus conventional farming and enjoy healthier lifestyles and a healthier environment.

SUMMARY

- Most scientifically valid research studies find little or no difference in the nutritional value between organic and conventional foods.
- Conventional foods contain measurably higher concentrations of synthetic pesticides than organic foods. However, the concentrations are far below those known to cause harm in experimental animals.
- The potential risks of long-term exposure to low levels of synthetic pesticides in conventional foods are difficult to prove.
- Organic farming techniques are probably better for the environment, but organic foods are more expensive.
- Conventional farming techniques are popular with farmers because they are more efficient and hence more cost-effective.

Stop for a moment and make a list of what you ate for lunch—a hamburger, fries, and a soft drink? Pizza and bottled water? A salad, apple, and milk?

Your digestive system is your body's sole source supplier of **nutrients**—substances in food that are required for growth, reproduction, and the maintenance of health. Every ounce of water, every calorie of energy, every essential amino acid and vitamin, every molecule of salt or sugar (and yes, some toxins, too!) enter your body only via the digestive system. We humans don't get our energy from the sun (like plants), and we can't absorb water through our skins (like frogs). Every living cell in every organ in the body relies totally on the digestive system for what it needs.

Most of what the digestive system takes in requires considerable processing before it can be absorbed and used by the cells. The digestive system is a highly efficient disassembly line. It takes in food, breaks it into small pieces, and digests the fragments with enzymes and strong chemicals. Then it passes the mixture on down the line, where the nutrients, vitamins, minerals, and water are absorbed.

When it's absorbed nearly everything that the body can use, the digestive system stores what's left (waste) until there's a convenient time to eliminate it from the body. So in addition to being a supplier and a disassembly line, the digestive system is also a waste collection and removal system. ■

Does what you eat matter? Yes! Poor nutrition is associated with diseases ranging from cancer to cavities. Good nutrition, on the other hand, improves overall health and lowers the risk of health problems. In this chapter, we follow the digestive process and the fate of your food. Then we discuss how you can apply this information to promote a healthy diet and body weight.

MJ's BlogInFocus What evidence is there to support the claim that organic foods contain more antioxidants? Visit MJ's blog in the Study Area in MasteringBiology and look under "Antioxidants in Organic Foods."

http://goo.gl/mNzofs

14.1 **The digestive system brings nutrients into the body**

The **digestive system** consists of all the organs that share the common function of getting nutrients into the body. It includes a series of hollow organs extending from the mouth to the anus: the mouth, pharynx, esophagus, stomach, small intestine, large intestine, rectum, and anus.

These organs form a hollow tube called the **gastrointestinal (GI) tract** ("gastric" is the adjective for stomach). The space within this hollow tube—the area through which food and liquids travel—is called the *lumen.*

The digestive system also includes four accessory organs—the salivary glands, liver, gallbladder, and pancreas. **Figure 14.1** summarizes the functions of the organs and accessory organs of the digestive system.

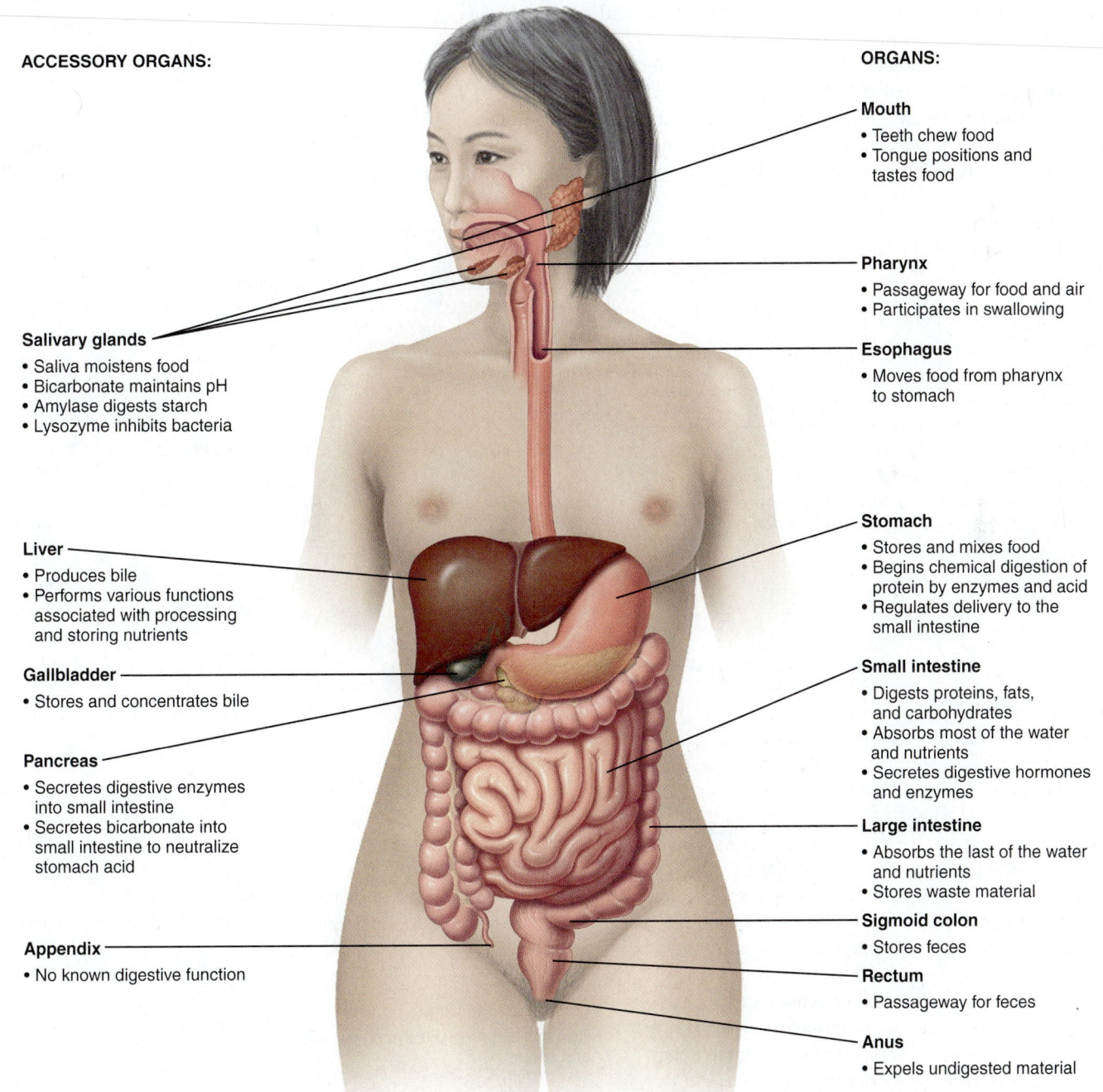

ACCESSORY ORGANS:

Salivary glands
• Saliva moistens food
• Bicarbonate maintains pH
• Amylase digests starch
• Lysozyme inhibits bacteria

Liver
• Produces bile
• Performs various functions associated with processing and storing nutrients

Gallbladder
• Stores and concentrates bile

Pancreas
• Secretes digestive enzymes into small intestine
• Secretes bicarbonate into small intestine to neutralize stomach acid

Appendix
• No known digestive function

ORGANS:

Mouth
• Teeth chew food
• Tongue positions and tastes food

Pharynx
• Passageway for food and air
• Participates in swallowing

Esophagus
• Moves food from pharynx to stomach

Stomach
• Stores and mixes food
• Begins chemical digestion of protein by enzymes and acid
• Regulates delivery to the small intestine

Small intestine
• Digests proteins, fats, and carbohydrates
• Absorbs most of the water and nutrients
• Secretes digestive hormones and enzymes

Large intestine
• Absorbs the last of the water and nutrients
• Stores waste material

Sigmoid colon
• Stores feces

Rectum
• Passageway for feces

Anus
• Expels undigested material

Figure 14.1 **Organs and accessory organs of the digestive system and their functions.**

The walls of the GI tract are composed of four layers

From the esophagus to the anus, the walls of the GI tract share common structural features (Figure 14.2). The walls of the GI tract consist of four layers of tissue.

1. *Mucosa.* The innermost tissue layer (the mucous membrane in contact with the lumen) is the **mucosa.** All nutrients must cross the mucosa to enter the blood.
2. *Submucosa.* Next to the mucosa is a layer of connective tissue containing blood vessels, lymph vessels, and nerves, called the **submucosa.** Components of food that are absorbed across the mucosa enter the blood and lymph vessels of the submucosa.
3. *Muscularis.* The third layer of GI tract tissue, called the **muscularis,** is responsible for motility or movement. The muscularis consists of two or three sublayers of smooth muscle. In general, the fibers of the inner sublayer are oriented in a circular fashion around

the lumen, whereas those in the outer sublayer are arranged lengthwise, parallel to the long axis of the digestive tube. The exception is the stomach, which has a diagonal (oblique) sublayer of muscle inside the other two.

4. *Serosa.* The outermost layer of the GI tract wall, or **serosa,** is a thin connective tissue sheath that surrounds and protects the other three layers and attaches the digestive system to the walls of the body cavities.

Some of the organs of the GI tract are separated from each other by thick rings of circular smooth muscle called *sphincters.* When these sphincters contract they can close off the passageway between organs.

Five basic processes accomplish digestive system function

In practical terms, the digestive system is a long disassembly line that starts with huge chunks of raw material (we call

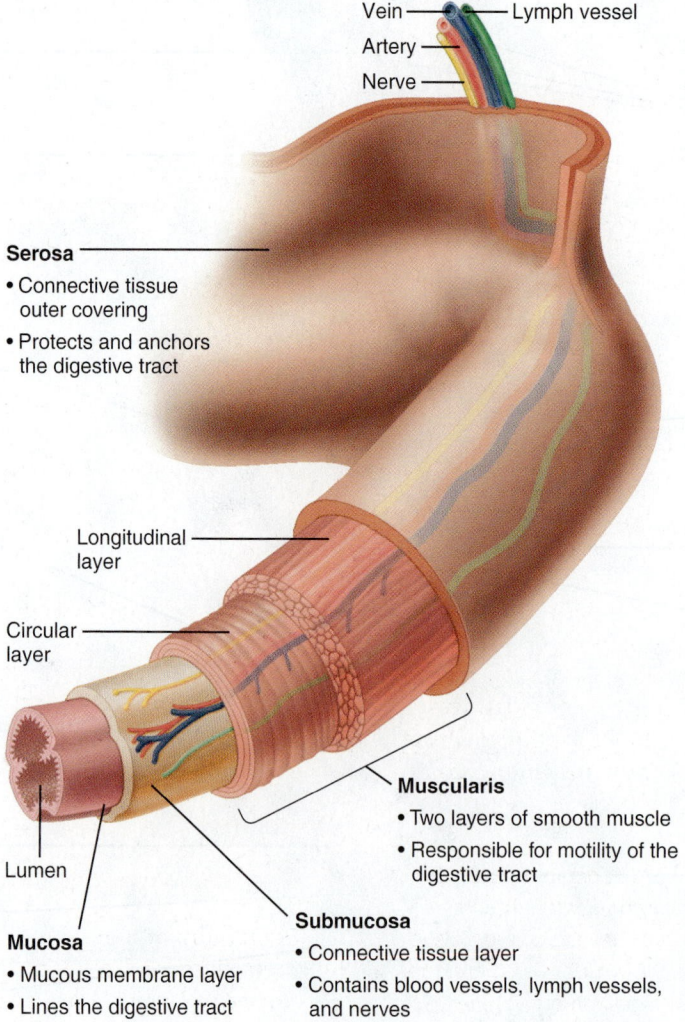

Vein —— Lymph vessel
Artery
Nerve

Serosa
• Connective tissue outer covering
• Protects and anchors the digestive tract

Longitudinal layer

Circular layer

Muscularis
• Two layers of smooth muscle
• Responsible for motility of the digestive tract

Lumen

Mucosa
• Mucous membrane layer
• Lines the digestive tract

Submucosa
• Connective tissue layer
• Contains blood vessels, lymph vessels, and nerves

Figure 14.2 The four tissue layers of the GI tract wall.

it food) and takes them apart so that the nutrients in the food can be absorbed into the body. The digestive system accomplishes this task with five basic processes.

1. *Mechanical processing and movement.* Chewing breaks food into smaller pieces, and two types of movement (motility) mix the contents of the lumen and propel it forward.
2. *Secretion.* Fluid, digestive enzymes, acid, alkali, bile, and mucus are all secreted into the GI tract at various places. In addition, several hormones that regulate digestion are secreted into the bloodstream.
3. *Digestion.* The contents of the lumen are broken down both mechanically and chemically into smaller and smaller particles, culminating in nutrient molecules.
4. *Absorption.* Nutrient molecules pass across the mucosal layer of the GI tract and into the blood or lymph.
5. *Elimination.* Undigested material is eliminated from the body via the anus.

Two types of motility aid digestive processes

The smooth muscle of the GI tract produces two kinds of motility, called *peristalsis* and *segmentation*. The functions of peristalsis and segmentation are quite different.

Peristalsis propels food forward (**Figure 14.3a**). Peristalsis begins when a lump of food (called a *bolus*) stretches a portion of the GI tract, causing the smooth muscle in front of the bolus to relax and the muscle behind it to contract. The contractions push the food forward, stretching the next part of the tube and causing muscle relaxation in front and contraction behind. The peristaltic wave of contraction ripples through the organs of the GI tract, mixing the contents of the stomach and pushing the contents of the esophagus and intestines forward. Peristalsis occurs in all parts of the GI tract but is especially prevalent in the esophagus, where it transports food rapidly to the stomach.

Segmentation mixes food (**Figure 14.3b**). In segmentation, short sections of smooth muscle contract and relax in seemingly random fashion. The result is a back-and-forth mixing of the contents of the lumen. Food particles are pressed against the mucosa, enabling the body to absorb their nutrients. Segmentation occurs primarily in the small intestine as food is digested and absorbed.

Recap The digestive system consists of organs and accessory organs that share the function of bringing nutrients into the body. The wall of the GI tract consists of four tissue layers: the mucosa, the submucosa, the muscularis, and the serosa. The five basic processes of digestion are (1) mechanical processing and movement, (2) secretion, (3) digestion, (4) absorption, and (5) elimination. ▪

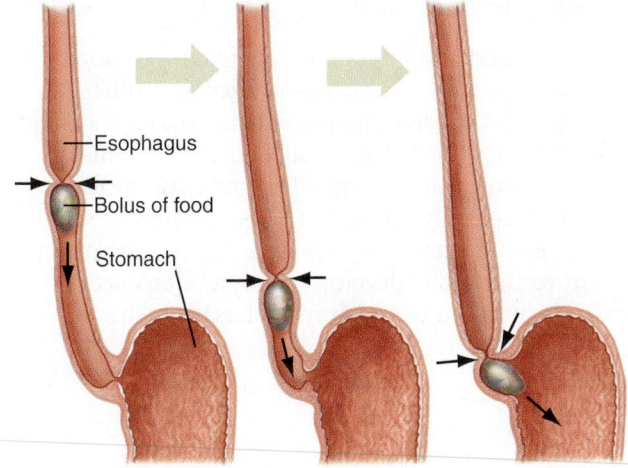

a) Peristalsis.

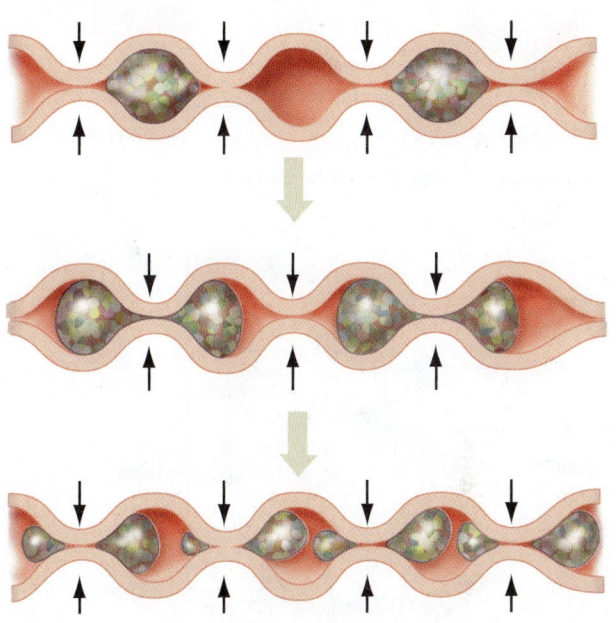

b) Segmentation.

Figure 14.3 Motility of the gastrointestinal tract.
a) Peristalsis in the esophagus. The three sequential diagrams show how peristalsis moves the contents of the lumen onward. b) Segmentation in the small intestine. Segmentation mixes the contents of the lumen.

14.2 The mouth processes food for swallowing

The mouth, or oral cavity, is the entrance to the GI tract. Digestion begins in the mouth with the process of chewing, which breaks food into smaller and smaller chunks. Essentially, the mouth functions as an effective food processor.

Teeth bite and chew food

The teeth chew food into pieces small enough to swallow. There are four types of teeth, each specialized for a different purpose (Figure 14.4a). The sharp-edged incisors cut food and the pointed canines tear it. The flat surfaces of the premolars and molars are well adapted to grinding and crushing food.

Children have only 20 teeth. Their teeth develop by about age 2 and are gradually replaced by permanent teeth. Permanent teeth usually develop by late adolescence, except for the wisdom teeth (third molars), which

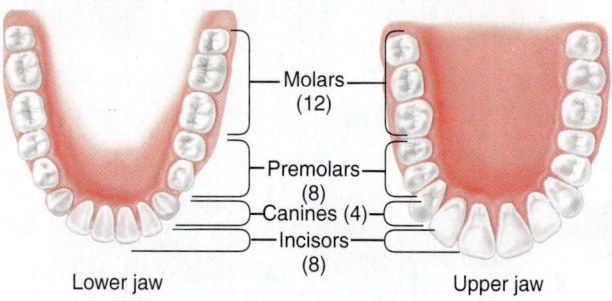

a) Locations and types of adult human teeth.

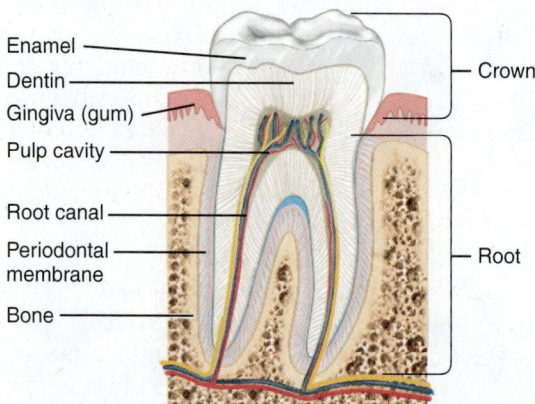

b) Anatomy of a tooth.

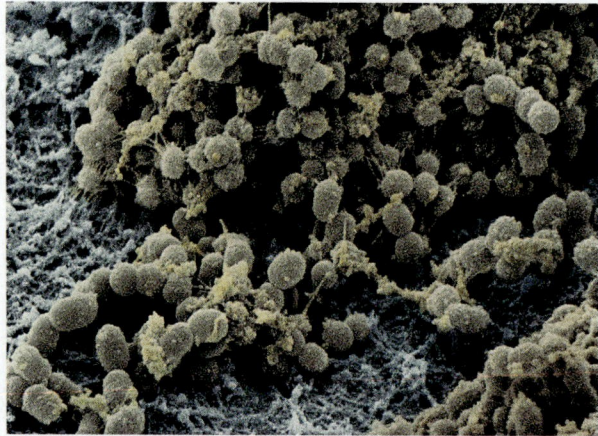

c) Bacteria on a tooth's surface (approx. ×8,000).

Figure 14.4 Teeth.

generally appear by age 25. Most adults have 32 permanent teeth.

Each tooth (Figure 14.4b) consists of a visible region called the *crown* and a region below the gum line called the *root*. The crown is covered by a layer of enamel, an extremely hard nonliving compound of calcium and phosphate. Beneath the enamel is a bonelike living layer called *dentin*. The soft innermost pulp cavity contains the blood vessels that supply the dentin, as well as the nerves that cause so much pain when a tooth is infected or injured. The entire tooth sits in a socket in the jawbone that is lined with periodontal membrane.

Our mouths contain large numbers of bacteria that flourish on the food that remains between the teeth (Figure 14.4c). During their metabolism, these bacteria release acids that can dissolve enamel, creating cavities or *dental caries*. If not treated, cavities deepen, eroding the dentin and pulp cavity and causing a toothache. Tooth decay may inflame the soft gum tissue (gingiva) around the tooth, causing *gingivitis*. Decay that inflames the periodontal membrane leads to *periodontitis*. However, good dental hygiene—including regular exams and teeth cleaning—can prevent most dental problems from becoming serious.

The tongue positions and tastes food

Chewing would be inefficient without the muscular tongue, which positions food over the teeth and mashes it against the roof of the mouth. The tongue consists of skeletal muscle enclosed in mucous membrane, so we have voluntary control over its movements. The tongue contributes to the sense of taste (Chapter 12) and is important for speech.

Saliva begins the process of digestion

Three pairs of **salivary glands** produce a watery fluid called *saliva*. The parotid gland lies near the back of the jaw, and the smaller sublingual and submandibular glands are located just below the lower jaw and below the tongue, respectively (Figure 14.5). All three glands connect to the oral cavity via ducts.

Saliva moistens food, making it easier to chew and swallow. Saliva contains four main ingredients, each with important functions. One is *mucin*, a mucus-like protein that holds food particles together so they can be swallowed more easily. An enzyme called *salivary amylase* begins the process of digesting carbohydrates. *Bicarbonate* (HCO_3^-) in saliva maintains the pH of the mouth between 6.5 and 7.5, the range over which salivary amylase is most effective. Salivary bicarbonate may also help protect your teeth against those acid-producing bacteria. Saliva also contains small amounts of an enzyme called *lysozyme*, which inhibits bacterial growth.

↩ Recap The four kinds of teeth (molars, premolars, canines, and incisors) mechanically digest chunks of food. Salivary glands secrete saliva, which moistens food, begins the chemical digestion of carbohydrates, maintains the pH of the mouth, and protects the teeth against bacteria. ∎

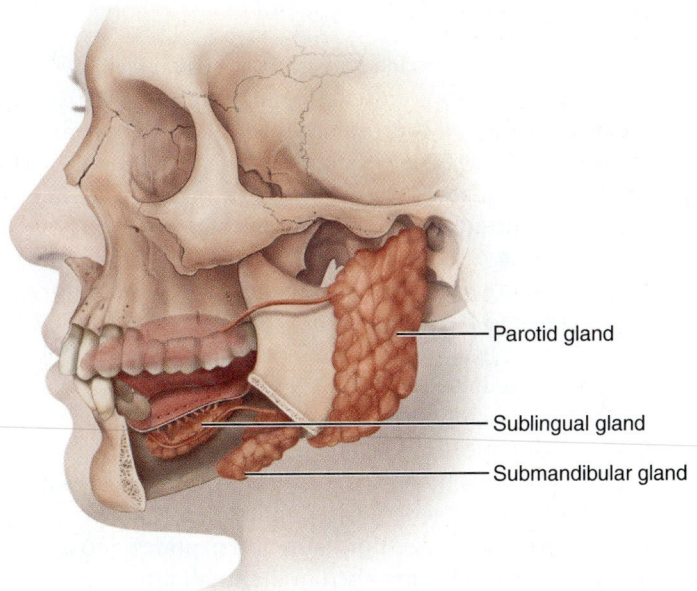

Figure 14.5 The salivary glands. The three pairs of glands deliver saliva to the mouth via salivary ducts.

14.3 The pharynx and esophagus deliver food to the stomach

After we have chewed our food and mixed it with saliva, the tongue pushes it into the **pharynx,** or throat, for swallowing. The act of swallowing (Figure 14.6) involves a sequence of events that is coordinated with a temporary halt in breathing.

First, ❶ voluntary movements of the tongue and jaws push a bolus of food into the pharynx. At rest and during this voluntary phase, ❷ muscles keep the esophagus closed. The arrival of the food in the pharynx stimulates receptors that initiate the "swallowing reflex," an involuntary act that cannot be stopped once it is started. The purpose of the swallowing reflex is to close off the

passageways for air to and from the pharynx while the bolus of food passes through the pharynx. Breathing is halted temporarily during the swallowing reflex. ❸ The soft palate rises to close off the passageway into the nasal cavity, and ❹ the epiglottis bends down to close the airway to the trachea. At the same time, the tongue pushes the food back even farther, sliding it past the epiglottis and into the esophagus.

Just beyond the pharynx is the **esophagus,** a muscular tube consisting of both skeletal and smooth muscle that connects the pharynx to the stomach. The lining of the esophagus produces lubricating mucus that helps food slide easily. Occasionally, thick foods such as mashed potatoes or peanut butter stick briefly to the mucosa of the esophagus, causing a painfully strong reflex contraction behind the food. If the contraction does not dislodge the food, drinking liquid may help.

Gravity assists peristalsis in propelling food. However, peristaltic contractions enable the esophagus to transport food even against gravity, such as when we are lying down.

The lower esophageal sphincter, located at the base of the esophagus, opens briefly as food arrives and closes after it passes into the stomach. The sphincter prevents reflux of the stomach's contents back into the esophagus. Occasionally, this sphincter malfunctions, resulting in the backflow of acidic stomach fluid into the esophagus. This condition, known as *acid reflux,* is responsible for the burning sensation known as heartburn. Acid reflux becomes more common with weight gain, pregnancy, and age. Occasionally, it indicates a *hiatal hernia,* a condition in which part of the stomach protrudes upward into the chest through an opening (hiatus) in the diaphragm muscle. Mild or temporary acid reflux is usually not serious and often improves with weight loss. However, prolonged acid reflux may cause esophageal ulcers because stomach acid can erode the mucosa of the esophagus.

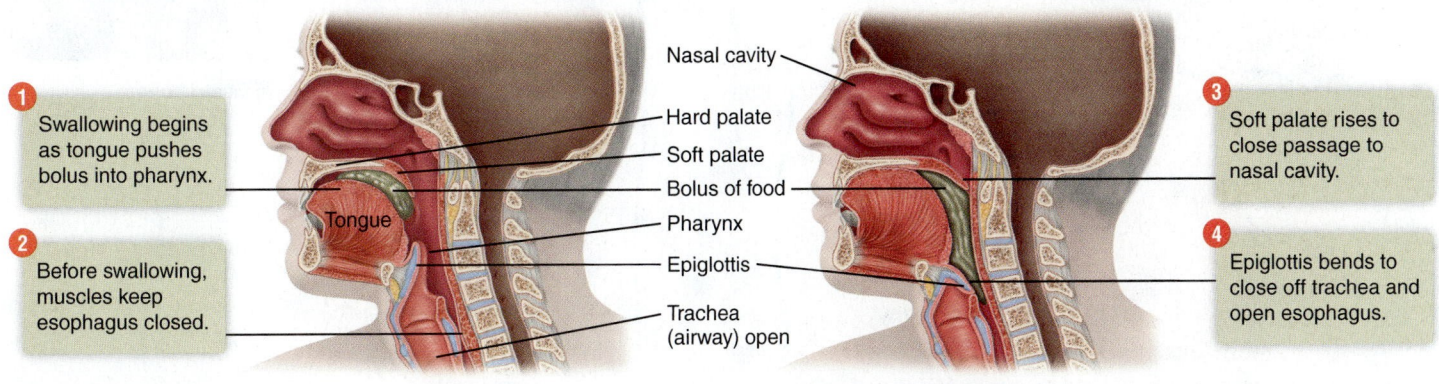

❶ Swallowing begins as tongue pushes bolus into pharynx.

❷ Before swallowing, muscles keep esophagus closed.

Nasal cavity
Hard palate
Soft palate
Bolus of food
Pharynx
Epiglottis
Trachea (airway) open

Tongue

❸ Soft palate rises to close passage to nasal cavity.

❹ Epiglottis bends to close off trachea and open esophagus.

a) Voluntary phase.

b) Involuntary phase.

Figure 14.6 Swallowing.

Recap Swallowing begins with voluntary movements of the tongue; the presence of food initiates an involuntary swallowing reflex. Peristalsis and gravity transfer food through the esophagus to the stomach. ∎

14.4 The stomach stores food, digests protein, and regulates delivery

The **stomach** is a muscular, expandable sac that performs the following three important functions:

- *Food storage.* Humans tend to eat meals several times a day. The stomach stores the food until it can be digested and absorbed. The stomach shrinks when empty and then expands to 1–3 liters of capacity when we eat.
- *Digestion.* The stomach is more than a storage bag, however. It also digests proteins, using strong acid and protein-digesting enzymes. The strong acid also kills most bacteria. Muscle contractions mix these secretions with food, assist in mechanically breaking apart food particles, and push the mixture into the small intestine.
- *Regulation of delivery.* The stomach regulates the rate at which food is delivered to the small intestine.

Gastric juice breaks down proteins

The walls of the stomach consist of the usual four layers: mucosa, submucosa, muscularis, and serosa (Figure 14.7a). A closer look at the mucosal layer reveals millions of small openings called *gastric pits* that lead to gastric glands below the surface (Figure 14.7b and c). Some of the cells lining the glands secrete either hydrochloric acid (HCl) or mucus, but most secrete pepsinogen, a large precursor molecule that becomes a protein-digesting enzyme called **pepsin** once it is exposed to HCl in the stomach. Collectively, the HCl, pepsinogen, and fluid that are secreted into the lumens of the glands are known as *gastric juice.*

The stomach typically produces 1–2 liters of gastric juice per day, most of it immediately after meals. The acid in gastric juice gives the stomach an acidic pH of approximately 2. The pepsin and acid in gastric juice dissolve the connective tissue in food and digest proteins and peptides into amino acids so they can be absorbed in the small intestine. The watery mixture of partially digested food and gastric juice that is delivered to the small intestine is called **chyme.** The pyloric sphincter, between the stomach and the small intestine, regulates the rate of transport of chyme into the small intestine.

If gastric juice is powerful enough to digest proteins, why doesn't it digest the stomach, too? The reason is that some of the cells lining the stomach and the gastric glands continuously produce a protective barrier of mucus. Normally, the stomach contents are in contact with mucus, not living cells. If the mucous layer

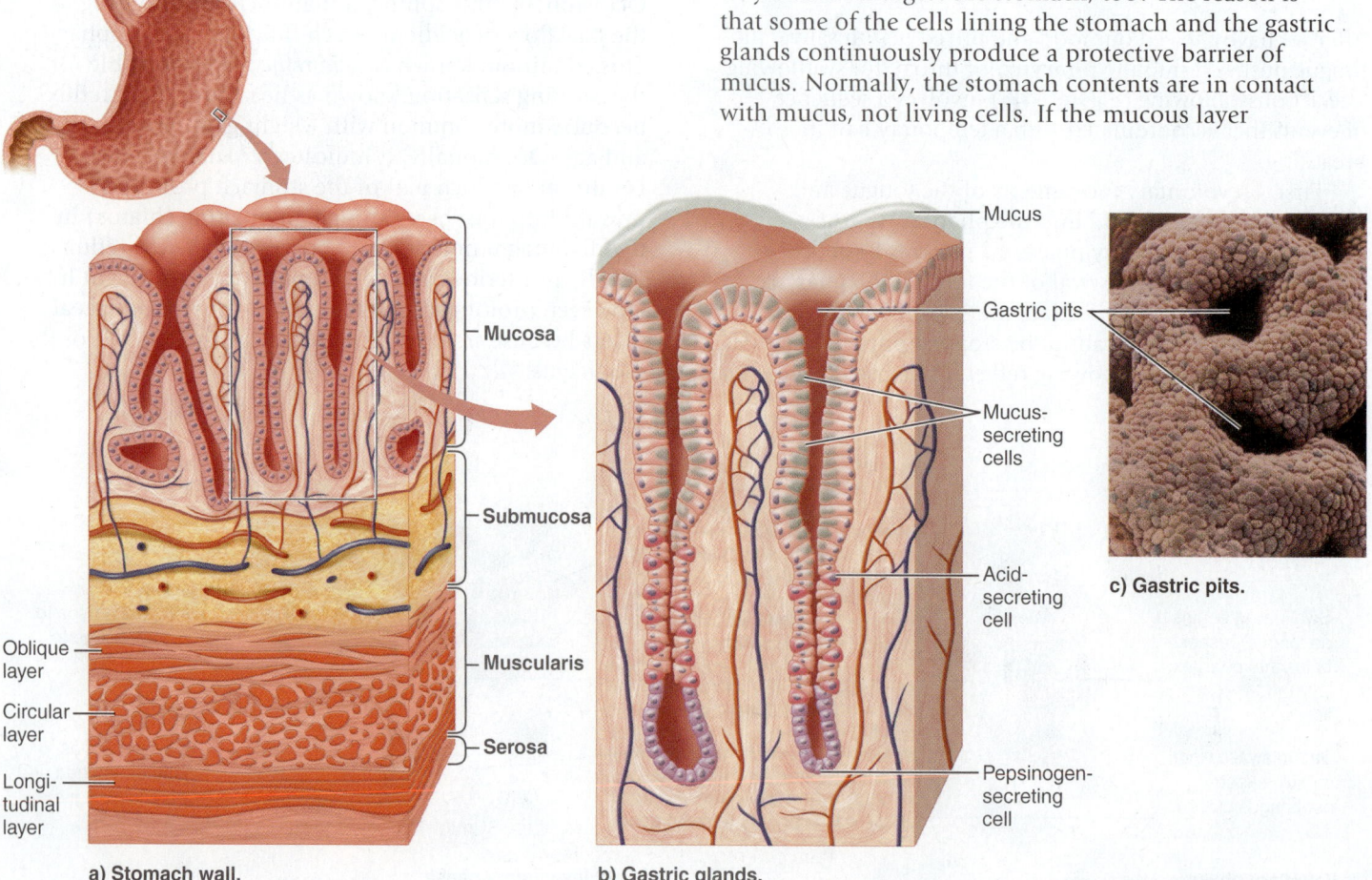

Oblique layer
Circular layer
Longi-tudinal layer

Mucosa
Submucosa
Muscularis
Serosa

a) Stomach wall.

Mucus
Gastric pits
Mucus-secreting cells
Acid-secreting cell
Pepsinogen-secreting cell

b) Gastric glands.

c) Gastric pits.

Figure 14.7 The stomach.

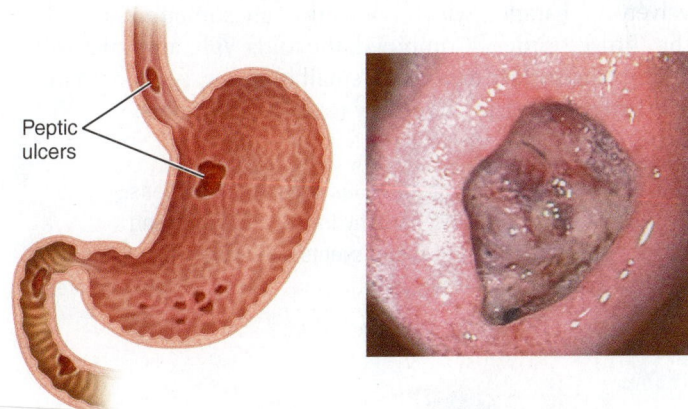

Figure 14.8 Peptic ulcers.

becomes damaged, however, this leaves the underlying tissue vulnerable. An open (sometimes bleeding) sore called a **pepticulcer** (from the Greek word *peptein*, meaning "digest") may form (**Figure 14.8**). Peptic ulcers occasionally occur in the esophagus and upper part of the small intestine as well.

In addition to pepsinogen, HCl, and mucus, some mucosal cells secrete *intrinsic factor*, a protein that binds to vitamin B_{12} so that it can be absorbed in the small intestine. Finally, certain cells in the gastric glands secrete a hormone called *gastrin* into the bloodstream. We will talk about gastrin when we discuss the regulation of the digestive process.

✅ Suppose the acid-secreting cells of the stomach were not working properly and did not secrete any HCl. Would the stomach still contain pepsin? Which of the stomach's functions would be affected?

Stomach contractions mix food and push it forward

While your stomach is empty, muscle contractions keep it small. When you eat a meal, contractions cease and the stomach relaxes and stretches to accommodate the food. Stretching signals peristalsis to increase.

Each wave of peristalsis starts at the lower esophageal sphincter and moves toward the pyloric sphincter, becoming stronger as it proceeds. The peristaltic wave pushes the chyme forward and then, when there is nowhere else for it to go, backward again in a squeezing, mixing motion (**Figure 14.9**). Each contraction propels about a tablespoon of chyme into the small intestine before the pyloric sphincter closes momentarily. A peristaltic contraction occurs every 15–25 seconds. Have you ever noticed your stomach gurgling at this rate after a meal? You're hearing peristalsis in action.

It takes two to six hours for the stomach to empty completely after a meal. Most forceful when the stomach is full, peristalsis declines as the stomach empties. Chyme with a high acid or fat content stimulates the release of hormones that slow stomach peristalsis, giving the small intestine more time to absorb the nutrients.

The stomach does not absorb nutrients because it lacks the required cellular transporting mechanisms and because its inner lining is coated with mucus. Exceptions to this rule are alcohol and aspirin, both of which are small lipid-soluble substances that can cross the mucus barrier and be absorbed into the bloodstream directly from the stomach. Nevertheless, this is not the primary absorptive pathway for alcohol; most alcohol is absorbed in the intestine along with the nutrients.

↩ **Recap** The stomach stores food, digests it, and regulates its delivery to the small intestine. Gastric juice dissolves connective tissue, large proteins, and peptides in food. The presence of food stretches the stomach and increases peristalsis. Peristaltic contractions mix the chyme and push it gradually into the small intestine. ■

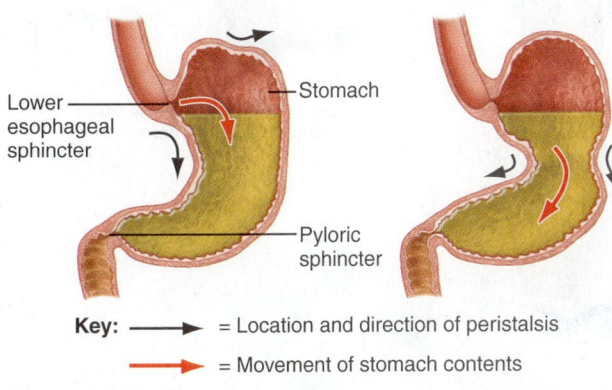
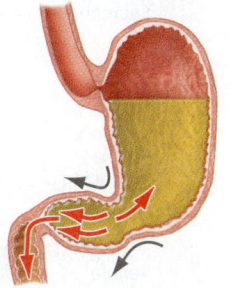
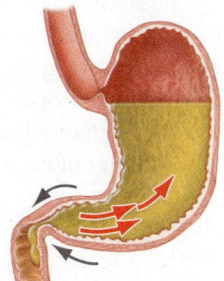

Key: ⟶ = Location and direction of peristalsis

⟶ = Movement of stomach contents

Figure 14.9 Peristalsis of the stomach. A peristaltic wave of contraction occurs in the stomach approximately every 15–25 seconds. Peristalsis mixes the contents of the stomach and forces a small amount of chyme into the small intestine with each contraction.

14.5 The small intestine digests food and absorbs nutrients and water

The process of digestion continues in the small intestine, so-named because it is smaller in diameter than the final segment of the digestive tract, the large intestine. The small intestine has two major functions.

1. *Digestion.* The stomach partially digests proteins to smaller peptides, under the influence of strong acids and pepsin. Protein digestion continues in the small intestine, but here we also digest carbohydrates and lipids. Digestion of proteins, carbohydrates, and lipids in the small intestine involves neutralizing the highly acidic gastric juice and adding additional digestive enzymes from the intestine and pancreas.
2. *Absorption.* Eventually, the proteins, carbohydrates, and lipids in food are broken down to single amino acids, monosaccharides, fatty acids, and glycerol, which are small enough to be transported across mucosal cells into the blood. Nearly 90% of the absorbable nutrients and water is absorbed in the small intestine.

The small intestine consists of three different regions. The first region, the **duodenum,** is only about 10 inches long, but it is here that most of the digestion takes place. The products of digestion are absorbed primarily in the other two segments, the *jejunum* and the *ileum,* which together are about 10 feet long.

The structure of the small intestine wall makes it well suited for absorption (**Figure 14.10a**). The mucosa contains large folds covered with microscopic projections called **villi** (singular: **villus**) (**Figure 14.10b**). At the center of each villus are blood capillaries and a small blind-ended lymphatic capillary called a *lacteal* (**Figure 14.10c**). Lacteals transport nutrients, some of which are too large to enter the blood capillaries, to larger lymph vessels and eventually back to the blood. Each epithelial cell of the villi has dozens of even smaller, cytoplasmic projections called *microvilli* (**Figure 14.10d**). The microvilli give the mucosal surface a velvety appearance, which is why they are sometimes called the "brush border." Combined, the folds, villi, and microvilli enlarge the surface area of the small intestine by more than 500 times, increasing its ability to absorb nutrients.

↩ **Recap** The small intestine has two major functions: (1) digesting proteins, carbohydrates, and lipids and (2) absorbing most of the nutrients and water we consume. Villi in the mucosa increase the small intestine's surface area. ■

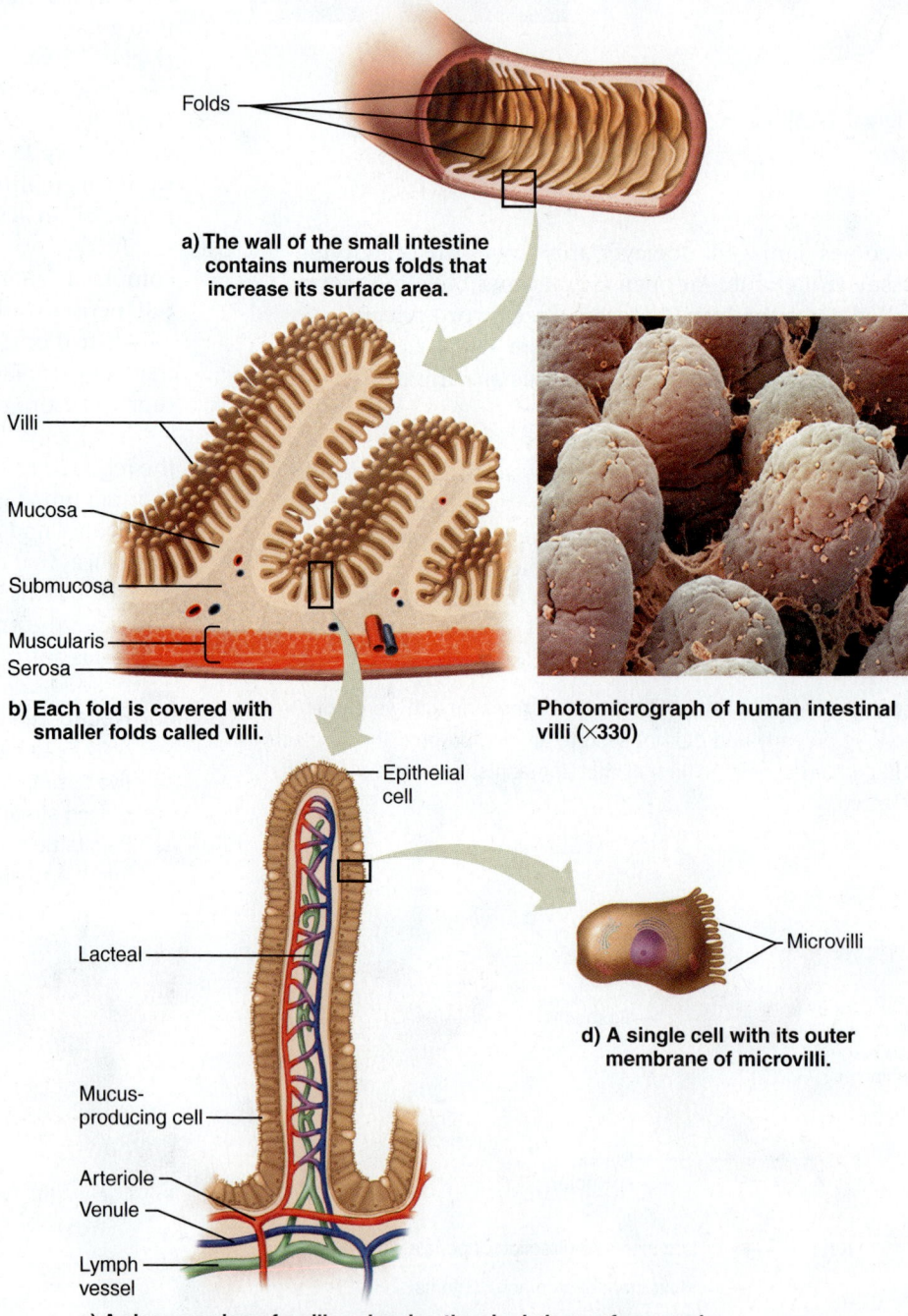

a) **The wall of the small intestine contains numerous folds that increase its surface area.**

b) **Each fold is covered with smaller folds called villi.**

Photomicrograph of human intestinal villi (×330)

d) **A single cell with its outer membrane of microvilli.**

c) **A close-up view of a villus showing the single layer of mucosal cells covering the surface and the centrally located lymph vessel (lacteal) and blood vessels.**

Figure 14.10 The small intestine.

14.6 Accessory organs aid digestion and absorption

The digestive system has four accessory organs: salivary glands, pancreas, gallbladder, and liver (Figure 14.11). We have discussed the salivary glands. Let's now look at the other accessory organs.

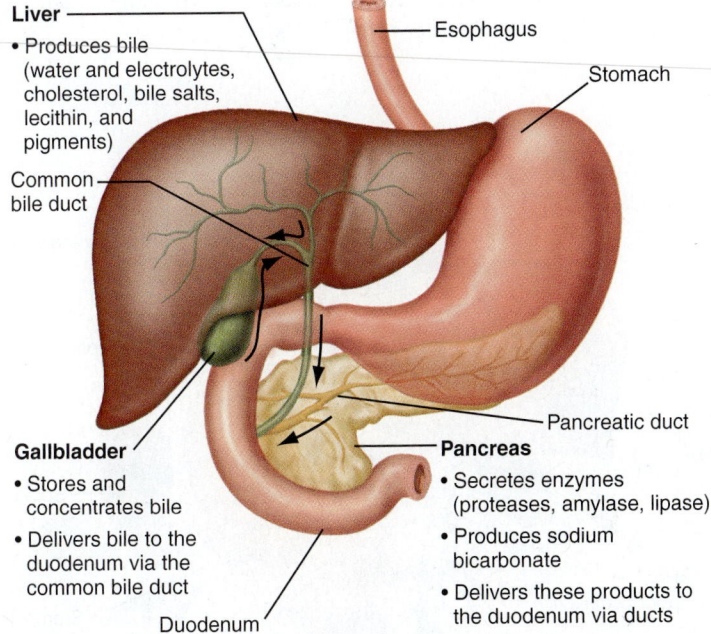

Liver
- Produces bile (water and electrolytes, cholesterol, bile salts, lecithin, and pigments)

Common bile duct

Gallbladder
- Stores and concentrates bile
- Delivers bile to the duodenum via the common bile duct

Esophagus

Stomach

Pancreatic duct

Pancreas
- Secretes enzymes (proteases, amylase, lipase)
- Produces sodium bicarbonate
- Delivers these products to the duodenum via ducts

Duodenum

Figure 14.11 Locations and digestive functions of the liver, gallbladder, and pancreas. These three accessory organs are connected to the small intestine via ducts. The common bile duct joins one of the two pancreatic ducts before it enters the duodenum. Arrows indicate the movement of secretory products.

The pancreas secretes enzymes and NaHCO₃

The **pancreas** is an elongated organ that lies just behind the stomach. In the chapter on the endocrine system (Chapter 13), we learned that the pancreas is an endocrine gland that secretes several hormones (most importantly insulin and glucagon) into the blood. But in addition, the pancreas is an exocrine gland that produces and secretes several products directly into ducts leading to the digestive tract. Table 14.1 lists major digestive enzymes in the gastrointestinal tract. The exocrine products of the pancreas are:

- *Sodium bicarbonate (NaHCO₃).* Except for pepsin, most digestive enzymes work best at a fairly neutral pH. Sodium bicarbonate from the pancreas neutralizes the stomach acid in the small intestine so that further digestion can proceed.
- *Digestive enzymes.* Most of the enzymes involved in digestion come from the pancreas. (The small intestine contributes only a limited amount of protein- and carbohydrate-digesting enzymes.) Digestive enzymes secreted by the pancreas include:
  - *Proteases* (enzymes that digest proteins), such as trypsin, chymotrypsin, and carboxypeptidase,
  - *Pancreatic amylase,* which continues the digestion of carbohydrates only partially accomplished by salivary amylase, and
  - *Lipase,* a lipid-digesting enzyme. The small intestine does not have lipase, so this product of the pancreas is particularly important.

The liver produces bile and performs many other functions

The liver is a large organ located in the upper-right abdominal cavity. The liver performs many significant functions, some of which are associated with digestion.

Table 14.1 Major enzymes of digestion

Enzyme	Source	Where active	Substance digested	Breakdown products
Carbohydrate Digestion				
Salivary amylase	Salivary glands	Mouth	Polysaccharides	Disaccharides
Pancreatic amylase	Pancreas	Small intestine	Polysaccharides	Disaccharides
Intestinal enzymes	Small intestine	Small intestine	Disaccharides	Monosaccharides (e.g., glucose)
Protein Digestion				
Pepsin	Stomach	Stomach	Proteins	Peptides
Trypsin	Pancreas	Small intestine	Proteins	Peptides
Chymotrypsin	Pancreas	Small intestine	Proteins	Peptides
Carboxypeptidase	Pancreas	Small intestine	Peptides	Amino acids
Intestinal enzymes	Small intestine	Small intestine	Peptides	Amino acids
Lipid Digestion				
Lipase	Pancreas	Small intestine	Triglycerides	Free fatty acids, monoglyceride

The liver's primary digestive function is to facilitate the digestion and absorption of lipids by producing **bile.** Bile is a watery mixture containing electrolytes, cholesterol, bile salts derived from cholesterol, a phospholipid called *lecithin*, and pigments (primarily bilirubin) derived from the breakdown of hemoglobin. The bile salts emulsify lipids in the small intestine; that is, they break them into smaller and smaller droplets. Eventually, the droplets are small enough to be digested by lipases (lipid-digesting enzymes) from the pancreas.

An important feature of the vascular anatomy of the GI tract is the **hepatic portal system** (**Figure 14.12**). In general terms, a *portal system* carries blood from one capillary bed to another: We have seen a portal system before, in the vascular connection between the hypothalamus and the anterior pituitary gland. In the digestive system, the hepatic portal system carries nutrient-rich blood directly from the digestive organs to the liver (*hepatos* is the Greek word for "liver") via the hepatic portal vein. Therefore, the liver is ideally located to begin processing and storing nutrients for the body just as soon as digestion and absorption have begun. After passing through the liver, the blood is returned to the general circulation.

The liver serves a number of other functions that maintain homeostasis:

- Storing fat-soluble vitamins (A, D, E, and K) and iron
- Storing glucose as glycogen after a meal, and converting glycogen to glucose between meals
- Manufacturing plasma proteins, such as albumin and fibrinogen, from amino acids
- Synthesizing and storing some lipids
- Inactivating many chemicals, including alcohol, hormones, drugs, and poisons
- Converting ammonia (NH_3), a toxic waste product of metabolism, into less toxic urea
- Destroying worn-out red blood cells

Because of its central role in so many functions, liver injury can be particularly dangerous. Overexposure to toxic chemicals, medications, or alcohol can damage the liver because it takes up these substances to "detoxify" them, killing some liver cells in the process. Long-term exposure, such as prolonged alcohol abuse, can destroy enough cells to permanently impair liver function, a condition known as *cirrhosis*.

☑ Many people eat the livers of domestic animals such as cows and chickens. Considering the contents of the liver (see earlier), do you think liver is generally a nutritious food choice? Can you think of any situations in which it might be dangerous?

The gallbladder stores bile until needed

The bile produced by the liver flows through ducts to the **gallbladder.** The gallbladder concentrates bile by removing most of the water and then stores the concentrated bile until it is needed. After a meal, bile is secreted into the small intestine via the common bile duct, which joins the pancreatic duct.

↻ Recap The pancreas secretes digestive enzymes and sodium bicarbonate. The sodium bicarbonate neutralizes stomach acid, making the digestive enzymes more effective. The liver produces bile, which is stored in the gallbladder until after a meal. The liver also produces plasma proteins; inactivates toxic chemicals; destroys old red blood cells; stores vitamins, iron, and certain products of metabolism; and performs other functions important for homeostasis. ∎

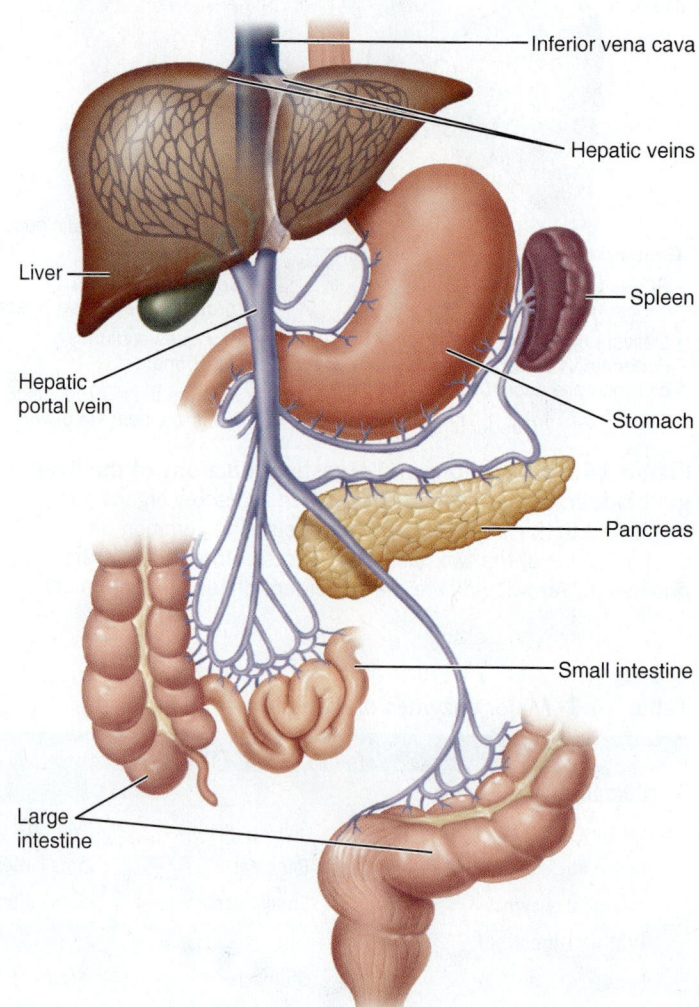

Figure 14.12 The hepatic portal system. The hepatic portal system carries nutrient-rich venous blood from the stomach, intestines, pancreas, and spleen directly to the liver before the blood is returned to the general circulation. Only the first part of the large intestine is shown.

14.7 The large intestine absorbs nutrients and eliminates wastes

By the time the contents of the digestive tract reach the large intestine, most of the nutrients and water have been absorbed. The **large intestine** absorbs most of the remaining nutrients and water and stores the now nearly solid waste material until it can be eliminated.

Although the large intestine is larger in diameter than the small intestine, it is only about half as long. It begins at a pouch called the *cecum*, which receives the chyme from the small intestine (Figure 14.13). A small fingerlike pouch, the **appendix,** extends from the cecum. The appendix has no known digestive function, but we become acutely aware of its presence if it becomes inflamed or infected, a condition called *appendicitis*.

Most of the large intestine consists of four regions collectively called the **colon.** The ascending colon rises along the right side of the body (the left side of Figure 14.13, viewed from the front), the transverse colon crosses over to the left side, and the descending colon passes down the left side to the short final segment, the sigmoid colon. Feces are stored in the sigmoid colon until defecation, when they pass through the rectum to the anus.

In addition to indigestible material, the feces contain about 5% bacteria by weight. Many strains of bacteria thrive on the leftover material in the colon that we cannot digest. Some of these bacteria release by-products that are useful to us, such as vitamin K, which is important for blood clotting. They also produce less helpful substances such as intestinal gas, a by-product of their metabolism as they break down food.

Defecation is controlled by a neural reflex. Normally, the anus is kept closed by contraction of a ring of smooth muscle called the *internal anal sphincter*. But when feces enter the rectum and the rectum is stretched, a neural reflex causes the internal anal sphincter to relax and the rectum to contract, expelling the feces.

MJ's BlogInFocus Did you know that there's a national human feces bank? What do you suppose it's used for? Visit MJ's blog in the Study Area in MasteringBiology and look under "A Human Feces Bank."

http://goo.gl/qmsILb

We can prevent defecation by voluntarily contracting the *external anal sphincter*, a ring of skeletal muscle under our conscious control. "Potty training" of children involves learning to control the external anal sphincter. Until they

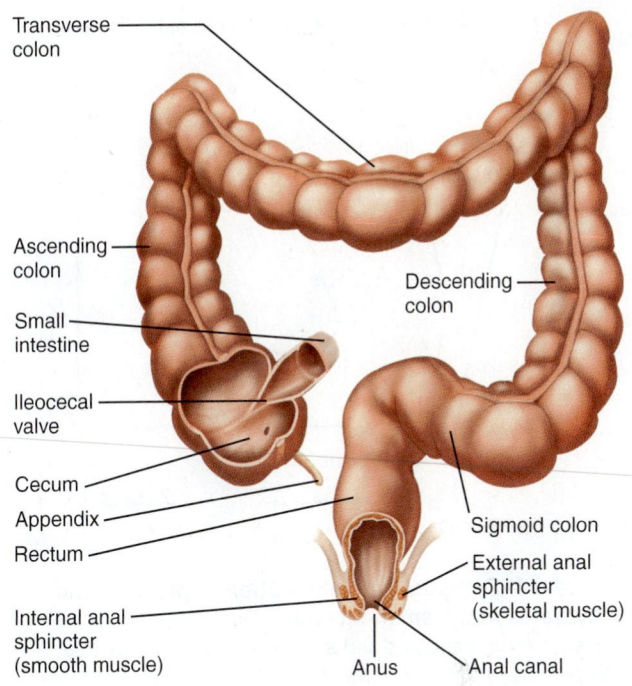

Figure 14.13 **The large intestine.** The large intestine begins at the cecum and ends at the anus.

learn how to control it consciously, children defecate whenever the defecation reflex occurs.

Recap The large intestine absorbs the last of the water, ions, and some nutrients and stores feces until defecation occurs. ∎

14.8 How nutrients are absorbed

Once food has been digested, how does your body absorb the nutrients? The mechanism depends on the type of nutrient.

Proteins and carbohydrates are digested, then absorbed

In the small intestine, enzymes from the pancreas and enzymes secreted by the mucosal layer of the stomach and from the small intestine break down proteins into amino acids, completing the digestion of proteins that began in the stomach. The amino acids are then actively transported into the mucosal cells. Eventually, they move by facilitated diffusion out of the mucosal cells and make their way into the capillaries.

Carbohydrate digestion begins in the mouth, where salivary amylase breaks down polysaccharides into disaccharides (review Table 14.1). It is completed in the

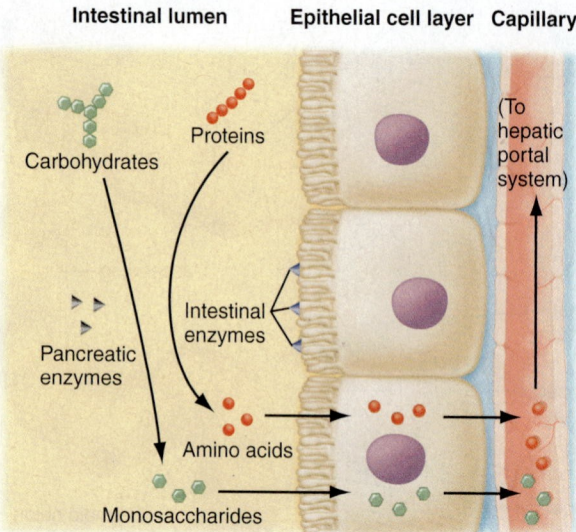

Figure 14.14 Digestion and absorption of proteins and carbohydrates in the small intestine. Digestion depends on enzymes from the pancreas and enzymes attached to the surface of the epithelial cells of the intestine. The amino acid and carbohydrate products of digestion cross the epithelial cell layer of the mucosa and enter a capillary for transport to the liver.

small intestine with the addition of pancreatic amylase and enzymes from the small intestine (**Figure 14.14**). Together these enzymes break down the remaining carbohydrates into monosaccharides (simple sugars such as glucose). Monosaccharides follow transport pathways that are similar to those for amino acids. However, they use different active transport proteins.

Lipids are broken down, then reassembled

Recall that bile salts emulsify lipids into small fat droplets, which are then digested by pancreatic and intestinal lipases. The products of lipid digestion are fatty acids and monoglycerides. Because they are nonpolar, the fatty acids and monoglycerides quickly dissolve in *micelles*—small droplets composed of bile salts and lecithin that have a polar outer surface and a nonpolar inner core. The function of micelles is to transport fatty acids and monoglycerides to the outer surface of the mucosal cells so that they can be absorbed into the cells (**Figure 14.15**).

Once inside the cells, the fatty acids and monoglycerides recombine into triglycerides. Clusters of triglycerides are then coated with proteins to form

water-soluble droplets called *chylomicrons*. Chylomicrons are released from the cell by exocytosis. However, they are too large to enter the capillaries directly. Instead, they enter the more permeable lacteals and travel in the lymph vessels until the lymph is returned to the venous blood vessels near the heart.

Water is absorbed by osmosis

As nutrients are absorbed in the small intestine (or when you drink pure water), the concentration of the water in the intestinal lumen becomes higher than in the intestinal cells or in the blood. A higher concentration of water in the lumen represents a strong driving force for the diffusion of water through the epithelial layer of cells of the small intestine and into the blood. The capacity for water absorption by the small intestine is nearly limitless, which explains why you do not experience diarrhea every time you drink a lot of fluid.

Water absorption continues in the large intestine, but here the capacity is limited. Conditions that cause the small intestine to deliver too much food residue to the large intestine or that speed up the rate of movement through the large intestine can lead to diarrhea (watery feces). A common cause of diarrhea is a bacterial infection in the small intestine.

The opposite problem is constipation, in which the food residue, now called *feces*, remains in the large intestine and colon so long that too much water is absorbed. The feces become dry and hard, making defecation difficult.

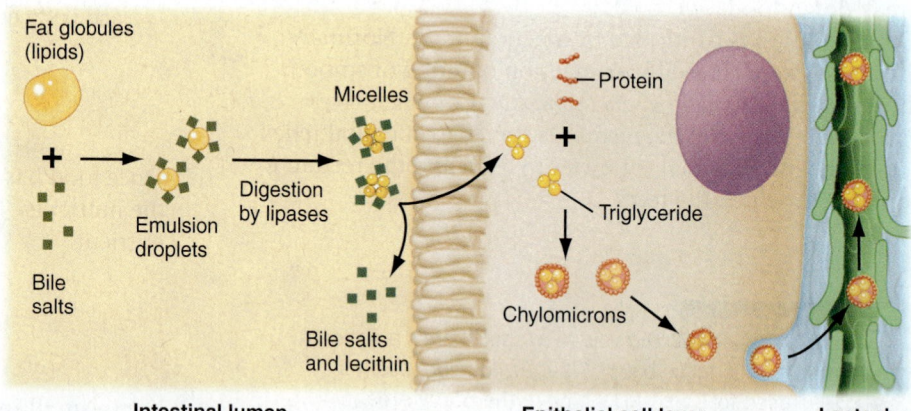

Figure 14.15 Digestion and absorption of fats in the small intestine. Bile salts emulsify large fat droplets into smaller droplets so that lipases can digest the fats into fatty acids and monoglycerides. The fatty acids and monoglycerides dissolve in micelles composed of bile salts and lecithin. Micelles near the cell's surface release their fatty acids and monoglycerides, which then diffuse across the cell membrane. In the cell, the fatty acids and monoglycerides are resynthesized to triglycerides and coated with protein. The protein-coated triglycerides are called *chylomicrons*. Chylomicrons exit the cell by exocytosis and enter a lacteal for transport to the circulation.

☑ Blood capillaries and lacteals are both present in the area directly beneath a small intestinal cell (review Figure 14.10c). Why do chylomicrons enter a lacteal, rather than a blood capillary?

HEALTH & WELLNESS

Should You Drink Raw Milk?

Drinking raw (unpasteurized) milk continues to gain in popularity. The primary reason some people choose to drink raw milk is because they are concerned that pasteurization kills probiotics (translation; "good bacteria") and degrades certain nutrients found in milk. On that point they are only partially right. The high heat of pasteurization does kill certain beneficial bacteria in milk such as *Lactobacillus,* which facilitate the digestion of lactose. However, a healthy person already has a colony of *Lactobacillus* in their digestive system. Drinking raw milk for the *Lactobacillus* probably doesn't represent much of a health gain. There are other less risky ways to recolonize the gut with *Lactobacillus* if necessary. And for the record, the concern that pasteurization degrades proteins (including enzymes) and carbohydrates in milk is based on a misunderstanding of biology; our own digestive system breaks them down anyway before they are absorbed.

Anyone choosing to drink raw milk should know that they are taking a serious risk. The point of pasteurization, after all, is to eliminate the bacteria, most notably the "bad" bacteria such as *Escherichia coli (E. coli)* and *Campylobacter.* These bacteria are not commonly found in milk but when they are, pasteurization kills them. And that's important, because it doesn't take many *E. coli* bacteria to cause bloody diarrhea, renal failure, and even death. People who drink raw milk are opting for the marginal health benefits of raw milk, while risking the chance, albeit rare, that the milk is contaminated with bad bacteria. And although the raw milk may be produced by licensed dairies and sold in some grocery stores, that doesn't mean it's as safe as pasteurized milk.

Consider the following true case: In 2006, a 7-year-old boy in California consumed raw milk purchased from a licensed dairy. He became sick with severe gastrointestinal symptoms, including bloody diarrhea. His condition deteriorated rapidly, and he was hospitalized with kidney disease, pancreatitis, and even cardiac involvement. He spent nearly two months in the hospital and incurred over $500,000 in hospital bills. Although he recovered, he is considered to be at risk of developing severe kidney disease in the future. The California Department of Public Health investigated the case and found that five other people had also become sick after consuming raw milk products from the same licensed dairy.

The federal government recognized the dangers of unpasteurized milk nearly 30 years ago. Since 1987, the Food and Drug Administration (FDA) has required that all milk sold across state lines be pasteurized. However, states are free to regulate the sale of milk that stays within their own borders. About half of the states require that all milk destined for human consumption be pasteurized. The others, including California, allow some form of commercial or private sale of raw milk. What is the law in your state?

Constipation can result from stress, lack of exercise, or insufficient fiber (indigestible material) in the diet.

Vitamins and minerals follow a variety of paths

How vitamins are absorbed depends on whether they are fat soluble or water soluble. Fat-soluble vitamins dissolve in the micelles and are absorbed by diffusion across the lipid membrane of the mucosal cell layer, just like the components of lipids. Water-soluble vitamins are absorbed either by active transport or diffusion through channels or pores. Minerals (ions) such as sodium, potassium, calcium, phosphate, sulfate, and magnesium are electrically charged and hence not lipid soluble. These are either actively transported or are absorbed by diffusion via specific transport proteins, pores, or channels.

In addition to the nutrients we ingest in our food, the body digests and reabsorbs the components of the digestive secretions themselves. Nearly 9 liters of gastric juice, pancreatic juice, digestive enzymes, and bile are produced every day. The water and minerals in the digestive secretions are reabsorbed by the normal mechanisms for these nutrients. Enzymes are digested to their component amino acids, and the amino acids are then reabsorbed. Bile salts are reabsorbed, returned to the liver, and used again.

↩ **Recap** Proteins and carbohydrates are transported into the cells lining the small intestine by active transport processes, and then diffuse into capillaries. The products of lipid digestion are transported to the mucosa in micelles, diffuse into the cell, and recombine into lipids within the cell. Then they are coated with protein to become chylomicrons that enter the lymph. The intestines also absorb water, vitamins, minerals, and digestive secretions. ■

14.9 Nerves and hormones regulate digestion

The digestive system is most active when food or chyme is present and fairly inactive when it is not. Regulation of digestion involves altering the motility and secretions of various organs so that each operates as efficiently as possible.

The endocrine system and nervous system regulate digestion according to both the volume and content of food. Because most digestion and absorption occurs in the stomach and small intestine, most regulatory processes involve those organs.

When the stomach stretches to accommodate food, neural reflexes increase stomach peristalsis and secretion of gastric juice. Stretching and the presence of protein stimulate the stomach to release the hormone **gastrin,** which triggers the release of more gastric juice.

When chyme enters the small intestine, neural reflexes within the duodenum increase the rate of segmentation to mix the chyme. The duodenum also secretes two hormones into the bloodstream. Acid in chyme triggers the secretion of **secretin,** which stimulates the pancreas to secrete water and bicarbonate to neutralize acid. Fat and protein stimulate the secretion of **cholecystokinin (CCK),** which signals the pancreas to secrete more digestive enzymes. CCK and stretching the duodenum also stimulate the gallbladder to contract and release bile.

☑ Do carbohydrates result in the release of any hormones from the duodenum? Where in the digestive system are carbohydrates first broken down?

Secretin, CCK, and stretching of the small intestine inhibit stomach motility and stomach secretions. In other words, if chyme flows too quickly from the stomach, the small intestine will slow stomach activity accordingly.

Shortly after eating, motility of the large intestine increases. This is why people often feel the urge to defecate after their first meal of the day. The mechanism involves the hormone gastrin and a neural reflex, both triggered by stretching of the stomach.

↩ **Recap** The endocrine and nervous systems regulate digestion based on content and volume of food. Regulatory mechanisms include neural reflexes involving organ stretching and release of the hormones gastrin, secretin, and cholecystokinin. ■

14.10 Nutrition: you are what you eat

Nutrition is the interaction between an organism and its food. Because nearly all nutrients enter the body through the digestive system, it is fair to say that "you are what you eat." What nutrients do we actually need, and why?

ChooseMyPlate.gov offers a personalized approach

Figure 14.16 presents one answer to these questions. Developed by the Center for Nutrition Policy and Promotion at the U.S. Department of Agriculture (USDA), *MyPlate* is a comprehensive, personalized approach that includes physical activity as well as healthy nutrition. It replaces the old food pyramid, which was thought to be too confusing for most consumers. MyPlate divides foods into five groups and gives you recommendations on what to eat from each group.

MyPlate provides many options to help you prepare a personalized nutrition plan. The system includes food plans appropriate to different calorie levels of food intake, based on the latest *Dietary Guidelines for Americans* issued by the USDA and the Department of Health and Human Services. If you enter your age, gender, height, weight,

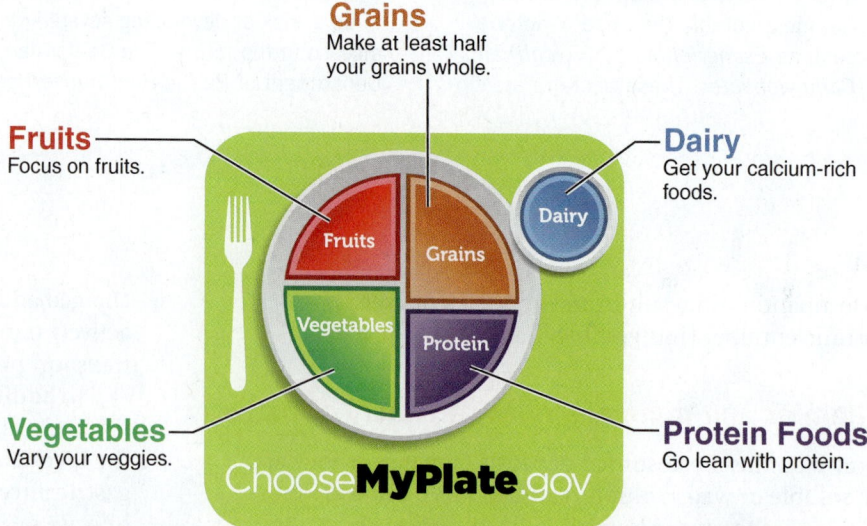

Figure 14.16 MyPlate. MyPlate is a visual representation of the recommended dietary guidelines of the USDA. The plate indicates the recommended consumption of foods from five different groups, taking into account both nutritional needs and concerns about the effects of diet on nutrition-related disease. MyPlate is designed to be highly interactive.

and physical activity level on the MyPlate Web site, the system will match you with the best plan for your needs. It will suggest how many calories you need per day and recommend specific amounts of each food group, as well as the upper limit of empty calories (extra fats and sugars) you should consume per day. You can print worksheets to help you set nutrition goals and monitor your progress. MyPlate even offers advice on foods to substitute if you choose not to eat specific foods, what to do if you are a vegetarian, for example.

Most nutritionists would agree with the general components of a healthy diet listed in MyPlate. However, some of the specific recommendations of MyPlate have been challenged by critics, who charge that the dietary guidelines on which it is based are influenced by intense pressure from lobbying groups such as the National Dairy Council and the National Cattleman's Beef Association. For example, current dietary guidelines are that for "improved bone health" (and presumably to prevent osteoporosis), adults should drink three cups of low-fat milk per day. That's over 300 calories just in milk. There are other ways to get the required calcium, of course, and not everyone is at risk of osteoporosis. MyPlate also lumps meat (beef, poultry, and fish) and beans together in the protein category, ignoring the mounting evidence that replacing red meat with a combination of fish, poultry, and beans is a healthier choice. Nevertheless, MyPlate is still a good place to start if you want basic information about a healthy diet overall.

General recommendations for a healthy diet include the following:

- Eat a variety of foods.
- Maintain a healthy weight.
- Eat plenty of fruits, vegetables, and whole-grain products.
- Choose a diet low in cholesterol and saturated fat.
- Use sugar only in moderation.
- Consume salt and sodium in moderation (less than 2,300 milligrams, or about 1 teaspoon of salt per day).
- If you drink alcoholic beverages, do so in moderation. How much is "moderate"? The equivalent of up to one drink per day for women and up to two drinks per day for men.

We consider the three basic nutrient groups first—carbohydrates, lipids, and proteins—before discussing vitamins, minerals, and the importance of fiber.

Carbohydrates: a major energy source

Carbohydrates are one of the body's main sources of energy, and many nutritionists recommend that approximately 45–65% of our calorie intake come from carbohydrates. Carbohydrates may be either simple or complex.

Simple carbohydrates (sugars) are found in many natural foods such as fruits, some vegetables, and honey. Refined sugars, such as granulated sugar and corn syrup,

have had most other nutrients removed during the refinement process, so they are less nutritious than their natural counterparts. It is difficult to tell exactly how much refined sugar we eat because so much is disguised on food labels as "corn sweeteners," "dextrose," or "fructose" (levulose). The USDA estimates the average North American consumes more than 142 pounds of sweeteners and refined sugar annually—almost 3 pounds each week.

Complex carbohydrates, such as starch or glycogen, consist of many sugar units linked together. Whole foods containing complex carbohydrates are better for us than refined sugars because they release sugars more slowly and also contribute fiber, vitamins, and minerals. In the body, stored starch and glycogen are broken down to glucose, one of the premier sources of energy. Good sources of complex carbohydrates include starchy vegetables (potatoes, corn, peas, and beans) and grains (wheat, rice, and oats).

Lipids: essential cell components and energy sources

Lipids (including fats) are essential components of every living cell. Phospholipids and cholesterol make up most of the cell membrane. Cholesterol also forms the backbone of steroid hormones and is used to synthesize bile. Fat stores energy and several vitamins, cushions organs, and insulates the body under the skin. Most of the fats in food are triglycerides, which consist of three fatty acids attached to a glycerol molecule. Recall (from Chapter 2) that fats are classified as either *saturated* or *unsaturated*, according to the ratio of hydrogen to carbon atoms in their fatty acids (**Figure 14.17**).

- Saturated fats have two hydrogen atoms for every carbon atom in their fatty acid tails. They tend to be solids at room temperature. Red meats and dairy products are higher in saturated fats than poultry and fish. Saturated fats are also found in a few plant sources such as coconut and palm kernel oil. They tend to raise blood levels of LDL cholesterol, the "bad" cholesterol that is associated with atherosclerosis and heart disease (Chapter 8).
- Unsaturated fats are missing one or more pairs of hydrogen atoms in their fatty acid tails. Every missing pair of hydrogen atoms leads to a double bond between

a) Animal fat, a saturated fat.   b) Olive oil, an unsaturated fat.

Figure 14.17 Fats.

adjacent carbons and a kink, or bend, in the tail, making it more difficult for adjacent molecules to form bonds with each other. Consequently, unsaturated fats are liquids (oils) at room temperature. In general, unsaturated fats are considered healthier than saturated fats because they tend to lower LDL cholesterol levels. Olive, canola, safflower, and corn oils are all unsaturated fats derived from plants. Certain cold-water fish (salmon, trout, and sardines) are rich in omega-3 fatty acids, polyunsaturated fatty acids that have been linked to a reduced risk of heart disease.

Vegetable oils can be turned back into solids at room temperature by partial hydrogenation of the fatty acid tails. However, the process of partial hydrogenation reconfigures the positions of some of the remaining unpaired hydrogen atoms, producing *trans fats*. Trans fats were once very popular with fast-food restaurants as a deep-frying oil because they increase the shelf life and flavor stability of foods. Trans fats are used in commercial baked goods (cookies, crackers, and snacks) and are also found in vegetable shortening and margarine. However, like saturated fats, trans fats tend to raise LDL cholesterol and thus may increase the risk of cardiovascular disease.

The liver can synthesize cholesterol and most of the lipids the body needs, but it cannot produce every last one of them. The few fatty acids that our bodies cannot make, called *essential fatty acids*, must be consumed as food. Two examples are linoleic and linolenic acids, present in corn and olives, both of which contribute to proper cell membrane structure. A teaspoon of corn or olive oil, or its equivalent in food, is enough to satisfy our daily requirements.

Although lipids are essential to human health, most of us consume far more than we need. Nutritionists recommend that lipids account for no more than 20–35% of calories consumed per day. In addition to watching the amount of fat we eat, though, it's important to pay attention to the type. Diets high in saturated fat, cholesterol, and trans fats place us at risk for cardiovascular disease and certain cancers. Less than 10% of calories consumed daily should come from saturated fats, with less than 300 milligrams per day of cholesterol and as little trans fat as possible.

✅ Why are fish living in cold waters particularly likely to have stores of unsaturated fats in their bodies, rather than saturated fats? (Hint: With few exceptions, fish are "cold-blooded," meaning their body temperature is the same as the temperature of the surrounding water.)

Complete proteins contain every amino acid

Like lipids, **proteins** are vital components of every cell. They make up the enzymes that direct metabolism, they serve as receptor and transport molecules, and they build our muscle fibers. A few are hormones.

Despite their variety, all proteins are composed of just 20 different amino acids. The body can make 12 of these amino acids. The 8 that the body cannot produce (isoleucine, leucine, lysine, methionine, phenylalanine, threonine, tryptophan, and valine) are called **essential amino acids** because they must be ingested in food. Two more that the body can make, histidine and arginine, are sometimes considered essential in children because their rapidly growing bodies cannot synthesize these amino acids fast enough.

A **complete protein** contains all 20 of the amino acids in proportions that meet our nutritional needs. Most animal proteins are complete, but nearly all plant proteins (with the exception of soybeans) lack one or more of the essential amino acids. Vegetarians must be careful to choose the right combinations of plant-based foods to obtain all amino acids. A combination of foods from any two of the three rightmost columns of Figure 14.18 will provide the necessary balance of all essential amino acids. Foods such as trail mix (roasted soybeans and nuts), hummus (chick peas and sesame seeds), a bean burrito in a corn tortilla, and red beans and rice do just that.

Approximately 15% of our calories should come from protein. In many parts of the world, protein deficiency is a serious health problem. Every enzyme has a unique amino

Complete proteins	Incomplete proteins: low in lysine	Incomplete proteins: low in methionine	Incomplete proteins: low in tryptophan
Legumes: Soybean Tofu Soy milk	**Legumes:** Peanuts	**Legumes:** Beans (dried) Black-eyed peas Chickpeas Lentils Lima beans Mung beans Peanuts	**Legumes:** Beans (dried) Garbanzos Lima beans Mung beans Peanuts
Cereal grains: Wheat germ	**Cereal grains:** Barley Corn Oats Rice Rye Wheat		**Fresh vegetables:** Green peas Mushrooms Swiss chard
Dairy products: Milk Cheese Yogurt Eggs	**Nuts, seeds:** Almonds Cashews Coconut English walnuts Hazelnuts Pecans Sunflower seeds Sesame seeds	**Fresh vegetables:** Asparagus Broccoli Green peas Mushrooms Potatoes	**Nuts, seeds:** Almonds English walnuts
Animal products: Meats Fish		**Nuts, seeds:** Hazelnuts	

Figure 14.18 Getting the essential amino acids. Most plants are low in one or more of the essential amino acids. A vegetarian should choose a combination of foods from any two of the three rightmost columns to ensure an adequate intake of all essential amino acids.

✅ Do you think eggs and milk contain complete proteins? Why or why not? (Hint: What are eggs and milk used for in nature?)

acid sequence, so if even one amino acid is missing in the diet, the body may be unable to produce crucial enzymes at the proper time in development. Protein deficiencies during pregnancy and childhood can retard growth and alter physical and mental performance. Unfortunately, foods with complete proteins such as meat and milk are often expensive in poor societies, and many people remain uninformed about ways to achieve a balanced diet of plant proteins.

Vitamins are essential for normal function

In addition to carbohydrates, lipids, and proteins, humans need **vitamins,** a group of at least 13 chemicals that are essential for normal functioning. The body can produce only a few vitamins; our skin synthesizes vitamin D when exposed to sunlight, and bacteria living in the colon manufacture vitamins K, B6, and biotin. We must obtain all others from our food and absorb them in the digestive tract. Table 14.2 lists the 13 vitamins along with their sources, functions, and signs of deficiency or excess.

Vitamins fall into two groups: fat soluble and water soluble. The distinction reflects how a vitamin is absorbed and stored and how steady a supply of it is needed. Fat-soluble vitamins are absorbed more readily if there is fat in the diet. They tend to be absorbed along with the components of fat, and they are stored in fat tissue

Table 14.2 Vitamins

Vitamin	Common sources	Primary functions	Signs of severe, prolonged deficiency	Signs of extreme excess
Fat-Soluble Vitamins				
A	Fortified dairy products, egg yolk, liver, yellow fruits, yellow and dark green leafy vegetables	Synthesis of visual pigments and the development of bones and teeth	Night blindness, dry scaly skin, increased susceptibility to infections	Damage to bone and liver, blurred vision
D	Fish oil, fortified milk, egg yolk; also produced by skin	Promotes absorption of calcium, required for healthy bones and teeth	Bone deformities in children, bone weakening in adults	Calcium deposits in tissues, leading to cardiovascular and kidney damage
E	Vegetable oils, nuts, whole grains	Thought to be an antioxidant that prevents cell membrane damage	Damage to red blood cells and nerves	Generally nontoxic
K	Green vegetables, also produced by bacteria in colon	Formation of certain proteins involved in blood clotting	Defective blood clotting, abnormal bleeding	Liver damage and anemia
Water-Soluble Vitamins				
B₁ (thiamine)	Whole grains, legumes, eggs, lean meats	Coenzyme in carbohydrate metabolism	Damage to heart and nerves, pain	None reported
B₂ (riboflavin)	Dairy products, eggs, meat, whole grains, green leafy vegetables	Coenzyme in carbohydrate metabolism	Skin lesions	Generally nontoxic
Niacin	Nuts, meats, grains, green leafy vegetables	Coenzyme in carbohydrate metabolism	Causes a disease called *pellagra* (damage to skin, digestive tract, and nervous system)	Skin flushing, potential liver damage
B₆	High-protein foods in general; also made by colonic bacteria	Coenzyme in amino acid metabolism	Muscle, skin, and nerve damage, anemia	Poor coordination, numbness in feet
B₉ (Folic acid)	Green vegetables, nuts, grains, legumes, orange juice	Coenzyme in nucleic acid and amino acid metabolism	Anemia, digestive disturbances	Masks vitamin B₁₂ deficiency
B₁₂	Animal products	Coenzyme in nucleic acid metabolism	Anemia, nervous system damage	Thought to be nontoxic
Pantothenic acid	Widely distributed in foods, especially animal products	Coenzyme in glucose metabolism, fatty acid and steroid synthesis	Fatigue, tingling in hands and feet, intestinal disturbances	Generally nontoxic except occasional diarrhea
Biotin	Widely distributed in foods; also made by colonic bacteria	Coenzyme in amino acid metabolism, fat and glycogen formation	Scaly skin (dermatitis)	Thought to be nontoxic
C (ascorbic acid)	Fruits and vegetables, especially citrus fruits, cantaloupe, broccoli, and cabbage	Antioxidant; needed for collagen formation; important for bone and teeth formation; improves iron absorption	A disease called *scurvy*, poor wound healing, impaired immune responses	Digestive upsets

and released as needed. Body stores of some fat-soluble vitamins may last for years, so it is rarely necessary to take supplements.

Water-soluble vitamins are absorbed more readily than fat-soluble vitamins, but they are stored only briefly and rapidly excreted in urine. Thus we need to consume foods containing water-soluble vitamins on a regular basis.

☑ Vegans—people who do not eat meat, dairy products, or eggs—are particularly vulnerable to a deficiency of a particular vitamin. Looking at Table 14.4, can you tell which one? Explain.

Minerals: elements essential for body processes

Minerals are the atoms of certain chemical elements that are also essential for body processes. They are the ions in blood plasma and cell cytoplasm (sodium, potassium, chloride, and many others). They represent most of the chemical structure of bone (calcium, phosphorus, and oxygen). They also contribute to the activity of nerves and muscles (sodium, potassium, and calcium), among many other functions. Twenty-one minerals are considered essential for animals. Nine of the 21 (arsenic, chromium, cobalt, manganese, molybdenum, nickel, selenium, silicon, and vanadium) are called *trace minerals* because they represent less than 0.01% of your body weight. Table 14.3 lists the 12 most important minerals.

How much do we need of each vitamin and mineral? The National Research Council publishes the current best estimate, the **Recommended Dietary Allowance (RDA).** Most healthy people can achieve the RDA without taking supplements if they eat a balanced diet of whole foods. That's a big "if" for some of us, considering our diets. Scientific studies suggest that supplements may also benefit certain groups, such as newborns, the elderly, or people taking medications that interfere with nutrient absorption. Calcium is advised for postmenopausal women to prevent the bone loss of osteoporosis.

Never take massive doses of any vitamin or mineral unless prescribed by your doctor. As the "excess" columns show, more is not always better.

Table 14.3 **Major minerals in the human body**

Mineral	Common sources	Primary functions	Signs of severe, prolonged deficiency	Signs of extreme excess
Calcium (Ca)	Dairy products, dark green vegetables, legumes	Bone and teeth formation, nerve and muscle action, blood clotting	Decreased bone mass, stunted growth	Kidney damage, impaired absorption of other minerals
Chloride (Cl)	Table salt	Role in acid-base balance, acid formation in stomach, body water balance	Poor appetite and growth, muscle cramps	Contributes to high blood pressure in susceptible people
Copper (Cu)	Seafood, nuts, legumes	Synthesis of hemoglobin, melanin	Anemia, bone and blood vessel changes	Nausea, liver damage
Fluoride (F)	Fluoridated water, tea, seafood	Maintenance of teeth and perhaps bone	Tooth decay	Digestive upset, mottling of teeth
Iodine (I)	Iodized salt, marine fish and shellfish	Required for thyroid hormone	Enlarged thyroid (goiter), metabolic disorders	Goiter
Iron (Fe)	Green leafy vegetables, whole grains, legumes, meats, eggs	Required for hemoglobin, myoglobin, and certain enzymes	Iron-deficiency anemia, impaired immune function	*Acute:* shock and death. *Chronic:* liver and heart failure
Magnesium (Mg)	Whole grains, green leafy vegetables	Coenzyme in several enzymes	Nerve and muscle disturbances	Impaired nerve function
Phosphorus (P)	Meat, poultry, whole grains	Component of bones, teeth, phospholipids, ATP, nucleic acids	Loss of minerals from bone, muscle weakness	Depressed absorption of some minerals, abnormal bone deposition
Potassium (K)	Widespread in the diet, especially meats, grains	Muscle and nerve function; a major component of intracellular fluid	Muscle weakness	*Acute:* cardiac arrhythmias and death. *Chronic:* muscle weakness
Sodium (Na)	Table salt; widespread in the diet	Muscle and nerve function; major component of body water	Muscle cramps	High blood pressure in susceptible people
Sulfur (S)	Meat, dairy products	Component of many proteins	None reported	None reported
Zinc (Zn)	Whole grains, meats, seafood	Component of many enzymes	Impaired growth, reproductive failure, impaired immune function	*Acute:* nausea, vomiting, diarrhea. *Chronic:* anemia and impaired immune function

Fiber benefits the colon

Fiber—found in many vegetables, fruit, and grains—is indigestible material. Even though our bodies cannot digest it, we need a certain amount in our diet. Fiber is beneficial because it makes feces bulky and helps them pass more efficiently through the colon.

A low-fiber diet can lead to chronic constipation, *hemorrhoids* (swollen veins in the lining of the anus, often caused by straining during defecation), and a disorder called *diverticulosis* (see section 14.12). A low-fiber diet has also been associated with a higher risk of colon cancer, perhaps because cancer-causing substances remain in the colon longer (see Chapter 18). Doctors recommend eating 20–35 grams of fiber every day—considerably more than the average 12–17 grams that most North Americans consume.

 **Recap** A healthy diet includes plenty of fruits, vegetables, and whole grains and adequate fiber. Saturated fat, sugar, salt, and alcohol should be ingested only in modest amounts. Some essential fatty acids and eight essential amino acids must be ingested to meet the body's nutritional requirements because the body cannot synthesize them. Complete proteins contain all 20 amino acids. Most animal proteins are complete, but most plant proteins are incomplete. All minerals and nearly all vitamins must be obtained from food. ■

MJ's BlogInFocus Will drinking bone broth improve your health? Visit MJ's blog in the Study Area in MasteringBiology and look under "Bone Broth."

http://goo.gl/VPxtSN

14.11 Food labels

By law, packaged foods must have a food label on their packaging that lists certain key pieces of nutritional information. Food labels can help you make nutritional choices if you take the time to read and understand them.

A typical food label offers nutritional facts and ingredients, as shown in **Figure 14.19**. This particular food label is from a box of crackers. At the top is information on serving size and serving per container. You may find that the serving size is quite small compared to the amount you might choose to eat and that it will vary between products, so keep that (and the calories per serving) in mind when comparing products.

Below the serving size and calories data is a breakdown of the percent of the recommended daily value of certain nutrients provided by just one serving. For some nutrients (fats, cholesterol, and sodium), the goal should be to keep your daily intake *under* 100% of the percent daily value;

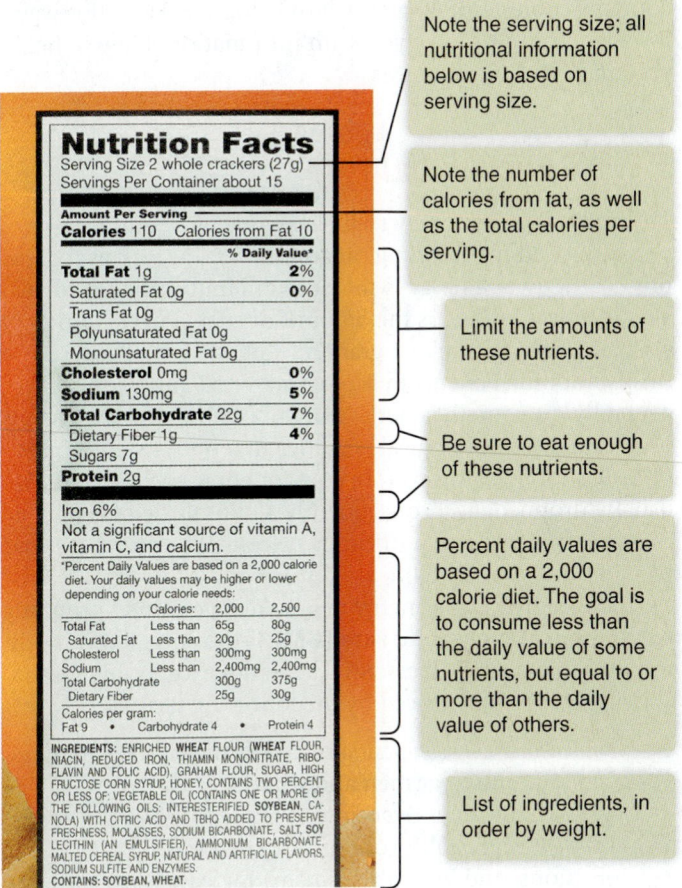

Figure 14.19 A food label.

for others (fiber, protein, and certain vitamins), the goal is to get *at least* 100% of the daily value each day. Next is a listing of the actual amounts of certain nutrients you should eat each day. Note that the amounts will depend on the total number of calories you aim to consume each day. Again, for some nutrients the goal is to stay under the listed amount; for others, it should be to get at least that amount.

Near the nutritional facts on the food label is another good source of information: the list of ingredients, which are always listed in order, top to bottom, by their weight. If you are trying to reduce your sugar intake, for example, you might be startled to see that sugar is among the first three ingredients in a product you regularly buy. You may also be surprised to see how many different ingredients there are, including flavorings, colorings, and preservatives. If you are allergic to any one potential food ingredient, here is where to look.

Reading and understanding the food labels on all the different products we buy is probably more work than most of us are willing to do. However, if you are trying to choose between two fairly similar products, it gets a lot easier; just compare the labels side by side. Which one has more calories in the serving size you would eat? Which one contains cholesterol, and which one doesn't? Used this way, food labels can provide you with information that will help you make healthier choices.

14.12 Energy balance

The body requires energy to fuel essential activities such as breathing and maintaining organ function, as well as for physical activity. Energy is measured in units called **calories.** Technically, a calorie is the amount of energy needed to raise the temperature of 1 gram of water by 1 °C. Because this is not much energy in biological terms, nutritionists generally measure the energy content of food and the energy used to perform biological activities in terms of kilocalories (1,000 calories). However, in subsequent usage the "kilo" is usually dropped, and consequently a kilocalorie of energy is referred to as just one nutritional calorie. We will follow that convention here as well.

Your daily caloric energy needs are determined primarily by your **basal metabolic rate (BMR),** the energy your body requires to perform all essential activities except physical activity. BMR is influenced by the following factors:

- *Gender and body composition.* BMR is higher in males and higher in muscular men and women.
- *Age.* BMR declines over time.
- *Health.* Some health conditions (such as fever, infections, and hyperthyroidism) increase BMR. Other conditions (such as hypothyroidism) lower it.
- *Food intake.* Eating increases BMR, whereas fasting and extreme dieting decrease it. This is why crash diets often fail to lower body weight permanently. Although weight may fall temporarily, a lower BMR makes it difficult to keep the pounds from returning.
- *Genetics.* Genetics plays a role in determining your BMR, although precisely how has not been determined.

By definition, BMR only includes a person's *basal* metabolic energy needs. Most people's energy expenditure is 50–100% higher than their BMR unless they spend all day in bed. The additional energy is expended in physical activity.

Energy balance, body weight, and physical activity

Healthy weight control involves balancing energy intake against energy expenditure. When we eat more calories than we use, the excess energy is stored in specialized cells as fat. The number of fat cells a person has is largely determined by the time he or she is an adult. In other words, fat cells do not multiply once adulthood is reached; they only decrease or increase in size. This may be why some people find losing weight so difficult and others do not. Research suggests that overweight people have two to three times more fat cells than normal-weight individuals, so when they diet and shrink their fat reserves in each cell, their bodies respond as if they are starving. Dieting is difficult for

chronically overweight persons because they are constantly fighting the body's own weight-control system, which responds as if their excess weight were normal.

Although your BMR stays fairly constant, the total number of calories you burn each day depends on your level of physical activity. Every 3,500 calories expended during physical activity equates to the loss of a pound of fat. Of course, physical activity has many benefits in addition to weight loss. It improves cardiovascular health, strengthens bones, tones muscles, and promotes a general sense of well-being. Your best strategy for good health is a well-balanced diet combined with moderate physical activity.

Healthy weight improves overall health

Numerous studies reveal a direct correlation between obesity and the incidence of heart disease, diabetes, cancer, arthritis, and other health problems. The connection is strong enough that government groups and insurance companies regularly publish charts that use a person's height and weight to define **body mass index (BMI).** Read your own BMI from the chart in Figure 14.20.

According to the government, a BMI between 18.5 and 25 is considered healthy, between 25 and 30 is considered overweight, and 30 or higher represents obesity. These numbers should be considered a general guide only because they are based on population statistics and do not take into account such factors as bone structure, fitness, or gender. BMI may overestimate the amount of body fat in individuals

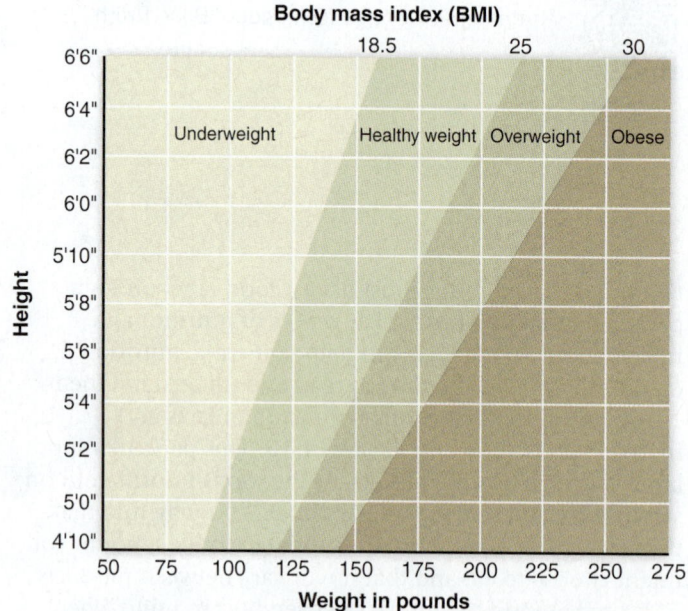

Figure 14.20 Find your BMI. Find your height in the left column of the chart and read right to find your weight. The government's weight categories (Underweight, Healthy weight, Overweight, or Obese) are indicated by colors. Their corresponding BMI ranges are shown at the top of the chart. The weight categories are based on large population groups. They are not necessarily an accurate measure of an individual's health or fitness.

who are very muscular or have a heavy bone structure. It may also underestimate the amount of body fat in people who have lost muscle mass. Thus, BMI charts do not provide definitive information about the appropriate weight status for any one individual. Within each range, the higher weights generally apply to men and the lower weights to women.

↩ **Recap** Weight control involves balancing energy consumed in food against energy spent. Calculating your BMR helps you estimate how many calories you need each day. Increasing physical activity is an efficient way to increase calorie expenditure. The best strategy for losing weight combines a healthy diet with moderate regular exercise. ■

Obesity

Obesity is on the rise. In the United States, the prevalence of obesity has increased from 13% of the population in 1990 to 36% in 2010. Figure 14.21 compares the figures state by state.

Body weight is determined by both internal factors (such as genetic makeup) and environmental factors (such as activity level and availability of food). There's a lot of interest in genetic factors these days, such as "fat genes" that might make it harder to maintain a healthy weight. But genetics cannot account for an entire population's weight gain, because genes don't evolve that quickly. It took us millions of years to become what we are, and our collective gene pool cannot change in just 20 years. The basic problem is that we now live in an environment that favors an imbalance between caloric intake and caloric expenditure. Computers, cars, desk jobs—all combine to produce a more sedentary lifestyle.

Physical exercise has become, for many of us, a leisure time activity rather than a requirement for survival. At the same time, food is relatively cheap, readily available, and highly caloric. We eat out (away from home) more often, we eat calorie-rich foods, and we eat more total calories per day—nearly 25% more than in 1970, according to government statistics. The rise in the prevalence of obesity is all too easy to understand; the question is what to do about it.

☑ Suppose two women start an exercise program. They are the same weight, they eat the same diet, and they both choose to do exactly 200 calories worth of physical activity per day. However, they choose different forms of exercise: One woman lifts weights to build stronger muscles, whereas the other person walks, which does not result in muscle growth. Which person is likely to lose more body fat? Why?

14.13 Eating disorders

Eating disorders are a group of conditions characterized by abnormal eating habits, to the point of damaging a person's physical or mental health. They are most common in women living in industrialized Western countries, especially young women with a history of extreme dieting. The three most common eating disorders are anorexia nervosa, bulimia nervosa, and binge eating disorder.

Anorexia nervosa (anorexia) is a condition in which a person diets excessively or stops eating altogether, even to the point of starvation and death. Untreated anorexia has the highest mortality rate of any mental illness. One of the most famous anorexics was singer Karen Carpenter, who died in 1983 at the age of 32.

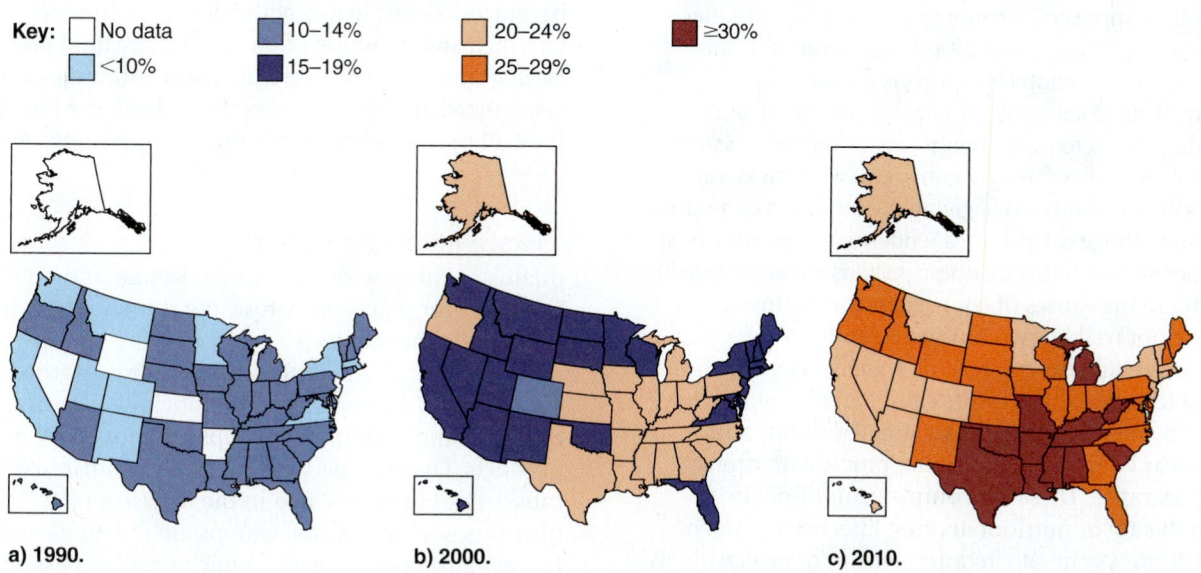

Key: ☐ No data ▦ 10–14% ▦ 20–24% ▦ ≥30%
      ▦ <10%   ▦ 15–19% ▦ 25–29%

a) 1990.   b) 2000.   c) 2010.

Figure 14.21 Obesity trends among U.S. adults, 1990, 2000, and 2010. These maps show the percentage of the population classified as obese in each state. Obesity is defined as a body mass index greater than 30.

Source: Behavioral Risk Factor Surveillance System, Centers for Disease Control and Prevention.

☑ Why do you think there are regional differences in the prevalence of obesity? Can you think of any local dietary or activity patterns that might be contributing factors?

Symptoms include the following:

- Refusal to maintain healthy body weight; people with anorexia often weigh less than 85% of their ideal weight.
- Intense fear of gaining weight, even though underweight.
- Distorted perception or preoccupation with body weight or shape.
- In premenopausal women, the absence of at least three consecutive menstrual cycles. Severe undernutrition interferes with the hormonal cycles of menstruation. Male anorexics also experience hormonal abnormalities.

Bulimia nervosa (bulimia) is a binge-and-purge condition in which someone eats and deliberately vomits or takes other steps to minimize the calories ingested. Symptoms of bulimia include the following:

- Recurrent episodes of binge eating. An episode of binge eating involves both (1) eating large amounts of food and (2) feeling a lack of control over eating.
- Taking recurrent inappropriate steps to prevent weight gain, such as self-induced vomiting; misusing laxatives, diuretics, enemas, or other medications; fasting; or exercising excessively.
- Binge eating and compensatory behaviors that occur, on average, at least twice a week for three months.
- Preoccupation with body shape and weight. However, unlike anorexics, some bulimics maintain a normal weight.

Binge eating disorder is like bulimia but without the purge. Binge eaters simply have no control over how much they eat on certain occasions. They are compulsive eaters, likely to eat even when they are not hungry.

All three forms of eating disorders play havoc with the body and mind. Anorexics become malnourished and suffer insomnia, hair loss, fatigue, and moodiness. Over time, they lose bone mass and develop osteoporosis (Figure 14.22). Bulemics are likely to suffer ulcers, chronic heartburn, and rectal bleeding due to repeated trauma to the digestive system. Recurrent vomiting also damages gums, erodes tooth enamel, and makes salivary glands swell, giving the bulimic a chipmunk-like appearance. Binge eaters have a tendency to become obese (but that does *not* mean that all obese persons are binge eaters).

The underlying causes of (or triggers for) eating disorders are not well known but seem to vary widely. They may include low self-esteem, fear, or anxiety; a dysfunctional home life; a feeling of helplessness and a desire for control where there is none; difficulty managing emotions and a tendency toward depression; or unhappiness with one's physical appearance. These are complex emotional issues rather than dietary or nutritional ones. Effective treatment of eating disorders generally requires a team of professionals who can address the patient's medical, psychiatric, dental, psychological, and nutritional needs.

14.14 Disorders of the digestive system

Although common, many digestive problems are not necessarily life threatening. One of the most common

Figure 14.22 Anorexia nervosa. French model Isabelle Caro posed for a controversial "No Anorexia" campaign in 2007. She died in 2010 at the age of 28.

conditions worldwide is *food poisoning,* caused by food and beverages contaminated with bacteria or their toxic products. Diarrhea and vomiting often accompany food poisoning. *Food allergies* can also cause diarrhea, vomiting, and generalized allergic responses throughout the body. Common food allergens include shellfish, wheat, peanuts, and eggs.

Disorders of the GI tract

Lactose intolerance: difficulty digesting milk Human infants are born with the enzyme lactase in their small intestines for digesting lactose, the primary sugar in milk. However, many adults gradually lose the enzyme, and with it their ability to digest lactose. The result is lactose intolerance. Symptoms of lactose intolerance include diarrhea, gas, bloating, and abdominal cramps after ingesting milk products. Diarrhea occurs because the undigested lactose causes fluid to be retained in the digestive tract. The gas, bloating, and abdominal cramps are due to bacterial fermentation of the lactose, which produces gases.

Lactose-intolerant people can eat cheese or yogurt because the lactose in these milk products has already been digested. Lactose-free milk is also available.

Peptic ulcers: sores in the stomach Peptic ulcers are painful erosions of the mucosal lining of the stomach or duodenum. Most peptic ulcers are associated with infection by one of the few bacteria that can live in the acidic environment of the stomach, called *Helicobacter pylori*. The bacterial infection leads to chronic inflammation, an increase in gastric acid secretion, and damage to the mucosal lining. Peptic ulcers can also be caused by excessive use of aspirin or nonsteroidal anti-inflammatory drugs (NSAIDs). NSAIDs block the production of the mucus that protects the gastric mucosa from gastric acid.

After treatment to eliminate any *H. pylori* infection, most peptic ulcers heal on their own. Antacids may also be prescribed as needed. The use of aspirin and NSAIDs should be discontinued if possible.

✅ Why do peptic ulcers typically occur in the stomach or duodenum, but not in the rest of the small intestine or the large intestine?

Celiac disease (gluten intolerance)
When people with celiac disease (also known as sprue) eat gluten, a protein found primarily in wheat, rye, and barley, their immune systems respond by damaging or destroying the villi that line the small intestine. The result is malabsorption of nutrients of all kinds, not just gluten. Celiac disease is an inherited disorder with a prevalence of about 1 in 130 persons. Symptoms vary widely, from acute abdominal pain and vomiting to chronic fatigue, depression, and eventually malnutrition, depending on a person's age, how much gluten they eat, and how sensitive they are to it.

Celiac disease often goes undiagnosed at first because its symptoms are so varied and so similar to other diseases and conditions. A definitive diagnosis is made by a blood test for a certain autoantibody in the patient's blood. The only effective treatment is a gluten-free diet.

Diverticulosis: weakness in the wall of the large intestine
Diverticulosis gets its name from its characteristic diverticula (small sacs produced when the mucosal lining of the large intestine protrudes through the other layers of the intestinal wall). Most people with diverticulosis do not experience discomfort, but sometimes the diverticula become infected or inflamed, in which case, it is called *diverticulitis*. Antibiotics usually resolve the problem.

Inadequate dietary fiber is thought to contribute to the development of diverticulosis. A low-fiber diet produces smaller feces, narrowing the colon and making its contractions more powerful. This increases the pressure on colon walls, forcing weak areas outward and forming diverticula. The disease is more prevalent with advancing age. Although it is rare in someone under 40, it may be present in up to two-thirds of all 85-year-olds.

Colon polyps: noncancerous growths
A polyp is a noncancerous growth that projects from a mucous membrane. Polyps can develop in many areas of the body, including the colon. The majority of colon polyps do not become cancerous, but because most colon cancers start as polyps, doctors recommend removing them. Polyps can be detected and removed in a *colonoscopy*, an outpatient procedure in which a flexible fiber-optic scope is inserted into the colon through the anus.

Disorders of the accessory organs

Hepatitis: inflammation of the liver Inflammation of the liver, referred to as **hepatitis,** is generally caused by viruses or toxic substances. Researchers have identified at least five viruses that cause hepatitis. The most common are hepatitis A, B, and C.

Hepatitis A is transmitted by contaminated food or water and causes a brief illness from which most people recover completely. A vaccine is available for hepatitis A. Although the vaccine is not considered necessary for most residents of industrialized nations, it may be a good idea for travelers to underdeveloped countries.

Hepatitis B travels in blood or body fluids, so it is usually passed via contaminated needles, blood transfusions, or sexual contact with infected individuals. Approximately 300,000 new cases are diagnosed each year. If not treated, hepatitis B can lead to liver failure. Symptoms include jaundice (the skin takes on a yellowish color due to accumulated bile pigments in the blood), nausea, fatigue, abdominal pain, and arthritis. A vaccine is available for hepatitis B. Federal law requires all health care workers to be vaccinated.

Hepatitis C is also transmitted in infected blood, usually through contaminated needles or blood transfusions. By 1992, U.S. blood banks began routine testing for hepatitis C, so the risk of contracting it from transfusions has fallen. However, researchers estimate that as many as 4 million Americans may already be infected, and many of them have no symptoms. Hepatitis C may remain dormant for years but still damage the liver. Severe cases can lead to chronic hepatitis, cirrhosis, or liver cancer. Doctors recommend testing for anyone who has had kidney dialysis (Chapter 15) or an organ transplant before 1992. Also at risk are those who have injected drugs or had sexual contact with a hepatitis C carrier.

Gallstones can obstruct bile flow The gallbladder normally concentrates bile about 10-fold by removing about 90% of the water. Excessive cholesterol in the bile may precipitate out of solution with calcium and bile salts, forming hard crystals called *gallstones* (**Figure 14.23**). Only about 20% of gallstones ever cause problems, but if

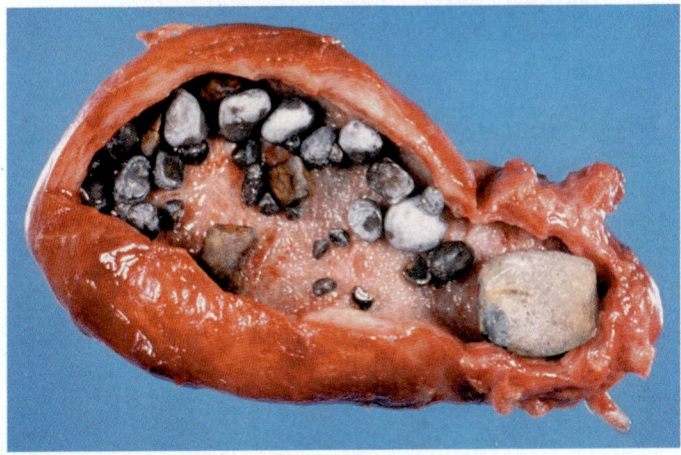

Figure 14.23 Gallstones.

the crystals grow large enough they can obstruct bile flow and cause intense pain, especially after a meal. Treatments include drugs to dissolve the crystals, ultrasound vibrations or laser treatments to break the stones apart, or surgery to remove the gallbladder.

Chapter Summary

14.1 The digestive system brings nutrients into the body *p. 325*

- The digestive system consists of the GI tract and four accessory organs: the salivary glands, liver, gallbladder, and pancreas.
- The five basic processes of the digestive system are motility, secretion, digestion, absorption, and excretion.
- Two types of motility in the GI tract are peristalsis, which propels food forward, and segmentation, which mixes the contents.

14.2 The mouth processes food for swallowing *p. 327*

- The 32 teeth of adults include incisors, canines, premolars, and molars. Teeth cut, tear, grind, and crush food.
- Saliva moistens food, begins the digestion of starch, and helps to protect against bacteria.

14.3 The pharynx and esophagus deliver food to the stomach *p. 329*

- Swallowing is a reflex that is initiated by voluntary movements of the tongue. Once started, swallowing is involuntary.
- The sole function of the esophagus is to get food from the mouth to the stomach.
- The lower esophageal sphincter prevents reflux of stomach contents.

14.4 The stomach stores food, digests protein, and regulates delivery *p. 330*

- The stomach stores ingested food until it can be delivered to the small intestine.
- Glands in the mucosa of the stomach secrete gastric juice into the lumen, beginning the process of protein digestion.
- Peristalsis of the stomach mixes the food and pushes it toward the small intestine.

14.5 The small intestine digests food and absorbs nutrients and water *p. 332*

- Digestion occurs primarily in the first part of the small intestine, called the *duodenum*.
- The jejunum and ileum of the small intestine absorb most of the products of digestion.
- The small intestine has a very large surface area because of its many folds, villi, and microvilli.

14.6 Accessory organs aid digestion and absorption *p. 333*

- The pancreas secretes fluid containing bicarbonate and digestive enzymes into the small intestine.
- The liver produces bile and participates in homeostasis in a variety of ways.

Recap Disorders of the GI tract and accessory organs include lactose intolerance, diverticulosis, colon polyps, gallstones, and hepatitis. ∎

- All of the venous blood from the GI tract is routed directly to the liver.
- The gallbladder stores bile from the liver and concentrates it by removal of most of the water.

14.7 The large intestine absorbs nutrients and eliminates wastes *p. 335*

- The large intestine absorbs most remaining nutrients and water and vitamin K produced by bacteria. It also stores wastes.
- Defecation is generally controlled by a neural reflex, but it can be overridden by conscious control.

14.8 How nutrients are absorbed *p. 335*

- Amino acids and simple sugars are actively transported into the mucosal cells that line the small and large intestines.
- The products of fat digestion enter the mucosal cells by diffusion and then reform into triglycerides. Triglycerides are coated with protein and move into lymph vessels for transport to the blood.
- Water is absorbed by osmosis.
- Vitamins and minerals follow a variety of specific pathways.

14.9 Nerves and hormones regulate digestion *p. 338*

- The volume and content of food play a large part in regulating digestive processes.
- Stretching of the stomach increases stomach peristalsis and the secretion of gastric juice.
- Stretching of the small intestine inhibits gastric motility, increases intestinal segmentation, and causes the secretion of two digestive enzymes, secretin and cholecystokinin.
- Acid in the small intestine triggers the secretion of pancreatic juice containing bicarbonate.

14.10 Nutrition: you are what you eat *p. 338*

- Good nutrition requires a variety of foods weighted toward fruits, vegetables, and whole-grain products.
- The human body needs certain nutritional components that it cannot make, including a few fatty acids, 8 amino acids, 13 vitamins, and all essential minerals.

14.11 Food labels *p. 343*

- Packaged foods are required to have food labels listing certain nutritional information.
- Food labels permit easy comparison of the nutritional value of similar packaged foods.

14.12 Energy balance *p. 344*

- BMR represents the daily energy needs of the body for all essential activities except physical activity.

- BMI takes into account a person's weight in relation to their height. A BMI over 30 is generally considered unhealthy.
- A person must use 3,500 calories more than he or she ingests to lose a pound of body fat.

14.13 Eating disorders p. 345

- Two serious eating disorders are anorexia nervosa, in which the person diets excessively or stops eating entirely, and bulimia, a binge-and-purge condition in which the person eats and deliberately vomits or takes other steps to minimize caloric intake.
- The most effective treatments are multidisciplinary, addressing medical, psychiatric, dental, psychological, and nutritional needs.

14.14 Disorders of the digestive system p. 346

- Lactose intolerance is caused by the lack of the enzyme (lactase) that normally digests lactose.
- The only effective treatment for celiac disease is to maintain a gluten-free diet.
- Hepatitis, or inflammation of the liver, can be caused by several different viruses.

Terms You Should Know

anorexia nervosa, *345*
basal metabolic rate (BMR), *344*
bile, *334*
bulimia nervosa, *346*
calories, *344*
cholecystokinin (CCK), *338*
chyme, *330*
colon, *335*
complete protein, *340*
esophagus, *329*
essential amino acids, *340*
gallbladder, *334*
gastrin, *338*
gastrointestinal (GI) tract, *325*
hepatic portal system, *334*
hepatitis, *347*
large intestine, *335*

mucosa, *326*
muscularis, *326*
nutrients, *324*
pancreas, *333*
pepsin, *330*
pepticulcer, *331*
peristalsis, *327*
Recommended Dietary
  Allowance (RDA), *342*
salivary glands, *328*
secretin, *338*
segmentation, *327*
serosa, *326*
submucosa, *326*
villus, *332*
vitamin, *341*

Concept Review

Answers can be found in the Study Area in MasteringBiology.

1. Describe the organs of the GI tract including the four accessory organs.
2. Name the five basic processes associated with carrying out the function of the GI tract.
3. Describe the role saliva plays in initiating the digestive process.
4. Compare and contrast *peristalsis* and *segmentation*.
5. Indicate how many teeth adult humans have.
6. Describe the events that occur in the stomach that contribute to the digestive process.

7. Indicate where in the GI tract most of the absorption of nutrients occurs.
8. Rank the major nutrient groups (carbohydrates, lipids, or proteins) according to calories per gram.
9. Indicate what fraction of our daily caloric expenditure is accounted for by our BMR.
10. Explain why it is harder to get an adequate supply of all amino acids from a vegetarian diet than from meat.

Test Yourself

Answers can be found in the Appendix.

1. Which of the GI tract tissue layers is most responsible for peristalsis and segmentation?
   a. serosa
   b. muscularis
   c. submucosa
   d. mucosa

2. All of the following are secreted into the lumen of the digestive tract during the process of digestion except:
   a. acid
   b. digestive enzymes
   c. hormones
   d. bile

3. All of the following are found in saliva except:
   a. mucin
   b. salivary amylase
   c. pepsin
   d. bicarbonate

4. Which of the following might be true of an individual unable to synthesize and secrete intrinsic factor?
   a. They would be unable to absorb adequate amounts of vitamin B_{12}.
   b. They would be unable to digest lactose found in dairy products.
   c. They would be unable to absorb amino acids in the small intestine.
   d. They would be unable to digest and absorb triglycerides.

5. Which of the following nutrients are digested within the small intestine?
   a. proteins
   b. lipids
   c. carbohydrates
   d. all of these choices

6. The region of the digestive tract most responsible for absorption of nutrients is the:
   a. stomach
   b. small intestine
   c. large intestine
   d. liver

7. Which of the following lists structures in order from smallest to largest?
   a. small intestine, folds, microvilli, villi
   b. villi, microvilli, folds, small intestine
   c. microvilli, folds, villi, small intestine
   d. microvilli, villi, folds, small intestine

8. What do the enzymes pepsin, chymotrypsin, trypsin, and carboxypeptidase have in common?
   a. They are secreted and active within the stomach.
   b. They are pancreatic enzymes.
   c. They are secreted and active within the small intestine.
   d. They are enzymes that participate in protein digestion.

9. Once nutrients are absorbed into the circulatory system, the next major organ they pass through is the:
   a. heart
   b. liver
   c. kidneys
   d. brain

10. Which of the following terms does not belong with the others?
    a. monoglycerides
    b. micelle
    c. peptide
    d. bile

11. Each of the following is absorbed into capillaries except:
    a. triglycerides
    b. amino acids
    c. monosaccharides
    d. water

12. Which of the following foods would be the best source of complex carbohydrates?
    a. strawberries
    b. honey
    c. orange
    d. oatmeal

13. Which of the following statements about trans fats is correct?
    a. Trans fats are found in animal products such as meat and dairy.
    b. Trans fats will lower LDL cholesterol levels.
    c. Trans fats are produced by the hydrogenation of unsaturated fats in vegetable oils.
    d. Salmon, trout, and sardines are rich in healthy trans fats.

14. Each of the following is a source of complete protein except:
    a. milk
    b. soy products
    c. peanuts
    d. eggs

15. Which of the following is associated with celiac disease?
    a. inability to digest and absorb proteins in dairy products
    b. damage to the villi resulting from the consumption of wheat products
    c. erosion of the wall of the stomach and duodenum due to hyperacidity
    d. chronic constipation due to low intestinal motility

Apply What You Know

Answers can be found in the Study Area in MasteringBiology.

1. A meal containing a lot of fat takes longer to be emptied out of the stomach than a low-fat meal. Why?
2. Procter & Gamble produces a calorie-free fat substitute called Olean (olestra) that is used to deep-fry foods such as potato chips. Olestra is a fat-soluble molecule that is neither digested nor absorbed. There is a concern that eating too much olestra might lead to certain vitamin deficiencies. What types of vitamins do you think would be affected by the use of olestra, and how might they be affected?
3. Other side effects of the consumption of too much olestra include gas, diarrhea, and a sudden need to defecate. What could be a cause of these?
4. Many prescription medications are supposed to be taken with a meal. Others are only to be taken on an empty stomach. Why would there be such differences?
5. Sometimes the cure for gallstones is to surgically remove the gallbladder. What effect would removal of the gallbladder have on digestion?
6. People with lactose intolerance cannot digest the common dairy sugar lactose. Besides being careful not to eat dairy foods, is there anything else a lactose-intolerant person can do?
7. Cholera is a dangerous infectious disease that is common in places that do not have a steady supply of clean drinking water. Severe untreated cholera can be fatal within a matter of days. Cholera is caused by a bacterial toxin that causes severe diarrhea. What do you think might cause death from cholera, if the only symptom is diarrhea?

MJ's BlogInFocus

Answers can be found in the Study Area in MasteringBiology.

1. What do those "Sell by" and "Best if used by" dates stamped on many prepared foods mean? Is a store allowed to sell food after its "Sell by" date? Visit MJ's blog in the Study Area in MasteringBiology and look under "Food 'Sell By' Dates." http://goo.gl/uTDgwv

MasteringBiology®

Students Go to MasteringBiology for assignments, the eText, and the Study Area with animations, practice tests, and activities.

Professors Go to MasteringBiology for automatically graded tutorials and questions that you can assign to your students, plus Instructor Resources.

Answers to ✓ questions are available in the Appendix.

The Urinary System

SEM of kidney glomeruli (red) and the blood vessels that supply them. The glomerular capsule and the nephron tubules have been removed.

Key Concepts

- **The primary organs of the urinary system are the kidneys.** They play a role in *homeostasis*, regulating the ionic composition and solute concentration of blood and contributing to the control of fluid volume and blood pressure.

- **The kidneys produce urine.** Urine is just a waste product. It consists of excess water and ions, metabolic wastes, and sometimes drugs, toxic chemicals, and vitamins.

- **The kidneys are endocrine organs.** They secrete a hormone called *erythropoietin* that regulates the production of new red blood cells. They also secrete an enzyme called *renin* that is part of the renin-angiotensin hormonal system.

- **The unit of function in a kidney is called a *nephron*.** Nephrons have the capability of producing either a diluted or a concentrated urine as needed to maintain proper fluid volume and composition.

- **When kidneys fail, the only option may be a kidney transplant** from another living person or from a cadaver.

CURRENT ISSUE
A Shortage of Kidneys

For years now, a serious imbalance has been developing between the numbers of kidneys available for transplantation and the number of patients who need them. In 2014, there were only 17,000 kidneys available for transplant and over 100,000 people on the waiting list for a kidney. Where do kidneys for transplant come from, and how is it decided who gets a kidney when one becomes available?

About one-third of donated kidneys come from living donors; the rest come from cadavers. Patient and donor must have the same blood type and six key tissue antigens. The more antigens that match, the greater the chance that the patient will not reject the organ. Living donors are usually close relatives, but even among close relatives the chances of a perfect match are not good. Matches between unrelated donors are rare. *The responsibility for finding a living donor rests with the patient.* There is no national registry of living unrelated potential donors.

If a patient cannot find his/her own living donor, the only source is a kidney from a deceased person (a cadaver). Organs may be harvested from a cadaver only when permission has been granted by the deceased or by the deceased's relatives. Persons who wish to donate their organs after death can sign and carry a donor card available through most state motor vehicle departments or on the Web. To make it easy, some states will indicate "organ donor" on your driver's license, with your permission.

Allocating Cadaveric Kidneys: A Changing Story

In 1984, when it became apparent that kidney transplants could save lives but that there was a severe shortage of kidneys for transplants, the Department of Health and Human Services (DHHS) established guidelines for how cadaveric kidneys would be allocated to patients. Three primary factors were considered in the guidelines: (1) the closeness of the immunological match, (2) the region of the country in which the kidney was donated, and (3) the length of time the recipient patient had been on a waiting list for a kidney. Here's how it worked: Patients who needed a kidney placed themselves on the waiting list of one of 11 regional Organ Procurement Organizations (OPOs). The OPOs then combined their lists, in effect creating a single national list of needy patients. When a cadaveric kidney became available in a particular OPO, it was tested for blood type and antigens. Perfect matches—the same blood type and the same six antigens—are so rare that patients anywhere in the country had first priority for a perfectly matched kidney.

In the case of partial matches, the kidney was first made available to all patients in the region of the OPO that procured the kidney. Within that OPO, patients who had been on the waiting list the longest would have the highest priority. Finally, if none of the patients in the regional OPO was a good partial match, the kidney was made available to patients in all other OPOs.

When it was first designed the allocation system seemed fair, but over time unintended consequences became apparent. Because the health of the cadaveric kidney was not taken into account and because the allocation guidelines favored the sickest patients (who had been on the transplant list the longest), nearly 30% of cadaveric kidney transplants failed

Questions to Consider

1 How do you think we should allocate cadaveric kidneys and other organs?

2 Would you be willing to donate your kidneys or other organs after death? Have you completed a donor card?

3 Should the government set up a national registry to match living donors with unrelated patients? Why or why not?

within five years. As a result, some patients received a second transplant, while other people on the transplant list died without ever receiving a transplant.

In an effort to fix this problem, the transplant guidelines were changed in 2014. Now, the 20% of donated kidneys with the greatest expected post-transplant longevity are allocated first to the 20% of transplant candidates with the best expected chance of long-term survival. In other words, the potential success of the transplant is being moved up the priority list, ahead of length of time on the transplant list. However, for most persons, time on the transplant list will still be a major factor. The idea is to maximize the benefit of what is admittedly a very limited resource, while still being as fair as possible. Whether the change will improve the overall longevity of kidney transplants remains to be seen.

Developments to Watch

Currently, it is against the law to sell a kidney—They can only be donated. Studies show that only about 50% of families of a deceased potential donor give their consent for kidney donation. To increase this number, the American Medical Association suggests that families be "compensated" when they donate the cadaveric kidneys of a recently diseased loved one. So far compensation isn't allowed.

Some countries have adopted *presumed consent* laws, wherein organs are presumed to be available for donation unless the donor (or family) explicitly states otherwise. Presumed consent laws have dramatically increased donations in countries that have adopted them. In the United States, however, presumed consent laws have not gained widespread acceptance, because they can be viewed as subtle government coercion or religious discrimination. →

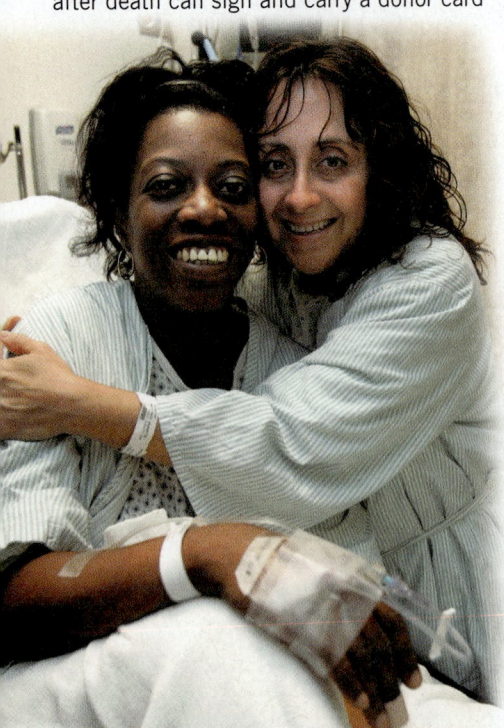

Kidney transplant recipient Valerie Beckford (left) with her donor, Jeanette Martinez.

Some patients find living donors on the Internet. The first commercial Web site to match living "altruistic donors" with patients was MatchingDonors.com. Patients who need an organ make an appeal on the site, and potential donors choose a patient to help. There are also Web sites that can increase your chances of finding a cadaveric kidney privately. LifeSharers (www.lifesharers.org) is a nonprofit national network of people who are willing to donate their organs to the network after death, in exchange for the chance of receiving an organ from the network should they need one during their lifetimes.

Some states offer income tax credit for expenses related to living-donor kidney donations or provide state employees with paid time off for donating a kidney. At least five states (Kentucky, Louisiana, Oklahoma, South Carolina, and Wisconsin) have passed laws mandating that organs donated in their states must be offered first to in-state patients. Whether such laws will stand up if challenged is an open question. The battle for scarce kidneys continues.

SUMMARY

- There are not enough donated cadaveric kidneys to meet the need for kidney transplants.
- The responsibility for finding a living donor rests with the patient.
- Under new federal allocation rules, patients with the greatest expected survival after transplant are more likely to be given a cadaveric kidney.
- Some patients are using Web sites to find unrelated living or cadaveric donors.

Drop by drop—slowly but steadily, the kidneys go about the business of making urine. Minutes or even hours pass as the urinary bladder slowly fills. Finally, although you hadn't been thinking about it before, the thought comes to your mind that you need to find a bathroom, and before long you can think of little else.

We are sometimes aware of our heartbeat, we can deliberately control our muscle movements, and we are often preoccupied by our stomach, especially when it's empty. But do we give our kidneys a second thought? They are forgotten organs, tucked away at the back of the abdominal cavity. With urine as their primary product, the kidneys may not serve a glamorous function. Urine itself serves no purpose; it is simply the end product of the regulation of the internal environment, the discarded waste. But do not underestimate the importance of your kidneys; they maintain the constancy of some of the most critical components of the internal environment. Without your kidneys, the cells that comprise your body would drown in a rising tide of toxic waste. It might take a week or two, but you'd be dead just the same. ▪

15.1 The urinary system regulates body fluids

As a recurrent theme in physiology, we have discussed homeostasis in terms of regulating body temperature, blood pressure, and other controlled variables in connection with the functions of organ systems. This chapter describes the central role of the urinary system in maintaining the constancy of the volume and composition of the body fluids.

Excretion refers to processes that remove wastes and excess materials from the body. Let's briefly review the excretory organs and systems involved in managing metabolic wastes and maintaining homeostasis of water and solutes (Figure 15.1).

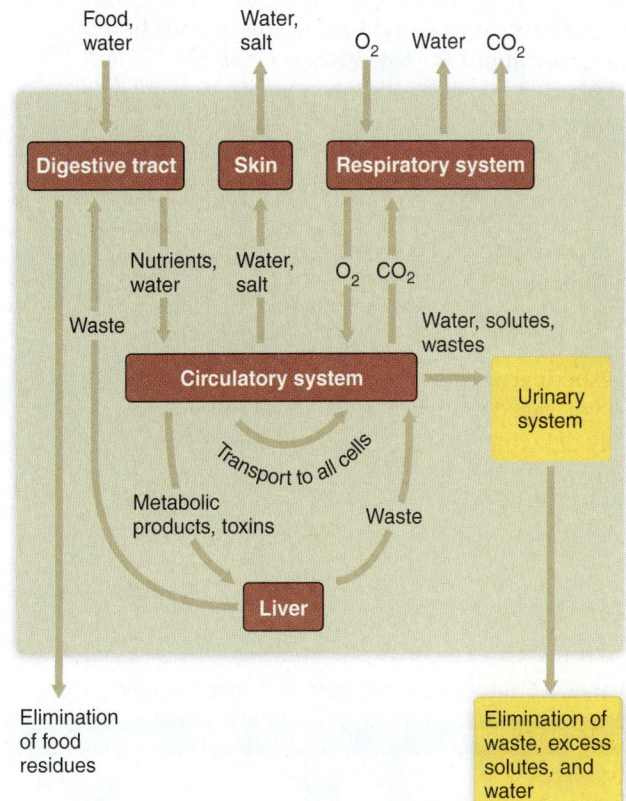

Figure 15.1 Organ systems involved in removing wastes and maintaining homeostasis of water and solutes. With the large tan box representing the body, this diagram maps the inflow and outflow of key compounds we consume.

The digestive system provides the body with nutrients and water, and excretes food residues. The lungs take in oxygen and excrete carbon dioxide gas. The skin gets rid of heat and is a site of water and salt loss, especially when we perspire. The liver destroys or inactivates numerous substances, some of which are excreted as bile in the feces. But the key organs in maintaining homeostasis of body fluids are the kidneys.

The **urinary system** consists of the kidneys, the ureters, the bladder, and the urethra. The two **kidneys** produce urine. The other components of the urinary system transport and store urine until it is eliminated from the body.

Urine is primarily water and solutes. Among the solutes are various ions, drugs, vitamins, toxic chemicals, and virtually every small waste molecule produced anywhere in the body. But the kidneys do much more than just get rid of liquid garbage. They also carefully regulate how much water and salt are excreted out of the body on a minute-to-minute basis in order to maintain homeostasis of fluid volume and composition.

Notably, substances not found in urine normally are the three classes of essential nutrients (carbohydrates, lipids, and proteins). The kidneys retain essential nutrients, leaving their management to other organs.

The kidneys regulate water levels

Water is the most abundant molecule in your body, representing about 60% of body weight. Normally, you exchange about 2½ liters of water per day with the external environment (Table 15.1). We consume most of our water in food and beverages, but we also produce about 300 milliliters as part of cellular metabolism. Meanwhile, we lose water steadily through evaporation from lungs and skin (not counting sweat) and through defecation.

Homeostasis is maintained only when water intake equals water output, as it does in Table 15.1. But you know from experience that water intake can vary tremendously. When you take in a lot of water or when you become severely dehydrated on a hot day, it's up to the kidneys to excrete the excess water or to conserve as much water as possible. The kidneys have a tremendous capacity to adjust water excretion as necessary, from a minimum of about 1/2 liter per day to nearly 1 liter per hour.

Table 15.1 Sources of water gain and loss per day

Water gain (ml/day)		Water loss (ml/day)	
Drinking fluids	1,000	Urine	1,500
Water in food	1,200	Evaporative loss (lungs)	500
Metabolic production	300	Evaporative loss (skin)	350
		Feces	150
Total	2,500	Total	2,500

The kidneys regulate nitrogenous wastes and other solutes

In addition to regulating water balance, the kidneys regulate a wide range of solutes. The primary solutes excreted by the kidneys are nitrogenous wastes, excess ions, and trace amounts of other substances.

The normal metabolism of proteins leaves us with excess nitrogenous wastes, which must be excreted by the kidneys. The metabolism of protein initially liberates ammonia (NH_3). Ammonia is toxic to cells, but the liver quickly detoxifies it by combining two ammonia molecules with a molecule of carbon dioxide to produce **urea** ($H_2N-CO-NH_2$) plus water. Most of the urea is excreted by the urinary system.

Dozens of different ions are ingested in excess with food or liberated from nutrients during metabolism. The most abundant ions in the body are sodium (Na^+) and chloride (Cl^-), both of which are important in determining the volume of the extracellular fluids, including blood. The volume of blood, in turn, affects blood pressure. Other important ions include potassium (K^+), which maintains electrical charges across membranes; calcium (Ca^{2+}), important in nerve and muscle activity; and hydrogen (H^+), which maintains acid-base balance. The kidneys regulate the urinary excretion of these ions to maintain homeostasis of each one.

Trace amounts of many other substances are excreted by the kidneys in direct proportion to the rate at which those substances are produced. Among them are creatinine, a waste product produced during the metabolism of creatine phosphate (an energy source in muscle), and various waste products that give urine its characteristic yellow color.

Recap The urinary system maintains a constant internal environment by regulating water balance and body levels of nitrogenous wastes, ions, and other substances. It filters metabolic wastes from the blood and excretes them in urine. The major nitrogenous waste product is urea. ∎

15.2 Organs of the urinary system

The urinary system is comprised of the kidneys, ureters, urinary bladder, and urethra (Figure 15.2a). The primary organs of function are the kidneys; the other three organs merely store and transport urine produced by the kidneys. The kidneys are responsible for excretion of metabolic wastes (especially urea), maintenance of water and salt balance, and control of the production of red blood cells. In addition, they contribute to the activation of vitamin D and to the regulation of acid-base balance and blood pressure. How they carry out these functions will be discussed in later sections after we review the structures of the urinary system.

The kidneys are located on either side of the vertebral column, near the posterior body wall. Each kidney is a dark reddish-brown organ about the size of your fist and shaped like a kidney bean. A renal artery and a renal vein connect each kidney to the aorta and inferior vena cava, respectively (*renal* comes from the Latin word *ren*, meaning "kidney").

Seen in a longitudinal section (Figure 15.2b), each kidney consists of inner pyramid-shaped zones of dense tissue (called *renal pyramids*) that constitute the **medulla** and an outer zone called the **cortex.** At the center of the kidney is a hollow space, the **renal pelvis,** where urine collects after it is formed.

A closer look at a section of the renal cortex and medulla reveals that it contains long, thin, tubular structures called *nephrons* (Figure 15.2c). Nephrons share a common final section called the *collecting duct*, through which urine is delivered to the renal pelvis.

Besides being the primary organs of the urinary system, the kidneys have many other functions related to maintaining homeostasis, which are discussed later in this chapter.

Ureters transport urine to the bladder

The renal pelvis of each kidney is continuous with a **ureter,** a muscular tube that transports urine to the bladder (Figure 15.2). Peristaltic waves of smooth muscle contraction, occurring every 10–15 seconds, move urine from the kidneys along the 10-inch length of the ureters to the bladder.

Urinary bladder stores urine

The **urinary bladder** stores urine (Figure 15.3). The bladder consists of three layers of smooth muscle lined on the inside by epithelial cells. Typically, the bladder can hold about 600–1,000 ml of urine, though volumes that large may make one feel uncomfortable. Women generally have a smaller bladder capacity than men because their bladders are slightly compressed by the nearby uterus.

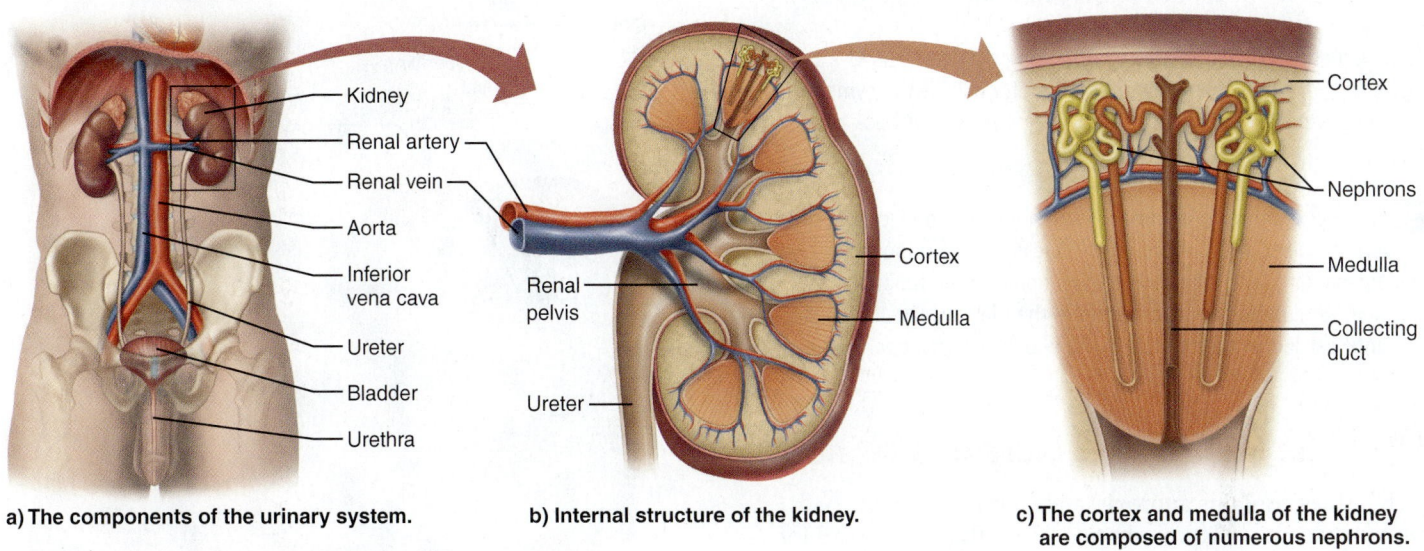

a) **The components of the urinary system.**

b) **Internal structure of the kidney.**

c) **The cortex and medulla of the kidney are composed of numerous nephrons.**

Figure 15.2 The human urinary system.

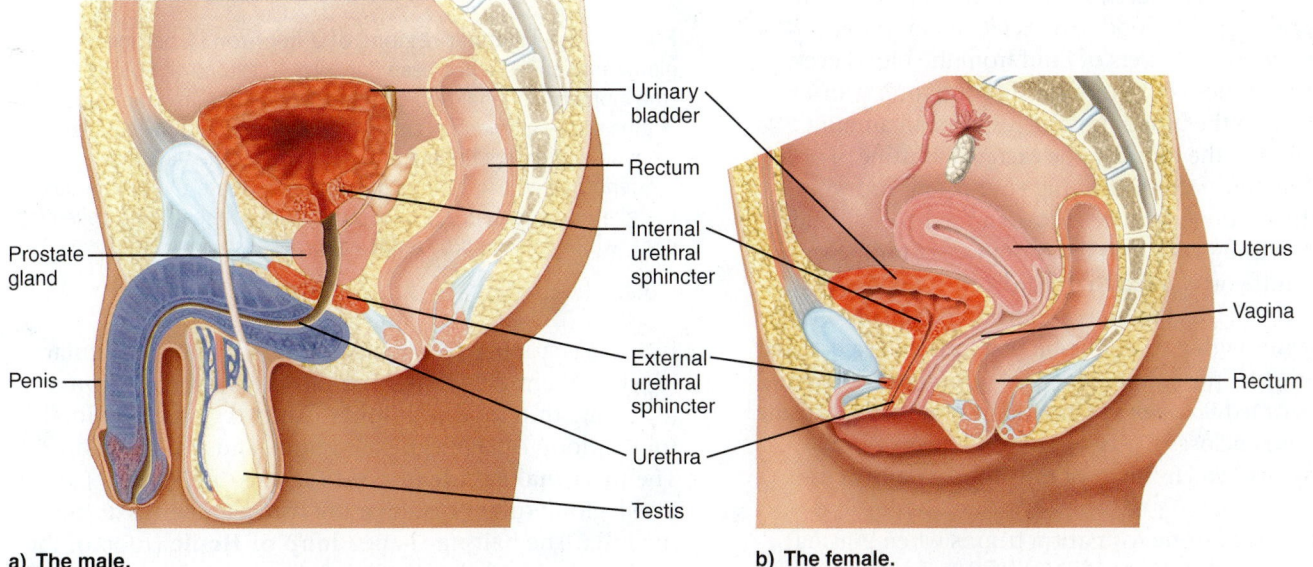

a) **The male.**

b) **The female.**

Figure 15.3 The positions of the bladder, the urethra, and associated organs in males and in females.

The urethra carries urine from the body

During urination, urine passes through the **urethra,** a single muscular tube that extends from the bladder to the body's external opening. Until then the bladder is prevented from emptying by the internal urethral sphincter, where the bladder joins the urethra, and the external urethral sphincter farther down the urethra. The urethra is about 8 inches long in men and about 1.5 inches long in women (Figure 15.3).

☑ A woman visits her doctor, complaining of symptoms of a burning sensation when she urinates. The doctor asks if the symptoms include any back pain. Why did the doctor ask about back pain?

↩ **Recap** Organs of the urinary system include the kidneys, ureters, bladder, and urethra. The kidneys are the principal urinary organs, although they have several homeostatic functions as well. The ureters transport urine to the bladder, where it is stored until carried by the urethra to the body's external opening. ■

15.3 The internal structure of a kidney

Each kidney contains approximately a million small functional units called **nephrons** (from the Greek word for "kidney," *nephros*). An individual nephron consists of a thin, hollow tube of epithelial cells, called a *tubule,* plus the blood vessels that supply the tubule. The function of the nephron is to produce urine. However, nephrons don't just pick the waste molecules out of blood and excrete them. Instead, they remove about 180 liters of fluid from the blood every day (about 2½ times your body weight) and then return almost all of it to the blood, leaving just a small amount of fluid behind in the tubule to be excreted as urine. If you took a similar approach to cleaning your room, you would take everything out of the room, dust and scrub it all, and then put everything back into the room except for the small amount of stuff you wanted to discard.

Figure 15.4 shows the tubular portion of a nephron. The nephron begins with a cup of tissue that looks like a deflated ball with one side pushed in, called the **glomerular capsule** (sometimes called *Bowman's capsule*). The glomerular capsule surrounds and encloses a network of capillaries called the **glomerulus,** which is part of the blood supply of the nephron.

The process of urine formation begins when plasma fluid is filtered out of the capillaries of the glomerulus and

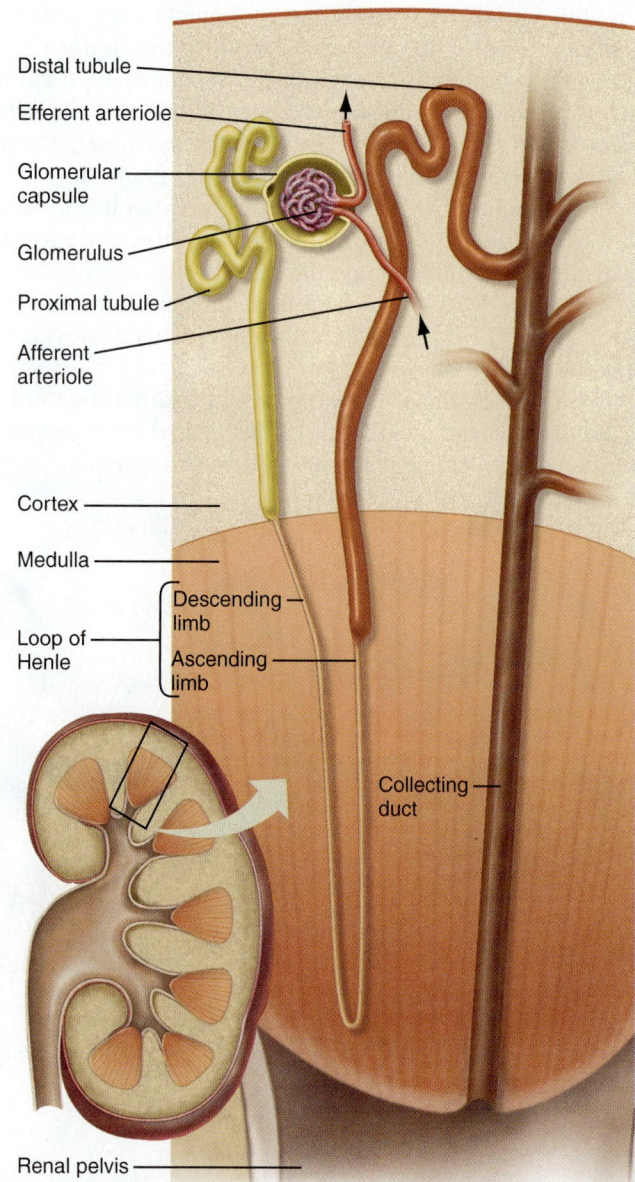

Figure 15.4 Tubular regions of a nephron. A portion of the glomerular capsule has been cut away to expose the blood vessels (the glomerulus) enclosed by the capsule. The proximal tubule begins at the glomerular capsule and ends where the tubule narrows and enters the medulla. The loop of Henle consists of descending and ascending limbs; it ends where the tubule passes the glomerular capsule. The distal tubule connects to a collecting duct, which is shared by many nephrons.

into the space between the two layers of the glomerular capsule. From the glomerular capsule, the tubule continues as a long, thin tube with four distinct regions: proximal tubule, loop of Henle, distal tubule, and collecting duct. The **proximal tubule** (*proximal* means "nearest to") starts at the glomerular capsule in the cortex and ends at the medulla. The hairpin-shaped **loop of Henle** (loop of the nephron) extends into the medulla as the *descending limb* and then loops back up as the *ascending limb*. After it passes the glomerular capsule, the tubule is called the **distal tubule** (*distal* means "more distant from"). Finally, the distal tubules of up to a thousand nephrons join to become a **collecting duct.** The collecting duct extends from the cortex through the medulla to reach the renal pelvis.

Special blood vessels supply the tubule

Figure 15.5 provides a closer look at the blood vessels serving the nephrons. The renal artery supplying a kidney branches many times to serve the million nephrons. Ultimately, every nephron is supplied by a single arteriole, called the **afferent arteriole** (*afferent* means "directed toward"). The afferent arteriole enters a glomerular capsule and then divides many times to become the network of capillaries that constitutes the glomerulus. Here, plasma fluid and solutes are filtered from the blood into the capsular space.

The glomerular capillaries rejoin to become the **efferent arteriole** (*efferent* means "directed away"), which carries filtered blood from the glomerulus. The efferent arteriole divides again into another capillary network that surrounds the proximal and distal tubules in the cortex, called the **peritubular capillaries** (*peri-* means "around"). The peritubular capillaries remove water, ions, and nutrients, which are reabsorbed by the proximal and distal tubules. The efferent arterioles of a few nephrons descend into the medulla and divide into long, thin capillaries called the **vasa recta** (Latin for "straight vessels") that supply the loop of Henle and collecting duct. Eventually, the filtered blood flows into progressively larger veins that become the single renal vein leading to the inferior vena cava.

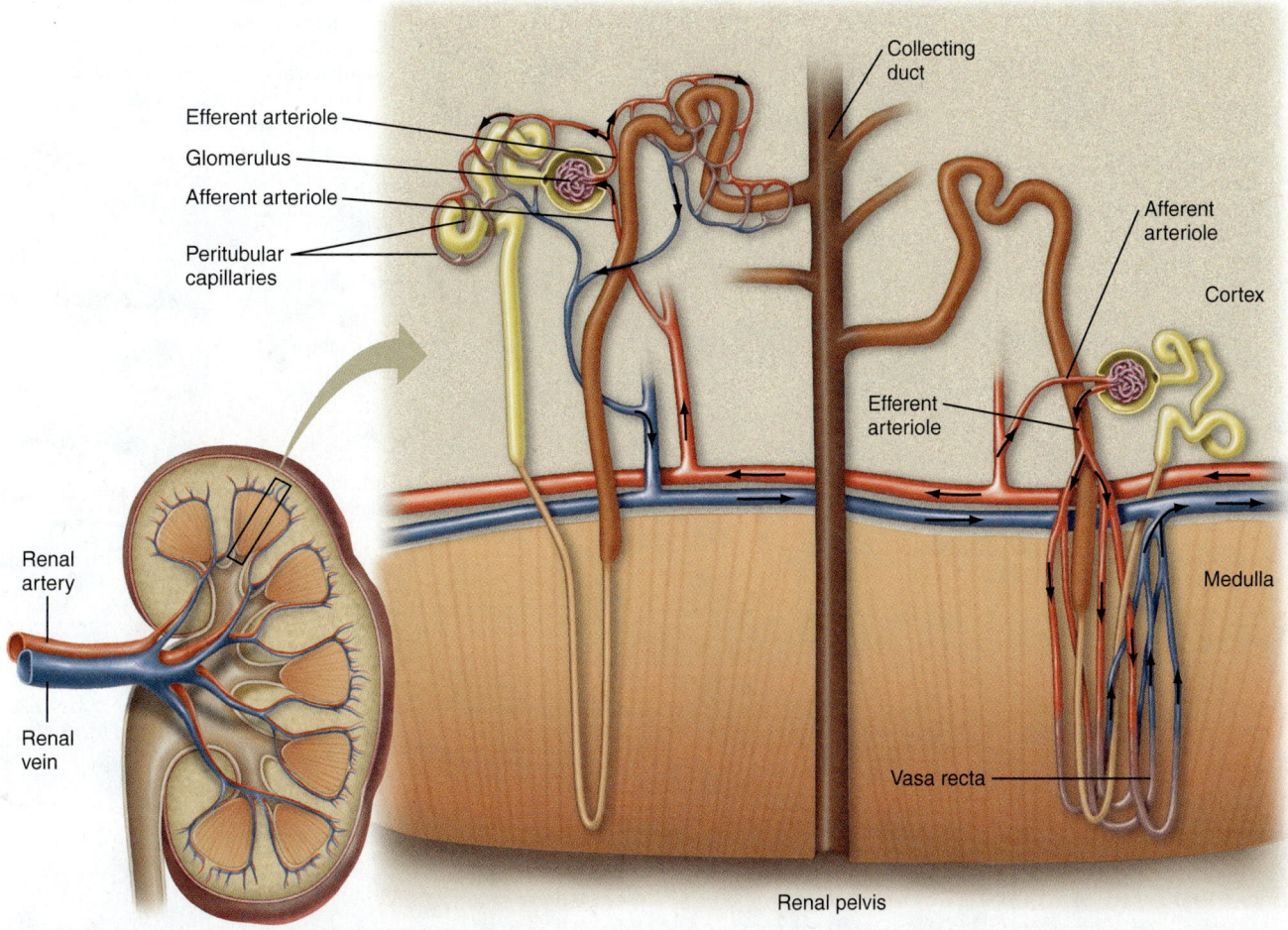

a) Main blood vessels in the kidney. The renal artery and renal vein branch many times to deliver blood to each glomerulus.

b) Post-glomerular blood vessels. The efferent arteriole of most nephrons, such as the one shown here on the left, divides to become the peritubular capillaries, which supply proximal and distal tubules in the cortex. In a few nephrons, such as the one on the right, the efferent arteriole descends into the medulla to become the vasa recta, which supply loops of Henle.

Figure 15.5 Relationship between the tubular and vascular components of nephrons.

↩ **Recap** A nephron is the functional unit of a kidney. A nephron tubule consists of a glomerular capsule, where fluid is filtered, and four regions, the proximal tubule, loop of Henle, distal tubule, and collecting duct, in which the filtered fluid is modified before it becomes urine. Blood flows to the glomerulus via the renal artery and afferent arterioles. Peritubular capillaries carry the blood to the proximal and distal tubules, and vasa recta supply the loops of Henle and collecting ducts. ∎

15.4 Formation of urine: filtration, reabsorption, and secretion

As we have noted, the urinary system regulates the excretion of water and ions to achieve homeostasis of fluid volume and composition. It excretes certain wastes while retaining precious nutrients. How do the kidneys select what to retain and what (and how much) to excrete in urine?

The formation of urine involves three processes (**Figure 15.6**):

❶ *Glomerular filtration.* The movement of a protein-free solution of fluid and solutes from the glomerulus into the space within the glomerular capsule.

❷ *Tubular reabsorption.* The return of most of the fluid and solutes back into the peritubular capillaries or vasa recta.

❸ *Tubular secretion.* The addition of certain solutes from the peritubular capillaries or vasa recta into the tubule.

The fluid and solutes that remain in the tubule constitute the urine, which is eventually excreted. Because matter is not being created or destroyed, the amount of any

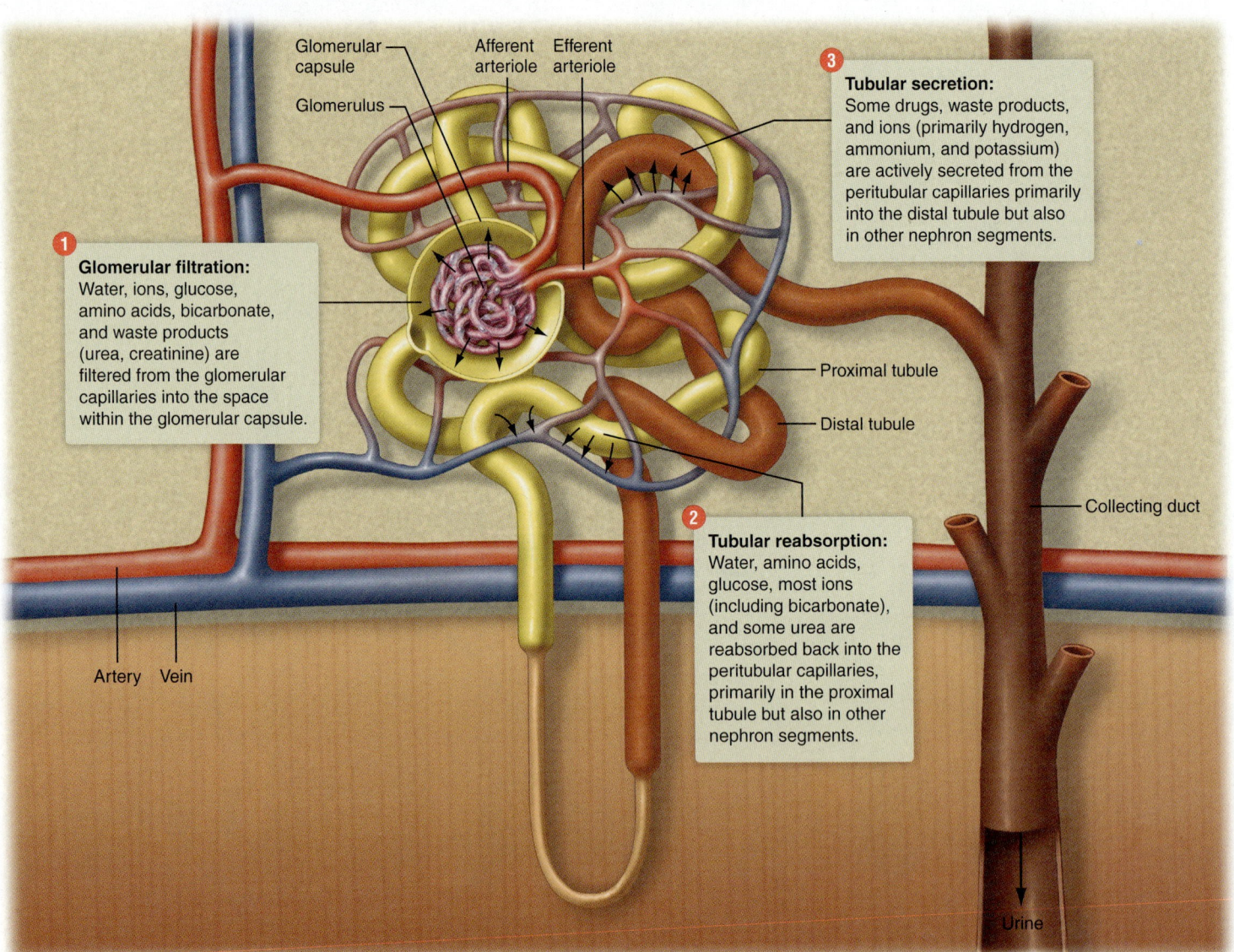

Glomerular capsule

Afferent arteriole **Efferent arteriole**

Glomerulus

❸ **Tubular secretion:** Some drugs, waste products, and ions (primarily hydrogen, ammonium, and potassium) are actively secreted from the peritubular capillaries primarily into the distal tubule but also in other nephron segments.

❶ **Glomerular filtration:** Water, ions, glucose, amino acids, bicarbonate, and waste products (urea, creatinine) are filtered from the glomerular capillaries into the space within the glomerular capsule.

Proximal tubule

Distal tubule

Collecting duct

❷ **Tubular reabsorption:** Water, amino acids, glucose, most ions (including bicarbonate), and some urea are reabsorbed back into the peritubular capillaries, primarily in the proximal tubule but also in other nephron segments.

Artery Vein

Urine

Figure 15.6 The three processes that contribute to the formation of urine.

☑ Explain why it makes sense to reabsorb nearly all of some of the substances that are filtered (such as water, salt, glucose, and amino acids) but very little, if any, of others (drugs, waste products, excess hydrogen ions, etc.).

substance excreted in urine must be equal to the amount filtered, minus the amount reabsorbed back into the blood, plus the amount secreted into the tubule:

Filtered − Reabsorbed + Secreted = Excreted

Glomerular filtration filters fluid from capillaries

Urine formation starts with **glomerular filtration,** the process of filtering a large quantity of protein-free plasma fluid from the glomerular capillaries into the glomerular space. The filtration barrier consists of two cell types: modified tubular epithelial cells called **podocytes** that cover and surround the outside surface of the capillaries and the capillary cells themselves (Figure 15.7a). Fluid first passes through pores in the capillary cells (Figure 15.7b) and then through tiny slits between cytoplasmic extensions of the podocytes in order to enter the glomerular space.

The podocytes and the glomerular capillary cells are like the sieve used to drain the water from spaghetti after cooking; they are highly specialized for filtration of a large volume of fluid. Together, they are nearly 100 times more permeable to water and small solutes than the capillaries in most other vascular beds. However, they are also highly selective; they are *less* permeable than other capillaries to large proteins and whole cells. As a result, the filtered fluid, called the *glomerular filtrate,* contains water and all of the small solutes in the same concentrations found in blood plasma, but it contains no large proteins or blood cells.

Together, the two kidneys produce about half a cup of glomerular filtrate per minute, a whopping 180 liters per day. Contrast that amount with the daily urine excretion of about 1.5 liters, and you can see how efficient the kidneys are at reabsorbing substances that are filtered.

Glomerular filtration is driven by high blood pressure in the glomerular capillaries—about twice as high as in any other capillary. The reason that glomerular capillary pressure is so high is that the efferent arteriole is much narrower than the afferent arteriole. The efferent arteriole is like a kink in a hose that is downstream from a very leaky section (the glomerular capillaries). The combination of a high pressure in glomerular capillaries and their leakiness allows large amounts of plasma fluid to be filtered out of the capillaries and into the glomerular space. The kidneys themselves do not have to expend any energy to produce the filtrate.

The rate of filtration is regulated in two ways:

- Under resting conditions, pressure-sensitive cells in the arterioles and cells in the tubule walls release chemicals that adjust the diameter of the afferent arterioles. These local feedback mechanisms maintain a relatively constant rate of glomerular filtration, allowing the kidneys to carry out their regulatory functions.

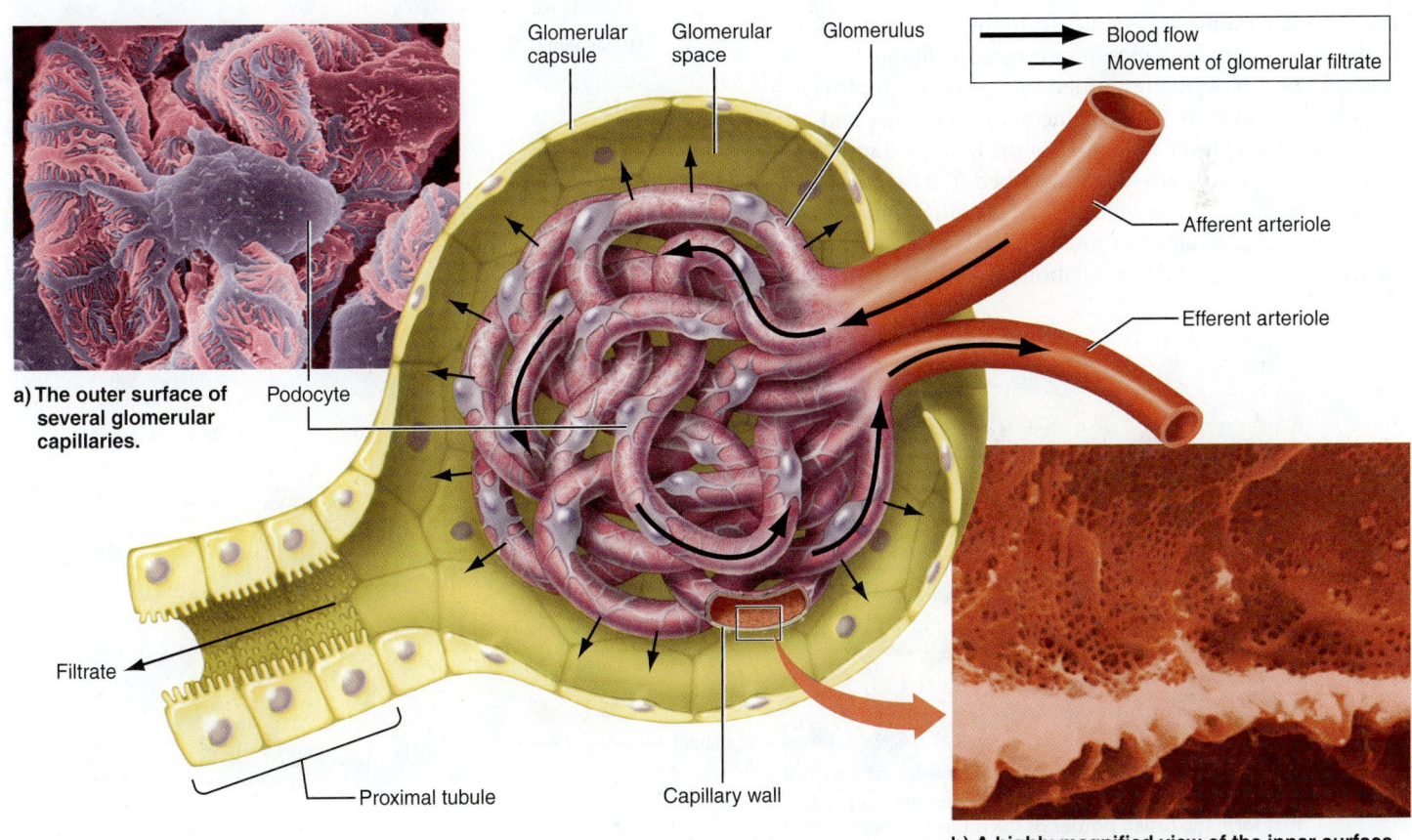

a) **The outer surface of several glomerular capillaries.**

Podocyte

Filtrate

Proximal tubule

Glomerular capsule

Glomerular space

Glomerulus

→ Blood flow
→ Movement of glomerular filtrate

Afferent arteriole

Efferent arteriole

Capillary wall

b) **A highly magnified view of the inner surface of a single glomerular capillary, revealing its porous sievelike structure.**

Figure 15.7 Glomerular filtration.

● During times of stress (such as after a loss of blood or while running a marathon), the sympathetic division of the autonomic nervous system constricts afferent and efferent arterioles, reducing blood flow and the rates of glomerular filtration and urine formation. This constriction is just part of normal mechanisms for redistributing blood flow to more critical organs during times of need. The kidneys are unharmed because they do not need a high blood flow to survive. In fact, a reduction in urine formation and water loss is just what is needed at such times.

If the delicate sievelike structure of the glomerular capillaries is disrupted, proteins may cross the filtration barrier into the tubule fluid. This condition is called *proteinuria*, the appearance of protein in urine. Persistent proteinuria is a sign of glomerular damage, perhaps by toxins or sustained high blood pressure. However, temporary proteinuria can occur after strenuous exercise, even in healthy people. The reason for this is unknown, but it may be that reduced renal blood flow during exercise allows blood to remain in contact with the glomerular barrier longer, letting more plasma proteins leak through. Exercise-induced proteinuria is not dangerous and goes away within a day or so.

Tubular reabsorption returns filtered water and solutes to blood

Tubular reabsorption, the second step in urine formation, returns filtered water and solutes from the tubule into the blood of the peritubular capillaries or vasa recta. As filtrate flows through the tubule, major nutrients are almost completely reabsorbed—all of the filtered glucose, amino acids, and bicarbonate, and more than 99% of the water and sodium. About 50% of the urea is also reabsorbed. The final product—urine—contains just enough water and sodium to balance the daily net gain from all other sources, the equivalent of all the urea produced in one day's metabolism, and trace amounts of other ions and wastes. Some waste products, such as creatinine, are not reabsorbed at all once they are filtered (Table 15.2).

Most tubular reabsorption occurs in the proximal tubule, leaving the fine-tuning and regulation of reabsorption to the more distal regions. A good example is water. The proximal tubule always reabsorbs about 65–70% of the water, and the loop of Henle reabsorbs another 25%, both completely unregulated. The distal tubule and collecting duct together reabsorb less than 10%, yet it is here that water excretion is effectively regulated.

To be reabsorbed, substances must cross the layer of epithelial cells of the proximal tubules to reach blood capillaries. Like the epithelial cells of the digestive tract, proximal tubular cells have a "brush border" of microvilli that increases surface area. In proximal tubule cells, the microvilli face the lumen (the inside) (Figure 15.8). The process of reabsorbing water, solutes, and nutrients from filtrate begins in the proximal tubule.

Table 15.2 Amounts of various substances filtered and excreted or reabsorbed

Substance	Amount filtered per day	Amount excreted in urine per day	Proportion reabsorbed
Water	180 L	1–2 L	99%
Sodium (Na^+)	620 g	4 g	99%
Chloride (Cl^-)	720 g	6 g	99%
Potassium (K^+)	30 g	2 g	93%
Bicarbonate (HCO_3^-)	275 g	0	100%
Glucose	180 g	0	100%
Urea*	52 g*	26 g	50%
Creatinine*	1.6 g*	1.6 g	0

*The plasma concentrations of these substances and the amounts filtered per day vary with diet, age, and level of physical activity.

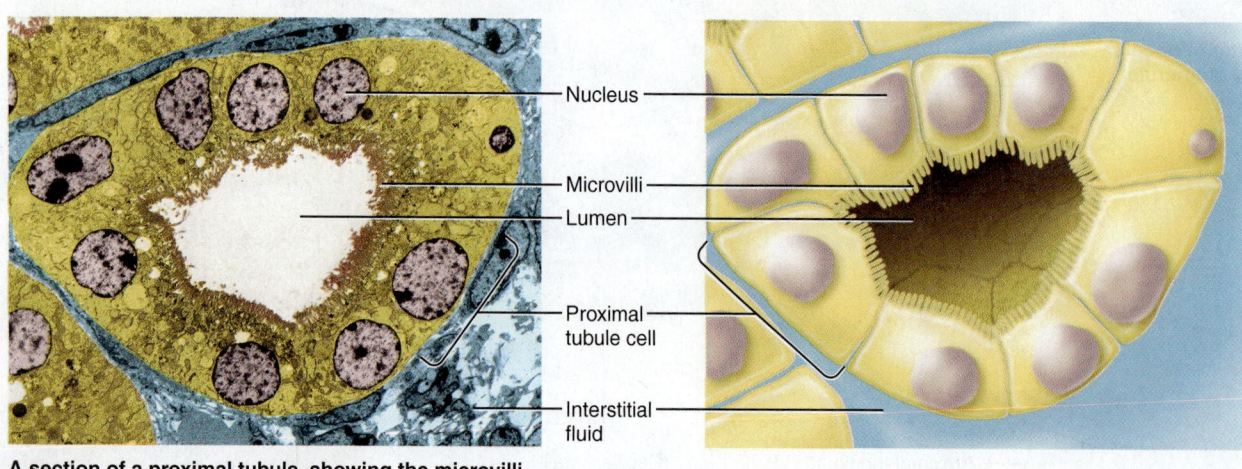

A section of a proximal tubule, showing the microvilli that form the brush border of the inner (luminal) surface.

- Nucleus
- Microvilli
- Lumen
- Proximal tubule cell
- Interstitial fluid

Figure 15.8 The proximal tubule.

Reabsorption in the proximal tubule (**Figure 15.9**) is a linked series of events, all dependent on just one active step: ❶ the active transport of sodium out of the proximal tubular cell on the capillary side. ❷ From there, the sodium ions diffuse into a capillary. In turn, that active transport of sodium out of the proximal tubule cell sets up ❸ a concentration gradient that favors the facilitated diffusion of sodium from the tubular lumen into the proximal tubule cell. Together, these steps result in the reabsorption of sodium from the filtrate back into the blood.

But that is only the beginning. Because sodium is a positively charged ion, its diffusion into the cell ❹ drives the diffusion of chloride across the tubular cell. (Negative charges must follow positive ones to maintain electrical neutrality.) And because sodium and chloride are also osmotic particles, ❺ their diffusion sets up a concentration gradient for water

that favors the reabsorption of water by diffusion as well. Finally, the facilitated diffusion of sodium inward across the luminal cell membrane is used as the energy source for ❻ the secondary active transport (also called co-transport) of glucose and amino acids into the cell. ❼ Once inside the cell, the glucose and amino acids diffuse into the interstitial fluid and eventually into the peritubular capillaries.

The key to the entire process is the active transport of sodium (Na^+) across the tubular cell membrane on the capillary side of the cell. This step requires energy in the form of ATP. The active transport of sodium keeps the intracellular sodium concentration low, which permits sodium to diffuse into the cell from the luminal side. Negatively charged chloride ions follow in order to maintain electrical neutrality, and water follows to maintain osmotic equilibrium. Finally, the diffusion of sodium across the luminal membrane is used as an energy source for the reabsorption of glucose and amino acids.

Note that the reabsorption of water, salt, glucose, and amino acids in the proximal tubule all depend on just one metabolic energy-using process: the primary active transport of sodium out of the tubular epithelial cell. Reabsorption in the more distal regions of the tubule, although less in amount, also relies on the active transport of sodium as the energy-utilizing step.

Tubular secretion removes other substances from blood

A few substances undergo movement in an opposite direction from tubular reabsorption. **Tubular secretion** is the movement of substances from the capillaries (either peritubular capillaries or vasa recta) into the tubule lumen. Tubular secretion is a mechanism for removing unwanted substances from the blood and excreting them in the urine, over and above the amounts that would already be in the urine by virtue of filtration. (Substances that undergo tubular secretion generally are not reabsorbed in the proximal tubules.) Tubular secretion may occur by either active transport or passive diffusion, depending on the substance being secreted.

Tubular secretion plays a critical role in removing nearly all of certain chemicals from the blood, including toxic and foreign substances and drugs. The list of substances secreted into the tubule includes penicillin, cocaine, marijuana, many food preservatives, and some pesticides. Note that the kidneys do not always determine which foreign substances are actually good for us. Most tubular secretion occurs in the distal tubule, rather than the proximal tubule.

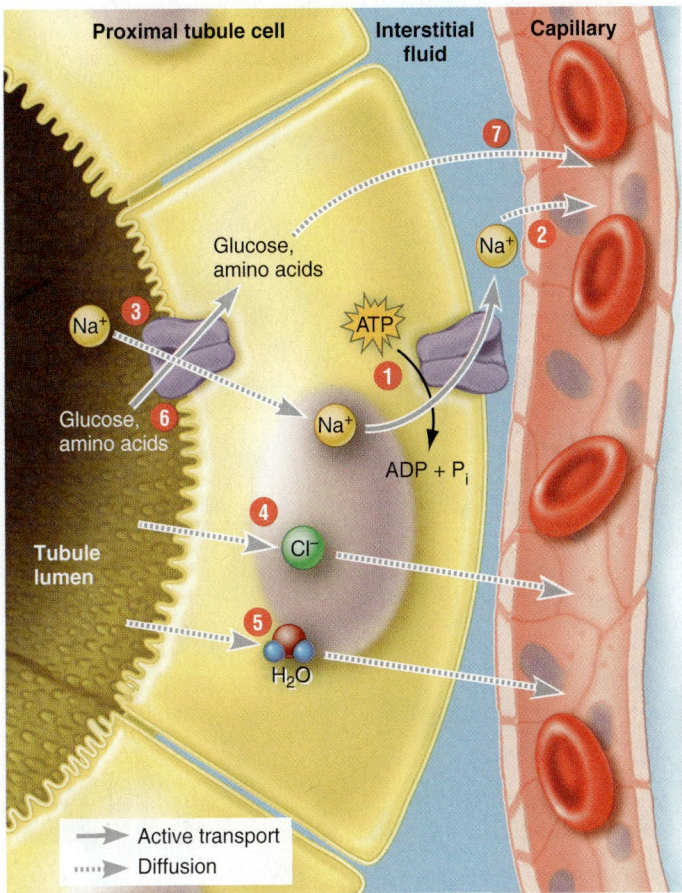

Figure 15.9 Basic mechanisms of reabsorption in the proximal tubule. The key to the entire process is the active transport of sodium (Na^+) across the tubular cell membrane on the capillary side of the cell. This step requires energy in the form of ATP. The active transport of sodium keeps the intracellular sodium concentration low, which permits sodium to diffuse into the cell from the luminal side. Negatively charged chloride ions follow in order to maintain electrical neutrality, and water follows to maintain osmotic equilibrium. Finally, the diffusion of sodium across the luminal membrane is used as an energy source for the reabsorption of glucose and amino acids.

 **MJ's BlogInFocus** How might the study of metabolites (products of metabolism) excreted in the urine assist in the diagnosis of medical diseases and conditions? Visit MJ's blog in the Study Area in MasteringBiology and look under "Metabolomics."

http://goo.gl/PNGK5s

Some substances that occur naturally in the body are controlled by tubular secretion, just to maintain normal homeostatic levels. The proximal tubule secretes hydrogen (H^+) and ammonium (NH_4^-) to regulate the body's acid-base balance, and the distal tubule secretes potassium (K^+) to maintain healthy levels of that mineral.

> ↩ **Recap** In the glomerulus, glomerular filtration separates some of the plasma fluid and small solutes from larger proteins and blood cells. Tubular reabsorption returns nearly all the filtered water and sodium and all the major nutrients back to the peritubular capillaries or vasa recta. Tubular secretion removes toxic, foreign, and excess substances from the peritubular capillaries or vasa recta. Tubular secretion is essential to the regulation of acid-base balance, potassium balance, and the excretion of certain wastes. ■

15.5 Producing diluted or concentrated urine

Your kidneys are capable of producing urine that is either more diluted or more concentrated than plasma. In other words, the kidneys can conserve water when it is in short supply and get rid of it when there is too much. The ability to do this depends on a high concentration of solutes in the renal medulla, coupled with the ability to alter the collecting ducts' permeability to water as needed. This is done by varying the concentration of the hormone **antidiuretic hormone (ADH)** from the posterior pituitary gland (see Chapter 13). ADH increases the permeability of the collecting duct to water. How this controls the volume and concentration of urine is described next.

Producing dilute urine: excreting excess water

If you drink a large glass of water quickly, the water is absorbed by your digestive system and enters your blood, increasing your blood volume and decreasing the concentration of ions in blood and body fluids. Most of the water enters your cells. To prevent osmotic swelling and damage to cells, the kidneys adjust the process of urine formation so that you reabsorb less water and produce dilute urine. Figure 15.10a illustrates this process.

The process of forming dilute urine begins at the descending limb of the loop of Henle. ❶ The descending limb is highly permeable to water. Because the solute

concentration is higher in the interstitial fluid than in the lumen, water diffuses out of the loop, into the interstitial fluid, and eventually into the blood. As a result, the remaining fluid becomes more concentrated.

❷ When fluid turns the hairpin corner of the loop, however, the permeability characteristics of the tubule change. The first part of the ascending limb is permeable to NaCl and urea but not to water. Because so much water was removed in the descending limb, sodium concentration in the ascending limb is now higher than in the interstitial fluid of the medulla. With the change in permeability, NaCl can now diffuse out of the ascending limb and into the interstitial fluid of the medulla.

Meanwhile, ❸ urea diffuses into the ascending limb of the loop from the collecting duct. The concentration of urea in the deepest (innermost) portion of the medulla is actually quite high because urea is being reabsorbed from the collecting duct. In other words, urea recycles—from

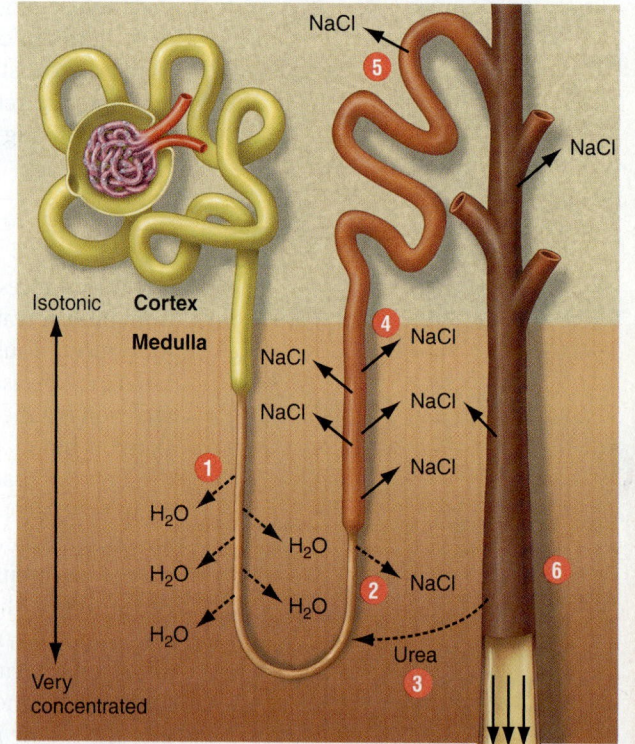

a) **Dilute urine.** In the absence of antidiuretic hormone (ADH), the distal tubule is impermeable to water. The tubular fluid has been diluted by the active reabsorption of salt (NaCl), and this fluid passes through the medulla in the collecting duct without being reabsorbed.

b) **Concentrated urine.** In the presence of ADH, the collecting duct is permeable to water. Consequently, most of the water in the collecting duct is reabsorbed by passive diffusion.

Figure 15.10 The formation of diluted or concentrated urine. For simplicity, in (b) common reabsorptive events that occur prior to the collecting duct are not shown.

☑ When a kidney is producing a lot of urine (Figure 15.10a), is it expending more energy (working harder) than when urine volume is low (Figure 15.10b)? Explain.

collecting duct, to ascending limb of the loop of Henle, and back to the collecting duct again. Because of this recycling pattern, urea concentrations can become much higher in medullary interstitial fluid than in blood plasma.

4 The final portion of the ascending limb of the loop of Henle becomes impermeable to salt and urea as well as water, so no passive diffusion occurs. However, this region of the ascending limb actively transports sodium chloride (salt) into the interstitial fluid by a process that requires energy in the form of ATP. Consequently, the interstitial fluid gains solutes while the fluid remaining in the tubule becomes more dilute.

5 Active salt reabsorption without the reabsorption of water continues in the distal tubule and collecting duct.

6 By the time tubular fluid reaches the end of the collecting duct, it is a very dilute solution of a lot of water, a little urea, and very little salt. You may need to urinate frequently when you are forming a very dilute urine because your kidneys are producing over a liter of urine per hour.

☑️ If the loop of Henle were somehow removed—that is, if the proximal tubule were connected directly to the distal tubule—would the kidney still be able to produce dilute urine?

Producing concentrated urine: conserving water

Sometimes our problem is too little water, perhaps due to perspiration or living in an arid climate. Less water in the blood means lower blood volume, declining blood pressure, and a risk of dehydration for body cells. Again your kidneys compensate, this time by reabsorbing more water. The result is a more concentrated urine.

We urinate less when we are dehydrated because the kidneys are reabsorbing more of the glomerular filtrate water than usual (**Figure 15.10b**). The formation of concentrated urine is similar to the formation of dilute urine until the tubular fluid reaches the collecting duct. In the presence of ADH, the collecting duct is highly permeable to water, and so water is reabsorbed by passive diffusion as the collecting duct passes through the inner medulla. Consequently, only a small volume of highly concentrated urine is excreted.

🔄 **Recap** Production of dilute urine requires the reabsorption of salt without the concurrent reabsorption of water in the ascending limb of the loop of Henle, the distal tubule, and the collecting duct. The formation of concentrated urine requires ADH. In the presence of ADH, most of the water is reabsorbed from the collecting duct, leaving a small volume of concentrated urine to be excreted. ■

15.6 Urination depends on a reflex

Urination depends on a neural reflex called the *micturition reflex*. This involves the two urethral sphincters and the bladder. The internal urethral sphincter, which consists of smooth muscle, is an extension of the smooth muscle of the bladder. It remains contracted unless the bladder is emptying. The external urethral sphincter is skeletal muscle and therefore under voluntary control.

Normally, the external urethral sphincter is kept closed by tonic activity of somatic motor neurons controlled by the brain. As the bladder fills with a cup (roughly 250 ml) or more of urine, it starts to stretch. Stretching stimulates sensory nerves, which sends signals to the spinal cord. Nerves in the spinal cord initiate an involuntary (parasympathetic autonomic) reflex that contracts the smooth muscle of the bladder and relaxes the internal urethral sphincter. Stretch receptor input also goes to the brain, which decreases the activity of somatic motor neurons to the external urethral sphincter, allowing it to relax so that urine can flow through.

Humans don't usually have the opportunity to urinate whenever the urge strikes. Fortunately, the brain can voluntarily override the micturition reflex by increasing the activity of the somatic nerves that control the external urethral sphincter. This delays urination, for extended periods if necessary. When we finally decide to urinate, the somatic nerve activity is shut off, the external urethral sphincter relaxes, and urination occurs. Voluntary prevention of urination becomes increasingly difficult as the bladder approaches maximum capacity.

🔄 **Recap** Urination depends on the neural micturition reflex; bladder stretching initiates involuntary relaxation of the internal urethral sphincter. The brain can override the reflex by voluntary contraction of the external urethral sphincter. ■

15.7 The kidneys contribute to homeostasis in many ways

The kidneys are much more than simply the organs that make urine. As mentioned earlier, they play other important roles in the body as well. The kidneys:

- contribute to the maintenance of water balance
- contribute to the maintenance of salt balance
- secrete an enzyme involved in the control of blood volume and blood pressure
- maintain acid-base balance and blood pH
- control production of red blood cells
- activate an inactive form of vitamin D

The kidneys are responsible for maintaining homeostasis (by excretion) of virtually every type of ion in the body, considering that ions cannot be created or destroyed by metabolic processes. Because the kidneys secrete hormones, they are also endocrine organs. And by chemically altering the vitamin D molecule so that it can function as a vitamin, they play a role in metabolism as well. The common theme in these activities is the maintenance of homeostasis.

ADH regulates water balance

As mentioned earlier, blood volume is determined in part by the amount of water in the body. Changes in blood volume are likely to cause parallel changes in blood pressure, so it is important that blood volume be maintained within a fairly narrow range. But it is also important that blood solute concentration be kept constant as well. If blood solute concentration can be kept constant under a wide variety of circumstances, then the control of blood volume can be determined by the regulation of salt balance, as we'll see below.

The kidneys, the hypothalamus of the brain, and the posterior pituitary gland of the endocrine system share the function of maintaining water balance. This balance is achieved by a negative feedback loop involving the hormone ADH (**Figure 15.11**).

When the concentration of blood solutes rises (too little water compared to the amount of solutes in the blood), the ADH-producing neurons in the hypothalamus cause ADH to be secreted from the posterior pituitary into the blood. ADH circulates to the kidney, where it increases the permeability of the collecting duct to water. This allows more water to be reabsorbed, reducing the amount of water excreted in the urine.

Note that increased reabsorption of water by the kidneys does not result in a net gain of water or a fall in blood solute

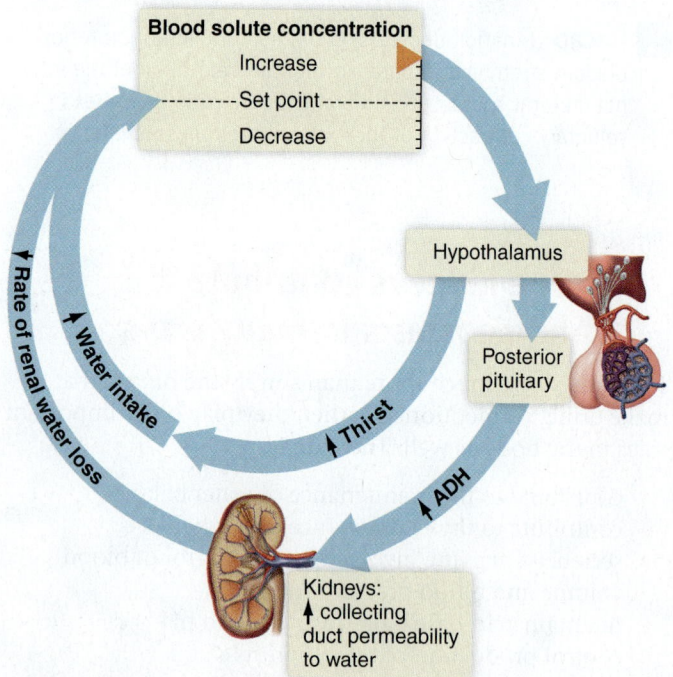

Figure 15.11 Negative feedback loop for the control of blood solute concentration. A rise in blood solute concentration triggers thirst and increases ADH secretion. ADH causes the kidneys to form a more concentrated urine and decrease their excretion of water. An increase in water intake coupled with a reduction in water loss from the kidneys dilutes the blood and returns blood solute concentration toward normal. Although only the responses to an increased solute concentration are shown here, the control loop responds equally well (with opposite responses) to a fall in blood solute concentration.

concentration. That's because all of the reabsorbed water was already in the blood before it was filtered in the glomerulus. However, in addition to stimulating ADH secretion, a rise in the blood solute concentration triggers a sensation of thirst, and so we drink more water. It's the increased water intake that eventually lowers the blood solute concentration toward normal again.

Normally, some ADH is present in the blood. This allows the control mechanism to respond to excess water, too. When there is too much water and blood solute levels fall, the hypothalamus signals the posterior pituitary to reduce its secretion of ADH. The fall in ADH concentration makes the renal collecting duct less permeable to water, so the excess water is excreted. We also feel less thirsty when we already have more than enough water.

The term *diuresis* refers to a high urine flow rate. A *diuretic* is any substance that increases the formation and excretion of urine. Diuretic drugs, such as furosemide (Lasix), are prescribed to reduce blood volume and blood pressure in certain patients, such as people with congestive heart failure or hypertension.

Caffeine is a mild diuretic because it inhibits sodium reabsorption, and as sodium is excreted it takes water with it. Alcohol is also a diuretic, although by a different mechanism: It inhibits ADH release. The result is that the renal tubules become less permeable to water, and more water is excreted. If you drink excessive amounts of alcohol, you may feel dehydrated and thirsty the next day.

Aldosterone regulates salt balance

The control of blood volume depends critically on maintaining the body's salt balance. Sodium (salt) excretion by the kidneys is regulated by **aldosterone,** a steroid hormone from the adrenal gland. Aldosterone causes more sodium to be reabsorbed from the distal tubule and collecting duct, back into the blood. High concentrations of aldosterone cause nearly all of the filtered sodium to be reabsorbed, so that less than 50 milligrams (0.05 gram) per day is excreted in the urine. Low levels of the hormone allow as much as 20–25 grams of sodium to be excreted each day. When we consider that the average North American consumes about 10 grams of sodium daily, we see that aldosterone provides more than enough control over sodium excretion.

Given what aldosterone does, we might think that its concentration would be controlled by blood sodium levels. This is partly true, because blood sodium concentration is a weak stimulus for aldosterone secretion. However, the real controller of aldosterone secretion is the *renin-angiotensin system.*

✅ You've just eaten a large pepperoni-and-cheese pizza that has an enormous amount of sodium. (You hardly drank anything, so your blood pressure does not change detectably.) Will this cause your ADH and aldosterone levels to increase or decrease? What will your kidneys do in response?

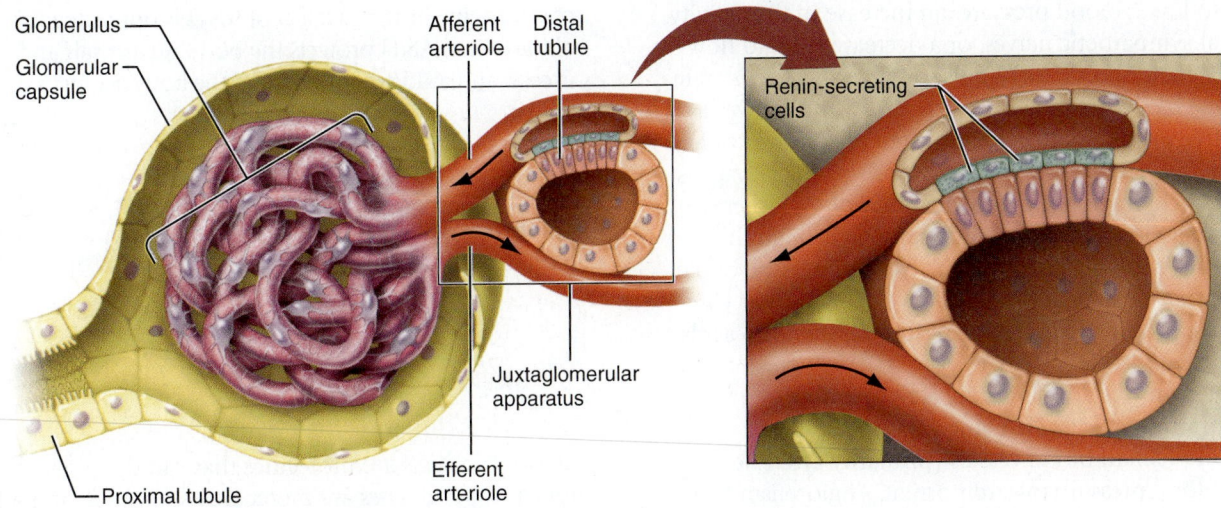

Figure 15.12 The juxtaglomerular apparatus. The juxtaglomerular apparatus is a region of contact between the afferent arteriole, the efferent arteriole, and the distal tubule. The afferent arteriolar cells of the juxtaglomerular apparatus synthesize and secrete renin.

The renin-angiotensin system controls blood volume and blood pressure

The **renin-angiotensin system** does not fit the classical definition of a hormone, because there is no one gland that releases a hormone into the blood. Rather, the production of a biologically active peptide (angiotensin II) that functions like a hormone requires three different organs (kidneys, liver, and lungs). The renin-angiotensin system is the primary hormonal controller of blood pressure and blood volume, both directly (via vasoconstriction) and indirectly (via the secretion of aldosterone).

The controlling component of the renin-angiotensin system is an enzyme called **renin.** Renin is synthesized and stored in specialized cells of the afferent arteriole in a region near the glomerulus called the *juxtaglomerular apparatus* (**Figure 15.12**), the region where the afferent and efferent arterioles are in close contact with the distal tubule. When blood pressure falls in the afferent arteriole or when the flow rate of fluid in the tubule in the region of the juxtaglomerular apparatus declines, renin is secreted into the blood of the afferent arteriole. Stimulation of the sympathetic nerves to the kidney also increases renin secretion.

How the secretion of renin ultimately leads to a return of a low blood volume and blood pressure to normal is illustrated in **Figure 15.13**. A fall in blood volume

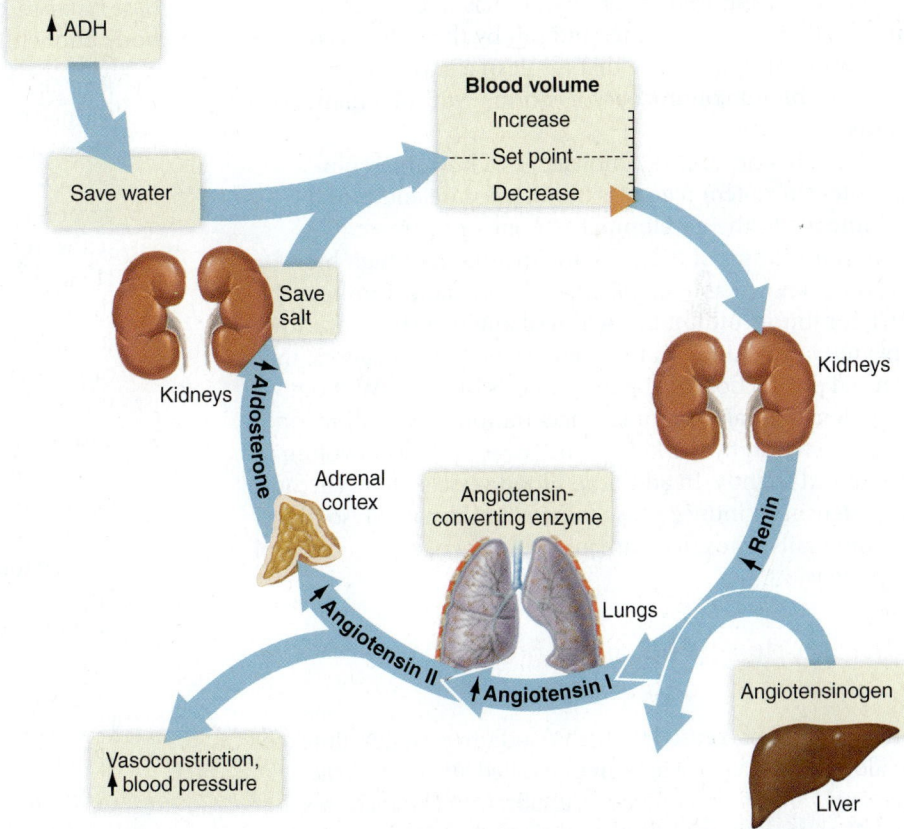

Figure 15.13 Regulation of blood volume by renin, aldosterone, and ADH. A decline in blood volume causes the kidneys to secrete renin. Renin activates a cascade of events that ultimately causes the kidneys to save salt and water, returning blood volume to normal.

✔ What makes the ADH increase? Does it increase before or after the other events shown in the diagram?

sufficient to lower blood pressure, an increase in the activity of the renal sympathetic nerves, or a decrease in fluid flow rate in the renal tubules are signals to the kidneys to secrete renin. In the bloodstream, renin attaches to and cleaves a large inactive protein molecule produced by the liver called *angiotensinogen*. The product of angiotensinogen cleavage is a 10-amino-acid fragment called *angiotensin I*. Angiotensin I is biologically inactive, but when it passes through the lungs another enzyme found primarily in lung tissue called *angiotensin-converting enzyme* (ACE) chops off another two amino acids, producing a peptide of just eight amino acids called *angiotensin II*.

Angiotensin II is a biologically active peptide with several important effects. First, angiotensin II causes constriction of small blood vessels (primarily arterioles), returning blood pressure toward normal. Angiotensin II–induced vasoconstriction can occur within minutes of the secretion of renin.

In addition, Angiotensin II is a powerful stimulus for the release of aldosterone from the adrenal glands. From this point on, you already know the actions of the other hormones involved; aldosterone causes the kidneys to save more salt, which raises blood solute concentration and triggers ADH-induced water retention as well. Overall, the increased retention of water and salt by the kidneys coupled with water and salt in the diet result in a slow but sure return of blood volume back to normal within a matter of hours.

On a historical note, knowledge of how the renin-angiotensin system regulates blood pressure and blood volume led to the development of one of the most important classes of drugs for the treatment of high blood pressure, known as *ACE inhibitors*. As the name implies, ACE inhibitors inhibit the action of angiotensin-converting enzyme in the lungs. In the presence of ACE inhibitors, the normal production of angiotensin II is blocked. Without angiotensin II, aldosterone concentration falls, sodium and water excretion by the kidneys increase, and blood volume is reduced slightly. In addition, blood vessels dilate (because angiotensin II–induced vasoconstriction declines), so blood pressure falls. The effects are just what are needed to control hypertension.

Atrial natriuretic hormone protects against blood volume excess

There is another controller of renal sodium excretion unrelated to aldosterone; a peptide hormone called *atrial natriuretic hormone (ANH)*. (A *natriuretic* is a substance that increases the excretion of sodium.) When the atria of the heart are stretched by an excessively high blood volume, they secrete ANH into the bloodstream. ANH inhibits sodium reabsorption in the distal tubules and collecting ducts of the kidneys, which leads to increased sodium excretion. Once again, ADH causes water to follow salt in order to keep the blood solute concentration

constant. In short, the effect of ANH is opposite to that of aldosterone; ANH protects the body against salt and water excess, and aldosterone protects the body against salt and water deficit.

☑ A man loses a large amount of blood in an accident. How will this affect the secretion of renin, and what will happen in response?

Kidneys help maintain acid-base balance and blood pH

Recall that *acids* are molecules that can donate hydrogen ions (H^+) and *bases* are molecules that can accept a hydrogen ion. Many of the body's metabolic reactions generate H^+. If not eliminated, these acids would accumulate—a dangerous situation, because blood pH must remain in a relatively narrow range of 7.35–7.45. Any change in blood pH by more than a few tenths of a unit beyond this range could be fatal.

The kidneys participate in the maintenance of acid-base balance, a role they share with various buffers in the body and with the lungs. The kidneys do this in two ways:

- *Reabsorption of filtered bicarbonate.* (**Figure 15.14a**). Kidney tubular cells secrete H^+ into the tubular lumen. Most of the secreted H^+ combines with bicarbonate (HCO_3^-) filtered by the glomerulus, creating water and carbon dioxide (CO_2). The CO_2 and water are then reabsorbed back into the cell, where they dissociate into H^+ and HCO_3^- again. The H^+ is secreted once again, but the HCO_3^- diffuses into the blood capillary. In effect, the secretion of H^+ is a mechanism for reabsorbing, or recovering, nearly all of the HCO_3^- that is filtered. When nearly all of the filtered HCO_3^- is recovered, any additional H^+ that is secreted is buffered by negatively charged ions in the urine (phosphate and sulfate) and eventually excreted in the urine.
- *Renal excretion of acid as ammonium.* (**Figure 15.14b**). Kidney tubule cells produce ammonium (NH_4^+) and HCO_3^- as part of their normal metabolism of the amino acid glutamine within the cell. The NH_4^+ is transported into the tubule lumen and then excreted in the urine. The new HCO_3^- diffuses into the blood to replace a HCO_3^- lost from the lung during respiratory control of CO_2 concentration.

Ordinarily, our diets provide us with a small amount of excess H^+ that must be excreted if we are to stay in balance. Most of our daily production of H^+ results from the creation of CO_2 during metabolism, which is handled by the lungs. The renal excretion of H^+ with sulfate, phosphate, and ammonia exactly offsets the excess acid intake per day.

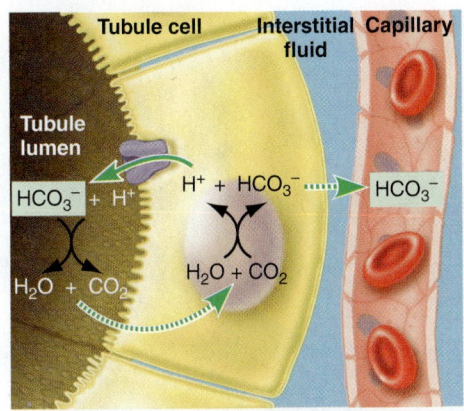

a) Reabsorption of filtered bicarbonate.
Hydrogen ions (H⁺) within the cell are secreted into the lumen, where they combine with filtered bicarbonate (HCO₃⁻) to form CO₂ and water. The CO₂ and H₂O diffuse into the cell and dissociate back into a H⁺ and HCO₃⁻. The H⁺ is secreted again, but the HCO₃⁻ diffuses from the cell back into the blood.

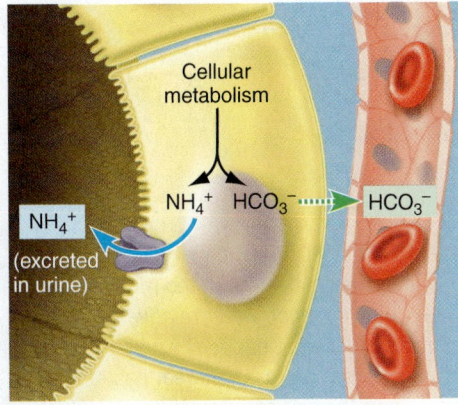

b) Excretion of ammonium. Kidney cells produce ammonium (NH₄⁺) and HCO₃⁻ during cellular metabolism of glutamine. The NH₄⁺ is secreted into the tubule and is excreted in the urine, but the new HCO₃⁻ diffuses into the blood.

Figure 15.14 Renal maintenance of acid-base balance.

☑ If you climb a very high mountain, the low air pressure of high altitude will probably cause you to hyperventilate slightly. This causes a condition called *respiratory alkalosis*, in which blood pH becomes too high. Should the kidney excrete more or fewer H⁺ ions in this situation? Explain.

Erythropoietin stimulates production of red blood cells

A very important function of the kidneys is to regulate the production of red blood cells in the bone marrow. They do this by secreting a hormone, **erythropoietin** (review Figure 7.6).

When certain oxygen-sensitive cells scattered throughout the kidneys sense a fall in the amount of oxygen available to them, they increase their secretion of erythropoietin, which then stimulates the bone marrow to produce more red blood cells. The presence of more red blood cells increases the oxygen-carrying capacity of the blood. Once oxygen is again being delivered normally, the oxygen-sensitive cells lower their rate of erythropoietin secretion, and RBC production slows. The regulation of erythropoietin secretion by oxygen-sensitive cells in the kidneys is part of the negative feedback loop for controlling oxygen availability to *all* cells, because it keeps the oxygen-carrying capacity of the blood within normal limits.

Kidneys activate vitamin D

Vitamin D is important for absorbing calcium and phosphate from the digestive tract and developing healthy bones and teeth. We absorb some vitamin D

from food, but the body also manufactures it in a three-step process involving the skin, liver, and kidneys.

Vitamin D synthesis begins when ultraviolet rays in sunlight strike a steroid molecule in skin that is similar to cholesterol, producing an inactive form of vitamin D. The inactive vitamin D is transported to the liver, where it is chemically altered, then carried to the kidneys, where it is converted to its active form under the influence of parathyroid hormone (PTH) from the parathyroid gland. During kidney failure, this final step may not occur. This is why people with long-standing kidney failure often suffer from vitamin D deficiency.

↩ **Recap** ADH controls blood total solute concentration, aldosterone controls blood sodium concentration, and the combination of renin-angiotensin, aldosterone, and ANF controls blood volume. The kidneys help maintain the body's acid-base balance and blood pH by reabsorbing all filtered bicarbonate and by excreting H⁺. Decreased oxygen delivery to the kidneys triggers the release of erythropoietin, which stimulates the production of red blood cells by bone marrow. The kidneys activate an inactive form of vitamin D. ▪

15.8 Disorders of the urinary system

The good news about urinary system disorders is that the kidneys have an enormous reserve of function. You can survive quite nicely on one kidney, and even on one kidney that is functioning only half as well as it should. Nevertheless, sometimes urinary system problems do occur. They include kidney stones, urinary tract infections (UTIs), and renal failure.

Kidney stones can block urine flow

Sometimes, minerals in urine crystallize in the renal pelvis and form solid masses called *kidney stones*. Most stones are less than 5 millimeters (1/5 inch) in diameter and are excreted in urine with no trouble. Others may grow larger and obstruct a ureter, blocking urine flow and causing great pain. Kidney stones can be removed surgically or crushed with ultrasonic shock waves, after which the pulverized fragments can be excreted with less pain.

HEALTH & WELLNESS

Water Intoxication

The kidneys are responsible for maintaining water balance, but sometimes they can't handle the job quickly enough. *Water intoxication* is a serious and potentially fatal condition of water imbalance. The most common cause of water intoxication is drinking a lot of water during or immediately after heavy exercise (even if it seems like the right amount) in order to stay hydrated or to rehydrate quickly. Other causes include water-drinking contests and psychiatric conditions in which a person deliberately drinks too much water.

During water intoxication, a low blood solute concentration favors the diffusion of water into cells, leading to cell swelling. The most serious consequences of cell swelling are in the brain, because the brain is encased in bone and has no room for expansion. It's not unlike what happens during a concussion, when brain injury causes swelling. Symptoms of water intoxication may include confusion, slurred speech, nausea and vomiting, and muscle cramps. Severe cases can lead to seizures, coma, and death. A blood sample would reveal the presence of hyponatremia (a low blood sodium concentration). For that reason, water intoxication is also called *dilutional hyponatremia*.

Water intoxication can occur even when a person appears to be in water balance. That's because the real problem in water intoxication is not too much water, but *too much water relative to salt*. Consider what happens when an athlete sweats profusely. Sweat has about half the salt concentration of blood. If an athlete loses four liters of sweat and replaces it by drinking four liters of water, he/she has actually lost the equivalent of all the salt in two liters of body fluid, potentially leading to a dangerously low sodium concentration in the blood. Dilutional hyponatremia is made worse if the athlete drinks more water than was actually lost.

Not surprisingly, marathon runners are particularly susceptible to water intoxication. A study published in the *New England Journal of Medicine*[1] found that 13% of the runners who finished the 2002 Boston marathon had symptoms of water intoxication, including hyponatremia. Even runners who drank sports drinks (containing salts) during the race were susceptible to hyponatremia. The one variable that correlated best with hyponatremia at the end of the race was a change in body weight; runners who gained weight during the race (that is, drank too much water or other fluids) were at greatest risk for hyponatremia.

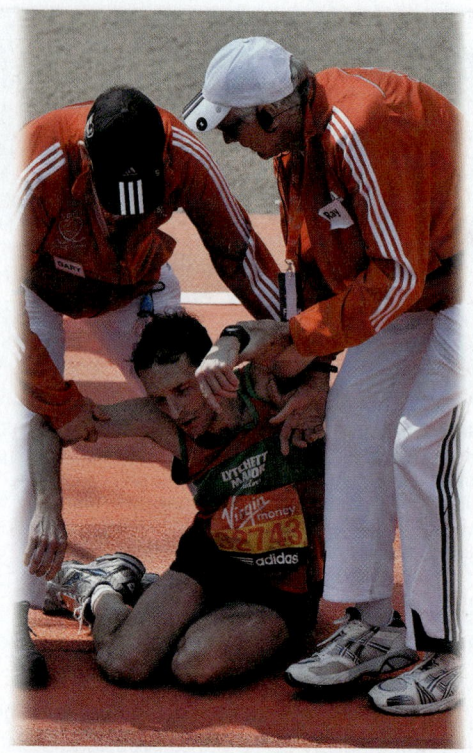

Water intoxication, also called dilutional hyponatremia, is caused by too much water relative to salt.

In mild cases of water intoxication, the only recommended treatments are rest and fluid restriction; given time, the kidneys will excrete the excess water. When symptoms include seizures or coma, diuretics (drugs that increase urine flow) may be necessary to remove the excess water quickly.

[1]Almond, C.S.D. et al. *New England Journal of Medicine* 352:1550–1556, Apr. 14, 2005.

Urinary tract infections are often caused by bacteria

A *urinary tract infection (UTI)* refers to the presence of microbes in urine or an infection in any part of the urinary system. Symptoms include swelling and redness around the urethral opening, a burning sensation or pain while urinating, difficulty urinating, bed-wetting, low back pain, and sometimes visible blood and pus in urine.

Most UTIs are caused by bacteria that make their way up the urethra from the body surface. Although UTIs can occur in men, they are more common in women because women's urethras are shorter, allowing organisms to reach the bladder more easily. From the urethra and bladder, microbes can travel up the ureters to the kidneys. This is why it is important to treat all UTIs immediately. Most cases can be cured with antibiotics.

Acute and chronic renal failure impair kidney function

Kidney impairments that are short term and possibly correctable are called *acute renal failure*. Conditions that might impair kidney function temporarily include sustained decreases in blood pressure to below the pressure required for kidney filtration, large kidney stones in the renal pelvis, infections, transfusion reactions, burns, severe injuries, and toxic drugs or chemicals.

Acute renal failure is fairly common after severe crush injuries or when blood flow to one or more limbs is interrupted for too long. The reason is that as muscle tissue dies it releases large amounts of protein into the blood. These proteins can essentially plug up the filtration barrier in the glomeruli, resulting in a loss of filtration and thus a complete or nearly complete cessation of urine flow.

Many common medications are toxic to the kidneys if they are taken for too long or at high doses. The kidneys are particularly vulnerable because of their reabsorptive efficiency, which tends to concentrate harmful substances in the urine. The toxicity problem may be prevented by drinking lots of water and excreting a more dilute urine.

If the renal tubular cells do not receive the nutrients they need or if glomerular filtration is blocked for too long, the tubules may become plugged with dead tubular cells and cellular debris. When that happens, the inability to produce urine may become permanent. *Chronic renal failure*, also known as *end-stage renal disease (ESRD)*, is defined as long-term irreversible damage leading to at least a 60% reduction in functioning nephrons and failure of the kidneys to function properly. People with ESRD may have less than 10% of the normal filtering capacity of the kidneys, and some essentially have no renal function at all.

MJ's **BlogInFocus** What's the connection between a rare, genetic kidney disease and a parasitic disease called African sleeping sickness? Visit MJ's blog in the Study Area in MasteringBiology and look under "Kidney Disease and African Sleeping Sickness."

http://goo.gl/GjXbTM

One of the most common causes of ESRD is diabetes. About 20–30% of people with diabetes eventually develop a mild form of renal impairment called *diabetic nephropathy*. The condition is first diagnosed as a modest but progressive loss of renal filtration capacity, along with the presence of protein in the urine; both are signs of glomerular damage. Diabetic nephropathy frequently progresses to ESRD, especially when blood sugar is not adequately controlled. In the United States, diabetic nephropathy now accounts for about 40% of all new cases of ESRD each year.

☑ A young man is in a hospital after an auto accident. He is still not producing any urine after three days. The physicians were not too concerned on the first day, but now they seem very concerned. Why the change in their attitude?

Dialysis cleanses the blood artificially

The body cannot replace irreparably damaged nephrons. This leaves two options for people with severe renal failure: dialysis or a kidney transplant.

Dialysis attempts to duplicate the functions of healthy kidneys. A dialysis technique that can be done at home without a kidney machine is called *continuous ambulatory peritoneal dialysis (CAPD)*. In this procedure, dialysis fluid (a fluid similar to extracellular fluid) is placed directly into the peritoneal cavity through an access port permanently implanted in the abdominal wall. The fluid is left in the peritoneal cavity for several hours, during which time it exchanges wastes and ions with the capillaries. Then the fluid is drained and discarded. If done on a regular basis, CAPD can be modestly effective. It is convenient and allows freedom of movement. However, there is a risk of infection because of the access port through the abdominal wall.

In another type of dialysis, called *hemodialysis*, the patient's blood is circulated through an artificial kidney machine consisting of semipermeable membranes in contact with a large volume of clean fluid (**Figure 15.15**). Metabolic wastes and excess ions diffuse from the blood into the dialysis fluid, which is discarded. The procedure is costly (over $50,000 per year) and time consuming. Typically, the patient must go to a medical site called a *dialysis unit* for treatment several times each week, with each treatment taking a whole morning or afternoon.

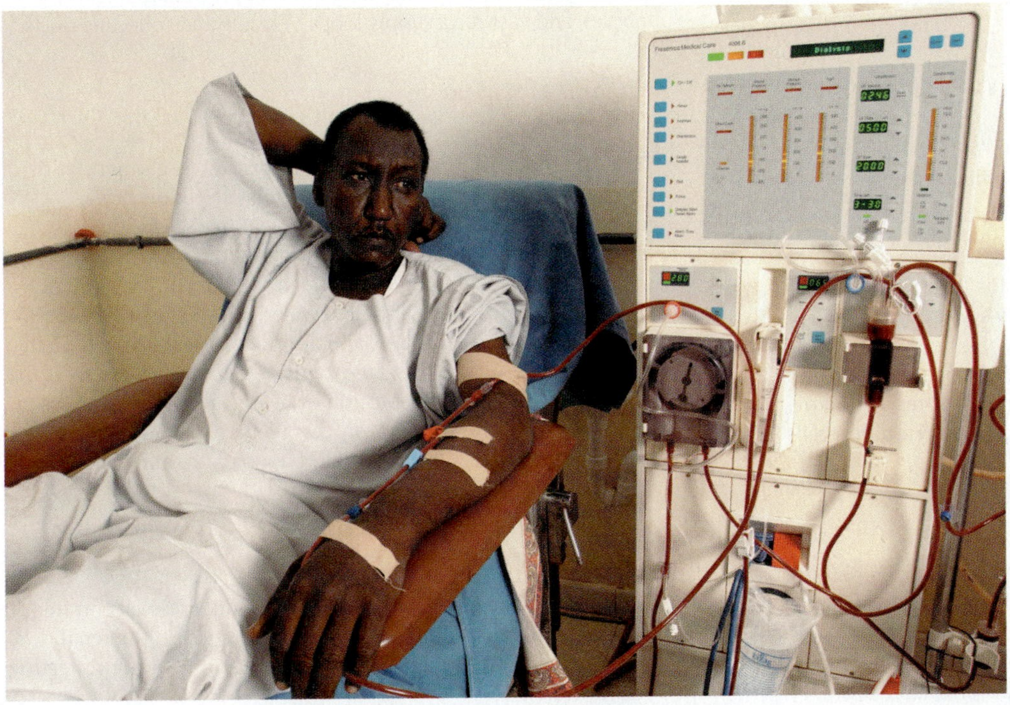

Kidney transplants are a permanent solution to renal failure

Many people currently on renal dialysis will remain on dialysis for the rest of their lives. Although dialysis is a life-saving technique, it is not the perfect solution. It is difficult, if not impossible, to achieve complete homeostasis of all ions and wastes by artificial means. Furthermore, dialysis does not replace the renal hormones.

Figure 15.15 A patient undergoing hemodialysis.

The best hope for chronic renal-failure patients is to receive a donated kidney, either from a living person or from a cadaver. When kidney transplants began in the 1960s, most donated kidneys came from living close relatives, because the biggest challenge was to find a good immunological match so that the recipient's body would not reject the foreign kidney. With the advent of better drugs to suppress the immune system, improved tissue-matching techniques, national data banks of patients, and rapid jet transport, it is now feasible for the donor kidney to come from a cadaver. The main bottleneck now is that not enough people have offered to donate their organs after death. Thousands of people currently on waiting lists will die before they can receive a kidney.

MJ's BlogInFocus What are some of the arguments against paying donors for kidneys, and is there any evidence for or against those arguments? Visit MJ's blog in the Study Area in MasteringBiology and look under "Paying Donors for Kidneys."

http://goo.gl/dUfH1J

Recap Acute or chronic renal failure can result from prolonged changes in blood pressure, disease, large kidney stones, transfusion reactions, burns, injuries, toxic substances, and other conditions such as diabetes. When kidneys fail, dialysis may be needed to keep the patient alive. Kidney transplants are the best hope for people in renal failure. Currently, there is a severe shortage of kidneys available for transplant. ■

Urinary incontinence is a loss of bladder control

Urinary incontinence (a loss of bladder control) is fairly common in older persons. It can be inconvenient and embarrassing, but it is certainly not life-threatening. Persistent urinary incontinence develops with age because with aging the bladder muscles do not stretch as well, leading to a decrease in bladder capacity and symptoms of an overactive bladder—the strong and sudden need to urinate due to contractions or spasms of the bladder (also called "urge incontinence"). Other gender-neutral causes include neurological disorders that interfere with neural control of urination and obstructions such as tumors or stones.

Women are more likely to develop urinary incontinence than men. During pregnancy, hormonal changes and the increased weight of the enlarging uterus can cause temporary incontinence. Pregnancy and childbirth may also predispose women to incontinence later in life because vaginal delivery may damage nerves or muscles involved in bladder control. Finally, a decline in estrogen in women after menopause leads to deterioration of the lining of the bladder. In men, urinary incontinence may be secondary to an enlarged prostate or prostate cancer.

Treatments for incontinence vary. Bladder training, management of fluid intake, and certain exercises may help. Anticholinergic drugs may help calm an overactive bladder. For women, a urethral insert may help. Several surgical techniques are also possible depending on the nature of the problem.

Chapter Summary

15.1 The urinary system regulates body fluids p. 353

- The kidneys, the lungs, the liver, and the skin all participate in the maintenance of homeostasis.
- The kidneys are the primary regulators of water balance and most excess solutes, especially inorganic ions and urea.

15.2 Organs of the urinary system p. 354

- The urinary system consists of those organs that produce, transport, store, and excrete urine. The urinary system includes the kidneys, the ureters, the bladder, and the urethra.
- Functions of the kidneys include regulation of the volume and composition of body fluids, excretion of wastes, regulation of blood pressure, regulation of the production of red blood cells, and the activation of vitamin D.

15.3 The internal structure of a kidney p. 356

- The functional unit of the kidneys is the nephron. Each nephron consists of a tubular component and the blood vessels that supply it.
- The tubular components of a nephron are the glomerular capsule, proximal tubule, loop of Henle, distal tubule, and collecting duct. The collecting duct is shared by many nephrons.

- A tuft of capillaries called the *glomerulus* is enclosed within each glomerular capsule. Peritubular capillaries supply proximal and distal tubules, and vasa recta supply loops of Henle and collecting ducts.

15.4 Formation of urine: filtration, reabsorption, and secretion p. 358

- The formation of urine involves three processes: glomerular filtration, tubular reabsorption, and tubular secretion.
- Approximately 180 liters per day of protein-free plasma fluid is filtered across the glomerulus and into the glomerular space. Filtration is driven by high blood pressure in the glomerular capillaries.
- Ninety-nine percent of all filtered water and salt and all of the filtered bicarbonate, glucose, and amino acids are reabsorbed in tubular reabsorption. The active transport of sodium provides the driving force for the reabsorption of nearly all other substances.
- Tubular secretion is a minor process relative to reabsorption but is critical for the regulation of acid-base balance and for the removal of certain toxic wastes.

15.5 Producing diluted or concentrated urine p. 362

- The ability of the kidneys to form either dilute or concentrated urine depends on the high solute concentration in the renal medulla and the ability to alter the permeability of the collecting duct to water.

- Dilute urine is formed in the absence of the hormone ADH. In the absence of ADH, reabsorption of salt without reabsorption of water continues in the collecting duct.
- Concentrated urine is formed when ADH increases the collecting duct's permeability to water, allowing water to diffuse toward the high solute concentration in the medulla.
- The kidneys regulate the rate of water loss from the body. The control of water gain resides with a thirst mechanism.

15.6 Urination depends on a reflex *p. 363*

- Urination is caused by the micturition reflex, a neural reflex initiated when the bladder is stretched.
- Urination can be prevented by higher (voluntary) neural signals from the brain.

15.7 The kidneys contribute to homeostasis in many ways *p. 363*

- Water balance is maintained by a negative feedback loop involving ADH. The main stimulus for the secretion of ADH is an increase in the solute concentration of the blood.
- Blood volume is regulated by maintenance of the body's salt balance, which is controlled primarily by a negative feedback loop involving the renin-angiotensin system and aldosterone.
- A hormone from the heart called *atrial natriuretic peptide* increases sodium excretion by the kidneys.
- The kidneys secrete (and excrete) H^+ and NH_4^+ in amounts equal to the net gain of acid per day (other than as CO_2). They also secrete H^+ as part of the mechanism for the reabsorption of all filtered HCO_3^-.
- The kidneys synthesize and secrete erythropoietin, the hormone responsible for the regulation of red blood cell production.
- The kidneys are required for the activation of vitamin D.

15.8 Disorders of the urinary system *p. 367*

- The kidneys are vulnerable to damage by toxic substances, infections, sustained decreases in blood pressure, or blockage of urinary outflow.
- Nephrons that are damaged beyond repair are not replaced.
- Hemodialysis is an artificial procedure for cleansing the blood of wastes and excess solutes.
- Renal transplantation is technically easy and usually quite successful, but there is a shortage of donor kidneys.
- Urinary incontinence is a loss of bladder control.

Terms You Should Know

afferent arteriole, 357
aldosterone, 364
antidiuretic hormone (ADH), 362
collecting duct, 357
dialysis, 369
distal tubule, 357
efferent arteriole, 357
glomerular capsule, 356
glomerular filtration, 359
glomerulus, 356

loop of Henle, 356
nephron, 356
peritubular capillaries, 357
proximal tubule, 356
renin, 365
tubular reabsorption, 360
tubular secretion, 361
urea, 354
ureter, 355
urethra, 356
vasa recta, 357

Concept Review

Answers can be found in the Study Area in MasteringBiology.

1. Describe in general terms the kidneys' role in maintaining homeostasis.
2. Describe the function of each of the organs of the urinary system.
3. Give the names and functions of each of the tubular sections of the nephron.
4. Indicate where vasa recta are located, and list the two tubular regions of the nephron they serve.
5. Describe the driving force responsible for filtration of fluid from the glomerular capillaries into the glomerular space.
6. Name the one region of the tubule where water permeability is *regulated*.
7. Indicate the primary stimulus for the secretion of the hormone ADH.
8. Describe the function of the hormone aldosterone.
9. Describe how chronic renal failure differs from acute renal failure.
10. Discuss the treatment options available to someone in chronic renal failure (end-stage renal disease).

Test Yourself

Answers can be found in the Appendix.

1. Which of the following organ systems is/are involved in excretion?
   a. urinary system
   b. integumentary system
   c. respiratory system
   d. all of these choices
2. Three of the choices below are organ systems directly involved in excretion of wastes. Which system is not directly involved in excretion but interconnects the other three?
   a. urinary system
   b. respiratory system
   c. circulatory system
   d. digestive system
3. The _____ detoxifies ammonia, producing _____, which is excreted by the _____.
   a. urinary system . . . urine . . . kidneys
   b. liver . . . urea . . . kidneys
   c. urinary system . . . urea . . . liver
   d. liver . . . carbon dioxide . . . respiratory system
4. The nitrogenous wastes excreted by the kidneys are produced through the metabolism of:
   a. carbohydrates
   b. proteins
   c. triglycerides
   d. cholesterol
5. Which of the following indicates the correct order of the structures through which urine would pass during formation and excretion?
   a. nephron, renal pelvis, ureter, bladder, urethra
   b. renal pelvis, ureter, nephron, urethra, bladder
   c. nephron, renal pelvis, urethra, bladder, ureter
   d. renal pelvis, nephron, ureter, bladder, urethra
6. Which structure is correctly paired with its function?
   a. ureter: formation of urine
   b. nephron: storage of urine
   c. urethra: transport of urine
   d. bladder: concentration of urine

7. Protein is generally not found in the urine because:
   a. It does not pass through the glomerular filter into the space within the glomerular capsule.
   b. It is reabsorbed in the ascending limb of the loop of Henle.
   c. It is reabsorbed within the collecting duct.
   d. It is metabolized by the cells that line the nephron.

8. Blood enters the glomerulus through the _____ and exits the glomerulus through the _____ .
   a. efferent arteriole . . . afferent arteriole
   b. efferent arteriole . . . distal tubule
   c. proximal tubule . . . efferent arteriole
   d. afferent arteriole . . . efferent arteriole

9. Which of the following would not be found in the glomerular filtrate of a healthy individual?
   a. water   c. glucose
   b. hydrogen ions   d. hemoglobin

10. Which of the following is most responsible for the production of the glomerular filtrate?
   a. diffusion
   b. active transport
   c. glomerular capillary blood pressure
   d. facilitated diffusion

11. In tubular absorption, which of the following processes essentially drives all of the others?
   a. active transport of Na^+ from the proximal tubular cell toward a capillary
   b. facilitated transport of Na^+ from the tubular lumen into the proximal tubular cell
   c. diffusion of water from the proximal tubular cell toward a capillary
   d. diffusion of Cl^- from the tubular lumen into the proximal tubular cell

12. All of the following would happen if an individual was perspiring heavily and becoming dehydrated except:
   a. ADH hormone would be secreted.
   b. The kidneys would reabsorb more water.
   c. The collecting ducts would become less permeable to water.
   d. Less urine would be produced.

13. How does ADH decrease the loss of water through the kidneys?
   a. ADH increases water reaborption in the collecting ducts.
   b. ADH decreases the amount of blood flowing through the kidneys.
   c. ADH increases water reabsorption in the proximal tubule.
   d. ADH decreases water excretion in the loop of Henle.

14. ACE inhibitors decrease blood pressure by:
   a. inhibiting the activity of ADH
   b. decreasing the amount of angiotensin II
   c. altering the permeability of the renal tubules to water
   d. inhibiting the activity of ANH

15. All of the following statements about UTIs are true except:
   a. Most UTIs are caused by bacteria.
   b. The shorter urethra predisposes women to more UTIs than men.
   c. UTIs are generally treated with antibiotics.
   d. UTIs usually start in the kidney and spread to the bladder.

Apply What You Know

Answers can be found in the Study Area in MasteringBiology.

1. Do your kidneys have to work harder (expend more metabolic energy) to get rid of the extra salt and water after you binge on salty snacks and liquid beverages? Explain your reasoning.
2. Explain the mechanism for why you feel thirsty after heavy exercise accompanied by sweating.
3. What does the production of a low volume of dark yellow urine indicate?
4. The kidneys filter a large volume of fluid, putting it on a path to ultimate disposal, and then reabsorb nearly all of it back into the body. Only a few waste products are actively secreted into the tubule. Why are the kidneys set up in this way? Wouldn't it be easier and more energetically efficient to selectively secrete only the waste molecules and the exact right amount of excess water and ions directly into an excretory tube or the bladder?
5. Why aren't children already potty trained—able to hold urination and defecation as well as adults can—when they are born?
6. A possible complication of a strep throat infection is post-streptococcal glomerulonephritis. In this disease, antibodies produced against the streptococcus invader also cause inflammation of glomeruli in the kidneys and damage to the glomerular filtration barrier. What might be the symptoms of damage to the glomerular filtration barrier, and what effect would you expect this to have on kidney function?

MJ's BlogInFocus

Answers can be found in the Study Area in MasteringBiology.

1. Based on what you know, do you think it would be legal to solicit for a kidney from a live donor on Craigslist or other similar Web site? Why or why not? To find out what happened in one such case, visit MJ's blog in the Study Area in MasteringBiology and look under "Kidneys on Craigslist?"
   http://goo.gl/a4uq07

MasteringBiology®

Students Go to MasteringBiology for assignments, the eText, and the Study Area with animations, practice tests, and activities.

Professors Go to MasteringBiology for automatically graded tutorials and questions that you can assign to your students, plus Instructor Resources.

Answers to ✔ questions are available in the Appendix.

Reproductive Systems

A scanning electron micrograph (SEM) of a human sperm on the surface of a human egg.

Key Concepts

- **The male reproductive system** can deliver over 100 million sperm at once. **The female reproductive system** usually releases just one egg each month.

- **In females, a cyclic pattern of changes controlled by hormones prepares one egg for release each month.** Another parallel cycle prepares the uterus to receive and nurture the egg if fertilization occurs.

- **In both males and females, sexual excitement leads to arousal and orgasm.** In males, orgasm leads to ejaculation, the expulsion of sperm.

- **Birth control methods and devices include** altering female hormones to disrupt maturation or fertilization of the egg, condoms or diaphragms that prevent sperm and egg from making contact, spermicides that kill sperm, intrauterine devices that prevent implantation of the egg, and surgical sterilization.

- **Sexually transmitted diseases range from mildly irritating to deadly.** Some are difficult or impossible to cure.

Would You Like a Boy or a Girl?

Your parents probably did not know whether you were a boy or a girl until the very moment you were born, unless ultrasound was performed sometime around 18–26 weeks of pregnancy. They certainly had no way of selecting a child for its gender. But with the advent of DNA technology and advanced fertility methods, all that is changing.

Early Gender Detection at Home: Fun but Not Accurate

In 2005, a biotech company called Acu-Gen Biolab, Inc., started offering a gender testing service and kit called Baby Gender Mentor™. According to the company, Baby Gender Mentor could determine the gender of an unborn child with 99.9% accuracy as early as five weeks after fertilization. For $275, the company would analyze a small blood sample from the mother and provide confidential results via e-mail, generally within 48 hours. Baby Gender Mentor was a big news story when it first came out. But within a year, parents who had used the service were suing the company, claiming that it was not nearly as accurate as claimed. The company filed for Chapter 11 bankruptcy in 2009, and the product was taken off the market.

The failure of Baby Gender Mentor™ didn't stop other companies from following in their footsteps. Two current products are "Baby Gender Predictor Test" and "Boy or Girl," both of which use a urine sample. Having learned a hard lesson from the experience with Baby Gender Mentor™, though, neither test claims to be accurate. Baby Gender is marketed as "Way more fun than a quiz and is a great Baby Shower Gift," and Boy or Girl calls its product "a fun pre-birth experience." The bottom line is that early gender detection tests should be considered fun products rather than medical tests. They're simply not accurate enough to be taken seriously.

Gender Selection Is Now Possible

If you would prefer to have a child of a specific gender without resorting to an elective abortion, there's a relatively easy way to at least increase the odds. A procedure called *sperm sorting* is based on the fact that the X (female) chromosome is larger and carries more DNA than the Y (male) chromosome. First, the sperm in a semen sample from the would-be father are labeled with a fluorescent dye that attaches to DNA. Because sperm with the female (X) chromosome have more DNA, they fluoresce more brightly than sperm with the male (Y) chromosome. The sperm are then sorted on the basis of light intensity by a machine called a *flow cytometer*. After sorting, either the "female-enriched" or the "male-enriched" sperm sample is used to artificially inseminate the mother. For couples who want a girl, sperm sorting produces girls 91% of the time; for boys the odds are a bit lower (only 76%), but that's still a lot better than random chance.

Could a couple choose their next child's gender with absolute 100% certainty? Yes, they could, but the techniques involved are expensive. According to the standard procedure at most fertility clinics, eggs and sperm would be collected from the prospective couple and mixed together. Once the early-stage embryos had begun to develop, each one would be tested for the presence or absence of the Y-chromosome

The Nash family. Linda and Jack Nash used genetic selection to choose an embryo that could donate umbilical cord blood to their daughter upon birth.

Questions to Consider

1 If you and your partner decided to have a child together, would you want to know the gender before the baby is born?

2 Would you try to select the gender of your next child by one of the methods described? Explain why or why not.

3 Would you be willing to select an embryo to have a child specifically to cure an ill child born earlier, as the Nashes did?

by a technique called *preimplantation genetic diagnosis (PGD)*. The presence of a Y-chromosome would indicate a male embryo, whereas the absence of a Y-chromosome would indicate a female embryo. A couple wanting only a boy, then, could choose to have only a boy-embryo implanted into the prospective mother.

The American Society of Reproductive Medicine does not endorse PGD for gender selection. Nevertheless, a number of fertility clinics will perform PGD for this purpose. To avoid controversy over the ethics of gender selection, the clinics may describe gender selection as a way to select against certain diseases. (Females don't get prostate cancer, for example, and males are at very low risk for breast cancer.)

Beyond Gender Selection

In addition to gender selection, PGD can be used to test for genetic disorders. It can also be used to select embryos for certain specific genotypes, including being a compatible tissue match for a sibling.

Molly Nash was born with Fanconi's anemia, a rare, incurable disease in which the bone marrow does not produce blood cells normally. Typically, people with this condition die in their early 20s. Fanconi's anemia is a genetic disorder. Although neither of Molly's parents have the disease, they both carry the gene that causes it. In a desperate attempt to save their daughter, Molly's parents turned to a fertility clinic that performed PGD. Eggs were harvested from the mother's ovaries, fertilized, and then tested for the abnormal Fanconi gene, among other genetic disorders. Embryos that lacked the fatal gene were screened further to determine whether they would be a safe tissue donor match for Molly. Only embryos that lacked the fatal gene and that were a good tissue donor match for Molly were implanted into the mother's uterus. After four attempts, Molly's mother →

gave birth to a healthy boy, whom the couple named Adam. Shortly after Adam's birth, blood-forming stem cells harvested from Adam's umbilical cord blood were infused into Molly (see the Current Issue on banking cord blood in Chapter 7). Apparently, enough of these cells became blood-forming stem cells in Molly that she was cured. Today, both children are healthy, and Molly has started school.

When the Nash story first hit the press, it caused a firestorm of controversy. The media labeled Adam the "savior sibling" and predicted dire effects on his future mental health from knowing that he had been brought into being to save his sister. Molly's parents insisted that they wanted more children anyway, so this procedure enabled them to have a healthy child and

also save their daughter's life. However, many people questioned whether it is ethical to have a child for the purpose of being a tissue donor. And of course when multiple eggs are fertilized *in vitro* but only one or two are implanted, there is always the issue of what happens to the embryos not selected for implantation.

There's no doubt that as reproductive technologies improve, our ability to screen for genetic abnormalities will also improve. But beyond just avoiding genetic diseases, what about using PGD to influence physical characteristics such as hair, skin, and eye color, intelligence, even musical ability? If we could pick and choose from a menu of specific traits for our offspring, would we want to?

SUMMARY

- At-home gender test kits are not accurate enough to be taken seriously.
- Sperm sorting can improve the odds of having a child of the desired gender.
- *In vitro* fertilization and preimplantation testing could guarantee having a child of the desired gender.
- As our ability to specify gender and other genetic traits develops, so does the possibility of being able to "customize" children genetically.

Most of the organ systems we have considered so far in this book are not exactly topics of daily interest to most people. How many songs have you heard praising the urinary system or digestive processes? How many movies show close-ups of people just breathing? Certainly some of the subjects in this chapter, such as differences in male and female anatomy and in patterns of sexual arousal, get more than their share of attention. The beauty and mystery of sex, however, are a source of endless fascination to us.

And yet, despite our fascination with sex, our reproductive system is the only organ system that we could almost do without. Our reproductive system doesn't serve to keep us alive for the next minute, or hour, or week, as the other organ systems do. We can choose to use it for its intended purpose or not, and if we choose not to (that is, choose not to have children), we won't die at a younger age. The primary function of our reproductive system, then, is not to contribute to our personal health and continued survival; it's to perpetuate the survival and evolution of humans as a species. In that sense, it's the only unselfish organ system we have. ■

Human reproduction requires both male and female **reproductive systems,** consisting of the tissues and organs that participate in creating a new human being. Each system (male and female) has highly specific structures and functions that allow it to complement the other perfectly. Ultimately, both systems share the same goal: to join a sperm and an egg in an environment (the woman's body) that enables the fertilized egg to become a new member of our species.

16.1 The male reproductive system delivers sperm

Figure 16.1 illustrates the organs of the male reproductive system. The male reproductive system evolved to deliver the male reproductive cells, called **sperm,** to the female. As we will see, the chance of any one sperm fertilizing a female reproductive cell (an **egg**) is extremely low. The reproductive strategy of the male, then, is to deliver huge numbers of sperm at a time to improve the odds that one will reach the egg.

Testes produce sperm

The sites of sperm production in the male are the paired **testes** (singular: testis). Shortly before birth the testes descend into the **scrotum,** an outpouching of skin and smooth muscle. The scrotum regulates the temperature of the developing sperm within the testes, because sperm develop best at temperatures a few degrees cooler than body temperature. When the temperature within the scrotum falls below about 95°F, the smooth muscle of the scrotum contracts, bringing the testes closer to the warmth of the body. When the temperature warms to above 95°F, the smooth muscle relaxes and the scrotum and testes hang away from the body.

Each testis is only about 2 inches long, but it contains over 100 yards of tightly packed **seminiferous tubules,** where the sperm are produced. The seminiferous tubules join to become the **epididymis,** a single coiled duct just outside the testis. The epididymis joins the long **ductus (vas) deferens,** which eventually joins the duct from the seminal vesicle to become the **ejaculatory duct.** The newly

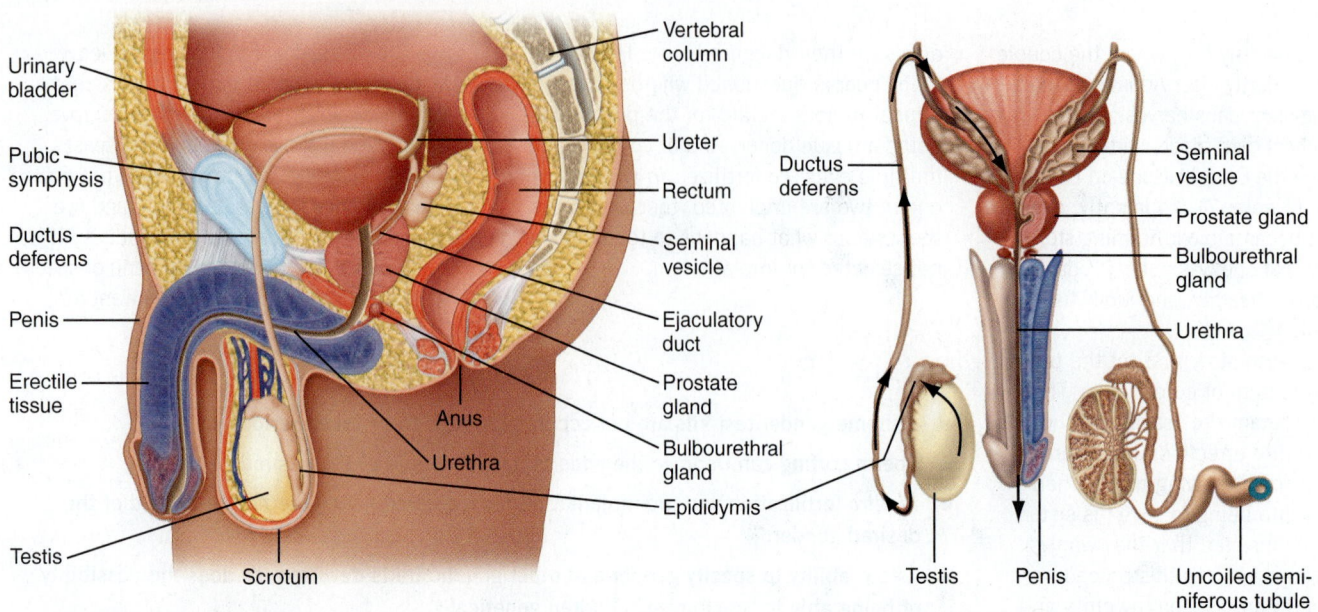

a) A sagittal cutaway view.

b) A posterior view of the reproductive organs and bladder.

Figure 16.1 **The male reproductive system.**

formed sperm are not fully mature (they cannot yet swim) when they leave the seminiferous tubules. Their ability to swim develops in both the epididymis and the ductus deferens. The epididymis and the ductus deferens also store the sperm until ejaculation.

When the male reaches sexual climax and ejaculates, rhythmic contractions of smooth muscle propel the sperm through the short ejaculatory duct and finally through the urethra, which passes through the penis.

The **penis** is the male organ of sexual intercourse. Its function is to deliver sperm internally to the female, safely away from the harsh external environment. The penis contains erectile tissues that permit **erection**—an increase in length, diameter, and stiffness of the penis that facilitates its entry into the vagina.

Accessory glands help sperm survive

Sperm must endure the rapid journey from epididymis to vagina and then continue to live within the woman for many hours. To improve their chances of survival, the male delivers sperm in a thick, whitish mixture of fluids called **semen.** In addition to sperm, semen contains secretions from the seminal vesicles, the prostate gland, and the bulbourethral gland.

The **seminal vesicles** produce seminal fluid, a watery mixture containing fructose and prostaglandins that represents about 60% of the volume of semen. Fructose, a carbohydrate, provides the sperm with a source of energy. The prostaglandins are thought to induce muscle contractions in the female reproductive system that help sperm travel more effectively. The **prostate gland** contributes an alkaline fluid. The vagina is generally fairly acid (pH 3.5–4.0), which

helps prevent infections but is too acidic for sperm. The prostate secretions temporarily raise the vaginal pH to about 6, optimal for sperm. Finally, the **bulbourethral glands** secrete mucus into the urethra during sexual arousal. The mucus washes away traces of acidic urine in the urethra before the sperm arrive and also provides lubrication for sexual intercourse. The sperm are not mixed with seminal and prostate fluids until the moment of ejaculation.

Table 16.1 summarizes the organs and glands of the male reproductive system.

Table 16.1 **Male reproductive organs and glands**

Organ	Function
Testis (2)	Produces sperm, testosterone, and inhibin
Scrotum	Keeps the testes at the proper temperature
Epididymis (2)	Site of sperm maturation and storage
Ductus deferens (2)	Duct for sperm maturation, storage, and transport
Ejaculatory duct (2)	Duct for transporting sperm and glandular secretions
Penis	Erectile organ of sexual intercourse
Accessory Glands	
Seminal vesicle (2)	Secretes fructose and most of the seminal fluid
Prostate gland	Secretes watery alkaline fluid to raise vaginal pH
Bulbourethral gland (2)	Secretes lubricating mucus

Sperm production requires several cell divisions

Let's take a closer look at the process of sperm production. As mentioned, it takes place in the seminiferous tubules within each testis (Figure 16.2a). Toward the outer surface of each tubule are undifferentiated cells (cells that have not yet become any kind of specialized cell) called *spermatogonia* (singular: spermatogonium) (Figure 16.2b). Spermatogonia are *diploid* cells, meaning they have 46 chromosomes, the normal human number.

The formation of the **gametes** (sperm and eggs) involves a series of cell divisions called *mitosis* and *meiosis*, discussed in more detail when we describe cell reproduction and differentiation (Chapter 17). The important point to remember is that during the two successive cell divisions of meiosis (meiosis I and II), the number of chromosomes in the developing gametes is reduced by half. Gametes are called *haploid* cells (from the Greek *haploos*, meaning "single") because they contain only 23 chromosomes, rather than the usual diploid number, 46. This ensures that when a sperm and egg unite, the embryo will again have the proper (diploid) number of chromosomes.

Spermatogonia divide several times during the course of sperm development. The first division, by mitosis, ensures a constant supply of *primary spermatocytes*, each with the diploid number of chromosomes. Primary spermatocytes then undergo two successive cell divisions during meiosis

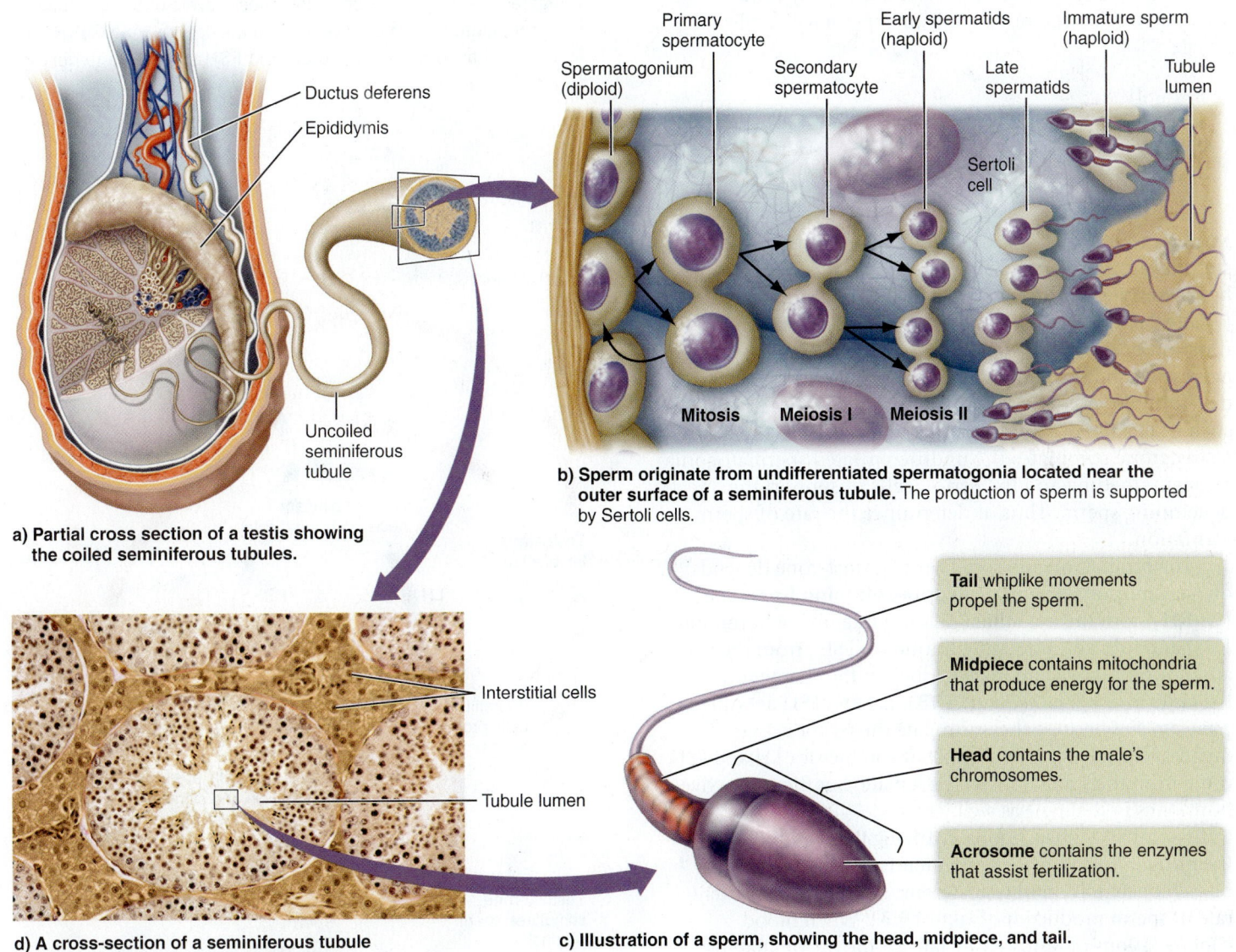

a) **Partial cross section of a testis showing the coiled seminiferous tubules.**

b) **Sperm originate from undifferentiated spermatogonia located near the outer surface of a seminiferous tubule.** The production of sperm is supported by Sertoli cells.

d) **A cross-section of a seminiferous tubule showing the location of the testosterone-producing interstitial cells between the tubules.** Mature sperm are barely visible in the tubular lumen.

c) **Illustration of a sperm, showing the head, midpiece, and tail.**

Tail whiplike movements propel the sperm.

Midpiece contains mitochondria that produce energy for the sperm.

Head contains the male's chromosomes.

Acrosome contains the enzymes that assist fertilization.

Figure 16.2 Sperm formation in a seminiferous tubule.

✓ Which gives rise to more sperm, a spermatogonium or a primary spermatocyte? Explain.

to become *secondary spermatocytes* and finally *spermatids.* Spermatids, which are haploid cells, mature slowly to become the male gametes, or sperm. The entire process of sperm formation and maturation takes about 9–10 weeks. During this time the cells are surrounded and nourished by the large **Sertoli cells** that make up most of the bulk of the seminiferous tubules.

A mature sperm has a head, midpiece, and tail (**Figure 16.2c**). The head contains the DNA, organized into chromosomes. The head is covered by an *acrosome*, a cap containing enzymes that help the sperm penetrate the egg. The sperm's whiplike tail moves it forward, powered by energy produced by mitochondria in the midpiece.

Tens of millions of sperm are produced every day after puberty in young men. A typical ejaculation may contain 100–300 million sperm, but only one sperm will fertilize an egg.

Testosterone affects male reproductive capacity

Male reproductive capacity depends on **testosterone,** a steroid hormone produced by **interstitial cells** located between the seminiferous tubules within the testes (**Figure 16.2d**). Testosterone controls the growth and function of the male reproductive tissues, stimulates aggressive and sexual behavior, and causes the development of secondary sexual characteristics at puberty, such as facial hair and deepening voice. Within the testes, testosterone stimulates the undifferentiated spermatogonia to begin dividing and the Sertoli cells to support the developing sperm. Thus, it determines the rate of sperm formation.

The production and secretion of testosterone depends on three hormones: **gonadotropin-releasing hormone (GnRH)** from the hypothalamus and luteinizing hormone (LH) and follicle-stimulating hormone (FSH) from the anterior pituitary gland. As described in the chapter on the endocrine system (Chapter 13), LH and FSH are called *gonadotropins* because they stimulate the reproductive organs—the gonads. Historically, the names for LH and FSH came from their functions in the female, but they are active hormones in the male as well.

Negative feedback loops involving the hypothalamus, the pituitary, and the testes maintain a fairly constant blood concentration of testosterone, and thus a consistent rate of sperm production (**Figure 16.3**). When blood concentrations of testosterone fall, ❶ the hypothalamus secretes GnRH, ❷ which stimulates the anterior pituitary gland to secrete LH and FSH. ❸ LH causes the interstitial cells of the testes to produce and secrete testosterone, and ❹ FSH enhances sperm formation indirectly by stimulating the Sertoli cells. ❺ Testosterone also stimulates sperm production.

Once the testosterone concentration is normal again or if it rises above normal, ❻ testosterone inhibits the secretion of GnRH and LH. In addition, ❼ when Sertoli cells are highly active they secrete a hormone called **inhibin** that inhibits the secretion of GnRH and FSH.

↩ **Recap** The male reproductive system comprises the testes, the penis, and associated ducts and glands. Semen consists of sperm and three glandular secretions that provide energy and the proper pH environment for the sperm and lubrication for sexual intercourse. Millions of sperm form every day throughout a man's life; a typical ejaculate contains up to 300 million. Testosterone stimulates the growth and function of the male reproductive system, and encourages aggressive and sexual behavior. Blood levels of testosterone are regulated by a negative feedback loop involving GnRH from the hypothalamus and LH and FSH from the anterior pituitary. ∎

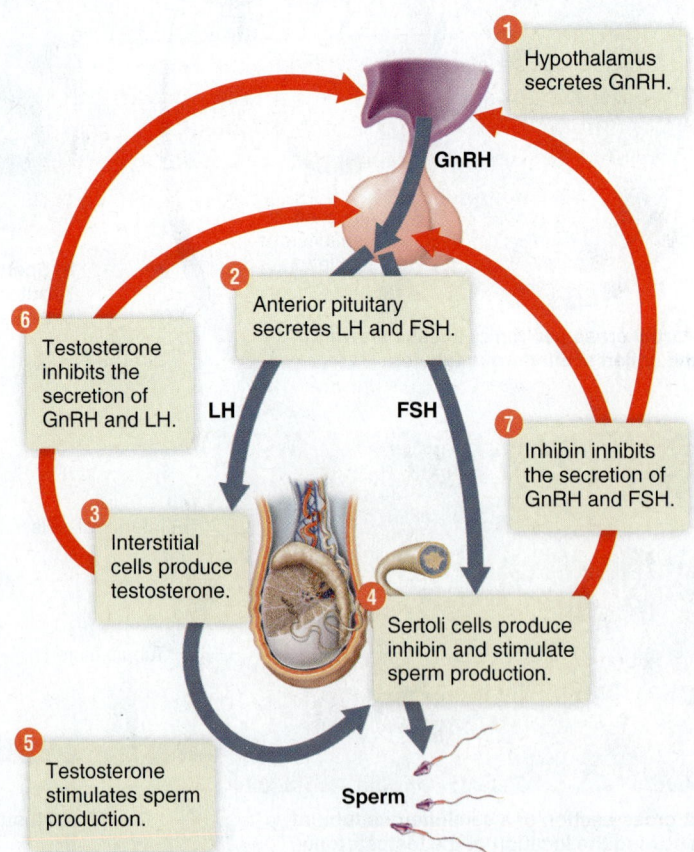

Figure 16.3 Feedback control of blood testosterone concentration and sperm production. Negative feedback control loops effectively maintain testosterone concentration (and hence sperm production) relatively constant over time. Negative feedback (red arrows) occurs at both the hypothalamus and the anterior pituitary.

16.2 The female reproductive system produces eggs and supports pregnancy

The female reproductive system evolved to perform more functions than that of the male. Eggs are precious, because the system releases only one or two every month. In addition, because it can never be known for certain if an egg will be fertilized, the female reproductive system must go through cyclic changes every month just to prepare for the possibility of fertilization. If fertilization does occur, the female system adjusts to pregnancy and supports the developing fetus.

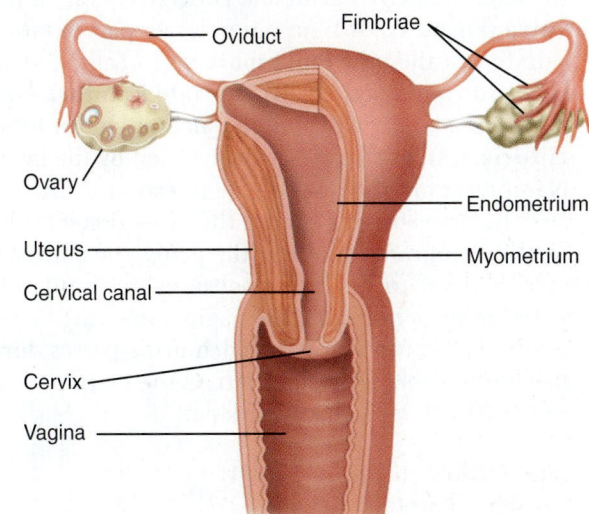

a) Sagittal section showing the components of the system in relation to other structures.

Ovaries release oocytes and secrete hormones

Figure 16.4a illustrates the organs of the female reproductive system. The primary female reproductive organs are the two **ovaries,** which release the female gametes, immature eggs called *oocytes,* at regular intervals during the reproductive years. The ovaries also secrete the female sex steroid hormones **estrogen** and **progesterone.** (The term *estrogen* is commonly used to refer to three hormones known collectively as *estrogens*:17b-estradiol, estrone, and estriol.)

Once released, the oocyte is swept into the open end of an **oviduct** (also called a *Fallopian tube*), which leads from the ovary to the uterus. The open end of the oviduct has fingerlike projections called fimbriae that help move the oocyte into the oviduct. If fertilization by a sperm occurs, it usually takes place in the upper third of the oviduct. The unfertilized oocyte or fertilized egg moves down the oviduct over the course of about 3–4 days to the uterus. If the egg is fertilized, it will embed itself in the lining of the uterus and fetal development will begin. An unfertilized egg will eventually degenerate and then either be reabsorbed by the body or discharged during menstruation.

✅ Embryos sometimes implant somewhere other than the uterus, most often in the oviduct before it reaches the uterus and sometimes in the abdominal cavity. How could an embryo end up in the abdominal cavity?

b) Anterior view of internal organs.

The uterus nurtures the developing embryo

The **uterus** is a hollow, pear-shaped organ where the embryo grows and develops. The walls of the uterus comprise two layers of tissue (Figure 16.4b).

The inner layer of the uterus, the **endometrium,** is a lining of epithelial tissue, glands, connective tissue, and blood vessels. A fertilized egg attaches to the endometrium in a process called *implantation.* After implantation, the endometrium helps to form the *placenta.* The placenta provides nourishment to the developing fetus and also provides for waste removal and gas exchange (see Chapter 21).

The outer layer of the uterus, or **myometrium,** consists of thick layers of smooth muscle. The myometrium stretches during

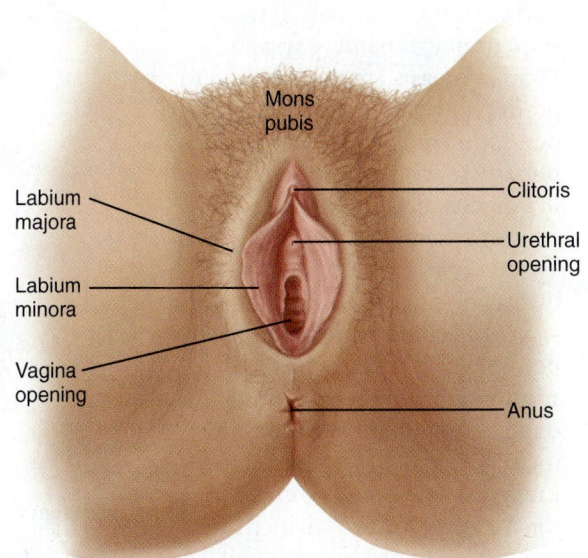

c) Vulva: the female external genitalia.

Figure 16.4 **The female reproductive system.**

pregnancy to accommodate the growing fetus. It also provides the muscular force to expel the mature fetus during labor.

The narrow part of the uterus that extends slightly into the vagina is the **cervix.** An opening in the cervix permits sperm to enter the uterus and allows the fetus to exit during birth.

The vagina: organ of sexual intercourse and birth canal

The cervix joins the **vagina,** a hollow muscular organ of sexual intercourse and also the birth canal. Glands within the vagina produce lubricating secretions during sexual arousal. A thin ring of tissue called the *hymen* may partially cover the opening to the vagina. The hymen is ruptured by the first sexual intercourse, or sometimes by the insertion of tampons or vigorous physical activity.

The vagina is continuous at the body surface with the female external genitalia, collectively called the **vulva** (Figure 16.4c). An outer, larger pair of fat-padded skin folds called the **labia majora** (in Latin, "major lips") surround and enclose the **labia minora** (singular: labium), a highly vascular but smaller pair of folds. The **clitoris,** a small organ partly enclosed by the labia minora, is important in the female sexual response. The clitoris and the penis originate from the same tissue during early fetal development, and like the penis, the clitoris contains erectile tissue that is highly sensitive to stimulation. Between the clitoris and the vaginal opening lies the urethral opening, through which urine passes during urination. Table 16.2 summarizes the main components of the female reproductive system.

Mammary glands nourish the infant

Human breasts contain **mammary glands,** modified sweat glands that technically are part of the skin, or integumentary system. Though breasts are not required to produce a viable fetus, they undergo developmental changes during puberty in women and they support the infant's development, so we consider them here.

A *nipple* at the center of each breast contains smooth muscles that can contract, causing the nipple to become erect in response to sexual stimulation, cold temperatures, or a nursing infant. Surrounding the nipple is the pigmented *areola.* Internally, the breasts contain the mammary glands, which are specialized to produce milk (Figure 16.5). The mammary glands

consist of hundreds of small milk-producing lobules. Contractile cells around each lobule allow the milk to be released, and ducts deliver it to the nipple. Most of the breast consists of adipose tissue (fat), so breast size does not indicate the potential for milk production. A network of fibrous connective tissue supports the glands and adipose tissue.

At puberty, female mammary glands enlarge under the control of the hormone estrogen. Both estrogen and progesterone prepare the glands for **lactation,** or the production of milk, late in pregnancy. At birth, prolactin stimulates lactation and oxytocin stimulates the contractions that release milk during nursing. In males the mammary glands are *vestigial*, meaning that although they may have had a function at some point in human evolution they no longer do.

Table 16.2 **Summary of the components of the female reproductive system**

Component	Function
Ovary (2)	Site of storage and development of oocytes
Oviduct (2)	Duct for transporting oocyte from ovary to uterus; also site of fertilization if it occurs
Uterus	Hollow chamber in which embryo develops
Cervix	Lower part of the uterus that opens into the vagina
Vagina	Organ of sexual intercourse; produces lubricating fluids; also the birth canal
Clitoris	Organ of sexual arousal

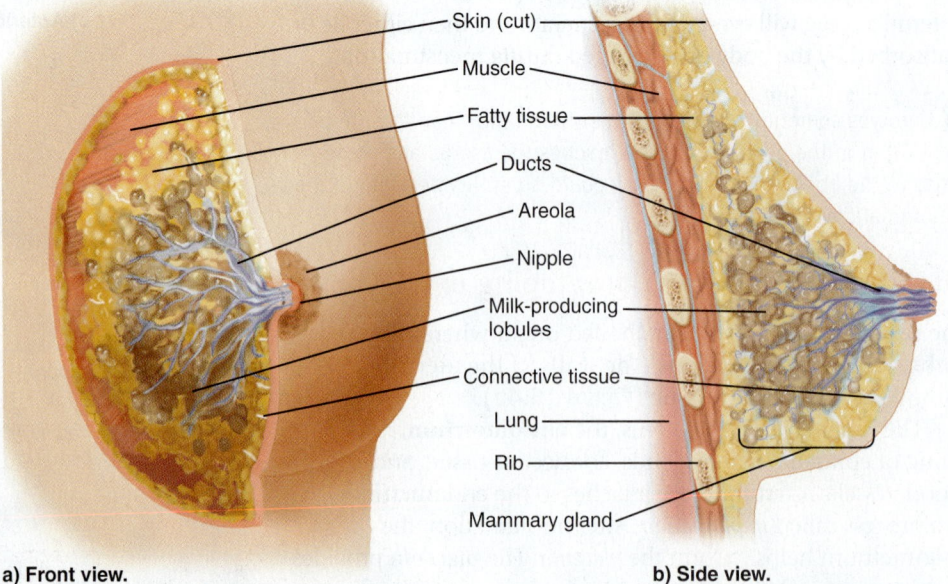

a) Front view.

b) Side view.

Figure 16.5 **The female breast.** Anatomical relationships between the milk-producing mammary glands, adipose tissue, and muscle.

↩ **Recap** The ovaries secrete estrogen and progesterone, store immature oocytes, and (usually) release one oocyte at a time at intervals of about 28 days. The oocyte travels through the oviduct to the uterus, where implantation occurs if the egg has been fertilized. The vagina is the female organ of sexual intercourse and the birth canal; around its opening are the structures of the vulva. Mammary glands are accessory organs that produce and store milk. ∎

16.3 The menstrual cycle consists of ovarian and uterine cycles

Every month, the ovaries and uterus go through a pattern of changes called the **menstrual cycle.** The menstrual cycle lasts about 28 days and is controlled by hormones. Menstrual cycles begin at puberty and generally continue throughout the reproductive years, except during pregnancy.

A complete menstrual cycle consists of two linked cycles, the *ovarian cycle* and the *uterine cycle.* Together, they periodically release oocytes and prepare the uterus to receive a fertilized egg.

The ovarian cycle: oocytes mature and are released

The **ovarian cycle** is a regular pattern of growth, maturation, and release of oocytes from the ovary. At birth, a female has approximately one million *primary oocytes* already formed and stored in each ovary, and no more are ever produced. Each primary oocyte has already developed partway through meiosis at birth, but at birth all further development halts until after puberty. By puberty, 85% of them have been resorbed, leaving only about 300,000 in both ovaries.

Each month perhaps a dozen primary oocytes start the development process, but typically only one completes it. Only about 400–500 oocytes are released during a woman's lifetime, a number that contrasts sharply with the 100–300 million sperm released in a single ejaculation.

Figure 16.6 illustrates the events of the ovarian cycle:

① A primary oocyte is surrounded by a layer of *granulosa cells* that nourish it. Together, a primary oocyte and its granulosa cells constitute an immature

follicle. At the beginning of the ovarian cycle, an increase in GnRH from the hypothalamus slowly increases the secretion of FSH and LH by the anterior pituitary. These two hormones stimulate the follicle to grow, enlarge, and change form. FSH and LH also stimulate the follicle (and surrounding ovarian cells) to secrete estrogen.

② Development begins. Stimulated by FSH, the granulosa cells begin to divide and to secrete a sugary coating of glycoproteins around the oocyte.

③ A fluid-filled space called the *antrum* develops within the follicle. The granulosa cells continue to divide; some of them begin to secrete estrogen.

④ The follicle matures. The primary oocyte completes stage I of meiosis to yield a *secondary oocyte* and a nonfunctional *polar body.* The mature follicle, sometimes called a Graafian follicle, consists of a secondary oocyte, a polar body, and numerous granulosa cells, all surrounded by ovarian connective tissue cells. The granulosa cells continue to secrete estrogen.

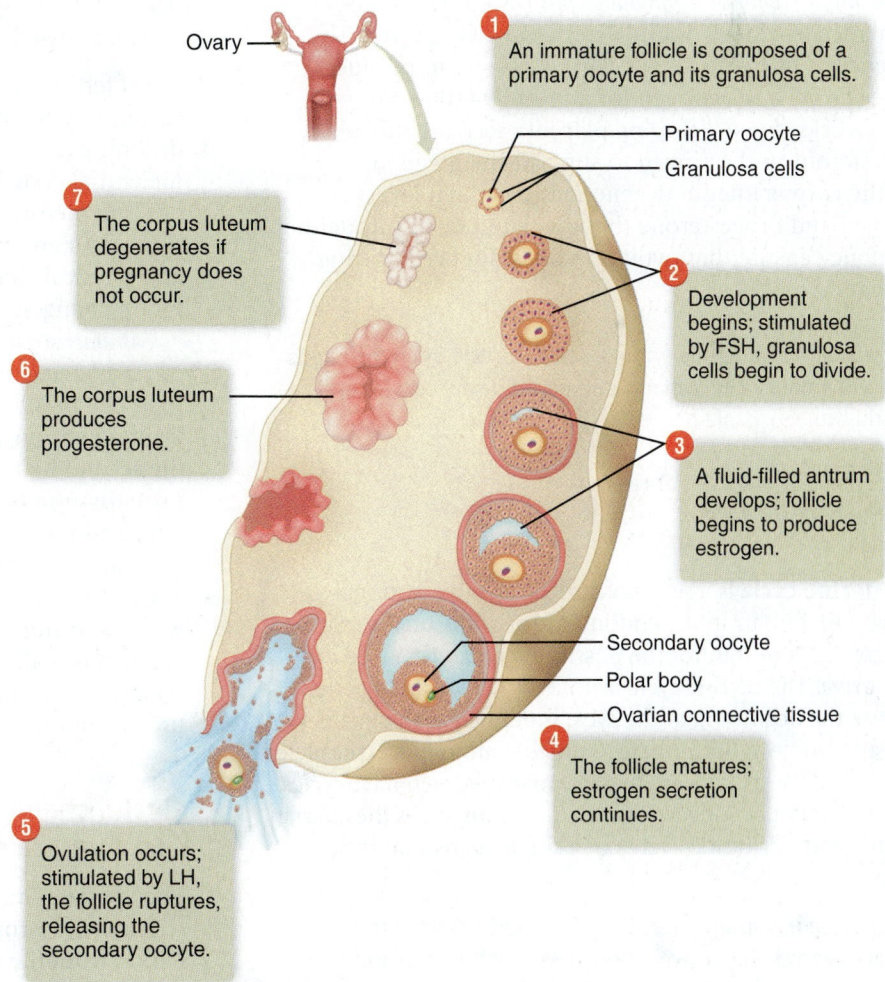

Ovary

① An immature follicle is composed of a primary oocyte and its granulosa cells.

Primary oocyte
Granulosa cells

② Development begins; stimulated by FSH, granulosa cells begin to divide.

③ A fluid-filled antrum develops; follicle begins to produce estrogen.

Secondary oocyte
Polar body
Ovarian connective tissue

④ The follicle matures; estrogen secretion continues.

⑤ Ovulation occurs; stimulated by LH, the follicle ruptures, releasing the secondary oocyte.

⑥ The corpus luteum produces progesterone.

⑦ The corpus luteum degenerates if pregnancy does not occur.

Figure 16.6 The ovarian cycle. Approximately one dozen primary follicles start this process each month, but generally, only one completes it. For any particular primary follicle the events take place in one location, but for clarity the events are shown as if they migrate around the ovary in a clockwise fashion.

5 Ovulation occurs. Midway through the ovarian cycle, a sharp rise in LH from the anterior pituitary triggers **ovulation.** The follicle wall balloons outward and ruptures, releasing the secondary oocyte with its polar body, zona pellucida, and surrounding granulosa cells into the extracellular fluid. Some women experience a brief sharp pain when the follicle ruptures.

6 A corpus luteum forms. Stimulated by LH, what is left of the ruptured follicle becomes the **corpus luteum** (the name "luteinizing hormone" comes from the fact that LH causes the corpus luteum to form). The corpus luteum has an important function: It secretes large amounts of progesterone, which prepares the endometrium of the uterus for the possible arrival of a fertilized egg.

7 If fertilization does not occur, the corpus luteum degenerates in about 12 days. At this point, estrogen and progesterone levels fall, and the ovarian cycle starts again.

If pregnancy does occur, a layer of tissue derived from the embryo, called the *chorion,* begins to secrete a hormone called *human chorionic gonadotropin (hCG).* Home-use pregnancy test kits work by detecting hCG in urine. The presence of hCG signals the corpus luteum to produce progesterone and estrogen for another 9–10 weeks. By 10 weeks, the developing placenta secretes enough progesterone and estrogen to support the pregnancy, and the corpus luteum degenerates. The high levels of estrogen and progesterone that are maintained throughout pregnancy ensure that ovulation does not occur during pregnancy.

☑ A woman has a pituitary tumor removed, and afterward she is unable to produce LH. Will she still have primary oocytes and can she still ovulate? Explain.

The uterine cycle prepares the uterus for pregnancy

The **uterine cycle** is a series of structural and functional changes that occur in the endometrium of the uterus as it prepares each month for the possibility that a fertilized egg may arrive. The uterine cycle is linked to the ovarian cycle. Because the beginning of the cycle is defined as the first day of menstruation, we begin there. Note that although a complete cycle generally lasts about 28 days, many women have cycles that are shorter or longer. Figure 16.7 summarizes the ovarian, uterine, and hormonal events of a single menstrual cycle of 28 days:

- *Days 1–5: menstrual phase.* If no fertilized egg from the previous cycle has arrived, body levels of estrogen and progesterone decline precipitously. Without these hormones, the uterus cannot maintain its endometrial lining, which has thickened to prepare for a potential pregnancy. **Menstruation** is the process by which the endometrial lining disintegrates and its small blood vessels rupture. The shed tissue and blood

pass through the vagina and are excreted. The period of visible menstrual flow, which lasts about 3–5 days, is called the *menstrual period.*
- *Days 6–14: proliferative phase.* Triggered primarily by rising levels of estrogen from the next developing follicle, the endometrial lining begins to proliferate. It becomes thicker, more vascular, and more glandular. In addition, the lining of the cervix produces a thin, watery mucus that facilitates the passage of sperm, if present, into the uterus from the vagina.
- *Ovulation.* Ovulation occurs midway through the cycle, around the 14th day.
- *Days 15–28: secretory phase.* Production of estrogen and especially progesterone by the corpus luteum causes the endometrium to continue proliferating, until it becomes almost three times thicker than it was after menstruation. Uterine glands mature and begin producing glycogen as a potential energy source for the embryo. The cervical mucus becomes thick and viscous, forming a "cervical plug" that prevents sperm and bacteria from entering the uterus. These structural changes in the endometrium prepare the uterus for a developing embryo.

If fertilization does not occur, degeneration of the corpus luteum at about 12 days after ovulation results in declining estrogen and progesterone levels, and another menstrual cycle begins.

The menstrual cycle is not a pleasant experience for some women, to put it mildly. It involves rapid hormonal and physical changes that can sometimes cause pain or mood changes.

Premenstrual syndrome (PMS) is a group of symptoms often associated with the premenstrual period, including food cravings, mood swings, anxiety, back and joint pain, water retention, and headaches. It generally develops in the second half of the cycle, after ovulation, and lasts until menstruation begins. Exercise and medications that reduce pain and water retention can help.

Painful menstruation is known as *dysmenorrhea.* The endometrium of the uterus produces prostaglandins, which can trigger contractions and cramping of the uterine smooth muscle. Aspirin and ibuprofen may offer some relief from dysmenorrhea because these drugs inhibit prostaglandin formation.

Cyclic changes in hormone levels produce the menstrual cycle

The ovarian and uterine cycles are regulated by cyclic changes in hormone levels. In the female (as in the male), GnRH from the hypothalamus causes the secretion of FSH and LH. FSH stimulates development of the oocyte and the secretion of estrogen, and LH stimulates the development of the corpus luteum and the secretion of progesterone. During most of the cycle, increases in estrogen and progesterone inhibit GnRH, FSH, and LH secretion, just as one would

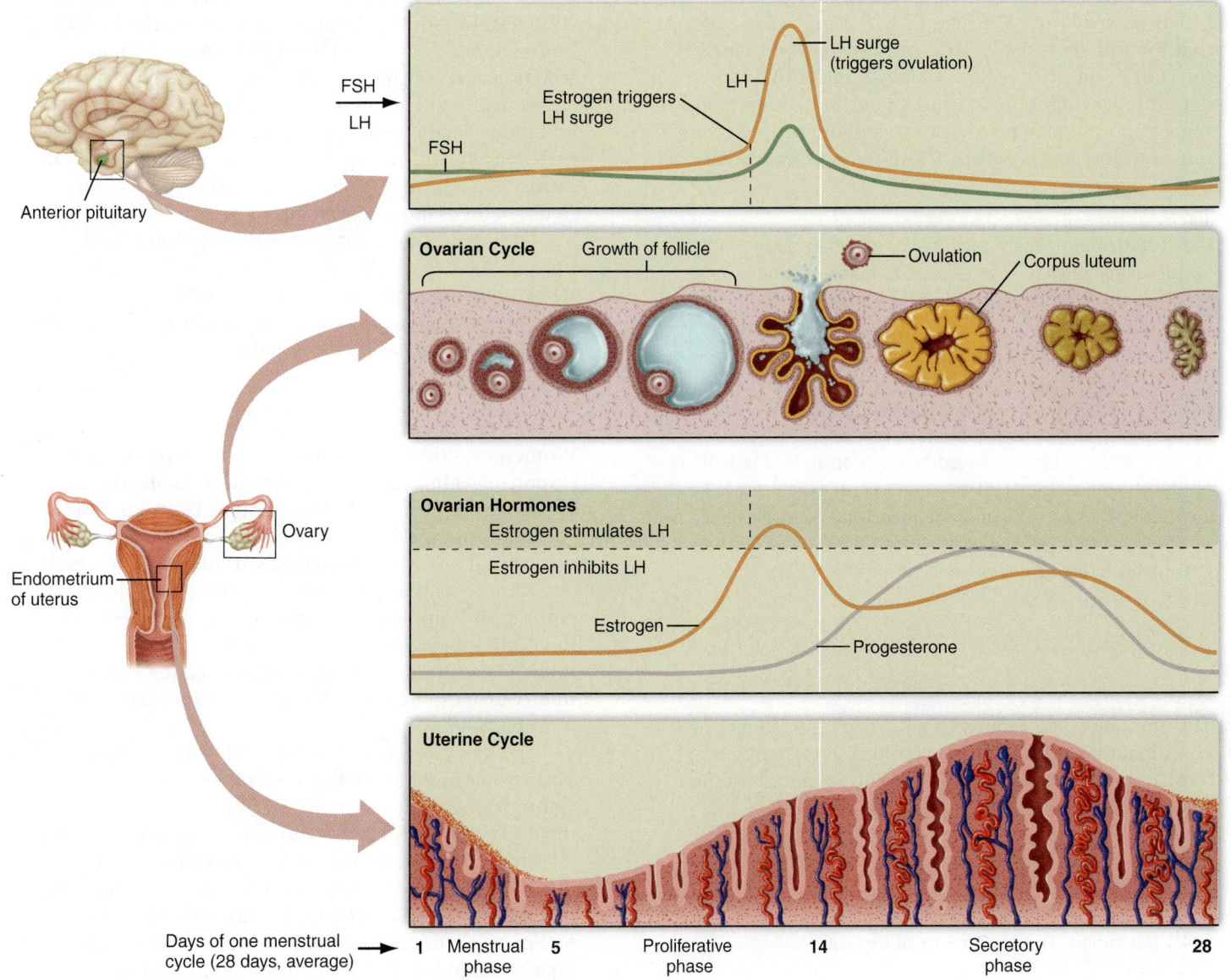

Anterior pituitary

Endometrium of uterus

Ovary

Figure 16.7 **The menstrual cycle.** The menstrual cycle consists of linked ovarian and uterine cycles. The ovarian cycle is a series of changes associated with oocyte maturation, which depends on the blood levels of FSH and LH. The uterine cycle involves changes in the endometrial lining of the uterus, which depend on the blood levels of estrogen and progesterone.

✅ What makes progesterone (purple line) rise during the second half of the menstrual cycle, and what makes it fall at the end of the cycle? (Hint: Where is this progesterone coming from?)

expect in a typical negative feedback control loop. However, a single exception to this pattern accounts for why the female hormonal pattern is cyclical. In females, there is a single *positive feedback* event just before the midpoint of the cycle. The positive feedback event is that a high and rapidly rising estrogen concentration briefly stimulates the secretion of LH rather than inhibiting it. The resultant **LH surge** (review Figure 16.7) triggers ovulation, which in turn leads to the formation of the corpus luteum and a rise in progesterone.

During the second half of the menstrual cycle, the hormonal control system returns to the typical negative feedback control. The LH surge is short-lived because high but steady levels of estrogen and progesterone from the corpus luteum once again inhibit LH and FSH secretion. Without LH and FSH, the next follicle does not develop. This prevents a second ovulation from taking place until the cycle is complete and estrogen and progesterone have returned to low levels again.

From a functional standpoint, hormonal regulation of the menstrual cycle makes perfect sense. Because only a mature follicle produces enough estrogen to trigger the LH surge, ovulation is triggered at the precise moment that the follicle is ready. And once ovulation occurs, the start of another cycle is delayed until the body determines whether or not pregnancy has taken place. If there is no pregnancy, then estrogen and progesterone return to low levels within about two weeks and the cycle begins again.

> ↩ **Recap** During the ovarian cycle, a primary oocyte within a developing follicle divides once to form a secondary oocyte. The follicle ruptures, releases the oocyte, and forms the corpus luteum that secretes progesterone and estrogen. Rising levels of estrogen cause the endometrium to proliferate. If pregnancy does not occur, hormone levels fall and the endometrial layer disintegrates and is shed, a process known as *menstruation.* Ovulation is triggered by a surge of LH, which in turn is caused by the positive feedback effect of a high concentration of estrogen from the maturing follicle. During the second half of the menstrual cycle, sustained high levels of estrogen and progesterone from the corpus luteum inhibit further ovulation. ∎

16.4 Human sexual response, intercourse, and fertilization

The human sexual response is a series of events that coordinate sexual function to accomplish intercourse and fertilization. The response can have up to four phases in both men and women:

1. *Excitement.* Increased sexual awareness and arousal
2. *Plateau.* Intense and continuing arousal
3. *Orgasm.* The peak of sexual sensations
4. *Resolution.* The abatement of arousal

The male sexual response

In men, excitement is accompanied by pleasurable sensations from the penis and by an erection. Erection occurs when neural activity relaxes (dilates) arterioles leading into vascular compartments within the penis. The compartments fill with blood, and as they fill they press on the veins draining the penis and reduce the rate of venous outflow. As more blood enters the penis and less blood leaves it, the penis swells, lengthens, and stiffens. Heart rate and breathing increase, as does the tone of skeletal muscle. Sights, sounds, memories, and other psychological stimuli; kissing or other physical contact; and especially physical contact with the penis all contribute to excitement. Arousal continues and becomes more intense in the plateau phase, which can last for many minutes.

At some point, sexual arousal becomes so great that **orgasm** occurs. An orgasm is a brief, intensely pleasurable reflex event that accomplishes **ejaculation,** or the expulsion of semen. During ejaculation, sympathetic nerve activity contracts smooth muscle in the seminal vesicles, epididymis, bulbourethral glands, and prostate to move glandular secretions into the urethra. The internal urethral sphincter closes tightly to prevent urine from leaking into the urethra, and the skeletal muscles at the base of the penis contract rhythmically, forcing semen out of the urethra in several spurts. In a typical ejaculation, the male expels about 3–4 ml (a teaspoon or more) of semen.

During the resolution phase, the erection subsides and breathing and heart rate return to normal. Generally, a refractory period follows, during which another erection and orgasm cannot be achieved. The refractory period may last from several minutes up to several hours.

The female sexual response

Women experience the same four phases of the sexual response. As in men, the excitement phase is triggered by sights, sounds, psychological stimuli, and physical stimulation, especially of the clitoris and breasts. During the arousal and plateau phases, stimulation initiates neural reflexes that dilate blood vessels in the labia, clitoris, and nipples. The tissues of the outer region of the vagina swell, and the area around the labia secrete lubricating fluids. Without these lubricating fluids, intercourse can be difficult and sometimes painful for both partners.

The female orgasm, like that of the male, consists of rhythmic and involuntary muscular contractions. Contractions of the uterus and vagina are accompanied by intense feelings of pleasure and sensations of warmth and relaxation. The rate of orgasm is more variable in females than in males. Some women do not experience orgasms frequently, whereas others experience multiple orgasms within one continuous period of sexual activity. Female orgasm is not required for fertilization or pregnancy, nor is arousal, for that matter.

Fertilization: one sperm penetrates the egg

During ejaculation, the male may deposit several hundred million sperm inside the vagina, near the cervix. The sperm have a long way to go, however, before one of them encounters an egg. Random swimming takes some sperm through the cervical opening, through the uterus, and up the two oviducts. The journey can take hours or even a day or more. Fertilization occurs when a sperm encounters an oocyte in an oviduct and penetrates the zona pellucida (**Figure 16.8**). Once one sperm penetrates the zona pellucida, all other sperm are excluded. Some sperm may be viable for up to five days in the female reproductive tract, although most do not last more than two days.

Now that we know how fertilization normally takes place, we turn to how to control fertility (avoid pregnancy) should you wish to and how to improve

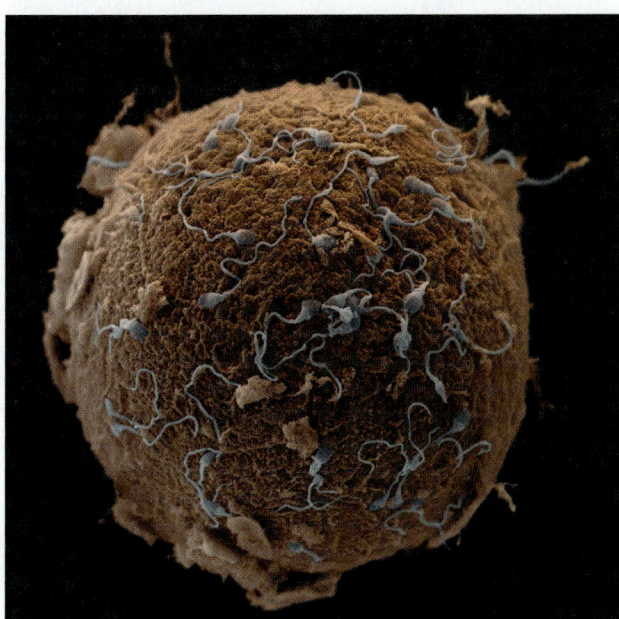

Figure 16.8 Colored SEM of several sperm in contact with an egg. Only one sperm will fertilize the egg.

Table 16.3 Approximate failure rates of various contraceptive methods

Contraceptive method	Annual number of pregnancies expected per 100 women
Spermicide Alone	28
Female Condom	21
Male Condom	18
Diaphragm with Spermicide	12
Hormonal Contraceptives	
Oral contraceptives	9
Skin patch	9
Vaginal contraceptive ring	9
Injected hormonal contraceptives	6
Intrauterine Devices	<1
Sterilization Surgery—Men	<1
Sterilization Surgery—Women	<1

Source: Adapted from *Birth Control Guide*, FDA Office of Women's Health, 2014.

fertility (increase the chances of becoming pregnant) if infertility becomes a problem. The details of human development from a single fertilized egg into a fully formed baby in just nine months are the subject of another chapter (Chapter 21).

🔄 **Recap** Women and men experience the same four phases of sexual responsiveness. Sexual arousal in the male results in penile erection that leads to orgasm and ejaculation. Females experience sexual arousal and pleasurable orgasms marked by rhythmic muscular contractions. During ejaculation, the male deposits several hundred million sperm in the vagina. Fertilization of the egg by a single sperm occurs within five days, if it occurs at all. ■

16.5 Birth control methods: controlling fertility

Couples or individuals might choose to control fertility for many reasons, and there are many ways to do it. Birth control (contraceptive) methods vary widely in effectiveness (Table 16.3). Some methods reduce the chance of pregnancy from just a single intercourse; others virtually eliminate the chances of pregnancy for a lifetime. An understanding of the advantages and the limitations of each method will help you make an informed choice, should you choose to control your own fertility for any period of time.

Abstinence: not having intercourse

The only completely effective method of birth control is to not have sexual intercourse at all.

Abstinence, as it is called, works for many people for short periods of time, and throughout life for others. Abstinence is natural in that no intervention or artificial method is employed. However, the idea of abstinence doesn't resonate with most people. When abstinence is not the option of choice, there are still many ways to reduce the probability of pregnancy.

Surgical sterilization: vasectomy and tubal ligation

Both men and women may opt for surgical sterilization (Figure 16.9). Male surgical sterilization is called *vasectomy*.

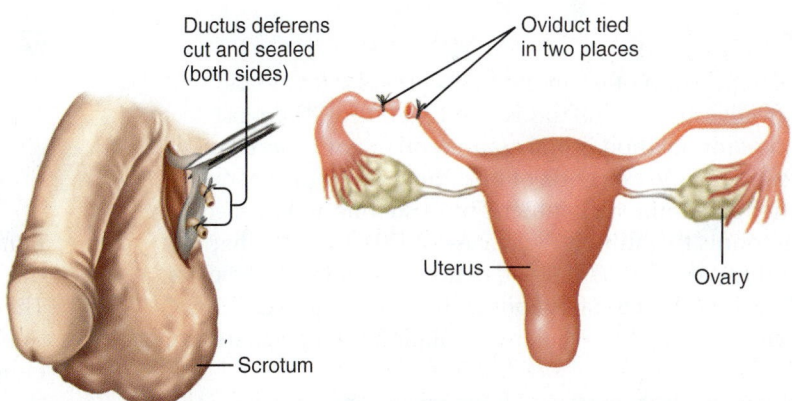

a) **Vasectomy.** Small incisions are made in the scrotum, and each ductus deferens is tied in two places and cut.

b) **Tubal ligation.** An incision is made in a woman's abdominal wall, the two oviducts are located, and each is tied in two places and cut.

Figure 16.9 Surgical sterilization.

It can generally be performed in a doctor's office under local anesthesia. The doctor makes small incisions in the scrotum, locates each ductus deferens, and ties them in two places. The segments between the two ties are then removed, preventing sperm from reaching the urethra.

After a vasectomy, the testes continue to produce testosterone and sperm, and sexual interest and all secondary sexual characteristics remain unchanged. With no access to the urethra, however, the sperm eventually are resorbed, just as they are if ejaculation never takes place. Generally, it is impossible to detect a change in semen volume per ejaculation. If done properly, surgical male sterilization is close to foolproof. However, to be on the safe side, the surgeon will probably recommend doing a sperm count several months after the operation.

Female sterilization involves a *tubal ligation*. The doctor makes a small incision in the woman's abdominal wall, locates each oviduct, and ties (ligates) each at two sites. The sections between the ties are cut away, leaving the oocyte no way to reach the uterus or sperm to reach the oocyte. In a newer version of a tubal ligation, called a *hysteroscopy*, a flexible tube is inserted through the vagina and uterus and into the oviduct. The tube emits electrical current that seals shut the oviduct. Both tubal ligation and hysteroscopy are highly effective but require surgery that is slightly more difficult than in the male. In addition, tubal ligation may leave two small but possibly visible scars.

A drawback to surgical sterilization in both men and women is that although it *may* be possible to reverse, sterilization generally is permanent. Sterilization should be carefully considered before it is chosen as a method of birth control.

✔ Would a man be sterile the day after a vasectomy and would a woman be sterile the day after a tubal ligation? Explain.

Hormonal methods: pills, injections, patches, and rings

Manipulation of hormone levels provides reasonably effective and safe methods of birth control. The most common method is oral contraceptives, or *birth control pills*. Although the exact mix of hormones in the pills varies, they all administer synthetic progesterone and estrogen in amounts that inhibit the release of FSH and LH. That way, follicles do not mature and ovulation does not take place. It is important to take a pill at the same time every day; if pills are not taken regularly, ovulation (and pregnancy) may occur.

Most birth control pills are taken for three weeks and then replaced with placebo pills for a week so that normal menstrual periods occur each month. However, many physicians have questioned the need for regular monthly periods. In 2003, the FDA approved the first *continuous birth control pill (Seasonale)* that reduces the number of menstrual periods to only four per year. In 2007, the FDA approved a pill that eliminates all periods for a full year (*Lybrel*). Continuous birth control pills are particularly attractive to women who routinely suffer from PMS.

Roughly one-third of all women in the United States use birth control pills for at least some period of their lives. Oral contraceptives have several side effects, some beneficial and others harmful. Some women report that oral contraceptives reduce cramps and menstrual flow, and they may offer some protection against cancers of the ovaries and uterus. On the negative side, oral contraceptives can cause acne, headaches, fluid retention, high blood pressure, or blood clots. Women who smoke have a greater risk of blood clots and vascular problems from birth control pills.

To avoid the drawbacks of taking—or forgetting to take—a daily pill, some women opt for hormone injections, skin patches, vaginal rings, or implants. *Depo-Provera* is a progesterone injection that lasts three months. *Ortho Evra* is a patch that slowly releases hormones through the skin into the bloodstream. The user applies a fresh patch every week for three weeks straight, followed by one week with no patch, during which she menstruates. *NuvaRing*, a flexible ring about 2 inches in diameter, is inserted into the vagina, where it slowly releases hormones. The ring remains in place for three weeks, and then is removed for one week to allow menstruation. Injections, patches, and rings can all have side effects similar to those of hormone pills, including weight gain, headaches, and irregular menstrual periods. *Implanon* and *Nexplanon* are matchstick-size rods containing synthetic progesterone that are implanted under the skin; both provide effective contraception for up to three years.

Hormonal methods of contraception have the advantage of not interfering with sexual activity or foreplay in any way. However, none of them protect against sexually transmitted diseases (STDs).

A variety of contraceptive methods are shown in Figure 16.10.

✔ What natural physiological state are these hormonal birth control methods most similar to? Explain.

IUDs are inserted into the uterus

Intrauterine devices (IUDs) are small pieces of plastic or metal that are inserted into the uterus by a health care provider. They should be removed only by a health care provider.

IUDs create a mild chronic inflammation that prevents either fertilization or implantation of the fertilized egg in the uterine wall. On the positive side, they are relatively effective and, like hormonal methods, do not interfere with sexual activity. However, risks include uterine cramping and bleeding, infection, and possible damage to the uterus. IUDs offer no protection from STDs.

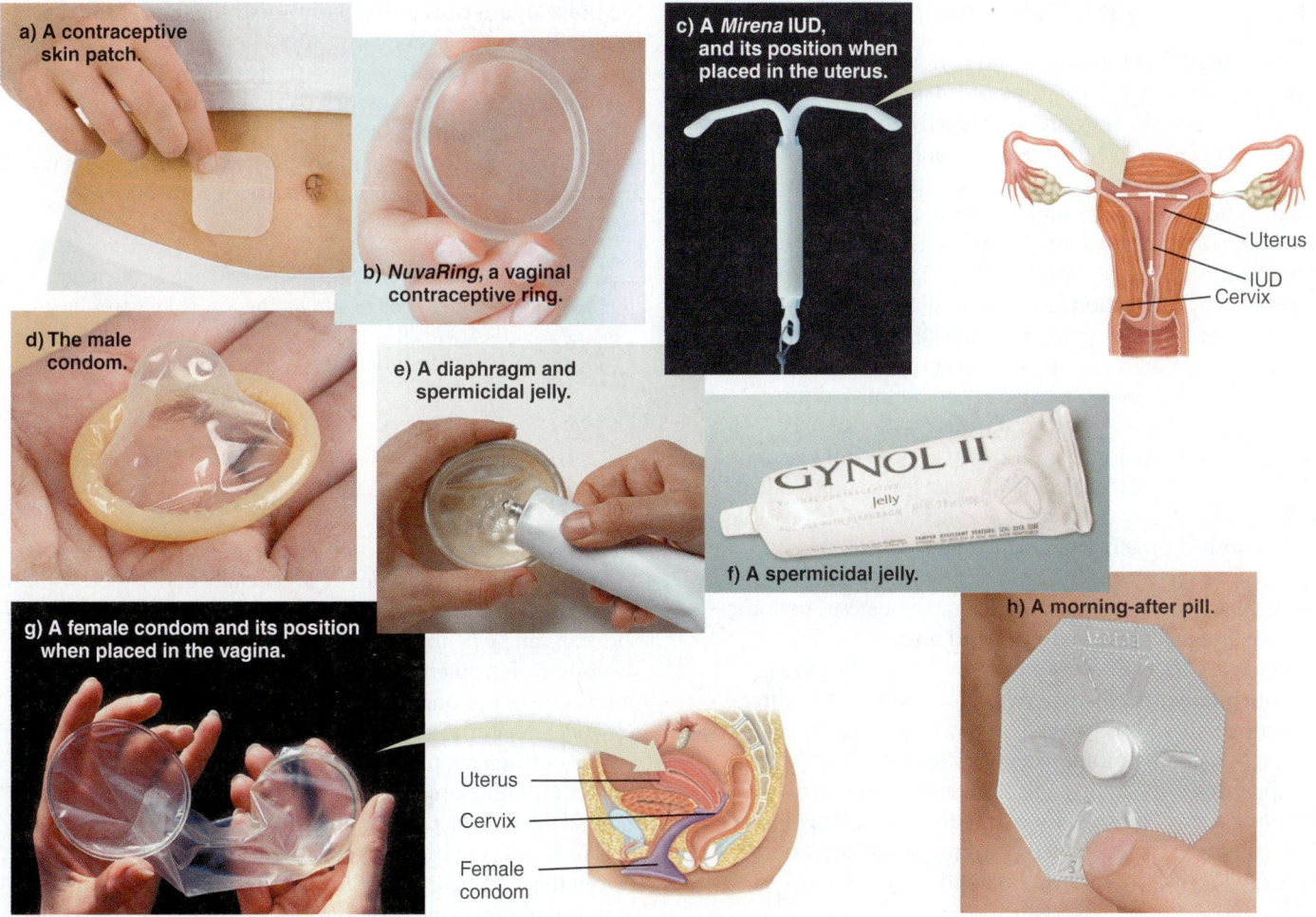

a) A contraceptive skin patch.

b) *NuvaRing*, a vaginal contraceptive ring.

c) A *Mirena* IUD, and its position when placed in the uterus.

Uterus
IUD
Cervix

d) The male condom.

e) A diaphragm and spermicidal jelly.

GYNOL II
Jelly

f) A spermicidal jelly.

g) A female condom and its position when placed in the vagina.

Uterus
Cervix
Female condom

h) A morning-after pill.

Figure 16.10   Some birth control methods.

An IUD called *Mirena* contains small amounts of a progesterone-like steroid. The steroid decreases the possibility of conception and also reduces menstrual flow, a significant benefit to women who normally experience heavy bleeding during their periods. *Mirena* is effective for up to five years.

Diaphragms and cervical caps block the cervix

Diaphragms and *cervical caps* are latex devices that a woman inserts into her vagina. They cover the cervical opening and prevent sperm from entering the uterus.

Both types of devices must be fitted initially to the user's cervix. A disadvantage is that they must be inserted shortly before intercourse and then removed sometime later, making sexual activity less spontaneous. However, they are fairly effective, especially if inserted correctly and used with chemical spermicides. Diaphragms and cervical caps do not protect against STDs.

Chemical spermicides kill sperm

Chemical *spermicides*, which destroy sperm, are available in a wide variety of forms, including foams, creams, and jellies.

Their failure rate as a contraceptive method is relatively high compared to other available methods. To be effective, they must be inserted into the vagina shortly before intercourse, not after. Some chemical spermicides also kill organisms that cause sexually transmitted diseases; others do not.

Condoms trap ejaculated sperm

Despite their decidedly low-tech approach (a condom is just a balloon, really), *condoms* remain one of the most popular forms of birth control. They are fairly effective if used properly. Combining them with a chemical spermicide makes them even more effective. They also offer some protection against STDs.

The male condom is a soft sheath, made of latex or animal membrane, that encloses the penis and traps ejaculated sperm. There is also a female condom that is essentially a polyurethane liner that fits inside the vagina. Condoms made of latex (but not animal membrane) have the tremendous advantage of offering some protection against diseases, including AIDS.

Withdrawal and periodic abstinence

Two other methods, withdrawal and periodic abstinence, are not very effective in preventing pregnancy. In fact, for a sexually active couple, they are so ineffective in the long run that some people believe they should not be classified as birth control at all. Couples who are serious about preventing pregnancy should consider more effective techniques.

In principle, the *withdrawal* method means that the man withdraws his penis from the vagina at just the right moment before ejaculation, so that ejaculation occurs outside the vagina. In practice, it takes skill and commitment on the part of the male and a lot of trust on the part of the female. In addition, some sperm can be flushed into the ejaculatory duct even before ejaculation, so there is a possibility of fertilization even if the man does not ejaculate.

People who practice *periodic abstinence* (sometimes called the *rhythm method*) rely on the fact that fertilization is possible for only a limited time in each menstrual cycle, defined by the life span of sperm and the time during which an oocyte can be fertilized after ovulation. As its name suggests, periodic abstinence requires the woman to avoid intercourse for about an eight-day span every month in the middle of her cycle, from approximately five days before ovulation to three days after ovulation. It is possible to determine the day of ovulation by carefully monitoring body temperature, because the temperature rises slightly—several tenths of a degree—at ovulation and stays elevated for about three days. However, it is difficult to determine which five days *precede* ovulation, because obviously it has not yet occurred.

Pills that can be used after intercourse

When intercourse has occurred in the absence of any birth control method, couples have several options that may either prevent pregnancy or terminate it early on. A product called *Ella* (ulipristal acetate) is a single pill that, if started within five days *after* intercourse, may prevent a pregnancy. Ella is thought to work by preventing ovulation if it has not already occurred. It may also prevent attachment to the uterus (implantation) even after ovulation, but this is not well documented. Another after-intercourse pill, called *Plan B*, contains progesterone. Both Ella and Plan B are commonly called "emergency contraception" or "morning-after" pills. Plan B is now available without a prescription to persons aged 17 and older; Ella requires a prescription.

More controversial is *Mifeprex* (mifepristone, formerly called RU-486), a drug that *blocks* the action of progesterone. By blocking the action of progesterone, Mifeprex can prevent the endometrial lining from proliferating. If taken in the first few days after intercourse, it can prevent pregnancy by preventing implantation of the pre-embryo in the uterine lining. However, Mifeprex is more commonly prescribed *after* a pregnancy has been identified, because its main advantage is that it is effective up to seven weeks after pregnancy. By causing the endometrial lining to regress to the point that it can no longer nourish the fetus, Mifeprex essentially causes a very early-stage chemical abortion. For this reason, Mifeprex has generally been labeled an "abortion pill" rather than an emergency contraception. Mifeprex first became available in France in 1988, but concern over its mode of action caused so much controversy that it was not approved for sale in the United States until 2000.

MJ's BlogInFocus Which state passed one of the most restrictive laws regarding the use of Mifeprex in 2014: Was it *your* state? Visit MJ's blog in the Study Area in MasteringBiology and look under "Mifeprex."

http://goo.gl/ZbWJcu

Elective abortion

An *elective abortion* is an elective termination of a pregnancy. In common usage, the term *elective* is generally understood without being stated. Aside from the very early-stage use of Mifeprex, an abortion can be performed in several ways, including vacuum suctioning of the uterus, surgical scraping of the uterine lining, or infusion of a strong saline solution that causes the fetus to be rejected. Sometimes abortions are performed in cases of incest or rape or when continuing the pregnancy would endanger the mother's health. Some women choose to have an abortion if medical tests reveal a nonviable fetus.

Abortions are highly controversial. Some people support a woman's right to choose her own reproductive destiny, whereas others support an embryo's right to life. The beliefs on both sides are often based on strongly held moral and religious convictions.

The future in birth control

Several new forms of birth control are still in the research and development stage. Possibilities include a male birth control pill that reduces sperm production and several vaccines for women. For instance, a vaccine is being developed to immunize women against hCG, the hormone that maintains the corpus luteum so implantation can occur. Also under development is a vaccine against sperm.

As it stands today, women have most of the choices. It's worth knowing what the options are so that one's choices can be well informed.

↩ Recap Surgical sterilization should be considered a permanent method of birth control. Hormonal methods—pills, injections, patches, and rings—are also relatively effective but can have side effects. Physical barriers (diaphragms, cervical caps, and condoms) and chemical spermicides are moderately effective; a few afford some protection against diseases. IUDs are fairly effective against pregnancy but do not protect against diseases. Withdrawal and periodic abstinence are not effective forms of birth control in the long term. Abortion is an elective but controversial procedure that terminates a pregnancy. ∎

16.6 Infertility: inability to conceive

If both partners are healthy, pregnancy is usually not difficult to achieve: The annual pregnancy rate without any birth control averages 85–90%. Timing can improve the odds. For example, if a couple wants to increase their chance of having a child, they might have intercourse near the middle of the menstrual cycle, around the time of ovulation.

Some couples find it difficult to become pregnant. Next, we describe conditions that can impair fertility and solutions that are available.

Infertility can have many causes

A couple is considered infertile if they fail to achieve pregnancy after a year of trying. By this definition, about 15% of all couples in the United States are infertile. Infertility is about equally attributable to men and women.

Male infertility is the insufficiency or lack of normal, healthy sperm. The chance of any one sperm achieving reproductive success is so low that men whose sperm count is below 60 million per ejaculation are, for all practical purposes, infertile. Reasons for a low sperm count include low testosterone levels, immune disorders that attack the sperm, radiation, certain drugs such as anabolic steroids, and diseases such as mumps and gonorrhea.

Sperm function best at temperatures slightly cooler than body temperature. One simple measure that may raise the viable sperm count is to switch from tight-fitting underwear, which hold the testes close to the body, to looser-fitting boxer shorts.

Causes of female infertility are more variable. Abnormal production of LH or FSH may limit the development of oocytes into follicles. Irregular menstrual cycles can make it difficult to time the period of ovulation. Some women produce strongly acidic vaginal secretions or thick cervical mucus that damage sperm or make it difficult for sperm to move toward the egg. Uterine tumors may interfere with implantation.

One of the most common reasons for female infertility is *pelvic inflammatory disease (PID)*, a general term for any extensive bacterial infection of the internal female reproductive organs. PID that involves the oviducts can result in scar tissue that seals the oviducts shut, preventing passage of either eggs or sperm (Figure 16.11).

Some women (1–3%) develop *endometriosis*, a condition in which endometrial tissue migrates up the oviduct during a menstrual period and implants on other organs such as the ovaries, the walls of the bladder, or the colon. There it grows, stimulated each month by the normal hormonal cycle. Endometriosis can cause pain and infertility, but can sometimes be corrected by surgery, drugs, or hormone therapy.

Female reproductive capacity is more affected by age than male reproductive capacity. By the mid-40s, women begin to run out of oocytes. In addition, the ovaries become less responsive to LH and FSH, leading eventually to menopause and the end of menstrual cycles. Oocytes ovulated later in life are also more likely to have been damaged by years of exposure to radiation, chemicals, and disease.

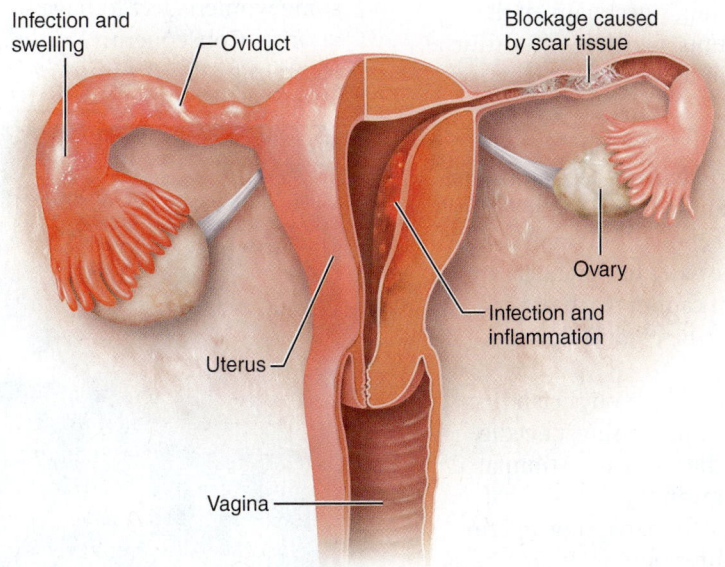

Figure 16.11 Pelvic inflammatory disease (PID) can cause infertility. Bacterial infections of the uterus and oviducts can lead to inflammation and scarring; in the oviduct, scarring can completely seal off the oviduct.

A common cause of failure to achieve reproductive success even when the couple is fertile is spontaneous abortion, or *miscarriage*, defined as the loss of a fetus before it can survive outside the uterus. Perhaps as many as one-third of all pregnancies end in miscarriages, some of them so quickly that a woman has no idea she ever conceived.

✅ A woman who has had several previous miscarriages will sometimes be given a prescription for progesterone pills if she becomes pregnant again. Why?

Enhancing fertility

Several options are available for infertile couples who wish to have a child. None guarantee success and most are expensive. It is not unusual for couples to spend tens of thousands of dollars, even $200,000 or more, to conceive a child by some of these methods. Some couples choose to hire a surrogate mother.

Artificial insemination By far the easiest, most common, and least expensive technique is *artificial insemination*. Sperm are placed, with a syringe, into the vagina or uterus, as close to the time of ovulation as possible. Artificial insemination is the method of choice when the man's sperm count is low, because his sperm can be collected over a period of time and all delivered at once. It is also a good option for men who produce no sperm, in which case the woman may receive sperm from an anonymous donor who has sold his semen to a "sperm bank," and for single women who make the personal choice to become a parent. However it is not an option when the problem is a woman's failure to conceive. The overall success rate (pregnancy) through artificial insemination is only about 20% per ovulatory cycle, but about 60–70% after six attempts or more. However, the percentages are somewhat misleading since many of the sperm donors have low sperm counts, which is why artificial insemination is being attempted in the first place.

Artificial reproductive technologies Artificial reproductive technologies (ARTs) refer to all techniques in which both sperm and eggs are handled outside the body. ART techniques are the method of choice for couples who are unable to conceive because the woman's oviducts are damaged in such a way that sperm cannot reach the egg. The most usual cause is scarring of the oviduct due to a prior PID.

In all of the various ART techniques, a physician first harvests immature eggs from the woman using a needle guided by ultrasound. Generally, the woman is stimulated with fertility-enhancing drugs in advance to ensure that a sufficient number of eggs are ready for harvesting. Sperm is collected from the man. From this point on the methods vary slightly.

The simplest and least expensive ART technique is *in vitro fertilization (IVF)*. In IVF, the eggs are allowed to mature in laboratory glassware under sterile conditions and then sperm are added. The method is called *in vitro* fertilization because the eggs are fertilized outside the body, that is, under glass (*vitreus* is Latin for "glass"). Two to four days later, when it is clear that fertilization has occurred because cells are dividing (Figure 16.12), the embryo is inserted into the woman's uterus via the vagina. IVF with vaginal insertion results in a live birth only about 20% of the time, so some couples try it several times before achieving pregnancy.

MJ's BlogInFocus What is an embryoscope, and how is it improving the efficiency and safety of IVF procedures? Visit MJ's blog in the Study Area in MasteringBiology and look under "The Embryoscope."

http://goo.gl/m4ei3T

Two variations on the IVF method that improve success rates are *gamete intrafallopian transfer (GIFT)* and *zygote intrafallopian transfer (ZIFT)*. In GIFT, unfertilized eggs and sperm are placed directly into an oviduct (Fallopian tube) through a small incision in the woman's abdomen, without waiting for an embryo to develop. In ZIFT, the egg is first fertilized outside the body and then placed into the oviduct (below any point of scarring, of course).

There are now over 450 ART clinics in the United States that can help infertile couples achieve their dream of having a child. The overall success rate at achieving a live birth from an ART cycle (a cycle of fertility drugs, egg collection, and insertion) is now about 30%. Because the chances of pregnancy in any one cycle are low, some women elect to have several embryos transferred at once. Consequently, among women undergoing ART techniques about a third of all live births are multiple

Figure 16.12 A human embryo at the eight-cell stage, approximately 2½ days after *in vitro* fertilization.

births. Over 60,000 babies are born each year in the United States as a result of ART techniques.

If the woman cannot produce healthy eggs because of an ovarian problem, ART techniques can be performed with eggs donated by another woman. It's also possible to freeze unused embryos from a cycle so they can be used later if the first GIFT or ZIFT attempt isn't successful.

Fertility-enhancing drugs Medications are available to boost the production of developing eggs. Sometimes these drugs are tried before IVF or given to women who are preparing for IVF so that several eggs may be harvested at once. Occasionally, multiple pregnancies can occur as a result of the medications themselves. When you hear news reports of multiple births of four or more babies, the mother was almost certainly taking fertility-enhancing drugs.

Surrogate motherhood Some couples choose to pay another woman, called a *surrogate mother*, to become pregnant and bear a baby for them. The prospective parents may contribute sperm, eggs, or both.

Surrogate motherhood raises a number of difficult legal issues that the couple and surrogate mother should explore before entering into an agreement. The surrogate must confirm legally that she will give up the baby for adoption to the parents. However, there have not yet been enough court cases to define legal precedents should conflicts arise between the surrogate mother and the prospective parents at a later date.

> ↻ **Recap** Male infertility is an insufficiency or lack of sperm. Causes of female infertility are variable and include failure to ovulate, damage to oviducts, PID, secretions that impair sperm function, uterine tumors, endometriosis, age-related changes, and miscarriages. The choice of options to improve fertility depends on the cause of the infertility. Options include artificial insemination, IVF, GIFT, ZIFT, fertility-enhancing drugs, and surrogate motherhood. ■

16.7 Sexually transmitted diseases

Disorders of the reproductive systems fall into four classes: (1) infertility, (2) complications of pregnancy, (3) cancers and tumors of the reproductive organs, and (4) sexually transmitted diseases. We have already considered infertility and approaches to overcoming it. We discuss potential complications of pregnancy in the chapter on human development (Chapter 21), and we cover cancer in a separate chapter (Chapter 18). In the rest of this chapter, we focus on STDs.

Sexually transmitted diseases (STDs) are transmitted by sexual contact, whether genital, oral-genital, or anal-genital. STDs are on the rise worldwide, apparently because young people are becoming sexually active earlier and birth control reduces the fear of pregnancy. In the United States

alone, over 12 million people contract an STD each year. A quarter of them are teenagers.

STDs are not just a social embarrassment. Some cause serious organ damage, and a few are deadly. Many don't actually affect the reproductive organs, even though they are transmitted sexually. Some STDs are difficult if not impossible to cure, so an ounce of prevention is definitely worth a pound of (attempted) cure.

Now that we have studied the human reproductive systems, you can see why STDs are so common. Most disease organisms cannot live long outside a warm, moist environment. That's why their most common sites of entry are those that are warm and moist, such as the digestive tract, respiratory tract, and, of course, the reproductive organs. Some organisms, such as the virus that causes AIDS, travel in body fluids. Others, like the pubic lice known as "crabs," simply take the opportunity afforded by intimate contact to hop from one human to another.

STD-causing organisms include viruses, bacteria, fungi, protozoa, and even arthropods. Table 16.4 summarizes common STDs, including causes, symptoms, potential complications, and treatments.

Bacterial STDs: syphilis, gonorrhea, and chlamydia

Syphilis The bacterium *Treponema pallidum* causes syphilis, one of the most dangerous STDs if left untreated. Fortunately it is no longer very common. Syphilis develops in three phases, often separated by periods in which the disease seems to disappear. One to eight weeks after infection, the disease enters its *primary phase*. A hard, dry, bacteria-filled sore, called a *chancre*, appears at the infection site. The chancre is not painful and disappears spontaneously after several weeks; infected women may not even notice it.

During the *secondary phase*, which can last several years, the bacteria invade lymph nodes, blood vessels, mucous membranes, bones, and the nervous system. Symptoms include a widespread but non-itchy rash, hair loss, and gray patches representing areas of infection on mucous membranes. Again, the symptoms may disappear, and sometimes the symptoms end there. However, some people progress to the *tertiary phase*, characterized by widespread damage to the cardiovascular and nervous systems, blindness, skin ulcers, and eventual death. A child of a syphilis-infected woman can be born blind, malformed, or dead because the bacteria cross the placenta into the developing fetus. Syphilis can be treated with penicillin.

Gonorrhea Caused by the bacterium *Neisseria gonorrhoeae*, gonorrhea is easily transmitted—roughly 50% of women and 20% of men will contract it after just a single exposure (sex with a gonorrhea-infected person). It can be transmitted to the mouth and throat by oral-genital contact or to the mouth or eyes through hand contact. An infected mother

Table 16.4 Sexually transmitted diseases

Disease	Organism	Symptoms	Complications	Comments
Syphilis	*Treponema pallidum* bacterium	Primary: Chancre at infection site	Secondary: Non-itchy rash, hair loss, gray areas of infection on mucous membranes Tertiary: Widespread damage to cardiovascular and nervous systems, blindness, skin ulcers	Symptoms may disappear spontaneously between phases; can be treated with antibiotics
Gonorrhea	*Neisseria gonorrhoeae* bacterium	Men: Pus from penis, painful urination Women: Painful urination, vaginal discharge	Men: Inflamed testes and sterility Women: Pelvic inflammatory disease, scarring, sterility	May be asymptomatic, or symptoms may disappear even though disease remains; can be treated with antibiotics
Chlamydia	*Chlamydia trachomatis* bacterium	Men: Discharge from penis, burning during urination Women: Vaginal discharge, burning, and itching	Pelvic inflammatory disease, urethral infections, sterility, complications of pregnancy	Can be treated with antibiotics
AIDS (also see Chapter 9)	Human immunodeficiency virus (HIV)	Fatigue, fever, chills, night sweats, swollen lymph nodes or spleen, diarrhea, loss of appetite, weight loss	Infections, pneumonia, meningitis, tuberculosis, encephalitis, cancer	Asymptomatic phase can last for years, even though disease is still contagious; no cure
Hepatitis B (also see Chapter 14)	Hepatitis B virus	Nausea, fatigue, jaundice, abdominal pain, arthritis	Cirrhosis, liver failure	Preventable with a vaccine
Genital herpes	Herpes simplex virus types 1 and 2	Genital blisters, painful urination, fever, swollen lymph nodes in groin	Symptoms disappear during remission, then reappear; contagious during active phases	Drugs available to control active phase
Genital warts	Human papillomavirus (HPV)	Warts on penis, labia, anus, or in vagina or cervix	Risk factor for cancers of cervix or penis	Warts can be removed by surgery, freezing, or drugs. Preventable with a vaccine
Yeast infections	*Candida albicans*	Men: Discharge from penis, painful urination Women: Vaginal pain, inflammation, cheesy discharge	Can develop in women without sexual contact, but is transmitted sexually to men	Can be treated with antifungal drugs
Trichomoniasis	*Trichomonas vaginalis* protozoan	Men: Discharge from inflamed penis Women: Frothy, foul-smelling vaginal discharge	Recurrent urethral infections	Can be treated with drugs
Pubic lice	*Phthirus pubis*	Intense itching, skin irritation in pubic area; lice are small but visible	Can also be transmitted through infected sheets, clothing	Can be treated with drugs

can pass gonorrhea to her child during birth, causing a potentially blinding eye infection.

In men, early symptoms of gonorrhea include discharge of pus from the penis and painful urination, but the symptoms may disappear even though the disease remains. Women may experience a burning sensation during urination and vaginal discharge that may not appear particularly abnormal. However, nearly 20% of infected men

and up to 80% of infected women show no symptoms, at least for periods of time, so they do not seek medical advice. Left untreated, gonorrhea can lead to inflammation, scarring, and sterility.

Gonorrhea is far less common than it was just 20 years ago (**Figure 16.13**). Although gonorrhea generally responds to antibiotics, in recent years antibiotic-resistant strains have emerged.

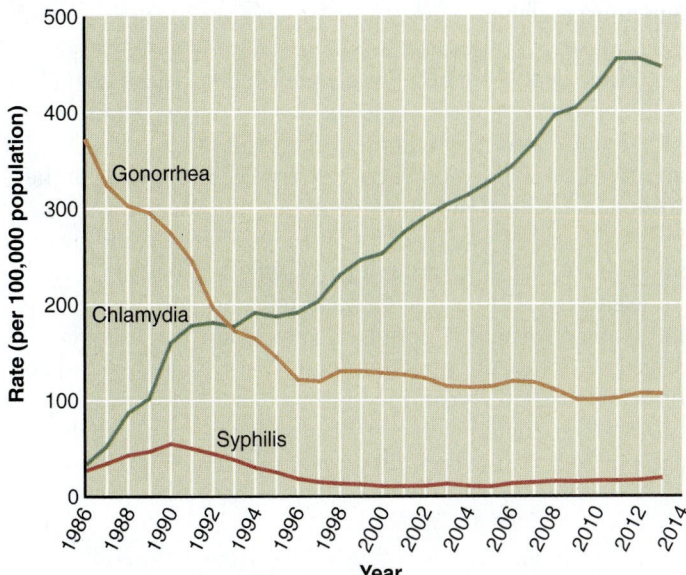

Figure 16.13 **Relative rates of infection for the three most common bacterial STDs.** Data are expressed in terms of numbers of infection per 100,000 population per year in the United States, since 1986. Syphilis rates refer to primary and secondary syphilis.

✔ What could explain the fact that the rate of chlamydia infection is increasing, whereas the rates of the other two STDs are declining?

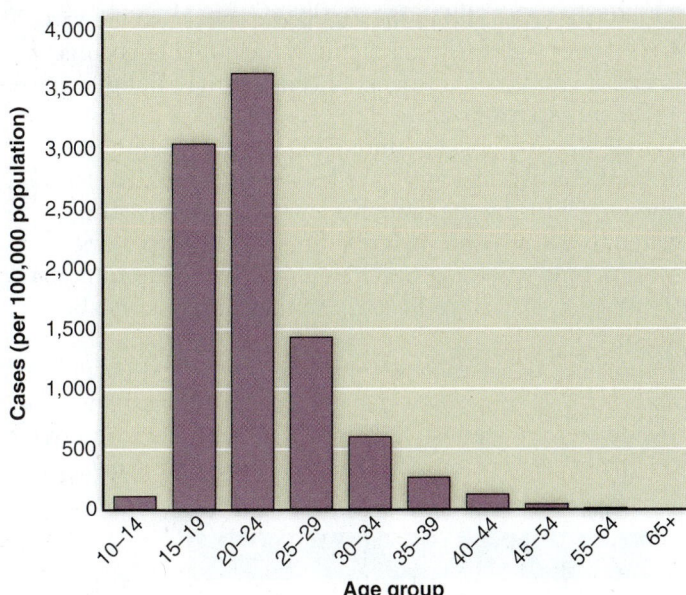

Figure 16.14 **Rates of chlamydia infection in women, by age.** Young women ages 15–24 are particularly likely to contract chlamydia.

Chlamydia

Chlamydia is a bacterial infection caused by *Chlamydia trachomatis*. Although it is easily cured with antibiotics, it often goes undiagnosed and untreated because the symptoms are so mild. Men may experience a discharge from the penis and a burning sensation during urination, and women may have a vaginal discharge and a burning and itching sensation. The overall rate of chlamydia infection seems to have stabilized in the past few years after rising steadily for decades (review Figure 16.13).

Women are diagnosed with chlamydia three times as often as men. Women between 15 and 24 years old are especially at risk, with reported rates of infection nearly 10 times that of the general population (Figure 16.14).

The eventual health consequences for women can be severe despite the apparent mildness of the symptoms. If left untreated, chlamydia can cause PID, permanent scarring of the Fallopian tubes, complications of pregnancy, and sterility. It can also infect a newborn's eyes and lungs at birth, leading to pneumonia. Health officials recommend that all sexually active women under the age of 26 be tested for chlamydia every year.

Viral STDs: HIV, hepatitis B, HPV, and genital herpes

The government does not keep accurate statistics on most of the viral STDs. Some of them are very dangerous; others are just very common.

HIV

Human immunodeficiency virus (HIV), the virus that causes *acquired immune deficiency syndrome (AIDS)*, is one of the most dangerous STDs of our time. One of the most dangerous features of AIDS is that it can be transmitted from person to person long before symptoms of the disease appear.

HIV slowly destroys the immune system over a period of years. Early symptoms include fatigue, fever and chills, night sweats, swollen lymph nodes or spleen, diarrhea, loss of appetite, and weight loss. As the immune system grows weaker, patients experience numerous infections and complications that lead eventually to death. As you learned previously in the chapter on immune system (Chapter 9), there are drugs that can keep HIV in remission, but it cannot be cured.

Hepatitis B

Several viruses can cause *hepatitis* (inflammation of the liver). The hepatitis B virus is transmitted in blood or body fluids during unprotected sex. It is more contagious than HIV but not as deadly. A vaccine for hepatitis B became widely available in 1991, and since that time the incidence of hepatitis B has declined. The government requires all health care workers to be vaccinated against hepatitis B because of their risk of being stuck with a needle on the job.

Human papillomavirus (HPV)

Infection with any one of about 40 HPV viruses can cause **warts** in the genital area, including the penis, the vulva (the area outside the vagina), around the anus, and in the vagina or cervix. Genital warts can be removed by several methods including laser surgery, freezing with liquid nitrogen, or certain drugs, but they have a tendency to reappear.

HPV is the most common STD worldwide. More than 80% of U.S. women will have been infected with at least one HPV virus by the time they are 50. HPV infections generally go away on their own in less than two years. Most people who are infected will have no symptoms at all (fewer than 5% will get warts), so they don't worry about it much.

The real danger of HPV infections is not the warts they may cause. We now know that at least two of the strains (types 16 and 18) are responsible for nearly all cases of cancers of the cervix, as well as many of the cancers of the penis, vulva, labia,

and anus; together, these cancers are responsible for about 5,000 deaths per year in the United States.

These deaths are now largely preventable. In 2006, the FDA approved Gardasil, a vaccine against the two strains of HPV that cause cervical cancer. Three years later, the FDA approved a second vaccine called Cervarix. If everyone in the United States were vaccinated with one of these two vaccines before the age of sexual activity, over 4,000 lives could be saved every year. (The drugs are not 100% effective.) However, to be effective, Gardasil or Cervarix must be given *before* HPV infection; neither drug is effective against an existing infection. For that reason, the Centers for Disease Control and Prevention (CDC) recommends that all children (girls *and* boys) be vaccinated with either Gardasil or Cervarix at about age 11 or 12, before their first sexual contact.

HEALTH & WELLNESS
Have You Had Your Gardasil?

Most women will be sexually active at some point in their lives. Nearly all will contract at least one HPV infection, whether they know it or not. A very few of those infections will lead to cancer. Fortunately, two vaccines are now available to protect against the cancer-causing strains of HPV: Gardasil was approved in 2006, and Cervarix was approved in 2009. An improved version of Gardasil that protects against more strains of HPV became available in 2014.

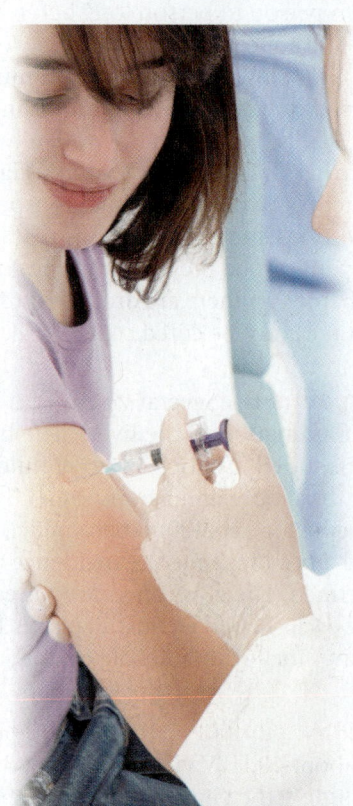

From the time it was first introduced, Gardasil (and now Cervarix) generated controversy. Some parents are concerned that immunizing young girls before they are sexually active, and perhaps even before sex has been discussed with them, might encourage sexual activity at

too young an age. Gardasil was quickly labeled the "promiscuity vaccine" by some people. There is also a public outcry against government intervention whenever states try to make the vaccine mandatory for school attendance. Currently only Virginia, Rhode Island, and the District of Columbia require immunization against HPV for girls in secondary school; Rhode Island also requires boys be vaccinated.

As a result of the controversy and perhaps because vaccination requires three shots over six months at a cost of about $400, not as many girls and young women have been vaccinated so far as health authorities might like. By 2013, only 36% of all girls aged 13–17

For maximum effectiveness, the HPV vaccine should be given to girls before they become sexually active.

had received all three required shots, according to the CDC. These vaccines should be given before girls become sexually active, because neither vaccine will protect against an existing infection. And according to the CDC, over one-third of all young women aged 14–19 are already infected with a high-risk strain of HPV.

The CDC is now recommending that women up to the age of 26 be vaccinated if they haven't already, because there's still a good chance that they have not yet been exposed to both of the two highest risk strains of HPV. Studies are under way to see if the vaccines might even be partially effective for older women as well, up to the age of 45.

Gardasil was approved in 2009 for boys as well, because of increasing evidence that they can protect against cancers of the penis and anus that are caused by HPV. However, from a public health standpoint, vaccinating all boys may not be cost-effective because cancers of the penis and anus are more rare than cervical cancer. Vaccinating boys would most likely provide an indirect benefit for girls by reducing HPV transmission rates, according to the American Academy of Pediatrics.

Have you had your Gardasil yet? Are you planning to?

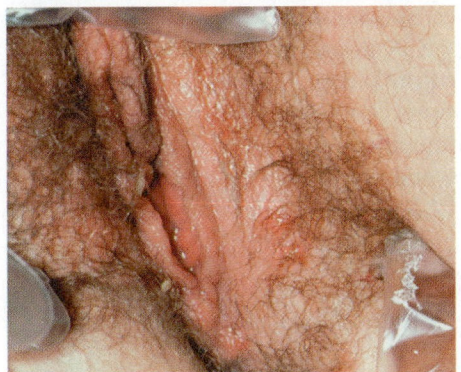

a) **Genital herpes on the external genitalia of the female.** Genital herpes are generally caused by HSV-2.

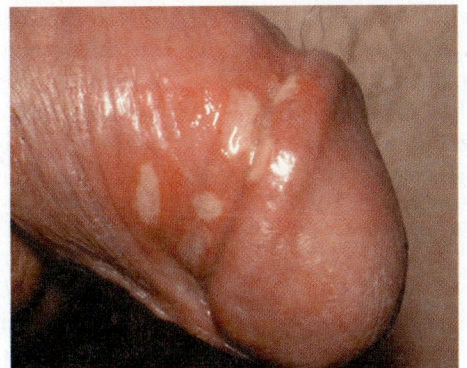

b) **Genital herpes on the penis of the male.**

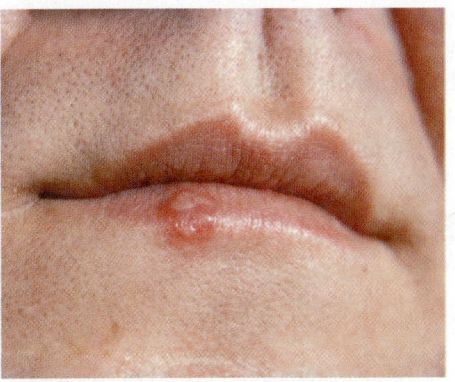

c) **Oral herpes.** Cold sores on the lips (also called fever blisters) generally are caused by HSV-1.

Figure 16.15 **Herpes simplex virus.**

Genital herpes Genital herpes is caused by *herpes simplex viruses*. Generally, it is caused by herpes simplex virus type 2 (HSV-2) as a result of sexual contact or skin-to-skin contact. Less commonly, the type 1 virus (HSV-1) that causes cold sores or blisters around the mouth may be spread to the genital area during oral sex.

Although the government does not keep accurate statistics on genital herpes, it is thought that over 8% of the population is infected. Most infected people have no or minimal symptoms. The primary symptoms are blisters around the genitals or rectum, which eventually break, leaving painful sores that are slow to heal (Figure 16.15). The blisters may recur from time to time, but usually their number and frequency decrease over a period of years. The virus can stay in the body indefinitely.

There is no cure for genital herpes, but drug treatments are available that can suppress the active (and contagious) phase. Like so many STDs, genital herpes can infect infants' eyes during birth and cause blindness if not treated.

MJ's BlogInFocus Does vaccination against the HPV virus change young women's attitudes toward sex or their sexual behavior, as some opponents of the vaccine contend? Visit MJ's blog in the Study Area in MasteringBiology and look under "Vaccination Against HPV."

http://goo.gl/r6YfGg

Other STDs: yeast infections, trichomoniasis, and pubic lice

Yeast infections *Candida albicans* is a yeast (a type of fungus) that is always present in the mouth, colon, and vagina. Normally, it is kept in check by competition from other organisms. Sometimes, however, the *Candida* population may grow out of control in the vagina, leading to pain, inflammation, and a thick, cheesy vaginal discharge. A woman with an active yeast infection can pass it to her

partner. Infected men sometimes develop a discharge from the penis accompanied by painful urination. It is easily treated by antifungal drugs.

Trichomoniasis Another condition that can cause *vaginitis* (inflammation of the vagina) is *trichomoniasis*, caused by the protozoan *Trichomonas vaginalis*. In women, it causes a frothy, foul-smelling vaginal discharge; in men, it inflames the penis and causes a discharge. Trichomoniasis can be transmitted even when there are no symptoms, and reinfection is common if partners are not treated simultaneously. It is treatable with drugs.

Pubic lice Pubic lice are tiny arthropods related to spiders and crabs (Figure 16.16). Commonly called "crabs," pubic lice live in hair but especially prefer pubic hair, jumping from one host to the next during sexual contact. They can also be transmitted by infected bed sheets or clothes. The adult lice lay

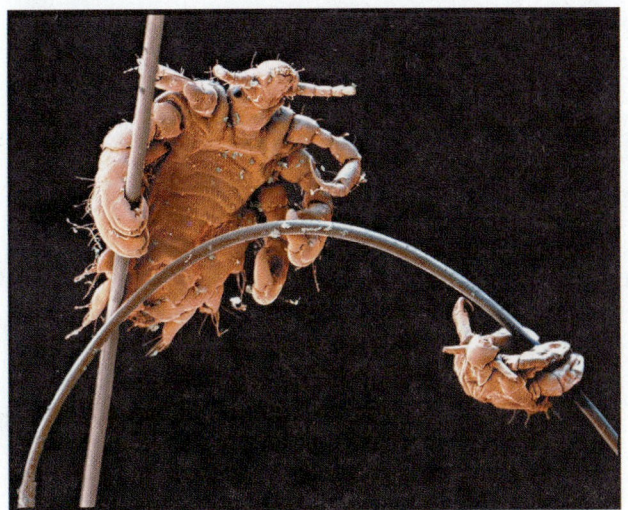

Figure 16.16 **The pubic louse, *Phthirus pubis*.** (approximately ×30)

eggs near the base of the hair, and the eggs hatch in a few days. Pubic lice nourish themselves by sucking blood from their host, causing intense itching and skin irritation; the adult lice are very small but visible with the naked eye. They can be killed with anti-lice treatments, but all clothes and bedding should be thoroughly washed in hot water to prevent reinfestation.

✅ Overall, which sex do you think is more likely to get STDs—women or men? Why?

Protecting yourself against STDs

We've just reviewed a group of diseases with symptoms ranging from painful urination to brain damage and outcomes ranging from full recovery to death. The good news is that you can probably avoid contracting an STD, but you need to know what to do and, more importantly, be willing to do it.

Previously (Chapter 9), we presented a number of safer-sex recommendations that can lower your risk of STDs. Many of them are just commonsense precautions:

- Select your partner wisely. A monogamous relationship with someone you know and trust is your best defense.
- Communicate with your partner. If you have questions or concerns, share them, and be willing to do the same yourself.

- Use suitable barriers, depending on the sexual activity in which you engage. Be aware, though, that barrier methods of birth control are not a cure-all for risky behavior. For example, condoms are only marginally effective in preventing the transfer of HIV, HPV, and genital herpes, and they will not protect against pubic lice.

Finally, know your own health. If you or your partner think you are at risk for (or may have) a certain disease, get tested. It's confidential, it's not difficult, and it can save you a lot of embarrassment. It may even save your life.

↻ **Recap** Major bacterial STDs include gonorrhea, syphilis, and chlamydia. The most dangerous viral STD is HIV. Hepatitis B can be prevented by a vaccine. Genital herpes is irritating but not particularly deadly. HPV can cause warts and is a risk factor for cervical cancer—it, too, can be prevented by a vaccine. Yeast, normally present in the vagina, can multiply and cause a yeast infection. Pubic lice are tiny arthropods that are transmitted during intimate contact with an infected person or by contact with infested clothes or bedding. You can reduce your risk of contracting an STD with a little effort. Choose your partner wisely, use a barrier method of birth control, and, if you suspect you have a disease, get tested promptly. ∎

Chapter Summary

16.1 The male reproductive system delivers sperm *p. 375*

- Sperm are produced in the male reproductive organs, called the *testes*, and stored in the epididymis and ductus deferens.
- Semen contains sperm and the secretions of three glands: the seminal vesicles, the prostate gland, and the bulbourethral glands.
- Tens of millions of sperm are formed every day throughout the male's adult life. The production of sperm is under the control of three hormones: testosterone, LH, and FSH.

16.2 The female reproductive system produces eggs and supports pregnancy *p. 379*

- The female reproductive organs (the ovaries) produce mature oocytes and release them one at a time on a cyclic basis.
- Fertilization of the oocyte (if it occurs) takes place in the oviduct.
- The fertilized egg makes its way to the uterus, where it implants and begins to develop into a fetus.
- The vagina contains glands that produce lubricating fluid during sexual arousal.
- The hormone estrogen causes the mammary glands to enlarge at puberty.

16.3 The menstrual cycle consists of ovarian and uterine cycles *p. 381*

- The cyclic changes in the female reproductive system are called the *menstrual cycle*. The menstrual cycle consists of an ovarian cycle that produces mature oocytes and a uterine cycle in which the uterus prepares for pregnancy.
- The menstrual cycle is controlled by the hormones estrogen, progesterone, LH, and FSH.
- Estrogen produced by the mature follicle causes a surge in LH secretion, triggering ovulation.
- Progesterone produced by the corpus luteum prepares the uterus for implantation.

16.4 Human sexual response, intercourse, and fertilization *p. 384*

- Both males and females are aroused by certain stimuli and respond in ways that facilitate intercourse and ejaculation.
- In males, the penis swells and hardens. In females, glandular secretions provide lubrication.
- Both males and females experience a pleasurable reflex event called an *orgasm*. Orgasm is accompanied by ejaculation, or the expulsion of semen, in the male.
- Sperm deposited in the vagina during intercourse make their way through the cervix and uterus and migrate up the oviducts to the egg. Only one sperm fertilizes the egg.

16.5 Birth control methods: controlling fertility *p. 385*

- The most effective methods of preventing pregnancy are abstinence and female or male surgical sterilization. Sterilization should be considered permanent, though it may be reversible.
- Other effective methods of birth control include IUDs and manipulation of hormone levels with pills, patches, injections, or vaginal rings.
- Moderately effective methods of birth control include condoms, cervical caps and diaphragms, and various spermicides. The "natural" methods (withdrawal and periodic abstinence) are the least effective.
- Morning-after pills are now available to be used as emergency contraceptives up to 72 hours after intercourse.

16.6 Infertility: inability to conceive *p. 389*

- Infertility is defined as the failure to achieve a pregnancy after a year of trying. Nearly 15% of all couples are infertile.
- For male infertility, the primary option is artificial insemination. When the male has a low sperm count but still some viable sperm, the sperm can be concentrated before insemination. In many cases, donor sperm are used.
- Female infertility may be overcome by several methods. In IVF, egg and sperm are combined under laboratory conditions. The fertilized egg is allowed to develop for several days and then inserted into the uterus. In GIFT, the egg and sperm are inserted directly into an oviduct immediately after collection. In ZIFT, the egg and sperm are combined first, and then the fertilized egg is inserted into the oviduct.

16.7 Sexually transmitted diseases *p. 391*

- The common feature of STDs is that they are transmitted during sexual contact. Their disease effects are not necessarily on the reproductive system.
- Bacterial STDs include gonorrhea, syphilis, and chlamydia. Syphilis is the most dangerous; chlamydia, the most common. All are treatable with antibiotics.
- Viral STDs include HIV (the virus that causes AIDS), hepatitis B, genital herpes, and genital warts (HPV). HIV is particularly deadly, and there is as yet no cure. There are now vaccines for hepatitis B and HPV.
- Avoiding the diseases caused by STDs is a matter of reducing your risk of exposure and paying attention to (and taking responsibility for) your own health.

Terms You Should Know

cervix, *380*	ovarian cycle, *381*
clitoris, *380*	ovaries, *379*
corpus luteum, *382*	oviduct, *379*
ductus (vas) deferens, *375*	ovulation, *382*
egg, *375*	penis, *376*
endometrium, *379*	progesterone, *379*
epididymis, *375*	semen, *376*
erection, *376*	sperm, *375*
estrogen, *379*	testes, *375*
follicle, *381*	testosterone, *378*
gametes, *377*	uterine cycle, *382*
menstrual cycle, *381*	uterus, *379*
menstruation, *382*	vagina, *380*
orgasm, *384*	

Concept Review

Answers can be found in the Study Area in MasteringBiology.

1. Name the three accessory glands of the male reproductive system.
2. Describe the development of a sperm cell.
3. Explain where fertilization takes place in the female reproductive system.
4. Describe the phases of the uterine cycle.
5. Describe what is happening in the uterus that leads to menstruation, and explain why menstruation does not occur during pregnancy.
6. Compare and contrast the male and female sexual responses.
7. Describe how IVF is done.
8. Name three viral STDs.
9. Describe birth control options, and list those that afford at least some protection against STDs.
10. Describe the various STDs of worldwide concern.

Test Yourself

Answers can be found in the Appendix.

1. Which of the following might interfere with a sperm's ability to swim?
   a. an acrosome with defective enzymes
   b. a defective flagella
   c. a defective midpiece lacking an appropriate number of mitochondria
   d. both choice (b) and (c)

2. Anabolic steroids are similar in structure and function to testosterone. Which of the following would be a likely outcome in a male taking anabolic steroids?
   a. Testosterone production by the interstitial cells would increase.
   b. Release of GnRH by the hypothalamus would increase.
   c. Testosterone production by the interstitial cells would decrease.
   d. Testosterone production would be unaffected.

3. Which of the following structures is correctly matched with its function?
   a. myometrium: contracts during labor and child birth
   b. cervix: transports oocyte from ovary to uterus
   c. oviduct: transports oocyte from uterus to vagina
   d. endometrium: birth canal

4. Which of the following statements about oocytes and ovulation is correct?
   a. The production of primary oocytes continues in women from the onset of puberty until the onset of menopause.
   b. Meiotic divisions are complete at the time of ovulation.
   c. At ovulation, the structure released from the follicle includes a secondary oocyte and polar body.
   d. All of these statements are correct.

5. Which of the following would provide the most accurate prediction of ovulation?
   a. a surge of progesterone
   b. a sudden drop in FSH
   c. a sudden drop in estrogen
   d. a surge of LH

6. What is the role of the corpus luteum during early pregnancy?
   a. It produces the hormones that will support the early pregnancy.
   b. The fertilized egg will implant in the corpus luteum.
   c. The corpus luteum nourishes the rapidly growing embryo.
   d. The corpus luteum protects the fertilized egg during its transport to the uterus.

7. Which of the following lists the female reproductive structures that sperm will pass through on their way to fertilize an oocyte, in the correct order?
   a. cervix . . . vagina . . . oviduct . . . ovary
   b. cervix . . . oviduct . . . vagina . . . uterus
   c. uterus . . . cervix . . . vagina . . . ovary
   d. vagina . . . cervix . . . uterus . . . oviduct

8. The prescription drugs used currently to treat erectile dysfunction primarily affect:
   a. testosterone levels
   b. blood flow to the penis
   c. contraction of skeletal muscles
   d. sympathetic nervous system

9. Which two means of birth control are most similar in the way they prevent pregnancy?
   a. tubal ligation and IUDs
   b. oral contraceptives and spermicides
   c. tubal ligation and condoms
   d. oral contraceptives and hormone patch

10. The most widely used hormonal methods of birth control:
    a. prevent fertilization by killing sperm
    b. prevent ovulation by inhibiting release of FSH and LH
    c. prevent implantation into the endometrium
    d. block transport of fertilized egg to uterus

11. If a woman is infertile as a result of oviducts blocked by the scarring of PID, which of the following methods would enable her to become pregnant?
    a. artificial insemination
    b. hormonal treatments
    c. *in vitro* fertilization
    d. all of the above

12. Which of the following STDs is most likely to infect a fetus and cause birth defects?
    a. syphilis
    b. gonorrhea
    c. chlamydia
    d. trichomoniasis

13. Which of the following STDs can be prevented by vaccination?
    a. genital herpes
    b. syphilis
    c. gonorrhea
    d. hepatitis B

14. Which of the following STDs is associated with the development of cervical cancer?
    a. genital herpes
    b. human papillomavirus infection
    c. hepatitis B infection
    d. trichomoniasis

15. PID may be associated with:
    a. chlamydia
    b. syphilis
    c. gonorrhea
    d. both (a) and (c)

Apply What You Know

Answers can be found in the Study Area in MasteringBiology.

1. Male sexual responsiveness and secondary sex characteristics remain unchanged after a vasectomy. Explain why this is so.
2. What keeps the ovarian and uterine cycles always exactly in phase with each other, even though the length of the menstrual cycle can vary slightly?
3. Why is it important for a new mother to breast-feed her baby immediately after birth if she wishes to breast-feed the baby once she gets home?
4. What provides the force that pushes the baby out of the uterus during childbirth?
5. Both women and men can have an STD and not have any symptoms. What might be the long-term consequences of an untreated STD?
6. If a woman does not have an intact hymen does that mean she's not a virgin?

MJ's BlogInFocus

Answers can be found in the Study Area in MasteringBiology.

1. In some countries (China and India are examples), there are reportedly more boys than girls born each year. What are some possible explanations for this phenomenon? Visit MJ's blog in the Study Area in MasteringBiology and look under "Gender Preference." http://goo.gl/KHxnUH

2. In Table 16.3, it is reported that the failure rate of condoms is 18%. Why do you think the failure rate is so high? (Hint: The withdrawal method is not 100% effective, either.) Visit MJ's blog in the Study Area in MasteringBiology and look under "Birth Control Method Failures." http://goo.gl/OaHGvE

MasteringBiology®

Students Go to MasteringBiology for assignments, the eText, and the Study Area with animations, practice tests, and activities.

Professors Go to MasteringBiology for automatically graded tutorials and questions that you can assign to your students, plus Instructor Resources.

Answers to  questions are available in the Appendix.

Cell Reproduction and Differentiation

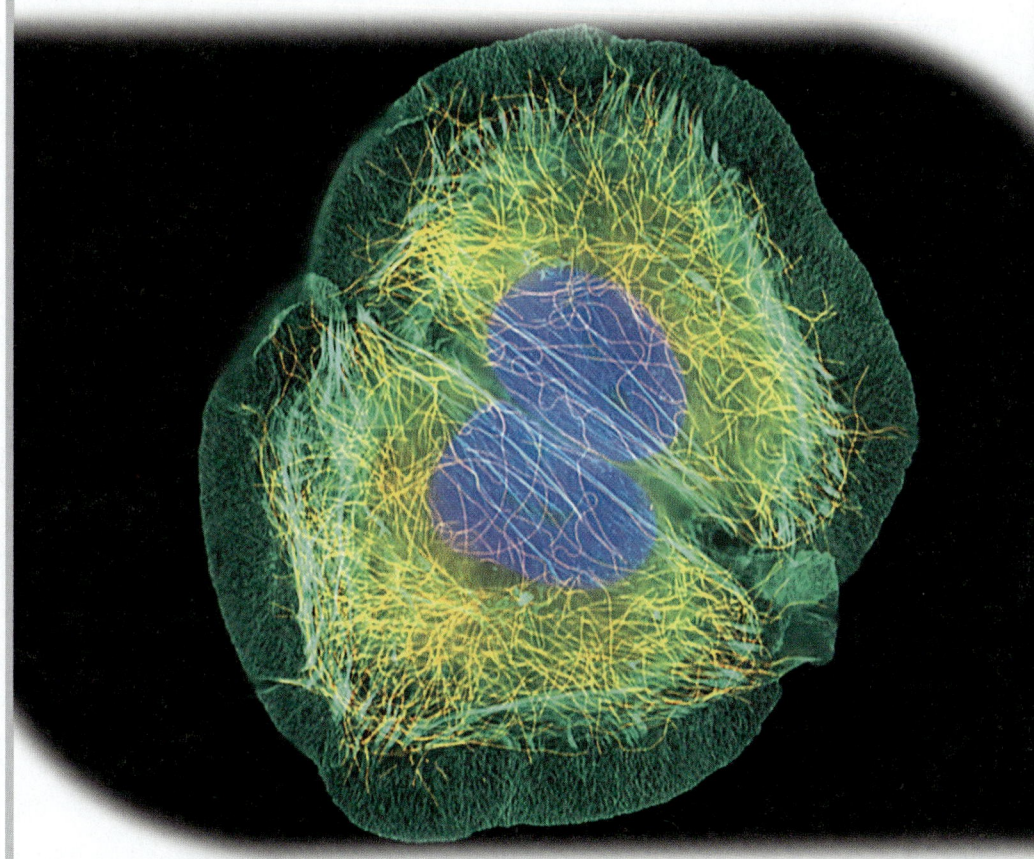

Immunofluorescence light micrograph of a human cell in the process of dividing into two daughter cells.

Key Concepts

- **All cells in every living organism were created by division of previously existing cells.** The exception is a fertilized egg, produced from the union of two cells (sperm and egg).

- **Before a living cell divides into two, its genetic material must be copied completely and accurately.**

- **In most multicellular organisms, cells *differentiate* during development**, taking on specialized forms and functions.

- **Cell division and differentiation are highly regulated.** Some cells rarely (if ever) divide in an adult organism; others continue to divide throughout life.

CURRENT ISSUE
Therapeutic Cloning

Imagine that you're in the hospital because you've just had a massive heart attack. The doctors have stabilized you for the moment, but there is a significant area of damage to heart muscle that was deprived of oxygen for too long. The damaged muscle cells have already died. If you live (and that's still in question), the dead cells will eventually be replaced by scar tissue, leaving your cardiac function forever compromised.

Now imagine that the year is 2040. As soon as you arrive at the hospital, a doctor from the new therapeutic cloning department takes a small scraping of skin cells from your arm. With cutting-edge therapeutic cloning techniques developed through research over the last 25 years, the doctor will turn back the clock on your mature skin cells, ultimately transforming them into cells as undifferentiated as the fertilized egg from which you arose. Then, with the right combination of enzymes and nutrients at just the right moment, the cells are coaxed along a pathway of growth, reproduction, and differentiation that ultimately turns the cells into immature (precursor) heart muscle cells. (Cell differentiation is the complex process whereby a single cell [the egg] divides and becomes specialized for a particular tissue, such as heart muscle, neurons, and skin.) Within a week of coaxing the cells to differentiate, the precursor cells are microinjected into the damaged area of your heart. There, the cells continue to grow, divide, and differentiate until they become mature heart muscle cells. Within a month, your heart muscle is completely repaired. After that, a surgeon replaces your partially blocked coronary arteries with new blood vessels made from smooth muscle cells grown on a tubular scaffold of collagen/

elastin fibers. The smooth muscle cells, too, are grown from your own skin cells. The tubular scaffolds are prepared from pig arteries after the muscle cells are dissolved away.

Sound too futuristic? Welcome to the brave new world of *therapeutic cloning*, defined as the use of undifferentiated, genetically identical cells to repair or replace damaged tissues or organs. Right now, therapeutic cloning is more a dream than a reality, but it's a dream that is likely to come true in your lifetime. Every year, we learn more and more about the mysteries of cell differentiation and how to replicate it.

There are signs of progress toward making therapeutic cloning a reality. The first step is cloning, creating undifferentiated cells that are genetically identical to the patient. We already know how to do this. However, the only practical technique for cloning at the moment involves fusing a mature cell from the patient with a human egg, growing that embryo to an 8- or 16-cell stage, and then sacrificing the embryo to harvest embryonic stem cells. To many people, this raises serious ethical issues about human life: At what point

Questions to Consider

1 How do you feel about the creation and then sacrifice of an early-stage human embryo to produce stem cells for a transplant patient? Should it be permitted?

2 How far should we go—to what lengths and at what expense—to prolong life? Are there limits?

in embryonic development is it unethical to sacrifice the embryo? It remains to be seen whether in the future embryonic sacrifice continues to be the primary means by which to clone undifferentiated cells. Scientists are working on ways to cause a mature cell from a patient to revert to an undifferentiated stage (as in the imagined scenario of 2040). But while scientists may be able to figure this out, it is not possible just yet. A report in 2014 of success in producing undifferentiated cells from mature skin cells proved to be premature, when other scientists were unable to duplicate the feat.

The second step of therapeutic cloning would be to coax an undifferentiated cell down a particular path of development so that it becomes the kind of cell needed by the patient. Although this step has proved to be difficult, it's not impossible.

A few cell types have been successfully produced from stem cells, including nerve cells, muscle cells, and pancreatic cells. Nerve cells produced from embryonic stem cells have been used to treat a mouse model of Parkinson's disease with partial success, leading to the hope for the treatment of that and other neurologic diseases, such as Alzheimer's, or neurological conditions, such as damaged spinal cords. Embryonic stem cells have also been used to create newly differentiated pancreatic cells capable of producing insulin, raising the possibility →

Nerve cells (red) produced from human embryonic stem cells, grown in tissue culture with mouse cells (green).

that therapeutic cloning techniques might someday cure Type I diabetes.

The third step of therapeutic cloning is an area of active research. Techniques are being developed for growing cloned cells on specially designed scaffolds that would allow us to produce functioning heart muscle, blood vessels, and skin. But so far scaffold techniques have not been tried on patients.

A long-term goal in therapeutic cloning is to create whole organs suitable for transplant. Meeting that goal, however, is still a long way off. But who knows? We may see the feat accomplished sooner than we think.

SUMMARY

- Therapeutic cloning is defined as the use of cloning techniques to produce cells that can be used to repair or replace damaged tissues or organs.
- Undifferentiated cells genetically identical to the patient currently must come from early-stage human embryos. Techniques are being developed to produce them from fully differentiated adult cells.
- Several newly differentiated cell types, including muscle, nerve, skin, and pancreatic cells, have been created in the laboratory.
- Routine treatment of human medical conditions by therapeutic cloning is still not feasible.
- The ability to produce whole organs for transplant is a long-term goal.

Cells reproduce by dividing into two. Cell *reproduction* is one of the defining features of life. For single-celled organisms, reproduction is accomplished by cell growth followed by cell division; a single-celled organism becomes two organisms. For multicellular organisms such as humans, reproduction is more complicated.

You started life as just a single cell created by the union of your father's sperm and your mother's egg. By the time you were born, you were already composed of over 10 trillion cells, each a result of cell reproduction. But something else was happening as well. As your cells continued to reproduce, they also began to undergo a process known as *differentiation*, whereby cells became different from each other and from their parent cell. For example, some cells differentiated into the ancestors of cells that eventually formed your heart, whereas changes in other cells caused them to become nerve cells or to give rise to cells that ultimately developed into the kidney or the liver. The subjects of this chapter—how cells reproduce and differentiate—are fundamental to the growth and development of every tissue and organ in the body. ▪

Cell division, and to a lesser extent cell differentiation, continues throughout childhood and adolescence as the body grows and matures. Even after adulthood, some cells continue to divide in order to replace cells that die or become damaged. The mechanisms of cell division are the same in all eukaryotes (organisms with cell nuclei) regardless of organism size. In this chapter, we see how the cell's genetic material is copied to prepare for cell division, how the cell divides into two, and how cell growth and division are regulated. Finally, we discuss how cells differentiate to take on special forms and functions.

17.1 The cell cycle creates new cells

The creation of new cells from existing cells involves a sequence of events known as the **cell cycle** (Figure 17.1).

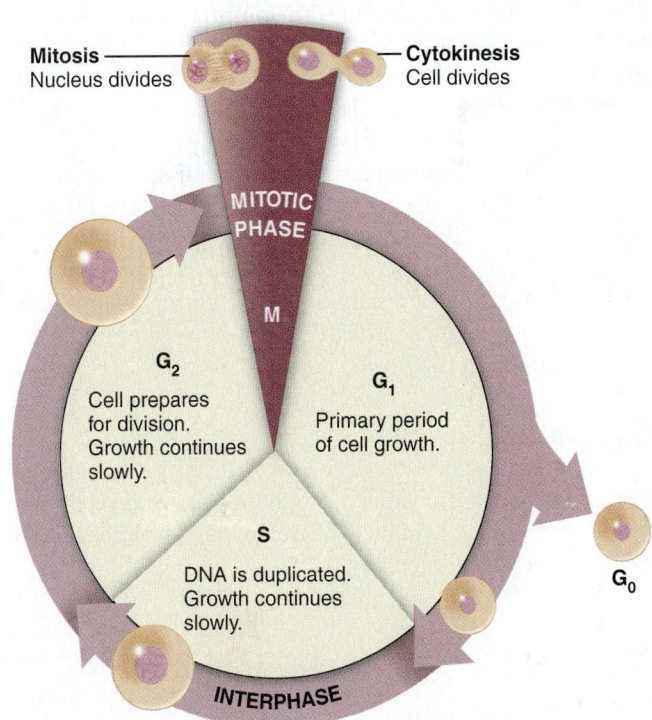

Figure 17.1 **The cell cycle.**

The cell cycle consists of two distinct periods, called *interphase* and the *mitotic phase*.

Interphase is a long growth period during which the cell grows and DNA is duplicated in preparation for the next division. Interphase is divided into three subphases, as follows:

- G_1 *phase.* The cell cycle begins with G_1 ("first gap"), the period between the last cell division and DNA synthesis. At the start of G_1, the cell is at its smallest size. G_1 is a period of very active cell growth.
- *S phase.* The S stands for "synthesis" of DNA. During the S phase, the cell's chromosomes are duplicated. Growth continues throughout the S phase, although at a slower pace than in the G_1 phase.
- G_2 *phase.* During G_2 ("second gap"), the cell continues to grow slowly as it prepares for cell division.

The **mitotic phase** is a much shorter period during which the nucleus and then the cytoplasm divide. The mitotic phase consists of (1) *mitosis*, during which the duplicated DNA is divided into two sets and the nucleus divides, and (2) *cytokinesis*, during which the cytoplasm divides and two new cells, called "daughter" cells, are formed. (The two new cells are called daughter cells regardless of the person's sex.) This cycle repeats itself over and over.

In mammals, a complete cell cycle takes about 18–24 hours if the cells are actively dividing. (Cell cycles as short as 2 hours have been recorded in some nonmammalian species.) DNA replication during the S phase takes about 7–8 hours, and mitosis and cytokinesis generally take only about 30–45 minutes. The remaining 12–18 hours compose G_1 and G_2.

Most of the cells in your body eventually enter a nongrowing, nondividing state called G_0. In tissues that do not need a steady supply of new cells, G_0 is a healthy condition that stops cells from proliferating unnecessarily. Cells in some tissues can be called back from G_0 to begin dividing again, but other tissues cannot restart division. Neurons, cells of the nervous system, and osteocytes, cells in bone, for example, generally remain in G_0 forever after adolescence.

↺ **Recap** Cell reproduction is required for growth and to replace cells throughout life. The cell cycle consists of a long growth phase (interphase), during which the cell's DNA is replicated, and a shorter phase (mitotic phase), during which the nucleus and then the cell cytoplasm divide. ∎

17.2 DNA structure and function: an overview

As described in the chapter on chemistry, **deoxyribonucleic acid (DNA)** is a double-stranded string of nucleotides intertwined into a helical shape. Three billion base pairs of human DNA are packed onto 46 separate structures called

chromosomes, which organize and arrange the DNA within the nucleus (**Figure 17.2**). Chromosomes also contain proteins called *histones* that confer a certain structure to the chromosome molecule. The number of chromosomes varies among species of organisms, but it is always the same for all the individuals of a species. Humans have 46 chromosomes in their cell nuclei.

Throughout most of the cell cycle, the chromosomes are not visible because they are so long and thin. During

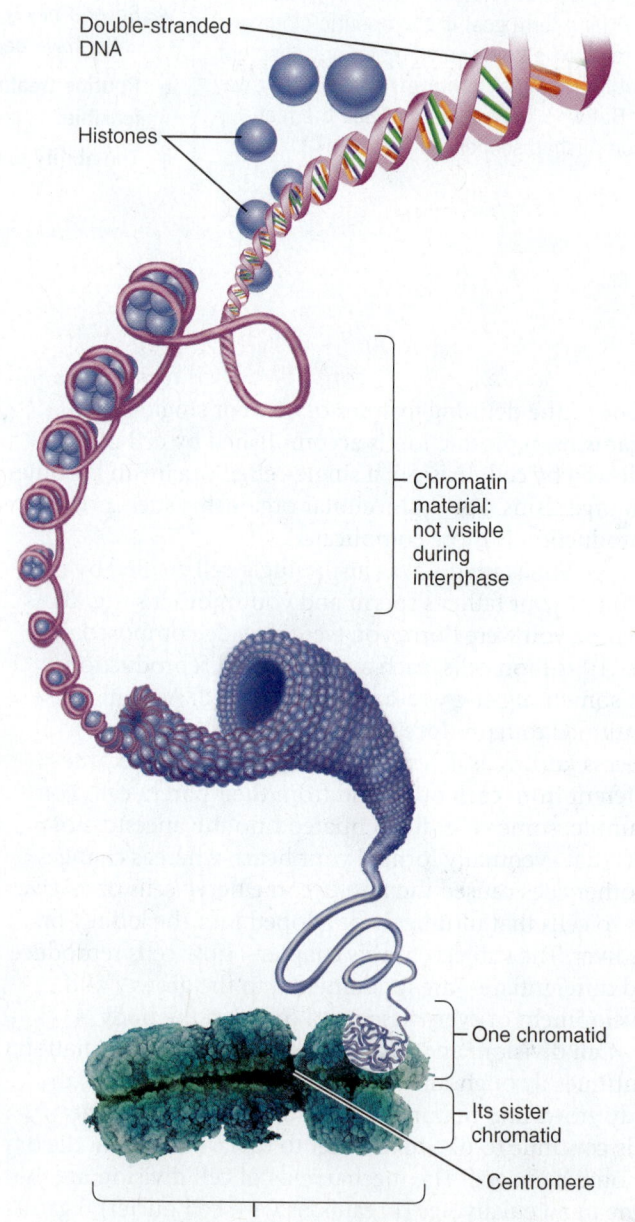

Double-stranded DNA

Histones

Chromatin material: not visible during interphase

One chromatid

Its sister chromatid

Centromere

Chromosome: visible during mitosis

Figure 17.2 The structure of a chromosome. DNA combines with histone proteins to become long, thin strands of chromatin material. During most of the cell cycle the chromatin material of a chromosome is not visible within the cell. During mitosis, the chromatin material condenses and the duplicated chromosome becomes visible as two identical sister chromatids held together by a centromere. The two sister chromatids separate during nuclear division.

interphase, they appear only as a diffuse, grainy substance known collectively as *chromatin* material. However, just before the nucleus divides, the duplicated chromosomes condense briefly into a shorter, compact shape that can be seen under the microscope. Chromosomes are moved around a lot during nuclear division, and their compactness at this stage helps prevent the delicate chromatin strands from breaking. Each visible duplicated chromosome consists of two identical **sister chromatids** held together by a *centromere*.

A **gene** is a short segment of DNA that contains the code, or recipe, for one or more proteins. A single gene is the smallest functional unit of DNA. There are approximately 20,000 genes on the 46 chromosomes, each one located at a precise position on a particular chromosome.

Let's look more closely at three processes essential to DNA function:

- **Replication:** The process of copying the cell's DNA prior to cell division.
- **Transcription:** The process of creating a coded message of a single gene that can be carried out of the nucleus.
- **Translation:** The process of converting the coded message into proteins useful to the cell.

☑ Explain in your own words the difference between genes, chromosomes, chromatin, and chromatids.

Replication: copying DNA before cell division

Before a cell can divide, all of its DNA must be duplicated so that the two new cells resulting from the division each have an identical set of genes. The process of duplicating the entire DNA of a cell, called replication, begins with the uncoiling of the DNA helices. The nucleotides in each strand of the double-helical DNA molecule consist of a sugar group, a phosphate group, and one of four bases (thymine, adenine, guanine, or cytosine). The double helix is formed as the bases pair up, thymine with adenine and guanine with cytosine. This precise pairing facilitates the replication of DNA, because upon the uncoiling and "unzipping" of the two strands, each single strand can serve as a template for the creation of a new complementary strand. The precise nature of the base pairing of the nucleotides (T–A and G–C) ensures that the new complementary strand is exactly like the original complementary strand (**Figure 17.3**). The new nucleotides are positioned and linked together by enzymes called *DNA polymerases*.

Mutations are alterations in DNA

Alterations in DNA are called **mutations.** Mutations may result from mistakes made during DNA replication. In addition, chemicals or physical forces can damage a segment of DNA before it is replicated. Unless these errors are detected and corrected before the next replication, they may be passed on to future cells or could even prevent the DNA from being copied at all. In other words, the DNA needs to be in as good a shape as possible before it is replicated.

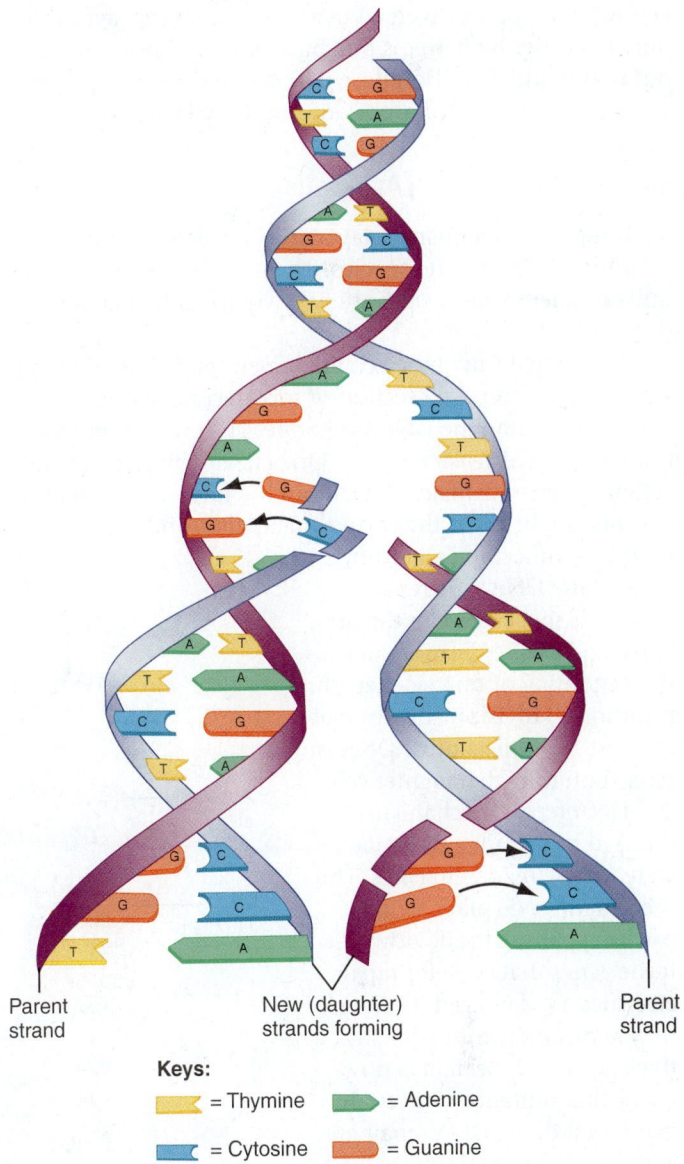

Parent strand New (daughter) strands forming Parent strand

Keys:

= Thymine = Adenine

= Cytosine = Guanine

Figure 17.3 How DNA replicates. The double strand of DNA unwinds, and each single strand serves as the template for a new complementary strand.

If the DNA of a somatic cell (a cell of the body other than a gamete) is severely damaged, it may not be possible to repair all the errors in it. In this case, substantial changes in cell function and even life-threatening cancers may result. For example, skin cancer can be caused by damage to DNA from excessive exposure to sunlight. Mutations of somatic cells, fortunately, are not passed to the offspring.

On the other hand, mutations that occur in the gametes (sperm or egg) may be passed to future generations. These are the changes that lead to evolutionary change, including changes that cause species to diverge. Generally speaking, heritable mutations that lead to species divergence occur very slowly. The DNA of humans and chimpanzees, for example, differs by only about 1% even though our evolutionary paths diverged over 5 million years ago. Two humans differ by perhaps one base pair in a thousand, but that is still 3 million base pair differences, more than enough to account for all of our individual variations.

Mechanisms of DNA repair

DNA repair mechanisms play a crucial role in the survival of an organism and its species. These mechanisms are quite efficient unless they are overwhelmed by massive damage.

DNA repair involves recognizing errors, cutting out and replacing the damaged section or incorrect nucleotide base, and reconnecting the DNA backbone. The process utilizes numerous *DNA repair enzymes*. One enzyme might recognize certain types of damage and cut out the impaired nucleotide, another might insert the correct nucleotide, and yet another might reconnect the adjoining ends of the DNA. The repair process is most active in the hour between DNA replication and the beginning of mitosis. Repair prior to cell division ensures that the best possible copy of DNA is passed on to each daughter cell.

DNA repair mechanisms are directed by certain genes that code for the repair enzymes. Thus, the genetic code plays a vital role in repairing itself. However, if the genes that code for repair enzymes are damaged, DNA errors may accumulate more quickly than normal. Researchers now know that mutations of certain genes that direct DNA repair are associated with an increased risk of several cancers, including forms of colon and breast cancer.

Transcription: converting a gene's code into mRNA

A gene is "expressed" when it is called into action and the protein for which it codes is produced. An essential step in this process is transcription, which involves using the DNA code of a single gene to create a complementary single strand of **ribonucleic acid (RNA)**, specifically a form of RNA aptly named **messenger RNA (mRNA)** because it carries the gene's message to the cytoplasm.

Transcription is different from the process of DNA replication in the following ways (**Figure 17.4a**):

- Only the segment of DNA representing a single gene unwinds, as opposed to the entire molecule.
- RNA is single-stranded, so that only one of the two strands of DNA actually carries the genetic code specifying the synthesis of RNA.
- One of the four complementary base pairs of RNA is different from those of DNA. (Uracil in RNA replaces thymine in DNA.)
- The sugar group of RNA is ribose rather than deoxyribose.

Once it is fully transcribed, the RNA molecule is released from the DNA strand and the two DNA strands entwine around each other again.

How is the specific gene to be transcribed identified on the chromosome? A unique base sequence called the

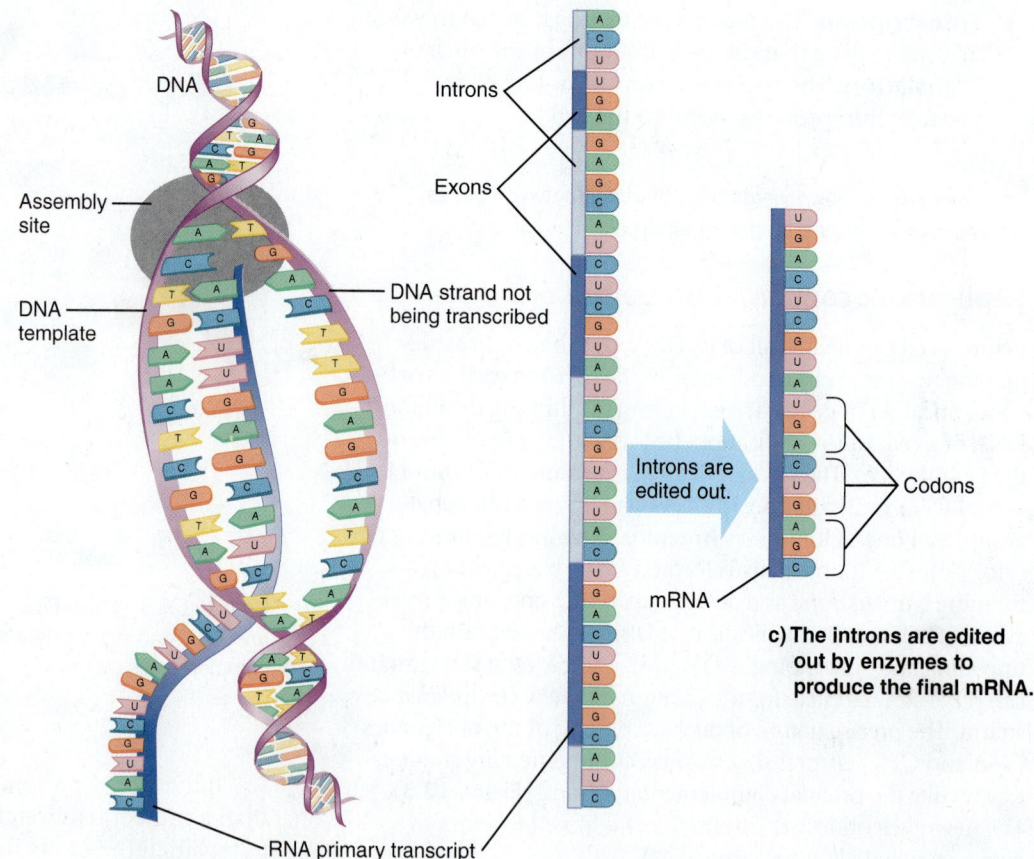

DNA

Assembly site

DNA template

RNA primary transcript

a) The portion of the DNA molecule corresponding to the gene unwinds temporarily, and a complementary strand of RNA is produced from one of the DNA strands.

Introns

Exons

DNA strand not being transcribed

b) The strand of RNA released from DNA is called a primary transcript. It contains sections that carry genetic information, called exons, and sections that may allow different combinations of genetic information, called introns.

Introns are edited out.

mRNA

Codons

c) The introns are edited out by enzymes to produce the final mRNA.

Figure 17.4 Transcription of a gene into mRNA. Transcription takes place entirely within the nucleus.

☑ Mark the place where you would expect to find the promoter. (*Hint:* Notice that transcription is already well under way.)

promoter marks the beginning of every gene. An enzyme called *RNA polymerase* attaches to the promoter, starts the DNA unwinding process, and assists in attaching the appropriate RNA nucleotides to the growing chain. The process ends at another base sequence that identifies the end of the gene.

The RNA molecule first transcribed from DNA is called the *RNA primary transcript* (Figure 17.4b). It is not yet functional because most of the DNA base sequence of a gene (and hence most of the RNA primary transcript) do not carry the code for the amino acid sequence of a protein. Before it can be used to make a protein, the RNA primary transcript must be "edited" by specific enzymes that snip out the sections that do not carry the genetic code, called **introns,** leaving only the sequences that do carry genetic code, called **exons** (Figure 17.4c). Nevertheless, the introns are not just "junk." Evidence indicates that they may allow the exon fragments to be joined in different ways so that the same gene could be used to make slightly different mRNA molecules (and hence slightly different proteins) in different cells.

The mRNA molecule produced from the RNA primary transcript is essentially a template that can be "translated" into a specific sequence of amino acids. The message is encrypted as a *triplet code,* so-called because three successive bases of mRNA, called a **codon,** each code for one of the 20 amino acids. Because there are four possible nucleotide bases and the code is a triplet code, there are precisely 64 different possible codons (4 × 4 × 4). We know exactly which amino acids are specified by the 64 possible codons of mRNA (Figure 17.5). Note the following two important points:

1. Several different codons can specify the same amino acid because there are more possible codons than there are amino acids. For example, both the UUU and UUC codons code for phenylalanine (Phe).
2. The codon AUG specifies a "start" code and three others specify "stop" codes. These are needed to specify where to begin and end the protein.

Some hormones act by binding to the promoters of particular genes in such a way that RNA polymerase can attach much more easily to that gene. What effect would this have on whether or not the gene is "expressed"?

Translation: making a protein from RNA

Translation is the process of using mRNA as a template to convert, or "translate," the mRNA code into a precise sequence of amino acids that composes a specific protein.

The genes of DNA actually produce at least three different kinds of RNA, each with a specific role in the production of proteins. Only mRNA carries the code for producing a new protein. The other two kinds of RNA are called *transfer RNA* and *ribosomal RNA*. Each has an important role in protein synthesis.

Transfer RNA (tRNA) are small molecules that carry the code for just one amino acid. They also carry an *anticodon,* a base triplet that is the complementary sequence to a codon of mRNA. The function of tRNA is to capture single amino acids and then bring them to the appropriate spot on the mRNA chain.

Once they have been captured and brought to the mRNA molecule, the amino acids must be connected together into one long chain in order to become a protein. This is the function of a ribosome. A **ribosome** consists of two subunits composed of **ribosomal RNA (rRNA)** and proteins. A ribosome has binding sites for both mRNA and tRNA. It also contains the enzymes that connect the amino acids together. The function of a ribosome, then, is to hold the mRNA and tRNA in place while joining the amino acids.

Second position

First position	U	C	A	G	Third position
U	UUU Phe	UCU Ser	UAU Tyr	UGU Cys	U
U	UUC Phe	UCC Ser	UAC Tyr	UGC Cys	C
U	UUA Leu	UCA Ser	UAA Stop	UGA Stop	A
U	UUG Leu	UCG Ser	UAG Stop	UGG Trp	G
C	CUU Leu	CCU Pro	CAU His	CGU Arg	U
C	CUC Leu	CCC Pro	CAC His	CGC Arg	C
C	CUA Leu	CCA Pro	CAA Gln	CGA Arg	A
C	CUG Leu	CCG Pro	CAG Gln	CGG Arg	G
A	AUU Ile	ACU Thr	AAU Asn	AGU Ser	U
A	AUC Ile	ACC Thr	AAC Asn	AGC Ser	C
A	AUA Ile	ACA Thr	AAA Lys	AGA Arg	A
A	AUG Met/start	ACG Thr	AAG Lys	AGG Arg	G
G	GUU Val	GCU Ala	GAU Asp	GGU Gly	U
G	GUC Val	GCC Ala	GAC Asp	GGC Gly	C
G	GUA Val	GCA Ala	GAA Glu	GGA Gly	A
G	GUG Val	GCG Ala	GAG Glu	GGG Gly	G

Phe = Phenylalanine	Pro = Proline	Gln = Glutamine	Cys = Cysteine
Leu = Leucine	Thr = Threonine	Asn = Asparagine	Trp = Tryptophan
Ile = Isoleucine	Ala = Alanine	Lys = Lysine	Arg = Arginine
Met = Methionine	Tyr = Tyrosine	Asp = Aspartic acid	Gly = Glycine
Val = Valine	His = Histidine	Glu = Glutamic acid	
Ser = Serine			

Figure 17.5 The genetic code of mRNA. Three successive bases represent a codon, and all but three codons encode for an amino acid. The letters correspond to the four RNA bases: uracil (U), cytosine (C), adenine (A), and guanine (G). To read the chart, start with the letter on the far-left side, then move across to one of the letters at the top, and then down the right side to the appropriate letter. AUG is the "start" codon and the only codon for the amino acid methionine; there are three "stop" codons.

A strand of DNA has the following sequence: TAC GGG CAT ATT. If it is transcribed, what will be the nucleotide sequence of the resulting mRNA, and what will be the amino acid sequence of the final peptide?

Translation occurs in three steps, as illustrated in **Figure 17.6**:

❶ Initiation A particular initiator tRNA binds to the smaller of two ribosome subunits and to the mRNA molecule. The tRNA and the ribosome subunit move along the mRNA until they encounter a "start" (AUG) codon. At this point, they are joined by a larger ribosomal subunit to form the intact ribosome, which holds the mRNA in place while the tRNAs bring amino acids to it.

❷ Elongation The chain of amino acids lengthens one amino acid at a time. A tRNA molecule carrying the next appropriate amino acid binds to the ribosome and to mRNA. As the mRNA passes between the two ribosomal subunits, the ribosome catalyzes the formation of the bond between the newest amino acid and the previous amino acid in the growing chain. The tRNA molecule is then released to find another amino acid.

❸ Termination There is no tRNA anticodon corresponding to a "stop" codon on mRNA. When a "stop" mRNA codon is encountered, the ribosomal subunits and the newly formed peptide chain detach from the mRNA.

To use a cooking analogy, the mRNA brings the recipe, the tRNAs find and deliver the ingredients, and the ribosome (with its rRNA) is the cook who creates the final product. What happens next to the newly formed peptide chain depends on its function. Some peptides enter the endoplasmic reticulum for further processing, packaging, and shipping, whereas others remain in the cytoplasm. Many become parts of new structural units, organelles, cell membrane components, and enzymes that the cell will need as it grows and then divides into two new cells.

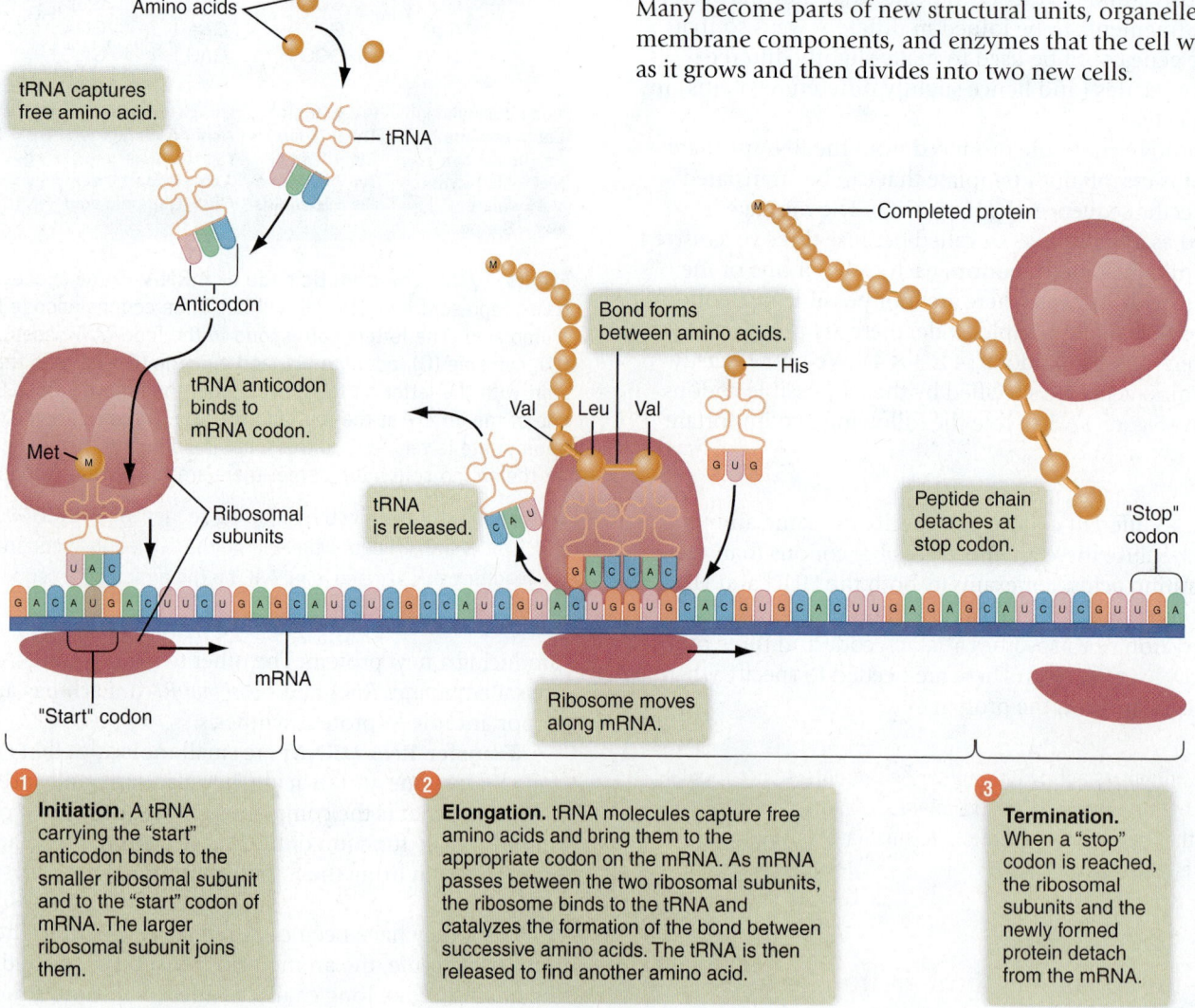

1 **Initiation.** A tRNA carrying the "start" anticodon binds to the smaller ribosomal subunit and to the "start" codon of mRNA. The larger ribosomal subunit joins them.

2 **Elongation.** tRNA molecules capture free amino acids and bring them to the appropriate codon on the mRNA. As mRNA passes between the two ribosomal subunits, the ribosome binds to the tRNA and catalyzes the formation of the bond between successive amino acids. The tRNA is then released to find another amino acid.

3 **Termination.** When a "stop" codon is reached, the ribosomal subunits and the newly formed protein detach from the mRNA.

Figure 17.6 The three steps of translation. Translation takes place in the cytoplasm.

✓ Suppose the gene that produced this mRNA happens to mutate so that one of the codons in the middle of the mRNA becomes a stop codon. What will happen the next time the gene is transcribed and translated? Will any protein be produced from the mutant mRNA? Explain.

↺ **Recap** Before a cell divides, its DNA is replicated. During *replication,* the two strands of DNA unwind and separate from each other. Enzymes called *DNA polymerases* add new nucleotides to each of the original strands, producing two complete molecules of DNA from one. Mutations may result from mistakes in DNA replication or from physical or chemical damage. Repair mechanisms remove and replace damaged DNA, if possible, before replication. During *transcription,* short portions of DNA representing single genes are converted into a readable and transportable mRNA code. *Translation* is a three-step process (initiation, elongation, and termination) by which proteins are assembled from amino acid building blocks according to an mRNA code. Assembly takes place on a ribosome. ∎

17.3 Cell reproduction: one cell becomes two

We turn now to the mitotic phase of the cell cycle (review Figure 17.1), when first the nucleus and then the cell cytoplasm divide. The division of the nucleus is called **mitosis.** It is followed by **cytokinesis,** the division of the cytoplasm (in Greek, *cyto-* means "cell" and -*kinesis* means "movement" or "activity").

Mitosis: daughter cells are identical to the parent cell

Mitosis is the process of nuclear division in which the sister chromatids of each duplicated chromosome are separated from each other (Figure 17.7). Following mitosis, each daughter cell has a complete set of DNA and is identical to the parent cell. Mitosis ensures that all cells of a complex organism have the same set of DNA.

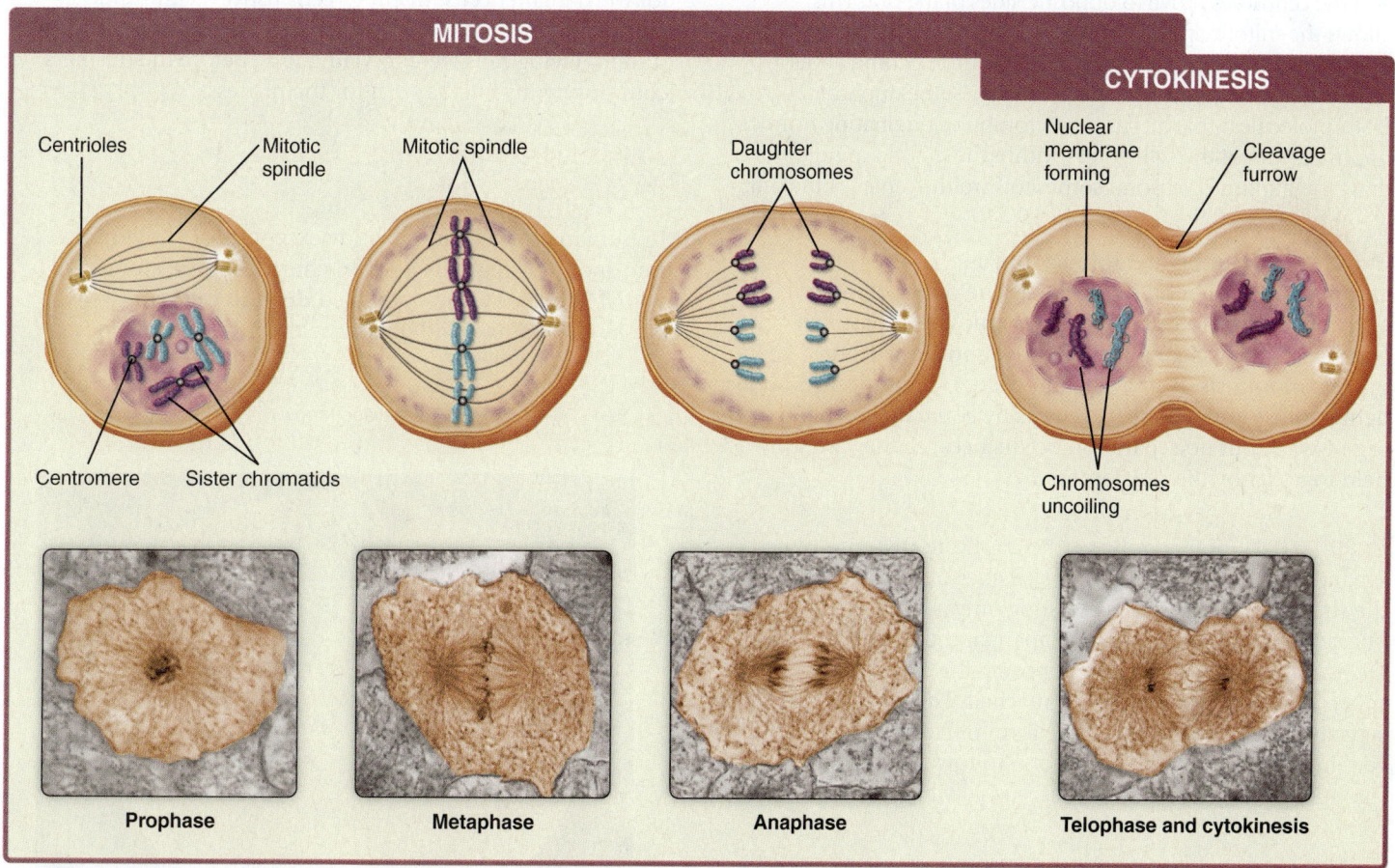

Figure 17.7 Mitosis. Mitosis accounts for most of the mitotic phase of the cell cycle, overlapping slightly with cytokinesis, the division of the cell into two cells. The duplicated chromosomes become visible in prophase, line up in metaphase, are pulled apart in anaphase, and uncoil and become surrounded by nuclear membranes in telophase. All 46 duplicated chromosomes undergo this process in humans, although only 4 are depicted in the drawings.

Mitosis includes a sequence of phases that compose most of the mitotic phase of the cell cycle. The phases are defined by the structural changes taking place in the cell as the DNA divides and two new nuclei form. Although each stage has a name and certain defining characteristics, mitosis is really a seamless series of events.

Prophase In late G_2 of the interphase, the strands of replicated DNA and associated proteins begin to condense and coil tightly. Eventually, they become visible as a duplicated chromosome, consisting of sister chromatids held together by the centromere. **Prophase** begins when the duplicated chromosomes first become visible.

Recall that the cell has a cytoskeleton that helps hold it together. During prophase, the tubular elements of the cytoskeleton come apart and are reassembled between the pairs of centrioles. This newly formed parallel arrangement of microtubules, called the *mitotic spindle*, will become the structure that causes the two identical sets of DNA to divide. The pairs of centrioles drift apart and the mitotic spindle lengthens between them. The nuclear membrane dissolves, and the centrioles move to opposite sides of the cell. This allows the mitotic spindle to cross through the middle of the cell. The centromere in each duplicated chromosome develops into two separate structures (one for each of the duplicate DNA molecules) that attach to microtubules originating from opposite sides of the cell. Forces within the mitotic spindle draw the duplicated chromosomes toward the center of the cell.

Metaphase During **metaphase,** the duplicated chromosomes align on one plane at the center of the cell. Cells often seem to rest at this stage, and metaphase can occupy as much as 20 minutes of the entire one-hour process of mitosis. In fact, the duplicate DNA molecules are being pulled in opposite directions by equal forces (like a tug-of-war), but no separation occurs because they are still held together by the centromeres.

Anaphase During **anaphase,** the duplicate DNA molecules separate and move toward opposite sides of the cell. This is the shortest phase of mitosis, lasting only a minute or so. The centromere abruptly comes apart, and the now-separated DNA molecules are pulled in opposite directions by the microtubules to which they are attached. The process of separation requires energy in the form of ATP and utilizes certain proteins that act as "motors" to tow the microtubules in opposite directions. Cytokinesis usually begins at this time.

Telophase When the two sets of chromosomes arrive at opposite poles of the cell, **telophase** begins. At this point, the mitotic spindle comes apart, and new nuclear membranes form around the chromosomes. The chromosomes uncoil and revert to the extended form in which they are no longer visible under a microscope.

☑ Suppose a chromosome in one of your skin cells has acquired a mutation in its centromere, such that its chromatids cannot separate from each other. During which stage of mitosis will this cause a problem, and what do you think will happen?

Cytokinesis divides one cell into two identical cells

Cytokinesis is the process by which a cell divides to produce two daughter cells. Certain physical properties of the cell facilitate this process. A contractile ring of protein filaments forms just inside the cell membrane. The contractile ring tightens, forming a cleavage furrow and pinching the cell in two (**Figure 17.8**). The ring is perpendicular to the long axis of the mitotic spindle, which ensures that one nucleus is enclosed in each of the daughter cells.

Cytokinesis is an example of the resourcefulness and efficiency of life's processes. The contractile ring is assembled just before it is needed from the remnants of the cytoskeleton that dissolved during mitosis. After the cell divides, the ring itself is disassembled, presumably to form new cytoskeletons for the daughter cells. In effect, cells use a "just-in-time" materials delivery system coupled to an efficient recycling system. Living cells were recycling long before the business community invented a term for the process.

Mitosis produces cells identical to the parent cell

In humans, all cells in the body except those developing into sperm or egg cells have 46 chromosomes. Human cells with 46 chromosomes are called **diploid** cells (in Greek, *diploos* means "twofold") because the 46 chromosomes

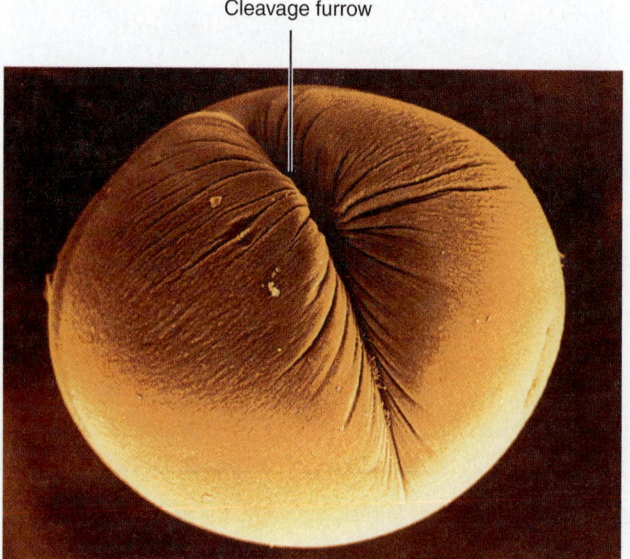

Cleavage furrow

Figure 17.8 Cytokinesis. A cleavage furrow forms as the ring of contractile filaments in the cell tightens.

actually represent 23 *pairs* of chromosomes. Diploid cells reproduce by first replicating the 46 chromosomes and then undergoing mitosis (one nuclear division), so that the two daughter cells end up with 46 chromosomes that are identical to those of the parent cell.

In diploid cells, one of each pair of chromosomes came from each parent. There are 22 pairs of **autosomes** (chromosomes other than the sex chromosomes) plus the sex chromosomes X and Y. The 22 pairs of autosomes plus the two sex chromosomes XX (in the female) are called **homologous chromosomes** because they look identical under the microscope and they have copies of the same genes in the same location (*homologous* means "same form and function"). Only the X and Y sex chromosomes in the male look different from each other as a pair.

Meiosis prepares cells for sexual reproduction

Sperm and eggs are **haploid** cells, meaning that they have only one set of 23 chromosomes (in Greek, *haploos* means "single"). Haploid sperm and eggs are created by **meiosis,** a sequence of two successive nuclear divisions in which the human genes are mixed, reshuffled, and reduced by half. Once fertilization occurs, the fertilized egg and all subsequent cells have the diploid number of chromosomes again. The two successive nuclear and cell divisions of meiosis are called *meiosis I* and *meiosis II*, both of which have the same four stages as mitosis: prophase, metaphase, anaphase, and telophase.

Meiosis I Before meiosis begins, the precursor reproductive cell undergoes a typical S phase of interphase in which the DNA is replicated. But then something very different happens, compared to mitosis. During prophase of meiosis I, duplicated homologous chromosomes pair up and swap sections of DNA, a process known as *crossing-over* (Figure 17.9). Consequently, the homologous chromosomes now contain an entirely new combination of genes derived from the genes of both parents. During the rest of meiosis I—metaphase through telophase—the homologous pairs of chromosomes separate (Figure 17.10).

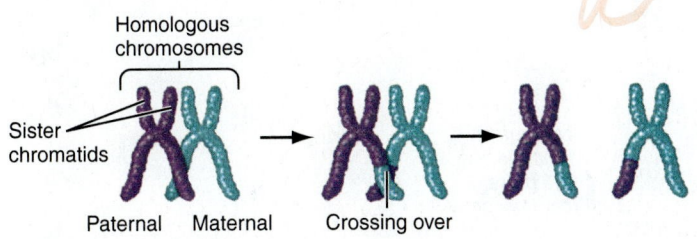

Figure 17.9 Crossing over. During prophase of meiosis 1 in the precursor cells of sperm and eggs, pairs of homologous chromosomes line up together and exchange sections of DNA.

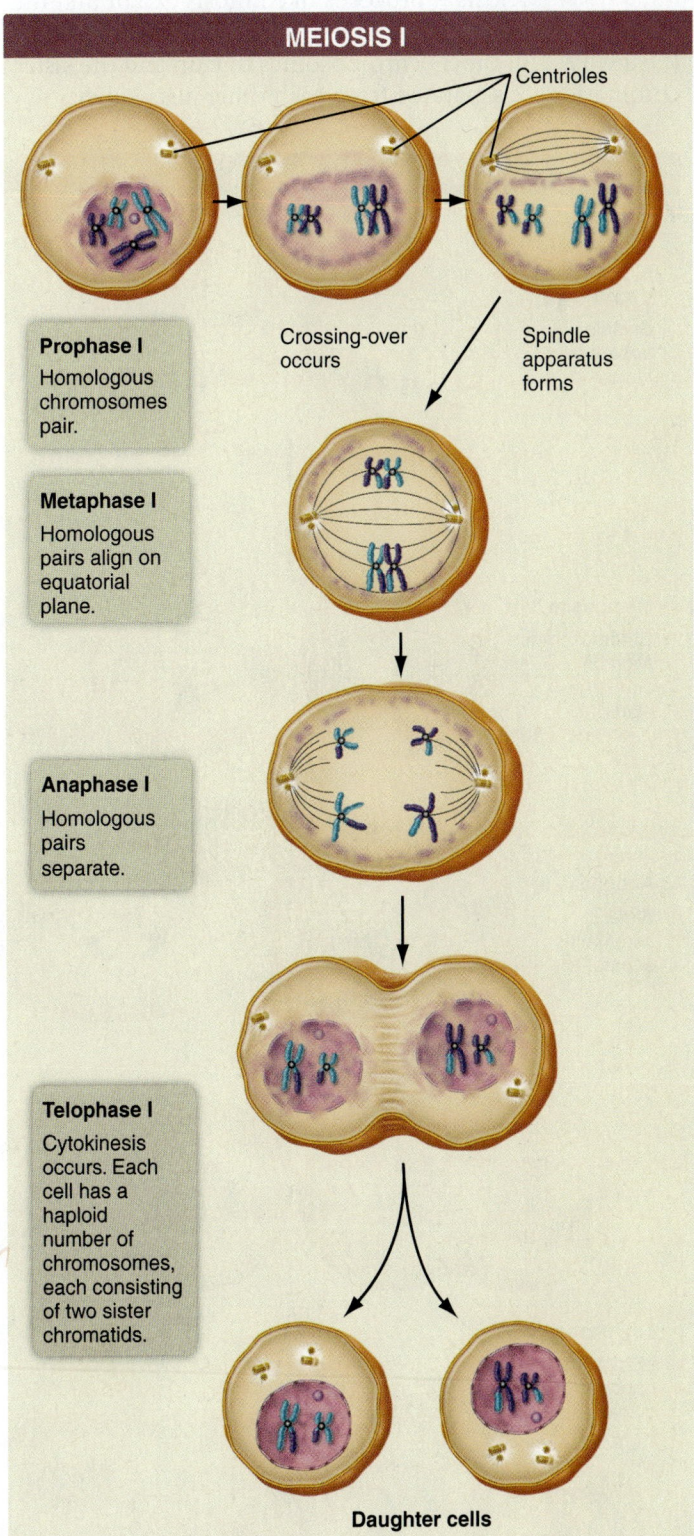

MEIOSIS I

Centrioles

Prophase I
Homologous chromosomes pair.

Crossing-over occurs

Spindle apparatus forms

Metaphase I
Homologous pairs align on equatorial plane.

Anaphase I
Homologous pairs separate.

Telophase I
Cytokinesis occurs. Each cell has a haploid number of chromosomes, each consisting of two sister chromatids.

Daughter cells

Figure 17.10 Meiosis I. The blue and purple colors indicate homologous chromosomes that were inherited from each of parent. During prophase of meiosis I, pairs of homologous chromosomes pair up and crossing-over occurs. During the rest of meiosis I, the homologous pairs are separated from each other.

Meiosis II Meiosis II proceeds like mitosis except that the chromosomes are not duplicated again. During meiosis II (Figure 17.11), the 23 chromosomes line up and the sister chromatids are separated from each other just as they are in mitosis. However, because of crossing-over during meiosis I, none of the four haploid daughter cells are exactly alike.

MEIOSIS II
Prophase II
Spindle apparatus forms.
Metaphase II
Chromosomes align on equatorial plane.
Anaphase II
Sister chromatids separate.
Telophase II
Nuclei form at opposite poles; cytokinesis occurs.
Four haploid daughter cells

Figure 17.11 Meiosis II. In meiosis II, the sister chromatids of each chromosome are separated from each other, as they are in mitosis. The result is four daughter cells, none of which are alike.

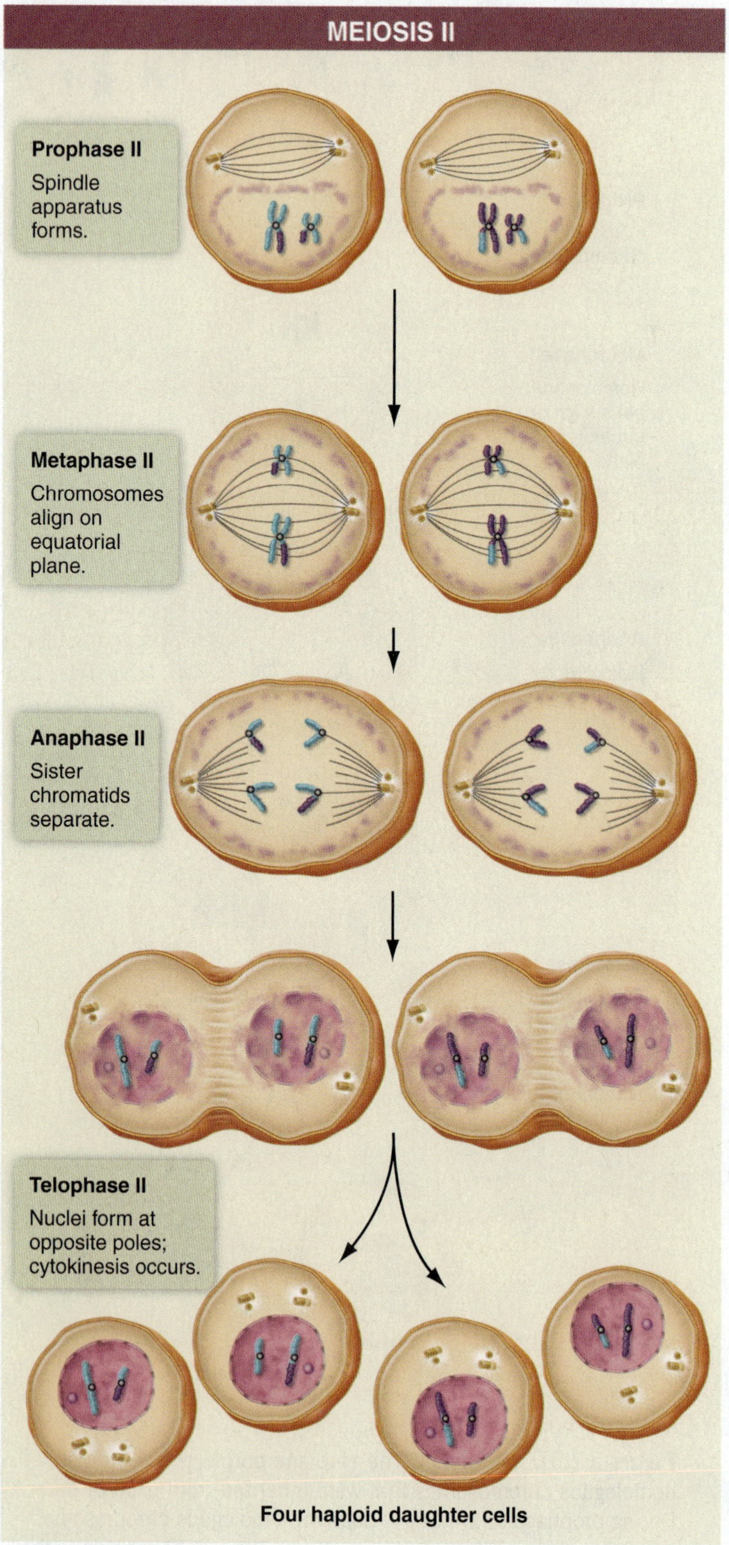

✓ A normal diploid cell from a chimpanzee has 48 chromosomes. How many chromosomes will be in a chimpanzee cell just before meiosis begins, after meiosis I, and after meiosis II? After which of these stages will the chromosomes have two chromatids?

Sex differences in meiosis: four sperm versus one egg

There is a fundamental difference in how meiosis proceeds in males and females, and it has to do with the different functions of sperm and eggs. There is only a slim chance that any one sperm will ever reach an egg. It is important, therefore, that males produce lots of sperm. Meiosis in males produces four equal-sized but genetically different sperm from every precursor cell (Figure 17.12).

If fertilized, an egg will need a lot of energy to grow and develop. A large egg with plenty of cytoplasm and lots of organelles has a better chance of surviving the early stages of development. In females, therefore, as much of the cytoplasm as possible is reserved for only one daughter cell at each meiotic division. The smaller cell produced at each division is called a *polar body*. The polar body produced during meiosis I may or may not divide again during meiosis II, but in any event the two (or three) polar bodies eventually degenerate.

Sperm
(23 chromosomes)
Polar bodies
Egg
(23 chromosomes)
Fertilization
Nucleus of fertilized egg
(23 pairs of chromosomes)

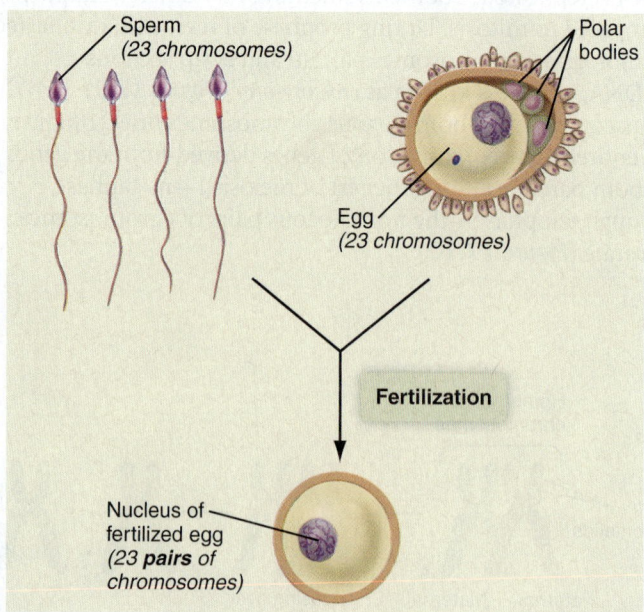

Figure 17.12 Sex differences in meiosis. In males, meiosis produces four different small sperm. In females, nearly all of the cytoplasm is reserved for the future egg at each stage of meiosis. Meiosis I in females thus results in a large oocyte and two or three polar bodies.

↩ **Recap** *Mitosis* is a four-phase process in which the cell nucleus divides into two nuclei. The duplicated chromosomes become visible in prophase, align along the center of the cell in metaphase, are pulled apart in anaphase, and move to opposite sides of the cell and become surrounded by new nuclear membranes in telophase. During *cytokinesis*, the cell divides into two cells, each containing a nucleus. *Meiosis* is a two-step process in which the nucleus and cell divide twice, producing sperm or eggs with the haploid number of chromosomes. Crossing-over ensures that the sperm (and also the eggs) are genetically different from each other. ∎

that the control mechanism can be stopped or started at certain checkpoints by signals from inside or outside the cell.

MJ's BlogInFocus  Are there any known natural biological factors that can restart cell growth and division in older animals? Visit MJ's blog in the Study Area in MasteringBiology and look under "An Anti-Aging Protein in Blood."

http://goo.gl/boRmHw

17.4 How cell reproduction is regulated

Although division is the norm for most cell types, not all cells in the body divide at the same rate. Table 17.1 lists examples of variations in the rates of cell division. Your 25–30 trillion red blood cells are completely replaced every 120 days, which works out to an astonishing 175 million cell divisions every minute. In contrast, most nerve and muscle cells do not divide at all in adulthood. Other cells reproduce at highly variable speeds, increasing or decreasing their rates of cell division in response to regulatory signals.

What causes such different and sometimes changing rates of cell divisions? Although we do not yet have all the answers, we do know that cells have an internal control mechanism that undergoes cyclic changes and

Consider a washing machine or a dishwasher. Once a wash cycle is started, a whole sequence of timed events occurs until the entire cycle is finished. The same thing happens in a living cell. Under normal circumstances, progression through the G_1, S, and G_2 phases is controlled by cyclic fluctuations in the concentrations of certain proteins called *cyclins*. Cyclins in turn activate certain regulatory proteins that initiate specific events within the cell, such as DNA replication or formation of the mitotic spindle.

Most washing machine control devices can stop the cycle at certain checkpoints, based on input received from internal sensors. If the washing machine does not fill with water, the cycle may stop right before agitation would normally begin. If a load of clothes is out of balance, the spin cycle may stop prematurely. Cells also have internal surveillance and control mechanisms to ensure that the cell is ready for the

Table 17.1 Variation in rates of cell division by human cells

Rate of division	Comments
Divide constantly and rapidly throughout life	
Skin cells	Skin is continually being formed from deep layers of cells. The outermost layer of dead cells is constantly being sloughed off.
Most epithelial cells	Epithelial cells lining the inner surfaces of body organs such as the digestive tract and the lungs are exposed to frequent damage and must be replaced.
Bone marrow cells	Stem cells in the bone marrow produce red blood cells (RBCs) and white blood cells (WBCs) throughout life. WBC production can be increased as part of the immune response.
Spermatogonia (after puberty)	Spermatogonia divide to produce sperm throughout life in the adult male. The rate declines with age.
Will divide under certain circumstances	
Liver cells	Liver cells don't normally divide in adulthood but will do so if part of the liver is removed.
Epithelial cells surrounding the egg	Called granulosa cells, they are quiescent for most of adult life. They begin to divide as the follicle matures.
Normally do not divide in adulthood	
Nerve cells	Although there are exceptions, most human nerve cells apparently do not divide in adulthood.
Osteocytes	Mature bone cells called *osteocytes* become trapped within the hard crystalline matrix of bone.
Muscle cells	The commonly held view is that most muscle cells do not divide in adulthood, or divide very slowly.

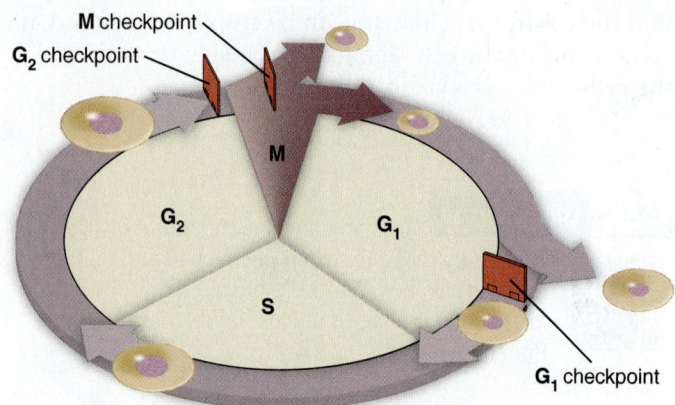

Figure 17.13 Cell cycle checkpoints. The cell cycle stops at several well-defined checkpoints unless the cell is ready for the next phase.

next phase before it proceeds (Figure 17.13). Has all of the DNA been replicated, and has it been checked for errors? Has the mitotic spindle been formed properly? Is the cell large enough to be divided in to two? Unless the cell surveillance system gives the go-ahead, the cell may stop at one of several checkpoints. The most important checkpoint seems to be near the end of G_1, but there are also checkpoints near the end of G_2 and at metaphase of mitosis.

Finally, the cell cycle can be influenced by conditions outside the cell. The cell cycle may stop at a checkpoint if certain nutrients are not available, if a particular hormone is not present, or if certain growth factors are not supplied by other cells. For example, a connective tissue cell called a *fibroblast* will not divide unless *platelet-derived growth factor* (derived, obviously, from platelets) is present. Even the presence of other cells affects the cell cycle. When cells come into contact with each other, for instance, they release a substance that inhibits cell division. Contact inhibition is an effective negative feedback mechanism to regulate tissue growth and organ size.

> ↺ **Recap** An internal cyclic control mechanism regulates the cell cycle. The cycle can be stopped at certain checkpoints by internal surveillance systems and is influenced by conditions outside the cell. ▪

17.5 Environmental factors influence cell differentiation

All of the cells in your adult body originated from a single cell, the fertilized egg, and all of those cells (except those destined to become your sperm or eggs) have exactly the same set of DNA. Yet your body is composed of lots of different kinds of specialized cells. If all your cells have the same exact set of genes, why don't all your cells stay

identical, like identical twins? What causes the cells to change form and function?

Differentiation is the process by which a cell becomes different from its parent or sister cell. At various developmental stages of life, cells differentiate because they begin to express different genes. The examples below show that differentiation is probably due to environmental factors. We now know that these environmental factors alter gene expressions by either modifying the histones that are associated with DNA or by methylating (adding methyl groups to) the DNA. The study of how gene expression is altered by factors other than the cell's underlying DNA sequence is called *epigenetics*.

Differentiation during early development

Let's examine the earliest stage of human development, starting from a fertilized egg. The fertilized egg is a relatively large cell. Soon after fertilization it begins to divide, becoming two cells, then four, then eight. During these early divisions, up to about 16 or 32 cells, the cells do not grow. Consequently, each cell gets progressively smaller as a ball of identical cells develops (Figure 17.14).

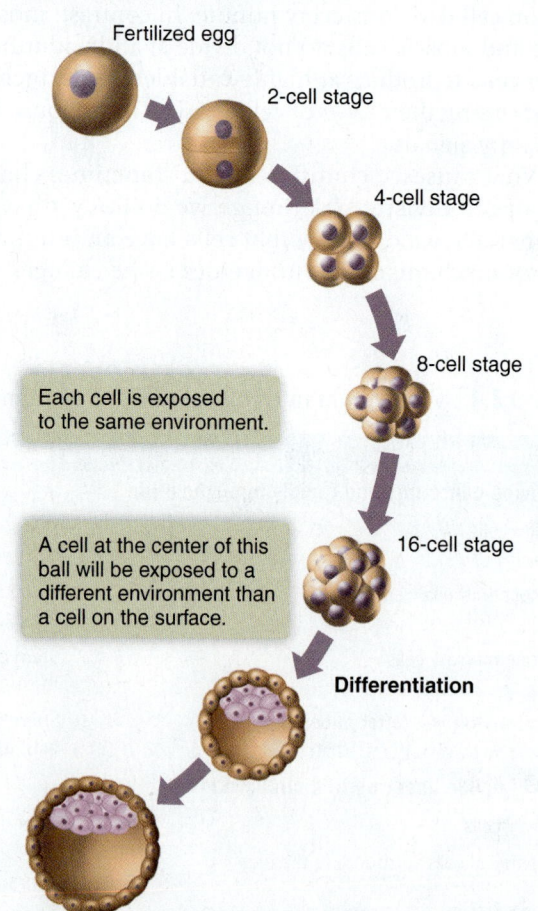

Fertilized egg

2-cell stage

4-cell stage

8-cell stage

Each cell is exposed to the same environment.

16-cell stage

A cell at the center of this ball will be exposed to a different environment than a cell on the surface.

Differentiation

Figure 17.14 How early differentiation might occur. Beyond about the eight-cell stage, the cells begin to be exposed to different environments. Differences in environments may trigger different gene expressions, leading to differentiation.

In a ball of eight or fewer cells each cell is exposed to the same environment—the surrounding fluid plus contact with cells just like themselves. Once the ball becomes more than about eight cells, the situation, and thus the environment, for some of the cells changes. From this point on, only the cells at the outer surface of the ball are exposed to the surrounding fluid; the rest are exposed only to the extracellular fluid between the cells. In other words, the environment that surrounds a cell at the center of the ball is different from the environment of a cell at the surface. Cells at the center may be exposed to a lower O_2 content, higher levels of CO_2, a different pH, or they may be influenced by products secreted by the other cells. Any one of these or other environmental factors could lead to methylation of DNA or alterations of histone structure that would alter which genes are being expressed, leading to different developmental pathways for the cells at the center versus the cells at the surface.

Notice that no feedback control mechanisms are needed to account for early differentiation; the process of differentiation during this time can be accounted for solely by differences in the physical and chemical environment of individual cells. Cell differentiation begins long before homeostatic feedback mechanisms ever develop in the organism.

Scientists have known for some time that it is possible to **clone** (make copies of) an organism by "embryo splitting;" that is, by splitting the eight-cell ball that develops from a fertilized egg into eight separate cells. At that stage, all eight cells are equally capable of becoming complete organisms, presumably because they have not yet differentiated. After about 16 cells, however, cloning by embryo splitting no longer works.

Differentiation later in development

Differentiation at later stages is similar in principle to that at earlier stages. Each cell is shaped by two factors: the developmental history of the cells that came before it and its local environment. Genes are "turned on" at certain stages of development and then "turned off," some never to be used again.

For example, early in development some cells begin to express the genes that cause them to become epithelial cells. Others become connective tissue, nerves, or muscle cells. Those that are destined to become muscle tissue differentiate even further, becoming either skeletal, smooth, or cardiac muscle.

Even cells that are closely related in function and structure can differ in hundreds of ways. Each difference in gene expression, added to those that occurred before, eventually results in a very specialized cell. For example, your smooth and skeletal muscle cells differ in their metabolic pathways (how they use energy), speed and strength of contraction, shape and size, and internal arrangement of muscle fibers. They got that way by an ever-increasing cascade of selective gene expressions.

The mechanisms of differentiation explain why the developing fetus is more vulnerable to genetic damage than an adult. Alteration of a gene that plays a critical role in early development can be so disruptive to embryonic differentiation that the embryo may not survive. Later in development, the embryo may be able to survive the damage but be left with marked physical deformities. In an adult, alteration of a single gene may be similarly dangerous, or it could have little or no visible effect. It all depends on what the proteins coded by the gene do.

These are some substances that can harm a fetus:

- *Cigarette smoke.* Smoking can retard fetal growth and may contribute to spontaneous abortions.
- *Alcohol.* Babies born to mothers who abuse alcohol may have fetal alcohol syndrome. These babies may have heart defects, limb abnormalities, and delays in motor and language development.
- *Prescription and over-the-counter medications.* If a woman becomes pregnant, she should review with her physician all the drugs she takes.
- *Illegal drugs.* Cocaine and heroin use during pregnancy can kill a fetus or cause abnormalities. The use of marijuana is associated with poor fetal weight gain and behavioral abnormalities in the newborn.
- *Chemicals in air, water, and soil.* Examples include lead, mercury, pesticides, and industrial solvents such as polychlorinated biphenyls (PCBs) and toluene.
- *Radiation, for instance, from radon gas or pelvic irradiation during cancer treatment.* Note that video display terminals and natural radiation from cosmic rays are not thought to pose any risk.

Intrauterine infections can also be dangerous. The most damaging are HIV infection, syphilis, and rubella (German measles).

↪ **Recap** Differentiation is the process whereby cells become different from each other. During early development, environmental influences trigger differentiation. Some genes are expressed only at certain stages of development, so genetic mutations during early development can be particularly damaging. In adults, hundreds of different genes may be expressed by various types of specialized cells. ∎

17.6 Reproductive cloning requires an undifferentiated cell

Making copies of (cloning) entire organisms, called **reproductive cloning,** has proven to be much more difficult than cloning specific molecules by genetic engineering techniques. Reproductive cloning requires a completely undifferentiated cell—a cell as close to a fertilized egg as possible—as the starting point. The two techniques for accomplishing reproductive cloning are called *embryo splitting* and *somatic cell nuclear transfer*.

Embryo splitting produces identical offspring

In *embryo splitting*, the undifferentiated cell is a fertilized egg (Figure 17.15). ❶ The egg is fertilized *in vitro* and ❷ allowed to divide to the eight-cell stage. At this stage, the cells are all still identical because they have not yet begun to differentiate. ❸ The cells are then separated, and ❹ each is implanted into a different surrogate mother. Each of the cells will develop normally, ❺ producing an organism that is an exact genetic copy of all the others. This type of reproductive cloning is currently quite feasible; indeed, it is being used effectively to produce valuable farm animals. The embryos can even be shipped to other countries and implanted in less valuable surrogate mothers, for herd improvement.

Although the offspring produced by embryo splitting are clones of each other, they are not clones of either parent. Genetically, they are like eight identical children, different from each parent because they received some of their genes from each one.

So far, no one has attempted to produce identical human octuplets by embryo splitting. The techniques required are not that difficult—it could be done if someone really wanted to.

☑ Suppose you're working at a zoo that has the last living pair of Arabian oryx—an endangered antelope. The female can only produce one calf per year, and you're worried that she can't produce enough calves for the species to survive. Outline a strategy using embryo splitting to try to save this species. What is one obvious problem with this approach?

Somatic cell nuclear transfer produces a clone of an adult

A *somatic cell* is any cell in the body except a germ (sex) cell. Somatic cells have the full set of instructions for creating a complete, unique individual, as opposed to germ cells (sperm and egg), which contain only half a set of DNA.

Cloning an organism by *somatic cell nuclear transfer* yields a true clone of an adult organism. This was the technique used to produce the first clone, Dolly the sheep, back in 1997. The feat of cloning a sheep through somatic cell nuclear transfer shocked the world, because at that time no one thought that a fully differentiated cell could be made to "start over." Basically, the technique involves combining a single somatic cell taken from an adult animal with an *enucleated* fertilized egg (an egg with its nucleus removed). The nucleus of the adult cell contains the set of instructions for making a copy (clone) of the adult, and the enucleated egg contains the proper environment for carrying out those instructions.

Figure 17.16 shows how somatic cell nuclear transfer is done in practice. ❶ An unfertilized egg is removed from a donor animal, and the nucleus is then carefully removed. ❷ A somatic cell is removed from the animal to be cloned and inserted into the enucleated egg. ❸ Electric current is used to fuse the two cells into one. ❹ The new cell is similar to a fertilized egg, but its nucleus is from the adult animal being cloned. ❺ The egg is then implanted into a surrogate mother and allowed to develop normally. ❻ Because the DNA is from the adult organism being cloned, the organism is a true clone of the adult organism from which the somatic cell was taken.

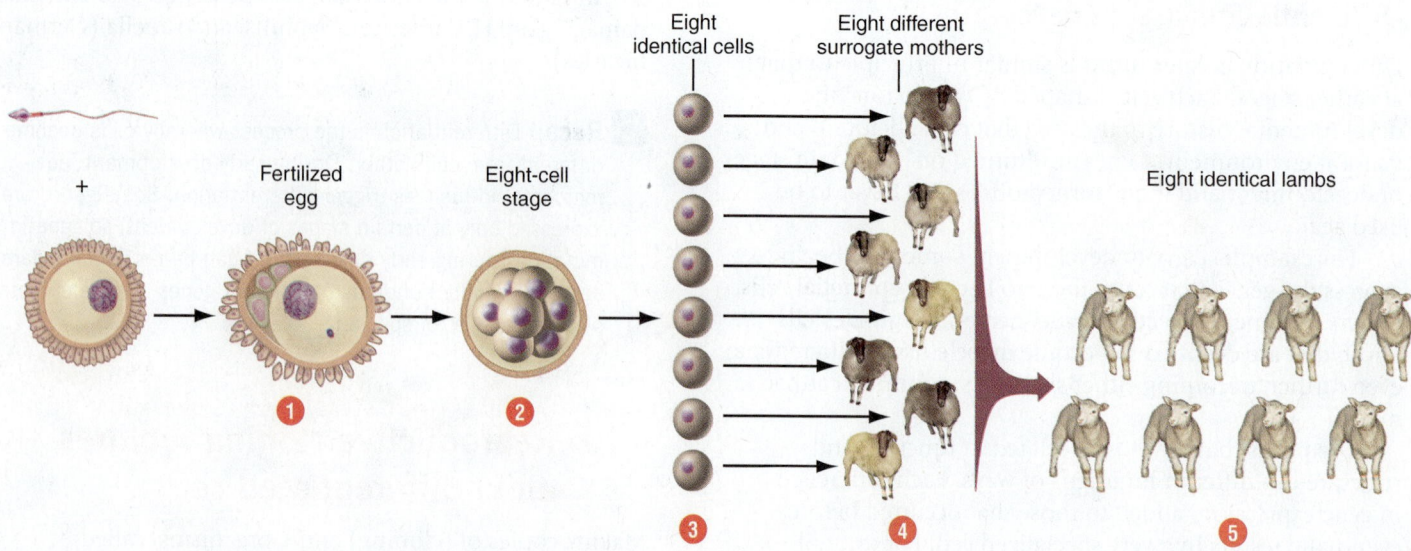

Figure 17.15 Cloning by embryo splitting. Clones produced by embryo splitting are genetically identical to each other but are not exact copies of either parent.

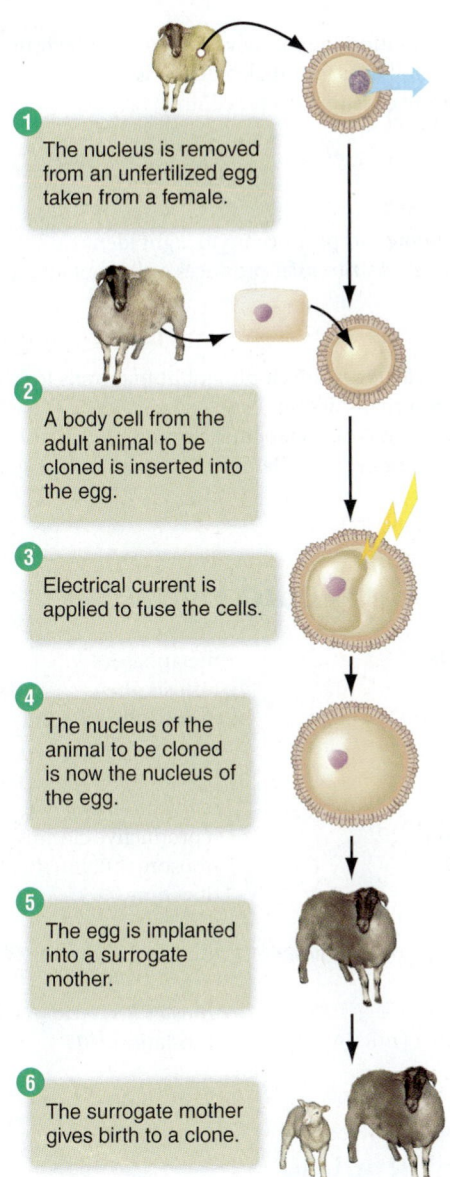

1. The nucleus is removed from an unfertilized egg taken from a female.

2. A body cell from the adult animal to be cloned is inserted into the egg.

3. Electrical current is applied to fuse the cells.

4. The nucleus of the animal to be cloned is now the nucleus of the egg.

5. The egg is implanted into a surrogate mother.

6. The surrogate mother gives birth to a clone.

Figure 17.16 Cloning by somatic cell nuclear transfer. The clone is genetically identical to the animal that provided the body cell.

MJ's BlogInFocus Is cloning animals for profit economically feasible yet? Visit MJ's blog in the Study Area in MasteringBiology and look under "Cloning Goes Commercial."

http://goo.gl/p9Vn5W

The popular press tends to view cloning by somatic cell nuclear transfer as science fiction, but it is not. It is feasible. It has been done successfully in animals, though not with the same success as embryo splitting. It could be done with humans, though so far it hasn't. It probably will be done in your lifetime unless legal sanctions are put in place to prevent it.

Dolly died at a relatively young age for a sheep. Scientists worry that somatic cells taken from an adult may already have suffered so much mutational damage that clones might routinely live shorter lives than the original adult. More experience over time with cloned farm animals may answer that question.

↩ **Recap** A clone is a copy. Reproductive cloning by *embryo splitting* produces genetically identical offspring that are different from either parent. Reproductive cloning by *somatic cell nuclear transfer* produces a genetic twin of the adult that was cloned. So far, humans have not been cloned by either method. ∎

17.7 Therapeutic cloning: creating tissues and organs

Whereas the purpose of reproductive cloning is to produce offspring that are genetically identical to the parent, the goal of **therapeutic cloning** is to create cells that can be used to treat specific diseases or conditions in human patients. Cells (and any tissues or organs derived from them) created by therapeutic cloning techniques would have the advantage that there would be no danger of tissue rejection since they would be genetically identical to the cells of the patient.

As a practical treatment for human disease, therapeutic cloning is a technique still in its infancy, as we learned in the Current Issue in this chapter. Nevertheless, therapeutic cloning has the potential to transform medical treatment in the future. But the more we learn about cell division and differentiation and the closer we get to therapeutic cloning, the closer we also get to being able to clone adults. We need to think about how to use and control cloning technology.

☑ Suppose that you were a researcher who wanted to grow heart muscle cells for repairing damaged hearts. Where might you look for cells to use in your research?

Chapter Summary

17.1 The cell cycle creates new cells *p. 401*

- Cells reproduce by a repetitive cycle called the *cell cycle* in which one cell grows and then divides into two.
- The cell cycle has two primary phases: interphase and the mitotic phase. During interphase, the cell grows and the DNA is replicated. During the mitotic phase, first the nucleus and then the cell divide into two.

17.2 DNA structure and function: an overview *p. 402*

- Human DNA consists of 46 separate molecules, packed with histones into structures called *chromosomes*.
- A gene is the smallest functional unit of a chromosome. A gene contains the code for making one or more specific proteins.
- DNA replication is a process in which the two strands of DNA separate and a new complimentary copy is made of each strand.
- DNA is repaired when it is damaged and is checked for errors after it is replicated.
- For a gene to be expressed, the strand of DNA with that gene must be transcribed to create a complementary strand of mRNA. The mRNA leaves the nucleus, attaches to a ribosome, and serves as the template for protein synthesis.
- Translation is the process of making a protein using the mRNA as a template. Three successive bases of mRNA, called a codon, code for a particular amino acid.
- The amino acid building blocks for the protein are captured in the cytoplasm by tRNA, brought to the mRNA, and attached to each other by ribosomal enzymes.

17.3 Cell reproduction: one cell becomes two *p. 407*

- Mitosis is a sequence of events in which the replicated chromosomes are separated to form two new genetically identical nuclei.
- Cytokinesis is the process whereby the cell divides into two new cells, each with one of the new nuclei produced by mitosis and roughly half of the cell's organelles and mass.
- Meiosis is a sequence of two cell divisions that produces haploid cells. Meiosis occurs only in cells destined to become sperm or an egg. Crossing-over during meiosis mixes the genes of homologous chromosomes, and subsequent cell divisions reduce the number of chromosomes by half.

17.4 How cell reproduction is regulated *p. 411*

- Cell reproduction is regulated in part by selective gene expression. Selective gene expression is controlled by regulatory genes.
- The cell cycle may be influenced by the physical and chemical environments both inside and outside the cell.

17.5 Environmental factors influence cell differentiation *p. 412*

- Differentiation is the process by which cells become different from each other, acquiring specialized forms and functions.
- Because all cells have the same set of genes, differentiation in the early embryo must be triggered by environmental influences.

- Cell differentiation later in development can be influenced by environmental cues, but it also depends on the developmental history of the cells that preceded it.

17.6 Reproductive cloning requires an undifferentiated cell *p. 413*

- *Embryo splitting* can produce up to eight identical offspring.
- *Somatic cell nuclear transfer* produces a clone of an adult animal.

17.7 Therapeutic cloning: creating tissues and organs *p. 415*

- *Therapeutic cloning* is the cloning of human cells for the purpose of treating human disease.
- The ultimate goal of therapeutic cloning is to be able to create cells, tissues, or even whole new organs for human patients.

Terms You Should Know

anaphase, *408*	metaphase, *408*
cell cycle, *401*	mitosis, *407*
chromosome, *402*	mitotic phase, *402*
clone, *413*	mutation, *403*
codon, *405*	prophase, *408*
cytokinesis, *407*	replication, *403*
differentiation, *412*	reproductive cloning, *413*
diploid, *408*	ribosomal RNA (rRNA), *405*
gene, *403*	ribosome, *405*
haploid, *409*	telophase, *408*
homologous chromosomes, *409*	therapeutic cloning, *415*
interphase, *402*	transcription, *403*
introns, *405*	transfer RNA (tRNA), *405*
messenger RNA (mRNA), *404*	translation, *403*

Concept Review

Answers can be found in the Study Area in MasteringBiology.

1. Describe how DNA is replicated before cell division.
2. Compare and contrast the processes of transcription and translation.
3. Explain what mutations are and the role of DNA repair mechanisms.
4. Name the four phases of mitosis and describe briefly what is happening to the chromosomes during each phase.
5. Explain why only one large egg is formed during meiosis in the female, whereas four sperm are formed during meiosis in the male.
6. Describe what is meant by *selective gene expression* and why it is important to how a cell functions.
7. Explain how factors present in the environment can influence cell differentiation.
8. Describe how ribosomes contribute to the formation of a protein.
9. Compare and contrast the roles of introns and exons.
10. Describe how the cell cycle is regulated.

Test Yourself

Answers can be found in the Appendix.

1. What would be the outcome if a cell completed mitosis but did not undergo cytokinesis?
   a. The cell would have two nuclei.
   b. The cell would have one nucleus but twice as many chromosomes.
   c. The cell would be cancerous.
   d. The cell would die.

2. Which of the following cell types are most likely to remain in G_0?
   a. neurons
   b. skin cells
   c. bone marrow cells
   d. cells lining the G_I tract

3. If the human genome was considered to be like a large cookbook, which of the following would represent the individual recipes?
   a. chromosomes
   b. genes
   c. nucleotides
   d. nitrogenous bases

4. During which stages of the cell cycle do chromosomes consist of two sister chromatids?
   a. G_1
   b. G_2
   c. prophase of mitosis
   d. both (b) and (c)

5. Which of the following are listed in order from largest, most inclusive, to smallest?
   a. genome—chromosome—nucleotide—gene
   b. gene—nucleotide—chromosome—genome
   c. chromosome—genome—nucleotide—gene
   d. genome—chromosome—gene—nucleotide

6. What might be a likely outcome if a mutation occurred in a promoter region of a gene, such that it was no longer recognized by RNA polymerase?
   a. The DNA would not be replicated.
   b. The protein encoded by that gene would not be synthesized.
   c. The cell would not be able to produce any proteins.
   d. The gene would be transcribed normally.

7. Which is likely to be the shortest chain of nucleotides?
   a. the DNA of a gene
   b. the primary transcript of the gene
   c. the mRNA
   d. DNA, the primary transcript of the gene, and mRNA all contain the same number of nucleotides.

8. How many different amino acids could be encoded if the genetic code was a doublet code (two bases) instead of a triplet code?
   a. 2   c. 8
   b. 4   d. 16

9. What is the most likely target of cell surveillance as a cell approaches the M (metaphase) checkpoint?
   a. Has all of the DNA been replicated?
   b. Has the DNA been checked for errors following replication?
   c. Is each chromosome properly joined to the mitotic spindle?
   d. Is the cell large enough?

10. Why do cells within an organism differentiate, such that one cell may eventually become a liver cell and another will become an epithelial cell?
    a. Cells differentiate because of differences in gene expression.
    b. Cells differentiate because different genes within them encode for different proteins.
    c. Cells differentiate because of crossing-over, a random process.
    d. Cells differentiate because of differences in genetic code.

11. Which method of cloning is most similar to the way identical (monozygotic) twins are formed?
    a. therapeutic cloning
    b. embryo splitting
    c. somatic cell nuclear transfer
    d. production of undifferentiated cells from mature cells

12. At which point in the cell cycle are cells the smallest?
    a. the beginning of G_1 c. G_2
    b. S   d. the beginning of mitosis

13. By the end of meiosis II, a cell that had entered meiosis I with 32 chromosomes would have produced _____ daughter cells, which are genetically _____ and which each has _____ chromosomes.
    a. 2 . . . identical . . . 32
    b. 2 . . . identical . . . 16
    c. 4 . . . nonidentical . . . 16
    d. 4 . . . identical . . . 32

14. Transcription occurs in/at the _____ and produces _____ from a _____ template.
    a. nucleus . . . RNA . . . DNA
    b. nucleus . . . DNA . . . RNA
    c. ribosomes . . . RNA . . . protein
    d. ribosomes . . . protein . . . RNA

15. How does the production of sperm differ from the production of eggs?
    a. Sperm production involves meiosis, whereas egg production involves only mitosis.
    b. Meiosis during sperm production produces four sperm cells, whereas meiosis during egg production produces only one functional egg cell.
    c. Sperm are haploid, whereas eggs are diploid.
    d. Sperm production begins during fetal development, whereas the meiotic process that begins egg production doesn't begin until puberty.

Apply What You Know

Answers can be found in the Study Area in MasteringBiology.

1. Recall that the base pairing in DNA is always A–T and C–G. Suppose that during a DNA replication, a C is accidentally paired with a T. Explain why it is important that the error be detected and corrected before the cell divides again, by illustrating what would happen at the next cell division if the error were not corrected.

2. The RNA code is a "triplet code," meaning that three bases code for a single amino acid. Would a "doublet code" be sufficient? Explain.

3. Are gene mutations that occur during fetal development always fatal? Why or why not?

4. Why do so many commercials for various pharmaceutical drugs contain warnings for women who are pregnant or are trying to become pregnant?

5. Mitochondria contain their own DNA that is inherited independently of nuclear DNA. From which parent did you inherit your mitochondrial DNA?

6. Bacteria can reproduce by simple cell division. Sexual reproduction is far more complicated and difficult; success is not guaranteed. What is the main advantage of sexual reproduction?

MJ's BlogInFocus

Answers can be found in the Study Area in MasteringBiology.

1. Could stem cells be made to differentiate into meat cells and then cloned to produce edible meat? Visit MJ's blog in the Study Area in MasteringBiology and look under "Cloning Meat." http://goo.gl/pivoOq

MasteringBiology®

Students Go to MasteringBiology for assignments, the eText, and the Study Area with animations, practice tests, and activities.

Professors Go to MasteringBiology for automatically graded tutorials and questions that you can assign to your students, plus Instructor Resources.

Answers to ✓ questions are available in the Appendix.

Cancer: Uncontrolled Cell Division and Differentiation

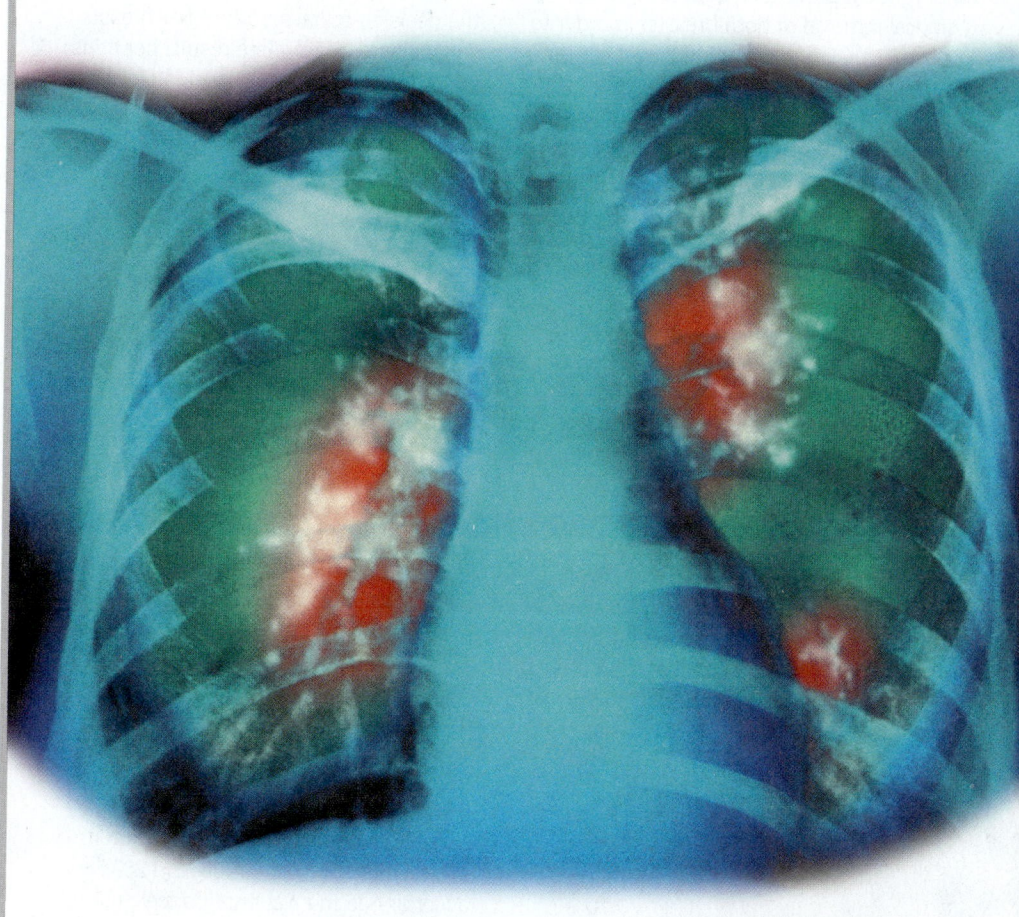

Color-enhanced X-ray showing lung cancer in both lungs (red/white).

Key Concepts

- **The essence of all cancers is a loss of control of cell division and differentiation.** As cancer cells divide out of control and even invade tissues, they may prevent the proper functioning of organs and organ systems.

- **Cell growth, division, and differentiation are controlled by specific regulatory genes.** When these regulatory genes mutate or are damaged, cancer may result.

- **Newer imaging and diagnostic testing techniques and sophisticated cancer treatments are slowly leading to gains against some cancers.** But no single cure for cancer is likely in the foreseeable future.

- **Many cancers could be prevented by changes in lifestyle.** Tobacco smoke accounts for over 30% of all cancer deaths.

Preventive Double Mastectomy to Reduce Breast Cancer Risk

In May of 2013, actress and director Angelina Jolie announced that she had undergone a preventive double *mastectomy* (surgical removal of both breasts) in order to reduce her risk of developing breast cancer. She was only 37 years old.

Ms. Jolie is one of a growing number of women who, with a higher than average risk for breast cancer, make the decision to have a double mastectomy. In each case, it's a very personal decision based on one's family history of breast cancer, the presence of one or more mutated genes known to be associated with breast cancer, an assessment on their particular risk of developing breast cancer, and feelings about self-image in the aftermath of a mastectomy. According to Ms. Jolie, tests revealed that she had a mutated allele of a gene, called BRCA1, known to be associated with a very high risk of developing breast cancer—in Ms. Jolie's case as high as 87%, according to her doctor. Her mother, actress and producer Marcheline Bertrand, died at age 56 after an eight-year battle with breast and ovarian cancer.

Angelina Jolie in 2001 with her mother, Marcheline Bertrand

With six children to raise and the desire to live her life with less fear of developing breast cancer, Ms. Jolie made the decision to have both of her breasts surgically removed. As a result, her odds go up in favor of a long life. Ms. Jolie's risk of developing breast cancer in the future is now only about 5%.

BRCA1 and BRCA2— The Breast Cancer Genes

Better genetic testing procedures and more accurate information about the risks associated with specific mutated genes means that women can find out whether they are at high risk for breast cancer while they are still young and healthy. Mutations in two genes in particular, called *BRCA1* (short for breast cancer 1) and *BRCA2*, are strongly associated with breast cancer. In their normal forms, these two genes suppress cell (and tumor) growth. Mutated forms of BRCA genes are relatively rare—fewer than 1% of all women have a mutated form—but for those who do, the risk of cancer goes up dramatically. In men, mutated BRCA genes are associated with a slight risk of breast and prostate cancers. For all women, the lifetime risk of developing breast cancer is 12%, according to the National Cancer Institute. For women with a mutated BRCA gene but no known family history of breast cancer, the risk of breast cancer is 45–65%. Add a family history of breast cancer, and that percentage range goes up. Armed with this knowledge, some women (including Ms. Jolie) have opted to have both breasts removed even *before* breast cancer develops.

Should *you* be tested for the mutated BRCA genes? That depends on whether you would really want to know the answer, and what (if anything) you would choose to do if you did know you had a mutated BRCA gene. If you are convinced that you could never have your breasts removed while you are still

Questions to Consider

1 Blood tests for BRCA1 and BRCA2 are readily available. Are you planning to be tested? Why or why not?

2 If you knew that you had a mutated BRCA gene, would it affect your attitude toward double mastectomy or how you live your everyday life?

3 What would you want a person in your life to do if they had a mutated BRCA gene?

healthy, then perhaps you would be better off not knowing. Your choice might also be influenced by whether or not you have a family history of breast cancer. In fact, mutated BRCA genes can be inherited from either of your parents.

Breast Reconstruction After Double Mastectomy

A woman's decision to have both breasts removed is likely to depend in some part on how she thinks she will look and feel after reconstructive surgery. Improvements in surgical techniques make a double mastectomy more appealing than in the past. Breast reconstruction is generally accomplished either with silicone implants and/or with tissue taken from elsewhere in the body. The availability of this technique may be one of the reasons women at high risk of breast cancer are willing to undergo a double mastectomy while they are still healthy. Ms. Jolie chose a "nipple-sparing mastectomy"—removal and reconstruction of the underlying breast tissue while leaving the skin, nipple, and areola (dark area around the nipple) intact. Because the nerves to the nipple are cut, however, generally no feeling remains in the nipple. Another option is to construct a new nipple and areola, as with breast reconstruction using tissue taken from elsewhere in the woman's body. Nipple construction is followed by tattooing to create the desired color. For women who opt to have a tattoo alone, a skilled plastic surgeon or tattoo artist generally can make a flat tattoo look three-dimensional.

What Is Peace of Mind Worth?

These days, it's not just women at high risk of developing cancer in both breasts who are opting for double mastectomies. Approximately 10% of women in their →

40s who need to have a single mastectomy to remove a breast cancer are now asking to have their healthy breast removed as well. For these cancer survivors, not having to worry about mammograms, biopsies, or breast surgery ever again may be the motivating factor. Some surgeons worry that these cancer survivors may not fully understand that if they are not carrying a mutated BRCA gene, removal of the healthy breast apparently has no effect on long-term survival. But who's to say what their motivation is or to question their choices?

We've been talking about probabilities of cancer and even death in an impersonal way, as if breast cancer were a game of numbers. But it is not a game. Each of us may have to someday make difficult choices because of the disease, for ourselves or a loved one. We'll be forced to balance personal feelings against impersonal odds. Ms. Jolie acknowledges that her decision to go ahead with a double mastectomy might not be right for every woman but adds that she is happy with her decision. She encourages all women, especially those with a family history of breast cancer, to seek out information and medical advice so that they can make their own informed choices.

SUMMARY

- Heritable mutations of two genes known as BRCA1 and BRCA2 are strongly associated with breast and ovarian cancer.
- Genetic tests for mutated BRCA1 or BRCA2 genes permit women to know whether they carry these genes.
- One option for women at high risk for developing breast cancer later in life is preventive double mastectomy (surgical removal of both breasts).
- When cancer is not yet present, improvements in reconstructive surgery have made it possible to save the skin, nipple, and areola during a mastectomy.
- Armed with the facts, some women who have an increased risk for breast cancer are opting for double mastectomies while they are still healthy, rather than waiting to see if cancer develops.

The word *cancer* may conjure up images of pain, suffering, and lingering death. We may associate the word with visits to the hospital with our parents when we were children; we may recall years of long-term care of a grandparent at home; we may remember a friend or loved one no longer with us. A diagnosis of cancer sounds like a death sentence.

And yet, people do recover from cancer. There are known risk factors for most types of cancers and things we can do to reduce our risk. Treatments are available for many cancers, some of which work rather well. The death rates for certain types of cancers are even declining as we learn more about the disease.

We know of at least 100 different types of cancers, but all are the same in one respect—they all are a disease of cell division and differentiation. Something goes wrong—very wrong—with normal physiological process within cells. Sometimes the cause is one or more mutated (altered) genes. Those same genes, when they are functioning properly, keep cell activities, including replication, under control. At other times, the cause of cancer may be environmental—something outside the body that we ingest or something we are exposed to that alters our cells and how they replicate. Cancer, therefore, is the manifestation of some cells out of control. Once cancer is understood in this way, as deviant change in cellular structure and function, it makes some sense. It makes sense why cancer can occur in 100 different guises in virtually every organ in the body. It makes sense why some cancers can kill us. ∎

In this chapter, we examine the biological basis for how cancer develops and spreads. Then we discuss how cancer is diagnosed and look at new advances in treatment and prevention.

To understand how cancer develops we must first understand two key characteristics of normal cells:

- Normal cells have regulatory mechanisms that keep their rates of cell division appropriately in check. They may divide more frequently at some times than at others, but their rates of cell division are carefully controlled by regulatory mechanisms that may involve an internal clock and various hormones. In addition, their rates of division are generally inhibited by signals from nearby cells.
- Normal cells generally remain in one location throughout their lifetime because they adhere to their nearby neighbors. (Blood cells are an exception, because they must travel throughout the body to carry out their normal function.)

18.1 Tumors can be benign or cancerous

Sometimes even normal cells increase their rate of cell division as part of their normal physiological function. Examples are the growth of the lining of the uterus during the menstrual cycle and the growth of breast tissue during pregnancy. In general, any substantial increase in the rate of cell division is called **hyperplasia** ("increased" plus Greek *plasis*, meaning "formation").

In the progression of normal cells toward cancer, however, hyperplasia is not necessarily a normal event.

Instead, it may signal a loss of control over cell division. Eventually, these rapidly dividing cells will develop into a discrete mass called a **tumor**, or *neoplasm* (meaning literally "new growth").

Not all tumors are cancers. Tumors that remain in one place as a single well-defined mass of cells are *benign tumors* (from Latin *benignus,* meaning "well disposed") (**Figure 18.1**). The cells in a benign tumor still have most of the structural features of the cells from which they originated, and in particular, they stay together and may be surrounded by a layer of connective tissue. Benign tumors tend to threaten health only if they become so large that they crowd normal cells, which can happen sometimes in the brain. In most cases, surgery can easily remove benign tumors because they are so well contained. Many moles on the skin are a form of a benign tumor so inconsequential that usually we don't bother to remove them. Occasionally, a previously benign tumor may change into something more damaging. For example, certain moles have the potential to develop into melanoma, a form of skin cancer.

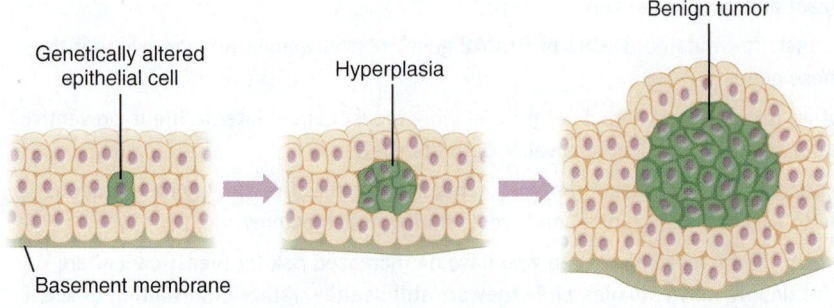

Figure 18.1 Development of a benign tumor. A benign tumor begins when a cell (in this case, an epithelial cell) becomes genetically altered and begins to divide more frequently than normal. The mass enlarges but stays well contained, generally by a capsule of normal cells.

☑ Should benign tumors be removed? Explain why or why not.

18.2 Cancer cells undergo structural and functional changes

If and when cells progress toward cancer, they undergo a series of structural changes in addition to dividing more frequently than normal (**Figure 18.2**). The nucleus often becomes larger, there is less cytoplasm, and the cells lose their specialized functions and structures. An abnormal structural change is called **dysplasia** (from *dys,* meaning

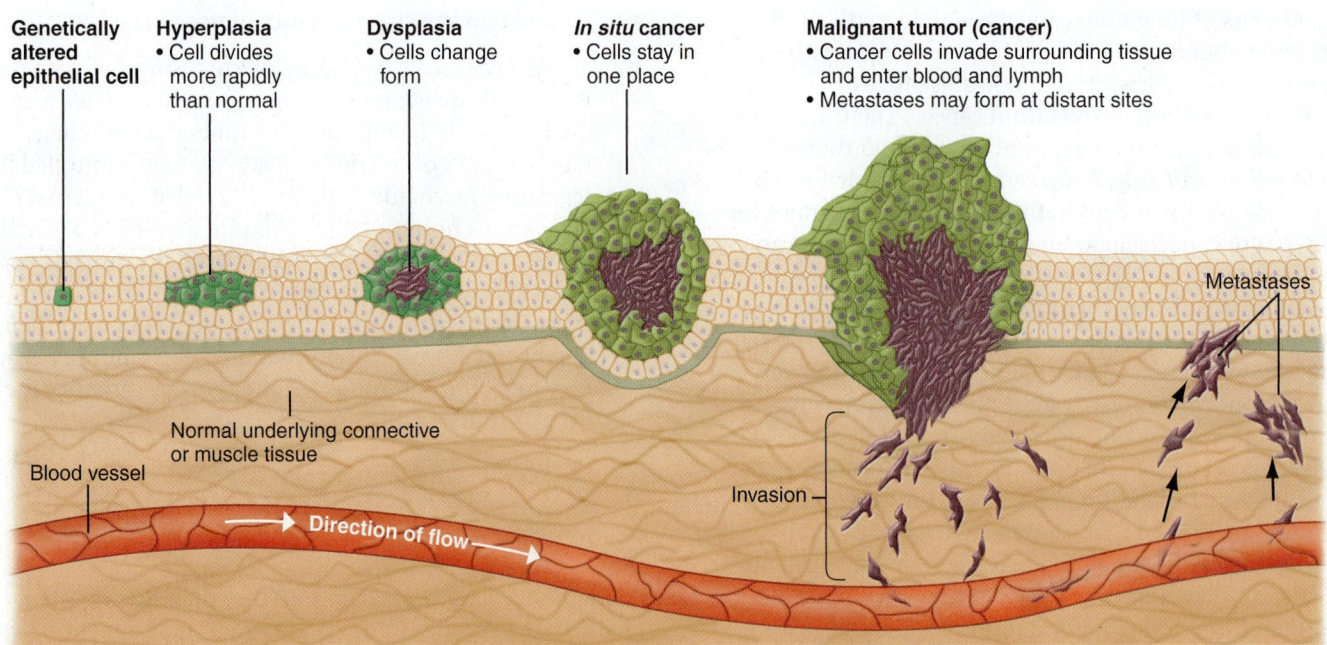

Figure 18.2 The development of a malignant tumor (cancer). A genetically altered cell begins to divide more frequently than normal, resulting in an increased number of cells (hyperplasia). Additional genetic alterations cause some cells to change form (dysplasia) and to lose all semblance of organization or control over cell function and division. As long as the mass stays in one place, it is called an *in situ* cancer. However, additional changes in the cells may cause them to break away from the tumor and invade the surrounding tissues or enter the blood or lymph. These invading cells may set up new colonies of cancer (called *secondary tumors*) at distant sites. A cancer that invades normal tissue and may produce secondary tumors is called a *malignant tumor.*

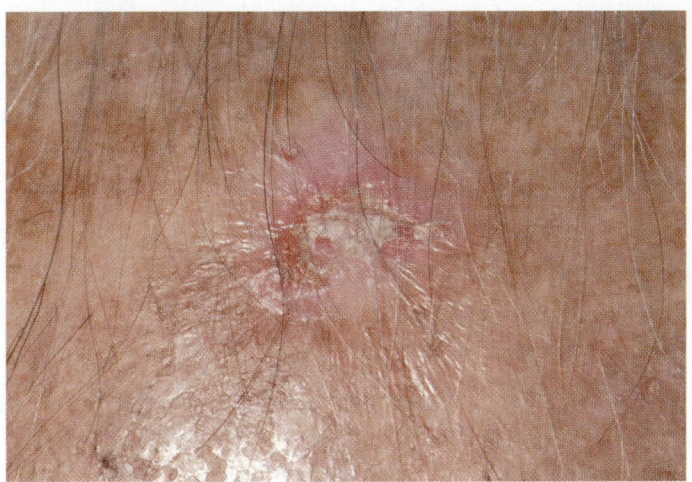

Figure 18.3 Actinic keratosis. Actinic keratosis is a precancerous skin lesion caused by repeated exposure to sunlight. Actinic keratosis lesions usually do not progress to cancer.

"abnormal," plus *plas,* meaning "growth"). Dysplasia may lead to cancer.

A pattern of changes leading to a lack of control

Doctors have learned to recognize characteristic patterns of changes that cells typically go through as they get more and more out of control. Dysplasia is often a sign that tumor cells are *precancerous,* meaning they are altering in ways that may herald the possibility of cancer. The tumor itself becomes more and more disorganized as cells pile on top of each other in seemingly random fashion.

An example of a precancerous skin lesion is *actinic keratosis* (Figure 18.3). Actinic keratoses generally appear as dry, scaly, or rough skin lesions on areas of the skin exposed to the sun. Often, they are felt more easily than seen, although sometimes they appear as reddish patches. They may grow slowly and even disappear for a time, only to return later. Only 10–20% of all actinic keratoses ever become cancerous. Fortunately, they are easily removed by chemical peeling, freezing, or scraping.

A tumor is defined as a **cancer** when at least some of its cells completely lose all semblance of organization, structure, and regulatory control. As long as the entire tumor remains in one place, it is known as an ***in situ* cancer.** *In situ* cancers can often be successfully removed surgically if they are detected early enough.

If cancer cells go through additional changes that cause them not to adhere to each other in the normal fashion, the result may be **metastasis**—the spread of cancer to another organ or region of the body (see Figure 18.2). Metastasis occurs when cancer cells break away from the main tumor, travel via the blood or lymph, and develop into new colonies of cancer cells at distant sites. Cancers that invade surrounding tissues until they compromise organ function are called **malignant tumors** (from Latin *malignus,* meaning "ill-disposed"). Many (but by no means all) malignant tumors also produce secondary tumors. Table 18.1 compares benign, *in situ,* and malignant tumors.

A malignant, metastasizing tumor is like a forest fire that is out of control and throwing off sparks. The sparks start new fires in sometimes distant locations, which also spread uncontrollably. Metastasizing cancer continues to spread until it completely overruns the functions of tissues, organs, and organ systems. Ultimately it may kill the entire organism, just as a forest fire may not die out until it has consumed the whole forest.

Cancer stages

For ease of communication with each other and with patients, health professionals classify most cancers by a scale of stages ranging from zero to roman numeral four (IV). The scale is intended to indicate how advanced the cancer is, based on several factors: the size of the tumor and whether it has spread to surrounding tissues, whether regional lymph nodes are involved, and whether the cancerous cells have spread to distant locations throughout the body. Each higher stage represents a potentially more problematic cancer, and each type of cancer uses slightly different criteria to define the stages. As a general rule, however, the stages and their meanings are:

- *Stage 0:* Also known as an *in situ* tumor. Although the cells in the tumor are abnormal, they are still all clumped together. A diagnosis of a Stage 0 tumor is good news, because generally the tumor can be removed successfully and completely.

Table 18.1 A summary of the characteristics of benign tumors and cancers

	Benign tumor	*In situ* tumor (cancer)	Malignant tumor (cancer)
Frequency of Cell Division	More rapid than normal	Rapid and completely out of control	Rapid and completely out of control
Cell Structure and Function	Slightly abnormal	Increasingly abnormal	Very abnormal
Tumor Organization	A single mass generally surrounded by a capsule	Still a single mass, but increasingly disorganized and not always surrounded by a capsule	Some cells break away to invade underlying normal tissues or to enter the blood or lymph. New tumor colonies (secondary tumors) become established at distant sites.

- *Stages I–III:* Each stage is defined for each type of cancer, according to the cancer's size, degree of spread to surrounding tissues, and whether or not lymph nodes are now involved. The higher the number, the larger and more extensive the cancer. Most cancers are still treatable at these three stages, especially at the lower stages.
- *Stage IV:* This is not a diagnosis that you want to hear. At stage IV, the cancer has definitely metastasized, spreading to distant sites and organs. Targeted therapies such as radiation will not work because there are just too many targets, even if they could all be located (which is doubtful). Treatment will have to be both broad and potentially dangerous to normal cells.

Not all cancers are identified by the stages described above. Cancers of blood and bone marrow do not have a clear-cut staging system, and cancers of the brain and spinal cord are identified by a name indicating the cell type (for example, astrocytoma, glioblastoma, meningioma) as well as a grade.

One in three people in the United States will experience cancer in their lifetime, and one in four will die of it. Cancer ranks second only to heart disease as a cause of death. The good news is that most cancers can be prevented. Later in this chapter, we discuss how you can lower your risk.

↺ **Recap** Some tumors are benign, but when tumor cells change form dramatically and divide uncontrollably, the tumor is called *cancer*. Cancer becomes malignant when cells invade and metastasize, starting new tumors at distant sites. ∎

18.3 Factors contributing to cancer development

For cancer to develop, at least two things must happen simultaneously:

- The cell must grow and divide uncontrollably, ignoring inhibitory signals from nearby cells to stop dividing.
- The cell must undergo physical changes that allow it to break away from surrounding cells. These events begin when some of the cell's genes become abnormal and no longer function properly.

Mutant forms of proto-oncogenes, tumor suppressor genes, and mutator genes contribute to cancer

Several types of genes control the various activities of a cell. *Structural genes* code for the proteins necessary for cell growth, division, differentiation, and even cell adhesion

(the adherence of cells in the same tissue to each other). *Regulatory genes,* on the other hand, code for proteins that either activate or repress the expression of these structural genes.

Some regulatory genes code for proteins secreted by cells in order to influence nearby cells. These proteins are called *growth factors* or *growth inhibitors,* depending on which it is that they do. Growth inhibitors are especially important in regulating cell division because they influence cells in a particular tissue to stop dividing when enough cells are present.

Under normal circumstances, two classes of regulatory genes control the cell's activities: proto-oncogenes and tumor suppressor genes. **Proto-oncogenes** are normal regulatory genes that *promote* cell growth, differentiation, division, or adhesion (*proto,* meaning "first"; *onco-* means "mass, bulk," that is, a tumor). Mutated or damaged proto-oncogenes that contribute to cancer are called **oncogenes** (**Figure 18.4**). Some oncogenes drive the internal rate of cell growth and division faster than normal. Others produce damaged protein receptors that fail to heed inhibitory growth signals from other cells. However, it is important to recognize that one oncogene alone is not sufficient to cause cancer. Because so many cellular processes must be disrupted

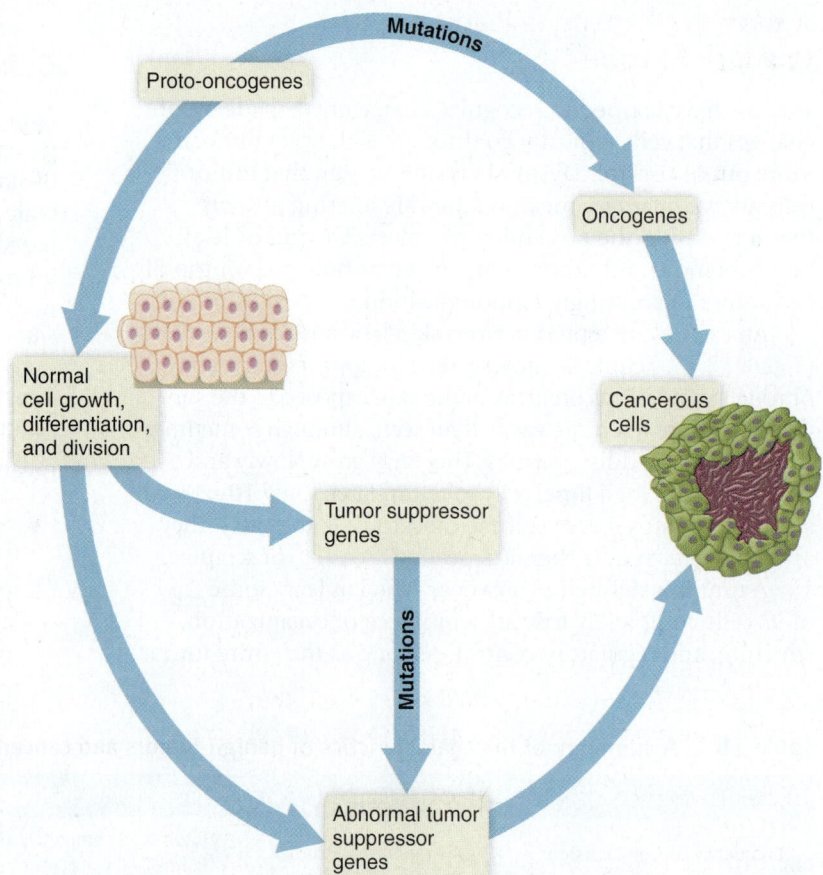

Figure 18.4 Proto-oncogenes and tumor suppressor genes. Proto-oncogenes promote cell growth, differentiation, and division. Tumor suppressor genes keep these activities in check. Mutations in either types of genes may contribute to cancer.

at once and each process may be controlled by multiple genes, cancer develops only when multiple oncogenes are present.

Tumor suppressor genes are regulatory genes that under normal conditions apply the brakes to unchecked cell growth, division, differentiation, or adhesion. They tend to *inhibit* these cell activities when cells are stressed or when conditions are not right for cell division. When they become damaged, tumor suppressor genes contribute to cancer because the cell activities they normally regulate are then likely to continue unchecked.

One tumor suppressor gene, called *p53*, has received a lot of attention lately. The p53 gene codes for a protein found in most tissues that prevents damaged or stressed cells from dividing. In fact, its primary role may be to inhibit division by cells that already show cancerous features. If the p53 gene becomes damaged, a variety of cancers may develop more easily. To date, mutated p53 genes have been identified in tumors of the cervix, colon, lung, skin, bladder, and breast.

A third class of gene may also contribute to cancer. In their unmutated form, **mutator genes** are involved in DNA repair during DNA replication. When these genes mutate, the cell becomes increasingly prone to errors in DNA replication. As a result, the cell may accumulate mutations in other genes at a faster rate than normal.

☑ A gene for a growth factor (a protein involved in normal cell growth) mutates so that it now produces growth factor constantly, triggering abnormal cell growth. What is the best term for the mutated gene: proto-oncogene, oncogene, tumor suppressor gene, or mutator gene? Why?

A variety of factors can lead to cancer

A car analogy demonstrates how a cell progresses toward cancer. At first, your new car runs very well. As time goes on, a few parts wear out, but you manage to keep it in good repair with regular service checkups. But as the car gets older, the need for repairs gets more frequent, and you aren't always able to keep up with them. Now the suspension needs replacing, the tires are worn out, the accelerator pedal sticks occasionally, and the brakes are thin. No single defect is enough to cause an accident, but together they spell trouble. There may come a moment when you may not be able to control the car, and you'll have an accident.

The same sort of progression occurs within cells. In fact, the single most important factor in the development of cancer may be age. One of the reasons that it has been so difficult to make any headway against cancer is that we have been so successful at preventing early deaths from other causes. The longer we live, the more likely we are to die of cancer. We all must die of something, some day.

The significance of aging can be seen quite clearly with skin cancer. Skin cancer is uncommon before adulthood, even though most severe sunburns occur during childhood.

The outer layers of your skin are continually being replaced by basal cells located at the base of the epidermis. The basal cells of a child's skin haven't been through very many cell divisions yet—they are still in good repair. But decades later, after thousands more cell divisions and a few more sunburns, those same basal cells are about worn out. Cellular repair mechanisms fail more frequently; mechanisms controlling cell division become less effective. First, there's dysplasia, then cancer. The cells are out of control.

In addition, your parents may carry a few of the cancer-promoting mutations that have accumulated in the precursor cells to human eggs and sperm during the course of human evolution. These rare heritable mutations may not be enough to guarantee cancer, but they do increase your *heritable susceptibility*. If six mutations are required to produce cancer, for example, you could inherit five mutations and still be cancer free. This is why, even in families with a history of cancer, some siblings develop the disease and others do not.

The multigene basis of cancer also explains why it is possible to develop cancer even if there is no family history. Remember that we all inherit a slightly different combination of our parents' genes, and if you happen to get mutated genes from both parents, you could be at greater risk than either of them. (We discuss genetic inheritance in more detail in Chapter 19.)

The process of transforming a normal cell into a cancerous one is called *carcinogenesis*. A **carcinogen** is any substance or physical factor that causes cancer. The word *carcinogen* comes from the recognition by early physicians that certain skin cancers had a crablike appearance (Greek *karkinos*, meaning "crab"; *genein*, meaning "to produce"). Some carcinogens are described below.

Certain viruses and bacteria Some viruses and bacteria may contribute to certain cancers. Viruses reproduce by inserting their DNA into the host's DNA and forcing the host to make the proteins needed by the virus. If the viral sequence of DNA impairs the function of a normal gene, perhaps by preventing a regulatory gene from being turned off after it is turned on, this will increase the cancer risk.

Viruses thought to contribute to cancer include:

- Human papillomavirus (HPV; cancers of the cervix and penis)
- Hepatitis B and hepatitis C viruses (liver cancer)
- HIV (Kaposi's sarcoma, non-Hodgkin's lymphoma)
- Epstein-Barr virus (Hodgkin's disease and non-Hodgkin's lymphoma)
- Human T-cell leukemia/lymphoma virus (HTLV-1; T-cell non-Hodgkin's lymphoma)

The *Helicobacter pylori* bacterium, which causes ulcers, may contribute to stomach cancer as well. Although this list may sound frightening, viruses and bacteria probably account for fewer than 15% of all cancers.

Figure 18.5 Environmental carcinogens. This agricultural worker is wearing protective clothing and an air-filtering mask while spraying pesticides. Some pesticides may increase the risk of cancer.

Toxic chemicals in the environment

Some chemical carcinogens damage DNA directly. Others do not damage DNA, but their presence increases the potency of still other carcinogens. The list of chemical carcinogens in the environment includes many chemicals used in (or produced by) industrial processes, such as coal tar, soot, asbestos, benzene, vinyl chloride, and some pesticides and dyes (Figure 18.5). Table 18.2 lists examples of viral, chemical, and radiation carcinogens, ranked by likelihood of exposure.

Tobacco

About 2% of all cancers are thought to be caused by industrial pollutants. In contrast, tobacco is by far the single most lethal carcinogen in the United States, accounting for over 30% of all cancer deaths. Smoking dramatically increases your risk of cancer of the lungs, mouth, pharynx, pancreas, and bladder. Smokers are not the only ones at risk; passive smoking (inhaling secondhand smoke) can cause lung cancer, too.

Smokeless tobacco (chewing tobacco or snuff) is also a carcinogen. It is a major risk factor for cancers of the mouth, pharynx, and esophagus.

Alcohol

The Department of Health and Human Services lists alcohol among the risk factors for cancer; the higher the alcohol consumption, the greater the risk. The primary risks are to throat, esophagus, and liver, but alcohol also increases the risk of breast cancer and cancers of the colon and rectum.

Radiation

Sources of radiation include the sun and radioactive radon gas, which is released by rocks, soil, and groundwater at intensities that vary in different geographic areas. Cellular telephones, household appliances, and electric power lines also emit trace amounts of radiation. All sources of radiation combined cause only about 2% of all cancer deaths, most of which are due to skin cancer caused by the sun's ultraviolet rays.

The high-frequency ultraviolet B rays in sunlight cause over 80% of all skin cancers, including a particularly dangerous cancer of the melanin-producing cells of the skin called **melanoma.** Frequent sunburns, especially

Table 18.2 Examples of carcinogens

Carcinogen	Source of exposure or types of persons exposed	Type of cancer	Exposure of general population
Tobacco (smoking and smokeless)	Smokers, chewers, people exposed to secondhand smoke	Lung, mouth, pharynx, bladder, cervix, colon, pancreas	Common
Diesel exhaust	Buses and trucks, miners, railroad yard workers	Lung	Common
Benzene	Paints, dyes, furniture finishes	Bone marrow (leukemias)	Common
Pesticides	Agricultural workers	Lung	Common
Ultraviolet light	Sunlight	Skin cancers	Common
Ionizing radiation	Radioactive materials, medical and dental procedures	Bone marrow (leukemias)	Common
Alcohol	Drinking alcoholic beverages	Throat, esophagus, liver	Common
Human papillomaviruses	Sexually transmitted	Cervix, penis	Common
Hepatitis B virus	Sexually transmitted	Liver	Less common
HIV	Sexually transmitted	Kaposi's sarcoma	Less common
Hydrocarbons in soot, tar smoke	Firefighters, chimney cleaners	Skin, lung	Uncommon
Asbestos	Shipyard, demolition, and insulation workers, brake linings	Lung, epithelial linings of body cavities	Uncommon
Radon	Mine workers, basements of houses	Lung	Uncommon

during childhood, increase the risk of melanoma. Fair-complexioned people who sunburn quickly are more vulnerable than those who tan easily (Figure 18.6). However, cumulative exposure to ultraviolet radiation, even without burning, raises the risk of skin cancer. Ultraviolet radiation from sunlamps and tanning booths is just as dangerous as radiation from natural sunlight.

> **MJ's BlogInFocus** Does radiation therapy for cancer treatment ever cause additional cancers? Visit MJ's blog in the Study Area in MasteringBiology and look under "Radiation and Cancer."
>
> http://goo.gl/nAnIuL

Diet and obesity Dietary factors and especially obesity may be involved in up to 30% of cancers, rivaling smoking. In particular, red meat and saturated animal fat raise the risk of cancers of the colon, rectum, and prostate gland, and a high salt intake may contribute to stomach cancer.

Figure 18.6 Risky behavior. Persons who were often sunburned at a young age have an increased risk for skin cancer later in life. Persons with lighter complexions are at greater risk.

Individuals who are obese or have Type 2 diabetes have a much higher risk of dying of cancer than people of normal weight. A current hypothesis is that in these individuals, insulin and a substance called *insulin-like growth factor* (IGF) may alter the metabolic environment of tumor cells, stimulating them to grow and divide.

Certain fungi and plants produce carcinogens. One such carcinogen is *aflatoxin*, a by-product of a fungus that may be present in ground nuts, including raw peanut butter. Commercial peanut butters are processed to eliminate the fungus.

In recent years, researchers have become interested in what might be *missing* from the diet that could contribute to cancer. Recent evidence suggests that fruits and vegetables in particular may be mildly protective. A compound in broccoli sprouts has received a lot of attention lately as a cancer risk–reducing agent. Analyzing the role of diet in cancer is complicated because there are literally thousands of compounds in the foods we eat.

Internal factors It is difficult to estimate how significant internal factors might be to the development of cancer. In theory, errors introduced during DNA replication and damage caused by certain chemicals within the body could contribute to the disease.

In particular, scientists are studying the effects of *free radicals*, highly reactive fragments of molecules that are produced by the body's biochemical processes. Free radicals are a normal by-product of cellular metabolism, and ordinarily they are detoxified by peroxisomes (Chapter 3). However, if detoxification processes become less efficient for any reason, free radicals may accumulate in the body and damage other molecules, including DNA, which could lead to cancer. Some researchers believe that free radicals play a role in the aging process as well.

Certain vitamins known as *antioxidants* appear to neutralize free radicals, although the mechanism by which they do this is still controversial. In particular, vitamins A, C, and E seem to exert antioxidant benefits. Whether antioxidants affect the development of cancer or the aging process, however, is still an open question. Antioxidant-containing foods include blueberries, spinach, and tomatoes.

☑ People who live in basements have a higher risk of lung cancer than people who live above ground. Which of the above influences on cancer might explain this?

The immune system plays an important role in cancer prevention

Under normal circumstances, the immune system plays a critical role in defending against damaged or defective cells, including cancer cells. Recall that all cells display proteins that identify them as "self" so that the immune system leaves them alone. As cells change in ways that make them cancerous, they may stop displaying "self" proteins, or they may display completely different proteins that mark them as abnormal. These scenarios make cells subject to attack and destruction by the immune system, at least as long as it is functioning normally.

Suppression of the immune system by drugs, viruses such as HIV, or even mental states of anxiety, stress, or depression may allow certain cancers to develop more easily. Some cancers apparently can suppress the immune system, and others may have mechanisms for disguising themselves from attack by the immune system. Figure 18.7 presents an overview of how cancer develops.

↺ Recap Proto-oncogenes and tumor suppressor genes normally control the rate of cell division. Mutator genes are involved in DNA repair. Mutations of any of these genes may contribute to cancer. Aside from heritable susceptibility, factors that may contribute to the development of cancer include viruses and bacteria, environmental chemicals, tobacco, radiation, dietary factors, and alcohol. Free radicals created during cellular metabolism may also contribute to cancer. The immune system normally protects us from cancer cells by killing them before they spread. Immune system suppression allows cancers to develop more easily. ■

18.4 Advances in diagnosis enable early detection

In 1971, President Nixon signed into law the National Cancer Act, declaring "war" on cancer. At that time, many researchers hoped we could overcome cancer with early diagnosis and better treatments.

The sobering fact is that the overall death rate from all cancers is still higher now than it was 80 years ago. The main reason has been a dramatic increase in lung cancer that closely follows the popularity of smoking, first in men and later in women (Figure 18.8). There was even a time when men in the military were given cigarettes for free, as part of their rations. We are now paying the price with much higher rates of cancer in older Americans. As these older generations are replaced with generations who grew up with anti-smoking campaigns, lung cancer rates are beginning to decline in younger age groups.

Although we have made significant progress against certain types of cancer, much more remains to be done. Early detection (diagnosis) of cancer is an important key to success. Early detection means that therapy can begin sooner,

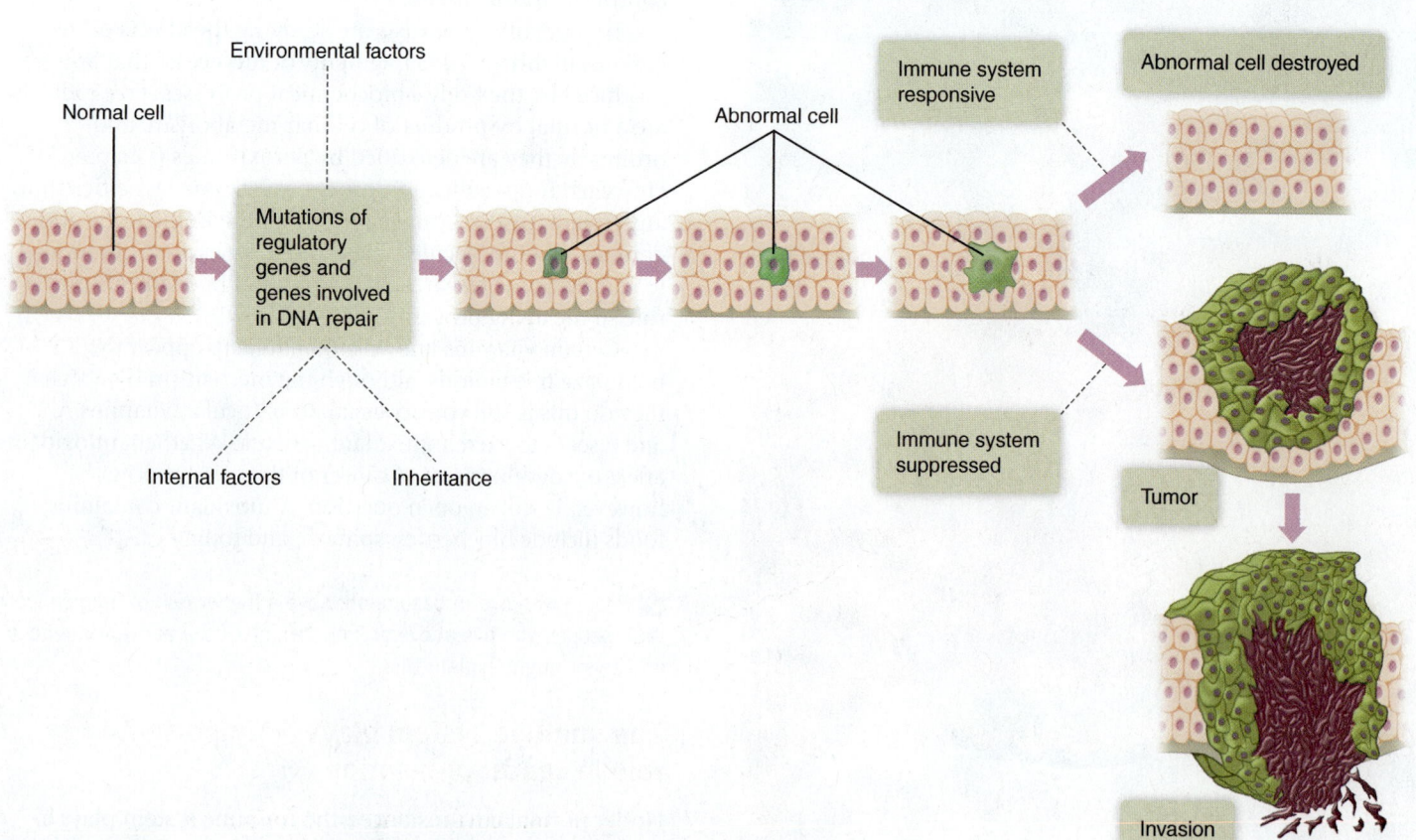

Figure 18.7 Overview of the development of cancer. The progression to cancer requires multiple genetic changes in regulatory genes, some of which may be inherited and some of which may be caused by environmental or internal factors. A cancer cell reproduces rapidly (develops into a cancer) and may invade and colonize distant tissues (metastasize). Suppression of the immune system appears to be involved in many cancers.

a) Ads promoting smoking were common in the 1940s and 1950s.

Figure 18.8 Smoking and lung cancer.

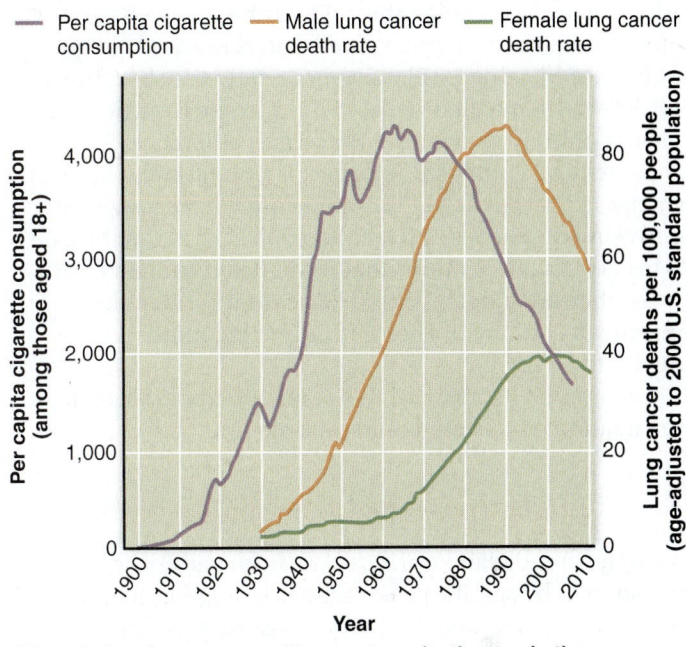

b) **Trends in tobacco use and lung cancer death rates in the United States.**

hopefully before the disease has had a chance to metastasize. Prompt treatment can even cure some cancers.

Tumor imaging: X-rays, PET, and MRI

A variety of imaging techniques are available for diagnosing tumors and for determining how advanced they may be.

Traditionally, an X-ray (Figure 18.9a) is most commonly used to provide an image of a tumor. A newer form of X-ray analysis called a computed axial tomography (CAT) scan (Figure 18.9b) creates multiple images that look as if one has taken slices of the body across the axial (transverse) plane.

Other advanced imaging techniques include ultrasound, also called sonography (Figure 18.9c), which uses sound waves above the audible range to create images of soft tissues. Two imaging techniques show physiological differences between tissues that cannot be diagnosed effectively with X-rays or ultrasound. Positron emission tomography (PET) scans (Figure 18.9d) employ radioactive substances to create three-dimensional images showing the metabolic activity of body structures. In

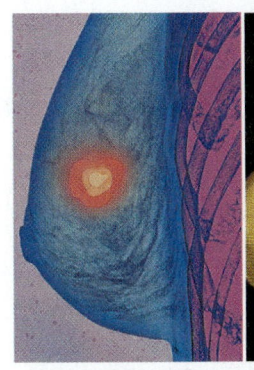

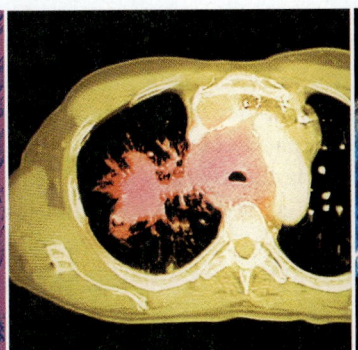

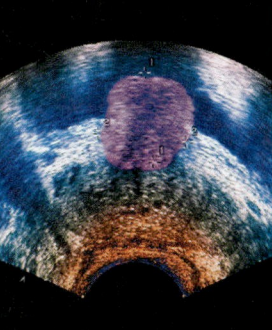

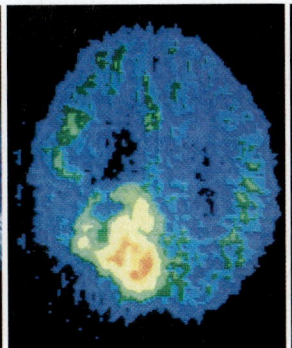

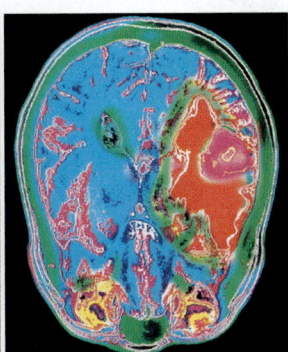

a) **An X-ray.** This is a color-enhanced X-ray of a breast with a tumor.

b) **A CAT scan** showing cancer in the left lung.

c) **An ultrasound scan.** This one shows a prostate tumor (purple).

d) **A PET scan** of the brain. The yellow/red area is a tumor.

e) **An MRI.** This is an MRI of a brain, showing a brain tumor (magenta and red area to the right).

Figure 18.9 Tumor imaging techniques.

addition to diagnosing cancer, PET scans are used to study blood flow, heart function, and brain activity. Magnetic resonance imaging (MRI) (**Figure 18.9e**) uses short bursts of a powerful magnetic field to produce cross-sectional images of body structures. MRI detects differences in water and chemical composition between tissues on the basis of changes in magnetic fields; tumors often have a slightly different consistency and composition than normal tissue. A big advantage of MRI is that it can detect tumors even when they are hidden behind bone, so it is often the imaging technique of choice for detecting tumors of the brain.

X-rays, CAT scans, and PET scans, all expose the patient to radiation. Ultrasounds and MRIs do not.

Genetic testing can identify mutated genes

As we learn more about which human characteristics are transmitted by specific genes, genetic testing for diseases, including cancers, becomes a real possibility. Already hundreds of new genes and their mutated counterparts have been identified, and tests are being devised to detect them.

Such tests are highly controversial, however, and society will have to choose how to use them. One of the concerns about genetic testing is that it may be possible to test for mutated genes that have no cure. Some people believe that such knowledge may not be in the patient's best interest. In addition, it is difficult to convince people that in many cases a mutated gene is nothing more than a risk factor. It does not guarantee development of a disease.

Enzyme tests may detect cancer markers

An enzyme called *telomerase* is rarely found in normal cells but is present in nearly all cancer cells. Some researchers propose that a sensitive test for telomerase in body fluids could detect early stages of cancer. An enzyme test could also be used to screen large populations or monitor the progression of the disease in known cancer patients. Researchers are exploring other potential cancer markers as well.

↩ **Recap** Early diagnosis and prompt treatment of cancer are vital. Diagnostic techniques include X-rays, positron-emission tomography (PET), magnetic resonance imaging (MRI), genetic testing, and enzyme markers. ∎

18.5 Cancer treatments

Many cancers are treatable, especially if discovered early. Although some forms of the disease have higher survival rates than others, current treatments cure approximately 50% of all cancer cases.

Conventional cancer treatments: surgery, radiation, and chemotherapy

Classic treatments for cancer include surgery, radiation, chemotherapy, or a combination of two or more of these. Better imaging techniques have benefited the surgical treatment of cancer, and techniques for focusing external sources of radiation directly onto the cancer have improved as well. These focused sources of intense radiation are designed to damage and kill cancer cells in a specific location. It is sometimes possible to implant radioactive materials directly into a tumor, so that the radiation will kill primarily tumor cells rather than surrounding healthy tissue.

 **MJ's BlogInFocus** What is a double mastectomy, and why are they becoming more common? Visit MJ's blog in the Study Area in MasteringBiology and look under "Double Mastectomies."

http://goo.gl/5tYCok

Healthy cells recover from radiation more readily than cancer cells, but radiation therapy generally does injure or kill some normal tissue, too. A drawback of both surgery and radiation is that they may miss small groups of metastasized cells, allowing the cancer to reappear later.

☑ Children with cancer who are treated with radiation treatments have an increased risk of developing other cancers later in life. Why?

Chemotherapy involves the administration of cytotoxic (cell-damaging) chemicals to destroy cancer cells. Chemotherapy addresses some of the limitations of surgery and radiation. Some chemotherapeutic drugs act everywhere in the body, and for that reason, they can sometimes inflict significant damage on normal cells. Some chemotherapy chemicals stop cells from dividing by interfering with their ability to replicate DNA. Chemicals that inhibit cell division act primarily on cancer cells because cancer cells divide very rapidly. But such chemicals also damage or kill many normal cells that divide rapidly, such as bone marrow cells and cells lining the digestive tract. Damage to cells in the digestive tract is the reason so many chemotherapy patients experience nausea.

In addition to nausea and vomiting, common side effects of chemotherapy include hair loss, anemia (a reduction in the number of red blood cells), and an inability to fight infections (due to fewer white blood cells). A major problem is that many tumors become resistant to chemotherapeutic drugs, just as bacteria become resistant to antibiotics.

Treatment then becomes a battle of trying to destroy abnormal cells with stronger and stronger chemicals without killing the patient in the process. Sometimes combination therapy, in which surgery is followed by radiation and combinations of chemotherapeutic drugs, works best.

☑ Why does chemotherapy cause anemia, increased risk of infections, and hair loss? Hint: What kind of tissues undergo a rate of cell division that would cause these symptoms, and what do these tissues have in common?

Magnetism and photodynamic therapy target malignant cells

To avoid the side effects of standard radiation and chemotherapy, researchers are developing techniques to target malignant cells more precisely while sparing healthy tissue. Promising developments include the following:

- *Magnetism:* Positioning a powerful magnet at the tumor site and injecting tiny metallic beads into the patient's bloodstream. The magnet pulls the beads, which have been coated with a chemotherapy drug, into the tumor, where they kill the cancer cells. Magnetism is undergoing clinical trials in liver cancer patients.
- *Photodynamic therapy:* Targeting cancer with light-sensitive drugs and lasers. The patient takes light-sensitive drugs that are drawn into the rapidly dividing cancer cells. Laser light of a particular frequency is focused on the tumor, where it triggers a series of chemical events that kill malignant cells. Photodynamic therapy has been approved for several years to treat tumors in the

esophagus and lung, and new refinements in lasers make it promising for other cancers, too (Figure 18.10).

Immunotherapy promotes immune response

Immunotherapy attempts to boost the general responsiveness of the immune system so that it can fight cancer more effectively. Recent efforts have focused on finding specific antigen molecules that are present on cancer cells but not on normal cells, and using them to produce antibodies that target cancer cells for destruction by the immune system. This approach is showing particular promise. Taking it one step further, researchers are working to attach radioactive molecules or chemotherapeutic drugs to the antibodies so that treatments can be delivered only to cancer cells and spare normal cells. Researchers are also working to create vaccines that prevent specific cancers.

Researchers are also testing other vaccines created from a patient's own cancer cells. When injected into the patient, the modified cancer cells appear to stimulate the person's immune defenses to recognize and combat the abnormal cells.

"Starving" cancer by inhibiting angiogenesis

Tumors grow and divide rapidly, so they require a great deal of energy. We now know that something within tumors promotes *angiogenesis* (the growth of new blood vessels) to serve their energy needs. Without angiogenesis, the tumor would reach about the size of a pea and stop growing. At least a dozen proteins stimulate angiogenesis, some of which are released

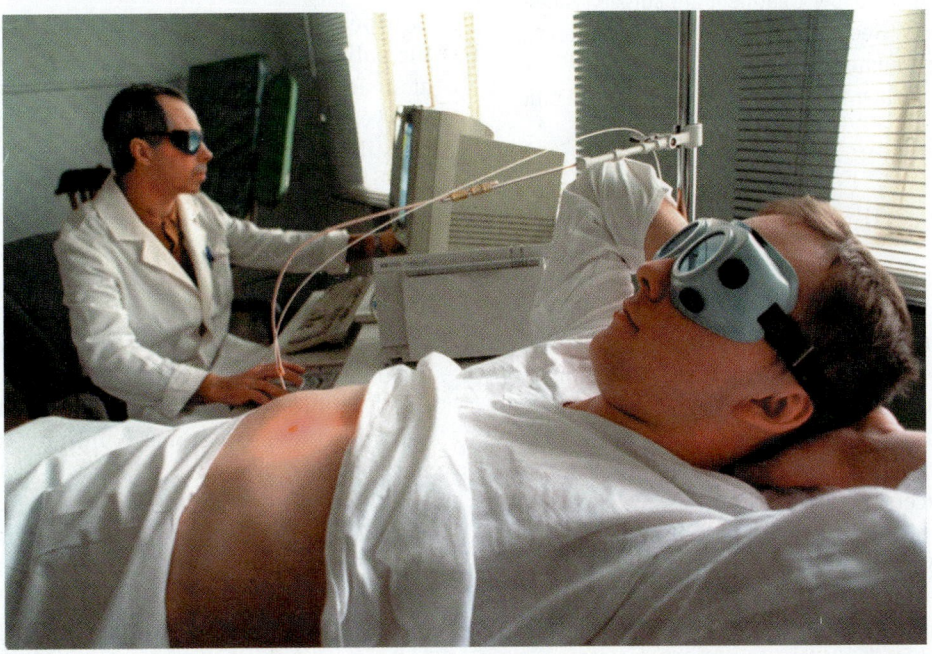

Figure 18.10 **Photodynamic therapy.** Here, it is being used to treat a skin cancer.

by tumor cells, and several other proteins are known to inhibit angiogenesis. Interest is very high in *anti-angiogenic* (against blood vessel growth) drugs, because they may be able to starve tumors by limiting their blood supply. Anti-angiogenic drugs now on the market include Avastin, a blockbuster drug now approved for the treatment of a variety of cancers, and Revlimid, recently approved for the treatment of multiple myeloma, a cancer of antibody-producing white blood cells in bone.

Interestingly, researchers have found that anti-angiogenic drugs may not work just by starving tumors, at least at first. As they begin to inhibit the growth of the abnormal tangle of blood vessels normally found in a tumor, they actually normalize the vasculature and improve blood flow for a time. And that time (perhaps a month) may be a window of opportunity for delivering anti-tumor chemicals to the tumor via the bloodstream.

☑ Patients on Avastin are advised to stop taking the drug temporarily if they are planning to have surgery. Why?

Molecular treatments target defective genes

Sophisticated molecular strategies target the defective genes that lead to cancer. Some researchers are attempting to inactivate specific genes, or the proteins they encode, to slow the runaway cell division. Future techniques may include *gene therapy*, in which defective genes are repaired or replaced with their normal counterparts. A key target for gene therapy is the p53 tumor suppressor gene that, when defective, contributes to many cancers.

↩ **Recap** The mainstays of cancer treatment are surgery, radiation, and chemotherapy. Recent advances include magnetism, photodynamic therapy, immunotherapy, drugs to inhibit angiogenesis, and molecular treatments that focus on specific defective genes. ■

18.6 The ten deadliest cancers

Table 18.3 lists the ten deadliest cancers in order, by numbers of deaths per year. Each has its own risk factors, warning signs, methods of detection, treatments, and (most important of all) survival rate. At the very top of the list is lung cancer. Lung cancer is responsible for as many deaths as the next four cancers combined. Cancers of the colon and rectum, breast, and pancreas are the second, third, and fourth most deadly cancers, respectively.

Note that deaths from cancers do not necessarily parallel the rates of cancer incidence (numbers of new cases diagnosed per year). The three most common cancers in terms of incidence are cancers of the lung, breast (primarily in women), and prostate (in men).

In the sections that follow, we'll describe each cancer in descending order starting with the most deadly, lung cancer.

Table 18.3 The top 10 cancers ranked by number of deaths per year

Cancer	New cases per year	Deaths per year
Lung	221,200	158,040
Colon and rectum	132,700	49,700
Breast	231,840 (women)	40,290 (women)
	2,350 (men)	440 (men)
Pancreas	48,960	40,560
Prostate	220,800	27,540
Leukemia	54,270	24,450
Lymphoma	80,900	20,940
Urinary bladder	74,000	16,000
Esophagus	16,980	15,590
Uterus (corpus and cervix)	67,770	15,270

Source: American Cancer Society

Lung cancer: smoking is leading risk factor

Most cases of lung cancer are preventable, because by far the greatest risk factor is cigarette smoking. All other risk factors, which include exposure to smoke, radiation, and industrial chemicals such as asbestos, are relatively minor by comparison.

Unfortunately, there is no simple screening test for lung cancer, and so it is generally not detected early. Most of the symptoms of lung cancer (persistent cough, recurrent pneumonia or bronchitis, voice change) are relatively nonspecific. By the time blood appears in the sputum and the patient experiences chest pain, the cancer may already be at an advanced stage. Early detection improves the odds. The five-year survival rate (percentage of persons diagnosed who are still alive after five years) is 54% when the disease is still localized to one spot in the lung. However, overall five-year survival rate for all persons diagnosed with lung cancer is still only 17%.

Treatment options depend on when lung cancer is detected. When it is detected early and is still localized, surgery is the treatment of choice. However, in most cases the disease has already spread by the time it is detected. For the more advanced stages, radiation therapy and chemotherapy, sometimes combined with surgery, are the only options.

Current research efforts are directed at finding earlier, more effective detection methods for lung cancer. A large clinical trial is currently under way to determine whether screening high-risk individuals with low-dose computed tomography (CT) scans can reduce lung cancer deaths. Such scans are not cheap, however. Other research is aimed at finding molecular markers for lung cancer in sputum so that inexpensive screening tests could detect the disease at an earlier stage.

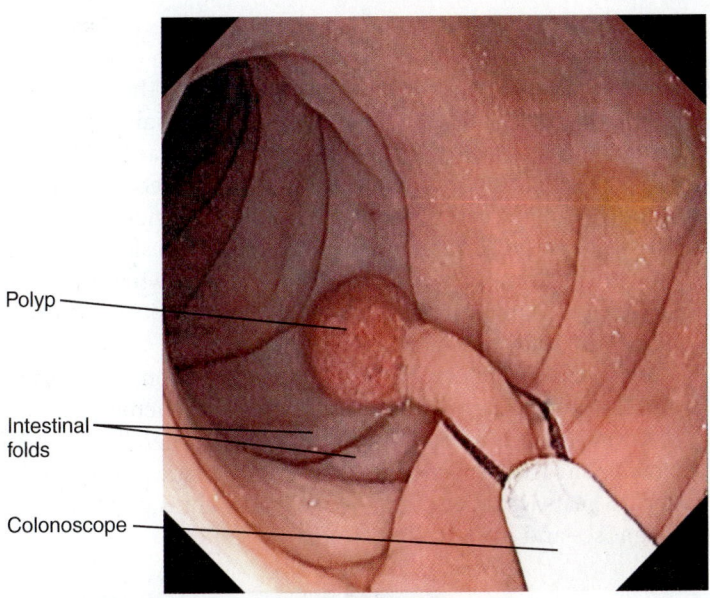

Polyp

Intestinal folds

Colonoscope

Figure 18.11 Polyps. A polyp in the colon being removed with a wire snare.

Cancers of colon and rectum: tests can detect them early

An obvious worrisome sign of colon or rectum cancer is blood in the stool or rectal bleeding, though bleeding can also result from noncancerous causes. Risk factors include sedentary lifestyle, obesity, smoking, a family history of colorectal cancer, and a low-fiber, high-fat diet with not enough fruits and vegetables.

Most colorectal tumors start as **polyps,** small benign growths that develop from the colon lining (Figure 18.11). The majority of colon polyps never turn malignant, and even when they do, the process usually takes years. This underscores the importance of regular exams to detect polyps and remove them before they can become dangerous. For everyone over age 50, the American Cancer Society recommends a *colonoscopy* (examining the interior of the complete colon with a flexible fiber-optic scope) every 10 years. People who have any risk factors should begin screening earlier and undergo testing more often. Polyps can generally be removed rather easily with a small wire snare or by electrocautery (electric heat).

The five-year survival rate for all cases of colorectal cancer has improved to 65%, largely because of early detection and treatment. For people with secondary tumors at the time of diagnosis, the five-year survival rate drops to just 13%. Here is another case where early detection is extremely beneficial. However, fewer than 40% of colorectal cancers are detected at an early stage because rates of screening are still low.

Breast cancer: early detection pays off

Breast cancer is almost exclusively a cancer of women (but note that it can occur in men; more than 2,000 new cases of male breast cancer are diagnosed each year). Breast cancer is usually first diagnosed by a mammogram (an X-ray of breast tissue) or when the patient or health care provider detects an abnormal lump during a physical exam.

Recent research suggests that there are at least two inherited susceptibility genes for breast cancer, called *BRCA1* and *BRCA2*. Fortunately, mutated forms of these genes are relatively rare. A major risk factor is age; only 1 in 200 women will get breast cancer before age 40, but 1 in 8 women who live to age 85 will get it in her lifetime. Other risk factors include early onset of menstruation or late onset of menopause, obesity after menopause, use of oral contraceptives, and hormone replacement therapy after menopause. Environmental and dietary factors may also play a role.

Early detection of breast cancer dramatically increases the chance of survival. The five-year survival rate for breast tumors that have not yet metastasized is 99%, but it falls to 25% when the cancer has already metastasized to distant tissues at the time of diagnosis. Overall, the five-year survival rate for all diagnoses is now 89%, up from 63% in the 1960s.

Current recommendations are that all women over the age of 50 should have an annual mammogram; women younger than 50 should have mammograms less frequently than every year until age 50, unless they are judged to be at high risk. In the past, women were strongly advised to also examine their breasts themselves. One change since 2009 is that the United States Preventive Services Task Force (USPSTF), the agency charged with making recommendations about preventive care services for people who do not have any signs or symptoms of disease, is now recommending against breast self-examination as a way to detect breast cancer. In reviewing the scientific literature, the USPSTF found that breast self-examinations do not lower the mortality rate from breast cancer and that women who perform them tend to have more imaging procedures and biopsies than women who don't, creating unnecessary expenses and risks. The American Cancer Society followed suit; it now calls breast self-examinations "optional," meaning that it is acceptable not to do them.

☑ Use the data in Table 18.3 to calculate breast cancer deaths per year, in men and then in women, as a percentage of newly diagnosed cases. In which gender is a newly diagnosed patient more likely to die? Can you explain these results?

Pancreatic cancer: rarely detected early enough

Cancer of the pancreas often goes undetected until an advanced stage because there are no obvious early signs. In its more advanced stages, symptoms may include weight loss and abdominal pain that radiates to the back, neither of which is very specific. As a result, the death rate from pancreatic cancer is the second highest of all cancers (after esophageal cancer).

Risk factors including smoking (about 20% of pancreatic cancers are attributable to smoking) and having one of the breast cancer genes, BRCA1 or BRCA2. Treatments such as surgery, chemotherapy, or radiation may extend survival but seldom produce a cure unless the cancer is caught at an early stage. The five-year survival rate for all pancreatic cancers is only 7%; even for the few patients diagnosed at an early stage it is only 26%.

Prostate cancer: most common after age 50

Early-stage prostate cancer may have no symptoms at all. As the cancer grows, symptoms may include difficulty urinating or inability to urinate, blood in the urine, and pain, usually in the pelvic area.

The biggest risk for prostate cancer is advancing age. In men with the symptoms listed above, prostate cancer is initially diagnosed (or ruled out) by a digital rectal exam or by a *prostate-specific antigen (PSA)* test, which detects elevated blood levels of a protein produced by the prostate gland when prostate cancer is present. The initial diagnosis is generally confirmed by a biopsy and examination of the prostate cells. However, physicians no longer recommend regular yearly screening for prostate cancer in men with no symptoms. Prostate tumors tend to grow so slowly that they are not generally the cause of death even when they are detected. Men are advised to consult a physician before being tested for prostate cancer.

Treatment options include surgery, radiation therapy, and hormones. Aggressive treatment in older men is generally discouraged, again because prostate cancer generally grows so slowly. The five-year survival rate is now 99%, up from 68% just 25 years ago. The dramatic improvement is due to earlier diagnoses and improved treatments.

HEALTH & WELLNESS
What If You Could Save Someone's Life?

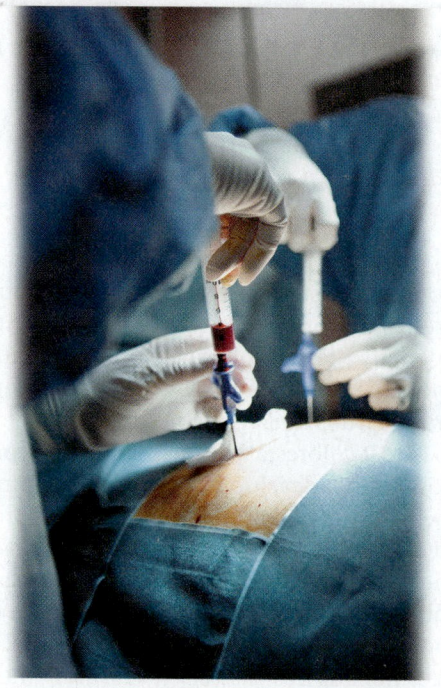

It sounds crazy, but it's entirely possible: You could be responsible for saving the life of someone with leukemia or lymphoma. That's because both cancers are curable, but only with the help of stem cells from a donor. And that donor could be you.

The standard cure for leukemias and lymphomas involves killing all of the patient's blood-forming stem cells with chemotherapy and radiation and then repopulating the patient's bone marrow with healthy stem cells from a donor. Some blood cancer patients are lucky enough to find a close relative who is a good immunological match, but for most patients the only source of donor stem cells is an unrelated, altruistic donor. But how is a patient to find an unrelated donor who is also a good immunological match?

The key is to have a lot of potential donors (like you) whose immunological match is known and who have agreed that they might like to donate if they ever match a patient in need. It's like agreeing to be an organ donor after your death, except in this case, you make a donation (and save a life) while you are still alive. Here's how it works. If you think that you might be willing to donate bone marrow or peripheral blood stem cells (PBSC), contact Delete Blood Cancer, Be The Match, or another nonprofit donor registry. The registry will determine your immunological match and "bank" that information—that is, make that information available (anonymously, of course) to doctors with patients in need. Then you wait.

If/when you are the best match for a patient (only one in 500 people who sign up will ever donate), you will be contacted and asked if you still want to donate. You will undergo further interviews and testing to ensure you are in good health, and then you will be invited to donate either bone marrow or stem cells, depending on the needs of the patient. If you agree to donate bone marrow, you will undergo a minor surgical procedure; under anesthesia, needles are used to collect a small amount (1–5%) of your bone marrow. If you're donating stem cells, you will receive a medication for five days that increases the number of stem cells in your blood. Then you will undergo a procedure in which blood is removed from one arm, passed through a machine that separates out the stem cells, and then is returned via your other arm.

Meanwhile, the patient who will receive your bone marrow or stem cells will undergo chemotherapy and perhaps radiation therapy to kill all of his/her cancerous bone-forming cells. Then the patient will be given your cells, and if the procedure works, your cells will re-colonize his/her bone marrow. The patient will be cured, and he/she will carry some of your cells for the rest of his/her life.

You cannot be paid directly for your efforts, and yes, it would take a special effort on your part. However, all expenses (including, perhaps, for travel to where the patient is located in order to donate) are paid. And best of all, you would know that you saved a life.

Leukemia: chemotherapy is often effective

Leukemias are cancers of immature white blood cells within bone marrow. In its more advanced stages, the bone marrow becomes completely filled with cancerous cells. Eventually, the production of normal blood cells declines, leading to anemia and a reduced resistance to infections. The causes of leukemia are largely unknown but may be linked to Down syndrome, excessive exposure to ionizing radiation, and benzene. Certain leukemias are caused by HTLV-1 (human T-cell leukemia/lymphoma virus-1).

Although leukemia is commonly thought of as a childhood disease, more adults than children develop leukemias. The symptoms of leukemia are fairly general (fatigue, weight loss, increased incidence of infections), and as a result, leukemias are not always diagnosed early. A firm diagnosis is made with a blood test and a bone biopsy.

The standard treatment for leukemias is chemotherapy, followed by a bone marrow transplant or a transplant of peripheral blood stem cells. The survival rates for certain types of leukemias have improved over the years, especially among children.

Lymphoma: cancers of lymphoid tissues

Lymphoma is the general term for cancers of the lymphoid tissues, which include Hodgkin disease and non-Hodgkin lymphoma. Symptoms include enlarged lymph nodes, intermittent fever, itching, weight loss, and night sweats.

Risk factors are not entirely clear but seem to relate to altered immune function. People at risk include organ transplant recipients, those with HIV or HTLV-1 infection, and perhaps workers with high occupational exposure to herbicides. Treatments generally are radiation or chemotherapy. Treatments involving specific antibodies to lymphoma cells can be effective. If treatment is insufficient, high-dose radiation therapy followed by a stem cell or bone marrow transplant may be necessary.

Urinary bladder cancer: surgery is often successful if done early

Blood in the urine is an important sign, although there may be reasons for it other than cancer. Smoking quadruples the risk of bladder cancer; in fact, smoking is responsible for about half of all bladder cancers. Living in an urban area, exposure to arsenic in the water supply, or occupational exposure to rubber, leather, or dye are also risks.

Bladder cancer is usually diagnosed by microscopic examination of cells in the urine and by direct visualization of the bladder wall with a cytoscope (a thin tube with a lens) inserted into the bladder via the ureter. Surgery combined with chemotherapy is generally successful if the cancer is detected while it is in the layer of cells in which it began (five-year survival rate is 96%) or still localized to the bladder (69%). However, the five-year survival rate declines to just 6% if the cancer has already metastasized to distant sites by the time it is detected.

☑ Propose a hypothesis that could explain why smoking increases the risk of bladder cancer, despite the fact that the bladder is never exposed to smoke.

Esophageal cancer: a high ratio of deaths to cases

Although esophageal cancer represents only 1% of all new cancer cases in the United States, it has the highest ratio of deaths to cases of the top ten deadliest cancers; there are 92 deaths for every 100 new cases. Most cases of esophageal cancer are diagnosed after age 50; the median age at diagnosis is 67. Risk factors include heavy alcohol use, tobacco use, obesity, and chronic gastric reflux, in which repeated exposure to stomach acid erodes the lining of the esophagus.

Like pancreatic cancer, esophageal cancer rarely is diagnosed at an early stage. The earliest symptoms are likely to be difficulty swallowing or pain while swallowing, but these do not generally appear until at least 60% of the circumference of the esophagus is already cancerous (Figure 18.12). If the cancer is still localized, an esophagectomy (surgery to remove a section of the esophagus) may be able to remove all of the cancer. For more advanced stages, the focus is on *palliative* care: care that focuses on relief of symptoms, pain, and physical and mental stress, with the goal of improving quality of life for whatever time remains. Chemotherapy can lengthen survival time, and placement of a stent may allow the patient to swallow. Experimental cancer drugs may offer some hope to patients who enroll in clinical trials, but the bottom line is that current treatment methods are generally not very effective against esophageal cancer.

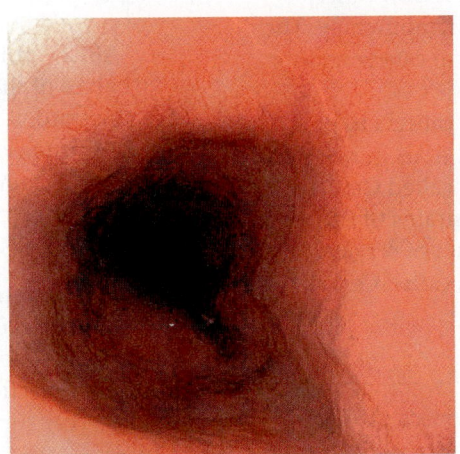

a) An endoscopic view of a normal esophagus.

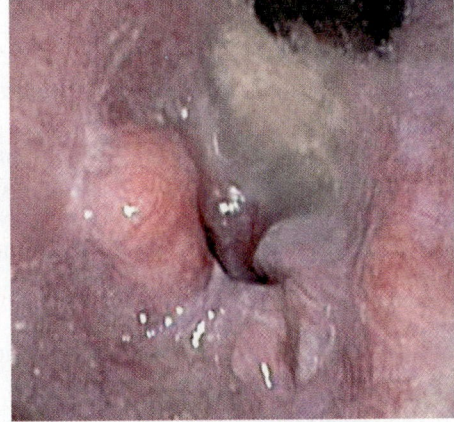

b) Esophageal cancer (upper right).

Figure 18.12 **Esophageal cancer.**

Cancer of the uterus: unusual uterine bleeding is major symptom

Uterine cancer includes cancers of the endometrium and cervix. The primary symptoms of endometrial cancer are abnormal bleeding or spotting. Pain is a late manifestation of the disease. High lifetime exposure to estrogen is the major risk factor for endometrial cancer. Reasons for high exposure include early onset of menstrual periods, late menopause, obesity, not having children, and/or estrogen replacement therapy after menopause. (In estrogen replacement therapy, progesterone added to estrogen is thought to offset the risk related to estrogen alone.) Pregnancy and the use of oral contraceptives seem to offer some protection from endometrial cancer.

Cervical cancer, in contrast, is strongly correlated to sexually transmitted infection with high-risk strains of the human papillomavirus (HPV). Women who have had many sexual partners (or whose partners have had many sexual partners) are at higher risk for cervical cancer. Cigarette smoking also increases the odds. Cervical cancer is usually detected by a Pap test (examination of a smear of cervical cells; the name comes from Papanicolaou, the doctor who devised the test). The Pap test is not very effective at detecting cancer of the endometrium, which is generally detected by a pelvic exam.

> **MJ's** **BlogInFocus** Why is it now recommended that young women only have a Pap test for cervical cancer every three years instead of every year? Are there any alternatives to the Pap test? Visit MJ's blog in the Study Area in MasteringBiology, and look under "Screening for Cervical Cancer."
>
> http://goo.gl/gRUZsL

Treatments for both endometrial and cervical cancer include surgery, radiation, hormones, or chemotherapy. The five-year survival rate is 82% for endometrial cancer and 68% for cervical cancer.

Cervical cancer may be one of the first cancers to be largely preventable by a vaccine. In clinical trials in more than 17,000 girls and women, the vaccine, called Gardasil, was nearly 100% effective in blocking the strains of the human papillomavirus that cause most cases of cervical cancer and genital warts. A second vaccine, called Cervarix, is also now available.

 **Recap** Cancers vary widely in terms of incidence, ease of detection and treatment, and death rate. For any cancer, the probability of survival is much higher if the disease is detected early, before it metastasizes. ■

18.7 Some other notable cancers

Kidney cancers

The kidneys are particularly susceptible to cancers caused by exposure to certain toxins, such as asbestos and cadmium. Smoking, too, is an important risk factor for cancers of the kidneys. Cancer-causing agents and toxic chemicals are more likely to affect the kidneys than many other organs because as the kidneys concentrate the urine, they also concentrate the toxic agents in it. Other risk factors that you can do nothing about are genetic factors, age (renal cancer occurs primarily after middle age), and gender (males are twice as likely as females to get it). The main treatment is removal of an affected kidney; one kidney is enough for good health.

Skin cancer

Cancers of the skin are classified as either melanoma or nonmelanoma cancers. A melanoma can be very serious.

- *Basal cell cancer* is the most common of the nonmelanomas. Recall that basal cells located at the base of the epithelium divide throughout life to produce the squamous epithelial cells that form the outer layer of the skin. Basal cell cancer occurs when the basal cells begin to divide out of control. Basal cell cancer usually appears as a pink or flesh-colored bump with a smooth texture. Sometimes, the bump bleeds or crusts. Basal cell cancer rarely metastasizes but should still be removed. If not treated, it can eventually spread to underlying tissues.
- *Squamous cell* cancer arises from the epithelial cells produced by the basal cells. It consists of pink scaly patches or nodules that may ulcerate and crust. Squamous cell cancer can metastasize, although slowly.
- *Melanoma* accounts for only 5% of all skin cancers, but it is the deadliest of the skin cancers because it is likely to metastasize fairly quickly. It arises from abnormal melanocytes, the cells that produce the dark brown pigment called *melanin*. Look for dark spots or patches on your skin and evaluate them according to the "ABCD" rule (Figure 18.13): A = *Asymmetry*—the two halves of the spot or patch don't match; B = *Border*—the border is irregular in shape; C = *Color*—color varies or is intensely black; and D = *Diameter*—diameter is greater than 6 millimeters, or about the size of a pea. Itchiness, scaliness, oozing, or bleeding may also be significant signs.

Early detection of skin cancers is important. Both basal and squamous cell cancers have a better than 95% cure rate if removed promptly because they metastasize slowly, if at all. Early detection is especially important for melanomas; the five-year survival rate is 98% if the melanoma is detected while it is still localized, but only 15% if it has already metastasized.

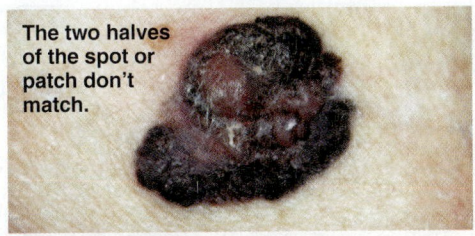

a) A: asymmetry.

The two halves of the spot or patch don't match.

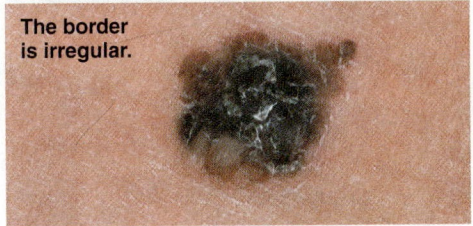

b) B: border.

The border is irregular.

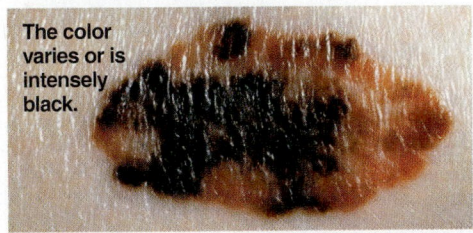

c) C: color.

The color varies or is intensely black.

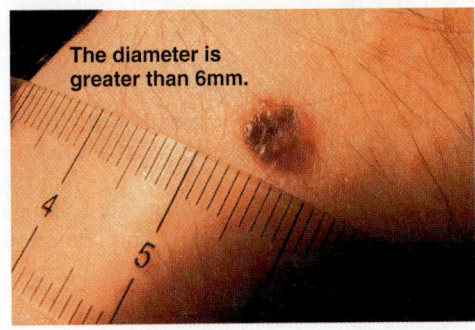

d) D: diameter.

The diameter is greater than 6mm.

Figure 18.13 Melanoma. Melanomas generally are diagnosed by the ABCD rule.

☑ An interesting fact about melanocytes is that, in embryonic life, they originate near the spinal cord and then migrate to reach their final locations in the skin. No other skin cells migrate in this way. How might this be related to the fact that melanomas are the most dangerous form of skin cancer?

Ovarian cancer

Cancer of the ovaries is another of those diseases with sometimes confusing signs and symptoms. However, it's important to know what they are, because if detected early ovarian cancer can be cured by surgical removal of one or both ovaries. Women with persistent symptoms of bloating and swelling of the abdomen, pelvic or abdominal pain, difficulty eating or feeling full quickly, or urinary urgency or frequency should consult their physician. For the 15% of cases diagnosed at an early stage, the five-year survival is 92%, but that drops to 27% once the cancer has spread.

Testicular cancer

Although not very common (fewer than 9,000 cases and 400 deaths per year), testicular cancer is noteworthy because it is so easy to detect, and therefore, it's worth taking the time to look for it. Checking a man's testicles is usually part of a general physical exam. In addition, the American Cancer Society provides information on how males can do a self-examination if they wish.

18.8 Most cancers could be prevented

Some cancers are linked to the genes we inherit from our parents, such as the BRCA genes that increase the risk of breast and ovarian cancer. For heritable cancers, early detection is not only vital but increasingly possible. Early detection of risk factors will allow treatment options to begin earlier. However, true prevention of heritable cancers will have to wait for advances in gene therapy and vaccines.

Nevertheless, most incidences of cancer are thought to be preventable. At least 60% of all cancer cases other than nonmelanoma skin cancers are thought to be caused by just two factors—smoking and poor diet. In addition, most nonmelanoma skin cancers could be prevented by reducing our exposure to the sun.

If we really want to make a dent in the cancer death rate, we need to work on public education: encouraging people to exercise, control their weight, eat a healthy diet, limit alcohol, avoid too much sun, and, above all, not smoke. Researchers are studying what constitutes a healthy diet and why certain dietary components are harmful or beneficial.

What can you do to reduce your own risk of cancer?

- Know your family history. Because cancer generally involves multiple genetic defects, you may have inherited one or more risk factors. Of course, this does not guarantee you will develop cancer. However, until we know of and can test for all such defects, you can at least assess your potential by examining your family's history of cancer and discussing it with your physician.
- Know your own body. Suspicious lumps in a breast or a testicle should be reported to your physician. Be on the lookout for changes in the skin that might represent a melanoma (refer to Figure 18.13).
- Get regular medical screenings for cancer. As we've seen, for most cancers early detection and treatment markedly improves the long-term survival rate. A good source of information on cancer screening is a document available on the World Wide Web entitled "American Cancer

Society Guidelines for the Early Detection of Cancer." Your physician may also suggest specific tests for you based on your family history, gender, and age.

- Avoid direct sunlight between 10 A.M. and 4 P.M., wear a broad-brimmed hat, and apply sunscreen with a sun protection factor (SPF) of 15 or greater. Also avoid sunlamps and tanning salons.

- Don't smoke. Here is a personal choice you can make that will dramatically lower your risk. Smoking is considered *the* single largest preventable risk factor for cancer. It accounts for fully 85% of all lung cancers and also contributes to cancers of the bladder, mouth, colon, cervix, pancreas, and pharynx. Ultimately, we may save more lives by persuading people not to smoke than by any amount of early detection.

- If you consume alcohol, drink in moderation. One to two alcoholic beverages per day *may* reduce the chance of cardiovascular disease, but the tradeoff is that alcohol consumption appears to raise the risk of certain cancers. It is particularly risky when combined with smoking.

- Watch your diet and your weight. There are no absolutes here, but the general recommendation is a diet high in fruits, vegetables, legumes (peas and beans), and whole grains, and low in saturated fats, red meat, and salt. Unsaturated fats, such as fish and olive oil, appear to be beneficial (Figure 18.14).

- Stay informed. There's a wealth of information on cancer in general, on specific types of cancer, and on the effects of nutrition and diet. Good resources include your

Figure 18.14   A healthy meal.

physician, the American Cancer Society, the National Cancer Institute, or any good library.

↻ Recap Smoking is by far the leading preventable risk factor for cancer. Strategies to reduce your cancer risk include knowing your family history and your own body, getting regular medical screenings, avoiding sunlight and sunlamps, watching your diet and weight, drinking alcohol in moderation if at all, and staying informed. ∎

Chapter Summary

18.1 Tumors can be benign or cancerous *p. 421*

- Normal cells have two key characteristics: (1) Their rates of division are kept under control and (2) most remain in one location in the body.
- A mass of cells that is dividing more rapidly than normal is called a *tumor*. Some tumors are benign.

18.2 Cancer cells undergo structural and functional changes *p. 422*

- Cancer develops when cells divide uncontrollably, undergo physical changes, and no longer adhere to each other.
- Eventually, some cancer cells invade nearby tissues. Others metastasize, colonizing distant sites.
- Cancer is the second leading cause of death in the United States.

18.3 Factors contributing to cancer development *p. 424*

- Mutations of proto-oncogenes, tumor suppressor genes, and/or mutator genes may contribute to cancer.
- Inheritance of one or more mutated genes from your parents may increase your risk of developing certain cancers.
- Known carcinogens include some viruses and bacteria, environmental chemicals, tobacco, radiation, and dietary factors

including alcohol consumption. Internal factors, such as faulty DNA replication and internally produced chemicals, can also contribute to cancer.

- Cancer is a multifactorial disease. Perhaps six or more mutations may need to be present before a cell becomes cancerous.

18.4 Advances in diagnosis enable early detection *p. 428*

- Most cancers are diagnosed when a tumor is detected by imaging techniques such as X-rays (including CAT scans), PET, or MRI.
- Enzyme tests may also identify cancer.
- Genetic testing can determine the presence of a specific known mutated gene, but the presence of the mutated gene does not necessarily mean that cancer will develop.

18.5 Cancer treatments *p. 430*

- Many cancers are treatable; early diagnosis is important.
- Conventional treatments for cancer include surgical removal of the tumor, radiation, and chemotherapy.
- Newer treatments include magnetism and photodynamic therapy to target malignant cells precisely, immunotherapy to activate the patient's immune system, drugs to inhibit angiogenesis and cut off the tumor's blood supply, and molecular treatments that target specific genes.

18.6 The ten deadliest cancers *p. 432*

- Lung cancer causes more deaths than any other type of cancer.
- Prostate cancer is common in men; breast cancer is common in women.
- The deadliest cancers (cancers that kill the most people per year) are cancers of the colon and rectum and cancers of the lung.
- Esophageal cancer is rare, but it has the highest ratio of deaths to cases.
- Leukemias and lymphomas can now be cured. Complete cure requires a stem cell or bone marrow transplant.

18.7 Some other notable cancers *p. 436*

- The kidneys are particularly susceptible to cancers caused by toxins.
- Among skin cancers, melanomas are by far the most dangerous.
- Testicular cancer is rare, but looking for it is easy.

18.8 Most cancers could be prevented *p. 437*

- Some cancers will occur despite our best efforts because we cannot control inherited risk factors.
- However, most cancers can be prevented. The single most effective way to reduce cancer deaths is to reduce the rate of smoking.
- A healthy diet will also decrease your cancer risk. Eat lots of fruits, vegetables, legumes, and grains; reduce your intake of saturated fat, red meat, and salt.
- Other preventive strategies include knowing your family's health history, getting regular medical screenings, maintaining a healthy weight, avoiding direct sunlight and sunlamps, drinking alcohol in moderation if at all, and staying informed.

Terms You Should Know

cancer, *423*	malignant tumor, *423*
carcinogen, *425*	melanoma, *426*
chemotherapy, *430*	metastasis, *423*
dysplasia, *422*	mutator gene, *425*
hyperplasia, *421*	oncogene, *424*
immunotherapy, *431*	proto-oncogene, *424*
in situ cancer, *423*	tumor, *422*
leukemia, *435*	tumor suppressor gene, *425*
lymphoma, *435*	

Concept Review

Answers can be found in the Study Area in MasteringBiology.

1. Compare and contrast a benign tumor and a malignant tumor.
2. Describe the key changes in cancer cells that allow them to metastasize.
3. Describe the impact a carcinogen has on a cell.
4. Explain why we have not yet made much progress against lung cancer, compared to other types of cancer.
5. Explain what causes most skin cancers.
6. List two dietary factors associated with an increased risk of cancers of the colon and rectum.

7. Describe the role the immune system plays in preventing different types of cancer.
8. Describe how tumors are diagnosed.
9. Describe how antibodies against cancer cells could improve the delivery of chemotherapeutic drugs.
10. List the things one can do to help prevent cancer.

Test Yourself

Answers can be found in the Appendix.

1. Which of the following represents the correct order of steps leading to the development of malignant cancer?
   a. dysplasia . . . hyperplasia . . . *in situ* cancer . . . metastasis and/or invasion
   b. hyperplasia . . . dysplasia . . . metastasis and/or invasion . . . *in situ* cancer
   c. dysplasia . . . *in situ* cancer . . . hyperplasia . . . metastasis and/or invasion
   d. hyperplasia . . . dysplasia . . . *in situ* cancer . . . metastasis and/or invasion

2. Metastasis is enabled by the:
   a. nervous system   c. circulatory system
   b. lymphatic system   d. both (b) and (c)

3. A mutation in which of the following types of genes would likely allow cell division to proceed even if conditions are not suitable or if the cells are already showing cancerous tendencies?
   a. mutator genes   c. tumor suppressor genes
   b. oncogenes   d. proto-oncogenes

4. Which of the following statements regarding skin cancer is true?
   a. Chemical carcinogens contribute more to the development of skin cancer than UV radiation.
   b. Sun exposure during childhood does not impact the development of skin cancer later in life.
   c. Tanning booths enable safer tanning because the UV exposure from lamps is not likely to contribute to the development of skin cancer.
   d. Skin cancer can develop from cumulative sun exposure, even in the absence of sunburns.

5. Which of the following foods is associated with an increased risk of cancer development?
   a. unprocessed peanut butter   c. broccoli, kale, cauliflower
   b. red and yellow peppers   d. none of these choices

6. Which of the following imaging techniques expose(s) the body to radiation?
   a. X-rays   c. PET scans
   b. MRIs   d. both (a) and (c)

7. Which of the following cancer treatments would be most effective at targeting both the original tumor as well as all secondary tumors?
   a. surgery   c. chemotherapy
   b. focused sources   d. photodynamic therapy
      of radiation

8. Hair loss, anemia, nausea, and depletion of white blood cells are side effects most often associated with:
   a. immunotherapy   c. radiation therapy
   b. chemotherapy   d. angiogenesis inhibition

9. Which of the following cancer treatments prevents growth of blood vessels that support growing tumors?
   a. immunotherapy
   b. anti-angiogenic drugs
   c. gene therapy
   d. radiation therapy

10. The ABCD rule refers to the evaluation of:
    a. melanoma
    b. squamous cell carcinoma
    c. basal cell carcinoma
    d. all types of skin cancer

11. The most common cause(s) of cancer deaths in the United States is/are:
    a. viruses
    b. radiation
    c. smoking
    d. poor diet and lack of exercise

12. Which of the following statements about breast cancer is true?
    a. Most women have a mutated form of either the BRCA1 or the BRCA2 gene.
    b. Post-menopausal hormone replacement therapy can reduce the risk of development of breast cancer.
    c. The risk of breast cancer declines with advancing age.
    d. Early onset of menstruation and/or late onset of menopause is a risk factor for the development of breast cancer.

13. Smoking increases the risk of developing:
    a. lung cancer
    b. bladder cancer
    c. cervical cancer
    d. all of these cancers

14. Which of the following would be useful in detecting a tumor within the urinary bladder?
    a. X-ray
    b. cytoscope
    c. colonoscope
    d. a blood test for a urinary bladder–specific antigen

15. Which cancer appears to be prevented by a vaccine?
    a. endometrial cancer
    b. cervical cancer
    c. ovarian cancer
    d. all of these cancers

Apply What You Know

Answers can be found in the Study Area in MasteringBiology.

1. Why do you suppose that the death rate from lung cancer has increased much more in women than in men over the past 35 years? What information would you like to have that might support or disprove your hypothesis?

2. If you can't do anything about the genes you inherit, does it do any good to know your family history for particular cancers? Explain your reasoning.

3. The overall death rate from cancer has dropped only 14% between 1950 and 2011. Why do you think we seem to have made so little progress against cancer?

4. From time to time, we hear stories in the press that cell phones may cause certain types of brain cancer. Then someone else says the evidence is not yet clear. From what you know of the scientific method, why do you think this issue has been so hard to resolve?

5. Common therapies for cancer include chemotherapy and radiation treatments. How effective are these therapies, and why do they both have such serious side effects?

6. The first cancer that can be nearly completely prevented by a vaccine is cervical cancer. How does the vaccine work? If a vaccine can be developed against cervical cancer, why hasn't someone developed a vaccine against the number one cancer killer—lung cancer?

MJ's BlogInFocus

Answers can be found in the Study Area in MasteringBiology.

1. Imagine that a new test has been developed to detect a particular cancer. It is inexpensive, and it can correctly identify the cancer 95% of the time and only rarely gives false positives (indicating cancer may be present when in fact it is not present). As a health administrator, would you recommend it as a screening test for a whole population? Is there anything else you would like to know before you decide? For help with this question, visit MJ's blog in the Study Area in MasteringBiology and look under "Screening Tests." http://goo.gl/en2lsC

MasteringBiology®

Students Go to MasteringBiology for assignments, the eText, and the Study Area with animations, practice tests, and activities.

Professors Go to MasteringBiology for automatically graded tutorials and questions that you can assign to your students, plus Instructor Resources.

Answers to ✓ questions are available in the Appendix.

Genetics and Inheritance

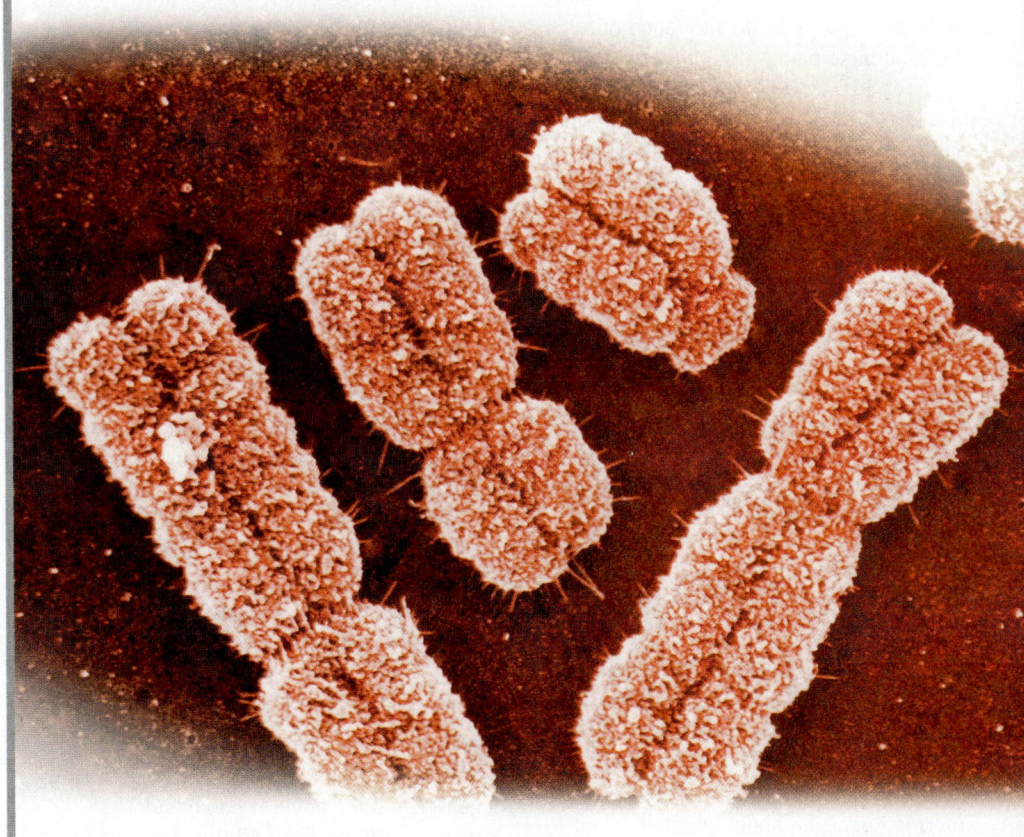

SEM (x13,000) of several human chromosomes.

Key Concepts

- **A *gene* is a segment of DNA that carries the code for making one or more proteins.** Humans have about 22,000 different genes on 23 pairs of chromosomes.

- **Alleles are two identical or nearly identical copies of a gene.** One of the alleles is inherited from each parent. Differences between alleles are the result of rare mutations over millions of years of evolution.

- **Your complete set of alleles, called your genotype, determines your particular physical and functional traits,** and even your susceptibility to certain diseases.

- **When two alleles in an individual are different, usually one is *dominant* and the other is *recessive*.** A trait expressed by a recessive allele will only be present when there is no dominant allele, that is, when both alleles are recessive.

- **The X and Y chromosomes, also called the *sex chromosomes*, carry the genes that determine the gender of the individual.** Females have two X chromosomes and males have an X and a Y chromosome.

CURRENT ISSUE

Should You Have Genetic Tests for Disease Risks?

Your entire set of genetic material, called your genome, is contained in the 3 billion base pairs in your DNA. It is now possible to determine the entire sequence of all 3 billion base pairs, although at a cost of about $10,000 it's too expensive for most people. However, for several hundred dollars there are companies that will determine the genetic sequence of only certain regions of your DNA. These genetic tests, called Genome-wide Arrays (GWAs), can identify thousands of gene mutations that signal an increased risk for certain genetic conditions or disorders. Should you have these tests so that you might know your risks of developing certain genetic diseases in the future? What are the benefits to you and your doctor, and are there any risks associated with having such tests?

Genetic Testing Offers Many Benefits

Genetic testing can help couples make informed reproductive choices. With genetic testing, couples can find out whether they are carriers of recessive disorders, enabling them to make an informed decision about whether or not to conceive a child together. For couples who are already expecting a child, prenatal genetic testing can be used to test fetal cells for approximately 40 genetic defects. Early genetic testing of the fetus gives parents time to prepare emotionally if a genetic defect is evident, or perhaps

to terminate the pregnancy if their beliefs allow for it.

Knowledge about your genetic makeup might also help you make better decisions about your health care. For example, the American Cancer Society recommends that most people have their first colonoscopy at age 50, but if you knew that you carried a gene that predisposes you to colon cancer, you and your doctor might decide that you should have your first colonoscopy at a younger age. If you're a known high risk, earlier testing could save your life.

But this is where the benefits of genetic testing get a little murky. Just because a mutation is found does not necessarily mean that a person will develop the disease associated with it. Testing for thousands of mutations and then trying to determine what the results mean for the future health of an individual is in part a guessing game, and a potentially scary one at that. It can frighten as well as inform.

Social, Medical, and Legal Issues

Privacy If you do decide to undergo genetic testing, who should have access to your genetic data? Should it be considered medical information and thus become part of your medical record, or should it

Questions to Consider

1 If you might be a carrier of a gene for an unpreventable condition, would you want to know? Why or why not?

2 Should genetic counseling be required before a person is allowed to purchase a genetic test? What about counseling afterward, to deal with any negative findings?

be considered personal information, not available to others without your permission? Your existing medical records are probably already available to the physicians and staff of the various medical facilities you have visited over the years and perhaps to your medical insurance company as well.

Health Insurance and Employment Discrimination Prior to 2008, the issue of privacy of genetic information was a serious consideration for anyone undergoing genetic testing. Before that time, insurers could (and did) deny health insurance coverage to persons with pre-existing conditions, which they interpreted as including genetic mutations that could lead to a health issue in the future. In other words, healthy persons were being denied health insurance solely on the basis of having one or more genetic mutations. Some employers, too, were making hiring decisions and even firing employees on the basis of genetic tests.

But before we condemn these insurers and employers too much for discriminating against people with certain genes, consider that no company can lose money consistently and stay in business for long. The costs of health care are soaring. To a health insurer deciding if they should cover you or to a company that offers you health insurance as part of your employment package, information about your likelihood of developing a disease, even if in the distant future, is information that could impact their bottom line.

To prevent genetically based discrimination in health insurance and employment, →

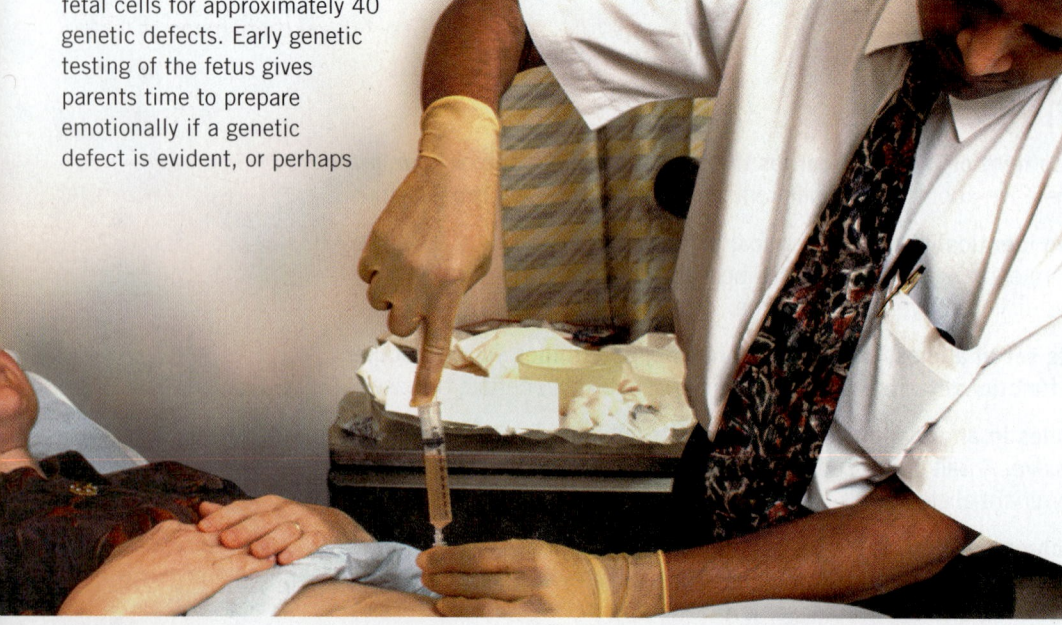

Fetal cells harvested during amniocentesis can now be tested for over 40 inherited genetic disorders.

the U.S. Congress passed the Genetic Information Nondiscrimination Act (GINA) in 2008. GINA prohibits insurers from setting higher premiums or denying coverage to healthy people based on the results of genetic tests. GINA also makes it illegal for employers to use genetic information in making decisions about hiring or firing a person. The basic tenets of GINA were subsequently included in the Affordable Care Act of 2010 (Obamacare). Under Obamacare, one cannot be denied health insurance on the basis of a pre-existing medical condition and that includes genetic abnormalities revealed by a genetic test.

Life, Disability, and Long-Term Care Insurance
In hindsight, GINA turned out to be a decidedly imperfect bill when it comes to genetic nondiscrimination. That's because when it was first proposed, GINA faced strong opposition from insurance companies and their lobbyists. To get GINA passed and signed into law, lawmakers compromised by not mentioning life, disability, or long-term care insurance in the bill. What that means is that if you choose to undergo genetic testing, you may *still* be denied life, disability, or long-term care insurance on the basis of the test results. It's something to think about if you're considering genetic testing.

Psychological and Emotional Impacts
Discovering that you have a genetic predisposition for a disorder or condition, for example, colon cancer, might motivate you to take preventive steps. On the other hand, some health professionals are cautious about recommending widespread genetic testing of healthy people as a future health-predicting tool. They worry that the discovery that a person is likely to develop a genetic disease in the future might do more harm than good, especially if there is nothing that can be done about it.

Ethnic or Racial Discrimination Sickle-cell anemia, an inherited disease of the blood, is most common in people of African heritage. Tay-Sachs disease, which causes progressive deterioration of nerve cells,

occurs most frequently in Ashkenazi Jews. People of Northern European descent are most likely to develop hemochromatosis, a genetic disorder that causes the body to absorb excess iron. Does this mean we should test all African Americans for sickle-cell anemia, all Jews for Tay-Sachs, or all Northern Europeans for hemochromatosis? Should we require members of high-risk groups to be tested if they are getting married and considering having children? Where do we want to draw the line between individual privacy and public health?

The science of genetic testing is progressing far more rapidly than our solutions to the social, medical, and legal issues that it raises. Expect continued debate on these issues in the future.

SUMMARY

- Genetic testing is a complex process for diagnosing genetic disorders by the direct examination of a person's DNA.
- Genetic testing is used to diagnose or rule out genetic disorders, help couples make informed reproductive choices, and prevent future health problems.
- Genetic testing raises difficult questions of privacy, eligibility for insurance, discrimination, and psychological or emotional harm.

An organism's genome is the sum total of all of its DNA, even though not all of that DNA codes for proteins. The information necessary for the development of a new human being is contained in the DNA within the nucleus of the fertilized egg. DNA sequences that code for one or more specific proteins are called **genes. Genetics** is the study of DNA and genes and their transmission from one generation to the next.

The word **inheritance** often conjures up notions of money or possessions. The broader definition of inheritance, however, is something received from an ancestor or another person. In terms of genetics, each of us inherits two sets of DNA, one set from our mother and one set from our father. As a result, we are not identical to either parent. But just as important, the genes of each parent are shuffled just before egg and sperm are formed so that the genes we inherit are also different from our siblings. With the exception of identical twins, then, no two persons on Earth are exactly alike. Slight variations in our genes are essential for the

process of evolution, which depends on there being at least some differences between individuals. ∎

In this chapter, we explore how genes are passed from generation to generation and how genes affect our traits, or distinguishing features, such as our appearance, health, and possibly even our thoughts and actions. Then we look at some of the things that can go wrong as genes are passed from one generation to the next.

19.1 Your genotype is the genetic basis of your phenotype

Chromosomes, the structures that contain most of an organism's DNA, are identifiable in cells only just before cell division. At this time, each chromosome can be identified by its characteristic size, centromere location, and distinct banding pattern. A composite display of all the chromosomes of an organism is called its **karyotype.**

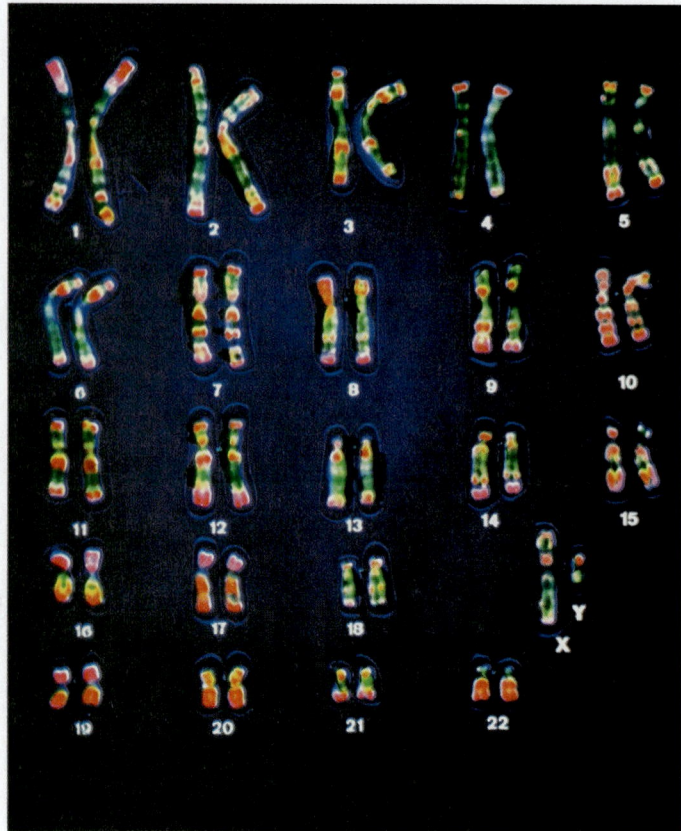

Figure 19.1   The human karyotype. Chromosomes are identified and paired according to their size, centromere location, and characteristic banding patterns. A numbered arrangement of all 23 pairs is called the *human karyotype.* This karyotype is from a male, as indicated by the dissimilar X and Y chromosomes of the unnumbered 23rd pair.

The karyotype in **Figure 19.1** shows the 23 pairs of human chromosomes. Of these, 22 are always matched pairs (called *autosomes*). The two chromosomes of the **23rd** pair are the sex chromosomes, X and Y. The sex chromosomes do not look alike in males because males have one X and one Y chromosome (females have two Xs). The sex chromosomes look different from each other and function differently because they carry different genes. They represent a pair in that you get one from each parent as a consequence of meiosis.

The 22 pairs of autosomes are called *homologous chromosomes* (homo- means "same") because they look alike and they each have a copy of the same gene at the same location, or *locus,* on the chromosome (**Figure 19.2**). Humans inherit one of each pair of autosomes and one sex chromosome from each parent, giving us two copies of each gene.

Although autosomes may look alike, closer inspection reveals that they are not completely identical. Small differences in DNA sequence may exist between any pair of autosomes. These sequence differences may occur within genes, and when they do, they produce alternative versions of genes called **alleles.** Because they are different, alleles may code for different proteins with slightly different structures. Protein structure can affect how the protein functions.

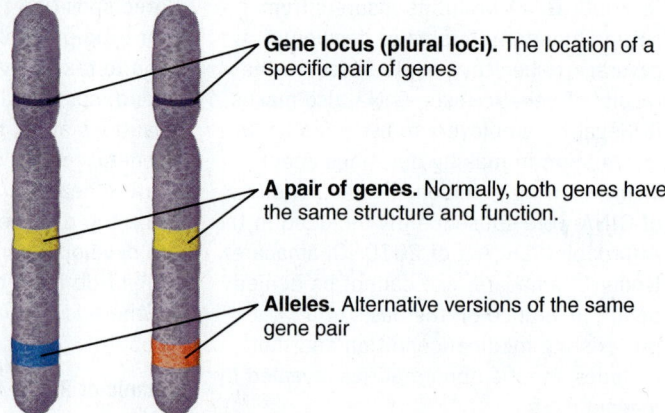

Pair of autosomes. Each autosome carries the same genes at the locus.

Gene locus (plural loci). The location of a specific pair of genes

A pair of genes. Normally, both genes have the same structure and function.

Alleles. Alternative versions of the same gene pair

Figure 19.2   Autosomes.

If an individual possesses two identical alleles of a particular gene, the person is said to be **homozygous** for that gene (*zygous* means "yoked," or joined together). A person who has two different alleles of a gene is **heterozygous** (*hetero-* means "different"). In many cases, there are more than two alleles of a particular gene in the human population, although a given individual can inherit only two of them. For example, there are three alleles for ABO blood type: the A, B, and O alleles.

Where did these different alleles come from? Most likely, different alleles resulted from millions of years of **mutations,** or random changes to the DNA sequence, of cells destined to become sperm or eggs. Because those mutations were not corrected before cell division and meiosis, and because they did not cause the sperm or egg to be nonviable, they were passed on to the next generation as slight variations of the original gene, and they remain in the human population to this day. Collectively, all the various genes and their alleles in the human population are known as the *human gene pool.* There are enough different alleles among our approximately 22,000 genes to account for the uniqueness of each individual.

MJ's BlogInFocus   Why did one genetic testing company stop marketing its genetic tests as a way to test for genetic diseases, and now sells genetic tests solely for the purpose of determining ancestry? Visit MJ's blog in the Study Area in MasteringBiology and look under "Genetic Testing."

http://goo.gl/FGqCHO

Your complete set of alleles is called your **genotype.** Your genotype has a profound influence on your **phenotype,** or the observable physical and functional traits that characterize you. Among your phenotypic traits are your hair, eye, and skin color; your height and body shape;

and your ability (or inability) to curl your tongue. Many phenotypic traits, such as blood type and susceptibility to disease, are less observable. Your phenotype is determined not only by the alleles you inherit from your parents (your genotype), but also by environmental factors and lifestyle choices such as how much you are exposed to the sun, whether you exercise, and how much and what you eat.

> ↺ **Recap** We inherit two alleles of every gene found on auto-somes; these alleles may be identical or slightly different. The sum of our alleles is our genotype, and the physical and func-tional expression of those alleles is our phenotype. ▪

19.2 Genetic inheritance follows certain patterns

One easy way to designate two alleles of the same gene is to assign them uppercase (such as A) and lowercase (a) letters. Recall that people with the same two alleles of a gene (either aa or AA) are homozygous for that gene and people with different alleles (Aa) are heterozygous.

Punnett square analysis predicts patterns of inheritance

When gametes (sperm and eggs) are formed in the parents, each gamete receives only one of the parents' two alleles. When sperm and egg combine, the fertilized egg again has two alleles, one from each parent. A **Punnett square** provides a simple way to represent patterns of inheritance

of alleles and to predict the probability that a particular genotype will be inherited (**Figure 19.3**).

To create a Punnett square, place the possible alleles of the male gametes on one axis and the possible alleles of the female gametes on the other. Bear in mind that although the parents are diploid—they contain two copies of each chromosome—they will each donate *only one* of each chromosome to their offspring. By combining the letters for the sperm and egg in the appropriate squares, which represent the possible combinations that might occur during fertilization, we can see all of the possible genotypes of the offspring and the ratios of each possible genotype.

Mendel established the basic principles of genetics

In the 1850s, Gregor Mendel, a university-educated Austrian monk who specialized in natural history, described many of the fundamental principles of inheritance through a number of experiments involving controlled breeding of garden peas. He chose pea plants as a "model" organism because they were cheap and easy to grow and matings between plants could be set up easily. Many biologists use model organisms to investigate the fundamental principles of biology because they can do experiments such as setting up matings that would be unethical to do with human subjects. Mendel followed the inheritance of traits such as flower color, pea color, pea texture, and plant height from one generation to the next. He collected an extensive set of data over a seven-year period, and proposed that there were discrete "factors" of heredity that united during fertilization and then separated

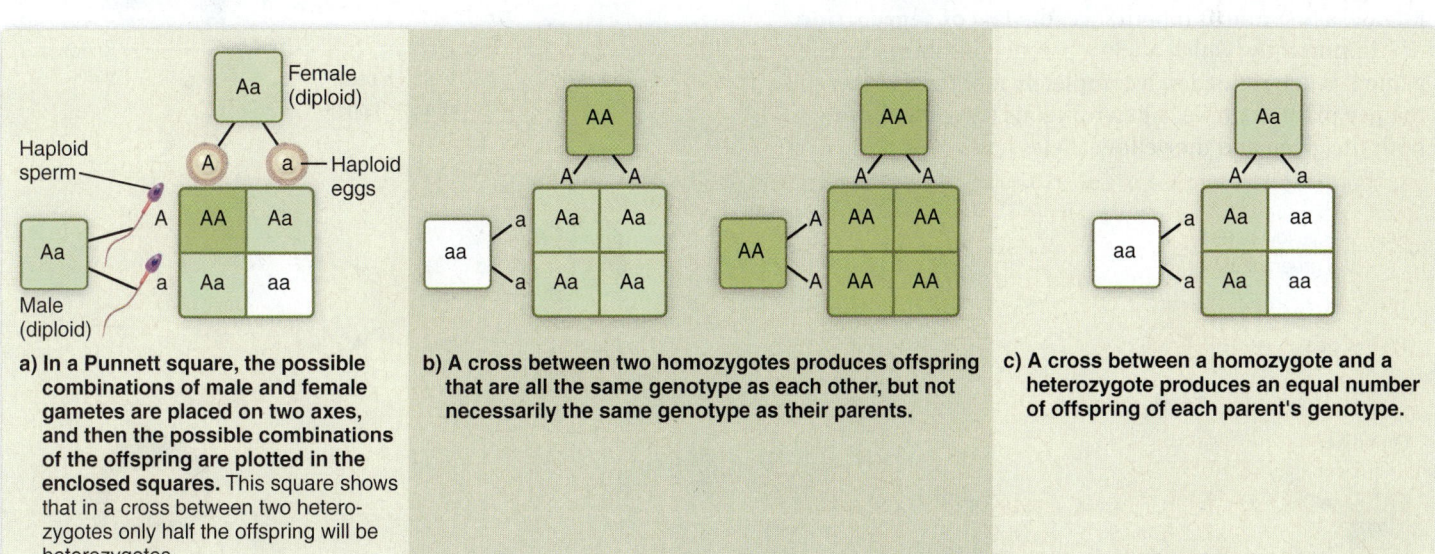

a) **In a Punnett square, the possible combinations of male and female gametes are placed on two axes, and then the possible combinations of the offspring are plotted in the enclosed squares.** This square shows that in a cross between two hetero-zygotes only half the offspring will be heterozygotes.

b) **A cross between two homozygotes produces offspring that are all the same genotype as each other, but not necessarily the same genotype as their parents.**

c) **A cross between a homozygote and a heterozygote produces an equal number of offspring of each parent's genotype.**

Figure 19.3 Punnett squares. Punnett squares are used to predict the outcomes of a particular combination of parental alleles.

☑ Suppose the parents shown in the first Punnett square have a baby. How likely percentage-wise is it that this baby will have the genotype AA? How about Aa? More generally, how likely is any one of the four cells in a Punnett square?

again in the formation of sperm and eggs. He concluded that pea plants inherit two units of each factor, one unit from each parent. Today, we call these factors *genes* and we know that they are found on chromosomes. Even though Mendel knew nothing about chromosomes, DNA, or meiosis, his pioneering studies with pea plants provided great insight into the principles of genetics in many complex organisms, and his rules of inheritance hold for humans as well as pea plants.

To better understand the laws that govern the passage of traits from generation to generation, consider one of Mendel's experiments. Mendel performed a one-trait cross, where he followed the inheritance of a single trait: pea color. Pea plants can produce either green or yellow peas; when Mendel mated, or crossed, a plant with green peas to a plant with yellow peas, he found that all of the offspring produced yellow peas. These peas were identical to those produced by the parent with yellow peas; it seemed that the green pea trait had disappeared completely in these offspring. But was the green trait really gone? To check, Mendel crossed two of these yellow pea–producing offspring together. To his surprise, Mendel found that only about 3/4 (75%) of this second group of offspring produced yellow peas, while the remaining 1/4 (25%) of these offspring produced green peas. Thus, the green pea trait hadn't disappeared; it was as if the green trait had simply "skipped" a generation. Mendel deduced that the information needed to produce green peas was still present in the first group of yellow pea-producing offspring but was somehow hidden. Based on this and other experiments, Mendel reasoned that each pea plant must have two factors (what we refer to as "alleles of the gene") for pea color and that each parent contributes only one factor (allele) to the offspring (Figure 19.4). This is now known as Mendel's first rule of inheritance, the **law of segregation.**

Importantly, which allele a parent contributes to a gamete is determined in a completely random manner. In the pea plant example, a heterozygous parent harboring both the green and the yellow alleles has a 50% chance of donating the green allele and a 50% chance of donating the yellow allele to each gamete. Today, we understand that gametes are haploid, with only half the number of chromosomes and genes of the parent. The primary reason for phenotypic variation among human individuals is that each parent randomly donates just one of their two alleles of each gene to each offspring, and there are 22,000 genes!

Dominant alleles are expressed over recessive alleles

In the experiment described previously, Mendel observed that the green pea trait "skipped" a generation. Why did this happen? Mendel reasoned that the yellow allele behaves in a *dominant* manner with respect to the green allele, meaning that when both alleles are present in the same plant, the plant will only show the yellow phenotype. The green allele is thus *recessive* to the yellow allele and only visible when no yellow alleles are present. It is customary to denote the dominant allele by an uppercase letter (in this case, Y) and the recessive allele by a lowercase letter (y).

Many human genes have dominant and recessive alleles as well. The gene that determines hairline pattern in humans has a dominant allele (W) that produces the widow's peak phenotype and a recessive allele (w) that causes the straight hairline phenotype (Figure 19.5). If a person homozygous

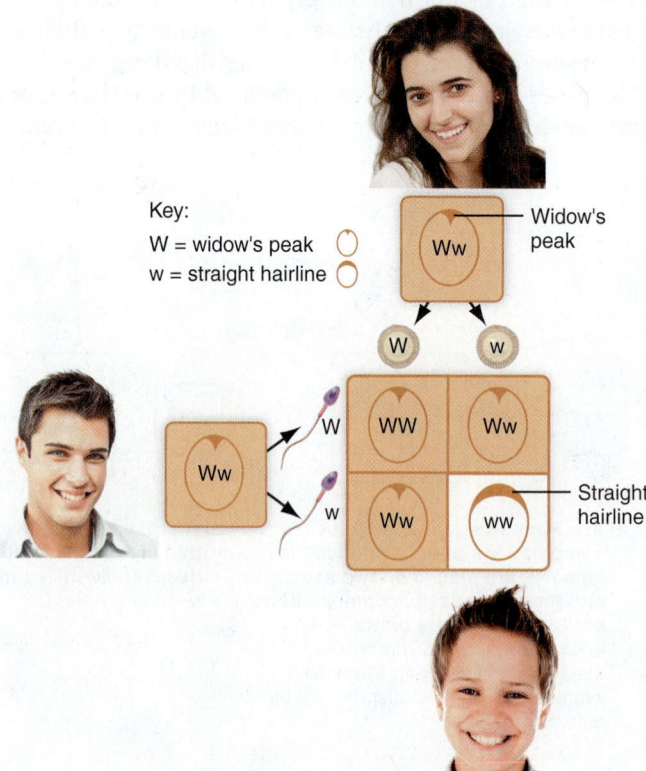

Figure 19.5　Hairline patterns in humans. Hairline pattern is determined by a gene with a dominant allele (W) for widow's peak and a recessive allele (w) for straight hairline. The person on the left (widow's peak phenotype) must be either WW or Ww genotype. Only the homozygous recessive genotype (ww) will have the straight hairline phenotype, shown on the right.

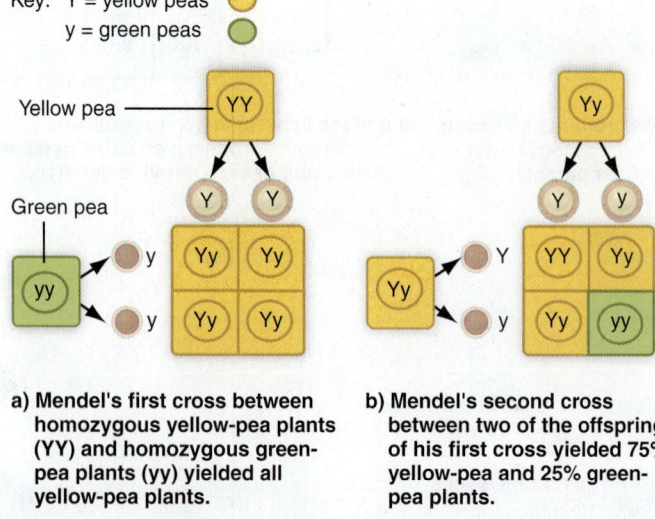

a) Mendel's first cross between homozygous yellow-pea plants (YY) and homozygous green-pea plants (yy) yielded all yellow-pea plants.

b) Mendel's second cross between two of the offspring of his first cross yielded 75% yellow-pea and 25% green-pea plants.

Figure 19.4　Mendel's one-trait crosses in pea plants.

for the widow's peak allele (WW) were to have children with a person homozygous for the straight hairline allele (ww), all of their children would be heterozygous (Ww) and would show the dominant widow's peak phenotype. But, if one of these heterozygous children were to mate with another person who was also heterozygous for the hairline pattern gene, the recessive straight hairline phenotype could reappear in their offspring. It would appear that the straight hairline trait had skipped a generation, just like the green pea trait in Mendel's experiment.

Most **recessive alleles**—such as the allele for a straight hairline and also the allele for attached earlobes (Figure 19.6)—do not confer a distinct advantage or disadvantage to the homozygous recessive person. They just originated as mutations at some point in our evolutionary history, and they remain in the human gene pool because they do no harm.

However, some recessive alleles result in the absence of a functionally important protein. An example is cystic fibrosis (see Health & Wellness feature), which afflicts only homozygous recessive individuals. These alleles no longer code for a specific essential protein because their nucleotide sequence has been altered slightly from the dominant form.

In principle, the frequency of truly harmful recessive alleles in a population is kept in check by the occasional premature death of homozygous recessives. Inheritance of two copies of a harmful recessive allele may result in miscarriage of a fetus, for example, or in death before reproductive age, both of which prevent the transfer of the allele to the next generation. Truly harmful recessive alleles tend to persist in the population primarily in heterozygotes, in which their presence is masked by the dominant normal allele.

Finally, the term **dominant allele** refers only to how an allele behaves in combination with a recessive allele in a heterozygote. Dominance has nothing to do with the frequency with which an allele is found in a population; in fact, some dominant alleles are quite rare. Consider the number of human fingers and toes. Most people are homozygous recessive (pp) for the gene controlling the number of fingers and toes, and they have four fingers and a thumb on each hand and five toes on each foot. But some people possess a dominant allele (Pp or PP), and the presence of this allele causes them to have extra fingers and toes on each hand and foot (Figure 19.7a), a condition known as *polydactyly*. Even though this allele behaves in a dominant fashion with respect to the allele for the normal number of fingers and toes, it is extremely rare in the population. Interestingly, there is a dominant allele in cats that causes affected individuals to have extra toes, as well (Figure 19.7b).

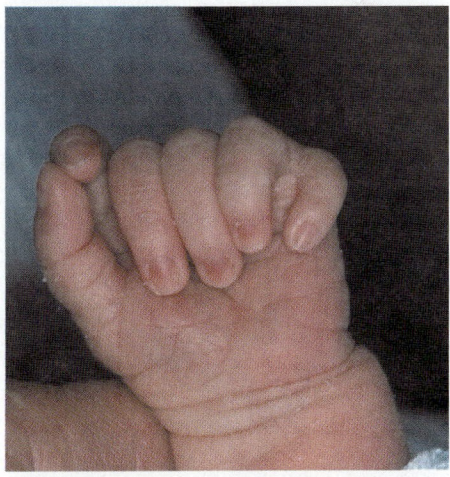

a) A human infant with polydactyly.

b) A polydactyl cat.

Figure 19.7 Dominant alleles are not always common in a population. Polydactyly, or the presence of extra digits, is an uncommon phenotype in both humans and cats caused by a dominant allele.

Attached earlobes (Johnny Depp, left) and unattached earlobes (George Clooney, right).

Figure 19.6 Harmless dominant/recessive traits and alleles. The allele for unattached earlobes is dominant over that for attached earlobes. Johnny Depp (on the left) is homozygous for the recessive allele, whereas George Clooney (on the right) is either homozygous for the dominant allele or is heterozygous.

Two-trait crosses: independent assortment of genes for different traits

So far we have followed the inheritance patterns of single genes throughout multiple generations. But what happens when we follow *two* genes? Are they inherited as a set, or can they be inherited independently of each other? To answer this question, let's consider the human hairline pattern gene and a second gene that controls whether a person's earlobes hang freely or are attached. For the earlobe gene, the allele for free earlobes (E) is dominant over the allele for attached earlobes (e). If a person with a widow's peak and free earlobes who is homozygous dominant for both genes (EEWW) mates with a person with a straight hairline and attached earlobes who is homozygous recessive for both genes (eeww), all of their children will be heterozygous for both genes (EeWw), and will have a widow's peak and free earlobs (Figure 19.8a).

But what happens if two people heterozygous for both genes mate? Would their children have either both dominant or both recessive phenotypes, or are other combinations possible? As shown in Figure 19.8b, a mating between two heterozygotes for both genes can indeed produce children with both dominant traits and children with both recessive traits, but other combinations are also possible. Some offspring may have the dominant widow's peak and the recessive attached earlobe phenotypes, and others may have the recessive straight hairline and the dominant free earlobe phenotypes.

When predicting the results of a two-trait cross, remember that each parent donates only one allele of each gene, and that the genes for different traits assort independently of each other during the formation of sperm and egg. That is, a parent can donate a dominant hairline pattern allele and a dominant earlobe allele, a recessive hair pattern allele and a recessive earlobe allele, or a mixture of the two. Each of these outcomes is equally probable because the hairline pattern and earlobe alleles are donated to egg and sperm independently of each other. The outcome of a two-trait cross can be predicted either by using a large

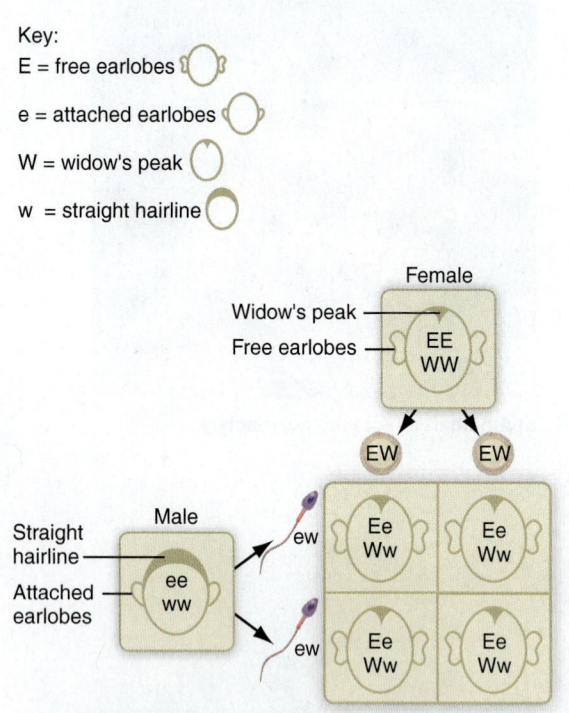

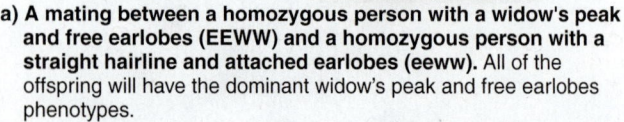

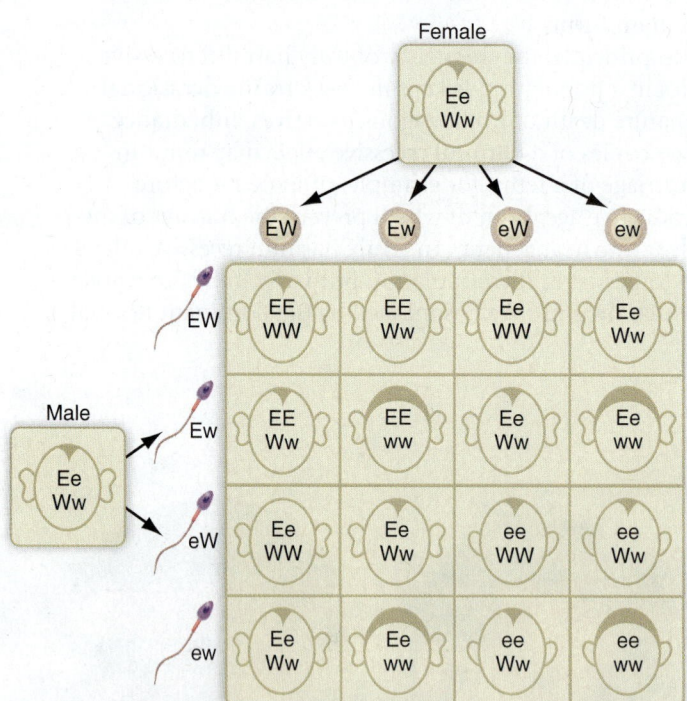

a) A mating between a homozygous person with a widow's peak and free earlobes (EEWW) and a homozygous person with a straight hairline and attached earlobes (eeww). All of the offspring will have the dominant widow's peak and free earlobes phenotypes.

b) A mating between two heterozygous people with widow's peaks and free earlobes (EeWw). Because the alleles for the two traits assort independently, some of the offspring show one dominant and one recessive trait.

Figure 19.8   A two-trait cross showing independent assortment of alleles for different traits.

☑ What is the ratio of the four possible phenotypes in the second mating (Figure 19.8b)? If you observed this same ratio in the offspring of an actual mating, what could you conclude about the parents?

Punnett square, as shown in Figure 19.8b, or simply by multiplying the ratios from small Punnett squares as shown in **Figure 19.9**. The latter works because the two events are independent of each other.

Mendel saw similar results in pea plants when performing two-trait crosses, and he formulated his **law of independent assortment,** which states that the alleles of different genes are distributed to egg and sperm cells independently of each other during meiosis. But, on this point, Mendel was only partly right. Recall that he knew nothing about chromosomes or meiosis. We now know that the 22,000 or so genes in the human genome do not simply float around independently of each other in the cell's nucleus. Instead, genes are structurally part of the 23 pairs of chromosomes found in the cell's nucleus. Only genes that are located on different chromosomes always assort truly independently. Alleles located on the same chromosome, called *linked alleles,* may or may not be inherited together, as discussed later in this chapter.

☑ In humans and other mammals, red hair is recessive to dark hair. Suppose a husband and wife are both heterozygous for the red hair color gene. What are the husband's and wife's own hair colors? If they have a child, what is the probability that he or she will have red hair?

↩ **Recap** A Punnett square is a useful method for predicting the ratios of possible offspring genotypes from a particular mating. Mendel's law of segregation holds that when sperm and eggs form, the two alleles of each gene separate from each other so that each sperm or egg receives only one allele of each gene. Where two alleles of each gene exist in the human gene pool, one is often dominant over the other and controls the phenotype of a heterozygote. According to Mendel's law of independent assortment, during the formation of sperm and egg, alleles for different traits are distributed independently of each other provided they are located on different chromosomes. ▪

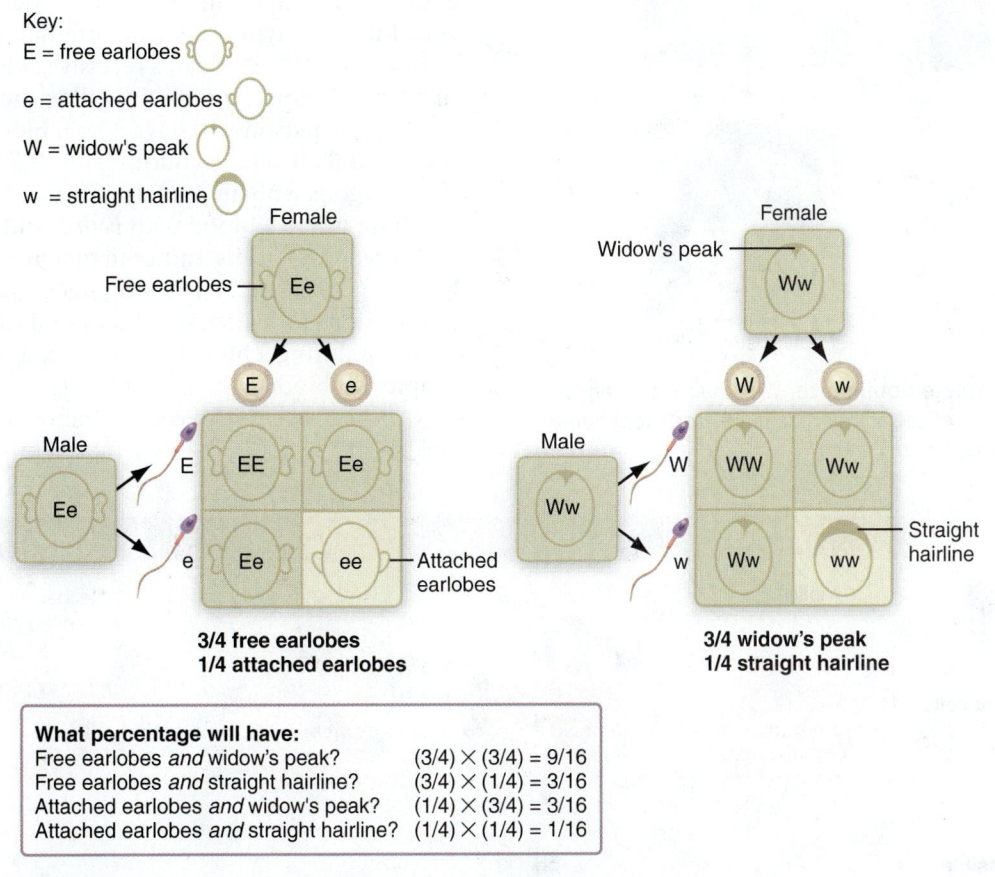

What percentage will have:

Free earlobes *and* widow's peak?	(3/4) × (3/4) = 9/16
Free earlobes *and* straight hairline?	(3/4) × (1/4) = 3/16
Attached earlobes *and* widow's peak?	(1/4) × (3/4) = 3/16
Attached earlobes *and* straight hairline?	(1/4) × (1/4) = 1/16

Figure 19.9 Another method of calculating phenotypic ratios from two-trait crosses. Instead of drawing the large Punnett square in Figure 19.8b, each trait can be considered individually. The phenotypic ratios obtained can then be multiplied together to get the same results shown in Figure 19.8b.

19.3 Incomplete dominance and codominance

Incomplete dominance: heterozygotes have an intermediate phenotype

Some alleles do not follow the dominant/recessive pattern described previously. In **incomplete dominance,** the heterozygous genotype results in a phenotype that is intermediate between the two homozygous phenotypes. An example of incomplete dominance in Caucasian humans is the trait of having straight, wavy, or curly hair (**Figure 19.10**). Another example is *familial hypercholesterolemia.* People homozygous for familial hypercholesterolemia may have blood cholesterol concentrations more than six times the normal level. The homozygous condition is quite rare (around one in a million people) because most homozygous individuals have heart attacks in childhood or adolescence and die by age 20. In contrast, the heterozygous condition is fairly common (about 1 in 500 people). People heterozygous for the disease may have blood cholesterol concentrations that are two to three times the normal level. With diagnosis and treatment, most can live fairly normal, healthy lives.

Figure 19.10 Incomplete dominance. Wavy hair (sc) is an intermediate phenotype between straight (ss) and curly (cc) hair because neither allele is dominant.

Codominance: both gene products are equally expressed

In **codominance,** the products of both alleles are expressed equally. One important example of codominance is the gene for blood type. There are three alleles for this gene: A, B, and O. The O allele is recessive; when combined in a heterozygous person with either the A allele or the B allele, the person will have A or B blood type. But the A and B alleles are codominant. If an individual is heterozygous with an A allele and a B allele, that person will have type AB blood with *both* A and B antigens on his or her red blood cells, rather than a mixed or intermediate antigen (**Figure 19.11**). Blood phenotype has important implications for which types of blood can be safely given to a patient during a blood transfusion, as you learned in the chapter on blood.

Another example of codominance is **sickle-cell anemia.** Sickle-cell anemia is a disease caused by one of two alleles

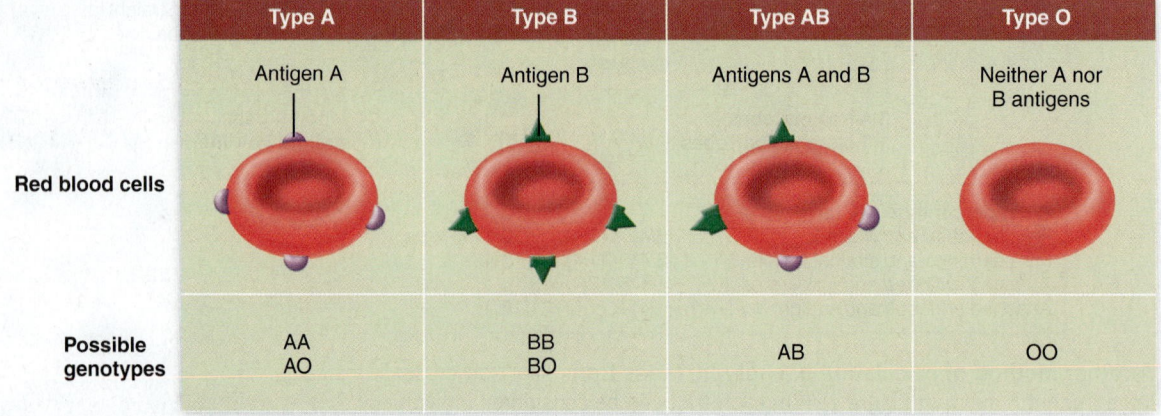

Figure 19.11 Blood type: an example of codominance. Characteristics of the four major blood types are shown, including their surface antigens and the genotypes that cause each blood type phenotype.

☑ A woman with blood type AB and a man with blood type O have a baby. What is the probability that the baby has blood type O, like his father? Explain your answer.

involved in the production of hemoglobin for red blood cells. One of the two alleles codes for a hemoglobin molecule (designated Hb^S) that differs from the normal hemoglobin molecule (designated Hb^A) by just one amino acid. People homozygous for *sickle-cell anemia* (Hb^{S}Hb^S) have only Hb^S hemoglobin in their red blood cells, whereas heterozygous individuals, who are said to have *sickle-cell trait* (Hb^{A}Hb^S), have equal amounts of each type of hemoglobin. As shown in Figure 19.12a, a mating between two individuals with the sickle-cell trait can result in children who are phenotypically normal (Hb^{A}Hb^A), children with sickle-cell trait (Hb^{A}Hb^S), and children with sickle-cell anemia (Hb^{S}Hb^S).

In people with sickle-cell anemia, breathing air with even slightly lower oxygen concentrations than normal can cause the Hb^S hemoglobin to crystallize within the red blood cells. When this happens, their red blood cells take on a crescent, or sickle, shape (Figure 19.12b). Sickled cells rupture easily because they are more fragile than normal red blood cells. They are also less flexible, so they tend to clog small blood vessels and block blood flow, and they carry less oxygen. The resultant oxygen deprivation of tissues and organs produces symptoms of fatigue, chest pain, bone and joint pain, swollen hands and feet, jaundice, and visual problems due to retinal damage. If sickle-cell anemia is left untreated, long-term complications can include stroke, blindness, pneumonia, and other health problems. Many people with sickle-cell anemia do not live into their 30s. In contrast, people with the sickle-cell trait rarely suffer

from any symptoms. The half of their hemoglobin that is normal is generally sufficient for them to lead normal lives. However, they may have mild episodes of red blood cell sickling when oxygen concentrations are lower than normal, such as at high altitudes or in airplanes in flight.

Sickle-cell anemia affects primarily Africans (and their descendants on other continents) and Caucasians of Mediterranean descent. In East Africa, sickle-cell anemia is fairly common and nearly 40% of the black population there has the sickle-cell trait. In contrast, only about 1 in 500 African Americans is born with sickle-cell anemia and only 8% have the sickle-cell trait. Why the big difference?

It turns out that the sickle-cell allele protects against malaria, a deadly parasitic disease common in tropical climates. Malaria kills about a million people every year worldwide. Where malaria exists—it is prevalent in Africa but rare in the United States—having sickle-cell trait (but not sickle-cell anemia) is a distinct advantage.

The malaria parasite spends part of its life cycle in humans (in the liver and red blood cells) and the other part in mosquitoes. When the malaria parasite enters the red blood cell of a person with sickle-cell trait or sickle-cell disease, the cell sickles and eventually ruptures, killing the parasite before it can reproduce and infect its host. Were it not for malaria, we could surmise that sickle-cell anemia and the sickle-cell trait would become rare in Africa, just as they are becoming rarer in the United States.

Because our phenotype includes all of our unique characteristics from our physical appearance to our biochemical makeup, the definition of an inheritance pattern as either incompletely dominant or codominant depends on the context. At the biochemical level, sickle-cell trait is an example of codominance because both the normal hemoglobin (Hb^A) and the sickle-cell hemoglobin (Hb^S) are produced in equal amounts. However, at the level of the whole person, the sickle-cell trait is an example of incomplete dominance, because the person with the sickle-cell trait exhibits a disease pattern that is intermediate between the two homozygous conditions. Understanding the concept is more important than an argument over whether the sickle-cell trait is one or the other.

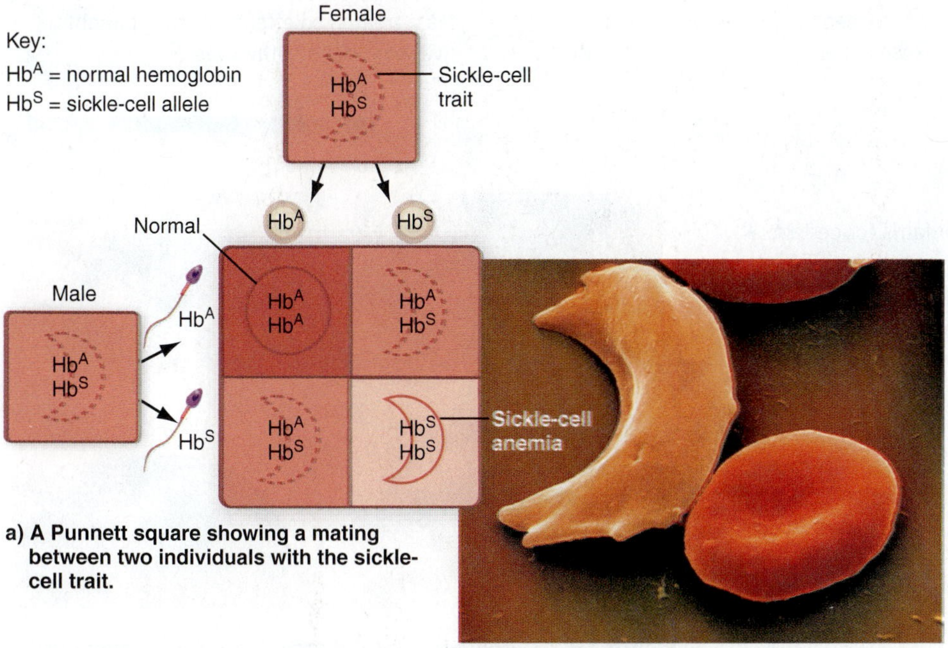

a) **A Punnett square showing a mating between two individuals with the sickle-cell trait.**

b) **A sickled red blood cell next to a normal red blood cell.**

Figure 19.12 Sickle-cell anemia: an example of codominance. If two people with the sickle-cell trait have children, each child will have a 25% chance of being phenotypically normal, a 50% chance of having the sickle-cell trait, and a 25% chance of having sickle-cell anemia.

Recap People heterozygous with incompletely dominant alleles have a phenotype that is intermediate between the phenotype of either homozygous genotype. People heterozygous with codominant alleles express the products of both alleles equally. Examples of codominance are AB blood type and sickle-cell anemia. ∎

HEALTH & WELLNESS

Cystic Fibrosis

Cystic fibrosis (CF), the most common potentially fatal genetic disease in North America today, is caused by a recessive allele of a gene on chromosome 7. Because it is recessive, parents may have no idea that they are both carriers for cystic fibrosis until they have a child with the disease. Approximately 1 in 23 Caucasians carries the recessive allele and about 1 Caucasian child in 2,000 has the disease. Among African Americans, only 1 child in 17,000 is born with the disease.

In cystic fibrosis, the recessive allele produces a defective version of a protein that normally aids the transport of chloride across the cell membrane of epithelial cells. Impaired chloride transport affects the secretion of salt and water across cells that line the ducts in several organs. As a result, organs that normally produce a thin, watery mucus secrete a thick, sticky mucus instead.

In the lungs, the thick mucus clogs the bronchioles and leads to chronic inflammation and infections that can be fatal. In the pancreas, the mucus blocks ducts and prevents pancreatic enzymes from reaching the intestines. The sweat glands are also affected, but ordinarily this does not cause major problems.

Typical symptoms of cystic fibrosis include persistent coughing, wheezing, or pneumonia; excessive appetite but poor weight gain; and foul-smelling, bulky feces. The disease is usually diagnosed by age 3 by testing for high salt levels in perspiration. A DNA-based blood test can screen for variations on the gene responsible.

Management of cystic fibrosis focuses on reducing the damage caused by the symptoms, often a constant battle throughout life. Regular, vigorous chest physical therapy helps to dislodge mucus in the lungs, and antibiotics are often needed to treat lung infections. Researchers are working on gene therapy techniques in which copies of the normal gene are inserted into viruses such as those that cause colds, and then the modified viruses are delivered into the airways. Improved treatments and better understanding of the disease have increased the average life expectancy for patients with cystic fibrosis from just five years before we knew much about the disease, to over 37 years today. In addition, genetic testing can now identify couples who are at risk of having a child with the disease. Do not be surprised if cystic fibrosis is cured in our lifetime, or at least treated so effectively that patients live long and healthy lives.

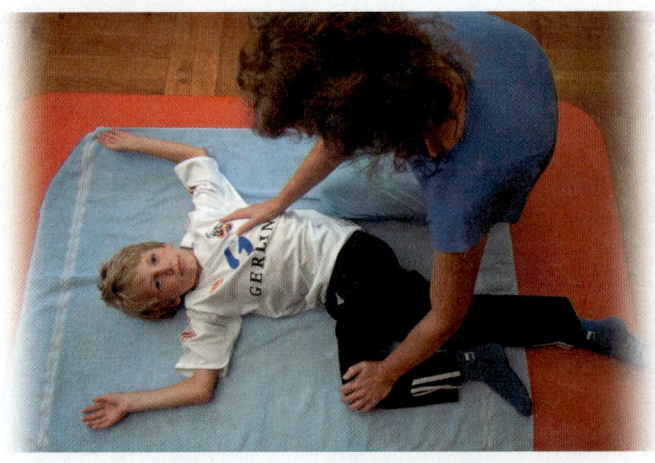

Cystic fibrosis. This child with cystic fibrosis is receiving vigorous percussion therapy ("clap therapy") to loosen mucus in the lungs.

 **MJ's BlogInFocus** Screening of newborn infants for certain known genetic disorders is required by law in all 50 states. What is required in your state? Visit MJ's blog in the Study Area in MasteringBiology and look under "Newborn Genetic Screening Tests."

http://goo.gl/nt6aa8

19.4 Other factors influencing inheritance patterns and phenotype

Polygenic inheritance: phenotype is influenced by many genes

Many traits result from not just one pair of genes, but many genes acting simultaneously. Inheritance of phenotypic traits that depend on many genes is called **polygenic**

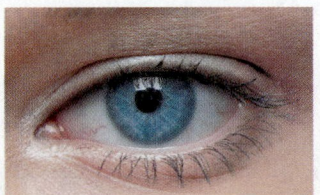
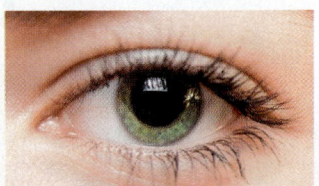
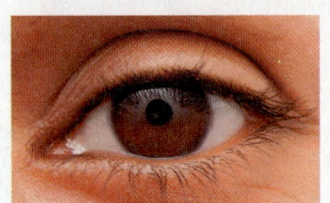
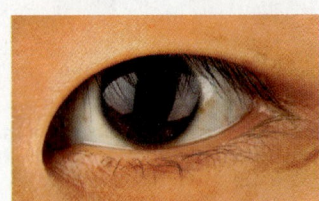

Figure 19.13 Polygenic inheritance. Eye color is determined by at least three genes, not just one. Different combinations of the alleles of the genes account for a range of eye colors.

inheritance (Figure 19.13). Eye color, for example, is controlled by at least three genes, yielding a range of different eye color phenotypes from nearly black to light blue. Both skin and eye color are determined by the amount of a light-absorbing pigment called melanin in the skin and iris. The striking differences in eye color occur because in people with relatively low amounts of melanin we see different wavelengths of light (green, gray, or blue) reflected from the iris.

Physical characteristics such as height, body size, and shape are also examples of polygenic inheritance. Traits governed by polygenic inheritance generally are distributed in the population as a continuous range of values, with more people at the middle (or median) and fewer people at the extremes. As an example of a trait influenced by multiple genes, let's consider height (Figure 19.14). We will simplify matters and assume that height is influenced by only three genes, each of which has two alleles. Each gene has a tall allele (A, B, or C) that behaves in an incompletely dominant manner with respect to the short alleles (a, b, or c). In this example, an individual homozygous for tall alleles for all three genes (AABBCC) would be very tall, whereas an individual homozygous for short alleles for all three genes (aabbcc) would be very short. Heterozygotes for all three genes (AaBbCc) would have an intermediate phenotype and would be of medium height. Now, consider a cross between individuals of medium height. Most of their offspring would

be similar in height to their parents, inheriting a mix of tall and short alleles, but by random chance, some offspring might inherit mostly tall or mostly short alleles, and a few might be very tall or very petite. This phenotypic distribution occurs throughout the human population, and a plot of the numbers of individuals at each value for the trait typically forms a normal (bell-shaped) curve.

We are just beginning to understand that many health conditions may be caused or at least influenced by polygenic inheritance. Examples include cancer, high blood pressure, heart disease, and stroke.

Both genotype and the environment affect phenotype

Phenotype is determined in part by our genotype and in part by environmental influences. A prime example of an environmental influence is the effect of diet on height and body size. Analysis of human populations reveals a clear trend toward increased height and weight in certain populations, particularly in developed countries. The changes in height and weight have sometimes occurred within one generation—too short a time to be due to changes in the gene pool. The trend is primarily due to improvements in diet and nutrition, especially early in childhood.

Except for a few rare genetic diseases such as Huntington disease(more on this later in the chapter), our genotype is not the sole determinant of whether we will actually develop a heritable disease (that is, have the disease phenotype). Usually, we inherit only a slightly increased *risk* of developing the disease, with those risks modifiable by environmental factors and our own actions. If you know that certain heritable conditions or diseases run in your family, it would be prudent to reduce any environmental risk factors associated with that disease. For example, if your family has a history of skin cancer, limit your exposure to the sun. If your family has a history of heart disease due to high cholesterol, have your cholesterol measured regularly and change your diet or begin other corrective measures if your cholesterol levels are above normal.

This is really the old nature-versus-nurture question: Are we the way we are because of our genes or because of our environment? The answer is both. Your genes carry the instructions for all your proteins, but the environment can influence how genes are expressed and how they contribute to your phenotype. There is no one answer to the nature-versus-nurture dichotomy. But since you cannot change your genes, your best defense against any possible future disease is to take proper care of yourself.

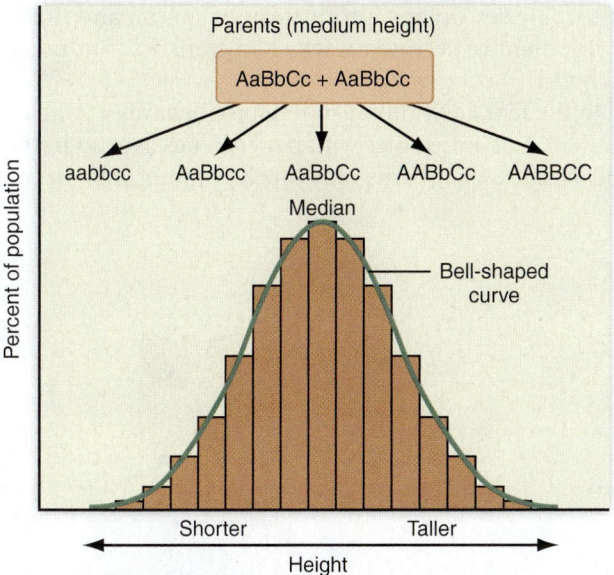

Figure 19.14 Continuous range of variation in traits governed by polygenic inheritance. Polygenic traits, such as height, appear in the population as a continuous range of phenotypes, with more people in the middle than at either extreme. In this hypothetical example, three genes influence height. Heterozygotes for all three genes have an intermediate (average) height and can produce a wide range of heights among their offspring, depending on how many tall alleles and short alleles each child inherits.

☑ Your father is taking medications to lower his cholesterol because it was found to be above normal. Does this mean that you will have high blood cholesterol levels later in life, too?

Linked alleles may or may not be inherited together

As previously mentioned, alleles of different genes are distributed to egg and sperm cells independently of each other during meiosis. But, because humans have only 23 pairs of chromosomes, it doesn't always work this way. Many alleles for different traits are often inherited together because they are physically joined on the same chromosome. These alleles are called **linked alleles,** and the closer they are located to each other on the chromosome the more likely they are to be inherited together. However, they are not *always* inherited together because of crossing-over, which partially reshuffles alleles across each pair of autosomes during meiosis. How often two linked alleles are inherited together is directly proportional to how close together they are and inversely proportional to how often crossing-over occurs (the more crossing-over occurs, the less likely the linked alleles will be inherited together).

To get a feel for how often linked alleles might be inherited together, consider that the average pair of chromosomes may have 2,000 or more genes, and that crossing-over occurs approximately 30 times per pair of chromosomes. Alleles located close to each other, then, may be inherited together quite often. Scientists have learned that they can map the positions of genes by studying how often particular linked alleles are inherited together, a technique called *linkage mapping.*

Adding it all up, there are three major sources of genetic variability as a result of sexual reproduction:

- Independent assortment of alleles located on different chromosomes
- Partial shuffling of linked alleles as a result of crossing-over between autosomes
- Random fertilization of an egg by a sperm

↩ **Recap** In polygenic inheritance, many different genes influence a single phenotype, such as height or eye color. In addition, some phenotypes can be influenced by environmental factors such as diet. Alleles located close together on the same chromosome are inherited together, except when crossing-over occurs. ▪

19.5 Sex-linked inheritance

As seen in **Figure 19.15,** the sperm from the father determines the gender of the offspring. Females have two X chromosomes and will donate one of them to the offspring, whereas males have an X and a Y chromosome, only one of which is donated to the offspring. If the male donates an X chromosome, the fertilized egg will have two X chromosomes and will develop into a female. If the male donates a Y chromosome, the fertilized egg has an X and a Y chromosome and will develop into a male.

Sex-linked inheritance depends on genes located on sex chromosomes

Gender determination by inheritance of sex chromosomes is not without cost, especially to males. The advantage of having pairs of autosomes is that you have two copies of each gene—a backup file, if you will. If something goes wrong with one copy, the other can still carry out its function.

In sexually reproducing animals in which the Y chromosome determines maleness, females have a homologous pair of X chromosomes but males do not. This means that males have a greater susceptibility for diseases associated with recessive alleles on the sex chromosomes. In fact, males are more likely to display any recessive trait found on the X chromosome, not just those that cause disease.

Sex-linked inheritance refers to inheritance patterns that depend on genes located on the sex chromosomes. Sex-linked inheritance is *X-linked* if the gene is located only on the X chromosome and *Y-linked* if the gene is located only on the Y chromosome. There are plenty of examples of X-linked inheritance because many of the genes on the X chromosome are not related to sex determination at all. However, there are relatively few well-documented examples of Y-linked inheritance. The reason is that the Y chromosome is relatively small and many of its genes relate to "maleness." Genes on the Y chromosome apparently influence differentiation of the male sex organs, production of sperm, and the development of secondary sex characteristics, but not much else.

In the female, X-linked inheritance behaves just like inheritance of autosomes with paired genes. Not so in the male, because he inherits only one X chromosome. In the

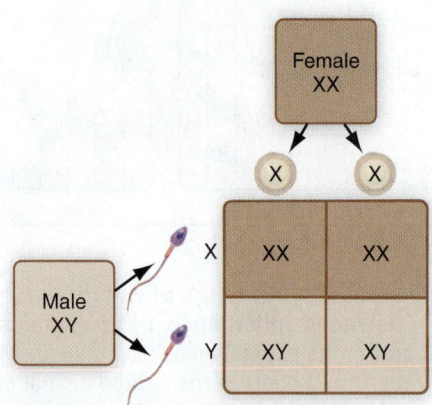

Figure 19.15 Sex determination in humans. The sex of the embryo is determined by the male's sperm, about 50% of which carry an X chromosome and 50% of which carry a Y chromosome. Maleness is determined by the inheritance of a Y chromosome.

male, X-linked genotype and phenotype are both determined solely by the single X chromosome he inherited from his mother.

The best known example of an X-linked disease is hemophilia, also known as *bleeder's disease*. Hemophiliacs lack a blood-clotting factor that is encoded by an X-linked gene with two alleles. The recessive allele (X^h) is a mutant that cannot produce the clotting factor, whereas the dominant allele (X^H) produces the clotting factor. Individuals have the disease only if they do *not* have at least one normal dominant allele (that is, if they are an X^hX^h female or an X^hY male).

The inheritance of hemophilia and other X-linked recessive diseases follows this pattern:

- Many more males have the disease than females. Females can be protected by inheriting at least one normal allele (X^H) out of two, whereas males inherit only one allele because they inherit only one X chromosome.

- The disease is passed to sons solely through their mothers even though the mothers may only be *carriers*. Carriers are heterozygous (X^HX^h); therefore, they can pass on the recessive hemophilia allele even though they do not suffer from the disease themselves. Statistically, half of the sons of carrier mothers will have the disease and half of the daughters will be carriers.

- A father cannot pass the disease to a son, but his daughters will all be carriers (unless the mother is also a carrier or has the disease, in which case, the daughters may also have the disease).

The inheritance pattern of a genetic disorder can be followed through many generations by using a pedigree chart to map out a family's lineage. **Figure 19.16** shows a pedigree following the hemophilia trait. Muscular dystrophy (Chapter 6) and red-green color blindness (Chapter 12) are also X-linked recessive conditions that follow this inheritance pattern. Nearly all X-linked diseases are caused by a recessive allele rather than a dominant one.

☑ Can a father ever pass an X-linked trait to his son(s)? Why or why not?

Sex-influenced traits are affected by actions of sex genes

A few genes that you might suspect are located on the sex chromosomes are actually located on one of the 22 autosomal pairs of chromosomes. A good example is a gene that influences baldness. The normal allele (b^+) and the allele for baldness (b) can be present in both men and women. However, the baldness allele is recessive in women. Women must be (bb) and probably have other baldness genes or factors as well before their hair loss is

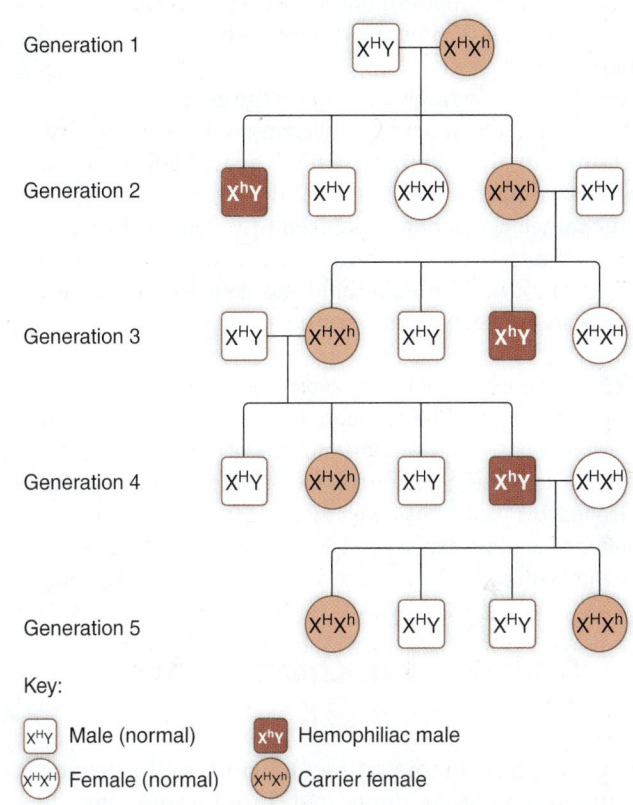

a) A pedigree chart following the passage of hemophilia for five generations. Female carriers pass the hemophilia allele to half their daughters and the disease to half their sons. Males with the disease pass the hemophilia allele to all their daughters (if they survive long enough to have children), but never to their sons.

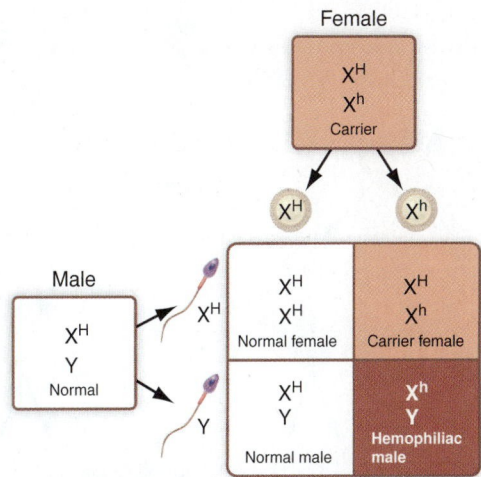

b) A Punnett square showing the possible outcomes of the mating in Generation 1.

Figure 19.16 A typical inheritance pattern for hemophilia, an X-linked recessive disease.

☑ Suppose that one of the women in Generation 5 marries and has a daughter (not a son) who has full-fledged hemophilia. How could this happen?

significant. Men become very bald if they are (bb) and develop significant pattern baldness even if they are heterozygous (bb⁺). Why the difference between men and women?

Testosterone strongly stimulates the expression of the baldness allele, in effect converting it from a recessive to a dominant allele in males. This is an example of a *sex-influenced* phenotype, one not inherited with the sex chromosomes per se but influenced by the actions of the genes on the sex chromosomes. In fact, a number of genes in addition to this particular baldness gene influence the presence and pattern of hair on the head.

🔙 **Recap** In humans, the sex chromosome carried by the male sperm determines the gender of the fertilized embryo. Males have only one X chromosome and are more susceptible to sex-linked recessive diseases than females. Sex-linked inheritance depends on genes located on the sex chromosomes; sex-influenced traits depend on genes located on autosomes that are influenced by the actions of sex-linked genes. ▪

19.6 Alterations in chromosome number or structure

During mitosis of diploid cells and meiosis of the precursor cells of sperm and eggs, duplicated chromosomes must separate appropriately into two different cells. Just as duplication of DNA is not always perfect, neither is separation of the chromosomes.

Failure of homologous chromosomes or sister chromatids to separate properly is called **nondisjunction.** Sometimes both sister chromatids go to one of the two cells formed during mitosis. Occasionally, homologous chromosomes or sister chromatids fail to separate during meiosis (Figure 19.17), resulting in an alteration in chromosome number of sperm or egg cells. Very rarely, a piece of a chromosome may break off and be lost, or it may reattach to another chromosome where it doesn't belong.

For the most part, we never see these chromosomal alterations. If mitosis goes awry, the two daughter cells probably just die and their place is taken by other dividing cells. Alterations that arise during meiosis are more serious because they have at least the potential to alter the development of an entire organism. However, it appears that we never see most of these either because so many of the genes on the chromosomes are essential for embryonic development. Any sperm, egg, or developing embryo with errors as great as an extra or missing chromosome is unlikely to survive. Most die early in embryonic development before we are even aware of their presence.

Down syndrome: three copies of chromosome 21

A few alterations of autosomal chromosome number do result in live births. The most common alteration of autosomal chromosome number is *Down syndrome*. There are actually three types of Down syndrome, but the overwhelming majority (95% of all cases) is caused by having three copies of chromosome 21 (trisomy 21) (Figure 19.18). Other disorders resulting from altered autosomal chromosomal number are Edwards syndrome (trisomy 18) and Patau syndrome (trisomy 13).

People with Down syndrome are typically short with round faces and other distinctive physical traits; most are friendly, cheerful, and affectionate. They generally are slow to

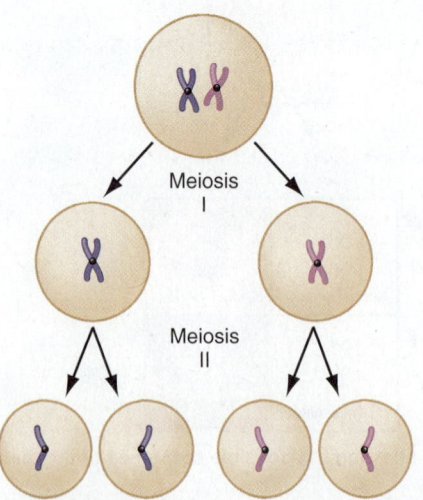
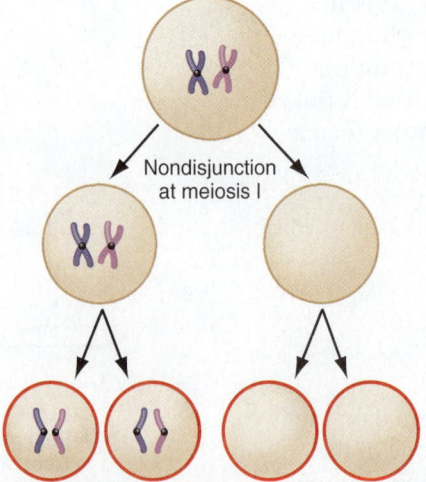
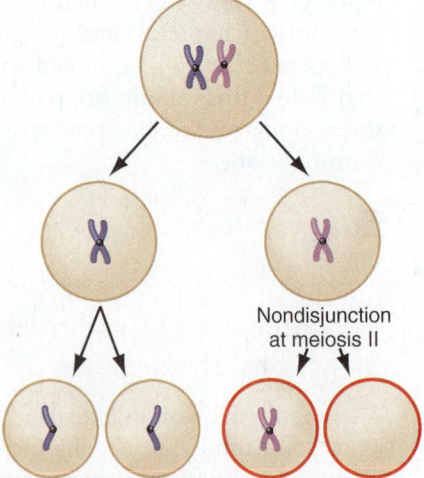

a) Normal meiosis. Duplicated homologous chromosomes separate during meiosis I, sister chromatids separate during meiosis II.

b) Nondisjunction during meiosis I. The duplicated homologous chromosomes fail to separate from each other.

c) Nondisjunction during meiosis II. Sister chromatids fail to separate from each other.

Figure 19.17 Nondisjunction during meiosis. Abnormal gametes have red outlines. This figure shows just one of the 23 pairs of human chromosomes.

Figure 19.18   A person with Down syndrome. Down syndrome is caused by inheriting an extra copy of chromosome 21. Note that alterations in chromosome number can happen for any of the chromosomes, but most alterations are not seen in the population because they are lethal during fetal development.

develop mentally and are prone to respiratory complications or heart defects that can lead to death at a young age. However, with improved care many now live well into their adult years.

Down syndrome affects approximately 1 in every 1,000 live births in the United States. The risk of giving birth to a Down syndrome baby rises markedly with the mother's increasing age. The chance is around 1 in 1,300 for mothers under 30, 1 in 100 at age 40, and 1 in 25 for mothers over age 45. Paternal age may also be a factor when both parents are 35 or older.

Some pregnant women elect to undergo fetal testing to detect Down syndrome and other disorders. In amniocentesis, a sample of amniotic fluid containing fetal cells is withdrawn and tested. Another option is chorionic villi sampling, in which a small sample of fetal tissue is suctioned from the placenta for examination. Fetal cells are examined for the presence of third copies of chromosome 21 (Down syndrome), for the presence of third copies of chromosomes 18 and 13, and for alterations of sex chromosome number.

✔️ Is Down syndrome caused by a gene mutation? Explain.

Alterations of the number of sex chromosomes

Nondisjunction of the sex chromosomes can produce a variety of combinations of sex chromosome number, several of which are fairly common in the human population. In general, an individual with at least one Y chromosome will be essentially a male phenotype. An individual lacking a Y chromosome will be a female phenotype, regardless of the number of X chromosomes. The four most common alterations of sex chromosome number are the following:

XYY—Jacob syndrome XYY individuals are males who tend to be tall but otherwise fairly normal, except that some show impaired mental function. At one time, it was thought that XYY males had a tendency toward violent criminal behavior, but careful study has shown this is not true. Whether their rate of minor, nonviolent criminality tends to be higher than normal is still a disputed issue.

XXY—Klinefelter syndrome People with Klinefelter syndrome are also tall male phenotype. They are sterile, may show mild mental impairment, and may develop enlarged breasts because of the extra X chromosome.

XXX—Trisomy-X syndrome Trisomy-X syndrome individuals are female phenotype. Typically, they are nearly normal except for a tendency toward mild mental retardation.

XO—Turner syndrome Individuals with only one X chromosome (XO) are phenotypically female. However, they tend to be short with slightly altered body form and small breasts. Most are not mentally retarded and can lead normal lives, though they are sterile and may have shortened life expectancies. Turner syndrome is relatively rare compared to the other three because the XO embryo is more likely to be spontaneously aborted.

Table 19.1 summarizes the most common alterations of chromosomal number.

✔️ Given the information above, what exactly causes an XY individual to be male? That is, is it the presence of the Y that causes maleness or the number of X chromosomes?

Table 19.1   Some common alterations of chromosome number

Chromosome genotype	Sexual phenotype	Common name	Approximate frequency among live births
Trisomy 21	Male or female	Down syndrome	1/1,000
Trisomy 18	Male or female	Edwards syndrome	1/6,000
Trisomy 13	Male or female	Patau syndrome	1/5,000
XYY	Male	Jacob syndrome	1/2,000
XXY	Male	Klinefelter syndrome	1/2,000
XXX	Female	Trisomy-X syndrome	1/2,000
XO	Female	Turner syndrome	1/10,000

MJ's **BlogInFocus** How accurate are the genome test results provided by companies that offer genome testing services? Visit MJ's blog in the Study Area in MasteringBiology and look under "Genome Testing Services."

http://goo.gl/mMqVYQ

Deletions and translocations alter chromosome structure

A **deletion** occurs when a piece of chromosome breaks off and is lost. In general, we cannot afford to lose genes, and therefore nearly all deletions are lethal to the sperm, egg, or embryo. There are a few rare conditions in which a chromosome deletion results in a live birth. An example is *cri-du-chat* (French, meaning "cat-cry") syndrome, caused by a deletion from chromosome 5. Babies born with this syndrome are often mentally and physically retarded, with a characteristic kittenlike cry due to a small larynx.

Translocation occurs when a piece of a chromosome breaks off but reattaches at another site, either on the same chromosome or another chromosome. Conceptually, you might think that translocations are not a problem because all of the genes are still present, just in different locations. However, translocations result in subtle changes in gene expression and therefore their ability to function. For instance, translocation appears to increase the risk of certain cancers, including one form of leukemia.

↩ **Recap** Nondisjunction leads to altered numbers of chromosomes in daughter cells. Down syndrome is usually caused by inheriting three copies of chromosome 21. When nondisjunction occurs in the sex chromosomes, usually people with at least one Y chromosome are male phenotype and people without a Y chromosome are female phenotype. A deletion occurs when part of a chromosome is lost; deletions rarely result in a live birth. ∎

19.7 Inherited disorders involving recessive alleles

People express inherited genetic disorders caused by defective recessive alleles only if they inherit two of the defective alleles. People who inherit only one defective allele can pass the gene to their children, but they do not get the disease caused by the defective allele themselves because their good allele is dominant. For this reason, recessive disease-causing alleles are likely to remain in a population more readily than dominant ones.

In some cases, more than one gene pair may contribute to a disease. Hypertension and heart disease are examples of diseases that apparently have a heritable component but aren't associated with any single gene pair.

One of the better known recessive genetic disorders is cystic fibrosis, discussed in the Health & Wellness feature

in this chapter. In this section, we discuss two other genetic disorders caused by recessive alleles (phenylketonuria and Tay-Sachs disease), as well as one (Huntington disease) that is caused by a dominant allele.

Phenylketonuria is caused by a missing enzyme

Phenylketonuria (PKU) is a human inherited disease in which homozygous recessive individuals are unable to make an enzyme that is needed for the normal metabolism of phenylalanine, an amino acid. If phenylalanine accumulates, some of it is used in other metabolic pathways, including one in which excess phenylalanine is converted to phenylpyruvic acid instead. Phenylpyruvic acid is toxic at high concentrations and can lead to mental retardation, slow growth rate, and early death.

Phenylketonuria occurs in about 1 in 12,000 births in Caucasians, so it is considerably less prevalent than cystic fibrosis. The disease is caused by mutation of the gene on chromosome 1 that is responsible for producing the enzyme phenylalanine hydroxylase.

Fortunately, people homozygous for the recessive allele are not affected before birth because the mother's enzymes metabolize phenylalanine for the fetus. After birth, the blood level of phenylalanine is very much dependent on the amount of phenylalanine in the diet. Given the seriousness of mental retardation, all states now require that newborns be tested for PKU by a simple blood test called the *Guthrie test*, which detects excess phenylalanine in the blood.

Treatment of PKU requires limiting the dietary intake of phenylalanine. This is not easy because most proteins contain phenylalanine, but it can be done by adhering to a diet consisting of fruits, vegetables, cereals, breads, and pastas (all low in phenylalanine) plus a medical food powder that contains all of the essential amino acids except phenylalanine. In addition, individuals with phenylketonuria cannot use the artificial sweetener NutraSweet because it contains aspartame, a chemical consisting of just two amino acids, phenylalanine and aspartic acid, linked together. When aspartame is broken down, people with PKU may experience a dangerous increase in phenylalanine.

☑ Suppose a mother and father have a child with PKU, though neither parent exhibits the disorder. What is the probability that their next child will also have PKU? Explain your answer.

Tay-sachs disease leads to brain dysfunction

Tay-Sachs disease is another enzyme deficiency disease, this one caused by a recessive gene located on chromosome 15. Individuals homozygous for the recessive allele are unable to make an enzyme that is responsible for the metabolism of a particular type of lipid, called a *sphingolipid*, found primarily in the lysosomes of brain cells. Without this enzyme, the sphingolipid accumulates in brain cells, causing cerebral degeneration. Infants seem unaffected at birth, but by 4–8 months of age motor function and brain function begin to decline. The child gradually develops seizures, becomes blind and paralyzed, and usually dies by age 3 or 4. There is no known cure or treatment.

Tay-Sachs disease is rare in the general population. However, it is fairly common among Ashkenazi Jews of Central European descent, with about 1 in 3,500 children having the disease. Carriers of the disease remain unaffected because they have sufficient enzyme activity for normal function. Tests are available to determine whether either member of a couple is a carrier of the disease.

Huntington disease is caused by a dominant-lethal allele

Huntington disease is marked by progressive nerve degeneration leading to physical and mental disability and death. Afflicted individuals begin to show symptoms in their 30s, and most die in their 40s or 50s.

Huntington disease follows the simple rules of inheritance of a single pair of alleles, one or both of which may be the dominant (HD) or recessive (hd) form. However, in this case, it is the *dominant* allele (HD) that causes Huntington disease. Anyone with even one HD allele will ultimately develop the disorder. The disease is fatal, and there is no known cure at the moment. For this reason, HD is called a **dominant-lethal allele.**

Dominant-lethal alleles are by their nature uncommon because they tend to eliminate themselves from the population, especially if they cause disease before the affected individual's reproductive years. Huntington disease is unusual in this regard. The disease remains in the human population because in the past, individuals who carried the HD allele had no way of knowing they would develop Huntington disease, and so they would pass the allele on to half their children before they developed any symptoms. Woody Guthrie, a famous folksinger/songwriter of the 1940s, and his mother both died of Huntington disease, and Woody could have passed it to his son, Arlo (**Figure 19.19**). Today, there is a test for the HD allele. Here is an example where genetic testing and counseling could have a dramatic effect on the prevalence of a disease in the future. (See the Current Issue box in this chapter for a discussion of the biological, social, and ethical issues raised by genetic testing.)

Genetic testing for Huntington disease and other inherited disorders is one of the great benefits of the *Human Genome Project.* Engaging scientists from around the world, this massive endeavor involved locating each of the approximately 22,000 genes in the human genome and identifying their functions. Researchers have already successfully identified the genes responsible for several disorders, including forms of breast cancer, skin cancer, and osteoporosis. This information holds promise for developing ways to prevent new cases of genetic diseases and treating existing cases with new technologies. One possibility might be to correct genetic defects with *gene therapy.* We discuss gene therapy in more detail in the chapter on DNA technology and genetic engineering.

Figure 19.19 Genetic testing revealed that folksinger Arlo Guthrie would not suffer from Huntington disease as his father Woody Guthrie did.

↩ **Recap** Recessive alleles such as those for phenylketonuria and Tay-Sachs disease cause disease only if the individual is homozygous recessive. Dominant disease-causing alleles are unusual because they tend to be eliminated from the population. Huntington disease is a rare example of a dominant-lethal disease. ∎

19.8 Genes code for proteins, not for specific behaviors

Genes represent the set of instructions for a human being. The actions of genes can be very specific, such as determining hair patterns or earlobe shape. In terms of behavior, the logical conclusion might be that specific genes determine specific behaviors. We hear of studies that show certain groups of criminals may share a common genetic defect, that some people may have a gene that predisposes them to depression, and that there may even be a "happiness gene." Is any of this true?

The bottom line is that genes code for specific proteins. Specific genes code for proteins that give you curly hair or freckles. Other genes (and their protein products) may strongly *influence* whole patterns of behavior or moods, such as a tendency toward anxiety or aggression. However, genes do not necessarily cause *specific* behaviors. They do not lead to specific thoughts, and there is no one gene that causes you to turn to the left and another one that causes you to turn to the right. Nor are there any genes that cause people to commit specific crimes, as far as we know.

The key to understanding how genes influence behaviors lies in understanding the many roles of proteins. Proteins may act as hormones, neurotransmitters, enzymes, or intracellular messengers. They make up many of the structural components of cells, and they control cell functions. Some proteins influence a broad range of other genes and proteins; others have very specific effects within only certain target cells. Together, groups of genes and their protein products apparently do influence broad patterns of behavior, such as feeding, mating, or learning. They may also influence moods. Nevertheless, there is no evidence that specific genes or their protein products cause depression, lead to complete happiness, or produce specific physical behaviors.

↩ **Recap** Genes influence patterns of behavior. They do not cause specific behaviors. ■

Chapter Summary

19.1 Your genotype is the genetic basis of your phenotype *p. 443*

- The DNA inherited from both parents represents the complete set of instructions for creating a human being.
- The two copies of every gene (one from each parent) can differ very slightly because of mutations. Different versions of a gene are called alleles.
- Our genotype is the complete set of alleles that we inherit; our phenotype refers to the physical and functional expression of those alleles.

19.2 Genetic inheritance follows certain patterns *p. 445*

- A Punnett square can be used to show patterns of inheritance for one or two pairs of alleles.
- A common pattern of phenotypic inheritance is one in which one allele exerts complete dominance over the other.
- Genes for different traits assort independently of each other.

19.3 Incomplete dominance and codominance *p. 450*

- In incomplete dominance, heterozygotes exhibit an intermediate phenotype.
- In codominance, heterozygotes express both phenotypes equally.

19.4 Other factors influencing inheritance patterns and phenotype *p. 452*

- In polygenic inheritance, multiple genes influence a single trait.
- Our phenotype is determined both by our genotype and by our environment.
- Alleles located on the same chromosome that may be inherited together are called linked alleles.

19.5 Sex-linked inheritance *p. 454*

- Maleness is determined early during development by genes on the Y (male) sex chromosome. In the absence of this chromosome, the individual becomes a female.
- Sex-linked inheritance refers to inheritance of genes located on the sex chromosomes. Most of these genes have nothing to do with sex determination.
- Diseases associated with sex-linked inheritance include hemophilia, color blindness, and muscular dystrophy.

19.6 Alterations in chromosome number or structure *p. 456*

- Nondisjunction refers to the failure of homologous chromosomes or sister chromatids to separate properly during mitosis or meiosis; this causes altered chromosome number in the daughter cells.
- Down syndrome is a condition usually caused by the inheritance of three copies of chromosome 21. The primary risk factor is the age of the parents, especially the mother.
- Deletion of even part of a chromosome is usually fatal to a sperm, egg, or embryo. Translocation of a chromosome segment may alter a gene's ability to function properly.

19.7 Inherited disorders involving recessive alleles *p. 458*

- Most autosomal heritable genetic diseases are caused by recessive alleles.
- Phenylketonuria, Tay-Sachs disease, and cystic fibrosis are diseases caused by recessive alleles.
- Huntington disease is caused by a dominant-lethal allele. It has persisted in the human population because people who inherit it do not show symptoms until after reaching reproductive age.

19.8 Genes code for proteins, not for specific behaviors *p. 459*

- Genes influence only general patterns of behavior or mood. They do not determine specific behaviors.

Terms You Should Know

allele, *444*
codominance, *450*
dominant allele, *447*
dominant-lethal allele, *459*
genotype, *444*
heterozygous, *444*
homozygous, *444*
incomplete dominance, *450*
inheritance, *443*
karyotype, *443*
law of independent assortment, *449*
law of segregation, *446*
mutation, *444*
nondisjunction, *456*
phenotype, *444*
polygenic inheritance, *452*
Punnett square, *445*
recessive allele, *447*
sex-linked inheritance, *454*
sickle-cell anemia, *450*
translocation, *458*

Concept Review

Answers can be found in the Study Area in MasteringBiology.

1. Where do we get the two copies we have of every gene?
2. What is the difference between a gene and an allele?
3. Distinguish between genotype and phenotype.
4. Describe the contributions of Mendel to the field of genetics.
5. Explain how alterations of chromosome number and structure can occur.
6. Distinguish between the concepts of dominance, incomplete dominance, and codominance.
7. Describe how genetic variability is accomplished via sexual reproduction.
8. Describe what is meant by sex-linked inheritance.
9. Explain why lethal diseases caused by dominant alleles are so rare.
10. Explain how genes affect mood or behavior.

Test Yourself

Answers can be found in the Appendix.

1. All of the following statements about homologous chromosomes are true except:
   a. One of each pair of homologous chromosomes comes from each parent.
   b. The alleles on homologous chromosomes are identical to each other.
   c. They have genes for the same traits at the same loci.
   d. Homologous chromosomes physically pair up and may exchange parts during meiosis.

2. How many different kinds of gametes (with respect to a particular gene locus) are formed by an individual who is homozygous at that locus?
   a. 1
   b. 2
   c. 4
   d. It can't be determined from this information.

3. Which of the following statements correctly expresses the relationship between dominant and recessive alleles?
   a. A dominant allele is always the preferred or most beneficial allele.
   b. A dominant allele is only expressed when it is homozygous.
   c. A dominant allele masks the expression of a recessive allele in a heterozygote.
   d. A dominant allele is always the most numerous allele in the population.

4. Which of the following would provide an exception to Mendel's law of independent assortment?
   a. adjacent genes on the same chromosome
   b. dominant genes
   c. recessive genes
   d. genes located on different chromosomes

5. Two parents are heterozygous for familial hypercholesterolemia, with blood cholesterol levels two to three times the normal level. What is the probability that their child will be unaffected by this disorder?
   a. 0%   c. 50%
   b. 25%   d. 100%

6. What is the likelihood that a parent with Type AB blood would have a child with Type O blood?
   a. 50%
   b. 25%
   c. 0%
   d. This can't be determined without knowing the blood type of the other parent.

7. All of the following human traits are determined by the interaction of multiple genes and alleles except:
   a. eye color   c. height
   b. hair color   d. sickle-cell anemia

8. Which of the following results in the separation (or unlinking) of linked genes?
   a. crossing-over during meiosis I
   b. independent assortment during mitosis
   c. segregation during meiosis
   d. separation of sister chromatids during mitosis

9. What is the likelihood that a man with hemophilia mating with a woman who is homozygous for the normal form of the blood clotting factor gene will have a son with hemophilia?
   a. 100%   c. 25%
   b. 50%   d. 0%

10. Which of the following could be detected by preparing a karyotype?
    a. sickle-cell anemia   c. Klinefelter syndrome
    b. cystic fibrosis   d. hemophilia

11. What tool is used to determine the probabilities of different genetic outcomes from various crosses?
    a. preparation of a karyotype
    b. chorionic villus testing
    c. biochemical testing of the amniotic fluid
    d. Punnett square analysis

12. Which of the following statements is/are true?
    a. Genotype determines phenotype.
    b. Phenotype determines genotype.
    c. Environment can affect phenotype.
    d. Both (a) and (c) are true.

13. What is the basis for the tremendous genetic diversity resulting from sexual reproduction?
    a. independent assortment during gamete production
    b. crossing-over and shuffling of linked genes during meiosis
    c. random fertilization
    d. all of these choices

14. Examination of a karyotype reveals 23 pairs of chromosomes, with each homologous pair illustrating similar size and centromere location. Which of the following can be concluded from this karyotype?
    a. The individual has Down syndrome.
    b. The individual is a female.
    c. The individual does not have cystic fibrosis.
    d. The individual does not have sickle-cell anemia.

15. Which of the following events or processes can result in Patau syndrome, Turner Syndrome, or Klinefelter syndrome?
   a. nondisjunction
   b. deletion of part of a chromosome
   c. crossing-over
   d. independent assortment

Apply What You Know

Answers can be found in the Study Area in MasteringBiology.

1. What fraction of the offspring of two wavy-haired Caucasians would have wavy hair? Explain using a Punnett square. *Hint:* The allele for curly hair exhibits incomplete dominance over the allele for straight hair.

2. Why is it that the range of resting blood pressures of humans is best represented by a bell-shaped curve covering a continuous range rather than by two populations, one with high blood pressure and one with normal blood pressure?

3. How is it that each parent is able to contribute only a single allele of each gene to his/her offspring, rather than both alleles?

4. Geneticists often study patterns of gene transfer in a variety of model organisms, including plants, fruit flies, and even worms. If they're really interested in patterns of inheritance in humans, why don't they use humans or at least larger animals more similar to humans, such as pigs or even primates?

5. Explain how a father who is blood type A and a mother who is blood type B can have a child who is blood type O.

6. Hemophilia is a sex-linked trait that mothers pass on to male offspring. Is it possible for females to have hemophilia, too? How would such a thing happen?

7. Nondisjunction during meiosis can lead to the formation of gametes (sperm or eggs) with extra copies of one or more chromosomes. Normally, fertilizations involving extra copies of chromosomes simply do not result in a live birth because too many developmental events are altered in some way. The most common extra-chromosomal condition that *does* lead to live births is Down syndrome (trisomy 21). Why do you suppose that trisomy 21 is more common than other trisomies?

MJ's BlogInFocus

Answers can be found in the Study Area in MasteringBiology.

1. Do identical twins have identical DNA? Explain the rationale behind your answer. To view a research paper on this question, visit MJ's blog in the Study Area in MasteringBiology and look under "DNA in Identical Twins." http://goo.gl/FGakie

MasteringBiology®

Students Go to MasteringBiology for assignments, the eText, and the Study Area with animations, practice tests, and activities.

Professors Go to MasteringBiology for automatically graded tutorials and questions that you can assign to your students, plus Instructor Resources.

Answers to ✔ questions are available in the Appendix.

DNA Technology and Genetic Engineering

Soybeans. Over 94% of the U.S. soybean crop is genetically modified.

Key Concepts

- **DNA can now be accurately sequenced in the laboratory.** The entire genomes of many species, including humans, have already been completely sequenced.

- **Individuals can be positively identified from just a small sample of their DNA.** The technique, called *DNA fingerprinting,* is now widely used in criminal investigations.

- **DNA from one species can be inserted into the genome of another species,** producing *transgenic* organisms. Human genes have been inserted into bacteria and large animals (goats, sheep, and cows) to produce valuable human proteins.

- **Transgenic plants can be produced with a wide variety of features,** including increased resistance to drought and disease, improved vitamin content, and even the ability to produce edible vaccines against certain human diseases.

- **The hope of the future is gene therapy**—the ability to insert human genes into human cells to treat or cure human diseases. Gene therapy techniques have not yet been perfected.

Genetically Modified Plants

By now, you have almost certainly eaten food products that came from *genetically modified* (GM) plants. GM plants are plants that have been genetically engineered to contain specific genes from other plants or even from animals or bacteria. They have been cultivated commercially for more than 20 years (since 1994).

Nearly all of the current commercially important GM plants contain genes that either allow the plant to produce its own insecticide (insect-killer) or that make the plant resistant to one of the most common herbicides (weed-killers). The latter allows a farmer to practice no-till farming because instead of tilling to uproot the weeds, they can spray herbicides to kill the weeds without harming the crop. Aside from insect and herbicide resistance, though, GM crops have been developed that taste better, can be stored longer, contain vitamins, are resistant to drought, and so on. There is almost no limit to the potential for tinkering with the genetic makeup of plants. Still, many people are worry that GM plants are not as beneficial as they might seem. They are concerned that GM plants may not have been tested rigorously enough for safety before they were widely adopted. Their concerns may be legitimate, even if the potential risks of GM crops have not yet been proven.

The subject of GM plants, therefore, tends to lead to a vigorous discussion, with some people in favor of them and others unyielding in their opposition. In fact, both sides have valid points. Let's take a look at some of them.

Are There Any Positive Benefits to GM Plant Crops?

Should we have allowed food plants to be genetically modified in the first place? Now that we have them, what do we know about them?

Now that GM crops have been around a while, some of the benefits of planting them are becoming clear. GM crops that are resistant to insects have reduced the cost of crop production for farmers because farmers can spray insecticides less often and use fewer toxic chemicals. This is a benefit to the environment, both in terms of the reduced use of toxic chemicals and in the reduced need for fuel for farm equipment. An even greater benefit to the environment may be in the reduced need to till fields to control weeds in fields of GM crops that are resistant to herbicides. Less tilling means less soil erosion, and therefore less soil and fertilizer runoff into streams and rivers. Farmers and food processors in the United States seem to have accepted the benefits of GM crops, as judged by the trends in crop production. In 2014, 93% of the U.S. corn crop and 94% of the soybean crop came from GM plants, according to the U.S. Department of Agriculture.

In theory, consumers should benefit from GM plants as well, as the reduced cost of crop production is passed on up the food production chain, ultimately resulting in lower prices at the grocery store. GM plants with more direct benefits to consumers are less common yet, though many are in the development stage. Scientists are working on plants that will produce edible vaccines, valuable human proteins, vitamins, and even drugs.

Are GM Foods Safe to Eat?

The answer is yes, at least so far. There is no documented evidence that any currently approved GM food product has ever caused a death, a medical condition, or even an allergic reaction. And that is significant, considering the fact that nearly all the corn and soybeans in this country are now genetically modified. Nevertheless, we would be wise to remain vigilant in evaluating GM foods for any adverse effects. The list of genetic modifications to various food plants grows almost daily. A recent report by the U.S. General Accounting Office concluded that the FDA's current regulation of GM foods is adequate but could be improved. Suggested improvements include making the FDA evaluation process more straightforward and less burdensome and randomly checking data from GM food-producing companies for any signs of an impending problem. →

Questions to Consider

1 Do you approve of the widespread planting of GM food crops? Do you feel comfortable eating them?

2 GM foods are created for a variety of reasons, from resistance to herbicides to the delivery of vaccines and vitamins. Should we permit the development of some GM foods and not others, and if so, which GM foods should we allow? Defend your position.

3 Should foods that contain genetically modified ingredients be labeled as such, so that consumers can decide whether they wish to buy them?

A field of genetically modified herbicide-resistant corn that has been sprayed with an herbicide to kill weeds.

Do GM Plants Endanger the Environment?

This is an area of legitimate concern. Certainly, GM plants have caused environmental changes; how damaging these will prove to be remains to be seen. Some of the currently known effects of GM plants on the environment include:

- Cross-pollination between GM crops and non-GM food crops nearby, rendering the latter no longer GM-free. This is a big problem for organic farmers and non-GM crop-producing farms that are near GM crop fields.

- The emergence of herbicide-resistant "superweeds"—weeds that have adapted to herbicides and thus are no longer killed by the chemicals. Superweeds are by-products of the increase in crops genetically modified for herbicide-resistance. This is no surprise, really; it's comparable to the emergence of antibiotic-resistant in bacteria due to the over-use of antibiotics.

- The appearance of insects able to overcome, or adapt to and survive, insecticides in GM crops. Again, this is no surprise. Genetically modifying plants to resist insects will always be a slow, losing battle because insects will inevitably adapt.

Opponents of GM crops have voiced a number of other concerns, not just about GM crops themselves but about how they have affected the culture of farming. The supply of GM plant seeds is tightly controlled by just a few seed companies. Backed by patent protection, these companies have taken an aggressive approach to farmers who try to save GM seeds for planting the following year, rather than buying GM seeds each year at top price. In addition, there is concern about the potential for a significant crop failure in the future, when nearly all of a food crop in the country is, in effect, a genetically identical monoculture. The problem with any monoculture, GMO or otherwise, is that all the plants are likely to respond the same way to an environmental challenge such as an excessively hot growing period or a drought. In most crops developed by traditional plant breeding techniques there is enough genetic variability for at least some plants to withstand such a challenge. And finally, little is known about the magnitude of the effect that insect-resistant GM plants have on non-harmful and often beneficial insects. For example, both the increased planting of insect resistant corn and the increased use of the herbicide glyphosate have caused a significant decline in the Monarch butterfly population in North America. All of these concerns are legitimate.

To add to the debate, opponents of GM farming techniques fear other problems related to GM crops may pop up that haven't even been considered yet. Nevertheless, the genie that is genetic modification cannot be put back in the bottle. For all intents and purposes, GM crops appear to be here to stay. Going forward, we'll need to learn to harness this new technology for the betterment of all and for the preservation of the environment. Simple but effective regulations will be needed to protect consumers and the environment from potentially harmful farming practices, while at the same time permitting farming practices that may be beneficial. Using new GM technology wisely is both an opportunity and a responsibility.

Should foods containing genetically modified ingredients be labeled?

SUMMARY

- Genetically modified (GM) plants contain genes from other plants, bacteria, or even animals.
- More than 93% of all of the corn and soybeans in the United States come from GM plants.
- The use of GM crops has improved crop yields, lowered crop production costs, and reduced some aspects of environmental damage.
- GM crops appear to be safe to eat.
- GM crops sometimes cross-pollinate other non-genetically modified crops. In addition, some insects are becoming resistant to insecticide-containing GM crops and weeds are adapting to survive herbicide-resistant GM crops.

Knowledge for the sake of knowledge alone has never been enough for humans—we have always felt the need to put our knowledge to good use. The technical application of biological knowledge for human purposes is called **biotechnology.**

The recent explosion of knowledge regarding DNA and the human genome has spawned a new field of biotechnology called **recombinant DNA technology.**

As the phrase implies, we now know how to take DNA apart, analyze its structure, and recombine it in new ways, producing molecules of DNA that did not exist previously in nature. We also know how to cut and splice genes into bacteria, plants, and animals, causing those organisms to make proteins they never made before. We are even beginning to understand how to insert, alter, or replace genes in humans to correct certain genetic diseases.

These are actions that have the potential not only to cure a genetic disease in the individual being treated but to prevent the occurrence of that genetic disease in all of his or her future descendants. In other words, recombinant DNA technology has the potential to affect the course of human evolution. ■

Manipulation of the genetic makeup of cells or whole organisms, including humans, is known as **genetic engineering.** As with all new technologies, genetic engineering comes with new risks and new responsibilities. If we tinker with the structure of DNA, could we accidentally produce harmful organisms? How will we even know what the potential risks are until something happens? What kinds of oversight and regulation are needed in the field of genetic engineering? On one hand, it would be tempting to prohibit or at least limit some of these new techniques because of their possible risks, but on the other hand, the potential benefits of genetic engineering may be just too great to ignore.

In this chapter, we discuss technical and practical aspects of recombinant DNA technologies so that we may make better judgments regarding their application. First, we look at how researchers determine the sequence of a piece of DNA. Then we look at how DNA can be combined in new ways and how to make more of it. We discuss how genes are located and identified and how they are being used to alter living organisms. Finally, we discuss the promises and pitfalls of genetic engineering.

20.1 DNA sequencing reveals the structure of DNA

As you have learned (review Figures 2.24 and 2.25), DNA consists of two strands of nucleotides joined together by complementary sequences of just four bases; cytosine, adenine, guanine, and thymine. Biologists have known for some time how to extract the DNA from a cell. But how do they determine the precise sequence of the bases that make up the individual strands of DNA? This is where the technique called *DNA sequencing* comes in.

DNA sequencing (**Figure 20.1**) begins when scientists ❶ extract a fragment of DNA from an organism, induce

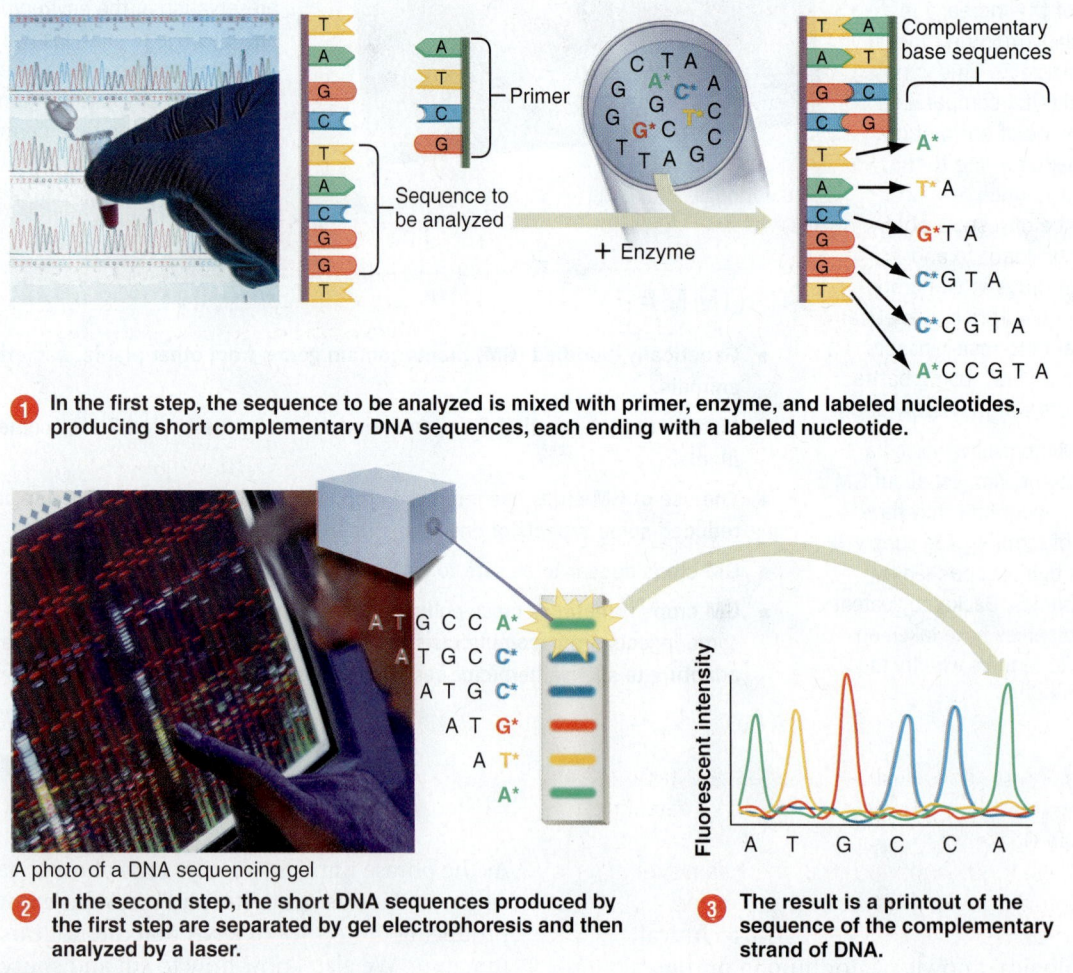

❶ In the first step, the sequence to be analyzed is mixed with primer, enzyme, and labeled nucleotides, producing short complementary DNA sequences, each ending with a labeled nucleotide.

A photo of a DNA sequencing gel

❷ In the second step, the short DNA sequences produced by the first step are separated by gel electrophoresis and then analyzed by a laser.

❸ The result is a printout of the sequence of the complementary strand of DNA.

Figure 20.1 DNA sequencing.

✅ A technician is preparing a sample for step 1, and he accidentally forgets to add the non-fluorescent nucleotides (the ordinary, unlabeled As, Ts, Gs, and Cs shown in the diagram for step 1). Will any DNA be produced? What will the printout in step 3 show? (Hint: Think about what will happen when the DNA polymerase adds the first nucleotide.)

replication of that DNA fragment, and place millions of identical copies of the fragment into a test tube. Then, they add short single-stranded pieces of DNA called **primers** that bind to one end of each DNA fragment in the test tube. The primer serves as the beginning point for synthesis of a new strand of DNA complementary to each DNA fragment in the text tube. Next, they add a mixture of the four nucleotides (T, A, G, and C) that normally constitute DNA, plus a mixture of four nucleotides designated T*, A*, G*, and C* that have been modified so that as soon as one of them is added to a growing DNA strand, further synthesis stops. Each of the four types of modified nucleotides has also been labeled with a different fluorescent color. Finally, they add an enzyme that facilitates the addition of nucleotides one by one to the growing strands of DNA. Synthesis of DNA begins.

During synthesis, either a normal or a modified nucleotide appropriate for a given point in a new strand is added. But because synthesis always stops as soon as a modified nucleotide is added, the final result is a mixture of pieces new stands of DNA of varying lengths, each ending with a single modified (and fluorescently-labeled) nucleotide.

Next ❷, the strands of new DNA are placed on a column (or a flat slab) of gel and subjected to a process called **gel electrophoresis.** Gel electrophoresis creates an electrical field that causes the DNA strands to migrate through the gel. Smaller strands move more quickly through the gel than larger strands.

Finally, a laser scans the gel, reading the locations of the four different fluorescent labels. ❸ The result is a graphic display of the labeled nucleotides in the new strands, arranged in order by the sizes of the strands. This represents the sequence of a single new strand of DNA complementary to the original fragment of DNA chosen for testing. The sequence of the nucleotides in the original unknown fragment of DNA can then be easily calculated on the basis of the known complementary base pairing in DNA.

↺ **Recap** DNA sequencing can determine the nucleotide sequence of fragments of DNA. Gel electrophoresis separates DNA segments by length. ∎

MJ's BlogInFocus Can commercial companies that identify potentially important variants of human genes patent those genes? Visit MJ's blog in the Study Area in MasteringBiology and look under "Patenting Human Genes."

http://goo.gl/FZILPE

20.2 DNA can be cloned in the laboratory

For billions of years, nature has been recombining DNA every time an organism reproduces. For thousands of years, humans have been domesticating and selectively breeding plants and animals to induce nature to recombine DNA in ways useful to us. In just the past few decades, we have learned enough about DNA to be able to cut, splice, and create huge quantities of it at will. In effect, we now have the capacity to develop organisms never before created in nature and perhaps even the ability to modify or fix defective human genes.

Recombinant DNA technology: isolating and cloning genes

Cutting, splicing, and copying (cloning) DNA and the genes it contains is *recombinant DNA technology*. The goal of recombinant DNA technology is to transfer pieces of DNA (and the genes the DNA contains) from one organism into another. Most commonly, it is used to insert specific genes (including human genes) into bacteria so that the bacteria can be induced to produce useful protein products. Recombinant DNA technology requires specialized tools and components, including restriction enzymes, DNA ligases, plasmids, and bacteria.

Restriction enzymes are naturally occurring enzymes in some bacteria that break the bonds between specific neighboring base pairs in a DNA strand. In nature, they protect bacteria from viral invasion by cutting up—restricting—the DNA of the invading virus, but they will cut DNA from any source. They are like a specialized editing tool that cuts a line of text after a specific word. There are many different restriction enzymes in nature, but the most useful for recombinant DNA technology are those that make their cut in *palindromic* nucleotide sequences (**Figure 20.2**). (A **palindrome** is a sequence of letters or words that reads the same backward as forward, such as the word, "racecar.")

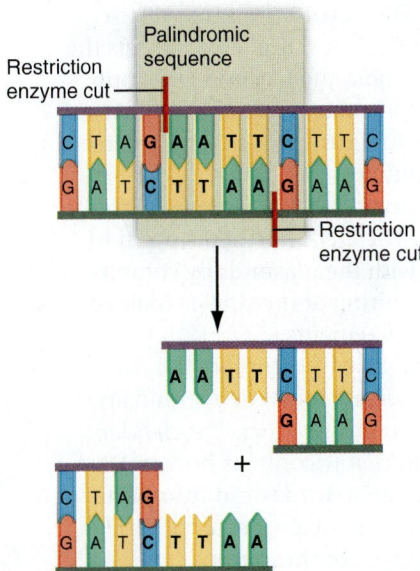

Figure 20.2 Cutting DNA with restriction enzymes. Restriction enzymes are used to cut DNA only at particular palindromic sequences, leaving single-stranded cut ends that are complementary to each other.

An example of a palindromic nucleotide sequence is CTTAAG in one strand of DNA paired with GAATTC in the other strand. A cut in a palindromic DNA sequence leaves two short single-stranded ends that are complementary to each other and also to any other DNA that is cut with the same enzyme.

DNA ligases are enzymes that bind fragments of DNA back together after the restriction enzymes have cut them.

Plasmids are small, circular, self-replicating DNA molecules found in bacteria. They are not part of the normal bacterial chromosome, but they are important to the bacterium because they contain certain genes needed for bacterial replication. Plasmids are useful to scientists because they can be extracted from a bacterium, made to take up (combine with) a foreign piece of DNA of interest, and then reinserted back into a bacterium. Once the plasmid is reinserted into a bacterium, it is copied every time the bacterium reproduces. By reproducing the bacteria in the laboratory, then, scientists can manufacture unlimited copies of a recombinant DNA plasmid.

Figure 20.3 illustrates the technique for producing recombinant DNA using human DNA:

❶ **Isolate DNA plasmids and the human DNA of interest.** The human DNA of interest is first isolated and purified from a sample of tissue. DNA plasmids are also prepared from bacteria.

❷ **Cut both DNAs with the same restriction enzyme.** The restriction enzyme cuts the DNA of interest and the plasmid DNA only at specific palindromic nucleotide sequences, leaving single-stranded ends that will match up with each other.

❸ **Mix the human DNA fragments with the cut plasmids.** The DNA fragments begin to join together with the plasmids by complementary base pairing of the single-stranded cut ends of the fragments.

❹ **Add DNA ligase to complete the connections.** DNA ligase joins the plasmid and human DNA strands together. The circular bacterial plasmid now contains human DNA.

❺ **Introduce the new plasmid into bacteria.** Relatively harmless bacteria, such as *E. coli* generally, are used for this purpose.

❻ **Select the bacteria containing the human gene of interest and allow them to**

reproduce. Many different plasmids will have been formed by the technique described above, each containing a different fragment of human DNA. Therefore, the bacteria that are carrying the gene of interest must first be identified and isolated before cloning the bacteria (and the human genes they are carrying) in large numbers. One way to do this is by adding labeled antibodies against the protein expressed

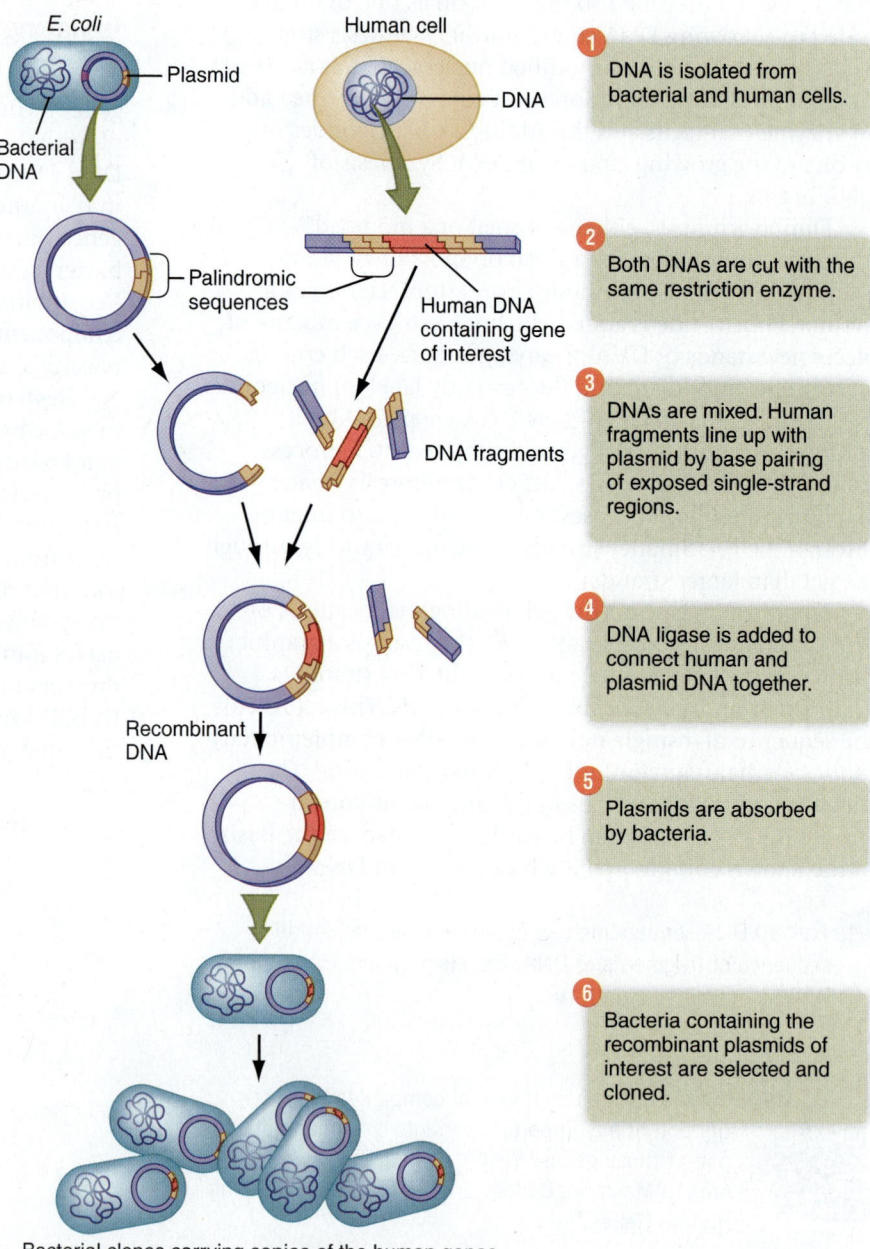

① DNA is isolated from bacterial and human cells.

② Both DNAs are cut with the same restriction enzyme.

③ DNAs are mixed. Human fragments line up with plasmid by base pairing of exposed single-strand regions.

④ DNA ligase is added to connect human and plasmid DNA together.

⑤ Plasmids are absorbed by bacteria.

⑥ Bacteria containing the recombinant plasmids of interest are selected and cloned.

Bacterial clones carrying copies of the human genes

Figure 20.3 Recombinant DNA technique for producing clones of a gene or the protein product of a gene.

✔ What would happen if the human DNA sample did not happen to contain any palindromic sequences? At the end of the procedure, would the bacteria have any plasmids, and would the plasmids have any human DNA?

by the gene to small colonies of bacteria. If the gene (and its protein product) is present in the bacterial colony, the antibody will bind to the protein and the bacterial colony of interest can be identified.

Recombinant DNA technology is, in essence, the technique for cloning genes. Taking the existing technology one step further, if the bacteria could be made to express (activate) the gene located in the DNA fragment, the bacteria could churn out virtually limitless quantities of a particular protein.

☑ Suppose you're trying to clone a particular human gene. You follow the recombinant DNA technique outlined, but to your disappointment, none of the resulting bacteria appear to have the complete human gene. Several of them have small pieces of the gene, however. What happened? What could you do next?

Cloning DNA fragments: the polymerase chain reaction

What if you only had a very small fragment of DNA and you wanted to determine the specific species or even the specific individual from which it came? Identifying the source of a sample of DNA requires a larger sample than is usually available from just a few cells or a small drop of blood. Fortunately, a technique called the **polymerase chain reaction (PCR)** can be used to make millions of copies of a small fragment of DNA very quickly. However, PCR is not a useful technique for copying (cloning) whole genes and the proteins they produce, because the copies of small segments of DNA produced by PCR lack the regulatory genes and proteins required to activate genes.

HEALTH & WELLNESS
DNA-Based Vaccines Against Viruses

Why are there vaccines for the prevention of nearly every significant bacterial disease but very few that work well against viral diseases? It's because of one important difference between bacteria and viruses: Bacteria are living cells, whereas viruses are not really alive, at least not in the traditional sense. Viruses have no organelles and can't reproduce on their own. They are composed of just the DNA or RNA needed to make their various proteins and surrounding protein coat.

Currently, most vaccines are produced from killed or weakened bacteria. The presence of foreign bacterial proteins in the vaccine triggers an immune response, priming the immune system for any future infection by live bacteria. Bacterial vaccines are safe because the bacteria used in the vaccines are either dead or no longer able to reproduce. However, the traditional method of producing bacterial vaccines doesn't work with viruses, because viruses aren't really alive in the first place. Viruses can't be killed or weakened in order to produce a vaccine and they certainly can't be injected whole as a vaccination.

An ideal viral vaccine would be one that contained just one viral protein, in order to trigger an immune response. But, how does one get exposed to just one viral protein? One way would be to get the body to make the viral protein itself. And the best way to do that would be to incorporate the gene for a single viral protein into human cells. Here's how it's done: Using recombinant DNA techniques, a snippet of viral RNA or DNA encoding for a single viral protein is incorporated into plasmids harvested from harmless bacteria. The modified plasmids containing the viral gene are injected into the person to be immunized. Some of the person's cells take up the plasmids and incorporate the viral gene in their DNA. The person's own cells then produce the viral protein, provoking an immune response that prepares the immune system for any future infection by that particular virus. The presence of just one viral protein is harmless to the person because none of the other proteins needed to make the whole virus are present.

On a practical level, DNA-based vaccines against viruses have not been very effective, at least yet. Preparing the plasmids with a viral protein is the easy part; the hard part is getting a person's cells to take up the plasmids, incorporate the viral DNA, and produce the viral protein. Three methods of plasmid delivery show some promise: One method is a noninvasive skin patch coated with plasmids linked to molecules that are taken up by the skin. A second method is microinjection of the plasmids into the skin as either dry powders or as tiny droplets of liquid. A third method is to inject the plasmids into muscle and then apply electrical pulses to the muscle tissue to alter the permeability of the muscle cells just enough that they take up the injected plasmids.

Scientists are currently working to develop DNA-based vaccines for some of the most important viral diseases, including HIV, swine flu, and West Nile Virus. Although none has been approved for use in humans yet, the vaccine against West Nile Virus has been approved for use in horses.

Conceptually, making multiple copies of DNA by PCR is fairly straightforward (**Figure 20.4**). First, the two strands of a short segment of DNA are unwound by gentle heating. Then, they are mixed with (1) primers that are complementary to one end of each strand, (2) nucleotides, necessary to create the new complementary strands of DNA, and (3) DNA polymerase, the enzyme that catalyzes the attachment of nucleotides to the growing complementary DNA strand. When the mixture is cooled slightly, the primers bind to the ends of the two single strands. The primers represent the starting point for replication of each strand. The nucleotides then attach to the growing complementary chain in sequence, assisted by DNA polymerase. Once both strands of DNA are completely replicated, the heating and cooling sequence is repeated again. Each heating and cooling cycle doubles the amount of the desired DNA sample. Repeat the reaction just 20 times and you have over a million identical copies of the original fragment of DNA.

☑ To apply PCR to a certain piece of DNA, there are two important things that you must already know (in advance) about the DNA's sequence. Can you figure out what two parts of the sequence you must know in advance, and why?

Identifying the source of DNA: DNA fingerprinting

DNA fingerprinting is a technique for identifying the source of a fragment of DNA after it has been sufficiently copied (cloned) by PCR. DNA fingerprinting is now commonly used for the positive identification of suspects in a criminal investigation and for paternity testing. The technique takes advantage of the fact that between the genes in DNA are long repeating sequences of base pairs consisting only of *short tandem repeats*, or STRs. An STR consists of 2–5 base pairs. An example of a repeating sequence would be the base pair sequence CATACATACATACATA, comprised of four consecutive repeats of the STR, CATA. The number of times an STR is repeated is highly variable between individuals; there might be 10 repeats of an STR between two specific genes in one individual and 60 in another.

DNA fingerprinting takes advantage of the fact that the lengths of the repeating sequences of STRs vary between individuals. In practice, here's how DNA fingerprinting is done. After a sample of the DNA of interest has been sufficiently copied by PCR, restriction enzymes are used to cut the DNA into pieces. Because the lengths of the repeating STR sequences between the cuts vary, the sizes of the pieces are different for every individual. These pieces are amplified by PCR and then labeled so that they can be visualized. Next, they are placed on a thin layer of gel and subjected to gel electrophoresis, which separates them according to size. A printout of the separation pattern of

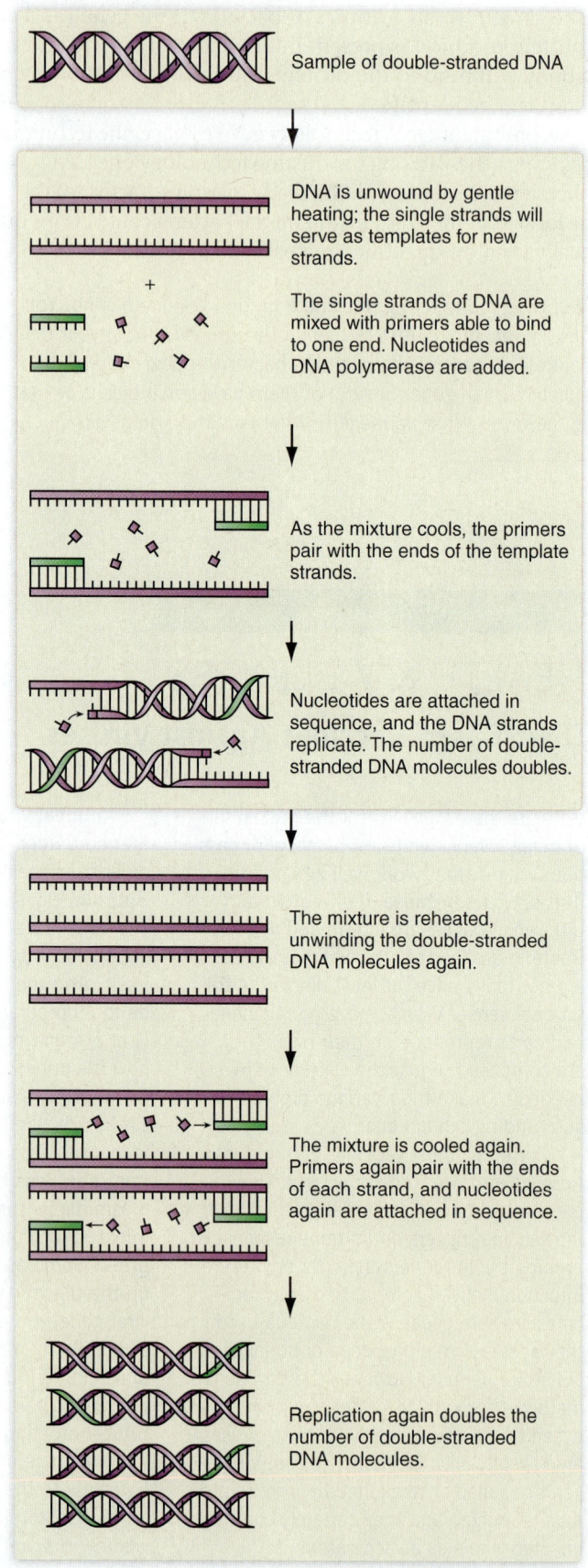

Sample of double-stranded DNA

DNA is unwound by gentle heating; the single strands will serve as templates for new strands.

The single strands of DNA are mixed with primers able to bind to one end. Nucleotides and DNA polymerase are added.

As the mixture cools, the primers pair with the ends of the template strands.

Nucleotides are attached in sequence, and the DNA strands replicate. The number of double-stranded DNA molecules doubles.

The mixture is reheated, unwinding the double-stranded DNA molecules again.

The mixture is cooled again. Primers again pair with the ends of each strand, and nucleotides again are attached in sequence.

Replication again doubles the number of double-stranded DNA molecules.

Figure 20.4   The polymerase chain reaction (PCR) for amplifying a small sample of a segment of DNA.

the DNA fragments on the gel is called an *electropherogram,* or a **DNA fingerprint.** In a DNA fingerprint, the pieces are arranged by size, from smallest to largest. When the separation pattern of an unknown sample matches that from a known individual, the sample is presumed to come from that individual (**Figure 20.5**).

The FBI maintains a database of DNA fingerprints known as CODIS (combined DNA index system). The DNA fingerprints in CODIS come from various law enforcement and forensic laboratories nationwide. The FBI returns the favor by allowing those same agencies access to CODIS where they can search for "hits," or matches, to samples from their criminal investigations. A standard DNA fingerprint requires analysis of only 13 different known STR sequences. If the tests are done properly and the samples are not contaminated, a DNA fingerprint will show a positive match about 99.99% of the time if the DNA samples are indeed from the same individual.

In addition to establishing the identity of individuals who were at a crime scene, DNA fingerprinting can be used to identify unknown deceased individuals following natural disasters, determine paternity, and trace ancestral relationships. Law enforcement agencies have used DNA fingerprinting to trace the illegal trade in endangered live animals, animal meat, and ivory. In human evolutionary studies, it has been used to establish evolutionary relationships between fossils, and in biology, it has been used to study the mating relationships between animals in a population.

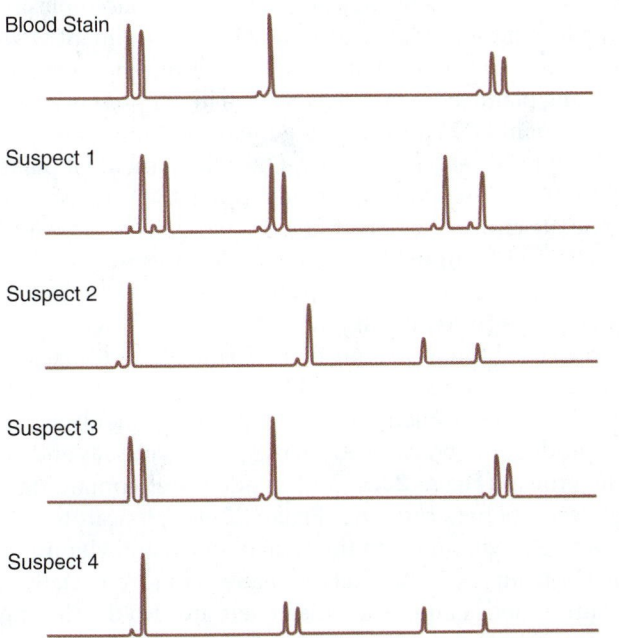

Figure 20.5 DNA fingerprints. This group of DNA fingerprints includes one from a blood sample found at a crime scene and four from different people. Can you identify the match?

Labels in figure: Blood Stain, Suspect 1, Suspect 2, Suspect 3, Suspect 4

↩ **Recap** Recombinant DNA technology is used to cut and splice DNA in order to clone entire functional genes in bacteria. Multiple copies of DNA fragments can be produced by the polymerase chain reaction (PCR) technique. DNA fingerprinting is a technique for identifying fragments of DNA as having come from a specific individual. ∎

20.3 Genetic engineering creates transgenic organisms

Spawning a new frontier for life scientists is the ability to produce **transgenic** organisms—organisms that have been genetically engineered so that they carry one or more foreign genes from a different species. Next, we describe some of the exciting applications of genetic engineering techniques in the twenty-first century.

Transgenic bacteria have many uses

In the past few decades, we have created new strains of transgenic bacteria specifically for our own purposes. Bacteria are rapidly becoming the workhorses of the genetic engineering industry because they readily take up plasmids containing foreign genes, and because their reproductive cycles are so short. Once a gene that codes for a desirable protein is introduced into bacteria by recombinant DNA techniques, it is only a matter of selecting those bacteria that express the gene (and thus make the protein) and then reproducing those bacteria endlessly to obtain nearly a limitless supply of the protein.

One of the very first applications of transgenic bacteria involved the manufacture of essential human proteins, most notably hormones. Recall that hormones are signaling molecules produced in very small amounts only by specific endocrine organs. The lack of a single hormone can be devastating, even life threatening. Take insulin, for example. People with insulin-dependent diabetes mellitus need to inject insulin daily just to survive. Prior to genetic engineering, insulin had to be purified by extraction from the pancreas glands of slaughtered pigs and cattle, making it expensive and difficult to obtain. Today, insulin is produced by transgenic bacteria grown in huge vats. Transgenic bacteria also produce human growth hormone and erythropoietin. Some nonhormone human proteins are also being manufactured, including tissue plasminogen activator (tPA) for preventing or reversing blood clots and human blood clotting factor VIII for treating hemophilia.

Transgenic bacteria can also be used to produce vaccines. Recall that vaccines prepare the immune system for a possible future attack by a disease-causing organism (bacterium or virus). Most vaccines are made from weakened or killed versions of the same organism that causes the disease, because in nature only these organisms carry the antigens that enable the immune system to recognize them. Unfortunately, a vaccine made from a weakened disease-causing organism carries a slight risk of transmitting the very disease it was designed to protect against. For example, a few people have

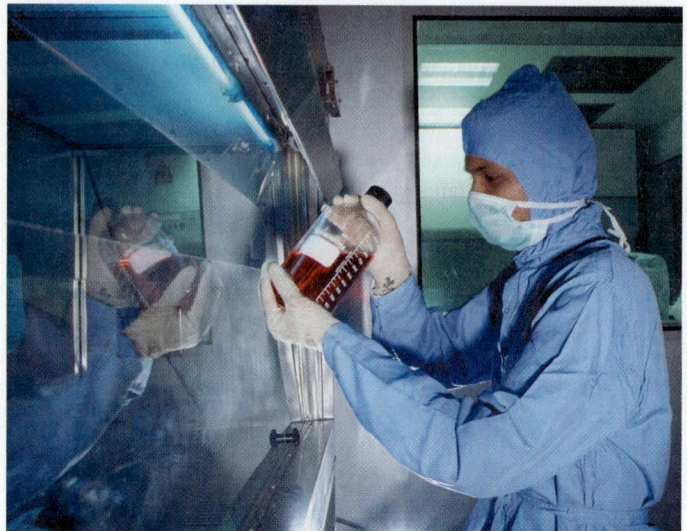

Figure 20.6 Transgenic bacteria can produce vaccines. The transgenic bacteria in this bottle will produce a vaccine against hepatitis B.

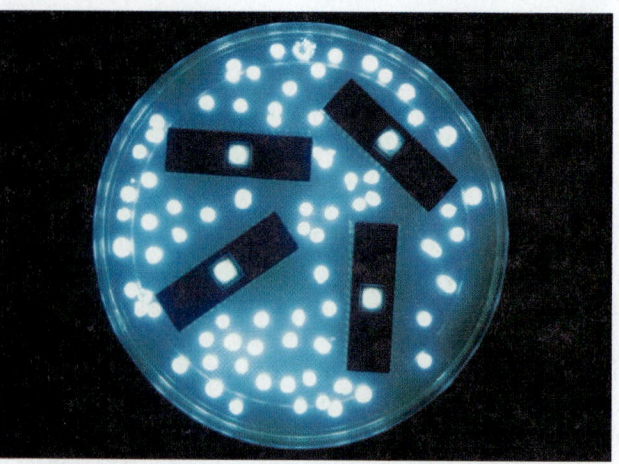

Figure 20.7 "Critters on a chip." The Pseudomonas fluorescens bacteria in this photo have been genetically engineered to emit light when they eat and digest naphthalene, a potential environmental pollutant. Some of the bacteria have been grown on small silicon chips (black rectangles) that are designed to convert the light to electrical current and send it to an electronic receiver. Bacteria and microelectronic devices such as this could be used to monitor hazardous waste sites.

developed polio from the polio vaccine. A second drawback is that vaccines are sometimes in short supply.

Vaccines are produced in genetically engineered bacteria by first inserting into a harmless bacterium a gene that encodes for a surface antigen protein of the disease-causing organism. The goal is to get the harmless bacteria to produce the surface antigen of the disease-causing organism for use as a vaccine. Because the vaccine is comprised solely of purified antigen—not the whole disease-causing organism—there is no chance vaccination can cause the disease.

Although producing a vaccine by genetic engineering sounds easy in principle, in practice it is quite time-consuming and expensive. Producing a genetically engineered vaccine requires successfully finding and transferring just the right gene into harmless bacteria, and then getting the bacteria to produce the desired protein in large amounts. To complicate matters, some disease-causing organisms evolve so rapidly that a vaccine produced one year may not work the next. Nevertheless, a genetically engineered vaccine against hepatitis B is now on the market and several other vaccines are in development including one against malaria (**Figure 20.6**). However, work on several potential vaccines against HIV had to be halted recently because the vaccines just didn't prevent infection by the virus. Successful production of vaccines by genetic engineering will continue to be a challenge.

Current practical uses of transgenic bacteria include producing enzymes (for example, amylase for breaking down starch to glucose in large-scale industrial processes), manufacturing citric acid and ethanol, and producing drugs for human use. Transgenic bacteria are also used to clean up toxic wastes and oil pollutants, remove sulfur from coal, and even monitor hazardous waste sites (**Figure 20.7**).

☑ When animal cells (including human cells) make proteins, they often modify the proteins in several ways. Bacteria cannot perform these modifications. How might this affect the use of transgenic bacteria to create human proteins?

Transgenic plants: more vitamins and better pest resistance

It is somewhat more difficult to create transgenic plants than transgenic bacteria. Typically, the DNA of interest is inserted into a plasmid in a particular bacterium that infects plants. When the bacterium is incubated with embryonic plant cells, a few of the plant cells may take up the DNA-laced plasmid, and a few of those may incorporate the recombinant DNA into their own genetic material. Other techniques include shocking plant cells with high voltage in the presence of recombinant DNA and shooting the recombinant DNA into plant cells at high velocity. Once the individual plant cells with the desired DNA are identified, they are used to regenerate entire plants with the desired traits.

The most common use of genetic engineering in plants has been to produce food crops that are resistant to insects or to herbicides (weed-killers). In addition, genetic engineering has produced tomato plants that resist freezing, tomatoes that last longer on the shelf, and apples and potatoes that resist browning. Genetic engineering has also yielded plants that produce larger leaves for better photosynthesis and faster growth (**Figure 20.8**) and "golden rice" containing high levels of beta-carotene (**Figure 20.9**). Beta-carotene is converted to vitamin A in the human body, and vitamin A is in short supply in the diets of many people, especially in Latin America and Asia. Researchers are also developing plants that mature earlier, develop more extensive root systems for better drought resistance; have short, stiff stems to better resist being blown down by the wind; and even are capable of nitrogen fixation, the process whereby nitrogen from the air is converted to a form that can be used by plants. Success in this latter area would reduce the reliance on nitrogen fertilizers.

Transgenic plants are also being used to produce edible vaccines against infectious diseases. Researchers have already produced a hepatitis B vaccine in raw potatoes, and they are investigating bananas as a source of vaccines against hepatitis B, cholera, and viruses that cause diarrhea. Bananas containing edible vaccines would be of great value because they are palatable to children, do not require cooking, and are easy to grow in tropical and subtropical climates.

✔ Explain in your own words why it is more difficult to create a transgenic plant than to create a transgenic bacterium.

Figure 20.8 Genetically engineered plants. The leaf on the right is from a poplar tree that was genetically engineered to grow larger. The leaf on the left is from a normal poplar tree.

✔ Do you think genetic engineering is fundamentally different from processes that occur in nature? Put another way, is it possible that a plant with the leaf on the right could evolve naturally? Explain your answer.

Figure 20.9 Transgenic plants can improve human nutrition. The "golden rice" on the right is high in beta-carotene, a nutrient in short supply in many areas of the world.

MJ's BlogInFocus Using genetic engineering techniques, scientists have created plant crops that are resistant to herbicides so that farmers can then spray their fields with herbicides during the growing season to kill unwanted weeds. What do you think might be the likely long-term effects (over many years) of this practice? Visit MJ's blog in the Study Area in MasteringBiology and look under "Herbicide Resistance in Plants."

http://goo.gl/kMq1Bb

Transgenic animals: a bigger challenge

Producing transgenic animals is even harder than producing transgenic plants. First, animal cells do not take up plasmids as do bacteria and plant cells, so introducing foreign DNA into an animal cell is a greater challenge. Second, the techniques for cloning animals from a single cell or group of cells are much more difficult than for cloning plants.

Producing a transgenic animal typically begins with inserting the DNA of interest into fertilized eggs. One way is to microinject DNA into each egg individually. Another is to agitate fertilized eggs in the presence of tiny silicon needles and DNA. The needles create holes in the eggs, allowing the DNA to enter. The eggs are then reimplanted into a female for gestation. The number of transgenic animals that can be produced at any one time is limited because usually only a few mature eggs are available at a time and fewer than 10% of the eggs incorporate the recombinant DNA into their own.

Despite these drawbacks, the production of transgenic animals is moving forward. To date, the gene for bovine growth hormone (bGH) has been inserted into cows, pigs, and sheep in order to create faster-growing and larger animals for food production. Transgenic animals are also used to study specific human diseases. For example, transgenic mice have been created that over-express the beta-amyloid protein thought to be responsible for Alzheimer's disease in humans.

MJ's BlogInFocus What are the alleged advantages of the genetically modified potato recently announced by its developer, Simplot? Visit MJ's blog in the Study Area in MasteringBiology and look under "A Genetically Modified Potato."

http://goo.gl/O4aM4o

Pharmaceutical companies are inserting human genes that code for useful proteins into goats, sheep, and cows. The process of producing proteins in farm animals for medical uses is called "gene pharming," aptly enough (Figure 20.10). Large mammals are preferred because it's possible to obtain large quantities of the desired protein from the female's milk, rather than having to repeatedly

Figure 20.10 **"Gene pharming."** These goats have been genetically engineered to produce milk containing antithrombin III, a human protein that can be used to prevent blood clots during surgery.

bleed or even sacrifice a transgenic animal to obtain the protein from its blood. One of the goats in **Figure 20.10**, for example, can produce as much antithrombin as the amount that could be extracted from 10,000 gallons of human blood.

The technique for producing a human protein in an animal is shown in **Figure 20.11**. After the human DNA of interest is incorporated into bacterial plasmids, ❶ plasmids with their human DNA are injected into a goat ovum and the ovum is fertilized *in vitro* (outside the body). ❷ The fertilized ovum is then implanted into a female goat. ❸ When the goat kid is born, its DNA is analysed to determine if it has indeed incorporated the human gene into its own. If it has and if the goat is a female, the human protein will (if all goes well) be produced in the goat's milk. ❹ The milk is collected, ❺ the proteins are separated out, and ❻ the human protein of interest is isolated and collected.

Note the level of uncertainty in this process and the time it takes to know if it has worked. First, only the female goat kids born by this technique are of any use, since the protein will appear in milk. Second, the technique is not always successful; only a fraction of the offspring will incorporate the human gene successfully. And third, it may be a year or more before the transgenic goat is old enough to produce milk. For these reasons, goats such as those shown in Figure 20.10 are (figuratively) worth their weight in gold.

🔄 **Recap** Bacteria are useful for genetic engineering because they readily take up plasmids containing foreign DNA and their reproductive cycles are short. Transgenic bacteria can produce human proteins, human hormones, and even vaccines. Transgenic plants can provide new agricultural crops and transgenic animals can replicate useful human proteins. ∎

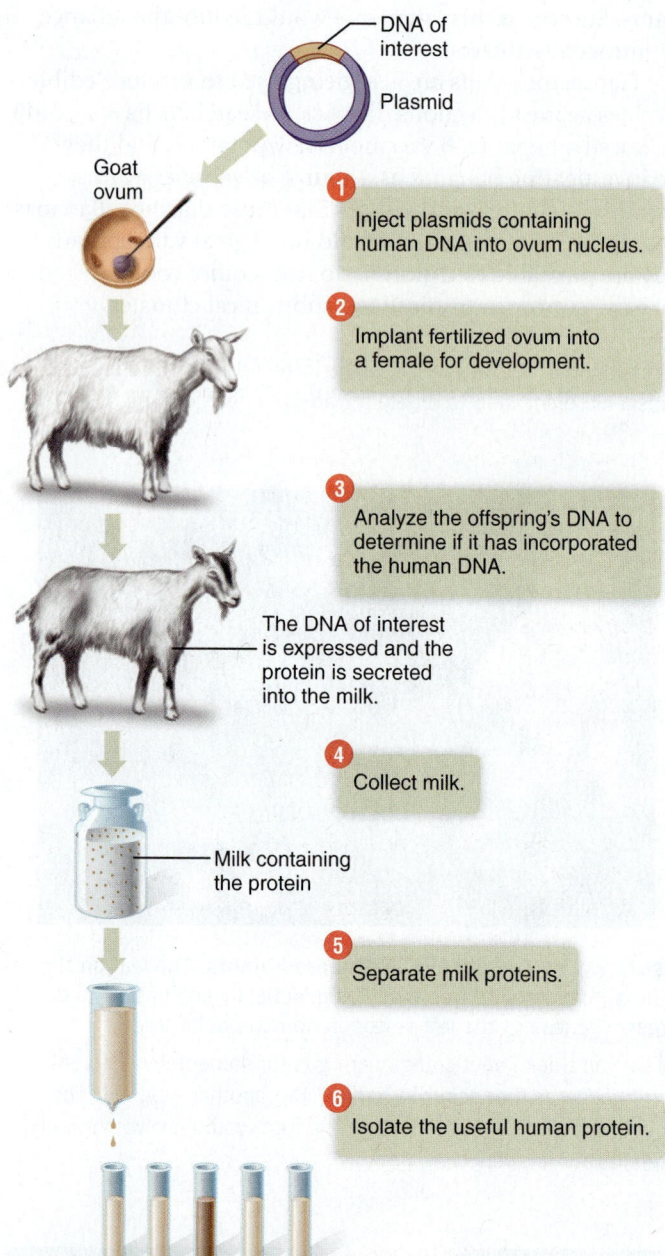

Figure 20.11 **Producing human proteins in transgenic animals.** Large animals such as goats are preferred because then the human protein of interest can be collected from the animal's milk once the transgenic animal matures.

20.4 Gene therapy: the hope of the future?

Nothing quite fires the imagination like the notion of **gene therapy**—the insertion of human genes into human cells to treat or correct disease. We now know the specific gene mutation and its chromosomal location for many genetic diseases, including cystic fibrosis, severe combined immunodeficiency disease, hemophilia, familial hypercholesterolemia, and others. The great hope of the future is that with gene therapy, these mutations could be fixed.

But a word of caution is in order: Despite considerable effort, successful gene therapy is still more of a dream than a reality. In this section, we consider some of the obstacles to success and describe what has been accomplished so far.

Gene therapy must overcome many obstacles

We have already discussed how difficult it is to produce an animal by genetic engineering. However, it is much easier to create an entire transgenic animal than to correct a genetic disease in a human adult, or even a child.

First, an adult is comprised of trillions of *somatic* cells—cells other than sperm and eggs—that have already differentiated and matured. It has proven exceedingly difficult to get the recombinant DNA containing the gene of interest into enough of the right somatic cells. Recall that transgenic animals are created by inserting the gene into a *single* cell (the fertilized egg) that has been isolated and is therefore available to the researcher for microinjection—and even then the procedure works less than 10% of the time. How would we ever insert the gene that codes for a single missing protein, for example, into every cell in the body, or at least into the somatic cells that normally produce that protein? How would we deliver the gene that codes for a specific missing hormone into the millions of cells in a particular endocrine gland that normally produces that hormone? Ideally, what we need are delivery systems that deliver recombinant DNA efficiently to all somatic cells or to specific tissues or cell types. No such delivery systems exist yet.

Second, even if we could correct a specific genetic disorder in an individual, it does not follow that the disease would also be corrected in his or her offspring. If we managed to cure cystic fibrosis in an individual by inserting the appropriate normal gene into enough lung cells to make a difference, that person's children could still inherit the disease—unless we were also able to insert the normal gene into the cells that become sperm or eggs, called *germ cells*. In fact, if we *do* become able to treat genetic diseases in individuals who might otherwise die before reproductive age, we may well increase the prevalence of genetic diseases in the population. We might be forced to use genetic engineering to correct more and more genetic disorders in the future. We should start thinking about the legal, ethical, and moral implications of germ cell gene therapy now, before they become real issues.

☑ Suppose a woman has a disease that is due to a single defective gene and that is certain to be passed on to her children. Yet she and her husband want to have children. In theory, might it be possible to do gene therapy on a human embryo from this woman and her husband, so that every cell in the resulting baby would be completely "cured"? If so, what techniques might work, and what might be some of the possible problems?

Vectors transfer genes into human cells

Although in principle it would be ideal if we could deliver a gene so that it replaces a damaged or missing gene in all of a person's cells, in practice this is not necessary. All we really need to do is get the gene into enough living cells to produce enough of the missing protein to prevent the disease. There are two strategies for doing this, both of which take advantage of transporters, or *vectors*, capable of delivering genes into human cells. The best vectors are a class of viruses called *retroviruses*. Retroviruses splice their own RNA-based genetic code permanently into the DNA of the cells they infect. This makes them ideal gene-transfer vectors as long as they are rendered harmless first.

In one gene-transfer method, shown in **Figure 20.12**, ❶ copies of the normal human gene that is missing from the patient with a genetic disease is incorporated into a retrovirus. Then, ❷ some of the patient's cells from a particular target

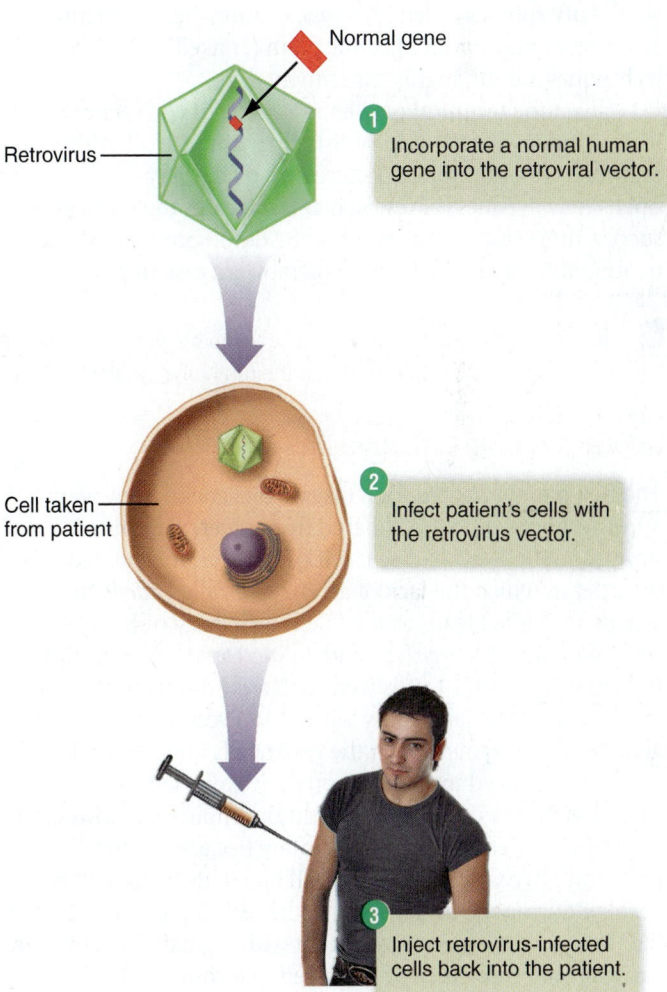

Normal gene

Retrovirus

1 Incorporate a normal human gene into the retroviral vector.

Cell taken from patient

2 Infect patient's cells with the retrovirus vector.

3 Inject retrovirus-infected cells back into the patient.

Figure 20.12   Gene therapy using a retroviral vector. Cells are removed from the human patient, exposed to retroviruses that contain the code for a human gene of interest, and then returned to the patient. The goal is for the infected cells to produce enough of the missing protein to correct symptoms of the disease.

tissue are removed from the patient and exposed to the retroviruses containing the human gene of interest. Finally, ❸ the retrovirus-infected cells are returned to the patient from whom they came. The hope is that the virus-infected cells will incorporate themselves back into the tissue from which they originated, except that now they will express the previously missing protein. In another technique, the retroviruses with their human gene payload are injected directly into the patient, or even directly into the desired tissue if possible. Again, the hope is that some of the cells will take up the retroviruses and incorporate the human gene into the cell's own DNA.

Retroviruses have their drawbacks as gene delivery vectors, however. One problem is that retroviruses generally insert foreign DNA into a cell's DNA only when the cell is dividing. Another is that they insert their genetic material randomly in the genome, so they might disrupt the function of other human genes. To get around these drawbacks, some researchers are experimenting with nonviral methods of gene transfer. These methods include embedding plasmids containing the genes in small fatty spheres called *liposomes,* coating them with amino acid polymers, or even injecting plain ("naked") DNA. Such techniques are still highly experimental.

Given the technical problems associated with gene therapy, it's a wonder we have accomplished anything at all. Although failures have outnumbered successes so far, every small advance opens new avenues for research and increases the chances of success in the future. The story of the development of the first treatment for a disease by gene therapy is a case in point.

✔️ Gene therapy that uses retroviruses tends to be more successful if patients are given drugs that suppress their immune systems. Why?

Success with SCID gives hope

The first patient treated for a disease by gene therapy was a four-year-old girl who suffered from severe combined immunodeficiency disease (SCID). SCID is an inherited disorder in which the lack of an enzyme called *adenosine deaminase (ADA)* leads to a marked deficiency of B and T cells of the immune system and an increased susceptibility to infections. In 1990, T cells were isolated from the girl's blood and grown in the laboratory with a viral vector containing the gene for the enzyme. When the genetically engineered T cells were reintroduced into her blood they expressed the gene, ADA was produced, and her condition improved. However, the improvement was only temporary because mature T cells don't live very long. Most SCID patients treated by gene therapy during the initial years of the technique's development had to supplement their treatment with regular doses of ADA (now produced by genetic engineering techniques).

As an improvement on the initial efforts, researchers tried isolating and treating blood-forming stem cells from the umbilical cords of infants with SCID and then reintroducing those treated cells into the infants' bloodstream. Researchers were hoping the treated stem cells would continue to produce corrected T cells throughout the patients' lives. The first results of this effort were only mildly encouraging: A few of the patients' T cells produced the enzyme, but not enough to prevent the disease without regular doses of the missing enzyme. However, as the techniques improved, eventually several SCID patients were treated successfully with gene therapy alone. The time span from the first gene therapy treatment in 1990 to real success was approximately 10 years.

✔️ Several severe combined immunodeficiency disease (SCID) patients in gene therapy trials have developed leukemia, a type of cancer. Why would gene therapy increase the risk of cancer?

Research targets cystic fibrosis and cancer

Recall that cystic fibrosis is a genetic disease caused by a mutant allele of a single gene. Because the disease primarily affects the respiratory system, researchers are experimenting with delivering the normal gene in a viral vector via a nasal spray. The results so far have been only slightly encouraging. Although about 5% of the cells of the respiratory system take up the normal gene, this is not enough to prevent the disease. Nor is it clear that the gene will be expressed by those cells. Nevertheless, the results are promising enough to warrant the search for more effective delivery vectors.

Some types of cancer may soon be treatable by gene therapy. One promising approach is to add genes for interleukins (which activate the immune system) to cancer cells in the laboratory. The cancer cells are then returned to the patient. As the cancer cells divide, the interleukins they produce stimulate the patient's own immune system to recognize and attack the cancer cells. Another method is to incorporate the gene for a protein of a foreign cell into cancer cells so that they are specifically targeted for destruction by the immune system.

Another promising approach is based on an understanding of why cells die when they become damaged. When normal cells become damaged, they undergo a process called *apoptosis* (programmed cell death). Apoptosis is nature's way of getting rid of damaged cells quickly. Scientists have discovered that certain proteins, including one called mda-7, are responsible for initiating apoptosis. However, cancer cells seem to avoid apoptosis. In one experiment, researchers inserted the mda-7 protein into a viral vector and injected it directly into malignant tumors. The tumors began to secrete mda-7, killing many cancerous cells and stimulating the patients' immune system.

↩️ **Recap** Gene therapy is still more of a hope than a reality. Even if we are able to correct genetic disease in an individual, that person's children could still inherit the disorder. Gene therapy depends on vectors to deliver corrected DNA into a patient's cells. Although there have been a few successes so far, gene therapy is still in its infancy. ∎

Chapter Summary

20.1 DNA sequencing reveals the structure of DNA *p. 466*

- DNA sequencing can now be done in the laboratory.
- The technique involves synthesizing a complementary new strand of DNA to a single strand that is to be sequenced and then determining its sequence.
- An enzyme called DNA polymerase is used to add new nucleotides to the growing strand.
- A technique called *gel electrophoresis* is used to sort DNA strands by size.

20.2 DNA can be cloned in the laboratory *p. 467*

- Recombinant DNA technology refers to techniques for splicing segments of DNA from one organism into the DNA of another organism.
- The technique involves cutting and splicing the DNA of interest into a plasmid and then delivering the plasmid into the organism of interest.
- The polymerase chain reaction makes multiple copies of short segments of DNA.
- DNA fingerprinting can identify fragments of DNA as having come from a specific individual.

20.3 Genetic engineering creates transgenic organisms *p. 471*

- Bacteria are commonly used to make human proteins and hormones because they readily take up recombinant DNA plasmids and can be grown easily in the laboratory.
- Creating transgenic plant crops has proven easy because many plants can be regenerated from individual transgenic plant cells.
- Many transgenic plant crops are already in use in the agricultural industry. Their long-term safety is largely untested.
- Transgenic animals are more difficult to produce, and they can be produced only one at a time.

20.4 Gene therapy: the hope of the future? *p. 474*

- Human gene therapy has proven difficult because it is hard to deliver the appropriate recombinant DNA to the right human cells.
- Gene therapy has shown some success in treating SCID.
- Some cancers may be treatable by gene therapy techniques in the near future.

Terms You Should Know

biotechnology, *465*
DNA fingerprinting, *470*
DNA ligase, *468*
gel electrophoresis, *467*
gene therapy, *474*
genetic engineering, *466*
palindrome, *467*

plasmid, *468*
polymerase chain reaction (PCR), *469*
primer, *467*
recombinant DNA technology, *465*
restriction enzyme, *467*
transgenic, *471*

Concept Review

Answers can be found in the Study Area in MasteringBiology.

1. Explain how DNA is sequenced.
2. Describe the enzymes used in recombinant DNA techniques.
3. Describe the process of producing recombinant DNA.
4. Discuss the polymerase chain reaction technique, and explain its advantages.
5. Describe how a human gene is inserted into bacteria.
6. Explain why bacteria are so frequently used in the genetic engineering industry, and discuss some of the products that can result.
7. Give two reasons why it is harder to produce large numbers of transgenic animals than it is to produce large numbers of transgenic plant crops.
8. Discuss the obstacles to successful gene therapy.
9. Discuss how gene therapy is presently used to treat genetic diseases.
10. List the diseases that gene therapy may be used to cure.

Test Yourself

Answers can be found in the Appendix.

1. Which of the following facilitates joining a piece of human DNA with bacterial plasmid DNA?
   a. DNA polymerase
   b. DNA ligase
   c. RNA polymerase
   d. restriction enzymes

2. Which of the following acts like a molecular copying machine, rapidly making large numbers of specific sequences of DNA?
   a. DNA fingerprinting
   b. DNA ligase
   c. DNA sequencing
   d. polymerase chain reaction

3. Which of the following is used to separate fragments of DNA by size?
   a. gel electrophoresis
   b. polymerase chain reaction
   c. DNA insertion into a plasmid
   d. DNA splicing

4. Genetically engineered human insulin, human growth hormone, and human clotting factor VIII are made by:
   a. gel electrophoresis
   b. polymerase chain reaction
   c. transgenic bacteria
   d. DNA fingerprinting

5. Which of the following vaccines is now produced by recombinant DNA technology?
   a. MMR vaccine
   b. hepatitis B vaccine
   c. polio vaccine
   d. tetanus vaccine

6. All of the following statements about transgenic plants are true except:
   a. Genes have been successfully introduced into plants to improve the nutritional value.
   b. Transgenic plants have been developed that are resistant to pests such as larval insects.
   c. Human genes cannot be incorporated into transgenic plants.
   d. Transgenic plants are being studied as a way to produce edible vaccines.

7. "Gene pharming" refers to:
   a. using transgenic farm animals to produce human proteins for medical uses
   b. producing transgenic plants with improved nutritional value
   c. producing drugs using transgenic plants
   d. using transgenic plants to produce vaccines

8. Which of the following poses the greatest challenge to gene therapy?
   a. identifying the chromosomal location of the defective gene
   b. obtaining a normal allele of the defective gene
   c. producing multiple copies of the normal allele to replace the defective allele
   d. delivering the normal allele into the cells of the recipient

9. What would be the advantage of introducing genes into germ cells rather than somatic cells?
   a. Genes can be more readily inserted into germ cells.
   b. Introduced genes are more often lost from somatic cells.
   c. Genes introduced into germ cells are more readily expressed.
   d. Genes introduced into germ cells would be passed on to gametes and future generations.

10. The role of restriction enzymes in bacterial cells is to:
    a. assist in DNA replication
    b. process primary transcripts into mRNA
    c. destroy invading viruses by cutting up their DNA
    d. join pieces of DNA

11. Which of the following is needed for sequencing DNA?
    a. DNA polymerase   c. primers
    b. gel electrophoresis   d. all of these choices

12. A tiny amount of blood was recovered from a crime scene. What technique will enable comparison of this DNA to that of a potential suspect?
    a. DNA fingerprinting   c. DNA sequencing
    b. DNA cloning   d. gene therapy

13. Which technique will allow amplification of the DNA from the crime scene, in order to provide a large enough specimen?
    a. gel electrophoresis   c. transgenic introduction
    b. polymerase chain reaction   d. DNA sequencing

14. A gene for a bacterial protein with insecticide properties has been introduced into a plant. This plant is now referred to as a(n):
    a. insectivorous plant   c. genetically modified plant
    b. transgenic plant   d. both (b) and (c)

15. Bacteria have been genetically modified to do all of the following except:
    a. produce human hormones   c. treat sickle cell anemia
    b. digest toxic wastes   d. produce vaccines

Apply What You Know

Answers can be found in the study area in MasteringBiology.

1. What do you think are the chances that someone will try to use genetic engineering techniques to produce a biological weapon of mass destruction? Should this affect our research into genetic engineering, and if so, how? Explain your reasoning.

2. Imagine a future in which the respiratory effects of cystic fibrosis, a genetic disorder, could be cured by a gene therapy technique in which the normal gene is delivered to the patient via a nasal spray. You are the patient's genetic counselor. Explain to the patient why the cure will not alter his or her future child's chances of having the disease.

3. Scientists in a Hollywood film used prehistoric dinosaur DNA to make real dinosaurs. Although re-creating a complete dinosaur is not yet within the realm of possibility, with present techniques it would at least be feasible to analyze a dinosaur's DNA (if some were available). What techniques would be used for such an analysis?

4. Pharmaceutical companies produce human insulin using genetic engineering techniques. The drug Humulin was one of the first recombinant DNA drugs ever approved by the FDA. How would such a drug be produced?

5. Why are vaccines produced by genetic engineering techniques (such as the Recombivax vaccine against hepatitis B) safer than vaccines produced from killed or weakened (attenuated) viruses?

6. There is interest in producing vaccines in foods. What would be the advantages of having a vaccine embedded in food?

7. Transgenic plants have the potential to improve food production around the world. They also have the possibility of being more nutritious than their naturally occurring cousins. Still, there are drawbacks. How could global climate change affect transgenic crops?

MJ's BlogInFocus

Answers can be found in the Study Area in MasteringBiology.

1. A distillery announced recently that it was going to produce a line of alcoholic beverages made from non-GMO grains. What do you think—will the product be any safer to drink than alcoholic beverages made from GMO grains? Visit MJ's blog in the Study Area in MasteringBiology and look under "GMO-Free Alcoholic Beverages." http://goo.gl/dkhINy

Answers to ✓ questions are available in the Appendix.

Development, Maturation, Aging, and Death

Staying fit may improve quality of life as we age.

Key Concepts

- **Humans develop from a single fertilized egg to a complex organism of over 10 trillion cells in just nine months.** A complex pattern of cell differentiation and cell division results in the formation of the various tissues, organs, and organ systems.

- **Birth represents the first step toward independence.** At birth the circulatory, respiratory, and digestive systems function independently for the first time.

- **Adolescence is the transition from childhood to adulthood.** Adolescence is marked by maturation of the reproductive systems and sexual responsiveness.

- **Aging is a process that begins in early adulthood and continues until death.** Aging is characterized by the slow decline in function of most tissues, organs, and organ systems. The aging process can be slowed, but it cannot yet be prevented.

- **Death, defined as the cessation of life, is the final transition.** The death of a person is not necessarily accompanied by the immediate death of all organs, tissues, or cells.

Death with Dignity (Brittany Maynard's Journey)

Brittany Maynard was only 29 years old, newly married, and the holder of a bachelor's degree in psychology and a master's degree in education. On January 1, 2014, after months of painful headaches, Brittany Maynard learned that she had a type of brain cancer called an astrocytoma. Surgery on January 10 to stop the growth of her cancer failed. In April, she was told that her cancer had shown signs of growing aggressively, and that she had about six months to live. Who among us wouldn't think, "It isn't supposed to end this way; at least not this soon!"

Brittany researched her options. She found that even if she underwent aggressive treatment that would include radiation and several surgeries, she faced months of steadily advancing disability, pain and suffering, and ultimately certain death. So she made a decision to have the option to control her own dying process "when the time was right." Fortunately for her, she could do so legally under what are called Death with Dignity laws, now in effect in several states.

Death with Dignity Laws

Death with Dignity laws, also called medical aid-in-dying laws, allow doctors to prescribe lethal drugs to a person certified to be terminally ill; the patient then chooses when (or if) to use the drugs. Only three

Brittany with her husband.

states (Oregon, Washington, and Vermont) had such laws at the time of Brittany's diagnosis. In addition, Montana and New Mexico have permitted the procedure in certain cases.

Death with Dignity laws do not permit someone to end their own life just because they are in pain at the moment or because they no longer have a desire to continue living. All current Death with Dignity laws require certification by qualified physicians that the patient is terminally ill and has no chance of long-term survival. Only then can a prescription for a lethal drug be written. After that, it is still the patient's choice when or whether to use the prescription. In Oregon, the first state to make the practice legal (in 1997), lethal prescriptions have been written for over 1,300 terminally ill patients, but fewer than 900 have actually used them.

Brittany Maynard's Journey

At the time of her diagnosis, Brittany Maynard lived in California, so in

Questions to Consider

1 Is your state considering a Death with Dignity law? If it considers one in the future would you support it or oppose it?

2 If you were in the same situation as Brittany Maynard was, would you make the same decision she did? Why or why not?

order to take advantage of Oregon's Death with Dignity law she first had to establish residency in Oregon. She found a new home, registered to vote, and obtained a driver's license. She also had to find physicians who could evaluate her and certify that she was terminally ill and therefore meet Oregon's Death with Dignity criteria.

Brittany was not one to waste the precious time she had left, so she put it to good use. She wrote articles for CNN and People.com about her experiences, and through a nonprofit organization called Compassion & Choices, she created thebrittanyfund.org to promote the legalization of medical aid-in-dying in other states. She also made sure to complete her bucket list—things she wanted to do before she died.

Perhaps because of her conviction that she had the right to control the circumstances of her death and because she was young (the average age of the people who have taken advantage of Oregon's death with dignity law is 71), for a brief time, Brittany became the face of a national debate over the need for (and the ethics of) Death with Dignity laws. According to Compassion & Choices, since Britanny's death over 50,000 people have written to their state's lawmakers urging them to pass death with dignity legislation, and by March 2015, at least 15 states and the District of Columbia had proposed such legislation. Brittany's home state of California passed its version of a Death with Dignity bill, "The End-of-Life Options Act" (SB-128), within one year of her death. How many states will eventually pass Death with Dignity legislation is unknown. Opponents of such laws include the Catholic Church and the National Right to Life Committee; the latter has referred to Death with Dignity laws as "doctor prescribed suicide."

Brittany received her prescription for a lethal drug in June of 2014. Her journey of life came to an end on November 1, →

2014, when she chose to end her life. According to *People* magazine, her final Facebook post read, "Today is the day I have chosen to pass away with dignity in the face of my terminal illness, this terrible cancer that has taken so much from me . . . but would have taken so much more." In her epitaph, she wrote, "It is people who pause to appreciate life and give thanks who are happiest."

SUMMARY

- **Death with Dignity laws permit a terminally ill patient to choose when to die.**
- **The laws require certification by physicians that the patient is indeed terminally ill.**
- **Five states—Oregon, Washington, Vermont, New Mexico, and Montana—now permit death with dignity. Other states are considering it.**
- **Opponents consider Death with Dignity laws to be a form of doctor-assisted suicide.**

Life is a journey that begins at fertilization and ends around the time of our last breath. It's an interesting and complicated process in which one developmental stage follows the next in a carefully choreographed dance.

The changes that occur during human development and maturation are nothing short of amazing: In just nine months, a single undifferentiated cell (the fertilized egg) develops into a baby comprising more than 10 trillion cells. Over the next 20 years, a human's organ systems grow and become fully functional. Along the way (sometime during adolescence), a human becomes capable of reproducing, and the cycle of life perpetuates. Then, starting at about age 40, there is a decline of function that we call *aging*. We don't fully understand what causes aging, but we do know that it ends in death. For all life-forms on Earth, life is finite; it begins, and it ends. ∎

The amazing journey of human life is what this chapter is all about. We begin at the beginning when sperm meets egg and end at the ending when life finally ceases.

21.1 Fertilization begins when sperm and egg unite

The possibility of fertilization begins with intercourse. Fertilization is a process that begins when a sperm and egg unite and ends when a **zygote** (a diploid cell) is formed.

The journeys of egg and sperm

It's remarkable that sperm and egg ever find each other at all, given the journey that each must take. An egg is released from one of the two ovaries at ovulation, and the sperm are deposited in the vagina near the cervix during intercourse. The egg moves slowly and passively down the oviduct, propelled by cilia that line the tube and sweep it gently downward.

Sperm are not much more than a head containing the all-important DNA, a midpiece containing mitochondria for energy production during the sperm's long journey, and a flagellum for movement (**Figure 21.1a**). The several

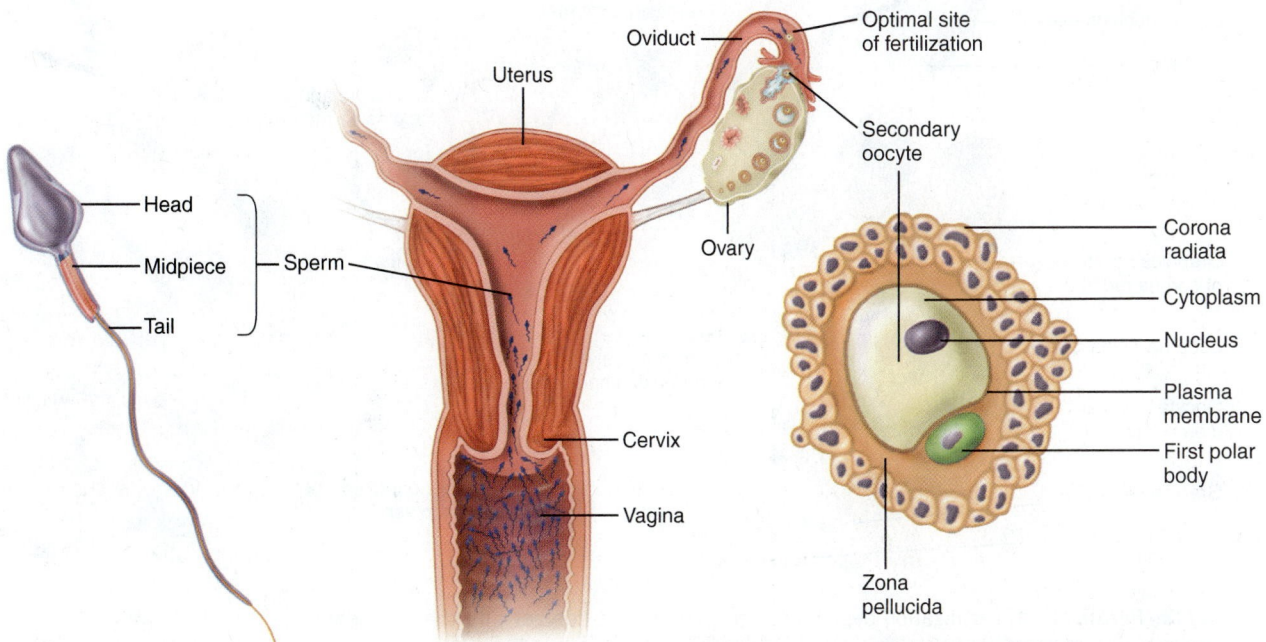

a) The male gamete, or sperm. In this illustration, the size of a sperm relative to the secondary oocyte has been greatly exaggerated.

b) Fertilization. Fertilization generally takes place in the upper third of the oviduct (Fallopian tube).

c) The female gamete. The female gamete is a secondary oocyte that is in an arrested state of stage II of meiosis.

Figure 21.1 Female and male gametes and fertilization.

hundred million sperm encounter many hazards on their journey toward the egg. First, they must pass through the mucus that blocks the cervical opening and cross the vast (to them) expanse of the uterus. Then, they must locate and enter the correct oviduct (half don't). Sperm swim randomly at a rate of ¼ inch per minute. Along the way they must tolerate the strongly acidic pH of the vagina, and avoid bacteria and the occasional white blood cells roaming the uterine lining.

A sperm's journey can take from several hours to several days. Only a few thousand or a few hundred make it successfully to the upper oviduct (fallopian tube); the rest are lost along the way and die. Fertilization typically takes place in the upper third of the oviduct approximately 6–24 hours after intercourse, provided that an egg is present (Figure 21.1b). If no egg is present, the sperm may live for a couple of days and then die. If no sperm arrive, the egg dies in about 24 hours.

One sperm fertilizes the egg

The odds of success for a single sperm are still not very good, for only one of the several hundred million sperm will fertilize the egg. This is important, for otherwise the zygote would end up with an abnormal number of chromosomes. The process of fertilization has evolved to ensure that only one sperm can succeed. Before fertilization, the egg is not completely developed; it is really only a secondary oocyte that has started but not finished the second stage of meiosis (Figure 21.1c). The second stage of meiosis is not completed until a sperm makes contact with the egg and fertilization begins. At this point, the secondary oocyte is surrounded by a protective covering called the *zona pellucida* and by a layer of granulosa cells derived from the follicle called the *corona radiata*. The egg is relatively large (nearly 2,000 times the mass of a sperm) because it contains a great deal of cytoplasm. The cytoplasm of the egg must support nearly two weeks' worth of cell divisions until the pre-embryo, as it is called, makes contact with the uterine lining and begins to receive nutrients from the mother.

When a sperm encounters the egg (Figure 21.2), the tip of the sperm head, called the *acrosome*, releases powerful enzymes. ❶ These enzymes digest a path for the sperm between the granulosa cells of the corona radiata and through the zona pellucida to the oocyte plasma membrane. Several sperm may be making this journey at the same time. ❷ When the first sperm makes contact with the oocyte plasma membrane, special protein "keys" of the sperm recognize receptor protein "locks" in the oocyte plasma membrane, ensuring that only human sperm can penetrate a human egg. The combination of lock and key causes the plasma membranes of egg and sperm to fuse so the nucleus of the sperm can enter the egg.

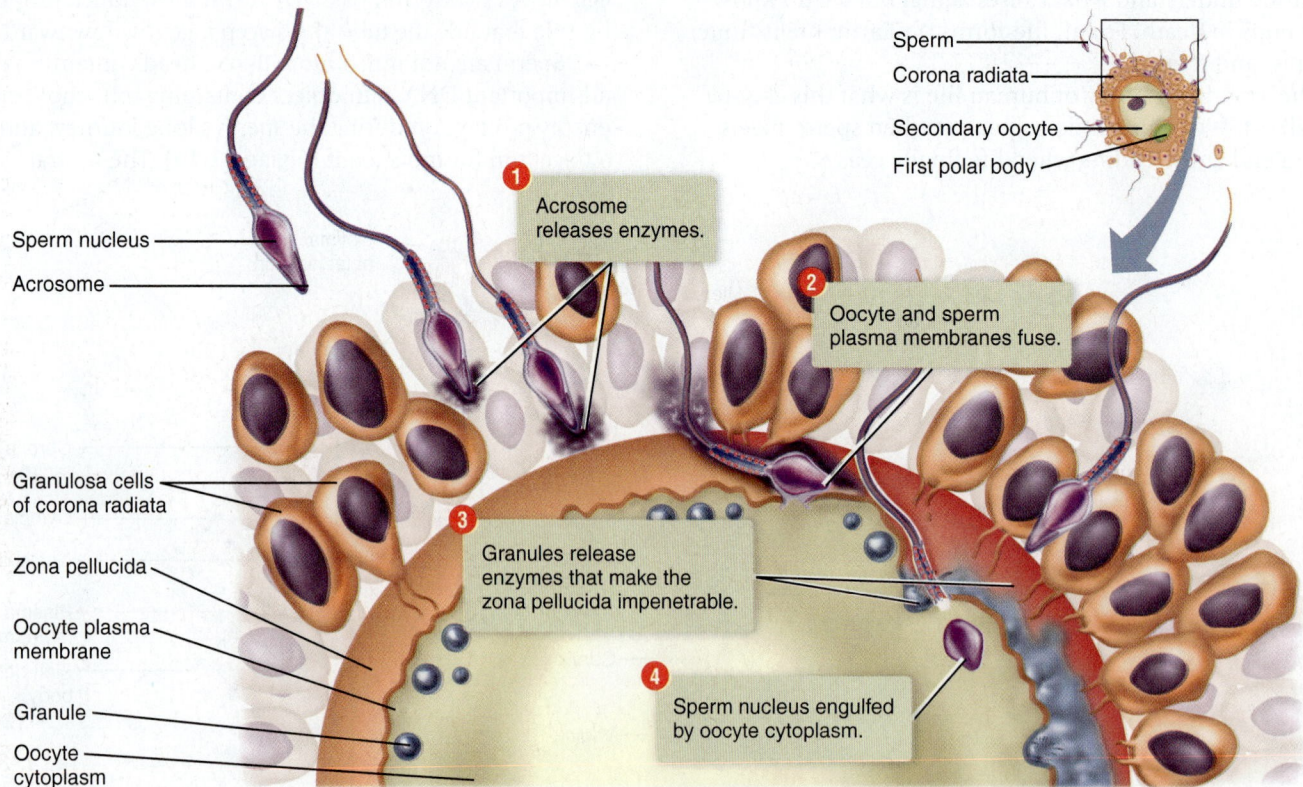

Figure 21.2 Fertilization. Fertilization begins when a sperm makes contact with the secondary oocyte and releases enzymes that digest a path through the zona pellucida. Once a sperm makes contact with the oocyte plasma membrane and enters the oocyte, granules are released from the oocyte that make the zona pellucida impenetrable to other sperm.

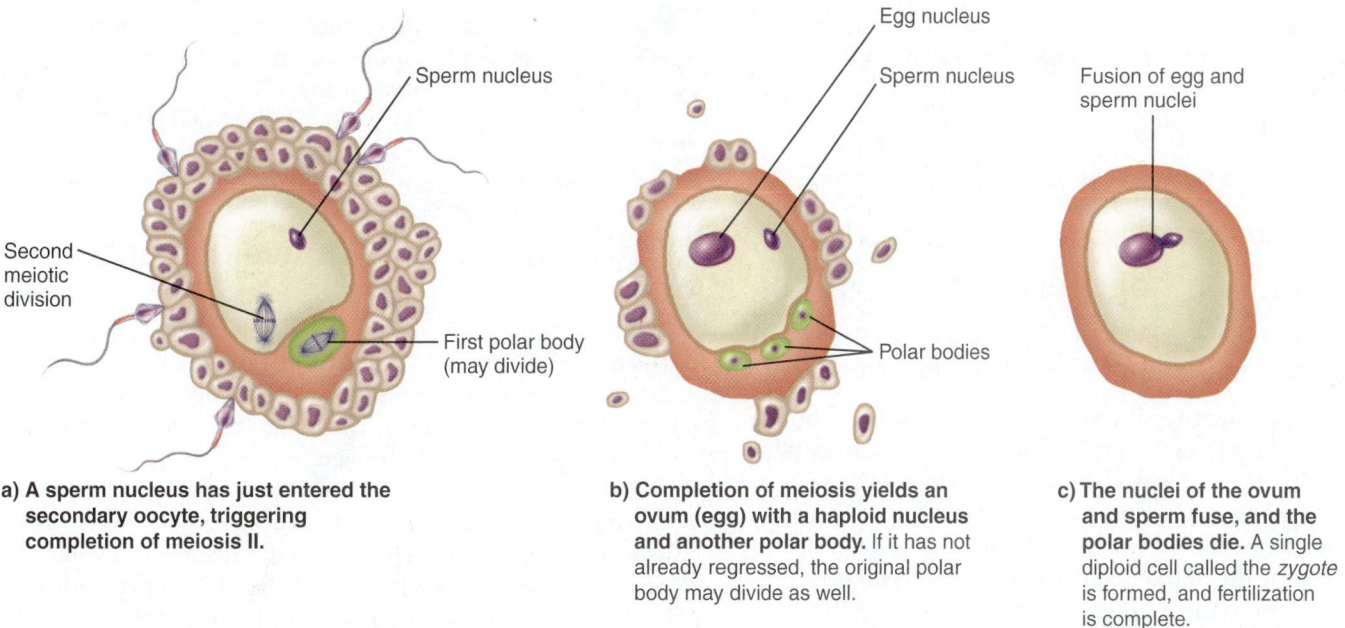

a) **A sperm nucleus has just entered the secondary oocyte, triggering completion of meiosis II.**

b) **Completion of meiosis yields an ovum (egg) with a haploid nucleus and another polar body.** If it has not already regressed, the original polar body may divide as well.

c) **The nuclei of the ovum and sperm fuse, and the polar bodies die.** A single diploid cell called the *zygote* is formed, and fertilization is complete.

Figure 21.3 Completion of fertilization.

Fertilization begins when the sperm's nucleus enters the egg. ❸ The entry of one sperm triggers the release of enzymes from granules located just inside the egg. These enzymes produce changes in the zona pellucida that make it impenetrable to all other sperm. ❹ One sperm nucleus, and only one, will enter the egg.

The process of fertilization is not yet complete, however, because the "egg" is still a secondary oocyte. Entry of the sperm nucleus triggers the completion of meiosis II and formation of the haploid **ovum,** or mature egg, and another polar body (**Figure 21.3**). Fertilization is considered complete when the haploid nuclei of sperm and mature ovum join, forming a single diploid cell (the zygote) with 46 chromosomes. Some people refer to the end of fertilization as *conception* and to the product of conception as the *conceptus*. The follicular cells that surrounded the secondary oocyte are shed and the polar bodies die, leaving just the fertilized cell, or zygote.

Twins may be fraternal or identical

Twins occur once in about every 90 births. Twins are the most common form of multiple births. Occasionally, we hear of sextuplets or septuplets (six or seven newborns at once), but these are rare and almost always result from the use of fertility drugs. Even natural triplets are rare (1 in 8,000 births).

Two different processes lead to twins. **Fraternal twins** arise from the ovulation of more than one oocyte in a particular cycle (**Figure 21.4a**). The oocytes are as different genetically as any two oocytes ovulated at different cycles, and each is fertilized by a different sperm. Fraternal twins are as different as any children by the same parents. Indeed, because they are derived from different sperm, they can be of different genders.

Identical twins, sometimes called maternal twins, arise from a single zygote (**Figure 21.4b**). Recall that up until about the 8-cell stage, all cells of the developing pre-embryo are identical. If the ball of cells breaks into two groups before differentiation has begun, two complete and similar individuals may be formed. Identical twins are always of the same gender and are usually closely alike in phenotypic appearance.

🔄 **Recap** Sperm deposited in the vagina must swim through the uterus and up the correct oviduct to meet the egg. Fertilization begins when one sperm's nucleus enters the oocyte and ends when the haploid nuclei of sperm and egg fuse, creating a new diploid cell. Fraternal twins result from the fertilization of two separate eggs. Identical twins occur when a single fertilized egg divides in two before differentiation has begun. ▪

a) **Fraternal twins arise when two eggs are ovulated and fertilized in the same monthly cycle.**

b) **Identical twins arise when a zygote divides in two during development.**

Figure 21.4 Twins.

☑ An egg can occasionally break out of its zona pellucida before the sperm reach it. If no zona pellucida is present when the sperm reach the egg, what might go wrong during fertilization?

21.2 Developmental processes: cleavage, growth, differentiation, and morphogenesis

During development, rapid and dramatic changes in size and form of the zygote take place. This involves four processes:

- *Cleavage.* **Cleavage** is a series of cell divisions without cell growth or differentiation during the first four days following fertilization. Cleavage produces a ball of identical cells that is about the same size as the original zygote. Growth does not occur because the ball of cells is traveling down the oviduct at this time. With no attachment to the mother, the only energy available to the zygote is that stored in the cytoplasm of the cells plus a little glycogen found within the oviduct.

- *Growth.* Starting about the time that the developing zygote becomes embedded in the endometrial lining of the uterus and begins to receive nutrients from the mother, the organism begins to grow in size. The **growth** of a human infant from fertilization to birth is truly spectacular—from a single cell too small to be seen to over 10 trillion cells with a combined weight of 6 or 7 pounds. Every time the cells divide, the two daughter cells double in mass to prepare for the next cell division. The only time cell division is not accompanied by cell growth is the initial period of cleavage.

- *Differentiation.* After about the eight-cell stage of development, cells begin to take on specialized forms and functions, a process we know as **differentiation.** Differentiation of cells is what leads to different tissue types and ultimately to different organs and organ systems. Cell differentiation, then, is the primary cause of morphogenesis.

- *Morphogenesis.* Shortly after cells begin to differentiate, the embryo undergoes a process of change in shape and form called **morphogenesis** (*morpho* meaning "form; shape" and *genesis* meaning "origin; production"). Starting as a ball of identical cells at day four, the organism becomes (in succession) several layers of different types of cells, a pre-embryo with a tail and head, an embryo with recognizable human features, and finally a fetus with a nearly complete human form.

Pregnancy is considered to comprise three periods of development called *trimesters*. Each trimester is approximately three months long, and characteristic events in embryonic and fetal development take place during each.

↩ **Recap** The four processes associated with development are (1) cleavage, a series of cell divisions producing a ball of identical cells; (2) growth in size; (3) differentiation, as cells assume specialized forms and functions; and (4) morphogenesis, a sequence of physical changes. Pregnancy is divided into three trimesters. ∎

21.3 Pre-embryonic development: the first two weeks

The period of prenatal ("before birth") human development can be divided into three stages called *pre-embryonic*, *embryonic*, and *fetal development*. Characteristic processes and changes take place in each stage.

Soon after fertilization, the fertilized egg, or zygote, also called the conceptus (product of conception) begins to make its way down the oviduct. Throughout the pre-embryonic period, the conceptus is known informally as a **pre-embryo** because many (if not most) of the cells are destined to become part of the placenta, not the embryo. While the conceptus is still in the oviduct, a series of successive cleavages yields a ball of about 32 identical cells called the **morula** (**Figure 21.5**). The term *morula* means "little mulberry," aptly describing its spherical, clustered appearance.

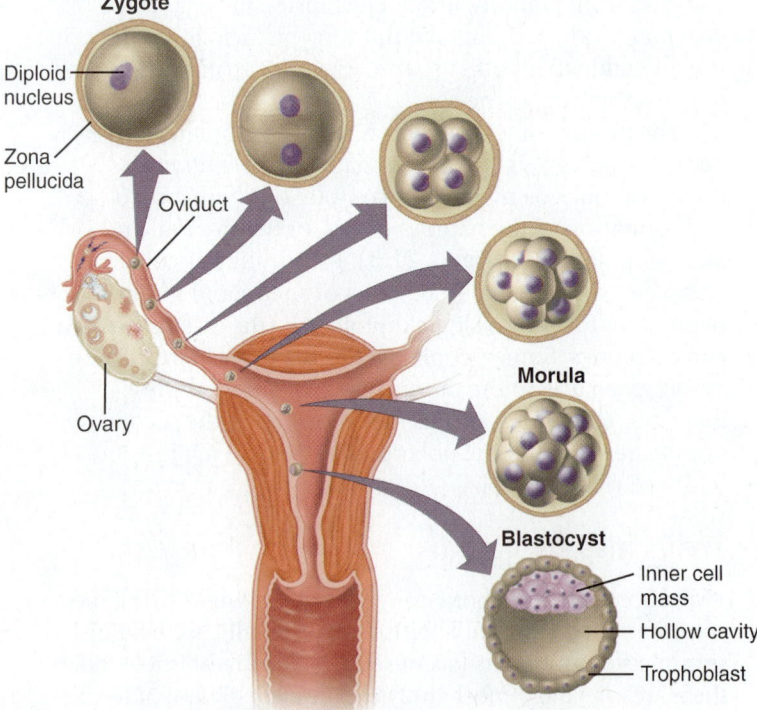

Figure 21.5 Pre-embryonic development leading up to implantation. After a series of cleavages, the morula enters the uterus and develops into a blastocyst comprising an outer layer of cells called the *trophoblast*, a hollow cavity, and an inner mass of cells. The blastocyst attaches to the uterine wall.

☑ How does cleavage differ from the normal cycles of cell division seen in adult tissues? (Hint: Compare the final size of the blastocyst with the original zygote.)

On about the fourth day, the morula is swept into the uterus, where it undergoes the first stages of differentiation and morphogenesis. Over the next several days, it becomes a **blastocyst,** a hollow ball comprising (1) an outer sphere of cells called a trophoblast, (2) a hollow central cavity, and (3) a group of cells called the inner cell mass. Only the inner cell mass is destined to become the embryo.

On about day six or seven, the trophoblast cells make contact with the endometrial lining and secrete proteolytic (protein-dissolving) enzymes that break down endometrial cells (**Figure 21.6**). ❶ The creation of a path for the blastocyst causes it to burrow inward. The process by which the blastocyst becomes buried within the endometrium is called **implantation.** At this stage, the conceptus begins to receive nutrients from the mother and starts to grow. Cells differentiate into several tissue layers, and the first stages of morphogenesis begin.

During the second week, ❷ the inner cell mass begins to separate from the surface of the blastocyst, creating a second hollow cavity that will become the amniotic cavity filled with amniotic fluid. At this point, the cell mass, now called the **embryonic disk,** ❸ differentiates into two cell types called *ectoderm* and *endoderm*. The appearance of an amniotic cavity and of ectoderm and endoderm in the embryonic disk marks the end of the pre-embryonic period.

Pre-embryonic development is a hazardous time because a lot can go wrong. Usually, a woman is not even aware that she is pregnant at this time, so she may continue risky behaviors such as smoking and drinking alcohol.

On rare occasions, the blastocyst implants outside the uterine cavity, resulting in an **ectopic pregnancy.** Most ectopic pregnancies occur in the oviduct, but they may also occur in the abdominal cavity, in an ovary, or near the cervix. Ectopic pregnancies usually are not successful because sites other than the uterus either are not large enough to accommodate a full-term baby or result in poor placental development. Occasionally, an ectopic pregnancy in an oviduct is successful, but only if the infant is delivered surgically as soon as it is able to survive with intensive care in a hospital.

↩ **Recap** During pre-embryonic development, successive cleavages yield a morula. Early stages of differentiation and morphogenesis cause the morula to become a blastocyst, which implants in the lining of the uterus. The embryonic disk is destined to become the embryo. ∎

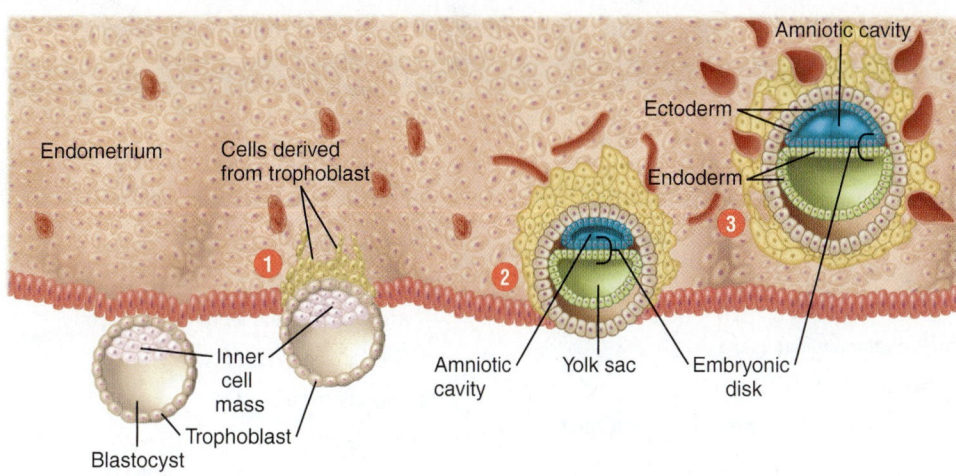

Figure 21.6 Implantation and the end of pre-embryonic development. The embryonic disk is destined to become the embryo.

✓ What does the trophoblast secrete that enables implantation to occur?

21.4 Embryonic development: weeks three to eight

From about the beginning of week three until the end of week eight, the developing human is called an **embryo.** During this time, growth, differentiation, and morphogenesis are especially rapid. All the organs and organ systems are established, though most of them are not fully functional. By the end of the embryonic period, the embryo has taken on distinctly human features but is still only 1 inch long.

Tissues and organs derive from three germ layers

The embryonic period begins when a third cell layer, called the *mesoderm*, appears between the other two layers in the embryonic disk (**Figure 21.7**). These three primary tissues,

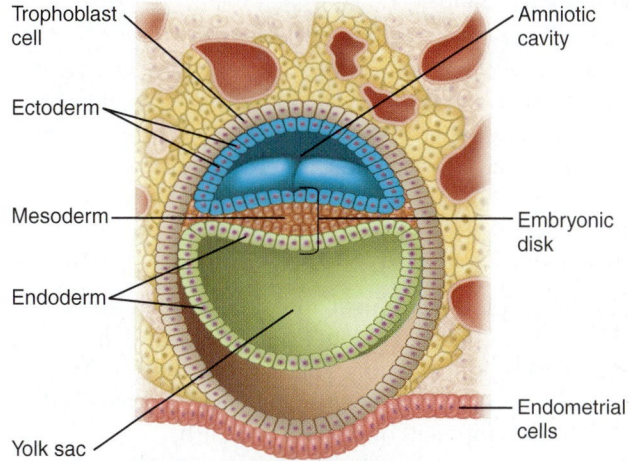

Figure 21.7 The three primary germ layers. The appearance of a third type of cell layer (mesoderm) between the ectoderm and endoderm in the embryonic disk marks the start of the embryonic period.

Table 21.1 Tissues, organs, and organ systems derived from the three primary germ layers

Ectoderm	Mesoderm	Endoderm
Epidermis of skin, including hair, nails, and glands	Dermis of skin	Epithelial tissues lining the digestive tract (except mouth and anus), vagina, bladder, and urethra
Mammary glands	All connective tissue, cartilage, and bone	Alveoli of lungs
The nervous system, including the brain, spinal cord, and all nerves	All muscle tissue (heart, skeletal, smooth)	Liver and pancreas
Cornea, retina, and lens of eye	Bone marrow (blood cells)	Thyroid, parathyroid, and thymus glands
Enamel of teeth	Kidneys and ureters	Anterior pituitary gland
Posterior pituitary gland	Testes, ovaries, and reproductive ducts	Tonsils
Adrenal medulla	Lining of blood vessels	Portions of the inner ear
Epithelial lining of nose, mouth, and anus	Lymphatic vessels	
	Adrenal cortex	

called the *germ layers*, represent the precursor (or germ) cells for the four basic tissue types (epithelial, muscle, connective, and nervous) and all organs and organ systems in the body. Differentiation and morphogenesis follow a predictable pattern, which is how we know which tissues and organs are derived from each germ layer. The three germ layers are:

- *Ectoderm.* The **ectoderm** is the outermost layer, the one exposed to the amniotic cavity. Tissues derived from the ectoderm become the epidermis of the skin, the nervous system, hair, nails, enamel of teeth, parts of the eye, and several other organs and tissues (Table 21.1).
- *Mesoderm.* The middle layer, or **mesoderm,** becomes muscle, connective tissue and bone, kidneys and ureters, bone marrow, testes or ovaries, the lining of the blood vessels, and other organs and tissues.
- *Endoderm.* The innermost layer is the **endoderm.** It gives rise to the liver and pancreas, the alveoli of the lungs, the linings of the urinary bladder, urethra and vagina, and several glands.

☑ A certain type of birth defect involves an absence of some of the nerves in various organs. In many of these patients, the skin and hair is unusually pigmented as well. Which germ layer is the likely cause of this birth defect? Explain your reasoning.

Extra-embryonic membranes

Early in embryonic development, four different extra-embryonic membranes form that extend out from or surround the embryo: the amnion, allantois, yolk sac, and chorion. Most of the components of these membranes are either resorbed during later development or discarded at birth.

The innermost layer is the **amnion,** also known as the "bag of waters" (Figure 21.8a). The amnion lines the amniotic cavity, which is filled with **amniotic fluid.** The amniotic fluid is derived from the mother's interstitial fluid and is in continuous exchange with it. Later in development when

the kidneys of the fetus form, the fetus will urinate into the amniotic fluid and the urine will be removed via exchange with the mother's blood. The amnion and amniotic fluid absorb physical shocks, insulate the fetus, and keep it from drying out.

The allantois is a temporary membrane that helps form the blood vessels of the umbilical cord. It degenerates during the second month of development.

The yolk sac forms a small sac that hangs from the embryo's ventral surface. In many species the yolk sac serves a nutritive function, but in humans that function has been taken over by the placenta, discussed shortly. Nevertheless, the yolk sac is important in humans because it becomes part of the fetal digestive tract. It also produces fetal blood cells until that job is taken over by other tissues, and it is the source of the germ cells that migrate into the gonads (testes and ovaries) and give rise to the gametes (sperm and eggs).

The outermost layer, the **chorion,** is derived primarily from the trophoblast of the early blastocyst. The chorion forms structures that will be part of the exchange mechanism in the placenta. It is also the source of **human chorionic gonadotropin (hCG),** a hormone that supports pregnancy for the first three months until the placenta begins producing enough progesterone and estrogen. The presence (or absence) of hCG in the urine is the basis of most home pregnancy tests.

The placenta and umbilical cord

As the embryo grows and develops, its relationship with the mother becomes more complex. The embryo cannot supply itself with nutrients or get rid of its own wastes, so exchange vessels form between the embryo and the mother for the exchange of nutrients and wastes without direct mixing of their blood. This is the function of the placenta and the umbilical cord (Figure 21.8b). The **placenta** is the entire structure that forms from embryonic tissue (chorion and chorionic villi) and maternal tissue (endometrium). The **umbilical cord** is the two-way lifeline that connects the placenta to the embryo's circulation.

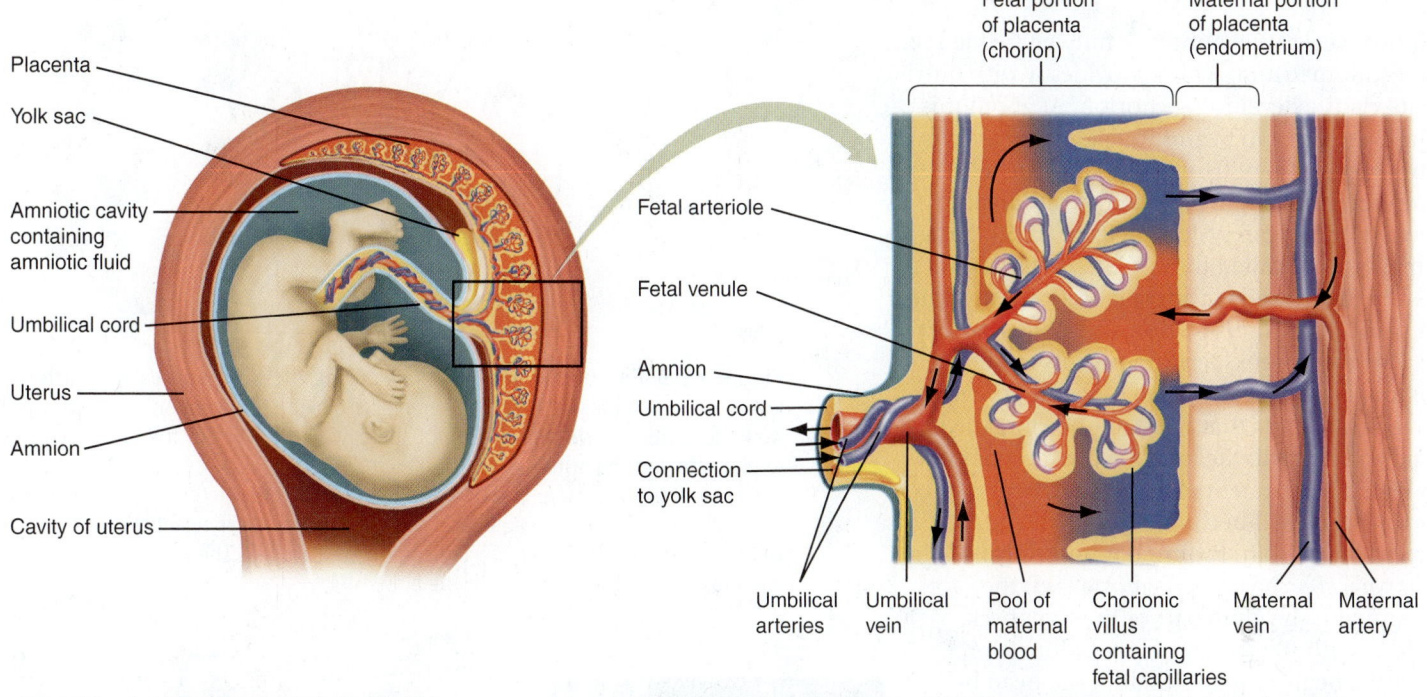

a) **The fetus is bathed in amniotic fluid within the amnion.** Its only connection to the mother is the umbilical cord.

b) **A closer view of portions of the placenta and umbilical cord, showing how nutrients and gases are exchanged between maternal and fetal blood without mixing.**

Figure 21.8 The placenta and umbilical cord.

The placenta develops because the cells of the chorion, just like the cells of the trophoblast from which they derive, secrete enzymes that eat away at the endometrial tissues and capillaries in the vicinity of the embryo. These enzymes rupture the capillaries, causing the formation of small blood-filled cavities. The developing chorion extends fingerlike structures called *chorionic villi* into these pools of maternal blood. Each chorionic villus contains small capillaries that are connected to umbilical arteries and veins that are part of the circulation of the fetus. In other words, the chorion damages the endometrium to cause local bleeding, and then it taps that bleeding as a source of nutrients and oxygen for the embryo and as a place to get rid of embryonic wastes, including carbon dioxide. In effect, an embryo behaves like a parasite within the mother.

The placenta is an effective filter. It permits the exchange of nutrients, gases, and antibodies between the maternal and embryonic circulations but not the exchange of large proteins or blood cells. The placenta allows the embryo to take advantage of the mother's organ systems until its own organs develop and become functional.

However, the placenta may permit certain toxic substances and agents of disease to cross over to the fetus as well. Examples include alcohol, cocaine, the HIV virus, and a wide variety of prescription and nonprescription drugs. Many of these substances can do a great deal of harm to the embryo when it is in the earliest stages of differentiation and morphogenesis, even though some of the prescription drugs may be therapeutic for

the mother (review Chapter 17). Upon learning she is pregnant, a woman should review all her prescription and nonprescription drugs with her physician.

The placenta is also an endocrine organ. Initially, the placenta secretes hCG that signals the corpus luteum, the remnant of the follicle that remains after ovulation, to continue its secretion of the progesterone and estrogen necessary to maintain the pregnancy (Chapter 16). Later, the placenta secretes its own progesterone and estrogen, and the corpus luteum regresses. Progesterone and estrogen promote the growth of the myometrium in preparation for intense contractions at birth, maintain the endometrial lining so menstruation does not occur, inhibit uterine contractions during pregnancy, and help form a thick mucous plug over the cervix to inhibit uterine infections.

The two umbilical arteries and single umbilical vein are considered part of the embryonic circulation, meaning the umbilical vein carries blood *toward* the embryonic heart and the umbilical arteries carry blood *away* from the embryonic heart and back to the placenta. Because the exchange of nutrients by the placenta requires a functional embryonic heart, the placenta and umbilical vessels do not become fully functional until the embryonic heart develops at about five weeks. Until then, the developing chorion supplies the nutrient needs of the embryo by diffusion.

✔️ Name the four extra-embryonic membranes and state the major function of each. Which one plays the largest role in the formation of the placenta?

The embryo develops rapidly

At two weeks, the embryo is fully embedded in the endometrium, which provides it with nutrients via the developing chorion. At about this time, the embryo begins to take shape as cells of the three primary tissue layers migrate to other locations and begin to form the rudimentary organs and organ systems (**Figure 21.9**). First, a small groove called the *primitive streak* appears in the flat, round embryonic disk, and the embryonic disk begins to elongate along one axis.

At three weeks, certain precursors to embryonic structures emerge. A neural groove of ectoderm forms that will later become the brain and spinal cord. Meanwhile, the mesoderm begins to separate into several segments called *somites* that will become most of the bone, muscle, and skin. Prominent bumps called the *pharyngeal arches* appear at one end; they are destined to become part of the face, neck, and mouth. By the end of week four, the heart is beginning to develop, the head begins to take shape, the position of the eyes becomes apparent, the neural groove has closed into a neural tube, and four limb buds and a tail appear (**Figure 21.10**).

The patterns of cell differentiation and organism morphogenesis are so consistent among vertebrates that even an expert would have a hard time telling whether a four-week-old embryo of unknown origin was destined to become a fish, a mammal, or a bird. Because early differentiation and morphogenesis follow such a common path, it is likely that these patterns developed early in the evolutionary history of the vertebrates.

Weeks five through eight mark the transition from a general vertebrate form to one that is recognizably human (**Figure 21.11**). The head grows in relation to the rest of the body, the eyes and ears are visible, and the four limbs are formed with distinct fingers and toes. A cartilaginous skeleton forms. The heart and circulatory system complete their development, and the umbilical cord and placenta become functional, so that now blood circulates throughout the fetus. However, the blood cells for the fetal blood are produced by the yolk sac because the blood-forming tissues of the fetus are not yet mature. Nutrients and wastes are now exchanged more efficiently and in greater quantities. The tail regresses. At the end of the embryonic period, the embryo is an inch long and weighs just 1 gram.

Although it is difficult to estimate how many embryos fail to survive to the end of the embryonic period, some estimates place the number as high as 20%. Spontaneous termination of pregnancy followed by expulsion of the embryo is called a **miscarriage,** or spontaneous abortion.

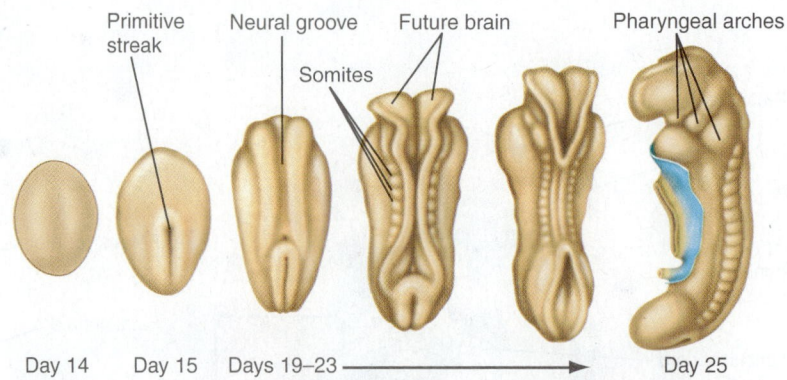

Day 14 Day 15 Days 19–23 ⟶ Day 25

Figure 21.9 Embryonic development during the third and early fourth week. Day 14 represents the flat embryonic disk; the illustration shows how it would look from above (from the amniotic cavity). On day 15, the embryo begins to elongate and the primitive streak of ectoderm appears. Days 19–23 mark the appearance of a neural groove of ectoderm. Somites develop in the mesoderm that will become bone, muscle, and skin. By day 25, pharyngeal arches that will contribute to structures of the head become visible. The illustration of the embryo on day 25 is a view from the side.

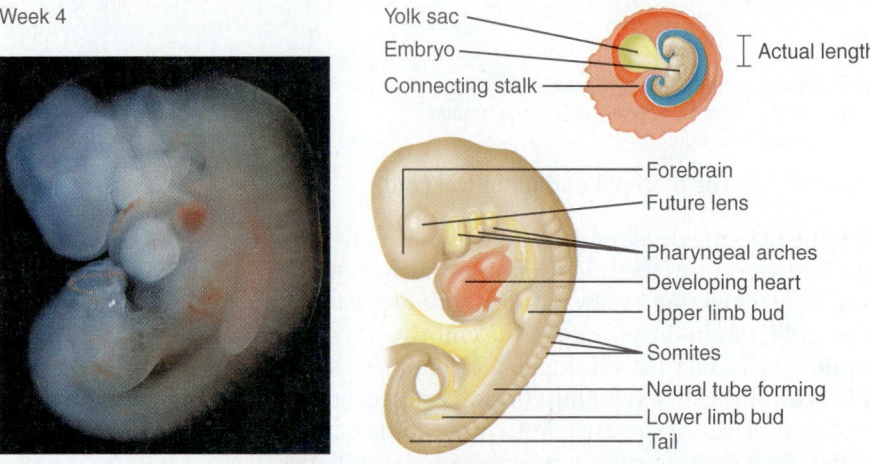

Figure 21.10 Week four of development.

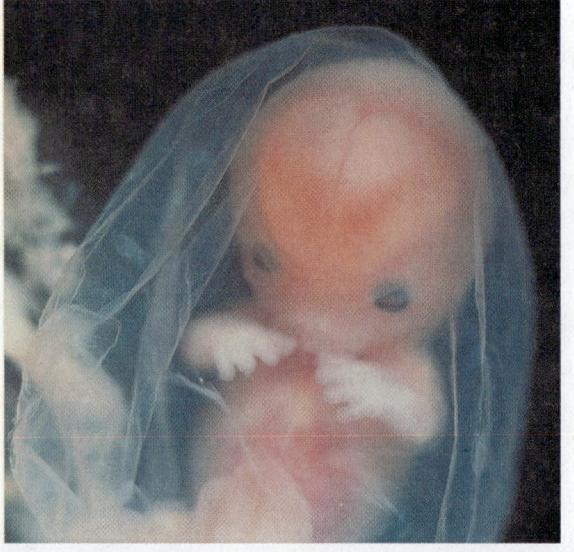

Figure 21.11 Week eight of development. The embryo is now distinctly human in appearance.

Sometimes a woman never even realizes she was pregnant. Miscarriages are probably nature's way of weeding out embryos with genetic disorders that might prevent them from developing normally.

> ↩ **Recap** By the beginning of embryonic development, the embryo comprises three primary germ layers called *ecto-derm*, *mesoderm*, and *endoderm* that ultimately give rise to fetal tissues and organs. Four extra-embryonic membranes (amnion, allantois, yolk sac, and chorion) serve varying supportive functions. The placenta exchanges nutrients and gases between embryo and mother and secretes hormones. The umbilical cord joins the embryo to the placenta. By the fifth week, the embryo is becoming distinctly human in form, and by eight weeks, it is 1 inch long. ∎

21.5 Gender development

Male and female external and internal genitalia do not begin to develop until about six weeks. Until that time, the embryo remains "sexually indifferent," meaning that it still has the capacity to develop into either a male or a female. However, the final outcome (male or female) is not in doubt; it was determined at the moment of conception by the single sperm that fertilized the egg. The development of either male or female external genitalia is shown in **Figure 21.12**. By six weeks, an undifferentiated urogenital groove has developed, topped by a bud and surrounded by labioscrotal swellings.

After six weeks, the presence of a Y chromosome causes the embryo to start developing male sexual characteristics. The process begins when a gene on the Y chromosome called *SRY* (for sex-determining region Y) is switched on. SRY encodes for a protein called *testis-determining factor* that directs the initial development of the testes internally. Shortly thereafter, the testes begin to secrete testosterone, which in turn, stimulates the development of the male internal ducts and external genitalia. (The embryonic testes also secrete a second hormone called *anti-Mullerian hormone* that suppresses the development of the female internal and external genitalia.) In males, the undifferentiated urogenital groove closes up to become the urethra within the penis. The penis elongates, and the previously undifferentiated bud becomes the head of the penis. The urogenital swellings develop into the scrotum into which the testes later descend.

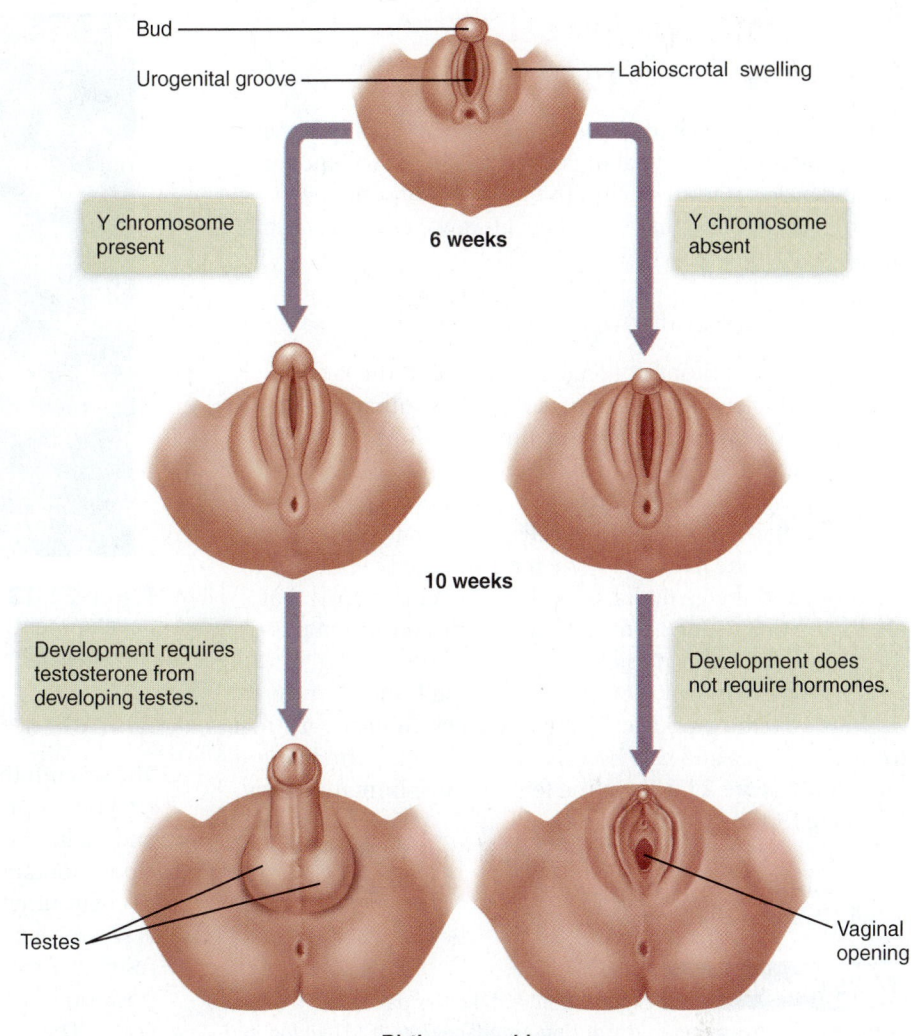

Figure 21.12 **Development of male or female external genitalia.**

☑ Occasionally an embryo has two X chromosomes and a Y chromosome (three sex chromosomes in total). What will happen to this embryo during sexual differentiation? Explain your answer.

In the absence of a Y chromosome, female internal and external genitalia develop and the embryo becomes a female. The urogenital groove expands to become the vagina, and the undifferentiated bud becomes the clitoris. Meanwhile the ovaries, fallopian tubes, and uterus all develop internally. The development of female internal and external genitalia is the "default" condition; it occurs in the absence of any hormones, whenever a Y chromosome is *not* present. In other words, gender is determined by the presence or absence of a Y chromosome, not by whether the embryo has one or two X chromosomes.

> ↩ **Recap** Gender development begins at about 6 weeks. The presence of a Y chromosome signals the embryo to develop into a male; the absence of a Y chromosome causes the embryo to develop into a female. ∎

21.6 Fetal development: nine weeks to birth

After the eighth week, the developing human is called a *fetus*. This is the time when fetal organ systems begin to function and the bones start to calcify. The mother, too, undergoes obvious changes in physical shape. Next, we examine each of these stages in detail.

Months three and four

During the third month of fetal development, the kidneys develop sufficiently for the fetus to begin eliminating some wastes as urine into the amniotic fluid. The limbs are well developed, the cartilaginous skeleton begins to be replaced with bone, and the teeth form. The spleen participates briefly in the production of red blood cells, a job that will soon be taken over by the liver and bone marrow. The liver begins to function, and the genitalia are well enough developed that sex can be determined. The end of the third month marks the end of the first trimester.

During the fourth month, the liver and bone marrow begin producing blood cells. The face takes on nearly its final form as the eyes and ears become fixed in their permanent locations (**Figure 21.13**). In the female fetus, immature eggs (ova) are forming in the ovaries.

Overall growth is rapid during this time. By the end of the fourth month, the fetus is already 6 inches long and weighs about 170 grams (6 ounces).

MJ's BlogInFocus Should women take the painkiller acetaminophen during pregnancy? Visit MJ's blog in the Study Area in MasteringBiology and look under "Acetaminophen and Pregnancy."

http://goo.gl/uWlaqe

Months five and six

By the fifth month, the nervous system and skeletal muscles are sufficiently mature for the fetus to begin moving. These movements are called *quickening,* and the mother may feel them for the first time. The fetus's skin, which is well formed although red and wrinkled, is covered by soft, downy hair. Its heartbeat is now loud enough to be heard with a stethoscope. Skeletal hardening in the fetus continues.

The sixth month marks the first point at which the fetus could, with the best neonatal care available, survive outside the uterus. At this stage, the fetus seems to respond to external sounds. Most importantly, the lungs begin to produce surfactant, the lipoprotein that reduces surface tension in the lungs and permits the alveoli to fill with air at birth.

The end of the sixth month marks the end of the second trimester. At this point, the fetus weighs nearly 700 grams, or approximately 1½ pounds.

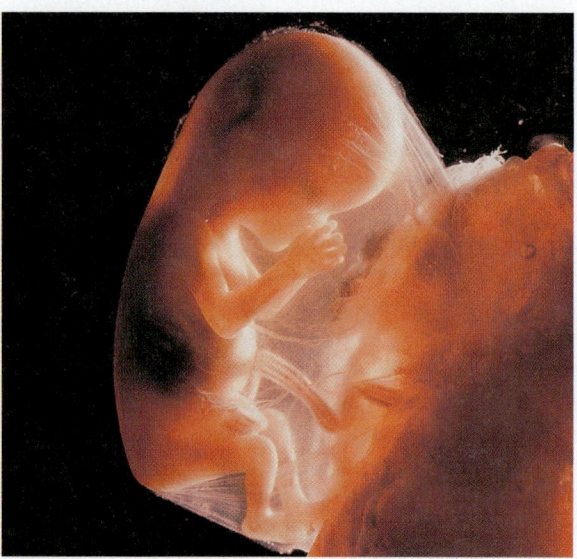

Figure 21.13 The fetus and placenta at four months of development. Note the amnion and the clear amniotic fluid surrounding the fetus. Actual length is 6 inches.

Months seven through nine

The seventh through ninth months (the third trimester) of fetal development are a period of continued rapid growth and maturation in preparation for birth. The eyes open and close spontaneously and can be conditioned to respond to environmental sounds. Activity increases, as if the fetus is seeking a more comfortable position in the increasingly restrictive space of the uterus. Usually, the fetus moves to a position in which the head is positioned downward, near the cervix. The skin begins to lose some of its reddish color and its coat of downy hair. In the male, the testes descend into the scrotal sac. Although neither the lungs nor the digestive systems have had a chance to function yet, both are now ready.

By nine months, the fetus is about 20 inches long and weighs approximately 6–7½ pounds. Birth usually occurs at about 38 weeks of development.

Recap The period of fetal development extends from nine weeks to birth at 38 weeks. Growth is rapid, with the mature fetus weighing approximately 6–7½ pounds at birth. The fetus begins to move at about five months, and life outside the womb is at least possible by about six months when the lungs begin to produce surfactant. ■

21.7 Birth and the early postnatal period

Other than perhaps conception and death, there is no period in our lives marked by greater developmental change than the hours during and immediately after birth. Within minutes, the newborn makes the shift from relying solely on the maternal circulation for nutrient and gas exchange to depending on its own circulation and lungs. The digestive tract, too, starts to function with the first swallow of milk.

Let's take a quick look at the changes that occur during birth and the postnatal ("after birth") period.

Labor ends in delivery

Birth involves a sequence of events that we call labor (it's hard work!), which ends in the delivery of the newborn into the world. Labor begins as a result of a series of events that is triggered by maturation of the fetal pituitary gland, which serves as the timing device to indicate that the fetus is now ready for birth.

The mature fetal pituitary gland begins to release ACTH, which stimulates the fetal adrenal gland to secrete steroid hormones that cause the placenta to increase its production of estrogen and decrease the production of progesterone. Estrogen, in turn, increases the number of oxytocin receptors and stimulates production of prostaglandins. Together, the increased estrogen, prostaglandins, and oxytocin receptors (along with the mother's oxytocin) stimulate the uterus to contract rhythmically.

A positive feedback cycle begins: rhythmic contractions of the uterus cause the release of still more oxytocin from the maternal pituitary, which increases contractions still further, and so on. As labor progresses, the periods of contraction get closer together, last longer, and become stronger, creating enough force to push the fetus toward the cervix and eventually through the vagina.

The period of labor and delivery lasts about 24 hours for the first birth and slightly less time for subsequent births. Labor and delivery are divided into three phases: dilation, expulsion, and afterbirth (**Figure 21.14**).

- *Stage 1—dilation.* The first phase of labor can last 6–12 hours. During this time, the rate, duration, and strength of contractions increase over time, pushing the head of the fetus against the cervix. The cervix itself is drawn back toward the uterus, widening the cervical opening and expelling the mucus plug. Continued pressure of the fetus against the cervix widens the cervical opening even further until eventually it is large enough to

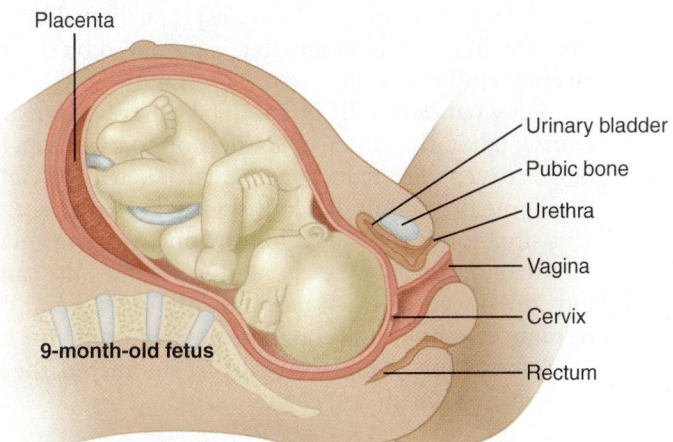

a) 9-month-old fetus. As birth approaches, the fetus usually is positioned with the head down and toward the cervix.

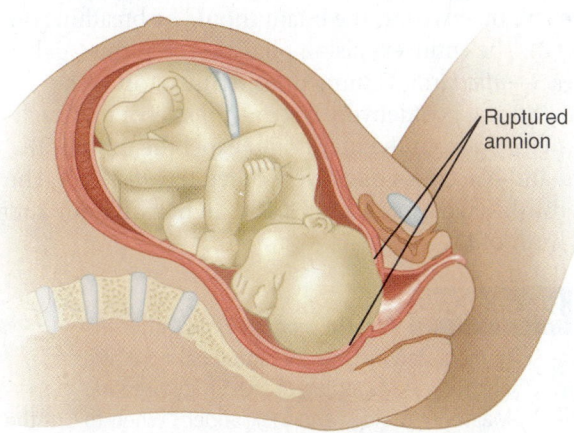

b) Stage 1: dilation. The cervical opening widens. The amnion may break at this stage.

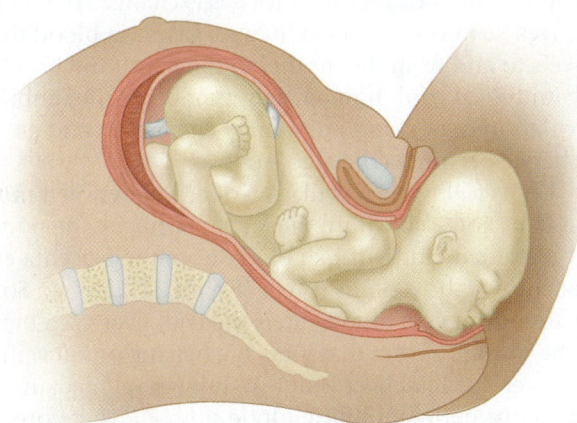

c) Stage 2: expulsion. The fetus passes headfirst through the cervical canal and the vagina.

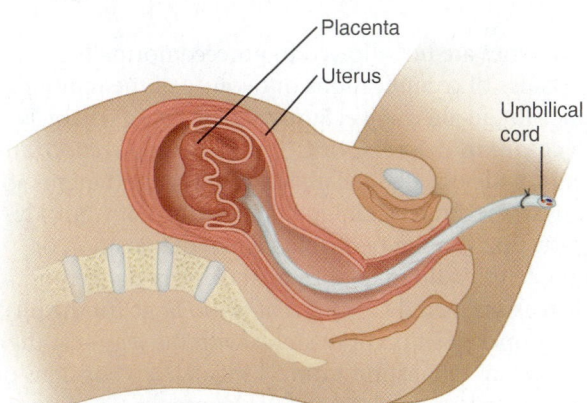

d) Stage 3: afterbirth. The placenta detaches from the uterus and is expelled along with the remainder of the umbilical cord.

Figure 21.14 The stages of birth.

accommodate the baby's head, or about 10 centimeters. At about this time, the pressure of the baby's head ruptures the amnion, releasing the amniotic fluid. This "breaking of the water" is a normal sign that delivery is proceeding. The physician keeps a close watch on the stage of dilation by measuring the degree of cervical dilation at regular intervals.

- *Stage 2—expulsion.* The period of expulsion extends from full cervical dilation through actual delivery. During this time, uterine contractions strengthen and the woman experiences an intense urge to assist the expulsion with voluntary contractions as well. To make the birth easier, some women opt for a surgical incision called an *episiotomy* to enlarge the vaginal opening. The intense contractions push the baby slowly through the cervix, vagina, and surrounding pelvic girdle. The appearance of the baby's head at the labia is called *crowning*. As soon as the baby's head appears fully, an attendant removes mucus from the baby's nose and mouth to facilitate breathing. The rest of the body emerges rather quickly. When the baby has fully emerged, the umbilical cord is clamped and then cut. By this time, the infant should be breathing on its own. The entire expulsion phase lasts less than an hour.

- *Stage 3—afterbirth.* Contractions do not stop with birth. Strong postdelivery contractions serve to detach the placenta from the uterus and expel the umbilical cord and placenta, collectively called the *afterbirth*. The afterbirth stage usually lasts less than half an hour after the birth of the infant.

MJ's BlogInFocus Is there a better time (other than immediately) to cut the umbilical cord after birth? Visit MJ's blog in the Study Area in MasteringBiology and look under "When to Cut the Umbilical Cord."

http://goo.gl/Y1fPYv

Cesarean delivery: surgical delivery of a baby

Some deliveries are not allowed to proceed normally, either because of complications or because of the mother's preference for an alternative. Surgical delivery of a baby is called a **cesarean delivery,** or C-section. In a C-section, an incision is made through the abdominal wall and uterus so that a physician can reach in and remove the baby quickly.

A C-section is often performed when the position or size of the fetus could make a vaginal delivery dangerous. Possible reasons include a fetus that is too large for the birth canal, improper position of the fetus with the legs near the cervix, maternal exhaustion from a long delivery, or signs of fetal distress such as an elevated fetal heart rate.

In the past, if a C-section was performed, it was common practice to do all subsequent deliveries by C-section as well. Today, most women who have a C-section can still opt for a vaginal delivery with the next child if they wish.

The transition from fetus to newborn

Somewhere between the point when its head emerges and the umbilical cord is clamped, the fetus must become capable of sustaining life on its own. The first necessity is for the newborn (also called the *neonate* for the first 28 days) to start breathing, preferably even before its umbilical connection to the mother is severed. Second, over the first few days of life, its cardiovascular system undergoes a series of remarkable anatomical changes that reflect the loss of the placental connection for gas and nutrient exchange. The newborn is now entirely on its own.

Taking the first breath When the infant emerges, the three hundred million alveoli in its lungs have never been inflated. After all, the infant essentially has been living underwater (more correctly, under amniotic fluid). The first inflation is critical and not all that easy. If you've ever tried to blow up a really stiff balloon, you know that it takes a lot of effort to start the inflation. The first inflation of the alveoli (representing small spheres) is facilitated by the lipoprotein called *surfactant* that is produced by some alveolar epithelial cells.

But what causes the infant to take that first breath? During labor, the placental connection to the mother begins to separate, reducing gas exchange with the fetus. Clamping the umbilical cord stops gas exchange entirely. This situation is equivalent to someone covering your mouth and nose completely. Within seconds, the carbon dioxide concentration in the fetus rises to the point that the respiratory centers in its brain stimulate respiration. With the enormous effort of one who is being smothered, the infant takes its first breath and begins to cry.

Changes in the cardiovascular system The fetal cardiovascular system is different from the cardiovascular system of an adult. While the fetus is still in the womb, the fetal lungs are not yet useful for gas exchange. The fetus receives its nutrients and exchanges gases via blood that travels to and from the mother to the fetus (and back) in the umbilical cord. Immediately after birth, the umbilical circulation ceases, and pulmonary circulation and gas exchange becomes crucial to survival.

Let us follow the circulation of blood between mother and fetus and also within the fetus, starting at the point that oxygen-laden, nutrient-rich blood from the placenta enters the fetus via the umbilical vein (**Figure 21.15a**). ❶ Some of the umbilical vein blood enters the fetal liver via a branch of the umbilical vein that joins the hepatic portal vein. However, the fetal liver is not yet fully developed and cannot handle the entire umbilical blood flow. ❷ Therefore, most of the blood bypasses the liver and is carried directly to the inferior vena cava by the ductus venosus. ❸ In the inferior vena cava, the nutrient-rich blood mixes with the venous blood of the fetus.

Most of the blood that enters the fetal heart must bypass the fetal lungs because they too are not fully developed. ❹ Some of the blood passes from the right atrium to the left atrium through the foramen ovale, and ❺ some is shunted from the pulmonary artery directly to the aorta via the ductus arteriosus.

Even though the fetal lungs are not yet functional and the digestive tract is not receiving nutrients, the aortic blood still has sufficient oxygen and nutrients (brought to the fetus via the umbilical vein) to supply all the fetal tissues. ❻ Some of the fetal arterial blood returns to the placenta via the umbilical arteries. In the placenta, the blood again picks up nutrients and oxygen and gets rid of carbon dioxide generated by fetal metabolism.

The situation changes dramatically at birth, because the umbilical circulation is immediately cut

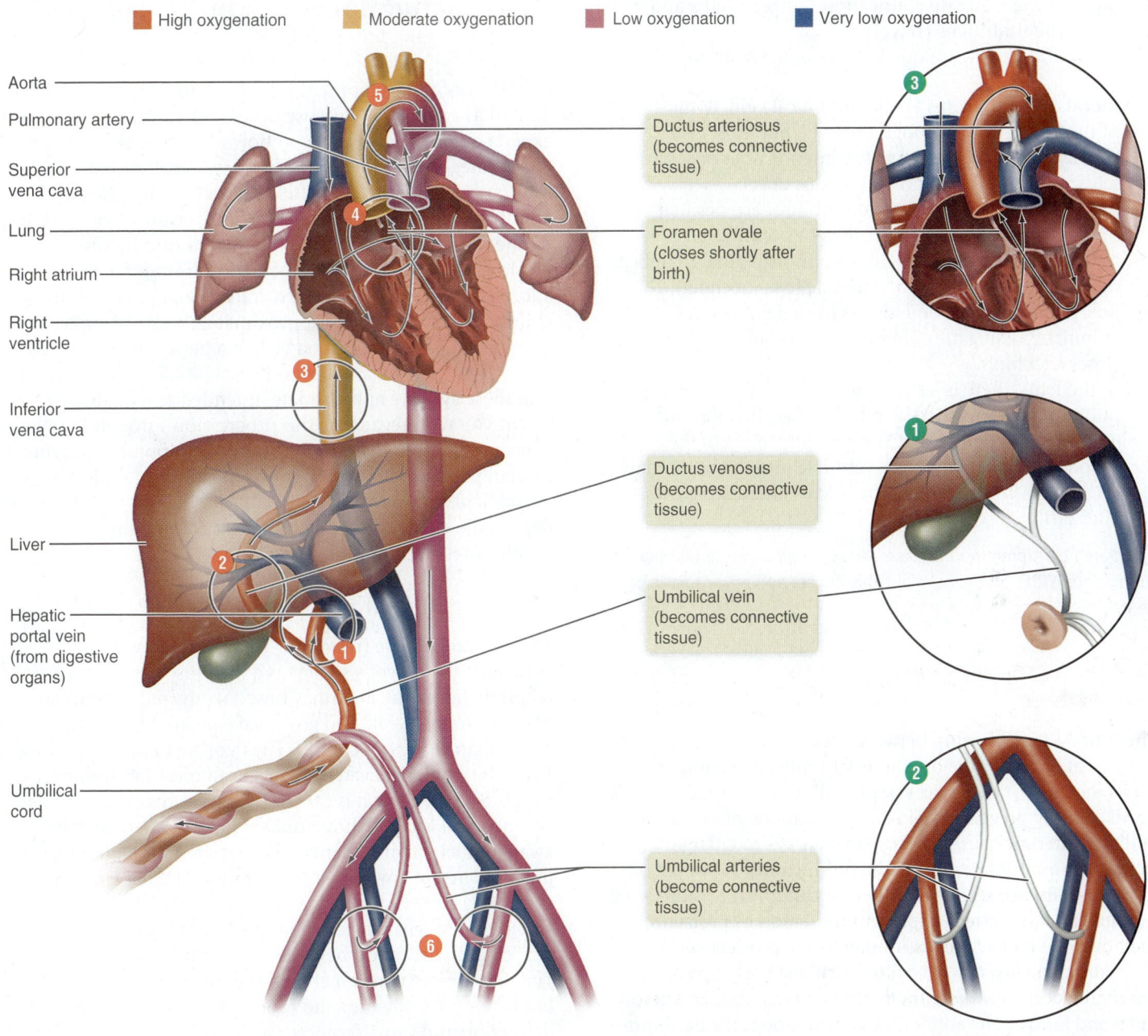

■ High oxygenation ■ Moderate oxygenation ■ Low oxygenation ■ Very low oxygenation

Aorta
Pulmonary artery
Superior vena cava
Lung
Right atrium
Right ventricle
Inferior vena cava
Liver
Hepatic portal vein (from digestive organs)
Umbilical cord

Ductus arteriosus (becomes connective tissue)
Foramen ovale (closes shortly after birth)
Ductus venosus (becomes connective tissue)
Umbilical vein (becomes connective tissue)
Umbilical arteries (become connective tissue)

a) Fetal circulation.

b) Circulation after birth.

Figure 21.15 The cardiovascular systems of the fetus and of the newborn. In the fetus, the immature lungs are not used for gas exchange, so most of the blood returning to the heart bypasses the lungs by flowing through either the foramen ovale or the ductus arteriosus. Gas exchange occurs via the placenta, so special fetal blood vessels connect the fetus to the placenta. These unique pathways for blood circulation in the fetus close shortly after birth.

✔ What would happen if either the ductus arteriosus or the foramen ovale did not close after birth?

off (Figure 21.15b). ❶ The umbilical vein, the ductus venosus, and ❷ the umbilical arteries all regress to become vestigial connective tissue. ❸ The foramen ovale and ductus arteriosus close off in the first few days or weeks after birth as well, so that all the cardiac output passes through the lungs for efficient gas exchange. Nutrients are now absorbed by the newborn's digestive tract and pass through the liver via the hepatic portal vein. The only reminder of the intimate connection between fetus and mother is the umbilicus (navel).

To summarize the unique features of the fetal circulation:

- Blood enters the fetus via the umbilical vein. Some of the blood flows through the liver, but most of it bypasses the liver and joins the inferior vena cava of the fetus via the ductus venosus (venous duct).
- Blood leaves the fetus via two umbilical arteries that originate from arteries in the lower extremities.
- In the fetus, there is a hole between the two atrial chambers of the heart called the foramen ovale (oval opening). The hole permits some blood to pass from right atrium to left atrium, bypassing the pulmonary circulation. The hole closes after birth.
- In the fetus, there is a shunt, or shortcut, from the pulmonary artery directly to the aorta called the *ductus arteriosus* (arterial duct). Consequently, most of the blood pumped by the right ventricle into the pulmonary artery bypasses the lungs. The ductus arteriosus closes after birth.

☑ Infants born prematurely often have not yet begun to produce surfactants. What problems might this cause, and how could it be treated?

Lactation produces milk to nourish the newborn

The intimate relationship between mother and child does not end at birth, for the human infant must rely entirely on its mother (or someone else) for all its nutritional needs. During pregnancy, high concentrations of estrogen and progesterone cause the woman's breasts to enlarge in preparation for lactation (milk production). However, the breasts do not secrete milk before childbirth because estrogen and progesterone prevent the action of prolactin, the hormone that actually stimulates milk production.

During the first day or so after birth, the breasts produce a watery milk called **colostrum** that is rich in antibodies but low in fat and lactose. The antibodies help to protect the newborn until its own immune system matures. Later, the high fat and lactose content of milk help the infant to grow rapidly.

A fourth hormone, oxytocin, is responsible for the contractions that deliver colostrum or milk only when needed. Oxytocin is released during childbirth and also as a result of a neural reflex every time the infant nurses. The expulsion of milk from the breast during suckling is called *milk ejection.*

↺ **Recap** The period of labor and delivery is composed of three stages: dilation, expulsion, and afterbirth. The dilation phase may last 6–12 hours. At birth, a sharp increase in carbon dioxide causes the newborn to take the first breath. Shortly after birth, anatomical changes in the newborn route all blood through the lungs. In the mother, prolactin stimulates lactation and oxytocin stimulates the release of milk during suckling. ■

21.8 Maturation: from birth to adulthood

The process of development from birth to adulthood is known as *maturation*. Because development is a continuum, experts do not always agree on how to categorize its stages. Some categories of life's stages may seem somewhat arbitrary, such as the distinction (if there is one) between adulthood and old age. At the risk of oversimplifying, we define the neonatal period as the first month, infancy as months 2–15, childhood as the period from infancy to adolescence, adolescence as the transitional period from childhood to adulthood (approximately 12 to 18 or 20 years of age), and adulthood (when a person is considered mature) as 20 years and beyond (Table 21.2). Bear in mind that these ages are approximate, intended as a guide to orient you with regard to your progression through life. We concentrate primarily on biophysical development, leaving human psychosocial and cognitive (learning) development to the fields of psychology and education. Figure 21.16 depicts the changes in body form that accompany human development, maturation, and aging.

The neonatal period: a helpless time

The human neonate (newborn) is fairly helpless. In particular, the nervous system and the muscular system are relatively immature at birth. Consequently, movements are uncoordinated, weak, and governed primarily by reflexes rather than conscious control. The neonate hardly has enough strength to hold its head up, so its head must be constantly supported whenever it is carried. The eyes are open, but often, they seem to wander about, unable to focus. Neonates are aware of their environment and can respond to it, but they are unable to retain any long-term memories. (What is *your* first memory?) They hear a narrower range of pitches than adults and are more sensitive to higher-pitched tones.

What neonates do best is suckle, urinate, and defecate. The digestive system can absorb the constituents of milk but is not yet ready for solid foods. The stools are soft, and defecation and urination occur as reflexes.

Infancy: rapid development and maturation of organ systems

Infancy is a time of rapid change. In just 15 months, the infant more than triples in weight to about 22–24 pounds. The bones harden gradually, and there is

Table 21.2 A summary of the stages of human maturation

Stage	Age
Prenatal (after fertilization)	

Pre-embryo
    Zygote
    Morula
    Blastocyst
	From fertilization to 2 weeks
	Days 1–3
	Days 4–7
	Days 8–14

Embryo	Weeks 3–8
Fetus	Weeks 9–40 (birth)
Postnatal (after birth)	
Neonate	The first month
Infant	Months 2–15
Child	16 months to 12 years
Adolescent	12–20 years
Adult	Beyond 20 years

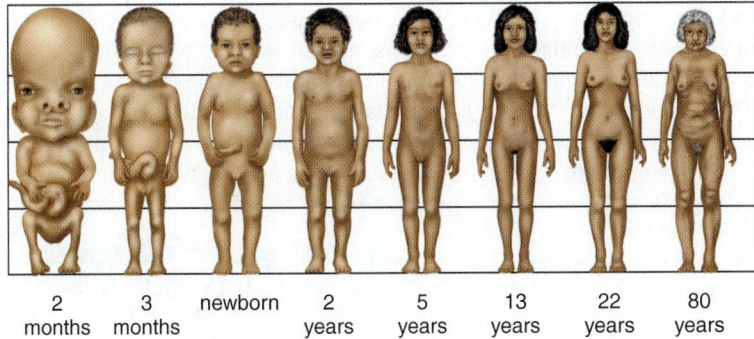

| 2
months | 3
months | newborn | 2
years | 5
years | 13
years | 22
years | 80
years |

Figure 21.16 Changes in body form and relative proportion throughout life.

a disproportionately large increase in muscle mass and strength. The brain also grows rapidly, with nearly half of all brain growth after birth occurring in infancy. Most of the growth occurs in the cerebral cortex, the area associated with sensory perception, motor function, speech, and learning. Most of the myelination of nerves occurs in infancy as well. The first teeth appear at about 6 months, and all 20 baby teeth are generally present by about 1 year. Most infants begin eating solid foods by about 6 months to a year.

By about 14 months, the combined maturation of the musculoskeletal and nervous systems allows the human infant to begin to walk on its own. A rapidly growing body needs lots of rest, and infants sleep a lot. Infants can easily sleep 10 hours a night and still take several naps during the day.

The immune system consistently lags behind most other systems in its development. Most vaccinations are ineffective in infants because their immune systems cannot produce the appropriate antibodies. The schedule of vaccinations in infants and children is designed so that each vaccine is given only after the immune system becomes able to respond to that vaccine.

Childhood: continued development and growth

The long period of childhood, from about 15 months to 12 years of age, involves continued growth of all systems. The brain grows to 95% of its final size during childhood. The immune system continues its slow process of maturation throughout childhood and even into adolescence. Body weight increases to an average of about 100 pounds in both boys and girls as the result of periods of slow growth interspersed with growth spurts. Both muscle strength and fine motor coordination improve.

In addition, body form alters as the long bones of the arms and legs lengthen. A lumbar curve develops in the small of the back, and abdominal musculature strengthens, so the childhood "pot belly" disappears.

By the end of childhood, the body form is distinctly adultlike, though still sexually immature. Most organ systems are fully functional or nearly so (although not necessarily their full adult size). The lone exceptions are the male and female reproductive systems, which have not yet matured.

Adolescence: the transition to adulthood

One of the most challenging times of life is the transition from childhood to adulthood. The final growth spurt, one of the largest, occurs in adolescence. The growth spurt begins at different times in different individuals, so there is a wide range of normal body weights and degrees of sexual maturity in the 12- to 14-year-old age group. Rapid growth can lead to awkwardness and a temporary loss of coordination.

Both the skeletal and muscular systems experience their greatest rates of growth during adolescence. Other organs increase in size as well. Notably, the lungs more than double in mass, and the kidneys and stomach increase by more than 50%. The brain, however, gains only about 5%.

Adolescence is marked by maturation of the reproductive systems and the human sexual response, an event known as **puberty.** Puberty is defined as the onset of menstrual periods (menarche) in the female at some time between 9½ and 15½ years and by reflex discharge of semen during sleep (nocturnal emissions) in the male sometime between 11½ and 17½ years.

The concentrations of all the reproductive hormones are low in children, but they begin to rise during the period of sexual maturation (known as the *pubertal period*). The pubertal period is initiated by maturation of certain neurons of the hypothalamus, which begin to secrete a releasing hormone called gonadotropin releasing hormone (GnRH). The initiation of GnRH secretion by the hypothalamus stimulates the secretion of luteinizing hormone (LH) and follicle-stimulating hormone (FSH) from the anterior pituitary, which in turn stimulate the production of sex hormones and the maturation of sex organs and secondary sexual characteristics.

The brain is one of the last organs to reach full maturity. The areas of the cerebral cortex that regulate impulse and emotion don't fully mature until about 18 years of age. This

HEALTH & WELLNESS
Prenatal Diagnostic Techniques

Fifty years ago, when a woman became pregnant there was little that she, the father, or a physician could do except hope that the baby would be born healthy. Late in pregnancy, a physician could hear the fetal heartbeat, but that was about it. Given that at the time nearly 7% of all newborns had a birth defect of some kind, pregnancy was a stressful period of wait-and-see. Today, fetal cells can be collected for genetic testing fairly early in fetal development, even early enough to permit a couple to terminate a pregnancy if that is their wish after reviewing the test results.

In **amniocentesis,** a needle is inserted through the abdominal wall and a sample of amniotic fluid containing fetal cells is collected. The harvested fetal cells are then grown in the laboratory until there are enough of them to examine the chromosomes and perform certain biochemical tests. Amniocentesis is relatively safe and may be recommended if there is a family history of a genetic disease or the risk of Down syndrome. Nearly 40 different fetal defects can be diagnosed using amniocentesis. A drawback is that amniocentesis cannot be performed until about the 15th week of gestation, and test results may not be available until the middle of the second trimester. This is close to the time that termination of pregnancy will no longer be an option.

In **chorionic villi sampling,** a thin, flexible tube attached to a syringe is inserted through the vagina and into the uterus, and a small sample of chorionic villi tissue is collected. Chorionic villi tissue originates from the fetus (review Figure 21.8), so most of the cells are fetal cells. The great advantage of chorionic villi sampling is that it can be done as early as five weeks of pregnancy. Because there is a slight but real risk of injury to the embryo, the technique is recommended only when a family history indicates a risk of genetic defects.

Percutaneous umbilical blood sampling (PUBS) is a technique for collecting fetal blood directly from the umbilical cord. Under ultrasound guidance, a needle is inserted through the abdominal wall and into a vein in the umbilical cord. Cord blood obtained by PUBS can be tested for sickle-cell anemia, hemophilia, anemia, and Rh problems, as well as genetic abnormalities. The risk of miscarriage is slightly higher than with amniocentesis or chorionic villi sampling; so again, this technique may be advisable only when there is a known risk of certain diseases.

---

may be why teenagers, who may seem to be responsible in most of their decisions, still sometimes make bad judgments when they are under intense peer pressure.

 Recap Maturation of organ systems occurs at different rates. At birth, the neonate is physically helpless and cannot form memories. Infants begin to eat solid foods and to walk. Brain growth is nearly complete by the end of childhood. Muscle strength and coordination continue to improve throughout childhood, and the body changes to a more adult form. Adolescence is accompanied by a growth spurt and by maturation of the reproductive systems. ◾

21.9 Aging

Physiologically, at least, humans reach their peak in early adulthood at about 20 years of age. Most body systems are fairly stable from 20 to about 40, and then the process that we call *aging* seems to accelerate. No matter whether we exercise regularly, eat right, and in general, take care of ourselves, there still seems to be at least some "natural" aging process, just like the carefully timed sequence of events that occurs in human development. To be sure, exercise and a proper diet do help, but in the end we all must face the fact that we will not be the same at age 80 as we were at 20. The big question, of course, is why we age and whether the aging process can be halted or at least slowed.

Aging is the process of change associated with the passage of time (**Figure 21.17**). Although the word does not necessarily imply deterioration (after all, some wines get better as they age), human aging invariably leads in that direction. *Aging,* as we use the term in biology, could be used synonymously with *senescence,* the progressive deterioration of multiple organs and organ systems over time. Human senescence (or aging) has no apparent or obvious cause—that is, it cannot be blamed on a particular disease or event.

The average life expectancy of a person born in the United States has increased from 47 years in 1900 to about 79 today (76 for males, 81 for females). Does that mean we are aging more slowly now? Not at all. **Longevity**—how long a person lives—does not depend solely on the aging process. Most experts believe that our increase in longevity is largely due to a reduction in deaths by accident or disease, not to a slowdown in the aging process itself.

In fact, it appears that nothing we have done (so far!) has decelerated the aging process. If we could solve every known disease and prevent every accidental death, would we still come face to face with a natural endpoint of life? Nobody knows for sure, but so far the answer seems to be a qualified yes. No one has beaten the aging process yet, and there are very few individuals with documented ages of over 100 years.

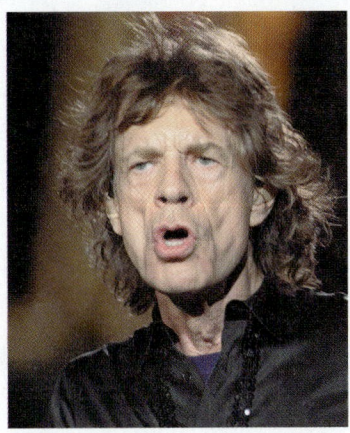

a) Rolling Stones' Mick Jagger at age 23, in 1967.

b) Mick Jagger at age 72, in 2015.

Figure 21.17 **Aging.**

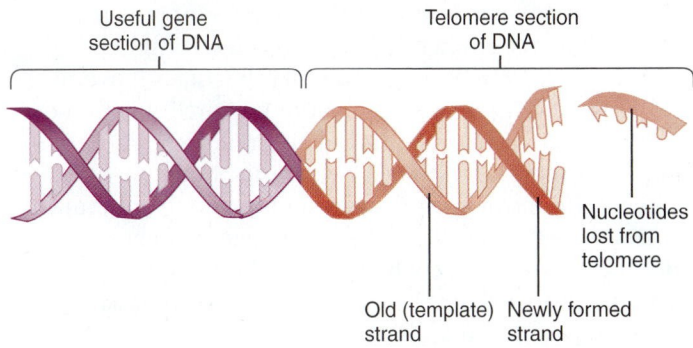

Useful gene section of DNA

Telomere section of DNA

Nucleotides lost from telomere

Old (template) strand

Newly formed strand

Figure 21.18 **Telomeres and aging.** Short terminal segments of DNA called *telomeres* are removed each time the DNA is replicated. As a result, the new strand will be shorter than the original strand. When all telomeres are gone, further cell divisions erode the useful genes.

What causes aging?

There are three hypotheses regarding the causes of aging. The first is that there is some sort of internal genetically determined program that dictates the timing of cell death. The second is that accumulated cell damage or errors eventually limit cells' ability to repair themselves. A third hypothesis is that the decline in function of a critical body system may lead to the parallel senescence of other systems. There is evidence to support each hypothesis.

An internal cellular program counts cell divisions

It has long been known that even when cells are grown under laboratory conditions that are ideal for cell growth and division, most cells will not continue to divide indefinitely. Most cells go through approximately 50–90 successive divisions and then just stop. Cancer cells are the exception, for they will continue to divide indefinitely.

Why do normal cells stop dividing? Every chromosome ends with a long tail of repetitive nucleotides called a **telomere** (Figure 21.18). A telomere protects the chromosome's genes from degradation over time. That's because the enzymes involved in replication can't replicate the chromosome all the way to the end; they must stop several base pairs short of the end. A telomere is like a roll of disposable tickets; every time the chromosome is replicated a ticket is removed (a small segment of the telomere is lost) and the chromosome gets shorter. When the chromosome's telomere is used up, cell division begins to eat away at vital genetic information. Eventually enough vital genetic information is lost that the cell just stops dividing, or even dies.

A curious fact about cancer cells is they manufacture an enzyme called *telomerase* that rebuilds the telomeres, replacing segments lost during cell division. There has been a lot of excitement about telomerase lately as a result of these findings. Some researchers have even suggested that telomerase may be the magic bullet that could reverse or prevent the aging process, or that an antitelomerase substance might offer the cure for cancer.

These ideas remain far-fetched at the moment; understanding why a single cell stops dividing does not tell us how to arrest the aging process at the level of the whole body. After all, aging is not just a decline in cell division but in cell function as well.

Cells become damaged beyond repair The second hypothesis holds that over time, accumulated damage to DNA and errors in DNA replication become so great that they can no longer be countered by the cell's normal repair mechanisms. The damage is thought to be caused by certain toxic by-products of metabolism, such as oxygen free radicals. Eventually, like a ticking time bomb, the damage becomes too great to allow survival.

According to proponents of this hypothesis, how long cells live is a function of the balance between how often damage occurs and how much of it can be repaired. Cells that undergo frequent injury and are slow to repair the damage have shorter lives than cells that are rarely damaged and repair themselves efficiently. Some people believe that taking antioxidant vitamins may slow the aging process and prolong life, but there is not yet enough long-term evidence to know for sure if this is true.

MJ's **BlogInFocus** Can severe caloric restriction prolong life? Visit MJ's blog in the Study Area in MasteringBiology and look under "Caloric Restriction and Longevity."

http://goo.gl/tghNjL

Aging is a whole-body process Why does aging seem to happen to virtually all body systems at once, instead of just one or two systems? The third hypothesis of aging is that because all the organ systems in a complex organism are interdependent, a decline in the function of any one of them eventually impairs the function of the others.

For example, a decline in the secretion of growth hormone by the endocrine system could well tip the balance away from growth and replacement and toward a net loss of cells or organ function. Impaired function of the cardiovascular system would affect nutrient delivery to and waste removal from every cell in the body. A critical system in terms of aging may turn out to be the immune system. Indeed, the elderly have a higher incidence of cancer than the young, which may indicate that the immune system is not recognizing damaged cells as efficiently as it once did.

It is also possible that aging is due to a specific type of tissue change that affects many organ systems at once. One hallmark of aging is that a certain type of protein cross-linking increases with age. This cross-linking could contribute to multiple aging processes, ranging from decreased elasticity of tendons and ligaments to functional changes in lungs, heart, and blood vessels. Of course, this hypothesis still leaves unanswered the question of what initiates the aging process at the cellular level.

☑ What might be one negative consequence of increasing the amount of telomerase in cells? (Hint: What cells often have high levels of telomerase?)

Body systems age at different rates

What actually happens during aging is fairly well documented, even if we don't know exactly why. Next, we describe important developments in each organ system.

Musculoskeletal system and skin On average, bone and muscle mass begin to decrease in most people in early adulthood. Bone loss is more marked in women than in men, especially after women hit menopause. The decline in muscle mass for either gender is a combination of a loss of muscle cells and a decrease in the diameter of remaining cells. By age 80, the average person has lost 30% of his or her muscle mass, and strength declines accordingly. Often, the joints become stiff and painful because they secrete less joint fluid. Ligaments and tendons become less elastic.

The loss of bone and muscle mass with age is not inevitable, however. Weight-bearing exercise (running rather than swimming) may reduce the rate of bone loss, and regular exercise of virtually any type may serve to maintain or even increase muscle mass. Increases in muscle mass are generally due to increased mass of the remaining muscle cells, rather than addition of new muscle cells.

The skin becomes thinner, less elastic, and wrinkled with age (**Figure 21.19**). There are fewer sweat glands, so it becomes harder to adjust to warm temperatures. Pigmentation may change as well, with the appearance of pigmented blotches called age spots or liver spots. Liver spots have nothing to do with the liver. They are caused by too much exposure to sunlight, which damages the skin and (over decades) causes the skin to produce excessive pigment in damaged areas.

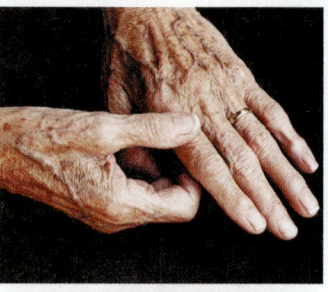

a) The hands of a young person have a smooth, elastic appearance.

b) With age, skin becomes thinner, wrinkled, and less elastic. Skin pigmentation may appear as blotches called liver spots.

Figure 21.19 Aging of skin. With age, skin becomes thinner, wrinkled, and less elastic. Skin pigmentation may appear as blotches called liver spots.

Cardiovascular and respiratory systems Lung tissue becomes less elastic, decreasing lung capacity. The heart walls become slightly stiffer due to an increase in the ratio of collagen to muscle, with the result that an aging heart generates less force. Blood vessels also lose elasticity, so the systolic pressure and pulse pressure may rise in elderly people. Overall, there is a reduced capacity for aerobic exercise as aging progresses.

Immune system The thymus decreases to only 10% of its maximum size by age 60, so there are fewer T cells. The activity, but not the number, of B cells declines as well. Wound healing slows, and susceptibility to infections increases—in the elderly, even the flu can become life-threatening. In addition the immune system fails to recognize "self" as well as it once did, leading to higher rates of autoimmune disorders.

Nervous and sensory systems Neurons in most areas of the brain do not divide in adults, so neurons that are lost throughout life generally are not replaced. The brain shrinks only slightly, however, because the neurons that remain tend to increase their connections with each other. Nevertheless, brain and sensory function do decline, in part because of degenerative changes that occur within the cell bodies of the remaining neurons. Virtually every neural activity is affected, including balance, motor skills, hearing, sight, smell, and taste. In addition to neural changes, sensory organs change as well. For example, hair cells in the inner ear stiffen, reducing hearing, and the lens of the eye stiffens, making it more difficult to focus on near objects. As a result of these many changes, elderly people become prone to injury. In addition, they may fail to eat properly because food seems less appealing.

Reproductive and endocrine systems Female reproductive capacity ceases when a woman reaches **menopause** (the cessation of ovulation and menstruation) in her late 40s or early 50s. Menopause occurs because the ovaries lose their responsiveness to LH and FSH and slowly reduce their secretion of estrogen. The onset of menopause may be gradual, with menstrual periods becoming irregular over several years.

The symptoms of menopause are variable. As a result of the decline in estrogen, some women experience "hot flashes" (periods of sweating and uncomfortable warmth). There is a tendency for vaginal lubrication to decline after menopause, making intercourse potentially less comfortable. Mood swings are not uncommon. However, female sexual desire (libido) may be undiminished because the source of female androgens is the adrenal glands, not the ovaries. After menopause, women are at increased risk for cardiovascular disease and osteoporosis. Hormone replacement therapy with estrogen and progesterone is sometimes prescribed for postmenopausal women to reduce menopausal symptoms and other health problems, although the therapy itself adds an additional health risk.

Men exhibit a long, slow decline in the number of viable sperm as they age. In addition, after age 50 men may take longer to achieve an erection. Nevertheless, both sexes can and do remain sexually active into old age.

The rate of decline of various components of the endocrine system varies widely. Notably, the secretion of growth hormone decreases substantially. However, there is not an appreciable decline in renin, TSH, ACTH, or ADH.

Digestion and nutrition The digestive system continues to function well except that some vitamins are not absorbed as efficiently. Nutritional requirements decline somewhat due to slower cellular metabolism, smaller muscle mass, and, often, less physical activity. Adequate nutrition can be a problem if the elderly fail to eat a balanced diet. Tooth loss varies widely among individuals, but people who have lost teeth may find it harder to eat fruit and other healthy but fibrous foods.

The liver's ability to detoxify and remove drugs from the body declines significantly. As a result, elderly people often require smaller doses of medications to achieve the appropriate therapeutic blood concentration.

Urinary system Kidney mass declines, and both blood flow and filtration rate may fall by as much as 50% by age 80. However, there is normally such a tremendous reserve of renal functional capacity that the decline is usually of little consequence. The elderly do lose some capacity to respond to extreme dehydration or overhydration, but otherwise the kidneys carry out their functions rather well.

The primary problem in men is usually prostate hypertrophy. The enlarged (hypertrophic) prostate presses against the urethra, making urination difficult. In women, the most common problem is incontinence (loss of bladder control) due to a weakened urethral sphincter.

☑ Which of the above aging-related declines do you think could be minimized, or even temporarily reversed, with regular exercise? Which specific type of exercise might benefit specific age-related declines?

Aging well

It would be easy to get depressed after reading the long list of declines in function that accompany aging. Bear in mind that not all of these developments occur to the same degree in everyone, nor do they all occur at the same age. The effects of aging are normal events in our long and exciting journey through life.

Will a healthy lifestyle slow the aging process? The answer is far from certain. However, regular exercise and healthy nutrition can certainly help us age better. Throughout human history, longevity has been determined primarily by the prevalence of disease, not by the "natural" length of a human life span (if indeed there is one). Exercise and a healthy diet improve cardiovascular and skeletomuscular fitness and reduce your risk of cancer and cardiovascular disease. A healthy lifestyle also improves the quality of life, providing a heightened sense of well-being and increased energy and vigor (**Figure 21.20**).

With these benefits, it really doesn't matter whether a healthy lifestyle can extend the "natural" human life span from 100 to, say, 105. If you can't live forever, you can at least age well and gracefully.

↩ **Recap** Aging is a complex process that is still poorly understood. The number of times a cell can divide may have an upper limit. In addition, damage may accumulate in cells until they no longer function properly. All organs and organ systems decline in function with advancing age, though not necessarily at the same rate. Even if the aging process cannot be stopped, it is certainly possible to improve human health and wellness throughout life. Regular exercise and healthy nutrition are important factors in aging well. ■

Figure 21.20 Aging well. Although it is not known for certain whether exercise and proper diet actually delay the aging process, they do improve performance and may provide an enhanced sense of wellness throughout life.

21.10 Death

The final transition in the journey of life is death. In terms of human biology, it is as natural as life itself.

Death is defined as the cessation of life. The death of a complex organism such as a human being is not necessarily accompanied by the immediate death of all its organs or cells. Death occurs when an organ system that is essential for life in the short term fails to function, making it impossible to sustain the life of other systems.

The critical organ systems whose failures lead to death very quickly are the brain (because it controls respiration), the respiratory system (without oxygen, life cannot be sustained), and the cardiovascular system (without oxygen and nutrient delivery to all cells, life ends). Failure of other organs with more long-term functions, such as the kidneys or digestive system, does not necessarily kill immediately. You can live without either of these systems for days as long as their loss does not cause one of the critical systems to fail.

The challenge of defining the precise moment of death, then, is that all of the organ systems don't necessarily die at the same time. Death is a process that begins with the failure of certain life-critical organ systems and then proceeds through the death of other organ systems, organs, and cells until the very last of the body's 100 trillion cells is no longer alive. That could take hours, days, or (with the assistance of life-support systems) even years.

Although death may occur gradually, our society requires a practical way to define the moment at which an individual dies. We need a definition of death so the living can begin to grieve and make plans, and so health care workers can stop the medical treatments that sustain life. We need a legal definition so that the police and courts can make a distinction between assault and murder, and so that estates can be settled. A precise definition of death is especially important because organs that might still be alive can be transplanted into another person and save a life.

Legal and medical criteria for declaring a person dead vary according to state law. Basically, all such criteria attempt to define the moment when, to the best of our knowledge and abilities, the person no longer has even a remote possibility of sustaining life. In general, they define death as either (1) *irreversible cessation of circulatory and respiratory functions* or (2) *irreversible cessation of all functions of the entire brain, including the brain stem.* Note that the definitions focus on critical systems that we cannot live without for more than a few minutes.

The key, of course, is the word *irreversible.* Usually, it is easy to tell when a person has had no respiration or blood flow for 10 minutes or so, long enough to cause brain death. Brain death alone, especially in a person on cardiopulmonary life support, is harder to define. Medical professionals generally run down a checklist of criteria that must be met in order to declare a person's brain nonfunctional, and even then they usually repeat the tests 6–12 hours later before they are willing to declare irreversible brain death. Even for the medical profession, the final transition in life can be difficult to define.

Recap Death is the termination of life. Death starts with the failure of one or more critical organ systems and ends ultimately with the inability to maintain an internal environment consistent with cellular life. ∎

Chapter Summary

21.1 Fertilization begins when sperm and egg unite *p. 481*

- Sperm and egg meet (and fertilization occurs) in the upper third of the oviduct.
- Enzymes in the head of the sperm create a pathway through the corona radiata and the zona pellucida to the egg cell membrane.
- Entry of the sperm nucleus triggers the completion of meiosis II by the secondary oocyte. Thereafter, the nuclei of sperm and egg fuse, creating a single diploid cell, the zygote.

21.2 Developmental processes: cleavage, growth, differentiation, and morphogenesis *p. 484*

- Development begins by a process called *cleavage* and then proceeds with morphogenesis, differentiation, and growth.
- The periods of development prior to birth are known as pre-embryonic (the first two weeks), embryonic (weeks three through eight), and fetal (weeks nine to birth).

21.3 Pre-embryonic development: the first two weeks *p. 484*

- During the first two weeks of development, the single cell develops into a hollow ball with an embryonic disk in the center that will eventually become the embryo.

- At about one week, the pre-embryo (now called a *blastocyst*) begins to burrow into the uterine wall, a process called *implantation.*

21.4 Embryonic development: weeks three to eight *p. 485*

- Embryonic development is marked by the presence of three primary germ layers (ectoderm, mesoderm, and endoderm) in the embryonic disk.
- During development, the embryo is completely surrounded by two membranes. The amnion contains amniotic fluid, and the chorion develops into fetal placental tissue. Two other membranes (the allantois and the yolk sac) have temporary functions only.
- By the fifth week, the embryo begins to take on distinctly human features.
- By the eighth week, the placenta and umbilical cord circulation are fully functional, and male or female gonads have begun to develop.

21.5 Gender development *p. 489*

- Until six weeks, an embryo is "sexually indifferent."
- The presence or absence of a Y chromosome determines whether the embryo will become phenotypically male or female.

21.6 Fetal development: nine weeks to birth *p. 490*

- Months three to nine are marked by rapid growth and development of the organ systems. By the fifth month, the fetus begins to move.
- By the sixth month, life outside the womb is possible with good medical care.

21.7 Birth and the early postnatal period *p. 490*

- Birth occurs at about 38 weeks of development (nine months).
- The three phases of labor are dilation, expulsion, and afterbirth.
- Shortly after the newborn takes its first breath, the newborn's cardiovascular system undergoes substantial changes. Within hours or a few days, the umbilical vessels regress and the ductus arteriosus and foramen ovale close, rerouting all blood through the pulmonary circulation.

21.8 Maturation: from birth to adulthood *p. 494*

- Human neonates are relatively helpless because the nervous and muscular systems are not yet mature.
- There is a disproportionate increase in the development of the musculoskeletal and nervous system during infancy. The immune system remains relatively immature.
- Body shape changes in childhood. The brain achieves 95% of its adult size.
- Adolescence is marked by the last spurt of rapid growth and the attainment of sexual maturity.

21.9 Aging *p. 496*

- Most cells have an internal mechanism that limits the number of times they can divide.
- Cumulative unrepaired cellular damage as a consequence of metabolic activity may limit the life of cells.
- Aging may affect all organ systems because a decline in one system will affect other systems.

21.10 Death *p. 500*

- Death is a process. Death begins with the failure of one or more critical organ systems, leading to the failure of other organ systems and eventually to the death of all cells.

Terms You Should Know

amnion, 486
amniotic fluid, 486
blastocyst, 485
cesarean delivery, 492
chorion, 486
colostrum, 494
ectopic pregnancy, 485
embryo, 485
embryonic disk, 485
endoderm, 486
fertilization, 483
fraternal twins, 483
human chorionic gonadotropin (hCG), 486
identical twins, 483
menopause, 498
miscarriage, 488
morphogenesis, 484
placenta, 486
puberty, 495
umbilical cord, 486
zygote, 481

Concept Review

Answers can be found in the Study Area in MasteringBiology.

1. Why does only one sperm fertilize the egg, and why is this important?
2. What determines whether an embryo will develop into a male or a female?
3. Describe the function of the placenta and the hormones that it secretes.
4. Describe the stages of labor and delivery.
5. What causes the newborn to take its first breath immediately after birth?
6. Why aren't all of a child's vaccinations given in the first year?
7. What causes the onset of puberty (sexual maturation)?
8. Summarize the three main hypotheses of the causes of aging.
9. What has caused the increase in longevity of the U.S. population during the past 100 years?
10. Why is it sometimes so hard to define the moment when an individual dies?

Test Yourself

Answers can be found in the Appendix.

1. Which of the following lists the correct order of structures through which a sperm must pass to accomplish fertilization?
   a. zona pellucida, corona radiata, secondary oocyte plasma membrane
   b. zona pellucida, secondary oocyte plasma membrane, corona radiata
   c. ovum plasma membrane, corona radiata, zona pellucida
   d. corona radiata, zona pellucida, secondary oocyte, plasma membrane

2. Which of the following will result in fraternal twins?
   a. fertilization of one egg by two different sperm
   b. division of a fertilized egg into two masses after differentiation has begun
   c. fertilization of two eggs by two different sperm
   d. division of a fertilized egg into two masses before differentiation has begun

3. Which of the following processes is the first to occur as a fertilized egg begins its journey down a uterine tube toward the uterus?
   a. differentiation   c. growth
   b. cleavage   d. all of these choices

4. A sperm with a defective acrosome would not be able to:
   a. swim
   b. penetrate the egg
   c. penetrate through the cervical mucus
   d. provide the energy for the sperm to swim

5. Which of the following indicates the correct order of pre-embryonic developmental stages?
   a. morula, blastocyst, embryonic disk
   b. blastocyst, morula, embryonic disk
   c. trophoblast, blastocyst, morula, embryonic disk
   d. embryonic disk, blastocyst, morula

6. Which of the following is a correct statement?
   a. The placenta is formed entirely from embryonic tissue.
   b. The umbilical cord brings maternal blood to the embryo.
   c. The embryonic/fetal blood and the maternal blood mix within the placenta.
   d. The placenta secretes hormones including HCG and progesterone.

7. Which of the following is mismatched?
   a. neural groove: spinal cord
   b. mesoderm: gastrointestinal tract lining
   c. urogenital groove: urethra
   d. placenta: chorion and endometrium

8. Which of the following events is correctly matched with the month of prenatal development?
   a. kidneys produce urine: month 3
   b. heart pumps to circulate blood: month 4
   c. gender development is completed: month 5
   d. bone marrow takes over blood cell production: month 7

9. Which fetal hormone plays a role in initiating labor?
   a. estrogen
   b. ACTH
   c. progesterone
   d. oxytocin

10. Which of the following would be detrimental to the health of a neonate?
    a. the ductus arteriosus stays open
    b. the foramen ovale closes
    c. the ductus venosus closes
    d. both (a) and (b)

11. _____ stimulates the production of milk, whereas _____ causes ejection of milk from the breast.
    a. Estrogen . . . prolactin
    b. Prolactin . . . oxytocin
    c. Estrogen . . . oxytocin
    d. Oxytocin . . . prolactin

12. Possible activities of telomerase could be:
    a. stopping the aging process by rebuilding the telomeres
    b. contributing to cancer by allowing tumor cells to continue dividing indefinitely
    c. blocking cancer by restoring telomeres after cell divisions
    d. both (a) and (b)

13. The aging of which organ or organ system is least problematic during the normal aging process?
    a. bones
    b. immune system
    c. nervous system
    d. kidneys

14. All of the following are theories regarding causes of aging except:
    a. Cells run out of telomeres and stop dividing.
    b. Cortisol production increases, and the stress response contributes to aging.
    c. Cells can no longer repair themselves properly, and damage accumulates.
    d. A decline in one critical body system induces parallel declines in others.

15. Which hormone plays a critical role both in childbirth as well as breast-feeding?
    a. oxytocin
    b. prolactin
    c. estrogen
    d. progesterone

Apply What You Know

Answers can be found in the Study Area in MasteringBiology.

1. Why is it biologically impossible for humans and gorillas to mate and produce offspring?

2. Conjoined twins are twins that are attached to each other. They may even share limbs and organs. From what you know of early human development, how do you think conjoined twins come about?

3. Why are many sperm but only one egg produced in the process of fertilization? Would humans be able to reproduce if the cell numbers were inverted?

4. Each year, it is estimated that over 40,000 newborn babies are impacted by the effects of alcohol consumed by the mother during pregnancy. Fetal alcohol syndrome is one of the leading causes of preventable birth defects. Why do you think alcohol is more likely to damage the fetus than the mother?

5. What exactly do home pregnancy tests test for as an indication that a woman is pregnant?

6. Babies born before a full nine months of gestation (premature babies) are often at a high risk of death. What is the earliest a baby can be born and stand a good chance of survival, and which organ system is generally the limiting one?

7. How is it possible that one's life could be prolonged without slowing the aging process?

MJ's BlogInFocus

Answers can be found in the Study Area in MasteringBiology.

1. There is evidence to suggest that fathers, particularly older fathers, pass on more genetic mutations to their offspring than mothers. Based on what you know of reproductive biology, can you think of any reason why that might be? Visit MJ's blog in the Study Area in MasteringBiology and look under "Older Fathers and Genetic Mutations." http://goo.gl/FQeZM9

MasteringBiology®

Students Go to MasteringBiology for assignments, the eText, and the Study Area with animations, practice tests, and activities.

Professors Go to MasteringBiology for automatically graded tutorials and questions that you can assign to your students, plus Instructor Resources.

Answers to ✓ questions are available in the Appendix.

Evolution and the Origins of Life

Skulls of human ancestors ranging from 3.0 million years ago (lower left) to 22,000 years ago (lower right).

Key Concepts

- The *theory of evolution* is the best explanation for the complexity and diversity of life on Earth. Although the theory has undergone some modification over time, it has withstood all scientific examination and challenge.

- Mutations coupled with natural selection cause slow evolutionary change in life-forms over time.

- Mass extinctions have dramatically affected the course of evolution at least five times during Earth's existence.

- Life originated on Earth about 3.5 billion years ago, only 1 billion years after Earth was formed. The first life-forms probably originated when self-replicating molecules of RNA and then DNA became enclosed within a lipid-protein membrane.

- Our human ancestors diverged from other primates about 4–5 million years ago. Chimpanzees are our closest living relatives.

- Modern humans are classified as genus *Homo*, species *sapiens*. *Homo sapiens* originated in Africa around 140,000 to 100,000 years ago.

Who Were the Flores People?

In 2004, archaeologists unearthed a female skeleton from Liang Bua Cave on the Indonesian island of Flores. They named the skeleton LB1, for Ling Bua #1. LB1 was barely 1 meter tall (about 3.3 feet) and weighed only about 25 kg (55 pounds). She was about 30 years old when she died.

Although similar in size and limb proportions to ancient species in the genus *Australopithecus* (early humans with ape characteristics), LB1 bore an eerie resemblance to more recent species in the genus *Homo* (modern humans). Like modern humans, LB1 exhibited posture for upright walking and a jaw for chewing, but her skull and brain were small, comparable to those of chimpanzees and early humans. So archaeologists were intrigued when her bones were dated to only 18,000 years ago—a time when *Homo sapiens* were thought to be the only species of the genus *Homo* remaining on Earth. Subsequently, the bone fragments and teeth of 14 others resembling LB1 were discovered in the same cave. The bones were dated between 95,000 and 12,000 years ago, indicating a long and successful history of LB1's kind in that area.

A New Human Species?

The discoverers proposed that the Flores people, as LB1 and her kind came to be called, represent a previously undiscovered human species, who, like *Homo sapiens*, descended from *Homo erectus*. Archaeologists named the new species, *Homo floresiensis*, in honor of the island where the species was discovered.

Almost immediately after she was discovered, anthropologists began arguing about whether LB1 actually represented a new human species. The most controversial aspect of LB1 was her brain size—only about a third the size of a modern human's. A large brain size, it has long been argued, is required for sophisticated toolmaking and complex social organizations, characteristics considered exclusive to the genus *Homo*. And yet, the Flores

Skulls of *Homo floresiensis* (left) and *Homo sapiens* (right).

people clearly were advanced in their activities and social structure. For instance, they crafted an impressive variety of stone tools. Though most of their implements were relatively simple, some were complex tools comparable to those honed by prehistoric *Homo sapiens*. Many tools of the Flores people were found mingled with the bones of *Stegodon,* a 2,200-pound dwarf relative of modern elephants that, although small for an elephant, would have posed a significant challenge to hunters only 3 feet tall. Bringing down such large prey not only points to use of sophisticated weaponry but it also indicates that the Flores people engaged in group hunting activities that required planning and cooperation.

One hypothesis to explain the discrepancy in brain size between the Flores people and other extinct species in the genus *Homo* is that the brain of the former became proportionally smaller only *after* they arrived on the island of Flores, not before. However, this runs counter to the observation that as humans become smaller from a larger ancestor, brain capacity doesn't necessarily decline. For example, there is not much difference in brain size between the present-day Masai people of east Africa, who are over 6 feet tall, and the present-day Bambuti people of the Congo, who are barely 4½ feet tall. How then, if the Flores people were human, could they have descended from larger-brained ancestors and lost so much brain size? The possible answers to this question have shaken up anthropologists' thinking about the importance of brain size in human evolution. →

Questions to Consider

1 Do you think the skeletons of Liang Bua Cave should be called a new species? Does it matter?

2 Is it important that we try to discover the evolutionary history of humans? Why or why not? Who should finance anthropological research, and why?

Liang Bua Cave, where the discovery of the Flores skeleton was made.

A Unique Body Form

LB1, the most complete Flores skeleton to date, has a number of unique physical characteristics. She has heavy brow ridges, a sloping forehead, and a small chin. Her arms are long but her lower extremities are short. But the real differences are in her hands and feet.

Modern humans have a complex of five bones in the wrist that absorb shock, such as when we pound with heavy tools, or provide stability, such as when we do precision work. The shape of the wrist bones of LB1 more closely resemble those of an ape, or of *Homo habilis*, who died out more than 1.6 million years ago.

The most striking feature of LB1 has to be her feet. Although her incredibly short big toes are in line with the rest of her toes, as toes are in modern humans, her feet are primitive, lacking a proper arch. Most remarkable is the length of her feet—fully 70% as long as her thigh bone, compared to about 55% in modern humans. The feet of the Flores people were so long that they would have walked with a high-stepping gate and been poor runners.

Taken together, the skull, hands, and feet of LB1 suggest that her species probably diverged from the line that became modern humans, sometime back about the time of *Homo erectus*. Current thinking, therefore, leans toward accepting LBI and her kind as sufficiently distinct to warrant their own species within the genus *Homo*.

The features of *Homo floresiensis* are primitive, indicating divergence from other humans, and yet *Homo floresiensis* managed to live on Earth with *Homo sapiens* until only 12,000 years ago. A remaining question is whether the two species actually interacted. They certainly could have, but we may never know. What we do know is that *Homo floresiensis* lived closer to the present day than any other extinct species of humans. Modern humans are the sole surviving human species.

SUMMARY

- Archaeologists recently discovered partial remains of up to 14 individuals of a small human species that became extinct only 12,000 years ago.
- Discoverers named this new species *Homo floresiensis.*
- The only skull found has a brain size comparable to that of a chimpanzee. However, these were intelligent people, as evidenced by their tools and hunting abilities.
- The discovery challenges anthropologists' current beliefs about the relationships between human brain size, mental capacity, and behavior.

Look around and you'll be impressed by the astounding variety of forms that life takes. From bacteria to beagles, hagfish to humans, living creatures come in a huge range of sizes, shapes, and colors. And yet, the evidence indicates that all living creatures on Earth are related. To the best of our knowledge, all of them descended from a single simple life-form that arose from the hot, steamy environment of Earth more than 3 billion years ago. Today, there are nearly 2 million different life-forms on Earth, according to estimates. Humans are just one of them. ■

The diversity of life on Earth is thought to result from **evolution,** defined as an unpredictable and natural process of descent over time with genetic modification. The formal definition of *evolution* has three key elements.

- *Descent over time.* As one generation of organisms gives rise to the next, populations of organisms undergo slow change that makes them different from their ancestors. Sometimes, a population becomes two populations that are so different from each other that they no longer interbreed, giving rise to new species.
- *Genetic modification.* The process of change depends on changes to the genes of the organisms.
- *Unpredictable and natural.* Evolution is affected by chance, *natural selection*, historical events, and changing environments.

The theory that evolutionary processes shaped life on Earth is associated with Charles Darwin, a British naturalist of the mid-1800s. The heart of Darwin's hypothesis is that life arose only once, probably in the primordial sea long ago, and all life as we know it descended from that early life-form. Darwin actually used the word *evolution* sparingly, preferring instead "descent with change" or "descent with modification." Key to Darwin's hypothesis is the idea that descent with modification resulted from "natural selection," which we describe later in this chapter.

Scientists reserve the word *theory* for major concepts that offer the best explanation to fit a broad range of established facts. They use the word very differently from opponents of evolution, who often say that evolution is "only a theory," as if the whole concept could be dismissed. Opponents of evolution also sometimes say "Darwin was wrong" by pointing out that certain specifics of his original explanation have had to be modified after extensive testing. Nevertheless, the *theory of evolution* remains the single most unifying concept in all of biology. There has never been a credible scientific challenge to the theory of evolution as the best explanation for how life on Earth came about.

In this chapter, we review the scientific evidence on which we base our understanding of evolution and discuss processes that affect evolutionary change. Then, we pick up the intriguing question of life's origin on Earth. Finally, we trace the relatively recent evolution of human beings and consider the features that make us distinctively human.

↺ **Recap** Evolution is a process of descent over time with genetic modification. First proposed by Charles Darwin in the 1800s, the theory of evolution remains the single most unifying concept in biology. ▪

22.1 The evidence for evolution

The evidence that humans and all other life-forms have evolved over time comes from several sources, including the fossil record, comparative anatomy and embryology, biochemistry, and biogeography.

The fossil record: incomplete but valuable

Fossils, the preserved remnants of organisms, are one of our richest sources of information about life-forms that lived in the past. Soft tissues decay quickly, so most of the fossil record consists of bones, teeth, shells, and occasionally spores and seeds. In general, remnants of organisms are preserved only if they are covered soon after death with layers of sediment or volcanic ash (**Figure 22.1**). Over time, the remains become mineralized with some of the same minerals that compose rocks, leaving a rocklike impression of the hard tissue of the organism, or a fossil.

Even though we have found fossils from over 200,000 species, there may be millions of species for which we will never have any record because they had no hard tissues. We have many fossils of ancient bony fishes, for example, but only a few fossils of jellyfish from the same time periods. We are more likely to discover fossils that were produced in great numbers. For example, we are more apt to find a seashell or spore from 100,000 years ago than to uncover the skeleton of an early human from the same time period. Fossils are found on land for the most part. However, there may be many more and possibly different fossils deep under the sea. By its nature, then, the fossil record tends to be incomplete and heavily weighted toward fossils that are easy to find.

One of the most important pieces of information in terms of evolution is a fossil's age, which places it at a certain point in history and shows how body structures have changed over time. Sediment and volcanic ash are laid down layer upon layer, a process known as *stratification*. Where fossils exist in the same location in different layers, it is usually possible to tell the chronology of the fossils, the older from the younger. This helps scientists establish relative relationships between fossils over time (**Figure 22.2**).

In addition, radiometric dating is often used to date rocks and fossils. Radiometric dating takes advantage of the fact that some atoms, called *radioactive isotopes*, are inherently unstable. Radioactive isotopes radiate, or give off, particles until they reach a more stable state, sometimes becoming a different, more stable atom. By measuring the radioactive isotopes in a mineral, scientist can determine the age of the mineral. The isotope most often used to date the oldest rocks (and by inference the fossils found near them) is radioactive potassium, which is slowly converted to argon, a gas. The half-life of radioactive potassium is 1.3 billion years, meaning that it takes 1.3 billion years for half of the radioactive potassium in a rock to become argon. It would take another 1.3 billion years for half of the remaining radioactive potassium to become argon, and so on.

a) A body will decompose unless it is covered by sediment or volcanic ash.

b) Hard elements of a body may be preserved from decomposition if they are covered quickly.

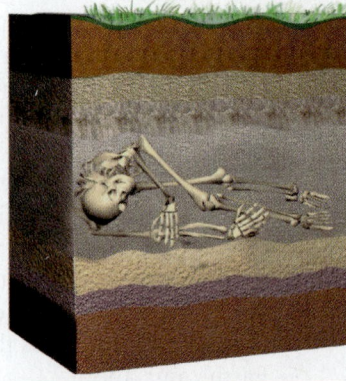

c) **Over time, more layers of sediment, ash, or soil are deposited.** The hard elements of the organism become mineralized by the same minerals that comprise rocks.

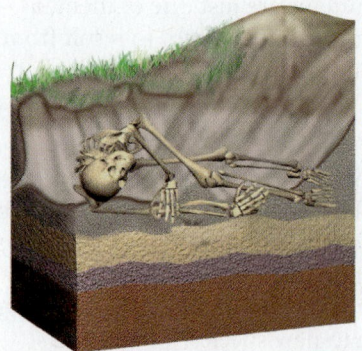

d) **Occasionally, erosion, uplifting of the earth's crust, or human excavation may expose fossils to the surface.**

e) **A fossilized human skeleton.**

Figure 22.1 How fossils are formed.

The argon formed from the decay of radioactive potassium remains trapped in the rock, so the ratio of radioactive potassium to argon in a rock can be used to estimate the rock's actual age. Because radioactive potassium decays so slowly, the radioactive potassium-to-carbon ratio is most useful for determining the age of rocks that are more than 100,000 years old. At the other end of the age spectrum, rocks and fossils less than 60,000 years old are dated using carbon-14, the radioactive isotope of carbon, which decays with a half-life of only 5,700 years.

☑ You find an interesting fossil of a tiny unicellular organism buried just underneath a layer of compacted ash. Measurement of radioactive potassium and argon in the ash shows that half of the radioactive potassium has decayed to argon. Approximately how old is the ash (and the fossil that it contains)?

Comparative anatomy and embryology provide more evidence

Additional evidence for evolution comes from comparing the anatomy of animals and the development of their embryos. When examining them in the context of evolution, scientists describe anatomical structures as *homologous, analogous,* or *vestigial.*

When different organisms share similar anatomical features, often it is because they evolved from a common ancestor. For example, the forelimbs of all vertebrates (animals with backbones) seem to share roughly the same set of bones arranged in similar patterns. This is true even though vertebrate forelimbs have undergone numerous modifications so animals can use them in a variety of specialized ways, such as flying, swimming, running, or grasping objects (Figure 22.3). **Homologous structures** are body structures that share a common origin. The degree to which homologous structures resemble each other can be used to infer the closeness of evolutionary relatedness.

In contrast, structures that serve the same function but do not arise from a common ancestor are called **analogous structures** (Figure 22.4). Good examples of analogous structures are the wings of birds and the wings of insects. Both types of wings evolved for flight. Nevertheless, they are not structurally similar, and therefore, birds and insects are not closely related.

Figure 22.2 Stratification of sedimentary deposits. The oldest layers at the bottom of the Grand Canyon are over 1.2 billion years old.

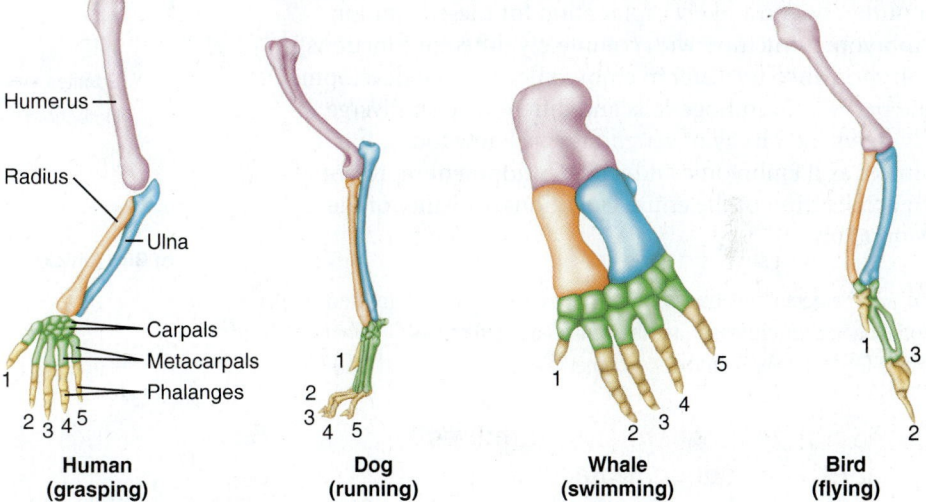

Figure 22.3 Homologous structures. These homologous forelimb structures share a common evolutionary origin. Numbers indicate digits. Notice how homologous bones (indicated by similar colors) have undergone modification in order to perform different functions in different vertebrates.

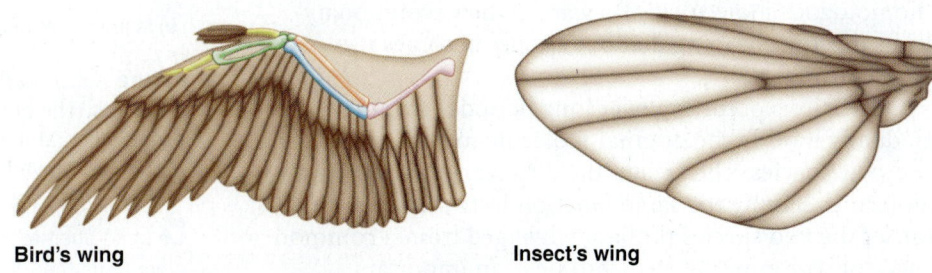

Figure 22.4 Analogous structures. The bird's wing and the insect's wing are both used for the same function, but they evolved from entirely different structures.

Finally, **vestigial structures** serve little or no function at all in an organism. However, vestigial structures may be homologous to body parts in other organisms, where the structures do still have a function. The human coccyx (tailbone) is the vestigial remnant of a tail, suggesting that we share a common ancestry with other vertebrates with tails. The muscles that enable you to wiggle your ears are vestigial in humans. Other animals, such as dogs and deer, still use these muscles to rotate their ears toward a sound so that they can hear it better.

Compelling evidence for evolution also comes from comparisons of the embryos of animals, especially vertebrates. The vertebrates are a varied group: fishes, amphibians, reptiles, birds, and mammals. Yet, despite their diversity, their early embryonic development follows the same general pattern (**Figure 22.5**). For example, all vertebrate embryos develop a primitive support structure called a *notochord*, which becomes the core of the intervertebral disks, and a series of folds called *somites* that will become bone, muscle, and skin. In addition, all vertebrate embryos develop a series of arches just below the head. In fishes and amphibians, these are called *gill arches* because they develop into gills. In humans, the homologous *pharyngeal arches* become parts of the face, middle ear, and mouth. The most likely explanation for these common embryonic structures with completely different functions is common ancestry. Later in embryonic and fetal development, the paths of morphogenesis and differentiation diverge, giving us the variety of vertebrates we know today. It is almost as if embryonic and fetal development represent an acceleration of the entire evolutionary history of the vertebrates.

☑ Are the dorsal fins (the stiff fins on the back) of dolphins and sharks homologous structures, or are they analogous structures? Explain. (Hint: Dolphins are mammals and sharks are cartilaginous fish.)

Comparative biochemistry examines similarities between molecules

Proteins and genes also provide clues to the relationships among species. When two species possess identical or nearly identical biochemical molecules, common ancestry may be indicated. In principle, the two similar molecules are akin to homologous structures—you might call them "homologous molecules." They arise when evolutionary paths diverge, meaning that one species becomes two different species.

Mutations occurring over long periods of time gradually modify the original molecule in each of the two new species. The greater the difference between two molecules serving the same function in two species, the earlier the two species probably diverged from a common ancestor. For example, cytochrome *c*, an important protein in the metabolic pathways for energy exchange, is found in organisms ranging from yeast to humans.

In humans, cytochrome *c* consists of 104 amino acids. A chimpanzee's cytochrome *c* is identical to that of a human, but both humans and chimpanzees differ by 1 amino acid from rhesus monkeys, 16 amino acids from chickens, and over 50 amino acids from yeast. From this data, you could hypothesize that humans are more closely related to chimpanzees than to rhesus monkeys and that humans are more closely related to other primates than to chickens or yeast.

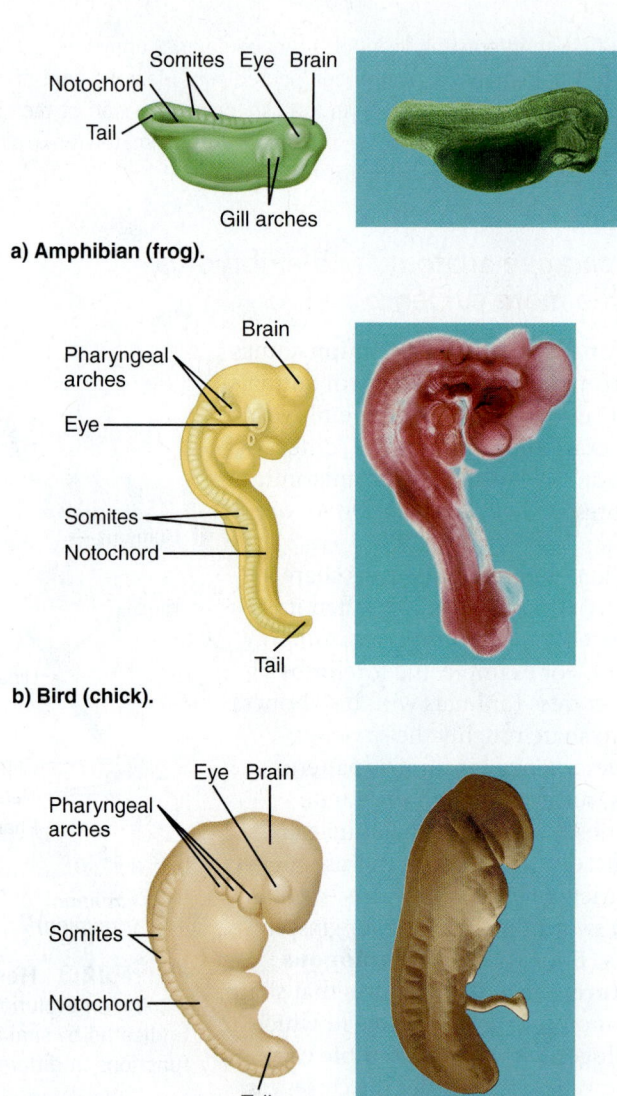

a) **Amphibian (frog).**

b) **Bird (chick).**

c) **Mammal (human).**

Figure 22.5 Early developmental similarity among vertebrate embryos. The embryos of all vertebrates tend to share certain developmental features, suggesting a common ancestry. In particular, all vertebrate embryos have either gill or pharyngeal arches. These arches ultimately have very different functions.

☑ In addition to the pharyngeal arches, human embryos have several other traits that disappear before birth. Three are visible in this diagram; can you find at least one of them? Propose an explanation for why we retain these traits as early embryos but lose them before birth.

Figure 22.6 Geographic barriers. Animal and plant migrations are inhibited by barriers such as these mountains and desert sands in Death Valley, California.

Biogeography: the impact of geographic barriers and continental drift on evolutionary processes

Biogeography is the study of the distribution of plants and animals around the world. Why do we find certain animals and plants in some locations on Earth and not in others? One explanation is that an organism's ability to migrate is inhibited if it lives in an isolated region such as an island, or a place with geographic barriers such as rivers, high mountain ranges, or deserts (**Figure 22.6**).

Where migrations of plants and animals are possible, the evidence indicates that closely related life-forms often evolve in one location and then spread to regions that are accessible to them. Eventually, their spread may be blocked by physical barriers or by environments in which they cannot survive, such as deserts or oceans. There are no snakes on the Hawaiian Islands, for example, because snakes evolved long before the volcanic Hawaiian Islands existed, and the Pacific Ocean represents an effective barrier to snake migration. When we see how easily some plants and animals adapt (adjust to new conditions) when placed in new locations, we can appreciate the impact of

geography on evolutionary processes and the distribution of life-forms.

Geologists tell us that Earth's continents are located on huge plates called *tectonic plates*, whose locations on the planet's surface are not fixed. The plates slowly move over time, a process called **continental drift.** About 200 million years ago, all the continents were joined in one interconnected landmass called *Pangaea* (**Figure 22.7**). The continents drifted apart after the first animals and plants appeared. As a result, related groups of organisms that were isolated from each other evolved separately, but along almost parallel paths. For instance, Australia separated from the other continents about 65 million years ago. Nevertheless, there are striking similarities between the unique marsupial (pouched) mammals of Australia and the predominant placental mammals (mammals that nourish their young with a true placenta) found on all other continents, even though marsupial and placental mammals are not closely related. Their similarities exist because the two species of mammals evolved from common ancestors and because evolution continued after separation. Evolution solved some of the same environmental challenges that both species faced in their different environments.

 **Recap** Comparisons of homologous anatomical structures, embryologic development, and DNA and protein structures can be used to estimate the closeness of relationships between different species. Fossils are the mineralized remains of the hard tissues of past life-forms. The fossil record provides compelling evidence for evolution. Radiometric dating can be used to date rocks and fossils. The slow drift of continents over the past 200 million years and physical barriers such as mountain ranges and deserts influence the evolution and distribution of life-forms. ■

MJ's BlogInFocus How is DNA analysis being used to help determine the evolutionary relationships between species? Visit MJ's blog in the Study Area in MasteringBiology and look under "Evolutionary Relationships."

http://goo.gl/4PTKJn

a) 200 million years ago.

b) 90 million years ago.

c) Present.

Figure 22.7 The formation of Earth's continents from a common landmass. The continents were once connected in one giant landmass called *Pangaea*.

22.2 Natural selection contributes to evolution

Recall that an organism's DNA represents the set of instructions for all of its life processes. If all creatures had an identical set of DNA, they would all *be* identical in form and function. Differences in DNA account for the differences among species and even among individuals of a species.

Random mutations underlie evolution

A species' gene pool changes very slowly over time (measured in thousands to millions of years) as a result of random mutations to the genes of individual members of the species. Mutations are rare accidental events that produce a different form of a gene, called an *allele*. The majority of mutations are neutral; a few are detrimental (even lethal) and a few are beneficial. Alleles that are not lethal may be passed on to future generations, to become part of the species' collective gene pool. Eventually these accumulating mutations may cause changes in the species' physical and functional traits. Ultimately, an accumulation of mutations may cause one species to diverge into two. Without random mutations, there could be no evolution.

Natural selection encourages changes in the gene pool

Gene mutations alone do not amount to evolution. Natural selection contributes as well.

Living organisms interact with their local environments, and as a consequence of that interaction, some live long enough to reproduce and some do not. The frequency with which any allele is found in the gene pool of a species depends critically on whether that allele confers survival. Darwin understood this principle even though he did not fully comprehend the genetic basis for it. He referred to it as "survival of the fittest" by means of natural selection. By **natural selection,** Darwin meant that individuals with certain traits are more fit for their local environment than other individuals without those traits, and therefore, the former are more likely than the latter to survive and reproduce. Mutations coupled with natural selection cause certain genes to increase and others to decrease from the gene pool of a population.

✅ A friend tells you that evolution occurs because of "good mutations." Explain what is wrong with this statement.

Genetic drift and gene flow alter populations

A **population** is a group of individuals of the same species that occupy the same geographical area. Darwin was convinced that the fittest members of a population were always the ones to survive. We now know that there are several other contributors to evolution besides natural selection. Other factors that affect the gene pool of a population include genetic drift, gene flow, and selective hunting.

Genetic drift (**Figure 22.8a**) refers to random changes in allele frequency because of chance events. Genetic drift is more likely to occur in smaller populations than in large ones because smaller populations are more vulnerable to change. One cause of genetic drift is the *bottleneck effect,* which occurs when a major catastrophe—such as a fire, a change in climate, or a new predator—wipes out most of a population without regard to any previous measure of fitness. Another cause of genetic drift, the *founder effect,* occurs when a few individuals leave the original group and begin a new population in a different location. In both cases the few genes left in the gene pool may no longer represent the original population, nor do they necessarily represent the most fit genes.

A good example of the bottleneck effect is found in the human population of the tiny island of Pingelap, in the Pacific Ocean. Pingelap Island was hit by a typhoon in 1775, leaving only 20 survivors. The ruler at the time (and one of the survivors) apparently was a carrier of a recessive allele for an extremely rare genetic disease characterized by total color blindness. Because the allele is recessive, the ruler himself was not color blind, but over the next few generations inbreeding led to the appearance of color blindness among the small population's descendants. Today, nearly 30% of the population of Pingelap Island are carriers of the colorblindness recessive allele and nearly 10% are color blind.

Differences in the gene pool of a particular population are also affected by movement of individuals into (immigration)

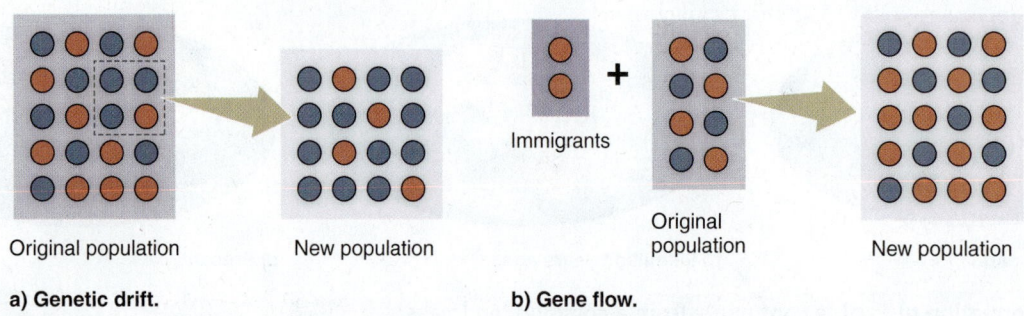

a) **Genetic drift.**   b) **Gene flow.**

Figure 22.8 Genetic drift and gene flow. a) Genetic drift may be due to either a bottleneck effect or a founder effect. b) Gene flow may be caused by either immigration (shown here) or emigration.

or out of (emigration) the population. This geographical redistribution of alleles, called **gene flow,** tends to mix pools of genes that might not otherwise mix (**Figure 22.8b**). The prevalence of international travel today has markedly increased the possibility of gene flow in the human population.

Certain human activities may also impact evolutionary processes. For example, the distinctly human practice of hunting big game animals for trophies specifically selects for larger, more fit males. How this practice might affect evolution is worth considering.

The slow processes of genetic drift and gene flow in complex organisms is quite different from the much faster evolution that occurs in simpler organisms, like viruses. **Antigenic shift** is a very rapid and dramatic change in a virus, brought about when two or more viruses combine or exchange genetic material to form a new virus, one with entirely different combination of genes The possibility of an antigenic shift resulting in a much more infective virus is what makes certain viruses, such as those causing bird flu and swine flu, so potentially dangerous.

☑ Recently, Florida panthers have been reduced to an extremely small population size of just a few dozen individuals. The few Florida panthers that survive today often carry an allele which causes bent tails, although bent tails were rare in the past. Explain which evolutionary process is most likely to have caused the change in the frequency of bent tails, and state whether it is a random or nonrandom process.

Mass extinctions eliminated many species

An **extinction** occurs when a life-form dies out completely. Throughout evolutionary history, major catastrophic events have sometimes wiped out whole groups of species regardless of fitness. In the past 530 million years, there have been at least five mass extinctions, each of which destroyed more than 50% of the species existing at that time.

The largest mass extinction, called the *Permian-Triassic,* occurred 251 million years ago at the boundary between the Permian and Triassic periods. It was accompanied by a carbon dioxide concentration nearly five times the current level, global warming of about 8 degrees Centigrade, and the extinction of approximately 95% of all species living at the time. The most recent and best-known mass extinction marked the end of the Cretaceous period and led to the extinction of the dinosaurs about 65 million years ago. Some scientists attribute the Cretaceous mass extinction to a dramatic change in climate after a comet or asteroid collided with Earth near what is now Cancun, Mexico. It appears that we may now be entering a sixth mass extinction, this one due to human activity.

MJ's BlogInFocus When exactly did Earth's largest mass extinction occur, and how quickly was it over? Visit MJ's blog in the Study Area in MasteringBiology and look under "Earth's Largest Mass Extinction."

http://goo.gl/LR24uo

Evolutionary trees trace relationships between species

An **evolutionary tree** (also called a *phylogenetic tree*) is an illustration of the evolutionary change and relationships among species. In an evolutionary tree, each branch represents a point of divergence between two species (the creation of a new species), and the length of the branches represents time (**Figure 22.9**). If a species becomes extinct, the branch for that species ends short of the present time.

When conditions are right, many new species may develop in a relatively short time from a single ancestor. Such short bursts of evolutionary activity are called **adaptive radiation** and are shown as numerous branches from a single point on an evolutionary tree. Periods of adaptive radiation most commonly follow changes in the environment that create new habitats.

Populations of living creatures are constantly undergoing evolutionary change. Sometimes it is necessary to further subdivide species into subspecies that look slightly different but can still interbreed. At any given time, some species are becoming extinct while others are in the process of diverging to become several new species. Evolutionary trees can be general, covering the entire history of all life-forms, or they can be detailed and focus on one species or one time period.

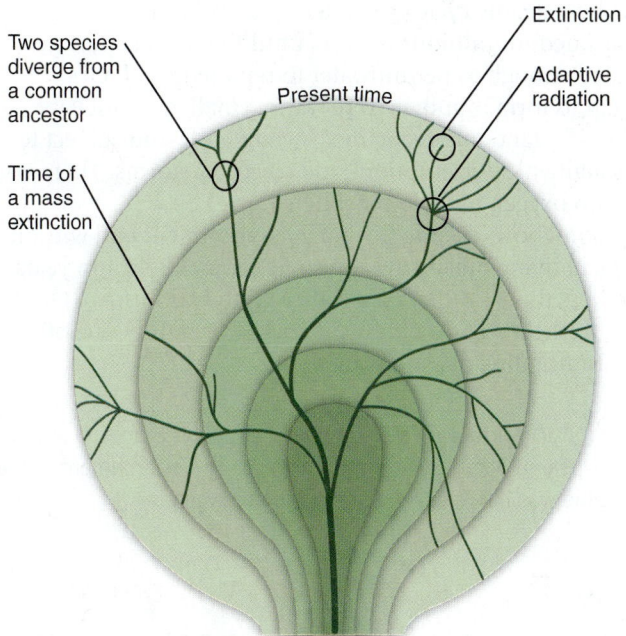

Figure 22.9 An evolutionary (phylogenetic) tree. Phylogenetic trees illustrate evolutionary change and relationships between species over time. Periods of mass extinctions are indicated by changes in color.

↩ **Recap** Mutations coupled with natural selection provide much of the basis for evolution. In addition, the course of evolution can be affected by genetic drift, gene flow, antigenic shift, and mass extinctions. An evolutionary tree visually depicts relationships between species. ■

22.3 In the beginning, Earth was too hot for life

To the best of our knowledge, Earth formed about 4.6 billion years ago. At one time, our sun was surrounded by a swirling cloud of gases, dust, and debris. As the outer region of the cloud cooled, it condensed into solid masses that today we know as the planets, one of which was Earth.

Four and a half billion years ago, Earth was an inhospitable place. Its tremendous heat melted the interior, which condensed as a core of liquid nickel, iron, and other metals. The solid but thin outer crust was constantly changing as frequent cracks in the surface and numerous volcanoes released molten rock and hot gases. The early atmosphere consisted primarily of carbon dioxide (CO_2), water vapor (H_2O), hydrogen (H_2), nitrogen (N_2), methane (CH_4), and ammonia (NH_3). There was no oxygen (O_2). Oceans did not exist because the tremendous heat vaporized all the water on the planet. Earth, not yet protected by an ozone layer, was constantly bombarded by ultraviolet radiation. The atmosphere crackled with electrical storms.

Over time, Earth cooled enough for water vapor at higher elevations to condense as rain, although most of the rain immediately vaporized again as it struck hot rock. The hot, steamy cycle of condensation and vaporization continued for millions of years until the planet's surface cooled enough to permit water to remain liquid. Oceans began to form, but they were warm, small, and not very salty. The land mass remained steamy, hot, and subject to frequent volcanic eruptions and electrical storms. There was still no oxygen in the atmosphere.

Somehow, despite the harsh environment and without oxygen, living organisms appeared about 3.8 billion years ago, less than 1 billion years after Earth was formed. Now we turn to the question of how life began, certainly one of the most intriguing questions in all of biology.

↩ **Recap** Earth was formed about 4.6 billion years ago and life appeared about 3.8 billion years ago. The environment at the time was hot, steamy, and devoid of oxygen. ■

22.4 The first cells lived without oxygen

We know from the fossil record that by 3.5 billion years ago single-celled creatures resembling bacteria were present. Once life appeared, evolutionary processes over billions of years produced the varied life-forms that we know today, as well as many more that have long since become extinct.

Organic molecules formed from atmospheric gases

Current scientific evidence indicates that the first step on the path to life was the creation of simple organic molecules from atmospheric gases (Figure 22.10). It was only *because* the early environment was so inhospitable (by our standards) that organic molecules were able to form. Apparently, the intense heat, ultraviolet radiation, and electrical discharges produced an extreme amount of energy, enough to combine molecules in the atmosphere into simple organic compounds, even without enzymes. These compounds (amino acids, simple sugars, fatty acids) dissolved in the sea, so that over time the oceans became a warm soup containing organic molecules and a little salt.

☑ Do you think a brand-new form of life could arise today on Earth's surface (that is, a form of life newly derived from inorganic molecules and not from preexisting life)? Why or why not?

Self-replicating RNA and DNA formed

The mere presence of organic molecules does not create a living organism that can grow and reproduce. How did these compounds join into molecules that could reliably reproduce themselves?

The only *self-replicating* molecules we know of are primitive single-stranded RNA and our modern double-stranded DNA, neither of which is likely to have formed spontaneously in the sea. Current thinking is that single-stranded self-replicating RNA first formed on templates of clay, in mudflats along the ocean's edge. Indeed, laboratory experiments have shown that, under the right conditions of moisture and heat, thin layers of clay can promote the formation of fairly complex molecules. Alternatively, the dry conditions and intense heat near volcanoes may have provided just the right environment for RNA formation.

If RNA was the first self-replicating molecule, the more stable DNA molecule must have developed from RNA very early in evolution. Modern cells use DNA as their self-replicating molecule and RNA to direct the synthesis of proteins, including enzymes. Only viruses use single-stranded RNA as their genetic material, and consequently viruses do not replicate on their own.

The first living cells were anaerobic

At some point, self-replicating molecules and small organic molecules became enclosed within a lipid-protein membrane. We do not know exactly how this happened. However, laboratory experiments have demonstrated that, under the right conditions, certain mixtures of amino acids and lipids will spontaneously form hollow, water-filled spheres.

The first simple cells relied on anaerobic metabolism, meaning metabolism without oxygen. Their ability to synthesize compounds was limited, so these simple cells relied on their immediate environment for the energy and raw materials they needed. When energy and raw materials were not available the cells died, or at least they did not replicate.

a) The primitive Earth was hot and steamy with small, warm oceans.

b) Simple organic molecules formed from atmospheric gases and dissolved in the oceans. The energy for their formation was provided by electrical storms, heat, and intense ultraviolet radiation.

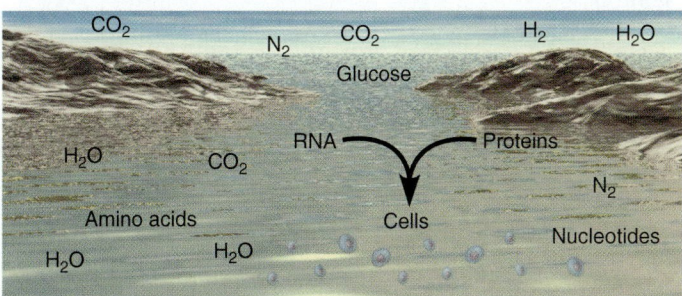

c) Self-replicating RNA probably formed on clay templates in shallow waters along the shorelines of oceans. RNA and other organic molecules became enclosed in a cell membrane and the first self-replicating cells were formed.

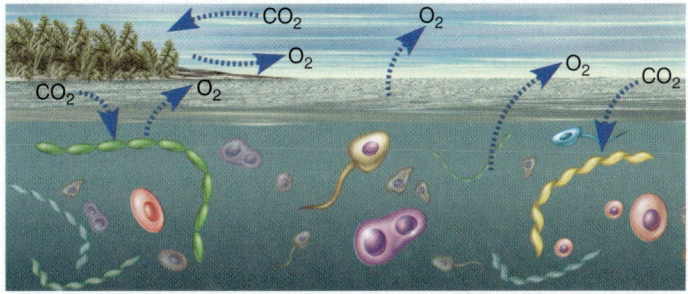

d) The development of photosynthesis created oxygen gas. The presence of oxygen permitted the later development of aerobic forms of metabolism.

Figure 22.10 The origin and early evolution of life.

☑ Outline an experiment that could test whether complex biological molecules could actually arise via the natural processes shown in this diagram. In your opinion, what might be a practical difficulty in replicating these processes in a laboratory experiment?

Could life also have arisen elsewhere in the solar system or universe? Given the manner in which it arose on Earth it's not hard to imagine that life could have formed elsewhere, but so far there is no proof that it did.

↩ **Recap** As best we can surmise, the sequence of events leading to self-replicating living cells was (1) the formation of organic molecules from atmospheric gases, (2) the formation of self-replicating RNA on templates of clay, and (3) the enclosure of RNA and organic molecules inside a cell membrane. All early cells used an anaerobic form of metabolism. ▪

22.5 Photosynthesis altered the course of evolution

By about 3 billion years ago, certain cells had mutated and developed the ability to produce, internally, some of the molecules they needed. They drew on the abundant CO_2 and water in their environment and the energy available in sunlight to create complex organic molecules containing carbon. This process is called *photosynthesis*, and its by-product is oxygen. As a result, oxygen began to appear in Earth's atmosphere. Note that oxygen appeared only after life began and as a direct consequence of life itself.

Aerobic organisms evolved

The availability of oxygen changed everything. Oxygen was toxic to anaerobic cells. In addition, because it is highly reactive, the oxygen began to break down the energy-containing molecules in the sea that the anaerobic organisms relied on for energy. Consequently, most of the anaerobic organisms that could not create their own organic molecules by photosynthesis died out.

In their place, new cells evolved that could harness the reactive nature of oxygen. These cells extracted energy from the abundant organic molecules stored within the cell itself, using the now readily available oxygen. As you may recall, the ability to use oxygen to extract energy from organic molecules is called *aerobic metabolism*.

The rise of animals and our human ancestors

Evolutionary processes are remarkably efficient at shaping life-forms to fit their environments. Some of the most significant of the evolutionary milestones include the following:

- DNA became enclosed within a nucleus about 1.7 billion years ago. Cells with nuclei, called *eukaryotes*, became the dominant cell type.
- The first multicellular organism, a type of seaweed, appeared approximately 1.3 billion years ago.
- Animals appeared about 600 million years ago.
- Dinosaurs became extinct about 65 million years ago.
- Our first distinctly human ancestors appeared only about 5 million years ago.

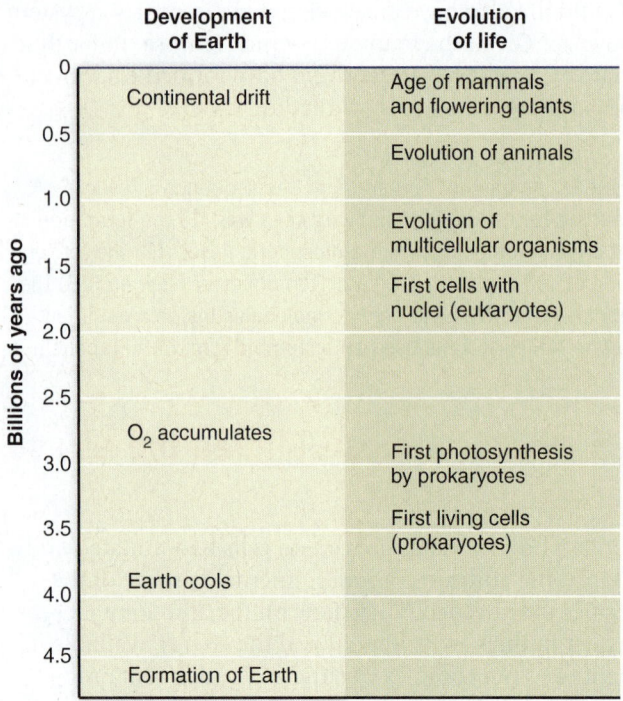

Figure 22.11 A time line of the evolution of life since Earth's formation. The entire span of human evolution lies within the horizontal line at the top of the figure.

Figure 22.11 presents a time line of biological milestones in the history of Earth so far. And how will it all end? The best estimate is that about 8 billion years from now, Earth will orbit closer and closer to an enlarged sun and be vaporized by the intense heat. If that is true, Earth is now about a third of the way through its period of existence.

↩ **Recap** Over time, photosynthesis evolved, resulting in increased atmospheric concentrations of oxygen. Some organisms subsequently developed an aerobic form of metabolism. A cell nucleus formed to enclose the DNA in most cells, multicellular organisms arose, and animals appeared. Our first recognizably human ancestors emerged about 5 million years ago. ■

22.6 Humans share a common ancestor with primates

Humans are primates

Taxonomy, the branch of science that focuses on classifying and naming life-forms, seeks a precise way to describe the relationships between all life-forms. The eight levels of the taxonomic classification system are *domain, kingdom, phylum, class, order, family, genus,* and *species.* A **species** is defined as a group of organisms that, under natural conditions, tend to interbreed and produce fertile offspring.

Life-forms are generally referred to by their genus and species names. Names are italicized, with the genus name capitalized and listed first. Table 22.1 shows the eight taxonomic categories as they apply to humans, or ***Homo sapiens.***

Humans belong to the class *Mammalia.* **Mammals** are vertebrates that have hair during all or part of their life and mammary glands that produce milk. Within the class Mammalia, we belong to the order *Primata,* more commonly called **primates.** Primates are mammals with five digits on their hands, fairly flat fingernails and toenails rather than hooves or claws, and forward-facing eyes adapted for stereoscopic vision. They include the lemurs (sometimes called pre-monkeys), monkeys, apes, and humans. The primate hand has a thumb that opposes four fingers.

Table 22.1 The taxonomic classification system as it applies to modern humans

Taxonomic category	Scientific name	First appearance*	Characteristics and examples of life-forms in each category
Domain	Eukarya	1.7 bya	Organisms with a membrane-bound nucleus within its cells.
Kingdom	Animalia	600 mya	Many-celled organisms with eukaryotic cells and a complex anatomy. Includes all animals: sponges, worms, insects, fish, birds, mammals, and many more.
Phylum	Chordata	550 mya	Animal with a nerve cord and a backbone. Fish, birds, and mammals are examples.
Class	Mammalia	120 mya	Members of the phylum Chordata with hair during part of their life cycle and mammary glands. Examples include mice, dogs, cows, whales, kangaroos, primates, and humans.
Order	Primata	60 mya	Members of the class Mammalia with five digits on the fore and hind limbs, flat fingernails, and stereoscopic vision. Lemurs, monkeys, apes, and humans are examples.
Family	Hominidae	5 mya	Primates that walk upright and that have slightly enlarged brains. Includes the genus *Australopithecus* and our *Homo* ancestors (now all extinct) and modern humans.
Genus	*Homo*	1.8 mya	Hominidae with an enlarged brain and a recognizable human skull and body form. Includes our extinct immediate *Homo* ancestors and modern humans.
Species	*sapiens*	0.1 mya	*Homo* with large brains and a complex cultural and social structure that includes spoken language. Modern humans.

*bya = billions of years ago; mya = millions of years ago.

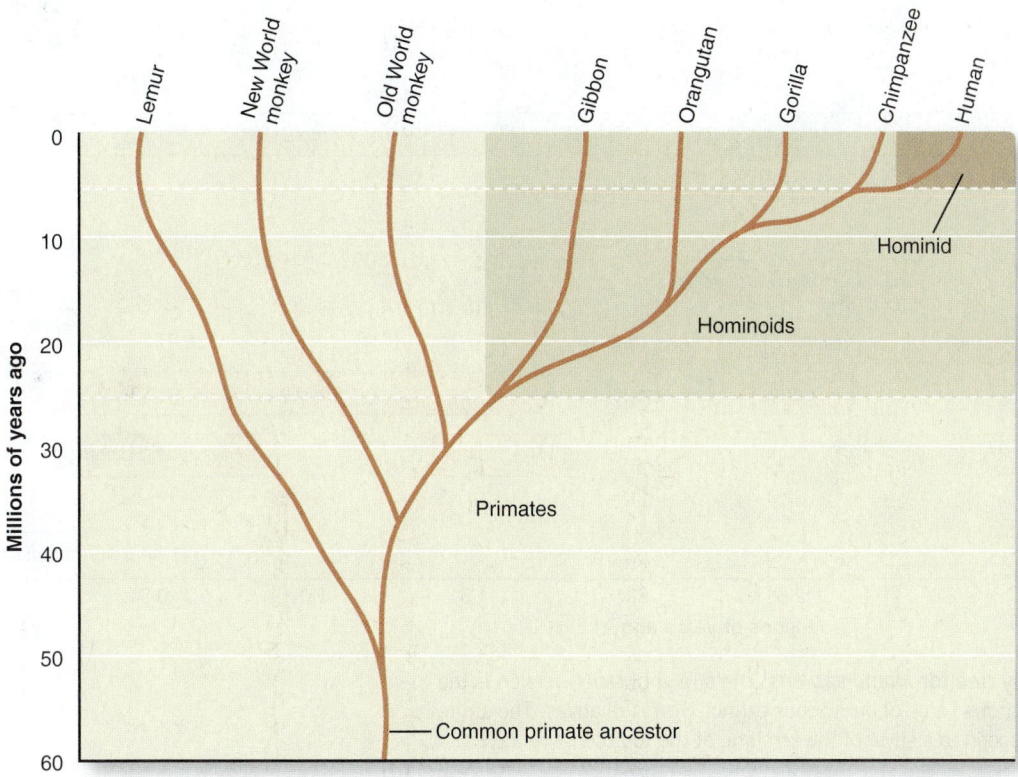

Figure 22.12 Evolution of modern primates from a common primate ancestor.

✔ Which are more closely related to humans: Old World monkeys or gibbons? (Hint: Look at the time since the last common ancestor.)

About 25 million years ago, a subgroup of primates diverged from a common ancestor of the old-world monkeys to become the present-day apes (gibbons, orangutans, gorillas, chimpanzees) and humans (Figure 22.13). **Hominoids** differ from the rest of the primates in that they are larger and have bigger brains. They also lack tails, and their social behaviors are more complex. Based on the number of differences in DNA, it appears that orangutans diverged first, followed by gorillas, chimpanzees, and humans. In other words, chimpanzees are our closest-living relatives among the hominoids. However, this does not mean that we descended from chimpanzees. It means that chimpanzees and humans apparently evolved from a common hominoid ancestor.

All of these features apparently evolved as adaptations for an arboreal lifestyle, that is, life in trees. The fossil record indicates that all present-day primates share a common ancestor that lived about 60 million years ago (Figure 22.12).

✔ Suppose you are at a paleontological dig site in Africa, and you find a fossil skeleton that is about 60 million years old. What traits could you examine to assess whether it might be a primate?

a) Gibbon.

b) Orangutan.

c) Gorilla.

d) Chimpanzee.

Figure 22.13 The hominoids. Modern humans (not pictured) are also hominoids, of the family *Hominidae*. Chimpanzees are our closest-living relatives.

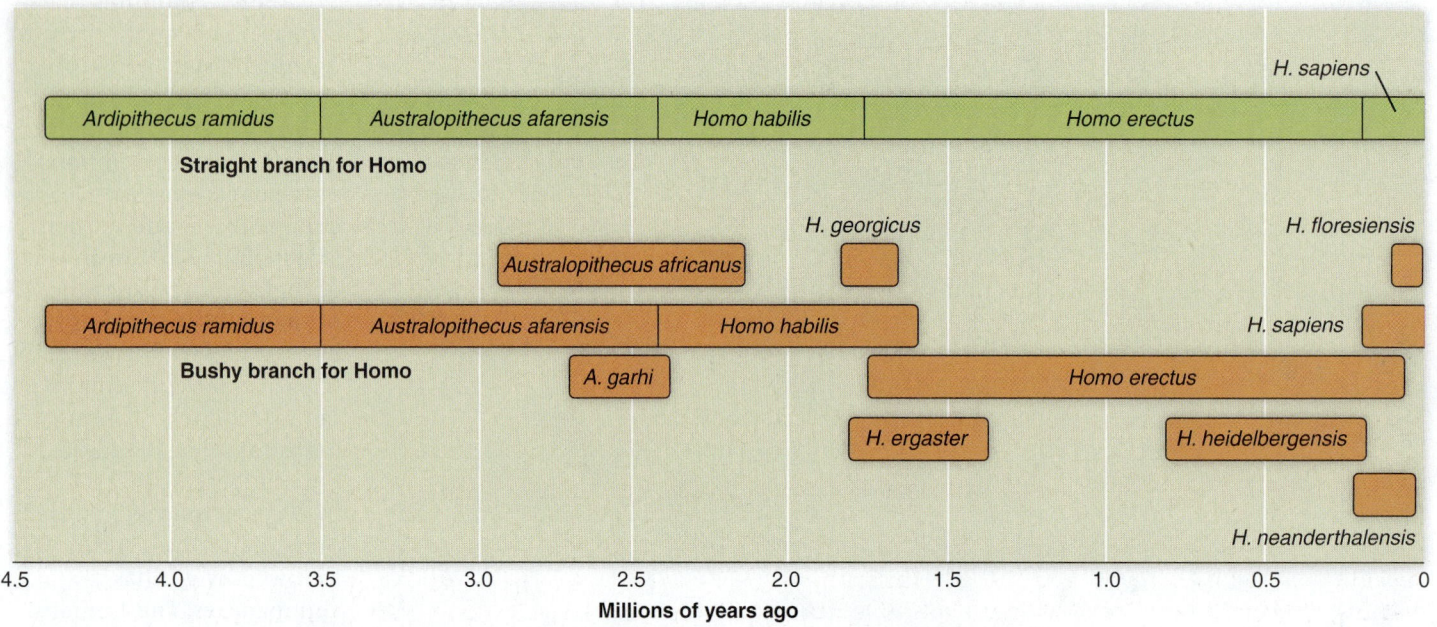

Figure 22.14 Two versions of the family tree for *Homo sapiens*. The straight branch version is the probable direct lineage. The bushy branch shows some of our various extinct *Homo* relatives. The entire 4.5 million years shown in this figure correspond to a sliver of the zero line at the top of Figure 22.11.

Evolution of *Homo Sapiens*

Although scientists disagree regarding specific dates and the classification of certain fossils, the general framework of human evolution is fairly well established. The story of the origins of modern humans begins about 4.4 million years ago in Africa (**Figure 22.14**). Anthropologists are still debating whether the pathway to modern humans was a single linear one or a branching, bushy one. Therefore, both pathways are represented in Figure 22.14.

Ardipithecus ramidus: walked upright some of the time

The fossil record indicates that the chimpanzee and human lines diverged from a common hominoid ancestor in Africa about 4–5 million years ago (review Figure 22.12). In the study of human origins, species of hominoids that are more like humans than chimpanzees are referred to as **hominids** (family Hominidae). The most likely first hominid, *Ardipithecus ramidus,* appeared about 4.4 million years ago in Africa. A nearly complete skeleton of a female *Ardipithecus ramidus* called "Ardi" was discovered in 1994 (**Figure 22.15**). Ardi had some body form and facial features in common with later hominids. She could walk upright according to the shape of her pelvis, but she also had long opposable toes for grasping branches with her feet and long fingers. In all likelihood, she did not walk upright very well, spending as much time in trees as walking upright on the ground.

Not all scientists are convinced that Ardi was a true hominid, at least not yet. Part of the debate is over whether habitual upright posture and bipedal walking are the defining features of a hominid. The other problem is that evolution occurs slowly, so it's hard to tell when

a) **The nearly complete skeleton of "Ardi," discovered in 1994.**

b) **An artist's depiction of Ardi.**

Figure 22.15 *Ardipithecus ramidus.*

a population is in the transitional phase to speciation and when it becomes a definitive species.

Australopithecus afarensis: most definitely walked upright

The genus *Australopithecus* contains several related species. One of them, most likely an African hominid named *Australopithecus afarensis*, is thought to be the direct ancestor of modern humans. Two other species, *Australopithecus africanus* and *Australopithicus garhi*, eventually became extinct.

A nearly complete 3.2-million-year-old skeleton of an *Australopithecus afarensis*, named Lucy by her discoverers, was found in Ethiopia in 1974. *Australopithecus afarensis* was partially apelike in its anatomy and social structure, with just the beginnings of what we would call human features. Their brains were still small relative to body size—more like chimpanzees than present-day humans. Their teeth had thick enamel, suggesting that they could grind foods; most likely they were vegetarian. The males were considerably larger than the females (*sexual dimorphism*), which suggests that *Australopithecus* had a social structure in which large, dominant males exerted control over small bands of females and less dominant males. Lucy, for example, was only about 1 meter tall. The arms of *Australopithecus* were long relative to their bodies, indicating that they still could travel from tree to tree (arboreal locomotion). But in terms of human evolution, the most important trait of *Australopithecus afarensis* was that they definitely walked upright.

Homo habilis: the first toolmaker

Our first distinctly human ancestor was **Homo habilis,** which first appeared about 2.4 million years ago in Africa (see Figure 22.14). Brain enlargement continued in *Homo habilis*, along with changes in teeth and facial features, a decline in sexual dimorphism, and conversion to a diet that included meat. As the first stone tools appear about this time, *Homo habilis* (which means "handyman") may have been the first toolmaker.

Homo erectus: out of Africa

The more human-looking **Homo erectus** (and its short-lived cousins *H. georgicus* and *H. ergaster*) apparently arose from *Homo habilis* ancestors about 1.8 million years ago (see Figure 22.14). *Homo erectus* migrated out of Africa to establish colonies as far away as Java and China. In the past, scientists thought that *Homo erectus* had been the direct ancestor of modern humans via populations that spread across Africa into Asia and Europe. However, *Homo erectus* died out in most locations by about 400,000 years ago. The very last *Homo erectus* lived in Java only 50,000 years ago, making them contemporary with modern humans.

With *Homo erectus* came even more brain enlargement, a longer period of infant development, a continued decline in sexual dimorphism, and the continuation and expansion of toolmaking. By 1.7 million years ago, *Homo erectus* had developed specialized stone tools such as hand axes and cleavers. There were probably changes in social structure as well, including male–female pair bonds, shared responsibilities for offspring, and the formation of hunting and gathering groups that shared food.

Other species of *Homo* apparently coexisted with *Homo erectus* but later became extinct (see Figure 22.14). Among them are *Homo heidelbergensis*, a species of archaic humans who created highly specialized stone tools, hunted cooperatively in groups, and may have had limited language skills. It is possible that some European descendants of *Homo heidelbergensis* became the Neanderthals, a controversial group of humans whose fossils were first discovered in the Neander Valley of Germany. Most anthropologists consider the Neanderthals to be a separate species (*Homo neanderthalensis*). Other anthropologists consider them a just subspecies of modern humans.

Homo sapiens: out of Africa again

The final phase of human development was marked by a substantial increase in brain size, the further development of spoken language, and the development of a physical structure that we would call thoroughly human. Although some of the details remain to be worked out, it appears that *Homo sapiens* evolved in Africa from *Homo erectus* between 200,000 and 140,000 years ago (see Figure 22.14). About 50,000 years ago a small band of *Homo sapiens* crossed the sea to what is now Yemen, on the Arabian Peninsula. From there, they migrated to all parts of the world, reaching the tip of South America about 10,000 years ago.

A key point is that wherever they went, the modern humans from Africa replaced all other existing *Homo* species. Ultimately *Homo heidelbergensis* and the Neanderthals vanished completely. Why? We don't know. Although some scientists speculate that there was interspecies violence, there is no evidence of it to date. It is just as likely that they fell victim to disease or were out-competed for resources. In addition, analysis of DNA recovered from Neanderthal bones shows that Neanderthals did interbreed with *Homo sapiens* to a limited extent, but because Neanderthals were so few in number compared to *Homo sapiens*, the Neanderthal distinctive physical features eventually disappeared. The last Neanderthals disappeared about 28,000 years ago.

MJ's BlogInFocus How many species of extinct humans are there? Visit MJ's blog in the Study Area in MasteringBiology and look under "Species of Extinct Humans."

http://goo.gl/KDmoSR

In what has been termed the greatest anthropological discovery in 50 years, archaeologists recently discovered evidence of what may have been yet another extinct human species, which they named *Homo floresiensis* (see Figure 22.14). This latter group apparently existed as recently as 12,000 years ago, making them fully contemporary with modern humans. (See the Current Issue: Who Were the Flores People?)

Homo sapiens is the sole surviving human (hominid) species. At a mere 200,000 years, modern *Homo sapiens* have not been around all that long. To put our existence in perspective, if the history of life on Earth were compressed into a 24-hour day, the entire history of *Homo sapiens* would represent less than the last three seconds.

☑ In the early twentieth century, before many hominid fossils had been found, most biologists thought that the first major human trait to evolve must have been an enlarged brain (much bigger than that of chimpanzees) and that upright walking must have evolved later. Explain whether this hypothesis turned out to be correct, and name the species in which these traits first appeared.

Differences within the human species

All humans belong to a single species. We are so similar that sometimes the sum of all humankind is called the human race. Nevertheless, there are certain distinct heritable physical differences within the human species that the dictionary defines as racial differences. How did these differences come about?

Throughout most of human history, the total human population existed mainly as smaller subpopulations that were geographically isolated from each other. Minor random mutations coupled with sexual reproduction within isolated subpopulations apparently led to slightly different human group phenotypes (referred to as *races*) that are still observable today. Human racial differences are a testament to the effectiveness of evolution over tens of thousands of years.

Some racial differences may have helped various subpopulations of humans adapt more effectively to their particular environment. For example, in a climate with intense sunlight, a dark skin color would have offered protection against excessive ultraviolet radiation. Whereas in a northern climate where sunlight was less intense, a light skin color would have been helpful, because small amounts of ultraviolet radiation are required to activate vitamin D, a vital nutrient. Facial features and even blood types may also have had their special advantages at the time. Other differences may simply have been neutral, meaning that they were the result of mutations that did no harm, and so were retained in the subpopulation in which they arose. In the final analysis, racial differences are nothing more than differences in phenotypes between subgroups of our common species.

Obviously, variation in human phenotypes does not stop at the subpopulation level. Each of us, every individual, has a distinct phenotype. In fact, DNA analysis reveals that there are as many differences between individuals of the same race as there are between individuals of different races. Given that there are no biological barriers to reproduction and that global travel and communication have recently become commonplace, human racial differences may disappear over the next 10,000 to 100,000 years, assuming we survive that long as a species.

↩ **Recap** A taxonomic system of eight categories classifies the relationship of any species to all others. Modern humans descended from a primate ancestor that began to walk upright nearly 5 million years ago. Over time, the brain enlarged and toolmaking skills and spoken language developed. The current best hypothesis is that modern humans first appeared in Africa about 140,000 years ago. All human beings belong to the same species. Racial differences most likely arose as adaptations to different environments. ∎

Chapter Summary

22.1 The evidence for evolution *p. 506*

- The theory of evolution is the single most unifying concept in biology today.
- The fossil record provides a physical record of life-forms that lived in the past.
- Patterns of embryonic development and structures with common evolutionary origins can be used to determine relationships between life-forms.
- Structural similarities between proteins and DNA can be used to define closeness of evolutionary relationships.
- Geographical separations can affect the distributions of evolving life-forms.

22.2 Natural selection contributes to evolution *p. 510*

- The pool of alleles in a population is affected by chance events such as genetic drift and gene flow.
- At least five mass extinctions have altered the course of evolution.
- An evolutionary tree visually depicts relationships between species.

22.3 In the beginning, Earth was too hot for life *p. 512*

- Earth was formed about 4.6 billion years ago. At first, it was a harsh environment with constant cycles of condensation and vaporization.
- Life began about 3.8 billion years ago.

22.4 The first cells lived without oxygen *p. 512*

- Organic molecules formed from gases in Earth's atmosphere.
- The first self-replicating molecule was probably single-stranded RNA.
- The first cell was formed when RNA and organic molecules became enclosed in a membrane.
- The early atmosphere was devoid of oxygen, so the first living cells survived by anaerobic metabolism.

22.5 Photosynthesis altered the course of evolution *p. 513*

- Oxygen appeared only after certain life-forms developed the process of photosynthesis as a way to produce organic molecules inside the cell.
- With oxygen available, other cells developed the process of aerobic metabolism as a way to extract energy from organic molecules.
- Once oxygen became plentiful in Earth's atmosphere, multicellular organisms evolved and Eukaryotes appeared.

22.6 **Humans share a common ancestor with primates** p. 514

- Humans are a relatively recent development on Earth. Humans and modern chimpanzees evolved from a common ancestor that lived about 5 million years ago.
- *Australopithecus afarensis* and *Homo habilis* were early hominids in human evolution.
- Modern humans apparently first appeared in Africa about 140,000 years ago. From there, they spread around the world.
- Heritable, observable, physical phenotype differences in human subpopulations are called *racial differences*. Racial differences are no more significant than differences between individuals. We all belong to the same human species.

Terms You Should Know

analogous structures, *507*
antigenic shift, *511*
evolution, *505*
evolutionary tree, *511*
extinction, *511*
fossil, *506*
gene flow, *511*
genetic drift, *510*
hominid, *516*
hominoid, *515*

Homo erectus, 517
Homo habilis, 517
Homo sapiens, 514
homologous structures, *507*
mammal, *514*
natural selection, *510*
population, *510*
primate, *514*
species, *514*
vestigial structures, *508*

Concept Review

Answers can be found in the Study Area in MasteringBiology.

1. Describe the three key elements of evolution.
2. Explain how a fossil is created.
3. List some of the main sources of evidence for evolution.
4. Describe how natural selection contributes to evolution.
5. Explain how genetic drift and gene flow affect populations.
6. Summarize the scientific theory of how life began on Earth.
7. Explain how rising atmospheric concentrations of oxygen affected life on Earth.
8. List the features that make Hominidae different from other families in the order Primata.
9. Describe the origins of *Homo sapiens*.
10. Explain why racial differences may disappear within the next 100,000 years.

Test Yourself

Answers can be found in the Appendix.

1. Which of the following statements about fossils is true?
   a. All body parts of an organism are equally likely to become fossilized.
   b. There are fossils of all species of organisms that have ever lived.
   c. Fossils in upper layers of rock are likely to be older than fossils in lower layers.
   d. Fossils are most likely to be found in sedimentary rock.

2. All the following would be considered homologous structures except a(n):
   a. elephant's foreleg
   b. bat's wing
   c. dolphin's flipper
   d. butterfly's wing

3. Examination of the skeleton of whales shows bones similar to the human pelvic girdle. In whales, these bones would be considered:
   a. vestigial structures
   b. anachronisms
   c. analogous structures
   d. heterologous structures

4. Vestigial structures, homologous structures, and biochemical similarities all provide evidence of:
   a. descent from a common ancestor
   b. barriers to migration
   c. random mating
   d. sexual selection

5. Which of the following statements about mutations is true?
   a. Mutations are always detrimental to survival.
   b. Mutations decrease the size of the gene pool and result in more similarities than differences.
   c. Mutations provide the genetic variation upon which evolution works.
   d. Whereas some mutations are random, many are caused by evolution.

6. Which of the following best describes evolutionary "fitness"?
   a. overall strength relative to other members of a population
   b. overall cardiovascular health relative to other members of a population
   c. ability to attract mates relative to other members of a population
   d. ability to survive and reproduce relative to other members of a population

7. Which of the following can affect the makeup of the gene pool within a population?
   a. natural selection
   b. mutation
   c. migration of individuals into or out of a population
   d. all of these choices

8. Genetic drift is most likely to affect:
   a. every population, regardless of size
   b. small populations
   c. large populations
   d. large and diverse communities

9. Which of the following would be described as genetic drift?
   a. the distribution of species resulting from the movement of the continents as they dispersed from Pangaea
   b. random mating that occurs in a large population
   c. the random changes that might occur in a gene pool as the result of a natural disaster that significantly reduced the size of a population
   d. the change in allele frequencies that result from sexual selection

10. Which of the following statements is true?
    a. Mass extinctions did not occur prior to the arrival of *Homo sapiens*.
    b. Increases in atmospheric CO_2 concentration and global warming have never occurred before.
    c. Environmental changes may create new habitats favoring the development of new species.
    d. Species can only become extinct; they can't diverge into more than one species.

11. In what order did events occur that led to life on Earth? (1) appearance of aerobic organisms, (2) appearance of photosynthetic organisms, (3) appearance and increase of O$_2$ in the atmosphere, (4) self-replicating RNA develops
    a. 4, 2, 3, 1   c. 3, 1, 4, 2
    b. 2, 3, 1, 4   d. 1, 2, 4, 3

12. Humans belong to all of the following taxonomic groups except:
    a. Primata   c. Prokaryotae
    b. Mammalia   d. Hominidae

13. Which of the following statements about the hominoids is false?
    a. Hominoids include gorillas, chimpanzees, and humans.
    b. Hominoids have larger brains than other primates.
    c. Hominoids lack tails.
    d. Hominoids typically are bipedal and walk upright.

14. Which of the following have coexisted with modern *Homo sapiens*?
    a. *Homo habilis*   c. *Homo floresiensis*
    b. *Homo neanderthalensis* d. both (b) and (c)

15. Which of the following is **not** believed to be a direct ancestor of *Homo sapiens*?
    a. *Homo erectus*   c. *Homo habilis*
    b. *Homo heidelbergensis*   d. *Australopithecus afarensis*

Apply What You Know

Answers can be found in the Study Area in MasteringBiology.

1. Because dogs are all one species, different breeds of dogs can mate and produce viable offspring. How, then, did different breeds of dogs with distinct characteristics evolve?
2. How would you expect an increase in gene flow in the human population to affect human phenotypes?
3. Critics of evolution often charge, correctly as it turns out, that the fossil record is incomplete and scientists cannot show how all living things evolved over time. Why don't we have a complete record?
4. Radiometric dating is a useful technique for identifying the age of fossils. Why does the suspected age of the fossil determine which isotope an investigator might use to try to date it?
5. There are species of fish that live in caves no longer connected to the surface. These fish have eye structures even though the fish are blind. Why do they have eyes even though the eyes don't function?
6. It is believed that humans evolved in Africa and then spread to all the other continents. If this is so, explain why the people of northern Europe do not have dark skin.

7. Throughout Earth's history, many species, and perhaps even most of the species that ever existed, have become extinct. The causes are numerous: climate change, volcanic activity, the actions of oxygen-producing organisms, changes in carbon dioxide concentration, and even asteroid impacts. Have humans ever caused extinctions?
8. Of all the great evolutionary advances of life on Earth, which organisms have contributed the most significant ones: simple organisms or complex multicellular organisms such as plants and animals? Defend your answer.

MJ's BlogInFocus

Answers can be found in the Study Area in MasteringBiology.

1. It has been argued that big-game trophy hunting, because of its emphasis on harvesting the largest males of a species, could have a negative impact on the evolution of those species. Based on what you know about biological processes, how do you think we could best minimize that impact and still allow big-game trophy hunting? To view a research paper on this subject, visit MJ's blog in the Study Area in MasteringBiology and look under "A Impact of Humans on Evolution." http://goo.gl/WcQsqN

2. Where (on which continent) would you guess that language originated? Visit MJ's blog in the Study Area in MasteringBiology and look under "Origins of Language." http://goo.gl/TgsjnK

MasteringBiology®

Students Go to MasteringBiology for assignments, the eText, and the Study Area with animations, practice tests, and activities.

Professors Go to MasteringBiology for automatically graded tutorials and questions that you can assign to your students, plus Instructor Resources.

Answers to ✓ questions are available in the Appendix.

Ecosystems and Populations

A school of sockeye salmon on the way to their spawning grounds.

Key Concepts

- **Under ideal conditions, a population of any species will grow** until growth is countered with *environmental resistance.*

- **Many species of organisms may live together in complex communities within an ecosystem.** Disruption of an ecosystem can have long-lasting consequences.

- **The energy for fueling life's complex chemical processes comes from the sun.** Producer organisms make their own organic molecules; consumer organisms get their organic molecules by consuming other organisms.

- **The molecules and elements that compose living organisms are recycled over and over again between living organisms and Earth.** The speed of recycling can vary from minutes to millions of years depending on the nature of the molecule and its location within its particular geochemical cycle.

- **The human population has grown rapidly since about 1700.** Currently, at 7.3 billion, it could increase another 25% by 2050.

Overharvesting Is Depleting the Oceans' Wildlife Populations

Humans have always harvested wildlife from the sea for food and other resources. In ancient times, ocean harvests were limited because of the crude way in which humans hunted, with hooks, spears, harpoons, and small nets. For centuries, the bounties to be harvested from the world's oceans were viewed as nearly inexhaustible. In reality, harvesting in the oceans was kept in check by the low-tech fishing methods of the day. Prior to 1950, most fishing boats had limited ranges and storage capacities and lacked refrigeration. Fishing tended to be limited to coastal areas, and catches were brought back to port each day while they were still fresh.

First Whales, Then Fish

Signs that humans could actually overharvest the ocean's vast resources first appeared in the early 1900s, when it became apparent that populations of the larger species of whales, such as the gray whale and the humpback, were being decimated by intense hunting. Then starting in the mid 1900s, humans began to harvest more fish as well.

In the year 1950, the worldwide annual fish harvest was 20 million metric tons. By 1987, the worldwide annual fish harvest had more than quadrupled to about 90 metric tons. A major reason for the increased harvest was the development of high-tech fishing vessels. Modern purpose-built super trawlers pull nets thousands of feet long, capable of capturing up to 400 tons of fish at once. The largest fishing vessels process and refrigerate their catch on site, staying at sea for months at a time.

The Tragedy of the Commons

Whale and fish overharvesting are made worse by what is known as the "tragedy of the commons." When there is free and unfettered access to a common resource (all the fish in the sea), everyone tries to reap as much of the benefit of that resource as possible. Unless there is joint agreement regarding who is responsible,

Commercial fishermen harvesting salmon in Alaska.

Questions to Consider

1 Do you think extending nations' ocean territorial limits is a good idea? Do you think it would have any effect on the world's fish populations?

2 Are there any marine protected areas (MPAs) or underwater national parks in your area? Have you visited them?

there is no incentive to use the resource in a sustainable manner. This is what is happening in the world's oceans. We are overharvesting a commonly held resource, but no nation is willing to stop when it knows that the other nations are not doing the same. Currently, the annual fish harvest remains at about 90 million tons per year, well over its pre-1950s level.

Attempts to limit harvests in the world's oceans by common agreement have so far not been very successful. International efforts in the 1930s to reduce the annual harvest of whales were met with stiff resistance by nations whose economies were tied to whaling. Commercial whaling was finally banned by an international agreement in 1986; however, Russia, Japan, and Norway rejected the ban. Since that time, Russia has discontinued whaling, but Japan continues to harvest whales under the guise of research and Norway still declares itself exempt from the ban. Today, the world's whale populations are nowhere near what they were several centuries ago. Agreements to limit fishing have similarly met with resistance.

Long-Lived Species Are More Susceptible to Decline

Because whales take so long to reach maturity and reproduce, their populations decline sharply when the individual death rates (in this case, from whaling) suddenly increase. Like whales, some of the most prized fish species are long-lived and late to mature and vulnerable to increased harvesting. Atlantic Bluefin tuna, for example, live for up to 50 years and may weigh more than 500 pounds. In 2009, the population of Atlantic Bluefin tuna in the Western Atlantic was estimated to be less than 20% of what it was 40 years previous.

→

Overall, the annual fish harvests of fully 30% of the species of fish preferred by humans have already declined by more than 90%. As the harvests of preferred fish decline, less desirable fish are being harvested instead. Sooner or later the less desirable fish will go the way of the more desirable, and the overall annual fish harvest will decline.

What Can Be Done About Overfishing?

Solutions to overfishing are difficult, especially where international cooperation is required. However, several actions could be taken:

- By international agreement, apportion the fish catch in commonly held waters. This will not be effective unless agreements are unanimous or nearly unanimous, as the experience with the whaling industry shows.

- Limit the annual catch of certain species of fish within each country's zone of exclusive fishing rights (generally 200 nautical miles from the shoreline). This does not require international cooperation, and it's already being done in some cases. In 2015, the New England Fishery Management Council cut the total allowable cod catch per fishing boat by 75%. This kind of drastic action has serious implications for the sustainability of the cod fishing industry, but it may be what it takes to keep the cod population from declining even further. In an effort to allow fishermen some latitude in deciding how to apportion the catch among themselves, the United States in 2010 established a "catch share

management" plan in which fishermen can sell their quota for a particular species of fish to another fisherman, as if it were a stock on Wall Street.

- Create marine protected areas (MPAs) where fish can live and breed undisturbed by fishermen. The Unites States now has more than 1,600 MPAs, most of them very small. Some allow limited fishing, while others do not. A few MPAs have been designated as underwater national parks. An example is Biscayne National Park in Florida.

- Clamp down on illegal fishing. Foreign vessels are allowed to come within 12 miles of a country's shoreline, well within the country's 200-mile exclusive fishing zone. With the vastness of the oceans making enforcement difficult, who's to stop foreign vessels from fishing a little while they're passing by? Nevertheless, as fish become a more valuable resource, nations and regulatory bodies are stepping up their enforcement to protect their fisheries industries. With better enforcement the illegal harvest of Chilean sea bass was reduced by more than 90% in the Southern Ocean between 1997 and 2010.

Will any of these efforts, if taken, be enough to allow fish populations to return to their pre-1980s levels? You may learn the answer in your lifetime. Keep an eye on the cost of fish in the market for a clue.

SUMMARY

- **The first sign of overharvesting of ocean wildlife was a decline in whale populations in the early 1900s.**

- **From 1950 to 1987, the worldwide ocean fish harvest more than quadrupled, from 20 million metric tons to about 90 million metric tons annually.**

- **Some fish populations are now below 10% of their peak levels. Annual fish harvests are holding steady but only because less desirable fish are being harvested.**

- **Unless something is done to increase ocean fish populations, a decline in the annual catch over the next 30 years is almost certain.**

- **Solutions include international agreements to limit the fish harvest in international waters and actions by individual nations to protect fish populations in waters they control.**

All living organisms, including humans, have certain features in common. All living organisms consist of one or more living cells. All have the capacity to reproduce. All share a common origin, and all have evolved over time. According to the latest estimates, there are nearly two million different species of living organisms on Earth. Some species are completely dependent on another species for survival. Other species compete with each other for the same food supply and space. Many species are engaged in intense predator-prey relationships. Every species relies on its environment for survival. How do all the species on Earth share the planet harmoniously? In fact, they don't. Sometimes a species flourishes for a while and then fades to relative obscurity. Sometimes a species is driven to extinction. ■

Ecology is the study of the relationships between organisms—including humans—and their physical environment. In this chapter, we begin by describing general principles of ecology and population dynamics. We also examine the rapid growth of the human population and the possible effects of this growth on our world in the future.

The dynamics of the human population on Earth are no different, conceptually, from those of a bacterial population in a laboratory dish—it's just that we are bigger and our generation times are longer.

23.1 Ecosystems: living organisms and their environment

The basic unit of study in ecology is an *ecosystem*. *Eco-* implies a connection to the environment, and *system* indicates a functional unit. An **ecosystem** consists of a community of organisms *and* the physical environment in which they live. An ecosystem comprises all the living things, all the matter, and all the energy in a particular area. In studying an ecosystem, we consider not only the relationships among the species that inhabit it but also how nonliving matter within it is recycled and how energy flows in and out of it. The components of an ecosystem form a hierarchy that includes populations, communities, and the physical environment.

A **population** is a group of individuals of the same species that occupy the same geographic area and interact with each other. A population of elk inhabits a mountain valley, and a population of frogs inhabits a small pond.

A **community** consists of the populations of all the species that occupy the same geographic area and interact. Insects feed on decaying wood. Woodpeckers feed on insects living in trees and make nests in those same trees. Deer browse on shrubs, cougars hunt the deer, and so on. A community is the living part of an ecosystem.

Together, all of the ecosystems on Earth compose the **biosphere,** or planetary ecosystem (refer back to Figure 1.7 on p. 10). The biosphere is so complex that it is difficult to study as one entity. Studies of simpler ecosystems have, however, led to some interesting conclusions that may apply to the entire biosphere as well. Most importantly, it appears that ecosystems are delicately balanced, self-sustaining units in which organisms interact with each other and their environment in complex ways. In the process, organisms can modify or control the physical and chemical nature of an ecosystem itself. Ecologists suggest that disturbing the delicate balance of the biosphere could endanger the survival of the human species—and all other species as well.

Recap An ecosystem consists of a community of organisms and their physical environment, and a community consists of all populations that interact within the same area. The biosphere comprises all ecosystems on Earth. ∎

23.2 The dynamic nature of populations

The population of any one species in a given geographic area is subject to a number of factors that may cause the population's size to change over time. Let's look at several factors that affect a population's size, including habitat, range, biotic potential, and ecological balance.

Where a species lives: habitat and range

A species' **habitat** is the place where it lives (Figure 23.1). Typically, this is determined by its tolerance for certain environmental conditions. For example, the habitat of the now rare northern spotted owl is mature evergreen forests, the habitat of the bison is grassland, and beavers prefer a habitat of mountain valleys with freshwater streams and small trees.

Each organism's habitat has certain chemical and physical characteristics that favor the organism's comfort and survival. Animals and plants that occupy the same or overlapping habitats form a community. Members of a community may benefit each other in some way, or they may compete with each other for resources, or they can even use each other for food and/or shelter.

Because the availability of ideal habitats varies, each species has a **geographic range**—the area over which it may be found. An organism's range

The habitat of these Springboks is the grasslands of Namibia, Africa.

The habitat of these students is a university campus.

These red-eyed tree frogs inhabit the tree canopy of the Costa Rican rain forest.

Figure 23.1 Habitats. An organism's habitat is the place where it lives.

is limited by several factors: (1) competition for resources such as sunlight or nutrients, (2) intolerable conditions such as extreme temperatures or altitude, or (3) physical obstacles, including mountain ranges, deserts, and bodies of water.

In the past few centuries, the development of worldwide transportation has enabled some animals to expand their ranges beyond physical barriers. For example, innovations in navigation and shipbuilding allowed European humans to cross the Atlantic Ocean and expand their range into North and South America.

Population growth rate tends toward biotic potential

One of the most important topics in ecology is how populations change in size over time. The maximum rate of growth of any population under ideal conditions is called its *biotic potential.* Biotic potential is a function of certain characteristics of the species itself, including:

- The number of offspring produced by each reproducing member (female, if reproduction is sexual);
- The time span before the offspring reach reproductive maturity;
- The sex distribution (ratio of males to females) if reproduction is sexual;
- The number of organisms at reproductive-age.

Typically, the biotic potential of any species follows an **exponential growth**—a repeated doubling of a population over similar time periods. In Figure 23.14, exponential growth is indicated by the curve with a characteristic J shape. For any species, the steepness of the growth curve, representing its biotic potential, is determined by the previous four factors. The J shape reflects the fact that under ideal conditions, populations would double again and again over similar time periods. If you know the percent change in the population size per year, you can calculate how long it will take the population to double in size by using the Rule of 72: Divide 72 by the percent change per year. For example, a population that is increasing at a rate of 4% per year will double in 18 years (72/4 = 18); at 3% it would take 24 years. The rule of 72 works for any variable with a known percent change per unit of time. (If your savings account is earning 4% a year, your money will double in 18 years.)

☑ If equal numbers of wolves and bears are introduced to a new area at the same time, the wolf population will often double in just a year, but the bear population can take a decade or more to double. Which species has greater biotic potential? What might explain this difference in biotic potential between wolves and bears?

Environmental resistance limits biotic potential

In the real world of finite resources and competition, an opposing force called **environmental resistance** limits any species' ability to consistently realize its biotic potential (**Figure 23.2**). Environmental resistance consists of factors that kill organisms and/or prevent them from reproducing. Examples include disease, predation by other organisms, environmental toxins, and limitations on nutrients, energy, and space.

Even when conditions are ideal to begin with, eventually every growing population reaches a point where environmental resistance begins to increase. For bacteria in a petri dish or humans living on an island, environmental resistance may be the food supply. For a population of mice, it may be disease brought about by overcrowding. For a woodpecker, it may be a lack

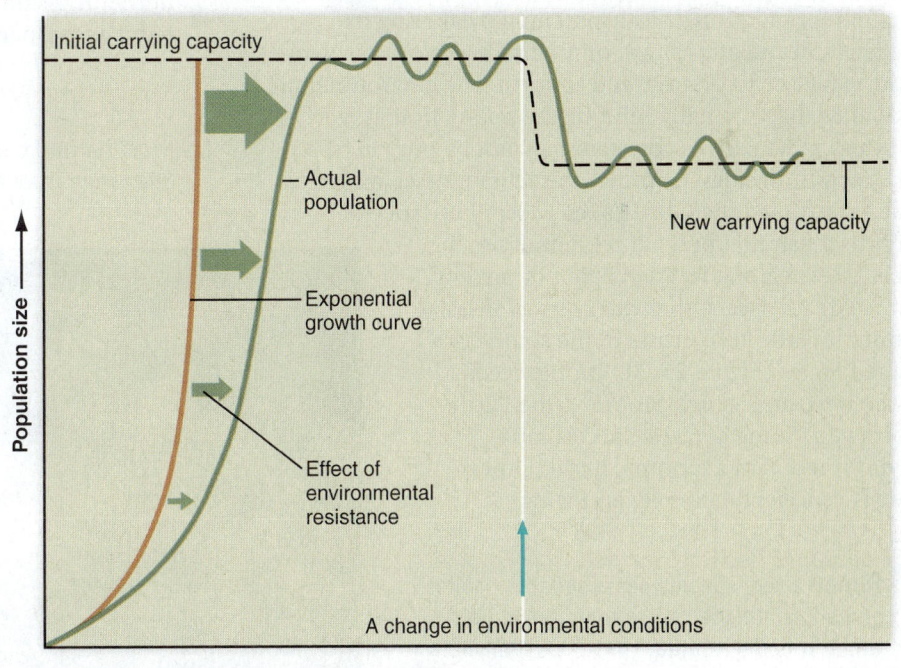

Figure 23.2 Effects of environmental resistance and carrying capacity on population size. As the population increases, the environmental resistance to continued exponential growth increases. Eventually, the population stabilizes near the carrying capacity. Changing environmental conditions can alter the carrying capacity.

☑ What would this graph look like if it depicted a population with a lower biotic potential? Would a lower biotic potential affect the carrying capacity?

of suitable nesting sites. Although we cannot always identify what specific environmental resistance is causing a population to stop growing or decline, we do know that no population grows at its full biotic potential indefinitely.

Ultimately, a population achieves an approximate balance between biotic potential and environmental resistance. At this point, it stabilizes, reaching a steady state. The population that the ecosystem can support indefinitely is called the **carrying capacity.** Population sizes may vary around their carrying capacity from time to time. In addition, the carrying capacity can change if environmental conditions change. However, in the long run, every population tends to stabilize near its carrying capacity. A wet spring might increase the carrying capacity of a prairie ecosystem for certain grasses and the animals that feed on those grasses. Conversely, three years of drought might reduce the carrying capacity to below its average value.

Invasive species alter the ecological balance

Humans sometimes alter the delicate balance of ecosystems inadvertently or on purpose through the introduction of **invasive species.** An invasive species is one that is not naturally found in an ecosystem; however, once the foreign species is introduced into an ecosystem it may overtake and dominate the native species. Invasive species challenge a natural population's biotic potential by altering its predator-prey relationships, omnivore/herbivore/carnivore ratio, exposure to disease, and so forth. The environmental resistance that normally inhibits the population growth of a species in its native ecosystem may not be present in an ecosystem it invades. Consider the zebra mussel, a species that has invaded the Great Lakes. Most of the species of birds and fish that prey on zebra muscles in their native habitat in Russia are not present in the Great Lakes. Other examples of invasive species include kudzu vine in the southeast United States (**Figure 23.3**), tumbleweeds in the west, and scotch broom in the Pacific Northwest. Some invasive species cause environmental or economic harm. Other invasive species are merely annoying.

↺ Recap Every species has a suitable habitat and an established geographic range over which it is distributed. Under ideal environmental conditions, a population would grow exponentially according to its biotic potential. However, environmental conditions generally are not ideal, and consequently, environmental resistance limits growth. The population of a particular species that an ecosystem can sustain indefinitely is called its *carrying capacity.* ∎

23.3 Communities: different species living together

Communities are composed of many different species. The relationships among the species may be quite complex, ranging from intense competition to mutual benefit. Oftentimes different species within a community rely on each other as sources of food, shelter, or protection.

Overlapping niches foster competition

An organism's *niche* is its role in the community—its functional relationship with all living and nonliving resources of its habitat. A woodpecker's habitat is woodland forests. Its niche is to rid trees of their insect infestations and to serve as a food supply for hawks or raccoons or ultimately for bacteria. A well-balanced ecosystem supports a wide variety of species, each with a different niche.

Although two species do not occupy exactly the same niche, niches can overlap enough that competition may occur between species for limited resources. For example, both grass and clover compete for sunlight in a lawn, and both foxes and hawks compete for mice as a food supply. Depending on their particular niches, one species may be able to completely out-compete another in a given location, a phenomenon known as *competitive exclusion*. The trees in a mature forest, for example, may outcompete grasses for sunlight and nutrients. As the human population expands into new habitats, humans are increasingly excluding other species.

Succession leads toward a mature community

Most natural communities undergo constant change. *Succession* is a natural sequence of change in terms of which

Figure 23.3 An invasive species. Scotch broom in the Pacific northwest region of the United States.

organisms dominate in a community. Succession is determined by population growth rates, the niches occupied by various species, and the kinds of competition that exist between them.

If you have ever seen an abandoned field grow first to weeds, then to shrubs, and finally to trees, you have watched succession in action. Organisms that can reproduce and grow quickly under the prevailing conditions are the first to arrive. These first organisms provide food and shelter for the next arrivals, which grow more slowly but occupy more specific niches. The number of species tends to increase as the succession advances and interrelationships between populations become ever more complex. If the area suffers no major shocks, such as a fire, succession eventually leads to a stable, **mature community** that changes very little over time.

Mature communities are the most efficient communities in terms of energy and nutrient utilization and the most varied in terms of numbers of species. A mature community may take hundreds of years to develop and may retain its character for many hundreds to thousands of years.

The type of mature community found in any region is largely determined by the physical environment but can also be influenced by the organisms living in surrounding areas. In the Northeast United States, the typical mature community is dominated by large but slow-growing deciduous trees (broad-leaved trees such as oaks that lose their leaves each winter, then grow new leaves each spring). In the Midwest, the mature community is prairie; in the Southwest, it is semiarid desert; and in the mountainous West, it is towering evergreen forests (**Figure 23.4**). Examples of mature communities in other parts of the world include the frozen treeless Arctic tundra and tropical rainforests near the equator.

Mature communities are not necessarily the only stable communities, however. The modern logging practice in some types of forests of selectively harvesting only a few mature trees on a regular basis creates a stable but immature community. Note that such a community is only maintained (kept in a perpetual state of immaturity) by regular, consistent intervention.

When mature communities are disturbed, they do not recover easily. Consider the introduction of the giant reed, *Arundo donax*, to coastal stream communities in Southern California. The mature community in this area includes coast live oak and laurel sumac. However, since *A. donax* was brought to the United States in the 1800s for erosion control, its rapid growth characteristics have allowed it to supersede slower-growing native species until today it dominates these communities. The very complexity and slow growth characteristics of mature communities make them vulnerable to disruption.

Figure 23.4 A mature community. This old growth temperate rainforest in Washington is a mature community. The species present are those best adapted to the environment once the environment reaches a stable condition.

☑ An abandoned field in the Northeast United States will eventually become a forest. However, a similar abandoned field in the Midwest will remain a field (eventually becoming a prairie—a natural field). Why won't trees colonize the midwestern field? More generally, why do different regions have different mature communities?

Ecosystems: communities and their physical environment

An ecosystem, as mentioned earlier, has both living and nonliving components. The total living component of an ecosystem is called that ecosystem's **biomass.** The nonliving components consist of the *chemical elements* (the physical matter that makes up Earth) and the *energy* that drives all chemical reactions.

A constant supply of energy is essential for continued life because the transfer of nutrients and energy from one organism to the next is always an energy-absorbing process. Nearly all of Earth's energy comes from the sun, with only a tiny amount supplied by the residual heat of Earth itself (thermal vents and volcanoes). In contrast to energy, the chemical matter of Earth is recycled over and over again between Earth and the biomass. The next two sections discuss these two key concepts in more detail.

↩ **Recap** Species with similar niches may be in competition with each other. Communities go through progressive changes over time, called succession, which increases overall efficiency and species variety. Left undisturbed, eventually, they become a mature community. Ecosystems consist of biomass, energy, and nonliving matter. ∎

23.4 Energy flows through living organisms

Nearly all of the energy needed by living organisms comes from the sun. Energy in sunlight is used to create the complex organic molecules required by living organisms. A constant supply of new energy from the sun is required to sustain life, because energy is constantly dissipated at each level of living systems.

The flow of energy through living organisms demonstrates two key principles of physics, called the *laws of thermodynamics:*

- The *first law of thermodynamics* states that energy is neither created nor destroyed. However, energy can change form, and it can be stored.
- The *second law of thermodynamics* states that whenever energy changes form or is transferred, some energy is wasted (converted to non-useful forms). Generally this wasted energy is lost as heat.

Producers capture and convert energy, consumers rely on stored energy

Energy flows in one direction through an ecosystem, from the sun to producers to consumers. **Producer** organisms use sunlight as a source of energy for combining nutrient matter found in the environment into the organic molecules they need. **Consumer** organisms get their organic molecules (and energy) by eating other organisms or dead and decaying material.

Most producers are capable of **photosynthesis** (**Figure 23.5**). A few producers that live entirely in the dark, such as bacteria located near deep-ocean hydrothermal vents, rely on a process called *chemosynthesis.* Producers are also sometimes called *autotrophs,* meaning "self-nutritive" or "self-growing." On land the producers are mainly green plants, whereas in aquatic ecosystems they are primarily species of algae.

The ability of producers such as plants to harness the energy of the sun depends on small organelles called *chloroplasts* within certain plant cells. Chloroplasts contain a green light-absorbing pigment called *chlorophyll* that gives plants their green color. In the first stage of photosynthesis, chlorophyll absorbs the energy of sunlight and then uses the energy to drive a chemical reaction in which water (H_2O) is disassembled and high-energy molecules of ATP and NADPH are produced. A by-product of the reaction is molecular oxygen (O_2), which is released to the atmosphere. In the second stage of photosynthesis, ATP and NADPH are used to fuel a biochemical cycle called the *Calvin cycle,* in which CO_2 from the atmosphere is used to produce high-energy carbohydrate molecules (sugars) consisting of multiple CH_2O units. Using the energy stored in the sugar and water and minerals obtained from the soil, plants are able to make all the other organic molecules they need, including proteins, lipids, and cellulose, a primary component of plant cell walls.

Notice how neatly the biochemical reactions of plants (producers) and animals (consumers) complement each other. Plants require CO_2 from the atmosphere for photosynthesis and produce O_2 as a by-product, whereas animals require O_2 for their metabolism and produce CO_2 as their waste product.

> **MJ's BlogInFocus** How is the rise in global atmospheric CO_2 concentration as a result of the burning of fossil fuels likely to affect plants? Visit MJ's blog in the Study Area in MasteringBiology and look under "CO_2 and Forest Growth."
>
> http://goo.gl/NtH3Xc

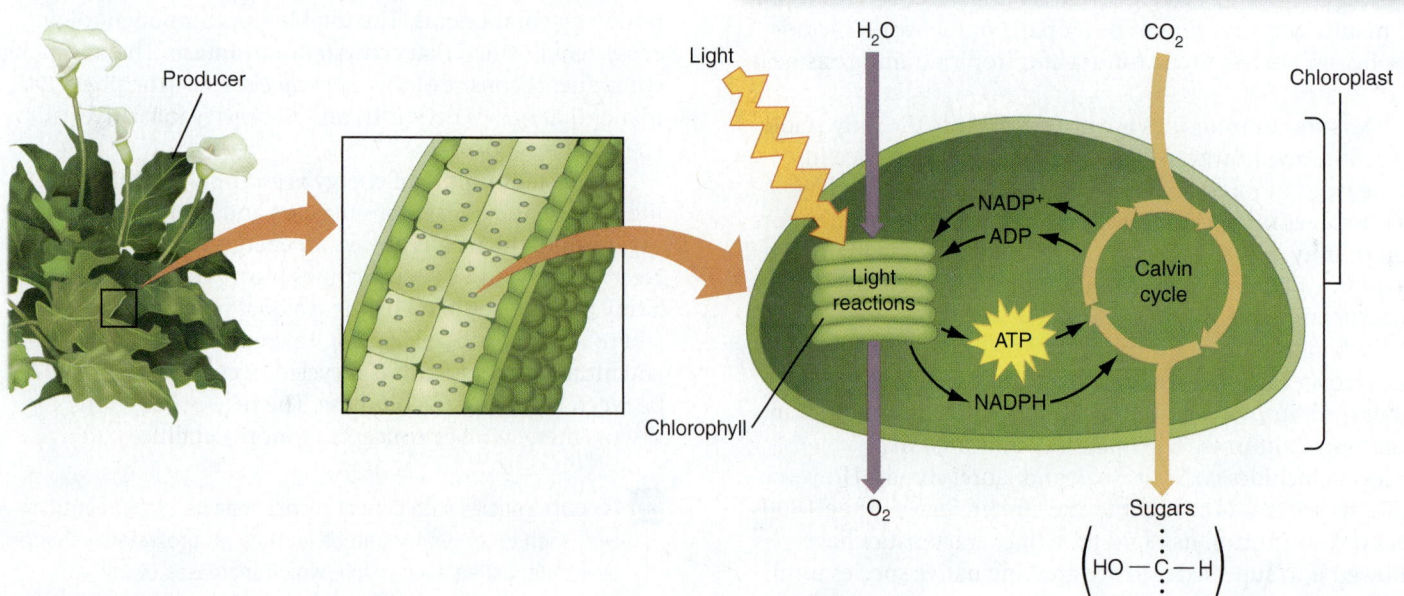

Figure 23.5 Photosynthesis. Producer organisms such as plants use the energy in sunlight, water, and CO_2 from the atmosphere to produce sugars. A by-product of photosynthesis is oxygen.

Most plants make more sugar than they need for their own use. The extra sugar is converted to starch, which is stored in the chloroplasts and in storage cells of roots, tubers, fruits, and seeds. On a global scale, it is estimated that photosynthesis accounts for the production of about 145 billion tons of carbohydrates per year, or nearly 23 tons for every person on Earth.

Animals, most bacteria, and fungi are *consumers* (also called *heterotrophs*). They cannot utilize the sun's energy to synthesize the molecules they need. Instead they must consume foods that already contain stored forms of energy. Consumers fall into four types, depending on what they use as a food source.

- **Herbivores** feed on green plants. They are also called *primary* consumers because they eat only producers. Sheep, goats, and cattle are herbivores.
- **Carnivores** feed on other animals, and thus they are *secondary*, *tertiary* (third-level), and sometimes *quaternary* (fourth-level) consumers. The primary forms of stored energy in animals are fats and carbohydrates such as glycogen. Wolves and hawks are carnivores.
- **Omnivores** feed on both animals and plants. Humans, pigs, and bears are omnivores.
- **Decomposers** feed on *detritus*, the disintegrated matter of dead organisms. Bacteria, fungi, earthworms, and certain small arthropods (a group of organisms that includes insects and spiders) are all decomposers.

Figure 23.6 shows a pond ecosystem, where energy enters as sunlight and is converted to biomass by algae and aquatic plants (producers). Fish (consumers) eat the plant material, and when they die, bacterial or fungal decomposers (such as chytrids) return the nutrients to the water and soil for recycling. Heat is lost at each step.

☑ In marine ecosystems, tiny arthropods called krill feed on photosynthetic algae. Baleen whales feed on krill, and killer whales prey on baleen whales. When whales die and fall to the bottom of the ocean, their bodies are consumed by microorganisms, including bacteria and fungi. Which of these species—algae, krill, baleen whales, killer whales, bacteria, and fungi—are producers and which are consumers? Which are herbivores, carnivores, or decomposers?

A food web: interactions among producers and consumers

Feeding relationships between producers and consumers are sometimes described by a simple *food chain*, an example of which is when A (grass) is eaten by B (a grasshopper), which is eaten by C (a sparrow), which is eaten by D (a fox). Although food chains are useful for presenting the relationships between types of consumers, they understate the degree of complexity in most ecosystems. Most organisms rely on more than one species for food and in turn may be eaten by more than one species.

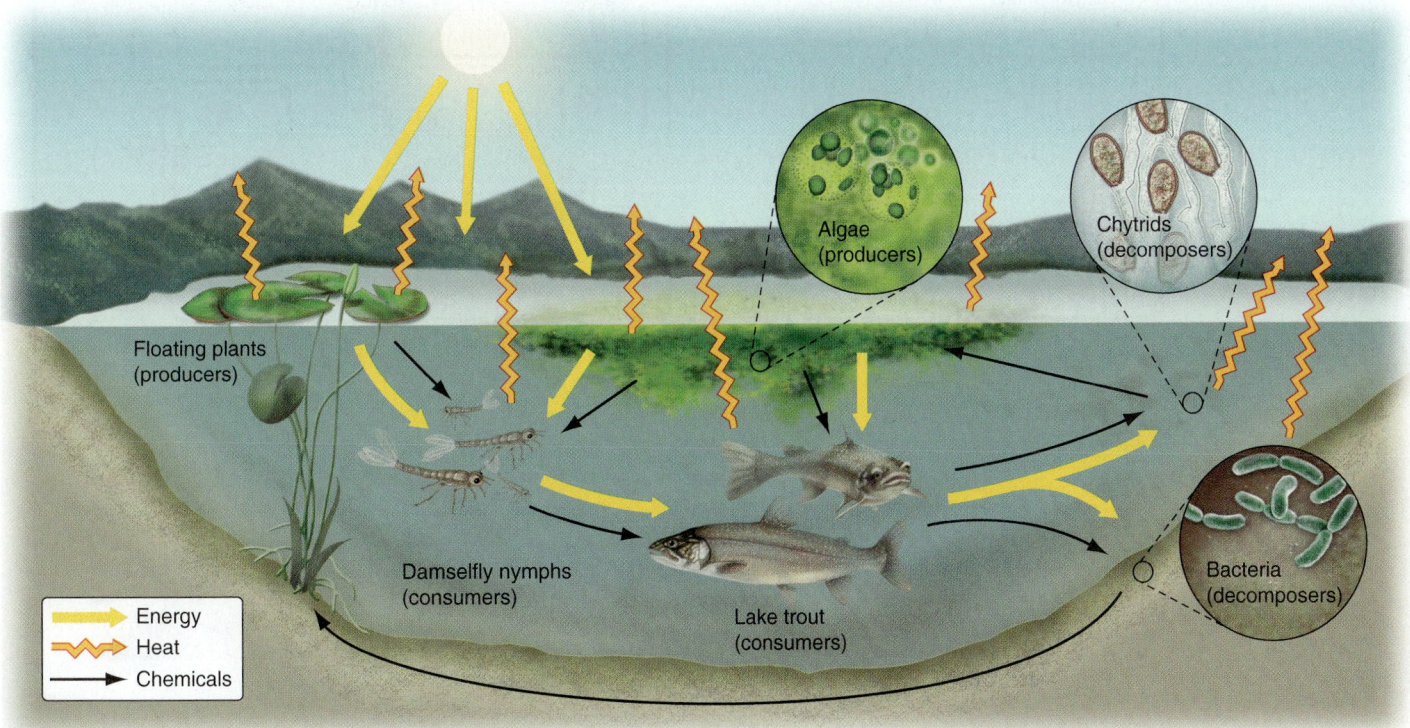

Figure 23.6 Energy flow and chemical cycling in a pond ecosystem. Energy flows in only one direction through ecosystems: from sunlight to producers and then to consumers. Most of the energy is eventually lost as heat. Chemicals are recycled continuously.

A more accurate depiction of the complex balanced nature of the feeding relationships in an ecosystem is shown in a **food web** (Figure 23.7). A particular consumer may eat different foods at different times depending on what is available. For example, as the population of a particular species of secondary consumer rises, it provides a more abundant food supply for a tertiary consumer. These tertiary consumers then rely less on other secondary consumer species whose populations are low at the moment. Consequently, no single population ever gets

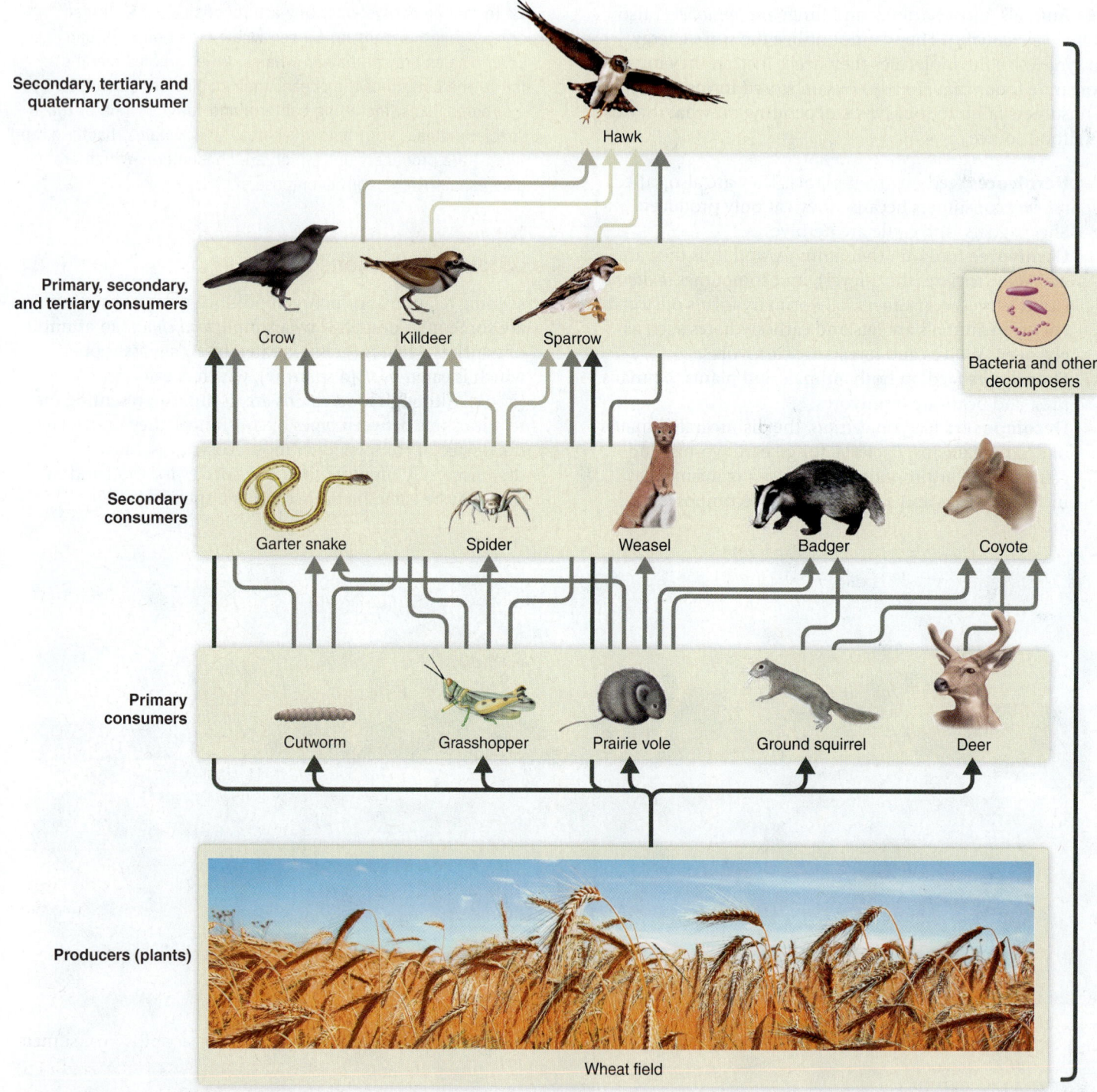

Figure 23.7 A food web. A food web shows the complexity of the feeding relationships among organisms in an ecosystem. Note that bacteria and other decomposers also play a role in the food web by feeding on dead organisms at all levels.

completely out of hand, and rarely does any species become extinct. In a stable ecosystem, every population rises and falls periodically as part of the normal ebb and flow of life. The variations in a species population are always near the carrying capacity of the ecosystem for that particular species.

The lower levels of an ecological pyramid support consumer populations

The processes of energy conversion, transfer, and storage are remarkably inefficient at every level. For example, photosynthetic processes utilize and store in the biomass only 2% of the total energy in sunlight. The rest is reflected as light or heat energy. Only about 10% of the energy at one level of a food web can be found in the tissues of the consumers at the next level in the web. In other words, only about 10% of the stored energy available in plants is stored in herbivores, and only about 10% of that (or 1%) is stored in secondary consumers.

An **ecological pyramid** (Figure 23.8) can depict either the total amount of energy stored at each level of an ecosystem or the total biomass at each level; this one depicts energy. The total amount of energy represented by tertiary consumers at the top of the food chain is very small indeed. To get a better image of the steepness of this pyramid, consider how many total pounds of large carnivores (wolves, bears, and eagles) there might be in a square mile of forest versus the total number of tons of grasses, shrubs, and trees.

☑ One of the most famous questions of ecology is: Why are the largest carnivores so rare? Explain why lions, bears, and wolves are relatively uncommon in natural ecosystems—even in places where they are not hunted by humans.

Human activities disrupt ecological pyramids

Ecological pyramids illustrate that the population of consumers at any level depends critically on the populations of consumers directly below it. Tertiary consumers are especially vulnerable to ecosystem disruptions because the small amount of energy available to them depends on the energy transfers at all levels below them.

Humans can disrupt ecosystems in balance rather easily with modern farming practices. When we create ecosystems with only one species of producer (mile after mile of wheat or corn, for example), we exclude other species from their natural food web and their place on the ecological pyramid. We do this to feed more people, of course, and it works. But the short-term high yield of one species comes with a long-term price in its effects on other species that no longer inhabit their natural ecosystem. We

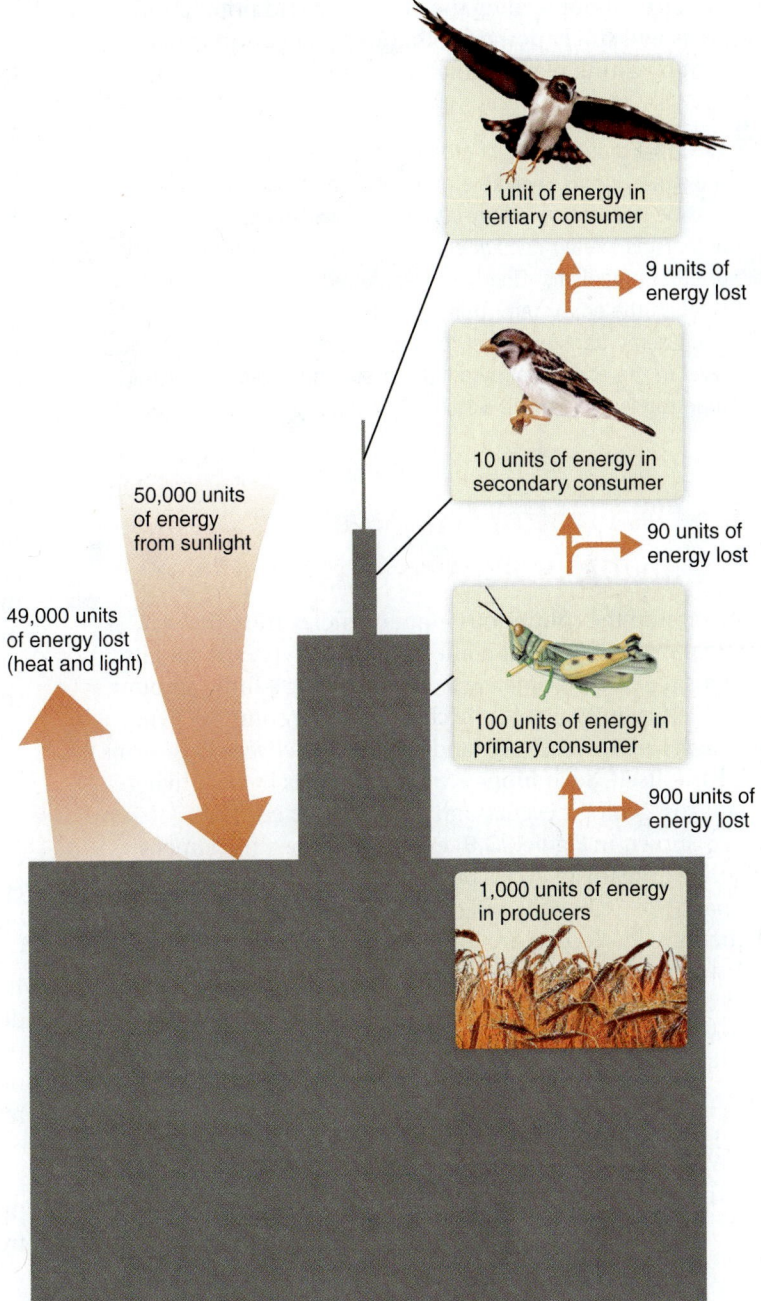

Figure 23.8 An ecological pyramid. This ecological pyramid depicts total energy stored in the biomass at each producer/consumer level. Energy transfer is inefficient; only about 10% of the energy at one level is transferred to the tissues of the consumer at the next highest level of the pyramid.

must recognize that our actions affect other species on Earth.

Humans can be either primary or secondary consumers. We are secondary consumers when we eat meat and primary consumers when we eat plants. Eating meat is energetically expensive. When we feed grain to cattle and then eat the beef, we utilize only 10% of the energy that would otherwise be available to us in the grain. At some point, if we face hard choices about feeding the growing world population, scientists will surely point out that we can be more energy-efficient by eating plants rather than meat.

↩ **Recap** Producers are photosynthesizing organisms that use the sun's energy, available nutrients, and atmospheric carbon dioxide to produce energy-containing molecules. All other organisms, called *consumers*, obtain their energy from plants or other organisms. Energy is transferred in only one direction through the ecosystem: from the sun to producers to consumers. Energy transfer is inefficient, so each producer/consumer level in the ecological pyramid can support fewer organisms than the level below it. ▪

23.5 The matter (material) comprising living organisms is recycled

In contrast to the continuous input of energy from the sun, the total mass of nutrient matter that makes up Earth and its biomass remains completely fixed. With only a finite amount of each element available, the chemicals that compose living organisms are recycled over and over again between organisms and Earth itself. Such **biogeochemical cycles** include living organisms, geologic events, and even weather events.

As shown in **Figure 23.9**, every molecule and element in a living organism cycles between three different pools: biomass, the exchange pool, and reservoir. Some of the elements or molecules are found in the *exchange pool* (water, soil, and atmosphere), from which the primary producers draw their nutrients. Using the energy of the sun, the primary producers incorporate nutrients into the *biomass*. The biomass is the only biotic (living) compartment in biogeochemical cycles; hence, molecules in the biomass are exchanged rapidly as organisms consume each other.

When an organism dies and decomposes, its nutrients return to the exchange pool, where they again become

available to primary producers. Coupled to the exchange pool is a large but hard-to-access pool of nutrients called the *reservoir*. Nutrients in reservoirs include minerals in solid rock and sediments deposited at the bottom of the oceans. Other important reservoirs are carbon-containing compounds in **fossil fuels** (coal, oil, and gas) that were created millions of years ago from the remains of living organisms and then covered by sediments. We turn now to the biogeochemical cycles for some of the most important molecules and elements of life—water, carbon, nitrogen, and phosphorus.

The water cycle is essential to other biogeochemical cycles

Water is probably more essential to life than any other molecule. It makes up about 60% of your body and covers approximately 75% of Earth's surface. However, freshwater—the kind that humans and most other organisms need in order to live—actually accounts for less than 3% of our planet's total water supply (and two-thirds of that is trapped in glaciers and polar ice caps). The remaining 97% of Earth's water is salty ocean water. Water is the universal solvent for many other elements—indeed, without the water cycle, all other biogeochemical cycles would cease.

Figure 23.10 illustrates the **water cycle.** Both the water cycle and the biomass are important to the biogeochemical cycles for minerals. Rain and surface water runoff slowly erode rocks and leach the minerals from them. Plants take up these minerals and incorporate them into the biomass. If there are no plants, the temperature is higher due to decreased evaporative cooling, and the minerals wash into the sea instead. When forests or grasslands are destroyed, nutrients lost from the soil may not be replenished from the reservoir pool for hundreds of years.

Water cycles between the atmosphere, oceans, and land. Although evaporation and precipitation occur primarily over oceans, on land evaporation is mostly from leaf surfaces and soil. Overuse of water can deplete underground stores, compromising deep-rooting, formerly drought-tolerant vegetation and decreasing available water for irrigation. A consequence of this overuse is the loss of wetlands worldwide: for example, Greek Macedonia in Northern Greece has lost 95% of its marshland since 1930, primarily due to diversion of water for irrigation. In addition, the combination of pollutants from human activities and changes in landcover due to urban development can degrade streams, rivers, and even oceans. (The impact of human activities on water is discussed in more detail in the chapter on Human Impacts, Biodiversity, and Environmental Issues.)

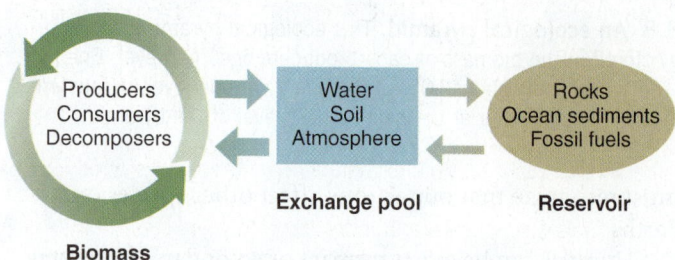

Figure 23.9 The flow of nutrients in biogeochemical cycles. The three components of a biogeochemical cycle are the biomass, the exchange pool, and the reservoir. The chemicals that compose living organisms are constantly recycled among the three components. The relative magnitudes of nutrient flows between the three components and within the biomass are represented by differences in arrow thickness.

☑ The most productive areas for human agriculture are river floodplains; the most productive areas for marine fisheries are the estuaries where rivers meet the sea. Why do the most productive areas occur near rivers?

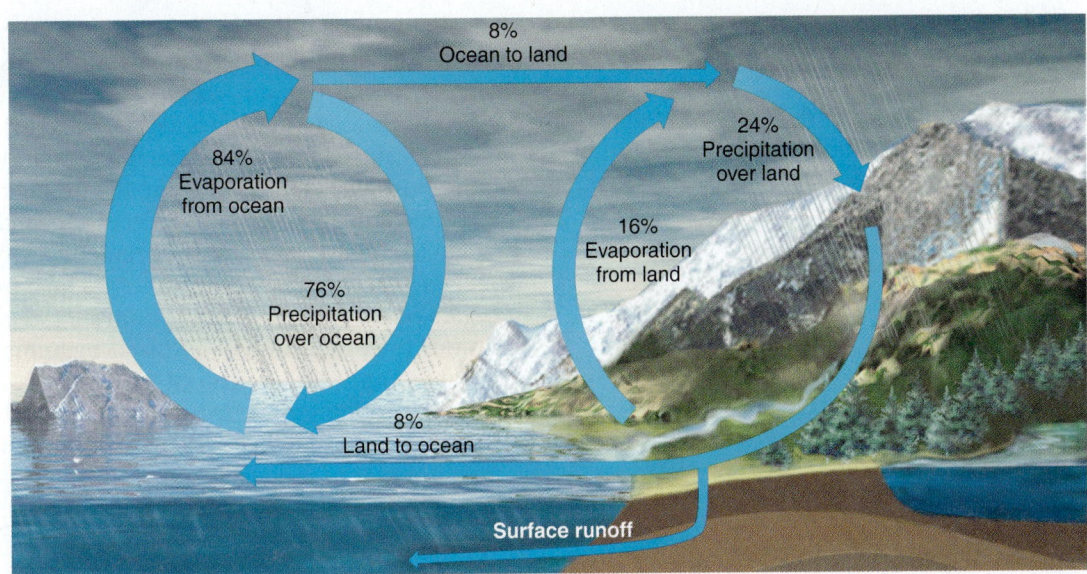

Figure 23.10 **The water cycle.** The amounts of water recycled relative to the total rates of evaporation and precipitation are indicated as percentages. Heat energy from the sun causes water to evaporate, primarily from oceans but also from leaf surfaces and soil on land. Most of the atmospheric water vapor precipitates as rain over the oceans, but approximately a quarter of it precipitates over land, replenishing surface water that was lost by evaporation and providing the extra water that becomes ponds, lakes, streams, and rivers.

The carbon cycle: organisms exchange CO_2 with the atmosphere

Carbon forms the backbone of organic molecules and the crystalline structure of bones and shells. Like the water cycle, the biogeochemical cycle for carbon (Figure 23.11) is called a *gaseous cycle* because carbon in living organisms is exchanged with atmospheric CO_2.

The carbon cycle is closely tied to photosynthesis by plants and aerobic respiration by both plants and animals. During photosynthesis, plants use the energy of sunlight to combine CO_2 with water, forming carbohydrates and releasing oxygen as a by-product. Both plants and animals then utilize aerobic respiration (metabolism in the presence of oxygen) to break down the carbohydrates, releasing the energy to make other complex molecules. In the process they produce CO_2 again. CO_2 is also released through the actions of decomposers such as fungi and bacteria.

In the absence of human interference, utilization and production of atmospheric CO_2 by the biomass would be in approximate balance. However, as shown in Figure 23.11, burning of vegetation and combustion of fossil fuels are tipping the balance toward increased CO_2 production, raising atmospheric CO_2 levels. The result may be global warming (Chapter 24).

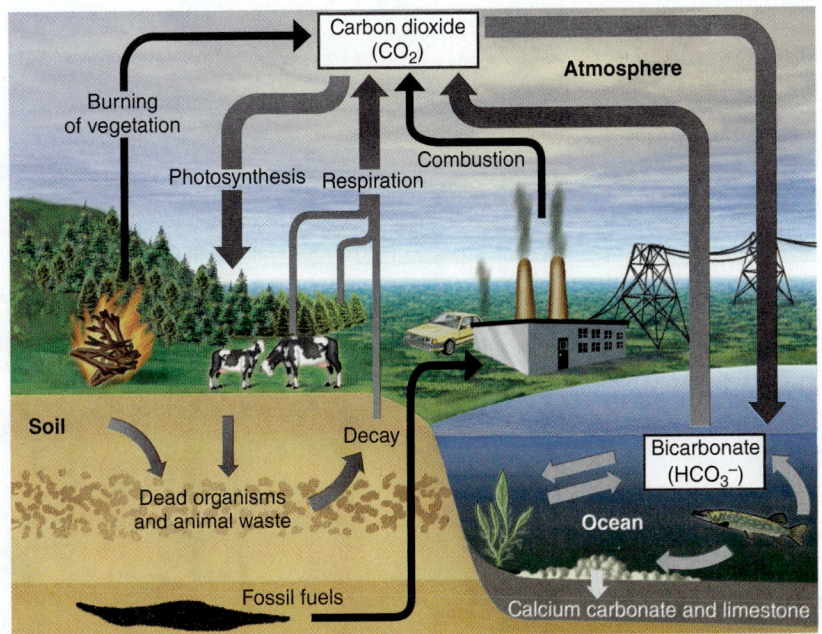

Figure 23.11 **The carbon cycle.** Carbon dioxide in the atmosphere and bicarbonate in the oceans constitute the exchange pool. The reservoir pool of carbon includes calcium carbonate and limestone from the shells of marine organisms and fossil fuels formed several hundred million years ago from decomposed plant material. Fossil fuels represent a reservoir for carbon that would not normally be accessible to the exchange pool, but over the past several hundred years humans have tapped it for energy. The darker-colored arrows indicate the effect of human activities.

☑ The amount of CO_2 in Earth's atmosphere shows a marked annual cycle—a decrease every July and August followed by an increase every December and January. Why? Which arrow in this figure represents a process that changes on an annual cycle?

Nitrogen: an essential component of nucleic acids and proteins

Nitrogen is an essential component of proteins and nucleic acids. By far, the largest reservoir of nitrogen is the nearly 79% of the air that is nitrogen gas (N_2). However, the nitrogen in air is unavailable to living organisms because the two nitrogen atoms are held tightly together by a triple covalent bond. Consequently, plant growth is more often limited by a shortage of usable nitrogen than a shortage of any other nutrient.

Figure 23.12 illustrates the nitrogen cycle. The process of converting nitrogen gas to ammonium (NH_4^+) is called **nitrogen fixation** because the nitrogen becomes trapped, or "fixed," in a form that ultimately can be used by plants. Nitrogen fixation is carried out by certain bacteria in the root nodules of plants such as peas, alfalfa, and clover and by modern fertilizer factories. Most of the ammonium is converted to nitrate (NO_3^-) before it is used by plants. The formation of nitrate is called **nitrification.** Modern fertilizer plants use fossil fuels to manufacture nitrate from air, and lightning provides the energy for a small amount of nitrate formation as well.

Plants take up ammonium and nitrates from the soil and assimilate them into their proteins and nucleic acids in the course of their cellular metabolism. All other organisms must rely on plants or other animals for their usable nitrogen. When organisms die and decompose, the nitrogen compounds in their proteins and nucleic acids are converted again to ammonia, aided by bacteria. Most of the ammonia undergoes nitrification again and recycles through the biomass of living organisms. Similar nitrogen cycles occur on land and in the sea.

Finally, **denitrification** by certain denitrifying bacteria converts some nitrate back to atmospheric nitrogen gas. In a balanced ecosystem, denitrification should equal nitrogen fixation. The production of fertilizers by humans has tipped the balance in favor of nitrogen fixation. Excessive fertilizer use leads to runoff of nitrates that pollute freshwater, upsetting the ecosystem's balance by stimulating excessive growth of plants and algae.

Phosphorus: a sedimentary cycle

Phosphorus is an essential element of life that most commonly exists as phosphate ions (PO_4^{3-} and HPO_4^{2-}). Producer organisms use phosphate to make ATP for energy, phospholipids for cell membranes, and nucleotides for RNA and DNA. Consumer organisms that feed on the producers use phosphate for these purposes too, but consumers also incorporate phosphate into bones, teeth, and shells.

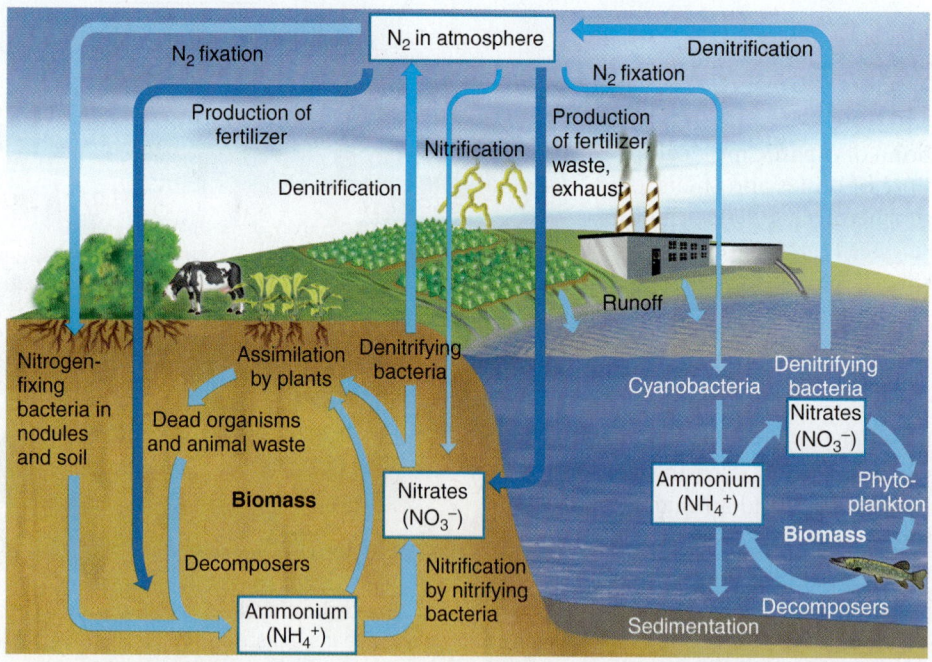

Figure 23.12 The nitrogen cycle. The nitrogen in nitrogen gas is not readily available to living organisms. Once it is "fixed" as ammonium or nitrified to nitrate, the nitrogen can be recycled over and over through the biomass. Under normal circumstances, the amount of nitrogen gas returned to the atmosphere by denitrification approximately balances nitrogen fixation. Humans alter this natural balance through the large-scale commercial production of nitrate and ammonium fertilizers (darker blue arrows).

✅ Circle the three types of bacteria that catalyze key steps in the nitrogen cycle on land, and then, without looking at the figure, write down the type of nitrogen conversion that is performed by each one.

Figure 23.13 shows the phosphorus cycle. It is called a *sedimentary cycle* because, unlike carbon or nitrogen, phosphorus never enters the atmosphere. Phosphate ions deposited in sediments over the course of millions of years are brought to Earth's surface by the upheaval of Earth's crust and the formation of new mountain ranges. The phosphates in sedimentary rocks are mined by humans and weathered by the forces of nature.

Without human intervention, phosphates from sedimentary rock are exchanged with water and soil, providing needed minerals for plant growth. With urbanization, an increase in covered surface areas (roads, parking lots, and buildings) leads to increased water runoff into aquatic ecosystems, with less water absorption by soil and plants. Excess phosphates leached from rock reservoirs and phosphate from fertilizers or from human or animal wastes may also enter aquatic ecosystems, causing excessive algae growth and eventually excessive algae death. The overabundance of dead algae translates to an overabundance of food for decomposers. Ultimately, the decomposers take up so much oxygen for metabolism that the oxygen availability plummets for other organisms. Consequently, those other organisms may suffocate.

In our discussion, we have focused on several major nutrients of life. There are similar biogeochemical cycles for all other nutrients, including oxygen, calcium, potassium, and sulfur. Pollutants and toxic substances such as heavy metals (mercury, cadmium, chromium, and lead), pesticides (DDT), and industrial chemicals (polychlorinated biphenyls, or PCBs) also cycle between living organisms and Earth. Environmentalists need to understand these cycles to advise us on how to minimize the effects of pollution.

MJ's BlogInFocus Global phosphorus supplies are being depleted by human mining activities. What could be done to prevent a future shortage? Visit MJ's blog in the Study Area in MasteringBiology and look under "Dwindling Phosphorus Supplies."

http://goo.gl/RL40GV

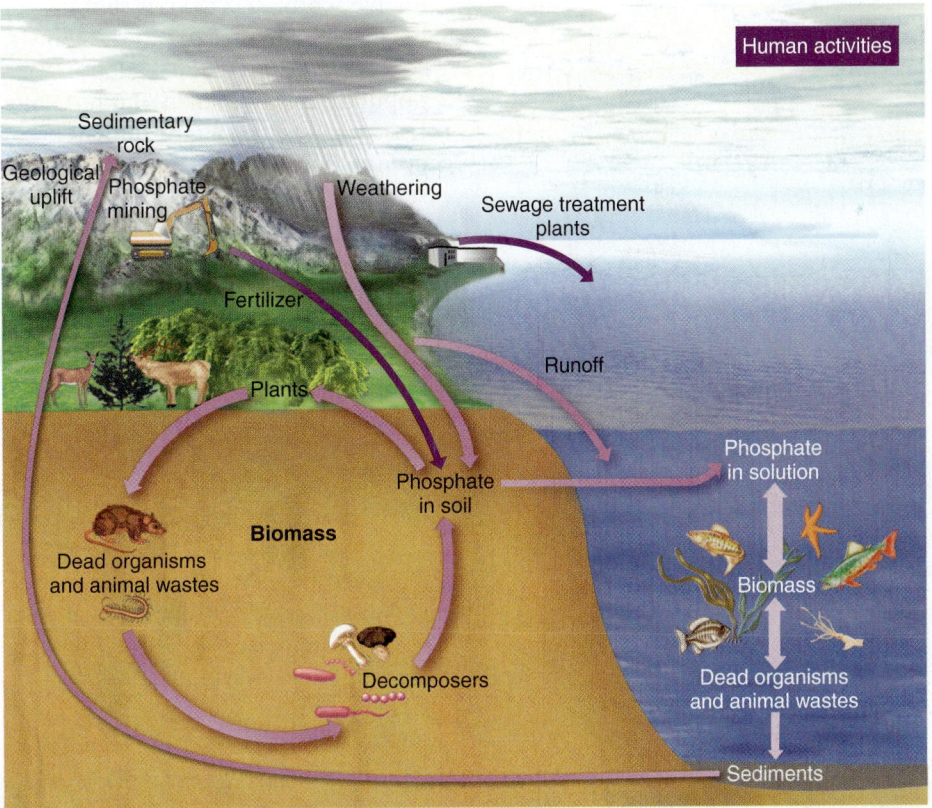

Figure 23.13 The phosphorus cycle. The reservoir pool for phosphorus is phosphate ions in sediments and sedimentary rocks. The slow weathering of rock makes "new" phosphate available to the exchange pool of phosphate in the soil. Some of this phosphate is taken up by plants and some runs off with the surface water to become the phosphorus supply for aquatic ecosystems. Eventually, the phosphate ends up in sediments, where it may remain for millions of years until geologic upheaval brings the sedimentary rocks to the surface and a weathering process can begin again.

↩ **Recap** The amount of matter in Earth is fixed. The chemicals and molecules that constitute all matter are recycled between the biomass, an exchange pool, and a reservoir pool. Most nutrients cycle rapidly within their biomass pool, less rapidly with their exchange pool, and rarely with their reservoir pool. Geochemical cycles can be described for virtually any chemical. Among the most important geochemical cycles for sustaining life are the cycles for water, carbon, nitrogen, and phosphorus. ∎

23.6 Human population growth

A graph of the global human population over just the past 4,000 years of human history shows that the human population is growing rapidly (**Figure 23.14**). How will this affect our world in the future?

For most of human history, the global population remained relatively low and apparently stable, never exceeding 10 million people. People lived in widely dispersed small groups whose numbers were held in check by disease, famine, and a lack of tools and technology with which to control the environment or even to grow food. Apparently, the inability of humans to overcome the environmental resistance kept human population at its carrying capacity.

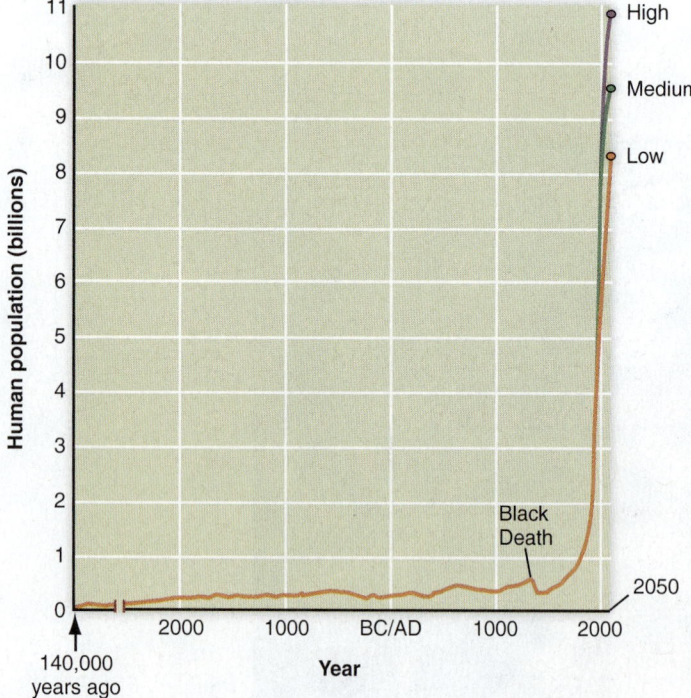

Figure 23.14 Past and projected growth of the human population. The human population currently is 7.3 billion. The dip in the population in the 1300s was caused by the Black Death, a bacterial plague that devastated the human population in Europe and Asia. The high, medium, and low world population projections are based on steady-state fertility rates of 2.6, 2.1, and 1.6 children per woman, respectively, by 2050.

Data after 1950 from Population Division of the Department of Economic and Social Affairs at the United Nations Secretariat.

About 4,000 years ago, however, the development of agriculture and the subsequent domestication of animals and plants for our own uses led to better conditions for our reproduction and survival (a reduction in environmental resistance and an increase in carrying capacity).

Rapid growth took hold in the 1700s (barely 300 years ago) with the Industrial Revolution. Advances in communication, transportation, technologies for feeding and housing more people, and medical care, including the discovery of antibiotics and vaccines, led to a marked increase in human births and survival. As a result, the human population has entered a phase of explosive growth. Note that the human population in Figure 23.14 resembles a J-shaped curve, typical of a population that is increasing according to its biotic potential, without any apparent environmental resistance.

We do not know how high the human carrying capacity could rise with future scientific discoveries. Some scientists believe that the concept of carrying capacity doesn't apply to humans, due to our ability to alter our environments to suit our needs. Humans may even have the capacity to limit our population growth voluntarily, before Earth's human carrying capacity is actually reached.

A more pessimistic view is that we have already inadvertently lowered our carrying capacity, through environmental degradation or other human actions. How soon environmental degradation would limit the human population is unknown. But no matter how you choose to look at it, eventually the human population must stabilize, even if that stabilization is just short of having one person standing on every square foot of Earth.

Zero population growth has not yet been achieved

The **growth rate** of a population is calculated as the number of births per year minus the number of deaths per year, divided by the total population. Currently the human population is growing at a rate of about 1.1% per year. Although that may not sound like much, at this rate the human population would double in the next 65 years (review the Rule of 72).

Zero population growth is the point at which births equal deaths. To reach that point, of course, we would have to either decrease the birth rate or increase the death rate. Given that no one would advocate increasing the death rate on purpose, let's look at how we might reach a zero population growth through changing the birth rate.

The human **fertility rate** is the number of children born to each woman during her lifetime. Fertility experts agree that the average fertility rate required to achieve long-term zero population growth, called the **replacement fertility rate,** is about 2.1 children per woman. Those 2.1 children would exactly replace the equivalent of the two people (a woman and a man) who produced them. The replacement fertility rate is slightly higher than 2.0 because some children die before reaching their reproductive years.

The United Nations Population Division estimates that if replacement fertility rate is achieved by 2050, the world population would stabilize at about 9.6 billion. However, if fertility rates stabilize at lower or higher levels, the global population could range from a low estimate of 8.3 billion to a high estimate of 10.9 billion (refer back to Figure 23.14). Note that the difference between the low and high projections is only one child per couple. (High, medium, and low population projections are based on fertility rates of 2.6, 2.1, and 1.6 children per couple.) Keep in mind that these graphs present global averages. Some areas, such as Africa and Asia, presently contribute more to population growth than other areas.

✅ The fertility rate in the United Kingdom today is approximately 1.8 children per woman, and yet the population is still growing. Explain why this combination of numbers is surprising, and propose an explanation.

Population age structure is linked to economic development

A population's *age structure* describes the number of people in each age group within the population; generally it is depicted by a graph showing the number of people in each age group. In a stable, mature human population the age structure would, in theory, be bullet-shaped (Figure 23.15a). In such a population, the number of births each year would exactly equal the number of deaths.

Although the world may achieve zero population growth overall by about 2050 or shortly thereafter, regional populations will by no means be stable. That's because economic development has been rapid and recent in some countries, and not consistent around the world. As a nation develops economically, the age structure of its population undergoes a *demographic transition*. The initial stage of a demographic transition is marked by a shift from a society dominated by poor living conditions and a high death rate to one of rapidly improving economic conditions, declining death rate, and higher birth rate. In the final stages of demographic transition, the birth rate declines until it equals the death rate, achieving zero population growth.

Demographers (scientists that study population growth) categorize nations according to their economic development. Nations with established industry-based economies are called **most developed countries (MDCs),** whereas nations only beginning to undergo economic development, or that have

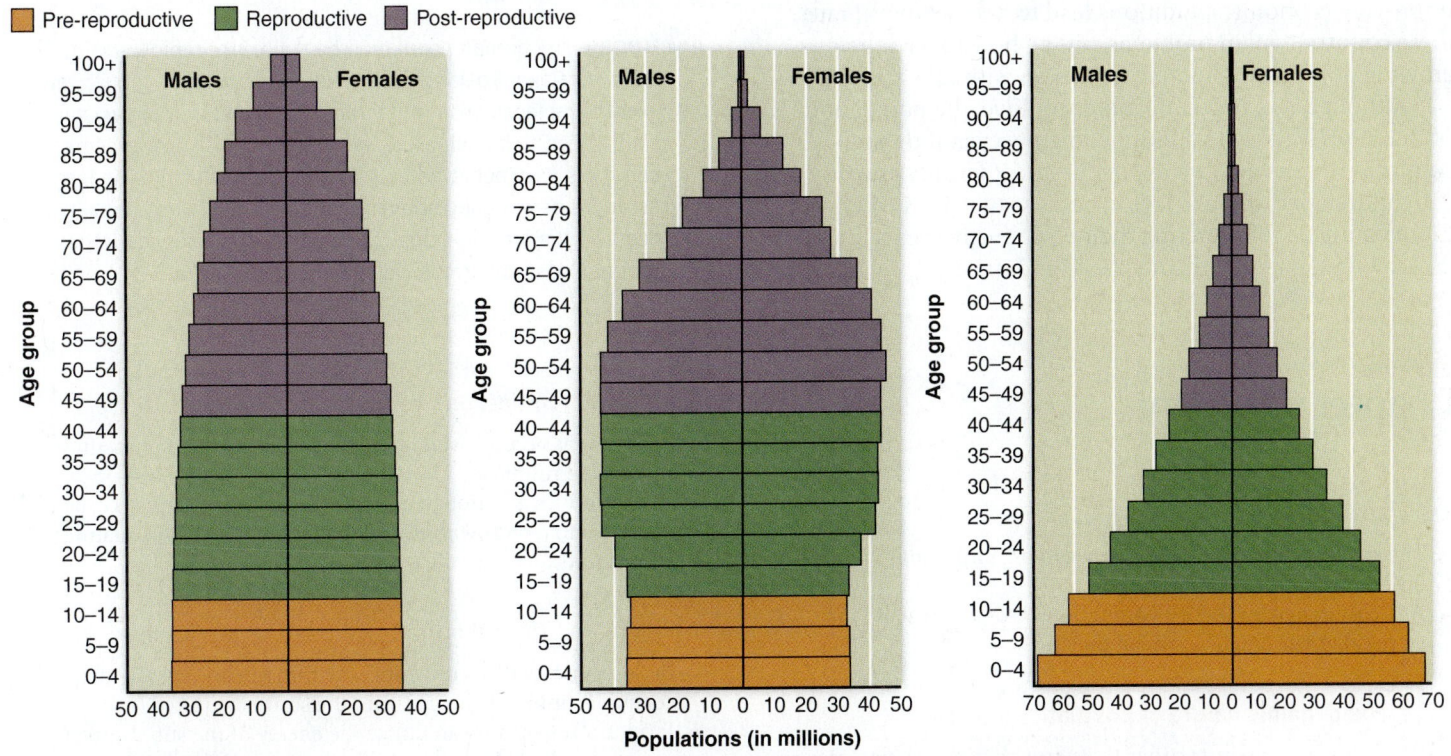

a) A stable population.   b) Most developed countries.   c) Least developed countries.

Figure 23.15 Population age structures. a) The age structure of a stable population would be bullet-shaped, with births equaling deaths. b) In the most developed countries (MDCs), populations have essentially stabilized, but a bulge in population persists in the middle age groups. c) In the least developed countries (LDCs), a large proportion of the population is younger than reproductive age, making it likely that the overall population will continue to expand.

Source for (b) and (c): U.S. Bureau of the Census, International Data Base.

not yet achieved a significant level of economic development, are called **least developed countries (LDCs).** The MDCs are primarily countries in Europe, North America, northeastern Asia, and Australia; the LDCs are located primarily in Africa, Latin America, and southwestern Asia. The demographic patterns suggest that economic development raises the human carrying capacity and allows the population to expand.

Figure 23.15b depicts the current population age structures of the MDCs. The population of the MDCs now shows an apparent stabilization in the younger age groups, suggesting a trend toward a stable, steady-state population overall in the next 50 years or so. However, because economic development has been rapid in these countries in the past 70 years, the MDCs have a distinct population bulge in the middle age brackets. These countries will be facing the problem of too few workers to support too many old people in the next several decades.

In contrast, the population of the LDCs (**Figure 23.15c**) has a distinct pyramid-like age structure, with a large proportion of young people still below reproductive age and a higher death rate in the middle years. When (or if) these countries begin to undergo significant development, they are likely to undergo a rapid rise in population overall, followed by a large bulge in the middle age segment, the same bulge now being experienced by the MDC. In the near term, as improving economic conditions lead to better survival rates, these countries' main problems will be how to feed their growing populations and provide them with jobs.

Although it may seem counterintuitive, the population of the LDCs is likely to continue to grow even if those countries were to achieve the replacement fertility rate of 2.1 children per woman. This is because the LDCs will have more young people entering their reproductive years than are leaving for at least the next several decades.

MJ's BlogInFocus What might cause a society collapse?
Visit MJ's blog in the Study Area in MasteringBiology and look under "Why Societies Collapse."

http://goo.gl/tCFcEZ

The problem for the world as a whole, therefore, is that the largest increases in population over the next 50 years are likely to occur in the very nations that are least able to provide for their citizens. To prevent poverty and starvation in these countries, we must seek solutions that promote their economic development while at the same time slowing their population growth rate. Family planning and other social programs that encourage delaying childbirth or reducing the number of children per couple would certainly help. Even government action has worked in some societies. In China, for example, it is against the law to have more than two children. In addition, substantial evidence supports the hypothesis that one of the most effective ways to reduce the birth rate is to improve the economic, social, and political standing of women in a society. Cultural norms and religious beliefs may conflict with this approach in some cultures, however.

Recap The human population began to rise rapidly about 300 years ago. To achieve zero human population growth, an average replacement fertility rate of 2.1 children per woman would need to be achieved and sustained worldwide. Some countries have achieved that goal; others have not. The largest increases in population over the next 50 years are likely to occur in the less industrialized countries, the very countries that are least able to provide for their citizens. ■

Chapter Summary

23.1 Ecosystems: living organisms and their environment *p. 524*

- The study of the relationships between organisms and their environment is called *ecology*.
- An ecosystem consists of a community of organisms and the physical environment in which they live.

23.2 The dynamic nature of populations *p. 524*

- When there are no restrictions to its growth, a population grows exponentially according to its biotic potential.
- Within an ecosystem, the stable size of the population of a particular species ultimately is determined by carrying capacity.
- Invasive species can cause economic or environmental harm to an ecosystem.

23.3 Communities: different species living together *p. 526*

- Species that occupy overlapping niches may be in competition for limited resources.
- Communities go through a natural sequence of changes called *succession* that ends with the establishment of a stable mature community.

23.4 Energy flows through living organisms *p. 528*

- Because energy flows only one way through an ecosystem, a constant supply of energy (from the sun) is required.
- Producer organisms can utilize the energy of the sun through a process called *photosynthesis*. Consumer and decomposer organisms must rely on other organisms for energy.
- Less energy is available at each higher consumer level of a food web.

23.5 The matter (material) comprising living organisms is recycled *p. 532*

- Chemicals (nutrients) recycle in an ecosystem, sometimes quickly, sometimes slowly.
- Water is constantly being recycled from the oceans to land and back to the oceans.
- In the carbon cycle, plants use CO_2 available in the atmosphere and animals obtain organic carbon molecules by consuming plants or other animals.
- Nitrogen gas in the atmosphere must be converted to NH_4^+ or NO_3^- ("fixed") before it is useful to living organisms.
- The phosphorus cycle is a sedimentary cycle. Phosphorus from decaying organisms is available for reuse.

23.6 Human population growth *p. 536*

- The human population has been growing rapidly since the Industrial Revolution, beginning in the 1700s.
- Currently, there are wide differences in fertility rates between the more industrialized countries and the less industrialized countries.
- Even when (or if) the replacement fertility rate of 2.1 children per woman is reached, the world population will continue to grow for several more decades.
- Humans may be able to increase the magnitude of their carrying capacity temporarily using technology, but these activities may ultimately reduce that capacity.

Terms You Should Know

biogeochemical cycle, *532*
biomass, *527*
biosphere, *524*
carnivore, *529*
carrying capacity, *526*
mature community, *527*
community, *524*
consumer, *528*
decomposer, *529*
denitrification, *534*
ecological pyramid, *531*
ecology, *523*
ecosystem, *524*
environmental resistance, *525*
fertility rate, *536*
food web, *530*
fossil fuels, *532*

geographic range, *524*
habitat, *524*
herbivore, *529*
invasive species, *526*
least developed countries (LDCs), *538*
most developed countries (MDCs), *537*
nitrification, *534*
nitrogen fixation, *534*
omnivore, *529*
photosynthesis, *528*
population, *524*
producer, *528*
replacement fertility rate, *536*
zero population growth, *536*

Concept Review

Answers can be found in the Study Area in MasteringBiology.

1. What are the components of an ecosystem?
2. Describe the concept of an ecosystem's carrying capacity.

3. Describe what happens to a country as it undergoes demographic transition. Give examples of nations in early and late stages of transition.
4. Define *zero population growth* and explain why it has not been achieved.
5. Describe the concept of a mature community, and explain why such communities are sensitive to disruption.
6. Why is a constant supply of energy needed to sustain life?
7. Discuss the interactions between producers and consumers in an ecosystem.
8. Explain why the water cycle is essential to all other biogeochemical cycles.
9. Discuss the importance of the carbon cycle.
10. Discuss the phosphorus cycle and its importance.

Test Yourself

Answers can be found in the Appendix.

1. Your college campus, including all the humans and other living things that occupy it as well as its physical environment, such as the buildings, soil, water, and air, is a(n):
   a. population
   b. community
   c. ecosystem
   d. biosphere

2. Which of the following will affect the biotic potential of a species?
   a. length of time it takes an individual to reach maturity
   b. number of members of the population at reproductive age
   c. number of offspring produced by each reproducing member
   d. all of the above

3. A population is growing at a rate of 2% per year. How many years will it take for this population to double in size?
   a. 16 years
   b. 20 years
   c. 72 years
   d. 36 years

4. The large intestine of humans is populated by a complex community of benign microorganisms (not harmful) which compete for nutrients and may prevent pathogenic (harmful) microorganisms from becoming established. Which term would best describe this?
   a. *competitive exclusion*
   b. *selective pressure*
   c. *succession*
   d. *environmental resistance*

5. Which of the following characterizes mature communities?
   a. They contain a limited number of different species.
   b. They are generally stable and long-lived.
   c. They contain equal numbers of producers and consumers.
   d. They can be reestablished quickly following significant environmental change.

6. _____ flow(s) one-way through ecosystems while _____ is/are recycled.
   a. Biomass . . . elements
   b. Energy . . . elements
   c. Elements . . . energy
   d. Producers . . . consumers

7. All of the following terms can appropriately describe humans except:
   a. *primary consumer*
   b. *autotroph*
   c. *heterotroph*
   d. *secondary consumer*

8. A human maintaining a vegan diet (containing no animal products) would be a:
   a. producer
   b. primary consumer
   c. secondary consumer
   d. decomposer

9. All of the following processes are involved in the carbon cycle except:
   a. photosynthesis
   b. cell respiration
   c. evaporation
   d. decomposition

10. The shape of an ecological pyramid, with each successive layer being smaller than the one below it, can be explained most readily by:
    a. the first law of thermodynamics
    b. inefficient geochemical cycles
    c. the second law of thermodynamics
    d. industrialization

11. Plant growth is most likely to be limited by the available supply of usable:
    a. nitrogen
    b. carbon
    c. oxygen
    d. phosphorus

12. Which biogeochemical cycle does not include exchange with the atmosphere?
    a. nitrogen
    b. carbon
    c. phosphorus
    d. water

13. The biogeochemical cycles involve the cycling of key molecules and elements between a reservoir, an exchange pool, and:
    a. the biomass
    b. producers
    c. consumers
    d. decomposers

14. Which of the following statements is false?
    a. The human population has grown steadily throughout human history.
    b. Human populations are now growing explosively.
    c. Human population growth can be represented by a J-shaped curve.
    d. Humans have been able to alter the carrying capacity of Earth.

15. Which statement about the human fertility rate is true?
    a. Fertility rates are the same in all human populations.
    b. Fertility rates can apply to both men and women.
    c. A sustained fertility rate of 2.1 would eventually lead to zero population growth.
    d. Death rates are factored in when calculating fertility rates.

Apply What You Know

Answers can be found in the Study Area in MasteringBiology.

1. Given absolutely no environmental resistance, how long would it take a population that was increasing by 8% a year to double?
2. Speculate as to what might eventually become the most important environmental resistance factor (or factors) determining the world's human carrying capacity.
3. Both hawks and owls feed on small mammals to sustain their respective populations without causing competitive exclusion. How can this work when both species live in the same geographic area?
4. The introduction of the zebra mussel into the Great Lakes by oceangoing vessels has had a significant impact on the lake ecosystem. How could such a small organism have had such a large impact?
5. What would happen in an ecosystem if suddenly there were no decomposers?
6. The term *algae bloom* refers to rapid, excessive growth of algae. What might cause an algae bloom, and what in turn would an algae bloom cause?
7. Least developed countries tend to be poorer than more developed countries, and at the same time, least developed populations are among the fastest growing. What sorts of interventions or assistance are likely to help least developed countries develop and prosper?

MJ's BlogInFocus

Answers can be found in the Study Area in MasteringBiology.

1. What should be done about an invasive species, especially one that dominates an ecosystem or causes economic damage? Should we try to eliminate the invasive species entirely? Should we try to contain it or reduce its influence? Should we just give up and learn to live with it? To view a research paper on this issue, visit MJ's blog in the Study Area in MasteringBiology and look under "Invasive Species." http://goo.gl/0UayjO

MasteringBiology®

Students Go to MasteringBiology for assignments, the eText, and the Study Area with animations, practice tests, and activities.

Professors Go to MasteringBiology for automatically graded tutorials and questions that you can assign to your students, plus Instructor Resources.

Answers to ✔ questions are available in the Appendix.

Human Impacts, Biodiversity, and Environmental Issues

A metallurgical plant in Chusovoy, Russia.

Key Concepts

- **Increased production of greenhouse gases is contributing to global warming and global climate change.** Certain air pollutants produced by human activities also contribute to acid rain, cause poor air quality, and deplete Earth's ozone layer.

- **Less than 1% of the water on Earth is freshwater.** Freshwater resources are declining in some regions of the world due to water pollution and inefficient use of water for agricultural purposes.

- **Some human activities damage ecosystems and lead to a loss of biodiversity on Earth.** The consequences of loss of biodiversity are not well understood.

- **Most of our energy comes from nonrenewable resources (coal, oil, and gas).** A shift to renewable energy sources is slowly getting under way.

- **Sustainability of life on Earth will require meeting our current needs without compromising the needs of future generations.** True sustainability will take into account the needs not only of humans but also all other species.

CURRENT ISSUE
Global Warming and Global Climate Change

Questions to Consider

1 Do you think we should take action to reduce the current levels of CO_2 in the atmosphere? Why or why not?

2 In what ways might your own lifestyle choices contribute to global warming? Would you be willing to support legislation that would affect your lifestyle choices in order to reduce global warming?

Earth's average surface temperature is rising, a phenomenon known as *global warming.* There is now general agreement among the world's most respected scientists that global warming is leading to *global climate change* (a change in weather patterns), and that global warming is due to rising atmospheric concentrations of greenhouse gases (primarily CO_2) brought about by the use of fossil fuels as a source of energy. To slow the rates of global warming and global climate change, we'll need to reduce the rate of rise of atmospheric CO_2. There are only two ways to do that: Either we reduce our dependence on fossil fuels or remove the resultant CO_2 from the atmosphere.

What Will Be the Effects of Global Warming and Global Climate Change?

To fully grasp the concept of global warming and why it may be a problem for future generations, you need to understand that global warming is a slow process; in the past 100 years the temperature has risen only about 0.8° Celsius (1.4° Fahrenheit). More disturbingly, the rate of rise has doubled in just the past 50 years. These small changes in temperature are too little for you to notice, but they are expected to have effects that last for hundreds,

thousands, or even tens of thousands of years.

What you will probably notice, however, is climate change. Current climate models suggest that if global warming continues, the number and intensity of hurricanes, typhoons, and tornadoes could increase. Precipitation patterns may change, causing some regions of the world to experience more rain and frequent floods, while other regions undergo prolonged droughts and become deserts. There could be widespread crop failures.

You will not experience other changes that are likely to occur due to climate change because they are very far off in the future and you will not live long enough. However, on a time scale of hundreds and even thousands of years, most of the polar ice cap is expected to melt. Rising sea levels will erode coastlines and flood low-lying coastal areas,

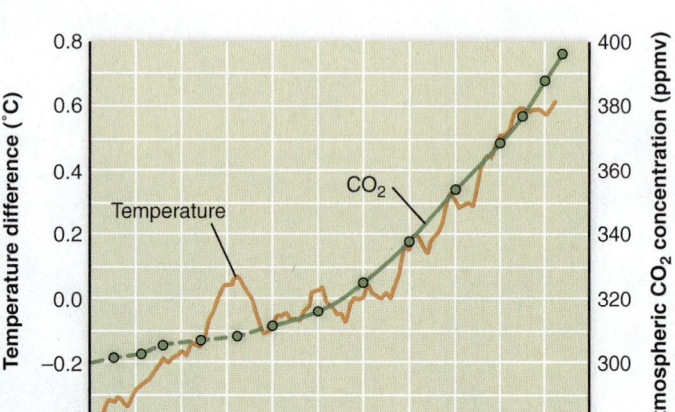

Global temperature differences (from a 1951–1980 base; five-year averages) and atmospheric CO_2 concentrations over the past 100 years. *Source:* NASA and the Carbon Dioxide Information Analysis Center.

leading to the abandonment of many of the world's major cities. Fragile ecosystems could unravel, leading to a mass extinction of many species of organisms. Life on Earth is expected to change subtly, slowly, and for a very long time if the global temperature continues on its current trajectory.

The International Community Begins to Respond

Once the potential threat of rising atmospheric CO_2 levels became clear, the international community began to come together to discuss what to do about it. The first joint effort to address global climate change came at a meeting of interested parties, called the Earth Summit, that took place in Rio de Janeiro, Brazil, during June of 1992. Under the sponsorship of the United Nations, attendees at the Earth Summit developed a nonbinding treaty, the United Nations Framework Convention on Climate Change (UNFCC). The stated goal of UNFCC was to "stabilize greenhouse →

Polar bears' habitat will be affected by a loss of arctic sea ice.

gas concentrations in the atmosphere at a level that would prevent dangerous anthropogenic [caused by humans] interference with the climate system." The treaty was signed initially by 154 nations; today there are 196 signatories.

Since 1995, the parties to UNFCC have met every year to try to hammer out the details of how specific goals could be met. Progress has been slow; in the meantime the atmospheric concentration of CO_2 continues to rise. Nevertheless, some progress has been made over the years. Three notable achievements stand out:

- 1997—The Kyoto Protocol set specific targets for developed countries to reduce their greenhouse gas emissions by certain dates and by certain amounts, from a benchmark year of 1992.
- 2010—The Cancun agreement concluded that any future increase in global temperature should be held to less than 3.6°F (2°C), relative to the preindustrial level.
- 2011—Attendees at UNFCC in Durban, South Africa, agreed in principle that any future agreement regarding how much each nation must contribute to reducing carbon emissions should be legally binding, not voluntary.

Both the Kyoto Protocol and the Cancun agreement require extensive data collection and analysis to be effective. In the Kyoto Protocol, for example, one of the first tasks was to determine what each nation's greenhouse gas emissions and recovery rates were in 1990, the benchmark year.

Going forward, under the Kyoto Protocol participating nations are required to inventory and report their CO_2 emissions annually—a task open to misinterpretation and manipulation. As for the Cancun agreement, the establishment of the future global warming limit of 2°C relies heavily on climate models of what might happen if we were to exceed that limit. Understandably, the proposed 2-degree limit has been controversial. Finally, as the Durban meeting revealed, the lack of legally binding, enforceable requirements for each nation to reduce its greenhouse gas emissions leaves the door wide open to do very little but talk.

In the absence of binding, enforceable agreements, some nations are taking action on their own. The two largest greenhouse gas emitters are China and the United States, in that order. China's greenhouse gas emissions are still increasing, and although it has not declared an actual target, China has said that its emissions will peak by 2030—that is, the country's emissions will hold steady or decline after that year. The United States reduced its net greenhouse

gas emissions by 9% between 2005 and 2013, and when he was in office, President George W. Bush pledged another reduction of 26–28% by 2025.

The actions nations have taken thus far to reduce greenhouse gas emissions are not enough, according to the most recent annual report of the Intergovernmental Panel on Climate Change (IPCC), the international body established by the United Nations for assessing climate change. The IPCC concludes that "without additional mitigation efforts beyond those in place today, and even with adaptation, warming by the end of the 21st century will lead to high to very high risk of severe, widespread and irreversible impacts globally (*high confidence*)."[1]

We have a choice. We can redouble our efforts to lower atmospheric CO_2 levels now, or we can wait, leaving the whole issue to future generations. It's up to us.

[1]IPCC, 2014: Climate Change 2014: Synthesis Report. Contribution of Working Groups I, II, and III to the Fifth Assessment Report of the Intergovernmental Panel on Climate Change [Core Writing Team, R.K. Pachauri and L.A. Meyer (eds.)]. IPCC, Geneva, Switzerland, page 151.

SUMMARY

- Earth's surface temperature is rising (global warming) and weather patterns are changing (global climate change).
- The primary cause of both global warming and global climate change is a rise in atmospheric CO_2 concentration, brought about by the burning of fossil fuels.
- The international community is in agreement that atmospheric CO_2 levels must be reduced and that global warming should be held to under 2°C if we are to avoid serious long-term consequences.
- So far, not enough action has been taken to achieve those two goals.

Humans live in communities with other species. However, unlike other species, humans tend to dominate the places where they live.

Our opposable thumbs and upright posture allow us to perform a surprising number of physical tasks. Our big brains can solve complex problems. Our mastery of complex language means that we can communicate our knowledge over time and distance. As a result, humans have the capacity to modify whole ecosystems. Only humans live in houses that may be hundreds of square feet per person and build skyscrapers thousands of feet high. Only humans irrigate lands that otherwise could not grow crops, level entire mountains in search of minerals and fossil fuels, and spread

herbicides, insecticides, and fertilizers over thousands of square miles. Human activities are causing global changes in biogeochemical cycles, altering the climate and affecting the populations of other living species. Whether we intended to or not, we have favored the proliferation of some species and driven others to extinction.

This chapter examines how we *Homo sapiens* are altering our environment on a global scale. We will look at the causes and impacts of pollution on air, water, and land. We will discuss how our energy requirements affect the environment. Finally, we'll describe the benefits of biodiversity and look at steps we can take to build a sustainable future for ourselves and other species.

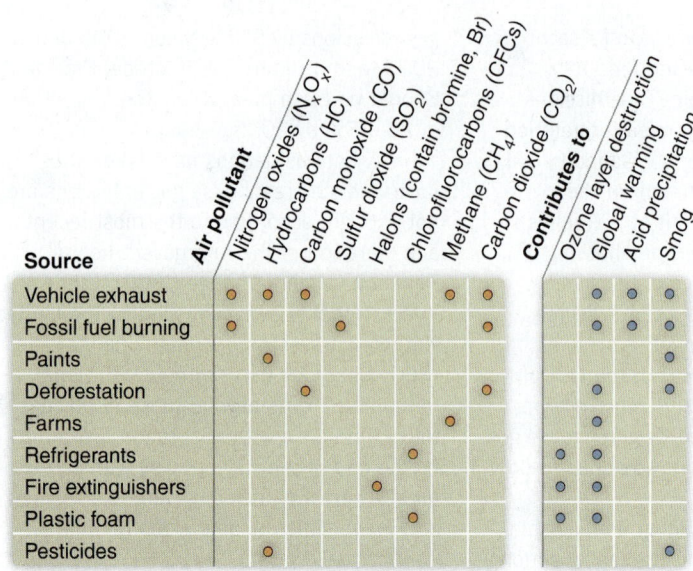

Figure 24.1 Major sources of air pollutants. A single activity (such as driving a car) may produce more than one type of air pollutant.

☑ Which source of air pollution releases the greatest number of different air pollutants? Which major concern regarding air pollution does that source not contribute to, and why not?

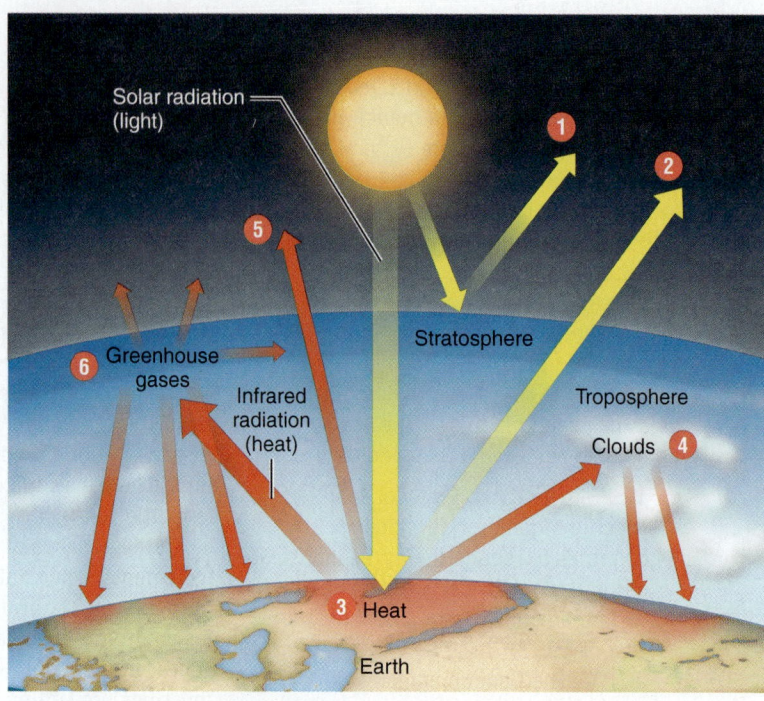

Figure 24.2 The greenhouse effect. A layer of stratospheric greenhouse gases consisting primarily of water vapor and carbon dioxide (CO_2) allows sunlight to pass but traps most of the heat. An overabundance of greenhouse gases may contribute to global warming.

24.1 Pollutants impair air quality

The air we breathe is a mixture of primarily nitrogen (nearly 79%) and oxygen (nearly 21%), with trace amounts of carbon dioxide (0.03%). Air also contains trace amounts of thousands of chemicals or particles that have adverse effects on living organisms, collectively known as **pollution.** The major concerns regarding the effects of air pollution fall into four areas:

- Global warming
- Destruction of the ozone layer
- Acid rain
- Smog

Figure 24.1 shows that for each of these issues there may be more than one contributing pollutant. Air pollution is difficult to remedy because it can be hard to determine who is responsible for it. Global problems require global solutions.

Excessive greenhouse gases are causing global warming

If you've ever been in a glass greenhouse on a sunny day you know that the glass roof of the greenhouse allows radiant energy from the sun in and then prevents the heat generated from escaping. As a result, the air in the greenhouse becomes warmer.

Certain gases in the upper layer of the atmosphere (the stratosphere), called **greenhouse gases,** also warm Earth, a phenomenon known as the *greenhouse effect* (**Figure 24.2**). When solar radiant energy in the form of light reaches Earth's

atmosphere, ❶ some of that solar radiant energy is reflected back out into space but most of it strikes Earth. After hitting Earth, ❷ another small fraction of solar radiant energy is reflected back out into space. The remainder either ❸ becomes heat at Earth's surface or is radiated back from Earth's surface as infrared radiation (heat). ❹ From there, some of the infrared radiation is reflected back to Earth by clouds, while ❺ some escapes back into space. ❻ Whatever infrared radiation remains is absorbed by greenhouse gases in the stratosphere, where it is reradiated in all directions including toward Earth. The result of the last step is an increase in Earth's surface temperature above what it would be in the absence of greenhouse gases. This is a natural and normal phenomenon. Without the greenhouse effect, most of the sun's radiant energy would radiate away from Earth and the average temperature would be below freezing. In other words, to maintain a normal surface temperature on Earth, a certain amount of greenhouse gases is needed in the stratosphere.

The most significant greenhouse gas is water vapor, accounting for approximately 60% of the total. The rest is mainly carbon dioxide (CO_2), plus small amounts of methane (CH_4), nitrous oxide (N_2O), and two air pollutants: the chlorofluorocarbons (CFCs), used primarily as refrigerants, and gases called *halons* that contain bromine, used in fire extinguishers.

Most of the CO_2 component of the greenhouse gases accumulated over millions of years as a result of respiration, volcanoes, fires, and plant decay. In modern times, however, human activities have increased the levels of greenhouse gases, especially CO_2. There is general agreement now among scientists that this excessive production of greenhouse gases is increasing the greenhouse effect and raising the average global temperature, a phenomenon known as **global warming.**

The main human activity that raises atmospheric CO_2 levels is the burning of fossil fuels for energy. The carbon in fossil fuels comes from decayed plant material that was buried underground by sedimentary processes over millions of years. When we burn fossil fuels, we release carbon into the air as CO_2.

The other main human activity that raises atmospheric CO_2 concentrations is **deforestation** (removing trees from large areas of land). Trees absorb CO_2 from the air during photosynthesis. A large growing tree will take up and store in its wood approximately 50 lb of CO_2 per year. Deforestation is due to indiscriminate logging for wood and widespread burning of forests to create new cropland. Deforestation by burning is doubly damaging—not only are there fewer trees to absorb CO_2 from the atmosphere but all of the carbon in the burned wood is immediately released back into the atmosphere.

☑️ Climate researchers try to model past weather patterns in order to better predict the effect of global warming on the average amount of cloud cover on Earth. Why is cloud cover of such interest to researchers?

CFCs deplete the ozone layer

Ozone (O_3) is found in two places in the atmosphere. In the atmospheric layer near the planet surface (the troposphere), O_3 is an air pollutant formed by the reaction of oxygen (O_2) with automobile exhaust and industrial pollution. O_3 is mildly toxic, causing plant damage and respiratory distress in animals, including humans.

Higher up in the atmosphere, O_3 is actually very beneficial. It forms a thin layer in the stratosphere that helps shield Earth from ultraviolet (UV) rays. UV rays damage DNA and contribute to skin cancer and may cause cataracts.

By the early 1980s, it had become apparent that the stratospheric ozone layer was suffering significant damage due to chlorofluorocarbons (CFCs), a group of chemicals used in refrigerators, air conditioners, and aerosol sprays. CFCs released near ground level migrate slowly upward toward the stratosphere and decompose, releasing chlorine (Cl) atoms. The Cl atoms combine with O_3 and destroy it, producing O_2 (**Figure 24.3**). Because a Cl atom can be used over and over in the reaction, a single Cl atom may destroy as many as 10,000 O_3 molecules.

By 1985, the stratospheric ozone layer had thinned noticeably, especially in the Southern Hemisphere over Antarctica. The United Nations predicted that rates of skin cancer would begin to increase as a result. Once the problem was recognized, quick action was taken. An international agreement to phase out the production of CFCs was signed by most industrial nations in 1987, and by the mid-1990s, the size of the hole in the ozone

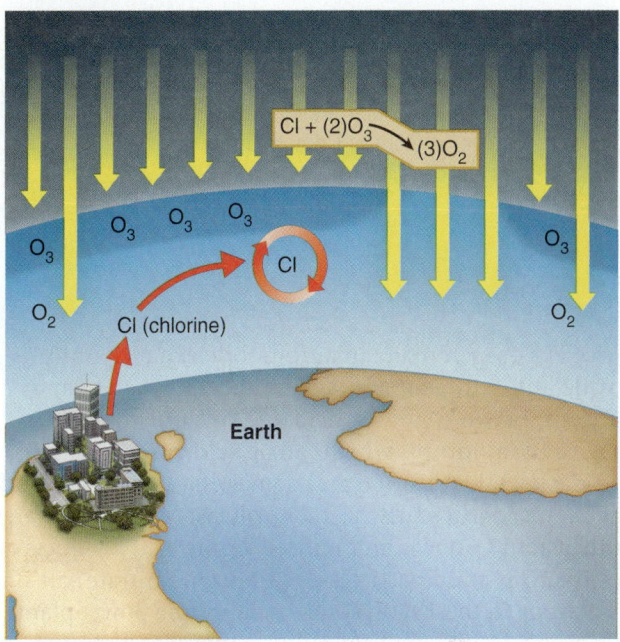

Figure 24.3 Destruction of the ozone layer. Chlorine released into the atmosphere reacts with ozone (O_3) to produce oxygen (O_2). Destruction of the ozone layer exposes Earth's surface to more ultraviolet radiation (yellow arrows).

layer had stabilized. It remains at about the 1996 level today (**Figure 24.4**).

Models of the stratosphere predict that it may be 100–150 years before the ozone layer recovers completely. Nevertheless, quick action on the part of the international community may have kept the problem from becoming even worse.

Pollutants produce acid rain

The major source of acid in the atmosphere is sulfur dioxide, which is released into the air as a result of burning high-sulfur coal and oil for power. A secondary

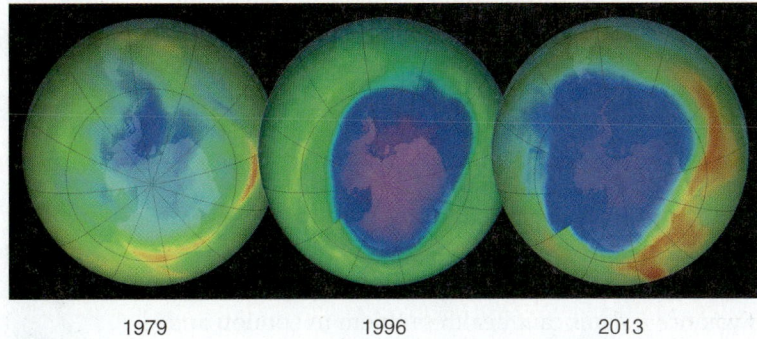

| 1979 | 1996 | 2013 |

Figure 24.4 Depletion of the ozone layer. Areas of depletion are indicated by purple and blue colors. The rate of depletion was halted by the mid-1990s, but it may be more than a hundred years before the ozone layer recovers completely. Source: NASA's Earth Observatory.

☑️ The country of New Zealand has the world's highest rate of death from skin cancer. Why? (Hint: There are two reasons.)

MJ's BlogInFocus How is global warming likely to change the climate in your region of the country? Visit MJ's blog in the Study Area in MasteringBiology and look under "Impact of Climate Change."

http://goo.gl/7WpflO

source is nitrogen oxide in automobile exhaust. Sulfur dioxide and nitrogen oxides combine with water vapor in the air, become sulfuric acid and nitric acid, and dissolve in raindrops, which fall as **acid rain.** Acid rain corrodes metal and stone and damages forests and aquatic ecosystems, particularly in the northeastern United States, southeastern Canada, and Europe. Some acid products also precipitate as solid particles of sulfate and nitrate salts.

Starting in the 1970s, some coal-burning power plants have been required to install sulfur removal and capture systems known as "scrubbers" to reduce their emissions of sulfur dioxide. As a result, since 1985, the deposition of sulfur in rainwater has declined by about 33% across most of the U.S. Northeast. Recent regulatory actions taken by the Obama administration are expected to eliminate most of the remaining sulfur emissions by 2014.

Smog blankets industrial areas

Several pollutants in air, most notably nitrogen oxides and hydrocarbons (chains of carbons linked to hydrogens), react with each other in the presence of sunlight and water vapor. They form a hazy brown or gray layer of **smog** (the term comes from "smoke" + "fog") that tends to hover over the region where it is produced. Most smog is caused by the burning of fossil fuels (coal and oil) and by automobile exhausts.

Smog also contains several chemicals that irritate the eyes and lungs and may lead to chronic respiratory illnesses such as asthma and emphysema. Depending on the source, smog may also contain small oil droplets and particles of wood or coal ash, asbestos, lead, dust, and animal waste. Smog can become especially troublesome during a **thermal inversion,** when a warm, stagnant upper layer of air traps a cooler air mass containing smog beneath it.

Like acid rain, the problem of smog has been reduced due to widespread cleanup efforts and reduced emissions of pollutants. For example, smog was once a significant health problem in London and Pittsburgh, but both of these cities have succeeded in major efforts to reduce smog (**Figure 24.5**). The state of California has passed strict automobile antipollution laws. Many regions of the United States now require the use of cleaner-burning, oxygenated automobile fuels containing ethanol or methanol

during certain winter months, and the use of lead as a fuel additive has been significantly reduced as well. Other areas, including cities such as Cairo and Mexico City, have not yet solved their smog problems.

Recap Air quality and air pollution are a worldwide concern. Human activities are destroying the protective layer of ozone around Earth, producing acid rain and smog, and this may be causing a slow global warming trend. Remedial efforts are effective, as shown by the reduction of smog in some cities and the application of new technologies for reducing acid rain. Global initiatives to slow global warming and reduce ozone layer destruction are already under way. ■

24.2 Pollution jeopardizes scarce water supplies

Human activities have three major detrimental effects on water quality and availability. First, humans use excessive amounts of water, depleting freshwater supplies. Second, the replacement of natural vegetation with buildings and roads, especially in urban areas, prevents rainwater from soaking in, causing runoff. Third, human activities sometimes pollute precious sources of water.

Water is scarce and unequally distributed

In one sense, water is a renewable resource because it evaporates continuously from the oceans and falls on land as rain or snow. However, all of the freshwater on the surface of the land and in **aquifers** (deep underground reservoirs) forms less than 1% of Earth's total water. Over 97% is salty ocean water, and 2% is frozen in glaciers and polar ice caps.

a) Pittsburgh at the height of the steelmaking era, before smog cleanup efforts began.

b) Pittsburgh after programs were instituted to reduce smog.

Figure 24.5 Smog.

MJ's BlogInFocus How is the High Plains Aquifer under eight western states being affected by current irrigation practices? Visit MJ's blog in the Study Area in MasteringBiology and look under "The High Plains Aquifer."

http://goo.gl/L6nVhl

In addition to general scarcity, water is not evenly distributed among human populations. Residents of the most developed countries use 10 to 100 times more water than people in the least developed countries. Some desert and semiarid regions of the world have already reached their human carrying capacity as a result of water shortage. When we divert water for our purposes, we take it away from other species or interfere with their migration patterns. Water rights are a controversial subject in the western United States, where water is limited. Already, choices have to be made between irrigating crops, supplying growing cities, and encouraging the reproduction of Pacific salmon, a species endangered because of drought.

Urbanization increases storm water runoff

In urban areas, the shift from woodlands and fields to impermeable roads and buildings has caused major stormwater runoff problems. In older cities on the U.S. East Coast, stormwater combines with sewage, leading to combined sewage overflow (CSO), which overwhelms receiving streams and flows into the nearby Atlantic Ocean. CSOs are a major source of pathogens in some areas, causing gastroenteritis, eye and ear infections, skin rashes, respiratory infections, and hepatitis in swimmers and kayakers. Furthermore, the use of pipes to transport stormwater runoff quickly from buildings and roads to streams leads to stream overflows during storms and insufficient water levels during dry periods. The result is erosion of stream beds and loss of aquatic life (Figure 24.6). Restoration of urban streams seeks to stabilize stream channels, reduce channel erosion, and restore aquatic wildlife populations.

Human activities pollute freshwater

Human activities tend to pollute freshwater. Untreated sewage, chemicals from factories, the runoff of pesticides and fertilizers, and rubber and oil from city streets all must go somewhere. Either these materials degrade and decompose chemically or they end up in water or soil.

Some water pollutants are *organic nutrients* that arise from sewage treatment plants, food-packing plants, and paper mills. When these nutrients are degraded by bacteria, the resulting rapid growth rate of bacteria can deplete the water of oxygen, threatening aquatic animals. Others are *inorganic nutrients* such as nitrate and phosphate fertilizers and sulfates in laundry detergents. These cause prolific growth of algae, which die and are also decomposed by bacteria.

Figure 24.6 Urban stream erosion and pollution. Urbanization has adverse effects on streams due to excess and polluted runoff.

Eutrophication refers to the rapid growth of plant life and the death of animal life in a shallow body of water as a result of excessive organic or inorganic nutrients (Figure 24.7). The rapid growth of single-celled plantlike organisms such as algae depletes the water of the oxygen, leading to the death of aquatic animals. Although eutrophication is part of the normal process whereby freshwater ponds first become marsh and then dry land, human activities can accelerate the process, upsetting the natural rhythm of the ecosystem. For example, excess nitrogen and phosphorus fertilizers in the water from

Figure 24.7 Eutrophication. Water pollution accelerates the normal process of eutrophication, encouraging prolific growth of algae and plants at the expense of animal life.

agribusinesses in the Midwest results in regular dead zones devoid of fish around the mouth of the Mississippi river.

A special problem is *toxic pollutants,* including polychlorinated biphenyls (PCBs), oil and gasoline, pesticides, herbicides, and heavy metals. Most of these cannot be degraded by biological decomposition, so they remain in the environment for a long time. In addition, they tend to become more concentrated in the tissues of organisms higher up the food chain, a phenomenon known as **biological magnification** (Figure 24.8). Biological magnification occurs because each animal in a food chain consumes many times its own weight in food throughout its lifetime.

A classic example of biological magnification occurs with the heavy metal mercury. The primary sources of mercury in the environment are emissions from coal burning, gold production, smelters for nonferrous metal production, and cement production. Mercury released into the air or on land often ends up in aquatic ecosystems, where it accumulates by biological magnification in tertiary consumers such as sharks, tuna, and whales. Exposure to mercury can result in diverse symptoms, including loss of coordination, decreased memory and intellect, and poor immune system function. Pregnant women and young children are especially at risk because mercury affects nervous system development. Many countries now have regulations governing the maximum allowable mercury concentrations in fish destined for human consumption.

Other important water pollutants worldwide include *disease-causing organisms* present in human sewage, *sediments* from soil erosion that clog waterways and fill in lakes and shipping canals, excess *nitrogen fertilizers* used on land to grow crops, and even *heat pollution* from power plants. Heat pollution reduces the amount of oxygen that water can carry while at the same time increasing the oxygen demand by aquatic organisms whose activity level is temperature dependent. As a result, heat pollution may suffocate aquatic life.

✓ Which of the following species do you think would have the very highest concentration of toxic pollutants: a swordfish (a short-lived secondary/tertiary consumer), a blue whale (a long-lived plankton feeder), a killer whale (a long-lived tertiary/quaternary consumer), or a sardine (a short-lived plankton feeder)?

Groundwater pollution may impair human health

Many of the same pollutants that taint surface water also pollute groundwater. However, groundwater pollution poses two additional concerns. First, groundwater is often used as drinking water, so its pollutants can quickly affect human health. Second, groundwater is a slowly exchanging pool. Once it becomes polluted, it may stay polluted for a long time.

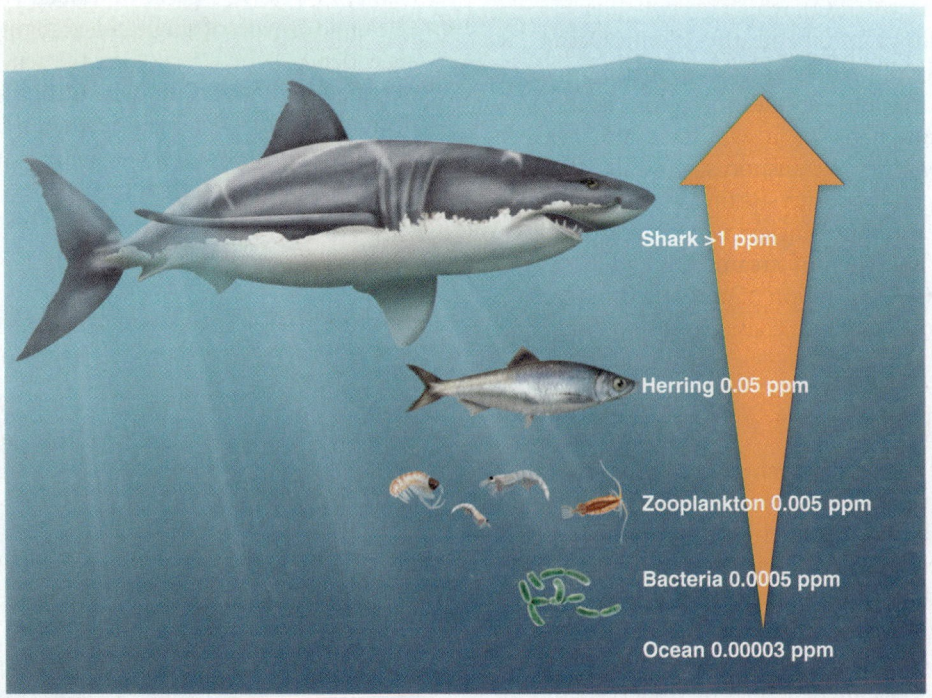

Shark >1 ppm

Herring 0.05 ppm

Zooplankton 0.005 ppm

Bacteria 0.0005 ppm

Ocean 0.00003 ppm

Figure 24.8 Biological magnification. The heavy metal mercury accumulates in body tissues as it moves up a food chain. Even though mercury may be present in only very trace amounts (about 0.00003 ppm, or parts per million) in water, it becomes highly concentrated (more than 1 ppm) by the time it reaches top predators such as sharks and whales.

The U.S. Environmental Protection Agency estimates that as many as 50% of all water systems and rural wells contain some type of pollutant. The most common are organic solvents such as carbon tetrachloride, pesticides, and fertilizers such as nitrates. We do not know the full effect of groundwater pollution because it is hard to separate from other possible causes of human disease. Public health officials suspect that some pollutants contribute to miscarriages, skin rashes, nervous disorders, and birth defects.

An issue of special concern regarding possible groundwater contamination is the disposal of radioactive wastes. Radioactive materials are used in nuclear power plants and in the diagnosis and treatment of human disease. Many radioactive wastes remain radioactive for thousands of years. They are usually stored in very dry environments in order to keep them from entering a water supply. Some radioactive wastes are so long-lived that for disposal they are incorporated into glass, which is then buried deep underground.

Figure 24.9 **Floating Garbage.** Discarded plastics may accumulate in certain regions of the oceans.

✔ Since 1997, New York City has spent over a billion dollars to purchase thousands of wooded acres in the watershed that supplies its drinking water. Why do you suppose the city has done that?

Oil pollution and garbage are damaging oceans and shorelines

A massive oil spill in the Gulf of Mexico in 2010 highlighted the potentially huge environmental costs associated with dependency on fossil fuels. The spill began after a British Petroleum drilling facility named Deepwater Horizon was destroyed by an explosion and fire. Ultimately, the Deepwater Horizon spill became the largest oil spill in U.S. history. Concerted efforts were made to burn the oil at sea, skim it off the surface, or disperse it with chemicals before it could reach shore. Despite these efforts, some of the oil ended up on pristine beaches and in fragile coastal estuaries and wetlands in the Gulf states. The full extent of the environmental and economic damage from just this one oil spill may not be known for decades.

In most years (2010 may have been an exception), several million tons of oil enter the world's oceans. About 50% comes from natural seepage; therefore, some oil pollution is a natural phenomenon. However, 30% of total oceanic oil pollution is caused by oil disposal on land that is washed to the sea in streams and rivers. The remaining 20% results from accidents at sea.

In general, when oil is spilled at sea about a quarter evaporates, nearly half eventually is degraded by bacteria, and the remaining quarter eventually settles to the ocean floor. In the short term, however, and especially if the spill is near shore, an oil spill can cause significant damage to marine and shoreline ecosystems. Before the oil dissipates it may coat living organisms, disrupting their ability to function and even choking and killing smaller organisms. Shoreline ecosystems may show signs of damage for years, including a loss of breeding grounds for shrimp and fish. Cleaning up an oil spill can sometimes save a shoreline ecosystem or help it recover more quickly, but it also shifts some of the pollution to land (if the oil is buried) or the air (if it is burned).

Another major source of ocean pollution is garbage, particularly plastic which does not degrade quickly. A Styrofoam cup may not degrade fully for 50 years; a plastic water bottle may take over 400 years. Floating garbage eventually ends up on shorelines or accumulates as giant rafts in certain regions of the oceans (**Figure 24.9**). An example is the Great Pacific garbage patch, also known as the Pacific Trash Vortex.

↺ **Recap** Freshwater is scarce and distributed unevenly. Freshwater can become polluted by surface water runoff from urban areas and by fertilizers, pesticides, and herbicides in rural areas. Pollution of rivers eventually reaches the oceans. Oil spills at sea are an example of how human activities directly pollute oceans and shorelines. ∎

24.3 Pollution and overuse damage the land

The main environmental issue with land may not be how we pollute it but how much of it we use. Humans consume a lot, and we tend to alter the landscape to suit our own purposes. We dam river valleys to produce hydroelectric power, strip mountaintops to find coal, and cut down forests for lumber or to clear space for crops.

The U.S. economy alone accounts for the direct consumption of 22 tons of fuels, metals, minerals, and biomass (food and forest products) for every person every year. Add to this number the amount of earth moved to build roads and waterways and to find energy, plus the erosion of soil by agricultural and forestry activities, and the total use of natural resources amounts to nearly 88 tons per U.S. citizen each year. It has been estimated that human activities have already altered a third of Earth's land mass, including the removal of nearly half of its forest cover.

As the human population grows, people tend to migrate toward cities. Cities often expand into nearby farmland because farmland is usually flat, and it is more economical to build where it is relatively flat. Cities also require large quantities of water and power and generate waste and pollution in a relatively small area.

The problems of land use in rural areas are quite different but just as significant as the land use issues in cities. More than half the people of the world live in rural poverty. Many rely almost entirely on their local environment to survive. It is not uncommon for impoverished rural communities to cut down nearly every tree in their area for fuel and shelter and overgraze communal lands with livestock. Stripping the biomass from fragile ecosystems leads to erosion and **desertification**—the transformation of marginal lands into near-desert conditions unsuitable for future agriculture (**Figure 24.10**). Every year, an estimated 15 million more acres of once-productive land become desert. Because the very survival of the rural poor is linked so closely to the use of the available resources, it is unlikely that we can halt land degradation in these regions without first doing something about the poverty there.

Figure 24.10 Desertification. Many poor residents of rural areas must depend on local resources to survive. To survive, they strip the biomass from their ecosystems causing desertification as once-productive land becomes eroded and barren. This photo was taken in Sudan.

Wars cause environmental damage that is often overlooked. In Iraq, the draining of the marshlands of the Euphrates/Tigris delta during the 1980s and 1990s resulted in the loss of valuable farmland. More recently, the oil spill at the Jiyyeh power station during the Israeli-Hezbollah conflict resulted in damage to 150 km of Lebanese and Syrian coastline, polluting beaches and coastal water.

And finally, there is the issue of how we should dispose of our garbage. Landfills are one solution, but there may be other ways to dispose of our waste that have less environmental impact, including recycling as much as possible. Landfills are not necessarily an environmental problem—as long as they are well designed and do not contribute to ground or water pollution. Nevertheless, we could all seek ways to generate less garbage and thereby reduce the risk of pollution. One way to reduce the impact of garbage is to recycle.

 Recap Land use and land pollution problems differ by region. In some areas, humans alter the landscape in search of fossil fuels and minerals. Cities expand into productive land and place a burden on resources. In rural regions, deforestation and desertification damage ecosystems and limit their productivity. ■

24.4 Energy: many options, many choices

Energy—we just can't get enough of it. We want inexpensive fuel for automobiles that weigh thousands of pounds. We heat our homes and even the water in our swimming pools. We need energy to refine raw materials, manufacture the products we use, and process and cook our food.

Energy use is tied to consumption, so we have a choice in determining how much energy to use. We also make choices about the sources of energy—choices that can affect the environment.

First, there are *fossil fuels*—coal, oil, and gas. We've already examined the environmental costs of retrieving and transporting fossil fuels, and we know that burning them may contribute to global warming, acid rain, and smog. But we must also consider that fossil fuels are *nonrenewable resources*, that is, they cannot be replenished. Their formation took place over millions of years, and we are on a path of consuming them all in a couple of centuries. Then what?

One way to reduce our dependence on fossil fuels would be to shift toward the use of renewable fuels or fuels with a seemingly infinite supply of energy. The possibilities include nuclear energy, biomass fuels, water, wind, geothermal energy, and solar power. Each has its advantages and its disadvantages.

Nuclear energy can supply a lot of power with the use of very little starting raw material. However, nuclear energy has proved to be relatively expensive because of the safeguards required to ensure that the core of the nuclear facility does not become overheated, suffer a meltdown, and release radioactivity into the environment. In addition, nuclear power plants generate wastes that remain radioactive for thousands of years. How to store radioactive wastes safely is a subject of current debate.

A growing trend is the production and use of *biomass fuels* (or biofuels)—fuels derived primarily from plants. Many rural poor have always used biomass fuels such as wood and the dung of herbivores. These are renewable if they are not depleted too fast, but they pollute the air and are not easily transported. In the most developed countries, ethanol (ethyl alcohol, an ingredient in alcoholic beverages), is used as a biofuel for automobiles, and biodiesel made from plant oils is a biofuel for some trucks and buses. Large-scale production of biofuels can only take place where biofuel crops can be grown efficiently (for example, in corn fields of the United States and sugarcane plantations in Brazil), so it is unlikely to be the fuel of choice of countries in cold climates. Biofuels production also removes valuable cropland from food production.

Where flowing water is readily available, hydroelectric power plants can generate electricity. Hydroelectric power plants use the kinetic energy of water to turn turbines that drive electric generators. However, hydroelectric power plants incur environmental costs because they require damming rivers, which disrupts valley and river ecosystems. In fact, instead of building more hydroelectric power plants, the opposite is happening. Some existing dams in the western United States are being dismantled because of the damage they have done to the Pacific salmon population.

Other types of renewable energy sources include wind farms that harness wind to generate electricity (Figure 24.11a) and geothermal power plants that extract heat from underground sources deep within Earth. Wind farms in particular have become popular lately, although not everyone wants one in their backyard. Wind and geothermal power sources have the advantage of providing renewable energy without the need for cooling water, as compared to fossil or nuclear power plants. Wind farms allow rural areas to retain jobs and to transmit energy to urban centers via utility grids. Geothermal energy is most plentiful near active volcanoes: 5% of California's energy and 25% of El Salvador's are from geothermal sources.

Ultimately, we may have to rely, just as plants do, on the one original and sustainable source of energy, *solar power*. Solar power can be harnessed to generate electricity in two ways. One way is through photovoltaic panels that convert light directly into electricity (Figure 24.11b). The panels absorb light units (photons) and release electrons as direct current. Photovoltaic panels are the method of choice where only a small amount of power is needed, such as for single homes. To create electricity on a commercial scale, powerful mirrors are used to focus sunlight on a fluid such as water, heating the water to boiling and producing steam. The steam is then used to run turbines that generate electricity for the electrical grid (Figure 24.11c).

At the moment, solar power technologies are less than 30% efficient at energy conversion, which makes them impractical to meet all our current energy demands. Nevertheless, the future of these technologies looks bright

a) A wind farm. Wind farms are increasingly being used to generate power in areas that have sustained winds.

b) Photovoltaic solar panels. Photovoltaic panels can provide enough electricity to power a home.

c) A solar power plant. This power plant is in Australia.

Figure 24.11 Alternatives to fossil fuel.

because the efficiency of photovoltaic panels and solar power plants improves with each passing year. With the use of high-efficiency photovoltaic panels on residences, solar tracking mirrors at solar power plants, and improved batteries to store electrical energy overnight, solar power could make a significant contribution to the world's energy needs. Right now, however, they contribute less than 1%.

MJ's BlogInFocus What are advanced biofuels, and why did the Environmental Protection Agency (EPA) lower its proposed 2022 advanced biofuels production targets? Visit MJ's blog in the Study Area in MasteringBiology and look under "Advanced Biofuels."

http://goo.gl/TJ3Onr

Using passive solar principles in building construction and site selection can reduce our energy consumption from other sources. South-facing windows collect solar energy passively during the day, especially if inside flooring material absorbs and holds a lot of heat. During the summer, heat-reflecting screens can keep the heat out. Using construction materials with a high thermal mass such as stone and concrete will reduce daily changes in building internal temperature.

↺ **Recap** Humans consume a disproportionate amount of energy compared to other species. Most of the energy we use comes from nonrenewable fossil fuels that contribute to pollution. Renewable energy sources include nuclear energy, biomass fuels, hydroelectric power, wind farms, geothermal energy, and solar energy. ∎

24.5 Environmental change and loss of biodiversity

Biodiversity refers to species richness, the assortment of living organisms on Earth. One measure of biodiversity is the number of species: To date, we have identified approximately 1.75 million species, but many scientists estimate that the total could be closer to 10, 30, or even 80 million. Ultimately, biodiversity can be considered as the variety of all forms of life, from species all the way to ecosystems. In this sense, biodiversity is a measure of ecological "health."

Human impacts on the biosphere contribute to a reduction in biodiversity worldwide. Indeed, some scientists estimate that the global extinction rate is 50 times higher now than at any time during the past 100,000 years.

Humans alter and destroy habitats

Pollution and overexploitation of natural resources are destroying whole habitats and driving many species of plants and animals toward extinction. The reasons are as varied as the places on Earth and the human cultures that inhabit them.

One cause of habitat destruction is farming, especially monoculture, the widespread practice of planting just one type of crop over a wide geographic area (for example, corn in the U.S. Corn Belt and wheat farther west). Monoculture substantially reduces the species variability within the local ecosystems. Intensive farming practices, monoculture or otherwise, can lead to soil erosion and loss of future productivity. The problem of erosion due to farming has been recognized and partially corrected with modern farming practices, which may include crop rotation and no-till farming.

In rain forest areas, deforestation is a major problem, as forests are cut down or burned to make way for subsistence farming, wherein farmers grow crops simply to feed themselves and their families (**Figure 24.12**). Despite their lush vegetation, rain forests do not make good farmland—they are fragile ecosystems with little topsoil and scant soil nutrients. Several years of intensive subsistence farming deplete the nutrients from the soil. To continue raising enough crops to survive, subsistence farmers must routinely clear a fresh section of rain forest, which starts another destructive cycle. The clear-cut areas subsistence farming leaves behind are slow to recover, if they recover at all.

Commercial logging is another factor in habitat destruction. European and Middle Eastern logging companies harvest 800-year-old hardwood trees in Africa and ship them to other continents, where they are processed into furniture, floors, and buildings. In recent years, the pace of logging has accelerated. The African nation of Ivory Coast once possessed 70 million acres of pristine tropical rain forest. Today, only 5 to 7 million acres remain, and even these may be gone in 10 years.

Exploitation of scarce resources poses another threat to habitats. In Brazil, poachers sneak into Itatiaia National Park

Figure 24.12 Deforestation. These loggers are harvesting old-growth timber in a slash-and-burn patch of the Amazon rainforest.

and cut down 100-year-old palm trees for the tender heart of palm, found only at the top of the tree. Hearts of palm are a popular gourmet product, and there is a growing market for them in the United States and other countries. Across Africa, illegal hunting of elephants for ivory and monkeys and antelopes for meat has soared. The United States has its own history of exploitation, including the complete extinction of the passenger pigeon and the near-extinction of the American bison in the late 1800s.

Clearly, habitat destruction and loss of biodiversity cannot be blamed on just one people, country, continent, or industry. After all, the logging industry is just responding to the demand from developed nations for paper, cardboard, and exotic woods. Subsistence farmers in the rain forests are just trying to feed themselves and their families. Halting the trend of habitat destruction and loss of biodiversity will require recognition of the problems and the willingness to do something about them.

Urbanization is a major force for environmental change

The worldwide shift to urban living has widespread effects on ecosystems, including biodiversity. Although cities cover less than 3% of Earth's surface, they are responsible for 78% of carbon emissions, 60% of residential water usage, and 76% of wood used for industrial purposes. The resource consumption and waste management requirements of a city typically expand well beyond the city limits. For example, the land area or "ecological footprint" required to serve London's resources and wastes is 120 times the size of the actual city.

Humans tend to plant specific trees, shrubs, and grasses in cities, resulting in fewer species overall, decreased native species, uneven plant coverage, and more invasive and exotic plants. Urbanization also shrinks the biodiversity of animal populations, with a shift to predominantly birds and insects. Furthermore, there is evidence that urbanization may reduce native species diversity on a regional or even global scale. For instance, urban sprawl, the spread of urban development into rural areas, in northern latitudes appears to reduce the populations of some migratory birds in southern latitudes.

✔ Considering the examples presented so far, where (in what types of ecosystems) do you think humans have affected biodiversity the most? How do you think this should impact decisions about land protection, such as which types of ecosystems to set aside?

Biodiversity is healthy for humans, too

Biodiversity is healthy for all organisms, including *Homo sapiens*, and large-scale disruptions to the biosphere threaten animal populations as well as human populations. For example, we depend on plants for a healthy atmosphere. Through photosynthesis, plants consume CO_2 and produce O_2, thereby not only producing O_2 for us to breathe but also removing CO_2 that might otherwise contribute to the greenhouse gases.

We depend on other species for clothing, shelter, food, and even medicine—25% of all prescription drugs sold in the United States are based on substances derived from fewer than 50 species of plants, many of which are found in rain forest areas that are being damaged by deforestation. It's estimated that there are 265,000 flowering plants in the world, and less than 1% have been tested so far for medicinal applications. By encouraging biodiversity, we maintain the potential to discover new medicines.

Biodiversity in the food supply may be important. Nutrition studies provide strong evidence that eating a varied diet, particularly a wide range of fruits and vegetables, is essential for optimal health. However, our current agriculture practices limit our food choices. Very few of the many known varieties of apples and tomatoes are typically available in U.S. supermarkets, and many leafy greens known to be high in antioxidants, such as arugula, watercress, and chard, often can only be found in specialty stores or farmers markets.

↩ **Recap** Human impacts are leading to a global biodiversity crisis. Urbanization, pollution, habitat destruction, and overexploitation of natural resources are destroying habitats. Loss of biodiversity threatens human populations, because humans depend on other species for oxygen, clothing, shelter, food, medicine, and other necessities. ▪

24.6 Toward sustainable development

In a sustainable world, humans would be an integral part of ecosystems so that those ecosystems (and we humans) could exist on the planet far into the future. This would mean practicing **sustainable development,** defined by the 1987 World Commission on Environment and Development as "development that meets the needs of the present without compromising the ability of future generations to meet their economic needs."

Measuring sustainability and quality of life

How can we determine whether development is sustainable? Before we can answer this question, we need to investigate the economics of sustainability. Economic progress is typically measured by the gross domestic product (GDP), which is defined as the total market value of all goods and services produced within a country in a year. An alternative measure is the genuine progress indicator (GPI). The GPI makes adjustments to the total market value of goods and services to account for environmental costs such as water and air pollution and loss of wetlands, as well as negative economic and social costs (crime, time spent commuting, loss of leisure time) associated with increased production. Quality of life is also taken into consideration by the GPI;

it weighs the value of all goods and services produced against enjoyment of life, including such factors as family breakdown and loss of parklands.

Strategies to support sustainable development

Given what we know today, let's focus on strategies that might help us reach a sustainable world.

Consume less This approach isn't popular, but by consuming less, you automatically use fewer resources, generate less garbage, and cause less pollution. If you use less, it leaves more for another person or even another species. *What* to consume less of may be a very personal choice, but unfortunately it's the big-ticket items that have the most impact. You could do a lot for the environment, for example, by living in a house that is no larger than your needs, keeping your car an extra year before replacing it (thus purchasing fewer cars in your lifetime), and living closer to work.

Recycle more This approach is more popular than consuming less. Recycling reusable materials such as paper, metal, and plastic reduces the manufacturers' need for raw materials and therefore the overall consumption of natural resources. It also keeps reusable materials from becoming either garbage or pollutants. We need to consider not only how we recycle substances back into our manufacturing processes but also how we recycle them back into the ecosystem. Some cities collect, compost, and then recycle leaves and organic material. Even landfills could be designed to encourage the breakdown of garbage and pollutants back into natural raw materials.

Encourage sustainable agriculture Current agriculture is based on an industrial model and is aptly referred to as agribusiness because it emphasizes high production. This approach has made food abundant and cheap but at the steep cost of degraded water and soil. In contrast, sustainable agriculture means developing systems to raise crops and livestock following the principles found in nature. With sustainable agriculture, we avoid depleting Earth's resources, or at least deplete them as little as possible. For example, sustainable agriculture on coffee plantations involves planting coffee plants under shade trees. Larger trees and greater diversity of vegetation enhance the growth of coffee plants, while also allowing native plants and birds to flourish (Figure 24.13). Rotating crops between nitrogen-fixers, such as clover, and nitrogen-users, such as corn, can reduce the use of synthetic fertilizers. By selling farm products locally, we reduce the environmental cost of transporting them to distant markets.

If you live in an area that still has small farms, consider participating Consumer Supported Agriculture (CSA) by connecting directly with a local farm. CSA members make a commitment to support a local farm, usually on a quarterly basis, by purchasing a share of the farm's harvest. Or, consider buying from a local farmers' market. The tradeoff

Figure 24.13 Sustainable agriculture. This shade-grown coffee plantation is in Ecuador.

for encouraging sustainable agriculture may be higher food prices, however, because practicing sustainable agriculture is not as efficient as modern agribusiness practices.

Support green roofs The replacement of conventional asphalt or tile on a flat roof with a living, vegetated roof system is a visually pleasing way to provide many benefits that support sustainability. Plants on green roofs typically consist of native species (for example, grasses in Scandinavia and succulents in California) and therefore require little irrigation or other maintenance (Figure 24.14). Green roofs increase the energy efficiency of buildings, primarily due to evaporative cooling from the plant material and soil. Green roofs help restore the natural water cycle by filtering and retaining stormwater. They also absorb airborne particulate pollutants, convert

Figure 24.14 A green roof.

harmful nitrogen oxides to organic nitrates, and use carbon dioxide for photosynthesis. One square meter of green roof absorbs the average amount of pollutants produced by an automobile driven 10,000 miles per year.

Lower fertility rates Fewer people translates to less resource consumption by humans and less environmental degradation. Fewer North and South Americans would mean fewer cars and houses in the United States and less deforestation in Brazil. A lower Asian population would mean fewer polluted rivers in India and less air pollution over China. Ending overpopulation in Africa would make reversal of desertification on the continent possible. A lower world population would mean a decrease in global warming trends. But, if the current fertility rate of the human species does not decline, eventually environmental degradation will reduce the human carrying capacity of Earth and the human death rate will rise to equal the birth rate. We can already see this happening in certain areas of Africa where life expectancy is short and infant mortality is high.

Reduce rural poverty People who are starving and without shelter think of their immediate needs, not future environmental impacts. It is virtually impossible to convey the long-term impact of desertification to someone who relies on a few goats and trees for day-to-day survival. More and more of the world will become uninhabitable due to deforestation and desertification unless the poor people in those areas are given a viable economic alternative to destruction of their own habitat.

Conserve energy in your home Opt for renewable energy at home, such as installing photovoltaic panels, and use energy-efficient appliances. Install low emittance (low-E) glass windows to reduce heat loss in winter and heat gain in summer. Switch incandescent bulbs to compact fluorescents to increase lighting efficiency.

Use environmentally preferable products Whenever possible, choose environmental products that are less harmful to the environment than their alternatives, such as paints that are low in volatile organic compounds (VOC), composite wood made from recycled materials, and carpets that are both low in VOC and high in recycled content.

Protect ecosystems that provide beneficial services From an economic point of view, it makes sense to protect our ecosystems, because if they are destroyed, we will lose crucial sources of food and medicines and many other benefits. For example, forests provide climate control, raw materials, and food. Because a forest is a natural asset that provides valuable services, it is a form of "natural capital." Similarly, the ozone layer provides protection from damaging ultraviolet radiation, and wetlands can protect adjacent areas from floods, support biodiversity, and filter pollutants.

Will we be able to achieve a sustainable world? Given our history so far, there is every reason for optimism. We humans have a remarkable capacity for making intelligent decisions and shaping our own environment. Already we can see cooperation on a global scale with the problem of global warming. More choices will be made in the future, and you will have a part in them. Your decisions will help to shape humankind's future.

> **Recap** A sustainable world will require stabilizing the human population growth rate and the global environment. Strategies for doing this include consuming less, recycling more, supporting sustainable agriculture and green building practices, lowering the global fertility rate and poverty rate, utilizing renewable energy sources, and protecting ecosystems. Achieving this goal will test our capacity to make intelligent decisions on a worldwide scale. ∎

Chapter Summary

24.1 Pollutants impair air quality p. 544

- Air pollution leads to excessive production of greenhouse gases, especially CO_2, and may be contributing to global warming.
- CFCs are thinning the ozone layer in the stratosphere. Although a number of nations are cooperating to reduce CFC emissions, ozone depletion remains a threat.
- Acid rain damages forests and aquatic ecosystems.
- Smog tends to hover over the area where it is produced and can cause chronic respiratory problems.

24.2 Pollution jeopardizes scarce water supplies p. 546

- Water is scarce and often distributed unequally. Freshwater forms less than 1% of the world's total water supply.
- Pollution of surface freshwater can contaminate groundwater and ultimately pollute the oceans.

- Pollution can accelerate the process of eutrophication.
- Because of biological magnification, toxic substances become more concentrated in organisms higher up the food chain.

24.3 Pollution and overuse damage the land p. 549

- Humans alter the landscape for their own purposes.
- Deforestation and desertification transform productive land into nonproductive land.

24.4 Energy: many options, many choices p. 550

- Humans use a great deal of energy compared to other species.
- Most of the energy we consume comes from nonrenewable fossil fuels—coal, oil, and gas—that contribute to pollution.
- Other possible sources of energy include biomass fuels, nuclear power, hydroelectric power, wind farms, geothermal power, and solar power. Solar power is renewable, but current solar power technologies cannot fulfill all our energy demands.

24.5 Environmental change and loss of biodiversity *p. 552*

- Pollution, urbanization, habitat destruction, and overexploitation of natural resources are damaging the biosphere and accelerating the global extinction rate. Human population growth, farming, logging, and exploitation of scarce resources are all factors.
- Humans depend on other species for clothing, shelter, food, medicine, oxygen and a healthy atmosphere, and other necessities of life. Loss of biodiversity places human populations at risk, as well as other species.

24.6 Toward sustainable development *p. 553*

- Strategies for achieving a sustainable future include consuming less, recycling more, supporting sustainable agriculture, planting green roofs, lowering the worldwide fertility rate to the replacement fertility rate, reducing rural poverty worldwide, conserving energy at home, using environmentally preferable products, and protecting ecosystems.
- Substituting the GPI (genuine progress indicator) for the GDP (gross domestic product) provides a measure of progress that takes into account our quality of life issues.
- Humans can shape the environment. It's up to us to shape it wisely.

Terms You Should Know

acid rain, *546*
aquifer, *546*
biodiversity, *552*
biological magnification, *548*
deforestation, *545*
desertification, *550*
eutrophication, *547*

global warming, *545*
greenhouse gases, *544*
ozone, *545*
pollution, *544*
smog, *546*
sustainable development, *553*
thermal inversion, *546*

Concept Review

Answers can be found in the Study Area in MasteringBiology.

1. Describe the importance of maintaining the ozone layer that surrounds Earth.
2. Describe the effects of acid rain and smog.
3. Discuss how human activities pollute Earth's water supply.
4. Explain how pollution can increase the greenhouse effect and how this might alter global temperatures.
5. Explain why tertiary consumers are more adversely affected by toxic pollutants than primary consumers.
6. Compare and contrast the various sources of energy that are available.
7. Describe the importance of biodiversity and how the loss of it impacts human health.
8. Describe why sustainable development makes sense from an economic point of view.
9. List strategies that support sustainable development.

Test Yourself

Answers can be found in the Appendix.

1. Global warming is most likely due to the impact of human activity on the:
   a. carbon cycle
   b. nitrogen cycle
   c. oxygen cycle
   d. phosphorus cycle

2. Which problem is attributed primarily to the use of chlorofluorocarbons (CFC) in refrigerators and aerosol cans?
   a. greenhouse gases leading to global warming
   b. depletion of the ozone layer in the stratosphere
   c. production of air pollution leading to acid rain
   d. development of smog over heavily populated areas

3. Which of the following is the primary source of the acid in acid rain?
   a. ozone produced by lightning
   b. carbon monoxide in automobile exhaust
   c. CO_2 from the burning of coal and oil for power
   d. sulfur dioxide from the burning of coal and oil for power

4. Which of the following are associated with the burning of gasoline and diesel fuel in automobiles and trucks?
   a. production of acid rain
   b. production of smog
   c. production of greenhouse gases that contribute to global warming
   d. all of the above

5. Which of the statements about freshwater distribution and use is correct?
   a. Freshwater supplies are evenly distributed throughout the world.
   b. About 10% of the world's water is freshwater.
   c. The supply of freshwater can limit the carrying capacity of a region.
   d. Freshwater supplies generally are unaffected by the water cycle.

6. All of the following statements about eutrophication are true except:
   a. Eutrophication is beneficial as it encourages growth of CO_2-consuming water plants and algae.
   b. Eutrophication is the result of sewage pollution and fertilizer runoff.
   c. Eutrophication can eventually result in the conversion of a wetland into dry land.
   d. Eutrophication is a natural process that can be accelerated by human activities.

7. Which of the following would be most affected by biological magnification of toxins?
   a. tertiary consumer   c. primary consumer
   b. secondary consumer   d. producer

8. All of the following are water pollutants except:
   a. fertilizers   c. heat
   b. ozone   d. disease-causing bacteria

9. Which environmental issue is most closely associated with rural poverty-stricken regions where people rely almost exclusively on their local environment for survival?
   a. air pollution
   b. groundwater pollution
   c. desertification
   d. eutrophication

10. All of the following are examples or sources of biomass fuels except:
    a. coal
    b. corn
    c. dung of herbivores
    d. sugarcane

11. The energy source raising the least concerns of pollution, environmental damage, or other risks is:
    a. nuclear energy
    b. hydroelectric energy
    c. solar energy
    d. biomass fuels

12. Which of the following exhibits the least biodiversity?
    a. a home vegetable garden
    b. a field of soybeans
    c. a tropical rain forest
    d. a northern pine forest

13. Which of the following contributes the most to the loss of biodiversity?
    a. water pollution
    b. air pollution
    c. habitat destruction
    d. mining and drilling for fossil fuels

14. Which of the following is likely to be a consequence of the burning of large areas of tropical rain forest to make way for croplands?
    a. a sustained increase in the economic well-being of the residents in the area
    b. an increase in the population of large game animals
    c. an increase in biodiversity
    d. an increase in atmospheric CO_2

15. All of the following are practices that promote sustainability except:
    a. planting large areas with the same crop
    b. reducing consumption
    c. recycling reusable materials
    d. reducing fertility rates worldwide

Apply What You Know

Answers can be found in the Study Area in MasteringBiology.

1. Your uncle in Iowa builds a farm pond and stocks it with bass and bluegill sunfish. At first, the pond water is fairly clear and the bass grow to large size, but within five years, the pond is covered with a thick green scum and all the fish are dead. Explain to your uncle what might have happened to the pond.

2. How can the seemingly simple choices a person makes in decorating a home on one continent affect the habitats of distant continents?

3. Most humans would agree that they wouldn't want to intentionally destroy something useful or valuable, and most agree that rainforests are valuable. If this is so, what is driving humans living in or near rainforests to destroy their own ecosystems?

4. In the 1970s, the Black Forest of Germany experienced a period when the number of trees dying in the forest increased substantially in a relatively short period of time. In addition, fish began to die in some of the lakes in the region. What could have caused these deaths to occur simultaneously in two very different habitats?

5. During the 2008 summer Olympic Games in Beijing, China, the government temporarily closed down many factories and mandated that people drastically reduce the use of their vehicles until the close of the Olympic Games. What was the government concerned about?

6. After Hurricane Katrina in 2005, many citizens of New Orleans were not able to return to their homes. Even those homes with only moderate water damage were declared condemned and ultimately torn down. Why were so many homes declared unfit and torn down rather than repaired?

7. Many different energy sources can be used to generate electricity. Which ones do you favor as the energy sources of the future, at least in your lifetime?

MJ's BlogInFocus

Answers can be found in the Study Area in MasteringBiology.

1. What is a carbon tax? Would you be in favor of a carbon tax as a way to reduce CO_2 emissions? For directions to an article on this issue, visit MJ's blog in the Study Area in MasteringBiology and look under "A Carbon Tax."
http://goo.gl/YHYhCx

MasteringBiology®

Students Go to MasteringBiology for assignments, the eText, and the Study Area with animations, practice tests, and activities.

Professors Go to MasteringBiology for automatically graded tutorials and questions that you can assign to your students, plus Instructor Resources.

Answers to ✔ questions are available in the Appendix.

Glossary

A

ABO blood typing Technique for classifying blood according to the presence or absence of two protein markers, A and B.

accommodation The adjustment of lens curvature in the eye so we can focus on either near or far objects.

acetylcholine (ACh) (AS-uh-teel-KOL-een) Neurotransmitter released by some nerve endings in both the peripheral and central nervous systems.

acid A substance that releases hydrogen ions when in solution. A proton donor.

acid precipitation Rain that has become acidic as the result of the airborne environmental pollutants sulfur dioxide and nitrogen oxide.

acquired immune deficiency syndrome (AIDS) Disease caused by human immunodeficiency virus (HIV). Symptoms include severe weight loss, night sweats, swollen lymph nodes, and opportunistic infections.

acromegaly (AK-roh-MEG-uh-lee) A condition of enlargement of the bones and muscles of the head, hands, face, and feet caused by an excess of growth hormone in an adult.

acrosome (AK-roh-sohm) A cap containing enzymes that covers most of the head of a sperm. Helps sperm penetrate an egg.

actin (AK-tin) A contractile protein of muscle. Forms the thin filaments in the myofibrils.

action potential A brief change in membrane potential of a nerve or muscle cell due to the movement of ions across the plasma membrane. Also called a nerve impulse.

active immunization The process of activation of the immune system in advance of the presence of a disease. Generally, active immunization is accomplished by administration of a vaccine.

active transport Process that transfers a substance into or out of a cell, usually against its concentration gradient. Requires a carrier and the expenditure of energy.

adaptation General term for the act or process of adjusting. In environmental biology the process of changing to better suit a particular set of environmental conditions. In sensory neurons, a decline in the frequency of nerve impulses even when a receptor is stimulated continuously and without change in stimulus strength.

adaptive radiation A short burst of evolutionary activity in which many new life-forms develop in a brief span of time from a single ancestor.

Addison's disease A condition caused by a failure of the adrenal cortex to secrete sufficient cortisol and aldosterone.

adenosine diphosphate (ADP) (uh-DEN-ohseen dy-FOSS-fate) A nucleotide composed of adenosine with two phosphate groups. When ADP accepts another phosphate group it becomes the energy-storage molecule ATP.

adenosine triphosphate (ATP) (uh-DEN-ohseen try-FOSS-fate) A nucleotide composed of adenosine and three phosphate groups. ATP is an important carrier of energy for the cell because energy is released when one of the phosphate groups is removed.

adipocyte (AH-di-po-syt) An adipose (fat) cell.

adipose tissue (AH-di-pohs) A connective tissue consisting chiefly of fat cells.

adrenal cortex (ah-DREE-nul KOR-teks) External portion of the adrenal gland. Its primary hormones, aldosterone and cortisol, influence inflammation, metabolism, interstitial fluid volume, and other functions.

adrenal glands (ah-DREE-nul) Hormone-producing glands located atop the kidneys. Each consists of medulla and cortex areas.

adrenal medulla (ah-DREE-nul meh-DUL-uh) Interior portion of the adrenal gland. Its hormones, epinephrine and norepinephrine, influence carbohydrate metabolism and blood circulation.

adrenocorticotropic hormone (ACTH) (uh-DREE-noh-kort-ih-koh-TRO-pic) Anterior pituitary hormone that influences the activity of the adrenal cortex.

aerobic metabolism (air-OH-bik) Metabolic reactions within a cell that utilize oxygen in the process of producing ATP.

afferent Directed to or toward a point of reference. In the kidney, an afferent arteriole carries blood to each glomerulus.

agglutination (uh-glue-tin-AY-shun) The clumping together of (foreign) cells induced by cross-linking of antigen-antibody complexes.

agranular leukocyte A white blood cell that does not contain distinctive granules. Includes lymphocytes and monocytes.

albumins (AL-byoo-mins) The most common group of proteins in blood plasma. Albumins have osmotic and transport functions.

aldosterone (al-DOS-ter-ohn or AL-do-STEER-ohn) Hormone produced by the adrenal cortex that regulates sodium reabsorption.

alleles (uh-LEELS) Genes coding for the same trait and found at the same locus on homologous chromosomes.

allergen (AL-ur-jen) Any substance that triggers an allergic response.

allergy (AL-ur-jee) Overzealous immune response to an otherwise harmless substance.

all-or-none principle The principle that muscle cells always contract completely each time they are stimulated by their motor neuron, and that they do not contract at all if they are not stimulated by their motor neuron.

alveolus (pl. alveoli) (al-VEE-oh-lus) One of the microscopic air sacs of the lungs.

Alzheimer's disease (AHLTZ-hy-merz) Degenerative brain disease resulting in progressive loss of motor control, memory, and intellectual functions.

amino acid (uh-MEE-no) Organic compound containing nitrogen, carbon, hydrogen, and oxygen. The 20 amino acids are the building blocks of protein.

amniocentesis A form of fetal testing in which a small sample of amniotic fluid is removed so that cells of the fetus can be examined for certain known diseases and genetic abnormalities.

amnionic fluid (am-nee-AHN-ik) A fluid similar to interstitial fluid that surrounds and protects the fetus during pregnancy. Amniotic fluid is in continuous exchange with maternal interstitial fluid.

ampulla (am-PULL-la) Base of a semicircular canal in the inner ear.

anabolic steroid Any one of over 100 synthetic and natural compounds, all of which are related to the male sex steroid hormone testosterone. Anabolic steroids increase muscle mass and contribute to the development and maintenance of masculine characteristics.

anabolism (ah-NAB-o-lizm) Energy-requiring building phase of metabolism, during which simpler substances are combined to form more complex substances.

analogous structures Structures that share a common function but not necessarily a common origin.

anaphase Third stage of mitosis, in which the two sets of daughter chromosomes move toward the poles of a cell.

androgen (AN-dro-jen) A hormone that controls male secondary sex characteristics, such as testosterone.

anemia (uh-NEE-mee-uh) A condition in which the blood's ability to carry oxygen is reduced because of a shortage of normal hemoglobin or too few red blood cells.

aneurysm (AN-yur-iz-um) A ballooning or bulging of the wall of an artery caused by dilation or weakening of the wall.

angina pectoris (AN-jin-uh PEK-tor-is) Sensation of severe pain and tightness in the chest caused by insufficient blood supply to the heart.

angiogram (AN-jee-oh-gram) An X-ray image of blood vessels.

angiotensin II (AN-jee-oh-TEN-sinTOO) A potent vasoconstrictor activated by renin. Also triggers the release of aldosterone from the adrenal gland.

anorexia nervosa (ah-nuh-REK-see-uh nur-VOH-suh) An eating disorder characterized by abnormally low body weight and unrealistic fear of becoming obese.

antibiotic (AN-tih-by-AH-tik) A substance that interferes with bacterial metabolism. Administered to cure bacterial diseases.

antibody A protein molecule released by a B cell or a plasma cell that binds to a specific antigen.

anticodon The three-base sequence in transfer RNA that pairs with a complementary sequence in messenger RNA.

antidiuretic hormone (ADH) (AN-tih-dy-yu-RET-ik) Hormone produced by the hypothalamus and released by the posterior pituitary. Stimulates the kidneys to reabsorb more water, reducing urine volume.

antigen (AN-tih-jen) A substance or part of a substance (living or nonliving) that the immune system recognizes as foreign. It activates the immune system and reacts with immune cells or their products, such as antibodies.

antigen-presenting cell (APC) A cell that displays antigen-MHC complexes on its surface. Triggers an immune response by lymphocytes.

antioxidant An agent that inhibits oxidation; specifically a substance that can neutralize the damaging effects of free radicals and other substances.

anus The outlet of the digestive tract.

aorta (ay-OR-tah) The main systemic artery. Arises from the left ventricle of the heart.

aortic bodies Receptors in the aorta that are sensitive primarily to the oxygen concentration of the blood. These receptors can also respond to H^+ and CO_2.

appendicular skeleton (ap-un-DIK-yuh-lur) The portion of the skeleton that forms the pectoral and pelvic girdles and the four extremities.

aquaporins Literally, water pores. Aquaporins are cell membrane proteins in certain specialized cells, especially the tubular cells in kidneys, that contain pores which permit the rapid diffusion of water.

appendix (uh-PEN-diks) A small finger-like pouch, with no known digestive function, that extends from the cecum of the large intestine.

aqueous humor (AY-kwee-us) Watery fluid between the cornea and the lens of the eye.

aquifer A deep underground reservoir of water.

Archaea (ar-KAY-uh) One of the three domains of life. The prokaryotes of Archaea have many of the features of eukaryotes, but they do not have true nuclei.

arrhythmia Irregular heart rhythm. Arrhythmias may be caused by defects in the cardiac conduction system or by drugs, stress, or cardiac disease.

arteriole (ar-TEER-ee-ohl) The smallest of the arterial blood vessels. Arterioles supply the capillaries, where nutrient and gas exchange takes place.

artery A blood vessel that conducts blood away from the heart and toward the capillaries.

arthritis Inflammation of the joints.

artificial insemination A technique to enhance fertility in which sperm are placed into the vagina or uterus with a syringe.

asthma (AZ-muh) A recurrent, chronic lung disorder characterized by spasmodic contraction of the bronchi, making breathing difficult.

astigmatism (uh-STIG-muh-tiz-um) An eye condition in which unequal curvatures in different parts of the lens (or cornea) lead to blurred vision.

atherosclerosis (ath-er-o-skler-OS-is) Buildup of fatty deposits on and within the inner walls of arteries. Atherosclerosis is a risk factor for cardiovascular disease.

atom Smallest particle of an element that exhibits the properties of that element. Atoms are composed of protons, neutrons, and electrons.

atrial natriuretic hormone (ANH) (AY-tree-ul NAH-tree-u-RET-ik) A hormone secreted by the atria of the heart. Helps regulate blood pressure by causing the kidneys to excrete more salt.

atrium (pl. atria) (AY-tree-um) One of the two chambers of the heart that receive blood from veins and deliver it to the ventricles.

auditory canal Part of the outer ear. The auditory canal channels sounds to the eardrum.

auditory tube Hollow tube connecting the middle ear with the throat. The auditory tube is important for equalizing air pressure in the middle ear with atmospheric pressure.

autoimmune disorder A condition in which the immune system attacks the body's own tissues.

autonomic division Division of the peripheral nervous system composed of the sympathetic and parasympathetic divisions. The autonomic division carries signals from the central nervous system that control automatic functions of the body's internal organs. Sometimes referred to as the autonomic nervous system (ANS).

autosomes (AW-tow-sohms) Chromosomes 1 to 22. The term *autosome* applies to all chromosomes except the sex chromosomes.

autotroph An organism that uses carbon dioxide and energy from its physical environment to build its own large organic molecules.

atrioventricular (AV) bundle Bundle of specialized fibers that conduct impulses from the AV node to the right and left ventricles.

atrioventricular (AV) node Specialized mass of conducting cells located at the atrioventricular junction in the heart.

atrioventricular (AV) valve Valve located between the atria and their corresponding ventricles which prevent blood from flowing back into the atria when the ventricles contract.

axial skeleton (AK-see-ul) The portion of the skeleton that forms the main axis of the body, consisting of the skull, ribs, sternum, and backbone.

axon (AK-sahn) An extension of a neuron that carries nerve impulses away from the nerve cell body. The conducting portion of a nerve cell.

B

B cell (B lymphocyte) (LIM-fo-syt) A white blood cell that matures in bone marrow and gives rise to antibody-producing plasma cells.

bacteria One of the three domains of life. Bacteria are prokaryotic microorganisms responsible for many human diseases.

balloon angioplasty Treatment to open a partially blocked blood vessel. The technique involves threading a small balloon into the blood vessel and then inflating the balloon, pressing the blocking material against the sides of the vessel.

baroreceptor Sensory receptor that is stimulated by increases in blood pressure.

basal metabolic rate (BMR) Rate at which energy is expended (heat produced) by the body per unit of time under resting (basal) conditions.

base A substance capable of binding with hydrogen ions. A proton acceptor.

basement membrane Layer of nonliving extracellular material that anchors epithelial tissue to the underlying connective tissue.

basophil (BAY-so-fll) White blood cell whose granules stain well with a basic dye. Basophils release histamine and other substances during inflammation.

bile Greenish-yellow or brownish fluid that is secreted by the liver, stored in the gallbladder, and released into the small intestine. Bile is important in emulsifying fat.

biodiversity The variety of living species on Earth.

biogeochemical cycle A cycle in which the chemicals that compose living organisms are recycled between organisms and Earth itself.

biogeography Study of the distribution of plants and animals around the world.

biological magnification Process by which non-excreted substances become more concentrated in organisms that are higher in the food chain.

biology The study of living things and life's processes.

biomass Total living component of an ecosystem.

biosphere (BY-oh-sfeer) The portion of the surface of Earth (land, water, and air) inhabited by living organisms.

biotechnology The technical application of biological knowledge for human purposes.

biotic potential (by-AHT-ic) Maximum growth rate of a population under ideal conditions.

birth control pill Oral contraceptive containing hormones that inhibit ovulation.

blastocyst (BLAS-toh-sist) Stage of early embryonic development. The product of cleavage.

blood A fluid connective tissue consisting of water, solutes, blood cells, and platelets that carries nutrients and waste products to and from all cells.

blood-brain barrier A functional barrier between the blood and the brain that inhibits passage of certain substances from the blood into brain tissues.

blood doping The practice of increasing a person's red blood cell production above normal by artificial means, for the purpose of increasing oxygen-carrying capacity.

blood pressure The force exerted by blood against blood vessel walls, generated by the pumping action of the heart.

blood type Classification of blood based on the presence of surface antigens on red blood cells, and the presence of antibodies to surface antigens other than one's own.

bone A connective tissue that forms the bony skeleton. Bone consists of a few living cells encased in a hard extracellular matrix of mineral salts.

bottleneck effect Evolutionary effect that occurs when a major catastrophe wipes out most of a population without regard to any previous measure of fitness.

botulism A form of poisoning caused by a bacterium, *Clostridium botulinum*, occasionally found in inadequately cooked or preserved foods.

Bowman's capsule Double-walled cup at the beginning of a renal tubule. Bowman's capsule, also called the glomerular capsule, encloses the glomerulus.

brain The command center of the body.

bronchiole (BRAHNG-kee-ohl) Small airways within the lungs that carry air to the alveoli. The walls of bronchioles are devoid of cartilage.

bronchitis (brahng-KI-tis) Inflammation of the bronchi.

bronchus (pl. bronchi) (BRAHNG-kus) Any one of the larger branching airways of the lungs. The walls of bronchi are reinforced with cartilage.

buffer Any substance that tends to minimize the changes in pH that might otherwise occur when an acid or base is added to a solution.

bulimia (byoo-LEE-mee-uh) Eating disorder involving episodes of binging and purging.

C

calcitonin (kal-suh-TOW-nin) Hormone released by the thyroid glands that promotes a decrease in blood calcium levels.

Calorie Amount of energy needed to raise the temperature of 1 kilogram of water 1° Celsius.

cancer A malignant, invasive disease in which cells become abnormal and divide uncontrollably. May spread throughout the body.

capillary (KAP-uh-lair-ee) The smallest type of blood vessel. Capillaries are the site of exchange between the blood and tissue cells.

carbaminohemoglobin (kar-buh-MEE-noh-HEE-moh-gloh-bin) Carbon dioxide bound form of hemoglobin.

carbohydrate (kar-boh-HY-drayt) Organic compound composed primarily of CH_2O groups. Includes starches, sugars, and cellulose.

carbonic anhydrase (kar-BAHN-ik an-HY-drase) Enzyme that facilitates the combination of carbon dioxide with water to form carbonic acid.

carcinogen (kar-SIN-o-jen) Cancer-causing agent.

cardiac cycle (KAR-dee-ak) Sequence of events encompassing one complete contraction and relaxation cycle of the atria and ventricles of the heart.

cardiac muscle Specialized muscle tissue of the heart.

cardiac output The amount of blood the heart pumps into the aorta each minute.

cardiovascular system The organ system composed of the heart and blood vessels.

carnivore (KAR-nih-vor) Organism that feeds only on animals.

carrying capacity The maximum number of individuals in a population that a given environment can sustain indefinitely

cartilage (KAR-til-ij) White, semiopaque, flexible connective tissue.

catabolism (kah-TAB-o-lizm) The process whereby larger molecules are broken down to smaller ones and energy is released.

catalyst (KAT-uh-list) A substance that changes the rate of a chemical reaction without itself being altered or consumed by the reaction.

cataracts (KAT-uh-rakts) Clouding of the eye's lens. Cataracts are often congenital or age-related.

cecum (SEE-kum) The blind-ended pouch at the beginning of the large intestine.

cell The smallest structure that shows all the characteristics of life. The cell is the fundamental structural and functional unit of all living organisms.

cell body Region of a nerve cell that includes the nucleus and most of the cytoplasmic mass, and from which the dendrites and axons extend.

cell cycle A repeating series of events in which a eukaryotic cell grows, duplicates its DNA, and then undergoes nuclear and cytoplasmic division to become two cells. Includes interphase (growth and DNA synthesis) and mitosis (cell division).

cell doctrine Theory consisting of three basic principles: (1) All living things are composed of cells and cell products. (2) A single cell is the smallest unit that exhibits all the characteristics of life. (3) All cells come only from preexisting cells.

cell junctions Various proteins that connect epithelial cells to each other.

cellular respiration The collective term for all cellular metabolic processes that result in the production of ATP.

central canal In bone, the hollow central tube of an osteon that contains nerves and blood vessels.

central nervous system (CNS) The brain and spinal cord.

centriole (SEN-tree-ohl) Small organelle found near the nucleus of the cell, important in cell division.

centromere (SEN-troh-meer) Constricted region of a chromosome where sister chromatids are attached to each other and where they attach to the mitotic spindle during cell division.

cerebellum (ser-ah-BELL-um) Brain region most involved in producing smooth, coordinated skeletal muscle activity.

cerebral cortex (suh-REE-brul KOR-teks) The outer gray matter region of the cerebral hemispheres. Some regions of the cerebral cortex receive sensory information; others control motor responses.

cerebral hemispheres The large paired structures that together form most of the cerebrum of the brain.

cerebrospinal fluid (suh-REE-broh-SPY-nul) Plasmalike fluid that fills the cavities of the CNS and surrounds the CNS externally. Protects the brain and spinal cord.

cerebrum (suh-REE-brum) The cerebral hemispheres and nerve tracts that join them together. Involves higher mental functions.

cesarean delivery (C-section) Surgical delivery of a baby.

chemical bonds Attractive forces between atoms that cause atoms to be bound to each other.

chemoreceptor (KEE-moh-rih-sep-tur) Type of receptor sensitive to various chemicals.

chemotherapy The use of therapeutic drugs to selectively kill cancer cells.

chlamydia (kla-MID-ee-uh) Sexually transmitted disease involving infection by the bacterium *Chlamydia trachomatis*.

cholecystokinin (CCK) (kohl-uh-sis-tuh-KYN-in) An intestinal hormone that stimulates contraction of the gallbladder and the release of pancreatic juice.

cholesterol (ko-LES-ter-ahl) Steroid found in animal fats as well as in most body tissues. Made by the liver.

chondroblast A cartilage-forming cell. In the fetus, chondroblasts produce the hyaline cartilage that forms the rudimentary models of future bones.

chorion (KOR-ee-ahn) Outermost fetal membrane. Helps form the placenta.

chorionic villi (kor-ee-AHN-ik VIL-eye) Extensions of the fetal chorion that project into maternal tissues. Samples of chorionic villi are sometimes taken in early pregnancy to obtain fetal cells for diagnosis of fetal disease.

choroid (KOR-oyd) The vascular middle layer of the eye.

chromatin (KROH-muh-tin) Threadlike material in the nucleus composed of DNA and proteins.

chromosome (KROH-muh-som) Rodlike structure of tightly coiled chromatin. Visible in the nucleus during cell division.

chyme (kym) Semifluid mass consisting of partially digested food and gastric juice that is delivered from the stomach into the small intestine.

cilium (pl. cilia) (SIL-ee-um) A tiny hairlike projection on cell surfaces that moves in a wavelike manner.

circulatory system The system consisting of the heart, blood vessels, and the blood that circulates through them.

cirrhosis (sir-ROH-sis) Chronic disease of the liver, characterized by an overgrowth of connective tissue (fibrosis).

citric acid cycle The metabolic pathway within mitochondria in which acetyl groups are completely disassembled into CO_2 and high-energy compounds. Also called the Krebs cycle.

clavicle The collarbone.

cleavage An early phase of embryonic development consisting of rapid mitotic cell divisions without cell growth.

clitoris (KLIT-uh-ris) In the female, a sensitive erectile organ located in the vulva.

cloning Production of identical copies of a gene, a cell, or an organism.

cochlea (KOK-lee-uh) Snail-shaped chamber in the inner ear that houses the organ of Corti.

codominance Pattern of inheritance in which both alleles of a gene are equally expressed even though the phenotypes they specify are different.

codon (KOH-dahn) The three-base sequence on a messenger RNA molecule that provides the code for a specific amino acid in protein synthesis.

coenzyme A small molecule that assists an enzyme by transporting small molecular groups. Most vitamins are coenzymes.

collagen fiber Strong and slightly flexible type of connective tissue fiber, made of protein.

colon (KOH-lun) The region of the large intestine between the cecum and the rectum. The colon includes ascending, transverse, descending, and sigmoid portions.

colostrum (koh-LAHS-trum) Milky fluid secreted by the mammary glands shortly before and after delivery. Contains proteins and antibodies.

community Array of several different populations that coexist and interact within the same environment.

compact bone Type of dense bone tissue found on the outer surfaces and shafts of bones.

complement system A group of bloodborne proteins which, when activated, enhance the inflammatory and immune responses and may lead to lysis of pathogens.

complete dominance Pattern of genetic inheritance in which one allele completely masks or suppresses the expression of its complementary allele.

complete protein A protein containing all 20 of the amino acids in proportions that meet our nutritional needs.

computed axial tomography (CAT scan) Computer analysis of a series of cross-sectional X-ray images made along a single axis of the body, used to construct three-dimensional images of internal body parts.

concussion An injury to the brain resulting from a physical blow that may result in a brief loss of consciousness.

cone cell One of the two types of photoreceptor cells in the retina of the eye. Cone cells provide for color vision.

congestive heart failure Progressive condition in which the pumping efficiency of the heart becomes impaired. The result is often an inability to adequately pump the venous blood being returned to the heart, venous congestion, and body fluid imbalances.

connective tissue A primary tissue; form and function vary extensively. Functions include support, energy storage, and protection.

consumer In ecosystems, an organism that cannot produce its own food and must feed on other organisms. Also called a heterotroph.

continental drift The slow movement over time of the Earth's tectonic plates.

control center In control system theory, a structure that receives input from a sensor, compares it to a correct, internally set value of the controlled variable, and sends an output to an effector. The most important control center in a living organism is the brain.

control group In scientific experiments, a group of subjects that undergoes all the steps in the experiment except the one being tested. The control group is used to evaluate all possible factors that might influence the experiment other than the experimental treatment. Compare *experimental group*.

controlled variable Any physical or chemical property that might vary from time to time and that must be controlled in order to maintain homeostasis. Examples of controlled variables are blood pressure, body temperature, and the concentration of glucose in blood.

core temperature The body's internal temperature. In humans, normal core temperature is about 98.6°F (37°C).

cornea (KOR-nee-uh) The transparent anterior portion of the eyeball.

coronary artery (KOR-uh-nair-ee) One of the two arteries of the heart leading to capillaries that supply blood to cardiac muscle.

coronary artery bypass graft (CABG) Surgery to improve blood flow. A piece of blood vessel is removed from elsewhere in the body and grafted onto a blocked coronary artery to bypass the damaged region.

corpus luteum (KOR-pus LOOT-ee-um) Structure that develops from cells of a ruptured ovarian follicle. The corpus luteum secretes progesterone and estrogen.

cortex (KOR-teks) General term for the outer surface layer of an organ.

Cortisol (KOR-tih-sahl) Glucocorticoid produced by the adrenal cortex, also called hydrocortisone.

covalent bond (koh-VAY-lent) Chemical bond created by electron sharing between atoms.

coxal bones The two large bones that connect the femur bones to the sacrum of the vertebral column. The coxal bones form our hips.

cranial nerve (KRAY-nee-ul) One of the 12 pairs of peripheral nerves that originate in the brain.

creatinine (kree-AT-uhn-een) A nitrogenous waste molecule excreted by the kidneys.

crossing-over Exchange of DNA segments between homologous chromosomes during prophase I of meiosis.

cystic fibrosis (SIS-tikfy-BRO-sys) Genetic disorder in which oversecretion of mucus clogs the respiratory passages. Cystic fibrosis can lead to fatal respiratory infections.

cytokines Signaling molecules secreted by helper T cells. Cytokines have a variety of functions, including promoting immune cell development, stimulating their activity, and attracting them to an area of infection.

cytokinesis (sy-tow-kih-NEE-sis) The division of cytoplasm that occurs after a cell nucleus has divided.

cytoplasm (SY-tow-plaz-um) The cellular material surrounding a cell nucleus and enclosed by the plasma membrane.

cytoskeleton A cell's internal "skeleton." The cytoplasm is a system of microtubules and other components that support cellular structures and provide the machinery to generate various cell movements.

cytotoxic T cell (sy-toh-TAHK-sik) A killer T cell that directly lyses foreign cells, cancer cells, or virus-infected body cells.

D

data Numerical information that can be organized and interpreted.

decomposer An organism that obtains energy by chemically breaking down the products, wastes, or remains of other organisms.

dehydration synthesis Process by which a larger molecule is synthesized by covalently bonding two smaller molecules together, with the subsequent removal of a molecule of water.

demographic transition A progression of changes followed by many nations, in which the society gradually moves from poor living conditions and a high death rate to improved economic conditions and a declining death rate.

denaturation (dee-nay-chur-AY-shun) Loss of a protein's normal shape and function produced by disruption of hydrogen bonds and other weak bonds.

dendrite (DEN-dryt) Branching neuron extension that serves as a receptive or input region. Dendrites conduct graded potentials toward the cell body.

denitrification (dee-nyt-ruh-fih-KAY-shun) Process in which nitrate is converted to nitrogen gas (N_2). Denitrification is part of the nitrogen cycle.

deoxyhemoglobin Hemoglobin that has given up its oxygen.

deoxyribonucleic acid (DNA) (dee-OX-ee-RY-bo-new-CLAY-ik) A nucleic acid found in all living cells. Carries the organism's hereditary information.

depolarization Loss of a state of polarity. In nerve and muscle cells, the loss or reduction of the inside-negative membrane potential.

dermis Layer of skin underlying the epidermis. Consists mostly of dense connective tissue.

desertification (deh-ZERT-uh-fuh-KAY-shun) Process by which marginal lands are converted to desertlike conditions.

detritus (dih-TRY-tus) Nonliving organic matter.

diabetes mellitus (dy-uh-BEE-teez MEL-ih-tus) Disease characterized by a high blood sugar concentration and glucose in the urine, caused by either deficient insulin production (Type 1 diabetes) or insufficient glucose uptake by cells (Type 2 diabetes).

dialysis (dy-AL-ih-sis) General term for several techniques that attempt to take the place of kidney function in patients whose kidneys have failed by letting the patient's blood exchange waste materials with artificial fluids. Two dialysis techniques are *continuous ambulatory peritoneal dialysis (CAPD)* and *hemodialysis*.

diaphragm (DY-uh-fram) A dome-shaped sheet of muscle that separates the thoracic cavity from the abdominal cavity. Also, a contraceptive device inserted into the vagina to cover the cervical opening.

diastole (dy-AS-toh-lee) Period of the cardiac cycle when a heart chamber is relaxed.

diastolic pressure (DY-uh-STAHL-ik) Lowest point of arterial blood pressure during a cardiac cycle. Diastolic pressure is the second of two numbers recorded by the health care provider, the first being systolic pressure.

dietary supplement A product not normally part of a diet, but can be taken to improve overall health or well-being.

differentiation Process by which a cell changes in form or function.

diffusion (dih-FYOO-shun) The movement of molecules from one region to another as the result of random motion. Net diffusion proceeds from a region of higher concentration to a region of lower concentration.

digestion Chemical or mechanical process of breaking down foodstuffs to substances that can be absorbed.

digestive system The system of organs that share the common function of getting nutrients into the body.

diploid The number of chromosomes in a body cell ($2n$), twice the chromosomal number (n) of a gamete. In humans, $2n = 46$.

disaccharide (dy-SAK-uh-ryd) Literally, a double sugar. A disaccharide consists of two monosaccharides linked together. Sucrose and lactose are disaccharides.

diuretic (dy-yoo-RET-ik) Any substance that enhances urinary output.

DNA fingerprinting A technique for identifying the source of a fragment of DNA after it has been sufficiently copied by PCR.

DNA ligases A class of enzymes that join fragments of DNA together.

dominant allele An allele that masks or suppresses the expression of its complementary allele.

Down syndrome A condition caused by inheriting three copies of chromosome 21.

duodenum (doo-oh-DEE-num) First part of the small intestine.

dysplasia (dis-PLAY-zhee-uh) Abnormal changes in the shape and/or organization of cells. Dysplasia may precede the development of cancer.

E

ecological pyramid Graph representing the biomass, energy content, or number of organisms of each level in a food web. An ecological pyramid includes both producer and consumer populations.

ecology (ee-KAHL-uh-jee) Study of the relationships between organisms and their physical environment.

ecosystem (EEK-oh-sis-tem) All living organisms, all matter, and all energy in a given environment.

ectoderm Embryonic germ layer. Ectoderm forms the epidermis of the skin and its derivatives, and nerve tissues.

ectopic pregnancy (ek-TOP-ik) A pregnancy that results from implantation outside the uterus. The most common form of ectopic pregnancy is a tubal pregnancy.

efferent Directed from or away from a point of reference. In the kidney, an efferent arteriole carries blood away from each glomerulus.

effector An effector takes the necessary action to correct an imbalance, in accordance with the signals it receives from the control center.

egg An ovum; a mature female gamete.

ejaculatory duct In the male, a short duct that carries sperm from the ductus deferens and seminal fluid from the seminal vesicle to the ureter.

elastic fibers Thin, coiled, fibrous connective tissues made primarily of the protein elastin, which can stretch without breaking.

electrocardiogram (ECG) Graphic record of the electrical activity of the heart, using sensors (electrodes) attached to the skin of the chest, wrists, and ankles.

electroencephalogram (EEG) Graphic record of the electric activity of the brain, using sensors (electrodes) attached to the skin of the head.

electron Negatively charged subatomic particle with almost no weight. Electrons orbit the atom's nucleus.

electron transport system A series of electron and energy-transfer molecules in the inner membrane of mitochondria that provide the energy for the active transport of hydrogen ions across the membrane. The electron transport system is essential to the ability of the mitochondria to produce ATP for the cell.

element One of a limited number of unique varieties of matter that composes substances of all kinds. Elements each consist of just one kind of atom.

embolus (EM-bo-lus) Material, such as a blood clot, that floats in the bloodstream and may block a blood vessel.

embryo (EM-bree-oh) An organism in an early stage of development. In humans, the embryonic period extends from the beginning of week 3 to the end of week 8.

embryonic disk A flattened disk of cells that develops in the blastocyst shortly after implantation. The embryonic disk will develop into the embryo.

emphysema (em-flh-SEE-muh) Chronic respiratory disorder involving damage to the bronchioles and alveoli. In emphysema, loss of elastic tissue in the bronchioles causes them to collapse during expiration, trapping air in the alveoli and eventually damaging them as well.

endocrine gland (EN-duh-krin) Ductless gland that secretes one or more hormones into the bloodstream.

endocrine system Body system that includes all of the hormone-secreting organs and glands. Along with the nervous system, the endocrine system is involved in coordination and control of body activities.

endocytosis (en-do-sy-TO-sis) Process by which fluids, extracellular particles, or even whole bacteria are taken into cells. In endocytosis, the materials become enclosed by a vesicle composed of cell membrane material and then internalized within the cell. Phagocytosis is an example of endocytosis.

endoderm (EN-do-derm) Embryonic germ layer. Endoderm forms the lining of the digestive tube and its associated structures.

endometrium (en-do-MEE-tree-um) Inner lining of the uterus. The endometrium becomes thickened and more vascular during the uterine cycle in preparation for pregnancy.

endoplasmic reticulum (ER) (en-do-PLAS-mik reh-TIK-yuh-lum) Membranous network of tubular or saclike channels in the cytoplasm of a cell. The endoplasmic reticulum is the site of most of the cell's production of proteins and other cell compounds.

end-stage renal disease A condition in which long-term irreversible damage to a kidney leads to a failure of the kidney to function properly. Also called chronic renal failure.

energy The capacity to do work. Energy may be stored (potential energy) or in action (kinetic energy).

environmental resistance Factors in the environment that limit population growth in a particular geographic area.

enzyme (EN-zym) A protein that acts as a biological catalyst to speed up a chemical reaction.

eosinophil (ee-oh-SIN-oh-fil) Granular white blood cell whose granules stain readily with a red stain called eosin. Eosinophils attack parasites and function in allergic responses.

epidermis (ep-ih-DUR-mus) Outermost layer of the skin. The epidermis is composed of keratinized stratified squamous epithelium.

epididymis (ep-ih-DID-ih-mis) Portion of the male reproductive system in which sperm mature. The epididymis empties into the vas deferens.

epiglottis (ep-ih-GLAHT-is) Flaplike structure of elastic cartilage at the back of the throat that covers the opening of the larynx during swallowing.

epilepsy (EP-ih-lep-see) Condition involving abnormal electrical discharges of groups of brain neurons. Epilepsy may cause seizures.

epinephrine (ep-ih-NEF-rin) Primary hormone produced by the adrenal medulla, also called adrenaline.

epithelial tissue (ep-ih-THEE-lee-ul) A primary tissue that covers the body's surface, lines its internal cavities, and forms glands.

erection An increase in the length, diameter, and stiffness of the penis that facilitates its entry into the vagina.

erythrocyte (red blood cell) Blood cell that transports oxygen from the lungs to the tissues and carbon dioxide from the tissues to the lungs.

erythropoietin (ih-rith-roh-PO-ih-tin) Hormone that stimulates production of red blood cells.

esophagus (ih-SOF-uh-gus) Muscular tube extending from the throat to the stomach. The esophagus is collapsed when it does not contain food.

essential amino acids Group of eight amino acids that the body cannot synthesize. Essential amino acids must be obtained from the diet.

estrogen (ES-tro-jen) Female sex hormone that stimulates the development of female secondary sex characteristics, helps oocytes mature, and affects the uterine lining during the menstrual cycle and pregnancy.

Eukarya One of the three domains of life. Eukarya includes all life-forms whose cells have nuclei.

eukaryote (yoo-KAIR-ih-ot) An organism composed of cells with nuclei and internal membrane-bound organelles. Animals, plants, protists, and fungi are all eukaryotes.

eutrophication (yoo-troh-flh-KAY-shun) Process by which mineral and organic nutrients accumulate in a body of water, leading to plant overgrowth and loss of oxygen. Eutrophication maybe accelerated by pollution.

evolution An unpredictable and natural process of descent over time with genetic modification. Evolution is influenced by natural selection, chance, historical events, and changing environments.

exocrine gland Specialized epithelial tissue that secretes a product directly into a hollow organ or duct.

exocytosis (ex-oh-sy-TO-sis) Mechanism by which substances are moved from the cell interior to the extracellular space. In exocytosis, a membrane-bound secretory vesicle fuses with the plasma membrane and releases its contents to the exterior.

exon A nucleotide sequence of DNA that specifies a useful informational sequence. Compare *intron*.

experiment A procedure designed to test a working hypothesis. In a controlled experiment, all conditions of the environment are controlled except the specific factor dictated by the working hypothesis.

experimental group In a controlled experiment, the group of subjects that receives the experimental treatment. Compare *control group*.

exponential growth Growth pattern of a population that follows a characteristic J-curve. A population undergoing exponential growth will double repeatedly over similar time periods.

external nose The visible portion of the nose.

extinction The loss of a particular life-form completely.

F

facilitated transport Passive transport of a substance into or out of a cell along a concentration gradient. Facilitated transport requires a carrier protein.

FAD A transport molecule within mitochondria that can accept hydrogen ions and electrons liberated by the citric acid cycle, forming high-energy $FADH_2$.

farsightedness Condition in which the eyeball is too short, causing nearby objects to be focused behind the retina; also called hyperopia.

fatigue A decline in muscle performance during exercise.

fatty acid Linear chain of carbon and hydrogen atoms (hydrocarbon chains) with an organic acid group at one end. Fatty acids are constituents of fats.

femur The large bone found in the upper leg. The femur is the longest and strongest bone in the body.

fertility rate Number of children that each female has during her lifetime.

fertilization Fusion of the nuclei of sperm and egg, thereby creating a zygote.

fever Body temperature that has risen above the normal set point (in humans, approximately 37°C, or 98.6°F). A fever often indicates illness.

fiber General term for a thin strand. A muscle fiber is a single muscle cell. A connective tissue fiber is a thin strand of extracellular material. In the digestive system, fiber refers to material that is indigestible but beneficial in the diet.

fibrin (FY-brin) Fibrous insoluble protein formed during blood clotting.

fibrinogen (fy-BRIN-oh-jen) A blood protein that is converted to fibrin during blood clotting.

fibroblast (FY-broh-blast) Young, actively mitotic cell that forms the fibers of connective tissue.

fibula The thinner of the two bones found in the lower leg.

fimbriae (FYM-bree-ay) Fingerlike projections at the open end of an oviduct that help move an oocyte into the oviduct.

flagellum (pl. flagella) (flah-JEL-um) Long, whiplike extension of the plasma membrane used by sperm and some bacteria for propulsion.

follicle (FAHL-ih-kul) Ovarian structure consisting of a developing egg surrounded by one or more layers of granulosa cells.

follicle-stimulating hormone (FSH) Hormone produced by the anterior pituitary that stimulates ovarian follicle production in females and sperm production in males.

food web Hierarchy consisting of numerous food chains. A food web depicts interactions between producers and consumers in an ecosystem.

forebrain Anterior portion of the brain that includes the cerebrum, the thalamus, and the hypothalamus.

fossil Preserved remnant of an organism. The fossil record is one of our richest sources of information about life-forms that lived in the past.

fossil fuel A carbon-containing compound (coal, oil, or gas) that was formed millions of years ago by living organisms, then covered by sediments.

fovea centralis (FOH-vee-uh-sen-TRA-lis) A small region at the center of the macula of the retina that is responsible for the highest visual acuity. The fovea centralis consists of densely packed cone cells.

fraternal twins Twins resulting from the ovulation and fertilization of more than one oocyte in a particular cycle. Fraternal twins are as different as any two children by the same parents.

free radicals Highly reactive chemicals with unpaired electrons that can disrupt the structure of proteins, lipids, and nucleic acids.

functional food A food or drink product that is said to have benefits beyond basic nutrition (also called a nutraceutical).

fungi (FUN-jy) One of the five kingdoms of life. The fungi include yeasts, molds, and mushrooms.

G

gallbladder Organ located beneath the right lobe of the liver. The gallbladder stores and concentrates bile.

gallstones Hard crystals containing cholesterol, calcium, and bile salts that can obstruct the flow of bile from the gallbladder.

gamete (GAM-eet) Haploid sex or germ cell. Sperm and eggs are gametes.

gamete intrafallopian transfer (GIFT) Technique to enhance fertility in which unfertilized eggs and sperm are placed directly into an oviduct (Fallopian tube) through a small incision in a woman's abdomen.

ganglion (pl. ganglia) (GANG-lee-on) Collection of nerve cell bodies outside the central nervous system.

gastric glands Glands located beneath the stomach lining. Gastric glands secrete gastric juice.

gastric juice Collective term for the hydrochloric acid, pepsinogen, and fluid secreted by gastric glands.

gastrin (GAS-trin) Hormone secreted into the blood by cells located in the gastric glands. Gastrin regulates the secretion of gastric juice by stimulating production of hydrochloric acid.

gastrointestinal (GI) tract A hollow tube that extends from the mouth to the anus. The GI tract includes mouth, pharynx, esophagus, stomach, small intestine, large intestine, rectum, and anus.

gel electrophoresis A technique for separating fragments of DNA or protein according to size.

gene (JEEN) The unit of heredity. Most genes encode for specific polypeptides, and each gene has a specific location on a particular chromosome.

gene flow The physical movement of alleles caused by movement of individuals into (immigration) or out of (emigration) a population. Gene flow tends to mix pools of genes that might not otherwise mingle.

gene therapy Technique in which defective genes are repaired or replaced with their normal counterparts.

genetic drift Random changes in allele frequency because of chance events.

genetic engineering The planned alteration of the genetic makeup of an organism by modifying, inserting, or deleting genes or groups of genes.

genital herpes (JEN-ih-tul HUR-peez) A sexually transmitted disease that results from infection by the herpes simplex type 2 virus.

genital warts A sexually transmitted disease caused by the human papillomavirus (HPV). Genital warts are a potential risk factor for cancers of the cervix or penis.

genome (JEE nohm) The complete set of DNA in the chromosomes of a particular organism.

genotype An individual's particular set of genes.

genus (JEE-nus) Level of taxonomic classification. Modern humans belong to the genus *Homo*.

geographic range The area of Earth over which a particular species is distributed.

gigantism Pituitary disorder caused by hypersecretion of growth hormone during childhood and adolescence.

gingivitis (jin-jih-VY-tis) Condition of inflamed gum tissue (gingiva), often caused by tooth decay and poor oral hygiene.

gland One or more epithelial cells that are specialized to secrete or excrete substances.

glaucoma (glaw-KO-muh) Eye condition caused by blockage of the ducts that drain the aqueous humor in the eye, resulting in high intraocular pressures and compression of the optic nerve. Glaucoma can result in blindness unless detected early.

glial cells (GLEE-ul) Cells in the nervous system that surround and protect neurons and supply them with nutrients. Glial cells do not generate action potentials.

global warming Increase in average global temperature.

globulins A group of proteins in blood plasma having transport and immune functions. Many antibodies are globulins.

glomerular filtration The process of filtering protein-free plasma fluid from the glomerular capillaries into Bowman's capsule.

glomerulus (glow-MAIR-yoo-lus) Cluster or tuft of capillaries inside the glomerular capsule in a kidney. The glomerulus is involved in the formation of the glomerular filtrate.

glottis (GLAHT-tis) Opening between the vocal cords in the larynx.

glucagon (GLOO-kah-gon) Hormone formed by alpha cells in the islets of Langerhans in the pancreas. Glucagon raises the glucose level of blood by stimulating the liver to break down glycogen.

glucocorticoids (gloo-kah-KOR-tih-koyds) Adrenal cortex hormones that increase blood glucose levels and help the body resist long-term stressors. The principal glucocorticoid is cortisol.

glycogen (GLY-ko-jen) Main carbohydrate stored in animal cells. A polysaccharide.

glycolysis (gly-KOL-ih-sis) Breakdown of glucose to pyruvic acid. Glycolysis is an anaerobic process.

goiter (GOY-tur) An enlarged thyroid, caused by iodine deficiency or other factors.

Golgi apparatus (GOHL-jee) Membranous system within a cell that packages proteins and lipids destined for export, packages enzymes into lysosomes for cellular use, and modifies proteins destined to become part of cellular membranes.

gonad (GOH-nad) An organ that produces sex cells. The gonads are the testes of the male and the ovaries of the female.

gonadotropins (goh-nad-o-TRO-pinz) Gonad-stimulating hormones produced by the anterior pituitary. The gonadotropins are FSH and LH.

gonorrhea (gahn-o-REE-uh) Sexually transmitted disease caused by the bacterium *Neisseria gonorrhoeae*.

graded potential A local change in membrane potential that varies directly with the strength of the stimulus. Graded potentials decline with distance from the stimulus.

granular leukocyte (GRAN-yoo-lur LOO-ko-syt) A type of white blood cell with granules that stain readily.

Graves' disease A disorder in which the thyroid gland produces too much thyroxine.

gray matter Area of the central nervous system that contains cell bodies and unmyelinated fibers of neurons. The gray matter is gray because of the absence of myelin.

greenhouse effect A natural phenomenon in which atmospheric gases allow sunlight to pass through but trap most of the heat that radiates outward from Earth's surface.

greenhouse gases Atmospheric gases, notably CO_2 and methane, responsible for the greenhouse effect.

growth hormone Hormone produced by the anterior pituitary that stimulates growth in general.

growth plate Cartilage plate located near the ends of bone. Bones become longer during childhood and adolescence because new cartilage is continually being added to the outer surface of the growth plates.

growth rate The percentage increase in a population per unit time. Growth rate is calculated as the number of births per year minus the number of deaths per year, divided by population number.

H

habitat The type of location where an organism chooses to live. An organism's habitat typically is determined by its tolerance for certain environmental conditions.

hair cells The mechanoreceptor cells of the ear.

haploid (HAP-loyd) Half the diploid number of chromosomes; the number of chromosomes in a gamete.

heart attack Condition characterized by the death of heart tissue caused by inadequate oxygen supply. The medical term for a heart attack is *myocardial infarction*.

heart failure A condition where the heart may become weaker and less efficient at pumping blood.

helper T cell Type of T lymphocyte that orchestrates cell-mediated immunity by direct contact with other immune cells and by releasing chemicals called lymphokines.

hematocrit (hee-MAHT-oh-krit) The percentage of blood that consists of red blood cells.

hemoglobin (HEE-moh-gloh-bin) Oxygen-transporting protein in red blood cells that gives the cells their characteristic red color.

hemophilia (hee-moh-FIL-ee-uh) Collective term for several different hereditary bleeding disorders with similar symptoms. Hemophilia is caused by a deficiency of one or more clotting factors.

hemostasis (HEE-mo-STAY-sis) Stoppage of bleeding or of the circulation of blood to a part.

hepatic portal system (heh-PAT-ik) The system of blood vessels connecting the organs of the digestive tract to the liver.

hepatitis (hep-uh-TY-tis) General term for inflammation of the liver. Hepatitis is sometimes caused by viral infections.

herbivore (HUR-buh-vor) An organism that feeds on plants, utilizing the energy stored in them. Also called a primary consumer.

heterotroph (HET-ur-oh-trof) An organism that cannot utilize the sun's energy to synthesize the molecules it needs. Instead, it must consume foods that contain stored forms of energy. Also called a consumer.

heterozygous (HET-ur-oh-ZY-gus) Having different alleles at the same location (on a pair of homologous chromosomes).

hindbrain Region of brain connected to the spinal cord. The hindbrain comprises the medulla oblongata, cerebellum, pons, and part of the reticular formation.

hippocampus (hip-po-KAM-pus) In the brain, a region of the limbic system that plays a role in converting new information to long-term memories.

histamine (HIS-ta-meen) Substance produced by basophils and mast cells that causes vasodilation and increases vascular permeability.

homeostasis (ho-mee-oh-STAY-sis) State of body equilibrium characterized by a relatively constant and stable internal environment.

hominids (HAHM-ih-nids) Common name for members of the family *Hominidae*. Hominids include present-day humans (*Homo sapiens*) and all of their extinct ancestors of the genus Australopithecus and the genus *Homo*.

hominoids (HAHM-ih-noyds) A sub-group of primates that diverged from a common ancestor of old-world monkeys to become the great apes (gibbons, orangutans, gorillas, chimpanzees) and modern humans.

Homo erectus (HOH-moh ih-REK-tus) A species of *Homo* that diverged from *Homo ergaster* and spread across Africa into Asia and Europe. *Homo erectus* died out only 50,000 years ago.

Homo ergaster (HOH-moh er-GAS-tur) A species of *Homo* that arose from *Homo habilis* ancestors about 1.9 million years ago and migrated out of Africa to establish colonies as far away as Java and China.

Homo habilis (HOH-moh HAB-ih-lus) First distinctly human ancestor. *Homo habilis* showed brain enlargement, changes in teeth and facial features, a decline in sexual dimorphism, and conversion to a diet that included meat. *Homo habilis* may also have used tools.

Homo heidelbergensis (HOH-moh HY-del-bur-GEN-sus) Sometimes called archaic *Homo sapiens*, this species descended from *Homo ergaster* and eventually gave rise to modern humans, *Homo sapiens*.

Homo sapiens (HOH-moh SAY-pee-enz) Genus and species of modern humans, thought to have evolved about 140,000 to 100,000 years ago.

homologous chromosomes (ho-MAHL-uh-gus) Chromosomes that look identical under the microscope. Homologous chromosomes have the same genes in the same locations.

homologous structures Body structures that share a common origin.

homozygous (hoh-moh-ZY-gus) Having identical alleles at the same location (on a pair of homologous chromosomes).

hormone (HOR-mohn) A chemical messenger molecule secreted by an endocrine gland or cell into the bloodstream that has effects on specific target cells throughout the body.

human immunodeficiency virus (HIV) Virus that destroys helper T cells, thus depressing cell-mediated immunity. Symptomatic AIDS gradually appears when lymph nodes can no longer contain the virus.

human papillomavirus (HPV) (pap-ih-LOH-ma-vy-rus) Virus that causes genital warts. HPV is a risk factor for cancers of the cervix or penis.

humerus The long bone of the upper arm.

hydrogen bond Weak bond that forms between a hydrogen atom with a partial positive charge and a nearby atom with a partial negative charge.

hydrolysis (hy-DRAHL-ih-sis) Process in which water is used to split a molecule into two smaller molecules.

hydrophilic A substance that is attracted to water. Most polar and charged compounds are hydrophilic, and therefore they dissolve easily in water.

hydrophobic A substance that is not attracted to water or is repelled by water. Nonpolar compounds such as oils are hydrophobic.

hyperopia (hy-per-OH-pee-uh) Condition in which the eyeball is too short, causing nearby objects to be focused behind the retina; also called farsightedness.

hyperplasia (hy-per-PLAY-zhee-uh) An increase in the number of cells in a tissue caused by an increase in the rate of cell division.

hyperpolarization Transient local change in the resting potential of a neuron that makes it even more negative than usual.

hypertension High blood pressure. Clinically hypertension is defined as a systolic blood pressure of 140 mm Hg or higher or a diastolic blood pressure of 90 mm Hg or higher.

hypertonic General term for above-normal tone or tension. In biology, a hypertonic solution is one with a higher solute concentration than plasma. Cells shrink when placed in a hypertonic solution.

hypertrophy (hy-PER-tro-fee) An enlargement or increase in size of an organ or tissue due to an increase in the size of its constituent cells, rather than an increase in cell number.

hypotension Low blood pressure.

hypothalamus Region of the brain forming the floor of the third ventricle. The hypothalamus helps regulate the body's internal environment by secreting releasing factors that affect the secretion of the hormones of the anterior pituitary.

hypothesis An explanation or description of a proposed truth. A hypothesis is expressed as a testable statement about the natural world.

hypotonic General term for below-normal tone or tension. In biology, a hypotonic solution is one with a lower solute concentration than plasma. Cells swell when placed in a hypotonic solution.

I

identical twins Twins that arise when a single fertilized ovum splits in two shortly after fertilization. Identical twins have identical genotypes.

immune response The activities of the immune system.

immune system A complex group of cells, proteins, and structures of the lymphatic system that work together to provide the immune response.

immunity Resistance to disease and pathogens.

immunization (im-yoo-nuh-ZAY-shun) A strategy for causing the body to develop immunity to a specific pathogen. Immunization may be active (the administraion of a vaccine) or passive (the administration of specific antibodies).

immunoglobulin (Ig) (im-yoo-noh-GLAHB-yoo-lin) Blood plasma proteins (gamma globulins) that play a crucial role in immunity. There are five classes.

immunosuppressive drugs Medications that block the immune response. Examples are corticosteroid drugs to suppress inflammation and cytotoxic medications to block activated lymphocytes.

immunotherapy Treatments that promote the general responsiveness of the immune system so that it can fight cancer more effectively.

incomplete dominance A pattern of genetic inheritance in which the heterozygous genotype results in a phenotype that is intermediate between the two homozygous phenotypes.

independent assortment Mendelian principle that genetic factors separate completely independently of each other during the formation of sperm and egg. Independent assortment applies fully only to genes located on different chromosomes.

inflammation (in-fluh-MAY-shun) A nonspecific defensive response of the body to tissue injury characterized by dilation of blood vessels and an increase in vessel permeability. The presence of inflammation is indicated by redness, heat, swelling and pain.

inheritance Characteristics or traits that are transmitted from parents to offspring via genes.

inner ear Region of the ear consisting of the cochlea and the vestibular apparatus. The cochlea sorts sounds by tone and converts them into nerve impulses. The vestibular apparatus helps maintain balance.

insertion The end of a muscle that attaches to another bone across a joint.

insulin (IN-suh-lin) A hormone secreted by the pancreas that enhances the uptake of glucose by cells, thus lowering blood glucose levels.

integumentary system The proper name for the skin and its accessory structures such as hair, nails, and glands.

interferon (in-tur-FEER-ahn) Chemical that is able to provide some protection against virus invasion of the body. Interferon inhibits viral reproduction.

interleukin (in-tur-LOO-kin) A substance that promotes development of immune cells.

internal environment The environment that surrounds the cells of a multicellular organism.

interneuron A neuron within the central nervous system located between two other neurons. Interneurons receive input from one or more neurons and influence the functioning of other neurons.

interphase One of two major periods in the cell life cycle. Includes the period from cell formation to cell division.

interstitial fluid (in-tur-STISH-ul) Fluid between body cells.

intervertebral disk (in-tur-vur-TEE-brul) A disk of fibrocartilage between vertebrae.

intrauterine device (IUD) Contraceptive device that is inserted into the uterus by a health care provider.

intron A noncoding nucleotide sequence of DNA. Compare *exon*.

in vitro fertilization (IVF) Fertilization outside the body. Eggs are harvested from a woman and fertilized, and the embryo is inserted into the woman's uterus through her cervix.

ion (EYE-ahn) An atom or molecule with a positive or negative electric charge.

ionic bond (eye-AHN-ik) Chemical bond formed by the attractive force between oppositely charged ions.

iris (EYE-ris) A colored disk-shaped muscle that determines how much light enters the eye.

isotonic General term for normal tone or tension. In biology an isotonic solution is one with the same solute concentration as plasma. Cells maintain their normal cell volume in isotonic solutions.

isotopes Different atomic forms of the same element, varying only in the number of neutrons they contain. The heavier forms tend to be radioactive.

J

jaundice (JAWN-dis) Yellowish color of the skin, mucous membranes, and sometimes the whites of the eyes. Jaundice is often caused by liver malfunction and high circulating levels of bilirubin.

jejunum (jeh-JOO-num) The part of the small intestine between the duodenum and the ileum.

joint The junction or area of contact between two or more bones; also called an articulation.

juxtaglomerular apparatus In the kidney, a region near the glomerulus where the afferent arteriole and the efferent arteriole make contact with the distal tubule. The hormone renin is secreted by specialized cells of the afferent arteriole of the juxtaglomerular apparatus.

K

karyotype (KAIR-ee-oh-typ) The diploid chromosomal complement in any species. The human karyotype typically is shown as a composite display of the 22 pairs of autosomes arranged from longest to shortest, plus the sex chromosomes X and Y.

keratinocyte (kair-uh-TIN-oh-syt) Type of cell in the epidermis that produces a tough, waterproof protein called keratin.

killer T cell T lymphocyte that directly lyses foreign cells, cancer cells, or virus-infected body cells; also called a cytotoxic T cell.

kinetic energy Energy in motion; energy actually doing work.

Klinefelter syndrome A condition caused by an XXY genotype. People with Klinefelter syndrome show a tall male phenotype.

Krebs cycle See *citric acid cycle.*

L

lactation Production and secretion of milk.

lacteal (LAK-tee-ul) Small lymphatic vessel in a villus of the small intestine that takes up lipids.

lactose intolerance (LAK-tohs) A common disorder of digestion and absorption caused by insufficient quantities of the enzyme lactase, which digests the lactose in milk and dairy products.

large intestine Portion of the digestive tract extending from the small intestine to the anus. Consists of the cecum, appendix, colon, rectum, and anal canal.

larynx (LAIR-inks) Cartilaginous organ containing the vocal cords, located between the trachea and pharynx. The larynx is also called the voice box.

law of independent assortment All alleles of different genes are distributed to egg and sperm cells independently of each other during meiosis.

law of segregation When gametes are formed in the parents, the alleles separate from each other so that each gamete gets only one allele of each gene.

lens A transparent, flexible, curved structure that focuses incoming light on the retina at the back of the eye.

leukemia (loo-KEE-mee-uh) Cancer of the cells that form white blood cells, resulting in overproduction of abnormal white blood cells.

leukocyte (white blood cell) One of several types of blood cells that are part of the body's defense system. Leukocytes are diverse in structure and specific function. They compose only about 1% of the volume of blood.

Leydig cell (LAY-dig) Cells located between the seminiferous tubules of the testes. Leydig cells produce testosterone.

ligament Dense fibrous connective tissue that connects bone to bone.

limbic system Functional brain system involved in emotional responses.

linked alleles Alleles physically joined on the same chromosome.

lipase (LY-pace) Any lipid-digesting enzyme.

lipid (LIP-id) Organic compounds formed of carbon, hydrogen, and oxygen. Fats, oils, and cholesterol are lipids. Lipids are not very soluble in water.

lipoprotein (ly-poh-PRO-teen) A compound containing both lipid and protein. Two medically important lipoproteins are the low-density lipoproteins (LDLs) and high-density lipoproteins (HDLs) that transport cholesterol.

liver Large, lobed organ that overlies the stomach in the abdominal cavity. The liver produces bile to help digest fat, and serves other metabolic and regulatory functions.

longevity How long a person lives.

loop of Henle (HEN-lee) Hairpin-shaped tubular structure that extends into the medulla of a nephron. The loop of Henle (also called the loop of the nephron) is important in the ability of the kidney to form either concentrated or dilute urine.

lumen (LOO-men) Cavity inside a tube, blood vessel, or hollow organ.

lupus erythematosis (LOO-pus air-uh-them-uh-TOW-sis) An autoimmune disorder in which the body attacks its own connective tissue.

luteinizing hormone (LH) (LOO-tee-in-eye-zing) Anterior pituitary hormone that aids maturation of cells in the ovary and triggers ovulation. In males, LH causes the interstitial cells of the testis to produce testosterone.

lymph (LIMF) Fluid derived from interstitial fluid transported by lymphatic vessels.

lymph nodes Small masses of tissue and lymphatic vessels that contain macrophages and lymphocytes, which remove microorganisms, cellular debris, and abnormal cells from the lymph before it is returned to the cardiovascular system.

lymphatic system System consisting of lymphatic vessels, lymph nodes, and other lymphoid organs and tissues. The lymphatic system returns excess interstitial fluid to the cardiovascular system and provides a site for immune surveillance.

lymphocyte (LIM-foh-syt) One of several types of white blood cells (T cells and B cells) that participate in nonspecific and specific (immune) defense responses.

lymphoma General term for cancers of the lymphoid tissues.

lysosome (LY-soh-sohm) Organelle originating from the Golgi apparatus that contains strong digestive enzymes.

lysozyme (LY-soh-zym) Enzyme in saliva and tears that kills certain kinds of bacteria.

M

macrophage (MAK-roh-fayj) Large phagocytic white blood cell derived from a monocyte; it engulfs damaged tissue cells, bacteria, and other foreign debris. Macrophages are important as antigen presenters to T cells and B cells in the immune response.

macula (MAK-u-luh) The central region of the retina, where photoreceptor density is the highest.

magnetic resonance imaging (MRI) Technique that uses short bursts of a powerful magnetic field to produce cross-sectional images of body structures.

major histocompatibility complex (MHC) proteins A group of proteins on the cell surface that identify an individual's cells as self and normally signal the immune system to bypass the individual's cells.

malignant tumor A mass of cells that has metastasized to become invasive cancer.

mammal Vertebrates that have hair during all or part of their lives, and mammary glands in the female that produce milk.

mammary gland Milk-producing gland of the breast.

mast cell Immune cell that detects foreign substances in the tissue spaces and releases histamine to initiate a local inflammatory response against them. Mast cells typically are clustered deep in an epithelium or along blood vessels.

mastectomy The medical term for surgical removal of a breast.

matter The material substance, composed of elements, that makes up the natural world. Matter takes up space and has weight.

mechanoreceptor (meh-KAN-oh-ree-sep-tur) Receptor sensitive to mechanical pressure or distortion, such as that produced by touch, sound, or muscle contractions.

medulla (meh-DUL-uh) General term for the inner zone of an organ. In the kidney, an inner pyramid-shaped zone of dense tissue that contains the loops of Henle. The term is also used for an area of the brain (medulla oblongata) and for the inner zone of the adrenal glands.

medulla oblongata (meh-DUL-uh ahb-long-GAH-tah) Portion of the brain stem. Contains reflex centers for vital functions such as respiration and blood circulation.

meiosis (my-OH-sis) Nuclear division process that reduces the chromosomal number by half. Meiosis results in the formation of four haploid

(n) cells. It occurs only in certain reproductive organs.

melanocyte (meh-LAN-oh-syt) Type of cell located near the base of the epidermis. Melanocytes produce a dark brown pigment called melanin that is largely responsible for skin color.

melanoma (mel-ah-NO-mah) Least common but most dangerous form of skin cancer.

melatonin (mel-ah-TON-in) A hormone secreted by the pineal gland that seems to synchronize our body's daily rhythms and may induce sleep. It may also play a role in determining the onset of puberty.

memory The act of storing information and retrieving it later, as needed.

memory cell Members of T cell and B cell clones that provide for immunologic memory.

meninges (meh-NIN-jeez) Protective coverings of the central nervous system.

meningitis (meh-nin-JY-tis) Inflammation of the meninges.

menopause (MEN-oh-pawz) Period of a woman's life when, prompted by hormonal changes, ovulation and menstruation cease.

menstrual cycle (MEN-stroo-ul) Pattern of changes in the ovaries and uterus. The menstrual cycle lasts about 28 days and is controlled by hormones.

menstruation (men-stroo-AY-shun) Process in which the endometrial lining disintegrates, its small blood vessels rupture, and the tissue and blood are shed through the vagina.

mesoderm (MEEZ-oh-durm) Primary germ layer that forms the skeleton and muscles of the body.

messenger RNA (mRNA) Long nucleotide strand that complements the exact nucleotide sequence of genetically active DNA. Messenger RNA carries the genetic message to the cytoplasm.

metabolism Sum total of the chemical reactions occurring in body cells.

metaphase Second stage of mitosis, during which the chromosomes align themselves on one plane at the center of the cell.

metastasis (meh-TAS-tuh-sis) The spread of cancer from one organ or location to another not directly connected to it.

microvilli (sing. microvillus) Tiny projections on the surface of some epithelial cells that function to increase a cell's surface area. Microvilli are often found on the luminal surface of cells involved in absorption of fluids, nutrients, and ions.

midbrain Region of the brain stem. The midbrain is a coordination center for reflex responses to visual and auditory stimuli.

middle ear Region of the ear consisting of an air-filled chamber within the temporal bone of the skull, bridged by the malleus, incus, and stapes. The middle ear amplifies sound.

mineral An inorganic chemical compound found in nature. Some minerals must be in the diet because they are necessary for normal metabolic functioning.

mineralocorticoid (MIN-er-al-oh-KOR-tih-koyd) Steroid hormone of the adrenal cortex that regulates salt and fluid balance.

mitochondrion (pl. mitochondria) (my-toh-KAHN-dree-ahn) A cytoplasmic organelle responsible for ATP generation for cellular activities.

mitosis (my-TOH-sis) Process of nuclear division during which duplicated chromosomes are distributed to two daughter nuclei, each with the same number of chromosomes as the parent cell. Mitosis consists of prophase, metaphase, anaphase, and telophase.

mitral valve (MY-trul) Left atrioventricular valve in the heart, also called the bicuspid valve.

molecule Particle consisting of two or more atoms joined together by chemical bonds.

Monera One of the five kingdoms of life, comprising the bacteria.

monoclonal antibodies Pure preparations of identical antibodies produced in the laboratory by a colony of genetically identical cells.

monocyte (MAHN-oh-syt) Large agranular leukocyte that can differentiate into a macrophage that functions as a phagocyte.

mononucleosis (mahn-oh-nu-klee-OH-sis) A contagious infection of lymphocytes in blood and lymph tissues, caused by the Epstein-Barr virus.

monosaccharide (mahn-oh-SAK-uh-ryd) Simplest type of carbohydrate with only one sugar unit. Glucose is a monosaccharide.

morphogenesis (mor-foh-JEN-ih-sis) Process involving dramatic changes in shape and form that an organism goes through during development.

motor neuron A neuron in the peripheral nervous system that conducts nerve impulses from the central nervous system to body tissues and organs.

motor unit A somatic motor neuron and all the muscle cells that it stimulates.

mucosa (myoo-KO-suh) The innermost tissue layer in the wall of the gastrointestinal tract.

mucus (MYOO-kus) A sticky, thick fluid secreted by mucous glands and mucous membranes. Mucus keeps the surfaces of certain membranes moist.

multiple sclerosis (skler-OH-sis) Progressive disorder of the central nervous system in which the myelin sheath around neuron axons disappears and is replaced by fibrous connective tissue.

muscle spindle Encapsulated receptor in skeletal muscles that is sensitive to stretch.

muscle tension The force exerted by a contracting muscle.

muscle tissue Tissue consisting of cells that are specialized to shorten or contract, resulting in movement of some kind. There are three types of muscle tissue: skeletal, cardiac, and smooth.

muscular dystrophy (DIS-trof-ee) A group of inherited muscle-destroying diseases.

muscularis One of the four tissue layers of the gastrointestinal tract wall. The muscularis consists primarily of smooth muscle.

mutation A change in the DNA base pair sequence of a cell.

mutator gene A class of gene that normally is involved in DNA repair during DNA replication. Mutated mutator genes may contribute to cancer because errors in DNA replication may not be corrected.

myelin sheath (MY-ih-lin) Fatty insulating sheath that surrounds the neuron axons of some types of neurons.

myocardial infarction (my-oh-KAR-dee-ul in-FAHRK-shun) Condition characterized by death of heart tissue caused by inadequate oxygen supply; also called a heart attack.

myocardium (my-oh-KAR-dee-um) Middle and largest layer of the heart wall, composed of cardiac muscle.

myofibrils Long cylindrical structures arranged in parallel inside a muscle cell.

myometrium The outer later of the uterus.

myopia (my-OH-pee-uh) A condition in which the eyeball is too long, causing distant visual objects to be focused in front of the retina. Also called nearsightedness.

myosin (MY-oh-sin) One of the principal contractile proteins found in muscle. Myosin is composed of thick filaments with cross-bridges.

N

NAD⁺ A transport molecule within mitochondria that can accept hydrogen ions and electrons liberated by the citric acid cycle, forming high-energy NADH.

nasal cavity The internal portion of the nose.

natural killer (NK) cells A type of lymphocyte that can lyse and kill cancer cells and virus-infected body cells before the immune system is activated.

natural selection The process by which, according to Darwin, individuals with traits that make them more fit for their local environments tend to survive and reproduce. Most changes in allele frequency in a population are the result of mutations coupled with natural selection.

natural world Matter and energy.

negative feedback A homeostatic control mechanism in which a change in a controlled variable triggers a series of events that ultimately opposes (negates) the initial change, returning the controlled variable to its normal value or set point.

neonate (NEE-oh-nayt) A newborn baby during its first 28 days after birth.

neoplasm (NEE-oh-plazm) Abnormal mass of proliferating cells; also called a tumor. Neoplasms may be benign or malignant.

nephron (NEF-rahn) Structural and functional unit of the kidney. A nephron consists of the renal tubule and the blood vessels that supply it.

nerve Cablelike bundle of many neuron axons, all wrapped together in a protective connective tissue sheath. Nerves carry nerve impulses to and from the central nervous system.

nerve impulse A self-propagating wave of depolarization, also called an action potential.

nervous tissue Tissue consisting of cells that are specialized for generating and transmitting nerve impulses throughout the body. Nervous tissue forms a rapid communication network for the body.

neuroendocrine cell A type of cell in the hypothalamus that essentially functions as both a nerve cell and an endocrine cell because it generates nerve impulses and releases a hormone into blood vessels.

neuroglial cells (noo-RAHG-lee-ul) Cells that provide physical support and protection to neurons and help maintain healthy concentrations of important chemicals in the fluid surrounding them.

neuromuscular junction Region where a motor neuron comes in close contact with a muscle cell. At the neuromuscular junction, neurotransmitter released by the motor neuron causes an electrical impulse to be generated in the muscle cell.

neuron (NYOOR-ahn, NOOR-ahn) Cell of the nervous system specialized to generate and transmit nerve impulses.

neurotransmitter A chemical released by a neuron that may stimulate or inhibit other neurons or effector cells.

neutron Uncharged subatomic particle found in the atomic nucleus.

neutrophil (NOO-truh-fil) Most abundant type of white blood cell. Neutrophils respond rapidly during an infection, surrounding and engulfing bacteria.

niche The role of an organism in its community.

nitrification The formation of nitrate (NO_3^-) by any of several means, including nitrifying bacteria, lightning, and nitrate-producing fertilizer factories.

nitrogen fixation The process whereby certain bacteria convert N_2 gas to ammonium (NH_4^+).

Nitrogen fixation is an important part of the nitrogen cycle.

node of Ranvier (RAHN-vee-ay) A short unmyelinated gap between adjacent Schwann cells where the surface of a neuron axon is exposed.

nondisjunction Failure of sister chromatids to separate during mitosis or failure of homologous pairs to separate during meiosis. Nondisjunction causes abnormal numbers of chromosomes in the resulting daughter cells.

nonsteroid hormone A hormone that consists of, or at least is partly derived from, the amino acid building blocks of proteins. Nonsteroid hormones generally are lipid insoluble, so they tend to act by binding to a receptor on the cell surface rather than by entering the cell cytoplasm.

norepinephrine (NOR-ep-ih-NEF-rin) A neurotransmitter and adrenal medullary hormone, associated with sympathetic nervous system activation.

nucleolus (pl. nucleoli) (noo-KLEE-oh-lus) A dense spherical body in the cell nucleus involved with ribosomal subunit synthesis and storage.

nucleotide A structural unit consisting of a nitrogen base, a five-carbon sugar, and one or more phosphate groups. Nucleotides are the structural units of DNA and RNA. ATP and ADP are also nucleotides.

nucleus General term for a central or essential part. The nucleus of an atom consists of neutrons and protons and contains most of the atomic mass. The nucleus of a eukaryotic cell contains the cell's DNA.

nutrients Substances in food that are required for cell growth, reproduction, and the maintenance of health.

O

obesity The condition of weighing more than 20% above ideal body weight.

olfactory receptor cell A modified sensory neuron, located in the upper part of the nasal passages, that detects odors.

oligodendrocyte (AHL-ih-goh-DEN-droh-site) A type of neuroglial cell that provides support and insulation to axons of neurons in the central nervous system.

oligosaccharide A short string of monosaccharides (single sugars) linked together by dehydration synthesis.

omnivore (AHM-nih-vor) An organism that can derive energy from either plants or animals. Humans, pigs, and bears are omnivores.

oncogene (AHNG-koh-jeen) A mutated or damaged proto-oncogene that contributes to cancer.

oocyte An immature egg in the female ovary or newly released from the ovary.

optic disk The area of the retina where the axons of the optic nerve exit the eye. The optic disk has no photoreceptors, so its presence creates a "blind spot" in each eye.

optic nerve One of the two cranial nerves that transmit nerve impulses from the retina to the brain.

organ A part of the body formed of two or more tissues and adapted to carry out a specific function. The stomach and the heart are examples of organs.

organ of Corti Collective term for the hair cells and tectorial membrane of the inner ear. The organ of Corti converts pressure waves to nerve impulses.

organelle (or-guh-NEL) One of many small cellular structures that perform specific functions for the cell as a whole.

organic molecule A molecule that contains carbon and other elements, held together by covalent bonds.

organ system A group of organs that together serve a broad function important to the survival of an organism or species.

orgasm (OR-gazm) A brief, intensely pleasurable reflex event consisting of rhythmic, involuntary muscular contractions.

origin The end of a muscle that is attached to the bone that does not move during muscular contraction.

osmosis (ahz-MO-sis) The net diffusion of water across a selectively permeable membrane, such as a cell membrane.

osteoblast (AHS-tee-oh-blast) A bone-forming cell.

osteoclast (AHS-tee-oh-klast) A cell that resorbs or breaks down bone.

osteocyte (AHS-tee-oh-syt) A mature bone cell.

osteodystrophy (AHS-tee-oh-DIS-troh-fee) Progressive disorder in which the bones become weak because of an abnormal mineral composition of bone.

osteon (AHS-tee-ahn) A cylindrical structure in bone composed of layers of living bone cells and hard extracellular material, arranged like the layers of an onion. In the center is a central canal through which nerves and blood vessels pass.

osteoporosis (AHS-tee-oh-poh-ROH-sis) Progressive disorder involving increased softening and thinning of the bone. Osteoporosis results from an imbalance in the rates of bone resorption and bone formation.

otoliths (OH-toh-lith) Hard crystals of bonelike material embedded in gel in the inner ear. Otoliths contribute to sensations of head position and movement.

outer ear Region of the ear consisting of the pinna (visible portion) and the auditory canal. The outer ear channels sound waves to the eardrum.

ovarian cycle (oh-VAIR-ee-un) Monthly cycle of follicle development, ovulation, and corpus luteum formation in an ovary.

ovaries (OH-vair-eez) The two female sex organs (gonads) in which ova (eggs) are produced.

oviduct (OH-vih-dukt) Tube that leads from an ovary to the uterus, also called the Fallopian tube or uterine tube.

ovulation (AHV-yoo-LAY-shun) Ejection of an immature egg (oocyte) from the ovary.

ovum (pl. ova) (OH-vum) Female gamete, also called an egg.

oxidative phosphorylation The process of producing ATP from ADP plus a phosphate group (P_i), using the energy derived from the electron transport system within mitochondria. The process is called *oxidative* because it requires oxygen and *phosphorylation* because a phosphate group is added (to ADP).

oxygen debt The amount of oxygen required after exercise to oxidize the lactic acid formed by anaerobic metabolism during exercise.

oxyhemoglobin (ahk-see-HEE-moh-gloh-bin) Oxygen-bound form of hemoglobin.

oxytocin (ahk-sih-TOH-sin) Hormone secreted from the posterior pituitary by neuroendocrine cells of the hypothalamus. Oxytocin stimulates contraction of the uterus during childbirth and the ejection of milk during nursing.

ozone In the stratosphere, a thin layer of O_3 gas that shields Earth from ultraviolet rays. Also an environmental pollutant near Earth's surface.

P

Pacinian corpuscle (puh-SIN-ee-un) Type of encapsulated mechanoreceptor located in the dermis that responds to deep pressure or high-frequency vibration.

pain receptors Receptors that respond to tissue damage or excessive pressure or temperature.

pancreas (PANG-kree-us) Organ located behind the stomach. The pancreas secretes digestive enzymes and bicarbonate into the small intestine and the hormones insulin and glucagon into the bloodstream.

pancreatic amylase (pang-kree-AT-ik AM-ih-lays) Pancreatic enzyme that continues the digestion of carbohydrates begun by salivary amylase.

parasympathetic division (par-uh-sym-puh-THET-ik) The division of the autonomic nervous system that generally promotes activities associated with the resting state, such as digestion and elimination.

parathyroid glands (par-uh-THY-royd) Small endocrine glands located on the thyroid gland that produce parathyroid hormone.

parathyroid hormone (PTH) Hormone released by the parathyroid glands that regulates blood calcium levels.

Parkinson's disease Progressive disorder of the nervous system involving death and degeneration of dopamine-containing nerve cells in the midbrain.Symptoms of Parkinson's disease include persistent tremor and rigid movements.

partial pressure The pressure exerted by one particular gas in a mixture of gases.

passive immunization The process of fighting an existing or anticipated infection by the administration of specific antibodies against the infective agent. The antibodies usually are isolated from a human or animal with immunity to the infective agent.

passive transport Membrane transport processes that do not require cellular energy. Diffusion is a form of passive transport.

pathogen (PATH-uh-jen) Disease-causing microorganism, such as a bacterium or virus.

pectoral girdle (PEK-tur-ul) Portion of the skeleton that attaches the upper limbs to the axial skeleton, composed of the clavicle and scapula.

pelvic girdle Portion of the skeleton that supports the weight of the upper body and attaches the lower limbs to the axial skeleton, composed of the two coxal bones and the sacrum and coccyx of the vertebral column.

pelvic inflammatory disease (PID) Infection of the internal female reproductive organs. PID can result in scarring that blocks the oviducts.

penis The male organ of sexual intercourse.

pepsin (PEP-sin) Protein-digesting enzyme secreted by the gastric glands of the stomach.

peptic ulcer An open, sometimes bleeding sore that forms in the inner lining of the stomach, the esophagus, or the upper part of the small intestine.

perforin (PUR-for-in) A chemical released by secretory vesicles of a cytotoxic T cell that perforates the membrane of an abnormal or foreign cell and kills it.

pericardium (pair-ih-KAR-dee-um) Tough fibrous sac that encloses and protects the heart and prevents it from overfilling.

peripheral nervous system (PNS) Portion of the nervous system that lies outside of the brain and spinal cord.

peristalsis (pair-ih-STAHL-sis) Progressive wavelike contractions of muscle in a tubular structure. Peristalsis moves food through the digestive tract and urine through the ureters.

peritubular capillaries (pair-ih-TOO-byoo-lur) Capillaries that form a network surrounding the proximal and distal tubules in the cortex of a nephron.

peroxisome (per-OX-ih-sohm) Membrane-bound vesicle in the cell cytoplasm containing powerful enzymes that detoxify harmful or toxic substances.

pH The measure of the relative acidity or alkalinity of a solution. Any pH below 7 is acidic and any pH above 7 is basic.

phagocyte (FAG-uh-syt) A white blood cell that destroys foreign cells through the process of phagocytosis.

phagocytosis (fag-uh-sy-TOH-sis) Process by which phagocytes surround, engulf, and destroy foreign cells.

pharynx (FAIR-inks) The region of the digestive and respiratory systems that extends from behind the nasal cavities to the esophagus, also called the throat. Both air and food pass through the pharynx.

phenotype (FEE-noh-typ) The observable physical and functional traits of an organism.

phospholipid A modified lipid that has a polar (water-soluble) region containing a phosphate group. Phospholipids are the main structural components of the cell membrane.

photodynamic therapy A therapy for cancer. First, the patient takes light-sensitive drugs that are drawn into the rapidly dividing cancer cells. Then laser light is focused on the tumor, triggering a series of chemical changes in the light-sensitive drugs that kill malignant cells.

photopigment Any one of several proteins located in the rods and cones of the eye that changes its shape when exposed to energy in the form of light. The change in shape causes the rod or cone to alter the amount of neurotransmitter it releases.

photoreceptor A specialized receptor cell that responds to light energy.

photosynthesis The process by which plants capture the energy in sunlight and convert it into chemical energy for their own use.

pineal gland (PY-nee-ul) Endocrine gland thought to be involved in setting the biological clock and influencing reproductive function. The pineal gland secretes melatonin.

pituitary gland (pih-TOO-ih-tair-ee) Neuroendocrine gland located just below the hypothalamus of the brain. Together, the anterior and posterior lobes of the pituitary gland secrete eight hormones.

pituitary portal system A system of small blood vessels between the hypothalamus and anterior lobe of the pituitary gland.

placebo (pluh-SEE-bow) A false treatment in a controlled experiment, given or performed to minimize the possibility of bias by suggestion.

placenta (plah-SEN-tuh) Temporary organ formed from both fetal and maternal tissues that provides nutrients and oxygen to the developing fetus, removes fetal metabolic wastes, and produces the hormones of pregnancy.

plasma (PLAZ-muh) The fluid component of blood in which the formed elements are suspended.

plasma cell A cell derived from a B cell lymphocyte specifically to mass-produce and release antibodies.

plasma membrane Membrane surrounding the cell, consisting of a phospholipid bilayer with embedded cholesterol and proteins. The plasma membrane regulates the passage of substances into and out of the cell.

plasma proteins The largest group of solutes in plasma which include albumins, globulins, and clotting proteins.

plasmid (PLAZ-mid) A small circular molecule of DNA found in the cytoplasm of some bacteria. Plasmids are used in recombinant DNA technology.

platelets (PLAYT-lets) Small cell fragments that are derived from certain cells in the bone marrow. Platelets are important in blood clotting.

pleural membranes (PLUR-ul) Membranes that line the thoracic cavity and cover the external surface of the lungs.

pneumonia (noo-MOHN-yuh) Respiratory condition involving inflammation and infection of the lungs. Pneumonia is usually caused by viruses or bacteria.

polar body Small nonfunctional cell with almost no cytoplasm that is formed when the primary oocyte completes stage I of meiosis.

pollution Chemicals or particles that have adverse effects on living organisms.

polygenic inheritance (pahl-ee-JEN-ik) Pattern of inheritance in which a phenotypic trait depends on many genes. An example is eye color, controlled by three or more genes.

polymerase chain reaction (PCR) (poh-LIM-ur-ays) A technique for making multiple identical copies of DNA in a test tube.

polypeptide A chain of 3–100 amino acids (*poly* means "many"). Chains of more than 100 amino acids are generally referred to as proteins.

polysaccharide A chain of monosaccharides. Starch and glycogen are polysaccharides.

population A group of individuals of the same species that occupy the same geographic area and interact with each other.

positive feedback A control system in which a change in the controlled variable sets in motion a series of events that amplifies the original change. The process of childbirth once labor has started is an example of positive feedback.

positron emission tomography (PET) Technique that employs radioactive substances to create three-dimensional images showing the metabolic activity of body structures.

potential energy Stored or inactive energy.

precapillary sphincter (SFINK-ter) Band of smooth muscle that controls blood flow into individual capillaries.

pre-embryo An organism during the first two weeks after fertilization. The term *pre-embryo* is used because many of the cells that result from cleavage and morphogenesis in the first two weeks will eventually not be part of the embryo at all, but will instead constitute extra-embryonic membranes and parts of the placenta.

premenstrual syndrome (PMS) Recurring episodes of discomfort during the menstrual cycle, generally beginning after ovulation and lasting until menstruation. PMS can include symptoms such as food cravings, mood swings, anxiety, back and joint pain, water retention, and headaches.

presbyopia (prez-bee-OH-pee-uh) Eye condition in which an increase in stiffness of the lens results in an inability to focus on near objects. Presbyopia typically begins after age 40.

primary motor area Region of the frontal lobe of the brain that initiates motor activity.

primary somatosensory area Region in the parietal lobe of the brain that receives sensory input from the skin.

primate (PRY-mayt) An order of mammals with five digits on their hands, fairly flat fingernails and toenails rather than hooves or claws, and forward-facing eyes adapted for stereoscopic vision. Primates include lemurs, monkeys, apes, and humans.

primer A short double-stranded piece of DNA used to initiate DNA synthesis during DNA sequencing.

prion (PREE-on) An infectious misfolded protein that replicates by causing a normal protein in the infected animal to misfold, producing another prion. Prions technically are not alive since they cannot reproduce on their own.

producer In ecosystems, an organism that makes its own organic molecules from inorganic compounds found in water, air, and soil, utilizing energy provided by the sun. Also called an autotroph. Green plants (on land) and algae (in water) are producers.

product In chemistry and biology, the end result of a chemical reaction.

progesterone (proh-JES-tur-ohn) Hormone partly responsible for preparing the uterus for the fertilized ovum. Progesterone is secreted by the corpus luteum of the ovary and by the placenta.

prokaryote (proh-KAIR-ee-oht) A single-celled organism that lacks the nucleus and the membrane-bound organelles characteristic of eukaryotes. Bacteria are prokaryotes.

prolactin (proh-LAK-tin) Hormone secreted by the anterior pituitary gland that stimulates the mammary glands to produce milk.

promoter A unique base sequence that marks the beginning of a gene.

prophase The first stage of mitosis. During prophase the chromosomes condense and thicken, the pairs of centrioles migrate to opposite sides of the cell, and the mitotic spindle forms.

prostaglandins (prahs-tuh-GLAN-dins) A group of lipid-based chemicals synthesized by most tissue cells that act as local messengers. Some prostaglandins dilate blood vessels and others constrict them.

prostate gland (PRAHS-tayt) Accessory reproductive gland in males that produces approximately one-third of the semen volume, including fluids that activate sperm.

protein One or more polypeptide chains of more than 100 amino acids. The three-dimensional structure of a protein is determined by the sequence of amino acids and by hydrogen bonds between different regions of the protein.

proteinuria (proh-tee-NYUR-ee-uh) Appearance of protein in urine. Persistent proteinuria may indicate glomerular damage.

prothrombin (proh-THRAHM-bin) A plasma protein that is converted to thrombin as part of the process of blood clot formation.

prothrombin activator Substance released by blood vessels and nearby platelets when blood vessels are damaged. Prothrombin activator activates the conversion of prothrombin into thrombin, facilitating the process of blood clotting.

Protista One of the five kingdoms of life, comprising unicellular and relatively simple multicellular eukaryotes such as protozoa, algae, and slime molds.

proton Positively charged subatomic particle in the nucleus of an atom.

proto-oncogene (proh-toh-AHNG-koh-jeen) Regulatory gene that promotes cell growth, differentiation, division, or adhesion. A proto-oncogene may contribute to cancer if it becomes mutated or damaged in such a way that it is turned on all the time.

proximal tubule Segment of a nephron that starts at Bowman's capsule and extends to the renal medulla.

puberty (PYOO-bur-tee) Period of life when reproductive maturity is achieved.

pulmonary circuit That part of the vascular system that takes deoxygenated blood to the lungs and returns oxygenated blood to the heart.

pulmonary embolism (EM-buh-liz-um) Condition that occurs when an embolus (blood clot) blocks an artery supplying blood to the lungs. A pulmonary embolism can cause sudden chest pain, shortness of breath, and even sudden death.

pulse Rhythmic expansion and recoil of arteries resulting from heart contraction that can be felt from the surface of the body.

Punnett square (PUN-et) A grid used for predicting patterns of inheritance and the probability that a particular genotype will be inherited.

pupil (PYOO-pul) Opening in the center of the iris through which light enters the eye.

Purkinje fibers (pur-KIN-jee) Modified cardiac muscle fibers that are part of the electrical conduction system of the heart.

pyloric sphincter (py-LOHR-ik SFINK-tur) A thickening of the circular layer of muscle at the distal end of the stomach that controls the rate at which the stomach empties.

pyrogen (PY-roh-jen) A fever-inducing agent. Macrophages and certain bacteria and viruses release pyrogens.

R

radiation High-energy waves or particles emitted by radioactive isotopes.

radius The smaller of the two bones in the lower arm.

rapid eye movement (REM) sleep Stage of sleep in which rapid eye movements occur. REM sleep is accompanied by an alert EEC pattern and dreaming.

receptor In sensory physiology, a specialized cell or nerve ending that receives a sensory signal such as touch, smell, light, or a chemical, and ultimately generates nerve impulses in a sensory neuron.

receptor protein A protein molecule of the cell membrane whose function is to transmit information across the cell membrane. Typically, when a particular molecule in the interstitial fluid binds to the receptor protein, a specific series of events is triggered within the cell.

recessive allele An allele that does not manifest itself in the presence of a more dominant allele. Two recessive alleles must be present in order for the recessive phenotypic trait to be expressed.

recombinant DNA technology Field of applied science that explores applications of cutting, splicing, and creating DNA.

Recommended Dietary Allowance (RDA) The U.S. National Research Council's current best estimate of how much of each vitamin and mineral we need to maintain good health.

recruitment Increasing the tone (or force) of a muscle by activating more motor units.

red blood cell (RBC) Blood cell that transports oxygen from the lungs to the tissues, and carbon dioxide from the tissues to the lungs. Also called an erythrocyte.

reflex An involuntary, automatic response to a stimulus.

refractory period (rih-FRAK-tuh-ree) Period immediately following a nerve impulse during which a neuron is unable to conduct another nerve impulse.

regulatory gene A gene that codes for a repressor or activator regulatory protein.

renin (REE-nin) An enzyme secreted into the bloodstream by the kidneys that leads to an increase in blood pressure and the secretion of aldosterone.

replacement fertility rate The fertility rate required to achieve long-term zero population growth. Replacement fertility rate is estimated to be an average of 2.1 children per woman.

replication The process of copying DNA prior to cell division.

repolarization In a nerve or muscle cell, return of the membrane potential to the initial resting (polarized) state after depolarization.

reproductive system Organ system that functions to produce offspring.

residual volume The amount of air remaining in the lungs even after a forceful, maximal expiration. Normal residual volume is approximately 1200 ml.

respiratory center Groups of nerve cells in the medulla oblongata near the base of the brain that are responsible for the cyclic nature of respiration.

respiratory system Organ system responsible for gas exchange. The respiratory system includes the nose, pharynx, larynx, trachea, bronchi, and lungs.

resting potential The slight difference in voltage (electrical potential) between the inside and outside of a cell.

restriction enzyme An enzyme found in bacteria that can be used to cut DNA at specific nucleotide sequences called restriction sites.

reticular activating system (RAS) (reh-TIK-yoo-lur) A group of neurons in the reticular formation that controls levels of sleep and wakefulness.

reticular fibers Thin, interconnecting fibers of collagen found in fibrous connective tissue. Reticular fibers provide an internal structural framework for soft organs such as the liver, spleen, and lymph nodes.

reticular formation Collective term for neurons in the pons that work with the cerebellum to coordinate the skeletal muscle activity required to maintain posture, balance, and muscle tone. The

reticular formation is also responsible for maintaining the level of wakefulness.

retina (RET-ih-nuh) Layers of tissue at the back and sides of the eye. The retina is composed primarily of photoreceptor cells, neurons, and a few blood vessels.

retrovirus (REHT-roh-vy-rus) A class of viruses that replicate by using their RNA to produce double-stranded DNA, which is inserted into the host cell's DNA. The host cell then produces RNA for new viruses. HIV is a retrovirus.

Rh factor A red blood cell surface antigen, first discovered in rhesus monkeys, that is a crucial consideration in blood transfusions.

rheumatoid arthritis (ROO-muh-toyd) A type of chronic arthritis involving inflammation of the synovial membrane that lines certain joints.

rhodopsin (rho-DOP-sin) The photopigment found in the rods of the eye.

ribonucleic acid (RNA) Nucleic acid that contains ribose and the bases A, G, C, and U. RNA carries out DNA's instructions for protein synthesis. Types of RNA include mRNA, rRNA, and tRNA.

ribosomal RNA (rRNA) The RNA component of a ribosome. Ribosomal RNA assists in protein synthesis.

ribosome (RY-boh-sohm) A cellular structure consisting of rRNA and protein at which amino acids are assembled into proteins. Some ribosomes float freely within the cytoplasm; others are attached to the endoplasmic reticulum.

RNA polymerase (poh-LIM-ur-ays) One of several enzymes that recognizes a promoter (the starting point of transcription). RNA polymerase attaches to the promoter, starts the DNA unwinding process, and assists in attaching the appropriate RNA segments to the growing chain.

rod cell One of the two types of photo-sensitive cells in the retina.

S

salivary amylase (SAL-ih-vair-ee AM-uh-layz) Enzyme in saliva that begins the process of digesting starch.

salivary glands Three pairs of glands that produce saliva to begin the process of digestion. The three pairs of salivary glands are the parotid, submandibular, and sublingual glands.

saltatory conduction Form of rapid transmission of a nerve impulse along a myelinated fiber, in which the nerve impulse leaps from node to node.

sarcomere (SAR-koh-mir) The smallest contractile unit of a muscle myofibril. A sarcomere extends from one Z-line to the next.

sarcoplasmic reticulum (sar-koh-PLAZ-mik reh-TIK-yuh-lum) Specialized endoplasmic reticulum of muscle cells that surrounds the myofibrils and stores the calcium needed for the initiation of muscle contraction.

saturated fat Type of fat with two hydrogen atoms for every carbon atom in the fatty acid tails. Saturated fats are found in meat and dairy products and in a few plant sources such as coconut and palm kernel oil.

scanning electron microscope (SEM) Microscope that bombards an object with beams of electrons to reveal what appears to be a three-dimensional view of the object's surface.

scapulas The shoulder blades.

Schwann cell A type of supporting cell in the peripheral nervous system. Schwann cells form the myelin sheaths around myelinated neurons and are vital to peripheral nerve fiber regeneration.

science The study of the natural world.

scientific method The process of science; the way scientific knowledge is acquired.

sclera (SKLEAR-uh) White, opaque outer layer of the eyeball.

scrotum (SKROH-tum) External sac of skin enclosing the testes.

second messenger Any intracellular messenger molecule generated by the binding of a chemical messenger (hormone or neurotransmitter) to a plasma membrane receptor on a cell's outer surface. A second messenger mediates intracellular responses to an extracellular (first) messenger.

secondary oocyte The haploid female reproductive cell that is released from an ovary and begins traveling down the oviduct during the ovarian cycle. A secondary oocyte has completed stage I of meiosis but not stage II.

secretion General term for the movement of a substance out of a cell of an endocrine or exocrine gland. Also, the substance secreted. The term is also used to denote the movement of any substance into the lumen of a tubule in the kidney.

segmentation Type of gastrointestinal motility in which short sections of smooth muscle contract and relax in seemingly random fashion, mixing the contents of the intestinal lumen.

segregation Mendelian principle that diploid organisms inherit two genes for each trait (on a pair of homologous chromosomes), and that during meiosis the two genes are distributed to different gametes.

selective gene expression The principle that, in any cell at any time, only a few of the genes are actually being expressed. Selective gene expression is responsible for differences in structure and function among different cells and for changes in cell function at different times.

selectively permeable The quality of allowing certain substances to pass while restricting the movement of others. The plasma membrane of a cell is selectively permeable.

semen (SEE-men) Thick, whitish fluid mixture containing sperm and the secretions of the male accessory reproductive glands.

semicircular canals Three fluid-filled canals in the vestibular apparatus of the inner ear that are important in sensing rotational movements of the head.

semilunar valves (sem-ee-LOO-nur) Heart valves located between the ventricles and the two major arteries of the body (pulmonary and aortic) that prevent blood from returning to the ventricles after contraction.

seminal vesicles (SEM-ih-nul) Male accessory reproductive structures that add fructose and prostaglandins to semen. Fructose serves as an energy source for sperm, and prostaglandins may induce muscle contractions in the female reproductive tract to help sperm travel more effectively.

seminiferous tubules (sem-ih-NIF-ur-us) Highly convoluted tubes within the testes in which sperm are formed.

senescence (seh-NES-ens) The progressive deterioration of multiple organs and organ systems over time; old age.

sensor In control system theory, any structure that monitors the current value of the controlled variable and sends the information to a control center. In biology, also called a receptor. Most receptors send information in the form of either action potentials in nerves or hormones in the bloodstream.

sensory neuron A neuron of the peripheral nervous system that is specialized to respond to a certain type of sensory stimulus, such as pressure or light. Sensory neurons conduct information about a stimulus to the central nervous system in the form of nerve impulses.

septicemia (sep-tih-SEE-mee-uh) Systemic disease caused by the spread of microorganisms or their toxins in blood. Also called blood poisoning or toxemia.

septum (SEP-tum) The muscular partition that separates the right and left sides of the heart.

serosa The outermost tissue layer of the gastrointestinal tract wall. It consists of a thin layer of connective tissue that surrounds the other three layers and attaches the gastrointestinal tract to the walls of the body cavities.

Sertoli cells Large cells that represent most of the bulk of seminiferous tubules in males. Also called *sustentacula cells,* they surround and nourish the developing sperm.

set point The internally set (expected, normal) value of a controlled variable.

sex chromosomes The chromosomes, X and Y, that determine gender (XX-female, XY-male). The 23rd pair of chromosomes.

sex-influenced trait A trait not inherited with a sex gene but influenced by the actions of the sex

genes. An example is the phenotypic trait of male baldness.

sex-linked inheritance Pattern of inheritance that depends on genes located on the sex chromosomes. Sex-linked inheritance is X-linked if the gene is located only on the X chromosome, and Y-linked if it is located only on the Y chromosome.

sickle-cell anemia An inherited disorder in which the red blood cells assume a sickle shape when the oxygen concentration is low.

sinoatrial (SA) node (sy-noh-AY-tree-ul) Group of specialized myocardial cells in the wall of the right atrium that initiates the heartbeat. The SA node is called the pacemaker of the heart.

sinus Mucous-membrane-lined, air-filled cavity in certain cranial bones.

sister chromatids The two identical chromosomes that remain attached at the centromere after replication. The centromere holds the sister chromatids together until they are physically pulled apart during the mitotic phase.

skeletal muscle Muscle composed of cylindrical multinucleate cells with obvious striations. Skeletal muscles attach to the body's skeleton.

skeleton The body's physical support system. The skeleton consists of bones and the various connective tissues that hold them together.

skepticism A questioning attitude.

skull Over two dozen bones that protect the brain and form the structure of the face.

sliding filament mechanism The mechanism of muscle contraction. Muscles contract when the thick and thin filaments slide past each other and sarcomeres shorten.

smog Hazy brown or gray layer of atmospheric pollution that tends to hover over the region where it is produced. The burning of fossil fuels (coal and oil) and automobile exhaust are the primary culprits.

smooth muscle Spindle-shaped muscle cells with one centrally located nucleus and no externally visible striations. Smooth muscle is found mainly in the walls of hollow organs.

sodium-potassium pump An active transport protein of the plasma membrane that simultaneously transports three sodium ions (Na^+) out of the cell and two potassium ions (K^+) in. The sodium-potassium pump is important for maintaining cell volume and for generating the resting membrane potential.

solute Any substance dissolved in a liquid.

solvent A liquid in which other substances dissolve.

somatic sensations Sensations that arise from receptors located at many points throughout the body. The somatic sensations include temperature, touch, vibration, pressure, pain, and awareness of body movements and position.

somatostatin (soh-mah-toh-STAHT-in) Hormone secreted from several locations in the body. In the pancreas, somatostatin inhibits the secretion of glucagon and insulin.

species The smallest classification category of life. A species is a group of organisms that under natural conditions tend to breed within that group.

sperm Male reproductive cell (gamete).

sphincter (SFINK-tur) General term for a ring of muscle around a hollow tube or duct that functions to restrict the passage of materials.

sphygmomanometer (SFIG-moh-mah-NOM-uh-tur) Device used to measure blood pressure.

spinal cord The portion of the central nervous system that lies outside the brain, extending from the base of the brain to about the second lumbar vertebra. The spinal cord provides a conduction pathway to and from the brain.

spinal nerve Any one of the 31 paired nerves that arise from the spinal cord.

spinal reflex An involuntary muscle response that is mediated at the level of the spinal cord, with little or no involvement of the brain.

spleen Largest lymphoid organ. The spleen removes old and damaged red blood cells and helps fight infections.

spongy bone Type of bone tissue characterized by thin, hard interconnecting bony elements enclosing hollow spaces. Red blood cells are produced in the spaces between the bony elements.

starch The storage polysaccharide utilized by plants.

stem cells General term for any cells that have not yet differentiated. Stem cells in bone marrow are the source of all blood cells and platelets.

steroids (STEER-oydz) Group of lipids having four interconnected rings that includes cholesterol and several hormones. Steroids are fat-soluble molecules.

sternum A flat blade-shaped bone composed of three separate bones that fuse during development and is found at the top of the chest cavity. Also called the breastbone.

stimulus (pl. stimuli) Sensory input that causes some change within or outside the body.

stomach Organ of the gastrointestinal tract where food is initially stored and where chemical breakdown of proteins begins.

stretch reflex Type of spinal reflex, important in maintaining upright posture and coordinating movement. Stretch receptors in a skeletal muscle stimulate sensory nerves when stretched, causing the sensory nerves to transmit nerve impulses to the spinal cord.

stroke Condition in which brain tissue is deprived of a blood supply, most often caused by

blockage of a cerebral blood vessel. Also called a cerebrovascular accident.

stroke volume Volume of blood pumped out of a ventricle during one ventricular contraction.

structural gene A gene that codes for an enzyme or a structural protein.

submucosa The middle layer of the gastrointestinal tract wall. The submucosa is a layer of connective tissue that contains blood vessels, lymph vessels, and nerves.

substrate In chemistry and biology, the starting material of a chemical reaction.

succession A natural sequence of change in terms of which organisms dominate in a community.

summation Accumulation of effects, especially those of muscular or neural activity.

suppressor T cell Regulatory T lymphocyte that suppresses the immune response.

surfactant (sur-FAK-tant) Secretion produced by certain cells of the alveoli that reduces the surface tension of water molecules, preventing the collapse of the alveoli after each expiration.

sustainable development Development that meets the needs of the present without compromising the ability of future generations to meet their economic needs.

sympathetic division The division of the autonomic nervous system that helps the body cope with stressors (danger, excitement, etc.) and with situations requiring high mental or physical activity.

synapse (SIN-aps) Functional junction or point of close contact between two neurons or between a neuron and an effector cell.

synaptic cleft (sih-NAP-tik) Fluid-filled space at a synapse.

synaptic transmission The process of transmitting information from a neuron to its target across a synapse.

synovial joint (sih-NO-vee-ul) A movable joint having a thin fluid-filled cavity between the bones.

syphilis (SIF-uh-lis) Sexually transmitted disease caused by infection with the bacterium *Treponema pallidum*.

systemic circuit The system of blood vessels that transport blood to all cells of the body except those served by the pulmonary circuit.

systole (SIS-toh-lee) Period when either the ventricles or the atria of the heart are contracting.

systolic pressure (sis-TOL-ik) The highest pressure reached in the arterial blood vessels during the cardiac cycle, and the first of two blood pressure numbers recorded by the health care

provider. Systolic pressure is achieved during ventricular systole.

T

T cells (T lymphocytes) Lymphocytes responsible for cell-mediated immunity, which depends on the actions of several types of T cells.

taste bud A cluster of taste receptor cells and supporting cells that respond to dissolved food chemicals in the mouth. The human tongue has about 10,000 taste buds.

taxonomy (tak-SAHN-uh-mee) The science of classifying and naming life-forms.

technology The application of science.

tectorial membrane (tek-TOR-ee-ul) The gelatinous noncellular material in which the mechanoreceptor hair cells of the inner ear are embedded. With the hair cells, the tectorial membrane forms the organ of Corti, which converts pressure waves to nerve impulses.

telomerase (tel-OH-mer-ays) An enzyme produced primarily by cancer cells and reproductive cells that builds new telomeres to replace telomeres lost during cell division.

telomeres (TEEL-oh-meer) Short segments of DNA at the end of each DNA molecule that do not encode for a protein. In most cells, telomeres are removed every time a cell divides until there are no telomeres left.

telophase In cell division, the last phase of mitosis in which the two new sets of chromosomes arrive at opposite poles of the cell, new nuclear membranes form around the chromosomes, and the chromosomes uncoil and are no longer visible under a microscope.

tendon A cord of dense fibrous connective tissue attaching muscle to bone.

teratogenic (TAIR-uh-toh-JEN-ik) Capable of producing abnormal development of a fetus. Teratogenic substances are one cause of birth defects.

testis (pl. testes) The male primary sex organ (gonad); produces sperm.

testosterone (tes-TAHS-teh-rohn) Male sex hormone produced primarily by the testes. (The adrenal cortex produces a small amount of testosterone in both sexes.) Testosterone promotes the development of secondary sexual characteristics and is necessary for normal sperm production.

tetanus (TET-uh-nus) In a muscle, a sustained maximal muscle contraction resulting from high-frequency stimulation. In medicine, an infectious disease also called lockjaw that is caused by an anaerobic bacterium.

thalamus (THAL-uh-mus) A region of the forebrain that receives and processes sensory information and relays it to the cerebrum.

theory A main hypothesis that has been extensively tested over time and that explains a broad range of scientific facts with a high degree of reliability. An example is the theory of evolution.

thermal inversion   A situation in which a warm stagnant upper layer of air traps a cooler air mass beneath it. Thermal inversions can cause air pollutants to accumulate close to the ground by preventing them from dissipating.

thermoreceptor   Receptor sensitive to temperature changes.

threshold   In an excitable cell such as a nerve or muscle, the membrane voltage that must be reached to trigger an action potential, or nerve impulse.

thrombin   Enzyme that facilitates the conversion of fibrinogen into long threads of fibrin. Thrombin promotes blood clotting.

thymus gland   Endocrine gland that contributes to immune responsiveness. T cells mature in the thymus.

thyroid gland   An endocrine gland, located in the neck, that produces the hormones thyroxine, triiodothyronine, and calcitonin.

thyroid-stimulating hormone (TSH)   Hormone produced by the anterior pituitary. TSH regulates secretion of thyroxine and triiodothyronine.

thyroxine   (thy-RAHK-sin) A hormone containing four iodine molecules that is secreted by the thyroid gland. Thyroxine and a closely related three-iodine hormone (triiodothyronine) accelerate cellular metabolism in most body tissues.

tibia   The thicker of the two bones found in the lower leg.

tidal volume   Volume of air inhaled and exhaled in a single breath. Normal tidal volume is approximately 500 milliliters, or about a pint.

tissue   A group of similar cells (and their intercellular substance) specialized to perform a specific function. Primary tissue types of the body are epithelial, connective, muscle, and nervous tissue.

tissue membrane   A layer of connective tissue and (with one exception) a layer of epithelial cells that line each body cavity and form our skin.

tonsils   Masses of lymphatic tissue near the entrance to the throat. Lymphocytes in the tonsils remove microorganisms that enter in food or in the air.

trachea   (TRAY-kee-uh) Tube extending from the larynx to the bronchi that is the passageway for air to the lungs. Also called the windpipe.

transcription   The production of a single strand of RNA from a segment (representing a gene) of one of the two strands of DNA The base sequence of the RNA is complementary to that of the single strand of DNA

trans fats   Vegetable oils that have been turned into solids by partial hydrogenation of their fatty acid tails. Trans fats increase the shelf life and

stability of foods, but they also increase the risk of heart disease.

transfer RNA (tRNA)   Short-chain RNA molecule that transfers amino acids to the ribosome.

transfusion reaction   The adverse reaction that occurs in a recipient when a donor's red blood cells are attacked by the recipient's antibodies. During a transfusion reaction the donated cells may clump together (agglutinate), blocking blood flow in blood vessels.

transgenic organism   A living organism that has had foreign genes inserted into it. Transgenic organisms are sometimes created to produce substances useful to humans, including human proteins.

translation   The process by which the genetic code of mRNA is used to string together the appropriate amino acids to produce a specific protein.

translocation   A change in chromosome location that occurs when a piece of a chromosome breaks off but reattaches at another site, either on the same chromosome or another chromosome. Translocations can result in subtle changes in gene expression and ability to function.

transmission electron microscope (TEM)   Microscope that bombards a very thin sample with a beam of electrons, some of which pass through the sample. A TEM provides images that are two-dimensional but highly magnified.

transport protein   A protein molecule of the cell membrane whose function is to transport one or more molecules across the cell membrane.

tricuspid valve   (try-KUS-pid) The right atrioventricular valve of the heart.

triglyceride   A neutral fat molecule consisting of a molecule of glycerol and three fatty acid tails. Triglycerides are the body's most concentrated source of energy fuel.

triiodothyronine   (try-i-oh-doh-THY-roh-neen) Hormone secreted by the thyroid gland, similar to thyroxine except that it contains three atoms of iodine instead of four. Like thyroxine, triiodothyronine accelerates cellular metabolism in most body tissues.

triplet code   The genetic code of mRNA and tRNA in which three successive bases encode for one of the 20 amino acids.

trisomy-X syndrome   (TRY-soh-mee) A condition caused by having three X chromosomes. Trisomy-X syndrome individuals are female phenotype, sometimes showing a tendency toward mild mental retardation.

trophoblast   Outer sphere of cells of the blastocyst.

tubal ligation   Procedure for female sterilization in which each oviduct is cut and tied.

tuberculosis   Infectious disease of the lungs caused by the bacterium *Mycobacterium tuberculosis.*

tubular reabsorption   The movement of fluid and solutes (primarily nutrients) from the renal tubules into the blood.

tubular secretion   The movement of solutes (primarily undesirable substances such as drugs, metabolic wastes, and excess ions) from the blood into the renal tubules.

tumor   A mass of cells derived from a single cell that began to divide at an abnormally high rate. The cells of a benign tumor remain at the site of origin, whereas those of a malignant tumor travel to distant sites (metastasize).

twitch   A complete cycle of contraction and relaxation in a muscle cell.

tympanic membrane   (tim-PAN-ik) Membrane between the outer ear and inner ear that receives sound waves and transmits the vibrations to the bones of the middle ear. The eardrum.

U

ulna   The larger of the two bones in lower arm.

ultrasound   Fetal imaging technique in which high-frequency (ultrasound) sound waves are aimed at the mother's uterus. The sound waves rebound when they hit hard tissues in the fetus. These reflecting waves are analyzed and converted to an image by a computer.

umbilical cord   (um-BIL-uh-kul) Cablelike structure connecting the fetus to the mother, through which arteries and veins pass.

unsaturated fat   Type of fat with fewer than two hydrogen atoms for every carbon atom in their fatty acid tails. Examples include plant oils—such as olive, safflower, canola, and corn oil—and fish oil.

urea   (yoo-REE-uh) Main nitrogen-containing waste excreted in urine.

ureter   (YUR-uh-tur) One of two tubes that transport urine from the kidneys to the urinary bladder.

urethra   (yoo-REE-thruh) Tube through which urine passes from the bladder to the outside of the body.

uterine cycle   (YOO-tur-in) A series of changes that occur in the uterus in preparation for the arrival of a fertilized egg. A complete cycle generally takes about 28 days.

uterus   (YOO-tur-us) Hollow, thick-walled organ that receives, retains, and nourishes the fertilized egg. The uterus is the site of embryonic/fetal development.

urinary system   Organ system that transports and stores urine until it is eliminated from the body.

V

vaccine Antigens prepared in such a way that when injected or taken orally they induce active immunity without causing disease.

vagina (vah-JY-nuh) Hollow muscular female organ extending from the cervix to the body exterior. The vagina functions as an organ of sexual intercourse and as the birth canal.

variable A factor that might vary during the course of an experiment.

varicose veins (VAIR-ih-kohs) Permanently dilated veins, often caused by improperly functioning venous valves. Varicose veins can appear anywhere but are most common in the legs and feet.

vasa recta Long, straight capillaries that perfuse the inner (medullary) portion of the kidney, supplying the loops of Henle and collecting ducts.

vasectomy (vuh-SEK-tuh-mee) Procedure for male surgical sterilization in which the vas deferens (ductus deferens) is cut and tied in two places.

vasoconstriction A narrowing of blood vessel diameter caused by contraction of the smooth muscle in the vessel wall.

vasodilation A widening of blood vessel diameter caused by relaxation of the smooth muscle in the vessel wall.

vein A thin-walled blood vessel that returns blood toward the heart from the venules.

ventricle (VEN-trih-kul) A general term for a cavity. In the heart, one of the two chambers that receives blood from an atrium and pumps it into an artery. In the central nervous system, one of four hollow cavities in the brain.

ventricular fibrillation (ven-TRIK-yoo-lur fib-rih-LAY-shun) A type of abnormal heart rhythm involving rapid irregular ventricular contractions.

venule (VEEN-yool, VEN-yool) A small blood vessel that transports blood from capillaries to a vein.

vertebral column (VUR-tuh-brul) Structure of the axial skeleton formed of a number of individual bones (vertebrae) and two composite bones (sacrum and coccyx). Also called the spine or backbone.

vertebrate (VUR-tuh-brayt) Any animal with a backbone composed of vertebrae.

vesicle (VES-ih-kul) A small membrane-bound, fluid-filled sac that encloses and contains certain substances within a cell.

vestibular apparatus (ves-TIB-yoo-lur) A system of fluid-filled canals and chambers in the inner ear. The vestibular apparatus consists of three semicircular canals for sensing rotational movement of the head, and the vestibule for sensing static (non-moving) position and linear acceleration and deceleration.

vestigial structure (ves-TIJ-ee-ul) A structure that may have had a function in some ancestor but which no longer serves any function. The human coccyx (tailbone) is a vestigial structure.

villus (pl. villi) (VIL-us) Fingerlike projection of the small intestinal mucosa that functions to increase the surface area for absorption.

virus A noncellular infectious agent that can replicate only within a host cell. Viruses consist only of one or more pieces of genetic material (DNA or RNA) surrounded by a protein coat.

vital capacity The maximum volume of air that can be expelled from the lungs by forcible expiration after the deepest inspiration.

vitamin Any one of more than a dozen organic compounds that the body requires in minute amounts but generally does not synthesize.

vitreous humor (VIT-ree-us) The clear fluid within the main chamber of the eye, between the lens and the retina.

vocal cords Two folds of connective tissue that extend across the airway. Audible sounds are heard when the vocal cords vibrate as air passes by.

vulva Female external genitalia that surround the opening of the vagina.

W

water intoxication A dangerous and possibly fatal condition caused by overhydration. Also called dilutional hyponatremia because its primary distinguishing feature is a low sodium concentration in the blood.

white blood cell (WBC) One of several types of blood cells that are part of the body's defense system. WBCs are diverse in structure and specific function. They compose only about 1% of the volume of blood. Also called a leukocyte.

X

X chromosome One of the two human sex chromosomes. An embryo that inherits two X chromosomes will develop into a female.

Y

Y chromosome One of the two human sex chromosomes. An embryo that inherits a Y chromosome will develop into a male.

yolk sac Extra embryonic membrane that serves as the first source for red blood cells for the fetus and as the source for primordial germ cells.

Z

zero population growth The condition that exists in a population when the birth rate equals the death rate.

zona pellucida (ZO-nuh peh-LOOS-ih-duh) A noncellular coating around an oocyte.

zygote (ZY-goht) The diploid cell formed by the union of the ovum and a sperm. The zygote is the product of fertilization.

zygote intrafallopian transfer (ZIFT) (ZY-goht IN-trah-fuh-LO-pee-un) A technique for enhancing fertility in which an egg is fertilized outside the body, then inserted into an oviduct through a small incision in the abdomen.

Answers to ✔ and Test Yourself Questions

Chapter 1

✔ Questions

✔ **p. 6:** Animals require energy for movement, of course, but *all* living things require energy to synthesize the molecules of life (proteins, carbohydrates, lipids, and nucleic acids). These molecules are not found naturally in the environment; they have to be synthesized from simpler atoms and molecules. Synthesizing these complex molecules requires energy. Animals get their energy from the food they eat (that is, other animals and/or plants). Plants get their energy from the sun. Ultimately, all the energy used by living systems is derived from the sun. ✔ **p. 7:** The cell without a nucleus is in either the domain Bacteria or the domain Archaea; you can't tell which from what you can observe, and neither is further subdivided into kingdoms. The cell with a nucleus is in the domain Eukarya (cells with nuclei), kingdom Protista (single and the simpler multicellular eukaryotes). Eukaryotes are more closely related to human cells because both have a nucleus.

✔ **p. 8:** Promising areas to study would be the skull (to estimate the size of the brain); the hands (to look for opposable thumbs); and the hips, knees, and feet (to look for signs of bipedalism). The fourth trait of humans, complex language, cannot be determined by examining skeletal material.

✔ **p. 11:** Hypothesis: "If regular exercise helps people sleep better … " Your prediction based on the hypothesis should specify the type and extent of exercise and the exact nature of "sleeping better," in as much detail as possible. For example, one prediction could be "… then people who jog for at least 40 minutes per day for at least two weeks will fall asleep more rapidly at night than people who do not do any exercise." Many other predictions are possible, but the main point is that the prediction must be testable and quantifiable and address the *if* statement that was your hypothesis.

✔ **p. 12:** No. A scientific hypothesis is any statement about the natural world that can lead to testable deductions. This statement is about the natural world, but it can't lead to a testable deduction because if Sasquatches leave no evidence of their existence, there is no way to test whether they exist! ✔ **Figure 1.8, p. 12:** Worse. The two groups of subjects would have different blood pressure even *before* receiving the drug, and this will make it very difficult to interpret the results. For example, the experimental group might respond to the drug, but their blood pressure, even after being lowered, may still be higher than the other group; this would make it appear as if the drug did not work at all. Alternatively, people with high blood pressure might simply respond *differently* to the drug than do people with normal blood pressure, making their results impossible to compare. Random assignment is a much better experimental design because it ensures that the only difference between the two groups is whether or not they receive the drug. ✔ **p. 13:** Several obvious problems are that the subjects were not randomly assigned to the jogging or non-jogging groups, and how "well" a person sleeps is too subjective to be quantifiable. He could have been more specific and asked, or better yet measured in a laboratory setting, the actual hours of sleep each subject got each night over a period of time. In addition, if he intended to study only joggers, perhaps the researcher should have reworded his hypothesis to "jogging helps people sleep better," because jogging is just one form of exercise. ✔ **Figure 1.9, p. 15:** Standard error bars give us an indication of how confident we should be about our estimate of the average value of the whole population. That's because standard error bars are determined by a mathematical formula that takes into account both the variability of the data and the size of the sample. The large standard error bar of the last data column, compared to the first one, means that we have much lower confidence in our estimate of average freshman enrollment in large universities as compared to small universities. (Note that there are fewer larger universities and their sizes are more variable.)

✔ **p. 16:** Both cold medications are relying on what is called anecdotal evidence. This type of evidence is notoriously unreliable because it is highly selected, so it will invariably give a very positive spin. It doesn't matter how many people give anecdotal evidence, or who that person is. It gives us no empirical information about the likelihood that the medications will work for other people nor about whether either supplement might be superior to the other (or, in fact, whether they're any better than a placebo).

Test Yourself

p. 19: 1. b; **2.** d; **3.** a; **4.** b; **5.** c; **6.** d; **7.** d; **8.** a; **9.** d; **10.** b; **11.** d; **12.** d; **13.** b; **14.** d; **15.** c.

Chapter 2

✔ Questions

✔ **p. 25:** An atomic number of 7 means a nitrogen atom always has 7 protons. Because in most atoms the number of protons equals the number of neutrons, it is a safe bet that a typical nitrogen atom also has 7 neutrons. If it's electrically neutral, it must also have 7 electrons—2 in the first shell and 5 in the second shell. ✔ **Figure 2.5, p. 27:** Your diagram should show one carbon atom in the center, surrounded by four hydrogen atoms. Each hydrogen atom should be sharing a pair of electrons (that is, forming a single covalent bond) with the carbon atom. ✔ **p. 28:** Carbon dioxide's structure is $O=C=O$; a carbon atom in the center connected by double covalent bonds to each of the two oxygen atoms. We can deduce this from our knowledge of the number of electrons in each atom's outer shell. Carbon has four electrons in its outermost shell and each oxygen has six. With double covalent bonds between each oxygen and the carbon, all three atoms have eight electrons in their outermost shell. And that is a very stable configuration. ✔ **p. 30:** Water is a polar liquid; it attracts polar and charged molecules and repels neutral ones. Oil is neutral; it attracts neutral molecules and repels polar and charged ones. So we can say that the dye is either polar or charged, but we cannot tell which. (It would also be called hydrophilic, or water-loving.) ✔ **p. 32:** Before the chemist added the solution, the concentration of hydrogen ions was 10^{-3} moles/liter (0.001 moles/liter); after, it was 10^{-5} moles/liter (0.00001 moles/liter). She must have added a base, which lowered the hydrogen concentration; that is, increased the pH. ✔ **p. 35:** Water. Each time a monosaccharide is added to a growing oligosaccharide chain, an H^+ is removed from one molecule and an OH^- is removed from the other. These will combine to produce a new water molecule. ✔ **Figure 2.15, p. 36:** Starch and cellulose. They are both made from glucose subunits. We can digest starch, but not cellulose, because of the way the glucose subunits are attached together. ✔ **p. 36:** Canola oil has more double bonds and fewer than two hydrogen atoms per carbon. When there are fewer than two hydrogens per carbon, double bonds form between adjacent carbons and that puts a kink in the tail. The kink prevents the molecules from packing closely together as a solid. ✔ **Figure 2.17, p. 38:** The phospholipid head is polar—meaning it is partially charged—and that means the head is electrically attracted toward the polar water molecules outside the membrane. The phospholipid tails, in contrast, are nonpolar and are not attracted toward water molecules; they're attracted only to the tails of other phospholipid molecules in the middle of the membrane. Any molecule of phospholipid that was oriented in the wrong direction would, on the basis of its polar head and neutral tail, simply flip around to the proper orientation. ✔ **p. 42:** It's probably a lipid. Lipids are generally insoluble in water and contain very few oxygen molecules. Some proteins are insoluble in water too, but proteins always contain nitrogen. Carbohydrates would contain a substantial amount of oxygen as well as carbon and hydrogen, and most are soluble in water.

Test Yourself

p. 45: 1. a; **2.** c; **3.** b; **4.** a; **5.** b; **6.** d; **7.** c; **8.** a; **9.** d; **10.** a; **11.** e; **12.** a; **13.** d; **14.** c; **15.** d

Chapter 3

✓ Questions

✓ **p. 56:** It's probably a vesicle. The nucleus and mitochondrion would have a double membrane, whereas ribosomes have no membranes at all; they're composed just of proteins and RNA It might contain products recently produced by the Golgi apparatus, or it might be a peroxisome or a lysosome. ✓ **p. 57:** Ribosomes are responsible for making proteins, so cells whose primary function is to secrete proteins typically have abundant ribosomes and rough ER but don't have much smooth ER ✓ **Figure 3.15, p. 60:** There will be a net movement of water from left to right, both because of diffusion (because initially the *concentration* of glucose will be higher on the right side) and also because of osmotic pressure, which will also be driving water to the right. Water will continue to move to the right until the height of the water columns is the same. ✓ **p. 61:** Na^+, Cl^-, and water molecules cannot cross a lipid bilayer on their own. Although they might be small enough, they also are either charged (the ions) or polar (water). Charged or polar molecules are not soluble in lipids, and thus, they cannot diffuse through lipid bilayers. ✓ **Figure 3.17, p. 62:** The fructose is moving "downhill," from high to low concentration, which is the natural direction that it would diffuse on its own. Therefore, the membrane protein is most likely allowing facilitated diffusion (a form of passive transport) and should not require any ATP. ✓ **p. 66:** The cell will tend to swell. A living cell always tends to accumulate a greater concentration of solutes inside than are present outside, which would (if uncorrected) result in a continual influx of water into the cell. Sodium-potassium pumps reduce this problem by removing sodium from the cell, thus reducing the concentration of solutes inside the cell. ✓ **Figure 3.21, p. 66:** Sweat is hypotonic. Therefore her extracellular fluid will probably become hypertonic—that is, it will have more solutes (and less water) than her cells. If it is extremely hypertonic, her red blood cells will start to shrivel. ✓ **Figure 3.28, p. 71:** The ATP synthase is "powered" entirely by the "downhill" flow of H^+ ions from one side of the inner mitochondrial membrane to the other. NADH and $FADH_2$ never interact with ATP synthase directly; they just cause more H^+ ions to be pumped across the membrane, making the concentration gradient stronger. ✓ **p. 71:** Yes, but not very much. Without ATP synthase, oxidative phosphorylation does not occur—that is, the electron transport chain cannot generate any ATPs—but four ATP molecules can still be produced from each glucose molecule by substrate-level phosphorylation (two from glycolysis and two more from the two acetyl CoA molecules that enter the citric acid cycle). ✓ **p. 74:** Aside from the obvious problem of becoming very hungry, a most important reason is that when you reduce caloric intake a little bit at each meal, the primary "extra" source of energy for the body (above daily caloric intake) is fat. But during prolonged fasting, the body also turns to utilization of protein. Protein is not primarily a storage depot for energy; it's there for a reason, meaning that it is of structural and functional significance. Prolonged periods without food result in muscle wasting, including the muscle of the heart.

Test Yourself

p. 75: 1. c; 2. b; 3. e; 4. a; 5. c; 6. b; 7. d; 8. e; 9. c; 10. b; 11. a; 12. e; 13. b; 14. a; 15. c

Chapter 4

✓ Questions

✓ **p. 81:** Columnar epithelium. Columnar epithelium contains the goblet cells that can secrete mucus. A single cell layer, such as simple epithelium, would also absorb the food molecules more efficiently than a stratified layer. ✓ **Figure 4.2, p. 81:** Gap junctions. The tiny pores in gap junctions allow exchange of ions, molecules, and other raw materials between cells; the other two types of junctions do not allow exchange between cells.

✓ **p. 84:** Ehler–Danlos syndrome (EDS) is a disorder of dense connective tissue caused by defective collagen. The connective tissue still has some normal elastin, but it has very little collagen, resulting in weak, stretchy ligaments that allow bones to dislocate easily and in weak, stretchy skin.

✓ **p. 87:** No. Every skeletal muscle cell is individually controlled by a nerve fiber, allowing for fine control and a wide range of muscle power. If there were gap junctions, then whenever one skeletal muscle cell started to contract, all of the other cells in the muscle would contract too, causing a forceful, simultaneous contraction of the entire muscle. A less forceful or a slow contraction would not be possible. ✓ **p. 90:** A serous membrane. Each pleural cavity contains a lung. The lubrication provided by a serous membrane reduces friction, allowing the lung to smoothly expand and contract without sticking to the inside of the pleural cavity. ✓ **Figure 4.10, p. 91:** He has a blister on the bottom of his right foot, on his heel. ✓ **p. 93:** If your keratinocytes began dividing more rapidly, your epidermis would simply become thicker (and might start flaking, as occurs in dandruff and psoriasis). Your skin should still be able to perform all of its major functions. In fact, it might improve the speed of skin healing after a minor injury. ✓ **p. 95:** It's a negative feedback system, because an elevation in the refrigerator's temperature will cause a response that brings the temperature back down, that is, negating the original change. The *controlled variable* is internal temperature; the *sensor* is a thermometer; the *control center* is a thermostat (wired to the thermometer); the *set point* is the temperature the thermostat is set at; and the *effector* is the refrigerator's compressor and refrigeration coils.

✓ **Figure 4.14, p. 95:** Your diagram should look like the diagrams in Figures 4.13 and 4.14. The sensors are probably in the brain, along with the control center. If fat stores increase above the set point, hunger decreases and less food is eaten, causing fat stores to return to the set point. If fat stores decrease, hunger increases and more food is eaten, again causing fat stores to return to the set point. The effector is most likely a "hunger center" in the brain that stimulates you to eat (or not). What isn't known precisely is what the sensors are actually sensing. ✓ **p. 96:** During a fever, the negative feedback system continues to work, but for some reason the set point is shifted to a higher-than-normal temperature. The "chills" and shivering represent the body's attempt to warm up to the new set point. When a fever "breaks," the set point returns to normal; this causes sweating to lower the temperature.

Test Yourself

p. 97: 1. a; 2. a; 3. d; 4. b; 5. b; 6. d; 7. d; 8. e; 9. a; 10. c; 11. d; 12. e; 13. c; 14. d; 15. c

Chapter 5

✓ Questions

✓ **p. 103:** Probably not; osteocytes not in direct contact with a blood vessel would most likely die from lack of oxygen and food and an accumulation of waste products. Osteocytes that are not near blood vessels rely on their gap junctions to pass materials to and from other osteocytes. ✓ **Figure 5.3, p. 104:** Osteoblasts produce a mixture of proteins called osteoid that lend structure to bone and enzymes that facilitate the deposition of hydroxyapatite, the hard crystals of bone. ✓ **p. 105:** Bone mass and density are increased not just by exercise but by weight-bearing exercise. Although swimming is great overall exercise in terms of muscles and cardiorespiratory fitness, it does not increase bone density and mass. If you want to increase or maintain bone density, lift weights or run. ✓ **p. 106:** His PTH levels would be high and his calcitonin levels would be low. This would stimulate his osteoclasts to break down bone in order to release the much-needed calcium into the bloodstream. ✓ **Figure 5.5, p. 106:** Breastbone: sternum; Collarbone: clavicle; Shoulder blade: scapula; Hip bone: coxal bone; Thighbone: femur; Shinbone: tibia. ✓ **Figure 5.10, p. 109:** The ribs and the sternum assist in respiration, and to do this they must expand the size of the cavity they surround. The flexible cartilage allows the ribs to move up and slightly away from the sternum. ✓ **p. 110:** Recall that one of the functions of the vertebral column is to serve as a site of attachment for limbs. Unlike most mammals, humans carry essentially all of their weight on just two legs, and we therefore need a strong site of attachment for the pelvis and legs. Fusing the sacral bones together results in an especially strong point of attachment. ✓ **Figure 5.15, p. 113:** Tendons attach bone to muscle; ligaments attach bone to bone. ✓ **Figure 5.16, p. 114:** Hinge joints generally only allow flexion and extension. Ball-and-socket joints can usually produce all of the motions shown in a, b, and c. Fibrous joints are not synovial joints and cannot move at all. ✓ **p. 116, first question on page:** A bone will generally heal more rapidly than a ligament. This is because living bone has a rich blood supply and many cells, whereas ligaments have a poor blood supply and few cells. ✓ **p. 116, second question on page:** A drug that increases chondroblast activity is most likely to help osteoarthritis patients. This is because chondroblasts are the cells that make new cartilage, and osteoarthritis begins with a wearing down of cartilage in the joints. A drug that increases osteoclast activity might, in theory, reduce the rate of bone thickening and bone spur development in more advanced patients, but it might also weaken the bones.

Test Yourself

p. 117: 1. b; 2. c; 3. a; 4. b; 5. d; 6. c; 7. b; 8. c; 9. d; 10. d; 11. a; 12. d; 13. b; 14. a; 15. b

Chapter 6

✓ Questions

✓ **p. 126:** Muscle cells are almost completely packed with myofibrils, which in turn consist almost entirely of two types of protein molecules (actin and myosin). So, foods comprised primarily of muscle tissue are mostly protein.

✓ **p. 128, first question on page:** This person would probably be completely unable to contract any skeletal muscles (that is, the muscles would always be relaxed). This is because the troponin-tropomyosin complex would always block myosin from binding to actin; calcium ions would be unable to move the troponin-tropomyosin complex out of the way. In fact, we're talking about a theoretical situation here, because if the diaphragm muscle (which is a skeletal muscle) couldn't function, the fetus would never draw a breath.

✓ **p. 128, second question on page:** Caffeine allows acetylcholine molecules to remain longer in the motor junction. The acetylcholine would stimulate the muscle cell for longer than normal, resulting in a prolonged series of contractions. ✓ **p. 129:** The calcium must be taken back into the sarcoplasmic reticulum, allowing the troponin-tropomyosin complex to move back over the myosin binding sites. Additionally, a new molecule of ATP must bind to each myosin head before it can "let go" of its current binding site on the actin filaments. ✓ **Figure 6.11, p. 132:** The increased force generated during summation and tetanus is produced by an increased rate of stimulation of a single motor unit; recruitment involves activation of additional motor units. They are similar in that they will both produce increased muscle tension.

✓ **p. 132:** The all-or-none principle means that all muscle cells in *one motor unit* contract simultaneously when stimulated by their motor neuron. Adjacent motor units controlled by other motor neurons may not contract. If your friend were correct, every muscle contraction would always be a maximum contraction, and we know that isn't true. ✓ **p. 133:** Marathons. The high ratio of red-to-white fibers indicates that her muscle cells have an unusually high concentration of fibers rich in myoglobin and/or blood vessels. In other words, she has a lot of slow-twitch muscle cells in her leg muscles, and she will be well suited for long-distance endurance events. ✓ **p. 135:** The pacemaker cells could still beat on their own, but they would no longer be able to drive the beat of the rest of the heart. Consequently, each heart cell would revert to its own intrinsic rhythm of beating, and the contraction of the heart as a whole would become uncoordinated. (This is known as fibrillation.) ✓ **p. 136:** The drug that blocks acetylcholine would likely be more effective, for it might block the action of acetylcholine sufficiently to allow the muscles to relax despite the nerves' continued release of acetylcholine. The most obvious problem is that if this drug is too strong or the dose used is too high, it might also paralyze the patient's respiratory muscles. (This was another "in theory" question designed to test your understanding of basic mechanisms; in truth, there aren't any drugs to effectively treat tetanus.)

Test Yourself

p. 137: 1. b; 2. c; 3. a; 4. c; 5. c; 6. a; 7. b; 8. b; 9. c; 10. d; 11. b; 12. d; 13. b; 14. c; 15. d

Chapter 7

✓ Questions

✓ **p. 144, first question on page:** Both of the conditions will make hemoglobin less likely to release oxygen to body tissues. Hemoglobin releases oxygen most readily under conditions of high body temperature and low (acidic) pH. ✓ **p. 144, second question on page:** This sample is probably from a man—the hematocrit is 44%, which is in the normal range for men but is slightly high for women. ✓ **p. 145:** Because humans have 4–6 liters of blood, the immediate effect of infusing a liter of plasma into the blood would be roughly a 20% expansion of blood volume and a 20% decrease in the concentration of all formed elements (RBC, WBC, and platelets). There would be a modest decline in oxygen carrying capacity per liter of blood because of the 20% reduction in hematocrit (the percentage of the blood that is RBC), but a decline of that magnitude is not life threatening. Twenty percent reductions in the concentrations of WBC and platelets would also be well within survivable limits. However, if *all* of the blood was replaced with plasma, the patient would die almost immediately. In the absence of RBCs, blood plasma alone cannot carry enough oxygen for survival. ✓ **Figure 7.6, p. 146:** If too few red blood cells are produced, the oxygen-carrying capacity of the blood declines. The decreased oxygen availability causes the kidneys to increase their production and secretion of erythropoietin, which (under normal circumstances) causes the bones to increase their production of RBC and bring oxygen-carrying capacity back to normal. But with bone disease, getting RBC production back even to normal might not be possible. The patient would be left with a decreased number of RBC (a low hematocrit) and an elevated blood concentration of erythropoietin. ✓ **p. 148:** These are symptoms of inflammation, and they are most likely caused by basophils, a type of granular leukocyte that is activated by injured body tissues. Basophils secrete histamine, which initiates the inflammatory process that brings in the nutrients, cells, and chemicals needed for tissue repair.

✓ **Figure 7.11, p. 151:** Native Americans are most able to accept blood transfusions from each other without reactions, because 91% of the people in this population share the same blood type (type O), which has neither A or B antibodies. However, it would still be risky to try this.

✓ **p. 151:** He can donate blood to people of any blood type—A, B, AB, or O. The fact that he does not have A or B antigens means that his red blood cells will not be attacked by the recipient's antibodies, no matter what their blood type. ✓ **p. 152:** No. First, the Rh-positive woman does not have any Rh antibodies (because the Rh antigen is a normal "self" protein for her); and second, the Rh-negative fetus doesn't have any Rh antigens anyway. ✓ **Figure 7.13, p. 153:** Your sketches should show that the Rh-positive blood will agglutinate when mixed with Rh antibodies; the Rh-negative blood will not. ✓ **p. 155:** Anemia due to renal failure is caused by low erythropoietin levels, so adding exogenous erythropoietin corrects the problem nicely. In the other four types, the low oxygen-carrying capacity has already caused erythropoietin secretion to increase in an attempt to increase RBC production and bring oxygen-carrying capacity back to normal. With endogenous erythropoietin levels already high, adding even more erythropoietin does little good.

Test Yourself

p. 157: 1. d; 2. a; 3. c; 4. d; 5. a; 6. b; 7. b; 8. c; 9. d; 10. b; 11. a; 12. d; 13. c; 14. d; 15. a

Chapter 8

✓ Questions

✓ **p. 164:** The walls of arteries and arterioles are too thick for gases and nutrients to be able to diffuse across them. Arteries must have thick muscular walls to withstand the high pressure of blood pumping out of the heart. Even the walls of arterioles, because they have a layer of muscle, are still too thick for effective diffusion. ✓ **Figure 8.4, p. 165:** Water is drawn back into the capillary at the distal end because there is a slightly higher concentration of dissolved molecules (especially proteins) in the capillary compared to the interstitial fluid and also because the blood pressure is lower at the distal end. Both of these forces favor movement of fluid back into the capillary. ✅ **p. 165:** If the lymphatic system in a limb is completely blocked, the excess fluid that leaves the capillaries every day cannot be drained from the limb. In that situation, the limb will become very swollen—sometimes to several times its normal size. ✅ **p. 169:** The pulmonary arteries are the only arteries in the body that carry deoxygenated blood. They are considered arteries because, by definition, arteries are thick-walled blood vessels that carry blood at high pressure away from the heart. ✅ **p. 170:** The problem is that the systemic circuit is almost completely isolated from the pulmonary circuit. Most of the blood in the systemic circuit would go to the body over and over again, without ever going to the lungs—and without picking up oxygen! Babies born with this condition can survive about a week without corrective surgery, because in the newborn there are open passages between the two atria and also between the pulmonary artery and aorta, allowing for some crossover of blood. But these passages close within 10 days. That's why this defect must be corrected surgically within a week or so of birth. ✅ **Figure 8.11, p. 171:** The AV and semilunar valves open and close passively due to changes in blood pressure. If blood pressure is higher "downstream" of the valve than "upstream," the valve automatically shuts. ✅ **p. 172:** The atria contract silently—no valves close when they contract, and the blood flows smoothly, so there is no noise. Contraction of the ventricles, however, makes the AV valves snap shut, which produce the first heart sound (the "lub" of the "lub-dub"). ✅ **Figure 8.16, p. 175:** The noise is heard only when the cuff pressure is between systolic and diastolic pressure because only then does the artery keep opening and shutting with every cardiac cycle. This repeated opening and shutting of the artery causes turbulence, which is what creates the audible tapping noise. When cuff pressure is above systolic pressure, the artery stays closed and so there are no sounds; when cuff pressure is below diastolic pressure, the artery stays open, flow is smooth and steady, and again there are no sounds. ✅ **p. 177:** Systolic blood pressure is 141 mm Hg, and diastolic blood pressure is 95 mm Hg. This indicates this patient may have Stage 1 Hypertension, but because one measurement is not considered sufficient to diagnose hypertension accurately (blood pressure does vary from moment to moment), the patient should be tested again on several different days before a final diagnosis is made.

✅ **p. 178:** Initially, the blood pressure will drop, causing the baroreceptors to send neural signals to the brain, which in turn will increase sympathetic nervous activity and decrease parasympathetic nervous activity. The increased sympathetic nervous activity will cause the heart to beat faster (both by direct nerve input and also via epinephrine), and the sympathetic nerves activation will cause blood vessels to constrict. Blood pressure will be restored to normal or close to it within minutes.

Test Yourself

p. 185: 1. d; 2. c; 3. a; 4. a; 5. c; 6. b; 7. c; 8. b; 9. b; 10. a; 11. d; 12. b; 13. c; 14. c; 15. b

Chapter 9

✅ Questions

✅ **p. 192:** It is most likely a virus. A bacterium would have a cell membrane (and usually a cell wall); a prion would consist only of protein without any nucleic acid. ✅ **p. 199:** Nasal congestion is due to histamine's vasodilation effect, which makes capillary walls leaky and causes swelling in the nasal tissues as part of the inflammation response to the common cold. When antihistamines prevent histamine from working, the capillaries in the nose return to their normal non-leaky state and the swelling diminishes.

✅ **p. 200, first question on page:** During pregnancy, the mother's immune system must somehow tolerate the presence of a fetus and a placenta that are genetically different from the mother, and that therefore have different (non-self) MHC proteins. (It is not understood why the mother's immune system does not attack the fetus and placenta during pregnancy.) ✅ **p. 200, second question on page:** The child should still be able to produce at least some antibodies, which are made by B cells. B cells arise from the bone

marrow, not the thymus. ✅ **p. 202:** A memory cell can make only a single type of antibody that has unique amino acid sequences in the variable regions of the light and heavy chains. Because of the unique variable regions, these antibodies will (most likely) only bind to a certain antigen on the surface of the hepatitis virus and will not be able to bind to the different antigens that occur on other pathogens. ✅ **Figure 9.13, p. 203:** Clonal expansion has to occur because typically only a very few T cells (sometimes just one) have a receptor that can bind to that antigen. With so few T cells able to recognize the antigen, very few cytokines will be produced unless clonal expansion occurs to produce many more identical T cells that all can recognize the same antigen. ✅ **p. 205:** Memory cells undergo clonal expansion when they come in contact with a specific antigen-presenting cell (usually a macrophage, but it can be an activated B cell as well). When a macrophage captures a foreign cell, it digests the cell and then displays a fragment of the cell (the antigen) on its surface as a sort of "red flag" that signals that a foreign cell has been found and captured. When the macrophage makes contact with a memory cell that can recognize that particular antigen, the memory cell undergoes clonal expansion. ✅ **Figure 9.16, p. 205:** A person lacking memory cells will not be able to mount a normal secondary immune response. Their first response will be normal (exactly like the green line), but their second response will be just like a first response (exactly like the green line again)—much smaller and slower than a normal secondary response. ✅ **p. 207:** This was passive immunization that provided only temporary protection. It was essentially a gamma-globulin shot from another species. ✅ **p. 210:** After each exposure, B cells produce more IgE antibodies, which bind to mast cells and basophils that were not previously involved in the allergy. On the next exposure, these mast cells and basophils will also react, causing a greater release of histamine, which causes worse symptoms.

Test Yourself

p. 217: 1. c; 2. a; 3. b; 4. c; 5. d; 6. c; 7. c; 8. b; 9. c; 10. c; 11. c; 12. a; 13. d; 14. c; 15. c

Chapter 10

✅ Questions

✅ **p. 223:** The upper respiratory tract is directly connected to the ears via the two auditory tubes. The microorganisms that cause respiratory tract infections can often spread up these tubes to the middle ears, where they can then cause ear infections. ✅ **Figure 10.3, p. 223:** Nose, nasal cavity, pharynx, trachea, bronchus, bronchiole, alveolus, and finally across the layers of cells that compose the alveolus and the capillary. Diffusional gas exchange occurs only across the cell layers of the alveolus and capillary. ✅ **p. 230, first question on page:** During exercise, tidal volume increases dramatically. Vital capacity remains the same because the vital capacity represents the maximum volume that can ever be exhaled under any circumstances (that is, the lungs do not suddenly grow larger; exercising just means that more of their vital capacity is being used). ✅ **p. 230, second question on page:** If the total atmospheric pressure at the summit of Everest is 260 mm Hg, the partial pressure of oxygen is 54.6 mm Hg (260 mm Hg × 0.21). ✅ **p. 231:** Oxygen will diffuse from the venous blood in the capillaries ($PO_2 = 40$ mm Hg) into the alveolar air (PO_2 of about 36 mm Hg)—that is, the passengers will actually start losing oxygen from their blood. This occurs because gases always diffuse from areas of high partial pressure to areas of low partial pressure. ✅ **Figure 10.11, p. 231:** PO_2 is lower in the tissues than in the aorta because the tissues are constantly using O_2 during cellular respiration. The difference in concentration between blood and interstitial fluid is what allows oxygen to diffuse into the interstitial fluid and ultimately into the cell, because oxygen (or any substance, for that matter) will only diffuse from an area of higher concentration to an area of lower concentration. ✅ **Figure 10.12, p. 232:** These people most likely do not have enough hemoglobin in their red blood cells (recall that iron is a necessary component of hemoglobin). The small amount of O_2 that is dissolved in the blood plasma is not enough to sustain life; we also need the much greater amount of O_2 that binds directly to hemoglobin. ✅ **p. 235:** A low concentration of H^+ ions in blood will result in low concentration of H^+ ions in cerebrospinal fluid as well. This will be detected by the receptor cells in the medulla oblongata, which will trigger a decrease in respiratory rate—resulting in less exhalation of CO_2 and causing H^+ concentrations to rise back toward normal. ✓ **p. 236:** Bronchitis is the most likely cause. The association with smoking suggests bronchitis or emphysema, but emphysema usually does not cause coughing; the coughing of asthma does not usually produce yellowish phlegm; and cystic fibrosis is a lifelong condition that would have been present since childhood.

✓ **p. 238:** Fever points to a probable infection. A chest X-ray would confirm an infection and most likely identify which one (TB and pneumonia X-rays look different to a physician). Finally, a tuberculin skin test or a sputum culture for *Mycobacterium tuberculosis* bacteria would confirm or rule out TB.

Test Yourself

p. 241: 1. a; 2. c; 3. a; 4. c; 5. d; 6. c; 7. d; 8. b; 9. a; 10. d; 11. c; 12. b; 13. c; 14. a; 15. b

Chapter 11

✓ Questions

✓ **p. 248:** The sodium-potassium pumps in this unusual scenario are producing a net movement of two (rather than the usual one) positive ions outward for every pumping cycle. This should make the resting potential more negative than the resting potential of a normal neuron.

✓ **Figure 11.4, p. 248:** The new action potential will look exactly the same as the first one (same height, same duration), because action potentials are all or none. Graded potentials summate, but action potentials do not. And because an action potential was triggered by three successive stimuli in the graph, that's what would happen if you tried it again with four stimuli: The action potential would be triggered after the first three stimuli and the fourth stimulus would not have any additional effect. ✓ **Figure 11.5, p. 249:** The green line represents *permeability* to K^+. When the green line goes up, that means K^+ channels have opened so that the membrane is now very permeable to K^+. K^+ ions then rush out of the cell, causing the interior of the cell to become *less positive* relative to the outside—which means the cell's membrane potential (blue line) falls. ✓ **p. 250:** No. Once the threshold voltage is reached, the neuron's voltage-gated Na^+ channels would open, but there would be no Na^+ ions to enter the cell. Without Na^+ ions entering the cell, the neuron would not depolarize and thus no action potential would be produced. ✓ **p. 253:** The presynaptic neuron would still be able to produce and propagate action potentials in the usual fashion, because an action potential only requires the presence of the usual concentrations of Na^+ and K^+. However, it would not be able to release neurotransmitters because a requirement for neurotransmitter release is that Ca^{2+} enter the neuron from the synaptic cleft. ✓ **p. 256:** Both reflexes are produced at the level of the spinal cord, without requiring the brain. Both trigger muscle contractions. Differences include (1) different types of receptors; the patellar reflex is triggered by stretch receptors and reflex responses to pain are triggered by pain receptors and (2) site and effect of response; the patellar reflex only causes contraction in the muscle being stretched, whereas the pain reflex to a foot causes contraction of flexor muscles in the leg with the painful stimulus (to withdraw the leg) and also contraction of extensor muscles in the opposite leg (to keep you from falling down).

✓ **p. 258:** This would reduce the activity of the sympathetic division but would not affect the parasympathetic or somatic divisions. Likely symptoms would include reduced heart rate and blood pressure, pupillary contraction, increased blood flow to the digestive tract, etc.—all the symptoms of *reduction* of sympathetic nervous activity. ✓ **p. 260:** The walls of the capillaries that produce cerebrospinal fluid (CSF) are fused tightly together with no slits between them, so it is more difficult for large and charged particles to gain access to the CSF than to the interstitial fluid. Compared to interstitial fluid, then, the CSF will have almost the same concentrations of electrolytes, gases, and small sugars, but it will have very few large protein molecules (including antibodies, which as you know are proteins). The CSF is also largely free of the bacteria and viruses that can so easily invade most

of the rest of the body. Although bacterial or viral infections of the CNS are rare, when they do occur, they are particularly difficult to fight because most antibiotics are also too large and too polar to gain access to the CNS.

✓ **p. 262:** Cessation of activity of the medulla oblongata function would cause death within minutes, because the medulla oblongata is required for the most basic of human functions, including respiration. Cessation of activity of the cerebellum would severely disrupt the ability to move normally, but if the person simply lies still, he would probably survive.

✓ **Figure 11.18, p. 263:** This can happen if the occipital lobe is damaged. The occipital lobe is at the back of the head, and its major function is processing visual information from the eyes.

Test Yourself

p. 271: 1. a; 2. d; 3. c; 4. c; 5. a; 6. d; 7. c; 8. b; 9. a; 10. d; 11. b; 12. c; 13. a; 14. b; 15. c

Chapter 12

✓ Questions

✓ **p. 277:** The person will most likely perceive action potentials from this neuron as some sort of light—even though the neuron was actually detecting sound waves. This is because the brain interprets sensory information according to the specific brain area being stimulated, and in this case the visual area of the brain would be stimulated. ✓ **p. 280:** Though your friend might briefly touch his toes during the "downward" part of the bounce, the muscle spindles will immediately trigger a stretch reflex that will cause the hamstrings to contract. After the bounce, the hamstrings will probably be tighter than before—so this is not an effective method for stretching muscles. ✓ **p. 282:** The tastants in foods must dissolve in saliva (or water) before they can bind to the chemoreceptor cells in the taste buds. The chemoreceptors must be stimulated for the brain to perceive any sensations of taste. ✓ **Figure 12.8, p. 283:** The new graph should have more waves per unit time (high frequency—that is, high pitch) than the "high" sounds illustrated in the original figure. The height of each wave (the "loudness") should be intermediate compared to the "loud" and "soft" sounds in the original figure. ✓ **p. 284:** Our sense of hearing would be much less sensitive—that is, all sounds would seem much softer, and we would probably be unable to detect very faint sounds. This is because the functions of the outer and middle ears are to channel and amplify sounds. The range of frequencies that we could hear would probably be unaffected, however.

✓ **Figure 12.10, p. 285:** A person with this condition would probably hear a very high pitched sound, such as a high-pitched hum or buzz. This is because activity of hair cells at the beginning of the cochlea normally generates impulses in sensory neurons that are interpreted by the brain as high-pitched sounds. ✓ **p. 287:** When you suddenly stop spinning, the fluid in the semicircular canals keeps moving under its own inertia for a few moments, flowing around the canals on its own. This will bend at least one of the cupulas, and the hair cells within it. The brain interprets this as if the head is moving in the other direction. ✓ **p. 290:** The absence of a lens in the eye is called *aphakia*. A person with aphakia would be able to see blurry images because most of the focusing of the light is done by the cornea. The lens is necessary only for fine-tuning, bringing the blurry images produced by the cornea into sharp focus. ✓ **p. 293:** Animals that are active only at night have poor or no color vision—they see the world in shades of grey. This is because their retinas contain mostly rods, which provide sharp vision in very dim light, and very few (if any) cones, which are required for color vision. ✓ **Figure 12.23, p. 295:** Yes. Though she lacks blue cones, she has normal green cones and red cones, which will allow her to distinguish between the colors green and red. The greens and reds will not look quite like the greens and reds a normal person sees, however.

Test Yourself

p. 296: 1. d; 2. c; 3. d; 4. a; 5. c; 6. d; 7. a; 8. b; 9. c; 10. d; 11. c; 12. c; 13. b; 14. b; 15. d

Chapter 13

✅ Questions

✅ **p. 304, first question on page:** In this example, 100,000 molecules of final product will be produced by a single hormone molecule. This example illustrates how hormones can be present at very low concentrations in the blood but still have a significant impact on the body. ✅ **p. 304, second question on page:** No. Steroid hormones generally don't use second messengers, because a second messenger is only needed for hormones that can't enter the cell. A second messenger is a molecule that relays the hormone's message into the interior of the cell. Steroid hormones simply enter the cell themselves. ✅ **p. 306:** The hypothalamus would detect that the blood has too much water, that is, low solute concentration, and would secrete less ADH, causing a decline in the ADH concentration in the blood. The reduction in ADH would cause the kidney to reabsorb less water from the urine, so that more water would be excreted in the urine. ✅ **p. 308:** If you answered that she would have trouble giving birth and difficulty nursing, you would have been on the right track, at least in theory. Interestingly, women who are oxytocin-deficient do undergo normal labor and delivery, indicating that there are other powerful forces driving childbirth besides oxytocin. She will still produce milk, because milk production is mainly under the influence of prolactin. However, a mother who is oxytocin-deficient will have trouble nursing normally because oxytocin is responsible for milk ejection.

✅ **p. 310:** Responding to a stressful situation usually requires energy (ATP). For example, healing a physical injury requires ATP (to rebuild the damaged tissues), and stressful emotions such as fear often indicate that we might need to run or fight, which also requires ATP. Cortisol helps raise blood glucose levels so that body tissues can use the glucose to produce more ATP. Put simply, cortisol gives you energy to do whatever needs doing. It also helps reduce inflammation after injury. ✅ **p. 312:** Thyroxine levels usually drop sharply during starvation. They also plummet during crash diets. This represents the body's attempt to save energy by reducing its metabolic rate, and it's one of the reasons that the harder you try to diet, the harder it is to lose weight. ✅ **Figure 13.14, p. 313:** The top (and starting point) of your diagram should be blood glucose concentration. Your diagram should show alpha and beta cells of the pancreas on the right side of the diagram, responding to an increase in blood glucose by reducing glucagon (alpha cells) and increasing insulin (beta cells). On the left side should be three major effectors: the liver, muscle, and fat tissue. The "increased insulin" arrow should go to all three effectors, whereas the "decreased glucagon" arrow should go to the liver alone. The liver and muscle both respond by converting the excess glucose into glycogen; fat cells respond by converting some of the excess glucose to fat. Arrows from these three effectors back to the starting point (glucose set point) should show that the changes caused by the three effectors will bring glucose back to normal. ✅ **p. 315:** People who are exposed only to dim office lights during the day often have elevated melatonin levels during the day as well as during the night, so they must not have a normal 24-hour sleep-wakefulness cycle. They might be lethargic during the day and have trouble falling asleep at night. High daytime levels of melatonin are thought to be one of the underlying causes of seasonal affective disorder, a syndrome of low energy and poor mood that is associated with wintertime.

✅ **p. 317:** It is almost always easier to treat failure to produce a hormone, because you can simply give the patient an injection of the missing hormone if it's available. Failure to respond to a hormone is a much more difficult condition to treat because neither giving the hormone nor blocking its action does very much of anything.

Test Yourself

p. 320: 1. c; 2. b; 3. a; 4. d; 5. a; 6. d; 7. a; 8. d; 9. c; 10. c; 11. b; 12. a; 13. d; 14. a; 15. d

Chapter 14

✅ Questions

✅ **p. 331:** The stomach would still produce pepsinogen, the precursor molecule to pepsin, but in the absence of an acidic environment it would not be converted to pepsin. The stomach could still perform its food storage and regulation-of-delivery functions and could still mechanically break up large pieces of food, but it would not be able to digest proteins effectively because of the lack of pepsin. Nor would bacteria ingested with the food be killed as readily. ✅ **p. 334:** Liver is generally a very nutritious food because it contains not only glucose stores and fat stores, but also iron and many vitamins. Liver is especially high in vitamin A, so pregnant women are advised to eat it in moderation, but it should not be fed regularly to young children in order to avoid vitamin A toxicity. Liver could also be dangerous if the animal it comes from has been exposed to environmental toxins that tend to accumulate in the organ, but that should be rare. Eaten in moderation, liver is definitely good for you. ✅ **Figure 14.15, p. 336:** Chylomicrons are too big to enter a blood capillary. The gaps between adjacent cells in a lacteal are wide enough to accommodate chylomicrons, however. ✅ **p. 338:** No—the duodenum releases the hormone cholecystokinin (CCK) in response to the arrival of proteins and fats but does not release any hormones in response to the arrival of carbohydrates. Carbohydrates are partially broken down in the mouth by the enzyme salivary amylase, and then are further broken down by enzymes from the pancreas and the small intestine. No hormones are involved. ✅ **p. 340:** Cold-water fish must use unsaturated fats so that their fat tissue will stay soft and pliable. If cold-water fish stored lipids as saturated fats, the saturated fats would be in solid form, making the fish too stiff to swim. Warm-water fish, in contrast, are free to use saturated fats—and they do. ✅ **Figure 14.18, p. 340:** Yes, eggs and milk both contain complete proteins. Unhatched bird embryos and baby mammals must assemble every protein in their bodies solely from the amino acids contained in the egg (for birds) and in milk (for mammals). ✅ **p. 342:** Vitamin B12. B12 is the only vitamin that is obtained only from animal products, and vegans do not eat any animal products. For this reason, many vegans take vitamin supplements and/or eat tofu that has been fortified with vitamin B12. ✅ **p. 345:** The woman who is lifting weights will probably lose more body fat. In addition to the 200 calories burned during the exercise itself, the extra muscle that she develops will increase her BMR throughout the rest of the day.

✅ **Figure 14.21, p. 345:** Obesity is most prevalent in southeastern states and the Appalachian mountains, especially West Virginia, Kentucky, and Tennessee. These areas are notable for a preponderance of fried, high-fat foods in people's diets. (Many other cultural factors may also contribute, including a lack of exercise.) ✅ **p. 347:** Peptic ulcers tend to occur only in the stomach and first portion of the duodenum because of the high acidity in those regions. The acid is neutralized by mid-duodenum by bicarbonate delivered from the pancreas.

Test Yourself

p. 349: 1. b; 2. c; 3. c; 4. a; 5. d; 6. b; 7. d; 8. d; 9. b; 10. c; 11. a; 12. d; 13. c; 14. c; 15. b

Chapter 15

✅ Questions

✅ **p. 356:** The woman almost certainly has a bladder infection, also called *cystitis*. The doctor is checking to see if the bladder infection might have spread up the ureters to the kidneys. The kidneys are located at the back of the abdominal cavity, so kidney infections can cause back pain. ✅ **Figure 15.6, p. 358:** Because of the way the kidneys' filtering mechanism is structured, large quantities of *all* small molecules are filtered; the filtering mechanism can't distinguish between essential and non-essential molecules. Consequently, the kidneys must then take back (reabsorb) all of the essential ones, leaving the non-essential ones to be excreted. It's as if in cleaning out your apartment or house, you first put *everything* outside and then brought back all of the things you wanted to keep (instead of just looking around for the things to discard). It sounds wasteful energetically, but in fact, the kidneys expend very little energy to reabsorb those essential molecules. ✅ **Figure 15.10, p. 362:** No. The ability of the kidneys to produce either a low volume of concentrated urine or a high volume of dilute urine depends on the transport activity of the loop of Henle, which doesn't vary that much. The only change that occurs between producing a dilute urine or a concentrated urine is a change in water permeability (and hence passive water diffusion) across the collecting duct. A change in water permeability does not require any additional energy. Rest assured that you are not overworking your kidneys when you have drunk a lot and are producing a lot of urine. ✓ **p. 363:** Yes. The distal tubule will still reabsorb sodium ions Na$^+$ and chloride ions (Cl$^-$) by active transport, leaving water in the urine. (However, the kidney would be unable to produce concentrated urine.) ✓ **p. 364:** The pizza has resulted in too much sodium and chloride in your blood (that is, a high concentration of solutes). Your kidneys must get rid of the excess sodium while losing as little water as possible. The high solute concentration will cause ADH to increase (to conserve water) and aldosterone will decrease somewhat (to allow more sodium to be excreted). Your kidney will produce very concentrated urine. ✓ **Figure 15.13, p. 365:** ADH increases after all of the other events have occurred. It increases because the action of aldosterone on the kidneys to save salt causes the blood solute levels to increase. The hypothalamus detects this and releases ADH in response. ✓ **p. 366:** The kidneys will sense the drop in blood pressure and will secrete renin, which will result in an increase in angiotensin II, which will stimulate aldosterone secretion from the adrenal gland. The angiotensin II will constrict arterioles within minutes, counteracting any fall in blood pressure caused by the blood loss. The aldosterone will cause an increase in sodium reabsorption, conserving as much salt as possible. Over time, the retained salt will raise the blood solute concentration, and then ADH will be secreted and water will be saved as well. Together, the retention of salt and water act to return blood volume toward normal. ✓ **p. 367:** The kidney should excrete fewer H$^+$ ions than usual, retaining H$^+$ in the blood to bring the pH back down to normal. (This is, in fact, exactly what the kidneys do, and this is why most people begin to feel better after about three days at high altitude.) ✓ **p. 369:** The young man initially had acute renal failure, meaning that his kidneys had stopped filtering and essentially shut down temporarily. Often that happens when blood pressure has fallen so low that filtration no longer is occurring. It's not a big problem if the kidneys begin functioning again within a day or so of getting his blood pressure back to normal; the body can do without renal function for several days at least. But if the lack of filtration is prolonged, then the tubules start to plug up and the acute phase may turn into permanent, irreversible renal failure. If that happens, the young man will need to be on dialysis and eventually will need a kidney transplant. And he may not get one before he dies.

Test Yourself

p. 371: 1. d; 2. c; 3. b.; 4. b; 5. a; 6. c; 7. a; 8. d; 9. d; 10. c; 11. a; 12. c; 13. a; 14. b; 15. d

Chapter 16

✓ Questions

✓ **Figure 16.2, p. 377:** A spermatogonium. A spermatogonium ultimately gives rise to millions of sperm, because it continually buds off new primary spermatocytes. Each primary spermatocyte gives rise to just four sperm. ✓ **p. 379:** Oocytes must pass from the ovary into the oviduct, where fertilization normally occurs. Embryos can sometimes escape into the abdominal cavity because the oviduct has an open end. ✓ **p. 382:** She will still have all of her primary oocytes, because those are all produced before birth. She won't be able to ovulate, though, because LH is the hormone that triggers ovulation. ✓ **Figure 16.7, p. 383:** The corpus luteum is the source of this progesterone. Progesterone levels rise when the corpus luteum forms from the remnants of the follicle just after ovulation. Progesterone levels fall when the corpus luteum degenerates approximately 12 days later—unless pregnancy occurs. ✓ **p. 386, first question on page:** The woman will be sterile immediately unless she ovulated right before the operation, in which case it might be 3–4 days before she cannot conceive. The man will not be sterile for several weeks. Maturing sperm are stored in the ductus deferens, so some viable sperm are usually present in the duct past the site of the vasectomy. It can take several weeks to get rid of these sperm. ✓ **p. 386, second question on page:** Pregnancy. In fact, most hormonal birth control methods are specifically designed to mimic the hormonal conditions of pregnancy. This

makes sense because pregnancy is the one physiological state during which ovulation is completely prevented for a period of several months. ✓ **p. 390:** Progesterone enables the uterus to thicken and maintain its inner lining, the endometrium—along with the embryo that is inside the endometrium. If progesterone is not high enough, the uterus will automatically lose its lining, just as it does during menstruation. A series of several miscarriages may indicate that the woman's progesterone is too low to maintain the uterine lining. If that's the case, then administering progesterone may help. ✓ **Figure 16.13, p. 393:** The symptoms of a chlamydia infection are very mild at first so the infection may go undiagnosed, especially in men. Undiagnosed infections are easily spread. Both gonorrhea and syphilis have symptoms that are severe enough to send a person to the doctor and easily diagnosed by the doctor. ✓ **p. 396:** Assuming equal levels of exposure, women would most likely be more susceptible to STDs than men. The vagina is lined by a warm, moist mucus membrane. The outer surface of the penis is covered with dry skin, which is a much less hospitable environment for microbes.

Test Yourself

p. 397: 1. d; 2. c; 3. a; 4. c; 5. d; 6. a; 7. d; 8. b; 9. d; 10. b; 11. c; 12. a; 13. d; 14. b; 15. d

Chapter 17

✓ Questions

✓ **p. 403:** A chromosome is a single long molecule of DNA along with its associated histone proteins. A gene is a short part of a chromosome that contains the instructions for making one or more proteins. Chromatin is simply a mixture of DNA with histone proteins (i.e., chromosomes are made of chromatin). Finally, a chromatid is one copy of a recently duplicated chromosome that is still attached to the other sister copy. ✓ **Figure 17.4, p. 404:** The promoter is most likely in the very bottom-most loop of the DNA shown in part A. The promoter is where transcription first began, and because transcription is already under way and the entire transcription complex is moving upward, the original starting point is now at the bottom of the figure. ✓ **p. 405:** If RNA polymerase can attach more easily, then transcription is much more likely to happen, producing mRNAs. The gene is now much more likely to be expressed, meaning that whatever protein it codes for will probably be produced. ✓ **Figure 17.5, p. 405:** The mRNA will have the sequence AUG CCC GUA UAA. The peptide will be only three amino acids long, with a sequence of met-pro-val; that is, methionine, then proline, then valine. (The "stop" codon UAA stops translation without adding a final amino acid.) ✓ **Figure 17.6, p. 406:** Initiation will occur normally. Elongation will proceed normally until the ribosome meets the new stop codon. At that point, termination will occur—too early! A protein will be produced, but it will be an abnormally short protein that is missing its second half. ✓ **p. 408:** This chromosome will probably complete prophase and metaphase normally, but as anaphase begins, the chromosome will still be stuck in the center of the cell with both of its chromatids still attached together. The chromatids will be unable to separate. Mitosis may stall completely, or possibly both chromatids may be tugged to the same side of the cell (most likely with devastating consequences for both daughter cells). ✓ **p. 410:** A chimpanzee cell just before meiosis will have 48 chromosomes, each with two chromatids. After meiosis I, each cell will have 24 chromosomes, each with two chromatids. After meiosis II, each cell will still have 24 chromosomes, but now each chromosome consists just of one copy. (Technically, the moment the chromatids separate from each other they graduate to the status of "chromosomes" and are no longer called chromatids.) ✓ **p. 414:** One strategy would be to collect eggs from the female, fertilize them with the male's sperm, split each embryo into eight identical embryos, and then implant all eight embryos in surrogate females. The main problem (besides the likely technical difficulties) is that the surrogate mothers would have to be of a different species because no other female Arabian oryx exists! Impregnating a female of a different species might make the pregnancies fail, and even if calves are born, they would have to be raised by mothers of the other species and might end up with abnormal behavior. However, if a closely related, but nonendangered antelope species could be used, this strategy could work.

An additional concern for any future reproduction of the species would be that all the offspring would be from just one pair of parents, so all of the off-spring would be nearly identical. ✓ **p. 415:** You could start by trying to grow cells you had harvested from a mature heart, but that probably wouldn't work because adult heart cells divide very slowly, if at all. Or you might try to grow cells taken from an eight-cell human embryo. But these are still undifferentiated cells, and we know very little about how to get an undifferentiated cell to start down the path to becoming a heart cell. Perhaps your best bet would be to take cells from an embryo's heart as soon as the developing heart could be identified. These are cells that have already partially differentiated into heart cells, so there would be fewer differentiation steps necessary to get them to become mature heart cells. Plus, they are still growing and dividing.

Test Yourself

p. 417: 1. a; 2. a; 3. b; 4. d; 5. d; 6. b; 7. c; 8. d; 9. c; 10. a; 11. b; 12. a; 13. c; 14. a; 15. B

Chapter 18

✓ **Questions**

✓ **Figure 18.1, p. 422:** Most doctors would say that benign tumors need not be removed if they are not pressing on nearby tissues. Because benign tumors rarely progress to become malignant tumors and because any surgical procedure inevitably has some risks, the risks of removal (of a tumor that is *known* to be benign) are often greater than the possible benefits.

✓ **p. 425:** The mutated gene is now an oncogene. Before it mutated, it was a proto-oncogene, because its normal function is to promote controlled cell growth. ✓ **p. 427:** The elevated risk of lung cancer is due to radioactive radon gas that is emitted naturally by the rocks and soil around basements in some areas. Because radon is a gas, it primarily affects the lungs.

✓ **p. 430:** Radiation is a carcinogen. It is used to kill cancer cells, but in the process it may also damage normal cells in tissues near the initial cancer, increasing the risk of other cancers in those tissues. ✓ **p. 431:** These three symptoms are all due to damage to cell types that normally undergo rapid cell division. Chemotherapy damages rapidly dividing stem cells in the bone marrow, which normally produces many new red blood cell and white blood cells every day. This causes the anemia and heightened infection risk. Hair follicles are also damaged because they are a part of the skin, and under normal conditions skin cells also divide rapidly.

✓ **p. 432:** Avastin slows healing of surgical incisions because it blocks angiogenesis, the growth of new blood vessels. After surgery, new blood vessels are needed to supply the new tissue being formed as wounds heal.

✓ **p. 433:** A slightly higher percentage of men with breast cancer die from the disease than women (19% versus 17%). The most likely explanation is that men are not examined regularly for breast cancer, as are women. Consequently, the disease is generally not diagnosed as early in men. By the time the cancer is discovered, it's more likely to be more advanced.

✓ **p. 435:** Cigarette smoke contains many carcinogens that don't just stay in the airborne smoke in the lungs; they dissolve into the blood plasma and circulate throughout the body. The leading hypothesis for the increase in bladder cancer is that many of these carcinogens are excreted in the urine, so the cells lining the bladder are exposed to high concentrations of these carcinogens while the bladder is holding urine. ✓ **p. 437:** The natural ability of melanocytes to migrate makes them much more likely to metastasize, spreading from their source to other regions or organs of the body. Though they usually do not move after embryonic life, their ability to move seems to become re-activated if and when they become cancerous.

Test Yourself

p. 439: 1. d; 2. d; 3. c; 4. d; 5. a; 6. d; 7. c; 8. b; 9. b; 10. a; 11. c; 12. d; 13. d; 14. b; 15. b

Chapter 19

✓ **Questions**

✓ **Figure 19.3, p. 445:** Each cell in a Punnett square has an equal probability of occurring. In this case, because there are four cells, each cell has a 25%

probability of occurring. Therefore, the baby has a 25% chance of having the AA genotype and a 50% chance of having Aa (50% = 25 + 25, because there are two different cells with this genotype). ✓ **Figure 19.8, p. 448:** The ratio is 9:3:3:1, which is to say, 9 out of 16 offspring have both dominant phenotypes (widow's peaks and free earlobes), 3 out of 16 offspring have widow's peaks and attached earlobes, another 3 out of 16 offspring have no widow's peaks and free earlobes, and 1 out 16 offspring has both recessive phenotypes (no widow's peak, attached earlobes). If you observed a similar 9:3:3:1 ratio in a natural mating that involved two traits, you could conclude that the parents were heterozygous for both traits.

✓ **p. 449:** The man and the woman both have dark hair. Each of their children will have a 25% probability of having red hair. That does not mean that if they have four children, one will have red hair, however. The 25% chance of each child having red hair is independent of all other children that the couple may have. ✓ **Figure 19.11, p. 450:** Zero. The baby must inherit either the B or the A allele from his mother, and thus cannot be blood type O. ✓ **p. 453:** Possibly, but not necessarily. Although it's possible that your father has a genetic makeup that predisposes him to high blood cholesterol and that you inherited that predisposition, it's also possible that his high cholesterol levels are due primarily to his diet. Knowing his history helps you, because you can change your diet if you wish. That alone may keep you from developing high cholesterol. In addition, you could choose to be tested for high cholesterol regularly, just in case it's primarily genetic. Then if your cholesterol ever rose above normal, you could begin treatment early enough to reduce your chances of cholesterol-related medical issues.

✓ **p. 455:** No. Sons always inherit their father's Y chromosome but never receive their father's X chromosome. (If they received their father's X chromosome, they would be daughters.) ✓ **Figure 19.16, p. 455:** The daughter must have inherited the hemophilia allele from *both* parents. This can only occur if the woman was a carrier and she married a hemophiliac man.

✓ **p. 457, first question on page:** No. Down syndrome does not involve any gene mutations. There are extra copies of some genes (the ones on chromosome 21), but each copy is a normal copy. ✓ **p. 457, second question on page:** It is the presence or absence of the Y chromosome that primarily determines gender. The number of Xs has only a minor effect, if any. That is, females are females not because they have two Xs, but because they lack a Y.

✓ **p. 458:** The baby must be a homozygous recessive (because only homozygous recessives have PKU), and therefore must have inherited a PKU allele from each parent. Since neither parent has the disorder, they can't be homozygous recessives themselves, so they must be heterozygotes. Any child of two heterozygous parents has a 25% chance of being homozygous recessive.

Test Yourself

p. 461: 1. b; 2. a; 3. c; 4. a; 5. b; 6. c; 7. d; 8. a; 9. d; 10. c; 11. d; 12. d; 13. d; 14. b; 15. a

Chapter 20

✓ **Questions**

✓ **Figure 20.1, p. 466:** Many copies of just a single green A* attached to the primer will be produced in the first step, but nothing else. This is because the only nucleotides available are the modified fluorescent ones that always terminate the growing strand. So the very first nucleotide used (the A*) will always be fluorescent *and* will always terminate the strand. The laser results in the third step will show only a single green peak. ✓ **Figure 20.3, p. 468:** The restriction enzyme would be unable to cut the human DNA into fragments. At the end of the procedure, the bacteria would have plasmids, but the plasmids would not contain any human DNA. But they could contain recombined fragments of bacterial plasmids. ✓ **p. 469:** The human gene most likely contains the palindromic sequence that is the target of the restriction enzyme—that is, the restriction enzyme is cutting at a location *within* the gene, as well as at a site outside the gene. Your best bet would be to try a different restriction enzyme that cuts a different palindromic sequence.

✓ **p. 470:** You must know the base pair sequences of the very beginning and the very end of the piece of DNA so that you can make primers with complementary sequences. To get the DNA copying started, the primers must be able to bind to one end of each single strand of DNA, and they will only

bind if they have complementary sequences. ✅ **p. 472:** It certainly makes it more difficult. Bacteria have been coaxed into making human proteins, but oftentimes it has taken extensive trial and error to find the right combination of plasmid and bacteria. ✅ **p. 473:** The major problem is that plant cells do not take up plasmids as readily as bacteria do, so a variety of methods may have to be tried to coax embryonic plant cells into taking up the desired DNA. Plants also grow much more slowly than bacteria, so once a plant actually does take up the DNA it will take longer to produce plant clones.
✅ **Figure 20.8, p. 473:** Yes, it's possible. Natural processes certainly do alter genomes through the process of random mutations. The result is evolutionary change over time. In addition, naturally occurring viruses sometimes transfer genes between species. So, a plant such as the one on the left in Figure 20.8 could indeed evolve through natural processes. The major differences between natural processes and genetic engineering are that natural processes are much slower than genetic engineering and that the direction evolution takes is random. ✅ **p. 475:** In theory, this could be done by using genetic engineering techniques on a human egg fertilized *in vitro*. For example, the desired DNA could be microinjected directly into fertilized human eggs, as described in Figure 20.11. However, this technique is successful only about 10% of the time. It might work eventually—but a lot of human embryos would probably have to be discarded. And, even if it did work, the infant might be born with serious birth defects. At this time, gene therapy on fertilized eggs is still in the realm of science fiction. ✅ **p. 476, first question on page:** Cells of the immune system attack viruses—including the retrovirus vectors that are carrying the valuable DNA! Immune suppression gives the retroviruses a better chance to deliver their payload successfully without being attacked. ✅ **p. 476, second question on page:** The added gene sometimes is inserted in an unfortunate place in the genome, and in several patients, it apparently was inserted next to or within one of the genes that controls cell division, altering its function. Even if this happens in only one of the patient's cells, a single cell with a disabled cell-division gene could become a cancer.

Test Yourself

p. 477: 1. b; 2. d; 3. a; 4. c; 5. b; 6. c; 7. a; 8. d; 9. d; 10. c; 11. d; 12. a; 13. b; 14. d; 15 . c

Chapter 21

✅ Questions

✅ **p. 484:** The zona pellucida normally becomes impenetrable after the first sperm has entered, thus preventing other sperm from entering. If there is no zona pellucida present, multiple sperm could fertilize the same egg (a fatal condition). ✅ **Figure 21.5, p. 484:** Cleavage has no period of cell growth between the rounds of cell division. The result is that individual cells get smaller and smaller—something that does not occur in the course of normal cell growth and division in adults. ✅ **Figure 21.6, p. 485:** The trophoblast secretes proteolytic enzymes that digest a path into the uterine tissue.
✅ **p. 486:** Ectoderm. A defect in ectodermal function will often cause defects in many different ectoderm-derived organs, such as the peripheral nerves and the skin. ✅ **p. 487:** The amnion cradles the embryo in a protective bag of fluid; the allantois helps form the blood vessels of the umbilical cord; the yolk sac forms part of the digestive tract, some blood cells, and the germ cells; and the chorion produces hCG and (together with maternal tissue) forms the placenta. ✅ **Figure 21.12, p. 489:** The XXY embryo will become male phenotype (though he might develop enlarged breasts at puberty as a result of the extra X chromosome). It is the presence or absence of the Y that determines gender, not the number of Xs. ✅ **Figure 21.15, p. 493:** If either the ductus arteriosus or the foramen ovale passageway did not close properly, some of the baby's venous blood would bypass the lungs and return directly to the body, without picking up any oxygen! Depending how much venous blood bypasses the lungs, a baby with either condition is likely to become oxygen deprived. Such conditions are correctable by surgery, if necessary, to close these passageways. ✅ **p. 494:** Without surfactants, the lungs will not inflate easily. Infants born prematurely will have difficulty breathing and can become severely oxygen-deprived. Treatment usually involves spraying a mist containing surfactants (often derived from

cow lungs) into the infant's lungs. ✅ **p. 498:** If telomerase enables cells to divide more times or even indefinitely, it could increase the risk of cancer. The fact that cancer cells often have high levels of telomerase is certainly cause for caution. ✅ **p. 499:** Regular exercise can slow or temporarily reverse several of the aging-related declines in the musculoskeletal system, cardiovascular, and respiratory systems. For example, weight-training exercise increases muscle mass and bone density; endurance exercises (walking, hiking, etc.) increase heart function; flexibility exercises (yoga, etc.) increase the flexibility and range of motion of ligaments and tendons.

Test Yourself

p. 501: 1. d; 2. c; 3. b; 4. b; 5. a; 6. d; 7. b; 8. a; 9. b; 10. a; 11. b; 12. d; 13. d; 14. b; 15. a

Chapter 22

✅ Questions

✅ **p. 507:** If 1/2 of the original potassium is still present, that means one radioactive potassium half-life has passed since the rock solidified. So the fossil is 1×1.3 billion = 1.3 billion years old. ✅ **p. 508:** These are analogous structures because they share a similar function but are not descended from a similar structure in a common ancestor. Dolphins descended from a mammalian ancestor that had no dorsal fin, and hence, the dolphins must have evolved their dorsal fin "from scratch," independently of the dorsal fins that evolved in sharks and their relatives. ✅ **Figure 22.5, p. 508:** In addition to the pharyngeal gill arches, the tail, the notochord, and the somites also disappear before birth in humans. All of these structures were inherited from our distant ancestors and appear to be "programmed into" our embryological development. Their persistence during development indicates that they may still play important roles in cuing the embryonic cells to give rise to other, more recently evolved, structures. After the latter structures have formed, the former structures seem to be unnecessary. ✅ **p. 510:** Mutations occur spontaneously and randomly—they are neither good nor bad. It just turns out that some mutations increase an organism's chances of survival, others decrease it. Those that decrease the chance of survival tend to be eliminated from the gene pool over time by natural selection.
✅ **p. 511:** The bent tails of Florida panthers are probably due to genetic drift, caused in this case by a population bottleneck. Genetic drift is a random process, so the bent-tail allele has become more common simply by chance, not because it is a superior allele. (Hunting probably played a role in that it further reduced population size, but in this case, hunting did not "select" for bent tails; that is, hunters were not specifically targeting cats with straight tails.) ✅ **p. 512:** Probably not. The same aspects that make today's atmosphere so hospitable to us also make it inhospitable for the formation of new kinds of life. The atmosphere today does not have enough energy (heat, UV radiation, and so forth) to produce complex macromolecules from simple precursors, and the abundant oxygen in today's atmosphere would quickly degrade any complex molecules that did manage to arise. (In addition, the abundant bacteria and other microorganisms that are present today would probably devour any fragile new life-forms.) ✅ **Figure 22.10, p. 513:** The simplest approach would be simply to replicate, in a sealed flask, the conditions of the atmosphere, the sea, and the surface (that is, the presence of clay), leave the flask for a while, and see whether the proposed molecules eventually appear. The flask would need to replicate the heat, electrical discharges, and UV radiation of ancient Earth. One of the many difficulties with such experiments is time: Some of the proposed steps may have taken millions of years to occur on ancient Earth. Typically, a scientific experiment can only run for a decade or two. Nevertheless, progress has been made in producing some of the molecules shown in panel b, and even some of the ones in panel c. ✅ **Figure 22.12, p. 515:** Old World monkeys and humans last shared a common ancestor 30 million years ago, whereas gibbons and humans last shared a common ancestor 25 million years ago. Hence, gibbons are more closely related to humans than are Old World monkeys. Note that the horizontal distance on the *x*-axis is not informative. It is the time from the last common ancestor that is the key for judging relationships. ✅ **p. 515:** You could look for forward-facing eyes on the skull, opposable thumbs, the presence of five digits on each foot, and the presence

of flat nails on the digits. There are other traits that could be used as well, such as the details of the teeth. ✓ **p. 518:** One of the great surprises of human paleontology is that this hypothesis has been completely disproven. Upright walking preceded the enlargement of the brain by several million years. Upright walking appears to a limited extent in *Ardipithecus*, with well-developed bipedal walking in *Australopithecus*, but the brain was still chimpanzee-sized in both of these species. A substantial increase in brain size (along with the ability to make stone tools) did not begin until *Homo habilis*.

Test Yourself

p. 519: 1. d; 2. d; 3. a; 4. a; 5. c; 6. d; 7. d; 8. b; 9. c; 10. c; 11. a; 12. c; 13. d; 14. d; 15. b

Chapter 23

✓ Questions

✓ **p. 525:** Wolves have a greater biotic potential than bears. This is not because of a predator-prey relationship, since both are near the top of the food chain. The most likely explanation for the biotic potential difference is in reproductive rates. A female wolf may have four to six pups every year, whereas a female bear is likely to have no more than two cubs every other year. Wolves also start breeding at a younger age. ✓ **Figure 23.2, p. 525:** If this species had lower biotic potential, both the orange and the green lines would have more gradual slopes. The species would take more time to reach carrying capacity. However, the carrying capacity itself would not necessarily be affected. Carrying capacity is determined by external environmental factors (such as food availability, predation, etc.), not by the species' biotic potential. ✓ **p. 527:** The mature communities in these two regions are different because of different environmental factors in the regions, particularly the climate. Specifically, prairies occur where there is not enough rainfall to support tree growth. ✓ **p. 529:** Algae are the producers; all the other species are consumers. (Technically, algae are not plants, but they gain their energy by photosynthesis, and thus are producers, like plants.) Among the consumers, the krill would have to be considered closer to an herbivore than a carnivore. The baleen whales and killer whales are both carnivores, and the bacteria and fungi are decomposers. ✓ **p. 531:** Large carnivores are generally upper-level consumers—that is, they are secondary, tertiary, or quaternary consumers that prey on other animals. They are rarer than their prey because of the inefficiency of energy conversion between levels of an ecological pyramid. For example, it takes many acres of western rangeland to feed just one elk, and about 20 elk per year to feed just one wolf.

✓ **p. 532:** Rivers contain minerals that have been eroded out of rocks in hills and mountains, and rivers deposit these valuable minerals in floodplains and estuaries. Hence plants and marine algae can grow abundantly in these areas, forming the base for a vibrant, productive ecosystem. In this way, the water cycle affects the productivity of ecosystems. ✓ **Figure 23.11, p. 533:** The annual cycle in atmospheric CO_2 occurs because of a strong cycle in the rate of photosynthesis. The vast forests of the large continents remove an enormous amount of CO_2 from the atmosphere every summer but then cease to do so during the northern winter. The southern continents are much smaller and less forested, so they have a much less noticeable effect. ✓ **Figure 23.12, p. 534:** Nitrogen-fixing bacteria perform N_2 fixation, converting N_2 to ammonium (NH_4^+); nitrifying bacteria perform nitrification, converting ammonium to nitrate (NO_3^-); and finally, denitrifying bacteria perform denitrification, converting nitrate back to N_2.

✓ **p. 537:** It's surprising because the population is still growing even though the fertility rate in the United Kingdom is below the replacement fertility rate of 2.1 children per woman. The most likely explanation for the United Kingdom's population growth is immigration. In theory, a population can still grow even when fertility rates are below the replacement rate, provided that a large fraction of the population is of reproductive age. While this would be a possible explanation in some countries, it is not as likely an explanation for population growth in the United Kingdom because this growth pattern is not typical of a developed nation, like the United Kingdom.

Test Yourself

p. 539: 1. c; 2. d; 3. d; 4. a; 5. b; 6. b; 7. b; 8. b; 9. c; 10. c; 11. a; 12. c; 13. a; 14. a; 15. c

Chapter 24

✓ Questions

✓ **Figure 24.1, p. 544:** Vehicle exhaust is the number one offender, producing five separate categories of pollutants that contribute to three of the four serious atmospheric problems. The only bright spot is that vehicle exhaust does not contribute to ozone destruction. Ozone destruction is caused by a single, very specific class of chemical (chlorine-containing gases, such as CFCs) not present in vehicle exhaust. ✓ **p. 545:** Any change in the amount of cloud cover could have a large effect on global warming. This is because clouds act as a reflective layer, sending infrared heat radiation coming from Earth back toward Earth again. More clouds mean more global warming. ✓ **Figure 24.4, p. 545:** The ozone layer is thinnest around the South Pole. New Zealand lies in a very southerly latitude close to the ozone hole, and thus receives much higher levels of ultraviolet radiation than most other countries. In addition, New Zealand has a higher percentage of light-skinned people than other nations at this latitude, and light-skinned people are especially vulnerable to ultraviolet radiation. ✓ **p. 548:** Killer whales have the highest contamination load. (In fact, many killer whales are so contaminated that if a killer whale carcass washes up on shore, it may have to be handled as "hazardous waste.") Next, would come swordfish, then blue whales, and last of all sardines. The most important factor affecting contamination load is the trophic level at which the animal feeds. Usually, the next most important factor is the animal's age, because contamination builds up over time. ✓ **p. 549:** Despite the high cost of buying the land, New York officials calculate that the city has *saved* money on its water supply. By owning and protecting almost the entire watershed, New York City can limit the degree to which agricultural and industrial pollutants enter the water, so the city's water is unusually clean. The financial benefit is that the city did not have to build a $6 billion water treatment plant, which would have been necessary if the watershed had not been protected. ✓ **p. 553:** Human activity tends to target grasslands (for agriculture), forests (for logging), and riversides and coastlines (for cities, towns, and near-shore fisheries). Least affected are high mountains, tundra, and taiga (northern coniferous forest). Thus, most biologists would say set aside ecosystems of grasslands, forests, and riverside/coastal areas.

Test Yourself

p. 556: 1. a; 2. b; 3. d; 4. d; 5. c; 6. a; 7. a; 8. b; 9. c; 10. a; 11. c; 12. b; 13. c; 14. d; 15. a

Credits

Photo Credits

Chapter 1
Opener: Gary I Rothstein/EPA/Newscom; **p. 2:** Dmitry Naumov/Shutterstock; **Fig. 1.1 top left:** NASA; **top right:** Peter Batson/Image Quest Marine; **bottom left:** Ammit/Alamy; **bottom right:** Michel Gunther/Science Source; **Fig. 1.2a**: Eye of Science/Science Source; **Fig. 1.2b:** Science Source; **Fig. 1.3 left:** Phil Schermeister/National Geographic Creative; **Fig. 1.3 top right:** SPL/Science Source; **Fig. 1.3 bottom right:** Hal Beral VWPics/SuperStock; **p. 11:** Scott Camazine/Alamy; **Fig. 1.8:** Stockbyte/Getty Images; **Fig. 1.11a:** Tyler Olson/Shutterstock; **Fig. 1.11b–c:** Science Source; **Fig. 1.12:** Maximilian Stock Ltd/Getty Images

Chapter 2
Opener: Perennou Nuridsany/Science Source; **p. 22 top:** Pearson Education, Inc.; **p. 22 bottom:** 68/Hill Street Studios/Corbis; **Fig. 2.2a:** CMSP/Custom Medical Stock Photo; **Fig. 2.2b:** Richard Megna/Fundamental Photographs; **Fig. 2.4a:** Julien Crosnier/DPPI/Icon SMI 547/Newscom; **Fig. 2.4b:** Chris Keane/Icon SMI 329/Newscom; **Fig. 2.11a:** CostinT/iStockphoto/Getty Images; **Fig. 2.11b:** JimD/iStockphoto/Getty Images; **Fig. 2.15c:** Don W. Fawcett/Science Source

Chapter 3
Opener: Dr. Torsten Wittmann/Science Source; **p. 48 top:** Chuck Kennedy/KRT/Newscom; **p. 48 bottom:** Partha Pal/Getty Images; **Fig. 3.1a:** Kallista Images/Custom Medical Stock Photo; **Fig. 3.1b:** Biology Pics/Science Source; **Fig. 3.2a:** Thomas Deerinck, NCMIR/Science Source; **Fig. 3.2b:** Science Source; **Fig. 3.2c:** Dr. Ed Reschke/Getty Images; **Fig. 3.4a–b:** Science Source; **Fig. 3.4c:** Kari Lounatmaa/Science Source; **Fig. 3.6:** Don W. Fawcett/Science Source; **Fig. 3.8:** Biology Pics/Science Source; **Fig. 3.10:** Science Source; **Fig. 3.12a:** Science Photo Library/Alamy; **Fig. 3.12b:** Susumu Nishinaga/Science Source; **Fig. 3.14:** Erik Schweitzer; **Fig. 3.18a:** Joseph F. Gennaro Jr./Science Source; **Fig. 3.18b:** Don. W. Fawcett/Science Source; **Fig. 3.21:** David M. Phillips/Science Source; **Fig. 3.29b:** ImageBroker/SuperStock

Chapter 4
Opener: Thomas Deerinck/NCMIR/Science Source; **p. 78 top:** David M. Phillips/Science Source; **p. 78 bottom:** BSIP SA/Alamy; **Fig. 4.2a:** Don W. Fawcett/Science Source; **Fig. 4.2b:** Don W Fawcett/Getty Images; **Fig. 4.2c:** David M. Phillips/Science Source; **Fig. 4.4a–b:** Ed Reschke; **Fig. 4.5c:** Pearson Education; **p. 85 top:** Keith Brofsky/Getty Images; **p. 85 bottom:** Neil McAllister/Alamy; **Fig. 4.6a:** Science Source; **Fig. 4.6b:** Ed Reschke; **Fig. 4.6c:** Pearson Education; **Fig. 4.7:** Ed Reschke/Getty Images; **Fig. 4.11, Fig. 4.12:** SPL/Custom Medical Stock Photo; **Fig. 4.14 left:** XiXinXing/Alamy; **Fig. 4.14 right:** Maridav/Fotolia

Chapter 5
Opener: Du Cane Medical Imaging Ltd./Science Source; **p. 100 top:** Kevin Beebe Custom Medical Stock Photo/Newscom; **p. 100 bottom:** BSIP SA/Alamy; **Fig. 5.1c:** Gene Cox/Science Source; **Fig. 5.1d:** Roseman/Custom Medical Stock Photo; **Fig. 5.3:** SPL/Custom Medical Stock Photo; **Fig. 5.9a:** J. Croyle/Custom Medical Stock Photo; **Fig. 5.9b:** Living Art Enterprises/Science Source; **p. 113:** Carolyn A McKeone/Science Source; **Fig. 5.16a:** Johannes Lehner/Fotolia; **Fig. 5.16b:** Lane Oatey/Blue Jean Images/Getty Images; **Fig. 5.16c:** Von Lieres/Fotolia; **Fig. 5.17a:** SPL/Science Source; **Fig. 5.17b:** Pietro M. Motta/Science Source; **Fig. 5.18:** Sally and Richard Greenhill/Alamy

Chapter 6
Opener: John Burcham/Getty Images; **p. 120 top:** Brian Baer/Sacramento Bee/ZUMAPRESS.com/Newscom; **p. 120 bottom:** Newscom; **Fig. 6.5b:** Science Source; **Fig. 6.6c:** Rosalind King/Science Source; **p. 130:** W. Heiber Fotostudio/Fotolia; **Fig. 6.10b:** Astrid & Hanns-Frieder Michler/Science Source; **Fig. 6.12a:** Andersen Ross/Getty Images; **Fig. 6.12b:** Newscom; **Fig. 6.13a:** Cultura RM/Alamy; **Fig. 6.14c:** Marshall Skla/Getty Images

Chapter 7
Opener: Juergen Berger/Science Source; **p. 140 top:** Jaclyn Albanese; **p. 140 bottom:** Science Source; **Fig. 7.2c:** Suthep/Shutterstock; **Fig. 7.3:** Science Source; **Fig. 7.7:** SPL/Science Source; **Fig. 7.9:** Steve Gschmeissner/Science Source; **Fig. 7.13:** John Scanlon; **p. 154:** LENEE/AGE Fotostock America Inc.; **Fig. 7.14:** Wayne Weinheimer

Chapter 8
Opener: SPL/Science Source; **p. 160 top:** Popperfoto/Getty Images; **p. 160 bottom:** B Busco/Getty Images; **Fig. 8.3a:** West Virginia University, School of Medicine (WVU); **Fig. 8.3b:** Science Source; **Fig. 8.6:** A & F. Michler/Getty Images; **Fig. 8.12a:** SPL/Science Source; **Fig. 8.12b:** Nathan Benn/Alamy; **Fig. 8.14a:** Eastphoto/AGE Fotostock America Inc.; **p. 176:** Biophoto Associates/Science Source; **Fig. 8.19:** Zephyr/Science Source; **Fig. 8.21a:** Newscom; **Fig. 8.21b:** Thomson Reuters (Markets) LLC; **Fig. 8.22:** Glamy/Fotolia

Chapter 9
Opener: Dr Olivier Schwartz, Institut Pasteur/Science Source; **p. 188 top:** Baz Ratner/Reuters; **p. 188 bottom:** Pascal Guyot/AFP/Getty Images; **Fig. 9.1a&c:** S. Lowry/Univ Ulster/Getty Images; **Fig. 9.1b:** Scimat/Science Source; **Fig. 9.4:** Steve Gschmeissner/Science Source; **Fig. 9.5:** Blend Images/SuperStock; **Fig. 9.6:** Lennart Nilsson/TT Nyhetsbyrån AB; **Fig. 9.7:** SPL/Science Source; **Fig. 9.15a:** Steve Gschmeissner/Science Source; **p. 206:** Michael Buckner/Getty Images; **Fig. 9.19a:** Southern Illinois University/Science Source; **Fig. 9.19b:** Science Source

Chapter 10
Opener: Prof. P.M. Motta/Dept. of Anatomy/University "La Sapienza", Rome/Science Source; **p. 220 top:** Pixelot/Fotolia; **p. 220 bottom:** Yuri Arcurs/iStock/Getty Images; **Fig. 10.4:** Custom Medical Stock Photo; **Fig. 10.6a:** Charles Daghlian/Science Source; **Fig. 10.6b:** Andrew P. Evan; Dale Sparks/AP Images; **Fig. 10.8b:** David M. Phillips/Science Source; **Fig. 10.10a:** Javier Larrea/age fotostock/Getty Images; **p. 237:** Dale Sparks/AP Images; **Fig. 10.14:** St Bartholomew's Hospital/Science Source

Chapter 11
Opener: Thomas Deerinck, NCMIR/Science Source; **p. 244:** Elaine Thompson/AP Images; **Fig. 11.7 bottom:** Don W. Fawcett/Science Source; **Fig. 11.8 top:** Omikron/Science Source; **Fig. 11.8 bottom:** SPL/Science Source; **Fig. 11.11:** Shutterstock; **Fig. 11.14:** SPL/Manfred Kage/Science Source; **p. 268:** KJ/Icon SMI/Newscom; **Fig. 11.21:** Robert Friedland/Science Source

Chapter 12
Opener: SPL/Science Source; **p. 275 top:** Sarah M. Golonka/Tetra Images/Alamy; **p. 275 bottom:** KMOX/UPI/Newscom; **Fig. 12.5 b:** SPL/Science Source; **Fig. 12.5c:** Astrid/Hanns Frieder Michler/Science Source; **Fig. 12.7:** Dmitriy Shironosov/123RF; **Fig. 12.10:** Susumu Nishinaga/Science Source; **Fig. 12.12:** Kresge Hearing Research Institute; **Fig. 12.19:** Ralph C. Eagle/Science Source; **p. 292:** Pascal Goetgheluck/Science Source; **Fig. 12.20:** Kresge Hearing Research Institute; **Fig. 12.21a:** Sue Ford/Science Source; **Fig. 12.21b:** Western Ophthalmic Hospital/Science Source; **Fig. 12.22:** National Eye Institute

Chapter 13
Opener: Biophoto Associates/Science Source; **p. 299 top:** Jose Gil/Fotolia; **p. 299 bottom:** David McNew/Getty Images; **Fig. 13.8:** ZUMA Press, Inc./Alamy; **Fig. 13.13:** Alison Wright/Science Source; **Fig. 13.15** Charles B. Wilson

Chapter 14
Opener: Susuma Nishina Calcula/Getty Images; **p. 323 top:** Inga Spence/Alamy; **p. 323 bottom:** Loomis Dean/The LIFE Picture Collection/Getty Images; **Fig. 14.4c:** SPL/Science Source; **Fig. 14.7c:** Steve Gschmeissner/Science Source; **Fig. 14.8:** David M. Martin/Science Source; **Fig. 14.10:** Steve Gschmeissner/Science Source; **p. 337:** Liz Hafalia/Corbis; **Fig. 14.17a:** lurs/Fotolia; **Fig. 14.17b:** Textu/Fotolia; **Fig. 14.19:** Martin Shields/Alamy; **Fig. 14.22:** Motte Jules/Abaca/Newscom; **Fig. 14.23:** Biophoto Associates/Getty Images

Chapter 15
Opener: Steve Gschmeissner/Science Source; **p. 352:** Craig Warga/NY Daily News Archive/Getty Images; **Fig. 15.7a:** Science Source; **Fig. 15.7b:** M.D Wilhelm Kriz; **Fig. 15.8:** Steve Gschmeissner/Science Source; **p. 368:** Ray Tang/Rex Features/AP Images; **Fig. 15.15:** Universal Images Group/Getty Images

Chapter 16
Opener: Eye of Science/Science Source; **p. 374:** Staff/Getty Images; **Fig. 16.2d:** Steve Gschmeissner/Science Source; **Fig. 16.8:** Eye of Science/Science Source; **Fig. 16.10a:** Tomasz Trojanowski/Fotolia; **Fig. 16.10b:** Garo/Phanie/Alamy;

Index

Citations followed by *b* refer to material in boxes; citations followed by *f* refer to material in figures or illustrations; and citations followed by *t* refer to material in tables.